U0172775

The Living Record of Science

《自然》学科经典系列

总顾问：李政道（Tsung-Dao Lee）

英方总主编：Sir John Maddox
Sir Philip Campbell
中方总主编：路甬祥

生命科学的进程 IV

PROGRESS IN LIFE SCIENCES IV

（英汉对照）

主编：许智宏

外语教学与研究出版社 · 麦克米伦教育 · 《自然》旗下期刊与服务集合

FOREIGN LANGUAGE TEACHING AND RESEARCH PRESS · MACMILLAN EDUCATION · NATURE PORTFOLIO

北京 BEIJING

图书在版编目（CIP）数据

生命科学的进程. Ⅳ ：英汉对照 / 许智宏主编. -- 北京 ：外语教学与研究出版社，2021.9

（《自然》学科经典系列 / 路甬祥等总主编）

ISBN 978-7-5213-2930-8

Ⅰ. ①生… Ⅱ. ①许… Ⅲ. ①生命科学－文集－英、汉 Ⅳ. ①Q1–53

中国版本图书馆 CIP 数据核字（2021）第 174604 号

地图审图号：GS（2021）4026 号

出 版 人　徐建忠
项目统筹　章思英
项目负责　刘晓楠　顾海成
责任编辑　刘晓楠
责任校对　王　菲　夏洁媛
封面设计　孙莉明　高　蕾
版式设计　孙莉明
出版发行　外语教学与研究出版社
社　　址　北京市西三环北路 19 号（100089）
网　　址　http://www.fltrp.com
印　　刷　北京华联印刷有限公司
开　　本　787×1092　1/16
印　　张　62
版　　次　2021 年 9 月第 1 版　2021 年 9 月第 1 次印刷
书　　号　ISBN 978-7-5213-2930-8
定　　价　568.00 元

购书咨询：（010）88819926　电子邮箱：club@fltrp.com
外研书店：https://waiyants.tmall.com
凡印刷、装订质量问题，请联系我社印制部
联系电话：（010）61207896　电子邮箱：zhijian@fltrp.com
凡侵权、盗版书籍线索，请联系我社法律事务部
举报电话：（010）88817519　电子邮箱：banquan@fltrp.com
物料号：329300001

《自然》学科经典系列

（英汉对照）

生命科学的进程

（英汉对照）

主编：许智宏

审稿专家（以姓氏笔画为序）

丁　梅	王　嵬	王社江	王敏康	文　波	方向东	石　磊
冯兴无	吕雪梅	刘佳佳	孙　军	杜　忆	杨崇林	肖景发
吴　琳	吴新智	沈　杰	张颖奇	陈继征	陈新文	林圣龙
赵凌霞	胡松年	胡卓伟	秦志海	崔　巍	崔娅铭	梁前进
彭小忠	董　为	焦炳华				

翻译工作组稿人（以姓氏笔画为序）

王耀杨　　刘　明　　关秀清　　李　琦　　何　铭　　蔡　迪

翻译人员（以姓氏笔画为序）

毛晨晖	田晓阳	刘皓芳	阮玉辉	苏怡汀	李　响	李　梅
张玉光	张锦彬	周平博	周志华	周晓明	郑建全	荆玉祥
侯彦婕	姜　薇	莫维克	高俊义			

校对人员（以姓氏笔画为序）

马　荣	马　亮	王　菲	王　敏	王帅帅	王丽霞	王晓敏
王晓蕾	王海纳	公　晗	石宇鹏	田晓阳	史明澍	邢路达
吕秋莎	朱晓霞	任　奕	任进成	任崤铭	刘　伟	刘　佩
刘若青	刘雨佳	刘俊杰	刘萍萍	许静静	阮玉辉	苏怡汀
李　四	李　芳	李　昆	李　娟	李　梅	李　景	李　婷
李盎然	李梦桃	李霄霞	杨　晶	吴兆军	邱珍琳	汪　涵
张　狄	张　晴	张玉光	张世馥	张欣园	张梦璇	张瑶楠
陈思婧	陈露芸	范艳璇	罗小青	周小雅	周少贞	周晓明
房秋怡	赵广宇	荣金诚	侯彦婕	侯鉴璇	姜　薇	贺乐天
夏洁媛	顾海成	黄　欢	黄　璞	黄元耕	黄雪嫚	第文龙
梁　瑜	葛聆沨	韩涛涛	焦晓林	谢周丽	蔡军茹	

Contents
目　　录

1983

1984

1985

1986

1987

1988

1995

1996

1997

Volume IV

New Hominoid Primates from the Siwaliks of Pakistan and Their Bearing on Hominoid Evolution

D. Pilbeam *et al.*

Editor's Note

The Miocene strata of the Siwalik Hills in Pakistan had yielded abundant fossils for decades, but the 1970s were especially rich. Pilbeam *et al.* document 86 new primate specimens from 18 localities, doubling the record then known. Of particular interest were specimens of the ape-like *Ramapithecus*, then thought of as an ancestor of man. Pilbeam and colleagues, however, add words of caution. *Ramapithecus*, along with another form, *Sivapithecus*, belonged to an increasingly appreciated diversity of Miocene apes worldwide, so drawing lines between these and modern forms would have been unwise. We now suspect that *Ramapithecus* and *Sivapithecus* are similar to each other, and closer to the orangutan than to man.

Siwalik deposits in the Punjab have yielded a rich collection of hominoid primate remains. Together with other recent finds they indicate the need for some changes in hominoid classification.

SINCE 1973 a team from the Geological Survey of Pakistan (GSP) and the Peabody Museum has collected fossils in the Potwar Plateau, Punjab Province, Pakistan (72°30′ E, 33°00′ N) as part of a joint research project aimed at a better understanding of the geological, floral and faunal history of South Asia. Neogene rocks of the Siwalik Group are widely exposed in India and Pakistan as part of the South Asian alpide system. The Soan synclinorium in the Potwar Plateau is the area in which all but one of the type localities for Siwalik formations and South Asian Land Mammal Ages are located (Figs 1 and 2 of preceding article). The lithostratigraphy and biostratigraphy of Siwalik deposits in the Plateau have been discussed in the preceding article[1].

巴基斯坦西瓦利克新发现的人猿超科灵长类动物及其在人猿超科进化上的意义

皮尔比姆等

编者按

在过去的数十年中，从位于巴基斯坦的西瓦利克山脉的中新世地层中已经出土了大量的化石，但是在20世纪70年代这里的出土量尤为突出。皮尔比姆等人从来自于18个发掘地点的化石中记录了86件新的灵长类动物标本，这使得当时已有的记录得以翻番。其中令人特别感兴趣的是似猿的腊玛古猿的标本，随后腊玛古猿被认为是人类祖先。但是皮尔比姆及其同事们提醒人们要谨慎对待这一问题。腊玛古猿和另一种类型的古猿——西瓦古猿，都属于多样性日渐丰富的中新世猿类，因此将它们与现代类型截然区分开是不明智的。现在我们怀疑腊玛古猿和西瓦古猿彼此类似，比起人类，它们更接近于猩猩。

旁遮普的西瓦利克沉积物中出土了大量人猿超科灵长类动物化石。结合其他的最新发现表明，有必要对人猿超科的分类做些改动。

自1973年以来，巴基斯坦地质调查所与皮博迪博物馆组成的考察队已经在巴基斯坦旁遮普省博德瓦尔高原（72° 30′ E，33° 00′ N）采集了一些化石，这是一个联合研究项目的一部分，旨在更好地了解南亚的地质、植物群及动物群的历史。作为南亚阿尔卑斯带系统的一部分，西瓦利克群的新近纪岩石在印度与巴基斯坦广泛出露。除了一个例外，其他所有的西瓦利克组及南亚陆地哺乳动物时代所在的典型地点都在博德瓦尔高原的索安复向斜区域（前面文章中的图1、图2）。在前面的文章[1]中已经对博德瓦尔高原西瓦利克的沉积物进行了岩石地层学与生物地层学方面的讨论。

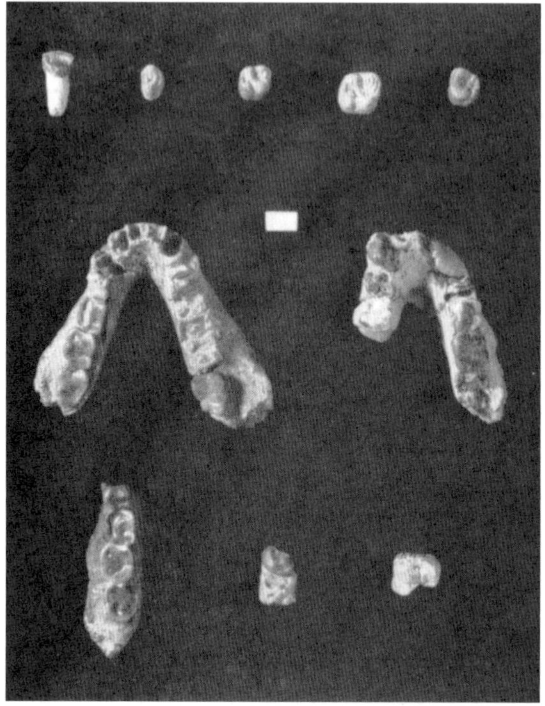

Fig. 1. Top row left to right: GSP 9903, 9906, 5019, 6206, 8702. Middle row: GSP 4622/4857, 9563/9902. Bottom row: GSP 6153, 7619, 7144.

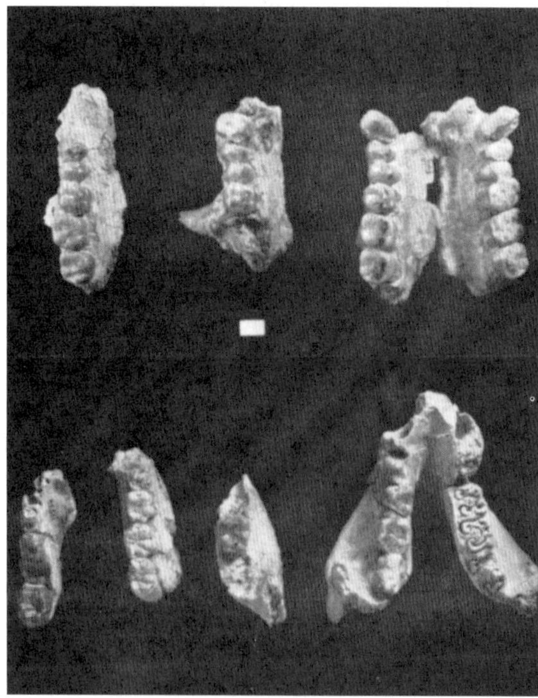

Fig. 2. Top row, left to right: GSP 11708, 11704, 9977/01/05/9564. Bottom row: 11536/7, 13163, 4230, 9977/01/05/9564.

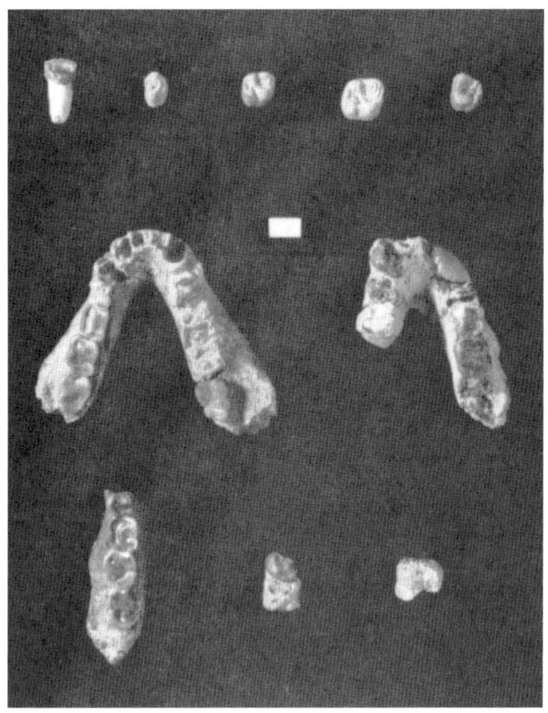

图 1. 上排从左到右：GSP 9903, 9906, 5019, 6206, 8702；中排从左到右：GSP 4622/4857, 9563/9902；下排从左到右：GSP 6153, 7619, 7144。

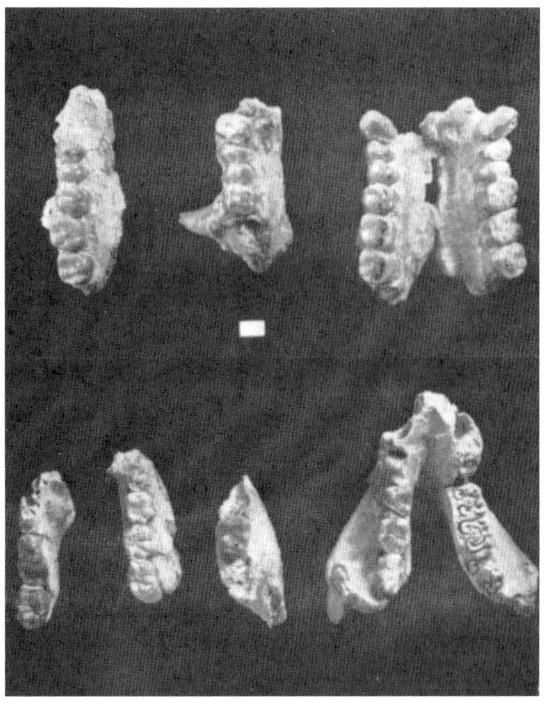

图 2. 上排从左到右：GSP 11708, 11704, 9977/01/05/9564。下排从左到右：11536/7, 13163, 4230, 9977/01/05/9564。

More than 13,000 catalogued fossil specimens representing a diverse vertebrate and invertebrate fauna have been collected from more than 300 localities; 18 localities have yielded 86 new hominoid primate specimens representing a minimum of 43 individuals (Tables 1–3). The hominoids are accurately located stratigraphically with well documented information concerning lithological context and faunal associations. They approximately double the previously known hominoid collections[2]. In addition, new stratigraphic information has facilitated further interpretation of earlier collections. Of particular interest are four localities (Table 2) in which remains of between five and nine hominoid individuals have been found together.

Table 1. Hominoid specimens collected from the Potwar Plateau, Punjab

Area	Locality	GSP no.	Specimen
Sethi Nagri	311	11536/7	Infant left mandibular corpus, $dP_{\overline{3}}$, $dP_{\overline{4}}$, associated $M_{\overline{2}}$
		10493	Left $\underline{C}$
		11534	Left $P^{\underline{4}}$ fragment
		11999	Right $M^{\underline{1}}$
		13162	Right $M^{\underline{1}}$ fragment
		12000	Right $M^{\underline{1}}$ fragment
		9986	Right $M^{\underline{1}}$
		10500	Left $M^{\underline{2}}$ fragment
		11533	Right $M^{\underline{2}}$ fragment
		7144	Left $M_{\overline{1}}$ fragment
		11998	Right $M_{\overline{2}}$
		13164	Central $\overline{I}$
		12648	Central $\overline{I}$
		6663	Distal humeral epiphyseal fragment
		6664	Distal pollicial phalangeal fragment
		6666	Distal phalangeal fragment
		6454	Intermediate cuneiform
		12271	Distal humeral fragment
Khaur	182	8928	Right $I^{\underline{1}}$
		8925	Right $\underline{C}$
		5019	Right $M^{\underline{1}}$
		5018	Left $M^{\underline{2}}$ fragment
		5067	Left $M^{\underline{3}}$
		8702	Left $M^{\underline{3}}$

已经从 300 多个地点采集到超过 13,000 个编入目录的化石标本，它们代表了不同的脊椎动物与无脊椎动物的动物群；在 18 个地点共出土了 86 件新的人猿超科灵长类动物标本，它们至少代表了 43 个个体（表 1~3）。利用记载详细的有关岩性背景与动物群的相关信息，准确地定位了人猿超科动物所在的地层。它们几乎是以往已知人猿超科动物标本的两倍 [2]。另外，新的地层信息促进了对早期标本的进一步解释。更为有趣的是，在其中 4 个地点（表 2）发现了聚集在一起的人猿超科动物个体，其个体数目介于 5 到 9 个之间。

表 1. 从旁遮普博德瓦尔高原采集到的人猿超科动物的标本

地区	地点	GSP编号	标本
塞西纳格里	311	11536/7	幼体左下颌共生体，$dP_{\bar{3}}$, $dP_{\bar{4}}$, 共生的 $M_{\bar{2}}$
		10493	左 $\underline{C}$
		11534	左 $P^{\underline{4}}$ 碎片
		11999	右 $M^{\underline{1}}$
		13162	右 $M^{\underline{1}}$ 碎片
		12000	右 $M^{\underline{1}}$ 碎片
		9986	右 $M^{\underline{1}}$
		10500	左 $M^{\underline{2}}$ 碎片
		11533	右 $M^{\underline{2}}$ 碎片
		7144	左 $M_{\bar{1}}$ 碎片
		11998	右 $M_{\bar{2}}$
		13164	中 $\bar{I}$
		12648	中 $\bar{I}$
		6663	肱骨末端骨骺残片
		6664	拇指指骨末端残片
		6666	指骨末端残片
		6454	中央楔状骨
		12271	肱骨末端残片
汉瓦	182	8928	右 $I^{\underline{1}}$
		8925	右 $\underline{C}$
		5019	右 $M^{\underline{1}}$
		5018	左 $M^{\underline{2}}$ 碎片
		5067	左 $M^{\underline{3}}$
		8702	左 $M^{\underline{3}}$

Continued

Area	Locality	GSP no.	Specimen
Khaur	182	4622/4857	Mandible, left $M_{\overline{1}}-M_{\overline{3}}$, right $M_{\overline{3}}$
		4230	Right mandibular fragment with $M_{\overline{2}}$
		5464	Right I-
		8679	Right $\overline{C}$
		5712	Left $P_{\overline{3}}$ fragment
		5020	Left $P_{\overline{4}}$
		4635	Right $M_{\overline{2}}$
		4735	Right $M_{\overline{3}}$
		8926	Right $M_{\overline{3}}$
		8927	Left $M_{\overline{3}}$
		4664	Left distal calcaneal fragment
Khaur	260	9977/01/05/	Maxilla, left and right $\underline{C}$ $M^{\underline{3}}$
		9564	right $I^{\underline{2}}$; mandible, right $\overline{C}$, left $P_{\overline{4}}-M_{\overline{3}}$; right condyle; Cranial and facial fragments
		9897	Left maxillary fragment
		9903	Right $I^{\underline{1}}$
		9898	Left $I^{\underline{1}}$
		13167	Left $\underline{C}$
		9906	Left $P^{\underline{4}}$
		13166	Left $P^{\underline{4}}$
		12647	Right $M^{\underline{1}}$
		9972	Right $M^{\underline{2}}$
		9896	Left $M^{\underline{2}}$
		9969	Right $M^{\underline{2}}$
		9895	Left $M^{\underline{3}}$
		9900	Right $M^{\underline{3}}$
		12709	Infant mandible, right $\overline{C}$, $dP_{\overline{3}}$
		9563/9902	Mandible, left $P_{\overline{3}}$, $M_{\overline{1}}$, right $M_{\overline{2}}$, $M_{\overline{3}}$
		13165	Right mandibular fragment, $M_{\overline{1}}-M_{\overline{3}}$
		9565	Left $\overline{C}$ fragment
		9899	Right $M_{\overline{3}}$
		12654	Femoral fragments
		9894	Femoral head fragment

续表

地区	地点	GSP编号	标本
汉瓦	182	4622/4857	下颌骨，左$M_{\bar{1}}$－$M_{\bar{3}}$，右$M_{\bar{3}}$
		4230	有$M_{\bar{2}}$的右颌骨碎片
		5464	右 I-
		8679	右$\overline{C}$
		5712	左 $P_{\bar{3}}$ 碎片
		5020	左 $P_{\bar{4}}$
		4635	右$M_{\bar{2}}$
		4735	右$M_{\bar{3}}$
		8926	右$M_{\bar{3}}$
		8927	左$M_{\bar{3}}$
		4664	左跟骨末端碎片
汉瓦	260	9977/01/05/	上颌骨，左右 $\underline{C}$ M^{3}
		9564	右 I^{2}；下颌骨，右$\overline{C}$，左 $P_{\bar{4}}$ － $M_{\bar{3}}$右骨节；颅骨与面部残片
		9897	左上颌骨残片
		9903	右 I^{1}
		9898	左 I^{1}
		13167	左 $\underline{C}$
		9906	左 P^{4}
		13166	左 P^{4}
		12647	右 M^{1}
		9972	右 M^{2}
		9896	左 M^{2}
		9969	右 M^{2}
		9895	左 M^{3}
		9900	右 M^{3}
		12709	幼体下颌骨，右$\overline{C}$，$dP_{\bar{3}}$
		9563/9902	下颚骨，左 $P_{\bar{3}}$，$M_{\bar{1}}$，右$M_{\bar{2}}$，$M_{\bar{3}}$
		13165	右下颌骨残片，$M_{\bar{1}}$－$M_{\bar{3}}$
		9565	左$\overline{C}$ 碎片
		9899	右$M_{\bar{3}}$
		12654	股骨残片
		9894	股骨头残片

Continued

Area	Locality	GSP no.	Specimen
Khaur	260	13168	Distal phalangeal fragment
Khaur	317	11708	Right maxilla, P^3–M^3
		11786	Right maxillary fragment, P^3–M^2
		11704	Right maxillary fragment, I^2 root–M^1
		7618	Right M^2
		11706	Right mandibular fragment
		11707/85	Mandible, left M_3
		7619	Left mandibular fragment, P_3
		9930	Left M_3
		7611	Radial diaphysis
		11867	Femoral head and neck
Khaur	137	3293	Left I^1
	191	12568	Left $\underline{C}$
	207	5001	Right M_2
	211	5260	Right M^1
	221	6758	Left M^3
		6759	Right M_3
	224	6153	Left mandibular fragment, P_4–M_3
		7308	Right M^1 fragment
		6178	Femoral head fragment
	226	6160	Right mandibular fragment, P_3–M_3
	227	6206	Right M^2
		13171	Left I^1
	230	6999	Right I^1
	251	8836	Right M^2
	261	9987	Right P^4
	309	11597	Left $\underline{C}$
		10232	Left $\underline{C}$ fragment
	310	10785	Talar fragment
	314	11003	Left $\underline{C}$
Chinji	38	780	Left $\underline{C}$ fragment

续表

地区	地点	GSP编号	标本
汉瓦	260	13168	指骨末端残片
汉瓦	317	11708	右上颌骨 P^3–M^3
		11786	右上颌骨残片 P^3–M^2
		11704	右上颌骨残片 I^2 残根 –M^1
		7618	右 M^2
		11706	右下颌骨残片
		11707/85	下颌骨，左 M_3
		7619	左下颌骨残片，P_3
		9930	左 M_3
		7611	桡骨骨干
		11867	股骨头与颈部
汉瓦	137	3293	左 I^1
	191	12568	左 $\underline{C}$
	207	5001	右 M_2
	211	5260	右 M^1
	221	6758	左 M^3
		6759	右 M_3
	224	6153	左下颌骨残片， P_4–M_3
		7308	右 M^1 残片
		6178	股骨头残片
	226	6160	右下颌骨残片， P_3–M_3
	227	6206	右 M^2
		13171	左 I^1
	230	6999	右 I^1
	251	8836	右 M^2
	261	9987	右 P^4
	309	11597	左 $\underline{C}$
		10232	左 $\underline{C}$ 残片
	310	10785	距骨残片
	314	11003	左 $\underline{C}$
成吉	38	780	左 $\underline{C}$ 残片

Table 2. Tentative listing of minimum numbers of hominoid individuals

Locality	No. of specimens	Minimum nos individuals			
		G	R	Si	Total
182	17	–	4	2	6
260	21	–	4	5	9
311	19	1	5	5	6
317	10	–	2	3	5
137	1	–	–	1	1
191	1	–	–	1	1
207	1	–	–	1	1
211	1	–	–	1	1
221	2	–	1	–	1
224	3	–	?2	–	2
226	1	–	?1	–	1
227	2	–	?1	–	1
230	1	–	–	1	1
251	1	–	1	–	1
261	1	–	–	1	1
309	2	–	–	2	2
310	1	–	?1	–	1
314	1	–	–	1	1
38	1	–	–	1	1
19	**86**	**1**	**17**	**25**	**43**

G, *Gigantopithecus*; R, *Ramapithecus*; Si, *Sivapithecus indicus*.

Table 3. Distribution of hominoid parts

Part preserved	Specimens
Cranial	
Associated maxillae and mandibles	1
Adult maxillae	4
Adult mandibles	9
Infant mandibles	2
Isolated teeth	
I^1	6

表 2. 人猿超科动物最小个体数的推测值

地点	标本编号	最小个体数			
		G	R	Si	合计
182	17	_	4	2	6
260	21	_	4	5	9
311	19	1	5	5	6
317	10	_	2	3	5
137	1	_	_	1	1
191	1	_	_	1	1
207	1	_	_	1	1
211	1	_	_	1	1
221	2	_	1	_	1
224	3	_	?2	_	2
226	1	_	?1	_	1
227	2	_	?1	_	1
230	1	_	_	1	1
251	1	_	1	_	1
261	1	_	_	1	1
309	2	_	_	2	2
310	1	_	?1	_	1
314	1	_	_	1	1
38	1	_	_	1	1
19	**86**	**1**	**17**	**25**	**43**

G，巨猿；R，腊玛古猿；Si，西瓦古猿印度种。

表 3. 人猿超科动物的部件分布表

保存的部件	标本
头颅部分	
共生的上颌骨与下颌骨	1
成年上颌骨	4
成年下颌骨	9
幼体下颌骨	2
单独的牙齿	
I^1	6

Continued

Part preserved	Specimens
$\underline{C}$	8
P^4	4
M^1	6
M^2	11
M^3	6
I-	3
$\overline{C}$	2
P_3^-	1
P_4^-	1
M_1^-	1
M_2	3
M_3	5
Postcranial	13

Detailed descriptions of the geology, revisions of major vertebrate and invertebrate groups, and more extensive descriptions and analyses of the hominoids will be published elsewhere as studies are completed. What follows is a preliminary announcement of the primate finds.

Geology and Palaeoecology

Most of the new hominoid sites are along the northern rim of the Soan synclinorium between Dhok Mila and Kaulial. One, 311, is at Sethi Nagri, another, 38, south of Chinji; both are along the southern margin of the synclinorium. All the northern sites, with two exceptions (251 and 261), are situated stratigraphically in the upper third of the Nagri Formation; localities 251 and 261 are somewhat lower in the section. Locality 311 at Sethi Nagri is probably situated somewhat below the level of the bulk of the northern sites. Locality 38 is in the Chinji Formation.

Approximate ages can be assigned to the localities on the basis of faunal comparisons to dated sequences elsewhere in the Old World, and these are apparently supported by palaeomagnetic surveys which will be published soon. Locality 38 may be about 12 Myr old; 251 and 261 about 9.5–10 Myr old; and the rest of the sites some 9.0 Myr old.

Abundant vertebrate and invertebrate faunal remains have been recovered from the Potwar Plateau both by us and by previous expeditions, and the biostratigraphic sequence is becoming clear. The upper Nagri levels are particularly well sampled, and new analyses of major taxonomic groups (rodents, carnivores, suids, bovids and equids) suggest close

续表

保存的部件	标本
$\underline{C}$	8
$P^{\underline{4}}$	4
M^1	6
M^2	11
M^3	6
I-	3
$\overline{C}$	2
$P_{\overline{3}}$	1
$P_{\overline{4}}$	1
$M_{\overline{1}}$	1
M_2	3
M_3	5
颅后骨骼	13

随着研究的完成，地质学方面的详细描述、对主要的脊椎动物与无脊椎动物的分类的修订，以及对人猿超科动物更为全面的描述与分析将在其他地方公开发表。接下来本文将公开有关灵长类动物的一些初步发现。

地质学与古生态学

大部分新的人猿超科动物遗址都沿着多克米拉与库里埃之间的索安复向斜的北缘分布。其中，编号为 311 的遗址位于塞西纳格里；而编号 38 的遗址则位于成吉之南；两个都是沿着复向斜的南缘分布。除了两个例外（251 号与 261 号），所有北面的遗址都位于纳格里组上部三分之一处；只有 251 号与 261 号地点在剖面上稍微低一点。311 号地点在塞西纳格里，可能稍微在大部分北面遗址的水平面之下。38 号地点则在成吉组内。

根据与旧大陆（东半球）的其他地方已测定年代的地层层序的动物群的对比，能够给出每一地点的近似年代值，并且这些数值明显得到了即将公开发表的有关它们的古地磁调查结果的支持。38 号地点的年代距今约 1,200 万年；251 号与 261 号地点的年代则距今约 950 万到 1,000 万年；其余的遗址距今约 900 万年。

我们和前人都从博德瓦尔高原的考察中挖掘出了大量属于脊椎动物与无脊椎动物动物群的化石，生物地层序列变得越来越清楚。特别在纳格里上层仔细采样，对于主要的分类组群（啮齿动物、食肉动物、猪科动物、牛科动物及马科动物）的新分析表明，

similarities to European and West Asian sites of late Vallesian and early Turolian age[3].

Earlier studies of Siwalik lithology[4], faunas[5-7] and floras[8,9] generated hypotheses about past habitats. These studies were large scale, involving the analysis of faunas from many localities and broad stratigraphic ranges, or used lithological or floral samples which could not be tied into firm stratigraphic or palaeogeographic frameworks. Only recently have more adequate studies started[10]. Our programme of detailed lithostratigraphical, magnetostratigraphical, sedimentological, taphonomical, faunal and floral analyses is aimed at a better understanding of Siwalik palaeogeography and past plant and animal associations.

Many primate localities are in pellet rock lithologies usually associated with minor grey-green sandstones within a red-bed sequence. In localities where several individuals are preserved, primates are often a relatively abundant component of the fauna. So far collections have been made in a taphonomically adequate way from few sites, nor have faunal analyses involving minimum numbers of individuals been completed. Possible faunal differences between primate and non-primate localities are being investigated. Preliminary faunal analyses from our "best-collected" localities suggest that the overall mammal community lacked a diversity of large mammals and had low numbers of grazing species, herbivores above 100 kg and arboreal species. By comparison with recent and Pleistocene faunas associated with riverine environments, this suggests woodland or bush habitats with open patches of grassland rather than extensive forests. The sedimentary features and lithologies of the upper Nagri indicate a braided fluvial regime of low sinuosity rather than a meandering one, and a rapidly aggrading multiple-channelled river would probably have been characterised by a mosaic of more and less open habitats rather than extensive riparian forest.

There have been too few palynological studies of Siwalik age sediments[8-10] to indicate clearly either the range of habitats at a particular time or habitat changes through time. Available evidence suggests a change during the deposition of the Chinji and Nagri Formations from mainly subtropical, forest habitats to more open, less low-lying habitats. Schaller[11] has summarised views on the modern vegetation of the area: evergreen and semi-evergreen forests have quite restricted distributions, most habitats being deciduous forest or thorn forest. Although South Asian Neogene climates may have been warmer and with more precipitation than Recent climates, it is possible that evergreen forests (including gallery forests) were never very widespread during the time of deposition of Siwalik group rocks. As noted, such limited information as we have suggests non-evergreen forest contexts for at least some of the major primate localities.

Primates

There is an extensive primary literature on Siwalik primates, and they have been mentioned and discussed many times[2]. Approximately 90 hominoid specimens representing more than 60 individuals are known from previous collections. The Potwar

这里和欧洲及西亚的属于晚瓦里西期与早图洛里安年代的遗址[3]非常相似。

对于西瓦利克岩石学[4]、动物群[5-7]及植物群[8,9]的早期研究形成关于过去生境的假说。这些研究规模很大，涉及许多地点的动物群分析，且涵盖了广泛的地层范围，或利用并未严格归入到确定的地层学或古地理学格架之中的岩石或者植物群样本。只是最近才开展了更为充分的研究[10]。我们对岩石地层学、磁性地层学、沉积学、埋藏学、动物群及植物群的详细分析，其目的是更好地理解西瓦利克的古地理学与过去动植物的相互联系。

许多灵长类动物的发掘地点都在球粒岩中，球粒岩通常与红层层序中的微小灰绿砂岩相关。在有几个个体被保存下来的地点上，灵长类动物常常是动物群中相当丰富的组成部分。迄今为止，只在少数几个遗址进行过具有埋藏学意义的充分采集，即使对涉及最小个体数的动物群，也没有完成对它的相关分析。正在对灵长类动物与非灵长类动物地点之间可能存在的动物群的差异性进行研究。我们对“采样采得最好的”地点的动物群的初步分析显示,整个哺乳动物群缺乏多样性的大型哺乳动物，且 100 千克以上的食草动物与树栖种类的数量也少。通过对现代及更新世与河岸环境相关的动物群的对比，显示出的是有着开阔的小块草地的林地或灌木生境而不是广阔的森林。纳格里上部的岩性与沉积特征显示，此处的河流系统是弯曲度低的辫状体系，而不是曲流体系，而且一条迅速淤高的多水道河流将可能以镶嵌状的或多或少的开阔生境为特征，而不是广阔的河滨森林。

对西瓦利克年代沉积物的孢粉学研究很少[8-10]，不能清楚地显示出一个特定时段的生境，也不能清楚地显示出生境随时间的变化。现有的证据显示，在成吉与纳格里组的沉积期间，从主要是亚热带森林的生境变为更开放、低洼更少的生境。沙勒[11]从现代植被的角度概述了这个地区的植被景观：常绿与半常绿林分布非常有限，大多数生境是落叶林或热带旱生林。虽然南亚新近纪气候可能是温暖一点，与全新世的气候相比降水量可能更多，但是在西瓦利克组群岩石沉积期间，可能常绿林（包括长廊林）从没有十分广泛地分布过。正如所述，我们所掌握的有限的信息表明，至少一些主要的灵长类动物地点是非常绿森林环境。

灵长类动物

关于西瓦利克灵长类动物的研究有大量的原始文献，它们多次被提及并讨论[2]。从以前搜集到的标本中分辨出大约 90 个人猿超科动物标本，代表着超过 60 个个体。博德瓦尔高原出土了一些单个的牙齿和少许下颌骨，大多数得自成吉的成吉组；几

Plateau yielded some isolated teeth and a few jaws, the bulk from the Chinji Formation at Chinji; several specimens came from around Hasnot, and a few individual teeth were known from other localities. Ramnagar in Kashmir was the source of a few hominoids, but the best collections came from Haritalyangar in India.

At least four hominoid species are represented in the Siwaliks, thought previously to range in age between about 14 Myr and 6 Myr. The four species are *Ramapithecus punjabicus, Sivapithecus sivalensis, S. indicus and Gigantopithecus bilaspurensis.* Until recently species of *Sivapithecus* have been classified in *Dryopithecus,* but it now seems preferable to separate them generically from *D. fontani* and other dryopithecines.

In several recent reviews of Miocene Hominoidea these Siwalik species and others from elsewhere in the Old World have been placed in either the Hominidae or Pongidae. *Ramapithecus* has often been included in Hominidae, while species of the other genera have usually been considered pongids[12]. This arrangement has not been accepted universally.

The new specimens from Pakistan discussed here (Tables 1–3), together with much new material from other Old World localities, facilitate a different view of earlier Neogene hominoid classification and evolution.

The new hominoids are discussed here taxonomically. Assignment of some individuals to species is difficult or impossible, and the scheme outlined is provisional. More comprehensive descriptions and discussions are being prepared. Measurements of the more complete specimens are given in Tables 4 and 5.

Table 4. Tooth measurements on hominoid maxillae

		9977/01		11704	11736	11708
		L	R	R	R	R
I^2	md	–	5.5			
	bl	–	7.5			
$\underline{C}$	max	14.2	13.4	14.6		
	tr	10.8	10.8	10.7		
P^3	md	9.3	9.3	9.2	9.2	9.5
	bl	11.8	14.3	11.2	–	11.6
P^4	md	8.7	7.5	7.7	8.5	8.3
	bl	12.4	12.4	11.7	12.2	12.0
M^1	md	11.5	12.1	11.1	12.0	12.4
	bl	13.3	13.3	12.8	–	13.1
M^2	md	13.5	13.6		12.5	13.7
	bl	14.6	14.2		13.4	14.0

个标本来自哈斯诺特附近，在其他地点也识别出来几个单个的牙齿。克什米尔的拉姆讷格尔是几种人猿超科动物的来源地，但是最好的标本来自印度的哈里塔尔扬加。

西瓦利克至少出现了 4 种人猿超科动物，以往认为它们的生存年代大约在 1,400 万年前到 600 万年前之间。这 4 个种分别是：腊玛古猿旁遮普种、西瓦古猿西瓦种、西瓦古猿印度种及毕拉斯普巨猿。直到最近西瓦古猿的各个种已经被划进森林古猿，但现在看来似乎将它们从种属上与方坦森林古猿及其他的森林古猿亚科分开更合适。

在最近的有关中新世人猿超科动物的几个讨论中，这些西瓦利克种类与其他的来自旧大陆（东半球）其他地方的种类不是被放入人科就是被放入猩猩科。腊玛古猿常常被包括在人科中，而其他属的种类通常被归为猩猩科 [12]。这种分类还未被普遍接受。

此处讨论的来自巴基斯坦的新标本（表 1~3），结合来自旧大陆（东半球）其他地点的许多新标本，都有助于促进关于新近纪早期人猿超科动物分类与进化的不同观点的形成。

在此，从分类学的角度讨论了新的人猿超科动物。要把某些个体指定为某个种类，是困难的或者是不可能的，因此，所概括出的方案也是暂时的。正在准备进行更为全面的描述与讨论。更完整的标本的测量结果列于表 4、表 5。

表 4. 对人猿超科动物上颌牙齿的测量结果

		9977/01		11704	11736	11708
		左	右	右	右	右
I^2	md	–	5.5			
	bl	–	7.5			
C	max	14.2	13.4	14.6		
	tr	10.8	10.8	10.7		
P^3	md	9.3	9.3	9.2	9.2	9.5
	bl	11.8	14.3	11.2	–	11.6
P^4	md	8.7	7.5	7.7	8.5	8.3
	bl	12.4	12.4	11.7	12.2	12.0
M^1	md	11.5	12.1	11.1	12.0	12.4
	bl	13.3	13.3	12.8	–	13.1
M^2	md	13.5	13.6		12.5	13.7
	bl	14.6	14.2		13.4	14.0

Continued

		9977/01		11704	11736	11708
		L	R	R	R	R
$M^{\underline{3}}$	md	12.7	12.7			12.5
	bl	13.7	13.5			13.8

Table 5. Measurements of hominoid mandibles

		9564/9905		13165	4622/4857	4230	9563/9902	6153	6160
$\overline{C}$	max	12.3	(R)						
	tr	10.1							
$P_{\bar{3}}$	max						11.2		11.0
	tr						7.0		7.0
$P_{\bar{4}}$	md	*9.4	(L)					8.1	8.5
	bl	–					–	9.5	9.8
M_1	md	*13.0	(L)	13.0	11.5		*10.5	10.3	11.1
	bl	*12.0		*10.9	9.5		*9.5	9.8	–
$M_{\bar{2}}$	md	14.5	(L)	14.0	12.7	14.7	12.7	12.4	12.7
	bl	12.7		11.7	10.5	12.7	10.7	10.9	10.7
$M_{\bar{3}}$	md	*15.5	(L)	*14.5	12.9	–	–	12.3	13.7
	bl	–		–	10.5	–	–	10.8	–
Breadth at $\overline{C}$		*35			22.5		*25		
Breadth at $M_{\bar{2}}$		*47			48.0		*40		
Symphysis									
Depth		52.5			*30.0		*34		
Thickness		20.0			15.0		*16		
At P4									
Depth		43.5		34.0	30.5		*28		
Thickness		15.5		15.0	13.0		*14		
At M3									
Depth		42.5			31.0	30.5			
Thickness		24.0			20.0	23.5			

*Estimated values.

At least one species of *Ramapithecus*, *R. punjabicus* (Fig. 1), is represented in Siwalik rocks; the genus is found in the Chinji and Nagri Formations, and possibly in rocks equivalent to the Dhok Pathan Formation. The probable age range is 13–8.5 Myr.

续表

		9977/01		11704	11736	11708
		左	右	右	右	右
M^3	md	12.7	12.7			12.5
	bl	13.7	13.5			13.8

表 5. 人猿超科动物下颌骨的测量结果

		9564/9905		13165	4622/4857	4230	9563/9902	6153	6160
$\overline{C}$	max	12.3	（右）						
	tr	10.1							
$P_{\overline{3}}$	max						11.2		11.0
	tr						7.0		7.0
$P_{\overline{4}}$	md	*9.4	（左）					8.1	8.5
	bl	–					–	9.5	9.8
M_1	md	*13.0	（左）	13.0	11.5		*10.5	10.3	11.1
	bl	*12.0		*10.9	9.5		*9.5	9.8	–
$M_{\overline{2}}$	md	14.5	（左）	14.0	12.7	14.7	12.7	12.4	12.7
	bl	12.7		11.7	10.5	12.7	10.7	10.9	10.7
$M_{\overline{3}}$	md	*15.5	（左）	*14.5	12.9	–	–	12.3	13.7
	bl	–		–	10.5	–	–	10.8	–
在$\overline{C}$的宽度		*35			22.5		*25		
在 $M_{\overline{2}}$的宽度		*47			48.0		*40		
骨联合部									
深度		52.5			*30.0		*34		
厚度		20.0			15.0		*16		
在 P_4									
深度		43.5		34.0	30.5		*28		
厚度		15.5		15.0	13.0		*14		
在 M_3									
深度		42.5			31.0	30.5			
厚度		24.0			20.0	23.5			

* 为估计值。

至少有一种腊玛古猿——腊玛古猿旁遮普种（图 1）出现在西瓦利克岩石中；在成吉与纳格里组中发现这个属，而且可能在相当于多克帕坦组的岩石中也有发现。其年代范围可能在 1,300 万年前到 850 万年前。

The new material from the Potwar Plateau helps considerably in understanding previous finds as well as adding significant new information. Particularly fine specimens are the adult mandibles GSP 4622/4857 from locality 182 and GSP 9562/9902 from locality 260, and an infant mandible, GSP 12709, also from locality 260. These specimens show that the incisor region was very narrow in *Ramapithecus*, canines small, anterior premolars and canines closely packed together, and postcanine tooth rows and mandibular corpora posteriorly divergent. Most teeth in the upper and lower dentitions are now known. Certain features are worth noting: $P_{\bar{3}}$ has a small but distinct lingual cusp, and its long axis is oriented at some 45° to the mesio-distal line of the tooth row. Unworn cheek teeth resemble those of *Australopithecus* and *Homo* quite strongly, particularly the maxillary molars. Occlusal surfaces are constricted and there is marked buccal (mandibular teeth) and lingual (maxillary) flare in unworn specimens (compare ref. 13). Occlusal surfaces broaden with wear and can be almost flat and still show no dentine. This is because occlusal enamel is very thick (between 2.5 and 3.0 mm on mandibular buccal cusps), as shown in a few broken specimens (GSP 8926) and by comparison of unworn and worn homologues of similar size. Mandibular rami are relatively robust; symphyses have marked superior and inferior transverse tori.

Gigantopithecus bilaspurensis (Fig. 1) is based on a complete mandible from the Haritalyangar area. It is likely to have an age of around 8.5 Myr[1,2,10], and is not significantly younger than the other hominoid primates from that area.

One partial molar from locality 311, GSP 7144, is tentatively assigned to *Gigantopithecus*, as is a previously described molar, GSI D175 from Alipur near Hasnot[14]. Both are probably 9.0-10 Myr old. Occlusal morphology is rather similar to that of *Ramapithecus*, and occlusal surface enamel is thick (about 3.5 mm on the mandibular buccal cusp of GSP 7144).

Specimens of *Sivapithecus indicus* (Fig. 2) are known from deposits ranging between about 13 and 8 Myr. Several relatively complete new specimens (especially GSP 9977/01/05/9564 from locality 260) facilitate mandibular and lower facial reconstructions. Tooth rows are subparallel with broad incisor regions and postcanine tooth rows that are markedly concave bucally.

Occlusal morphology in *S. indicus* resembles the other genera, although molars are somewhat broader and there are other minor differences. Occlusal surfaces are constricted on cheek teeth with marked buccal and lingual flare; enamel is thick (about 3 mm on mandibular buccal cusps).

Canines are projecting and moderately dimorphic and exhibit mesial, distal and apical wear; $P_{\bar{3}}$s are closely approximated to the canines, their long axes rotated to lie about 45° to the mesio-distal line of the tooth row. Mandibular rami are deep; the symphysis is long with a prominent inferior transverse torus and a relatively small superior transverse torus.

S. sivalensis is the most enigmatic of the Siwalik hominoids and it is not absolutely clear that it exists as a separate species.

来自博德瓦尔高原的新材料，对理解以前的发现大有帮助，还增加了新的重要信息。来自 182 号地点的 GSP 4622/4857 与来自 260 号地点的 GSP 9562/9902 的成年下颌骨，还有来自 260 号地点的 GSP 12709 的一块幼体下颌骨，都是保存很好的标本。这些标本显示，腊玛古猿的门齿区十分窄，犬齿小，前面的前臼齿与犬齿紧紧地挤在一起，后犬齿齿列与下颌体向后分开。现在已知上下齿列的大部分牙齿。其某些特征值得注意：P_3的舌侧齿尖虽然小但明显，其长轴与齿列中远端线大约 45°。未受磨损的颊齿，特别是上颌的臼齿，与南方古猿及人属的颊齿极其相似。上下齿咬合面缩小，未受磨损的标本其颊（下颌齿）与舌侧（上颌）倾斜明显（对比参考文献 13）。上下齿咬合面因磨损而变宽，几乎是平的，却仍未露出齿质。正如几个破损的标本（GSP 8926）所表现出来的以及通过对比未受磨损及磨损的程度类似的同源结构所看到的，这是因为上下牙咬合面的釉质非常厚（下颌骨颊侧齿尖，厚度介于 2.5~3.0 毫米之间）。下颌支相当粗壮；骨联合部的上下横向隆凸明显。

毕拉斯普巨猿（图 1）的确认是基于来自哈里塔尔扬加地区的完整的下颌骨。其可能的年代距今约 850 万年[1,2,10]，并不比来自那个地区的人猿超科灵长类动物年轻多少。

一颗来自 311 号地点的 GSP 7144 的部分臼齿被暂时定为是巨猿的，因为这与此前曾被描述的，来自靠近哈斯诺特的阿里布尔的 GSI D175 的臼齿类似[14]。两者都可能生存于 1,000 万年前到 900 万年前之间。上下齿咬合面的形态与腊玛古猿上下齿咬合面的形态相当类似，且上下齿咬合面的牙釉质厚（在 GSP 7144 的下颌颊侧齿尖处厚大约 3.5 毫米）。

从大约 1,300 万年前到 800 万年前的沉积物中识别出了西瓦古猿印度种的标本（图 2）。几个相对完整的新标本（特别是来自 260 号地点的 GSP 9977/01/05/9564）使下颌及下面部复原变得容易。齿列近似平行于宽的门牙区而颊齿列在颊侧有明显的凹入。

虽然臼齿稍微有点宽而且也存在其他小的差别，西瓦古猿印度种的上下齿咬合面的形态仍类似于其他属。颊齿的上下咬合面缩小，有明显的颊侧与舌侧倾斜；牙釉质厚（下颌骨颊侧齿尖处厚约 3 毫米）。

犬齿突出，适度二态，中、末端及顶上有磨损；P_3非常靠近犬齿，其长轴旋转，其与齿列近中—远中线大约 45°。下颌支深；骨联合部长，其下横隆凸突出，上横隆凸相对较小。

西瓦古猿是最令人困惑的西瓦利克人猿超科动物，甚至不能清楚的确定其能作为一个单独的种类而继续存在。

Parts of 13 hominoid postcranial bones (Fig. 3) have been collected so far and fit into three size groups. The largest specimen is a partial distal right humerous, GSP 12271, with the lateral supracondylar ridge, lateral epicondyle, capitulum and radial fossa preserved entire, and the coronoid and olecranon fossae and trochlear surface preserved in part. In its size and morphological features this specimen is similar to the distal humerus of adult female gorillas. Eight specimens are close in size to the corresponding bones of adult pygmy chimpanzees. This group includes a partial distal right humerus, GSP 6663, with parts of the capitular and trochlear surfaces preserved. Hindlimb specimens include a femoral head, GSP 6178, a femoral head plus part of the neck, GSP 11867, the proximal, mid-shaft, and distal non-articular parts of a right femur, GSP 12654, a left intermediate cuneiform, GSP 6454, and the distal end of a proximal hallucial phalanx, GSP 6664. These specimens, together with two distal ends of phalangeal bones, GSP 6666 amd GSP 13168, show some morphological similarities to extant hominoid species, although these are not as marked as in the case of the large specimen. Three fossils are smaller. One of them, a partial left radius, GSP 7611, is clearly juvenile. A partial right talus, GSP 10785, which includes most of the trochlear surface, the lateral process, and most of the posterior calcaneal articular surface, and a partial femoral head, GSP 9894, come from approximately adult macaque-sized animals. It is not certain that they are from adults. A partial calcaneus, GSP 4664, shows a number of equivocal morphological features and is not included in any of the three groups.

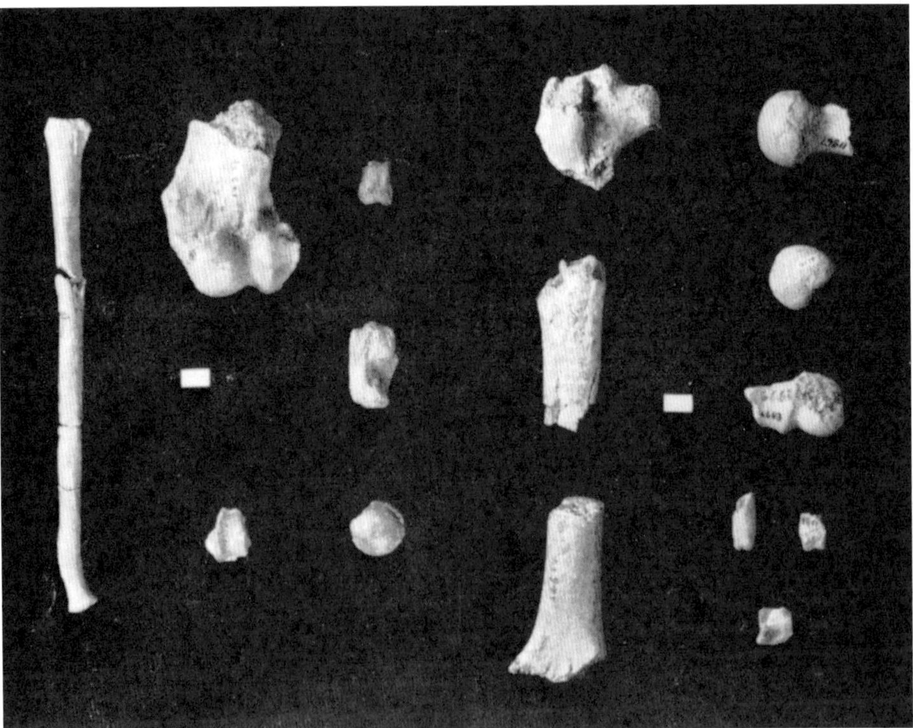

Fig. 3. Left to right, first column: GSP 7611; second column: top, GSP 12271, bottom, GSP 10785; third column: top, GSP 6664, middle, GSP 4664, bottom, GSP 9894; fourth column, GSP 12654; fifth column, top to bottom, GSP 11867, 6178, 6663, 6666 and 13168, 6454.

迄今为止，已经采集到13个人猿超科动物的颅后骨骼（图3）部分，按大小刚好放入3组。最大的标本是不完整的右肱骨远端，GSP 12271，带有保存完整的侧髁上脊、侧上髁、肋骨小头与桡骨窝，喙突窝与鹰嘴窝以及滑车面部分保存。在其大小与形态特征上，这个标本类似于成年雌性大猩猩的肱骨远端。8个标本大小上接近于成年矮小黑猩猩的对应骨头。这一组包括不完整的右肱骨远端，GSP 6663，其小头的部分与滑车面被保存下来。下肢标本包括股骨头，GSP 6178；股骨头加部分股骨颈，GSP 11867；右股骨近端、骨干中部及远端非关节部分，GSP 12654；左中间楔状骨，GSP 6454；以及大拇趾骨的远端，GSP 6664。这些标本及其他两段指骨远端一起，GSP 6666与GSP 13168，都在形态上显示出了与现生的人猿超科种类的类似，虽然这些类似并不如在大标本的情形下那样明显。3块化石较小。其中一块明显是青少年个体的不完整的左桡骨，GSP 7611；GSP 10785是右距骨的局部，包括大部分滑车面、外侧突、大部分后跟骨关节面；它和一个不完整的股骨头，GSP 9894，都来自近似于成年猕猴大小的动物。不能肯定它们是来自于成年个体。不完整的跟骨，GSP 4664，表现出大量意义不明确的形态特征，并不包括在上述3个组的任一组中。

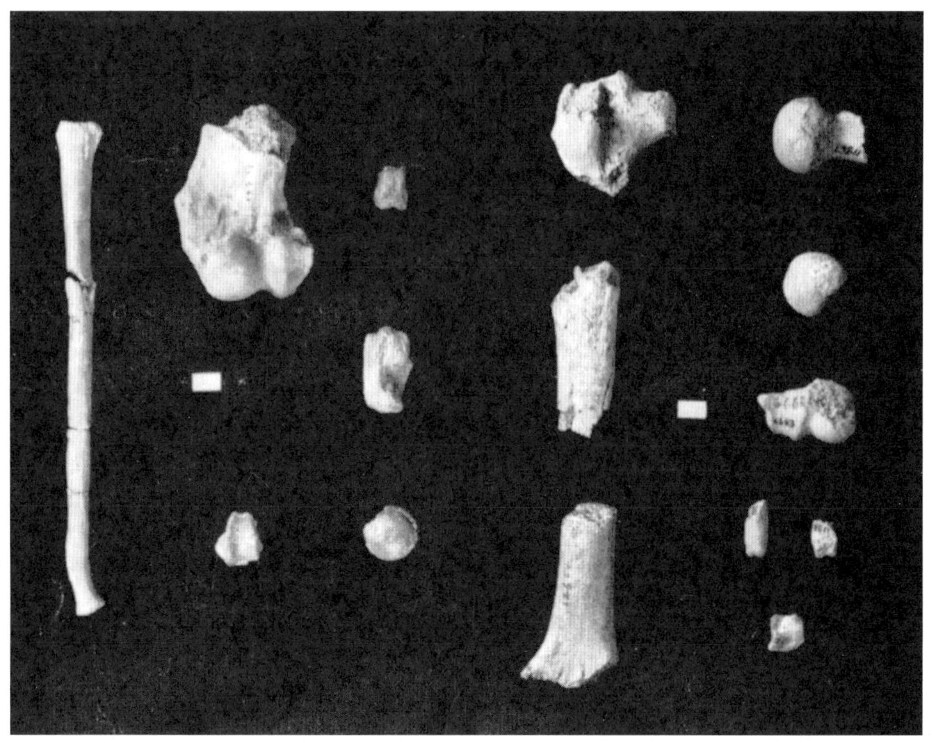

图3. 从左到右，第1列：GSP 7611；第2列：上，GSP 12271，下，GSP 10785；第3列：上，GSP 6664，中，GSP 4664，下，GSP 9894；第4列：GSP 12654；第5列：从上到下，GSP 11867，6178，6663，6666及13168，6454。

None of these specimens was found in direct association with cranial or dental material. Because only three different sized primate species are definitely represented by cranial and dental specimens, however, the largest postcranial specimen is provisionally assigned to *Gigantopithecus* cf. *bilaspurensis*, the intermediate sized group to *Sivapithecus indicus*, and the smaller specimens to *Ramapithecus punjabicus* (remembering that some of these specimens may represent juveniles of the middle size group). It is significant that the *Gigantopithecus* humeral fragment and many of the *Sivapithecus* specimens come from locality 311; dental remains from that locality include our only *Gigantopithecus* specimen and five *Sivapithecus indicus* (Table 2).

If the postcranial remains are correctly subdivided and if they are associated with the three size groups based on gnathic and dental remains, some interesting conclusions can be drawn.

First, all hominoid species in the Siwaliks apparently have cheek teeth with very thick occlusal enamel. Second, all are truly megadont, in that cheek teeth are very large relative to body size. In these two features, the Siwalik hominoids resemble Plio-Pleistocene hominids and differ from pongines and the Miocene hominoids now assigned to Dryopithecinae (*sensu strictu,* see below). Third, these species share a basically similar occlusal morphology.

The Siwalik hominoids subdivide into two groups, mainly on the basis of anterior tooth size. *S. indicus* has relatively large incisors and, like *S. sivalensis,* large dimorphic canines; *Ramapithecus* and *Gigantopithecus* species have relatively small incisors, (probably) non-projecting and (probably) moderately dimorphic canines.

Other Hominoid Material

New ideas and new hominoid fossil remains recovered during the past decade suggest that earlier, simpler schemes of hominoid evolution need to be modified. These changes can be summarised briefly as follows.

(1) Important dental differences between later hominids and living pongids lie less in arcade shape and incisor size as often stated in the past, but in relative tooth size and occlusal enamel thickness. Living apes have U-shaped dental arcades, living humans parabolic ones. Such arcade shapes are rarely found in other Neogene hominoids, the predominant form being some variant of a V-shape. Incisor size also seems to have been rather variable at all phases of hominoid evolution. More important, the apes, like almost all other non-human higher primates, have relatively small cheek teeth with thin enamel, perhaps an adaptation to predominantly browsing diets. Pliocene and earlier Pleistocene hominids in contrast have large cheek teeth with thick enamel (*Homo sapiens* has evolved small cheek teeth relatively recently). Apes have large tusk-like sexually dimorphic canines; hominids have small, somewhat incisiform canines exhibiting considerably less sexual dimorphism.

并没有发现这些标本中的任何一个与颅骨或牙齿材料直接共生。因为仅由颅骨与牙齿标本明确表示出了有 3 种不同大小的灵长类动物。然而，最大的颅后标本被暂时归入毕拉斯普巨猿相似种，中间大小的组被归入西瓦古猿印度种，较小的标本则被归入腊玛古猿旁遮普种（值得提醒的是，这些标本中的某些可能是中等大小组的青少年个体）。巨猿的肱骨残片与许多来自 311 号地点西瓦古猿的标本意义重大；来自该地点的牙齿标本包括我们唯一的巨猿标本与 5 个西瓦古猿印度种的标本（表 2）。

如果将颅后骨骼化石再进行恰当的细分，或能将它们与根据颌与牙齿的化石划分出的 3 个大小组别相关联，就可能得出一些有趣的结论。

第一，西瓦利克的所有人猿超科动物有特征明显的颊齿，其上下齿咬合面的牙釉质十分厚。第二，都是真正的巨型牙，相对其身体尺寸，颊齿显得很大。在这两个特征上，西瓦利克人猿超科动物与上新世 – 更新世人科成员类似，而与猩猩科动物以及现在被归入森林古猿亚科（严格意义上的，见下文）的中新世人猿超科动物不同。第三，这些种类的上下齿咬合面在形态上都基本类似。

主要根据前部齿大小，把西瓦利克人猿超科动物再细分为两组。西瓦古猿印度种拥有相对较大的门牙，与西瓦古猿西瓦种相像，犬齿二态性大；腊玛古猿与巨猿种的门齿相对较小，犬齿（可能）不突出并且（可能）适度二态。

其他人猿超科动物材料

新的概念以及过去十年间所发现的新人猿超科动物的化石表明，需要对过去关于人猿超科动物进化的较早的、较简单的方案进行修改。这些修改简要概述如下：

（1）在后来的人科动物与现生猩猩科类人猿之间，其牙齿的重要差异，并不像过去常常所说的那样仅仅在于齿弓形状与门牙的大小，而更在于牙齿的相对大小及其上下齿咬合面牙釉质的厚度。现生猿其齿弓呈 U 形，现生人类呈抛物线形。在其他新近纪人猿超科动物上这样的齿弓形状罕见，其主要形状是少许变形了的 V 形。在人猿超科动物进化的所有阶段，门齿的大小也似乎有相当大的变异。更重要的是，与几乎所有其他非人的高等灵长类动物一样，猿的颊齿相当小，其牙釉质薄，可能是为了适应其以草食为主的食性。相反的，上新世与早更新世的人科动物已经具备了大的颊齿和厚的牙釉质（在相对更晚近的时期，智人已经进化出小的颊齿）。猿的犬齿大，像獠牙，性别上二态；人科动物犬齿小，稍稍似门齿的形状，表现出来极少的性别二态现象。

(2) New discoveries in east Africa, and new analyses of earlier finds in Africa and Europe, suggest that early apes were considerably more diverse than previously believed[15]. Classifying them in only one genus, *Dryopithecus*, obscures this diversity. Thus at least three genera (*Proconsul*, *Rangwapithecus* and *Limnopithecus*) should probably be recognised in east Africa during the early and middle Miocene, separated from *Dryopithecus*, at least two species of which are found in middle Miocene deposits in Europe. These species can all be placed conveniently in the Dryopithecinae (or Dryopithecidae). All have dentitions basically like those of the living African pongids, with thin enamel and, if postcranial remains are correctly allocated, relatively small cheek teeth. Canine-premolar complexes are like those of modern apes.

Postcranial material of this group has proved difficult to interpret in a framework heavily dependent on comparisons with modern primate groups[16]. The fossil species were probably arboreal, and seem to be qualitatively different from living apes, being more "monkey-like" in certain ways; they are probably best viewed as truly primitive relative to modern hominoids.

Where palaeoecological contexts can be inferred plausibly, dryopithecine species seem to have been associated with predominantly forest floras and faunas[15]. Adaptively this rather diverse group was probably more like living ceboids or cercopithecoids than the low-diversity modern hominoids.

(3) Besides dryopithecines, other kinds of hominoids are present in the Old World middle Miocene. New material from Hungary[17], Greece[18], Turkey[19], Kenya[20] and China together with the specimens described here from Pakistan as well as earlier fossils from Europe and Africa show that thick-enamelled and (at least for those with postcranial remains) megadont hominoid species were widely distributed between 14 and 8 Myr ago, probably in predominantly non-forested habitats. During this time cercopithecoid monkeys seem to have been absent from Eurasia, and not particularly diverse in Africa[21].

This thick-enamelled middle Miocene cluster of species can be divided into two groups. One, consisting of species variously described as *Sivapithecus*, *Bodvapithecus*, *Ankarapithecus* and *Ouranopithecus* contains forms that are rather more ape-like, with large and sexually dimorphic canines. More than one generic name is probably necessary to reflect adequately the diversity of this group, which would be termed Sivapithecinae (or Sivapithecini). The other group would contain species described as *Gigantopithecus*, *Ramapithecus* and *Rudapithecus* (at least), forms with canines smaller and less dimorphic than those of pongines, dryopithecines or *Sivapithecus*-group species, although perhaps more so than australopithecines. Canine-premolar complexes resemble those of "primitive" australopithecines[22]. This cluster would be termed Ramapithecinae (or Ramapithecini).

(4) Recent discoveries suggest very strongly that the story of the Plio-Pleistocene hominids was more complex than previously thought. Between about 3.75 and 1 Myr ago several hominid lineages seem to have coexisted, certainly in Africa and possibly in Asia too[23].

（2）在东非的新发现，以及对非洲与欧洲早期发现的最新分析表明，早期猿类的种类多样性要远胜于人们以往所认为的[15]。把它们仅仅划分为森林古猿这一个属，就使这种多样性变得模糊。因而，在东非中新世早期与中期，至少应该可能识别出 3 个可以从森林古猿属中独立出来的属（原康修尔猿、腊玛古猿及湖猿），其中至少两种发现于欧洲中新世中期的沉积物中。这些种类全都可以被方便地放入森林古猿亚科(或森林古猿科)。所有的齿系基本上都像现生非洲猩猩类的齿系，牙釉质薄且颊齿相对较小（如果对保存的颅后骨骼定位准确的话)。犬齿 – 前臼齿复合体与现代猿的相像。

已经证明，要在十分依赖于与现代灵长类动物组群对比的框架内[16]对这一组的颅后材料进行解释是很困难的。化石种可能是树栖的，而且从性质上而言，似乎不同于现生猿，在某些方面更“像猴”；相对于现代人猿超科动物，它们可能最好被视为是真正原始的。

森林古猿种类似乎一直主要是与森林植物群及动物群相联系[15]，由此似乎可以推测出古生态环境来。与此相适应的是，这个变化相当大的组群可能更像现生的卷尾猴科或猕猴科动物，而不太像分化程度低的现代人猿超科。

（3）除了森林古猿类之外，其他种类的人猿超科动物出现在中新世中期的旧大陆（东半球)。来自匈牙利[17]、希腊[18]、土耳其[19]、肯尼亚[20]及中国的新标本，跟在此所描述的出自巴基斯坦的标本，还有来自欧洲与非洲的早期化石一起显示出，在 1,400 万年前到 800 万年前，牙釉质厚的与（至少对那些有颅后化石的）巨型牙的人猿超科动物可能在以非森林占优的生境中是广泛分布的。在此期间，猕猴科的猴子似乎已经不存在于欧亚大陆，而且在非洲种类变化也不是特别多[21]。

这一牙釉质厚的中新世中期的种类组群，能被分为两个组。其中一组，包括更像猿的形式，拥有大的、具性别二态犬齿的种类，由被描述成为西瓦古猿、波德瓦古猿、安卡拉古猿及乌朗诺古猿的不同种类组成。可能需要不止一个属名才能充分反映这个组的多样性，这个组被称为西瓦古猿亚科（或西瓦古猿族)。另一组将包括被描述为巨猿、腊玛古猿及鲁达古猿（至少）的种类，与猩猩科动物、森林古猿亚科或西瓦古猿组种类相比，这些种类的犬齿较小，性别二态较少，与南方古猿亚科动物相比，更是如此。犬齿 – 前臼齿复合体类似于“原始”南方古猿的犬齿 – 前臼齿复合体[22]。这个组群被称为腊玛古猿亚科（或腊玛古猿族)。

（4）最近的发现强有力地表明上新世—更新世的人科动物的发展历史比以往所认为的要复杂得多。大约在 375 万年前到 100 万年前，几个人科动物的谱系似乎共

These species have been classified in both *Australolithecus* and *Homo,* but are characterised by certain shared features: reduced canines exhibiting moderate size dimorphism, enlarged and thick-enamelled cheek teeth, relatively enlarged brains (compared with dryopithecines and pongines at least), and a postcranial skeleton showing numerous adaptations to habitual bipedalism. At least one of these species, by at least 2 and perhaps 2.5–3 Myr ago, made stone tools.

It has been increasingly realised that these socalled Plio-Pleistocene hominids are not to be regarded merely as "diminutive humans" but as creatures, although recognisably hominid, qualitatively different from middle and late Pleistocene hominids. They are a diverse group of truly primitive species.

The oldest specimens with teeth similar to those of *Australopithecus* or early *Homo* species (and distinguishable from those of the *Sivapithecus* or *Ramapithecus* groups) come from Lukeino and Lothagam in Kenya and are between 5 and 7 Myr old[24,25].

The period after about 7 Myr ago is one during which australopithecines and early hominines diversified; *Sivapithecus*-group and *Ramapithecus*-group species are so far unknown. In both Africa and Eurasia cercopithecoid monkeys, especially the more open-country types, become abundant for the first time.

Synthesis

Current knowledge of hominoid evolution suggests a tentative taxonomic and phylogenetic scheme that reflects the important advances of the past decade. These are summarised here by D.P.

First, Neogene hominoids were, during most of their evolution, a relatively diverse superfamily. Second, extinct hominoids were not identical with, nor, in some cases, particularly similar to living hominoids, and to interpret extinct hominoids as though they were very "modern" is potentially misleading. Third, it is very difficult to draw exact ancestor-descendant relationships given the complexity of the picture at any given time, the differences between descendants and available ancestral candidates, and the still substantial gaps in the fossil record. Rather, each radiation should be studied for its own sake in order to understand it as an adapted and successful group. Fourth, clearly the two taxonomic categories into which living large hominoids are normally subdivided, Pongidae and Hominidae, cannot be imposed on earlier Neogene species without suppressing important information and obscuring evolutionary concepts.

A minimum of six clusters of species is needed to describe the diversity of Neogene large hominoids, and these can be classified as tribes or subfamilies.

(1) Ponginae: the living great apes. (2) Dryopithecinae: species from Africa and Europe, ranging in age from about 23 to 9 Myr, sharing dental features with pongines but differing

同存在，在非洲肯定是如此，在亚洲也可能是这样[23]。这些种类已被分类为南方古猿与人属，但是它们仍有某些共同特征：简化的犬齿在尺寸上表现出适度的二态，颊齿增大、其牙釉质厚，脑量相对增大（至少是与森林古猿及猩猩科动物对比），颅后骨骼显示出众多适应两足行走的改变。这些种类中至少有一种，在至少 200 万年前甚至可能是 250 万到 300 万年以前，就可以制造石器。

现在已经越来越多地认识到这些通常所谓的上新世－更新世人科动物，虽然可被识别为人科，但不能仅仅被认为是“小型人类”，而应该从定性的方面上被看作是不同于中、晚更新世的人科动物的生物。它们是和真正原始种类不同的组。

最古老的标本，其牙齿类似于南方古猿或早期人属种类的牙齿（显然不同于西瓦古猿或腊玛古猿组的牙齿），它们出自肯尼亚的路奇诺与洛萨加姆，年代距今约 500 万年至 700 万年[24，25]。

南方古猿与早期人亚科是在大约 700 万年前之后的这个时期内发生了分化；而迄今为止尚不明确西瓦古猿组与腊玛古猿组的种类。在非洲与欧亚大陆，猕猴，尤其是生活在更为开阔的空间的类型，第一次变得繁盛。

综合分析

依据对人猿超科进化的现有认识，尝试提出了一个反映过去十年在分类和系统上的重要进展的方案。这些由戴维·皮尔比姆在此总结。

第一，新近纪的人猿超科动物，在其大部分进化过程中，是一个种类相对多样的超科。第二，灭绝了的人猿超科与现生人猿超科不会完全相同，也不会在某些情形下特别类似于现生人猿超科，而把灭绝了的人猿超科阐述得仿佛十分“现代”，这可能会产生误导。第三，考虑到情况的复杂性、后代类型与可选择的祖先类型之间的差异以及化石记录上仍然存在的很大的空白，要准确地勾画出祖先与后代的关系，是十分困难的。当然，为了认定它是一个适合的、成功的组，应该研究它的每一支辐射。第四，无疑，猩猩科与人科作为两个分类学单元，其中现生的大人猿超科通常再细分，这样的分类方法是不能强加于新近纪早期种类上的，那就会干扰重要信息，并使进化概念模糊不清。

需要最少 6 个种类集群来描述新近纪人猿超科的多样性，它们能被分为族或亚科。

（1）猩猩亚科：现生大猿。（2）森林古猿亚科：来自非洲与欧洲的种类，年代大约在 2,300 万年前至 900 万年前，其牙齿与猩猩科动物有共同特点，但许多颅后特征不同；（3）西瓦古猿亚科或西瓦古猿族：这是来自欧亚大陆与非洲的种类，年代范

in many postcranial characters; (3) Sivapithecinae or Sivapithecini: species from Eurasia and Africa, ranging in age from around 15 to 8 Myr, sharing dental features with both pongines and dryopithecines on the one hand and with australopithecines and early hominines on the other; (4) Ramapithecinae or Ramapithecini: Eurasian and African species resembling sivapithecines and australopithecines about equally in dental features; (5) Australopithecinae: species from Africa and possibly Asia ranging in age from at least 3.75 (and perhaps as much as 7) to 1 Myr; (6) Homininae: essentially similar to Australopithecinae in dental, cranial and postcranial morphology.

Figure 4 shows these groups distributed along temporal and qualitative morphological axes; the latter axis is shown as unidimensional although it is, of course, multidimensional. Three alternative phylogenies are indicated on the diagram in order of decreasing probability (in my opinion). Although definite relationships between the middle Miocene and Plio-Pleistocene hominoids are unclear because of gaps in the hominid record between 8 and 4 Myr and the pongid record after 9 Myr, I believe that the major groups as defined here represent at least some of the grades through which modern hominoids evolved. Phylogeny *A* is the most probable; I regard phylogenies *B* and *C* as less probable, although plausible.

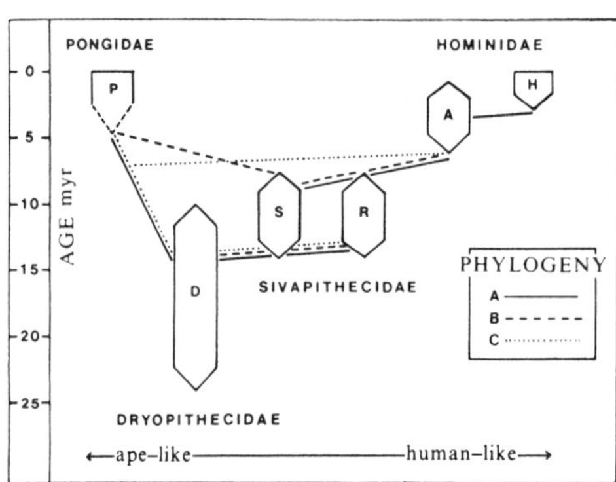

Fig.4. Tentative stratigraphic-morphological distribution of large hominoid groups with possible phylogenies. P, Pongidae; D, Dryopithecidae; S, Sivapithecinae, R, Ramapithecinae; A, Australopithecinae; H, Homininae. (Sivapithecidae should be Ramapithecidae; *note added in proof.*)

I suggest the following classification of large Hominoidea, a compromise of "vertical" and "horizontal" philosophies[26]. (1) Pongidae; (2) Dryopithecidae: including Dryopithecinae and older, more primitive hominoids; (3) Ramapithecidae: Sivapithecinae (or Sivapithecini) and Ramapithecinae (or Ramapithecini); (4)Hominidae: Australopithecinae and Homininae.

The classification is flexible in that it is compatible with any of the three phylogenies suggested in Fig 4. Whether or not the four groups are classified as families, subfamilies or tribes, they should be coordinate. As the fossil record improves I expect that certain (perhaps known) members of Ramapithecidae and Hominidae will prove more conclusively to be ancestors and descendants (that is phylogeny *A* in Fig. 4 becomes more probable), in

围大约在 1,500 万年前到 800 万年前，其牙齿一方面与猩猩科动物及森林古猿亚科有共同特点，另一方面与南方古猿亚科及早期人亚科有共同特点；(4) 腊玛古猿亚科或腊玛古猿族：欧亚大陆与非洲的种类，在牙齿特征上，与西瓦古猿亚科及南方古猿亚科的类似程度相当；(5) 南方古猿亚科：这是来自非洲的种类，也可能来自亚洲，年代范围从至少 375（可能差不多是 700）万年前至 100 万年前；(6) 人亚科：在牙齿、头骨及颅后形态上基本上类似于南方古猿亚科。

图 4 是这些组沿时间轴与定性形态轴的分布图示；虽然定性形态轴毫无疑问是多维的，但这里用一维表示。3 个备选系统发育史在图上依概率降序排列（依我看来）。虽然，由于 800 万年前至 400 万年前人科动物记录的缺失，以及 900 万年前以后猩猩科类人猿记录的缺失，还不清楚中新世中期的人猿超科与上新世－更新世的人猿超科之间的确切关系，但是我相信这里详细说明的主要的组至少代表了现代人猿超科进化的一些阶段。生物系统发育史 A 最为接近；我认为生物系统发育史 B 与 C 可能性小些，虽然它们也似乎有可能。

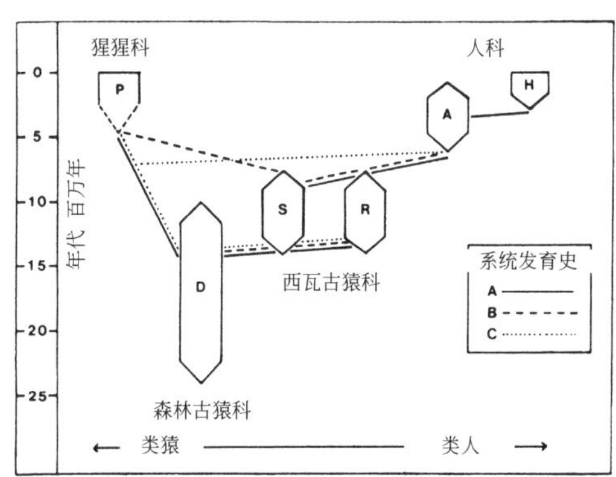

图 4. 反映可能的系统发育史的超大人猿超科群的尝试性地层－形态学分布图。P，猩猩科；D，森林古猿科；S，西瓦古猿亚科；R，腊玛古猿亚科；A，南方古猿亚科；H，人亚科。（附加说明：图中“西瓦古猿科”应为“腊玛古猿科”）

对“垂直”与“水平”基本原理进行折中 [26]，我提出如下大的人猿超科分类。(1) 猩猩科；(2) 森林古猿科：包括森林古猿亚科与更古老、更原始的人猿超科；(3) 腊玛古猿科：西瓦古猿亚科（或西瓦古猿族）与腊玛古猿亚科（或腊玛古猿族）；(4) 人科：南方古猿亚科与人亚科。

此分类是很灵活的，它与图 4 提出的 3 个生物的种系发展史中的任何一个都相容而不矛盾。无论 4 个组是否被分为科、亚科或族，它们都应该是并列的。随着化石记录的改善，我希望某些（可能已知）腊玛古猿科与人科的成员能最后被证明是祖先和后代（即图 4 中的生物种系发展史 A 变得更有可能），这种情形下，一些研

which case some workers might wish to include those (or all) ramapithecids in Hominidae. Finally, two important points should be re-emphasised. A rigidly dichotomous classification of Neogene large hominoids is not only inappropriate but potentially misleading. And the living hominoids are rather aberrant forms when the total array of Neogene large hominoids is considered.

This work was supported by NSF, SFCP, the Government of Pakistan, the Geological Survey of Pakistan, Yale University and NERC. Special thanks go to E. L. Simons and B. Lipschutz.

(**270**, 689-695; 1977)

David Pilbeam*, Grant E. Meyer*, Catherine Badgley*, M. D. Rose†, M. H. L. Pickford‡, A. K. Behrensmeyer§ and S. M. Ibrahim Shah¶

* Departments of Anthropology and Geology and Geophysics and Peabody Museum of Natural History, Yale University, New Haven, Connecticut 06520

† Section of Gross Anatomy, Department of Surgery, Yale Medical School, New Haven, Connecticut 06520

‡ Department of Geology, Queen Mary College, London, UK

§ Division of Earth Sciences, University of California, Santa Cruz, California 95064

¶ Geological Survey of Pakistan, Quetta, Pakistan

Received 8 August; accepted 11 October 1977.

References:

1. Pilbeam, D. *et al. Nature* **270**, 684-689 (1977).
2. Pilbeam, D. *Les Plus Anciens Hominidés* (eds Tobias, P. V. and Coppens, Y.) 39-59 (Centre National de la Recherche Scientifique, Paris, 1976).
3. Berggren, W. A. & Van Couvering, J. *Palaeogeogr. Palaeoclimatol. Palaeoecol.* **16**, no. 1/2 (1974).
4. Krynine, P. *Am. J. Sci.* **34**, 422-446 (1937).
5. Lewis, G. E. *Am. J. Sci.* **33**, 191-204 (1937).
6. Tattersall, I. *Nature* **221**, 451-452 (1969); **224**, 821-822 (1969).
7. Prasad, K. N. *Nature* **232**, 413-414 (1971).
8. Banerjee, D. *Rev. Palaeobotan. Palynol.* **6**, 171-176 (1968).
9. Nandi, B. *Himalayan Geol.* **5**, 411-424 (1975).
10. Johnson, G. D. *Geol. Rundschau.* **66**, 192-216 (1977).
11. Schaller, G. *The Deer and the Tiger* (Chicago University Press, 1967).
12. Simons, E. L. *J. Hum. Evol.* **5**, 511-528 (1976).
13. Simons, E. L. *Proc. Natl. Acad. Sci. U.S.A.* **51**, 528-535 (1964).
14. Pilgrim, G. E. *Rec. Geol. Surv. India* **45**, 1-74 (1915).
15. Andrews, P. & Van Couvering, J. *Approaches to Primate Paleobiology* (ed. Szalay, F.), 62-103 (Karger, Basel, 1975).
16. McHenry, H. & Corruccini, R. *Folia Primat.* **23**, 227-244 (1975).
17. Kretzoi, M. *Nature* **257**, 578-581 (1975).
18. de Bonis, L. & Melentis, J. *C. r. hebd. Séanc. Acad. Sci.* **284**, 1393-1396 (1977).
19. Tekkaya, I. *Bull. Min. Res. Expl. Inst. Turkey* no. **83**, 148-165 (1974).
20. Andrews, P. & Walker, A. in *Human Origins* (eds Isaac, G. & McCown, E.) 279-304 (Benjamin, New York, 1976).
21. Delson, E. *Approaches to Primate Paleobiology* (ed. Szalay, F.) 167-217 (Karger, Basal, 1975).
22. Johanson, D. & Taieb, M. *Nature* **260**, 293-297 (1976).
23. Leakey, R. & Walker, A. *Nature* **261**, 572-574 (1976).
24. Behrensmeyer, A. *Earliest Man and Environments in the Lake Rudolf Basin* (eds Coppens, Y., Howell, F., Isaac, G. & Leakey, R.) 163-170 (Chicago University Press, 1976).
25. Pickford, M. *Nature* **256**, 279-284 (1975).
26. Simpson, G. G. *Principles of Animal Taxonomy* (Columbia University Press, New York, 1961).

究人员可能希望列入人科中的那些（或所有的）腊玛古猿。最后，要再次强调两个重点。把新近纪人猿超科动物严格地分成两类，不仅不合理，还有可能产生误导。如果考虑到新近纪人猿超科动物的总序列，则现生人猿超科是相当的畸形的。

本研究由美国国家科学基金会、史密森外币项目、巴基斯坦政府、巴基斯坦地质调查所、耶鲁大学以及英国自然环境研究理事会资助。特别感谢西蒙斯与利普许茨。

（田晓阳 翻译；林圣龙 审稿）

Pliocene Footprints in the Laetolil Beds at Laetoli, Northern Tanzania

M. D. Leakey and R. L. Hay

Editor's Note

Since the 1930s, Louis Leakey and his colleagues had been persuaded that the Olduvai Gorge and the surrounding region would be productive of fossil human remains. This remarkable paper describes the discovery in ancient volcanic tuffs—volcanic ash consolidated by the action of rain—of the footprints of a great variety of animals and even insects and a striking series of imprints of human feet. One set of human prints appears to have been left by two human beings walking northwards alongside each other for more than 100 metres. M. D. Leakey is Mary Leakey, the wife of Louis Leakey.

Recent excavation of the tuffs of the Laetolil Beds in Tanzania has revealed the presence of a large variety of footprints from the Pliocene. Many of these prints can be correlated with fossilised remains of Pliocene animals found in the same area.

IT was stated previously[1] that the name Laetolil would be used in preference to either Garusi or Vogel River for the area where the Laetolil Beds are exposed. Laetolil, as stated then, is an anglicisation of the Masai word Laetoli and was first used by Kent[2]. The Tanzanian authorities have now asked that the term Laetolil should be dropped in favour of Laetoli. The name Laetoli will be used for the area, but the Pliocene deposits will continue to be known as the Laetolil Beds, as established in 1976[1].

The Laetolil Beds (Figs 1, 2) are dominantly tuffs which have a maximum known thickness of 130 m and are divisible into upper and lower units. Nearly all the fossils have come from the upper unit which is 45–60 m thick.

In 1975 three marker tuffs had been identified in the upper unit of the Laetolil Beds[1]. Since then, more than a dozen widespread air-fall tuffs (Fig. 3) have been recognised, permitting detailed correlations.

坦桑尼亚北部莱托利尔层中的上新世足迹

利基，海

编者按

从20世纪30年代开始，路易斯·利基和他的同事们就相信在奥杜威峡谷和周围的区域中保存有大量的人类化石。这篇值得关注的文章记述了在古老火山凝灰岩（在雨水的作用下加固的火山灰）中发现的大量不同种类的动物足迹（甚至是昆虫）和一系列惊人的人类脚印。其中一列人类的脚印有100多米长，看起来是由并排向北行走的两个人所留下。文章的第一作者是玛丽·利基，她是路易斯·利基的妻子。

最近在坦桑尼亚莱托利尔层的火山凝灰岩中发现了大量上新世时期的足迹。这些足迹中很多能够与在同一区域发现的上新世动物化石进行对比。

如前所述[1]，“莱托利尔”原先是用来指加鲁西或沃格尔河的区域，因为这些区域又是莱托利尔层出露的地方。正如之后所表述的，莱托利尔是马赛语莱托里英语化后确定下来的，肯特首次使用了它[2]。现在坦桑尼亚官方已经要求用“莱托里”替代“莱托利尔”。所以“莱托里”这个名称将用来指这个地区，但上新世的沉积物将沿用世人熟知的、于1976年确定的“莱托利尔层”[1]。

莱托利尔层（图1、2）以凝灰岩为主要岩性，目前已知的最大厚度有130米，而且被分为上、下两个层位。几乎所有的化石都是从45~60米厚的上部层位中发现的。

1975年在莱托利尔层的上部层位中发现了三个标志凝灰岩层[1]。其后，又辨识出了12处以上分布广泛的空落凝灰岩（图3），它们使不同化石点的地层的详细对比成为可能。

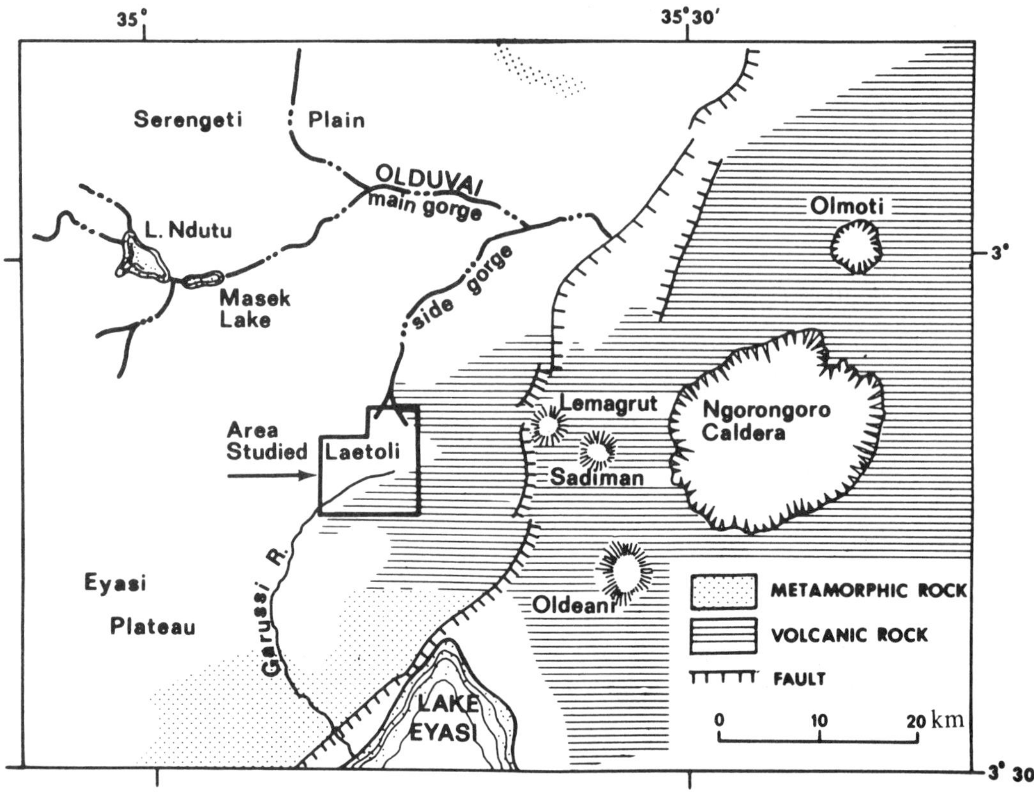

Fig. 1. Map of the southern Serengeti and volcanic highlands showing the position of the Laetoli area.

Description of the Footprints Tuff

The footprints tuff is divisible into two units of differing lithology and structure. The lower unit, 7–8.6 cm thick, is relatively uniform in thickness and is characterised by widespread ash layers of even thickness, commonly with rainprinted surfaces. The upper unit, generally 5–7 cm thick, thins over the higher areas of the lower unit and thickens in depressions to eliminate the undulations preserved by the mantle bedding of the lower unit.

Several unusual features of these tuffs can be explained by composite ash falls of natrocarbonatite ash and melilitite lava globules. The ash must have been cemented rapidly to have prevented erosion of the sand-sized lava globules by wind in this semi-arid climate in which 80–85% of the sediment was wind-worked. Natrocarbonatite ash would have dissolved incongruently in rainfall to yield soluble carbonates, which would have crystallised under the heat of the Sun to cement the ash layer in a few hours.

Footprints were made on at least six different surfaces but are by far the most common at two levels (Fig. 4). Prints of birds and hares are common to all levels, but prints more than about 10 cm in diameter have been found only in the upper unit. Particularly striking is the number of elephant prints in the higher levels compared to their apparent absence below.

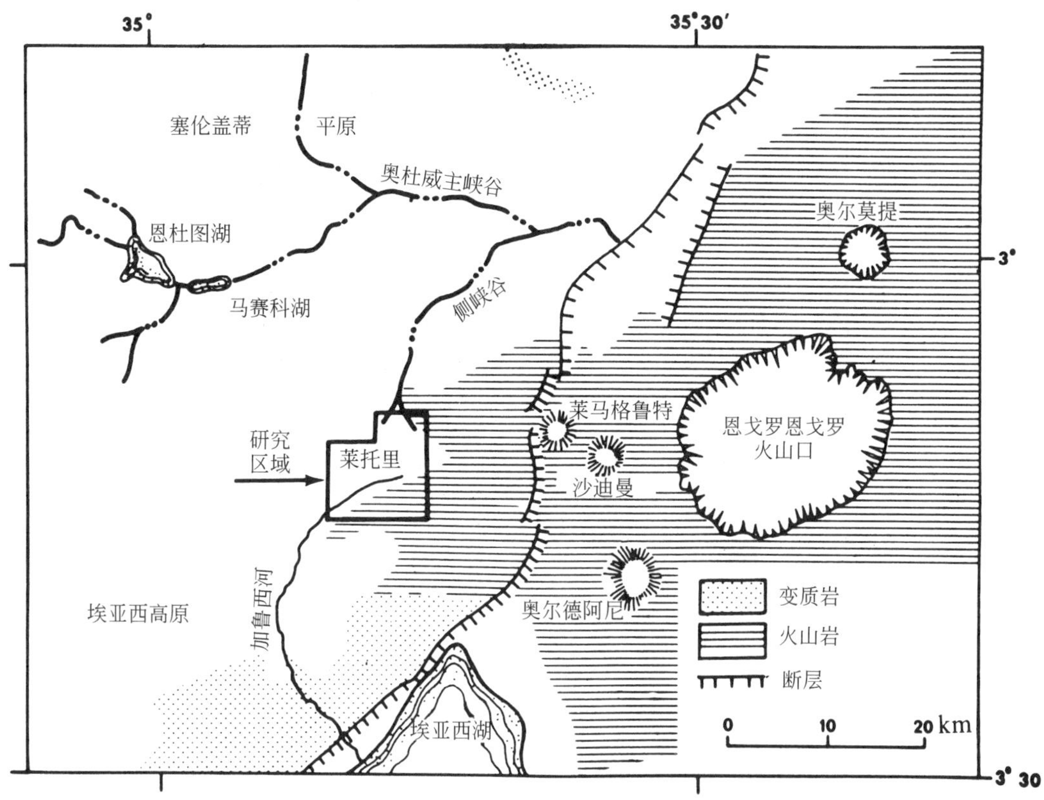

图 1. 莱托里在南塞伦盖蒂和火山高地中的地理位置图

足迹凝灰岩描述

足迹凝灰岩可以分为两种不同的岩性和结构。在下部层位中有 7~8.6 厘米厚的凝灰岩，它们厚度相对统一，而且还以广泛分布的相同厚度的火山灰层为特征，这些火山灰层的表面上普遍有雨痕。在上部层位，凝灰岩通常有 5~7 厘米厚，在下部层位较高的区域较薄，同时在低洼处变厚，以抵消下部层位地幔基岩原有的起伏。

这些凝灰岩表现出来的几个不同寻常的特征能够用钠碳酸岩质火山灰和黄长岩熔岩球形成的复成分火山灰降落来解释。在这种有 80%~85% 的沉积是由风成作用形成的半干旱气候下，火山灰必须迅速凝结以防止被砂粒大小的火山熔岩球所风蚀。钠碳酸岩质火山灰会不均匀地溶解在降雨中产生可溶性碳酸盐。可溶性碳酸盐在太阳辐射的加热下会结晶，并在几小时内凝结成火山灰层。

至少在 6 个不同层面发现了足迹，但目前为止只在两个层位中密集分布（图 4）。鸟类和野兔的足迹普遍存在于各个层位中，但直径大于 10 厘米的足迹只发现于上部地层中。特别引人注目的是在较高层位中的大象足迹的数量，这与低层位中大象足迹的明显缺失形成鲜明对比。

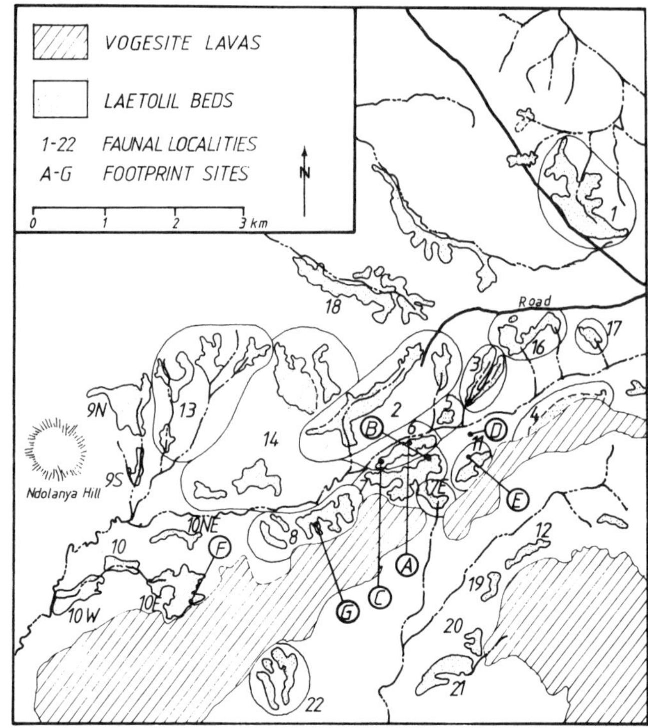

Fig. 2. Map of the fossil localities and footprint sites at Laetoli.

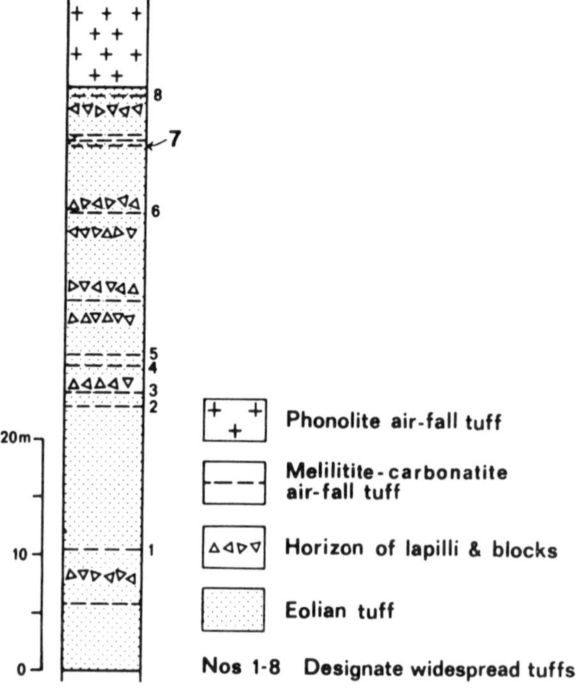

Fig. 3. Columnar section of the upper unit of the Laetolil Beds at Locality 1 showing air-fall tuffs and horizons of lapilli and blocks. The present report is based on studies of the lower part of Tuff 7, termed the Footprint Tuff. Tuffs 6, 7 and 8 were designated Tuffs a, b and c by Leakey *et al.*[1]. The pholite tuff was designated Tuff d.

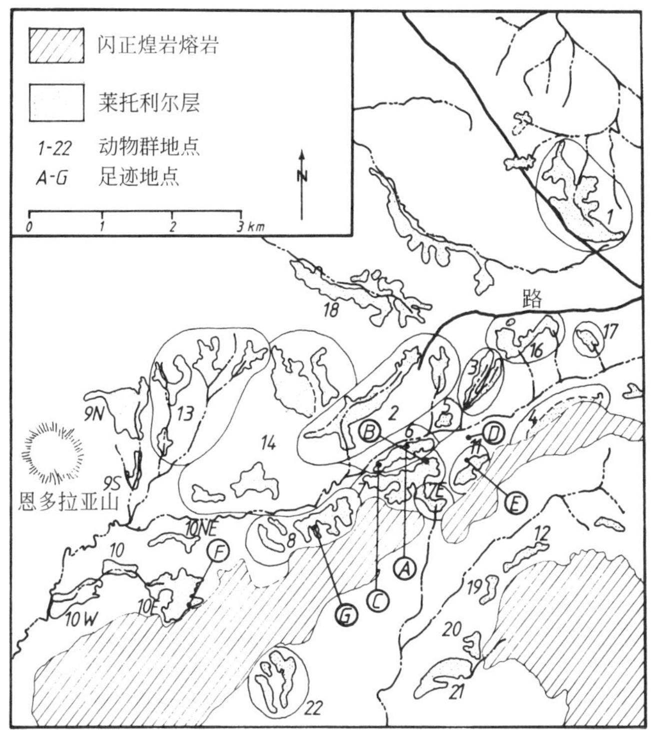

图 2. 莱托里地区的化石产地和足迹地点位置图

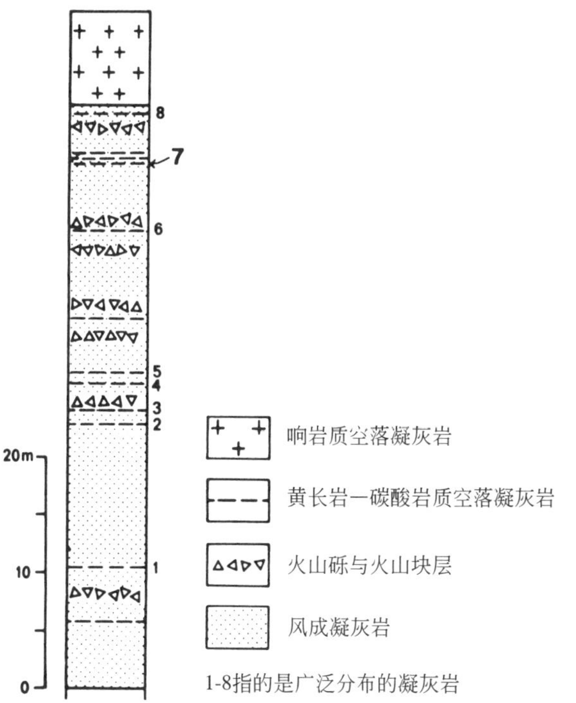

图 3. 地点 1 处的莱托利尔层上部的柱状剖面图，显示空落火山凝灰岩和火山砾、火山块层。本报告针对火山凝灰岩 7（又称为足迹凝灰岩）下部的研究。凝灰岩 6、7 和 8 被利基等人 [1] 称为凝灰岩 a、b 和 c。而响岩质凝灰岩则为凝灰岩 d。

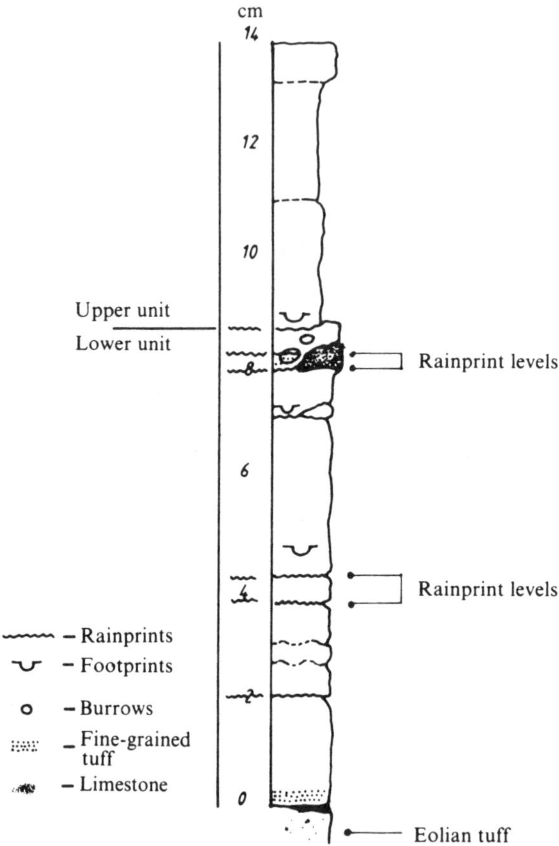

Fig. 4. Generalised columnar section for footprint tuff at site *A*.

On the basis of the available evidence a tentative history of the footprint tuff begins near the end of the dry season and continues into the rainy season. The first showers of ash fell on a relatively bare, nearly flat landscape with scattered *Acacia* trees. The ash layers which constitute the lower unit were cemented by intermittent showers near the beginning of the rainy season. A few times between showers extruded eolian sediments, together with some of the air-fall ash, were redeposited by wind. The sharp contact at the base of the upper unit may mark the onset of the rainy season. Stratification in the upper unit is compatible with sheetwash produced by heavy showers, but is unlike that of the lower unit in which individual ash layers vary little in thickness over a distance of 5 km. The smaller amount of calcite in the upper unit may have resulted from either heavy rains or a smaller original content of carbonatite ash, or both. Thus, the abrupt appearance of footprints of elephants and other large animals in the upper unit may represent, at least in part, their migration which accompanies the rainy season.

The 1975 and 1976 field seasons at Laetoli were devoted to study of the geology of the area and the collection and excavation of fossil vertebrates and molluscs from the Laetolil Beds and later deposits. The age of the upper, fossiliferous part of the Laetolil Beds was

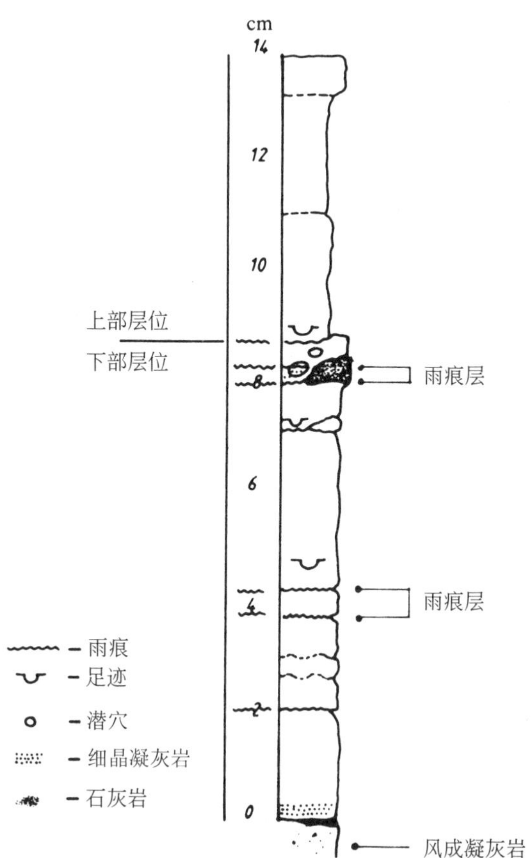

图 4. 遗址 *A* 处的足迹凝灰岩的综合柱状剖面图

基于现有证据，推想足迹凝灰岩的历史始于干旱季节几近结束时并持续到雨季。首波火山灰洒落在相对裸露、零星分散着金合欢树的近乎平坦的地方。组成下部层位的火山灰被雨季开始时的间歇性阵雨所凝结。在数次阵雨间歇内一些被挤压出的风成沉积物和空落的火山灰一起被风卷起混合后再沉积下来。上部层位底部的锐角不整合的接触可能标志着雨季的开始。上部层位的层理与暴雨产生的片状冲积层相一致，但与下部层位的层理不同，后者在 5 千米范围内，其所包含的单独的火山灰层厚度上几乎没有变化。上部层位中含有的少量方解石，可能是在大雨或者火山灰中原有的少量碳酸盐岩，或二者的共同影响下形成的。因此，在上部层位中突然出现大象和其他大型动物的足迹可能意味着，至少是某种程度上意味着，它们是伴随着雨季迁徙的。

在 1975~1976 年莱托里的野外季中，研究人员致力于该区域地质情况的研究和从莱托利尔层及其上部沉积物中采集和发掘脊椎动物、软体动物化石。莱托利尔层

established at 3.6–3.75 Myr (ref. 1).

While visiting the Laetoli camp during 1976 Dr. Andrew Hill observed a number of depressions in the surface of a fine-grained tuff exposed in a river bed. These proved to be footprints of birds and mammals ranging from elephant and rhinoceros to carnivores and hares, which had been exposed by natural erosion and weathering.

The first site where footprints were observed (site *A* in Fig. 2) lies just south of the Garusi River in fossil Locality 6. An area of ~490 m^2 has been exposed by natural erosion and by excavation. To the south-west, at a second site (*C*) there are ~156 m^2 of the footprint-bearing tuff exposed. Both these localities were studied in detail in 1977. Five further areas where the footprint-bearing tuff is well exposed are also known but have not yet been completely studied.

Deeply worn game tracks or pathways can be seen crossing the footprint areas at two sites. These were clearly used repeatedly by animals to reach some objective and on modern behavioural patterns it is likely that they were made by the game going to and from water holes.

Only a proportion of the animals represented in the fossil prints have been identified so far. Investigation of present-day game tracks in National Parks is underway to provide comparative information. In general, however, the fossil record agrees well with the footprints. A Musukuma tracker was employed to assist in identifications. He assisted considerably in identifying to family and generic levels, particularly in the case of Bovidae, which are commonly represented in both the footprint and fossil records.

A partial breakdown of the mammalian specimens recovered from the Laetolil Beds in 1975 was published[1] and may usefully be given here. Additional material was recovered in 1976 and 1977, but the overall proportions of various groups remain close to the first figures (Table 1).

Table 1. Mean percentages of mammalian specimens from various sites

<table>
<tr><td>Bovidae
(of which 15.1% are Madoqua, dik-dik)</td><td>43%</td><td>Equidae</td><td>4.4%</td></tr>
<tr><td>Lagomorpha</td><td>14.4%</td><td>Suidae</td><td>3.6%</td></tr>
<tr><td rowspan="2">Giraffidae
(including both a large and small species, as well as Sivatherium)</td><td rowspan="2">11.2%</td><td>Proboscidea</td><td>3.4%</td></tr>
<tr><td>Rodentia</td><td>3.3%</td></tr>
<tr><td>Rhinocerotidae</td><td>9.7%</td><td>Carnivora</td><td>3.1%</td></tr>
</table>

Avifauna, Cercopithecidae, Hominidae and Pedetinae were omitted from this list.

上部含化石部分的时代已确定为360万到375万年前（参考文献1）。

安德鲁·希尔博士在1976年访问莱托里营地期间曾观察到出露于河床且具有细晶的凝灰岩表面有很多的凹陷。这些被证明是鸟类及大象、犀牛、食肉动物还有野兔等哺乳动物留下的足迹，它们由于受到自然的侵蚀和风化作用，这些足迹已经暴露出来。

发现足迹的第一处遗址（在图2的位置*A*）位于加鲁西河的南边的第6处化石地点。由于自然侵蚀和挖掘，约490平方米的区域已经出露。在位于西南的第二处遗址（*C*）则有156平方米的含足迹凝灰岩出露。1977年曾对这两个遗址进行过详细研究。还有5处已知区域有出露很好的含足迹凝灰岩，但还未得到彻底研究。

在两处遗址的足迹区可以观察到印痕很深的游走足迹或路径。很显然，这些路径被动物屡次使用以到达某个目标，从现代行为学模型来看，它们很可能是动物游走时去水坑和离开水坑时留下的印痕。

在以化石足迹反映出的动物中，迄今只有一部分的分类地位得以确定。用来作为比较信息的现生动物游走足迹正在国家公园中进行研究。然而整体而言，化石记录与足迹十分吻合。我们聘请了一位穆苏库玛追踪者辅助鉴定工作。他在很大程度上辅助了科级与属级的鉴定工作，尤其是在牛科动物的足迹鉴定中，而牛科动物的足迹和化石材料都很常见。

1975年在莱托利尔层中出土的哺乳动物清单已经部分发表[1]，在此再复述一下也许有用。1976年和1977年又出土了一些新材料，但不同种类间的总体比例与第一次出土的化石数据相近（表1）。

表1. 不同产地哺乳动物种类的平均百分比

<table>
<tr><td>牛科
(15.1%是犬羚属)</td><td>43%</td><td>马科</td><td>4.4%</td></tr>
<tr><td>兔形目</td><td>14.4%</td><td>猪科</td><td>3.6%</td></tr>
<tr><td rowspan="2">长颈鹿科
(包括大和小物种，以及西洼兽属)</td><td rowspan="2">11.2%</td><td>长鼻目</td><td>3.4%</td></tr>
<tr><td>啮齿目</td><td>3.3%</td></tr>
<tr><td>犀牛科</td><td>9.7%</td><td>食肉目</td><td>3.1%</td></tr>
</table>

该表中省略了鸟类、猴科、人科和跳兔亚科

The footprints that have been recorded are briefly described below, with notes on the fossil material where it seems to be related.

Diplopoda

A single track approximately 20 cm long is known at site (*A*).

Fossil record: a small fragment of fossilised centipede was found at Locality 4.

Avifauna

(1) *Struthio* sp. Two isolated prints at site (*A*).

(2) *Phasianidae,* cf. Guinea fowl. Numerous tracks occur at all sites, generally in trails of four or more. They compare closely with tracks of the living helmeted Guinea fowl, common in the Laetoli area today. Average length of eight fossil prints 62 mm, of nine modern prints 60 mm.

(3) Similar but smaller tracks, averaging 45 mm in length, possibly of francolin.

Fossil record: numerous fragments of ostrich eggshell are known but no skeletal remains. Clutches of eggs comparable in size to those of modern Guinea fowl have been found at Locality 10.

Primates

Cercopithecidae: Tracks are known at three localities.

At site (*C*) there is a single trail comprising six hind foot prints with a digit protruding to one side. Each of these prints is accompanied by a roughly circular impression, always to the left (Fig. 5). When first discovered these prints were interpreted as knuckle impressions, but more thorough cleaning has revealed traces of the palms of the hands and they are undoubtedly prints of the forefeet. In the hind feet the longest digit is central and the prints range in length from 20.1 to 14.7 cm with an average of 17 cm. The width varies from 10.9 to 8.1 cm with an average of 9.9 cm (excluding the great toe). Stride length varies from 34 to 46 cm with an average of 41 cm. (Stride is here interpreted as the distance between the posterior margin of successive heel prints of the same foot.)

The second trail is at site (*D*). It was made by a single animal and is 4 m long. There are prints of both hind and forefeet. All are lightly imprinted on a surface which was clearly wet and slippery when the animal walked over it .The average length of the hind prints is 14.5 cm and of the forefeet prints 11 cm. Stride length averages 27.7 cm. These prints are not only smaller than those at site (*C*) but are relatively broader, with very narrow heel impressions.

下面简要地描述记录下来的足迹化石，同时也标注了这些足迹可能对应的化石材料。

倍 足 纲

一条长约 20 厘米的单独的足迹见于遗址（*A*）。

化石记录：出土于第 4 化石地点的一小段蜈蚣化石。

鸟 类 动 物

（1）鸵鸟属未定种。两个单独的脚印见于遗址（*A*）。

（2）雉科，珍珠鸡相似种。多个足迹见于所有遗址，一般每处有 4 列以上的足迹。它们与莱托里地区常见的现生盔珠鸡足迹非常相似。8 个化石脚印的平均长度为 62 毫米，9 个现生脚印的平均长度为 60 毫米。

（3）相似的但小一点的足迹，脚印平均长度 45 毫米，可能是鹧鸪。

化石记录：有很多鸵鸟蛋壳碎片，但没有骨骼化石。在第 10 化石地点发现了一窝蛋，蛋的大小与现在珍珠鸡蛋的相似。

灵 长 目

猴科：其足迹见于 3 处。

在遗址（*C*）有个单列足迹，由 6 个后脚印组成。脚印中有个趾头向侧面伸出。每个脚印的左侧都伴有一个大致呈圆形的印痕（图 5）。这些脚印在发现的初期被认为是指背行走的印痕，但经深入清理后发现手掌的痕迹，因此无疑是前掌的印痕。后脚的最长脚趾在中间，脚印的长度在 20.1~14.7 厘米之间，平均值为 17 厘米。脚宽在 10.9~8.1 厘米之间，平均值为 9.9 厘米（不包括大拇趾）。步幅长度在 34~46 厘米之间，平均值为 41 厘米（步幅在此的定义为两个相邻的同一只脚的脚印后跟之间的距离）。

第二列足迹在遗址（*D*）。这是一个个体留下的脚印，长 4 米。前后肢的脚印均有。这些脚印均为较浅的印痕，当动物经过时，地面显然是湿滑的。后脚印平均长 14.5 厘米，前脚印平均长 11 厘米。步幅平均长 27.7 厘米。这些脚印不仅比遗址（*C*）的要小，而且还相对较宽，后跟的印痕非常窄。

Fig. 5. Print of cercopithecoid fore and hind foot at site C.

At site (*F*), in fossil Locality 10, there are at least four sets of tracks going in slightly different directions, as do the tracks of present-day baboons when they move in a troop. The average measurements of the hind feet in each of the four trails range from 15.2 to 10 cm and of the forefeet prints from 7.6 to 4 cm (excluding the great toe).

Fossil record: a number of cercopithecoid mandibles and teeth and a few postcranial fragments have been provisionally attributed to *Papio* sp. and a colobine. Most are as small or even smaller in size than those of a living female baboon, but two mandibular fragments, a calcaneum and the distal ends of a humerus and femur are considerably larger than in any living baboon and compare in size with *Theropithecus oswaldi*. The difference in size between the prints described above is compatible with that of the known fossils, although age and sex differences are factors to be considered.

Hominidae: Three trails believed to be hominid are known at sites (*A*) (Fig. 6) and (*G*) (Figs 7, 8), in fossil Localities 6 and 8.

(1) At site (*A*) there are five prints in a trail 1.5 m long (Fig. 6). Natural erosion has almost entirely exposed two of the prints, but the remaining three are still filled with matrix of the overlying deposits. The exposed prints are short and broad, 15.5 cm long and 10.5 cm wide. The stride is also short with an average length of 31 cm. The gait was somewhat shambling, with one foot crossing in front of the other. Unlike the cercopithecoid prints, the longest digit is the great toe, situated as in the human foot.

图 5. 猴科动物在遗址 *C* 的前后脚印。

在遗址（*F*），第 10 化石地点，至少有 4 列去向略有不同的足迹，就如现生狒狒结队行走时会踩出的路径那样。每列后脚印都算出一个平均尺寸，这 4 列足迹平均尺寸在 10~15.2 厘米之间，前脚印的则为 4~7.6 厘米（不包括大拇指）。

> 化石记录：一些猕猴科动物的下颌骨和牙齿，以及少量颅后骨骼的碎块被暂时归入狒狒属未定种和一种疣猴。在大小方面，大部分与现生雌性狒狒的一样小或比它更小，但是有两件下颌骨破片、1 件跟骨、1 件肱骨远端和 1 件股骨远端显然大于任何现生狒狒，而与奥斯华狮尾狒的个体相近。尽管需要考虑年龄和性别差异的因素，上述脚印大小的差异与所出土的化石材料指示的个头差异是对应的。

人科：在化石地点 6 和 8 中的遗址 *A*（图 6）和 *G*（图 7 和图 8）中可见 3 列被认为是人科动物的足迹。

（1）在遗址（*A*）的一列长 1.5 米的足迹中有 5 个脚印（图 6）。自然侵蚀使其中的两个脚印几乎完全出露，但是其余 3 个仍然被上覆沉积物的围岩填充。出露的脚印又短又宽，长 15.5 厘米，宽 10.5 厘米。步幅也小，平均长度为 31 厘米。步态有点蹒跚，一只脚要绕到另一只脚的前面。与猴类的脚印不同，最长的脚趾是拇趾，位置与人类的脚相同。

Fig. 6. Hominid footprints at site *A*.

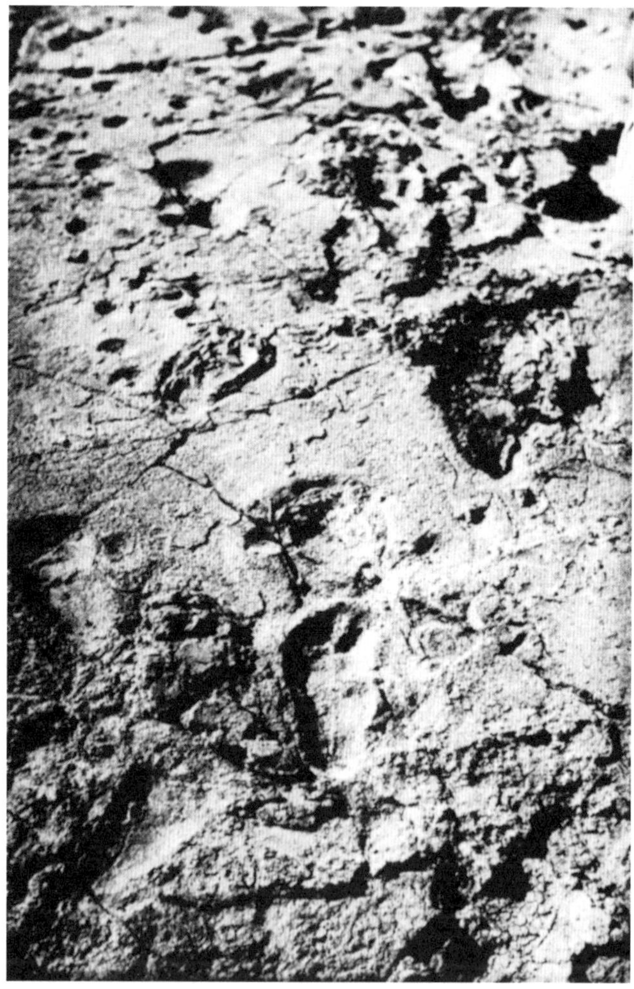

Fig. 7. Dual trail of hominid footprints at site *G*.

图 6. 遗址 *A* 的人科动物足迹。

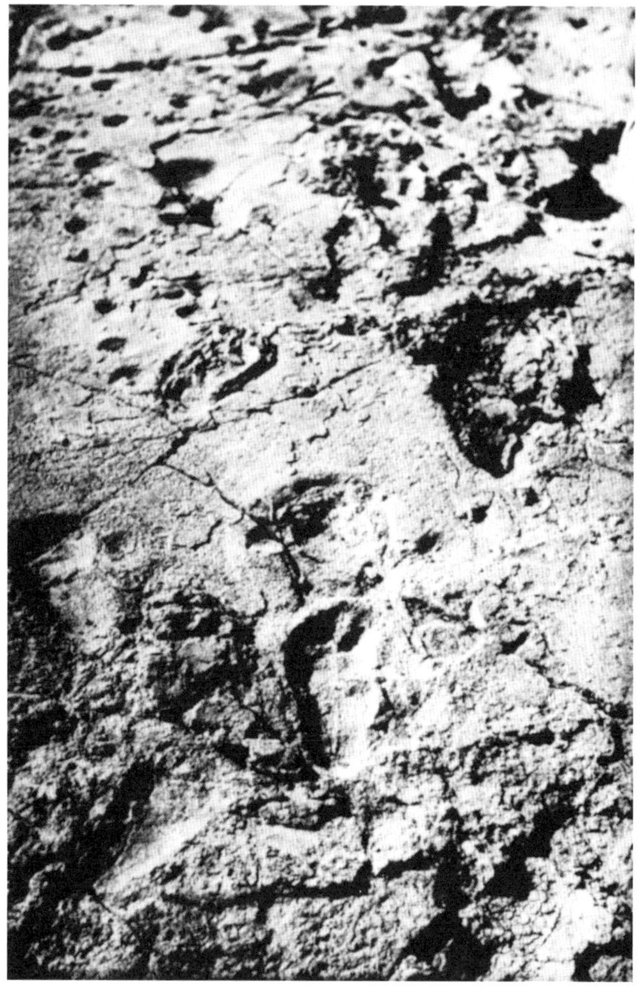

图 7. 遗址 *G* 的人科动物的双列足迹。

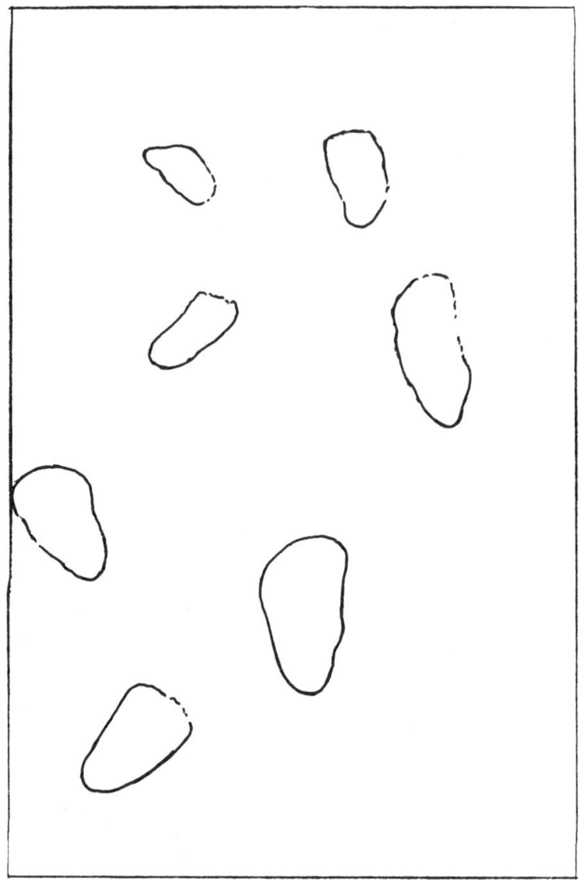

Fig. 8. Outline of footprints shown in Fig. 7.

It has been suggested that these prints might have resulted from the superposition of a hind on forefoot impression, or vice versa, either of a quadrupedal animal or a knuckle-walking primate. Careful examination of the original prints *in situ* reveals no indication of superposition, while the last interpretation invokes the hypothetical existence of an animal which is not present in over 5,000 fossil specimens.

(2) At site (*G*) there are trails left by two individuals travelling north (Figs 7–9). The trails are parallel and ~25 cm apart, too close for the hominids to have walked abreast. They followed the same line or pathway but it is possible that they did not pass by at the same time as there is a noticeable difference in the conditions of the two sets of prints. Those of the smaller individual are sharp and well defined, indicating a firm, compact surface, whilst those of the larger individual, with one notable exception, are blurred at the edges and enlarged, as would be the case if the surface had been dry and dusty. At one point the smaller individual appears to have paused and made a half-turn to the left before continuing in a northerly direction.

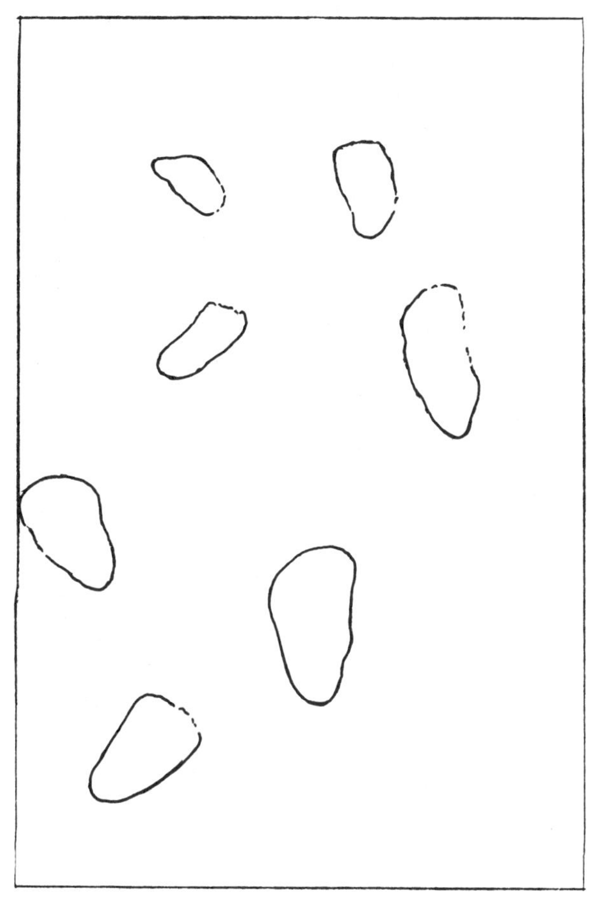

图 8. 图 7 中足迹的轮廓。

有人解释这样的脚印是由于后脚踩到前脚印上形成的或者有可能是反过来的，因此这些脚印要么是四足行走的动物要么是指背行走的灵长类的脚印。对原始脚印的现场细致的观察表明没有任何叠覆迹象，然而最后一种解释援引了一种在五千多件化石标本中都未出现的物种的假设性存在。

(2)在遗址(G)有两个个体向北行走留下的足迹(图 7~9)。这两列足迹互相平行，间距为 25 厘米左右；对于并排行走的人科动物来说，这样的间距太小了。这两列脚印沿着相同的路线或路径，但可能是在不同的时间里留下的，因为这两列脚印是在明显不同的情况下留下的。其中个体较小的脚印清晰，轮廓分明，指示一种稳重而紧密的接触状况；而个体较大的脚印，除了其中的一个明显的特例外，其余的都是边缘模糊、印痕扩大，就如在干燥而尘土多的状态下出现的情况一样。较小的个体在行走中似乎还在某处停顿下来向左转弯，然后又继续向北行走。

Fig. 9. Left footprint of larger individual in dual trail at site *G*.

In a number of prints the original surface in which they were made has been eroded, leaving only indentations in the underlying deposits. Thus, the measurements of these prints do not accurately reflect the dimensions of the original impressions and are not included.

The site is now being excavated and the trails have so far been found to extend for a distance of 23.54 m. Trail 1 contains 22 prints and Trail 2 contains 12 prints. The average length/breadth of the prints in Trail 1 is 18.5×8.8 cm and in Trail 2 is 21.5×10 cm. Average stride length in Trail 1 is 38.7 cm and in Trail 2 is 47.2 cm. Note that the longitudinal arch of the foot is well developed and resembles that of modern man, and the great toe is parallel to the other toes (Figs 7, 8 and 9).

Fossil record: two mandibles, parts of 2 maxillae, partial deciduous and permanent dentions as well as a partial infant skeleton and a number of isolated teeth have been recovered from the Laetolil Beds. They bear considerable resemblances to the material collected from the Afar in Ethiopia, although the Ethiopian material is substantially later.

图 9. 遗址 G 的双列足迹中较大个体的左侧足迹。

还有一些留下脚印的原始层面受到了侵蚀，仅在层面下的沉积物中留下一些凹痕。因此这些脚印的测量无法准确反映动物原有脚印的大小而没有列入统计中。

这处遗址目前还在发掘中，清理出的足迹长度已延续到 23.54 米。第 1 列足迹有 22 个脚印，第 2 列足迹有 12 个。第 1 列足迹的脚印长宽平均值为 18.5 厘米 ×8.8 厘米，而第 2 列足迹的为 21.5 厘米 ×10 厘米。第 1 列足迹的平均步幅长为 38.7 厘米，第 2 列足迹的平均步幅长为 47.2 厘米。请注意其纵向足弓和现代人一样很发育，拇趾和其他脚趾平行（图 7~9）。

化石记录：2 件下颌骨，2 件上颌骨残段，若干不完整的乳齿系和恒齿系，1 个不完整的幼体骨架，还有一些单独的牙齿，均出土于莱托利尔层。它们与埃塞俄比亚阿法尔地区出土的材料有很多相似之处，尽管埃塞俄比亚材料要晚得多。

Leporidae

Innumerable tracks of hares occur at all sites.

Fossil record: mandibles and other remains attributed to *Serengetilagus* sp. are abundant.

Carnivora

(1) Viverridae, indet. Series of small carnivore prints with non-retractile claws occur at all sites. They are generally lightly imprinted and rather faint. Average length/breadth 34×27 mm. Other small, rounded prints without claw marks, suggest genets. Average length/breadth 28×27 mm.

Fossil record: a variety of Viverridae are known among the fossils, some were described by Dietrich[3], those from the recent collections are being studied by Mme G. Petter.

(3) Hyaenidae. Hyenas have left a series of well-defined and relatively long trails containing large numbers of prints. Eight trails are known and two gaits are represented. Subject to research on living hyaenas, these appear to be a walk and a slow canter or lope. There is uniformity in size, depth and stride within each of the four trails measured, although it is evident that the animals varied in individual size. Average length/breadth/depth measurements:

Trail *C.*22 (19 prints) 122×94×12 mm
Trail *C.*27 (30 prints) 125×102×22 mm
Trail *C.*31 (11 prints) 102×87×13 mm
Trail *C.*39 (18 prints) 111×100×18 mm

Fossil record: in a preliminary report, M. G. Leakey noted the existence of three species of hyena from the Laetolil Beds: (1) *Hyaena bellax* (Ewer 1954); (2) *Hyaenictis* cf. *preforfex* (Hendey 1974); and (3) *Lycyaena* sp.

(3) Felidae. One trail of 12 prints at site (*A*), and several isolated prints appear to correspond in size to those of a serval cat. Average length/breadth/depth 65×52×13 mm. Average length of stride 29 mm.

Fossil record: remains comparable to *F. serval* and *F. caracal* have been noted in the 1974–76 collections.

(4) *Machairodontinae.* Two prints at site (*A*), measuring 134×115 mm and 110×115 mm probably belong to a large sabre-tooth cat, as no other felid of comparable size occurs in the fossil material.

Fossil record: The presence of a large sabre-tooth cat, cf. *Homotherium* is indicated by two teeth.

兔　科

所有遗址中都有不计其数的野兔足迹。

化石记录：可以归入塞伦盖蒂兔属未定种的下颌骨和其他部位的化石非常多。

食　肉　目

（1）灵猫科(属种不定)：带有不可伸缩的爪的一系列小型食肉类的脚印在所有遗址均有发现。它们的脚印较浅，印痕相当模糊，长宽平均值为 34 毫米 ×27 毫米。其他没有爪痕的小圆脚印可能属于麝猫。长宽平均值为 28 毫米 ×27 毫米。

化石记录：化石材料中有多种灵猫科成员，迪特里希曾记述过一些[3]，彼得夫人正在研究最近采集到的标本。

（2）鬣狗科：鬣狗留下了一系列清晰的、含有大量脚印的、相对较长的足迹。已知有 8 列足迹，表现出两种步态。据对现生鬣狗的研究，这些足迹像是散步、慢跑或大步慢跑。尽管鬣狗的个体大小显然有差异，但这 4 列足迹的测量显示，它们在大小、深度和步幅上都具有一致性。脚印的长、宽、深平均测量值如下：

列 *C*.22（19 个脚印）122 毫米 ×94 毫米 ×12 毫米
列 *C*.27（30 个脚印）125 毫米 ×102 毫米 ×22 毫米
列 *C*.31（11 个脚印）102 毫米 ×87 毫米 ×13 毫米
列 *C*.39（18 个脚印）111 毫米 ×100 毫米 ×18 毫米

化石记录：米芙·利基在一个初步报告中提到莱托利尔层中存在 3 种鬣狗（1）好斗鬣狗（尤尔，1954），（2）前剪刀鼬鬣狗相似科（亨戴，1974）和（3）狼鬣狗未定种。

（3）猫科：1 列 12 个脚印见于遗址（*A*），并且好几个单独的脚印在大小上似乎符合薮猫的脚印。平均长宽为：65 毫米 ×52 毫米 ×13 毫米。平均步幅长 29 毫米。

化石记录：可以归入薮猫和狞猫的材料见于 1974~1976 年采集到的化石。

（4）剑齿虎亚科：两个脚印见于遗址（A），大小为 134 毫米 ×115 毫米和 110 毫米 ×115 毫米，可能属于一种大型剑齿虎，因为化石材料中没有出现其他同样大小的猫科动物。

化石记录：有两枚大型似剑齿虎的牙齿。

Proboscidea

The proboscidean prints appear to belong to *Loxodonta exoptata* with the exception of one particularly well-preserved print in a trail of four at site (*C*) in which the phalanges and metapodials were more nearly vertical than in other known proboscidean prints and may be of a *Deinotherium*. The *Loxodonta* prints include a number made by juvenile animals as well as some that are usually large by present-day standards. The average for 15 measured prints, mostly adult, is 420×346×34 mm.

Fossil record: both *Loxodonta exoptata* and *Deinotherium* cf. *bozasi* occur in the Laetolil Beds, but the latter is relatively scarce. A high proportion of the *Loxodonta* teeth are from juvenile animals.

Equidae

Only two equid trails are known. They have only recently been discovered and have not yet been fully studied. Both are at site (*G*), one on either side of the hominid trails but travelling in an opposite direction, to the south. The best preserved trail is 4.97 m long and contains 15 prints, nine of which can be measured. They range in length from 9.9 to 7.8 cm with an average of 8.6 cm and in width from 10.6 to 8.3 cm with an average of 8.8 cm. The animal appears to have changed gait during this trail.

Fossil record: numerous teeth, postcranial material and some incomplete mandibles have been found. All can be attributed to *Hipparion* sp.

Rhinocerotidae

Both *Ceratotherium* and *Diceros* must be represented among the many prints of rhinoceros, but no distinguishing features have been observed to date except on the basis of size. A trail at site (*A*) is one of the longest known, measuring 22 m in length. It contains 31 prints, 23 of which are double, with the hindfoot superimposed on the front. At the western end of the trail, where the animal changed gait, the prints become single and are irregularly spaced. Comparison with modern prints of *Diceros bicornis* at Olduvai show very close similarity. Average length/breadth/depth of double prints in trail: 416×273×31 mm. Single prints, forefoot 246×248×25 mm, hindfoot 256×228×37 mm. Isolated single fore and hind prints in the adjacent game trail are unusually large and may be of *Ceratotherium*. They measure 285×310×42 mm and 400×270×33 mm respectively.

Fossil record: on the basis of the early collections only *Ceratotherium* was believed to be present in the Laetoli fauna. But a skull of *Diceros* has now been found, as well as numerous teeth and some postcranial material.

Chalicotheriidae

Two Chalicothere prints were found at site (*C*). They are deeply indented and measure 250×155×63 mm and 243×110×46 mm. The prints comprise impressions of three digits and of the palm. The digits have left symmetrical rounded grooves and the emplacement

长 鼻 目

长鼻类的脚印似乎属于古非洲象，只有一个例外是在遗址（*C*）的一列 4 个脚印中有一个保存得非常好的脚印，与其他已知长鼻目动物脚印相比，其趾骨和掌骨更加近似平直，有可能是恐象。非洲象的脚印包括幼年个体的和以现在标准看的大型个体的。15 个已测脚印（其中大多为成年个体的脚印）的平均值为 420 毫米 × 346 毫米 × 34 毫米。

化石记录：古非洲象和博氏恐象相似种均在莱托利尔层中出现，但后者相对稀少。有很大比例的非洲象牙齿来自幼年个体。

马 科

仅发现两列马科动物的足迹。马的脚印是最近才发现的，尚未得到充分研究。两列脚印均发现于遗址（*G*），有一列在人科动物足迹的一侧，但是行进方向相反，向南方前行。保存最好的足迹有 4.97 米长，包含 15 个脚印，其中 9 个可以测量。其长度在 9.9~7.8 厘米之间，平均值为 8.6 厘米；宽度在 10.6~8.3 厘米之间，平均值为 8.8 厘米。在这列足迹中，这匹马似乎在中途改变了步态。

化石记录：已经发现许多牙齿、颅后骨骼及一些不完整的下颌骨。所有材料可归入三趾马属未定种中。

犀 科

在这么多犀牛脚印中肯定有白犀和黑犀，但迄今除了根据大小来区分外还没有发现其他鉴定特征。在遗址（*A*）发现的一列足迹是目前已知的最长的足迹之一，有 22 米长，包括 31 个脚印，其中有 23 个脚印是重叠的，即后脚踩到前脚的脚印上。在这列足迹的西端，这头犀牛改变了步态，脚印变成单个并呈不规则隔开。它与奥杜威现生的黑犀脚印非常相似。这列足迹中脚印长宽深平均值为：重叠脚印 416 毫米 × 273 毫米 × 31 毫米；单独脚印中前脚印 246 毫米 × 248 毫米 × 25 毫米；后脚印 256 毫米 × 228 毫米 × 37 毫米。在相邻的一个游走足迹中的单独的前后脚印通常较大，并可能是白犀的。前后脚印测量值分别为 285 毫米 × 310 毫米 × 42 毫米和 400 毫米 × 270 毫米 × 33 毫米。

化石记录：根据早期发掘结果，一般认为在莱托利尔动物群中只有白犀，但最近出土了一件黑犀的头骨、许多牙齿和颅后骨骼材料。

爪 兽 科

在遗址（*C*）发现了两个爪兽的脚印。它们的印痕很深，测量值为 250 毫米 × 155 毫米 × 63 毫米和 243 毫米 × 110 毫米 × 46 毫米。脚印由三个脚趾和一个脚掌组成。这

for a claw can be seen on one. The prints are 1.30 m apart.

Fossil record: a few specimens which can be referred to *Ancylotherium* cf. *hennigi* have been recovered. They include a calcaneum, astragalus and some phalanges.

Suidae

Thirteen suid prints at site (*A*) have been measured. They are comparable in size to prints of the living warthog. Average length/breadth/depth measurements are 55×47×11 mm.

Fossil record: only two Suidae are known from the Laetolil Beds, *Potamochoerus* sp. and *Notochoerus euilus*. *Hylochoerus* and *Phacochoerus* were previously believed to be present in the early fauna, but during 1974–77 have only been found in the more recent deposits. On size, the prints are probably of *Potamochoerus*, as *Notochoerus* was a larger animal than the living warthog.

Giraffidae

(1) Three trails made by single animals are known. They consist of 6, 7 and 11 prints respectively. There are also 19 other prints attributed to giraffe. In one trail, the animal has dragged its feet after each step and left scuffed grooves up to 96 cm long. Average length/breadth/depth measurement for the prints in the trails are 190×151×29 mm, 202×144×10 mm, 211×150×17 mm and 205×150×28 mm. Averages for the isolated prints are 208×155×19 mm.

Fossil record: the prints are of similar size to those of the modern giraffe and can be attributed to *G. jumae* which occurs as a fossil.

(2) Small giraffe cf. *G. stillei*. There are no prints which can positively be allocated to this animal, although its fossil remains are by far the most common giraffid. However, there are many prints, including three trails, which were identified as 'eland' by the tracker. No eland is known in the fossil fauna, nor is there any other bovid in the collection of a size suitable to have made these prints. For the present, it seems justifiable to assign these prints to the small giraffe. They are abundant at site (*A*), where there are three trails consisting of 21, 14 and 5 prints, as well as 72 isolated prints. The average length/breadth/depth measurements for the prints in the three trails are 140×107×14 mm, 139×104×13 mm and 128×94×16 mm.

Fossil record: if the prints are correctly identified, they would belong to the animal originally named *Okapia stillei*, now known as *Giraffa stillei*.

(3) *Sivatherium*. No prints could be identified. The fossils consist mostly of isolated teeth and foot bones.

些脚趾留下了对称的圆沟，其中一个可以看到脚趾上的爪。脚印的间距为 1.30 米。

化石记录：已经出土了一些可归入钩爪兽相似种的标本，包括 1 个跟骨、距骨和一些趾骨。

猪　科

已经测量了遗址(*A*)的 13 个猪科动物脚印。它们与现生疣猪的脚印具有可比性。长宽深平均测量值为 55 毫米 × 47 毫米 × 11 毫米。

化石记录：在莱托利尔层中只有两种猪科成员，河猪未定种和南方猪兽。林猪和疣猪曾被认为出现在较早的动物群中，但是在 1974~1977 年的发掘中，只在较晚的沉积物中发现了它们。从大小上判断，脚印很可能是河猪的，因为南方猪是一种比现生的疣猪还要大的动物。

长颈鹿科

（1）有 3 列同一个种的动物留下的足迹。它们分别由 6、7、11 个脚印组成。还有其他 19 个脚印被认为是长颈鹿的。在一条足迹中，这头长颈鹿每走一步都拖沓着脚趾，在地面划出一道长达 96 厘米的沟痕。这几列足迹的长宽深平均值分别为 190 毫米 × 151 毫米 × 29 毫米，202 毫米 × 144 毫米 × 10 毫米，211 毫米 × 150 毫米 × 17 毫米，205 毫米 × 150 毫米 × 28 毫米。那些单独脚印的平均值则为 208 毫米 × 155 毫米 × 19 毫米。

化石记录：脚印与现生长颈鹿的大小相似，可以归入朱玛长颈鹿化石。

（2）小型的施氏长颈鹿相似种。没有可以明确归入这种动物的脚印，尽管迄今为止这个种的化石在长颈鹿科动物中最常见。但是有很多脚印，包括 3 列足迹，被跟踪员鉴定为“大羚羊”。而在出土的化石中既没有大羚羊，也没有其他个头合适的牛科动物可以留下那么大的脚印。暂时将这些脚印归入小型长颈鹿无可非议。在遗址（*A*）有很多脚印，那里有分别由 21、14 和 5 个脚印组成的 3 列足迹，还有 72 个单独的脚印。3 列足迹的长宽深平均值分别为 140 毫米 × 107 毫米 × 14 毫米、139 毫米 × 104 毫米 × 13 毫米和 128 毫米 × 94 毫米 × 16 毫米。

化石记录：如果脚印的鉴定是正确的，那么它们属于施氏长颈鹿，这个种原先被命名为施氏獾㹢狓。

（3）西洼兽：没有可鉴定的脚印。化石主要包括一些单独的牙齿和足部骨骼。

Bovidae

(1) Bovini cf. *Simatherium kohllarseni.* These prints consist of large, rounded and generally deeply indented tracks resembling those of the living African buffalo. They are clearly made by Bovini and can be attributed to *S. kohllarseni,* as this is the only bovine in the fossil record. The prints are represented by four trails containing 16, 12, 10 and 5 tracks, as well as 40 additional prints, either single or in pairs. The average length/breadth/depth measurements for the 43 prints in the trails are 185×149×25 mm.

Fossil record: a number of horn cores and teeth have been found, including a cranium associated with both horn cores.

(2) *Hippotragus* sp. Seven single prints from sites (*A*) and (*C*) were identified by the tracker as 'roan antelope'. They are characterised by widely splayed, elongate hoof-marks. The average length/breadth/depth measurements are 109×95×16 mm.

Fossil record: three horn cores and some teeth, collected during 1975 are believed to be a hippotragine and have provisionally been attributed to *Praedamalis deturi.* A horn core and teeth in the earlier collections may also belong to this species which is smaller than the living roan antelope.

(3) Alcelaphini. Eighteen prints at site (*A*) and two at site (*C*) were identified by the tracker as hartebeest. The prints at site (*A*) are both smaller and shallower than those at site (*C*). Average length/breath/depth for the former are 80×60×14 mm and for the latter 103×80×23 mm.

Fossil record: *Parmularius* sp. has been identified on a frontlet with horn cores, teeth and other fragmentary horn cores. There is also a cranium collected in 1959 which has been attributed to a larger species of alcelaphine. A third species may also be represented.

(4) Small antelopes and gazelles. Prints of dik-dik (*Madoqua*) are rare although dik-dik are the most abundant single species of fossil Bovidae. This anomaly may be explained by the fact that dik-dik are one of the Bovidae who do not require to drink water, subsisting on moisture from vegetation, while tracks of other Bovidae were probably made going to or from water holes.

Fossil record: *Neotragini*? and *Raphicerus* sp. (Steenbuck) are provisionally identified. *Madoqua* is very abundant. *Antilo-pini*: the gazelle appears to be *G. janenschi.*

Conclusions

The greater part of the fossil fauna from the Laetolil Beds is recorded in the fossil tracks. In all, over 20 taxa are represented. The preservation of the footprints can be attributed to an unusual and possibly unique combination of climatic, volcanic and mineralogic conditions. The available evidence indicates that the episode took place during a brief

牛　科

（1）牛族柯氏司马牛相似种。这些脚印组成大的、圆的、通常印痕很深的足迹，与现生的非洲水牛很像。它们显然是牛亚科动物留下的，可以认为是柯氏司马牛的，因为在化石记录中这是唯一的牛科动物。脚印有 4 列，分别由 16 个、12 个、10 个和 5 个脚印组成；另外还有 40 多个单独或成对的脚印。上述足迹中 43 个脚印的长宽深平均值为 185 毫米 × 149 毫米 × 25 毫米。

化石记录：已经发现一些牛角角心和牙齿，包括一件带有一对角心的颅骨。

（2）马羚未定种。在遗址（*A*）和（*C*）发现了 7 个单独的脚印，跟踪员把它们鉴定为“马羚”。它们的特点是每个脚印分成两瓣拉长的蹄印。长宽深平均测量值 109 毫米 × 95 毫米 × 16 毫米。

化石记录：采集于 1975 年的 3 件角心和一些牙齿被认为是一种马羚，并被暂时归入狄氏原转角牛羚。早些时候采集到的一件角心和一些牙齿也可以归到这个种，它比现生的马羚要小一些。

（3）狷羚族。遗址（*A*）的 18 个脚印和遗址（*C*）2 个脚印被跟踪员鉴定为狷羚的脚印。遗址（*A*）的脚印比遗址（*C*）的脚印更小更浅。前者的长宽深平均值 80 毫米 × 60 毫米 × 14 毫米，后者的为 103 毫米 × 80 毫米 × 23 毫米。

化石记录：依据一件带有角心的额骨、一些牙齿及角心碎块鉴别出了斗士羚未定种。有一件 1959 年采集到的颅骨被认为属于一种较大的狷羚类。可能还有第三种狷羚。

（4）小型羚羊和瞪羚。尽管犬羚是牛科化石中最丰富的一个种，但是犬羚的脚印非常少。这一异常现象可以用以下事实来解释，犬羚是牛科中几乎不需要饮水的种类之一，它们依赖植物中的水分生存，而其他牛科动物的足迹路径很可能是在前往水坑或从水坑返回时留下的。

化石记录：暂定的种类有岛羚族（尚未确定）和小岩羚未定种（斯廷巴克）。犬羚非常丰富。羚羊族：该瞪羚似乎为杰南齐瞪羚。

结　论

在莱托利尔层中出土的绝大多数化石种类都有化石足迹的记录。总共有超过 20 个种类。足迹的保留可以归因于气候、火山和矿物条件的异常且独特的组合。现有的证据表明这些足迹是在一个短暂的时期内留下的，大概是在某个雨季开始时恰逢

period, probably during the onset of a single rainy season which happened to coincide with the eruption of light ash showers from the nearby volcano Sadiman.

The locomotor pattern displayed by the trails of hominid footprints is still under examination but it is immediately evident that the Pliocene hominids at Laetoli had achieved a fully upright, bipedal and free-striding gait; a major event in the evolution of man which freed the hands for tool-making and eventually led to more sophisticated human activities. Moreover, evidence supplied by cranial parts of the somewhat later but related hominid fossils from the Afar in Ethiopia (dated between 2.6 and 3 Myr) indicates that bipedalism outstripped enlargement of the brain. To have resolved this issue is an important step in the study of human evolution, as it has long been the subject of speculation and debate.

With the hands free and available for purposes not connected with locomotion it is perhaps surprising that no form of artefact has been found. But the concept of tool-making may well have been beyond the mental ability of these small-brained creatures. Any "tools" or weapons used must have been solely of perishable materials as the Laetolil Beds are devoid not only of artefacts but of all stones other than volcanic ejecta.

Further exploration of sites and analysis of material will continue in 1979, but it is evident that Laetoli will give an unique perspective into hominid environment during Pliocene times.

We thank the United Republic of Tanzania for permission to continue research at Laetoli, the National Geographic Society, Washington, D. C. for financial support, A. A. Mturi, Director of Antiquities, Tanzania and A. J. F. Mgina, Conservator, Ngorongoro Conservation Authority, for their help, Philip Leakey and Peter Jones for organising field seasons, Drs A. W. Gentry, J. Harris and M. G. Leakey for identifying the bovid, giraffid, primate and carnivore fossils and all those who participated in the fieldwork.

(**278**, 317-323; 1979)

M. D. Leakey* and R. L. Hay†

* P. O. Box 7, Ngorongoro, Tanzania

† Department of Geology and Geophysics, University of California, Berkeley, California 94720

Received 28 September 1978; accepted 2 February 1979.

References:

1. Leakey, M. D. *et al. Nature* **262**, 460-466 (1976).
2. Kent, P. E. *Geol. Mag.* **78**, 173-184 (1941).
3. Dietrich, W. O. *Palaeontographica* **94**, 44-133 (1942).

附近的沙迪曼火山喷发出轻质的火山灰。

人科动物的足迹所表现出的运动模式尚无定论，但显而易见的是，莱托里上新世的人科动物已经具备了完全直立、两足行走并且随意跨步的步态。这在人类进化过程中是一个重大事件，这样就将人类的双手解放出来用以制造工具，并最终导致更加复杂精细的人类活动。此外，由稍晚但相关的来自埃塞俄比亚阿法尔的人科动物化石的部分颅骨（时间在距今 260 万到 300 万年间）提供的证据表明，两足行走的意义胜过人类脑容量的扩增。由于这个问题在很长时间内都是思考和争论的主题，所以解决这个问题是研究人类进化的重要步骤。

尽管解放出的双手可以用于非行走的目的，但或许出乎意料的是还没有发现任何形式的人工制品。制造工具的想法可能已经完全超过了这些小型脑量生物的智力水平。（这些人科动物）使用过的任何“工具”或武器肯定是完全用易腐材料制成的，因为在莱托利尔层中不仅缺乏人工制品，而且除火山喷出物外也没有其他石头。

遗址的进一步勘探和材料的分析将持续到 1979 年，但莱托里显然将为认识上新世人科动物的生存环境提供独特的视角。

感谢坦桑尼亚联合共和国允许我们继续在莱托里地区从事研究，感谢国家地理学会、华盛顿政府的经济支持，感谢坦桑尼亚文物部主管姆图里和管理员姆吉纳，感谢恩戈罗恩戈罗自然保护区管理局的帮助，感谢菲利普 · 利基和彼得 · 琼斯组织的野外季，感谢金特里博士、哈里斯和利基帮助鉴定牛科、长颈鹿科、灵长类动物和食肉动物的化石，感谢所有参加野外考察的人员。

（张玉光 翻译；董为 审稿）

Single Strands Induce recA Protein to Unwind Duplex DNA for Homologous Pairing

R. P. Cunningham *et al.*

Editor's Note

Homologous recombination, where genes swap over between DNA strands, plays a key role in generating genetic diversity and repairing DNA damage. Here biochemist Charles M. Radding and colleagues elucidate the function of a key protein in the process, the bacterial enzyme recA. The paper describes how single DNA strands stimulate recA to unwind double-stranded DNA for homologous pairing. The mechanism has since been fine-tuned: it is now known that recA forms a helical filament with single-stranded DNA which then binds to double-stranded DNA and samples for homology. Once found, the filament causes an exchange of strands that yields recombined DNA. Given the ability of short DNA fragment to stimulate unwinding, Radding correctly speculates that recombination may be inducible.

Single-stranded DNA, whether homologous or not, stimulates purified *Escherichia coli* recA protein to unwind duplex DNA. This helps to explain how recA promotes a search for homology in genetic recombination. As oligodeoxynucleotides also stimulate unwinding, a common mechanism may relate the function of recA protein in recombination to other functions (SOS) induced by oligonucleotides.

BECAUSE breakage of DNA by various means stimulates general genetic recombination, some investigators have suggested that recombination begins with the interaction of a single strand with duplex DNA (see ref. 1 for review). As a model for this kind of interaction, Holloman *et al.*[2] and Beattie *et al.*[3] studied the uptake of homologous single strands by superhelical DNA [replicative form I (RFI) or form I], a reaction that produces a D-loop (see Fig. 1). By transfection of *Escherichia coli*, Holloman and C.M.R.[5] implicated the *recA* gene in the homologous interaction of a single-stranded fragment with superhelical DNA. More recently, Shibata *et al.*[4] observed that stoichiometric amounts of highly purified recA protein promoted the rapid formation of D-loops by superhelical DNA and homologous single-stranded fragments (Fig. 1).

The properties of the uncatalysed formation of D-loops, which occurred rapidly only at non-physiological temperatures, suggested that the rate was limited by the initial unstacking of a small number of base pairs in superhelical DNA[3]. As recA protein bypassed this rate-limiting step, we suggested that it acted in part by unstacking base pairs in duplex DNA. We also observed that a preparation of nicked circular DNA (form II) was one-third as active as superhelical DNA in forming complexes with single-stranded fragments[4]. Subsequently, we

单链DNA诱导recA蛋白解旋DNA以利于同源配对

坎宁安等

编者按

同源重组即基因在DNA分子之间的交换，在遗传多样性的产生和DNA损伤的修复方面具有关键的作用。本文中，生化学家查尔斯·雷丁和他的研究小组阐明了细菌来源的酶recA这一关键蛋白在此过程中的功能。本文描述了单链DNA是如何触发recA将双链DNA解旋以进行同源配对的。这一机制是精密可控的：已知recA可以和单链DNA形成一种螺旋细丝，然后这种单链DNA就可以和双链DNA结合并进行同源配对。这种结合一旦形成，这段单链就会导致链之间发生交换并产生重组DNA。如果短的DNA片段具有触发解旋的作用，那么雷丁关于重组可能是被诱导的推测就是正确的。

无论是否同源，单链DNA都可以触发从大肠杆菌中纯化的recA蛋白解旋双链DNA。这有助于解释recA蛋白是如何促进遗传重组中（DNA的）同源配对的。由于寡脱氧核苷酸也可以触发DNA解旋，因此，这种共同的机制可能使recA蛋白在遗传重组中的功能与其他由寡核苷酸诱导的功能（如SOS）建立联系。

由于各种各样的因素所导致的DNA链的断裂普遍会触发遗传重组，因此，一些研究人员认为重组起始于单链DNA与双链DNA的相互作用（见参考文献1）。作为研究此类相互作用的一种模型，霍洛曼等人[2]以及贝蒂等人[3]研究了同源单链DNA被超螺旋DNA［I复制型（RFI）或I型］结合并形成D环（见图1）的过程。通过转染大肠杆菌，霍洛曼和雷丁[5]提出*recA*基因参与单链DNA片段与超螺旋DNA的同源相互作用。最近，柴田等人[4]观察到化学计量定量的高度纯化的recA蛋白可以促进超螺旋DNA和同源的单链DNA片段迅速形成D环（图1）。

非催化的D环的形成仅在非生理温度下迅速发生，表明其反应速率受超螺旋DNA少量碱基对的起始解离的限制[3]。由于recA蛋白不受这一限速步骤的影响，因此我们认为它在某种程度上通过解开双链DNA的配对碱基来发挥作用。我们还观察到，带切口的环状DNA（II型）在与单链DNA片段形成复合物时，其活性只有超螺旋DNA的三分之一[4]。接下来，我们证实了II型DNA是recA蛋白形成D

have confirmed that form II DNA is a substrate for the formation of D-loops by recA protein (see below). The latter observation strengthened the inference that recA protein has an unstacking or unwinding activity, which is the subject of this article.

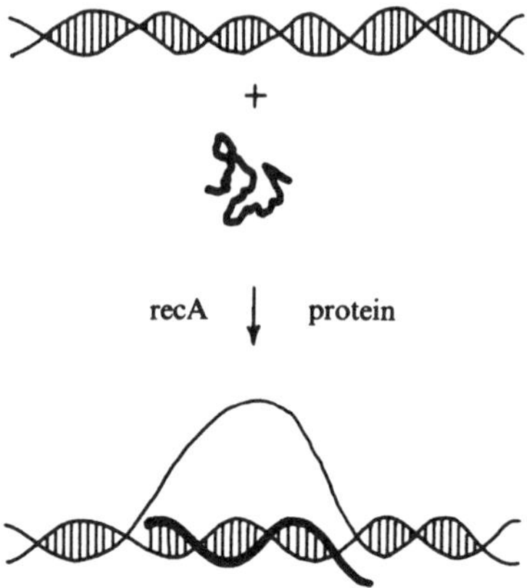

Fig. 1. The formation of a D-loop catalysed by recA protein[4]. The actions of recA protein apparently include (1) unwinding double-stranded DNA, (2) unfolding single-stranded DNA, (3) homologous pairing, and (4) uptake of a single strand rather than rewinding of the original duplex DNA.

Unwinding of Duplex DNA by recA Protein in the Presence of ATPγS and Single Strands

To obtain direct evidence that recA protein unwinds DNA, we incubated it with fd form II DNA, added DNA ligase, and examined the product by gel electrophoresis and isopycnic centrifugation in CsCl plus propidium diiodide[6,7].

In the absence of recA protein, ligase sealed more than half of the form II DNA as indicated by gel electrophoresis (Fig. 2*a*, *b*) and by the assay of Kuhnlein *et al.*[8] for covalently closed DNA. At the ionic strength of the gel, these covalently closed molecules (form I′) migrated as a set of bands of positively superhelical DNA, which overlapped a band due to linear DNA (form III) present in the preparation (Fig. 2*b*). When recA protein and adenosine 5′-*O*-(3-thiotriphosphate) (ATPγS) were incubated with form II DNA before ligation, the observed distribution of products was similar; ligase worked well but there was no apparent change in the average linking number of the population of sealed molecules (Fig. 2*c*). However, when we added fragments of single-stranded ΦX174 DNA during the incubation with recA protein, we observed a very different result. Ligase still closed about one-third to one-half of the fd form II DNA, but all the closed DNA migrated as a single band of more compact DNA, at about the position of form I (Fig. 2*d*). In three experiments, when recA protein and ATPγS were incubated with form II DNA in the absence of any added single-stranded DNA before ligation, we observed the formation of a small amount of material with the electrophoretic mobility of form I (see Table 1, lines 1 and 2). Controls included the following variations of the mixture represented in Fig. 2*d*: no recA protein; no form II DNA; no ATPγS; ATP in place of ATPγS; recA protein in the absence of ligase (Fig. 2*a*). In no case was DNA detected in the position of the rapidly

环时的一个底物（见下文）。这一观察结果进一步加强了关于 recA 蛋白具有解配对或解旋活性的推论，这也是本文的主题。

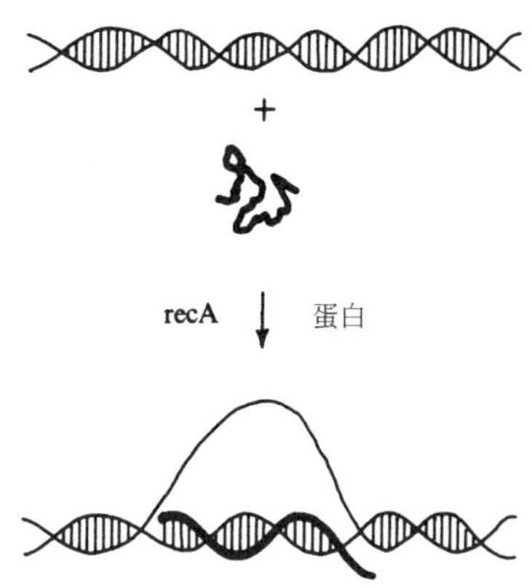

图 1. recA 蛋白催化 D 环的形成[4]。recA 蛋白的作用显然包括：(1) 促进双链 DNA 解旋；(2) 促进单链 DNA 的伸展；(3) 促进同源配对；(4) 促进解旋的 DNA 链与单链结合，而不是促进原始双链 DNA 重新形成螺旋。

在 ATPγS 和单链 DNA 存在时 recA 蛋白促使双链 DNA 解旋

为了获得 recA 蛋白能够解旋 DNA 的直接证据，我们将 recA 蛋白与 fd II 型 DNA 共同孵育，并加入 DNA 连接酶进行反应，然后使用凝胶电泳和加入二碘化丙啶的 CsCl 等密度离心的方法来检测反应产物[6,7]。

正如凝胶电泳的结果（图 2*a* 和 2*b*）所示，在没有 recA 蛋白存在时，DNA 连接酶可以将一半以上的 II 型 DNA 封闭。库恩莱因等人[8]关于共价闭合 DNA 的类似的实验进一步证实了这个结果。在凝胶的离子强度下，这些共价闭合的分子（I′型）在凝胶中以一系列正超螺旋 DNA 条带的形式迁移，且与样品中存在的一个线性 DNA（III 型）的条带重叠（图 2*b*）。如果把 recA 蛋白与硫代三磷酸腺苷（ATPγS）以及 II 型 DNA 在进行连接反应前共同孵育，也可以观察到相似的连接反应产物条带分布。此时连接酶可以正常工作，但是封闭分子的平均链环数似乎没有发生明显的改变（图 2*c*）。然而，当我们向 recA 蛋白与 DNA 的共同孵育体系中加入单链 ΦX174 DNA 片段时，我们观察到了一个非常不同的结果。此时，连接酶仍然可以闭合 1/3～1/2 的 fd II 型 DNA，不过所有这些闭合的 DNA 在电泳实验中以包装更为紧密的 DNA 所对应的单独的一条条带进行迁移，并最终位于 I 型 DNA 大概所处的位置处（图 2*d*）。在三次实验中，当把 recA 蛋白、ATPγS 和 II 型 DNA 在进行连接反应前共孵育并且不加入任何单链 DNA 的话，可以观察到形成了少量具有 I 型 DNA 电泳迁移率的物质（见表 1，第 1 和第 2 行）。我们设计了如下各种对照组（混合物成分见图 2*d*）：不含 recA 蛋白；不含 II 型 DNA；不含 ATPγS；ATP 代替 ATPγS；加入 recA 蛋白但不加连接酶（图 2*a*）。在这些对照试验中，快速泳动的条带处均检测不

migrating band.

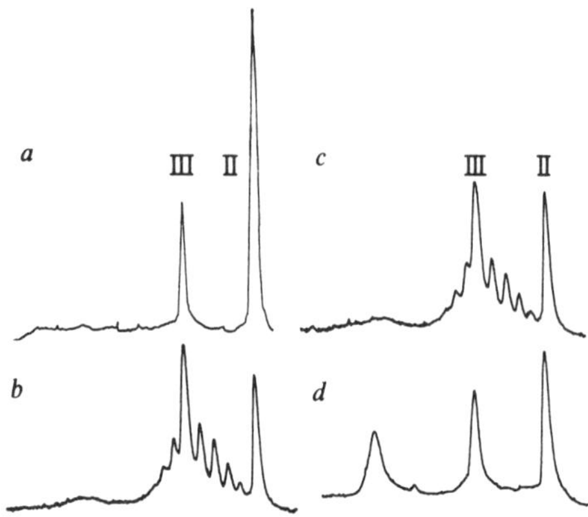

Fig. 2. The linking number of circular DNA sealed in the presence of recA protein and ATPγS (Boehringer–Mannheim). Microdensitometer tracings of electrophoretic gels prepared as described in Fig. 4, legend. *a*, Control, fd form II DNA incubated with recA protein but no ligase. The preparation of DNA contained about 30% form III DNA as indicated. *b*, Ligase only (New England BioLabs). *c*, Ligase plus recA protein. *d*, Ligase, recA protein and 12-μM fragments of heterologous single-stranded DNA from ΦX174. The various forms of circular DNA from phages ΦX174 and fd were prepared as cited previously[4]. Single-stranded fragments were prepared by boiling 840 μM DNA for 15 min in 0.1 mM EDTA and 10 mM Tris-HCl at *p*H7.5. All concentrations of DNA are expressed as mol of nucleotide. The reaction mixture was the same as that described in Table 2 legend, except that heterologous single-stranded fragments were substituted for homologous fragments, the volume was doubled and ATP was replaced by 0.5 mM ATPγS. After incubation at 37°C for 30 min, the mixture was chilled on ice, KCl, $(NH_4)_2SO_4$ and NAD^+ added to make their concentrations 25 mM, 10 mM, and 25 μM, respectively, and incubated at 18°C for 5 min. 0.3 units of *E. coli* DNA ligase were then added and the sample, 50 μl total volume, incubated for 90 min. The reaction was stopped by adding EDTA, 28 mM, and SDS, 0.17%. The sample was transferred to a bath at 37°C, proteinase K, 0.2 mg ml^{-1} added and the sample incubated for 15 more minutes.

To determine whether the compact closed DNA had a lower or a higher linking number than the DNA closed by ligase alone (form I′), we centrifuged the products of a similar experiment in an isopycnic gradient of CsCl and propidium diiodide (Fig. 3 and Table 1). In the absence of recA protein, half of the recovered DNA was found in a single peak at the dense end of the gradient, as expected for form I′ DNA derived from the closure of form II by ligase (Fig. 3*b*, Table 1, line 1). Incubation of form II DNA with recA protein and ATPγS before ligation had little effect on the distribution (Fig. 3*a*, Table 1, line 2). However, when recA protein, fd form II DNA, ΦX174 single-stranded fragments and ATPγS were incubated before ligation, the peak of form I′ DNA at the dense end of the gradient disappeared almost entirely and was replaced by DNA with a broad distribution of densities between those of form I′ and linear or form II DNA (Fig. 3*a*, Table 1, line 3), that is, DNA with lower linking numbers than form I′. The peak of material of intermediate density corresponded to the position of negatively superhelical fd DNA (form I DNA). The total recovery of radioactivity was similar in the three parts of this experiment; the fraction of closed circular DNA relative to linear or form II DNA (Fig. 3*a*, Table 1, line 1) was the same as that estimated by the assay of Kuhnlein *et al.*[8], which indicates that the distribution of recovered material was not biased.

到 DNA。

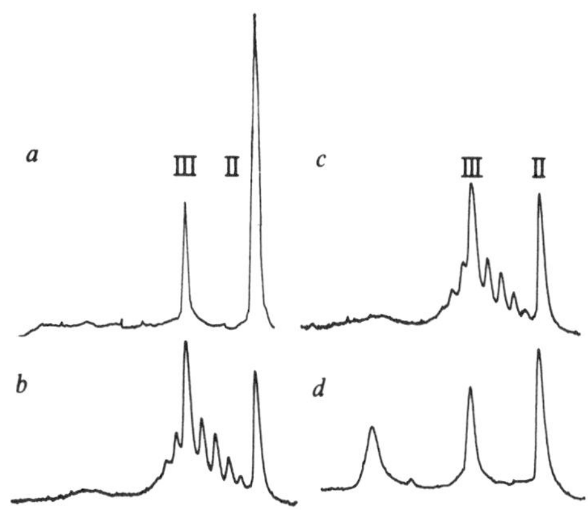

图 2. recA 蛋白和 ATPγS 存在条件下封闭的环状 DNA 的链环数（伯林格－曼海姆）。采用图 4 图注所述的方法制备凝胶电泳的显微密度计测定线。*a*，对照组，将 fd II 型 DNA 与 recA 蛋白共同孵育，但不加入连接酶。如图所示，样品中含有约 30% 的 III 型 DNA。*b*，只含有连接酶（购自新英格兰生物实验室公司）。*c*，只含有连接酶和 recA 蛋白。*d*，含有连接酶，recA 蛋白以及 12 μM 来自 ΦX174 的异源单链 DNA 片段。采用前文所引用的方法 [4] 制备源于 ΦX174 噬菌体及 fd 噬菌体的各种形式的环状 DNA。将 840 μM 的 DNA 在含有 0.1 mM EDTA 和 10 mM pH 7.5 的 Tris-HCl 缓冲液中煮沸 15 分钟即可得到单链 DNA 片段。所有的 DNA 浓度都使用核苷酸的摩尔数来表示。反应体系与表 2 注中所描述的一致，只不过异源单链 DNA 片段被换成了同源 DNA 片段，溶液的体积也扩大了一倍，另外 ATP 被换成了 0.5 mM 的 ATPγS。在 37℃孵育 30 分钟后，将反应体系在冰上冷却，然后向其中加入 KCl、$(NH_4)_2SO_4$ 和 NAD^+ 使其终浓度分别为 25 mM、10 mM 和 25 μM。然后，在 18℃孵育 5 分钟。向样品中加入 0.3 U 的大肠杆菌连接酶，并将反应液总体积调整到 50 μl，孵育 90 分钟。向反应体系中加入 EDTA 和 SDS，至终浓度分别为 28 mM 和 0.17%，使反应终止。将产物转移到 37℃水浴中，加入 0.2 mg/ml 的蛋白酶 K，再孵育 15 分钟。

为了证明紧密闭合 DNA 的链环数比经连接酶闭合的 DNA（I′ 型）低还是高，我们将一个类似实验的反应产物使用 CsCl 和二碘化丙啶进行了等密度梯度离心分离（图 3 和表 1）。当反应体系中没有 recA 蛋白时，约有一半回收的 DNA 在梯度的高密度区形成的一个单峰里，这与预期中的连接酶将 II 型 DNA 闭合而形成 I′型 DNA 的结果吻合（图 3*b*，表 1，第 1 行）。在连接反应前将 II 型 DNA 与 recA 蛋白和 ATPγS 孵育并不会对这种分布产生什么影响（图 3*a*，表 1，第 2 行）。然而，如果在进行连接反应前把 recA 蛋白、fd II 型 DNA、ΦX174 单链 DNA 片段以及 ATPγS 一起孵育，则位于梯度中最高密度区的 I′型 DNA 的峰几乎完全消失，取而代之的是密度介于 I′型 DNA 与线性或 II 型 DNA 之间的广泛分布的峰（图 3*a*，表 1，第 3 行），也就是说，这些 DNA 的链环数比 I′ 型 DNA 要低。位于中等密度区域的峰代表负超螺旋 fd DNA（I 型 DNA）。在这个实验的三个部分中，放射性的总回收率都很接近。闭合环状 DNA 与线性或 II 型 DNA（图 3*a*，表 1，第 1 行）的比值也与库恩莱因等人 [8] 的实验中测得的结果一致，这表明在这些实验中，回收的物质在分布上并不存在偏倚。

Table 1. Unwinding of DNA

	Total recovery (%)	Distribution (%)		
		I′	Intermediate	II + III
Ligase only	22	46	4	50
+recA protein	29	39	7	54
+recA protein and fragments	28	8	21	71

Data tabulated from Fig. 3. The DNA labelled "Intermediate" has a lower linking number than form I′ DNA, the product of sealing by ligase alone.

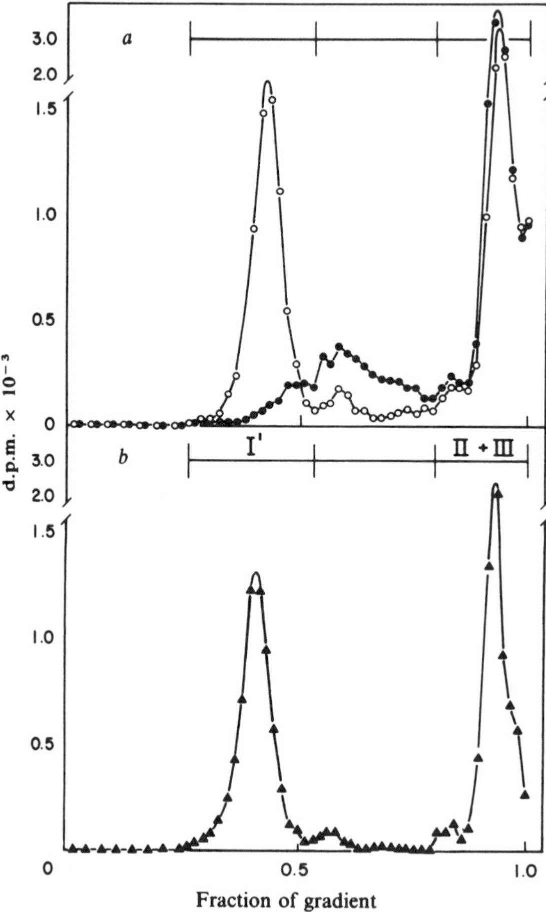

Fig. 3. Decreased linking number of circular DNA sealed in the presence of recA protein and ATPγS; isopycnic centrifugation in CsCl and propidium diiodide[7]. *a*, fd form II DNA sealed in the presence of recA protein only (○), and in the presence of recA protein plus heterologous fragments of single-stranded DNA from ΦX174 (●). *b*, Control. DNA sealed in the absence of recA protein (▲). The left end of the figure corresponds to the dense end of the gradient. We summed values as indicated by the brackets (Table 1). All values were corrected for quenching. Solutions, 7.4 ml, contained CsCl to make the density 1.565 g cm^{-3}, 500 μg ml^{-1} propidium diiodide, 10 mM Tris-HCl and 1 mM EDTA at *p*H 8.0. These solutions were centrifuged at 44,000 r.p.m. and 20°C for 69 h in a Beckman 75 ti rotor. The resulting gradients were collected from the bottom of the tube and counted in a scintillation fluor containing Triton X-100.

表 1. DNA 的解旋

	总回收率（%）	分布（%）		
		I′ 型	中间体	II 型 +III 型
只加连接酶	22	46	4	50
连接酶 +recA 蛋白	29	39	7	54
连接酶 +recA 蛋白 +DNA 片段	28	8	21	71

根据图 3 制成表的数据。被标记为“中间体”的 DNA 与仅由连接酶封闭的产物 I′ 型 DNA 相比具有更低的链环数。

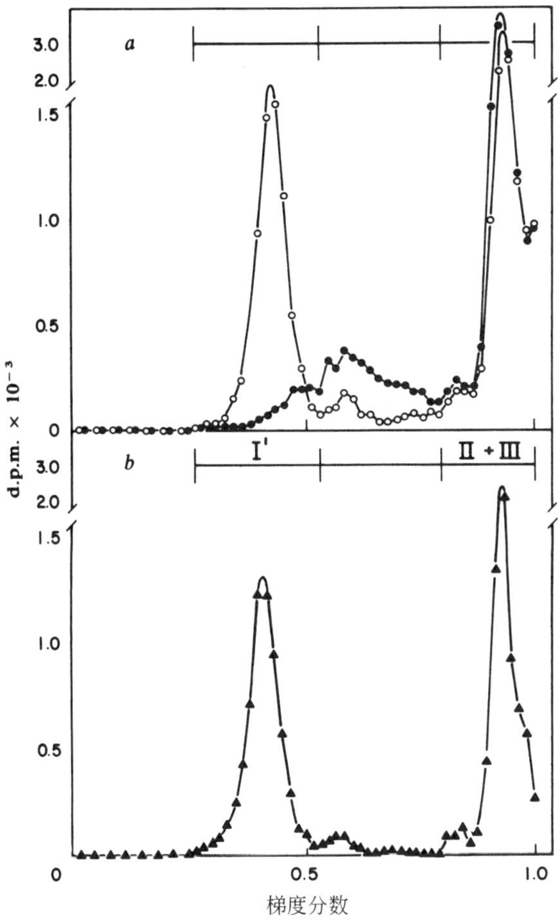

图 3. CsCl 和二碘化丙啶的等密度离心实验[7]表明，在 recA 蛋白和 ATPγS 存在时，环状 DNA（由连接酶闭合的，译者注）的链环数减少。*a*，在只含有 recA 蛋白（○）和同时含有 recA 蛋白及异源 ΦX174 噬菌体单链 DNA 片段（●）时，闭合的 fd II 型 DNA 的量。*b*，对照组。不含有 recA 蛋白时（▲），闭合的 DNA 的量。图的左端对应着密度梯度的高密度端。我们将各部分的数据相加，如表 1 所示。所有的数据都经过校正。溶液体积 7.4 ml，含有 CsCl，密度为 1.565 g/cm^3，二碘化丙啶浓度为 500 μg/ml，另外含有 10 mM pH8.0 的 Tris-HCl 和 1 mM pH 8.0 的 EDTA。这些溶液使用贝克曼 75ti 转子在 20℃和 44,000 转 / 分钟转速下离心 69 小时。从离心管底部逐层收集不同密度梯度的组分，并使用含有 Triton X-100 的荧光闪烁来对 DNA 定量。

Lack of Topoisomerase Activity of recA Protein

The unwinding observed in the experiments described above could have occurred before or after ligation, but three experiments show that recA protein lacks detectable topoisomerase activity in the presence of either ATPγS or ATP. We incubated recA protein with relaxed closed circular DNA (form I′) in conditions identical to those of the experiments shown in Figs 2 and 3, including a sample which contained heterologous fragments and ATPγS. When we observed the product by gel electrophoresis, we found no DNA in the position of form I DNA, and no perceptible shift in the distribution of form I′ DNA (Fig. 4). With the following additions or omissions, we examined, by gel electrophoresis, the fate of fd RFI incubated for 30 min at 37°C with recA protein and fragments of single-stranded ΦX174 DNA: ATP; ATPγS; ATPγS without single-stranded fragments; ATPγS without recA protein. We observed little or no change in the amount of form I DNA (data not shown).

In further experiments we made D-loops by incubating recA protein with fd form I DNA and fragments of fd single strands in the presence of ATP[4]. Measurement by electron microscopy of 22 D-loops with one or two single-stranded tails showed that the mean length was 416±45 nucleotides. (We were unable to distinguish the double-stranded arm of the D-loop from the single-stranded one. As there was little difference in the lengths of the two arms of any D-loop, we simply measured both and calculated the average of all values.) In the conditions of the reaction, fd form I DNA should contain about 40 superhelical turns which corresponds to a D-loop about 400 nucleotides long. Previous observations have shown a rough equivalence between the length of D-loops measured by electron microscopy and the length predicted from the superhelix density in solution[9]. Observation of the same relationship in these experiments suggests that when recA protein catalysed the formation of D-loops from form I DNA, the linking number of the closed circular DNA was not markedly changed.

Table 2. Homologous pairing of double-stranded DNA and single-stranded fragments

Fragments	% fd double-stranded [^{3}H] DNA retained	
	form I	form II
None	1	4
fd	34	22
ΦX	2	6

The reaction mixture, 20.5 μl in volume, contained 31 mM Tris-HCl at *p*H 7.5, 25 mM $MgCl_2$, 1.8 mM dithiothreitol, 88 μg ml^{-1} bovine serum albumin, 1.3 mM ATP, 8.8 μM RF[^{3}H]DNA, 12 μM single-stranded DNA fragments and 98 μg ml^{-1} recA protein[4]. The mixture was incubated at 37°C for 30 min, chilled and 18 μl transferred to 0.5 ml of cold 25 mM EDTA. Aliquots were taken from this mixture to measure total counts, D-loops and closed circular DNA, exactly as described before[4].

recA 蛋白缺乏拓扑异构酶活性

上述实验中所观察到的解旋现象有可能是在连接反应之前发生的，也可能是在反应完成后才发生的。但是，在这三个实验中，无论是加入 ATP 或是 ATPγS，recA 蛋白都没有可检测到的拓扑异构酶的活性。我们将 recA 蛋白与松弛闭合环状 DNA（I′ 型）在与图 2 和图 3 所示的实验相同的条件下孵育，并在一个样品中加入含有异源单链 DNA 片段和 ATPγS 的样品反应一段时间。当我们使用凝胶电泳实验来分析反应产物时，我们发现在 I 型 DNA 的位置上没有 DNA，而 I′ 型 DNA 的分布也没有显著移动（图 4）。于是我们又以下述实验条件的增删继续做了一些实验，把 fd RFI、recA 蛋白和 ΦX174 单链 DNA 片段分别与 ATP、ATPγS、ATPγS 但不含单链 DNA 片段或 ATPγS 但不含 recA 蛋白在 37℃孵育 30 分钟，并通过凝胶电泳实验来分析 DNA 迁移率的变化。结果发现 I 型 DNA 的含量几乎没有或仅有微小的改变（数据未展示）。

在进一步的实验中，我们将 recA 蛋白与 fd I 型 DNA 以及 fd 单链 DNA 一起孵育，并加入 ATP，以产生 D 环[4]。通过电子显微镜测量 22 个带有一个或两个单链 DNA 尾巴的 D 环，我们发现其平均长度为 416±45 个核苷酸。（我们无法区分 D 环的双链臂与单链臂。由于任一 D 环的这两种臂的长度都差不多，因此我们简单地采取了测量两种环的长度，取平均值的方法。）在这个反应条件下，fd I 型 DNA 应该含有大约 40 个超螺旋转弯，相当于一个约 400 个核苷酸组成的 D 环。之前的研究表明使用电子显微镜测得的 D 环的长度与通过溶液中超螺旋密度预测的长度大致相等[9]。这些实验中所观察到的与上述相同的等量关系表明当 recA 蛋白催化 I 型 DNA 形成 D 环时，闭合环状 DNA 的链环数并没有发生显著的改变。

表 2. 双链 DNA 和单链 DNA 片段的同源性配对

片段	残留的 [³H] 标记的 fd 双链 DNA 的比例（%）	
	I 型	II 型
无	1	4
fd	34	22
ΦX	2	6

反应混合物体积 20.5 μl，含有 31 mM pH7.5 的 Tris-HCl，25 mM $MgCl_2$，1.8 mM 二硫苏糖醇，88 μg/ml 牛血清白蛋白，1.3 mM ATP，8.8 μM RF[^{3}H]DNA，12 μM 单链 DNA 片段以及 98 μg/ml recA 蛋白[4]。反应混合物在 37℃孵育 30 分钟，冷却后取 18 μl 加到 0.5 ml 预冷的 25 mM 的 EDTA 溶液中。将样品等量分装后严格按照文献描述的方法[4]分别测量 DNA 总数、D 环数以及闭合环状 DNA 的数量。

Nature of The Cofactor for Unwinding by recA Protein

Unwinding is supported either by heterologous or homologous single-stranded fragments (Fig. 4*a*, *b*). Circular single-stranded DNA, which does not form D-loops, promoted unwinding (Fig. 4*d*). Double-stranded restriction fragments were inactive, but their boiled products were active (Fig. 4*f*, *g*). This observation indicates that the cofactor was not some fortuitous ingredient of the DNA preparations. A digest, consisting entirely of acid-soluble material produced by treatment with pancreatic DNase and boiling, also supported the unwinding activity very well (Fig. 4*e*). The following, added singly, promoted little or no unwinding: deoxyadenosine, dAMP, dGMP, dCMP, dTMP, AMP and CMP (data not shown).

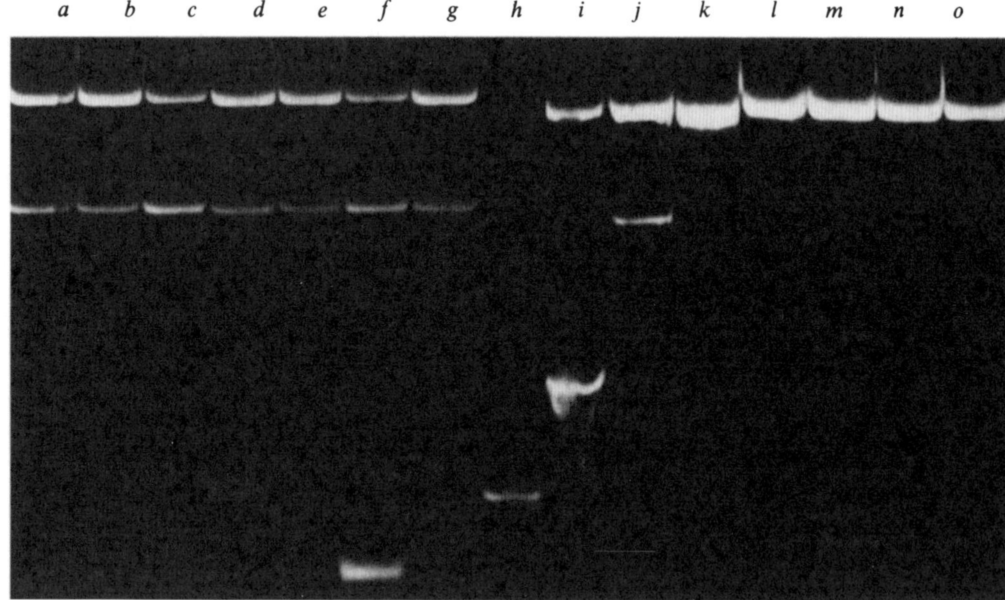

Fig. 4. Nature of the cofactor that stimulates unwinding by recA protein in the presence of ATPγS (*a–g*); absence of topoisomerase activity in recA protein (*l–o*): agarose gel electrophoresis. Lanes *a–g* contained fd form II DNA sealed in the presence of recA protein, ATPγS and 12 μM DNA cofactors: *a*, fragments of single-stranded ΦX174 DNA; *b*, fragments of single-stranded fd DNA; *c*, no addition; *d*, fd phage DNA; *e*, boiled DNase I limit digest of ΦX174 RFI DNA; *f*, restriction endonuclease *Hae*III digest of fd RFI; *g*, the preceding digest boiled. Lanes *h–k* contained markers: *h*, fd phage DNA; *i*, fd RFI DNA; *j*, fd form II DNA (~70%) and fd form III DNA (~30%), *k*, fd form I′ DNA. Lanes *l–o* contained fd form I′ DNA incubated with recA protein and ATPγS: *l*, no addition; *m*, fragments of single-stranded ΦX174 DNA; *n*, preceding sample minus recA protein; *o*, fragments of single-stranded ΦX174 DNA plus ATP instead of ATPγS. To obtain the limit digest indicated above 24 μM ΦX174 RF DNA was treated with 20 μg ml^{-1} pancreatic DNase (Worthington) for 30 min at 37°C in 5 mM $MgCl_2$ and 10 mM Tris-HCl at *p*H 7.5, which rendered all the DNA acid soluble. The product was boiled for 4 min to inactivate the pancreatic DNase and denature all the fragments. Nicked circular DNA, sealed by polynucleotide ligase was slightly positively superhelical under the conditions of gel electrophoresis (for example lane *c*) whereas closed circular DNA relaxed by calf thymus topoisomerase had a titrable superhelix density close to zero (for example lane *k*). We used 1% agarose gels, 16 cm×16 cm×0.3 cm in E buffer, which contained 40 mM Tris-acetate, 5 mM sodium acetate and 1 mM EDTA at *p*H 8.0. Current was applied at 35 V for 16–18 h, and the buffer was recirculated between the reservoirs, following which we stained the gels for 2 h in E buffer containing 0.5 μg ml^{-1} ethidium bromide. We illuminated the gels from below with a short-wavelength UV lamp (Ultra-Violet Products) and photographed them with Polaroid Tyffe 55 Land film. The negatives were scanned with a Joyce-Loebl recording micro-densitometer.

recA 蛋白解螺旋作用的辅因子的性质

解旋的过程需要异源或同源单链 DNA 片段的协助（图 4*a*，*b*）。环状单链 DNA 虽然不能够形成 D 环，但是可以促进双链 DNA 的解旋（图 4*d*）。双链限制性酶切片段不具有这种活性，但是其煮沸后的产物则可以促进双链 DNA 的解旋（图 4*f*，*g*）。这些观察结果表明，解旋的辅因子不是 DNA 样品的某些偶然性成分。使用胰 DNA 酶消化并煮沸处理产生的全部酸溶性产物也具有很好的促进解旋活性（图 4*e*）。而脱氧腺苷、dAMP、dGMP、dCMP、dTMP、AMP 和 CMP 单独加入反应体系时，不具有或仅具有很低的促进解旋的活性（数据未展示）。

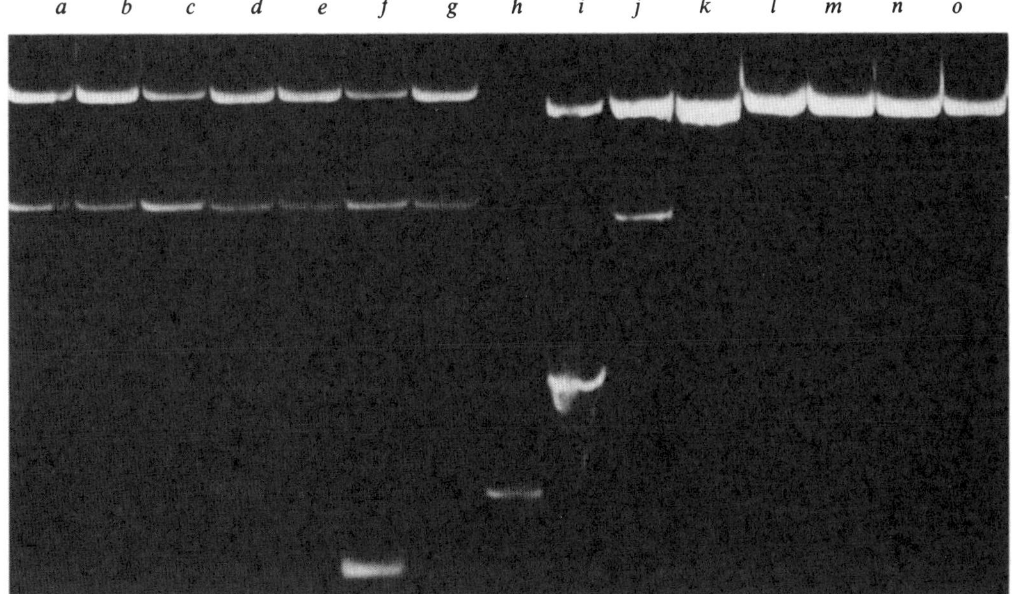

图 4. 琼脂糖凝胶电泳实验结果。当 ATPγS 存在时，触发 recA 蛋白发挥解旋作用的辅因子的性质（*a*~*g*）；recA 蛋白不具有拓扑异构酶的活性（*l*~*o*）。*a*~*g* 泳道含有在 ATPγS、12 μM DNA 辅因子以及 recA 蛋白作用下闭合的 fd II 型 DNA：(*a*)，ΦX174 单链 DNA 片段；(*b*)，fd 单链 DNA 片段；(*c*)，空白对照；(*d*)，fd 噬菌体 DNA；(*e*)，DNA 酶 I 不完全消化 ΦX174 RFI DNA 后经煮沸处理的产物；(*f*)，fd RFI 的 *Hae*III 限制性内切酶酶切产物；(*g*)，fd RFI 被 *Hae*III 限制性内切酶酶切后，经煮沸处理的产物。*h*~*k* 泳道为各种分子标记物：(*h*)，fd 噬菌体 DNA；(*i*)，fd RFI DNA；(*j*)，fd II 型 DNA（约 70%）和 fd III 型 DNA（约 30%）；(*k*)，fd I′ 型 DNA。*l*~*o* 泳道为 fd I′ 型 DNA 与 recA 蛋白和 ATPγS 共同孵育后的产物：(*l*)，空白对照；(*m*)，ΦX174 单链 DNA 片段；(*n*)，ΦX174 单链 DNA 片段但不加入 recA 蛋白；(*o*)，ΦX174 单链 DNA 片段，且用 ATP 代替 ATPγS。为了得到上述不完全消化的 DNA，我们用 20 μg/ml 的胰 DNA 酶（购自沃辛顿公司）在含有 5 mM $MgCl_2$、10 mM Tris-HCl 的 pH 7.5 的缓冲液中酶切 24 μM 的 ΦX174 RF DNA，37℃反应 30 分钟，最终所有的核酸都处于酸溶状态。将反应产物煮沸 4 分钟，以灭活其中的胰 DNA 酶，并使所有的 DNA 片段变性。由多核苷酸连接酶闭合的带切口环状 DNA 在凝胶电泳的实验条件下会形成轻微的正超螺旋（例如 c 泳道），而经小牛胸腺拓扑异构酶松弛的闭合环形 DNA 的超螺旋滴定密度则接近于零（例如 *k* 泳道）。凝胶电泳实验方法如下：1% 琼脂糖凝胶，大小为 16 cm × 16 cm × 0.3 cm，使用 E 缓冲液（40mM Tris– 乙酸，5 mM 乙酸钠，1 mM EDTA，pH 8.0）。恒定电压 35 V，电泳 16~18 小时，其间缓冲溶液在电泳槽之间不断地循环。电泳结束后，使用含有 0.5 μg/ml 溴化乙啶的 E 缓冲液染色 2 小时，然后用短波紫外灯观察凝胶，并使用 Polaroid Tyffe 55 胶片拍照。阴性结果使用 Joyce-Loebl 微量密度计进行扫描。

Homologous Pairing of Form II DNA and Single-stranded Fragments

Table 2 compares form II and form I DNA as substrates for recA protein in the homologous pairing with single-stranded fragments. In the conditions of these experiments (see Table 2, legend), which differ slightly from previously described conditions[4], the preparation of form II DNA was one-half to two-thirds as active as form I DNA in making complexes that were retained by nitrocellulose filters. As the preparation of form II DNA contained less than a few per cent of form I DNA, these complexes must have been made with DNA that was not superhelical. The synthesis of complexes with form II DNA required homologous single-stranded fragments, and the properties of the product were similar to those of D-loops made from RFI; the complexes were not dissociated by treatment with 0.2% SDS nor by heating at 50°C for 4 min in 1.5 M NaCl and 0.15 M Na citrate at *p*H 7. Preliminary electron microscopic observations suggest that a significant fraction of complexes made with form II DNA contain D-loops (data not shown). Recently, McEntee, Weinstock and Lehman also described the synthesis of D-loops from non-superhelical DNA, in that case, from linear phage P22 DNA[10].

The formation of D-loops by form II DNA suggests that recA protein unwinds DNA in the presence of ATP as well as in the presence of the analogue, ATPγS.

Discussion

Other experiments have shown that ATPγS competitively inhibits the ATPase activity of recA protein and blocks the formation of D-loops, but promotes the formation of complexes of protein, double-stranded DNA and single-stranded DNA (unpublished observations). The competitive inhibition shows that ATPγS occupies the same binding site as ATP; this, together with the other observations cited, suggests that the accumulation of partially unwound molecules reflects a step in the reaction that normally produces a D-loop. These effects of ATPγS on recA protein are reminiscent of observations on DNA gyrase which also catalyses several partial reactions in the presence of another non-hydrolysable analogue of ATP[11]. From the observed synthesis of D-loops with form II DNA, we infer that recA protein also unwinds DNA in the presence of ATP. The hydrolysis of ATP alters or dissociates the protein–DNA complex and accounts for the failure of partially unwound molecules to accumulate in the presence of ATP and DNA ligase (unpublished observations).

At first glance, the activation of unwinding by heterologous single strands seems incongruous for a mechanism that promotes homologous pairing, but further thought suggests that the opposite is true, that nonspecific stimulation of unwinding by any single strand explains in part how recA protein promotes a search for homology. As recombination in *E. coli* depends strongly on the function of recA[12], the action of recA protein by this mechanism implies that recombination often begins with the interaction of a single strand with duplex DNA.

Mutants of recA are pleiotropic, they affect not only recombination, but also a set of

II 型 DNA 与单链 DNA 片段的同源性配对

表 2 比较了 II 型 DNA 和 I 型 DNA 分别作为 recA 蛋白的底物与单链 DNA 片段进行同源配对的差异。这些实验的反应条件（见表 2，表注）与之前几个实验的条件[4]稍有不同，硝酸纤维素过滤器截留到的复合物的量上显示，反应物中 II 型 DNA 在形成复合物方面的活性只有 I 型 DNA 的二分之一到三分之二。由于制备的 II 型 DNA 中仅含有很少量的 I 型 DNA（百分之几），因此这些复合物应该是由非超螺旋的 DNA 形成的。II 型 DNA 在形成复合物时需要同源单链 DNA 片段，其产物的性质与 RFI 形成的 D 环很类似。使用 0.2% 的 SDS 处理，或者在 pH 7 的 1.5 M NaCl、0.15 M 柠檬酸钠混合溶液中 50℃ 加热 4 分钟，都不会引起该复合物的解离。初步的电子显微镜观察结果显示，由 II 型 DNA 形成的复合物中有很显著的一部分含有 D 环（数据未展示）。最近，麦肯蒂、温斯托克和莱曼也报道了非超螺旋 DNA 可以形成 D 环的现象。不过在他们的实验中，使用的是线性 P22 噬菌体 DNA[10]。

II 型 DNA 可以形成 D 环的现象表明，recA 蛋白可以在 ATP 或其类似物 ATPγS 存在的情况下，促进 DNA 解旋。

讨　论

其他的实验表明 ATPγS 可以竞争性抑制 recA 蛋白的 ATP 酶活性，从而抑制 D 环结构的形成。但是 ATPγS 会促进由蛋白质、双链 DNA 以及单链 DNA 组成的复合物的形成（未发表的观察结果）。竞争性抑制的现象表明 ATPγS 可以占据 ATP 的结合位点；加上本文引用的其他人的观察结果，我们认为部分解旋的 DNA 分子的积累反映了 recA 蛋白促进双链 DNA 解旋的整个过程中的某个步骤，这个步骤本应该会导致 D 环的形成。ATPγS 对 recA 蛋白的这种作用使我们想到过去关于 DNA 促旋酶的一些实验结果。DNA 促旋酶也可以在另一个不能够被水解的 ATP 类似物存在时催化部分反应的进行[11]。根据 II 型 DNA 可以形成 D 环这一观察结果，我们推断 recA 蛋白在 ATP 存在时也可以促进 DNA 的解螺旋。ATP 的水解会引起蛋白质 – DNA 复合物的改变或者解离，并在 ATP 和 DNA 连接酶存在的情况下，抑制部分解旋的 DNA 的积累（未发表的观察结果）。

初看上去，由异源单链 DNA 激活的解旋似乎并不适合作为促进同源配对的机制。但是进一步想来，由任意单链 DNA 非特异性地触发双链 DNA 解旋可以部分地解释 recA 蛋白如何促进对同源配对序列的搜寻，因此这一观点是有道理的。由于在大肠杆菌中，同源重组高度依赖于 recA 蛋白的功能[12]，因此 recA 蛋白在这种机制下的行为意味着同源重组的过程常常是从单链 DNA 与双链 DNA 的相互作用开始。

recA 蛋白的突变是多效性的，它们不仅会影响同源重组，还会影响一系列由此

inducible functions, the so-called SOS functions[13]; these include error-prone repair, inactivation of phage repressors and the synthesis of recA protein itself (for reviews see refs 14, 15). Damaged DNA or blocked replication are inducing stimuli, and oligonucleotides have been implicated as intermediates in the induction[16,17]. The *tif*-1 mutation, which is located in the *recA* structural gene, makes the synthesis of recA protein thermo-inducible[18-22], indicating that a suitable form of the protein stimulates its own synthesis. The interaction of oligonucleotides with wild-type protein presumably stimulates synthesis by the same mechanism. However, observations on the *tsl* mutation indicate that abundant synthesis of recA protein may not be sufficient to induce fully all of the SOS functions[19,23]. Do oligonucleotides have a direct role in the inducible functions? Roberts *et al.* have studied one of the SOS functions *in vitro*, namely the inactivation of λ repressor by proteolytic cleavage[24,25]. In spite of the difficulty of reconciling proteolysis and unwinding of DNA, there are marked similarities between our observations on the unwinding of DNA by wild-type recA protein and those of Roberts *et al.* on inactivation of λ repressor by *tif* recA protein. Both processes require ATP or ATPγS, and oligonucleotide or single-stranded poly-nucleotide. The experiments described in this article show that both oligonucleotides and single-stranded DNA affect the functional properties of recA protein. The stimulation of unwinding by oligonucleotides, as well as by single strands, suggests that a common mechanism relates the function of recA protein in recombination to the other functions that are induced by oligonucleotides. Accordingly, oligonucleotides or single strands may trigger a conformational change of recA protein that can both unwind DNA and alter the interaction of recA protein with certain other proteins that bind to DNA.

A further inference from the similar effect of single strands and oligonucleotides on recA protein is that recombination may be inducible to some extent. Induction of synthesis of recA protein by single strands would provide the requisite amounts of protein to form D-loops, amounts which are large *in vitro*[4]. On the other hand, the induction of recA protein synthesis by oligonucleotides would not necessarily produce an increase in the number of genetic exchanges if any other factor, such as single strands, were limiting.

T.S. is a visiting fellow from The Institute of Physical and Chemical Research, Saitama, Japan. We acknowledge the technical assistance of Lynn Osber. This research was sponsored by grant no. NP 90E from the American Cancer Society and grant no. PHS CA 16038-05 from the NCI.

(**281**, 191-195; 1979)

Richard P. Cunningham, Takehiko Shibata, Chanchal DasGupta and Charles M. Radding

The Departments of Internal Medicine, and Molecular Biophysics and Biochemistry, Yale University School of Medicine, New Haven, Connecticut 06510

Received 4 May; accepted 18 June 1979.

诱导的功能，比如所谓的SOS反应[13]，其中包括易错修复、噬菌体抑制子的失活以及recA蛋白自身的合成（参见参考文献14和15）。DNA的损伤或复制的阻断会引发SOS反应，寡核苷酸也作为媒介参与其中[16,17]。位于*recA*结构基因位点的*tif*-1突变会导致recA蛋白的合成变成热诱导型[18-22]，这表明recA蛋白的某种合适的形式会促进其自身的合成。寡核苷酸与野生型recA蛋白的相互作用很可能也通过相同的机制来触发recA蛋白的合成。然而，对*tsl*突变体的观察表明，recA蛋白的大量合成并不一定足以完全诱导所有的SOS反应[19,23]。那么寡核苷酸是否具有直接诱导SOS反应的功能呢？罗伯茨等人在体外研究了SOS反应的一个功能，即通过水解切割使λ抑制子失活[24,25]。尽管很难使蛋白水解和DNA解旋这两个过程相一致，但我们关于野生型recA蛋白触发DNA解旋的观察结果与罗伯茨等人关于*tif*突变的recA蛋白导致的λ抑制子失活的观察结果有显著的相似之处；以上两个过程均需要ATP或ATPγS以及寡核苷酸或单链DNA片段的参与。本文中所述的实验表明寡核苷酸以及单链DNA片段都可以影响recA蛋白的功能。寡核苷酸以及单链DNA片段可以触发双链DNA解旋的现象表明，recA蛋白在促进同源重组和由寡核苷酸诱导的其他功能之间存在某种共同的作用机制将两者相关联。相应地，寡核苷酸或单链DNA片段有可能引起recA蛋白发生构象改变，从而既能触发DNA解旋，又能改变recA蛋白与某些其他可以与DNA结合的蛋白的相互作用。

从寡核苷酸与单链DNA片段在对recA蛋白的作用方面的相似性中得到的进一步推论是，同源重组的过程在某种程度上或许是可诱导的。单链DNA片段诱导的recA蛋白的合成将产生规定数量的蛋白以形成D环，在体外实验中产生的蛋白的产量很高[4]。另一方面，如果有其他的因素，如单链DNA片段的数量有限，那么由寡核苷酸引发的recA蛋白的合成未必导致遗传交换数量的增加。

柴田是来自日本崎玉物理与化学研究所的访问学者。我们感谢林恩·奥斯伯在技术方面的协助。本研究受美国癌症协会NP 90E项目以及美国国家癌症研究所PHS CA 16038-05项目的资助。

（张锦彬 翻译；陈新文 陈继征 审稿）

References:

1. Radding, C. M. *Ann., Rev. Biochem.* 47, 847-880 (1978).
2. Holloman, W. K., Wiegand, R., Hoessli, C. & Radding, C. M. *Proc. Natl. Acad. Sci. U.S.A.* 72, 2394-2398 (1975).
3. Beattie, K. L., Wiegand, R. C. & Radding, C. M. *J. Molec. Biol.* **116**, 783-803 (1977).
4. Shibata, T., DasGupta, C., Cunningham, R. P. & Radding, C.M. *Proc. Natl. Acad. Sci.U.S.A.* **76**, 1638-1642 (1979).
5. Holloman, W. K. & Radding, C. M. *Proc. Natl. Acad. Sci. U.S.A.* **73**, 3910-3914 (1976).
6. Liu, L. F. & Wang, J. C. *Proc. Natl. Acad. Sci. U.S.A.* 75, 2098-2102 (1978).
7. Anderson, P. & Bauer, W. *Biochemistry* 17, 594-601 (1978).
8. Kuhnlein, U., Penhoet, E. E. & Linn, S. *Proc. Natl. Acad. Sci. U.S.A.* **73**, 1169-1173 (1976).
9. Wiegand, R. C., Beattie, K. L., Holloman, W. K. & Radding, C. M., *J. Molec. Biol.* **116**, 805-824 (1977).
10. McEntee, K., Weinstock, G. M. & Lehman, I. R. *Proc. Natl. Acad. Sci. U.S.A.* **76**, 2615-2619 (1979).
11. Sugino, A., Higgins, N. P., Brown, P. O., Peebles, C.L. & Cozzarelli, N. R. *Proc, Natl. Acad. Sci. U.S.A.* 75, 4838-4842 (1978).
12. Clark, A. J. *A. Rev. Genet.* 7, 67-86 (1973).
13. Radman, M. in *Molecular Mechanisms for Repair of DNA* Part A (eds Hanawalt, P.C. & Setlow, R. B.) 355-367 (Plenum, New York, 1975).
14. Devoret, R. *Biochimie* **60**, 1135-1140 (1978).
15. Witkin, E. M. *Bact. Rev.* **40**, 869-907 (1976).
16. Oishi, M. & Smith, C. L. *Proc. Natl. Acad. Sci. U.S.A.* **75**, 3569-3573 (1978).
17. Oishi, M., Smith, C. L. & Friefeld, B. *Cold Spring Harb. Symp. Quant. Biol.* **43**, 897-907 (1979).
18. Castellazzi, M., Morand, P., George, J. & Buttin, G. *Molec. gen. Genet.* **153**, 297-310 (1977).
19. McEntee, K. *Proc. Natl. Acad. Sci. U.S.A.* **74**, 5275-5279 (1977).
20. Gudas, L. J. & Mount, D. W. *Proc. Natl. Acad. Sci. U.S.A.* **74**, 5280-5284 (1977).
21. Little, J. W. & Kleid, D. G. *J. Biol. Chem.* **252**, 6251-6252 (1977).
22. Emmerson, P. T. & West, S. C. *Molec. gen. Genet.* **155**, 77-85 (1977).
23. Mount, D.W., Kosel, C.K. & Walker, A. *Molec. gen. Genet.* **146**, 37-41 (1976).
24. Roberts, J. W., Roberts, C. W. & Craig, N. L. *Proc. Natl. Acad. Sci. U.S.A.* **75**, 4714-4718 (1978).
25. Roberts, J., Roberts, C. W., Craig, N. L. & Phizicky, E. *Cold Spring Harb. Symp. Quant. Biol.* **43**, 917-920 (1979).

Selfish Genes, the Phenotype Paradigm and Genome Evolution

W. F. Doolittle and C. Sapienza

Editor's Note

In 1976, biologist Richard Dawkins popularized in his book *The Selfish Gene* the growing notion in evolutionary biology that genes are the autonomous agents of Darwinian evolution, which compete for replicative success. In this review article, W. Ford Doolittle and Carmen Sapienza at Dalhousie University in Canada consider the existence of "junk DNA", which proliferates in the genome without an observable effect on an organism's phenotype. They argue that it is this DNA, the only function of which is to replicate itself within the genome, that is truly "selfish". The nature of "junk DNA", and the issue of whether it truly has no role integral to the genome, has become a hot topic of research in the new era of genomics.

Natural selection operating within genomes will inevitably result in the appearance of DNAs with no phenotypic expression whose only "function" is survival within genomes. Prokaryotic transposable elements and eukaryotic middle-repetitive sequences can be seen as such DNAs, and thus no phenotypic or evolutionary function need be assigned to them.

THE assertion that organisms are simply DNA's way of producing more DNA has been made so often that it is hard to remember who made it first. Certainly, Dawkins has provided the most forceful and uncompromising recent statement of this position, as well as of the position that it is the gene, and not the individual or the population, upon which natural selection acts[1]. Although we may thus view genes and DNA as essentially "selfish", most of us are, nevertheless, wedded to what we will call here the "phenotype paradigm"—the notion that the major and perhaps only way in which a gene can ensure its own perpetuation is by ensuring the perpetuation of the organism it inhabits. Even genes such as the segregation-distorter locus of *Drosophila*[2], "hitch-hiking" mutator genes in *Escherichia coli*[3,4] and genes for parthenogenetic reproduction in many species[4]—which are so "selfish" as to promote their own spread through a population at the ultimate expense of the evolutionary fitness of that population—are seen to operate through phenotype.

The phenotype paradigm underlies attempts to explain genome structure. There is a hierarchy of types of explanations we use in efforts to rationalize, in neo-darwinian terms, DNA sequences which do not code for protein. Untranslated messenger RNA sequences which precede, follow or interrupt protein-coding sequences are often assigned a phenotypic role in regulating messenger RNA maturation, transport or translation[5-7].

自私的基因，表型模式和基因组进化

杜利特尔，萨皮恩扎

编者按

1976年，生物学家理查德·道金斯的著作《自私的基因》使得进化生物学中日益兴起这样的观点——基因是达尔文进化论的自主元件，它们为自身的成功复制相互竞争。在这篇综述中，加拿大达尔豪西大学的福特·杜利特尔和卡门·萨皮恩扎认为“垃圾DNA”是存在的，它们在基因组中扩增，并且不会对生物体的表型产生可观测的效应。他们认为，这种DNA唯一的功能是在基因组内自我复制，是十足的“自私”。“垃圾DNA”的性质及其是否真的对基因组没有任何作用这个问题，在基因组的新时代中成为研究热点。

基因组内进行的自然选择将不可避免地导致无任何表型表达的DNA的出现，其唯一的“功能”就是存在于基因组内。原核生物的转座元件和真核生物的中度重复序列就可以被看作是这种DNA，它们因此被认为没有表型或者进化的功能。

认为生物体只是DNA复制更多DNA的途径的观点是如此普遍，以至于我们都想不起来是谁首先提出了这一观点。毫无疑问，道金斯对这个观点以及自然选择作用于基因本身而不是作用于个体或者种群的观点做出了最强有力和最坚定的最新陈述[1]。尽管我们可能因此将基因和DNA在本质上看作是“自私的”，但是无论如何，大多数人坚持本文我们称为“表型模式”的观点——基因能够确保长期存在的主要或者可能唯一的方法就是确保其寄存的生物体一直存活。尽管诸如果蝇的分节突变位点[2]、大肠杆菌的“搭乘”增变基因[3,4]以及很多物种的孤雌生殖基因[4]等基因都是这样自私，即通过牺牲种群的进化适应性而促进自身的传播，但是它们也是通过表型实现的。

在表型模式的基础上尝试解释基因组结构。从新达尔文主义的角度，我们试图从各个层次给出某些DNA不编码蛋白质的合理解释。这些先于、紧随或者干扰蛋白编码序列的非翻译信使RNA常常在信使RNA的成熟、转运或翻译的调控方面被赋予一些表型作用[5-7]。转录子的部分去除被认为是处理过程中必需的[8]。未转

Portions of transcripts discarded in processing are considered to be required for processing[8]. Non-transcribed DNA, and in particular repetitive sequences, are thought of as regulatory or somehow essential to chromosome structure or pairing[9-11]. When all attempts to assign to a given sequence or class of DNA functions of immediate phenotypic benefit to the organism fail, we resort to evolutionary explanations. The DNA is there because it facilitates genetic rearrangements which increase evolutionary versatility (and hence long-term phenotypic benefit)[12-17], or because it is a repository from which new functional sequences can be recruited[18,19] or, at worst, because it is the yet-to-be eliminated by-product of past chromosomal rearrangements of evolutionary significance[9,19].

Such interpretations of DNA structure are very often demonstrably correct; molecular biology would not otherwise be so fruitful. However, the phenotype paradigm is almost tautological; natural selection operates on DNA through organismal phenotype, so DNA structure must be of immediate or long-term (evolutionary) phenotypic benefit, even when we cannot show how. As Gould and Lewontin note, "the rejection of one adaptive story usually leads to its replacement by another, rather than to a suspicion that a different kind of explanation might be required. Since the range of adaptive stories is as wide as our minds are fertile, new stories can always be postulated" (ref. 20).

Non-phenotypic Selection

What we propose here is that there are classes of DNA for which a "different kind of explanation" may well be required. Natural selection does not operate on DNA only through organismal phenotype. Cells themselves are environments in which DNA sequences can replicate, mutate and so evolve[21]. Although DNA sequences which contribute to organismal phenotypic fitness or evolutionary adaptability indirectly increase their own chances of preservation, and may be maintained by classical phenotypic selection, the only selection pressure which DNAs experience directly is the pressure to survive within cells. If there are ways in which mutation can increase the probability of survival within cells without effect on organismal phenotype, then sequences whose only "function" is self-preservation will inevitably arise and be maintained by what we call "non-phenotypic selection". Furthermore, if it can be shown that a given gene (region of DNA) or class of genes (regions) has evolved a strategy which increases its probability of survival within cells, then no additional (phenotypic) explanation for its origin or continued existence is required.

This proposal is not altogether new; Dawkins[1], Crick[6] and Bodmer[22] have briefly alluded to it.

However, there has been no systematic attempt to describe elements of prokaryotic and eukaryotic genomes as products of non-phenotypic selection whose primary and often only function is self-preservation.

录 DNA 尤其是重复序列被认为具有调节作用或者在某种程度上对染色体的结构或配对是必不可少的[9-11]。我们将特定的序列或者一类 DNA 的功能认定为有利于生物体的直接表型，当上述所有尝试失败以后，我们只能诉诸进化方面的解释。DNA 的存在是因为它促进了基因重排，有利于增加进化灵活性（因此有利于长期的表型获益）[12-17]，或者因为它可以作为提供新功能性序列的储存库[18,19]，或者最糟的解释——因为它是过去具有进化意义的染色体重排后即将被清除的副产物[9,19]。

这种对 DNA 结构的解释常常被证实是正确的；否则分子生物学研究不会如此硕果累累。然而，表型模式几乎总是冗赘的；自然选择通过生物体的表型作用于 DNA，因此，DNA 的结构必定具有即时或者长期（进化）的表型优势，尽管我们无法解释其原因。正如古尔德和列万廷所指出的："抵制一个适应性假说常常导致其被另一个假说替代，而不是怀疑是否需要换种方式来解释。由于适应性假说的范围与我们的思维一样广阔，因此总能推导出新的假说"（文献 20）。

非表型选择

本文我们假设有各种各样需要"不同解释"的 DNA。自然选择不仅仅通过生物体的表型作用于 DNA。细胞自身就是 DNA 序列能够复制、突变从而进化的环境[21]。尽管那些有助于生物体表型适应性或者进化适应性的 DNA 序列间接地增加了其本身被保留的机会，并且可能被经典的表型选择所保留，但是 DNA 直接面对的唯一选择压力是它们在细胞中存活的压力。如果存在某些方式，使 DNA 突变能够增加其在细胞内的存活可能性并且对生物体表型没有影响，那么这种将自我保护作为唯一"功能"的序列必然会增加，并以我们称之为"非表型选择"的方式被保留。此外，如果能够证明一个特定的基因（一个 DNA 片段）或者一组基因（多个片段）以增加其存活于细胞内可能性的策略来进化，那就不再需要针对其起源或持续存在的额外（表型）解释了。

这个提议并不完全新奇；道金斯[1]，克里克[6]和博德默[22]都简要地提及过。

但是，并没有系统的尝试将原核生物和真核生物基因组的元件描述为非表型选择的产物，其首要和常规的功能仅是自我保存。

Transposable Elements in Prokaryotes as Selfish DNA

Insertion sequences and transposons can in general be inserted into a large number of chromosomal (or plasmid) sites, can be excised precisely or imprecisely and can engender deletions or inversions in neighbouring chromosomal (or plasmid) DNAs[12-16]. These behaviours and, at least in some cases, the genetic information for the enzymatic machinery involved, must be inherent in the primary sequences of the transposable elements themselves, which are usually tightly conserved[12-16,23]. Most speculations on the function of transposable elements concentrate on the role these may have, through chromosomal rearrangements and the modular assembly of different functional units, in promoting the evolution of plasmid and bacterial chromosomes, and thus in promoting long-term phenotypic fitness[12-16]. Most assume that it is for just such functions that natural selection has fashioned these unusual nucleic acid sequences.

Although transposable elements may well be beneficially involved in prokaryotic evolution, there are two reasons to doubt that they arose or are maintained by selection pressures for such evolutionary functions.

First, DNAs without immediate phenotypic benefit are of no immediate selective advantage to their possessor. Excess DNA should represent an energetic burden[24,25], and some of the activities of transposable elements are frankly destructive[12-16]. Evolution is not anticipatory; structures do not evolve because they might later prove useful. The selective advantage represented by evolutionary adaptability seems far too remote to ensure the maintenance, let alone to direct the formation, of the DNA sequences and/or enzymatic machinery involved. A formally identical theoretical difficulty plagues our understanding of the origin of sexual reproduction, even though this process may now clearly be evolutionarily advantageous[1,4].

Second, transposability itself ensures the survival of the transposed element, regardless of effects on organismal phenotype or evolutionary adaptability (unless these are sufficiently negative). Thus, no other explanation for the origin and maintenance of transposable elements is necessary. A single copy of a DNA sequence of no phenotypic benefit to the host risks deletion, but a sequence which spawns copies of itself elsewhere in the genome can only be eradicated by simultaneous multiple deletions. Simple translocation (removal from one site and insertion into another) does not provide such insurance against deletion. It is significant that recent models for transposition require retention of the parental sequence copy[26,27], and that bacterial insertion sequences are characteristically present in several copies per genome[16]. The assumption that transposable elements are maintained by selection acting on the cell does not require that they show these characteristics. The evolutionary behaviour of individual copies of transposable elements within the environment represented by a bacterial genome and its descendants can be understood in the same terms as organismal evolution. Replicate copies of a given element may diverge in sequence, but at least those features of sequence required for transposition will be maintained by (non-phenotypic) selection; copies which can no longer be translocated

原核生物的转座元件是自私 DNA

插入序列和转座子一般来说都能被插入到大量的染色体（或质粒）位点内，能被精确或者不精确地剪切，并能引起临近染色体（或者质粒）DNA 的缺失或者倒位[12-16]。至少在某些情况下，这些行为以及涉及这些行为的酶体系的遗传信息必须是转座元件本身的原始序列中固有的，并且通常是高度保守的[12-16,23]。绝大多数关于转座元件功能的假设均集中于它们可能起到的作用，如通过染色体的重排和不同功能单位的模块化组装促进质粒和细菌染色体的进化，从而促进长期的表型适应性[12-16]。大部分人认为正是因为这个功能才使得自然选择塑造出了这些不寻常的核酸序列。

尽管转座元件可能在原核生物的进化中有利，但还是有两个理由认为它们未必是在这种进化功能的选择压力下产生或保持的。

首先，没有直接表型优势的 DNA 对于其拥有者没有直接的选择优势。过多的 DNA 则代表能量的负担[24,25]，还有转座元件的一些活动是具有直接破坏性的[12-16]。进化是不可预期的；结构没有进化是因为它们可能在将来被证明有用。进化适应性所代表的选择优势似乎对于 DNA 序列和（或）涉及的酶体系的保持作用都微乎其微，更别说指导其形成。一个形式上一致的理论难题妨碍了我们对于有性生殖起源的理解，尽管现在很清楚这个过程是具有进化优势的[1,4]。

其次，可转移性本身确保了已转座元件的存活，无论其对于有机体的表型或者进化适应性是否有作用（除非有充分的负面作用）。因此，没有必要再对转座元件的起源和保留做更多解释。对宿主表型没有益处的单拷贝 DNA 序列存在被删除的风险，但是在基因组内多个位置存在自身拷贝的序列只能通过多个位点同时缺失才能从基因组中彻底删除。单纯的易位（从一个位点移除插入到另一个位点）不能确保序列免于被删除。很显然，近来的转座模型均需要保留亲本的序列拷贝[26,27]，而且细菌的插入序列特征性地在每个基因组内存在多个拷贝[16]。假设转座元件通过作用于细胞的选择而得以保留，这个假设的成立并不需要它们具备这些特征。一个细菌基因组及其后代为代表的环境内的单拷贝转座元件的进化行为同样可以用生物体进化的方式来理解。特定元件的复制拷贝可能分散在序列上，但是至少转座所需序列的特征会通过（非表型）选择保留下来；再也不能易位的拷贝最终会被删除。一些分散的拷贝可能更容易转座；它们会以牺牲其他拷贝为代价增加其转座的频率。依靠宿

will eventually suffer elimination. Some divergent copies may be more readily transposed; these will increase in frequency at the expense of others. Transposable elements which depend on host functions run the risk that host mutants will no longer transpose them; it is significant that at least some transposition-specific functions are known to be coded for by the transposable elements themselves[26-29]. It is not to the advantage of a transposable element coding for such functions to promote the transposition of unrelated elements; the fact that given transposable elements generate flanking repeats[16,30] of chromosomal DNAs of sizes characteristic to them (that is, 5, 9 or 11–12 base pairs) may indicate such a specificity in transposition mechanism. It is to the advantage of any transposable element to acquire genes which allow independent replication (to become a plasmid), promote host mating (to become a self-transmissable plasmid) or promote non-conjugational transmission (to become a phage like Mu).

It is certainly not novel to suggest that prokaryotic transposable elements behave in these ways, or to suggest that more frankly autonomous entities like phages have arisen from them[12-16,31]. However, we think it has not been sufficiently emphasized that non-phenotypic selection may inevitably give rise to transposable elements and that no phenotypic rationale for their origin and continued existence is thus required.

Transposable Elements in Eukaryotes

There has long been genetic evidence for the existence in eukaryotic genomes of transposable elements affecting phenotype[32]. These have been assigned roles in the regulation of eukaryotic gene expression and in evolution, but would have escaped genetic detection had they not had phenotypic effect. More recent evidence for transposable elements whose effects are not readily identified genetically has come fortuitously from studies of cloned eukaryotic DNAs. For instance, the Ty-1 element of yeast (which has no known phenotypic function) is flanked by direct repeats (like some prokaryotic transposons) and is transposable[33]. It is present in some 35 dispersed copies and comprises some 2% of the yeast genome (like a higher-eukaryotic middle-repetitive DNA). The directly repeated δ-sequence elements flanking it are found at still other sites (just as prokaryotic insertion sequences can be found flanking transposons or independently elsewhere in the genome). Cameron *et al.* suggest that "Ty-1 may be a nonviral 'parasitic' DNA" but then go on to suggest, we think unnecessarily, that transposition "allows adaptation of a particular cell to a new environment" (ref. 33). The repetitive elements *412, copia* and *297* of *Drosophila* are physically similar to Ty-1 (and to bacterial transposable elements) and are transposable[34-37]. Strobel *et al.* suggest "it is possible that the sole function of these elements is to promote genetic variability, and that their gene products may only be necessary for the maintenance and mobility of the elements themselves, rather than for other cellular processes" (ref. 37). But if maintenance and mobility mechanisms exist, then no cellular function at all need be postulated.

A large fraction of many eukaryotic genomes consists of middle-repetitive DNAs, and the variety and patterns of their interspersion with unique sequence DNA make

主功能的转座子有不再被宿主突变体转座的风险；很显然，至少已知一些转座特异性的功能是由转座元件自身编码的[26-29]。对于转座元件本身而言，编码这些功能来促进无关元件的转座是无益的；事实上，特定的转座元件产生特异性大小（即 5，9 或者 11~12 个碱基对）的染色体 DNA 两侧的重复序列[16,30]，这可能说明了转座机制的特异性。对于任何转座子来说，获得一些基因，使其可以独立复制（变成质粒）、促进宿主配对（成为可自我传递的质粒）或者促进非接合传递（成为噬菌体，如 Mu 噬菌体），是非常有利的。

提出原核生物转座元件以这种方式作用或者提出更直接自主的实体（比如噬菌体）就是由上述方式衍生而来的观点并不新颖[12-16,31]。但是，我们认为还未充分强调的是，非表型选择可能不可避免地产生转座元件，而且我们不需要用表型理论去解释其起源和持续存在。

真核生物的转座元件

很早以前就有遗传学证据证明真核生物基因组内存在影响表型的转座元件[32]。人们认为这些转座元件在真核基因表达的调控以及进化中起作用，但是如果它们没有表型效应，它们可能会被基因检测遗漏。很偶然地，在最近对真核生物 DNA 克隆的研究中，发现了更多不便于从遗传学角度确认功能的转座子的证据。比如，酵母的 Ty-1 元件（没有已知的表型功能）两侧都是正向重复序列（类似某些原核生物转座子），而且可以转座[33]。这种元件大约以 35 个分散的拷贝存在，构成了约 2% 的酵母基因组（类似高等真核生物的中度重复 DNA）。在其他位点也可以找到两侧正向重复的 δ 序列（就像可以在转座子的侧翼或者基因组中其他独立的位置找到原核生物的插入序列一样）。卡梅伦等指出“Ty-1 可能是一种非病毒的‘寄生性’DNA”，他们又进一步指出转座“使得特定细胞适应新环境”（文献 33），但我们认为后续的论述没有必要。果蝇的重复元件 *412*，*copia* 和 *297* 在结构上都类似于 Ty-1（以及细菌的转座元件），并且可以转移[34-37]。斯特罗贝尔等指出“这些元件的唯一功能可能就是促进遗传变异，它们的基因产物可能仅仅对这些元件本身的保留和移动是必需的，而对其他细胞活动不起作用”（文献 37）。但是如果存在保留和移动的机制，那就无须再假设其他细胞功能了。

很多真核生物基因组含有很大比例的中度重复 DNA，它们与特定的 DNA 序列穿插分布的多样性和形式并没有特别的系统发育或者表型功能的意义。布里滕和戴

no particular phylogenetic or phenotypically functional sense. Britten, Davidson and collaborators have elaborated models which ascribe regulatory functions to middle-repetitive DNAs, and evolutionary advantage (in terms of adaptability) to the quantitative and qualitative changes in middle-repetitive DNA content observed even between closely related species[17,38-40]. Middle-repetitive DNAs are more conserved in sequence during evolution than are unique-sequence DNAs not coding for protein, and Klein *et al.* suggest that "restraint on repetitive sequence divergence, either within the repeat families of a given species, or over evolutionary time spanning the emergence of different species, could be due to [phenotypic] selective pressures which prevent free sequence change in large fraction of the repeat family members. Or perhaps repetitive sequences diverge as rapidly as do other sequences, but the type sequence of the family is preserved by frequent remultiplication of the 'correct' surviving sequences" (ref. 41). The evidence for a phenotypically functional role for middle-repetitive sequences remains dishearteningly weak[40-43], and if the calculations of Kimura[44] and Salser and Isaacson[45] are correct, middle-repetitive DNAs together comprise too large a fraction of most eukaryotic genomes to be kept accurate by darwinian selection operating on organismal phenotype. The most plausible form of "remultiplication of the 'correct' surviving sequences" is transposition. If we assume middle-repetitive DNAs in general to be transposable elements or degenerate (and no longer transposable and ultimately to be eliminated) descendants of such elements, then the observed spectra of sequence divergence within families and changes in middle-repetitive DNA family sequence and abundance can all be explained as the result of non-phenotypic selection within genomes. No cellular function at all is required to explain either the behaviour or the persistence of middle-repetitive sequences as a class.

The Rest of the Eukaryotic Genome

Middle-repetitive DNA can comprise more than 30% of the genome of a eukaryotic cell[46]. Another 1–40% consists of simple reiterated sequences whose functions remain unclear[10], and Smith has argued that "a pattern of tandem repeats is the natural state of DNA whose sequence is not maintained by selection" (ref. 47). Even unique-sequence eukaryotic DNA consists in large part of elements which do not seem to be constrained by phenotypic selection pressures[45]. Some authors have argued that the intervening sequences which interrupt many eukaryotic structural genes are insertion sequence-like elements[6,48,49]. If they are, they are likely to be the degenerate and no-longer-transposable descendants of transposable sequences whose insertion was rendered non-lethal by pre-existing cellular RNA:RNA splicing mechanisms. Such elements, once inserted, are relatively immune to deletion (since only very precise deletion can be non-lethal), and need retain only those sequence components required for RNA splicing. The rest of the element is free to drift and one expects (and observes) that only the position and number of intervening sequences in a family of homologous genes remain constant during evolution. Although evolutionary and regulatory phenotypic functions have been ascribed to intervening sequences[6,49-51], it is unnecessary to postulate any cellular function at all if these elements are indeed degenerate transposable elements arising initially from non-phenotypic selection. Another explanation for the origin and continued existence of intervening sequences, which also does not require phenotypically or evolutionarily advantageous roles, has been suggested elsewhere[50,51].

维森及同事精心构建了一个模型，该模型将调节功能归因于中度重复 DNA，同时将进化优势（在适应性方面）归因于中度重复 DNA 量和质的改变，这些变化甚至在相近物种之间也曾被观察到[17,38-40]。与单一序列的非编码蛋白质的 DNA 相比，中度重复 DNA 序列在进化过程中更加保守，并且克莱因等指出“对重复序列差异的限制，无论是在特定物种的重复序列家族内还是经过很长的进化时间出现不同的物种，都可归因于 [表型] 选择压力，这种压力防止了大部分重复家族成员中的自由的序列改变。也可能是因为重复序列和其他序列一样很快发生变化，但是家族的特征性序列通过频繁的‘正确’存活序列的再复制而得以保留”（文献 41）。证明中度重复序列具有表型功能的证据仍然不足[40-43]，而且如果木村[44]以及萨尔瑟和艾萨克森[45]的计算是正确的，中度重复 DNA 总体在大部分真核生物基因组中所占比例过大，以致作用于生物体表型的达尔文选择不能保持其正确无误。理论上，“正确存活序列的再增殖”最合理的形式就是转座。如果我们假设中度重复 DNA 通常都是转座元件或者这类元件的退化产物（不再具有转座能力并且最终被清除），那么观察到的家族内部序列变异范围以及中度重复 DNA 家族序列和丰度的改变都可以解释为基因组内非表型选择的结果。作为一个群体，中度重复序列的行为或者其持续存在均无需用细胞功能去解释。

真核生物基因组的其他组分

中度重复 DNA 可以构成超过真核细胞基因组的 30%[46]。另外 1%~40% 由功能未知的简单重复序列组成[10]，并且史密斯认为“串联重复的形式是 DNA 的天然状态，其序列并不通过选择来保持”（文献 47）。甚至组成一大部分元件的真核 DNA 的单一序列似乎也不受表型选择压力约束[45]。一些作者认为那些使很多真核生物结构基因中断的间隔序列都是类似于插入序列的元件[6,48,49]。如果确实如此，那么它们很可能是转座序列退化并且不再具备转座能力的产物，其插入通过预先存在的细胞 RNA–RNA 剪接机制呈现非致死性。这些元件，一旦插入后，相对难以被删除（因为只有非常精确的删除才是非致命性的），并且仅需保留那些 RNA 剪接所需的序列成分。元件的剩余部分可以自由移动，并且可以预期（并观察到），只有同源基因家族内间隔序列的位置和数目在进化中是保持恒定的。尽管也有人认为间隔序列具有进化和调节表型的功能[6,49-51]，如果这些元件确实是初始通过非表型选择退化的转座元件产物，那么任何有关细胞功能的假设都没有必要。另外一种有关间隔序列起源和持续存在的解释另有叙述，这种解释同样无需表型上或者进化方面的优势作用[50,51]。

Why Do Prokaryotes and Eukaryotes Differ?

It is generally believed that prokaryotic genomes consist almost entirely of unique-sequence DNA maintained by phenotypic selection, whereas the possession of "excess" unique and repetitive DNA sequences whose presence is at least difficult to rationalize in phenotypic terms is characteristic of eukaryotes. However, it is more accurate to say that there is continuum of excess DNA contents; at least 1% of the *E. coli* genome can be made up of copies of six identified insertion sequences alone[16]. Yeast, whose genome is no larger than that of some prokaryotes, has few repeated sequences other than those coding for stable RNAs, and *Aspergillus* may have none[52,53]. There is in general (but with many exceptions) a positive correlation between excess DNA content, genome size and what we anthropocentrically perceive as "evolutionary advancement". Many interpret this as the cause and/or consequence of the increasing phenotypic complexity which characterizes organismal evolution, and attribute to excess DNA a positive role in the evolutionary process[17-19,40]. The interplay of phenotypic and non-phenotypic forces, and the importance of understanding both in attempts to restore the "*C*-value paradox" are discussed more thoroughly by Orgel, and Crick in the following article.[54]

There is another, simpler and perhaps obvious explanation. Non-phenotypic selection produces excess DNA, and excess DNA logically must be an energetic burden; phenotypic selection should favor its elimination[24,25]. The amount of excess (and hence total) DNA in an organism should be loosely determined by the relative intensities of the two opposing sorts of selection. The intensity of non-phenotypic pressure on DNA to survive even without function should be independent of organismal physiology. The intensity of phenotypic selection pressure to eliminate excess DNA is not, this being greatest in organisms for which DNA replication comprises the greatest fraction of total energy expenditure. Prokaryotes in general are smaller and replicate themselves and their DNA more often than eukaryotes (especially complex multicellular eukaryotes). Phenotypic selection pressure for small "streamlined" prokaryotic genomes with little excess DNA may be very strong.

Necessary and Unnecessary Explanations

We do not deny that prokaryotic transposable elements or repetitive and unique-sequence DNAs not coding for protein in eukaryotes may have roles of immediate phenotypic benefit to the organism. Nor do we deny roles for these elements in the evolutionary process. We do question the almost automatic invocation of such roles for DNAs whose function is not obvious, when another and perhaps simpler explanation for their origin and maintenance is possible. It is inevitable that natural selection of the special sort we call non-phenotypic will favour the development within genomes of DNAs whose only "function" is survival within genomes. When a given DNA, or class of DNAs, of unproven phenotypic function can be shown to have evolved a strategy (such as transposition) which ensures its genomic survival, then no other explanation for its existence is necessary. The search for other explanations may prove, if not intellectually sterile, ultimately futile.

为什么原核生物和真核生物不同?

一般认为原核生物基因组几乎全部由表型选择所保留的单一序列 DNA 组成，而存在难以用表型合理解释的“冗余的”单一和重复 DNA 序列则是真核生物的特征。但是，更加准确地说，冗余 DNA 成分的连续统一体是存在的；至少大肠杆菌 1% 的基因组由六种已被确认的插入序列拷贝单独组成[16]。酵母的基因组不比一些原核生物的大，但是除了那些编码稳定 RNA 的序列之外，几乎没有重复序列，而曲霉菌可能并不含有重复序列[52,53]。一般来说（但是有很多例外），冗余 DNA 成分、基因组大小以及以人类为中心的所谓高级进化产物之间存在正相关性。很多人将这解释为标志生物体进化的表型复杂程度增加的原因和（或）结果，并认为冗余 DNA 在进化过程中具有正面作用[17-19,40]。奥格尔和克里克在后面的文章中更加详细地讨论了表型和非表型影响的相互作用，以及弄清楚二者对试图修复“*C* 值悖论”的重要性[54]。

有另外一个更简单，可能更显而易见的解释。非表型选择产生了过量的 DNA，理论上过量的 DNA 必然增加了能量负担；表型选择更倾向于清除这些 DNA[24,25]。生物体内冗余 DNA（以及总 DNA）的量应该不严格地由这两种对立选择力量的相对强度决定。非表型压力作用于 DNA，使得即使无功能的 DNA 也得以保存，其强度应该与生物体的生理无关。而清除冗余 DNA 的表型选择压力的强度则不然，该强度在 DNA 复制占据总能量消耗比例最高的生物体内最大。原核生物一般来说更小，并比真核生物更频繁地复制自身及其 DNA（尤其是复杂得多细胞真核生物）。小“流线型”原核生物基因组含有很少冗余 DNA，作用于这种基因组的表型选择压力可能是非常大的。

必要和不必要的解释

我们并不否认原核生物转座元件或者真核生物中不编码蛋白质的重复和单一序列 DNA 可能对于生物体具有直接的表型优势。我们也不否认这些元件在进化过程中的作用。如果对上述元件的起源和保留有可能存在其他的或更为简单的解释，那我们就要质疑，将这些作用赋予那些功能并不明显的 DNA，是否合理。不可避免地，我们称之为非表型的自然选择的一种特殊形式帮助 DNA 在基因组中发展，这些 DNA 的唯一“功能”就是在基因组中存活。当一个或者一组未经证实表型功能的特定 DNA 可以被证明进化出一种策略以确保其基因组的存活（比如转座），那么就不再需要其他解释其存在的理由。寻找其他解释的过程，即使不是毫无结果，也是没有意义的。

We thank L. Bonen, R. M. MacKay and M. Schnare for help in development of the ideas presented here, and C. W. Helleiner, M. W. Gray, C. Stuttard, R. Singer, S. D. Wainwright and E. Butz for critical discussions.

We are especially grateful to C. E. Orgel and F. H. C. Crick for discussing with us the ideas presented in the following article before publication and for encouragement and support.

(**284**, 601-603; 1980)

W. Ford Doolittle and Carmen Sapienza
Department of Biochemistry, Dalhousie University, Halifax, Nova Scotia, Canada B3H 4H7

References:

1. Dawkins, R. *The Selfish Gene* (Oxford University Press, 1976).
2. Crow, J. F. *Scient. Am.* **240**, 134-146 (1979).
3. Cox, E. C. & Gibson, T. C. *Genetics* 77, 169-184 (1974).
4. Maynard Smith, J. *The Evolution of Sex* (Cambridge University Press, 1978).
5. Darnell, J. E. *Prog. Nucleic Acid Res. Molec. Biol.* **22**, 327-353 (1978).
6. Crick, F. H. C. *Science* **204**, 264-271 (1979).
7. Murray, V. & Holliday, R. *FEBS Lett.* **106**, 5-7 (1979).
8. Sogin, M. L., Pace, B. & Pace, N. R. *J. Biol. Chem.* **252**, 1350-1357 (1977).
9. Fedoroff, N. *Cell* **16**, 697-710 (1979).
10. John, B. & Miklos, G. L. G. *Int. Rev. Cytol.* **58**, 1-114 (1979).
11. Zuckerkandl, E. *J. Molec. Evol.* **9**, 73-122 (1976).
12. Cohen, S. N. *Nature* **263**, 731-738 (1976).
13. Starlinger, P. & Saedler, H. *Curr. Topics Microbiol. Immun.* **75**, 111-152 (1976).
14. Nevers, P. & Saedler, H. *Nature* **268**, 109-115 (1977).
15. Kleckner, N. *Cell* **11**, 11-23 (1977).
16. Kopecko, D. in *Plasmids and Transposons: Environmental Effects and Maintenance Mechanisms* (eds Stuttard, C. & Rozee, K. R.) (Academic, New York, in the press).
17. Britten, R. J. & Davidson, E. H. *Q. Rev. Biol.* **46**, 111-138 (1971).
18. Ohno, S. *Evolution by Gene Duplication* (Springer, New York, 1970).
19. Hinegardner, R. in *Molecular Evolution* (ed. Ayala, F. J.) 160-199 (Sinauer, Sunderland, 1976).
20. Gould, S. J. & Lewontin, R. C. *Proc. R. Soc.* B**205**, 581-598 (1979).
21. Orgel, L. *Proc. R. Soc.* B**205**, 435-442 (1979).
22. Bodmer, W. in *Human Genetics: Possibillities and Realities*, 41-42 (Excerpta Medica, Amsterdam, 1979).
23. Johnsrod, L. *Molec. gen. Genet.* **169**, 213-218 (1979).
24. Zamenhof, S. & Eichorn, H. H. *Nature* **216**, 456-458 (1967).
25. Koch, A. L. *Genetics* **72**, 297-316 (1972).
26. Shapiro, J. A. *Proc. Natl. Acad. Sci. U.S.A.* **76**, 1933-1937 (1979).
27. Arthur, A. & Sherratt, D. *Molec. gen. Genet.* **175**, 267-274 (1979).
28. Gill, R., Heffron, F., Dougan, G. & Falkow, S. *J. Bact.* **136**, 742-756 (1978).
29. MacHattie, L. A. & Shapiro, J. A. *Proc. Natl. Acad. Sci. U.S.A.* **76**, 1490-1494 (1979).
30. Huberman, P., Klaer, R., Kühn, S. & Starlinger, P. *Molec. gen. Genet.* **175**, 369-373 (1979).
31. Campbell, A. in *Biological Regulation and Development* Vol. 1 (ed. Goldberger, R. F.) 19-55 (Plenum, New York, 1979).
32. McClintock, B. *Cold Spring Harb. Symp. Quant. Biol.* **16**, 13-47 (1952).
33. Cameron, J. R., Loh, E. Y. & Davis, R. W. *Cell* **16**, 739-751 (1979).
34. Finnegan, D. J., Rubin, G. M., Young, M. W. & Hogness, D. S. *Cold Spring Harb. Symp. Quant.Biol.* **42**, 1053-1063 (1977).

感谢博南、麦凯和施纳尔帮助我们形成了文中的观点，并感谢赫莱纳、格雷、斯图塔特、辛格、温赖特和巴茨对此进行的关键讨论。

特别感谢奥格尔和克里克和我们讨论他们未发表的文章观点，以及给予我们的鼓励和支持。

（毛晨晖 翻译；崔巍 审稿）

35. Carlson, M. & Brutlag, D. *Cell* **15**, 733-742 (1978).
36. Potter, S. S., Borein, W. J. Jr, Dunsmuir, P. & Rubin, G. M. *Cell* **17**, 415-427 (1979).
37. Strobel, E., Dunsmuir, P. & Rubin, G. M. *Cell* **17**, 429-439 (1979)
38. Britten, R. J. & Davidson, E. H. *Science* **165**, 349-357 (1969).
39. Davidson, E. H., Klein, W. H. & Britten, R. J. *Devl. Biol.* **55**, 69-84 (1977).
40. Davidson, E. H. & Britten, R. J. *Science* **204**, 1052-1059 (1979).
41. Klein, W. H. *et al. Cell* **14**, 889-900 (1978).
42. Sheller, R. H., Constantini, F. D., Dozlowski, M. R., Britten, R. J. & Davidson, E. H. *Cell* **15**,189-203 (1978).
43. Kuroiwa, A. & Natori, S. *Nucleic Acids Res.* 7, 751-754 (1979).
44. Kimura, M. *Nature* **217**, 624-626 (1968).
45. Salser, W. & Isaacson, J. S. *Prog. Nucleic Acid Res. Molec. Biol.* **19**, 205-220 (1976).
46. Lewin, B. *Cell* **4**, 77-93 (1975).
47. Smith, G. P. *Science* **191**, 528-534 (1976).
48. Tsujimoto, Y.G. & Suzuki, Y. *Cell* **18**, 591-600 (1979).
49. Gilbert, W. *Nature* **271**, 501 (1978).
50. Doolittle, W. F. *Nature* **272**, 581-582 (1978).
51. Darnell, J. E. *Science* **202**, 1257-1260 (1978).
52. Roberts, T.M., Lauer, G. D. & Klotz, L. *CRC Crit. Rev. Biochem.* **3**, 349-451 (1976).
53. Timberlake, W. F. *Science* **702**, 973-974 (1978).
54. Orgel, G. E. & Crick, F.H.C. *Nature* **284**, 604-607 (1980).

Selfish DNA: the Ultimate Parasite

L. E. Orgel and F. H. C. Crick

Editor's Note

In the second of two consecutive review articles on "selfish DNA", Leslie Orgel and Francis Crick at the Salk Institute in California consider how this form of DNA—which proliferates in the genome without significantly affecting an organism's phenotype—might arise during evolution. Such DNA is in effect a kind of relatively harmless parasite, which is insufficiently costly to produce (in metabolic terms) to be efficiently purged from the genome. Orgel and Crick point out that some of this apparently "useless" DNA might occasionally acquire a useful function—but most would not. Not only did these ideas help to dismantle the popular belief that natural selection ensures maximal genomic efficiency, but they also presaged the modern recognition that there is a great deal of DNA in the human genome with unknown function.

The DNA of higher organisms usually falls into two classes, one specific and the other comparatively nonspecific. It seems plausible that most of the latter originated by the spreading of sequences which had little or no effect on the phenotype. We examine this idea from the point of view of the natural selection of preferred replicators within the genome.

THE object of this short review is to make widely known the idea of selfish DNA. A piece of selfish DNA, in its purest form, has two distinct properties:

(1) It arises when a DNA sequence spreads by forming additional copies of itself within the genome.

(2) It makes no specific contribution to the phenotype.

This idea is not new. We have not attempted to trace it back to its roots. It is sketched briefly but clearly by Dawkins[1] in his book *The Selfish Gene* (page 47). The extended discussion (pages 39–45) after P. M. B. Walker's article[2] in the CIBA volume based on a Symposium on Human Genetics held in June 1978 shows that it was at that time already familiar to Bodmer, Fincham and one of us. That discussion referred specifically to repetitive DNA because that was the topic of Walker's article, but we shall use the term selfish DNA in a wider sense, so that it can refer not only to obviously repetitive DNA but also to certain other DNA sequences which appear to have little or no function, such as much of the DNA in the introns of genes and parts of the DNA sequences between genes. The catch-phrase "selfish DNA" has already been mentioned briefly on two occasions[3,4]. Doolittle and Sapienza[5] (see the previous article) have independently arrived at similar ideas.

自私的DNA：最终的寄生物

奥格尔，克里克

编者按

在有关“自私的DNA”的两篇连续综述的第二篇中，加利福尼亚索尔克研究所的莱斯利·奥格尔和弗朗西斯·克里克思考了这种形式的DNA——在基因组中扩增，但不明显影响生物体的表型——是如何在进化中产生的。实际上，这种基因是一种相对无害的寄生物，机体不足以以高代价产生（在代谢方面）将其从基因组中有效清除的机制。奥格尔和克里克指出，这种明显“无用的”DNA有些可能碰巧会获得一种有用的功能——但是大部分不会。这些观点不仅有助于消除人们关于自然选择会保证最大的基因组效率的普遍观念，而且预示人类基因组中有大量未知功能DNA的新认识。

高等生物的DNA通常分为两种，特异的和相对非特异的。情况似乎是这样：非特异性序列源于某些序列的扩散，这些序列对表型影响很小或没有影响。我们从基因组中优先复制子的自然选择的观点出发来检验这个想法。

这篇简短的综述的目的是普及自私DNA的概念。一段处于最简单形式的自私DNA，具有两种截然不同的特性：

（1）它是一段DNA序列在基因组中复制更多拷贝时产生的。

（2）它对生物的表型没有特定的作用。

这种想法并不新颖，我们还没有尝试追溯它的根源。道金斯[1]曾经在他的著作《自私的基因》（第47页）中简要清楚地叙述过这个观点。基于1978年6月举行的人类遗传学研讨会文集CIBA卷中，在沃克文章[2]后的扩展讨论（第39~45页）中显示，那个时候这种观点已经被博德默、芬彻姆和我们两位作者之一所熟知。这场讨论主要集中于重复DNA，因为这是沃克的文章主题，但是我们应该在更广泛的意义上使用自私的DNA这个概念，它不仅仅指明显的重复DNA，还包括其他的没有或有很少功能的DNA序列，比如内含子内的大多数DNA和基因之间的部分DNA序列。“自私的DNA”这一说法已经在两个场合简要提及[3,4]。杜利特尔和萨皮恩扎[5]（见上篇文章）已经不约而同达成了相似观点。

The Amount of DNA

The large amounts of DNA in the cells of most higher organisms and, in particular, the exceptionally large amounts in certain animal and plant species—the so-called *C* value paradox—has been an unsolved puzzle for a considerable period (see reviews in refs 6–8). As is well known, this DNA consists in part of "simple" sequences, an extreme example of which is the very large amounts of fairly pure poly d(AT) in certain crabs. Simple sequences, which are situated in chromosomes largely but not entirely in the heterochromatin, are usually not transcribed. Another class of repetitive sequences, the so-called "intermediate repetitive", have much longer and less regular repeats. Such sequences are interspersed with "unique" DNA at many places in the chromosome, the precise pattern of interspersion being to some extent different in different species. Leaving aside genes which code for structural RNA of one sort or another (such as transfer RNA and ribosomal RNA), which would be expected to occur in multiple copies (since, unlike protein, their final products are the result of only one stage of magnification, not two), the majority of genes coding for proteins appear to exist in "single" copies, meaning here one or a few. A typical example would be the genes for α-globin, which occur in one to three copies and the human *β*-like globins, of which there are four main types, all related to each other but used for slightly different purposes. Notable exceptions are the proteins of the immune system, and probably those of the histocompatibility and related systems. Another exception is the genes for the five major types of histone which also occur in multiple copies. Even allowing for all such special case, the estimated number of genes in the human genome appears too few to account for the 3×10^9 base pairs found per haploid set of DNA, although it must be admitted that all such arguments are very far from conclusive.

Several authors[8–13] have suggested that the DNA of higher organisms consists of a minority of sequences with highly specific functions plus a majority with little or no specificity. Even some of the so-called single-copy DNA may have no specific function. A striking example comes from the study of two rather similar species of *Xenopus*. These can form viable hybrids, although these hybrids are usually sterile. However, detailed molecular hybridization studies show that there has been a large amount of DNA sequence divergence since the evolutionary separation of their forebears. These authors[13] conclude "only one interpretation seems reasonable, and that is that the specific sequence of much of the single-copy DNA is not functionally required during the life of the animal. This is not to say that this DNA is functionless, only that its specific sequence is not important".

There is also evidence to suggest that the majority of DNA sequences in most higher organisms do not code for protein since they do not occur at all in messenger RNA (for reviews see refs 14, 15). Nor is it very plausible that all this extra DNA is needed for gene control, although some portion of it certainly must be.

We also have to account for the vast amount of DNA found in certain species, such as lilies and salamanders, which may amount to as much as 20 times that found in the human genome. It seems totally implausible that the number of radically different genes needed

DNA 的数量

大多数高等生物的细胞中存在的大量 DNA，特别是某些动物和植物物种中存在异常大量的 DNA，所谓的 *C* 值悖论已经在相当长的一段时期内成为一个未解的谜团（见参考文献 6~8 的综述）。众所周知，这种 DNA 存在于部分“简单”的序列中，一个极端的例子就是某种蟹中存在大量非常纯的多聚 d(AT)。大量位于染色体中但不完全位于异染色质中的简单序列通常不被转录。另外一种重复序列即所谓的“中度重复序列”，包含更长但更不规则的重复序列。这些序列与“独特的”DNA 穿插分布到染色体的很多位置，这种分布的精确模式在不同物种中有一定程度的不同。撇开编码各种结构 RNA（例如转运 RNA 和核糖体 RNA）的基因不谈，它们通常存在多个拷贝（与蛋白质的两步合成不同，它们的最终产物是一步合成的结果）。大多数编码蛋白质的基因以“单”拷贝形式存在，这里的“单”拷贝指一个或几个拷贝。一个典型的例子就是 α–球蛋白基因，存在一到三个拷贝；人类 β–样球蛋白有四种主要类型，彼此相互关联但用途稍有不同。值得注意的例外是免疫系统的蛋白质，比如那些组织相容性及其相关系统的蛋白质。另一个例外是编码五种主要类型组蛋白的基因，同样存在多个拷贝。即使算上所有这些特例，相对于每组单倍体 DNA 拥有的 3×10^9 个碱基对，对人类基因组内基因数量的估计还是太少了，虽然不可否认所有这些争议离下结论还为时过早。

几位作者[8-13]曾经提出这样的观点：高等生物的 DNA 是由少量具有高度特异性功能的序列和大量特异性程度低或没有特异性的序列组成。甚至一些所谓的单拷贝 DNA 可能也不具有特异性功能。一个突出的例子来自对爪蟾属两种特别相似物种的研究。它们可以繁殖可成活的杂交后代，虽然这些杂交后代通常是不育的。然而精细的分子杂交研究显示，从它们祖先的进化分离开始就出现了大量的 DNA 序列差异。这些作者[13]得出结论“只有一种解释较为合理，在动物的生命过程中许多单拷贝 DNA 的特异性序列在功能上并不是必需的。这并不是说这些 DNA 没有功能，只是它的特异序列并不重要。”

同样有证据表明大多数高等生物的大部分 DNA 序列并不编码蛋白质，因为它们根本不出现在信使 RNA 中(综述见参考文献 14,15)。尽管一部分必定会起到作用，但是要说这些多余的 DNA 都是基因调控所必需的，也并不合理。

我们还必须解释的是，在某些物种（例如百合和蝾螈）中发现了惊人数量的 DNA，其数量约为人类基因组的二十倍。蝾螈所需的完全不同基因的数量是人类的二十倍是完全不合理的假设；而且没有证据表明蝾螈的基因大都存在二十个相当类

in a salamander is 20 times that in a man. Nor is there evidence to support the idea that salamander genes are mostly present in about 20 fairly similar copies. The conviction has been growing that much of this extra DNA is "junk", in other words, that it has little specificity and conveys little or no selective advantage to the organism.

Another place where there appears to be more nucleic acid than one might expect is in the primary transcripts of the DNA of higher organisms which are found in the so-called heteronuclear RNA. It has been known for some time that this RNA is typically longer than the messenger RNA molecules found in the corresponding cytoplasm. Heteronuclear RNA contains these messenger RNA sequences but has many other sequences which are never found in the cytoplasm. The phenomenon, has been somewhat clarified by the recent discovery of introns in many genes (for a general introduction see ref. 4). Although the evidence is still very preliminary, it certainly suggests that much of the base sequence in the interior of some introns may be junk, in that these sequences drift rapidly in evolution, both in detail and in size. Moreover, the number of introns may differ even in closely related genes, as in the two genes for rat preproinsulin[16]. Whether there is junk between genes is unclear but it is noteworthy that the four genes for the human β-like globins, which occur fairly near together in a single stretch of DNA, occupy a region no less than 40 kilobases long[17]. This greatly exceeds the total length of the four primary transcripts (that is the four mRNA precursors), an amount estimated to be considerably less than 10 kilobases. There is little evidence to indicate that there are other coding sequences between these genes (although the question is still quite open) and a tenable hypothesis is that much of this interspersed DNA has little specific function.

In summary, then, there is a large amount of evidence which suggests, but does not prove, that much DNA in higher organisms is little better than junk. We shall assume, for the rest of this article, that this hypothesis is true. We therefore need to explain how such DNA arose in the first place and why it is not speedily eliminated, since, by definition, it contributes little or nothing to the fitness of the organism.

What is Selfish DNA?

The theory of natural selection, in its more general formulation, deals with the competition between replicating entities. It shows that, in such a competition, the more efficient replicators increase in number at the expense of their less efficient competitors. After a sufficient time, only the most efficient replicators survive. The idea of selfish DNA is firmly based on this general theory of natural selection, but it deals with selection in an unfamiliar context.

The familiar neo-darwinian theory of natural selection is concerned with the competition between organisms in a population. At the level of molecular genetics it provides an explanation of the spread of "useful" genes or DNA sequences within a population. Organisms that carry a gene that contributes positively to fitness tend to increase their representation at the expense of organisms lacking that gene. In time, only those organisms that carry the useful gene survive. Natural selection also predicts the spread of

似的拷贝。所以可以确信，很多这些冗余的 DNA 都是“垃圾”；换句话说，它们几乎没有任何特异性，而且对传递生物的选择优势几乎没有任何贡献。

另一处比预想中有更多核酸的地方在高等生物 DNA 的初级转录物中，这是在所谓的异核 RNA 中发现的。已知这种 RNA 通常比在相应的细胞质中发现的信使 RNA 要长。异核 RNA 包含信使 RNA 的序列，同时还包含许多细胞质中未发现的其他序列。基于近期对基因内含子的一些发现，这种现象在一定程度上得到了解释（简介见参考文献 4）。虽然这些证据还很初步，但是可以肯定一些内含子内部的很多碱基序列可能是垃圾，因为这些序列在进化中会迅速漂变，无论是在内容还是长度方面都是如此。而且在一些非常相似的基因（例如大鼠前胰岛素原的两个基因）中，内含子的数目也有可能不同[16]。基因之间是否有垃圾序列还不是很清楚，但是值得注意的是人类 β–样球蛋白的四个基因一起紧密地分布在 DNA 的同一条单链上，占据了一段不小于 40kb 的区域[17]。这大大超过了四个初级转录物（四个 mRNA 前体）的总长度，据估算这些初级转录产物的总长肯定小于 10kb。鲜有证据表明这些基因之间存在其他编码序列（虽然这个问题仍未解决），而一种可靠的假说就是大量分散在其中的 DNA 几乎都没有特异性的功能。

总而言之，大量的证据表明，高等生物中的很多 DNA 和垃圾差不多，但尚未证实。我们可以在下面的叙述中假定这个假设是正确的。那么我们首先需要解释的是这样的 DNA 是怎样产生的，以及既然像所定义的那样，它对生物的适应性几乎不起作用，那它为何最初没有被迅速地淘汰。

什么是自私的 DNA？

自然选择学说在其更为广义的阐述中，论述了复制实体间的竞争。该理论认为，在这种竞争中，可以看到能力较强的复制个体以牺牲能力较弱的竞争者为代价增加个体数量。经历了足够长的时间后，只有能力最强的复制子得以存活。自私的 DNA 的理论牢固地建立在自然选择这个普遍学说上，但它解决的是一个不常见的背景中的选择问题。

我们熟知的新达尔文自然选择学说涉及的是种群中生物体间的竞争问题。它在分子遗传学的层面上给出了“有用的”基因或 DNA 序列在种群中扩增的解释。携带一个促进其适应性的基因的生物以牺牲缺乏这种基因的个体为代价提高它们自身的存活率。经过一段时间，只有那些携带这种有用基因的生物才得以存活。自然选

a gene or other DNA sequence within a single genome, provided certain conditions are satisfied. If an organism carrying several copies of the sequence is fitter than an organism carrying a single copy, and if mechanisms exist for the multiplication of the relevant sequence, then natural selection must lead to the emergence of a population in which the sequence is represented several times in every genome.

The idea of selfish DNA is different. It is again concerned with the spread of a given DNA within the genome. However, in the case of selfish DNA, the sequence which spreads makes no contribution to the phenotype of the organism, except insofar as it is a slight burden to the cell that contains it. Selfish DNA sequences may be transcribed in some cases and not in others. The spread of selfish DNA sequences within the genome can be compared to the spread of a not-too-harmful parasite within its host.

Mechanisms for DNA Spreading

The inheritance of a repeated DNA sequence in a population of eukaryotes clearly requires that the multiplication which produced it occurred in the germ line. Furthermore, any mechanism that can lead to the multiplication of useful DNA will probably lead to the multiplication of selfish DNA (and vice versa). Of course, natural selection subsequently discriminates between multiple sequences of different kinds, but it does not necessarily prevent the multiplication of neutral or harmful sequences.

Multiplication in the germ-line sequence can occur in nondividing cells or during meiosis and mitosis (within lineages that lead to the germ line). In the former case, the mechanisms available resemble those that are well documented for prokaryotes, that is, multiplication may occur in eukaryotes through the integration of viruses or of elements analogous to transposons and insertion sequences. Doolittle and Sapienza[5] have discussed these mechanisms in some detail, particularly for prokaryotes. They are likely to lead to the spreading of DNA sequences to widely separated positions on the chromosomes.

During mitosis and meiosis, multiplication (or deletion) is likely to occur by unequal crossing over. This mechanism will often lead to the formation of tandem repeats. It is well documented for the tRNA "genes" of *Drosophila* and for various other tandemly repeated sequences in higher organisms.

The Amount and Location of Selfish DNA

Natural selection "within" the genome will favour the indefinite spreading of selfish preferred replicators. Natural selection between genotypes provides a balancing force that attempts to maintain the total amount of selfish DNA at an equilibrium (steady state) level—organisms whose genomes contain an excessive proportion of selfish DNA would be at a metabolic disadvantage relative to organisms with less selfish DNA, and so would be eliminated by the normal mechanism of natural selection. Excessive spreading of functionless replicators may be considered as a "cancer" of the genome—the uncontrolled

择也预言，如果满足所需条件，一个单基因组内的基因或其他DNA序列将在其中得以扩增。如果携带了一段序列的多个拷贝的生物比只携带一个拷贝的生物的适应性要好，并且此机制由于相关序列的增殖而继续存在，那么自然选择将导致一类种群的出现，在这个种群中，该序列在每个基因组中多次重复。

自私的DNA的概念并不相同。它也与基因组中一段给定的DNA的扩增有关。然而，对于自私的DNA来说，扩增的序列对于生物体的表型没有任何贡献，而且在这种情况下对于包含它的细胞来说是个轻微的负担。自私DNA的序列只在某些情况下才可能被转录，其他情况下都不会。自私DNA的序列在基因组中的扩增就像害处不大的寄生虫在宿主中的繁殖。

DNA扩增的机制

在真核生物种群中，一段重复DNA序列若要遗传下去，必须保证产生该序列的增殖发生在生殖细胞系中。进一步说，任何产生有用DNA的增殖的机制同样可能产生自私DNA的增殖（反之亦然）。当然，自然选择接下来会区别对待不同种类的多种序列，但未必会阻止中性或有害序列的扩增。

生殖细胞系中的序列增殖也会发生在非分裂的细胞中或者减数分裂（在产生生殖细胞系的株系中）和有丝分裂过程中。在前一种情况下，现有的机制类似于已在原核生物中得到很好证明的机制，真核生物中的基因复制可能是通过病毒或与转座子和插入序列类似的元件的整合发生的。杜利特尔和萨皮恩扎[5]已经在某些细节上讨论了这些机制，特别是对于原核生物。它们很可能导致DNA序列扩散到染色体广泛的分散位点上。

在减数分裂和有丝分裂过程中，增殖（或缺失）有可能通过不等交换发生。这种机制经常导致串联重复序列的形成。果蝇的tRNA“基因”和高等生物其他各种串联重复序列就是很好的证明。

自私DNA的数量和位点

基因组内的自然选择倾向于自私的优先复制子的无限扩增。基因型间的自然选择提供了一种平衡力使自私DNA的总量保持平衡（稳定状态）——基因组中含过量比例的自私DNA的生物比含较少自私DNA的生物，在新陈代谢方面将处于劣势，因此将在自然选择的正常机制下被淘汰。没有功能的复制子的过度扩增可以看作是基因组的“癌症”——基因组中一个片段不受控制的扩增将最终导致允许这种扩增

expansion of one segment of the genome would ultimately lead to the extinction of the genotype that permits such expansion. Of course, we do not know whether extinction of genotypes in nature even occurs for this reason.

It is hard to get beyond generalities of this kind. To do so we would, at least, need to know how much selective disadvantage results from the presence of a given amount of useless DNA. Even this minimal information is not easily acquired, so we cannot produce other than qualitative arguments.

It seems certain that the metabolic energy cost of replicating a superfluous short DNA sequence in a genome containing 10^9 base pairs would be very small. If, for example, the selective advantage were equal to the proportion of the genome made up by the extra DNA, a sequence of 1,000 base pairs would produce a selective disadvantage of only 10^{-6}. If the selective disadvantage were proportional to the extra energy cost divided by the total metabolic energy expended per cell per generation, the disadvantage would be much smaller. The selective disadvantage might be greater in more stringent conditions, but it is still hard to believe that a relatively small proportion of selfish DNA could be selected against strongly.

On the other hand, when the total amount of selfish DNA becomes comparable to or greater than that of useful DNA, it seems likely that the selective disadvantage would be significant. We may expect, therefore, that the mechanisms for the formation and deletion of nonspecific DNA will adjust, in each organism, so that the load of DNA is sufficiently small that it can be accommodated without producing a large selective disadvantage. The proportion of nonspecific DNA in any particular organism will thus depend on the lifestyle of the organism, and particularly on its sensitivity to metabolic stress during the most vulnerable part of the life cycle.

We can make one prediction on the basis of energy costs. Selfish DNA will accumulate to a greater extent in non-transcribed regions of the genome than in those that are transcribed. Of course, selfish DNA will in most cases be excluded from translated sequences, because the insertion of amino acids within a protein will almost always have serious consequences, even in diploid organisms (but see the suggestion by F.H.C.C.[18]).

At first sight it might seem anomalous that natural selection does not eliminate all selfish DNA. Since the suggestion that much eukaryotic DNA is useless distinguishes the selfish DNA hypothesis from many closely related proposals, it may be useful to take up this point in some detail.

First, the elimination of disadvantaged organisms from a population, by their more favoured competitors, takes a number of generations several times larger than the reciprocal of the selective disadvantage. If the selective disadvantage associated with a stretch of useless DNA in higher organisms is only 10^{-6}, it would take 10^6–10^8 years to

的基因型的消失。当然，我们甚至不知道自然界中基因型的消失是否是出于这个原因。

在这类问题上得到超出一般性的概括是很困难的。如果要研究，我们至少要知道一定数量无用 DNA 的存在将会产生多少选择性的劣势。即使这个最基本的信息都很难得到，所以我们只能进行一些定性的讨论。

似乎可以确定的是，在一个拥有 10^9 个碱基对的基因组中复制一段多余的短 DNA 序列所消耗的代谢能量是很低的。例如，如果选择优势与冗余 DNA 组成的基因组的比例是均等的，那么一段 1,000 个碱基对的序列产生的选择劣势仅仅为百万分之一。如果这种选择劣势与每一代每个细胞消耗的总代谢能量分配的额外能量消耗成比例，这种劣势将会更微不足道。这种选择劣势在较苛刻的条件下可能会变得更加重要，但是仍然很难让人相信相对较小比例的自私 DNA 会面临激烈淘汰。

另一方面，当自私 DNA 的总量与有用 DNA 的量相当或者比其更大的时候，选择性的劣势可能会变得很显著。因此，我们可以预见，非特异性 DNA 形成和消除的机制会在每个生物体中得以调节，使 DNA 的负荷足够小来适应环境，而不至于产生大的选择性劣势。因此任何特定生物中非特异 DNA 的比例将取决于这种生物的生存方式，特别是取决于它在整个生命周期最脆弱时期对代谢压力的敏感性。

我们可以基于能量消耗进行一番假设。自私 DNA 在基因组非转录区域将比转录区域积累的数量多。当然，自私 DNA 在大多数情况下被排除在翻译序列之外，因为蛋白质内氨基酸的插入往往会产生严重后果，即使在二倍体生物中也是如此（见克里克[18]的建议）。

自然选择并没有淘汰所有的自私 DNA，最初看起来有些不合常理。由于真核生物的多数 DNA 无用的观点使自私 DNA 假说与许多相近的提议区别开来，那么对这个观点进行一些深入研究可能是有用的。

首先，处于劣势的生物被处于相对优势的竞争者从种群中淘汰，需要经历的代数比选择劣势的倒数大很多倍。在高等生物中如果与一段无用 DNA 相关联的选择劣势只有 10^{-6}，那么通过竞争淘汰这段基因大概需要 10^6~10^8 年。对一些典型的

eliminate it by competition. For typical higher organisms this is a very long time, so the elimination of a particular stretch of selfish DNA may be a very slow process even on a geological time scale. Second, the mechanisms for the deletion of short sequences of DNA may be inefficient, since there is no strong selective pressure for the development of "corrective" measures when the "fault" carries a relatively small selective penalty. Taken together, these arguments suggest that the elimination of a particular piece of junk from the genome may be a very slow process.

This in turn suggests that the amount of useless DNA in the genome is a consequence of a dynamic balance. The organism "attempts" to limit the spread of selfish DNA by controlling the mechanism for gene duplication, but is constrained by imperfections in genetic processes and/or by the need to permit some duplication of advantageous genes. Selfish DNA sequences "attempt" to subvert these mechanisms and may be able to do so comparatively rapidly because mutation will affect them directly. On the other hand, the defence mechanisms of the host are likely to depend on the action of protein and therefore may evolve more slowly. Once established within the genome, useless sequences probably have a long "life expectancy".

For any particular type of selfish DNA, there is no reason that a steady state should necessarily be reached in evolution. The situation would be continually changing. A particular type of DNA might first spread rather successfully over the chromosomes. The host might then evolve a mechanism which reduced or eliminated further spreading. It might also evolve a method for preferentially deleting it. At the same time, random mutations in the selfish DNA might make it more like ordinary DNA and so, perhaps, less easy to remove. Eventually, these sequences, possibly by now rather remote from those originally introduced, may cease to spread and be slowly eliminated. Meanwhile, other types of selfish DNA may originate, expand and evolve in a similar way.

In short, we may expect a kind of molecular struggle for existence within the DNA of the chromosomes, using the process of natural selection. There is no reason to believe that this is likely to be much simpler or more easy to predict than evolution at any other level. At bottom, the existence of selfish DNA is possible because DNA is a molecule which is replicated very easily and because selfish DNA occurs in an environment in which DNA replication is a necessity. It thus has the opportunity of subverting these essential mechanisms to its own purpose.

The Inheritance of Selfish DNA

Although the inheritance of selfish DNA will occur mainly within a mendelian framework, it is likely to be different in detail and more complex than simple mendelian inheritance. This is due both to the multiplication mechanisms, which in one way or another will produce repeated copies (see the discussion by Doolittle and Sapienza[5]), and to the fact that these copies are likely to be distributed round the chromosomes rather than being located in a single place in the genome as most normal genes are. For both these reasons,

高等生物来说这是一段很长的时间，所以淘汰一段特定的自私 DNA 即使在地质年代表中也可能是一段漫长的过程。其次，短序列 DNA 的消除机制效率很低，因为当这种“错误”带来的选择惩罚相对较小时，对“矫正”措施的发展就没有很大的选择压力。总之，这些争议提示，从基因组中淘汰一段特定的垃圾序列可能是一段很漫长的过程。

这也反过来提示基因组中大量无用 DNA 的存在可能是一种动态平衡的结果。生物“试图”通过控制基因复制的机制来限制自私 DNA 的扩增，但是受到一些限制,包括遗传过程的缺陷和（或）一些有利基因复制的需求。自私 DNA 序列“试图”破坏这些机制，而且由于突变的直接影响而可能相对较快地实现。另一方面，宿主的防御机制很可能是依靠蛋白质的功能,所以进化得更加缓慢。一旦在基因组中形成，无用序列很可能拥有一段很长的“期望寿命”。

对任何一种特定类型的自私 DNA 而言，它们都没有理由要在进化过程中达到一种稳定状态。情况会一直变化。一种特定类型的 DNA 会首先在染色体范围成功扩增。宿主会随之进化出相应机制来减少或消除其进一步的扩增，也有可能进化出某种方法优先删除这段序列。与此同时，自私 DNA 内部的随机突变很可能使它与普通 DNA 更加类似，从而更加难以去除。最终，这些序列与当初产生时相比已经有了很大差异，而且可能会逐渐停止扩增并慢慢被淘汰。同时，其他类型的自私 DNA 也以相似的方式产生、扩散和进化。

总之，我们可以认为这是染色体 DNA 内的分子为了生存，利用自然选择过程而进行的一种战斗。没有理由相信这个过程比其他水平的进化过程更简单或更容易。从根本上说，自私 DNA 的存在是可能的，因为 DNA 是一种很容易复制的分子，而且自私的 DNA 所处的环境中，DNA 复制是一种必需。因此它有机会破坏这些基本机制，以达到自己的目的。

自私 DNA 的遗传

虽然自私 DNA 的遗传方式基本处在孟德尔定律的框架中，但是与单纯的孟德尔定律相比，其在细节上有所不同并且更加复杂。这不仅仅是因为增殖机制会以某种形式产生重复序列（见杜利特尔和萨皮恩扎[5]的讨论），还因为这些拷贝很可能分散到染色体各处，而不会像大多数普通基因那样只定位于基因组某个固定的位置。基于这两个原因，一种特殊类型的自私 DNA 很可能比低选择优势的普通基因在种

a particular type of selfish DNA is likely to spread more rapidly through a population than would a normal gene with a low selective advantage. It will be even more rapid if selfish DNA can spread horizontally between different individuals in a population, due to viruses or other infectious agents, although it should be remembered that such "infection" must affect the germ line and not merely the soma. If this initial spread takes place when the additional DNA produced is relatively small in amount, it is unlikely to be seriously hindered by the organism selecting against it. The study of these processes will clearly require a new type of population genetics.

Can Selfish DNA Acquire a Specific Function?

It would be surprising if the host organism did not occasionally find some use for particular selfish DNA sequences, especially if there were many different sequences widely distributed over the chromosomes. One obvious use, as repeatedly stressed by Britten and Davidson[19,20], would be for control purposes at one level or another. This seems more than plausible.

It has often been argued (see, for example, ref. 21) that for the evolution of complex higher organisms, what is required is not so much the evolution of new proteins as the evolution of new control mechanisms and especially mechanisms which control together sets of genes which previously had been regulated separately. To be useful, a new control sequence on the DNA is likely to be needed in a number of distinct places in the genome. It has rarely been considered how this could be brought about expeditiously by the rather random methods available to natural selection.

A mechanism which scattered, more or less at random, many kinds of repeated sequences in many places in the genome would appear to be rather good for this purpose. Most sets of such sequences would be unlikely to find themselves in the right combination of places to be useful but, by chance, the members of one particular set might be located so that they could be used to turn on (or turn off) together a set of genes which had never been controlled before in a coordinated way. A next way of doing this would be to use as control sequences not the many identical copies distributed over the genome, but a small subset of these which had mutated away from the master sequence in the same manner.

On this picture, each set of repeated sequences might be "tested" from time to time in evolution by the production of a control macromolecule (for example, a special protein) to recognize those sequences. If this produced a favourable result, natural selection would confirm and extend the new mechanism. If not, it would be selected against and discarded. Such a process implies that most sets of repeated sequences will never be of use since, on statistical grounds, their members will usually be in unsuitable places.

It thus seems unlikely that all selfish DNA has acquired a special function, especially in those organisms with very high *C* values. Nor do we feel that if one example of a particular sequence acquires a function, all the copies of that sequence will necessarily do

群中扩增得更快。如果自私 DNA 能通过病毒或其他感染因子在一个种群的不同个体中水平扩散，速度将会更快，应当记住，尽管这种“感染”肯定要影响整个生殖细胞系而不是仅仅影响体细胞。如果这种初始扩增在产生的冗余 DNA 数量相对较小时发生，则不太可能受到生物体选择对其的严重阻碍。对于这些过程的研究显然需要一种新型的种群遗传学。

自私 DNA 能获得一种特异性功能吗?

如果宿主生物没能偶然地发现特定自私 DNA 序列的某些功能，这会很令人惊奇，特别是有很多不同序列广泛分布于染色体的时候。如布里滕和戴维森[19,20]反复强调的，一个明显的功能就是在一定水平上实现调控的目的。这似乎是非常合理的。

我们经常讨论（例如见参考文献 21），对于复杂的高等生物的进化，与其说需要新蛋白的进化，不如说需要新的调控机制的进化，特别是集中调控几组基因的机制，而这几组基因之前是被分别调控的。更为有用的是，一段 DNA 上新的调控序列可能被基因组中很多不同的位置需求。很少有人思考，非常随机的自然选择是如何使这些事件快速发生的。

将多种重复序列或多或少任意地分散到基因组各处的机制对达成这个目标是很有利的。大多数这样的序列组不太可能位于正确的位置而发挥作用，但是偶尔也有可能某组特定序列的成员正好处在合适的位置可用来共同开启（或关闭）一组以前从未以协同方式调控的基因。接下来发生的是，这些序列被用作调控序列，它们并不是遍布基因组的很多相同的拷贝，而只是这些拷贝中以相同方式从主序列变异而来的一小部分序列。

在这个框架下，每组重复序列在进化过程中都会时常受到一个调控大分子（如一种特殊的蛋白质）的“检测”而得到识别。如果产生了一个有利的结果，自然选择就会认可并推广这种新机制。如果没有，这种机制就会被排斥并被淘汰。这样一个过程表明大多数的重复序列毫无用处，因为从统计学角度来看，它们的成员通常会位于不适当的位置。

因此，所有自私 DNA 都具有特殊功能是不太可能的，特别是对于那些具有高 *C* 值的生物而言。我们认为，如果一段特殊序列的一个拷贝获得了一种功能，其他

so. As selfish DNA is likely to be distributed over the chromosomes in rather a random manner, it seems unlikely that every copy of a potentially useful sequence will be in the right position to function correctly. For example, if a specific sequence within an intron were used to control the act of splicing that intron, a similar sequence in an untranscribed region between genes would obviously not be able to act in this way.

In some circumstances, the sheer bulk of selfish DNA may be used by the organism for its own purpose. That is, the selfish DNA may acquire a nonspecific function which gives the organism a selective advantage. This is the point of view favoured by Cavalier-Smith in a very detailed and suggestive article[12] which the reader should consult. He proposes that excess DNA may be the mechanism the cell uses to slow up development or to make bigger cells. However, we suspect that both slow growth and large cell size could be evolved just as well by other more direct mechanisms. We prefer to think that the organism has tolerated selfish DNA which has arisen because of the latter's own selective pressure.

Thus, some selfish DNA may acquire a useful function and confer a selective advantage on the organism. Using the analogy of parasitism, slightly harmful infestation may ultimately be transformed into a symbiosis. What we would stress is that not all selfish DNA is likely to become useful. Much of it may have no specific function at all. It would be folly in such cases to hunt obsessively for one. To continue our analogy, it is difficult to accept the idea that all human parasites have been selected by human beings for their own advantage.

Life Style

The effect of nonspecific DNA on the life style of the organism has been considered by several authors, in particular by Cavalier-Smith[12] and by Hindergardner[8]. We shall not attempt to review all their ideas here but instead will give one example to show the type of argument used.

Bennett[22] has brought together the measurements of DNA content for higher herbaceous plants. There is a striking connection between DNA content per cell and the minimum generation time of the plant. In brief, if such an angiosperm has more than 10 pg of DNA per cell, it is unlikely to be an ephemeral (that is, a plant with a short generation time). If it is a diploid and has more than 30 pg of DNA, it is highly likely to be an obligate perennial, rather than an annual or an ephemeral. The converse, however, is not true, there being a fair number of perennials with a DNA content of less than 30 pg and a few with less than 10 pg. A clear picture emerges that if a herbaceous plant has too much DNA it cannot have a short generation time.

This is explained by assuming that the extra DNA needs a bigger nucleus to hold it and that this increases both the size of the cell and the duration of meiosis and generally slows up the development of the plant. An interesting exception is that the duration of meiosis,

所有的拷贝未必也会获得该功能。由于自私 DNA 可能是以相当随机的方式遍布于染色体中，所以一个潜在有用序列的每一个拷贝不太可能都位于正确的位置发挥正常功能。例如，如果一段位于内含子内部的特异性序列起到调控内含子剪接的作用，那么显然另一段位于基因间非转录区的相似序列并不能发挥这种作用。

在一些情况下，生物体可能会利用自私 DNA 的绝对数量来达到自己的目的。即自私 DNA 可能获得一种非特异性的功能而使生物体具有选择优势。这是卡弗利尔－史密斯在一篇详尽且很有启发性的文章[12]中所支持的观点，读者可以作为参考。他提出冗余的 DNA 可能是细胞用于减缓发育或者产生更大细胞的机制。然而，值得我们怀疑的是减缓生长与增大细胞尺寸方面的进化都可以通过其他更直接的方式实现。我们更倾向于认为，生物体包容了因为（自私 DNA）自身选择压力而产生的自私 DNA。

因此，一些自私 DNA 可能获得有用的功能而赋予生物体选择优势。与寄生的过程相类似，轻微有害的感染最终可能转化为共生。我们应当强调的是并非所有的自私 DNA 都可能变成有用的。它们中的大多数根本没有任何特异性功能。在这些情况下，执意寻求一个功能是愚蠢的。继续我们的类比，这就好像很难让人接受所有的人类寄生虫都是由人类为自身利益而选择出来的。

生命方式

已有多位作者考虑到非特异性 DNA 对生物生命方式的影响，特别是卡弗利尔－史密斯[12]和欣德加德纳[8]。在这里我们不再试图复述他们所有的观点，而是举例来说明用到的论据的类型。

本内特[22]归纳了高等草本植物 DNA 含量的测量方法。每个细胞内的 DNA 含量和植物最短世代时间之间有显著相关性。简而言之，如果一种被子植物每个细胞含有超过 10 pg 的 DNA，它就不可能是一种短生植物（指世代时间很短的植物）。如果它是二倍体并含有超过 30 pg 的 DNA，那它极有可能是专性多年生植物，而不是一年生或短生植物。然而逆命题并不成立，有相当一部分多年生植物的 DNA 含量低于 30 pg，甚至有几种低于 10 pg。可以得到一个清晰的结论是如果一种草本植物含有大量 DNA，那么它的世代时间不会很短。

这种现象可以用一种假设解释，那就是冗余 DNA 需要更大的细胞核来容纳，这就增加了细胞的大小和减数分裂持续的时间，通常也减缓了植物的生长。但有一

is, if anything, shorter for polyploid species than for their diploid ancestors[23]. This suggests that it is the ratio of good DNA to junk DNA rather than the total DNA content which influences the duration of meiosis.

An analogous situation may obtain in certain American species of salamander. These often differ considerably in the rapidity of their development and of their life cycles, the tropical species tending to take longer than the more temperate ones. Drs David Wake and Herbert MacGregor (personal communication) tell us that preliminary evidence suggests that species with the longer developmental times often have the higher *C* values. This appears to parallel the situation just described for the herbacious plants. It remains to be seen if further evidence will continue to support this generalization. (See the interesting paper by Oeldorfe *et al.*[25] on 25 species of frogs. They conclude that "genome size sets a limit beyond which development cannot be accelerated".)

Testing the Theory

The theory of selfish DNA is not so vague that it cannot be tested. We can think of three general ways to do this. In the first place, it is important to know where DNA sequences occur which appear to have little obvious function, whether they are associated with flanking or other sequences of any special sort and how homologous sequences differ in different organisms and in different species, either in sequence or in position on the chromosome. For example, it has recently been shown by Young[24] that certain intermediate repetitive sequences in *Drosophila* are often in different chromosomal positions in different strains of the same species.

Second, if the increase of selfish DNA and its movement around the chromosome are not rare events in evolution, it may be feasible to study, in laboratory experiments, the actual molecular mechanisms involved in these processes.

Third, one would hope that a careful study of all the nonspecific effects of extra DNA would give us a better idea of how it affected different aspects of cellular behaviour. In particular, it is important to discover whether the addition of nonspecific DNA does, in fact, slow down cells metabolically and for what reasons. Such information, together with a careful study of the physiology and life style of related organisms with dissimilar amounts of DNA, should eventually make it possible to explain these differences in a convincing way.

Conclusion

Although it is an old idea that much DNA in higher organisms has no specific function[8–12], and although it has been suggested before that this nonspecific DNA may rise to levels which are acceptable or even advantageous to an organism[8,12], depending on certain features of its life style, we feel that to regard much of this nonspecific DNA as selfish DNA is genuinely different from most earlier proposals. Such a point of view is especially useful in thinking about the dynamic aspects of nonspecific DNA. It directs attention

个有趣的例外，多倍体物种减数分裂所持续的时间比它们的二倍体祖先更短[23]。这种现象说明，影响减数分裂时间的是有用DNA与垃圾DNA的比值，而不是全部DNA的含量。

在美国的某种蝾螈中也可以发现类似的情况。它们的发育和生长周期的速率往往相差很大，热带品种通常比温带品种需要更长的生长发育时间。戴维·韦克博士和赫伯特·麦格雷戈博士（个人交流）告诉我们，有初步证据显示发育期较长的物种通常具有较高的*C*值。这个结论与前面描述的草本植物的情况相类似。是否有更进一步的证据来继续支持这个推断，还要拭目以待。（请参阅厄尔德费等[25]关于25种蛙类的有趣文章。他们得出结论"基因组的大小为发育设置了上限，导致其不能加速发育。"）

检验此学说

自私DNA的学说并没有模糊到不可检验。我们可以想到三种普适的检验方法。首先，重要的是弄清楚那些看似没有显著功能的DNA序列位于什么位置，它们是否与侧翼或其他任何特殊类型的序列相关，以及同源序列在不同生物不同物种间差别如何，包括序列本身及其在其染色体中所处位置。例如近期扬[24]发现果蝇中某些中度重复序列在同一物种的不同种系中常常位于染色体的不同位置。

第二，如果在进化过程中，自私DNA的增加及其在染色体内的移动并不罕见，那么在实验室中研究这个过程中实际的分子机制是可行的。

第三，我们希望对冗余DNA所有非特异性作用的深入研究可以帮助我们更好地理解它如何影响细胞活动的不同方面。特别重要的是，去探索这些非特异性DNA的加入是否的确减缓了细胞的新陈代谢，以及是什么原因造成了这种现象。这些信息，加上对不同含量DNA的相关生物的生理和生命方式的深入研究，最终可能会有力地解释这些差异。

结　论

虽然高等生物的大多数DNA都没有特异性功能[8-12]这一想法不再新奇，以前也有人认为这种非特异性的DNA会升级到一定程度，使其不但被生物体接受，甚至会使生物具有一定优势[8,12]。但是基于它的生命方式的某些特点，我们认为把大量非特异性DNA视为自私DNA的观点与早期的大多数提议有切实的差异。这个观点在

to the mechanisms involved in the spread and evolution of such DNA and it cautions one against looking for a special function for every piece of DNA which drifts rapidly in sequence or in position on the genome.

While proper care should be exercised both in labelling as selfish DNA every piece of DNA whose function is not immediately apparent and in invoking plausible but unproven hypotheses concerning the details of natural selection, the idea seems a useful one to bear in mind when exploring the complexities of the genomes of higher organisms. It could well make sense of many of the puzzles and paradoxes which have arisen over the last 10 or 15 years. The main facts are, at first sight, so odd that only a somewhat unconventional idea is likely to explain them.

We thank W. Ford Doolittle and C. Sapienza for showing us their article before publication, and Drs D. Wake and H. MacGregor for allowing us to quote some of their unpublished conclusions about salamanders. This work was supported by the Eugene and Estelle Ferhauf Foundation, the J. W. Kieckhefer Foundation, the Ahmanson Foundation and the Samuel Roberts Noble Foundation.

(**284**, 604-607; 1980)

L. E. Orgel and F. H. C. Crick
The Salk Institute, 10010 N. Torrey Pines Road, La Jolla, California 92037

References:

1. Dawkins, R. *The Selfish Gene* (Oxford University Press, 1976).
2. Walker, P. M. B. in *Human Genetics: Possibilities and Realities,* 25-38 (Excepta Medica, Amsterdam, 1979).
3. Crick, F. H. C. in *From Gene to Protein: Information Transfer in Normal and Abnormal Cells* (eds Russell, T. R., Brew, K., Faber, H. & Schultz, J.) 1-13 (Academic, New York, 1979).
4. Crick, F. H. C. *Science* **204**, 264-271 (1979).
5. Doolittle, W. F. & Sapienza, C. *Nature* **284**, 601-603 (1980).
6. Callan, H. G. *J. Cell Sci.* **2**, 1-7 (1967).
7. Thomas, C. A. *A. Rev. Genet.* **5**, 237-256 (1971).
8. Hinegardner, R. in *Molecular Evolution* (ed. Ayata, F. J.) 179-199 (Sinauer, Sunderland, 1976).
9. Commoner, B. *Nature* **202**, 960-968 (1964).
10. Ohno, S. *J. Hum. Evolut.* **1**, 651-662 (1972).
11. Comings, D. E. *Adv. Hum. Genet.* **3**, 237-436 (1972).
12. Cavalier-Smith, T. *J. Cell Sci.* **34**, 274-278 (1978).
13. Galan, G. A., Chamberlin, M. E., Hough, B. R., Britten, R. J. & Davidson, E. H. in *Molecular Evolution* (ed. Ayala, F. J.) 200-224 (Sinauer, Sunderland, 1976).
14. Bishop, J. O. *Cell* **2**, 81-86 (1974).
15. Lewin, B. *Cell* **4**, 11-20 (1975); *Cell* **4**, 77-93 (1975).
16. Lomedico, P. *et al. Cell* **18**, 545-558 (1979).
17. Bernards, R., Little, P. F. R., Annison, G., Williamson, R. & Flavell, R. A. *Proc. Natl. Acad. Sci. U.S.A.* **76**, 4827-4831 (1979).
18. Crick, F. H. C. Eur. *J. Biochem.* **83**, 1-3 (1978).
19. Britten, R. J. & Davidson, E. H. *Science* **165**, 349-358 (1969).
20. Davidson, E. U. & Britten, R. J. *Science* **204**, 1052-1059 (1979).
21. Wilson, A. C. in *Molecular Evolution* (ed. Ayala, F. J.) 225-236 (Sinauer, Sunderland, 1976).

思考非特异DNA的动态变化方面特别有帮助。它将人们注意力导向这种DNA扩增和进化的机制上，而且告诫我们不要在基因组上的序列或位点中试图寻找每段快速漂变DNA的特异功能。

然而在给每段功能还未直接显现的DNA贴上自私DNA的标签以及在引用看似合理但未经证实的关于自然选择细节的假说的过程中，还应仔细斟酌。在探索高等生物基因组的复杂性时，参考这个想法或许会起到一些作用。它能很好地解释过去10到15年间提出的很多难题和矛盾。主要的事实起初看来非常奇怪，以至于只有些许非常规的想法才能解释它们。

感谢福特·杜利特尔和萨皮恩扎允许我们拜读他们未发表的文章，以及韦克博士和麦格雷戈博士允许我们引用一些关于蝾螈尚未发表的结论。此项工作受到尤金和埃斯特尔·费尔哈夫基金会，基克希弗基金会，阿曼森基金会以及塞缪尔·罗伯茨荣誉基金会的支持。

（李响 翻译；崔巍 审稿）

22. Bennett, M. D. *Proc. R. Soc.* **B181**, 109-135 (1972).

23. Bennett, M. D. & Smith, J. B. *Proc. R. Soc.* **B181**, 81-107 (1972).

24. Young, M. W. *Proc. Natl. Acad. Sci. U.S.A.* **76**, 6274-6278 (1979).

25. Oeldorfe, E., Nishioka, M. & Bachmann, K. *Sonderdr. Z. F. Zool. System. Evolut.* **16**, 216-24 (1978).

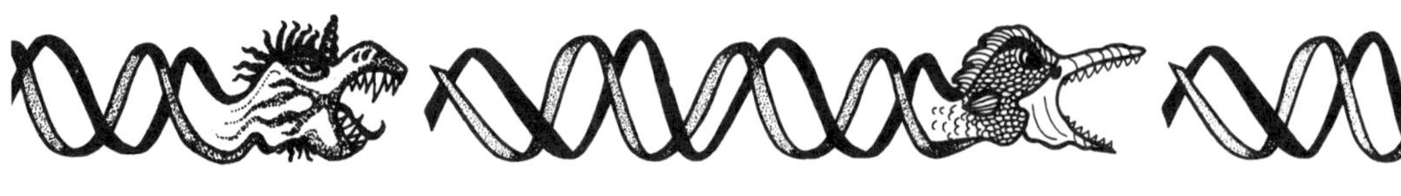

Selfish DNA

L. E. Orgel *et al.*

Editor's Note

In April of this year, the authors of this article published reviews examining how "selfish DNA" that has no phenotypic function can "live" within the genomes of organisms. Their views drew much comment, and here they reassess the earlier papers in the light of the ensuing debate. They clarify their terminology and acknowledge that there are several classes of DNA in genomes, including "dead" DNA that has lost its coding function, and DNA that functions much like a symbiotic entity.

In two review articles in the 17 April issue Doolittle and Sapienza (p. 601) and Orgel and Crick (p. 604) separately suggested that much of the DNA in the genome of higher organisms could be described as "selfish". They argued that such DNA has no appreciable phenotypic effect and functions only to ensure its own self-preservation within the genome. This view point stimulated a great deal of comment, some of which was published in the issue of 26 June (p. 617). Now the original authors have joined up with one of their critics and reassessed their ideas in the two articles below. A further comment is added by H. K. Jain.

DIFFICULTIES have been caused by the words "selfish," "junk," "specific" and "phenotype" that were used in the two reviews of selfish DNA[1,2].

Many people dislike the term "selfish DNA" and a more acceptable alternative might be "parasitic DNA". The word "parasitic" does not imply that the DNA can move between individuals, though certain viral DNAs might do this. It does imply that such DNA can usually move between different chromosomes in the same cell.

The word "junk" also seems to arouse strong feelings. The idea behind it can be clarified by considering what is meant by "specific". We consider a sequence highly specific if the change of any one of its bases almost always has a considerable effect on the organism.

The "selfish DNA" design at the top of the page was created by Linda Angeloff-Sapienza of Halifax, Canada and originally appeared on the cover of the issue of 17 April, 1980.

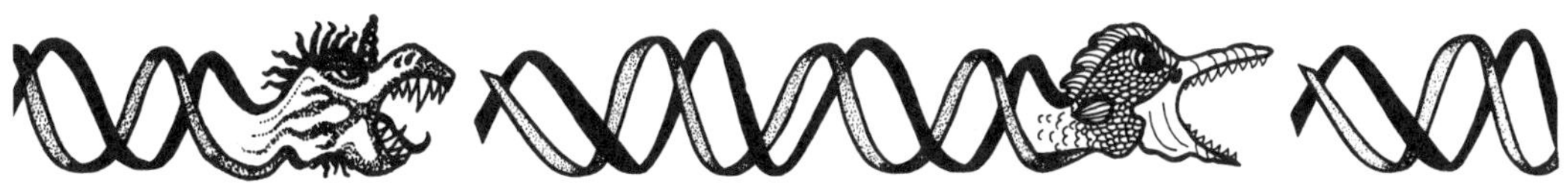

自私的DNA

奥格尔等

编者按

当年4月，本文的作者发表了两篇综述，探究了没有表型功能的“自私DNA”是怎样“生活”在生物体的基因组内的。他们的观点引起了众多评论，在此他们根据后续争论重新评价了之前的论文。他们澄清了他们的术语并认可基因组中有多种DNA，包括已经失去编码功能的“死”DNA和那些功能类似于共生实体的DNA。

4月17日那一期刊登了两篇分别由杜利特尔和萨皮恩扎（第601页）及奥格尔和克里克（第604页）发表的综述文章，这两篇文章指出高等生物基因组中的许多DNA可以被描述为“自私的”。他们认为这种DNA没有明显的表型效应和功能，它们仅能保证自身在基因组中的自我保存。这一观点激起了许多争论，其中有部分评论发表在了6月26日版上（第617页）。现在原创作者与其中的一位评论者一起，在下面两篇文章中重新评估他们的观点。贾因做了一些补充评论。

在关于自私DNA的两篇综述里，对“自私的”“垃圾”“特异的”“表型”这些词的使用引起了一些麻烦[1,2]。

许多人不喜欢“自私DNA”这个术语，而更倾向于接受“寄生DNA”这一替代词语。尽管某些病毒的DNA可以在个体之间移动，但是“寄生”这一词语并不意味着DNA可以在个体之间转移。不过这个词语的确暗示这种DNA可以在同一细胞的不同染色体间发生移动。

“垃圾”一词似乎也具有强烈的感情色彩。其隐藏的含意可以通过考虑“特异”这一含义而得以阐明。如果一条序列任意一个碱基的变化几乎都会对生物产生明显影响的话，我们就将其视为一条高度特异性的序列，例如一个与生理相关的限制性

本页最上方的“自私DNA”的设计图案是由加拿大哈利法克斯市的琳达·安杰洛夫–萨皮恩扎创作的，最早出现于1980年4月17日版的封面上。

An example would be the recognition site for a physiologically relevant restriction enzyme. At the other extreme are sequences whose deletion or extensive alteration would produce a negligible effect. Such sequences could reasonably be called junk. However, there is probably a continuum between these two extremes, including fairly specific sequences, where the alteration of most bases will produce some effect (many sequences coding for protein, and the different signals for starting and stopping transcription are likely to be of this type) and sequences whose deletion or extensive alteration will usually produce a small effect, such as a change in the local rate of recombination. In some cases close similarity of two sequences may be important, while the base sequences themselves may matter hardly at all—for example, within the introns of two neighbouring versions of a gene. The word "junk" is perhaps too broad to cover all those cases for which the effect of sequence on the phenotype of an organism is small or zero. We hope a more precise terminology will evolve as the facts become better known.

The word "phenotype" has also caused difficulties in spite of Doolittle and Sapienza's careful use of "organismal phenotype" to make their meaning clear. We obviously need two words: one to refer to the phenotype of the organism and the other to apply solely to the "phenotype" of the parasitic DNA, a distinction we would certainly make in the case of a true parasite. For the former we would suggest "organismal phenotype" and for the latter, following Cavalier-Smith[3], "intragenomic phenotype", but we would allow the word "phenotype" alone to be used when the context makes the meaning clear.

In our original definition we said that selfish DNA had two distinct properties: (1) It arises when a DNA sequence spreads by forming additional copies of itself within the genome. (2) It makes no specific contribution to the phenotype. By "phenotype" we meant organismal phenotype. We intended "specific" to be understood as "highly specific" or "fairly specific" in the discussion above. However it has been pointed out to us by R. Pritchard[4] that "no... contribution" is unnecessarily strict. It would have been more useful to include also DNA which made a small contribution to the organismal phenotype, either positive or negative. An example of the latter might be a viral DNA which became part of the genome.

There is obviously a continuum of possible selective advantages (positive or negative) to the organism. We had excluded from our definition of selfish DNA those cases where the selective advantage is very high. To decide whether a repeated sequence is parasitic or not, one must determine whether the presence of the repeated sequence in the population is mainly due to the efficiency with which the sequence spreads intragenomically or mainly due to the reproductive success of those individuals in the population who possess repeated copies of the sequence. Only in the former case do we consider it useful to use the term selfish or parasitic DNA, as opposed to useful or symbiotic DNA—the borderline between the two may not be sharp.

内切酶的识别位点。另一个极端是有些序列的缺失或广泛变异所产生的影响都微乎其微，这样的序列被称为垃圾 DNA 还是很合理的。然而，这两个极端之间可能存在着一个连续集合，这其中包括那些大部分碱基的变化会产生某种影响的相当特异的序列(许多编码蛋白质及转录起始和终止的不同信号的序列可能就是这种类型的)，以及缺失或者广泛变化通常产生微弱效应的那些序列，例如局部重组率的变化。某些情况下，两条序列的密切相似性可能很重要，而碱基序列本身则几乎无关紧要——例如一个基因的两个相邻译本间的内含子。“垃圾”一词可能含义太宽泛而不能囊括所有对生物表型影响很小或者没有影响的那些序列。我们希望随着事实变得越来越明朗，能够总结出一个更准确的术语。

尽管杜利特尔和萨皮恩扎谨慎地用了“生物表型”一词来让他们表达更明确，但是“表型”这一词语还是引起了一些麻烦。显然我们需要两个词语：一个用来表示生物的表型，另一个则只用来表示寄生 DNA 的“表型”，这样当我们遇到真正的寄生生物时就能有一个明确的区分。对于前者我们建议使用“生物表型”，而对于后者，我们想沿用卡弗利尔－史密斯[3]的“基因组内表型”，但是当上下文意思很明确时，我们也允许单独使用“表型”一词。

在最开始的定义中，我们认为自私 DNA 有两种独有的特征：(1) 当一条 DNA 序列通过形成自身的额外拷贝在基因组内得以传播时，就会产生自私 DNA。(2) 自私 DNA 对表型没有特异性的贡献。我们使用的“表型”是指生物表型。在上述的讨论中，我们认为应该将“特异的”一词理解成“高度特异的”或者“非常特异的”。然而，普理查德[4]给我们指出“没有……贡献”的说法未必严谨。将对生物表型有着很小贡献的 DNA 也包含进去可能更有意义，无论这种贡献是积极的还是消极的。后者的一个例子可能是成了基因组一部分的病毒 DNA。

很显然，生物存在一个合理的选择优势（积极的或消极的）的连续集合。我们已经从自私 DNA 的定义中排除了那些选择优势非常高的情况。要确定一条重复序列是否是寄生的，就必须确定种群中重复序列的存在主要是序列在基因组间扩散的效率造成的还是主要由于种群中具有重复拷贝序列的那些个体的成功复制产生的。我们认为只有前一种情况使用“自私的”或者“寄生的 DNA”一词才有意义，正如反对使用“有用的”或“共生的 DNA”这种词语的理由一样——二者的界限可能并不明确。

In considering the spread of parasitic DNA one should not underestimate the power of natural selection. For example, if a particular transposon was inserted at random, it would run the risk of inactivating many genes and thus be selected against. A transposon which usually inserted at sites between genes would be at a selective advantage. Sites very near essential genes (as pointed out by Bruce Grant[5]) may be harder to delete than those in the middle of long stretches of junk and so parasitic DNA in the former positions is likely to survive longer. Effects of this type would lead to the selection of selfish DNA sequences that inserted preferentially at special sites in the genome.

Competing theories differ in their analysis of the factors determining the amount of non-specific DNA and of the way in which it comes into existence. Although we cannot at present decide on the quantitative contribution of the different types of non-specific DNA to the genome, it is still helpful to classify the various theories.

We proposed[2] that the amount of non-specific DNA present in a given genome is often determined by the balance between the intragenomic spreading of selfish sequences and phenotypic selection against excess DNA—the weaker the phenotypic selection against non-specific DNA the larger the DNA content of the genome. In another group of theories it is proposed that there is an optimal DNA content for each organism, which may be substantially greater than the amount of specific DNA that is needed to define the phenotype. The amount of non-specific DNA is then principally determined by the difference between the optimal DNA content and the essential content of specific DNA. The theories are not mutually exclusive, but differ substantially in emphasis in their explanation of C-values.

Cavalier-Smith's proposal[3,6] is an interesting example of an "optimal DNA content" theory. One of his ideas, which we misinterpreted in our previous paper[2], is that in large cells, particularly in oocytes, the transport of messenger RNA across the nuclear membrane may become a limiting factor and that the only way to increase the rate of transport is by increasing the number of nuclear pores by extending the surface of the membrane. If the area of nuclear membrane is determined by the DNA content of the nucleus, it follows that selection for a larger cell must lead to an increase in the DNA content of the genome. Thus, rather surprisingly, extra non-specific DNA is selected for because it allows such a cell to grow *faster*. While we do not question the logic of the argument, given the various assumptions, we do not find all the assumptions particularly plausible. It may be that there is sometimes selection for increased cell volume and increased nuclear volume. In cells so selected, non-specific DNA can accumulate. Whether it does so because large cells with large nuclei require such accumulation, or because they simply permit it remains to be seen. We feel that more experimental work is needed to unravel the complexities of the situation. In particular, we should like to know in which stages and in which organisms the surface of the nuclear membrane is saturated with nuclear pores.

Cavalier-Smith[3] also cites the widely different DNA contents of germ cells and somatic cells in some invertebrates as evidence against the selfish DNA hypothesis. However these

当考虑寄生 DNA 的传播时，不应该低估自然选择的力量。例如，如果一个特定的转座子随机插入后，将会经历让许多基因失活的风险，因此它会被选择性地排斥。插入到基因间位点的转座子一般具有选择优势。非常靠近重要基因的位点（布鲁斯·格兰特[5]指出的那些位点）可能比那些位于长的垃圾 DNA 中间的那些位点更难发生缺失，所以位于前者中的寄生 DNA 可能存活的更久一些。这种影响会导致对优先插入到基因组特异位点的自私 DNA 序列的选择作用。

与此相对立的学说在分析哪些因素能够决定非特异 DNA 的数量及其产生方式等方面有所不同。尽管目前我们还不能对不同类型的非特异 DNA 对基因组的贡献加以定量，但这仍然有助于对各种不同的学说进行分类。

我们提出[2]存在于某一特定基因组中的非特异 DNA 的数量通常由自私序列在基因组内的传播和对冗余 DNA 的表型选择间的平衡来确定——非特异性 DNA 的表型选择越弱，基因组的 DNA 含量越大。另一组学说提出每种生物都有一个最优的 DNA 含量，这一含量可能比用来定义表型所需的特异 DNA 的数量要大得多。非特异 DNA 的数量主要由最优 DNA 含量和特异 DNA 的基本含量之间的差异来确定。这些学说并不是相互排斥的，但是它们在解释 C 值时所强调的重点不同。

卡弗利尔－史密斯的观点[3,6]是“最优 DNA 含量”学说中的一个有趣的例子。我们在之前的论文[2]中对他的一个观点有些误解，那就是他认为在大细胞中，尤其是在卵母细胞中，信使 RNA 穿过核膜的转运可能变成一个限制因素，唯一可以提高转运速率的方式是通过扩展膜表面来增加核孔的数目。如果核膜面积由细胞核的 DNA 含量决定，那么就会遵循如下原则，即自然选择倾向于大细胞，这必然导致基因组 DNA 含量的增加。因此，令人非常惊讶的是，冗余的非特异 DNA 由于其允许这种细胞生长得**更快**而被选择。尽管我们不会质疑这个论点的逻辑性，但考虑到多方面的假设，我们发现并不是所有的假设都是特别可信的。有时环境可能会选择增大的细胞体积和增大的细胞核体积，在这样选择出来的细胞中，非特异性 DNA 能够积累下来。上述行为的原因，可能是具有大细胞核的大细胞需要这种累积，或者细胞只是简单地允许其能够存在而最终被观察到。我们认为有必要做更多的实验来揭开这一情况的复杂性。尤其我们想知道在哪个阶段以及在何种生物体内核膜表面的核孔能够达到饱和。

卡弗利尔－史密斯[3]也引用了某些无脊椎动物的生殖细胞和体细胞的 DNA 含量间的巨大差异作为反对自私 DNA 假说的证据。然而这些结果也可以用自私 DNA

observations can also be explained in terms of the selfish DNA theory. Such DNA "needs" only to remain in the germ line to function parasitically. On the other hand, organismal selection might sometimes be stronger against surplus DNA in the soma than in the germ line. Thus representation in the germ line but not in the soma may sometimes be an optimal strategy for parasitic DNA. As for B chromosomes, in many cases the evidence appears to us to give some support to the idea (originally proposed by Östergren[7] in 1945) that they are largely parasitic, but there is certainly evidence that they sometimes have phenotypic effects which may possibly be useful[8,9].

Smith[10] has pointed out that the DNA of vertebrates usually has about 42 percent GC whereas the GC content of invertebrate and prokaryotic DNA varies over a much wider range. The theory of parasitic DNA has rather little to say on this point. There are many factors which might affect the GC content of an organism's DNA. If much of the parasitic DNA has descended rather recently from insertion elements which themselves originally coded for proteins, then it would not be surprising if their present GC content were similar to that of genes which still code for protein. This may, perhaps, explain the constancy of GC in vertebrates.

As for our own ideas, we now feel that there may perhaps be reasons why too *little* DNA can in some cases produce a selective disadvantage. For example, Zuckerkandel[11] has suggested that there may be a minimum size for a "domain" necessary for stability of the chromatin in the folded state. Thus a domain containing only a few genes might benefit from having some non-specific DNA as "padding". This would mean that there is indeed an optimal amount for total DNA.

In our original paper[2] we feel that we did not put enough emphasis on the distinction between sequences which are repeated, exactly or nearly exactly, in many tandem repetitions and sequences which are more widely dispersed over the chromosomes and which occur in only one or a few copies in any one place. It seems plausible that these two types of sequence evolved different mechanisms. It is possible that the mechanisms generating the tandemly repeated type are usually more "ignorant" (in Dover's sense[12]) than the more dispersed type. If the latter have any specific function it is likely to be that of the control, at one level or another, of gene expression, whereas the tandemly repeated type seem more likely to influence chromosome mechanics.

One possibility to which we feel we should have given more weight is that of "dead genes", also called "pseudogenes"[13,14]; that is, sequences which can no longer code for a protein (or a structural RNA) but which appear to have descended from a sequence that did. Whether these conform to our definition of parasitic DNA remains to be seen, but we suspect this is unlikely, since they usually exist in only a single copy, or as multiple tandem copies in only one place.

In our recent experience most people will agree, after discussion, that ignorant DNA, parasitic DNA, symbiotic DNA (that is, parasitic DNA which has become useful to

学说来解释。自私DNA只“需要”待在生殖细胞系中以寄生方式发挥功能。另一方面，有时生物在体细胞中对过剩DNA的选择作用可能比其在生殖细胞中的选择作用更强烈。因此对于寄生DNA来说，相比存在于体细胞中，存在于生殖细胞系中有时也许是最佳策略。至于B染色体，我们看到很多情况下的证据是支持存在大量寄生DNA这一观点（最初由奥斯特格伦[7]于1945年提出），但是也有证据明确表明这些DNA有时是有表型效应的，而且这些表型效应可能是有用的[8, 9]。

史密斯[10]指出脊椎动物DNA的GC含量通常为42%，而无脊椎动物和原核生物DNA的GC含量变化范围很大。寄生DNA学说对于这一点几乎没做任何解释。影响生物DNA中GC含量的因素有很多。假设很多寄生DNA是近期从插入元件产生而来，而这些插入元件最初是编码蛋白质的，那么它们现在的GC含量与那些仍然编码蛋白质的基因的GC含量相似一点也不令人感到意外。这也许可以解释脊椎动物的GC含量为何具有稳定性。

至于我们自己的观点，目前我们认为，之所以只有极**少**的DNA会在个别情况下产生选择劣势是有原因的。例如，楚克尔坎德尔[11]提出，为了保证染色质在折叠状态下的稳定性，维持这一稳定性所必要的“结构域”大小可能有一个最低限度的要求。因此只含有几个基因的结构域可能受益于某些作为“填料”的非特异DNA。这意味着，对于总DNA确实具有一个最优数量。

在我们原来的文章[2]中，我们认为对于许多串联重复序列中严格重复或接近严格重复的序列与更广泛地分散在染色体上的或者仅在某一位点产生单个或几个拷贝的序列之间的差异，我们强调得还不够。这两种序列有着不同的进化机制，这似乎是有道理的。产生串联重复型序列的机制通常比产生更分散型序列的机制更加“无意识”（在多弗所使用的意义上[12]）。如果后者有特定功能的话，那么其可能是对基因某一水平的表达进行控制，然而串联重复型序列的作用更有可能是影响了染色体的结构。

我们认为可能更应该侧重的一个方面是“死基因”，也称为“假基因”[13,14]，即不再编码蛋白质（或者结构RNA），但是似乎这些序列是从某些可以编码蛋白质或者结构RNA的序列产生来的。这些与我们对寄生DNA的定义是否吻合还有待验证，但是我们推测这种一致性不太可能存在，因为它们通常是以单拷贝存在的，或者只在一个位点作为多串联拷贝存在。

经过讨论，大部分人都同意无意识的DNA、寄生DNA、共生DNA（即对机体有用的寄生DNA）和“死”DNA中的任何一种都可能存在于高等生物的染色体中。

the organism) and "dead" DNA of one sort or another are all likely to be present in the chromosomes of higher organisms. Where people differ is in their estimates of the relative amounts. We feel that this can only be decided by experiment. We expect that due to the recent advances in genetic engineering and related techniques much sequence information will accrue in the near future. This should help to decide between the different alternatives.

(**288**, 645-646; 1980)

L. E. Orgel*, F. H. C. Crick* and C. Sapienza†
* The Salk Institute, San Diego, California
† Department of Biochemistry, Dalhousie University, Halifax, Canada

---

References:

1. Doolittle, W. F. & Sapienza, C. *Nature* **284**, 601 (1980).
2. Orgel, L. E. & Crick, F. H. C. *Nature* **284**, 604 (1980).
3. Cavalier-Smith, T. *Nature* **285**, 617 (1980).
4. Pritchard, R. (personal communication).
5. Grant, B. (personal communication).
6. Cavalier-Smith, T. *J. Cell. Sci.* **34**, 247 (1978).
7. Östergren, G. *Bot. Notiser* **2**, 157 (1945).
8. Jones, R. N. *Int. Rev. Cytol.* **40**, 1 (1975).
9. Ames, A. & Dover, G. *Chromosoma* (in the press).
10. Smith, T. F. *Nature* **285**, 620 (1980).
11. Zuckerkandel, E. (personal communication).
12. Dover, G. *Nature* **285**, 618 (1980).
13. Loomis, W. *Devl. Biol.* **30**, F3-F4 (1973).
14. See, for example, Proudfoot, N. *Nature* **286**, 840 (1980).

人们的分歧之处在于，他们对相对数量的估计不尽相同。我们认为这点只能通过实验来确定。我们希望随着基因工程和相关技术的进步，在不久的将来可以获得并积累更多序列信息。这将有助于我们在不同的学说中做出更科学的选择。

（刘皓芳 翻译；王嵬 审稿）

Modes of Genome Evolution

G. Dover and W. F. Doolittle

Editor's Note

Following on from the previous addendum to two pieces on "selfish DNA" published in April of this year, here the author of one of those pieces, biochemist Ford Doolittle, collaborates with one of his critics, geneticist Gabriel Dover at Cambridge University, to modify and sharpen his earlier arguments. Doolittle and Dover say that some DNA is better seen as "ignorant" rather than selfish, in that it is prone to generating meaningless rearrangements.

OUR original articles[1,3] presented antagonistic positions and perhaps obscured many areas of agreement. In essence, we are approaching similar phenomena from different perspectives and it might be useful to clarify where our views agree and differ and to indicate the sorts of evidence which would allow one to decide whether the origin of non-coding DNA in eukaryotic cells is more precisely viewed as "ignorant"[3] or "selfish"[1,2].

The eukaryotic genome is constantly in flux, and this plasticity results from a variety of known and as yet mysterious mechanisms which amplify and disperse segments of DNA throughout a set of chromosomes[4–7]. Replication and recombination are complex processes requiring many enzymes that have evolved by natural selection. What we both wish to stress is that, in establishing these processes, evolution has inadvertently endowed the genome with built-in mechanisms for irregular and recurrent random sequence rearrangements and created an environment in which elements capable (to varying extents) of promoting their own amplification and dispersion will inevitably arise[1,2,7]. We acknowledge that there is evidence which suggests that some proportion of genome rearrangements may have effects on the biology of an organism (for instance on chromosome behaviour, on nuclear RNA processing, on cell-cycle times, on recombination frequencies and on gene expression; see refs 1–3, 8). Where this is the case, the change in frequency of a sequence rearrangement and the behaviour of elements which promote their own amplification and dispersion will of course depend on the effects they have on fitness. Hence, the net accumulation of these particular families of sequences reflects a balance between the intrinsic rate of accumulation (intra-genomically) and the effect of natural selection on phenotypic differences.

We agree that the amplification and dispersion of segments may occur either at random (sequence-independent or "ignorant") or with preference for certain sequences (sequence-dependent or "selfish"). Although both "ignorant" and "selfish" are unfortunate terms, they should be understood in the spirit in which they are used and defined: sequence-independent and sequence-dependent processes of amplification and dispersion respectively.

基因组的进化模式

多弗，杜利特尔

编者按

上一篇文章是当年四月发表的两篇关于“自私的DNA”综述的补篇，之后，其中一篇综述的作者——生物化学家福特·杜利特尔，与他的观点批判者——剑桥大学的遗传学家加布里埃尔·多弗进行合作，在本文中修正并深化了他早期的论点。杜利特尔和多弗认为，某些DNA更应该被视为“无意识的”，而不是自私的，因为它们更倾向于产生无意义的重排。

我们最初的文献中[1,3]介绍了一些对立的观点，这可能使很多存在共识的领域变得模糊。其实，我们正在从不同的角度解释相似的现象，这将有助于阐明我们观点中相同或相异的地方，并有助于指出有关真核细胞中非编码DNA起源的各种证据，来判断这种起源该被视为“无意识的”[3]，还是“自私的”[1,2]才更为精确。

真核基因组一直处于变化状态，这种可塑性是由种种已知的和至今仍未可知的机制导致的，这些机制使DNA在一整套染色体中扩增和分散[4-7]。复制和重组都是复杂的过程，此过程需要许多经由自然选择进化而来的酶。我们想要强调的两点是，在建立这些过程时，进化在无意间赋予了基因组对不规则和重复的随机序列进行重排的内在机制，并且创造出了使元件必然能够（在不同程度上）促进自身扩增和分散的环境[1,2,7]。我们承认，已有证据表明一定比例的基因组重排可能对生物体产生生物学影响（例如染色体行为、核RNA加工、细胞周期时序、重组频率及基因表达等；详情可见参考文献1~3、8）。在这种情况下，序列重排频率的改变，及那些能促进自身扩增和分散的元件的行为，都必将依赖于它们对适应性的影响。因此，这些特定序列家族的净积累反映了固有累积速率（基因组内部的）与自然选择对表型差异影响之间的一种平衡。

我们赞同以下观点，即片段的扩增和分散可能是随机发生的（序列无关的或“无意识的”），或者也可能更加偏好特定序列（序列相关的或“自私的”）。尽管“无意识的”和“自私的”都是不太贴切的术语，但是应该本着如下应用和定义的初衷对它们进行理解：它们分别用于修饰序列无关和序列相关的扩增和分散过程。

The "selfish process" produces DNAs which are preferentially chosen either by virtue of their nucleotide sequence or by virtue of the fact that they may code for gene products for their own amplification and dispersion. Mobile elements such as bacterial insertion sequences and transposons, the "*Ty-1*-like" elements of yeast, the "*copia*-like" elements of *Drosophila*, and vertebrate retroviruses, which show surprising similarities in structure and (perhaps) dispersal mechanisms[8] are almost certainly the self-perpetuating products of this sort of process. An alternative term "self-selection" might usefully describe the process of accumulation of these sequences. Interestingly, there may be an element of self-selection in the process of accumulation and dispersion of some sequence-independent segments. Extensive sharing of similar sequence patterns of repetitive DNAs between chromosomes (see refs 9 and 10), and occasionally between species[11,12], suggests that an arbitrarily accumulated sequence arrangement preferentially enjoys further rounds of amplification and dispersion[10]. Simulation of unequal recombination by computer appears to show a degree of self-perpetuation of sequences initially chosen at random[13] and also that amplification and dispersion can be part and parcel of the same recombinational irregularity[14].

A process of recurrent amplification can explain the frequent observation of a greater within-species than between-species sequence homogeneity of shared families[6,9,12]. Similarly, polymorphisms and variations in sequence patterns of ribosomal genes[15], histone genes[16] and some non-coding families[17–19] in several diverse organisms can be interpreted as the most recent, and hence localized, amplifications of these sequences.

We wish to emphasize, however, that "ignorant" and "selfish" self-perpetuation are terms that uniquely apply to these particular DNA processes and cannot be used, meaningfully, to describe changes in frequencies of other elements that are totally dependent on the natural selection of phenotypes. "Selfish genes"[20] and "replicator selection"[21] are not synonymous terms and the evolutionary processes to which they allude are unknown. A very limited accumulation of supernumerary B chromosomes is the only other process that can be described as self-accumulation, often the result of meiotic and mitotic non-disjunction. We doubt if this term can be used to describe the mis-named "meiotic drive" mechanisms of regular A chromosomes in the rare instances where this occurs. For example, in the case of segregation distortion (SD) in *Drosophila melanogaster*, the preferential recovery of the SD chromosome is not so much due to its accumulation *per se* but is the outcome of dysgenesis of cells carrying the non-SD homologue[22]. This, and other cases, are analogous to the relative changes in frequency of alleles at a gene locus that are the outcome of natural selection; and "selfish" and "replicator selection" are misleading descriptions of this process of differential accumulation of alleles and chromosomes.

We do not agree upon the relative contributions of randomly accumulated and preferentially accumulated DNAs to the evolution of eukaryote genomes. It is clear that much of the non-coding DNA of yeast could be of the sort one could call "selfish"[4,8] whilst non-coding elements of the *D. melanogaster* genome are made up of varying proportions of essentially "ignorant" repetitive DNA (for example the satellite DNAs[23]) and "selfish"

“自私的过程”会产生由于某些原因被优先选择的DNA，或者由于这些DNA的核苷酸序列，或者由于它们可以编码用于自身扩增和分散所需的基因产物。几乎可以肯定的是，在结构和（可能）分散机制上[8]具有惊人相似性的可移动元件（例如细菌的插入序列和转座子，酵母的“*Ty-1*样”元件，果蝇的“*copia*样”元件，以及脊椎动物的逆转录病毒），它们多是这类过程自我保存的产物。另一个术语——“自我选择”——可能有助于描述这类序列的积累过程。有意思的是，在某些序列无关片段的积累和分散过程中，可能也存在一个自我选择元件。在染色体之间（见参考文献9和10），偶尔也在不同的物种[11,12]之间，大量的DNA重复序列具有相似的序列模式，表明一个随意积累的序列排布优先享有下一轮扩增和分散的机会[10]。计算机对于不对等重组的模拟结果显示，原本通过随机选择产生的序列具有一定程度的自我保存能力[13]。同时，扩增和分散可能是相同的无规则重组中不可缺少的一部分[14]。

周期性的扩增过程可以用来说明为什么经常能在种内观察到比种间更高的共有基因家族的序列同质性[6,9,12]。同样地，可以将各种不同生物体的核糖体基因[15]、组蛋白基因[16]和一些非编码家族[17-19]在序列模式上的多态性和多样性解释为这些序列在时间上最近的，也是局部范围内扩增的结果。

然而，我们希望强调的是，所谓的“无意识的”和“自私的”自我保存都是专门用于说明这类特殊DNA过程的术语，不能被用作描述其他元件（此类元件完全依赖于对表型的自然选择）的频率变化之用。“自私基因”[20]和“复制子选择”[21]并不是同义的术语，至于它们暗指的进化过程，尚不得而知。多余B染色体的一种非常有限的积累过程是仅有的另外一个可以被描述为自我积累的过程，它往往是减数分裂和有丝分裂不分离的结果。我们怀疑，这个术语是否可以用于描述极少数情况下，发生在标准A染色体上，被误称为“减数分裂驱动”的机制。以黑腹果蝇的偏分离（SD）为例，SD染色体的优先恢复并不完全取决于它的积累本身，而是携带非SD同源物的细胞发育不良的结果[22]。这种情况及其他情况类似于等位基因在一个基因位点上频率的相对改变，这种改变也是自然选择的结果。“自私的”和“复制子选择”都是对等位基因和染色体的这种差异性积累过程的误导性描述。

我们并不认同随机积累和优先积累的DNA对真核基因组进化的相关贡献。很明显，酵母的很多非编码DNA都可能是被称为“自私的”的类型[4,8]；而黑腹果蝇基因组的非编码元件则是由不同比例的本质上“未知的”重复DNA（例如卫星DNA[23]）和“无意识的”转座DNA（例如“*copia*样”元件[5]）组成。温辛克和他的同事们[24]

transposable DNAs (for example the "*copia*-like" elements[5]). The scrambled arrangements of repetitive elements in *D. melanogaster* observed by Wensink and co-workers[24] could reflect either sequence-independent shuffling processes or sequence-dependent insertion of transposable elements, similar to *copia*, at adjacent sites. The available data on repetitive DNA families of species of sea urchins, Graminea, rodents, primates and insects (see ref. 3) do not permit a clear assessment of the mechanisms that gave rise to them, for no direct sequence data are available. The multiple-copy "Alu I" family of mammalian genomes may have a cellular function[25,26], but it is not impossible that these are the descendants of a family of transposable elements[27].

Finally, we suggest that the accumulation of sequences might be affected by constraints on the mechanisms themselves. For example, sequence-dependent amplification and dispersion will favour sequences that accurately contain the required sequence for transposition and that have not drifted too far into unacceptable divergent sequences. Similarly, sequence-independent mechanisms may be constrained to particular lengths of sequences[10]. Such constraints impose a type of selection on sequences, not necessarily as a result of their phenotypic effects but more as a consequence of the molecular mechanisms of replication and recombination.

We do not know what proportions of the repetitive DNAs of "higher" organisms are amplified and dispersed by either sequence-dependent or sequence-independent mechanisms. Similarly, we do not know to what extent each of these mechanisms is constrained nor do we know the extent to which the frequencies of sequence patterns are an outcome of their possible effects on individual fitness. It is clear, however, that there are several modes of sequence rearrangement within rapidly evolving genomes. The problem now, as with most scientific debates, is one of quantification.

(**288**, 646-647; 1980)

Gabriel Dover* and W. Ford Doolittle†
* Department of Genetics, University of Cambridge, UK
† Department of Biochemistry, Dalhousie University, Canada

References:

1. Doolittle, W. F. & Sapienza, C. *Nature* **284**, 601 (1980).
2. Orgel, L. E. & Crick, F. H. C. *Nature* **284**, 604 (1980).
3. Dover, G. A. *Nature* **285**, 618 (1980).
4. Cameron, J. R. *et al. Cell* **16**, 739 (1979).
5. Potter, S. S. *et al. Cell* **17**, 424 (1979).
6. Flavell, R. B. *et al.* in *Genome Organization and Expression in Plants* (Plenum, 1980).
7. Dover, G. A. *Chromosomes Today* **6**, 105 (1977).
8. *Cold Spring Harb. Symp. Quant. Biol.* **45** (1980).
9. Dover, G. A. *Nature* **272**. 123 (1978).
10. Brown, S. D. M. & Dover, G. A. *Nucleic Acids Res.* **8**, 781 (1979)
11. Brown, S. D. M. & Dover, G. A. *Nature* **285**, 47 (1980).
12. Donehower, L. & Gillespie, D. *J. Molec. Biol.* **134**, 805 (1979).

在黑腹果蝇中观察到的重复元件的无序排列可能反映了序列无关的交换过程，也可能反映了在相邻位点上序列相关的转座元件（类似于*copia*）的插入。由于无法获得直接的序列数据，凭借现有的关于海胆类、禾本科、啮齿目动物、灵长类和昆虫等生物在重复DNA家族方面的数据（见参考文献3），我们还不能对其产生机制进行确切估计。哺乳动物基因组的多拷贝“Alu I”家族可能具备一种细胞功能[25,26]，但要说它们是转座元件家族的后代，也不是不可能[27]。

最后，我们认为序列的积累可能受到它们自身机制约束的影响。例如，序列相关的扩增和分散将更倾向以下两种序列：精确包含了转座必需序列的序列和那些没有过分漂移至不可接受的相异序列的序列。同样地，序列无关的机制可能受制于序列的特定长度[10]。这些约束强加给序列一种选择，这种选择更多的是一种复制和重组的分子机制作用的后果，而没有必要将其视为表型效应的结果。

我们不知道“高等”生物体有多大比例的重复DNA是通过序列相关或者序列无关的方式进行扩增和分散的。同样，我们不知道这些机制各自在多大程度上受到约束；我们也不知道，序列模式的频率在多大程度上可能是它们作用在个体适应性上的结果。然而，很明显的是，在快速进化的基因组中，确实存在数个序列重排模式。伴随着大多数科学辩论，如今这只是一个量化的问题。

（阮玉辉 翻译；王嵬 审稿）

13. Smith, G. P. *Science* **191**, 528 (1976).

14. Smith, T., Brown, S. D. M. & Dover, G. A. (unpublished results).

15. Wellauer, P. K. *et al. J. Molec. Biol.* **105**, 487 (1976).

16. Cohn, R. H. & Kedes, L. J. *Cell* **18**, 855 (1979).

17. Cooke, H. J. & Hindley, J. *Nucleic Acids Res.* **6**, 3177 (1979).

18. Christie, N. T. & Skinner, D. M. *Proc. Natl. Acad. Sci. U.S.A.* 77, 2786 (1980).

19. Donehower, L. *et al. Proc. Natl. Acad. Sci. U.S.A.* 77, 2129 (1980).

20. Dawkins, R. *The Selfish Gene* (Oxford Univ. Press, 1976).

21. Dawkins, R. A. *Tierpsychology* **47**, 61 (1978).

22. Crow, J. F. *Scientific American* **104**, 1 February (1979).

23. Brutlag, E. E. *et al. Cold Spring Harb. Symp. Quant. Biol.* **42**, 1137 (1979).

24. Wensink, P. C. *et al. Cell* **18**, 1231 (1977).

25. Jelinek, W. R. *et al. Proc. Natl. Acad. Sci. U.S.A.* 77, 1398 (1980).

26. Rubin, C. M. *et al. Nature* **284**, 372 (1980).

27. Bell, G. L. *et al. Nucleic Acids Res.* **8**, 4091 (1980).

Mutations Affecting Segment Number and Polarity in *Drosophila*

C. Nüsslein-Volhard and E. Wieschaus

Editor's Note

How do higher organisms get their shape? It was recognized that they are generally composed of similar repeating units, the segmented body of the fruit fly *Drosophila melanogaster* being a prime example. But the genetics of the patterning process was still obscure. Here Christiane Nüsslein-Volhard and Eric Wieschaus of the European Molecular Biology Laboratory in Heidelberg identify the key genes responsible for *Drosophila* segmental development, mutations of which alter the body plan. It later became clear that body shapes in a wide variety of organisms are controlled by just a few so-called homeobox or *Hox* genes, of which these are examples. This work earned the two researchers the Nobel Prize in Physiology or Medicine in 1995.

In systematic searches for embryonic lethal mutants of *Drosophila melanogaster* we have identified 15 loci which when mutated alter the segmental pattern of the larva. These loci probably represent the majority of such genes in *Drosophila*. The phenotypes of the mutant embryos indicate that the process of segmentation involves at least three levels of spatial organization: the entire egg as developmental unit, a repeat unit with the length of two segments, and the individual segment.

THE construction of complex form from similar repeating units is a basic feature of spatial organisation in all higher animals. Very little is known for any organism about the genes involved in this process. In *Drosophila*, the metameric nature of the pattern is most obvious in the thoracic and abdominal segments of the larval epidermis and we are attempting to identify all loci required for the establishment of this pattern. The identification of these genes and the description of their phenotypes should lead to a better understanding of the general mechanisms responsible for the formation of metameric patterns.

In *Drosophila*, the anlagen for the individual segments arise as equally sized subdivisions of the blastoderm, each segment represented by a transverse strip of about three or four cell diameters[1]. A cell lineage restriction between neighbouring segments is established at or soon after this stage[2]. Two basic types of mutation have been described which change the segmental pattern of the *Drosophila* larva. Maternal effect mutations like *bicaudal* lead to a global alteration of the embryonic pattern[3]. Bicaudal embryos develop two posterior ends arranged in mirror-image symmetry, and lack head, thorax and anterior abdomen. The *bicaudal* phenotype suggests that the initial spatial organisation of the egg established

影响果蝇体节数量和极性的突变

尼斯莱因－福尔哈德，维绍斯

编者按

高等生物的形状是如何获得的？公认的是，它们一般由相似的重复单元组成，黑腹果蝇分节的身体就是一个典型的例子。但是，模式形成过程中涉及的遗传学内容仍然模糊不清。在本文中，海德堡欧洲分子生物学实验室的克里斯蒂亚娜·尼斯莱因－福尔哈德和埃里克·维绍斯确定了负责果蝇体节发育的关键基因，这些基因的突变会改变机体规划。后来知道，在很多生物体中，体型仅仅由少数所谓的同源框或者 *Hox* 基因控制，这些就是例子。这项工作使得两位研究者在 1995 年获得了诺贝尔生理学或医学奖。

在对黑腹果蝇胚胎致死突变体的系统研究中，我们鉴定出 15 个突变后可以改变幼虫体节模式的基因座。这些基因座可能代表了果蝇中大多数这类基因。突变胚胎的表型表明果蝇分节过程至少涉及空间结构的三个层次：作为发育单位的整个卵、一个具有两体节长度的重复单元以及单个体节。

由相似的重复单位构成的复杂结构是所有高等动物空间结构的基本特征。而参与这一过程的基因在任何生物体中都知之甚少。果蝇中这种类型的体节性质在幼虫表皮胸节和腹节上体现得最明显，并且我们正试图鉴定控制这种模式建成所需的所有基因座。对这些基因的确定和表型的描述可以帮助我们更好地理解分节模式形成的普遍机制。

在果蝇中，发育成单个体节的原基起源于胚盘上同样大小的细化区域，每个体节是一个长约三到四个细胞直径的横向条状区域 [1]。在这个时期或稍后会在相邻体节之间形成一个细胞系界线 [2]。本文描述了改变果蝇幼虫分节模式的两个基本突变类型。类似“双腹”突变的母体效应突变会导致胚胎形态的完全改变 [3]。“双腹”突变的胚胎发育出两个尾部，成镜像对称分布，没有头部、胸部和前腹部。“双腹”突变表型说明在卵子发生时建立起来的最初的空间结构涉及成形素的梯度分布，成形

during oogenesis involves a morphogen gradient that defines antero-posterior coordinates in early embryonic pattern formation[3,4]. The subdivision of the embryo into segments is thought to occur by a differential response of the zygotic genome to the maternal gradient. Homeotic mutations (for example, *bithorax*[5,6]) seem to be involved in a final step of this response process. These mutations change the identity of individual segments; for example, *Ultrabithorax* transforms the metathoracic and first abdominal segments into mesothoracic segments. However, the homeotic loci do not affect the total number, size or polarity of the segments, nor do they point to any other step which might intervene between the maternal gradient and the final pattern of segments.

We have undertaken a systematic search for mutations that affect the segmental pattern depending on the zygotic genome. We describe here mutations at 15 loci which show one of three novel types of pattern alteration: pattern duplication in each segment (segment polarity mutants; six loci), pattern deletion in alternating segments (pair-rule mutants; six loci) and deletion of a group of adjacent segments (gap mutants; three loci) (Table 1, Fig. 1).

Table 1. Loci affecting segmentation in *Drosophila*

Class	Locus	Map position*	No. of alleles†	Ref.
Segment-polarity	*cubitus interruptus*^{D} (ci^D)	4-0	(2)	20
	wingless (*wg*)	2-30	6	9
	gooseberry (*gsb*)	2-104	1	This work
	hedgehog (*hh*)	3-90	2	This work
	fused (*fu*)‡	1-59.5	(9)	8, 20
	patch (*pat*)	2-55	8	This work
Pair-rule	*paired* (*prd*)	2-45	3	This work
	even-skipped (*eve*)	2-55	2	This work
	odd-skipped (*odd*)	2-8	2	This work
	barrel (*brr*)	3-27	2	This work
	runt (*run*)	1-65	1	This work
	engrailed (*en*)	2-62	6	11, 20
Gap	*Krüppel* (*Kr*)	2-107.6	6	12, 20
	knirps (*kni*)	3-47	5	This work
	hunchback (*hb*)	3-48	1	This work

* For the new loci (see last column) the map positions are based on recombination between the markers *S*, *Sp*, *Bl*, *cn*, *bw* for the second chromosome, and *ru*, *h*, *th*, *st*, *cu*, *sr*, e^S, *ca* for the third chromosome. For description of markers see ref. 20. The loci *runt*, *Krüppel* and *knirps* were further mapped using the breakpoints of deficiencies and duplications for the respective regions. All mutants were mapped by scoring the embryonic progeny of single recombinant males backcrossed to heterozygous females from the original mutant stocks.

† The numbers in parentheses refer to the alleles listed in Lindsley and Grell[20]. All other alleles, except the *runt* allele,

素浓度梯度在胚胎早期形态建成中决定头部和尾部的分化方向[3,4]。一般认为胚胎分化成体节是通过合子基因组对母体生理梯度的不同反应所产生的。同源异型突变（如“双胸”突变[5,6]）似乎参与了这一反应过程的最后一步。这些突变改变了单个体节的特征；比如，“超双胸”基因把后胸和第一腹节转变成了中胸体节。然而，同源异型基因座并不影响体节总数、大小或极性，也无法表明它们具体是哪一步在母体生理梯度和最终体节形态之间起了干扰作用。

我们对依赖合子基因组而影响体节模式的突变展开了系统的研究。本文我们描述的突变体涉及 15 个基因座，每个突变体表现出以下三种新型模式改变中的某一种：每个体节都存在重复模式（体节极性突变体；6 个基因座）、体节交替缺失（成对控制突变体；6 个基因座）和一组邻近体节缺失（间隙突变体；3 个基因座）（表 1，图 1）。

表 1. 影响果蝇体节形成的基因座

类型	基因座	图谱位置 *	等位基因数目 †	参考文献
体节 – 极性	尺骨中断 (ci^D)	4–0	（2）	20
	无翅 (*wg*)	2–30	6	9
	鹅莓 (*gsb*)	2–104	1	本文
	刺猬 (*hh*)	3–90	2	本文
	融合 (*fu*) ‡	1–59.5	（9）	8，20
	斑点 (*pat*)	2–55	8	本文
成对控制	配对 (*prd*)	2–45	3	本文
	偶数 – 遗漏 (*eve*)	2–55	2	本文
	奇数 – 遗漏 (*odd*)	2–8	2	本文
	桶状 (*brr*)	3–27	2	本文
	侏儒 (*run*)	1–65	1	本文
	锯齿状 (*en*)	2–62	6	11，20
间隙	跛子 (*Kr*)	2–107.6	6	12，20
	折叠扇 (*kni*)	3–47	5	本文
	驼背 (*hb*)	3–48	1	本文

* 对于新的基因座（见上表），图谱定位基于二号染色体的 *S*,*Sp*,*Bl*,*cn* 和 *bw* 标记之间的重组以及三号染色体上 *ru*,*h*,*th*,*st*,*cu*,e^s 及 *ca* 标记之间的重组。对标记的描述见参考文献 20。运用各自区域的缺失和复制断点，对“侏儒”“跛子”和“折叠扇”基因座进行了进一步定位。所有突变都是通过计算重单重组体雄性与来自最初突变体品系的杂合体雌性回交后产生的胚胎后代数量来定位的。

† 圆括号内的数字参考了林斯利和格雷尔所列的等位基因[20]。除“侏儒”等位基因，三个“跛子”等位基因和一个

three *Kr* alleles and one *knirps* allele, were isolated in screen for embryonic lethal mutants on the second chromosome. 5,800 balanced stocks were established from individual males heterozygous for an ethyl methane sulphonate-treated *cn bw sp* chromosome using the DTS-procedure suggested by Wright[21]. 4,500 of the stocks had one or more new lethal mutations. Unhatched embryos from 2,600 putative embryonic lethal stocks were inspected for cuticular abnormalities[22]. Third chromosomal mutants discovered in the second chromosomal balanced lines were recovered after selection through individual females by balancing individual third chromosomes over TM3. Complementation tests were carried out between mutants with similar phenotypes whereby the occurrence of mutant embryos among the progeny of the crosses served as the criterion for allelism. Three new *Kr* alleles were isolated in a screen for lethals over the original *Kr* of Gloor[12], and one *knirps* allele of presumably spontaneous origin was discovered on a TM1 chromosome. The *runt* allele was isolated in a screen for X-linked lethals.

‡ *fused* is a male-rescuable maternal-effect locus[8]. Thus, the segment polarity reversal is observed in *fu/fu* embryos from *fu/fu* mothers. The progeny of *fu/+* females show a normal embryonic pattern regardless of embryonic genotype.

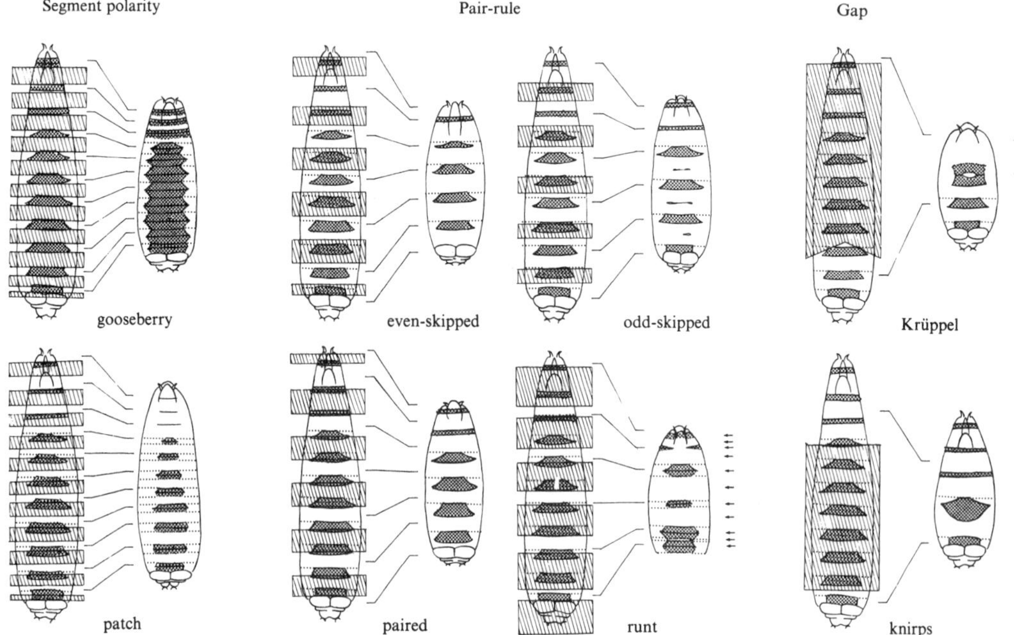

Fig. 1. Semi-schematic drawings indicating the regions deleted from the normal pattern in mutant larvae. Dotted regions indicate denticle bands, dotted lines the segmental boundaries. The regions missing in mutant larvae are indicated by the hatched bars. The transverse lines connect corresponding regions in mutant and normal larvae. Planes of polarity reversal in *runt* and *Krüppel* are indicated by the arrows. The two segment polarity loci *patch* and *gooseberry* are represented at the left. For indication of the polarity of the patterns, see Fig. 3. The patterns of *fused* and ci^{D} (not shown) look similar to the *gooseberry* pattern, whereas in *hedgehog* and *wingless* the deleted regions are somewhat larger, cutting into the denticle bands at either side. Four pair-rule mutants are shown in the centre. The interpretation of their phenotypes is based on the study of weak as well as strong alleles, combinations with *Ubx* (see text) and, in the case of *runt*, on gynandromorphs (unpublished). They probably represent the extreme mutant condition at the respective loci. The phenotypes of all known *barrel* and *engrailed* alleles (not shown) are somewhat variable and further studies are needed to deduce the typical phenotype. At the right, the two gap loci *Krüppel* and *knirps* are shown. Both patterns represent the amorphic phenotype as observed in embryos homozygous for deficiencies of the respective loci. The only known *hunchback* allele (not shown) deletes the meso- and metathorax.

“折叠扇”等位基因外，所有其他等位基因都是在筛选二号染色体上的胚胎致死突变体时分离到的。用赖特提出的DTS方案[21]，用甲基磺酸乙酯处理筛选 *cn bw sp* 染色体，用上述方法从单个雄性杂合体中，建立了5,800个平衡品系。其中4,500个品系含有1个或多个新的致死突变。从2,600个假定的胚胎致死品系中取未孵化的胚胎，检查其表皮异常[22]。在二号染色体的平衡品系中发现了三号染色体突变体，该突变体在通过三号染色体对TM3的雌性平衡个体中筛选后，得到了恢复。对表型相似的突变体进行互补测验，以杂交后代中出现突变胚胎为这些突变的等位性标准。在从格洛尔发现的原始“跛子”突变体中筛选致死胚胎时，分离出了三个新的“跛子”等位基因[12]，其中，在TM1染色体上发现了一个可能是自然发生的折叠扇等位基因。在筛选X–连锁致死突变体时分离出了“侏儒”等位基因。‡“融合”基因是一个雄性可以补救的母体效应基因座[8]。因此在“融合”纯合体母代产生的纯合体胚胎中可观察到反转的体节极性。不管胚胎的基因型如何，杂合体母本的后代皆表现正常的胚胎模式。

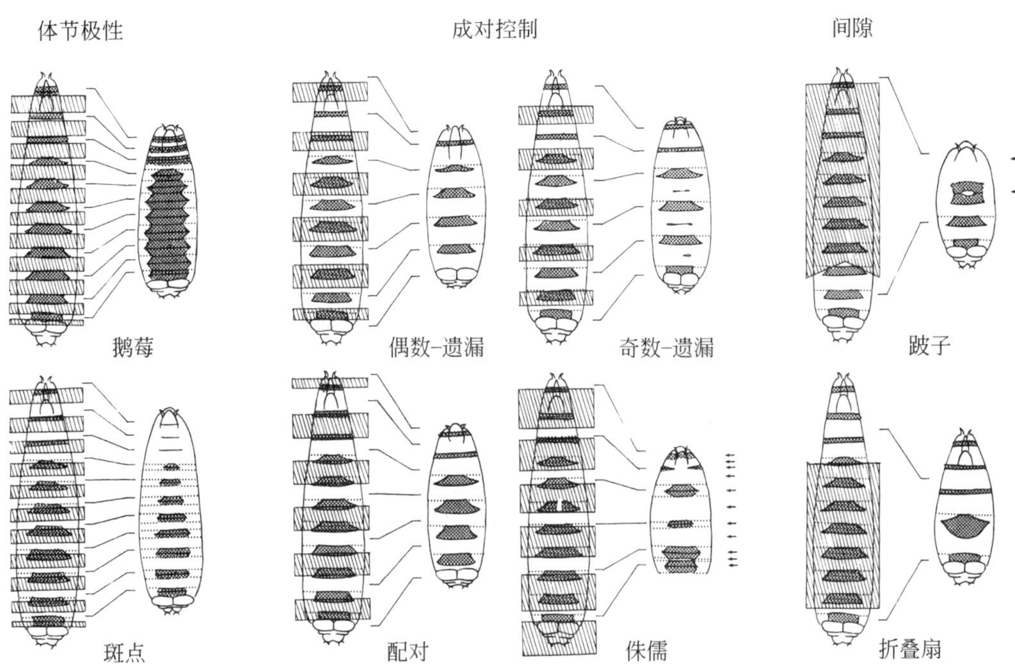

图 1. 半示意图表示突变体幼虫从正常模式中缺失的区域。点状区域表示齿状突起带，点线表示体节边界。阴影框所示为突变体幼虫中的缺失区域。横线连接突变体与正常幼虫中相对应的区域。箭头所示为“侏儒”和“跛子”突变体中极性反转面。图左侧为两个体节极性基因座“斑点”和“鹅莓”的突变体。对这种模式的极性的说明见图3。“融合”和“尺骨中断D”（未显示）的突变体形态与“鹅莓”相似，而“刺猬”和“无翅”突变体的缺失区域略大一些，从每侧切入齿状突起带。图中间所示为四个成对控制突变体。对它们表型的解释基于对弱、强等位基因的研究以及超双胸突变的结合(见正文部分)；至于“侏儒”突变，则基于雌雄嵌合体（未发表）。这或许代表了相应基因座的极端突变情况。所有已知的“桶状”和“锯齿状”等位基因突变的表型（未显示）都有些许多样性，需要深入研究以推断其典型表型。图右侧是两个间隙基因座“跛子”和“折叠扇”的突变体。在缺失相应基因座的纯合胚胎上观察到了这两种基因座所代表的无定形表型。唯一已知的“驼背”等位基因突变(未显示)缺失中胸和后胸。

The Segmental Pattern of the Normal *Drosophila* Larva

Figure 2 shows the cuticular pattern of a normal *Drosophila* larva shortly after hatching. The larval body is comprised of three thoracic and eight abdominal segments. Although differences are observed in different body regions, all segments have certain morphological features in common. The anterior of each segment is marked with a band of denticles, most of which point posteriorly. The posterior part of each segment is naked. The segment borders run along the anterior margins of the denticle bands[7], they have no special morphological features. The polarity of the pattern is indicated by the orientation of the dentcles and, in the abdomen, by the shape of the bands (Fig. 3). In the thoracic segments the bands are narrow with fine denticles whereas those in the abdominal segments are broader and comprised of thick pigmented denticles (for a detailed description of the cuticular pattern see ref. 1).

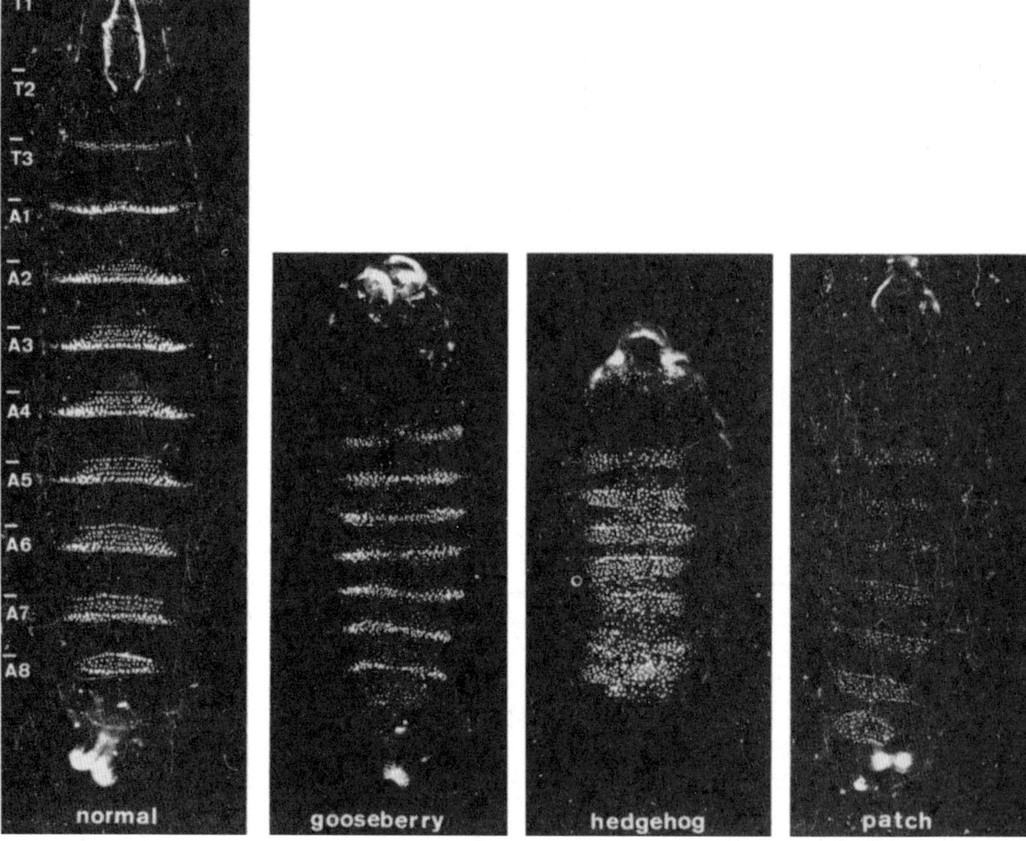

Fig. 2. Ventral cuticular pattern of (from left to right) a normal *Drosophila* larva shortly after hatching, and larvae homozygous for *gooseberry*, *hedgehog* and *patch*. The mutant larvae were taken out of the egg case before fixation. All larvae were fixed, cleared and mounted as described in ref. 22. A, abdominal segment; T, thoracic segment. For further description see text and Fig. 3. ×140.

正常果蝇幼虫的分节模式

图 2 显示了一只刚孵化不久的正常果蝇幼虫的表皮模式。幼虫身体由三个胸节和八个腹节组成。虽然可观察到不同身体部位的区别，但所有的体节都有某些共同的形态特征。每个体节的前端都有一条齿状带，它们中的大部分都指向后部。每个体节的后部是裸露的。体节的边界沿着齿状带的前缘延展[7]，它们没有特别的形态特征。小齿的方向及腹部条带的形状都表明了其模式的极性（图 3）。在胸节上条带很窄，且小齿细小，然而在腹节上条带较宽，由具厚重色素的小齿组成（对表皮模式的详细描述见参考文献 1）。

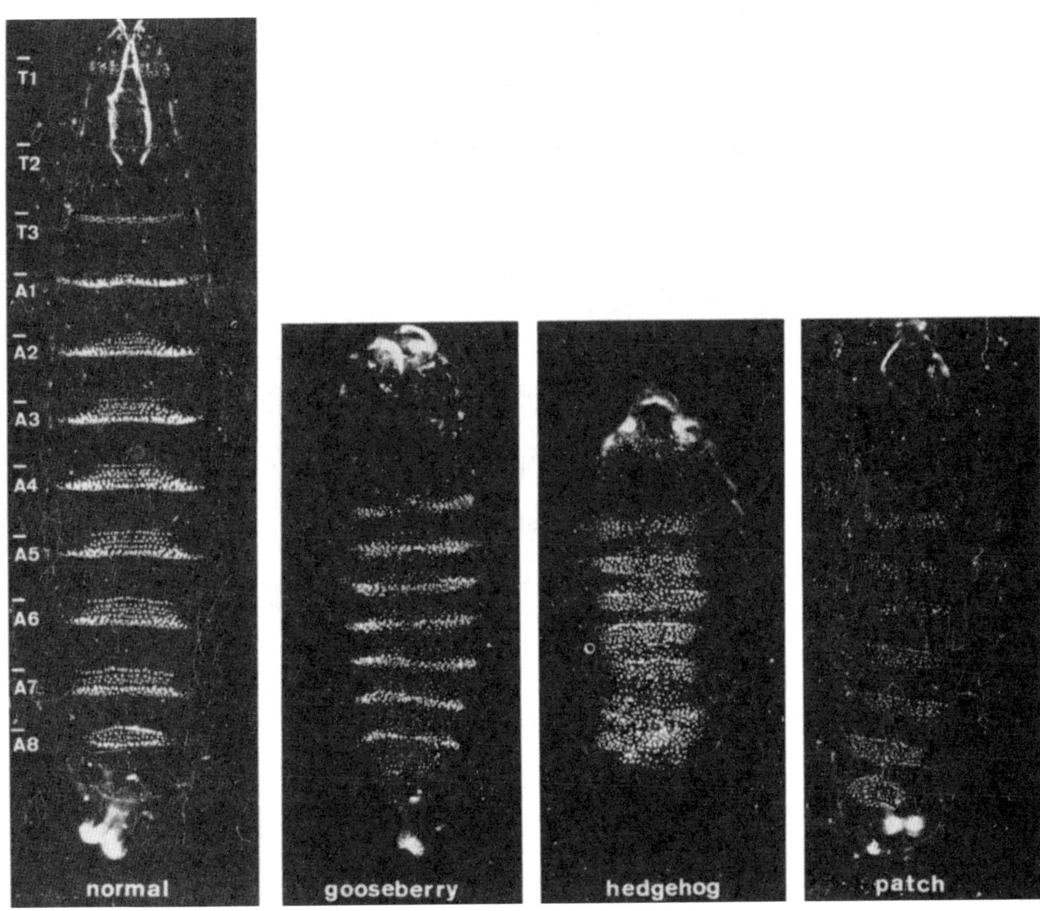

图 2. 从左到右依次是刚孵化不久的正常果蝇幼虫、“鹅莓”“刺猬”和“斑点”纯合体幼虫的腹侧表皮模式。在固定前将突变幼虫从卵壳中取出。按照参考文献 22 的方法将所有幼虫固定、清洗并制成标本。A，腹节；T，胸节。更多描述见正文及图 3。（×140）。

Segment Polarity Mutants: Deletions in Each Segment

Mutants in this class have the normal number of segments. However, in each segment a defined fraction of the normal pattern is deleted and the remainder is present as a mirror-image duplication. The duplicated part is posterior to the "normal" part and has reversed polarity (Figs 1–3).

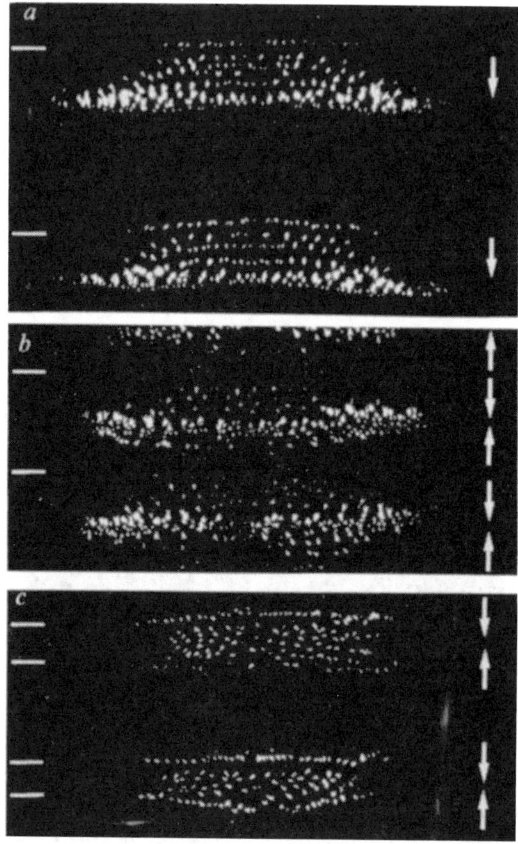

Fig. 3. Details from the ventral abdomen of a normal (*a*), a *gooseberry* (*b*), and a *patch* (*c*) larva. The positions of the segment boundaries are indicated at the left by the transverse lines. The arrows at the right indicate the polarity of the pattern as judged by the orientation of the denticles as well as the shape of the denticle bands.

Six such loci have been identified. Three loci, *fused*[8], *wingless* (ref. 9 and G. Struhl, personal communication) and *cubitus interruptus*^{D}, were previously known whereas *gooseberry*, *hedgehog* and *patch* are new (Table 1). All the mutations in this class are zygotic lethals and the phenotypes are only produced in homozygous embryos. One of the loci, *fused*, also shows a maternal effect in that a wild-type allele in the mother is sufficient to rescue the mutant phenotype of the homozygous embryos.

In all mutants except *patch* the region deleted includes the naked posterior part of the pattern and the duplication involves a substantial fraction of the anterior denticle band. In

体节极性突变：每个体节都有缺失

这个类型的突变体体节数目正常。然而，每个体节都缺失了正常模式中特定的一部分，剩余的部分呈镜像重复。重复的部分位于“正常部分”之后并具有相反的极性（图 1~3）。

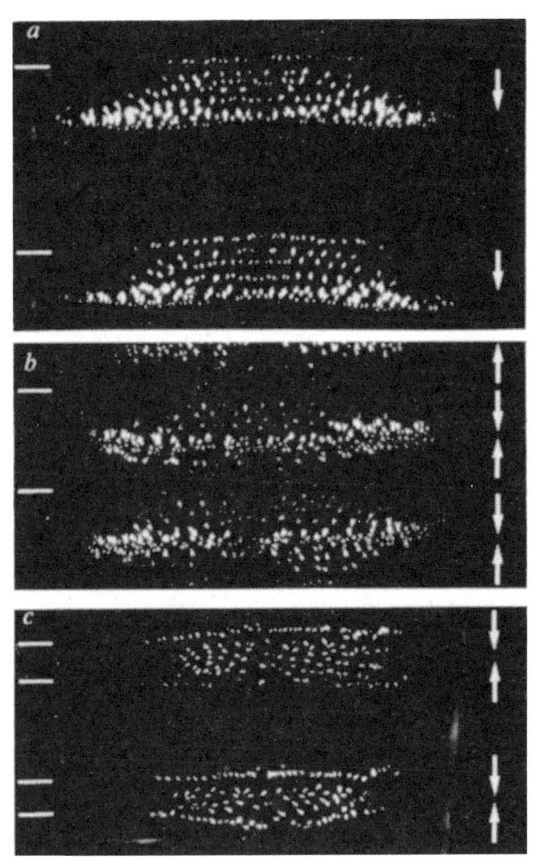

图 3. 幼虫腹节腹侧详图，正常（*a*），“鹅莓”突变体（*b*），“斑点”突变体（*c*）。左侧横线标出了体节边界的位置。右侧箭头指示依据小齿方向和齿状突起带形状所确定的模式的极性。

已经鉴定出了六个这一类型的基因座。其中，“融合”[8]“无翅”（参考文献 9 以及与斯特鲁尔的个人交流）和“尺骨中断 D”这三个基因座是已知的，而“鹅莓”“刺猬”和“斑点”是新基因（表 1）。这一类型的所有突变都是纯合致死的，并且突变表型只在纯合胚胎中产生。其中“融合”这个基因座还具有母体效应，因为母体中一个野生型等位基因足以挽救其纯合胚胎的突变体表型。

除“斑点”外，所有突变体的缺失部分都包括体节模式中裸露的后部区域，而其重复部分则都包含前部齿状带的大块区域。突变幼虫每个体节的腹侧都几乎完全

mutant larvae, the ventral side of each segment is almost entirely covered with denticles, the posterior fraction of which point anteriorly. The segment identity seems to be normal, the denticles of the abdominal segments being large and pigmented whereas those of the thorax are short and pale (Figs 1, 2). The anterior margin of the region duplicated in these mutants coincides with the segment boundary in only two cases, *fused* and *gooseberry* (Fig. 3). In *wingless* and *hedgehog* it lies posterior to the boundary, such that these larvae apparently lack all segment boundaries.

The phenotype of embryos homozygous for *patch* contrasts with that produced by the five loci described above in that the duplicated region includes some naked cuticle anterior to each denticle band. The duplicated unit thus involves structures of two adjacent segments. *Patch* larvae, despite the normal number of denticle bands, have twice the normal number of segment boundaries (Figs 1, 3).

Despite these differences, the common feature of all mutants in this class is that a defined fraction of the pattern in each segment is deleted. This deletion is associated with a mirror image duplication of the remaining part of the pattern. We suggest that these loci are involved in the specification of the basic pattern of the segmental units.

Pair-rule Mutants: Deletions in Alternating Segments

In mutants of this class homologous parts of the pattern are deleted in every other segment. Each of the six loci is characterized by its own specific pattern of deletions (Table 1, Figs 1, 4). For example, in *even-skipped* larvae, the denticle bands and adjacent naked cuticle of the pro- and metathoracic, and the 2nd, 4th, 6th and 8th abdominal segments are lacking. This results in larvae with half the normal number of denticle bands separated by enlarged regions of naked cuticle (Figs 1, 4). In *paired* larvae the apparent reduction in segment number results essentially from the deletion of the naked posterior part of the odd-numbered and the anterior denticle bands of the even-numbered segments (Figs 1, 4). The double segments thus formed are composites of anterior mesothorax and posterior metathorax, anterior first abdominal and posterior second abdominal segment, etc. The identification of the regions present or deleted in mutant larvae is based on the phenotypes produced by alleles with lower expressivity. In such embryos the deletions are in general smaller and variable in size. In a "leaky" allele of *even-skipped*, the denticle bands of the even-numbered segments are frequently incomplete, whereas in *paired*^{2} (*prd*^{2}) the pattern deletions often involve only part of the naked cuticle of the odd-numbered segments, leading to a pairwise fusion of denticle bands (Fig. 4).

被小齿覆盖，小齿后部指向前部。体节特性似乎正常，腹节的小齿大且色素化而胸节的小齿短且颜色淡（图 1 和 2）。只有在“融合”和“鹅莓”这两种突变体中重复部分的前缘才与体节边界一致（图 3）。在“无翅”和“刺猬”突变体中，重复部分的前缘位于体节边界之后，以至于这些幼虫明显缺少了所有体节边界。

“斑点”纯合突变胚胎重复区域的每个齿状突起带前部都包含一些裸露表皮，其表型与前面描述的五个基因座突变表型截然不同。所以其重复单位涉及两个相邻体节的结构。因此，尽管“斑点”突变幼虫齿状带的数目正常，但其体节边界数是正常数目的两倍（图 1 和 3）。

虽然有这些区别，但这一类型所有突变体的共同特征是每个体节中都缺失特定的区域，这种缺失与体节模式中剩余部分的镜像重复有关。我们认为这些基因座都参与了对体节单位基本模式的特化。

成对控制突变：交替体节的缺失

在这个类型的突变体中，每隔一个体节缺失模式中的同源部分。六个基因座中的每一个都有其独特的缺失方式（表 1，图 1 和 4）。例如，在“偶数 – 遗漏”突变幼虫中，不仅其前、后胸的齿状带和相邻裸露的表皮缺失，同时其第二、四、六、八腹节也缺失了。这导致突变幼虫中半数于正常数目的齿状带被裸露表皮的扩展区域分隔开（图 1 和 4）。在“配对”突变幼虫中，体节数目的明显减少关键是由奇数体节裸露的后部和偶数体节前部的齿状带的缺失造成的（图 1 和 4）。这样就形成了由中胸前部、后胸后部、第一腹节前部和第二腹节后部等组成的双体节。突变幼虫一个区域的存在或缺失是根据低表现度等位基因产生的表型鉴定的。这些胚胎中的基因缺失一般较小且大小变化多样。在“偶数 – 遗漏”的一个“遗漏”等位基因突变体中，偶数体节的齿状带往往不完整；然而在“配对 ^{2}”（*prd*^{2}）突变体中，缺失往往只涉及奇数体节裸露表皮的一部分，导致齿状带两两融合（图 4）。

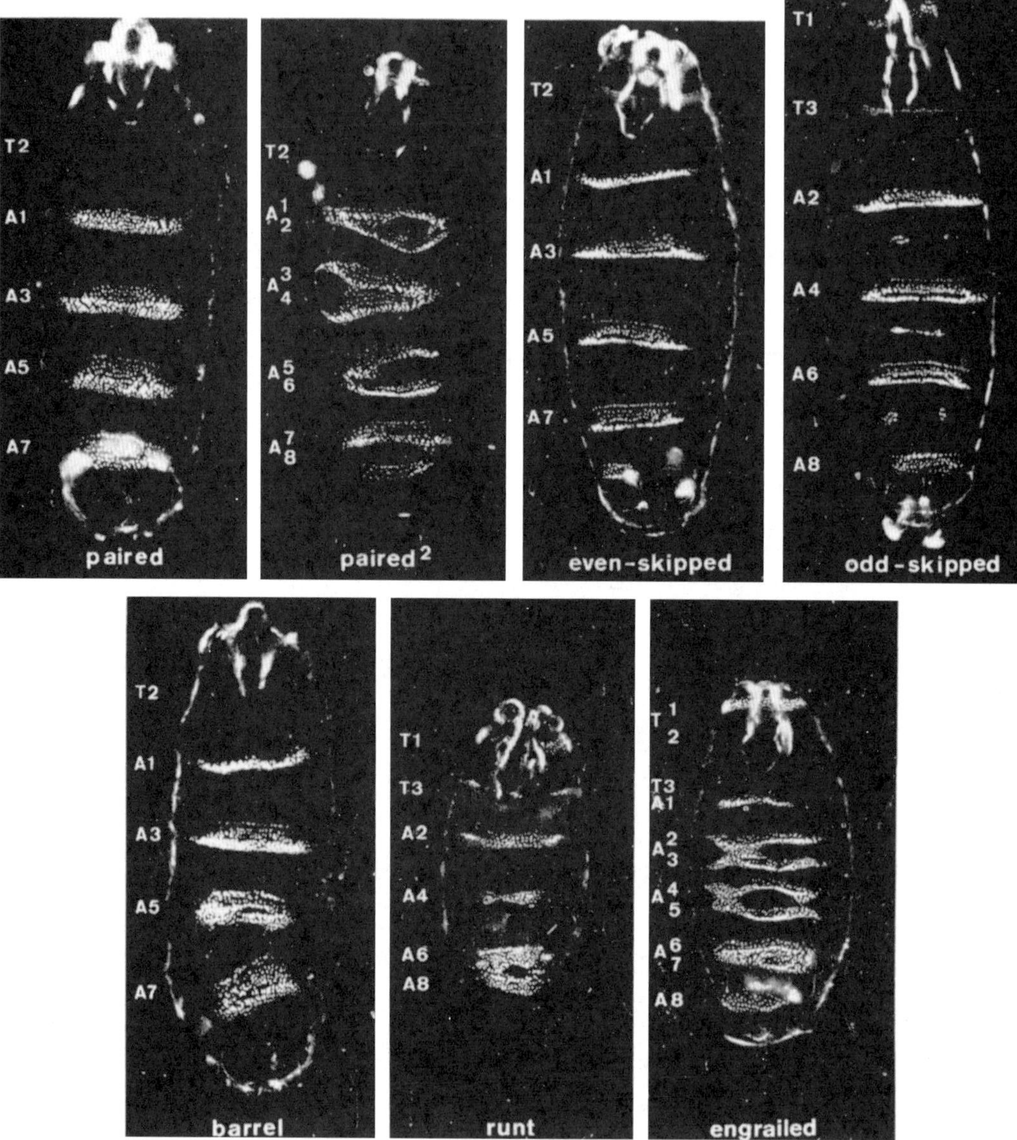

Fig. 4. Larvae homozygous for mutations at the six pair-rule loci. The segmental identity of the denticle bands is indicated at the left of each picture. A, abdominal band; T, thoracic band. For comparison with the normal pattern see Figs 1 and 2.

Further support for the composite nature of the segments in *prd* larvae was obtained in combinations with *Ultrabithorax* (*Ubx*), a homeotic mutation which when homozygous causes the transformation of the first abdominal segment into mesothorax[5]. In such *prd*; *Ubx* larvae, only the *denticle band* of the first double segment in the abdomen is transformed. The posterior margin of this band and the naked cuticle which follows remain abdominal in character and lack, for example, the typical mesothoracic sense organs. The composite nature of the segments in *paired* shows that the establishment of segmental identity does not require the establishment of individual segments as such.

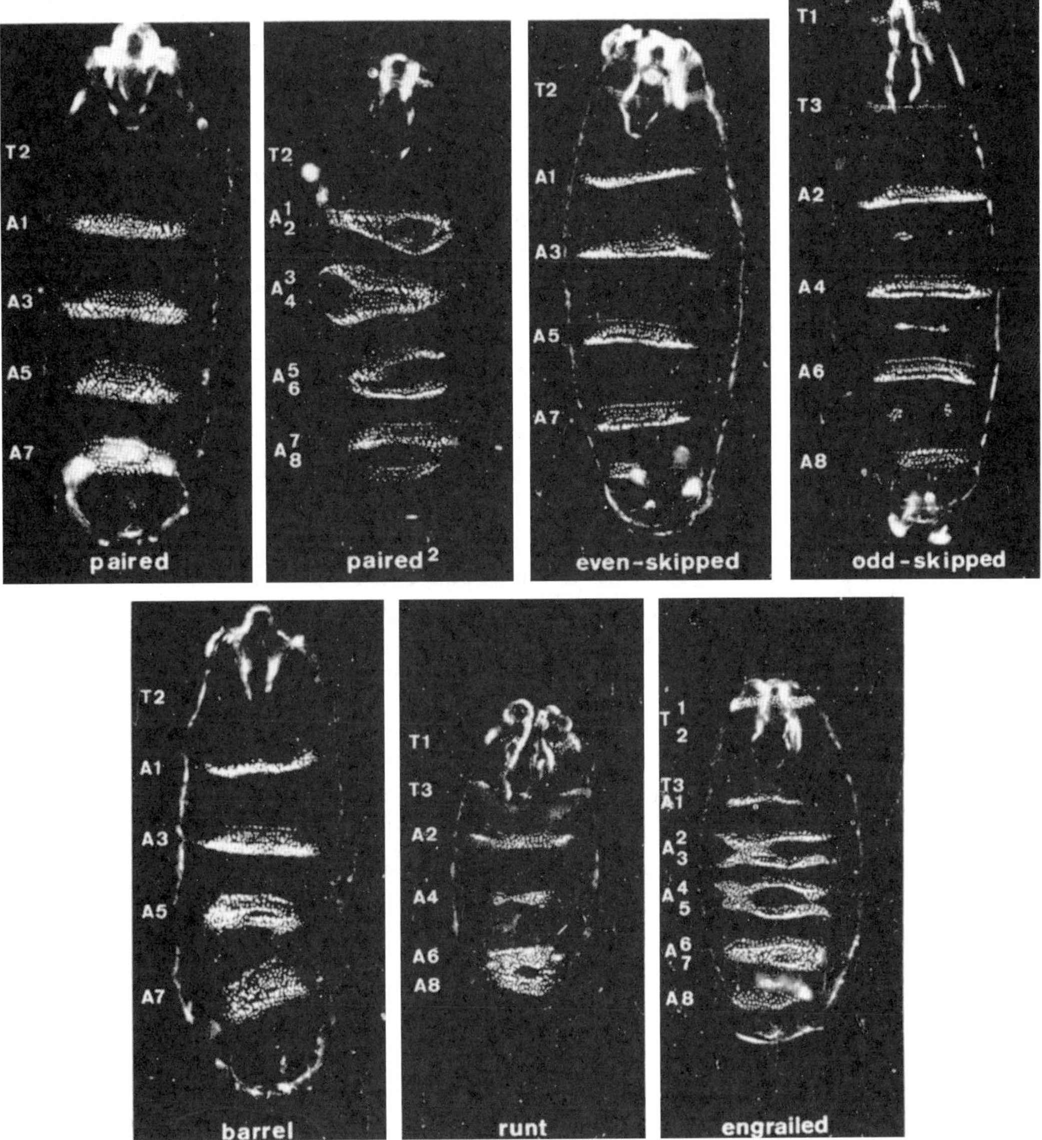

图 4. 六个成对控制基因座纯合突变的幼虫。在每个图的左侧注明了其齿状带的体节特性。A，腹带；T，胸带。其与正常模式的比较见图 1 和图 2。

“超双胸”(*Ubx*) 突变是一种纯合时可以将第一腹节变为中胸的同源异型突变[5]，它与“配对”突变结合进一步支持了配对突变幼虫中体节的复合性质。在这种“配对 – 超双胸”双突变幼虫中，只有腹部第一个双体节的齿状带转变了。这个条带的后缘和紧接着的裸露表皮保留了腹部的特征并缺失(例如)典型的中胸感觉器官。“配对”突变体中体节的复合性质表明，体节特性的建立并不需要个别体节本身特性的建立。

The segmentation pattern in *prd*; *Ubx* larvae is typical of that of *paired* alone and is not affected by the *Ubx* homeotic transformation. Similarly, in *prd*; *Polycomb* larvae the naked cuticle of alternating segments is deleted just as it is in *paired* alone, even though in *Polycomb* embryos all the thoracic and abdominal segments have an 8th abdominal character[5]. These combinations, and similar ones with *even-skipped*, indicate that the observed grouping of segments in pairs depends on the position of segments within the segmental array rather than the segmental identity. These combinations thus provide evidence that mutations such as *paired* and *even-skipped* affect different processes from those altered in homeotic mutants.

Other mutants in this class show different deletion patterns. The phenotype of *odd-skipped* is similar to that of *even-skipped*. However, in this case it is the odd-numbered denticle bands that are affected. In *odd-skipped*, the deleted region is smaller and is restricted to a more posterior part of the segment than in *even-skipped* (Figs 1, 4). *Barrel* has a phenotype similar to *paired* although the pattern is often less regular (Fig. 4). *Runt*, on the X chromosome, is the only pair-rule mutant showing mirror-image duplications. *Runt* embryos have half the normal number of denticle bands, each a mirror-image duplication of the anterior part of a normal band (similar to the duplications found in *patch*). The bands in *runt* embryos, as well as the region of naked cuticle separating them, are of unequal sizes (Figs 1, 4).

To the list of pair-rule loci we have added the *engrailed*[10] locus. Lethal *engrailed* alleles[11] lead to a substantial deletion of the posterior region of even-numbered segments. In addition, the anterior margin and adjacent cuticle of each segment are affected. Thus, the defect pattern in *engrailed* shows repeats which are spaced at both one- and two-segment intervals.

Each of the six different pair-rule loci affects a different region within a double segmental repeat. In no case does the margin of the deleted region coincide with a segment boundary. When the deleted region corresponds in size up to one entire segment (*paired*, *even-skipped*) it includes parts of two adjacent segments.

The phenotypes of the pair-rule mutants suggest that at some stage during normal development the embryo is organized in repeating units, the length of which corresponds to two segmental anlagen.

Gap Mutants: One Continuous Stretch of Segments Deleted

One of the striking features of the mutations of the first two classes is that the alteration in the pattern is repeated at specific intervals along the antero-posterior axis of the embryo. No such repeated pattern is found in mutants of the third class and instead a single group of up to eight adjacent segments is deleted from the final pattern. Three loci have been identified which cause such gaps in the pattern (Table 1, Figs 1, 5). *Krüppel* (*Kr*) was originally described by Gloor[12]. Embryos homozygous for *Kr* lack thorax and anterior abdomen. The posterior-terminal region with the abdominal segments 8, 7 and 6 is normal, although probably somewhat enlarged. Anterior to the 6th abdominal segment is a plane of mirror-image symmetry followed by one further segment band with the

“配对–超双胸”双突变幼虫具有“配对”单突变幼虫体节模式的典型特点，且不受“超双胸”同源异型转变的影响。同样地，虽然在“多梳”单突变胚胎中所有胸节和腹节都具第八腹节的特征[5]，但是，在“配对–多梳”双突变幼虫中，交替体节裸露表皮的缺失和“配对”单突变中的一样。这些组合及相似的“偶数–遗漏”组合表明，实验中观察到的成对体节组的形成是由体节在序列中的位置而不是体节特性决定的。因此，这些组合证明，诸如“配对”和“偶数–遗漏”的这类基因突变影响的过程不同于同源异型突变体中发生变化的过程。

这个类型的其他突变体表现出不同的缺失模式。“奇数–遗漏”突变表型与“偶数–遗漏”相似。然而，这时受影响的是奇数齿状带。与“偶数–遗漏”突变相比，在“奇数–遗漏”突变中缺失区域更小并局限于体节上更靠后的区域(图1,4)。“桶状”突变与“配对”表型相似但其模式往往较不规律（图4)。“侏儒”位于X染色体上，是唯一一个具有镜像重复的成对控制突变体。“侏儒”突变体胚胎有半数于正常数目的齿状带，每个带都有正常突起带前部的镜像重复（与“斑点”突变中的重复类似)。“侏儒”突变胚胎中的突起带及分隔它们的裸露表皮的尺寸大小不均一（图1,4)。

在成对控制基因座的列表中我们增加了“锯齿”基因座[10]。致死的“锯齿”等位基因[11]能造成偶数体节后部区域的大量缺失。此外，每个体节的前缘和相邻表皮也会受到影响。因此，“锯齿”缺陷模式表现出在一个或两个体节间隔间的重复。

六个不同的成对控制基因座分别影响双体节重复中的一个不同区域。缺失区域的边缘从不与体节边界一致。当缺失区域大小达到一个完整体节时（“配对”，“偶数–遗漏”)，它也会包含两个相邻体节的部分区域。

成对控制基因突变体的表型暗示，在正常发育的某个阶段胚胎被划分为重复单位，每个重复单位的长度相当于两个体节原基。

间隙突变体：一组连续体节缺失

前两类突变的一个显著特征是，沿着胚胎的前后轴模式的改变以特定的间隔重复出现。在第三类突变中不存在这种重复特性，取而代之的是在最终模式中最多连续八个相邻体节一起缺失。已经鉴定出三个引起这类间隙突变的基因座（表1，图1和5)。“跛子”（*Kr*）突变最早是由格洛尔描述的[12]。“跛子”纯合胚胎没有胸部和前腹部。尾部末端及第八、七、六腹节正常，但是有些许变大。在第六腹节的前面是一个镜像对称的平面，该平面连着另一个具有第六体节特征但极性相反的体节带。对称平面的精确位置是变化的，通常不与体节边界一致。“跛子”突变的大部分模式

character of a 6th segment oriented in reversed polarity. The exact position of the plane of symmetry varies and does not usually coincide with a segmental boundary. A large part of the *Krüppel* pattern is reminiscent of the pattern observed in embryos produced by the maternal-effect mutant *bicaudal*, although no maternal component is involved in the production of the *Krüppel* phenotype (in preparation).

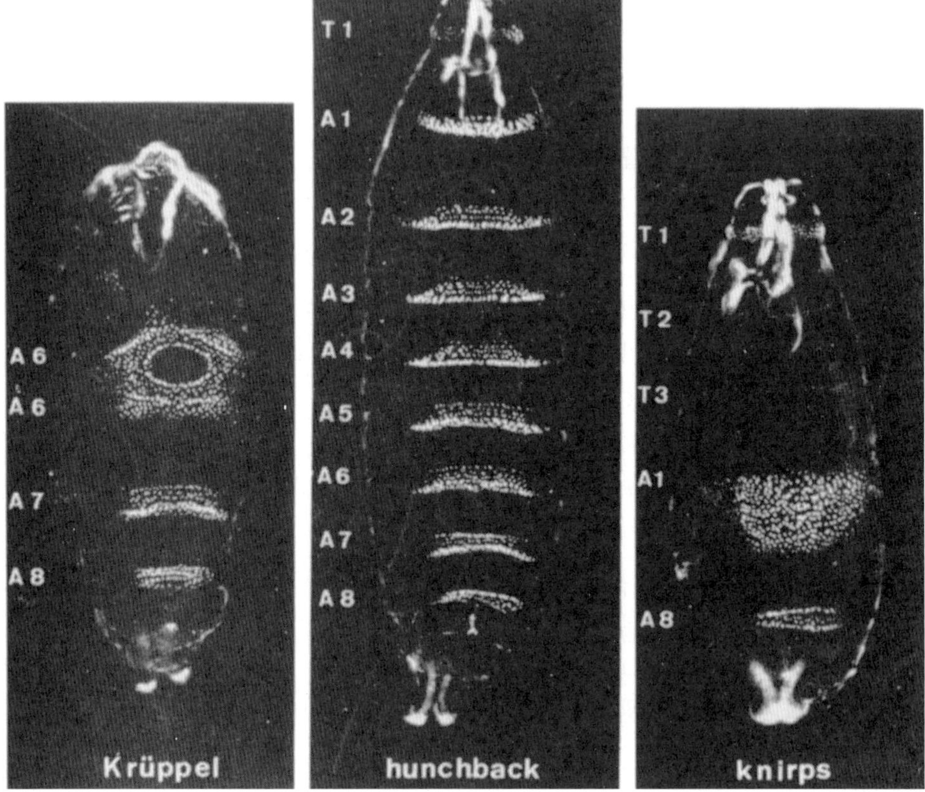

Fig. 5. Larvae homozygous for mutations at the three gap loci. A, abdominal segment; T, thoracic segment.

In the other two loci of this class, the gap in the pattern occurs in specific morphologically defined subregions of the larval pattern, in the thorax and in the abdomen, respectively. In *hunchback*, the meso- and metathoracic segments are deleted. In embryos homozygous for *knirps*, only two rather than eight denticle bands are formed in the abdomen. The posterior-terminal region including the 8th abdominal segment seems normal whereas the anterior abdominal denticle band is considerably enlarged. The anterior margin of the denticle band is morphologically similar to the first abdominal segment but combinations with *Ubx* show that the band is a composite with more than one segmental identity.

All three loci are required for a normal segmental subdivision of one continuous body region. The lack of a repeated pattern of defects suggests that the loci are involved in processes in which position along the antero-posterior axis of the embryo is defined by unique values.

都使人联想到母体效应“双尾”突变体产生的胚胎模式，虽然母体成分并未参与“跛子”表型的产生过程（文章准备发表）。

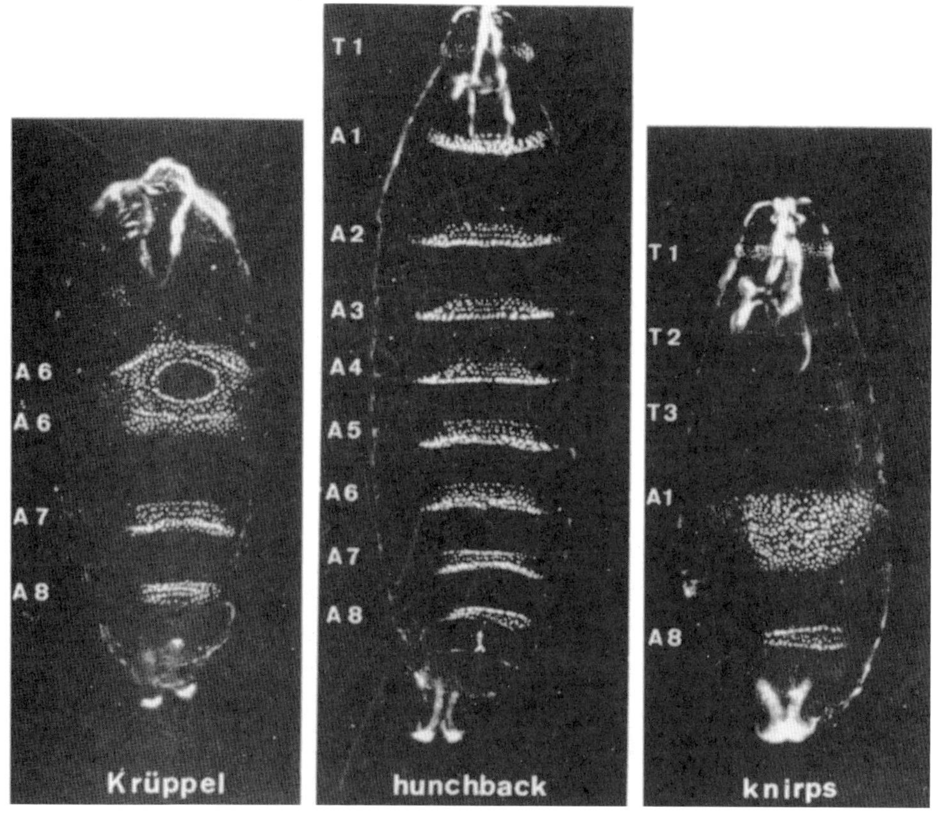

图 5. 三个间隙基因座突变的纯合幼虫。A，腹节；T，胸节。

这一类型的其他两个基因座中，幼虫模式上的间隙分别产生于胸部和腹部的形态特化的亚区域。“驼背”突变幼虫的中胸和后胸体节缺失了。“折叠扇”突变纯合胚胎的腹部只形成了两条而非八条齿状突起带。尾部末端区域包括第八腹节看起来正常，但其前腹部齿状带增大了很多。齿状带的前缘形态上与第一腹节相似，但与“超双胸”突变的组合表现出这条突起带是具有多种体节特性的复合体。

上述三个基因座都是身体上一个连续区域细分成正常体节所必需的。这些缺陷缺少重复性模式，表明这些基因座参与了由固定值划定的胚胎前后轴模式的形成过程。

When Are Genes Affecting Segment Number Active?

The phenotypes described above are observed only in homozygous embryos, indicating that the loci identified by these mutations are active after fertilization and are crucial for the normal segmental organisation of the embryo. We have described the mutations in terms of their effect on the differentiated pattern. However, in many instances their effect can be observed much earlier in development. In normal embryos segmentation is first visible 1 h after the onset of gastrulation as a repeated pattern of bulges in the ventral ectoderm (unpublished observations). In *paired* and *even-skipped* embryos the number of bulges is reduced and corresponds to the number of segments observed in the differentiated mutant embryos. *Krüppel*, *runt* and *knirps* embryos can be identified 15 min after the onset of gastrulation. All three mutations cause shorter germ bands, a phenomenon clearly related to the strong reduction in segment number observed in the differentiated larvae. Further evidence for an early activity at the *paired* locus was obtained using a temperature-sensitive allele. The extreme phenotype is only obtained when the embryo is kept at the restrictive temperature during the blastoderm stage. All these results indicate that the wild-type genes defined by the mutations are active before the end of gastrulation during normal development.

Discussion

Segmentation in *Drosophila* proceeds by the transition from a single field into a repeated pattern of homologous smaller subfields. Mutant alleles at the 15 loci we have described interfere with this process at various points. Although each locus has its own distinct phenotype, we were able to distribute the mutations in three classes. In one class only a single large subregion of the embryo is affected, whereas in mutants of the other two classes a reiteration of defects is produced with a repeat length of one or two segments, respectively. This suggests that the process of segmentation involves three different units of spatial organization.

The organization of the egg is thought to be controlled by a monotonic gradient set up during oogenesis under the control of the maternal genome[3,4]. All the mutants we have described depend on the embryonic rather than the maternal genome. None of them alters the overall polarity of the embryo, the head always being at the anterior and the telson at the posterior end of the egg. The most dramatic alterations of the pattern are produced in the gap mutants and involve one large subregion of the egg. *Hunchback* and *knirps* affect the development of thorax and abdomen, respectively, and might be involved in the establishment of these large morphologically unique subregions of the embryo. The large mirror-image duplications found in the posterior pattern of *Krüppel* embryos are similar to those of *bicaudal*. The *bicaudal* phenotype has been interpreted as resulting from an instability in the maternal gradient[3,13], and thus *Krüppel* might be involved in the maintenance of elaboration of this gradient in the posterior egg region after fertilization.

The smallest repeat unit, the individual segment, is affected in mutants at the six segment-polarity loci. The pattern alteration consists of a mirror-image duplication of part of the normal pattern with the remainder deleted. In these mutants the deleted region

基因在何时有效地影响体节数目?

上述表型只有在纯合基因突变胚胎中才能得到，说明通过这些突变体鉴定的基因座在受精之后起作用，而且是胚胎形成正常体节所必需的。我们已经描述了这些突变体对分化模式的影响。然而，在许多情况下，它们的影响在更早的发育时期就已出现。正常胚胎中最早观察到体节分化是在最初原肠胚形成后一个小时，表现为外胚层腹侧重复模式的突起（未发表的观察结果）。在“配对”和“偶数－遗漏”突变胚胎中突起数目有减少，并且与已分化的突变体胚胎中观察到的体节数目相同。在最初原肠胚形成后 15 分钟后就可以鉴定出“跛子”“侏儒”“折叠扇”突变体的胚胎。这三种突变都可引起胚带变短，这一现象与在已分化的幼虫中观察到的体节数目大量减少有明显关系。通过温度敏感等位基因得到了关于配对基因座早期活性更深入的证据。只有在囊胚层期将胚胎置于限制温度，才能得到极端突变表型。所有结果表明，在正常发育中，通过突变定义的野生型基因作用于原肠胚形成结束以前。

讨　论

果蝇的分节是一个从单域向同源的较小亚区的重复模式的转变过程。我们已经描述的 15 个基因座的突变体等位基因在不同时间点干扰这个过程。虽然每个基因座都有其独特的表型，但我们还是可以将这些突变体划分为三个类型。在一类中，只有胚胎上一个大的亚区受影响。而在另两类突变体中，分别产生了重复长度为一个体节或两个体节的重复性缺陷。这暗示了分节过程涉及三种不同的空间组织单元。

卵的结构形成被认为是由单一梯度决定的，该梯度是在母体基因组控制下在卵子发生时建立的[3,4]。我们描述的所有突变体都是以胚胎而不是以母体基因组为基础的。它们都未改变胚胎的整体极性，头仍然在卵的前部，尾节仍在卵的后端。间隙突变体产生的模式改变最明显，涉及卵的一个大的亚区。“驼背”和“折叠扇”分别影响胸部和腹部的发育，可能参与胚胎上大的形态独特的亚区的建立。发生在“跛子”突变胚胎后部的大面积镜像重复与“双尾”相似。“双尾”突变表型可以被理解为母体效应不稳定的结果[3,13]，因此，“跛子”可能参与维持了受精后卵的后部区域梯度的细分。

最小的重复单位，即单个体节，在六个体节－极性基因座突变中受到了影响。其模式改变包括部分正常模式的镜像重复及其剩余部分的缺失。在这些突变体中缺

corresponds in size to more than half a segment. Mutations causing smaller deletions in the segmental pattern do not lead to polarity reversals (unpublished observations). The mirror-image duplications produced by the pair-rule mutation *runt* are also associated with a deletion of more than half a repeat unit, that is, more than one segment. The tendency of partial fields containing less than half the positional values to duplicate is well described for imaginal disks in *Drosophila*[14], as well as for the larval epidermis in other insects[15,16]. On the other hand, the same types of pattern duplication are also produced in conditions which do not involve cell death and regeneration, but rather a reorganization of the positional information in the entire field[3,17]. More detailed studies of early development in mutant embryos may reveal which mechanism is responsible for the different mutant phenotypes.

Given the evidence for a homology between segments, the existence of a class of mutants affecting corresponding regions in each individual segment is perhaps not surprising. The discovery of a mutant class affecting corresponding regions in every other segment was not expected and suggests the existence at some time during development of homologous units with the size of two segmental anlagen. It is possible that the double segmental unit corresponds to transitory double segmental fields which are established early during embryogenesis and are later subdivided into individual segments. At the blastoderm stage the epidermal primordium giving rise to thorax and abdomen is only about 40 cells long[1], An initial subdivision into double segments would avoid problems of accuracy encountered in a simultaneous establishment of segment boundaries every three to four cells. A stepwise establishment of segments implies that the borders defining double segments be made before the intervening ones. The mutant phenotypes do not definitely show which, if any, of the segment borders define a primary double segmental division in normal development. The mutant phenotypes which come closest to a pattern one would expect if the transition from the double segment to the individual segment stage were blocked are the patterns of *paired* and *even-skipped*. Both suggest the frame meso- and methathorax, 1st and 2nd, 3rd and 4th abdominal segment, etc.

It is also possible that the double segmental units are never defined by distinct borders in normal development. The existence of a double segmental homology unit may merely reflect a continuous property such as a wave with a double segmental period responsible for correct spacing of segmental boundaries (see, for example, refs 18, 19). We have not found any mutations showing a repeat unit larger than two segments. This may indicate that the subdivision of the blastoderm proceeds directly by the double segmental repeat with no larger intervening homology units. However, the failure to identify such larger units may reflect the incompleteness of our data.

Drosophila has been estimated to have about 5,000 genes and only a very small fraction of these when mutated result in a change of the segmental pattern of the larva. Some of the loci described here were known previously but only in the case of *Krüppel* has the embryonic lethal phenotype been recognized as affecting segmentation.[12] The majority of the mutants described here have been isolated in systematic searches for mutations affecting the segmentation pattern of the *Drosophila* larva. These experiments are still incomplete. Most of the alleles on the second chromosome were isolated in one experiment which yielded an average allele frequency of four or five alleles per locus

失区域大于半个体节。引起体节模式中较小缺失的突变并不会导致极性反转（未发表的观察结果）。成对控制突变“侏儒”产生的镜像重复也与大于半个重复单位（也就是大于一个体节）的缺失有关。包含少于一半位置值的部分区域发生重复的趋势已被果蝇成虫盘[14]和其他昆虫幼虫表皮[15,16]为例很好地描述。另一方面，在不涉及细胞死亡和再生的条件下也会产生相同类型的重复模式，但不是完整区域内位置信息的重组[3,17]。对突变胚胎早期发育更细致的研究也许会揭示不同突变表型的形成机制。

鉴于体节之间的同源性，存在一组影响各个体节相应区域的突变体也许并不奇怪。发现一组影响间隔体节相应区域的突变体却出乎意料，这暗示了在某个发育阶段存在大小为两个体节原基的同源单位。很可能双体节单位对应胚胎发生早期形成的短暂存在的双体节区域，这些区域之后再分化成了单个体节。在囊胚期，将要分化为胸部和腹部的上皮原基只有 40 个细胞的长度[1]。最初分化为双体节可避免每三到四个细胞同时形成各体节边界时所遇到的准确性问题。体节的逐步形成意味着双体节边界是在中间边界之前形成的。突变体表型并未确切地显示确立正常发育中的双体节分区的体节边界。如果从双体节到单体节的转变过程被打断，与预期模式最接近的突变表型是“配对”和“偶数–遗漏”突变的表型。两者都提示中胸和后胸，第一、第二、第三和第四腹节等结构。

也可能双体节单位在正常发育中从未被明显的边界所划分。双体节同源单位的存在可能仅仅反映了一种连续性质，就像一种以双体节为周期的波，这种波与体节边界的正确划分有关（举例说明可以见参考文献 18, 19）。我们未发现重复单位大于两个体节的突变体。这可能说明了胚盘直接通过没有更大的中间同源单位的双体节重复进行分化。然而未能确定这种较大的同源单位，可能反映出我们的数据不足。

据估计，果蝇约有 5,000 个基因，其中只有一小部分突变后会引起幼虫体节模式改变。本文描述的基因座中有些是已知的，但只有“跛子”突变体的胚胎致死表型被认为影响分节[12]。本文描述的大多数突变体都已在影响果蝇幼虫分节模式突变的系统研究中分离出来。这些实验仍不完整。二号染色体上绝大多数等位基因是在一个实验中分离出来的，在这个实验中，每个基因座的平均等位基因频率为 4~5 个(基于 42 个胚胎致死基因座)。根据这个结果及三号、一号染色体相似的计算结果，我

(based on 42 embryonic lethal loci). From this yield and similar calculations for the third and first chromosomes, we estimate that we have identified almost all segmentation loci on the second chromosome and about 50% each of those on the third and first chromosome. Our sample of 15 loci should therefore represent the majority of the loci affecting segmentation in the *Drosophila* genome. Thus, in *Drosophila* it would seem feasible to identify all genetic components involved in the complex process of embryonic pattern formation.

We thank Hildegard Kluding for excellent technical assistance, Adelheid Schneider, Maria Weber and Gary Struhl for help during various parts of the mutant screens, Gerd Jürgens for stimulating discussion and our colleagues from the *Drosophila* laboratories in Cambridge, Freiburg, Heidelberg and Zürich for critical comments on the manuscript, and Claus Christensen for the photographic prints. Thomas Kornberg and Gary Struhl provided us with lethal alleles of *engrailed* and *wingless* respectively which facilitated the identification of our alleles. All mutants are available on request.

Note added in proof: All known *barrel* alleles fail to complement *hairy*[20], suggesting that the *barrel* mutations are alleles at the *hairy* locus.

(**287**, 795-801; 1980)

Christiane Nüsslein-Volhard and Eric Wieschaus
European Molecular Biology Laboratory, PO Box 10.2209, 69 Heidelberg, FRG

Received 26 June; accepted 29 August 1980

References:

1. Lohs-Schardin, M., Cremer, C. & Nüsslein-Volhard, C. *Devl. Biol.* 73, 239–255 (1979).
2. Wieschaus, E. & Gehring, W. *Devl. Biol.* **50**, 249–263 (1976).
3. Nüsslein-Volhard, C. in *Determinants of Spatial Organisation* (eds Subtelney, S. & Konigs-berg, I. R.) 185–211 (Academic, New York, 1979).
4. Sander, K. *Adv. Insect Physiol.* **12**, 125–238 (1976).
5. Lewis, E. B. *Nature* **276**, 565–570 (1978).
6. Garcia-Bellido, A. *Am. Zool.* **17**, 613–629 (1977).
7. Szabad, J., Schüpbach, T. & Wieschaus, E. *Devl. Biol.* **73**, 256–271 (1979).
8. Counce, S. *Z. Induktive Abstammungs-Vererbungslehre* **87**, 462–81 (1958).
9. Sharma, R. P. & Chopra, V. L. *Devl. Biol.* **48**, 461–465 (1976).
10. Lawrence, P. A. & Morata, G. *Devl. Biol.* **50**, 321–337 (1976).
11. Kornberg, T., in preparation.
12. Gloor, H. *Arch. Julius-Klaus-Stift. VererbForsch*, **25**, 38–44 (1950).
13. Meinhardt, H. *J. Cell Sci.* **23**, 117–139 (1977).
14. Bryant, P. J. *Ciba Fdn Symp.* **29**, 71–93 (1975).
15. Wright, D. & Lawrence, P. A., in preparation.
16. Lawrence, P. A. in *Developmental Systems: Insects* (eds Counce, S. & Waddington, C. H.) 157–209 (Academic, London, 1973).
17. Jürgens, G. & Gateff, E. *Wilhelm Roux Arch.* **186**, 1–25 (1979).
18. Meinhardt, H. & Gierer, A. *J. Cell. Sci.* **15**, 321–346 (1974).
19. Kaufmann, S. A., Shymko, R. M. & Trabert, K. *Science* **199**, 259–270 (1978).
20. Lindsley, D. & Grell, E. H. *Genetic Variations of Drosophila melanogaster* (Carnegie, Washington, 1968).
21. Wright, T. R. F. *Drosoph. Inf. Serv.* **45**, 140 (1970).
22. Vander Meer, J. *Drosoph. Inf. Serv.* **52**, 160 (1977).

们估计我们已经鉴定出了二号染色体上几乎所有的分节基因座，并且完成了对三号、一号染色体上各 50% 的分节基因座的鉴定。所以我们鉴定的 15 个分节基因座可以代表果蝇基因组中影响分节的大部分基因座。因此，鉴定出参与果蝇分节模式形成复杂过程的所有遗传组分是可行的。

我们感谢希尔德加德·克鲁丁出色的技术支持，感谢阿德尔海德·施奈德、玛丽亚·韦伯、加里·斯特鲁尔在突变体筛选各阶段的帮助，感谢格尔德·于尔根斯富有启发性的讨论和来自剑桥、弗赖堡、海德堡、苏黎世等地的果蝇实验室的同事们对手稿的关键性评论，感谢克劳斯·克里斯坦森在图像处理方面的工作。托马斯·科恩伯格和加里·斯特鲁尔分别赠予“锯齿”和“无翅”的致死等位基因便于我们鉴定等位基因。如有需求，所有突变体均可提供。

附加说明：所有已知的“桶状”等位基因都不能互补“多毛”突变[20]，这意味着“桶状”突变体是“多毛”基因座的等位基因。

（李梅 翻译；沈杰 审稿）

Establishment in Culture of Pluripotential Cells from Mouse Embryos

M. J. Evans and M. H. Kaufman

Editor's Note

British biologists Martin Evans and Matt Kaufman were the first to isolate and culture stem cells. Here they describe the process, which yielded mouse stem cells capable of self-renewal and differentiation into other cell types. Careful culture conditions and the use of dormant early embryos aided their success, and the study's impact has been enormous. Since researchers demonstrated that the DNA changes in genetically modified stem cells could be passed through the germline of chimaeric animals, thousands of genes have been "knocked out" in mouse models, shedding light on development and disease. Stem cells enabled reproductive cloning to become reality, and are currently being studied for regenerative medicine. In 2007 Evans received a Nobel Prize for his work.

Pluripotential cells are present in a mouse embryo until at least an early post-implantation stage, as shown by their ability to take part in the formation of chimaeric animals[1] and to form teratocarcinomas[2]. Until now it has not been possible to establish progressively growing cultures of these cells *in vitro*, and cell lines have only been obtained after teratocarcinoma formation *in vivo*. We report here the establishment in tissue culture of pluripotent cell lines which have been isolated directly from *in vitro* cultures of mouse blastocysts. These cells are able to differentiate either *in vitro* or after inoculation into a mouse as a tumour *in vivo*. They have a normal karyotype.

PREVIOUS attempts to obtain cultures of pluripotential cells directly from a mouse embryo have been unsuccessful[3,4] although cells with a similar appearance have been reported to be present transiently[5,6]. We considered that success might depend on three critical factors: (1) the exact stage at which pluripotential cells capable of growth in tissue culture exist in the embryo; (2) explantation of a sufficiently large number of these precursor cells from each embryo; and (3) tissue culture in conditions most conducive to multiplication rather than differentiation of these embryonic cells. These considerations have been discussed at greater length elsewhere[7]. An indication of the optimal stage of embryonic development might be gained by a comparison of the properties of embryonic cells at various stages with established cultures of embryonal carcinoma (EC) cells. Cell-surface antigen expression and the patterns of protein synthesis revealed by two-dimensional electrophoresis have suggested that neither the cells of the $6\frac{1}{2}$-day ectoderm nor those of the $3\frac{1}{2}$-day inner cell mass show homology with EC cells, but that epiblast

小鼠胚胎多能细胞培养体系的建立

埃文斯，考夫曼

编者按

英国生物学家马丁·埃文斯和马特·考夫曼率先分离并培养干细胞。在本文中，他们描述了这个产生小鼠干细胞的过程，该小鼠干细胞能够进行自我更新并分化成其他细胞类型。严格的培养条件和休眠的早期胚胎的使用促成了他们的成功，而且这项研究非常有影响力。自从研究人员证实在遗传修饰的干细胞中发生的DNA改变可以通过嵌合动物的生殖细胞系来遗传,此后数以千计的基因都在小鼠模型中被“敲除”，这些为发育和疾病的研究提供了便利。干细胞使得生殖性克隆成为现实，并且目前人们在再生医学方面正在进行干细胞研究。2007年，埃文斯凭借他的研究获得了诺贝尔奖。

多能细胞在小鼠胚胎中一直存在，至少到着床后阶段的早期才消失，这是因为实验显示它们具有参与形成嵌合动物[1]和畸胎癌[2]的能力。迄今为止还不能在体外建立这些细胞的持续培养体系，而且只有在体内形成畸胎癌后才能获得细胞系。本文中我们报道了直接从体外小鼠囊胚培养物中分离出来的多能细胞系组织培养体系的建立。这些细胞能够在体外分化，或者作为一个肿瘤接种到小鼠体内后发生分化。它们具有正常的核型。

尽管已有报道称具有类似多能细胞表型的细胞能够短暂存活于体外[5,6]，但是以前试图直接从小鼠胚胎中获得多能细胞的培养都没有成功[3,4]。我们认为成功培养可能取决于三个关键因素：(1) 胚胎中能够在组织培养中生长的多能细胞所处的确切阶段；(2) 从每个胚胎中外植足够多的前体细胞；(3) 组织培养的条件要最适合于诱导这些胚胎细胞的增殖，而不是分化。这些观点在其他文献中有篇幅更长的讨论[7]。通过比较不同阶段的胚胎细胞与已经建立的胚胎癌（EC）细胞培养物的特性就可以获得胚胎发育最佳阶段的提示。通过二维电泳显示的细胞表面抗原的表达和蛋白合成的模式表明，6.5天的外胚层细胞与3.5天的内细胞团都不与EC细胞具有同源性，但是交配后5.5天的早期着床后胚胎的上胚层细胞可能与EC细胞具有同源性[8]（发现交配栓的时间定义为0.5天）。早期着床后胚胎的细胞似乎是用于多能细胞培养的

cells of the early post-implantation embryo at $5\frac{1}{2}$ days post coitum may do so[8] (the day of finding coital plug is termed day $\frac{1}{2}$). Cells from embryos of an early post-implantation stage seem to be the best candidates for direct progenitors of pluripotential cells in culture. As these embryos are difficult to isolate, and as the cell number in the isolated epiblast is small, we chose an alternative route to obtain embryo cells at this stage of development.

Mouse blastocysts may be induced to enter a state of diapause just before implantation. This delay in implantation depends on the maternal hormonal conditions, and may be induced experimentally by ovariectomy at an appropriate stage[9]. Embryos in implantational delay hatch from the zona but remain free-floating in the uterine lumen. A gradual increase in cell number occurs[10], and the primary endoderm may be formed but no further development takes place until implantation occurs, under the control of hormonal stimuli.

129 SvE mice were caged in pairs and examined for mating plugs each morning. They were ovariectomized on the afternoon of day $2\frac{1}{2}$ of pregnancy, injected subcutaneously with 1 mg Depo–Provera (Upjohn), and delayed blastocysts were recovered 4–6 days later. The blastocysts were cultured intact in groups of about six embryos in small drops of tissue culture medium under paraffin oil on tissue culture plastic Petri dishes for 4 days. The blastocysts attached within 48 h and the trophectoderm cells grew out and differentiated into giant trophoblast cells. The inner cell mass cells subsequently developed into large egg cylinder-like structures, with a group of small round cells surrounded by endodermal cells growing attached to the Petri dish. The egg cylinder-like structures were picked off the dish, dispersed by trypsin treatment and passaged on to gelatin-pretreated Petri dishes containing mitomycin C-inactivated STO fibroblasts. All culture was carried out in Dulbecco's modified minimal essential medium supplemented with 10% fetal calf serum and 10% newborn calf serum. The cultures were examined daily and passaged by trypsinization every 2–3 days. Actively proliferating colonies of cells closely resembling EC cells were apparent from an early stage. These colonies were picked out, passaged and mass cultures grown. The cell cultures had the appearance and general growth characteristics of feeder-dependent EC cells (Fig. 1).

直接前体细胞的最佳选择。由于这些胚胎很难分离，而且分离出的上胚层细胞非常少，我们选择了另一种方法来获得这个发育阶段的胚胎细胞。

小鼠的囊胚可以在即将着床时被诱导进入休眠状态。这种着床延迟取决于母体激素水平。而且可以在合适的阶段通过卵巢切除，实现着床延迟的实验性诱导[9]。着床延迟的胚胎从透明带中孵出，但是在子宫腔内仍然保持自由漂浮状态。在激素刺激的控制下，细胞的数量逐渐增多[10]，原内胚层可能会形成，但是着床之前不会有进一步发育。

将 129 SvE 小鼠成对关在笼子里，每天早晨查看是否有交配栓形成。在妊娠 2.5 天的下午切除它们的卵巢，皮下注射 1 毫克狄波－普维拉醋酸甲羟孕酮注射液（普强公司），延迟发育的囊胚在 4~6 天后恢复。以大约 6 个胚胎为一组将囊胚完整地在少量组织培养基中培养 4 天，培养基盛放在组织培养用的塑料培养皿中，上面覆盖石蜡油。囊胚在 48 小时内贴壁，滋养外胚层细胞长出并分化成巨大的滋养层细胞。内细胞团细胞随后发育成大的近似卵筒结构，并伴随一群贴在培养皿上生长的由内胚层细胞包围的小圆形细胞。将近似卵筒结构挑出培养皿，用胰蛋白酶处理使其分散，并在明胶预处理的含有丝裂霉素 C 灭活的 STO 成纤维细胞的培养皿中进行传代。所有的培养都使用杜尔贝科改良的最低必需培养基加上 10% 的胎牛血清和 10% 的新生牛血清。每天检查培养物，每 2~3 天用胰蛋白酶消化进行传代培养。从早期阶段开始，非常类似于 EC 细胞的活跃增殖的细胞集落就很明显了。将这些集落挑出、传代并大规模培养。这些细胞培养物具备饲养层依赖的 EC 细胞的外观和一般生长特征（图 1）。

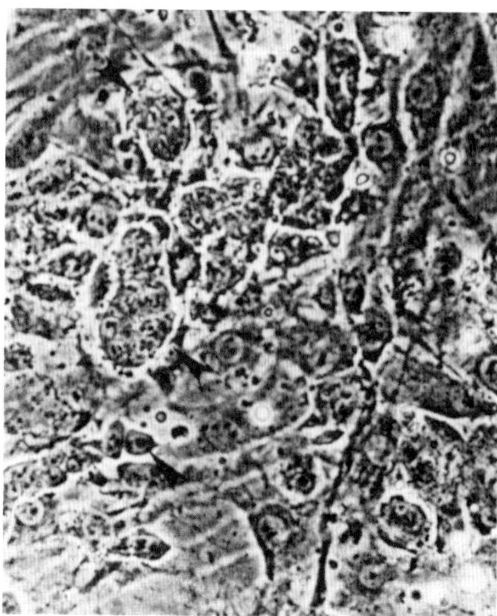

Fig. 1. Groups of pluripotential embryo cells (arrowed) growing in monolayer culture on a background of mitomycin C-inhibited STO cells. The isolation of a definite cell line from a blastocyst takes only ~3 weeks and the pluripotential cell colonies are visible within 5 days of passage. We have had 30% yield of lines from blastocysts in one experiment. Two of the lines have been rigorously cloned by single-cell isolation but most were only colony-picked—this makes no difference.

The embryos used to initiate these cultures are from normal 129 SvE strain mice, that is, from the same strain of mice as many EC cell lines, in particular those grown in this laboratory. Therefore it was important to exclude any possibility of contamination of these cultures with EC cells from established cell lines. Cell cultures were established from different embryos in three separate experimental series, but the best indication of their separate identity came from their karyotype. Cultures were initiated from 6–12 embryos, thus it might be expected that both male and female cells should be present. None of the 129 embryonal carcinoma cell lines in this laboratory have a normal karyotype, and, in particular—in common with most available embryonal carcinoma cell lines—they do not contain a Y chromosome. These embryo-derived cells have a completely normal karyotype. An XY karyotype is shown in Fig. 2. Three additional cell lines have been analysed; two of these are normal 40XX and one is normal 40XY. We have termed these directly embryo-derived cells EK to distinguish them from EC cells. EK cells grow rapidly in culture and have been maintained for over 30 passages *in vitro*.

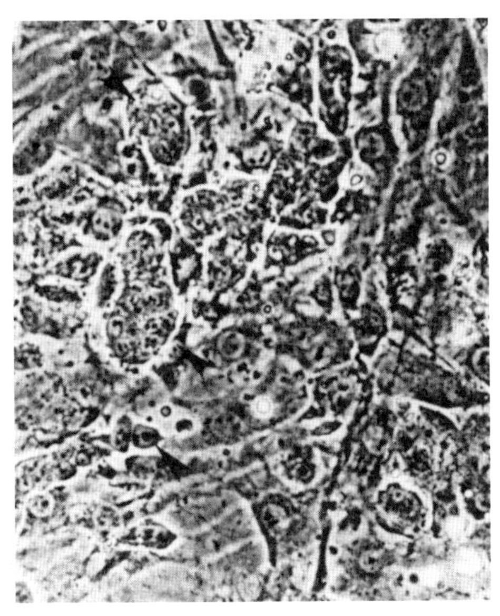

图 1. 在丝裂霉素 C 抑制的 STO 细胞背景中，呈单层生长的多能胚胎细胞群（箭头）。只需要大约 3 周就可从囊胚中分离到一种特定的细胞系，而且在传代 5 天之内就能观察到多能细胞集落。在一次实验中，我们已经从囊胚中获得了 30% 的细胞系。其中两个细胞系已经通过单细胞分离严格地克隆出来，但是多数细胞系仅从集落中挑出——这没有太大的差别。

用于起始培养的胚胎均来自正常的 129 SvE 种系小鼠，也就是说，这些胚胎像许多 EC 细胞系尤其是本实验室培养的那些细胞系一样，来源于同一种系的小鼠。因此排除任何来自已建立的细胞系的 EC 细胞污染是非常重要的。在三个单独进行的实验系列中细胞培养物都是从不同的胚胎中建立的，但是核型才是鉴别它们各自特性的最好指标。培养物起始于有 6~12 个胚胎，因此可以预计，雄性和雌性细胞同时存在。本实验室使用的 129 个胚胎癌细胞系都没有正常的核型，而且尤其是，与大部分可获得的胚胎癌细胞系一样，它们都没有 Y 染色体。这些胚胎来源的细胞具有完全正常的核型。图 2 中显示了 XY 核型。我们还分析了另外三个细胞系，其中两个是正常的 40XX，另一个是正常的 40XY。我们将这些直接胚胎来源的细胞称为 EK 细胞，以区别于 EC 细胞。EK 细胞在培养基中生长非常迅速，而且在体外能够维持传代达 30 代以上。

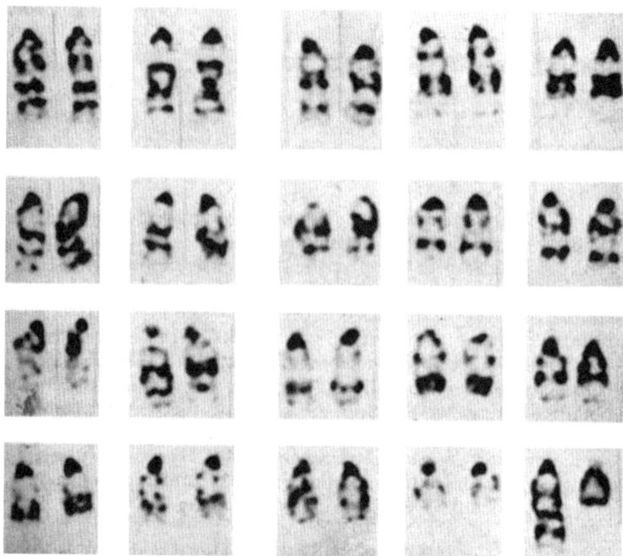

Fig. 2. Karyotype of an embryo-derived pluripotential cell line, 40XY. Over 80% of the spreads of this clonal line possessed 40 chromosomes and had a clearly identifiable Y chromosome.

Cultures of EK cells were collected by trypsinization, and $\sim 10^6$ cells injected subcutaneously into the flank of syngeneic male mice. Tumours grew in all cases, and histological examination of these revealed that they were teratocarcinomas. When the EK cells were passaged without feeder cells they formed embryoid bodies which, when kept in suspension, became cystic. Embryoid bodies allowed to attach to a Petri dish spread out and differentiated in the usual way into a complex of tissues. Preliminary observations indicate that, like early ectoderm cells of the mouse embryo and EC cells, EK cells carry the cell-surface antigens recognized by M1-22-25 (Forssman)[8,11] and anti-I Ma (lacto-*N*-iso-octaosyl ceramide)[12,13] and also that two dimensional gel electrophoretic separations of their proteins very closely resemble those of the EC cell line PSMB.

We have demonstrated here that it is possible to isolate pluripotential cells directly from early embryos and that they behave in a manner equivalent to EC cells isolated from teratocarcinomas. The network of inter-relationships between the mouse embryo and pluripotential cells derived from it has previously lacked only the direct link between the embryo and cells in culture for completion. We have now demonstrated this (Fig. 3).

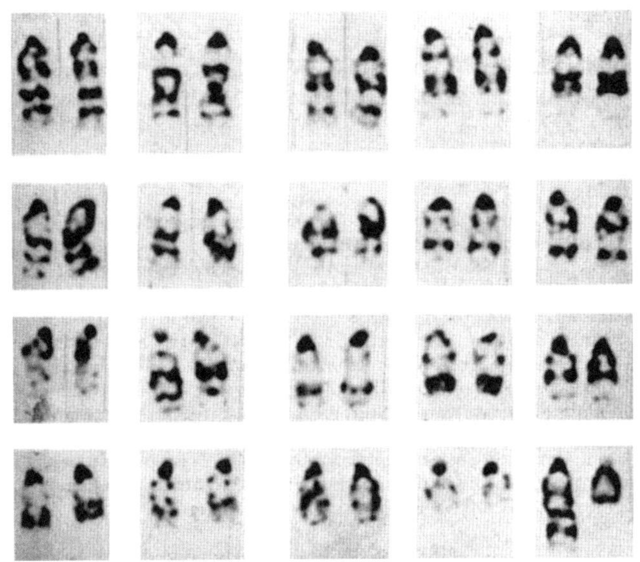

图 2. 胚胎来源的多能细胞系的核型，40XY。这种克隆细胞系超过 80% 的细胞涂片都具有 40 条染色体，并具有清晰可辨的 Y 染色体。

通过胰蛋白酶消化收集培养的 EK 细胞，然后将大约 10^6 个细胞皮下注射到同系雄性小鼠的侧腹部。所有实验小鼠均长出肿瘤，而且组织学检查显示它们都是畸胎癌。如果在没有饲养细胞的情况下对 EK 细胞进行传代，它们就会形成拟胚体，如果保持悬浮培养就会成为囊性胚体。能够黏附到培养皿上的拟胚体就会迅速铺展并按照正常的方式分化成复杂的组织。初步观测结果表明，就像小鼠胚胎的早期外胚层细胞和 EC 细胞一样，EK 细胞携带的细胞表面抗原能够被 M1–22–25（福斯曼）[8,11] 和抗 I–Ma（乳酰–*N*–异 –二十八烷基神经酰胺）[12,13] 识别，而且 EK 细胞蛋白质的二维凝胶电泳分离结果与 EC 细胞系 PSMB 的结果非常接近。

本文中我们已经证明有可能直接从早期胚胎中分离出多能细胞，而且它们的行为方式与从畸胎癌中分离出来的 EC 细胞相同。之前，关于小鼠胚胎和其来源的多能细胞之间完整的相互关系网络仅缺乏胚胎和培养细胞之间的直接联系，我们现在展示这种联系（图 3）。

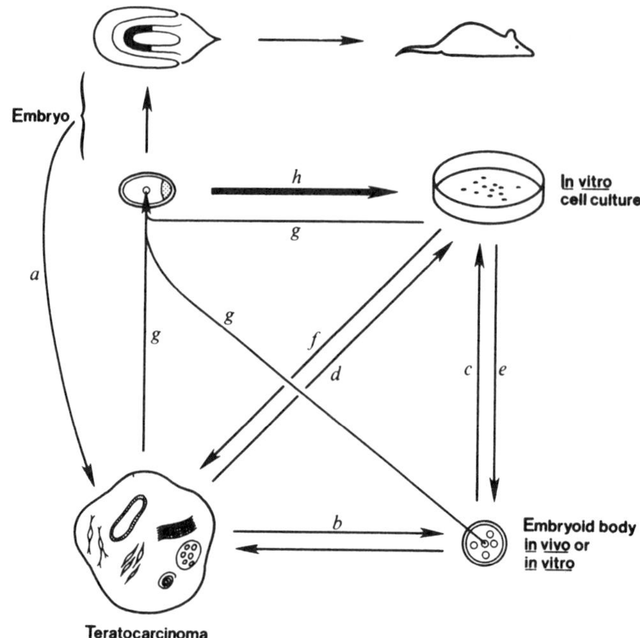

Fig. 3. Inter-relationships of cell lines, teratocarcinomas and embryoid bodies with normal mouse embryos. Arrows indicate routes of cell transfer: *a*, formation of teratocarcinoma by ectopic implantation of embryos; *b*, formation of embryoid bodies from teratocarcinoma and vice versa; *c*, derivation of cell culture from embryoid bodies; *d*, cell culture obtained directly from solid tumours; *e*, differentiation to embryoid bodies from culture; *f*, formation of solid tumours on reinjection of cells from culture; *g*, transfer of embryonal carcinoma cells either from cell culture or from the core of an embryoid body or from a solid tumour back to a blastocyst. All these procedures may result in chimaerism of the resulting mouse; *h*, the missing link supplied here.

Teratocarcinoma cells are now being widely used as a model for the study of developmental processes of early embryonic cell commitment and differentiation. Their use as a vehicle for the transfer into the mouse genome of mutant alleles, either selected in cell culture or inserted into the cells via transformation with specific DNA fragments, has been presented as an attractive proposition. In many of these studies the use of pluripotential cells directly isolated from the embryos under study should have great advantages. We have now shown that these EK cell lines are readily established from cultures of single blastocysts and so far have 15 lines of independent embryonic origin, some of which have been isolated from non-129, outbred mouse stocks. We are now studying the chimaeric mice formed from these cells.

We thank Mrs A. Burling for technical assistance and Dr E. P. Evans for advice regarding karyotype analysis. M.J.E. and M.H.K. were supported by the MRC; M.J.E. also received support from the Cancer Research Campaign.

(**292**, 154-156; 1981)

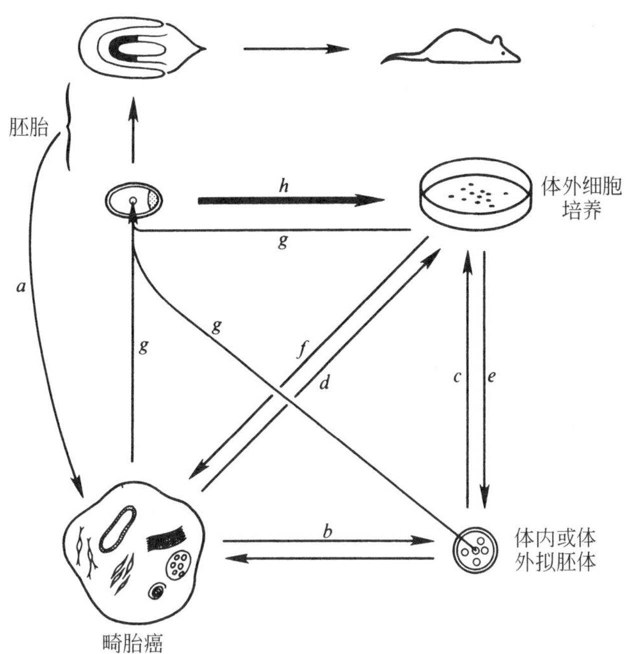

图 3. 细胞系、畸胎瘤和拟胚体与正常小鼠胚胎之间的相互关系。箭头表示细胞移植的途径：*a*，胚胎的异位植入形成畸胎瘤；*b*，从畸胎瘤形成拟胚体，反之亦然；*c*，从拟胚体衍生出细胞培养物；*d*，直接从实体瘤获得细胞培养物；*e*，培养细胞分化成拟胚体；*f*，重新注射培养细胞形成实体瘤；*g*，将细胞培养物、拟胚体核心或者实体瘤来源的胚胎癌细胞移植回囊胚中，所有的这些步骤都可能导致小鼠的嵌合型；*h*，本文获得的先前未发现的联系。

畸胎癌细胞目前作为模型广泛用于研究早期胚胎细胞定向和分化的发育过程。它们的用途是作为载体将突变的等位基因转移到小鼠的基因组中，无论是从细胞培养物中选择突变的等位基因还是通过特定 DNA 片段的转化将其插入细胞，这已经是目前具有吸引力的主题。在这一领域的很多课题中，使用从胚胎中直接分离出来的多能细胞进行研究具有很大的优势。我们已经表明从单个囊胚的培养物中能够容易地建立 EK 细胞系，而且迄今为止已经拥有 15 个独立胚胎来源的细胞系，其中的一些已经从非 129 的远交小鼠品系中分离出来了。我们目前正在研究这些细胞形成的嵌合小鼠。

感谢伯林女士的技术支持和埃文斯博士关于核型分析的建议。埃文斯和考夫曼都得到了医学研究理事会的资助；埃文斯还得到了英国癌症研究运动的支持。

（毛晨晖 翻译；梁前进 审稿）

M. J. Evans* **and M. H. Kaufman**†
* Department of Genetics, University of Cambridge, Downing Street, Cambridge CB2 3EH, UK
† Department of Anatomy, University of Cambridge, Downing Street, Cambridge CB2 3EH, UK

Received 6 February; accepted 14 April 1981.

References:

1. Gardner, R. L. & Papaioannou, V. E. in *The Early Development of Mammals* (eds Balls, M. & Wild, A. E.) 107-132 (Cambridge University Press,1975).
2. Stevens, L. C. *Devl. Biol.*, **21**, 364-382 (1970).
3. Cole, R. J. & Paul, J. in *Preimplantation Stages of Pregnancy* (eds Wolstenholme, G. E. W. & O'Connor, M.) 82-122 (Churchill, London,1965).
4. Sherman, M. I. *Cell* **5**, 343-349 (1975).
5. Atienza-Samols, S. B. & Sherman, M. I. *Devl. Biol.* **66**, 220-231 (1978).
6. Solter, D. & Knowles, B. *Proc. Natl. Acad. Sci. U.S.A.* 72, 5099-5102 (1975).
7. Evans, M. J. *J. Reprod. Fert.* **62**, 625-631 (1981).
8. Evans, M. *J.*, Lovell-Badge, R. H., Stern, P. L. & Stinnakre, M. -G. INSERM *Symp.* **10**, 115-129 (1979).
9. McLaren, A. *J. Endocr.* **50**, 515-526 (1971).
10. Kaufman, M. H. in *Progress in Anatomy* Vol. 1 (eds Harrison, R. J. & Holmes, R. L.) 1-34 (Cambridge University Press, 1981).
11. Stern, P. L. *et al. Cell* **14**, 775-783 (1978).
12. Feizi, T. *Blood Transfusion Immunohaemat.* **23**, 563-577 (1980).
13. Kapadia, A., Feizi, T. & Evans, M. J. *Expl Cell Res.* **131**, 185-195 (1980).

Test-tube Babies, 1981

R. G. Edwards

Editor's Note

Robert Edwards, in collaboration with Patrick Steptoe in England, enabled the first birth of a baby by *in vitro* fertilization (IVF), Louise Brown, in 1978. Their research had been considered controversial, and the birth was met with a mixture of surprise, excitement and dismay. As the title of this article by Edwards indicates, infants born by IVF were almost immediately dubbed "test-tube babies", even though only the fertilization of the egg by sperm and very initial growth of the embryo were conducted "in glass". Edwards implies that by 1981 the technique was already becoming routine, with 15–20 IVF babies born in the UK that year. In that sense, this paper records the "normalization" of this form of assisted conception.

Between fifteen and twenty babies will be born this year after the *in vitro* fertilization of human eggs. Many of the essential steps now have high rates of success, including the recovery of preovulatory oocytes, and fertilization and embryo cleavage *in vitro*. Implantation of the embryo following its replacement in the mother remains the major difficulty. Some implications of the work are discussed.

THIS year should prove a turning point for the birth of children by the fertilization of human eggs *in vitro*. Between 15 and 20 such babies will be born in approximately equal numbers in the United Kingdom and Australia, and there will be one or two elsewhere. These methods will be introduced in many countries, primarily to alleviate human infertility. Fundamental aspects of human conception will be analysed and increasing debate will presumably be given to genetic engineering. This is, therefore, an appropriate time to assess the relevant clinical and scientific issues raised by this work[1].

The First Essential: Timing of Ovulation

I will first discuss the methods involved in timing ovulation for the collection of preovulatory oocytes.

Harvesting preovulatory oocytes is the first of several steps essential for obtaining human embryos. They must be collected during their final stages of maturation just before ovulation occurs, when meiosis is advanced and cortical granules have established the defence against polyspermy. Follicular growth and ovulation must be regulated by endocrine therapy, or natural ovulation must be closely predicted during the menstrual cycle.

试管婴儿，1981

爱德华兹

编者按

1978 年，在英国科学家罗伯特·爱德华兹及其同事帕特里克·斯特普托的努力下，第一例体外受精（IVF）的婴儿——路易丝·布朗诞生了。他们的研究曾被认为具有争议性，人们对试管婴儿的诞生表现出惊喜、兴奋和惊慌交加的复杂感情。正如爱德华兹的这篇文章标题所示，尽管只有精子和卵细胞的受精作用和非常早期的胚胎的生长是在“玻璃（试管）”中进行，但通过体外受精诞生的婴儿却几乎立即就被称为“试管婴儿”。爱德华兹暗示，截至 1981 年，这项技术已经成为常规技术，当年在英国有 15~20 例试管婴儿诞生。从这个意义上考虑，本文记录了这种辅助受孕形式的“常态化”。

这一年将有 15 到 20 名婴儿通过人类卵细胞的体外受精方式诞生。很多关键步骤目前都有很高的成功率，包括排卵期前的卵母细胞的收集以及体外的受精及胚胎卵裂。主要的困难仍然在于胚胎移入母体后的着床。本文还对这项研究的一些应用进行了讨论。

这一年应该是通过人类卵细胞体外受精诞生婴儿的转折点。在英国和澳大利亚，将有 15 到 20 名婴儿通过这种方式来到人世，两个国家的数目相近，其他国家也有 1~2 名。很多国家会引入这种技术，主要用于减轻人类不育症。影响人类怀孕的基本因素将得到分析，而且针对基因工程的争论想必也将越来越多。因此，是时候来评价这项技术带来的相关临床和科学问题了 [1]。

第一项要素：排卵时间

首先我将讨论计算排卵时间以便收集排卵期前的卵母细胞的方法。

获得人类胚胎的几个关键步骤中首要步骤是收集排卵期前的卵母细胞。卵母细胞必须刚好在排卵前成熟的最后阶段进行收集，这时候减数分裂还在进行，皮质颗粒已经形成了抵御多精入卵的屏障。卵泡生长以及排卵必须通过激素疗法来调节，或者必须准确预测月经周期中的自然排卵。

The regulation of ovulation is undoubtedly easier. Several follicles can be primed using human menopausal gonadotropin (HMG) or clomiphene early in the menstrual cycle. An endogenous surge of luteinizing hormone (LH) will then induce ovulation. Alternatively, a single injection of human chorionic gonadotropin (HCG, 5,000 IU) can be given between days 11 and 14, according to the follicular response of each patient. Levels of urinary oestrogens of 80–100 μg per day (refs 2, 3), or follicular diameters of 1.5–2 cm measured by ultrasound[4–6] are believed to be appropriate indications to inject HCG. Ovulation can be induced at any desired time of day or night, a considerable help in organizing laboratory or surgical teams for oocyte recovery. Two or more preovulatory oocytes can be collected from many patients, another advantage of stimulating the ovary.

There may be problems to offset these advantages. Wide variations exist in the rate of growth of individual follicles in each patient, revealed by the different levels of steroids in follicular fluids[7] and by variations in embryonic growth when superovulation techniques are applied to animals. Some patients fail to respond to clomiphene. Others produce increasingly large amounts of oestrogens as several follicles grow, and their endogenous LH surge stimulates ovulation. A difficult situation occurs in patients with moderate or high levels of oestrogens and no endogenous surge of LH (Fig. 1). As in other tissues[8–11], clomiphene may have depleted cytoplasmic oestrogen receptors in the pituitary gland of such patients over several days, so preventing the LH surge in response to rising levels of oestrogens. HCG must be given at some arbitrary time, before follicles become atretic, yet while it is uncertain if the patient will have her own endogenous LH surge. Ovarian stimulation can also distort the menstrual cycle, inducing a short luteal phase and a disorganized endometrium, both incompatible with establishing pregnancy[12]. An average of eight cycles of treatment with clomiphene is needed for oligomenorrhoeic women to conceive naturally[3]. Such disadvantages may be greater with HMG than with clomiphene.

毫无疑问，调节排卵更容易些。在月经周期早期，使用人绝经期促性腺激素（HMG）或者克罗米芬能够促发多个卵泡。随后内源性黄体生成素（LH）峰就能诱导排卵。另一种方法就是，根据每个患者的卵泡反应，在月经周期的第 11 到 14 天期间单次注射人绒毛膜促性腺激素（HCG，5,000 IU）。尿雌激素水平达到每天 80~100 微克（参考文献 2、3）或者超声波测量卵泡直径达到 1.5~2 厘米时 [4-6]，被认为是注射 HCG 的合适时间。诱导排卵可以安排在白天或者晚上任何理想的时间，这对组织实验室或者手术小组进行卵母细胞收集非常有利。刺激卵巢的另一个优点是，很多患者都可以采集到两个或者更多排卵期前的卵母细胞。

但仍然可能会有一些问题抵消这些优点。个体卵泡液中类固醇水平的不同 [7] 以及对动物使用超排卵技术时胚胎生长的差异显示，每个患者各个卵泡的生长速率差异很大。有一些患者对克罗米芬没有反应；另一些人随着数个卵泡生长产生越来越多的雌激素，并且其内源性 LH 峰刺激了排卵。一种很难处理的情况是患者产生中到高水平的雌激素，但是没有内源性的 LH 峰（图 1）。正如在其他组织中一样 [8-11]，克罗米芬可能在数天内耗尽这些患者脑垂体细胞质中的雌激素受体，从而阻止响应雌激素水平增高的 LH 峰的出现。HCG 必须在卵泡闭锁之前某个任意时间使用，但是很难确定患者是否会有自发产生的内源性 LH 峰。卵巢刺激也会使月经周期紊乱，导致黄体期很短以及子宫内膜紊乱，这两者均对妊娠建立不利 [12]。通常使月经过少的女性自然怀孕平均需要 8 个周期的克罗米芬治疗 [3]。与使用克罗米芬相比，使用 HMG 时这种缺点更为明显。

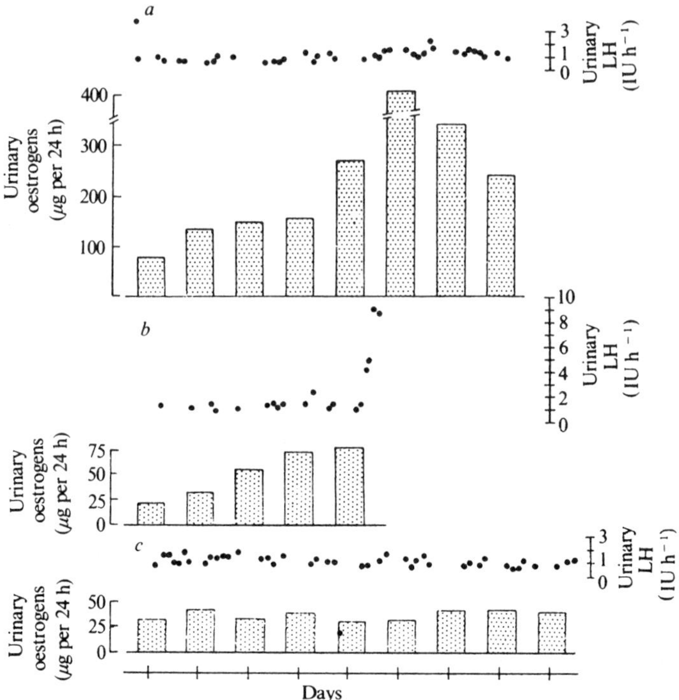

Fig. 1. Response of patients to clomiphene. *a*, Levels of urinary oestrogens rose considerably, and then fell in the absence of an endogenous LH surge. *b*, Rising levels of urinary oestrogens followed by surge of LH in urine. *c*, No response in urinary oestrogens and no surge □, Urinary oestrogens; ●, urinary LH, measured by the Hi-Govanis kit.

The alternative is to monitor the approach of ovulation during the natural menstrual cycle, and then aspirate the single preovulatory oocyte. This method was used for the conception of the first child by fertilization *in vitro*[12], and is widely used in our practice today. Plasma or urinary oestrogens are used to assess follicle growth, and the onset of the LH surge at mid-cycle provides a warning of the approach of ovulation.

The disadvantages are obvious. There is usually only one preovulatory oocyte, and any pathological conditions in the ovary or abdomen will limit the chances of collecting it. The LH surge is less predictable than an injection of HCG, hence more difficulty arises in organizing laboratory and clinical staff to collect the oocyte. Some difficulties have proved less than feared. Repeated blood sampling may provide a reliable guide for assaying LH[6,13–15], but is unnecessary, because urine samples are suitable for rapid assays of LH using the kit Hi-Gonavis (Mochida Pharmaceuticals)[12] and for oestrogens. Fortunately, the LH surge begins in early morning in almost three-quarters of the women (Fig. 2), which is most convenient for the collection of preovulatory oocytes 24–28 h later. This "critical period" in the LH surge in women resembles the situation in rats[16,17], and its distinct diurnal component must modify the feedback effects of ovarian oestrogens and LH releasing hormone on the pituitary gland[18]. If women discharge their LH at other times of day, their oocytes must sometimes be aspirated less than 24 h later and cultured *in vitro* to complete maturation.

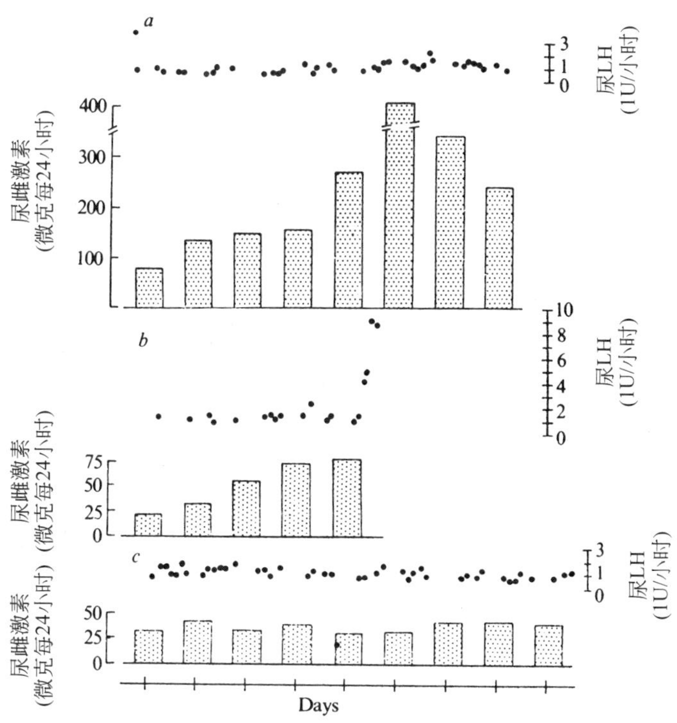

图 1. 患者对克罗米芬的反应。*a*，尿雌激素水平明显升高，然后无内源性 LH 峰出现时尿雌激素水平下降。*b*，尿雌激素水平逐渐增高，随后尿中出现 LH 峰。*c*，尿雌激素对克罗米芬没有响应，没有 LH 峰。□，尿雌激素；●，尿 LH，用 Hi-Govanis 试剂盒测量。

另一种方法是监测自然月经周期中即将排卵的时期，然后吸取出单个排卵期前的卵母细胞。第一个体外受精婴儿的孕育就是使用了这种方法 [12]，如今广泛运用于我们的临床实践。血浆或者尿中的雌激素水平被用于评价卵泡的生长状况，并且月经周期中期 LH 峰的出现提供了即将排卵的信号。

这种方法的缺点是很明显的。通常只有一个排卵期前的卵母细胞，卵巢或者腹腔的任何病理情况都将降低其采集的成功率。相对于注射 HCG 来说，LH 峰更难预测，因此在组织实验室和临床人员收集卵母细胞时将面临更多的困难。实际上，一些困难比想象中的要小一些。重复采血能可靠地检测 LH 的水平 [6,13-15]，但这没有必要，因为尿液样本很适合使用 Hi-Gonavis 试剂盒（日本持田制药）进行 LH 的快速检测 [12] 以及雌激素的快速检测。幸运的是，几乎四分之三的女性 LH 达峰的时间都在清晨（图 2），这非常有利于 24~28 小时后收集排卵前的卵母细胞。女性到达 LH 峰的这个“关键时期”与大鼠的情况非常相似 [16,17]，其独特的昼间成分必定改变了卵巢雌激素和促 LH 释放激素对垂体的反馈效应 [18]。如果女性在其他时间释放 LH，那么必须在 24 小时内吸取其卵母细胞并体外培养直至完全成熟。

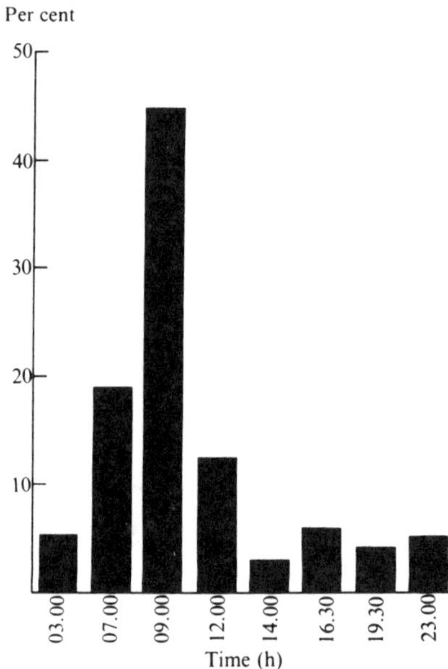

Fig. 2. Diurnal rhythm in the onset of the urinary LH surge in women. The time shows the initial increase in levels of LH which was followed by a sustained rise (see Figs 1*b* and 3).

Fertility may be higher during the natural cycle than after ovarian stimulation, an obvious advantage in establishing pregnancy. Several surveys have revealed a 1 in 4 chance of pregnancy during unprotected intercourse in any menstrual cycle, but even this low rate may be higher than during induced cycles. Clomiphene may be indicated in patients with irregular or prolonged cycles.

Aspiration of Oocytes

A double aspirating needle or two separate needles are used, one channel being used for aspiration and the other to flush out the follicle if the oocyte is not collected[19,20]. The flushing solution may contain heparin, to prevent clotting within the follicle.

The highest rates of collection are achieved 32 h after the injection of HCG, or 26 h after the rise of LH in urine (Table 1). It is essential to time the onset of ovulation correctly. Should HCG be given after an endogenous surge of LH, ovulation may occur before aspiration begins. Diurnal rhythms in tonic LH release, sometimes reaching low surge levels over a few hours, can confuse the correct timing of the LH surge (Fig. 3). Pelvic adhesions, hydrosalpinx, cystic follicles, endometriosis and other ovarian conditions impair the collection of oocytes, and preliminary laparoscopy may be required to alleviate these conditions. If the ovary is accessible, oocytes can be collected during the natural cycle from almost 90% of patients, laparoscopy being completed within a few minutes (Table 1). None of our patients had ovulated when laparoscopy was performed, and each of them had a large preovulatory follicle.

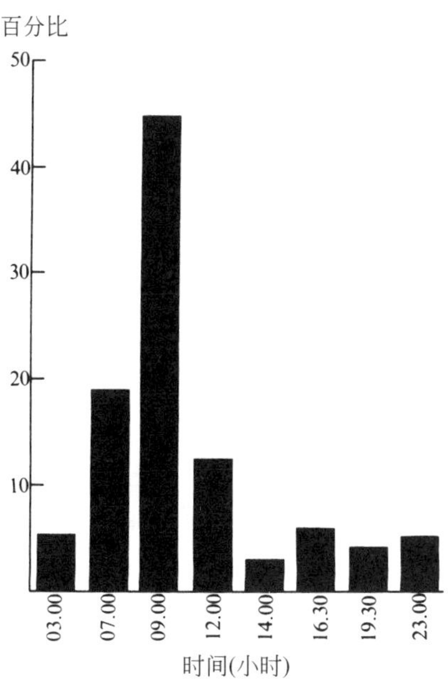

图 2. 女性尿 LH 达峰的昼夜节律。时间轴显示了 LH 水平开始增高并且保持持续增长（见图 1b 和 3）。

自然月经周期下的生殖力要比卵巢刺激后的更高，这在妊娠建立中是一个显著的优势。数个调查显示在月经周期的任何阶段进行无保护性交而怀孕的概率为 1/4，但即便是这个看似很低的概率也要高于诱导周期中的怀孕概率。月经不规律或者周期延长的患者可以使用克罗米芬。

卵母细胞的吸取

使用双腔吸引管或者两根单独的针管，一个用于吸取细胞，另一个用于在没有采集到卵母细胞时冲洗掉卵泡[19, 20]。冲洗液可以含有肝素，可以防止卵泡内凝血。

采集成功率最高的时间是在 HCG 注射后的第 32 小时，或者在尿液中 LH 含量升高后的第 26 小时（表 1）。准确计算排卵起始的时间是非常重要的。如果在内源性 LH 峰后注射 HCG，开始吸取卵母细胞之前排卵就可能发生。紧张性 LH 释放的昼夜节律性有时会在数小时内达到较低的峰值，而混淆了 LH 达峰的准确时间（图 3）。盆腔粘连、输卵管积水、囊状卵泡、子宫内膜异位以及其他卵巢疾病都会影响卵细胞的采集，可能需要用腹腔镜预处理以减轻这些症状。如果卵巢可直接到达的话，几乎 90% 的患者都可以在自然周期中采集到卵母细胞，利用腹腔镜在数分钟内就可以完成这个过程（表 1）。没有一个患者在腹腔镜操作过程中排卵，而且所有患者都有较大的排卵前卵泡。

Table 1. Aspiration of oocytes from preovulatory follicles

	Natural cycle*	Natural and clomiphene cycle[20]	Clomiphene cycle[6]
No. of patients	122	–	–
No. with accessible ovaries	109	–	–
No. of preovulatory follicles aspirated	109	172	107
No. (and %) of oocytes collected	95 (88%)	110 (64%)	96 (89%)

* Recent series of 122 patients in Bourn Hall, Cambridge. Their average age was 33.9 yr.

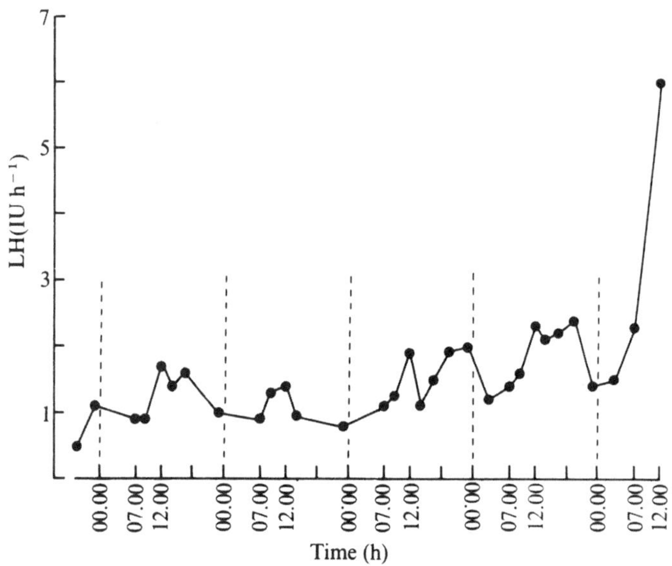

Fig. 3. Diurnal variations in the levels of urinary LH in women. "Tonic" levels rise each morning, then decline. The LH surge began during the morning in this patient.

Preovulatory oocytes can be identified quickly. They are embedded in a viscous follicular fluid, to a diameter of 0.5 cm or more and can be seen by eye[2]. The viscosity of the cumulus mass serves as a guide to the stage of maturity of the oocyte. A few hours in culture in the presence of some follicular fluid helps to complete maturation, especially in those deliberately collected less than 26 h after the LH rise, which appear "unripe".

Fertilization and Cleavage *In Vitro*

Provided there are no pathological conditions in the husband's spermatozoa, almost 90% of the preovulatory oocytes aspirated during the natural cycle can be fertilized *in vitro* (Table 2). The conditions of culture are rapidly becoming standardized[3,6,7,12]. Simple media are used, including Earle's solution with pyruvate, a medium designed to support mouse embryos *in vitro*[21,22], and Ham's F10, each supplemented with serum albumin or homologous human serum (7.5–8.6%). We prefer Earle's solution containing pyruvate,

表 1. 从排卵前卵泡内吸取卵母细胞

	自然周期 *	自然和克罗米芬周期[20]	克罗米芬周期[6]
患者总数	122	–	–
可到达的卵巢数	109	–	–
吸取的排卵前卵泡数	109	172	107
采集卵母细胞数目（和 %）	95（88%）	110（64%）	96（89%）

* 剑桥波恩诊所最近的 122 名患者，平均年龄 33.9 岁。

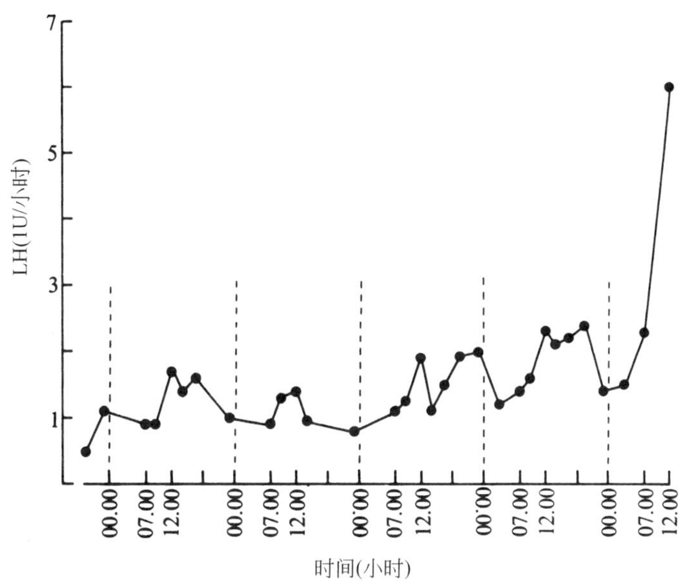

图 3. 女性尿 LH 水平的昼夜变化。“紧张性” LH 水平在每天早上升高，随后又降低。该患者 LH 峰出现在早晨。

排卵前卵母细胞可以被迅速识别。它们包被在黏性卵泡液内，直径达到 0.5 厘米或更大并可用肉眼观察到[2]。周围聚集的卵丘细胞的黏稠度可以指示卵母细胞成熟的阶段。在卵泡液存在的条件下对卵母细胞进行数小时的培养有助于其完全成熟，尤其是那些特意在 LH 升高 26 小时以内时采集的看起来“未成熟”的卵母细胞。

体外受精和体外卵裂

假设丈夫的精子没有任何病理状况，在自然周期内吸取的 90% 排卵前卵母细胞都可以在体外完成受精(表 2)。培养条件正在迅速标准化[3,6,7,12]。使用简单的培养液，包括含丙酮酸盐的厄尔氏溶液，一种设计用于体外培养小鼠胚胎的培养液[21,22]，以及汉姆氏 F10，每种培养液中均添加血清白蛋白或者同源的人类血清（7.5%~8.6%）。我们更倾向于采用含丙酮酸钠的厄尔氏溶液，添加各患者自己的血清（表 3）[12]。精

with serum from each patient (Table 3)[12]. Sperm numbers vary between 10^5 and 10^6 per ml, or between 1 and 5×10^5 living spermatozoa per ml. Droplets of medium held under paraffin oil or small culture tubes can be used for fertilization. Our present practice is to begin fertilization in droplets of medium under oil, adding $1.5–2\times10^5$ living spermatozoa per ml between 3 and 4 h after aspiration of oocytes; the oocytes are placed in tubes some hours later when most of their cumulus cells have been shed[12].

Table 2. Fertilization of human eggs *in vitro*

	Patients with occluded oviducts		Patients with idiopathic infertility[6]
	Bourn Hall*	Melbourne[6]	
Total no. patients	95	–	–
No. with pathological conditions:			
Spermagglutination cells or debris in semen	5	–	–
Abnormal movement of spermatozoa	3	–	–
Viscous seminal plasma	1	–	–
Remainder	86	40	–
No. (and %) fertilized	76 (88%)	37 (92%)	35%

* Recent survey of 122 patients (See Table 1).

Table 3. Composition of the media used for fertilization and cleavage of human eggs *in vitro* (g l^{-1})

	Modified Earles[12]	Media for mouse eggs [21,22]	Ham's F10*
$CaCl_2 \cdot 2H_2O$	0.2649	–	0.441
KC1	0.400	0.356	0.285
$MgSO_4 \cdot 7H_2O$	0.200	0.294	0.1527
NaCl	6.800	5.140	7.400
$NaH_2PO_4 \cdot 2H_2O$	0.1583	–	–
Na_2HPO_4	–	–	0.156
KH_2PO_4	–	0.162	–
$NaHCO_3$	2.1	1.9	1.2
Ca lactate	–	0.527	–
Na pyruvate	0.011	0.025	0.110
Glucose	1.000	1.000	1.100
Na lactate (60% syrup)	–	3.7 ml l^{-1}	–

Embryos growing most rapidly are two-cell by 21 h after insemination, four-cell at 40 h, and eight-cell at 44–54 h (refs 3, 23).

* Also contains $CuSO_4$, $FeSO_4$, $ZnSO_4$, amino acids, etc.

子的数目波动范围在每毫升 10^5 到 10^6 个，或者每毫升（1~5）× 10^5 个活精子。覆盖石蜡油的培养液小滴或者小的培养管都可以用于受精。我们目前采用的是在覆盖石蜡油的培养液小滴中开始受精，在吸取卵母细胞后 3~4 个小时之内，于每毫升培养液中加入（1.5~2）× 10^5 个活精子。数小时后当大部分卵丘细胞都脱落时，就把卵母细胞转移到培养管内 [12]。

表 2. 人卵细胞体外受精

	输卵管闭塞的患者		原发性不育的患者 [6]
	波恩诊所 *	墨尔本 [6]	
患者总数	95	–	–
具有病理状况的患者数			
精子凝集或精液内有碎片	5	–	–
精子运动异常	3	–	–
精浆黏稠	1	–	–
其他情况	86	40	–
受精数目（和百分比）	76（88%）	37（92%）	35%

* 最近调查的 122 名患者（见表 1）。

表 3. 人类卵细胞体外受精和卵裂使用的培养液组分（克 / 升）

	改良的厄尔氏溶液 [12]	小鼠受精卵用的培养液 [21,22]	汉姆氏 F10*
$CaCl_2 \cdot 2H_2O$	0.2649	–	0.441
KCl	0.400	0.356	0.285
$MgSO_4 \cdot 7H_2O$	0.200	0.294	0.1527
NaCl	6.800	5.140	7.400
$NaH_2PO4 \cdot 2H_2O$	0.1583	–	–
Na_2HPO_4	–	–	0.156
KH_2PO_4	–	0.162	–
$NaHCO_3$	2.1	1.9	1.2
乳酸钙	–	0.527	–
丙酮酸钠	0.011	0.025	0.110
葡萄糖	1.000	1.000	1.100
乳酸钠（60% 糖浆）	–	3.7 毫升 / 升	–

胚胎成长最快的时候是授精后 21 小时的二细胞阶段、40 小时的四细胞阶段以及 44~54 小时的八细胞阶段（参考文献 3、23）。

* 也含有 $CuSO_4$、$FeSO_4$、$ZnSO_4$、氨基酸等。

The oocytes are fertilized within a few hours. Some have been fertilized following a re-insemination 48 h later and grew normally to the eight-cell stage (J. M. Purdy and R. G. E., unpublished). Conditions such as spermagglutination, viscous seminal plasma and erratic sperm movement reduce the chances of fertilization (Table 2). Almost all eggs are monospermic, indicating that the block to polyspermy is efficient[3,6,23].

Most embryos cleave normally. They tolerate a wide variety of culture media, similar to those used for fertilization. Serum is almost universally added; in our work, we use 15% v/v of homologous serum[12]. Cleavage times are remarkably similar despite the varying concentration of energy sources in different media (Table 3). There have been occasional reports of cell fragments in cleaving eggs, although similar conditions have been found in embryos flushed from the reproductive tract and may not be pathological[24,25]. Many embryos grow to morulae and blastocysts, and escape from the zona pellucida *in vitro*[3,6], although some apparently develop abnormally during these later stages of growth[6]. The fastest-growing embryos are apparently the most successful in establishing pregnancy[6,26]. Cine films of cleavage with IR photography may enable embryos to be examined for their nuclear structure before they are replanted in the mother[27].

Replanting Embryos

Implantation is the most difficult and unpredictable stage. In animals, rates of implantation are high when embryos are replaced surgically (>70%), but lower if replaced non-surgically via the cervix (~50%)[28,29]. Almost all human embryos have been replaced transcervically, and approximately one-quarter of them have implanted in some series (Table 4). Embryos might be best replaced during the evening[12], because the spontaneous contractility of the human uterus is lower at night[30,31]. Disorders in the luteal phase, including progesterone deficiency and a short luteal phase, could preclude implantation (Table 4). Some implanted embryos live for only a short time after the expected return of menstruation, and are identified by a transitory rise in HCGβ (ref. 3) (Table 4).

Table 4. Pregnancies after replacing human embryos transcervically (same group of patients as in Tables 1 and 2)

	Single catheter	Double catheter
No. of patients	51	23
No. with deficient luteal phases:		
Short luteal phase*	2*	0
Progesterone deficiency†	1	3
No. with delayed return to menstruation‡:		
Brief elevated levels of HCGβ	2 } (9.8%)	0
Brief elevated levels of HCG/LH	3 }	3
No. pregnant 4 weeks after replacement	12 (23.1%)	2 (8.7%)

* Luteal phase of 12 days or less; these patients had luteal phases of 5 and 12 days, respectively, one also having progesterone deficiency.

† These patients had between 0.4 and 2.7 ng ml^{-1} on days 7 or 8 of the luteal phase.

‡ Some other patients showed a delayed return to menstruation but no evidence was found of elevated HCGβ or HCG/LH.

卵母细胞在数小时内完成受精。有些卵母细胞会在48小时后再授精时完成受精，并正常成长到八细胞阶段（珀迪和爱德华兹，未发表）。诸如精子凝集、精浆黏稠以及精子运动不稳定等状况都会降低受精的概率（表2）。几乎所有的卵细胞都是单精受精，说明卵细胞可有效地阻止多精受精[3,6,23]。

大多数胚胎都正常分裂。它们能够耐受多种培养液，这些培养液与受精时使用的培养液类似。它们几乎都加入了血清；在我们的工作中，我们使用体积比为15%的同源血清[12]。尽管不同培养液中能源物质的浓度各不相同（表3），但胚胎的分裂次数却非常相似。偶有报道称分裂的受精卵中有细胞碎片，但类似的情况也会发生在从生殖道中排出来的胚胎中，因此可能并非病理性的[24,25]。很多胚胎在体外长到桑椹胚和囊胚阶段，并可在体外培养中从透明带逸出[3,6]，尽管有些胚胎在发育的后期阶段会出现明显的发育异常[6]。很明显，生长最快的胚胎是最容易成功建立妊娠的[6,26]。红外摄影得到的分裂期照片有助于在胚胎重新植入母体之前对其进行核结构的检查[27]。

胚 胎 移 植

着床是最困难且无法预测的阶段。在动物中，手术移植胚胎着床的成功率较高（>70%），但是如果用非手术方式经宫颈移植则着床成功率较低（~50%）[28,29]。几乎所有的人类胚胎都是经宫颈移植的，一些研究中报告大约四分之一能够成功着床（表4）。胚胎着床最好在夜间[12]，因为夜间人类子宫的自发收缩较弱[30,31]。黄体期的异常，包括孕酮缺乏以及黄体期缩短都能影响着床（表4）。一些移植的胚胎只能在预期的月经复潮之后存活很短时间，这可以通过 HCGβ 的暂时升高来判断（参考文献3）（表4）。

表4. 人类胚胎经宫颈移植后的妊娠情况（与表1和2是同一批患者）

	单导管	双导管
患者总数	51	23
黄体期不全的数目		
黄体期短 *	2*	0
孕酮缺乏 †	1	3
月经复潮延迟的数目 ‡		
短暂的 HCGβ 水平升高	2 } (9.8%)	0
短暂的 HCG/LH 水平升高	3	3
着床4周后妊娠的数目	12（23.1%）	2（8.7%）

* 黄体期12天或更短；两位患者的黄体期分别是5和12天，其中一位还有孕酮缺乏。

† 这些患者在黄体期的第7天或者第8天体内孕酮的浓度是0.4 ng/ml 到 2.7 ng/ml 之间。

‡ 一些其他患者表现出了月经复潮的延迟，但是没有发现 HCGβ 或者 HCG/LH 升高的证据。

Why do so few embryos implant? Perhaps 20% or thereabouts is all that can be expected, since it is similar to the incidence of natural human fertility reported in serval surveys. This view seems to be pessimistic. Some embryos may be lost or infected if there are difficulties in passing a catheter through the internal os. An outer metal cannula can be useful in such cases, but pregnancy rates are lower (Table 4). A Teflon catheter with a smooth tip and side aperture evidently passes easily through the internal os[6], but may require the use of a tenaculum on the cervix. Implantations have occurred after replacing all stages between the two-cell and blastocyst, but there are insufficient data as yet on the optimal stage for replacement[26].

Physiological disorders might prevent implantation. Vaginal distension during replacement could invoke discharges of prolactin, and a catheter passed through the uterine cavity may invoke a premature decidualization of the endometrium. Slight myometrial contractions could expel embryos soon after their replacement and might be inhibited using β_2 -mimetics[30,31]. Restricting the volume of medium might encourage implantation (0.03–0.10 ml are used at present). Surgical transfer might circumvent these problems, but a second anaesthesia will be needed soon after that used for oocyte recovery, and the myometrium and endometrium will undoubtedly bleed when a needle is passed through them. In rhesus monkeys, 11 out of 15 embryos replaced surgically in the oviduct, and 2 of 8 replaced in the uterus developed to advanced stages of gestation[32].

Growth of Human Fetuses after Replacement in the Mother

At least 12 children have been born after fertilization *in vitro*, 3 in the initial series in Oldham[33], 7 in Australia[6,34], and 2 recently in the United Kingdom. Two of these were born prematurely at approximately 20 weeks, soon after amniocentesis. The children have been healthy, although a male twin required corrective surgery, and several current pregnancies appear to be progressing normally[6,26]. Details of all pregnancies are being recorded in an international register to evaluate any risks to the fetuses and the causes of abortion or fetal death.

Between one-quarter and two-fifths of the fetuses arising from fertilization *in vitro* die *in utero*, mostly in the first trimester. Maternal age (late thirties and early forties) may be a disposing factor in some cases. One dead fetus was triploid[33], the others have not been karyotyped. Triploidy may not be serious quantitatively, as the vast majority of fertilized eggs are monospermic. Most triploid fetuses arising during fertilization *in vivo* are caused by dispermy, fewer being due to fertilization by a diploid spermatozoon or cleavage errors in the embryo[35]. Nor should trisomy be more frequent after fertilization *in vitro*. Most human trisomies arise during the first meiotic division of the oocyte[36,37], a stage which is virtually complete before the oocytes are aspirated. Others arise during the second meiotic division, that is, after fertilization, or during cleavage of the embryos, but there is no record of their incidence *in vitro*. Amniocentesis should be performed on all fetuses arising through fertilization *in vitro*, if the parents agree, until the risks of trisomy have been assessed.

为什么只有少数胚胎能着床？或许预期着床成功的概率只有 20% 左右，这个数据类似于数个调查中所报告的人类自然生育率。这个观点似乎有些悲观。如果导管进入宫颈内口遇到困难，那么有些胚胎可能丢失或者被感染。这种情况下使用外层金属套管是有帮助的，但是怀孕率更低（表 4）。带有光滑尖端和侧孔的特氟龙导管明显很容易通过宫颈内口 [6]，但是可能需要在宫颈内使用宫颈钳。从二细胞到囊胚期的各个阶段的胚胎都能够着床，但是没有充足的数据说明哪个阶段是移植的最佳时期 [26]。

生理异常可能妨碍着床。胚胎移植过程中的阴道扩张可能促发催乳素释放，导管通过子宫腔可能促发子宫内膜的过早蜕膜化。子宫肌层的轻度收缩可能将移植后的胚胎排出，但是可以用 β_2–类似物抑制收缩 [30,31]。限制培养液的体积可能有助于着床（目前使用 0.03~0.10 毫升）。手术移植可能可以避免这些问题，但是卵母细胞采集之后需要马上进行二次麻醉，而且用针穿刺子宫肌层和内膜时无疑会流血。在恒河猴中，通过手术移植到输卵管中的 15 个胚胎中的 11 个以及移植到子宫内的 8 个胚胎中的 2 个都发育到了妊娠后期 [32]。

人类胚胎移植后胎儿在母体内的生长

已有至少 12 名体外受精的婴儿诞生，3 名诞生于奥尔德姆的最初实验 [33]，7 名出生在澳大利亚 [6,34]，另外 2 名最近出生在英国。其中 2 名在大约 20 周羊膜穿刺术后不久早产。孩子们都很健康，尽管一对男性双胞胎需要矫正手术，另外几个进行中的妊娠都进展正常 [6,26]。所有的妊娠细节都记录在一个国际登记处，以评价对胎儿产生危险的因素以及流产或者胎儿死亡的原因。

体外受精后四分之一到五分之二的胎儿在子宫内死亡，大部分都死于妊娠早期。在一些案例中母亲的年龄（年近四十或者四十出头）可能是一个决定性因素。一个死亡的胎儿是三倍体 [33]，其他胎儿的染色体组型还没有分型。三倍体不会太多，因为绝大多数受精的卵细胞都是单精受精。体内受精期间大部分三倍体胎儿的形成是因为双精入卵，很少的情况是因为与二倍体精子的受精或者胚胎卵裂异常 [35]。体外受精后三倍体的形成也应该不多。大部分三倍体出现在卵母细胞的第一次减数分裂 [36,37]，事实上在卵母细胞被吸出以前这个阶段已经完成了。另一些出现在第二次减数分裂，也就是受精以后或者卵裂期间，但是没有体外发生这些情况的记录。如果双亲同意，所有体外受精的胎儿都要实施羊膜穿刺术，直到评估了三倍体的风险为止。

Some fetuses may die *in utero* through physiological or embryological factors associated with fertilization *in vitro*. These could include damage to the uterine wall during replacement of the embryos, the introduction of cervical flora into the uterine cavity and the implantation of the embryo low in the uterine canal[3]. Short or deficient luteal phases arising after ovarian stimulation, or through the withdrawal of too many granulosa cells from the preovulatory follicle occur in very few patients (see Table 4)[38]. Abnormal embryonic differentiation *in vitro* might have led to the transitory secretion of HCG in some patients, as the fetus developed into a "blighted ovum" or trophoblastic vesicles, resulting in its early death. Such conditions arise following conception *in vivo*[39,40].

Genetic Engineering

Arguments sometimes raised against the introduction of fertilization *in vitro* into clinical practice involve the possibility of genetic engineering—chimaeras, hybrids and cloning embryos. These issues are quite distinct from the alleviation of infertility. Three cloned mice have been born, using nuclei from blastocysts[41], and others after the use of nuclei from the embryonic ectoderm and proximal endoderm of 7-day mouse embryos, showing that the developmental potential of nuclei from earlier embryos is retained in these cell types[42].

The thought of cloned human embryos identical to a pre-existing individual is not attractive. Yet, in a sense, uniparental human embryos largely similar to the father already exist. Hydatidiform moles are the remains of diploid androgenetic fetuses, in which the trophoblast proliferates over several weeks to form large grape-like vesicles. Such fetuses evidently arise due to the expulsion of the female pronucleus from the fertilized egg[43–46]. The sperm chromosomes are doubled and the embryo expresses paternal chromosomes only. Rare hydatidiform moles may even retain much of the father's heterozygosity, since they could arise from a diploid spermatozoon. The reason androgenetic embryos undergo hydatidiform changes is not understood. Perhaps the embryonic cells die early, or the mother "rejects" a fetus lacking her own antigens[46]; nevertheless, androgenetic mouse embryos develop normally to full term[47]. Human gynogenones could also arise through similar processes involving expulsion of the male pronucleus from the egg, although no searches have yet been made for them before or after birth. Uniparental embryos could arise through delayed syngamy and an enlarged pronucleus, seen in two human eggs fertilized *in vitro*.

Subtle forms of genetic engineering have been introduced. DNA fragments containing the gene for human insulin were injected into pronucleate mouse eggs, and the gene was identified in tissues of fetuses at the 18th day of pregnancy[42]. The DNA evidently replicated and was inserted into the genome of the fetus, but there is no information as to whether it was transcribed. There is apparently no knowledge of the potential value of such treatments in preventing the expression of inherited conditions such as diabetes.

一些胚胎会因为与体外受精有关的生理或者胚胎因素而胎死宫内。这些因素可能有：胚胎移植过程中损伤了子宫壁，将宫颈菌群带到了子宫腔内以及胚胎低位着床在宫颈管内[3]。在极少部分患者中，由于卵巢的刺激或者从排卵期前卵泡内吸走了太多的粒状细胞而导致黄体期过短或者缺失（见表4）[38]。在一些患者中，体外胚胎的异常分化可能导致暂时性的HCG分泌，比如胚胎发育成了"萎缩性胚囊"或者滋养层囊泡，并导致其早期死亡。体内受孕中也会出现这样的情况[39, 40]。

基 因 工 程

有时会出现反对将体外受精运用于临床的争论，它们涉及对基因工程——嵌合体、杂交以及克隆胚胎的可能性的担忧。这些问题与治疗不孕症截然不同。用囊胚来源的细胞核已经诞生了3只克隆鼠[41]，其他还有使用7天小鼠胚胎外胚层细胞核和近端内胚层细胞核而诞生克隆鼠，这说明这些细胞类型中早期胚胎的细胞核保留着发育的潜质[42]。

克隆一个与已经存在的个体完全一样的人胚胎是没有吸引力的。但是，从某种意义上说，非常类似于父本的单性人类胚胎已经存在了。葡萄胎就是雄核发育二倍体胎儿的残余，其中滋养层细胞经过数周的增殖形成了巨大的葡萄样水泡。显然这种胎儿的形成要归因于受精卵中雌原核被排除[43-46]。精子染色体复制成双倍，胚胎仅仅表达了父源性染色体。少数葡萄胎能保留较多父亲的杂合度，因为它们可以来源于二倍体的精子。雄核发育胚胎变成葡萄胎的原因尚不清楚。可能胚胎细胞于早期死亡，或者母亲"拒绝"缺乏其自身抗原的胎儿[46]；不管怎样，雄核发育的小鼠胚胎能够正常发育至临产[47]。人类的雌核生殖可能也通过类似的过程即从受精卵中排除雄原核而形成，尽管还未在产前或者产后的胎儿中进行寻找。通过配子融合的延迟以及原核的扩大就能形成单性生殖的胚胎，这在两个体外受精的人卵细胞中观察到。

基因工程的精细技术已被引入。将含有人类胰岛素基因的DNA片段注射到小鼠的原核卵细胞内，这些基因随后可以在妊娠18天的胎儿组织中找到[42]。显然DNA已经过复制并被插入到了胎儿的基因组中，但是没有信息证明基因是否被转录。显然，有关这种治疗方法在预防遗传性疾病比如糖尿病的基因表达方面的潜在价值还没有相关知识。

Prospects

Many children will soon be born after the fertilization of human eggs *in vitro*. We have established more than 40 pregnancies since resuming work during the past 9 months, and the majority are surviving. This is most encouraging for those couples who could not be offered any other form of corrective surgery and have so far been without effective treatment (Table 5). The method can be carried out several times on the same patient, and success rates should soon exceed some forms of oviductal surgery. If ovarian stimulation is used, "spare embryos" may be available for embryological studies and one embryo has grown for 9 days *in vitro* until stage 5a (ref. 3). The frozen storage of human embryos still appears to be distant. Some of the fathers were oligospermic, and patients with idiopathic (unexplained) infertility, hostile cervical mucus, incompetent cervix, antibodies against the zona pellucida, might also be helped. Complex disorders leading to the abnormal growth of pronuclei can be investigated[48].

Table 5. Results on a recent series of 122 patients with tubal occlusion

No. of patients	122	
Failure to collect oocyte:		
Method failure	14	27
Adhesions, endometriosis, etc.	13	
(Table 1)		
No. with preovulatory oocytes:	95	
Failure of fertilization:		
Method failure	9+1?	19
Pathological spermatozoa	9	
(Table 2)		
Embryos not replaced	2	
No. of embryos replaced:	74	
Failure of implantation:		
Short luteal phase/progesterone deficiency	6	55
Method failure*	49	
Indications of pregnancy:		
Delayed RTM: early abortion?	5	
Pregnant 4 weeks after replacement	14	

The natural menstrual cycle was monitored. RTM, return to menstruation.
* See Table 4.

I thank Jean Purdy and Patrick Steptoe for their help at all stages of this work, and Simon Fishel for his comments on the manuscript.

(**293**, 253-256; 1981)

展　望

越来越多的婴儿将会通过人卵细胞体外受精的方式诞生。自恢复体外受精的研究后，在过去的 9 个月内我们已经为 40 多位患者建立妊娠，且绝大多数胎儿仍然存活。对于那些无法获得其他任何形式的矫正手术以及目前为止没有得到有效治疗的夫妇来说，这是非常令人鼓舞的（表 5）。这种方法在同一个患者身上可以进行多次，很快成功率将超过一些其他类型的输卵管手术。如果使用卵巢刺激，“多余的胚胎”还可以用于胚胎学研究，其中一个胚胎在体外一直成长了 9 天直到第 5a 阶段（参考文献 3）。人类胚胎的冷冻储存似乎还很遥远。一些少精父亲、先天性不育患者（尚不能解释原因的）和宫颈黏液不良、宫颈内口松弛、抗透明带抗体的患者也可以得到帮助。同时还可以研究导致原核异常生长的各种复杂疾病[48]。

表 5. 最近 122 名输卵管堵塞患者的研究结果

患者总数	122
卵母细胞采集失败：	
方法失败	14 } 27
粘连、子宫内膜异位症等	13
（表 1）	
排卵期前卵母细胞数目：	95
受精失败	
方法失败	9+1? } 19
病态精子	9
（表 2）	
未移植的胚胎	2
移植的胚胎数目	74
未着床：	
黄体期短 / 孕酮缺乏	6 } 55
方法失败 *	49
妊娠的指征：	
RTM 延迟：早期流产?	5
移植 4 周后的妊娠	14

监测自然的月经周期。RTM：月经复潮。
* 见表 4。

感谢琼 · 珀迪和帕特里克 · 斯特普托在工作的各个阶段给予的帮助，以及西蒙 · 菲谢尔对手稿的意见。

（毛晨晖 翻译；王敏康 审稿）

R. G. Edwards
Physiological Laboratory, Cambridge University, Cambridge CB2 3EG, UK, and Bourn Hall, Cambridge CB3 7TR, UK

References:

1. *3rd World Congress of Human Reproduction,* Berlin (Excerpta Medica, Amsterdam, 1981).
2. Edwards, R. G. & Steptoe, P. C. *J. Reprod. Fert. Suppl.* 22, 121 (1975); *Lancet* i, 683 (1970).
3. Edwards, R. G. *Conception in the Human Female* (Academic, London, 1980).
4. Kratochwil, A., Urban, G. & Friedrich, F. *Ann. Chir. Gynaec. Fenniae* **61**, 211 (1972).
5. de Crespigny, L. J. Ch., O'Herlihy, C., Hoult, I. J. & Robinson, H. P. *Fert. Steril.* **35**, 25 (1981).
6. Trounson, A. O. *et al. 3rd World Congress of Human Reproduction*, Berlin (Excerpta Medica, Amsterdam, 1981).
7. Fowler, R. E., Edwards, R. G., Walters, D. E., Chan, S. T. H. & Steptoe, P. C. *J. Endocr.* 77, 161 (1978).
8. Baudendistel, L. J., Ruh, M. F., Nadel, E. M. & Ruh, T. S. *Acta Endocr.* **89**, 599 (1978).
9. Katzellenbogen, B. S. & Ferguson, E. R. *Endocrinology* **97**, 1 (1975).
10. Watson, C. S., Medina, D. & Clark, J. H. *Endocrinology* **108**, 668 (1981).
11. Adashi, E.Y., Hsueh, A. J. W. & Yen, S. S. C. *J. Endocr.* **87**, 383 (1980).
12. Edwards, R. G., Steptoe, P. C. & Purdy, J. M. *Br. J. Obstet. Gynec.* **87**, 737 (1980).
13. Frydman, R., Testart, J. & Feinstein, M. C. *3rd World Congress of Human Reproduction*, Berlin (Excerpta Medica, Amsterdam, 1981).
14. Plashot, M., Mandelbaum, J. & Cohen, J. *3rd World Congress of Human Reproduction*, Berlin (Excerpta Medica, Amsterdam, 1981).
15. Mettler, L. *3rd World Congress of Human Reproduction*, Berlin (Excerpta Medica, Amsterdam, 1981).
16. Everett, J. W. & Sawyer, C. H. *Endocrinology* **47**, 198 (1950).
17. Everett, J. W. *A. Rev. Physiol.* **31**, 383 (1969).
18. Knobil, E. *Recent Prog. Horm. Res.* **36**, 53 (1980).
19. Lopata, A., Johnston, I. W. H., Houalt, I. J. & Speirs, A. L. *Fert. Steril.* **33**, 117 (1980).
20. Renou, P., Trounson, A. O., Wood, C. & Leeton, J. F. *Fert. Steril.* **35**, 409 (1981).
21. Brinster, R. L. in *Reproductive Biology* (eds Balin, H. & Glasser, S.) (Excerpta Medica, Amsterdam, 1972).
22. Hoppe, P. C. & Pitts, J. *Biol. Reprod.* **8**, 420 (1973).
23. Edwards, R. G., Purdy, J. M. & Steptoe, P. C. *Am. J. Obstet. Gynec.* (in the press).
24. Sundström, P., Nilsson, O. & Liedholm, P. *Acta Obstet. Gynec. Scand.* **60**, 109 (1981).
25. Wramsby, H. & Liedholm, P. *3rd World Congress of Human Reproduction,* Berlin (Excerpta Medica, Amsterdam, 1981).
26. Edwards, R. G., Steptoe, P. C. & Purdy, J. M. *3rd World Congress of Human Reproduction*, Berlin (1981).
27. Hamberger, L. *3rd World Congress of Human Reproduction,* Berlin (Excerpta Medica, Amsterdam, 1981).
28. Rowson, L. E. A. (ed.) *Egg Transfer in Cattle* (Commission of European Communities, Brussels, 1976).
29. Sreenan, J. N. *Theriogenology*, **9**, 69 (1978).
30. Lundström, V., Eneroth, P., Granström, E. & Swahn, K.-L. *3rd World Congress of Human Reproduction,* Berlin (Excerpta Medica, Amsterdam, 1981).
31. Akerjund, M., Andersson, K.-E. & Ingermarsson, I. *Br. J. Obstet. Gynec.* **83**, 673 (1976).
32. Marston, J. H., Penn, R. & Sivelle, P. C. *J. Reprod. Fert.* **49**, 175 (1977).
33. Steptoe, P. C., Edwards, R. G. & Purdy, J. M. *Br. J. Obstet. Gynec.* **87**, 757 (1980).
34. Lopata, A. *Nature* **288**, 642 (1980).
35. Jacobs, P. A. *et al. Ann. Hum. Genet.* **42**, 49 (1978).
36. Hassold, T. J. & Matsuyama, A. *Hum. Genet.* **46**, 285 (1978).
37. Niikawa, N., Merotto, E. & Kajii, T. *Hum. Genet.* **40**, 73 (1977).
38. Feichtinger, W., Kemeter, P., Szalay, S., Beck, A. & Janisch, H. *3rd World Congress of Human Reproduction,* Berlin (Excerpta Medica, Amsterdam, 1981).
39. Hertig, A. T. in *Progress in Infertility* (eds Behrman, J. & Kistner, R.W.) (Little Brown, Boston, 1975).
40. Batzer, F. R. Schlaff, S., Goldfarb, A. F. & Carson, S. L. *Fert. Steril.* **35**, 307 (1981).
41. Illmensee, K. & Hoppe, P. C. *Cell* 23, 9 (1981).
42. Illmensee, K. *3rd World Congress of Human Reproduction,* Berlin (Excerpta Medica, Amsterdam, 1981).
43. Kajii, T. & Ohama, K. *Nature* **268**, 633 (1977).
44. Jacobs, P. A., Wilson, C. M., Sprenkle, J. A., Rosenheim, N. B. & Migeon, B. *Nature* **286**, 714 (1980).
45. Lawler, S. D. *et al. Lancet* ii, 580 (1979).
46. Surti, U., Szulman, A. E. & O'Brien, S. *Hum. Genet,* **51**, 153 (1979).
47. Hoppe, P. C. & Illmensee, K. *Proc. Natl. Acad. Sci. U.S.A.* 74, 56-57 (1977).
48. Trounson, A. O., Leeton, J. F., Wood, C., Webb, J. & Kovacs, G. *Fert. Steril.* **34**, 431 (1980).

试管婴儿，1981

Enzymatic Replication of *E.coli* Chromosomal Origin is Bidirectional

J. M. Kaguni *et al.*

Editor's Note

Here molecular biologist Arthur Kornberg and his colleagues at Stanford University in California shed new light on how bacterial genomes are replicated. It was already known that bacterial chromosomes consist of circular double strands of DNA that are replicated in both directions at once, starting from a site denoted *oriC*. Kornberg and colleagues identify an enzyme that will replicate any plasmid (a circular stretch of bacterial DNA) into which the *oriC* segment from *E. coli* has been inserted. This confirmed the supposed replication mechanism, isolated the replicase enzyme, and suggested a way to replicate any arbitrary strand of bacterial DNA for biotechnological purposes.

A soluble enzyme system has been discovered which specifically recognizes and replicates plasmids containing the *Escherichia coli* chromosomal origin, *oriC*. Electron microscopy has shown that plasmid replication begins at or near *oriC* from which it progresses bidirectionally to completion. Control of initiation of a cycle of chromosomal replication and mechanisms of priming and fork movement can now be explored using this system.

REPLICATION of the *Escherichia coli* K-12 chromosome, as shown by genetic and biochemical analysis *in vivo*, begins at a unique site (*oriC*) and proceeds bidirectionally[1–3]. The DNA fragment containing *oriC* has been isolated by its ability to confer autonomous replication on plasmids or phage whose own replication origin has been inactivated[4–7]. Deletion analysis has localized *oriC* to a sequence of 232–245 base pairs (bp)[8]; insertions, deletions or substitutions in this essential region can inactivate *oriC*.

Initiation of a cycle of chromosomal replication in the cell requires the activities encoded by the genes *dnaA*, *dnaI* and *dnaP*[9–11]. In addition, *dnaB* and *dnaC* proteins, whose activities are essential for priming the synthesis of nascent chains during replication[12,13], are also required during or shortly after initiation[14,15]. RNA polymerase has also been implicated[16].

A soluble enzyme system has been discovered which specifically recognizes and replicates *oriC* plasmids[17]. The reaction requires *dnaA* protein, RNA polymerase and numerous replication proteins including *dnaB*, single-stranded DNA-binding protein and DNA gyrase. Recently, *dnaC* protein has been shown to be required for *in vitro* replication of *oriC*

大肠杆菌染色体起始位点的酶促复制是双向进行的

卡古尼等

编者按

在本文中，分子生物学家阿瑟·科恩伯格与其在加州斯坦福大学的同事对细菌基因组的复制方式提出了新的观点。已知，细菌的染色体由环状双链 DNA 构成，其复制可向两个方向同时进行，该复制起始位点是一个被称为 *oriC* 的位点。科恩伯格及其同事发现了一种酶，它可以复制任何一种插入了大肠杆菌 *oriC* 位点的质粒（一段环状细菌 DNA）。这一发现证实了关于复制机制的假设，分离到了复制酶，并且提供了一种出于生物技术目的而复制任何细菌 DNA 链的方法。

我们发现了一个可溶性的酶系统，它可以特异地识别并复制含有大肠杆菌染色体复制起始位点（*oriC*）的质粒。电子显微镜观察结果显示，质粒的复制从 *oriC* 位点或接近 *oriC* 位点的地方起始，双向进行，直至复制完成。通过这个酶系统，可以研究一个染色体复制周期中的起始调控、引发机制以及复制叉移动机制。

体内的遗传学和生物化学分析表明，大肠杆菌 K-12 菌株染色体的复制起始于一个独特的位点（*oriC*），并且是双向进行的[1-3]。含有 *oriC* 位点的 DNA 片段可以使自身复制失活的质粒或噬菌体 DNA 获得自主复制的能力，人们根据这一特点将其分离出来[4-7]。通过缺失分析，人们将 *oriC* 位点定位于一段含有 232~245 个碱基对(bp)的序列[8]，在这段必需序列中发生碱基插入、缺失或替换都会使 *oriC* 位点失活。

细胞中染色体复制周期的起始需要 *dnaA*、*dnaI* 和 *dnaP* 基因编码产物的活性[9-11]。此外，*dnaB* 和 *dnaC* 蛋白的活性既是复制过程中引发新生 DNA 链合成所必需的[12,13]，也是复制起始过程中或复制起始后不久所必需的[14,15]。RNA 聚合酶也在其中发挥作用[16]。

我们发现了一个可溶性酶系统，它可以特异地识别并复制含有 *oriC* 位点的质粒[17]。这一反应需要 *dnaA* 蛋白、RNA 聚合酶以及包括 *dnaB* 蛋白、单链 DNA 结合蛋白和 DNA 促旋酶在内的多种复制有关的蛋白质参与。最近，又发现 *dnaC* 蛋白是含有 *oriC* 的质粒在体外复制所必需的（未发表的结果）。对于酶促合成的复制中间体

plasmids (unpublished observations). Biochemical analysis of enzymatically synthesized replicative intermediates showed initiation occurring at or near *oriC* and was consistent with bidirectional progress from that point, but such evidence is essentially a statistical average. Is replication of an individual molecule, once initiated at *oriC*, then extended bidirectionally to completion? We present here the results of an electron microscopic study which indicate that with few exceptions this enzyme system creates an "eye" or replication "bubble" in a plasmid molecule at or near the *oriC* region which is extended in both directions to generate two complete molecules.

Formation of Replicative Intermediates

Two classes of supercoiled, *oriC* template DNA were examined. One, pSY317, is a 13.5-kilobase (kb) plasmid which contains a 5.6-kb *Eco*RI, *oriC* fragment (from pSY221)[5] and a 7.9-kb *Eco*RI kanamycin-resistance fragment (from pML21)[18]. The *oriC* plasmid contains *oriC* intact, together with extensive flanking sequences. The second template, M13*oriC*26, is a 12.2-kb, chimaeric M13 phage DNA containing *oriC* and its adjacent *asnA* gene[19]. In constructing M13*oriC*26, the *Xho*I site immediately to the right of *oriC* was interrupted. Although interruptions in and around this site do not alter the ability of this and similar *oriC* plasmids to replicate[8,19,20], the directionality of replication was reported to be affected, becoming unidirectional rather than bidirectional[21,22].

The extent of replication was limited to obtain a significant number of replicative intermediates for electron microscopic study. The chain terminator, 2′,3′-dideoxythymidine 5′-triphosphate (ddTTP) was used. This inhibitor, which prevents chain growth when incorporated into DNA, was present at concentrations relative to dTTP that generated replicative intermediates showing different exents of replication[23]. It is assumed that incorporation of a ddTMP residue in the leading or lagging strand at a replication fork stops fork movement. DNAs were purified, spread on parlodion-coated grids by the formamide spreading technique[24], and rotary shadowed at a low angle with platinum-tungsten vapour. Replicative intermediates generated with an enzyme fraction (fraction II)[17] from wild-type cells constituted 10–20% of the molecular forms observed, a value consistent with the fraction of template molecules used in comparable uninhibited reactions. It is possible that the fortuitous arrangement of a DNA fragment overlapping a circular molecule or unit-length linear fragment can give rise to structures that are indistinguishable from true replicative intermediates, but such events are expected to occur only rarely. Circular replicative intermediates of pSY317 and M13*oriC*26 appeared in every case as theta-like structures (see Fig. 2).

的生物化学分析结果显示，复制的起始发生在 *oriC* 位点或 *oriC* 位点附近，并且分析结果与从此位点开始双向进行的复制过程一致，但这些证据是一个平均的统计结果。对于一个单独的分子来说，一旦其复制从 *oriC* 位点开始，是否就双向延伸直到复制完成呢？本文展示了一些电子显微镜的研究结果，这些结果表明，这个酶系统在绝大多数的情况下能够在质粒分子的 *oriC* 位点或其邻近区域形成“眼”或“复制泡”样结构，然后这种结构从两个方向向两端延伸并最终产生两个完整的分子。

复制中间体的形成

我们检测了两种类型的含有 *oriC* 位点的超螺旋 DNA 模板。一种模板是 pSY317，它是一个 13.5 kb 的质粒，含有一个大小为 5.6 kb，具有 *Eco*RI 酶切位点的 *oriC* 片段（来自 pSY221）[5] 和一个大小为 7.9 kb，具有 *Eco*RI 酶切位点的卡那霉素抗性片段（来自 pML21）[18]，该 *oriC* 质粒含有完整的 *oriC* 位点以及大量的侧翼序列。另一种模板是 M13*oriC*26，它的大小为 12.2 kb，是含有 *oriC* 位点及其邻近 *asnA* 基因的 M13 噬菌体 DNA 嵌合体 [19]。在构建 M13*oriC*26 质粒的过程中，右侧紧临 *oriC* 位点的 *Xho*I 酶切位点被切断。尽管该位点及其周边区域的破坏并不影响该 *oriC* 质粒及类似质粒的复制 [8,19,20]，但据报道，复制的方向性会受到影响，会从双向复制变成单向复制 [21,22]。

通过控制复制的进程可以得到大量的可供电子显微镜观察的复制中间体。我们使用了可以使 DNA 链延伸过程终止的 2′, 3′ – 双脱氧胸腺嘧啶 –5′ – 三磷酸（ddTTP）来达到这一目的。ddTTP 进入到 DNA 链后，可以阻止 DNA 链的延伸。通过在反应体系中加入不同比例的 ddTTP 和 dTTP，可以得到各种不同复制程度的复制中间体 [23]。人们据此推断整合到复制叉的前导链或后随链中的 ddTMP 残基可以使复制叉停止移动。使用甲酰胺铺展技术 [24] 将纯化的 DNA 铺展在火棉胶片包被的载网上，然后用铂 – 钨蒸汽进行小角度旋转投影电镜观察。使用从野生型细胞中分离到的酶组分（组分 II）[17] 所产生的复制中间体在观察到的分子形式中占 10%~20%，这一比例与未受抑制的对照反应中的模板分子含量一致。环状 DNA 分子或单位长度的线性片段与一段 DNA 片段重叠的偶然排列可能形成与真正的复制中间体类似的结构，不过这种情况发生的概率很低。pSY317 和 M13*oriC*26 质粒的环状复制中间体在每个实验中都呈现出 θ 样的结构（见图 2）。

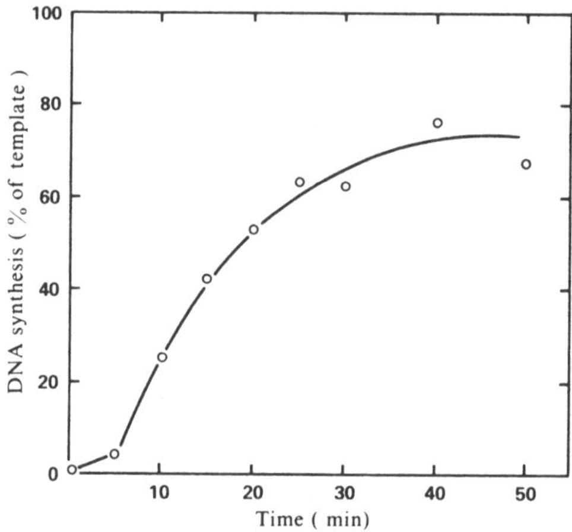

Fig. 1. Time course of *oriC* plasmid in *dnaA*-complementing conditions. Reactions were performed as described previously[17] in a 250 μl volume containing 40 mM HEPES *p*H 7.6, 2 mM ATP, 0.5 mM each of GTP, UTP and CTP, 43 mM creatine phosphate, 100 μM each of dGTP, dATP, dCTP and 5-methyl-[^{3}H]dTPP (85 c.p.m. per pmol of total deoxynucleotide), 6% (w/v) polyvinyl alcohol 24,000, 11 mM magnesium acetate, 100 μg ml^{-1} creatine kinase (Sigma), 8.6 μg ml^{-1} supercoiled pSY317 DNA, 2 mg of protein (fraction II, prepared from *E. coli* WM433 *dnaA*204 as described elsewhere[17]) and 230 units of *dnaA*-complementing activity (fraction III). Incubation was at 30°C. Aliquots (25μl) were TCA-precipitated and counted in a liquid scintillation counter. One unit is equal to one pmol of nucleotide incorporated per min.

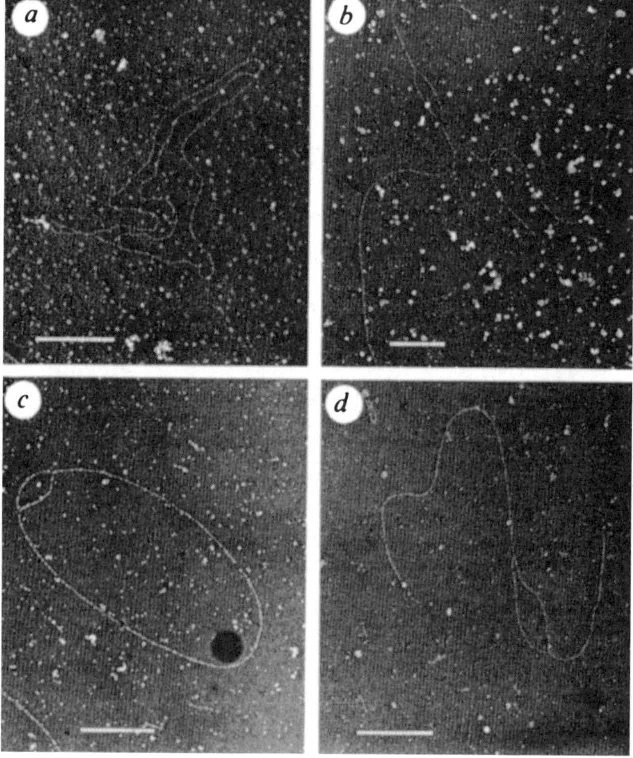

Fig. 2. Replicative intermediates of pSY317 and M13*oriC*26. The reaction with 8.6 μg ml^{-1} of supercoiled pSY317 was as described in Fig. 1 legend in a 125 μl volume with 50 μM ddTTP, 1,100 μg of *E. coli*

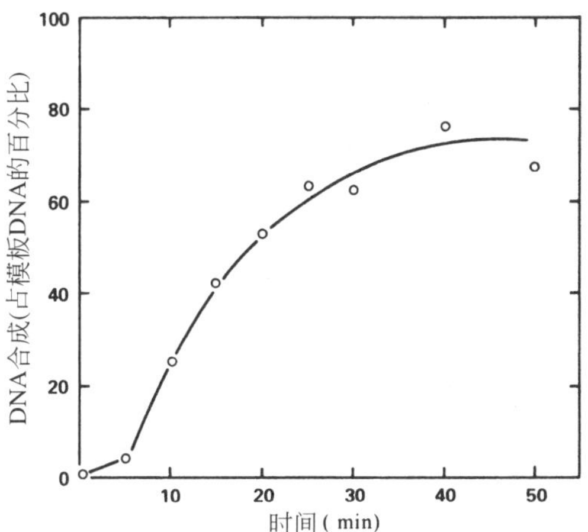

图 1. 在 *dnaA* 互补条件下，含有 *oriC* 位点的质粒复制随时间变化的曲线。反应按照之前文献报道的条件进行 [17]。反应体积为 250 μl，含有 40 mM pH 7.6 的 HEPES，2 mM ATP，GTP、UTP 和 CTP 各 0.5 mM，43 mM 磷酸肌酸，dGTP、dATP、dCTP 和 5– 甲基 –[^{3}H]dTPP（总脱氧核苷酸 85 c.p.m./pmol）各 100 μM，6%（w/v）的聚乙烯醇 24,000，11 mM 乙酸镁，100 μg/ml 肌酸激酶（购自西格玛公司），8.6 μg/ml 超螺旋 pSY317 DNA，2 mg 蛋白质（组分 II，采用文献报道的方法 [17] 从大肠杆菌 WM433 *dnaA*204 中分离）以及 230 单位的互补活性 *dnaA*（组分 III）。反应在 30℃进行。将反应物分装（25 μl 每管）后使用 TCA 沉淀并用液闪计数器计数。一个活性单位相当于每分钟掺入 1 pmol 核苷酸所需的 *dnaA* 量。

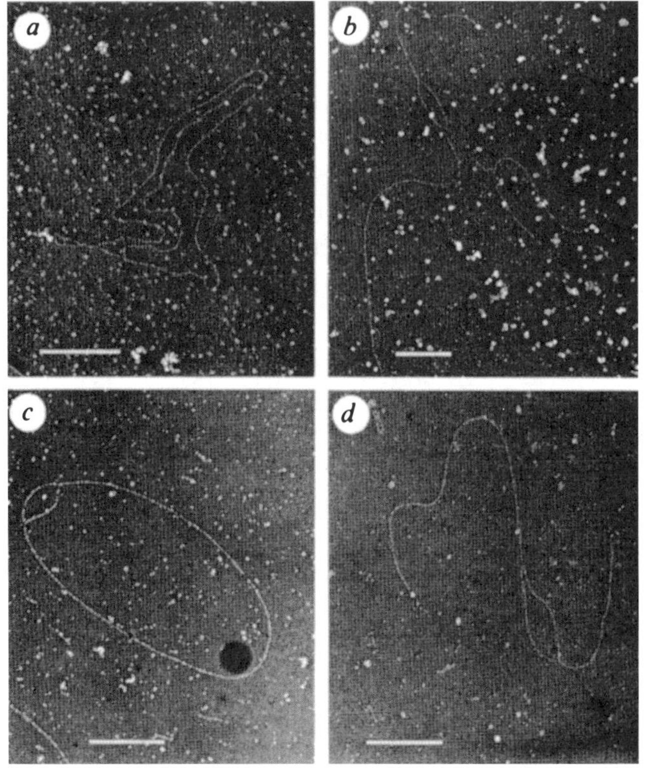

图 2. pSY317 和 M13*oriC*26 的复制中间体。8.6 μg/ml 的 pSY317 超螺旋质粒的反应条件如图 1 注，反应体系为 125 μl，含有 50 μM ddTTP，1,100 μg 提取自大肠杆菌 WM433 菌株的 *dnaA*204 组分 II 以及 275 μg

WM433 *dnaA*204 fraction II and 275 μg of *E. coli* HB101 (pBF101) fraction II, known to be enriched for *dnaA*-complementation activity[17]. Incubation was at 30°C for 10 min after which reactions were terminated by addition of SDS to 1% and EDTA to 50 mM. Following incubation at 65°C for 3 min, aliquots were removed for TCA precipitation and counted in a liquid scintillation counter. The sample was then incubated with 0.5 mg ml^{-1} of proteinase K (Merck) at 37°C for 60 min, phenol-extracted, ether-extracted, and centrifuged in a Beckman airfuge into a CsCl shelf of 1.42 g cm^{-3}. The shelf was recovered and DNA was ethanol-precipitated and resuspended in a small volume of 10 mM Tris-HCl *p*H 8.0 and 1mM EDTA. Recovery at this step was 60–70% of the initial reaction product. The sample was then incubated with 5 μg ml^{-1} of pancreatic RNase at 37°C for 60 min to digest contaminating RNA and either relaxed with *Sal*I endonuclease (given by K. Burtis) in the presence of ethidium bromide[31] (*a*), or restricted to completion with *Sal*I endonuclease in 10 mM Tris-HCl *p*H 8.0, 10 mM $MgSO_4$ and 100 mM NaCl at 37°C for 30 min (*b*). DNA was then spread on to parlodion-coated grids by the formamide technique[24], shadowed at a low angle with platinum-tungsten vapour, and viewed in a Philips 300 electron microscope. Replicative intermediates of M13*oriC*26 were prepared for electron microscopy as described above except that the sample was either relaxed with *Eco*RI endonuclease (given by C. Mann) (*c*) or linearized by digestion to completion with *Eco*RI endonuclease in 100 mM Tris-HCl *p*H 7.2, 5 mM $MgCl_2$, 2 mM 2-mercaptoethanol and 50 mM NaCl at 37°C for 60 min (*d*). Scale bars, 0.5 μm.

Defining the Initiation Site and Direction of Replication

Replication of an *oriC* plasmid (for example pSY317) requires *dnaA* protein[17]. Complementation of a crude enzyme fraction (fraction II)[17] prepared from a *dnaA* mutant provides an assay for purifying *dnaA* protein overproduced in cells bearing a plasmid carrying the *dnaA* gene. In a *dnaA* protein preparation enriched at least 200-fold over wild-type levels, the template molecules were almost completely used in a reaction (Fig. 1).

In conditions which complement *dnaA* activity, replicative intermediates of pSY317 were accumulated by inhibiting the reaction by 60% with 50 μM ddTTP. To determine the site of initiation and direction of replication, molecules were linearized by cleavage with *Sal*I endonuclease and examined by electron microscopy (Figs 2,3). In orienting the molecules in Fig. 3, *oriC* is located at a point 45% of the genome length from the right end of the linearized pSY317. With few exceptions, the replicated segment of the intermediate overlaps *oriC*. Although the slight asymmetry of the *Sal*I restriction cleavage relative to *oriC* introduces some ambiguity in orienting the more extensively replicated intermediates, it is clear that for most of them, replication proceeds bidirectionally. Inspection of the less extensively replicated molecules reveals that initiation occurs at or near *oriC*. Thus, we have shown that in these conditions, replication of the *oriC* plasmid pSY317 proceeds bidirectionally from a start in the *oriC* region.

提取自大肠杆菌 HB101 菌株（pBF101）的组分 II（已知其可以通过 *dnaA* 互补活性被富集[17]）。将反应体系在 30℃孵育 10 min，然后加入 SDS 和 EDTA 使其终浓度分别为 1% 和 50 mM 以终止反应。随后，将反应体系在 65℃孵育 3 min，分装后用 TCA 沉淀（去除蛋白质）并用液闪计数器计数。将样品加入 0.5 mg/ml 的蛋白酶 K（购自默克公司）在 37℃孵育 60 min，然后进行酚抽提和醚抽提，并用贝克曼公司的 airfuge 离心机进行氯化铯密度梯度离心。样品在氯化铯密度梯度中，最终沉降到密度为 1.42 g/cm^3 的一个薄层上。将这一薄层回收，并用乙醇沉淀其中的 DNA，然后用小体积的 10 mM pH 8.0 的 Tris-HCl 和 1 mM EDTA 溶液重悬 DNA。与初始反应产物相比，这一步骤的回收率为 60%~70%。将样品与 5 μg/ml 的胰 RNA 酶 37℃孵育 60 min，以消化污染的 RNA，并分别进行如下两种不同的操作来松弛 DNA 超螺旋：(*a*) 在溴化乙啶存在时用 *Sal*I 内切酶（由伯蒂斯馈赠）处理[31]；(*b*) 在含 10 mM pH 8.0 Tris-HCl、10 mM 硫酸镁、100 mM 氯化钠的溶液中用 *Sal* I 内切酶 37℃消化 30 min。使用甲酰胺铺展技术[24]将纯化的 DNA 铺展在包覆了火棉胶片的载网上，然后用铂－钨蒸汽进行小角度旋转投影，并使用菲利普斯 300 电子显微镜观察。M13*oriC*26 的复制中间体也用上述方法制备，唯一不同的是样品使用 *Eco*R I 内切酶（由曼馈赠）松弛（*c*）或使用 *Eco*RI 内切酶在含 100 mM pH 7.2 Tris-HCl、5 mM 氯化镁、2 mM 2–巯基乙醇、50 mM 氯化钠的溶液中，于 37℃下完全消化 60 min 使 DNA 线性化（*d*）。标尺，0.5 μm。

复制起始位点及复制方向的确定

含有 *oriC* 位点的质粒（如 pSY317）的复制需要 *dnaA* 蛋白的参与[17]。从 *dnaA* 突变体中制备的酶粗提物（组分 II）[17] 与 *dnaA* 蛋白的功能互补性为纯化 *dnaA* 蛋白提供了方法，该蛋白在含有携带 *dnaA* 基因的质粒的细胞中是过量表达的。当 *dnaA* 蛋白的浓度富集到野生型水平的 200 倍以上时，反应中几乎所有的模板分子都被利用了（图 1）。

在有互补 *dnaA* 活性的条件下，向反应中加入终浓度为 50 μM 的 ddTTP，反应被抑制 60%，并引起 pSY317 复制中间体的积累。为了确定复制的起始位点以及方向，我们用 *Sal* I 限制性内切酶消化切割使之线性化，然后用电子显微镜观察（图 2，3）。在图 3 中，对这些分子进行定向后发现，*oriC* 位点位于距离线性化的 pSY317 质粒右端 45% 基因组长度的位置。除了极少数的例外，绝大部分复制中间体的复制片段都与 *oriC* 位点重叠。尽管相对于 *oriC* 位点，*Sal*I 的限制性内切会造成轻微的不对称性，从而在更长片段的复制中间体的定向中引入了不确定性，但是可以肯定的是对于多数复制中间体来说，复制是双向进行的。对较短的复制分子的观察表明，复制起始于 *oriC* 位点或 *oriC* 位点附近的区域。因此我们已经证明了，在这些条件下，含有 *oriC* 位点的 pSY317 质粒的复制起始于 *oriC* 位点并且是双向进行的。

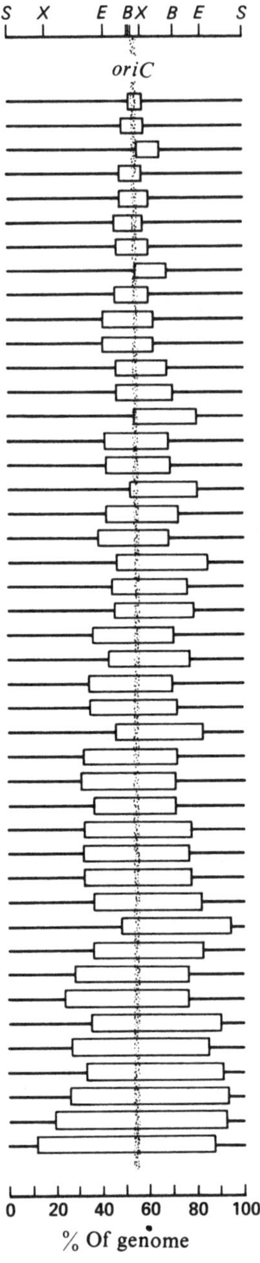

Fig. 3. Replication of the *oriC* plasmid pSY317 initiates at *oriC* and progresses bidirectionally. Replicative intermediates generated with 50 μM ddTTP, then purified and linearized with *Sal*I endonuclease (as described in Fig. 2 legend), were randomly selected and photographed. Molecules were analysed with a Hewlett-Packard 9810A calculator by measuring the non-replicated and both replicated segments of an individual molecule on a Hewlett-Packard 9864A digitizer board, averaging the replicated segments, and expressing each portion as a percentage of the total length. Only molecules in which both replicated segments were identical in length (±5%) and for which the total length was within ±10% of the length expected were included in the analysis. Molecules are aligned so that the longer unreplicated segment is to the left, and they are arranged according to increasing extents of replication. The position of *oriC* is indicated by the stippled area. The open boxes represent the replication "bubble". Restriction endonuclease sites are: S, *Sal*I; X, *Xho*I; E, *Eco*RI; and B, *Bam*HI.

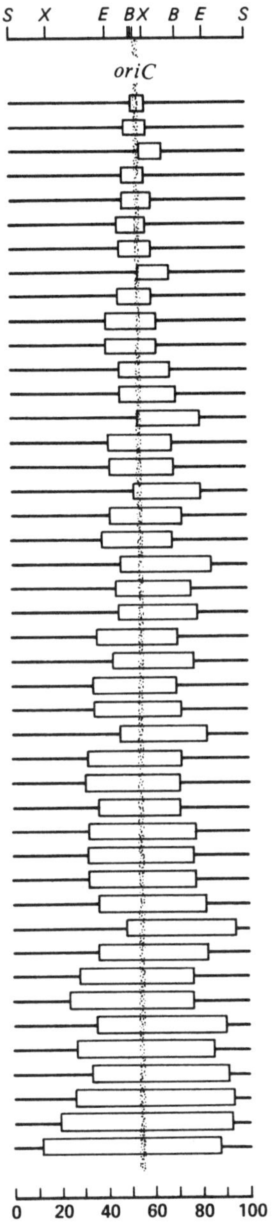

图 3. pSY317 质粒的复制从 *oriC* 位点起始并双向进行。使用 50 μM ddTTP 来制备复制中间体，纯化后使用 *Sal*I 内切酶消化使之线性化（如图 2 注所述）。然后随机选取这些复制中间体进行拍照。通过惠普 9864A 数字转换板测量单个分子未复制的片段和复制后的两个片段的长度，计算复制片段的平均值，将每一部分换算成总长度的百分比，然后，对这些分子进行分析（使用惠普 9810A 计算机）。只有那些两个复制片段的长度相同（±5%），并且分子总长度与预期值的差异不超过 10% 的分子才会列入分析范围。将这些分子排列对齐，使较长的未复制的片段位于左端，并按复制区长度由小到大进行排列。图中点带区域所示为 *oriC* 的位置。方框所示为“复制泡”。限制性内切酶的位点分别是：S，*Sal*I；X，*Xho*I；E，*Eco*RI；B，*Bam*HI。

Replicative intermediates of M13*oriC*26 DNA were generated with an enzyme fraction (fraction II)[17] from wild-type cells at several levels of ddTTP inhibition and analysed after cleavage with *Eco*RI endonuclease (Figs 2,4). In uninhibited reactions, no replication occurred from the M13 origin[17] due to the absence of the M13-encoded gene 2 protein required for replication of supercoiled M13 DNA[25–27]. Inhibition of replication as measured by nucleotide incorporation correlated well with the effect of ddTTP in decreasing the average length of the replicated segment (Fig. 5). These findings and similar results for pSY317 (data not shown) confirm that the "eye" and "Y" forms (molecules with a single fork) are replication intermediates.

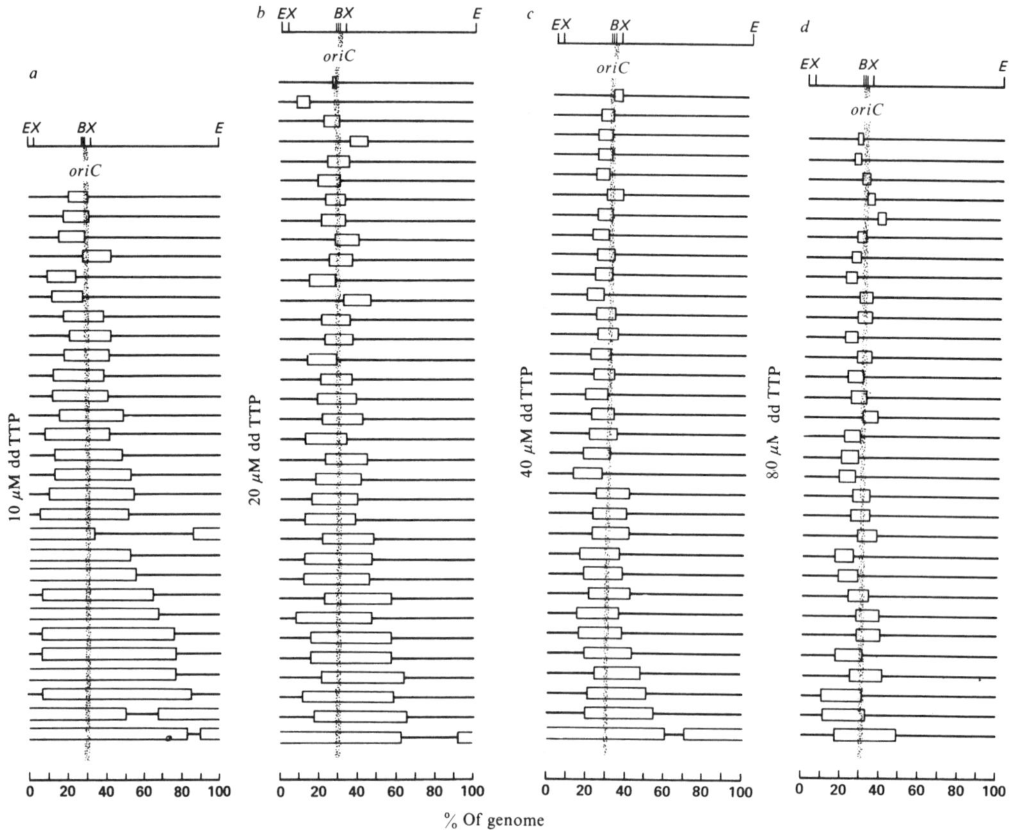

Fig. 4. Replication of M13*oriC*26 DNA initiates at *oriC* and progresses bidirectionally. Reactions were performed as described in Fig. 1 legend in a volume of 125 μl containing ddTTP (as indicated) and 1 mg of enzyme fraction II prepared from *E. coli* C600 as described elsewhere[17]. Incubation was at 30°C for 15 min. Replicative intermediates produced with *a*, 10 μM; *b*, 20 μM; *c*, 40 μM, and *d*, 80 μM ddTTP from samples prepared as described in Fig. 2 were restricted, after treatment with pancreatic RNase, with an excess of *Eco*RI endonuclease, spread on to parlodion-coated grids and examined. Replicative intermediates at each ddTTP concentration were selected randomly, photographed and measured. Only molecules having a correct total length (±10%) in which both replicated segments were the same length (±5%) were included in the analysis. Molecules are aligned so that the longer unreplicated segment is to the right, and they are arranged according to extent of replication. The position of *oriC* is indicated by the stippled area. The open boxes represent the replication "bubble". Restriction endonuclease sites are the same as for Fig. 3.

为了制备 M13*oriC*26 DNA 的复制中间体，我们在反应体系中使用了分离自野生型细胞的酶组分（组分 II）[17]，并且用不同浓度的 ddTTP 终止反应，然后用 *Eco*RI 限制性内切酶进行切割，最后对生成的产物进行分析（图 2，4）。在没有抑制的反应中，由于缺乏 M13 超螺旋 DNA 复制所必需的 M13 编码的基因 2 的蛋白质产物[25-27]，在 M13 的复制起点[17]没有观察到复制的发生。使用掺入核苷酸的方法检测的复制抑制情况与使用 ddTTP 后引起的平均复制片段长度减小的情况非常吻合（图 5）。这些发现以及使用 pSY317 质粒时得到的类似的实验结果（数据未展示）证实"眼睛"样结构和"Y"型结构（只含有一个复制叉的分子）就是复制中间体。

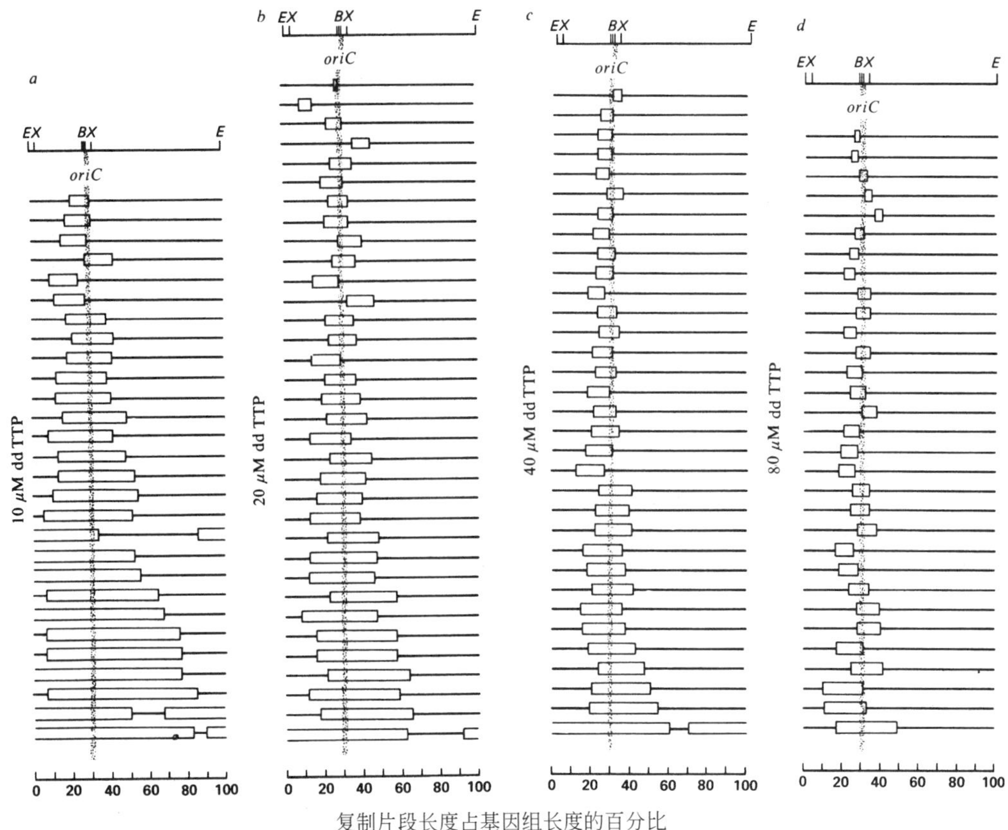

图 4. M13*oriC*26 DNA 的复制起始于 *oriC* 位点并且双向进行。反应按照图 1 注中所述的方法进行，反应体积 125 μl，含有 ddTTP（如图中所标注的浓度）以及 1 mg 酶组分 II（按照其他文献所述的方法[17]从大肠杆菌 C600 中制备）。30℃孵育 15 min。按照图 2 中所述的方法，分别用（a）10 μM、（b）20 μM、（c）40 μM、（d）80 μM 的 ddTTP 处理细胞，来制备复制中间体。用胰 RNA 酶消化样品中的 RNA 后，用过量的 *Eco*RI 限制性内切酶消化样品，然后将纯化的 DNA 铺展在包被了火棉胶片的载网上，使用电子显微镜观察。随机挑选在各种浓度的 ddTTP 作用下产生的复制中间体，拍照并测量长度。只有具有正确总长度（±10%）且两条复制片段长度相等（±5%）的分子才会被列入分析范围。将这些分子按照复制片段的长度进行排列，并使较长的未复制的片段位于右端。图中点带区域所示为 *oriC* 的位置。方框所示为"复制泡"。限制性内切酶的位点与图 3 中相同。

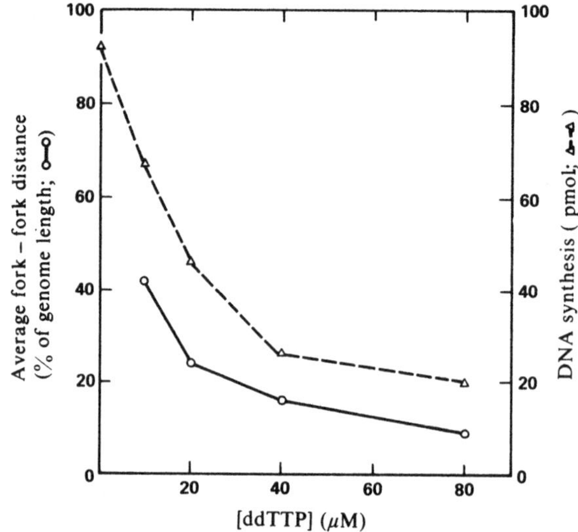

Fig. 5. Inhibition of DNA synthesis by the chain terminator ddTTP correlates with the fork-to-fork distance of replicative intermediates of M13*oriC*26. After incubation, aliquots (25 μl) from the reactions described in Fig. 4 legend were removed for TCA precipitation and counted to determine the amount of DNA synthesis. The replicative intermediates of Fig. 4 were then analysed individually for the extent of replication and averaged. Each point represents the average fork-to-fork distance of 28–34 replicative intermediates generated at the corresponding ddTTP concentration.

The minimal *oriC* sequence is ~30% of the genome length from one end of M13*oriC*26 DNA linearized by *Eco*RI endonuclease (Fig. 4). Recombinant M13 phage DNAs from which *oriC* has been deleted do not replicate from *oriC* either *in vitro* or *in vivo*[17,28]. Molecules aligned in Fig. 4 with the longer replicated segment to the right are consistent with replication initiating from the *oriC* segment. More extensively replicated molecules clearly indicate that replication proceeds bidirectionally in most cases, and the less extensively replicated molecules show initiation to be at or near *oriC*.

Replicative intermediates of M13*oriC*26 (Fig. 4) were analysed individually with respect to the extent of replication to the right and left of *oriC*. These values were averaged for each ddTTP concentration and plotted as the distance rightwards or leftwards relative to *oriC* (Fig. 6). The results indicate that for a population of molecules, replication from *oriC* is bidirectional.

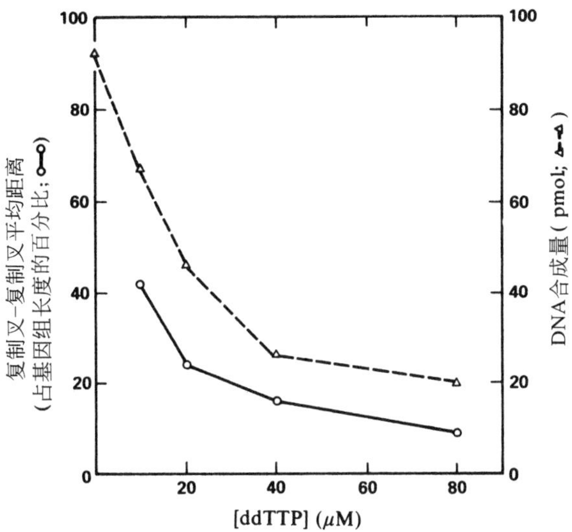

图 5. 链终止子 ddTTP 对 DNA 合成的抑制作用与 M13*oriC*26 复制中间体的复制叉之间的距离相关。反应体系按图 4 注描述的方法配制，孵育后取 25 μl，TCA 沉淀后计数，以确定 DNA 合成量。逐个分析图 4 中所述的复制中间体的复制进行的程度，并计算平均值。图中的每一个点代表在相应浓度的 ddTTP 作用下，28 到 34 个复制中间体的复制叉之间的距离的平均值。

最短的 *oriC* 序列位于距离 *Eco*RI 限制性内切酶消化的线性 M13*oriC*26 DNA 的一端大约 30% 基因组长度的地方（图 4）。敲除掉 *oriC* 位点的重组 M13 噬菌体 DNA 不管是在体外还是体内都无法从 *oriC* 位点起始复制 [17,28]。图 4 中排列对齐的那些在右侧具有较长的复制片段的分子与起始于 *oriC* 片段的复制一致。复制区较长的复制中间体分子表明，在多数情况下复制都是双向进行的，复制区较短的复制中间体分子表明复制起始于 *oriC* 位点或其附近的区域。

为了研究复制向 *oriC* 位点左右两端进行的情况，我们逐一分析了 M13*oriC*26 的复制中间体（图 4）。我们分别计算了各个 ddTTP 浓度下测得的数据的平均值，并用点图标示复制叉距离 *oriC* 位点左侧或右侧的长度（图 6）。结果表明，对于我们研究的这些分子群体来讲，从 *oriC* 位点起始的复制是双向进行的。

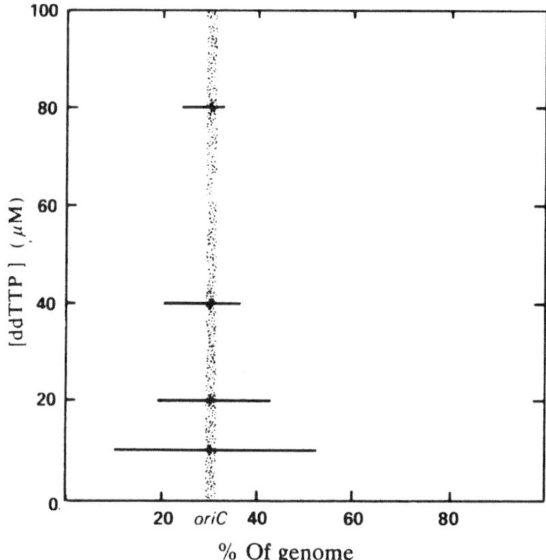

Fig. 6. Replication of M13*oriC*26 DNA is bidirectional for a population of molecules. Replicative intermediates at each ddTTP concentration shown in Fig. 4 were analysed individually for the extent of replication rightwards and leftwards from *oriC*. The values were averaged and plotted relative to the position of *oriC*, indicated by the stippled area, and to the physical map of M13*oriC*26 linearized at the single *Eco*RI site. Molecules (4 of 126) which did not seem to have initiated from *oriC* were excluded from the analysis.

Of 126 M13*oriC*26 replicative intermediates examined, 4 were judged not to have been initiated at *oriC* as the replicated segment did not overlap *oriC* within the error of measurement (5% of the genome length; Fig. 4). These discrepancies may be due to aberrant initiation or improper breakage of the duplex template. For several molecules of both template DNAs, the bidirectional progress of replication appears asymmetric or even unidirectional, which may be due to asynchrony in the initiation of the two replication forks or in chain terminations by ddTTP. In this regard, examination of the replicative intermediates of phage λ and F plasmids, which are known to replicate bidirectionally, also reveals molecules that replicate unidirectionally[29,30].

The bidirectional replication of M13*oriC*26, a template interrupted at the *Xho*I site, contrasts with the analysis of replicative intermediates of the *oriC* plasmids[20], pOC24, produced *in vivo*, in which predominantly unidirectional replication was attributed to an interruption at the same *Xho*I site[21,22]. This discrepancy may be due to differences in either the plasmid sequences or the experimental conditions; in the *in vivo* study, replicative intermediates were detected at a frequency of only 1 in 10^4 plasmid molecules and thus may not represent the principle mode of replication[21].

This electron microscopic study of individual plasmid molecules replicated in a soluble enzyme system is thus consistent with initiation of replication from *oriC*. Fork movement from that point progresses bidirectionally. The soluble enzymatic system used here and elsewhere[17] should allow a biochemical approach to the events of priming, fork movement,

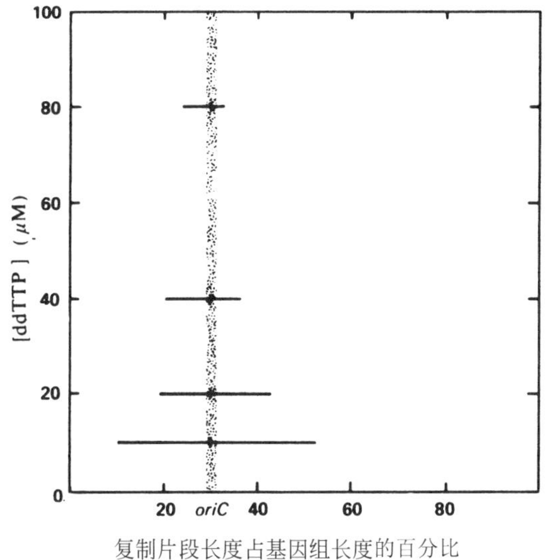

图 6. 对一个质粒分子群体的分析表明，M13*oriC*26 DNA 的复制是双向进行的。我们逐一计算了图 4 中所示的在各个浓度的 ddTTP 作用下产生的复制中间体中，复制从 *oriC* 位点分别向左侧和右侧进行的程度。我们计算了这些数据的平均值，根据与 *oriC* 位点的距离，并对照在 *Eco*RI 单一位点线性化后的 M13*oriC*26 质粒的物理图谱，标注出 *oriC* 的位置，即图中的点带区域。在 126 个分子中有 4 个似乎并不是从 *oriC* 位点起始复制的，我们未分析这些分子。

在我们分析的 126 个 M13*oriC*26 复制中间体中，有 4 个分子被判定为不是从 *oriC* 位点起始复制的，因为它们的复制片段在测量误差范围内（基因组长度的 5%，图 4）没有与 *oriC* 位点重合。这一差异可能是由于复制的异常起始或者双链模板的不正常断裂导致的。对于一些分子的两条 DNA 模板链来说，双向复制过程似乎是不对称的甚至可能是单向的，这有可能是由于两个复制叉的启动是不同步的或由于 ddTTP 对链延伸的终止不是同时的。就这一点而言，在对 λ 噬菌体和 F 质粒（已知它们的复制是双向进行的）的复制中间体进行分析时，也发现了一些单向复制的分子 [29,30]。

M13*oriC*26 是一种在 *Xho*I 位点被打断的双向复制模板，与体内产生的 *oriC* 质粒 [20] pOC24 复制中间体的分析相反，后者主要为单向复制，这是由于该质粒在同样的 *Xho*I 位点被打断 [21,22] 造成的。这一差异可能是由于质粒的序列不同或实验条件的不同而引起的；在体内实验中，只有万分之一的频率检测到复制中间体，因此无法反映复制的主要模式 [21]。

对在可溶性酶系统中复制的单个质粒分子的电子显微镜的研究也表明复制是从 *oriC* 位点起始的。复制叉从这一位点开始的移动是双向的。在本文或其他文献 [17] 中使用的可溶性酶系统，应该能够应用生物化学技术来研究复制的起始、复制叉移动、

regulation of initiation and other events occurring at the *E. coli* origin.

We thank Darrell Dobbertin for assistance in the electron microscopy. This work was supported in part by grants from the NIH and NSF. J.M.K. is a Fellow of the Damon Runyon–Walter Winchell Cancer Fund.

(**296**, 623-627; 1982)

Jon M. Kaguni, Robert S. Fuller and Arthur Kornberg
Department of Biochemistry, Stanford University School of Medicine, Stanford, California 94305, USA

Received 21 December 1981; accepted 8 March 1982.

References:

1. Masters, M. & Broda, P. *Nature New Biol.* **232**, 137-140 (1971).
2. Bird, R. E., Louarn, J., Martuscelli, J. & Caro, L. *J. Molec. Biol.* **70**, 549-566 (1972).
3. Prescott, D. M. & Kuempel, P. L. *Proc. Natl. Acad. Sci. U.S.A.* **69**, 2842-2845 (1972).
4. Hiraga, S. *Proc. Natl. Acad. Sci. U.S.A.* **73**, 198-202 (1976).
5. Yasuda, S. & Hirota, Y. *Proc. Natl. Acad. Sci. U.S.A.* **74**, 5458-5462 (1977).
6. von Meyenburg, K., Hansen, F. G., Nielsen, L. D. & Riise, E. *Molec. gen. Genet.* **160**, 287-295 (1978).
7. Miki, T., Hiraga, S., Nagata, T. & Yura, T. *Proc. Natl. Acad. Sci. U.S.A.* **75**, 5099-5103 (1978).
8. Oka, A., Sugimoto, K. & Takanami, M. *Molec. gen. Genet.* **178**, 9-20 (1980).
9. Hirota, Y., Mordon, J. & Jacob, F. *J. Molec. Biol.* **53**, 369-387 (1970).
10. Beyersmann, D., Messer, W. & Schlicht, M. *J. Bact.* **118**, 783-789 (1974).
11. Wada, C. & Yura, T. *Genetics* 77,199-220 (1974).
12. Lark, K. G. & Wechsler, J. A. *J. Molec. Biol.* **92**, 145-163 (1975).
13. Wechsler, J. A. *J. Bact.* **121**, 594-599 (1975).
14. Zyskind, J. W. & Smith, D. W. *J. Bact.* **129**, 1476-1486 (1977).
15. Carl, P. *Molec. gen. Genet.* **109**, 107-122 (1970).
16. Lark, K. G. *J. Molec. Biol.* **64**, 47-60 (1972).
17. Fuller, R. S., Kaguni, J. M. & Kornberg, A. *Proc. Natl. Acad. Sci. U.S.A.* **78**, 7370-7374 (1981).
18. Hershfield, V., Boyer, H. W., Yanofsky, C., Lovett, M. A. & Helinski, D. R. *Proc. Natl. Acad. Sci. U.S.A.* **71**, 3455-3459 (1974).
19. Kaguni, J. M., LaVerne, L.S. & Ray, D. S. *Proc. Natl. Acad. Sci. U.S.A.* **76**, 6250-6254 (1979).
20. Messer, W. *et al. Cold Spring Harb. Symp. Quant. Biol.* **43**, 139-145 (1979).
21. Meijer, M. & Messer, W. J. *J. Bact.* **143**, 1049-1053 (1980).
22. Messer, W., Heimann, B., Meijer, M. & Hall, S. *ICN-UCLA Symp. Molec. Cell. Biol.* **19**, 161-169 (1980).
23. Conrad, S. E., Wold, M. & Campbell, J. L. *Proc. Natl. Acad. Sci. U.S.A.* **76**, 736-740 (1979).
24. Davis, R. W., Simon, N. M. & Davidson, N. *Meth. Enzym.* **21**, 413-428 (1971).
25. Lin, N. S.-C. & Pratt, D. *J. Molec. Biol.* **72**, 37-49 (1972).
26. Fidánian, H. M. & Ray, D. *J. Molec. Biol.* **72**, 51-63 (1972).
27. Meyer, T. F. & Geider, K. *Single-stranded DNA Phages* (eds Denhardt, D. T., Dressler, D. & Ray. D. S.) 389-392 (Cold Spring Harbor Laboratory, New York, 1978).
28. Kaguni, L. S., Kaguni, J. M. & Ray, D. S. *J. Bact.* **145**. 974-979 (1981).
29. Eichenlaub, R., Figurski, D. & Helinski, D. R. *Proc. Natl. Acad. Sci. U.S.A.* **74**, 1138-1141 (1977).
30. Schnöss, M. & Inman, R. B. *J. Molec. Biol.* **51**, 61-73 (1970).
31. Shortle, D. & Nathans, D. *Proc. Natl. Acad. Sci. U.S.A.* 75, 2170-2174 (1978).

复制起始的调控以及在大肠杆菌复制起点发生的其他事件[31]。

感谢达雷尔·多贝廷在电子显微镜实验中的帮助。本文部分工作受美国国立卫生研究院和美国国家科学基金会的资金的支持。卡古尼是戴蒙·鲁尼恩–沃尔特·温切尔癌症基金会的研究员。

（张锦彬 翻译；梁前进 审稿）

Human EJ Bladder Carcinoma Oncogene is Homologue of Harvey Sarcoma Virus *ras* Gene

L. F. Parada *et al.*

Editor's Note

Here US cancer researcher Robert Weinberg and colleagues reveal that two seemingly very different types of cancer-causing "oncogene" are in fact very similar at the DNA level. One of the oncogenes, *ras*, was previously identified as the cancer-causing agent in the Harvey sarcoma virus. The other triggers cancerous growth of bladder cells in an entirely non-viral context. Their homology suggested that both types of oncogene might be derived from the same benign, cellular genetic element or "proto-oncogene", and that a single proto-oncogene can be activated by different molecular processes. The concept that proto-oncogenes can become cancer-causing when mutated is now well accepted, and all it takes for the normal *ras* gene to become activated is mutation of a single base.

Examination of homologies between retroviral oncogenes and transforming sequences defined by transfection reveals that the human bladder carcinoma (EJ) oncogene is homologous to the Harvey sarcoma virus oncogene (*ras*). Structural analysis limits the region of homology to a 3.0-kilobase *Sac*I fragment of the EJ oncogene. Both EJ and *ras* DNA probes detect similar transcripts in transfectants derived from bladder carcinoma cell lines.

TWO groups of cellular oncogenes have been discovered during the past decade. The first consists of genes that were characterized by virtue of their association with retroviruses. The prototype of this class is the *src* gene of avian sarcoma virus. Several experiments have indicated that this gene was acquired from the chicken genome by an avian retrovirus, and has been exploited by the chimaeric virus to transform cells[1,2]. Once incorporated into the viral genome, expression of the *src* gene is driven by viral controlling elements and is no longer responsive to control mechanisms that governed its expression while in the cellular chromosome. In addition to the *src* gene, this group includes at least 12 other gene sequences, each associated with a different chimaeric retrovirus[3,4]. These genes are conserved over great evolutionary distances[2,5] implying that they mediate essential cellular or organismic functions.

A second class of cellular transforming genes has been detected by the experimental route of DNA transfection. Recent reports have indicated that the DNAs of some non-virally induced tumour cell lines can induce transformation when applied to mouse

人类EJ膀胱癌癌基因是哈维肉瘤病毒*ras*基因的同源基因

帕拉达等

编者按

美国癌症研究人员罗伯特·温伯格和他的同事指出，两种看似非常不同的引发癌症的“癌基因”在DNA水平上实际上是非常相似的。*ras*作为一种癌基因，之前就已被鉴定为哈维肉瘤病毒的致癌因子；而另外一种癌基因则在完全无病毒的环境下仍可以触发膀胱细胞的癌变。它们的同源性说明两个类型的癌基因可能起源于同一良性基因，即相同的细胞遗传元件或者“原癌基因”，且单一的原癌基因可以通过不同的分子过程激活。当突变发生时，原癌基因就变为致癌性的，这个观点现已广为接受；且仅仅只是某个单一碱基的突变就足以导致正常*ras*基因被激活。

逆转录病毒癌基因和通过转染确定的转化序列的同源性研究显示人类膀胱癌（EJ）癌基因是哈维肉瘤病毒癌基因（*ras*）的同源基因。结构分析将该同源区域限定在EJ癌基因的一个3,000个碱基对的*Sac*I片段上。在膀胱癌细胞系内，用EJ和*ras* DNA探针检测到了相似的转录物。

在过去的十年里，人们发现了两组不同的细胞癌基因。第一组由与逆转录病毒关联为特点的基因组成。这一类的原型是禽肉瘤病毒的*src*基因。多个实验表明该基因是通过禽类逆转录病毒从鸡基因组中获得的，并被这个嵌合型病毒利用来转化细胞[1,2]。一旦整合到病毒基因组中，*src*基因的表达就由病毒的控制元件驱动，而不再受细胞染色体中支配其表达的控制机制的影响。除了*src*基因之外，这一组基因还包括至少12个其他基因序列，每一个都与不同的嵌合型逆转录病毒相关[3,4]。这些基因即使在较远的进化距离上也非常保守[2,5]，表明它们介导着非常重要的细胞学或者生物学功能。

通过DNA转染的实验方法检测到了第二类细胞转化基因。最近的报告指出，一些非病毒诱导的肿瘤细胞系的DNA在导入小鼠单层成纤维细胞后能够诱导其转

fibroblast monolayers. These tumour cell lines are derived from chemically induced animal tumours[6–9] and from human tumours of spontaneous origin[7,10–12].

The two classes of oncogenes have many properties in common, the most striking of which is the apparent origin of both types of genes from normally benign, cellular genetic elements. Because of this and other parallels, we undertook a search to determine whether the two groups of genes shared any members in common. Such overlap would have far-reaching consequences for our understanding of the mechanisms of viral and non-viral carcinogenesis.

Detection of Homologies by Nucleic Acid Hybridization

The search for relatedness between the two groups of genes depended on detection of nucleic acid sequence homologies between individual members of each group. One group consisted of a series of seven retrovirus-associated *onc* genes known to be unrelated or only distantly related to one another[4]. The other group studied was a collection of seven tumour oncogenes that had been defined by transfection[7,10]. Nucleic acid sequence probes have been derived for the retrovirus-associated genes, whereas only a few of the transfection-derived genes have been isolated in the form of molecular clones. We therefore used the virus-derived *onc* probes to survey the DNAs of cells which had acquired, via transfection, copies of the second class of genes. The survey was performed using the Southern gel-filter transfer procedure[13]. Virus-derived *onc* probes were from several sources (see Table 1).

Table 1. Oncogene probes

Oncogene ‡	Related virus or tumour	Clone	Sequence complexity	Source	Ref.
v-*abl*	Murine Abelson leukaemia virus	pAblsub9	3.0kbp	J. Y. J. Wang and D. Baltimore	*
v-*erb*	Avian erythroblastosis virus	PAE-*Pvu*II	2.5kbp	T. Gonda and J. M. Bishop	28
v-ST-*fes*	Snyder-Theilen feline sarcoma virus	PST-3	0.6kbp	C. Sherr	29
c-Ha-*ras*l (rat)	Murine Harvey sarcoma virus	LHXB-3	2.3kbp	R. Ellis and E. Scolnick	30
v-Ha-*ras*	Murine Harvey sarcoma virus	BS9	0.45kbp	R. Ellis and E. Scolnick	21
c-*mos*(Human)	Murine Moloney sarcoma virus	pHM1	2.75kbp	G. Vande Woude	†
v-*myc*	Avian myelocytomatosis virus	puMyC3-*Pst*	1.5kbp	T. Gonda and J. M. Bishop	31
v-*src*	Avian sarcoma virus	SRA-2	0.8kbp	T. Gonda and J. M. Bishop	32
	Human EJ bladder carcinoma	pEJ6.6	6.6kbp	C. Shih	18

* J. Y. J. Wang and D. Baltimore, in preparation. † G. Vande Woude, personal communication. ‡ v-indicates viral gene; c-indicates cellular gene.

We used each retrovirus-related *onc* probe to search for novel cross-reacting fragments in each of several transfectants. For example, an analysis using the v-*abl* probe is shown in Fig. 1: lane *a* represents the endogenous mouse sequences detected in the DNA of

化。这些肿瘤细胞系来自化学诱导的动物肿瘤[6-9]和人类自发形成的肿瘤[7,10-12]。

这两类癌基因有许多共同的特性，其中最突出的就是两种基因显然皆起源于正常良性的细胞遗传元件。由于这个以及其他类似的原因，我们试图探究确定这两种基因是否具有共同的组分。这种基因间的重叠可能对我们了解病毒以及非病毒致癌过程的机制具有深远的意义。

通过核酸杂交检测同源性

探求两组基因之间的相关性要依靠每组个体成员之间核酸序列同源性的检测。其中一组由七个与逆转录病毒相关的 *onc* 基因组成，并且已知这七个基因之间彼此无关联或者关系甚远[4]。另一组用于研究的基因经转染已被确定为七个肿瘤癌基因的集合[7,10]。逆转录病毒相关基因的核酸序列探针已经获得，但是仅有少量转染得到的基因以分子克隆的形式被分离出来。因此，我们使用病毒来源的 *onc* 探针进行 DNA 印迹分析来检测那些通过转染获得的第二类基因拷贝的细胞的 DNA[13]。病毒 *onc* 探针有多种来源（见表 1）。

表 1. 癌基因探针

癌基因‡	相关病毒或肿瘤	克隆	序列复杂性	来源	参考文献
v-*abl*	鼠艾贝尔森白血病病毒	pAblsub9	3.0 kbp	王和巴尔的摩	*
v-*erb*	鸟类成红细胞增多症病毒	PAE-*Pvu*II	2.5 kbp	贡达和毕晓普	28
v-ST-*fes*	猫科肉瘤病毒	PST-3	0.6 kbp	谢尔	29
c-Ha-*ras*1(大鼠)	鼠哈维肉瘤病毒	LHXB-3	2.3 kbp	埃利斯和斯科尔尼克	30
v-Ha-*ras*	鼠哈维肉瘤病毒	BS9	0.45 kbp	埃利斯和斯科尔尼克	21
c-*mos*(人)	鼠莫洛尼肉瘤病毒	pHM1	2.75 kbp	范德伍德	†
v-*myc*	鸟类成髓细胞增生症病毒	puMyC3-*Pst*	1.5 kbp	贡达和毕晓普	31
v-*src*	鸟类肉瘤病毒	SRA-2	0.8 kbp	贡达和毕晓普	32
	人类 EJ 膀胱癌	pEJ6.6	6.6 kbp	施	18

* 王和巴尔的摩，论文撰写中；† 范德伍德，个人交流；‡ v- 表示病毒基因，c- 表示细胞基因。

我们使用每个逆转录病毒相关 *onc* 基因探针在众多转染子中逐个搜索新的交叉反应片段。比如，图 1 显示了用 v-*abl* 探针进行的分析：泳道 *a* 表示用 v-*abl* 探针在未转染的小鼠成纤维细胞 NIH 3T3 细胞 DNA 中检测到的内源性小鼠序列。这些

untransfected NIH 3T3 cells by the v-*abl* sequence probe. These "background" bands of 28, 10 and 6 kilobases (kb) were also found in the DNAs of all transfectants (Fig. 1, lanes *b*, *d-g* and *i*). Figure 1*c* shows detection of the human homologue of the v-*abl* probe; lanes *d-g* show an analysis of DNAs of mouse cells transfected with four different human oncogenes. None of the DNAs contains any fragments beyond those present in the untransfected mouse control (lane *a*). We concluded that none of these transfected cells acquired the human homologue of the v-*abl* sequence. From a similar analysis in lanes *h* and *i*, we concluded that the rabbit bladder carcinoma oncogene is also not related to the *abl* gene. A more equivocal interpretation came from analysis of the DNA of a mouse cell transfected with a mouse fibroblast oncogene (Fig. 1*a*, *b*). Due to the lack of species-specific fragment markers, we were unable to rule out the identity of the *abl* and fibroblast oncogene. However, knowledge of their restriction enzyme cleavage sites excludes identity[14,15].

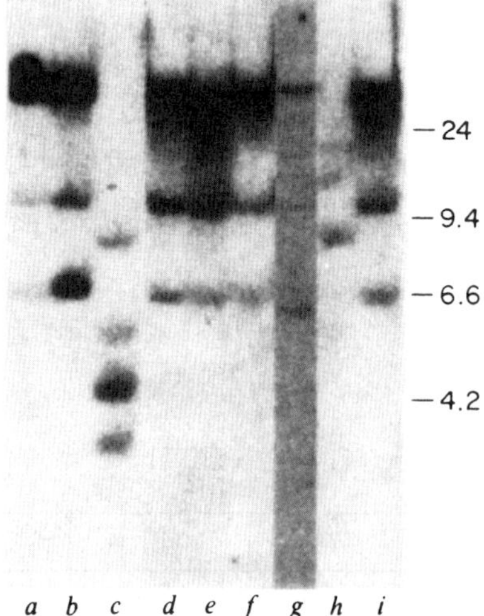

Fig. 1. Southern blot analysis of digested cellular DNAs from various transfectants probed with v-*abl* specific probe (see Table 1). 10 μg of each DNA were digested with endonuclease *Eco*RI, fractionated by electrophoresis through a 1% agarose gel and transferred to nitrocellulose paper[13]. The filters were incubated with 5×10^6 c.p.m. of nick-translated[39] ^{32}P-labelled Abelson virus specific probe. The DNAs analysed were from the following cell lines: *a*, NIH 3T3; *b*, Y5-1; *c*, HeLa; *d*, A5-2; *e*, SH-1-1; *f*, SW-2-1; *g*, EJ-6-1; *h*, rabbit embryo fibroblast; *i*, RBC-1. Lanes *a-f* and *h*, *i* are from one filter; *g* is from a different filter. The transfected cell lines are described in Table 2. Migration of *Hin*dIII-digested λ DNA fragments are shown at the right (in kilobase pairs, kbp).

Figure 2 shows a further Southern blot analysis of DNAs that were prepared from several cell lines derived by transfection of NIH 3T3 cells with DNAs of human tumour cell lines. The probe used was the rat cellular homologue of Harvey sarcoma virus, c-Ha-*ras*1, given by R. Ellis and E. Scolnick[16,17]. A novel fragment of 9.5 kb was found in the DNA of a mouse cell transfected with DNA of the EJ human bladder carcinoma cell line (Fig. 2*c*). In contrast, no novel fragments were present in the transfectants derived from other tumour cell lines (Fig. 2*d-g*).

28kb、10kb、6kb 的背景条带在所有的转染子 DNA 中都能找到（图 1，泳道 *b*、*d*~*g* 和 *i*）。图 1*c* 为使用 v-*abl* 探针对人类同源基因的检测结果；泳道 *d*~*g* 为四种不同的人类癌基因转染的小鼠细胞 DNA 的分析结果。结果显示这些 DNA 仅含有未转染的对照小鼠中含有的 DNA 片段（泳道 *a*）。因此我们的结论是：这些转染的细胞都没有获得 v-*abl* 序列的人类同源基因。对泳道 *h* 和 *i* 进行类似的分析，我们得出结论：兔膀胱癌癌基因也与 *abl* 基因无关。但是，对转染了小鼠成纤维细胞癌基因的小鼠细胞的 DNA 进行分析只能得到更为模棱两可的解释（图 1*a*, *b*）。由于缺乏物种特异性的片段标记物，我们不能排除 *abl* 和成纤维细胞癌基因是同一个基因的可能性。但是，对它们限制性内切酶酶切位点的分析结果排除了这种同一性[14,15]。

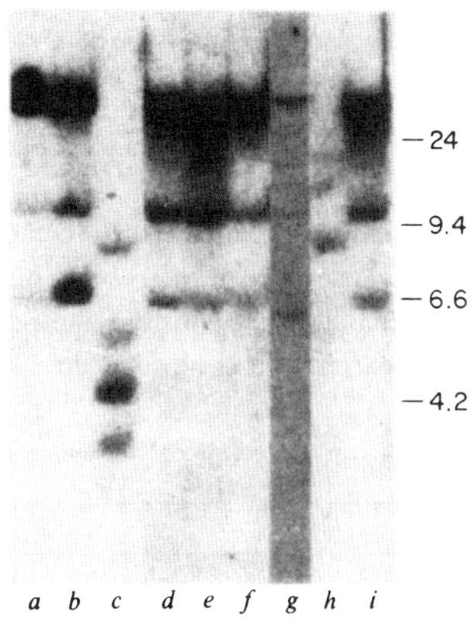

图 1. 用 v-*abl* 特异性探针对被消化的多个转染子来源的细胞 DNA 的 DNA 印迹分析结果（见表 1）。每种 DNA 取 10 μg，用核酸内切酶 *Eco*RI 进行酶切，经 1% 琼脂糖凝胶电泳分离后转移到硝酸纤维素膜上[13]。滤膜与 5×10^6 c.p.m. 切口平移法[39]制备的 ^{32}P 标记的艾贝尔森病毒特异性探针一起温育。所分析的 DNA 来源于以下细胞系：*a*，NIH 3T3；*b*，Y5-1；c，HeLa；*d*，A5-2；*e*，SH-1-1；*f*，SW-2-1；*g*，EJ-6-1；*h*，兔胚胎成纤维细胞；*i*，RBC-1。泳道 *a*~*f* 和 *h*，*i* 来源于同一滤膜，*g* 来源于另一张滤膜。表 2 描述了转染的细胞系，右侧显示了 *Hin*dIII 消化的 λ DNA 片段的迁移情况（以千碱基对，即 kbp 为单位）。

图 2 显示了用人类肿瘤细胞系的 DNA 转染 NIH 3T3 细胞获得的多个细胞系的 DNA 的进一步 DNA 印迹分析结果。实验使用的探针是埃利斯和斯科尔尼克提供的哈维肉瘤病毒的大鼠细胞同源基因 c-Ha-*ras*1[16,17]。我们在人类 EJ 膀胱癌细胞系 DNA 转染的小鼠细胞 DNA 中找到了一个长度为 9.5 kb 的新片段（图 2*c*）。与此相反的是，在其他肿瘤细胞系获得的转染子中则没有发现新的片段（图 2*d*~*g*）。

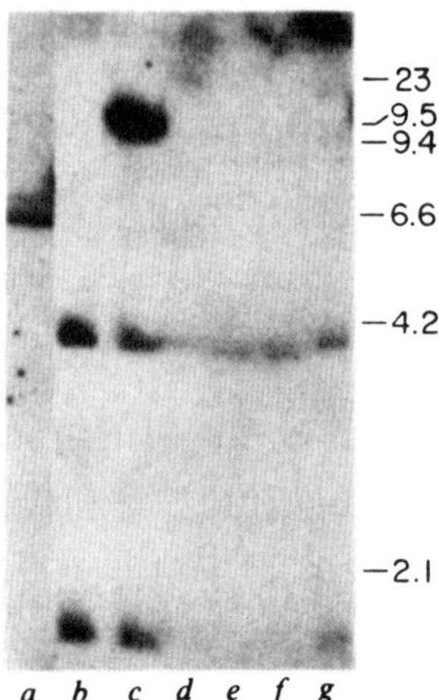

Fig. 2. Southern blot analysis of DNA from transfected cells with rat cellular *ras* probe (c-Ha-*ras*1). ^{32}P-labelled c-Ha-*ras*1 DNA was prepared as described in Fig. 1 legend and incubated with a filter carrying *Bam*HI-digested DNAs from the following cell lines: *a*, HeLa; *b*, NIH 3T3; *c*, EJ-6-1; *d*, A5-2; *e*, SH-1-1; *f*, SW-2-1; *g*, HL-60-9. Lanes *b-g* are from the same filter. The cell lines are described in Table 2.

Of the seven *onc* probes used in this survey, only one detected the presence of novel DNA fragments in transfection-derived cell lines. The results of this comparative oncogene survey are summarized in Table 2. Below we consider the relationship between the c-Ha-*ras*1 gene and the human bladder carcinoma gene suggested by Fig. 2.

Table 2. Comparison of transfected oncogenes with retrovial *onc* probes

onc Probes						Transfected celll lines		
	Y5-1-1	EJ-6-1	A5-2	SH-1-1	SW-2-1	HL60-1-9	B104-1-1	RBC-1
v-*abl*	*	–	–	–	–	–	NT	–
v-*erb*	*	–	–	–	–	–	NT	NT
v-*fes*	*	–	–	–	–	–	–	NT
c-Ha-*ras*1(rat)	*	+	–	–	–	–	–	NT
c-*mos* (Human)	*	–	NT	NT	–	–	NT	NT
v-*myc*	*	–	–	–	–	–	–	NT
v-*src*	*	–	–	NT	–	–	NT	–

The oncogene probes shown are described in Table 1. The transfected cell lines listed are of the following origins: Y5-1-1 is derived from two serial passages of the oncogenic DNA from a 3-methylcholantherene-induced mouse fibroblast cell line, MCA-16 (refs 6, 33); EJ-6-1 is a secondary transfected cell line derived from DNA of a human bladder carcinoma cell line, EJ[7]; A5-2 is a primary transfected cell line derived from human lung carcinoma cell line, A549 (ref. 34); SH-1-1 is a secondary transfected cell line derived from DNA from human neuroblastoma cell line, SK-N-SH (J. Føgh, personal communication); SW-2-1 is a secondary transfected cell line derived from human colon carcinoma cell line, SW-480 (refs 10, 35); HL60-1-9 is a secondary transfected cell line derived from human leukaemia cell line, HL60 (refs 10, 36); B104-1-1 is a secondary transfected cell line derived from rat neuroblastoma cell line, B104 (refs 7, 37); RBC-1 is a primary transfected cell line derived from rabbit bladder carcinoma cell line, RBC[7,38]. + Indicates hybridization of the probe to novel bands in addition to hybridization with NIH 3T3 cellular sequences; – indicates a well controlled negative correlation. * Apparent negative correlation that is not based on species-specific

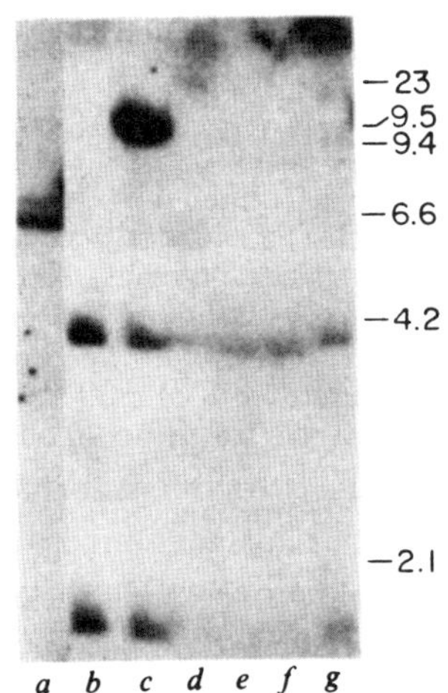

图 2. 用大鼠细胞 *ras* 探针（c-Ha-*ras*1）对转染细胞 DNA 进行 DNA 印迹分析。将如图 1 注中所述方法制备的 ^{32}P 标记的 c-Ha-*ras*1 DNA 与含有 *Bam*HI 消化的来源于如下细胞系的 DNA 的滤膜放在一起进行孵育：*a*，HeLa; *b*，NIH 3T3; *c*，EJ-6-1; *d*，A5-2; *e*，SH-1-1; *f*，SW-2-1; *g*，HL-60-9。泳道 *b*~*g* 来自同一滤膜。表 2 描述了相关细胞系。

这个研究中使用的七个 *onc* 探针中，只有一个在转染细胞系中检测到了新的 DNA 片段。表 2 总结了癌基因检测比较的研究结果。下面我们讨论图 2 揭示的 c-Ha-*ras*1 基因与人膀胱癌基因的关系。

表 2. 逆转录病毒 *onc* 探针对转染癌基因的比较

onc 探针						转染细胞系		
	Y5-1-1	EJ-6-1	A5-2	SH-1-1	SW-2-1	HL60-1-9	B104-1-1	RBC-1
v-*abl*	*	–	–	–	–	–	NT	–
v-*erb*	*	–	–	–	–	–	NT	NT
v-*fes*	*	–	–	–	–	–	–	NT
c-Ha-*ras*1（大鼠）	*	+	–	–	–	–	–	NT
c-*mos*（人）	*	–	NT	NT	–	–	NT	NT
v-*myc*	*	–	–	–	–	–	–	NT
v-*src*	*	–	–	NT	–	–	NT	–

表中显示的癌基因探针在表 1 中描述过。所列出的转染细胞系来源如下：Y5-1-1 是用 3 – 甲基胆蒽诱导的小鼠成纤维细胞系 MCA-16（参考文献 6,33）致癌 DNA 两次连续传代得到的细胞系；EJ-6-1 是从人膀胱癌细胞系 EJ[7] DNA 得到的二级转染细胞系；A5-2 是从人肺癌细胞系 A549（参考文献 34）得到的一级转染细胞系；SH-1-1 是从人神经母细胞瘤细胞系 SK-N-SH（福格，个人交流）DNA 得到的二级转染细胞系；SW-2-1 是从人结肠癌细胞系 SW-480（参考文献 10，35）得到的二级转染细胞系；HL60-1-9 是从人白血病细胞系 HL60（参考文献 10, 36）得到的二级转染细胞系；B104-1-1 是从大鼠神经母细胞瘤细胞系 B104（参考文献 7, 37）得到的二级转染细胞系；RBC-1 是从兔膀胱癌细胞系 RBC[7,38] 得到的一级转染细胞系。+ 表示除了与 NIH 3T3 细胞序列杂交外，探针还能与新的条带杂交；– 表示完全匹配的阴性对照。* 表示不基于可以区分受体细胞中固有染色体 DNA 与供体 DNA 的物种特异性 DNA 片段大小的阴性对照。NT 指未检测。

DNA fragment sizes that distinguish donor DNAs from resident chromosomal DNAs of the recipient cells. NT, not tested.

Homology of the Two Oncogene DNAs

A series of tests was performed to further substantiate the relationship between the EJ bladder oncogene and the c-*ras* oncogene. Figure 3*A* shows that nine cell lines derived by transfection of EJ bladder carcinoma DNA have all acquired novel DNA fragments reactive with the c-Ha-*ras*l probe. DNA from untransfected mouse cells (Fig. 3*A*, lane *a*) does not exhibit any of these novel fragments. The *Bam*HI-digested DNAs of the various transfectants exhibit differently sized novel fragments because of rearrangements occurring during the transfection process. Note that the oncogene of the T24 human bladder carcinoma is closely related to that of the EJ bladder carcinoma[18,19]. Figure 3*A*, lane *m*, indicates that a mouse transfectant carrying the T24 oncogene also contains in its DNA a novel acquired fragment reactive with the probe.

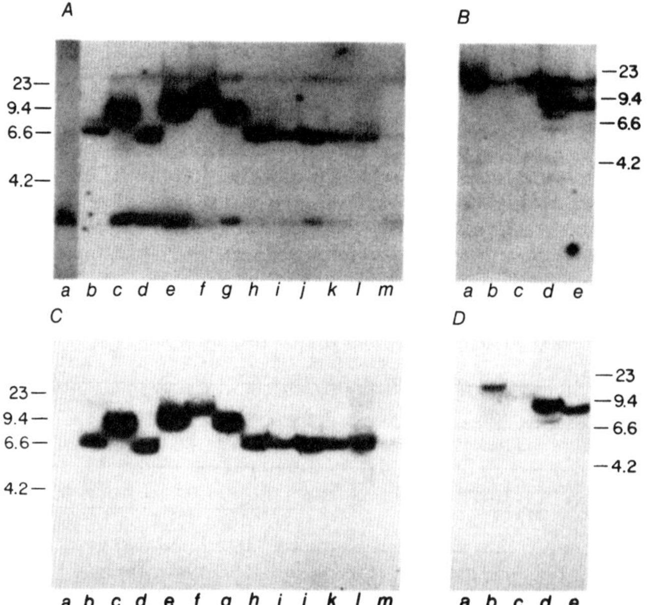

Fig. 3. Analysis of DNAs from EJ and T24 transfectants using the c-Ha-*ras*1 and EJ oncogene probes. DNAs were digested with endonuclease *Bam*HI (*A*, *C*) or *Eco*RI (*B*, *D*) and analysed as described in Fig. 1 legend. The filters shown *A* and *B* were incubated with 5×10^6 c.p.m. of c-Ha-*ras*1 DNA probe (2×10^8 c.p.m. μg^{-1}) and exposed for autoradiography for 16 h. The adsorbed ^{32}P-labelled probe was removed from the filters by washing in 0.1 M NaOH, 0.5 M NaCl, 1.0 mM EDTA. The washed filters were exposed to film for 48 h to verify that all the radioactive signal had been removed (not shown). The filters were then incubated with 5×10^6 c.p.m. of nick-translated pEJ6.6 DNA (6×10^7 c.p.m. μg^{-1}). *C* and *D* represent autoradiography after 16 h. The DNAs analysed in panels *A* and *C* were from the following cell lines: *a*, NIH 3T3; *b*, HeLa; *c*, EJ-6-1; *d*, EJ-2-R1; *e*, EJ-2-R5; *f*, E-4-R4-B; *g*, EJ-6-1; *h*, EJ-6-2-R; *i*, EJ-6-3; *j*, EJ-1-2; *k*, EJ-3-2; *l*, T-24 human cells[19]; *m*, T24-8-5. *B* and *D* display DNAs from: *a*, NIH cells; *b*, HeLa cells; *c*, EJ-6-2 (*Bam*)-1 cells; *d*, EJ-6-2(*Bam*)-2 cells; *e*, EJ-4(*Bam*)-1 cells. In *A* and *C*, lanes *c-f* and *h-k* represent DNA from cell lines that were derived from independent serial transfections of the EJ tumour cell oncogene; lanes *g* and *c* are duplicates. Lane 1 contains DNA from human bladder carcinoma cell line T24. Lane *m* contains DNA from a secondary transfectant of T24. Lanes *c-e* of *B* and *D* show an analysis of DNAs from tertiary transfectants induced by exposure to *Bam*HI-cleaved secondary transfectant DNA.

两个癌基因 DNA 的同源性

人们进行了一系列的检测来进一步证实 EJ 膀胱癌癌基因与 c-*ras* 癌基因之间的相关性。图 3*A* 显示了 EJ 膀胱癌 DNA 转染的 9 个细胞系都获得了能与 c-Ha-*ras*1 探针反应的新 DNA 片段。而未转染的小鼠细胞 DNA（图 3*A*，泳道 *a*）不存在任何新片段。由于在转染过程中发生了重排，不同转染子的 DNA 在被 *Bam*HI 消化以后得到了大小不同的新片段。我们注意到 T24 人膀胱癌癌基因与 EJ 膀胱癌癌基因高度相关 [18,19]。图 3*A*，泳道 *m* 显示了一个含有 T24 癌基因的小鼠转染子 DNA 中也含有能与探针反应的新片段。

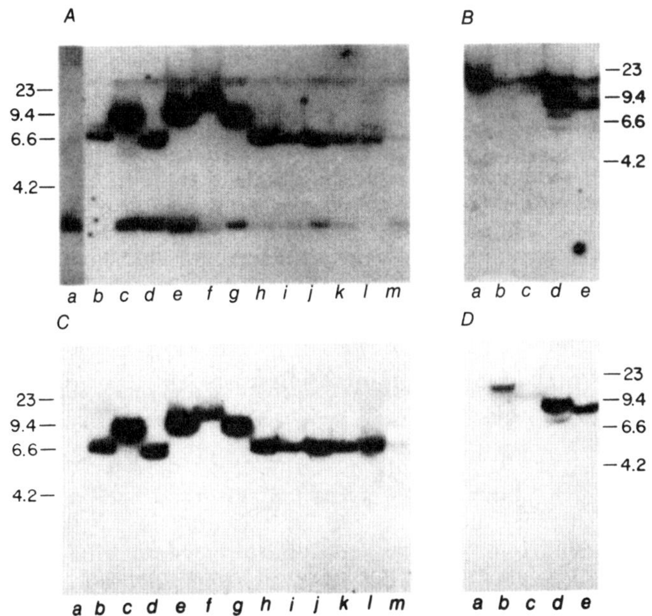

图 3. 用 c-Ha-*ras*1 和 EJ 癌基因探针分析 EJ 和 T24 转染子的 DNA。DNA 用核酸内切酶 *Bam*HI (*A*, *C*) 或者 *Eco*RI (*B*, *D*) 酶切，并用图 1 注所述的方法进行分析。*A* 和 *B* 所示的滤膜与 5×10^6 c.p.m. 的 c-Ha-*ras*1 DNA 探针（2×10^8 c.p.m. μg^{-1}）温育并曝光进行放射自显影 16 小时。用含 0.1 M NaOH，0.5 M NaCl，1.0 mM EDTA 的溶液清洗滤膜以洗脱吸附的 ^{32}P 标记探针。为确保所有的放射性标记都已被去除，将洗过的滤膜在胶片上曝光 48 小时（图中未显示）。然后将这些滤膜与经切口平移法制备的 5×10^6 c.p.m. 的 pEJ6.6 DNA（6×10^7 c.p.m. μg^{-1}）放在一起进行温育。*C* 和 *D* 代表 16 小时后的放射自显影成像。图 *A* 和 *C* 中分析的 DNA 来自如下细胞系：*a*，NIH 3T3；*b*，HeLa；*c*，EJ-6-1；*d*，EJ-2-R1；*e*，EJ-2-R5；*f*，E-4-R4-B；*g*，EJ-6-1；*h*，EJ-6-2-R；*i*，EJ-6-3；*j*，Ej-1-2；*k*，EJ-3-2；*l*，T-24 人细胞 [19]；*m*，T24-8-5。*B* 和 *D* 显示的 DNA 来自：*a*，NIH 细胞；*b*，HeLa 细胞；*c*，EJ-6-2(*Bam*)-1 细胞；*d*，EJ-6-2(*Bam*)-2 细胞；*e*，EJ-4(*Bam*)-1 细胞。在 *A* 和 *C* 中，泳道 *c*~*f* 和 *h*~*k* 代表独立连续转染 EJ 肿瘤细胞癌基因的细胞系的 DNA；泳道 *g* 和 *c* 是完全一样的。泳道 1 含有人膀胱癌细胞系 T24 的 DNA。泳道 *m* 含有 T24 二级转染子的 DNA。*B* 和 *D* 的泳道 *c*~*e* 为 *Bam*HI 酶切后的二级转染子 DNA 诱导的三级转染子的 DNA 的分析结果。

To further define the linkage between the EJ oncogene and the c-Ha-*ras*1 homologous sequences, we analysed the DNAs of three EJ transfectants derived by transfer of *Bam*HI-cleaved DNA (Fig. 3*B*). The donors of these DNAs were secondary transfectants derived previously by two serial passages of the bladder carcinoma oncogene. As endonuclease *Bam*HI does not inactivate the EJ oncogene[18], the transfection of *Bam*HI-cleaved secondary DNA ensured that almost the only human fragment present in resulting tertiary transfectants was the 6.6-kb *Bam*HI fragment bearing the EJ oncogene. The three DNAs of the tertiary transfectants were analysed after *Eco*RI cleavage as it was thought they might have lost some *Bam*HI sites during the second transfection. The DNAs of the three transfectants (Fig. 3*B*, lanes *a-c*) all showed acquired fragments reactive with the c-Ha-*ras*1 probe. This demonstrated that the linkage between the EJ oncogene and the c-Ha-*ras*1 homologous sequences could not be broken by *Bam*HI cleavage.

A further comparison between the genes depended on the fact that the EJ bladder carcinoma oncogene was one that we have recently isolated as a molecular clone[18]. The EJ human bladder oncogene has been cloned as a biologically active *Eco*RI fragment of 16 kb carried by a Charon 4A λ phage vector and termed Φ631. A biologically active 6.6-kb *Bam*HI fragment subclone has been inserted into plasmid vector pBR322 and termed pEJ6.6. All endonucleases shown to cleave within this 6.6 kb insert (Fig. 4) inactivate the focus-inducing activity of this DNA (ref. 18 and C. Shih, unpublished results).

Using the EJ 6.6-kb *Bam*HI fragment and the c-Ha-*ras*1 oncogene clone as sequence probes, we analysed the DNA fragments homologous to these genes in normal human DNA. Both probes detected a 6.6-kb *Bam*HI fragment (Fig. 3*A*, *C*, lanes *b*) and a 23-kb *Eco*RI fragment (Fig. 3*B*, *D*, lanes *b*) in human DNA (see also ref. 20). Furthermore, the EJ bladder oncogene probe detected the same novel fragments in transfected mouse lines (Fig. 3*C*, lanes *c-k*) that were previously detected using the c-Ha-*ras*1 probe (Fig. 3*A*, lanes *c-k*).

Endonuclease-cleaved Φ631 DNA was immobilized on a cellulose nitrate filter and probed with the c-Ha-*ras*1 sequences. Figure 4*a* indicates that homology between c-Ha-*ras*1 and the EJ clone is limited to the 6.6-kb *Bam*HI fragment of the bladder carcinoma oncogene; lane *b* further reduces the domain of homology between the two oncogenes to a 3.0-kb *Sac*I fragment within the 6.6-kb *Bam*HI fragment. Figure 4*f-i* shows a similar experiment to that of lanes *b-e* but in this case BS-9, a v-Ha-*ras* probe, was used (provided by Drs D. Lowy and E. Scolnick). BS-9 is a *ras* specific subclone of the Harvey sarcoma virus genome[21] and is ~450 base pairs (bp) long. This probe includes the 5′ half of the v-*ras* gene. Comparison of left and right panels of Fig. 4 confirms that the viral probe and c-Ha-*ras*1 crosshybridize with identical fragments of the EJ oncogene DNA. Double digests (Fig. 4*g*, *i*) with *Sac*I + *Kpn*I or *Sac*I + *Xba*I indicate that the v-Ha-*ras* homology straddles the *Kpn*I and *Xba*I cleavage sites indicated at the top of Fig. 4. Most of the reactivity of the c-Ha-*ras*1 probe lies in the larger of the two fragments created by these digests (Fig. 4*c*, *e*). The deduced alignments between the EJ oncogene and the v-Ha-*ras* and c-Ha-*ras*1 probes are shown at the top of Fig. 4. The direction of transcription, deduced from the results of the present study and previous data[16,17], is from right to left on the map.

为了进一步确定 EJ 癌基因和 c-Ha-*ras*1 同源序列之间的联系，我们分析了 *Bam*HI 酶切后 DNA 转染而来的三个 EJ 转染子的 DNA（图 3*B*）。这些 DNA 的供体是之前两次连续传代膀胱癌癌基因产生的二级转染子。由于核酸内切酶 *Bam*HI 不会使 EJ 癌基因失活 [18]，因此可以保证，经 *Bam*HI 酶切的二级 DNA 的转染产生的三级转染子中，几乎唯一存在的人类基因片段就是 6.6 kb 的含有 EJ 癌基因的 *Bam*HI 片段。我们分析了经 *Eco*RI 酶切后三个三级转染子 DNA 的产物，因为我们认为在二次转染的过程中它们可能丢失了一些 *Bam*HI 的位点。三个转染子的 DNA（图 3*B*，泳道 *a*~*c*）都显示产生的片段能与 c-Ha-*ras*1 探针反应。这表明 *Bam*HI 酶切并不会破坏 EJ 癌基因与 c-Ha-*ras*1 同源序列之间的关联。

我们最近以分子克隆的形式分离得到了 EJ 膀胱癌癌基因，基因之间的进一步比较正是基于此 [18]。EJ 人类膀胱癌基因已被克隆为 Charon 4A λ 噬菌体载体携带的、具有生物学活性的、16 kb 的 *Eco*RI 片段，并被命名为 Φ631。一个具有生物学活性的 6.6 kb 的 *Bam*HI 片段亚克隆被插入到质粒载体 pBR322 中，被命名为 pEJ6.6。所有能切开这个 6.6 kb 插入子的核酸内切酶（图 4）都能使该 DNA 失去中心诱导活性（参考文献 18 和施，未发表的结果）。

用 EJ 6.6 kb 的 *Bam*HI 片段和 c-Ha-*ras*1 癌基因克隆作为序列探针，我们分析了正常人 DNA 中与这些基因同源的 DNA 片段。两个探针都在人类 DNA 中检测到了一个 6.6 kb 的 *Bam*HI 片段（图 3*A*,*C*,泳道 *b*）和一个 23 kb 的 *Eco*RI 片段（图 3*B*,*D*,泳道 *b*）（见参考文献 20）。此外，EJ 膀胱癌癌基因探针在转染的小鼠细胞系中检测到了与之前用 c-Ha-*ras*1 探针检测到（图 3*C*，泳道 *c*~*k*）的相同的新片段（图 3*A*，泳道 *c*~*k*）。

经核酸内切酶酶切的 Φ631 DNA 被固定到硝酸纤维素膜上后，用 c-Ha-*ras*1 探针进行杂交。图 4*a* 显示了 c-Ha-*ras*1 和 EJ 克隆之间的同源序列仅限于膀胱癌癌基因上 6.6 kb 的 *Bam*HI 片段；泳道 *b* 进一步将两种癌基因的同源域缩小到 6.6 kb *Bam*HI 片段内部 3.0 kb 的 *Sac*I 片段上。图 4*f*~*i* 显示了与泳道 *b*~*e* 类似的实验，只是所使用的探针为一种 v-Ha-*ras* 探针——BS-9（由洛伊博士和斯科尔尼克博士提供）。BS-9 是哈维肉瘤病毒基因组的 *ras* 特异性亚克隆 [21]，大约 450 bp。该探针包含了 v-*ras* 基因的 5′ 端。图 4 的左右泳道的比较证实了病毒探针及 c-Ha-*ras*1 与 EJ 癌基因 DNA 的同一片段发生了交叉杂交反应。*Sac*I+*Kpn*I 或 *Sac*I+*Xba*I 的双酶切结果（图 4*g*，*i*）显示，v-Ha-*ras* 同源片段跨过了图 4 上方所示的 *Kpn*I 和 *Xba*I 酶切位点。c-Ha-*ras*1 探针的反应活性主要位于上述酶切反应形成的两个片段中较大的那个（图 4*c*，*e*）。推断出的 EJ 癌基因与 v-Ha-*ras* 及 c-Ha-*ras*1 探针之间的序列比对结果显示在图 4 的上方。从本研究以及先前的数据 [16,17] 推断出的转录方向在图谱上是从右到左的。

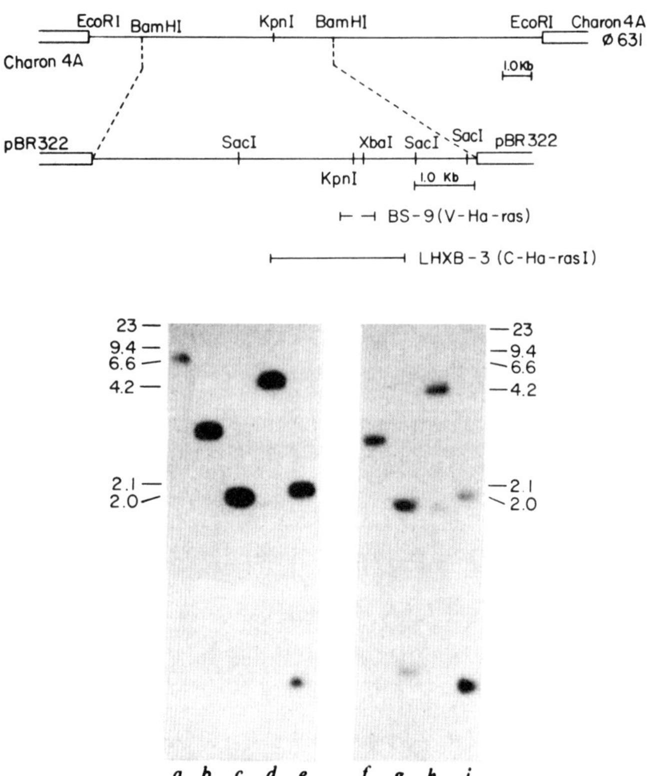

Fig. 4. Alignment of c-Ha-*ras*1 and of v-Ha-*ras* (BS-9) with the physical map of the EJ oncogene. DNA from the EJ-Charon 4A clone Φ631 was digested with several restriction enzymes. DNA (0.5 μg) was loaded onto each lane before electrophoresis and blot transfer. Identical nitrocellulose filters were prepared and incubated with ^{32}P-labelled DNA of c-Ha-*ras*1 (lanes *a-e*) or with HaSV subclone BS-9 (lanes *f-i*). Φ631 DNA was cleaved with *Bam*HI (*a*); *Bam*HI + *Sac*I (*b, f*); *Sac*I + *Kpn*I (*c, g*); *Bam*HI + *Xba*I (*d, h*); *Sac*I + *Xba*I (*e, i*). The alignment shown is accurate to within 200 nucleotides.

Taken together, these results indicate that the EJ bladder oncogene is closely linked to the human homologue of the rat c-Ha-*ras*1 sequences. Although the limits of the c-Ha-*ras*1 structural sequences have been well defined[16,17,20], the corresponding sequences of the EJ gene have not yet been mapped. Thus, the data above cannot exclude the possibility that the two genetic elements were adjacent to one another rather than congruent.

Analysis of Transcripts Homologous to the Clones

The transcripts encoded by these genes were analysed to further establish their relationship to one another. We examined the RNAs of transfected cells for molecules reactive with the two oncogene probes. As shown in Fig. 5, the two probes each detected transcripts of 1.2 and 5.1 kb in both the parental tumour cell line and in EJ- and T24-transfected mouse cell lines (see also ref. 22). These transcripts were not detected in untransfected NIH 3T3 cells. Thus, introduction of the EJ oncogene into mouse cells results in synthesis of RNAs that are homologous with the rat c-Ha-*ras*1 gene. As discussed below, these data support a congruency between the functionally active region of the EJ gene and that of the c-*ras* gene.

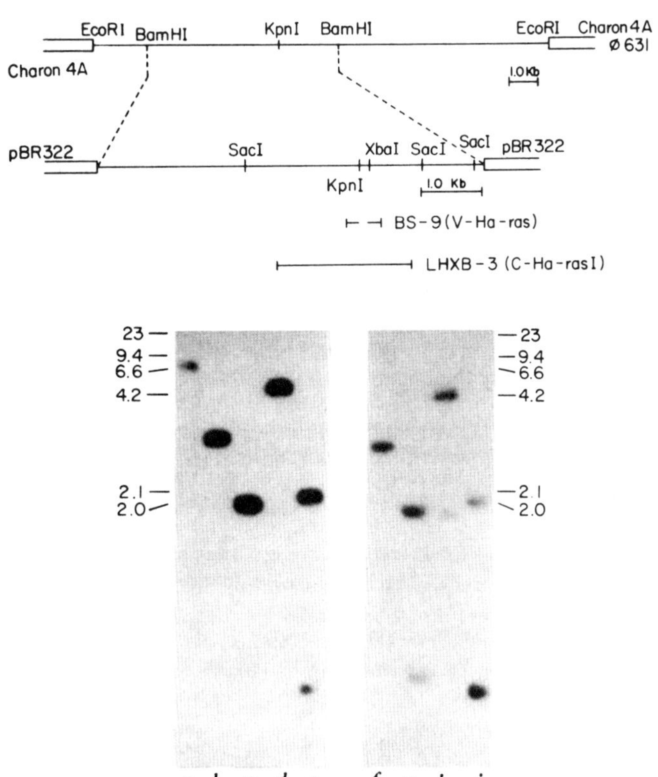

图 4. EJ 癌基因物理图谱及 c-Ha-*ras* 1 和 v-Ha-*ras* (BS-9) 的比对。用多种限制性内切酶对 EJ-Charon 4A 克隆 Φ631 的 DNA 进行酶切。将 DNA (0.5 μg) 点到每个泳道上再进行电泳和印迹转移。将准备好的相同的硝化纤维素膜与用 ^{32}P 标记的 c-Ha-*ras*1（泳道 *a*~*e*）或者 HaSV 亚克隆 BS-9（泳道 *f*~*i*）的 DNA 放在一起温育。Φ631 DNA 分别用 *Bam*HI (*a*); *Bam*HI+*Sac*I (*b*, *f*); *Sac*I+*Kpn*I (*c*, *g*); *Bam*HI+*Xba*I (*d*, *h*); *Sac*I+*Xba*I (*e*, *i*) 酶切。图中显示的比对精确到 200 个核苷酸以内。

总体来讲，这些结果表明 EJ 膀胱癌癌基因与大鼠 c-Ha-*ras*1 序列的人类同源序列关系密切。尽管 c-Ha-*ras*1 结构序列的范围已经明确 [16,17,20]，但 EJ 基因中的相应序列图谱却仍未绘出。因此，以上数据并不能排除这两个遗传元件是彼此相邻而非完全相同的可能性。

分析与克隆同源的转录物

为了进一步明确这些基因彼此之间的关系，我们对这些基因编码的转录物进行了分析。我们用两种癌基因探针检验了转染细胞的 RNA 的分子反应。如图 5 所示，这两个探针分别在亲代肿瘤细胞系和 EJ、T24 转染的小鼠细胞系中都检测到了 1.2 kb 和 5.1 kb 的转录物（见文献 22）。在未转染的 NIH 3T3 细胞中没有检测到这些转录物。因此，转染到小鼠细胞中的 EJ 癌基因导致了与大鼠 c-Ha-*ras*1 基因同源的 RNA 的合成。如下面所讨论的一样，这些数据表明 EJ 基因和 c-*ras* 基因的功能活性区域具有一致性。

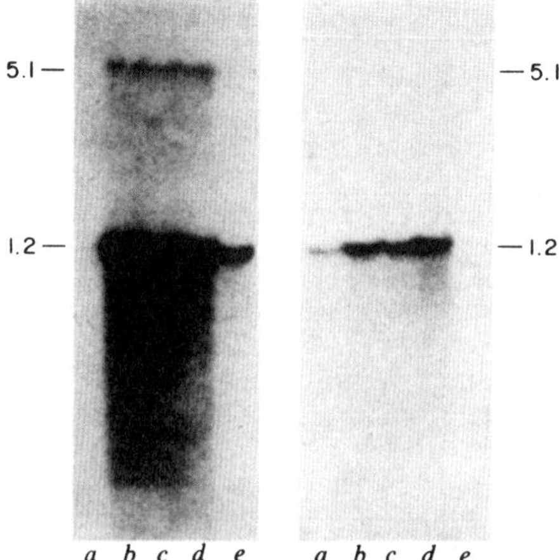

Fig. 5. Cellular polyadenylated RNAs analysed using pEJ6.6 and c-Ha-*ras*l ^{32}P-labelled probes. In the left-hand panel, nick-translated c-Ha-*ras*l DNA (2×10^{8} c.p.m. μg^{-1}) was used to probe RNA isolated from the following cell lines: *a*, NIH 3T3; *b*, T24-8-1 (a secondary transfectant derived from T24 human cell line); *c*, EJ-4(R1)-2; *d*, EJ-6-2; *e*, EJ human bladder. In the right-hand panel, nick-translated pEJ6.6 (6.6×10^{7} c.p.m. μg^{-1}) was used to probe the following cell lines: *a*, EJ; *b*, EJ-6-2; *c*, EJ-4(R1)-2; *d*, T24-8-1; *e*, NIH 3T3. The polyadenylated RNAs were prepared by the technique of Varmus *et al.*[40]. The RNA was then fractionated by electrophoresis through formaldehyde-containing gels and transferred to nitrocellulose (B. Seed and D. Goldberg, in preparation). ^{32}P-labelled probes, prepared as described in Fig. 1 legend, were annealed to the immobilized RNAs[41]. Bands which represent sequences homologous to the probes were visualized by autoradiography. Molecular weights were determined by comparison with markers obtained from *in vitro* run-off transcription of the adenovirus late promoter[42] and are shown in kilobases.

Discussion

The present data strongly suggest an evolutionary homology between the EJ human bladder carcinoma oncogene and the rat c-Ha-*ras*1 gene. We now consider the experimental basis for this conclusion and its implications.

The relatedness between the EJ and c-Ha-*ras*1 genes was first noted when we demonstrated that a cell acquiring the EJ oncogene also carried a novel DNA fragment reactive with the c-Ha-*ras*1 probe. As the linkage between the c-Ha-*ras*1 homologous sequence and the EJ oncogene was not broken by endonuclease *Bam*HI, we concluded that the two genetic elements lay within the same 6.6-kb *Bam*HI-generated fragment. These data alone were consistent with the two elements being either physically adjacent or congruent with one another. To resolve this ambiguity, we analysed the transcripts encoded by the two oncogenes. The EJ parental tumour and its derived transfectants, all express 5.1- and 1.2-kb transcripts that react with the oncogene probe. The fact that these transcripts are also present in cells transfected with *Bam*HI-cleaved DNA (data not shown) implies that the entire transcriptional unit of the 5.1-kb RNA is found within the confines of the 6.6-kb *Bam*HI-generated fragment (Fig. 3*B*, *D*).

It could be argued that the transcripts detected in transfected mouse cells are of murine origin. In this case, their synthesis would be induced indirectly in the mouse cells by the

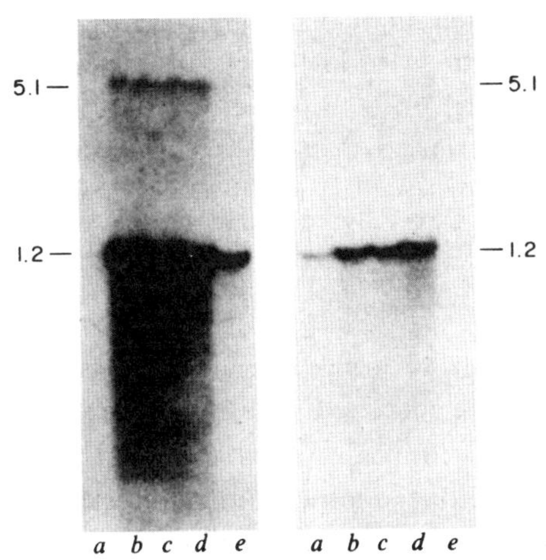

图 5. 用经 ^{32}P 标记的 pEJ6.6 和 c-Ha-*ras*1 探针分析细胞多聚腺苷酸 RNA。在左侧图中，用切口平移法制备的 c-Ha-*ras*l DNA (2×10^8 c.p.m. μg^{-1}) 作为探针，检测从以下细胞系中分离出的 RNA：*a*，NIH 3T3；*b*，T24-8-1(T24 人类细胞系产生的二级转染子)；*c*，EJ-4(R1)-2；*d*，EJ-6-2；*e*，EJ 人类膀胱。在右侧图中，用切口平移法制备的 pEJ6.6 (6.6×10^7 c.p.m. μg^{-1}) 作为探针，检测了从以下细胞系中分离出的 RNA：*a*，EJ；*b*，EJ-6-2；*c*，EJ-4(R1)-2；*d*，T24-8-1；*e*，NIH 3T3。用瓦默斯等描述的方法 [40] 制备多聚腺苷酸 RNA。使用含有甲醛的凝胶电泳分离这些 RNA 并转移到硝酸纤维素膜上（锡德和戈德堡，论文撰写中）。采用图 1 注所描述的方法制备 ^{32}P 标记的探针，并退火到已固定的 RNA 上 [41]。通过放射自显影就能看到与探针具有序列同源性的条带。分子量是通过与腺病毒晚期启动子 [42] 体外失控转录所获得的标记物进行比较确定的，以千碱基表示。

讨　论

目前的数据有力地表明 EJ 人类膀胱癌癌基因和大鼠 c-Ha-*ras*1 基因之间的进化同源性。我们接下来讨论这一结论和意义的实验基础。

当证实了获得 EJ 癌基因的细胞同时也携带能与 c-Ha-*ras*1 探针反应的新的 DNA 片段时，我们才首次注意到了 EJ 和 c-Ha-*ras*1 基因之间的相关性。由于核酸内切酶 *Bam*HI 并不能破坏 c-Ha-*ras*1 同源序列与 EJ 癌基因之间的关联，我们认为这两个遗传元件位于同一个由 *Bam*HI 产生的 6.6 kb 的片段内。单就这些数据而言，它们符合两个元件在物理位置上相邻或者是同一个基因的假设。为了明确这一问题，我们分析了由这两个癌基因编码的转录物。EJ 亲代肿瘤及其产生的转染子均表达了与癌基因探针反应的 5.1 kb 和 1.2 kb 的转录物。在 *Bam*HI 酶切过的 DNA 转染的细胞内也存在这些转录子（未显示数据），这个事实表明整个 5.1 kb 的 RNA 转录单元位于由 *Bam*HI 产生的 6.6 kb 的片段内部（图 3*B*，*D*）。

可以认为在转染小鼠细胞中检测到的转录物是鼠源性的。假若这样，这些转录物的合成就是在小鼠细胞内由其获得的人类癌基因间接诱发的，而不是由人类基因

acquired human oncogene rather than being encoded directly by the human gene. We consider this unlikely, as the rat c-Ha-*ras*1 probe used had a three-fold higher specific radioactivity than the human EJ probe, but yielded a 3–4-fold lower signal intensity on autoradiography (Fig. 5). This must reflect the relatively lower affinity of the rat probe for human transcripts present in the EJ transfectants. Thus, we conclude that the two probes both detect RNAs transcribed largely, if not entirely, from the human EJ oncogene template.

A similar, if not identical, pair of transcripts has been found by colleagues working with the c-Ha-*ras*1 gene and its RNAs[22]. As both the EJ and c-Ha-*ras*1 homologous 5.1-kb transcriptional units lie within the 6.6-kb *Bam*HI fragment, we conclude that the two genes are congruent with one another rather than adjacent. Consistent with this structural homology is a functional analogy in that both genes are able to induce fibroblast transformation.

The *ras* genes encode proteins of molecular weight 21,000 (ref. 23). Immunoprecipitation of metabolically labelled lysates of transfected cells has detected a protein of this size (R. Finkelstein and R. A. W., unpublished observations). However, the association of this protein with the bladder oncogene will only be well established after its detailed peptide structure has been analysed. Knowledge of the function of this protein may elucidate the important steps in human bladder carcinogenesis.

This p21 polypeptide represents a strong candidate for the protein mediating transformation of a human tumour cell. It is one of the first proteins implicated directly in the oncogenic conversion of a cell following its transformation by non-viral agents, and has been localized at the inner surface of the plasma membrane in Harvey sarcoma virus (HaSV)-transformed cells[24]. If this localization applies also to the EJ bladder carcinoma cells, then the transforming protein of these cells, the p21, should not display extracellular antigenic determinants. In this case, any tumour-specific surface antigens displayed by the bladder carcinoma cell should be encoded by genetic elements other than the oncogene itself.

The present work has several other implications. Perhaps the most apparent is that a single proto-oncogene can be activated by different molecular processes. The c-Ha-*ras*1 oncogene of the rat became activated via its affiliation with retrovirus sequences, forming the chimaeric Harvey sarcoma virus[21,25]. As demonstrated in the accompanying article[26], the human c-Ha-*ras*1 gene is also capable of oncogenic activation after it becomes linked *in vitro* to retrovirus promoter sequences. The mode of activation of the EJ oncogene is different, but not yet understood. Presently evidence suggests that the EJ oncogene and its normal human allelic counterpart sequence are indistinguishable by restriction enzyme site mapping[18]. It is possible that its activation depends on minor structural alterations, such as point mutations.

The relatedness between the EJ oncogene and that of a transforming retrovirus represents

直接编码的。我们认为这不太可能，因为使用的大鼠 c-Ha-*ras*1 探针的放射性比活度比人类 EJ 探针高出三倍，但是在放射自显影图像上其信号强度却低 3~4 倍（图 5）。这可能反映了大鼠探针对 EJ 转染子中存在的人类转录物的亲和力相对较低。因此，我们得出结论：这两个探针都检测到的 RNA 大部分（即便不是全部）是由人类 EJ 癌基因模板转录出来的。

研究 c-Ha-*ras*1 基因及 RNA 的同行们也发现了类似的（即便不是相同的）一对转录物 [22]。因为 EJ 以及 c-Ha-*ras*1 同源的 5.1 kb 的转录单元均位于 6.6 kb 的 *Bam*HI 片段内部，所以我们认为这两个基因就是同一个基因，而不是相邻的两个基因。与这种结构同源性一致的是其功能的相似性，即二者都能诱导成纤维细胞的转化。

ras 基因编码了分子量为 21,000 的蛋白质（参考文献 23）。对转染细胞内代谢标记的裂解产物进行的免疫沉淀实验检测到了类似大小的蛋白质（芬克尔斯坦和温伯格的未发表数据）。但是，只有解析了这个蛋白质具体的肽链结构以后，其与膀胱癌癌基因之间的关系才能确定。对这个蛋白质功能的认识可能有助于阐明人膀胱癌发生的关键步骤。

这种 p21 多肽很可能是蛋白介导人类肿瘤细胞转化的参与者。这是首批与通过非病毒元件转化发生癌变过程直接相关的蛋白质之一，并被定位在哈维肉瘤病毒 (HaSV) 转化的细胞质膜内表面上 [24]。如果该定位也适用于 EJ 膀胱癌细胞的话，那么这些细胞的转化蛋白 p21 就不应该是细胞外抗原决定簇。假若这样，膀胱癌细胞所表现的任何肿瘤特异性表面抗原应该是由遗传元件，而不是由癌基因本身编码。

这项工作具有许多其他意义。最显著的也许就是，一个单独的原癌基因能被不同的分子过程激活。大鼠的 c-Ha-*ras*1 癌基因通过逆转录病毒序列的加入而被激活，形成嵌合型哈维肉瘤病毒 [21,25]。正如附录文章所描述的 [26]，人类 c-Ha-*ras*1 基因在体外连接上逆转录病毒启动子序列后其致癌性也会被激活。EJ 癌基因的激活模式是不同的，且是未知的。目前的证据显示 EJ 癌基因及其正常的人类等位基因对应序列用限制性内切酶位点图谱法尚无法区分 [18]。其活化有可能依赖于微小的结构改变，比如点突变。

EJ 癌基因和转化的逆转录病毒基因之间的联系表明我们对人类膀胱癌的分子

an advance in our understanding of the molecular basis of human bladder carcinoma. This stems from the fact that the structure and function of the *ras* genes and their gene products have been extensively studied[16–18,20–23].

We have been unable to demonstrate other homologies between retrovirus *onc* genes and the transfection-derived tumour genes, but this may merely reflect the small repertoire of tumour genes presently available in cloned form. As other genes become available for study, additional connections will probably be found.

Two paradoxes seem to be raised by the unexpected association of a rat sarcoma oncogene with a human bladder oncogene. First, this work implies the ability of the c-Ha-*ras*1 gene to act in unrelated tissue environments. The rat gene, when carried in Harvey sarcoma virus, can induce sarcomas and erythroleukaemias[27] while its human counterpart is now implicated in the genesis of bladder carcinomas[10–12,18,19]. We consider it possible that the *ras* oncogene of either species is capable of transforming a wide range of target tissues, only a small portion of which has been studied experimentally.

Second, we have suggested that the precursor of the EJ oncogene, now identified as c-Ha-*ras*1, represents a preferred target for activation during bladder carcinogenesis[18]. It seems unlikely that the bladder urothelium was the site of acquisition of the *ras* gene during the events that let to the creation of the chimaeric HaSV genome. The two routes of oncogene activation must involve different molecular mechanisms which probably occur at different sites in the organism. Each mode of activation may be favoured by different predisposing factors present in different tissues. For example, in the bladder, the c-Ha-*ras*1 gene may be in a configuration particularly susceptible to mutational activation whereas in certain other tissues it may be expressed in a manner favouring the recombinational events that lead to creation of chimaeric retroviruses.

We thank our colleagues for providing the *onc* probes used in this study. This research was supported by US National Cancer Institute grants CA17537 and CA26717 to R.A.W.

(**297**, 474-478; 1982)

Luis F. Parada, Clifford J. Tabin, Chiaho Shih and Robert A. Weinberg
Center for Cancer Research and Department of Biology, Massachusetts Institute of Technology, Cambridge, Massachusetts 02139, USA

Received 8 April; accepted 17 May 1982.

References:

1. Spector, D.H., Varmus, H.E. & Bishop, J.M., *Proc. Natl. Acad. Sci. U.S.A.* 75, 4102-4106 (1978).

2. Stehelin, D., Varmus, H.E., Bishop, J.M. & Vogt, P. K. *Nature* **260**, 170-173 (1976).

3. Klein, G. (ed.) *Advances in Viral Oncology: Cell Derived Oncogenes* (Raven, New York, 1981).

4. Coffin, J. M. *et al. J. Virol.* **40**, 953-957 (1981).

5. Shilo, B-Z. & Weinberg, R.A. *Proc. Natl. Acad. Sci. U.S.A.* **78**, 6789-6792 (1981).

基础的认识有了提高。这主要基于对 *ras* 基因及其基因产物的结构和功能的广泛研究 [16-18, 20-23]。

我们尚不能阐明逆转录病毒 *onc* 基因和转染产生的肿瘤基因之间的其他同源性，但是这可能仅仅反映了目前以克隆形式获取的肿瘤基因的小部分内容。随着更多基因投入研究，也可能找到更多的联系。

随着大鼠肉瘤癌基因和人膀胱癌癌基因之间意外联系的发现，出现了两个矛盾。第一，这个研究表明 c-Ha-*ras*1 基因能够在不相关的组织环境中发挥功能。插入到哈维肉瘤病毒上的大鼠基因能够导致肉瘤和红白血病的发生 [27]，而其与人类相应的基因在膀胱癌的发生中起重要作用 [10-12,18,19]。我们认为有可能这两个物种之一的 *ras* 癌基因能够转化大范围的靶点组织，而正在进行实验研究的只是其中一小部分。

第二，我们已经指出 EJ 癌基因的前体，即现在所认为的 c-Ha-*ras*1，是膀胱癌致癌作用中偏好激活的靶位 [18]。膀胱上皮细胞似乎不太可能是产生嵌合型 HaSV 基因组过程中获取 *ras* 基因的位点。两种癌基因的激活途径肯定涉及可能发生在生物不同部位的不同分子机制。不同诱发因素的激活模式可能存在于不同的组织中。例如，在膀胱中，c-Ha-*ras*1 基因可能处于一种特别容易受到突变激活的构型，而在某些其他组织中，该基因的表达可能在一定程度上促进重组从而产生嵌合型逆转录病毒。

感谢我们的同事提供本研究使用的 *onc* 探针。美国国家癌症研究所基金 CA17537 和 CA26717 向温伯格赞助了此项研究。

（毛晨晖 翻译；彭小忠 审稿）

6. Shih, C., Shilo, B-Z., Goldfarb, M., Dannenberg, A. & Weinberg, R. A. *Proc. Natl. Acad. Sci. U.S.A.* **76**, 5714-5718 (1979).
7. Shih, C., Padhy, L. C., Murray, M. & Weinberg, R. A. *Nature* **290**, 261-264 (1981).
8. Hopkins, N., Besmer, P., DeLeo, A. B. & Law, L. W. *Proc. Natl. Acad. Sci. U.S.A.* **78**, 7555-7559 (1981).
9. Lane, M. A., Sainten, A. & Cooper, G. M. *Proc. Natl. Acad. Sci. U.S.A.* **78**, 5185-5189 (1981).
10. Murray, M. J. *et al. Cell* **25**, 355-361 (1981).
11. Krontiris, T. G. & Cooper, G. M. *Proc. Natl. Acad. Sci. U.S.A.* **78**, 1181-1184 (1981).
12. Perucho, M. *et al. Cell* **27**,467-476 (1981).
13. Southern, E. M. *J. Molec. Biol.* **98**, 503-517 (1975).
14. Shilo, B-Z. & Weinberg, R. A. *Nature* **289**, 607-609 (1981).
15. Goff, S. P., Gilboa, E., Witte, O. N. & Baltimore, D. *Cell* **22**, 777-785 (1980).
16. DeFeo, D. *et al. Proc. Natl. Acad. Sci. U.S.A.* **78**, 3328-3332 (1981).
17. Ellis, R. W. *et al. Nature* **292**, 506-511 (1981).
18. Shih, C. & Weinberg, R. A. *Cell* (in the press).
19. Goldfarb, M., Shimizu, K., Perucho, M. & Wigler, M. *Nature* **296**, 404-409 (1982).
20. Chang, E. H., Gonda, M. A., Ellis, R. W., Scolnick, E. M. & Lowy, D. R. *Proc. Natl. Acad. Sci. U.S.A.* (in the press).
21. Ellis, R. W. *et al. J. Virol.* **36**, 408-420 (1980).
22. Ellis, R. W., DeFeo, D., Furth, M. & Scolnick, E. M. *Cell* (in the press).
23. Shih, T. Y., Weeks, M. O., Young, H. A. & Scolnick, E. M. *Virology* **96**, 64-79 (1979).
24. Willingham, M. C., Pastan, I., Shih, T. Y. & Scolnick, E. M. *Cell* **19**, 1005-1014 (1980).
25. Harvey, J. J. *Nature* **204**, 1104-1105 (1964).
26. Chang, E. H., Furth, M. E., Scolnick, E. M. & Lowy, D. R. *Nature* **297**, 479-483 (1982).
27. Chesterman, F. C., Harvey, J. J., Dourmashkin, R. R. & Salaman, M. H. *Cancer Res.* **26**, 1759-1768 (1966).
28. Vennstrom, B., Fanshiev, L., Moscovici, C. & Bishop, J. M. J. *Virol.* **36**, 575-585 (1980).
29. Sherr, C. J., Fedele, L. A., Oskarsson, M., Maizel, J. & Vande Woude, G. *J. Virol.* **34**, 200-212 (1980).
30. Chang, E. H., Gonda, M. A., Ellis, R. A., Scolnick, E. M. & Lowy, D. R., *Proc. Natl. Acad. Sci. U.S.A.* (in the press).
31. Vennstrom, B., Moscovici, C., Goodman, H. & Bishop, J. M. *J. Virol.* **39**, 625-631 (1981).
32. DeLorbe, W. J., Luciw, P. A., Goodman, H. M., Varmus, H. E. & Bishop, J. M. *J. Virol.* **36**, 50-61 (1980).
33. Rapp, V. R., Nowinski, R. C., Reznikoff, C. A. & Heidelberger, C. *Virology* **65**, 329-409 (1975).
34. Giard, D. J. *et al. J. Natl. Cancer Inst.* **51**, 1417-1421 (1973).
35. Leibovitz, A. *et al. Cancer Res.* **36**, 4562-4569 (1976).
36. Collins, E. J., Gallo, R. C. & Gallagher, R. E. *Nature* **270**, 347-349 (1977).
37. Schubert, D. *et al. Nature* **249**, 224-226 (1974).
38. Summerhayes, I. C. & Franks, L. M. *J. Natl. Cancer Inst.* **62**, 1017-1021 (1979).
39. Rigby, P. W., Dieckmann, M., Rhodes, C. & Berg, P. *J. Molec. Biol.* **13**, 237-251 (1977).
40. Varmus, H. E., Quintrell, N. & Ortiz, S. *Cell* **25**, 23-36 (1981).
41. Wahl, G. M., Stern, M. & Stark, G. R. *Proc. Natl. Acad. Sci. U.S.A.* **76**, 3683-3687 (1979).
42. Manley, J. L., Fire, A., Cano, A., Sharp, P. A. & Gefter, M. L., *Proc. Natl. Acad. Sci. U.S.A.* 77, 3855-3859 (1980).

Neurone Differentiation in Cell Lineage Mutants of *Caenorhabditis elegans*

J. G. White *et al.*

Editor's Note

The use of the nematode worm *Caenorhabditis elegans* for studying gene function and development was pioneered by John Sulston. The British biologist developed techniques to study the cell divisions that transform fertilized egg into adult animal, proving that every worm undergoes the same program of cell division and differentiation. Here Sulston and colleagues describe two *C. elegans* mutants that have particular cell divisions blocked. The blocked cells, they show, yield only one of the two daughter cells that would normally be produced, and this is always a neuron. The study shed light on the processes controlling cell fate. Sulston, his coauthor H. Robert Horvitz, and nematode researcher Sydney Brenner later shared a Nobel Prize for their work on *C. elegans*.

The nematode *Caenorhabditis elegans* develops by an essentially invariant sequence of cell divisions[1–3] leading to an adult complement of 959 somatic cells. In this organism cell fate is correlated with cell lineage, suggesting that genealogy may be a determining factor for the differentiated state of a cell. The study of mutants with altered cell lineages may help elucidate the precise mechanisms by which cell fate is decided. Several cell lineage mutants have been isolated and characterized[4,5], some having more and some fewer cell divisions than wild type. We have now investigated the cell types produced by two cell lineage mutants; these mutants exhibit blocks in certain terminal or near terminal cell divisions, which in normal animals generally give rise to daughter cells that differentiate into distinctly different cell types. We find that the blocked cells in the mutants generally exhibit the differentiated characteristics of only one of the two daughter cells that normally would be produced. The differentiated state of the blocked precursors may be due to an intrinsic dominance of one cell type over another in what is essentially a fused cell, and/or it may reveal the state of commitment of the precursor in wild-type animals.

WE have studied two very similar mutants *unc-59* (*e*1005) and *unc-85* (*e*1414), both of which are variably blocked in some of the later divisions of the lineages that produce the adult complement of motoneurones in the ventral nerve cord[4,5]. These lineages have been characterized in wild-type animals by following the development of living animals in the light microscope[1]. The nuclei of 12 precursor cells (designated P1–P12) migrate into the ventral cord during the first larval stage; each cell then divides

秀丽隐杆线虫细胞谱系突变体的神经元分化

怀特等

编者按

约翰·萨尔斯顿开创了利用秀丽隐杆线虫来研究基因功能和发育的先河。这位英国生物学家探索出了一套方法来研究动物个体通过细胞分裂实现从受精卵到成体这一转化过程，并得出了每条线虫都经历相同的细胞分裂和分化程序的结论。在本文中，萨尔斯顿和他的同事们描述了两种细胞分裂在特定步骤被阻断的秀丽隐杆线虫的突变体。他们揭示了在这两种突变体中，细胞分裂阻断的细胞仅产生一个子细胞——正常情况下会产生两个子细胞，并且细胞分裂阻断的细胞往往成为神经元。这项研究为阐明细胞命运的调控过程开辟了道路。后来，萨尔斯顿、他的共同作者罗伯特·霍维茨以及线虫研究者西德尼·布伦纳凭借他们在秀丽隐杆线虫方面的工作共享了诺贝尔奖。

秀丽隐杆线虫通过一种基本上程序不变的细胞分裂过程[1-3]，发育形成由959个体细胞组成的成体。在这种生物中，细胞的命运往往与其细胞谱系相对应，这表明了种系可能是细胞分化状态的决定因素。研究细胞谱系发生改变的突变体可能有助于阐明决定细胞命运的确切机制。有数个细胞谱系突变体已经被分离并鉴定[4,5]，与野生型相比，有些突变体的细胞分裂次数增多，而有些减少。目前我们研究了两个细胞谱系突变体的细胞类型；这些突变体在一些细胞分裂的末期或临近末期的时候发生细胞分裂的阻断，而在正常情况下这些细胞能够分裂，并分化成不同的子细胞。我们发现：突变体中分裂阻断的细胞通常表现出正常情况下产生的两个子代细胞之一的分化特征。如果将细胞分裂受阻的细胞理解为两个子细胞融合在一起的状态，那么前体细胞的分化状态可能取决于内在的，其中一种细胞类型相对于另一细胞类型的显性表现，而且（或者）这也可能显示了野生型情况下该前体细胞应该呈现的本来状态。

我们研究了两种非常相似的突变体 *unc-59*（*e*1005）和 *unc-85*（*e*1414），这两者都在产生成虫腹神经索运动神经元的细胞分裂的较晚阶段发生了不同程度的细胞分裂阻断[4,5]。在野生型个体中用光学显微镜跟踪活体动物的发育过程，已经明确了这些细胞的分裂谱系[1]。12 个前体细胞（称为 P1~P12）的细胞核在第一龄幼虫期迁移到腹侧索部位，然后每个细胞分裂产生一个神经母细胞和一个上皮细胞。12 个神经

to produce a neuroblast and a hypodermal cell. The 12 neuroblasts (together with an additional neuroblast present at hatching) undergo identical sequences of divisions (Fig. 1), each producing 5 cells that intercalate with the pre-existing juvenile motoneurones to form a single row of cells along the ventral cord. All cleavages have longitudinal spindle axes, and descendant cells maintain their relative antero-posterior positions throughout development. Certain cells derived from some of the neuroblasts die soon after their formation; the pattern of cell death is invariant.

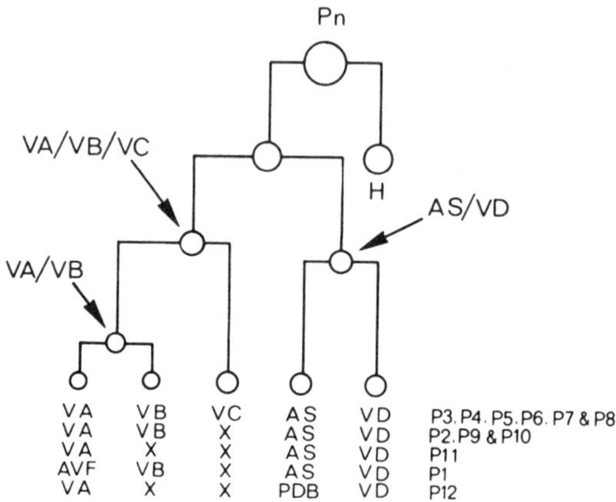

Fig. 1. The ventral cord precursor cells P1–P12 are hypodermal cells (J. G. W., unpublished observations). Their nuclei migrate into the ventral cord and each divides to produce a ventral hypodermal cell (H) (which functionally replaces the mother) and a neuroblast. All 12 neuroblasts then undergo identical series of divisions to produce five descendants[1]. All spindle axes are longitudinal; in the diagram anterior daughters are drawn to the left and posterior daughters to the right. The fates of the descendants from each precursor are shown. Cell types have been assigned on the basis of cell morphology and synaptic connectivity[6]. Precursors P3–P8 produce five classes of ventral cord motoneurone (VA, VB, VC, AS, VD). In the P1–P2 and P9–P12 lineages, cells lineally equivalent to VC neurones undergo programmed cell death (X), as do cells lineally equivalent to VB neurones derived from P11–P12. Alternative fates are seen for some of the cells derived from the two precursors at the ends of the cord (P1,P12): P1 produces an AVF interneurone instead of a VA, and P12 produces a PDB interneurone instead of an AS. (AVF and PDB are two distinct classes of interneurone which are quite dissimilar to the motoneurone classes.) These alternative fates are probably specified by local interactions[1,14]. In the *unc-59* and *unc-85* animals, certain cells which divide in the wild-type fail to do so. Such cells are labelled according to the fate of their normal descendants; for example, an AS/VD cell should have divided to produce an AS and a VD neurone.

The structure of the ventral cord of wild-type animals has been deduced by reconstructions from electron micrographs of serial sections[6]. In neuroblast lineages that do not contain cell deaths the five cells produced differentiate into five distinct classes of motoneurone designated VA, VB, VC, AS and VD (Fig. 1). Classes are defined on the basis of patterns of synaptic connections (see Pn, Table 2) and morphology (Fig. 2). Either criterion is sufficient to assign a cell to a class.

母细胞将（与孵化时就存在的另一个神经母细胞一起）经历相同顺序的分裂（图 1），其中每一个都会产生 5 个细胞，这 5 个细胞将并入到已经存在的，未成熟的运动神经元当中，一起形成沿着腹侧索排列的一排细胞。每次分裂都产生纵向的纺锤体轴，并且在整个发育过程中子代细胞都保持相对的前后位置关系。一些来源于神经母细胞的细胞在形成后不久即死亡，这些细胞死亡的模式都是固定不变的。

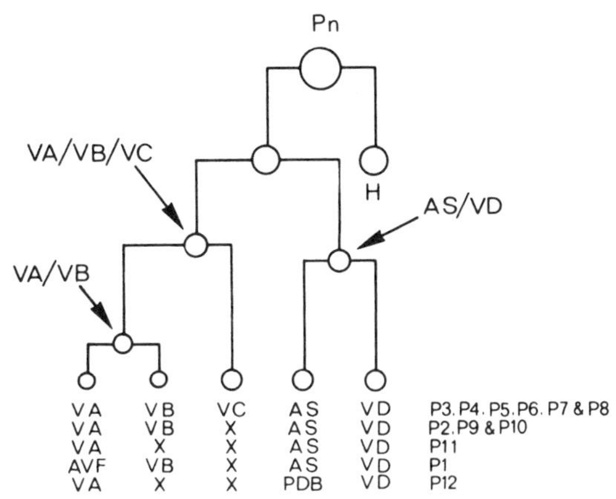

图 1. 腹侧索前体细胞 P1~P12 是上皮细胞（怀特，未发表的结果）。它们的细胞核迁移到腹侧索，而后每个细胞分裂产生一个腹侧上皮细胞（H）（其功能是替代母细胞）和一个神经母细胞。所有的 12 个神经母细胞经过相同的分裂程序，产生 5 个后代 [1]。每次分裂的纺锤体轴都是纵向的。靠前端的子细胞在图中标注于左侧，靠后端的标注于右侧。从每个前体细胞产生的子代细胞的细胞命运标注如图。基于细胞形态和突触的连接性对细胞类型进行了划分 [6]。前体细胞 P3~P8 产生五种腹侧索运动神经元（VA、VB、VC、AS、VD）。在 P1~P2 和 P9~P12 细胞系，相当于 VC 神经元位置的细胞经历程序性细胞死亡（X），而在 P11~P12 细胞谱系中相当于 VB 神经元位置的细胞也发生了程序性死亡。来源于腹侧索末端两个前体细胞（P1，P12）的部分子细胞有不同的命运：P1 产生 AVF 中间神经元，而不是 VA；P12 产生 PDB 中间神经元而不是 AS。（AVF 和 PDB 是两种不同类型的中间神经元，它们都与运动神经元非常不同。）这些另类的分化方式很可能是由局部的相互作用决定的 [1,14]。在 *unc-59* 和 *unc-85* 动物中，一些在野生型中能够分裂的细胞失去了分裂能力。根据这些细胞正常子细胞的分化方向对它们进行标记；例如，AS/VD 细胞应该分裂成一个 AS 神经元和一个 VD 神经元。

野生型个体的腹侧索结构已通过连续切片的电子显微图像重建而推断出来了 [6]。在没有发生子细胞死亡的神经母细胞系中，其产生的五个细胞分化成五种不同类型的运动神经元，分别是 VA、VB、VC、AS 和 VD（图 1）。分类依据是突触连接的形式（见 Pn，表 2）和细胞形态（图 2）。二者之中任一标准都足以将一个细胞归到一种类型中。

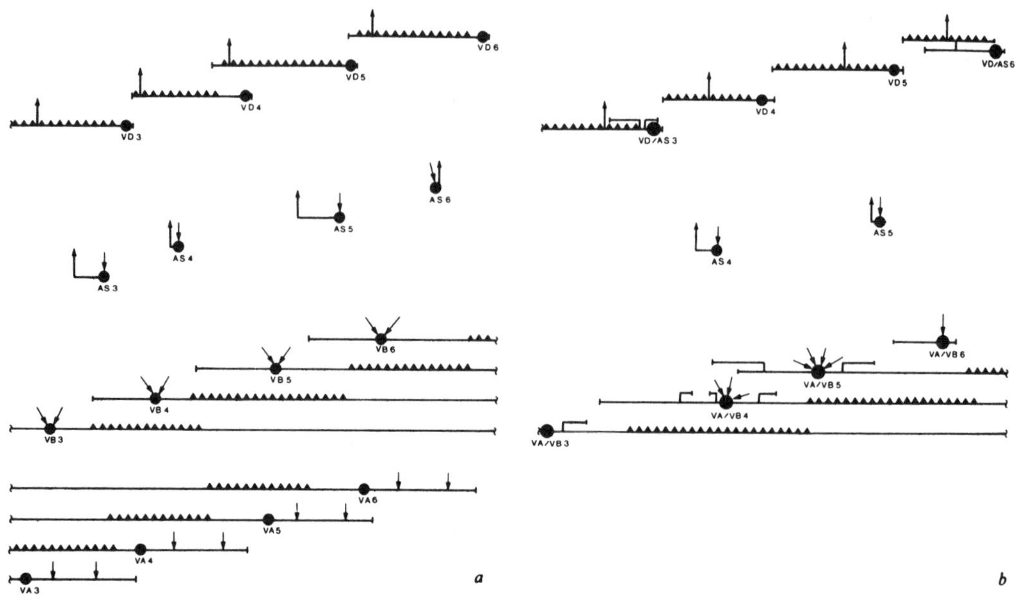

Fig. 2. The morphologies of the motoneurones derived from the precursors P3–P6 are shown for wild type (*a*) and *unc-85* (*b*). The wild-type structures are as described in ref. 6, and the mutant structures were derived, using similar techniques, by reconstruction from 3,000 serial section electron micrographs from an animal of known lineage. Four of the five motoneurone classes derived from the P cells are shown: VA, VB, AS and VD (class VC has been omitted, because it has few distinguishing characteristics). The extent of the processes in the ventral cord and the positions of the cell bodies (–●–), neuromuscular junctions (NMJs) (▲▲▲), commissures (| →) and synaptic input (| ←) are shown. The horizontal axis is equivalent to the longitudinal axis of the animal (anterior is drawn to the left); the vertical axis has been used simply to separate the neurones into their respective classes. VA neurones , which have anteriorly directed axons, form a regular sequence of regions of motor activity with little or no overlap between adjacent members of a class. VB neurones are similar but have posteriorly directed axons. VA neurones receive their synaptic input in a dendritic region near the cell body on a branch opposite the axon; VB neurones receive their synaptic input predomiantly at the cell body. The axons from class AS motoneurones lead to the dorsal cord via commissures and innervate dorsal muscles. They have a short dendritic region in the ventral cord. The axons from VD neurones end abruptly in gap junctions to axons from adjacent VD neurones. The dendritic regions behave in a similar fashion in the dorsal cord (not shown) and connect to the ventral cord via commissures. In the *unc-85* animal, the VA/VB-blocked precursors from P3–P5 had morphologies similar to wild-type VB neurones, forming NMJs from posteriorly directed axons. The blocked VA/VB precursor from P6 had failed to grow an axon. The AS/VD precursors from P4 and P5 had divided normally, and each produced apparently normal AS and VD daughters. The blocked AS/VD cells from P3 and P6 had the characteristic morphologies of VD motoneurones with their processes forming NMJs and ending abruptly in gap junctions with adjacent normal VD motoneurones. The only differences between the morphologies of these blocked precursors and those of their normal posterior daughters is that the former have larger nuclei, a few extra branches and more synaptic inputs.

The post-embryonic development of ventral cord cells was observed in several *unc-59* and *unc-85* animals. In these mutants, P-cell-derived neuroblasts produced fewer progeny cells than normal (Fig. 1), because some cell divisions failed. The set of affected cells varied between animals. Failures resulted in polyploid cells; sometimes the nuclei failed to divide after DNA replication, and sometimes apparently normal nuclei formed but then fused to

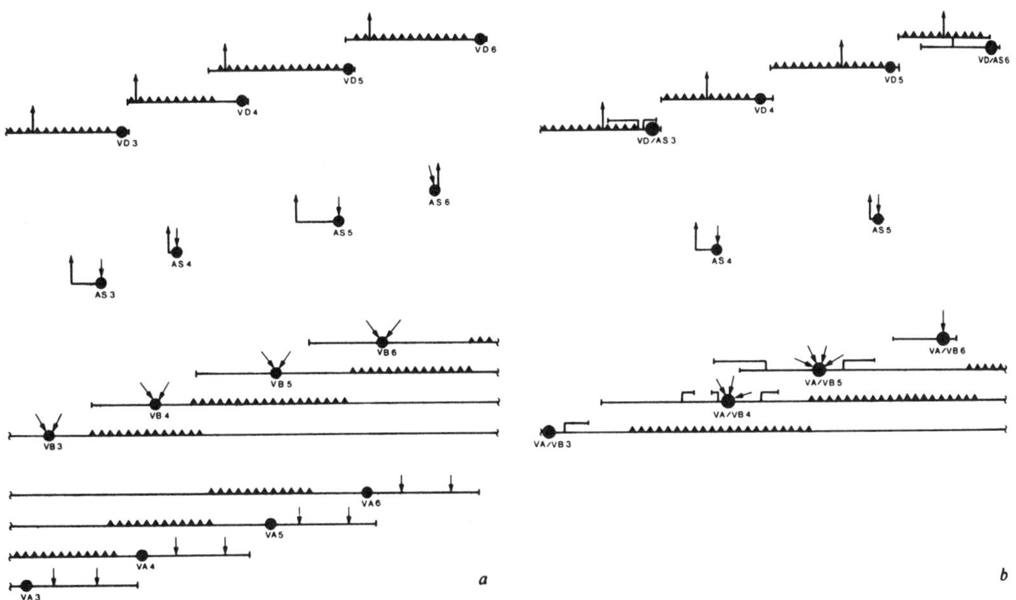

图 2. 图中显示了分别由野生型（*a*）和 *unc-85*（*b*）的前体细胞 P3~P6 来源的运动神经元的形态。野生型的结构见文献 6 中的描述，突变体结构是用类似的技术从一个细胞谱系已知的物种的 3,000 份连续切片电子显微镜图中重建出来的。图中显示了来源于 P 细胞的 5 个运动神经元类型中的 4 种：VA、VB、AS 和 VD（VC 类被略去，因为它的鉴别特征非常少）。图中显示了细胞在腹侧索内延伸的范围和细胞胞体（-●-）、神经肌肉接头（NMJ）（▲▲▲）、神经连合（|→）和突触传入（|←）的位置。水平轴相当于动物的纵轴（头端在左侧）；垂直轴用来将神经元划分到它们各自的类型中。VA 神经元向前延伸出轴突，形成常规顺序的运动活性区域，它们在相邻的同类细胞间几乎没有重叠。VB 神经元也是如此，但是轴突延伸方向朝后。VA 神经元通过与轴突相反方向的细胞胞体延伸出来的树突区域接受突触传入信息；VB 神经元主要通过细胞胞体接受突触传入。AS 运动神经元的轴突通过神经连合延伸至背侧索，支配背部的肌肉。它们在腹侧索有短的树突区域。VD 神经元的轴突以间隙连接的方式突然地中止于邻近 VD 神经元的轴突末端。背侧索的树突区域也是如此（数据未显示），并通过接合处与腹侧索相连。在 *unc-85* 动物中，P3~P5 来源的、细胞分裂受阻的 VA/VB 前体细胞与野生型 VB 神经元的形态类似，通过朝向尾端延伸的轴突形成 NMJ。P6 来源的、细胞分裂受阻的 VA/VB 前体细胞不能长出轴突。P4 和 P5 来源的 AS/VD 前体细胞正常分裂，每一个都产生了正常的 AS 和 VD 子代细胞。P3 和 P6 来源的、细胞分裂受阻的 AS/VD 细胞具有与 VD 运动神经元相似的形态特征，它们形成 NMJs 并以间隙连接的方式突然中止于相邻的正常 VD 运动神经元。细胞分裂受阻的前体细胞与它们正常的子代细胞之间形态的差别仅在于前者具有较大的核、一些额外分叉以及更多的突触传入。

（我们）观察了多个 *unc-59* 和 *unc-85* 个体中腹侧索细胞的胚后发育情况。由于一些细胞不能进行分裂，这些突变体的 P 细胞来源的神经母细胞比野生型产生的子代细胞少（图 1）。受影响的细胞系在不同的个体中有差异。细胞分裂失败导致多倍体细胞的产生；有时 DNA 复制后核没有分裂，有时虽然形成了正常的细胞核但随后

form a tetraploid nucleus or remained separated in a binucleate cell[5]. *unc-59* and *unc-85* individuals that displayed several examples of undivided ventral cord nuclei were selected for reconstruction from electron micrographs of serial sections.

In both *unc-59* and *unc-85* animals all blocked cells were found to have differentiated into neurones. Most of these neurones could be unequivocally identified as belonging to one of the motoneurone classes normally produced from these lineages; specifically, the blocked cells acquired the differentiated characteristics of one of the daughter cells that would normally be produced. These results are summarized in Table 1. All blocked VD/AS precursors differentiated into cells that had all the distinguishing characteristics of VD motoneurones (Fig. 2, Tables 1, 2). VA/VB or VA/VB/VC precursors differentiated into VB-like cells in the anterior ventral cord and into VA-like cells in the posterior ventral cord (Tables 1, 2). The choice of VA versus VB for the blocked precursors may be related to the regional preference exhibited by the progeny of these precursors in wild-type development: a VB but no VA is produced by the anterior P1 lineage, whereas a VA but no VB is produced by the posterior P11 and P12 lineages (Fig. 1).

Table 1. Post-embryonic development of P cells from *unc-85* and *unc-59* animals

Mutant	Blocked precursor		Cell type
unc-85	P3	VA/VB/*	VB
	P3	AS/VD	VD
	P4	VA/VB/*	VB
	P5	VA/VB	VB
	P6	VA/VB/*	VB‡
	P6	AS/VD	VD
unc-59			
Animal 1	P3	VA/VB/VC	VB
	P3	AS/VD	VD
Animal 2	P8	VA/VB	VA
	P8	AS/VD	VD
	P9	VA/VB/†	VA
	P9	AS/VD	VD
	P10	VA/VB/†	VA
	P10	AS/VD	VD
	P11	VA/†/†	VA‡
	P11	AS/VD	VD

* In these lineages a VC nucleus was produced but was not present when the adult was reconstructed. These nuclei probably either fused to make a VA/VB/VC or died.

† The cells lineally equivalent to VC in the P1–P2 and P9–P12 lineages and to VB in the P11–P12 lineages normally undergo programmed cell death[1].

融合成四倍体核或者在双核细胞中仍然保持分离状态[5]。（我们）选取了几个腹侧索细胞核不能分裂的 *unc-59* 和 *unc-85* 突变体进行了连续切片的电子显微镜图像重建分析。

在 *unc-59* 和 *unc-85* 个体中，所有细胞分裂受阻的细胞都分化成了神经元。这些神经元大部分都能够明确地归类到正常情况下由这些细胞系形成的运动神经元类型中去。特别要说明的是，这些细胞分裂受阻的细胞往往获得了其正常情况下产生的某一子代细胞所具有的分化特征。这些结果总结在表 1 中。所有被阻断的 VD/AS 前体细胞都分化成了具有 VD 运动神经元特征的细胞（图 2，表 1 和表 2）。VA/VB 或者 VA/VB/VC 前体细胞在腹侧索前段分化成了类似于 VB 的细胞，在后段分化成了类似于 VA 的细胞（表 1 和表 2）。阻断的前体细胞选择分化成 VA 还是 VB 可能与这些前体细胞在野生型动物的发育中子代细胞所表现出来的区域偏好有关：位于前端的 P1 细胞系产生 VB 而不是 VA，而位于后端的 P11 和 P12 细胞系产生 VA 而不是 VB（图 1）。

表 1. *unc-59* 和 *unc-85* 个体中 P 细胞的胚胎后发育

突变体	阻断的前体细胞		细胞类型
unc-85	P3	VA/VB/*	VB
	P3	AS/VD	VD
	P4	VA/VB/*	VB
	P5	VA/VB	VB
	P6	VA/VB/*	VB‡
	P6	AS/VD	VD
unc-59			
个体 1	P3	VA/VB/VC	VB
	P3	AS/VD	VD
个体 2	P8	VA/VB	VA
	P8	AS/VD	VD
	P9	VA/VB/†	VA
	P9	AS/VD	VD
	P10	VA/VB/†	VA
	P10	AS/VD	VD
	P11	VA/†/†	VA‡
	P11	AS/VD	VD

* 在这些细胞系中，VC 核能够产生，但是在成虫电镜重建时没有出现。这些核很可能融合以形成 VA/VB/VC 或者消亡了。

† 这些细胞在 P1~P2 和 P9~P12 系相当于 VC，在 P11~P12 系相当于 VB，它们正常情况下会经历程序性细胞死亡[1]。

‡ Axons failed to grow out of these cells although they had the synaptic contacts characteristic of the cell types indicated.

Table 2. Number of synapses formed by blocked precursors and typical synapses of wild-type ventral cord neurones (Pn)[6]

		AVA	AVB	VA	VB	VC	AS	VD	NMJ	
unc-85										
P3	VA/VB/*	‡	‡						15	
P3	AS/VD				§ 1			= 1	19	
P4	VA/VB/*		= 6						20	
P5	VA/VB	= 1	= 9		= 5				8	
P6	VA/VB/*		= 1		= 5					No axon
P6	AS/VD				§ 3			= 1	14	
unc-59 (1)										
P3	VA/VB/VC		= 15					# 1	5	
P3	AS/VD							= 6	14	
unc-59 (2)										
P8	VA/VB	§ 6 = 6							‡	
P8	AS/VD			§ 3				= 2	11	
P9	VA/VB/†	§ 3 = 5						# 3	22	
P9	AS/VD			§ 3				= 2	13	
P10	VA/VB/†	§ 9 = 5						# 3	16	
P10	AS/VD							=2	10	
P11	VA/†/†	§ 7								No axon
P11	AS/VD							= 2	12	
Wild type										
Pn	VA	§ 4 = 3					= 1	# 1	16	
Pn	VB		= 2		= 2			# 2	9	
Pn	VC					§ 3 = 5		# 7	2	
Pn	AS	§ 2 = 1	§ 2	= 1						
Pn	VD			§ 1	§ 2	§ 7		= 1	26	

Values for wild type are the mean values from precursors P3 to P5 obtained from a reconstruction of a wild-type animal. #, Presynaptic; §, postsynaptic; =, gap junctions.

* In these lineages a VC nucleus was produced but was not present when the adult was reconstructed. These nuclei probably either fused to make a VA/VB/VC or died.

† The cells lineally equivalent to VC in the P1–P2 and P9–P12 lineages and to VB in the P11–P12 lineages normally undergo programmed cell death[1].

‡ Processes in regions that normally would display these synapses were out of the region of reconstruction.

‡ 这些细胞不能长出轴突，尽管它们具备这些细胞类型显示出的突触联系特征。

表 2. 阻断的前体细胞形成的突触数目以及野生型腹侧索神经元（Pn）形成的典型突触数目 [6]

		AVA	AVB	VA	VB	VC	AS	VD	NMJ	
unc-85										
P3	VA/VB/*	‡	‡						15	
P3	AS/VD				§ 1			= 1	19	
P4	VA/VB/*		= 6						20	
P5	VA/VB	= 1	= 9		= 5				8	
P6	VA/VB/*		= 1		= 5					无轴突
P6	AS/VD				§ 3			= 1	14	
unc-59 (1)										
P3	VA/VB/VC		= 15					# 1	5	
P3	AS/VD							= 6	14	
unc-59 (2)										
P8	VA/VB	§ 6 = 6							‡	
P8	AS/VD			§ 3				= 2	11	
P9	VA/VB/†	§ 3 = 5						# 3	22	
P9	AS/VD			§ 3				= 2	13	
P10	VA/VB/†	§ 9 = 5						# 3	16	
P10	AS/VD							= 2	10	
P11	VA/†/†	§ 7								无轴突
P11	AS/VD							= 2	12	
野生型										
Pn	VA	§ 4= 3					=1	#1	16	
Pn	VB		= 2		= 2			# 2	9	
Pn	VC					§ 3 = 5		# 7	2	
Pn	AS	§ 2= 1	§ 2	= 1						
Pn	VD			§ 1	§ 2	§ 7		= 1	26	

野生型的突触数值是从野生型个体的电镜重建获得的 P3 至 P5 前体细胞突触数值的平均值。#，突触前；§，突触后；=，间隙连接。

* 在这些细胞系中，VC 的细胞核曾经产生，但是在成虫电镜重建时已不存在。这些核可能融合以形成 VA/VB/VC 或者消亡了。

† 这些细胞在 P1~P2 和 P9~P12 系相当于 VC，在 P11~P12 系相当于 VB，正常情况下会经历程序性细胞死亡 [1]。

‡ 这些区域内的突起正常时能够显示这些突触，但处于重建区域之外。

Several differences were apparent between the blocked precursors and the corresponding cell types in wild-type animals: the blocked cells had more synaptic inputs (Table 1), larger nuclei and extra branches (Fig. 2). These extra branches were generally devoid of synapses and had none of the characteristics of the other cell type. Even more extensive branching has been seen from precursors that are blocked earlier in ventral cord lineages in the mutant *lin-5* (ref. 7). The supernumerary branches may be a consequence of the polyploid nature of these cells; perhaps the total process length produced from a given neurone is proportional to its ploidy. The embryonically produced neurones in this region (that is, the DA, DB and DD motoneurones[6,8]) were normal in morphology and connectivity.

The observation that characteristics of some cell types were expressed normally in the blocked cells implies that the normal complement of wild-type cell divisions is not a necessary prerequisite for the differentiation of these cell types; however, some other cell cycle event (such as DNA replication[15]) may be required. Blast cells that are blocked earlier in a lineage either by drugs[9,10] or in the mutant *lin*-5 (ref. 7), also exhibit characteristics of their normal descendants, but in these cases characteristics of several cell types may be present in one polyploid cell. This may be a result of either localized differentiation or excessive dilution of regulatory elements within these abnormally large, highly polyploid cells. The isolation and characterization of mutants that are blocked at intermediate levels may shed some light on these differences.

Not all late divisions failed in these mutants, and those that did not fail produced normal cells. This observation indicates that the suppression of the differentiated characteristics of certain cell types seen in the blocked precursors does not simply reflect the inability of the mutants to produce these cell classes but implicates the failure to divide as the cause of their suppression.

The lack of expression of one cell type in the undivided, probably tetraploid precursors of *unc-59* and *unc-85* may be because there is an intrinsic mutual exclusivity in the expression of cell type such as has been described for certain fused cells[11] and cultured sympathetic neurones[12]. An alternative interpretation of these observations is that the differentiated state of the precursors reflects the state of commitment of these cells in wild-type animals; thus, one of the daughters would normally inherit the state of commitment of the mother. Some support for this notion is provided by observations of mutants in the genes *unc-86* and *lin-4*; these mutants undergo extra cell divisions in specific lineages. In these cases one of the daughters reiterates the characteristic divisions of the mother in a stem cell manner[13], indicating that it has the same state of commitment as its mother cell. These two interpretations are not necessarily incompatible if cells may only express one state of commitment at any time because of an intrinsic mutual exclusivity of cell states.

The electron microscopy was done by N. Thomson and Marilyn Anness. We thank S. Brenner and W. Fixsen for helpful discussions. H.R.H. was supported by postdoctoral fellowships from the Muscular Dystrophy Associations of America and the USPHS and by

细胞分裂受阻的前体细胞与野生型中与之相对应的细胞类型之间有几个明显的差别：阻断的细胞具有更多的突触传入（表 1），更大的细胞核和额外的分叉（图 2）。这些额外的分叉通常缺乏突触，而且不具备其他细胞类型的特征。在突变体 *lin-5* 中，腹侧索前体细胞的分裂在更早期被阻断，在这些前体细胞中观察到更大规模的分叉（文献 7）。这些过度的分叉可能是细胞多倍体性的结果；可能一个特定神经元的总突起长度是与其倍性成比例的。在这些区域中胚胎发育时期形成的神经元（即 DA、DB 和 DD 运动神经元 [6,8]）在形态和连接性上都是正常的。

细胞分裂受阻的细胞能够表现出某些细胞类型的特征，这个发现提示野生型中的正常细胞分裂过程并不是细胞分化的必要前提。但是，其他的细胞周期事件（比如 DNA 复制 [15]）可能是必需的。药物处理 [9,10] 或者在 *lin-5* 突变体中（文献 7），细胞分裂在较早时期已经阻断的母细胞是会体现其正常子细胞特征的，但这些情况下往往是多种细胞类型的特征同时出现在同一个多倍体细胞中。这一现象可能是这些超大的，染色体倍数特别多的细胞中的局部分化或者调节元件的过度稀释造成的。细胞分裂被部分阻断的突变体的分离和鉴定可能有助于区分这两种可能性。

（这两种）突变体中并非所有的晚期细胞分裂都不能进行，那些没有被阻断的细胞同样能够产生正常的细胞。这个发现表明：分裂受阻的前体细胞不能出现某些特定细胞类型的分化特性，不仅仅简单地反映了突变体不能产生这些细胞类型，而且暗示分裂失败可能是它们被抑制的原因。

unc-59 和 *unc-85* 中未分裂的、可能为四倍体的前体细胞不能显示某种细胞类型特征的原因可能是细胞类型间的内在排他性（这种现象在某些融合细胞 [11] 和培养的交感神经元 [12] 中也曾描述过）。上述现象的另一解释就是前体细胞的分化反映了野生型个体中这些细胞的本来状态；所以子代细胞之一自然就会遗传母细胞本来的分化状态。在突变体 *unc-86* 和 *lin-4* 的研究观察中得到了一些支持这一观点的证据；这些突变体在特定的细胞系中会经历额外的细胞分裂。在这种情况下，子代细胞之一以干细胞的方式反复进行母代特有的细胞分裂 [13]，这表明其具有与母代细胞相同的功能状态。如果由于细胞状态的内在排他性而使得细胞在任何时候都只能表达一种功能状态的话，这两种解释应该是可以共存的。

电子显微镜工作是由汤姆森和玛丽莲 · 安尼斯完成的。我们感谢布伦纳和菲克森提出的宝贵意见。霍维茨得到了美国肌萎缩协会和美国公共卫生署博士后研究基

USPHS grants GM24663 and GM24943.

(**297**, 584-587; 1982)

J. G. White*, H. R. Horvitz† and J. E. Sulston*
* MRC Laboratory of Molecular Biology, Hills Road, Cambridge CB2 2QH, UK
† Department of Biology, Massachusetts Institute of Technology, Cambridge, Massachusetts 02139, USA

Received 25 January; accepted 28 April 1982.

References:

1. Sulston, J. E. & Horvitz, H. R. *Devl. Biol.* **56**,110-156 (1977).
2. Kimble, J. & Hirsh, D. *Devl. Biol.* **70**, 396-417 (1979).
3. Deppe, U. *et al. Proc. Natl. Acad. Sci. U.S.A.* **75**, 376-380 (1978).
4. Horvitz, R. & Sulston, J. *Genetics* **96**, 435-454 (1980).
5. Sulston, J. & Horvitz, R. *Devl. Biol.* **82**, 41-55 (1981).
6. White, J., Southgate, E., Thomson, N. & Brenner, S. *Phil. Trans. R. Soc.* **B275**, 327-348 (1976).
7. Albertson, D., Sulston, J. & White, J. *Devl. Biol.* **63**, 165-178.
8. White, J., Albertson, D. & Anness, M. *Nature* **271**, 746-766 (1978).
9. Whittaker, J. R. *Proc. Natl. Acad. Sci. U.S.A.* **70**, 2096-2100 (1973).
10. Laufer, J., Bazzicalupo, P. & Wood, W. *Cell* **19**, 569-577 (1980).
11. Davidson, R. L. in *Somatic Cell Hybridization* (eds Davidson, R. L. & de la Cruz, F.) 131-150 (Raven, New York,1974).
12. Reichardt, L. & Patterson, P. *Nature* **270**, 147-151 (1977).
13. Chalfie, M., Horvitz, R. & Sulston, J. *Cell* **24**, 59-69 (1981).
14. Sulston, J. E. & White, J. G. *Devl. Biol.* **78**, 577-598 (1980).
15. Satoh, N. & Susumu, I. *J. Embryol. exp. Morph.* **61**, 1-13 (1981).

金以及美国公共卫生署基金 GM24663 和 GM24943 的支持。

（毛晨晖 翻译；丁梅 审稿）

Evidence on Human Origins from Haemoglobins of African Apes

M. Goodman *et al.*

Editor's Note

Fossil evidence for human antiquity was tempered in the 1970s by comparison of protein sequences, which provide a "molecular clock". Rather than supporting the fossil-based idea of a human lineage stretching back more than 10 million years, the molecular data suggested a contradictory story: that humans and chimpanzees diverged around 1.5 million years ago. Here Morris Goodman and colleagues used more refined techniques to show how the rate of change in protein sequences of apes and humans has decelerated over the past few million years, giving more accurate and consistent molecular-clock estimates of this divergence time.

Molecular data have influenced views concerning human origins, first, by supporting the genealogical classification of *Pan* (chimpanzee) and *Gorilla* with *Homo* rather than with *Pongo* (orangutan)[1,2] and, second, by suggesting that only a few million years separate humans and chimpanzees from their last common ancestor[3,4]. Indeed, the cladistic distances in phylogenetic trees constructed from amino acid sequence data, on detecting many superimposed mutations, yielded a "molecular-clock" divergence date between *Homo* and *Pan* of only 1–1.5 Myr BP[5]. This date, which is even more recent than that (4.2–5.3 Myr BP)[6] calculated using phenetic distances from immunological and DNA-hybridization comparisons (Table 1), is too near the present considering the existence of 3–4 Myr-old fossils of bipedal human ancestors[7] (and a 5.5 Myr-old jaw fragment assigned to *Australopithecus*[8]). Perhaps decelerated sequence evolution occurred; alternatively, hominoid distances could have been underestimated, because chimpanzee and gorilla were represented mostly by sequences inferred from peptide amino acid compositions, as was the case for their haemoglobins[9,10]. To help rectify this situation we report here the rigorously determined α- and β-haemoglobin amino acid sequences not only of chimpanzee (*Pan troglodytes*) and *Gorilla gorilla* but also pygmy chimpanzee (*Pan paniscus*). Our findings favour the explanation of decelerated evolution and point to selection preserving perfected haemoglobin molecules.

THE pygmy chimpanzee sequence is shown in Fig. 1, with supporting data in Fig. 2. There is no difference in this sequence from that of either chimpanzee or man, and only two differences from gorilla. At position $\alpha 23$, gorilla has aspartic acid instead of glutamic acid and at $\beta 104$, lysine instead of arginine. When these sequences are used with other known haemoglobin sequences in phylogenetic reconstructions by the maximum parsimony method, evidence is provided for cladistically joining *Pan* and *Gorilla* to *Homo* in

来自非洲猿血红蛋白的人类起源证据

古德曼等

编者按

20 世纪 70 年代，通过比较提供“分子钟”的蛋白质序列，古人类的化石证据得到一些修正。基于化石证据，人类谱系可以回溯到一千多万年前；然而分子数据非但不支持这一观点，反而提出了一个与其对立的情形：人类和黑猩猩大约在 150 万年前分化。本文中莫里斯·古德曼和他的同事们采用更精确的技术，展示了过去几百万年间类人猿和人类的蛋白质序列变化速率是怎样减缓的，并给出了更准确、一致的分离时间的分子钟估计值。

分子数据已经影响到了有关人类起源的观点，这种影响体现在两个方面：首先，分子数据支持黑猩猩属和大猩猩属的系统分类与人属相同，而与猩猩属不同 [1,2]；其次，分子数据提出，人类和黑猩猩从其最近的共有祖先分离出来只有几百万年的时间 [3,4]。实际上，根据检测许多叠加突变的氨基酸序列数据构建的系统演化树的进化支距离，我们可以得到人属和黑猩猩属的“分子钟”分离时间距今仅 100 万到 150 万年 [5]；比起利用免疫学和 DNA 杂交比较得到的表型距离计算出的时间（距今 420 万年到 530 万年）[6]，这一时间更近（表 1）。考虑到存在 300 万到 400 万年前的两足行走的人类祖先化石 [7]（以及一个被认定为 550 万年前的南方古猿的下颌骨碎片 [8]），这一时间似乎离现在太近了。这有可能是因为序列进化速率减缓了，或者因为黑猩猩和大猩猩的序列主要是以多肽的氨基酸组成推导出来的序列呈现的，就像他们的血红蛋白的情况一样 [9,10]，导致了对人科动物进化距离的过低估计。为了帮助修正这一情况，我们在此报道了经过严密方法确定的黑猩猩、大猩猩以及倭黑猩猩的 α–血红蛋白和 β–血红蛋白的氨基酸序列。我们的发现支持进化减速的解释，并且指出存在某种选择作用，该选择倾向于保留最好的血红蛋白分子。

图 1 所示为倭黑猩猩的序列，其支持数据在图 2 中列出。倭黑猩猩的这一序列与黑猩猩和人类都没有区别，与大猩猩只在两个位点有所不同。在位点 α23，大猩猩是天冬氨酸，而倭黑猩猩是谷氨酸；β104 位点处，大猩猩中是赖氨酸，而倭黑猩猩是精氨酸。当采用最大简约法使用这些序列和其他已知血红蛋白序列一起重建系统发育时，就可以得到相关证据证明在进化关系上黑猩猩属和大猩猩属跟人属聚

Homininae rather than to *Pongo* in Ponginae, these two subfamilies being sister groups (Fig. 3*a*). Specifically, substitutions at β87 (lysine → threonine) and β125 (glutamine → proline) group *Pan, Homo* and *Gorilla* into Homininae. When all relevant amino acid sequence data are considered, breaking up the African ape–human clade with orang-utan adds at least 10 nucleotide replacements (NRs) over the most parsimonious score (2 NRs apiece for fibrinopeptides A and B, myoglobin and β-haemoglobin, and 4 NRs for carbonic anhydrase[11]). Further parsimony evidence for a monophyletic chimpanzee–human–gorilla clade has been obtained from mitochondrial DNA nucleotide sequence data, representing close to 900 aligned positions of human[12], chimpanzee[13], gorilla[13], orang-utan[13], gibbon[13], mouse[14] and ox[15]. Taking mouse and ox as outgroups of Hominoidea and calculating NR lengths for each possible dichotomous branching order among the five hominoid lineages, it was found[16] that breaking up the human–chimpanzee–gorilla clade adds at least 19 NRs to the most parsimonious score.

Table 1. Comparison of "clock" dates from different sets of molecular data

Ancestral node	Immunological		DNA hybridization (Myr BP)	Amino acid sequence (Myr BP)	
	Albumin (Myr BP)	Transferrin (Myr BP)			
Theria	125	X	X	117	200
Eutheria	90–100	X	X	<u>90</u>	154
Primates	73.5	63	91	51	87.1
Anthropoidea	<u>35</u>	<u>35</u>	<u>35</u>	20.5	<u>35</u>
Catarrhini	20.3	18.6	21.4	13.4	22.9
Homo-Pan-(Gorilla)	4.2(4.2)	4.6(4.6)	5.3(5.3)	1.3(1.8)	2.2(3.1)

Times for the Theria (Metatheria–Eutheria) and Eutheria ancestral nodes for albumin immunological distances are from Sarich and Cronin[6] as described in the text of their article; the other divergence times from albumin and transferrin immunological distances and from DNA hybridization distances are calculated from Table 2 of their article using 35 Myr BP[6] for the Anthropoidea (platyrrhine–catarrhine) ancestral node as the setting of the "clock". The divergence clock dates shown in the last two columns are from Table 9 of ref. 5; these dates are based on calculations using phylogenetic trees constructed from amino acid sequence data on up to 10 polypeptide chains (α- and β-haemoglobins, myoglobin, lens α-crystallin A, cytochrome *c*, fibrinopeptides A and B, and carbonic anhydrases I, II and III). In the second to last column, the setting of the clock is 90 Myr BP for Eutheria ancestral node. In the last column, 35 Myr BP for Anthropoidea ancestral node is used to set the clock, thus allowing more direct comparison with values from Sarich and Cronin[6]. Note that the so-called molecular clocks are not in markedly good agreement with one another. The value underlined for each set of clock dates is the setting for these calculations. "X" Means that no clock date could be calculated because no corresponding distance value was given in Table 2 of Sarich and Cronin[6].

在人亚科的进化支下，而不是同猩猩属聚在猩猩亚科中。这两个亚科是姊妹群（图3*a*）。发生在β87（赖氨酸→苏氨酸）和β125（谷氨酸→脯氨酸）两个位点的替换将黑猩猩属、人属和大猩猩属归属到人亚科中。如果考虑到所有相关的氨基酸序列信息，将非洲猿类—人类分支与猩猩所在的分支打断至少会增加超过最大简约度10个NR（血纤肽A和B、肌红蛋白和β–血红蛋白各有两个NR，而碳酸酐酶有四个NR[11]；NR：nucleotide replacement，核苷酸的替代位点）。目前已经由线粒体DNA的核苷酸序列数据得到了更多关于黑猩猩－人类－大猩猩进化谱系是单源进化支系的简约性证据，这里所使用的线粒体DNA核苷酸序列数据代表了人类[12]、黑猩猩[13]、大猩猩[13]、猩猩[13]、长臂猿[13]、小鼠[14]和牛[15]的将近900个位点的比对信息。将小鼠和牛作为人猿超科的外群，在这五种人科家族各种可能的二歧分支次序上计算的核苷酸置换长度，结果发现[16]：打断人类－黑猩猩－大猩猩进化支，最大简约度至少会增加19个NR。

表1. 来自不同分子数据"分子钟"时间的比较

祖先节点	免疫学		DNA 杂交 (Myr BP)	氨基酸序列 (Myr BP)	
	白蛋白 (Myr BP)	转铁蛋白 (Myr BP)			
兽亚纲	125	X	X	117	200
真兽次亚纲	90~100	X	X	90	154
灵长目	73.5	63	91	51	87.1
类人猿亚目	35	35	35	20.5	35
狭鼻类	20.3	18.6	21.4	13.4	22.9
人属－黑猩猩属－(大猩猩属)	4.2 (4.2)	4.6 (4.6)	5.3 (5.3)	1.3 (1.8)	2.2 (3.1)

表1. 兽亚纲（后兽次亚纲－真兽次亚纲）和真兽亚纲的白蛋白免疫距离的祖先节点引用自萨里奇和克罗宁[6]的文章中的描述；其余白蛋白和转铁蛋白的免疫距离的分离时间以及DNA杂交距离的分离时间是由他们文章中表2给出的数据计算出来的，将类人猿亚目（阔鼻次目－狭鼻次目）的祖先节点距今3,500万年[6]作为"分子钟"设定值。最后两栏列出的分子钟日期来自参考文献5的表9；根据多达10条多肽链（包括：α－血红蛋白和β－血红蛋白、肌红蛋白、α－晶状体球蛋白A、细胞色素c、血纤肽A和B、碳酸酐酶I、II和III）的氨基酸序列信息构建系统进化树，利用该进化树计算得出这些日期。从第二栏到最后一栏，真兽次亚纲的祖先节点距今9,000万年作为分子钟设定值。最后一栏中，将类人猿亚目的祖先节点距今3,500万年作为分子钟设定值，因此可以与萨里奇和克罗宁文章中的数值进行更直接的比较[6]。注意：所谓的分子钟彼此并不十分一致。表示每组分子钟日期的数值都用下划线标出，这些数值是为进行这些计算而设定的。"X"表示由于萨里奇和克罗宁文章[6]的表2中没有给出相应的距离值，因而无法计算出其分子钟时间。

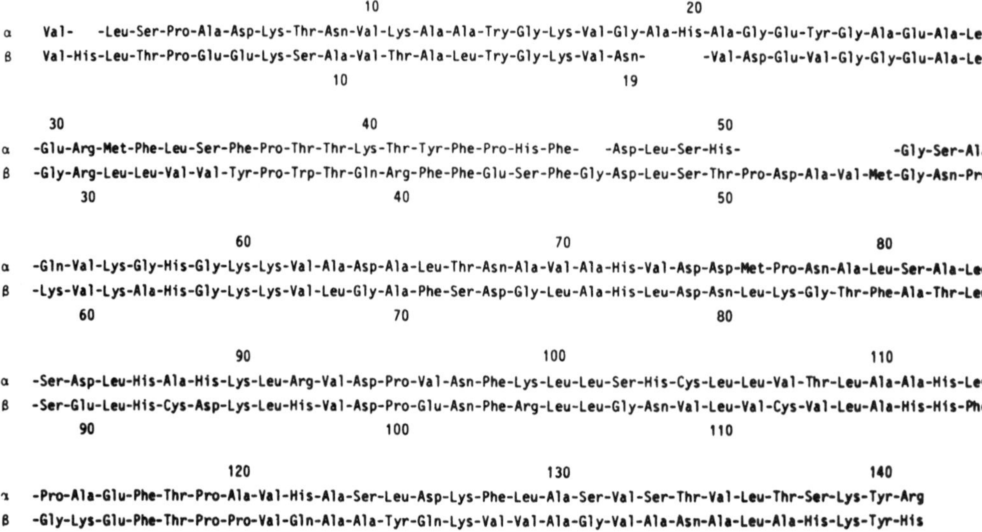

Fig. 1. Complete amino acid sequence of pygmy chimpanzee haemoglobin. The sequence is deduced in the liquid phase sequenator using the Quadrol program[26] and the *N,N*-diethylaminopropyne program[27]. The amino acid analysis of the most important peptides is given in Fig. 2. The amino acid sequences of human haemoglobin A and of the haemoglobins of pygmy chimpanzee and chimpanzee are all identical.

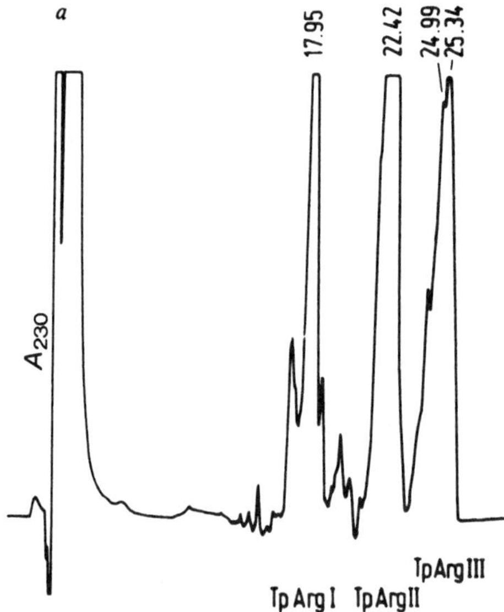

Fig. 2. *a*, Amino acid analysis of most important peptides for sequencing pygmy chimpanzee haemoglobin. Tp Arg, tryptical peptides after blocking lysine side chains. Pro Pep., C-terminal peptide after hydrophilic splitting of Asp–Pro bond[28]. The separation of most peptides was performed by HPLC chromatography. Separation of the "Arginine" peptides from the α-chains by HPLC (apparatus: Beckman Instruments, RP 2 column acetonitrile gradient; ammonium acetate buffer *p*H 6.0).

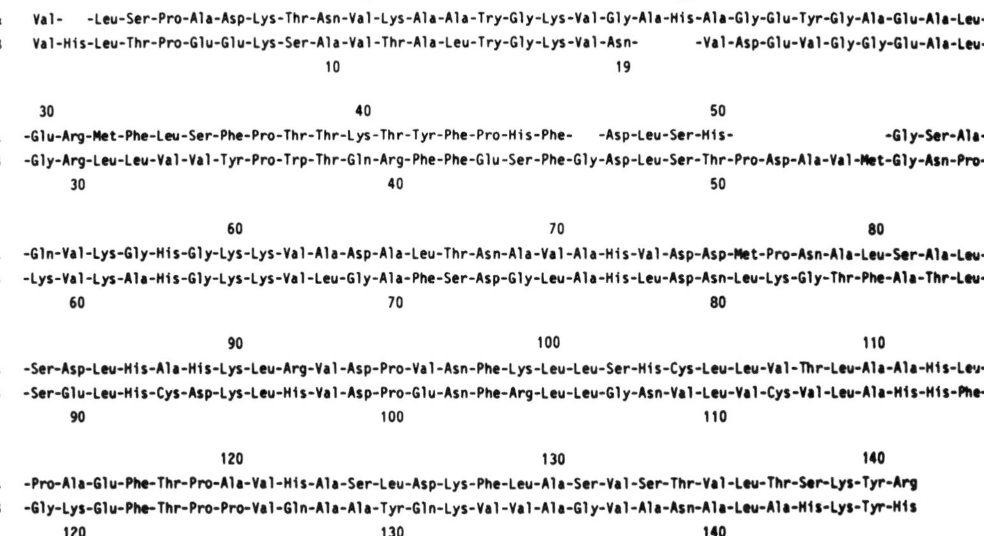

图 1. 倭黑猩猩血红蛋白的氨基酸全序列。该序列是使用乙二胺程序 [26] 和 *N*,*N*– 二乙基丙炔胺甲酸盐程序 [27] 通过固相序列分析仪推导出来的。图 2 给出了最重要的肽段的氨基酸分析。人类血红蛋白 A 的氨基酸序列与倭黑猩猩和黑猩猩的血红蛋白的氨基酸序列都是一样的。

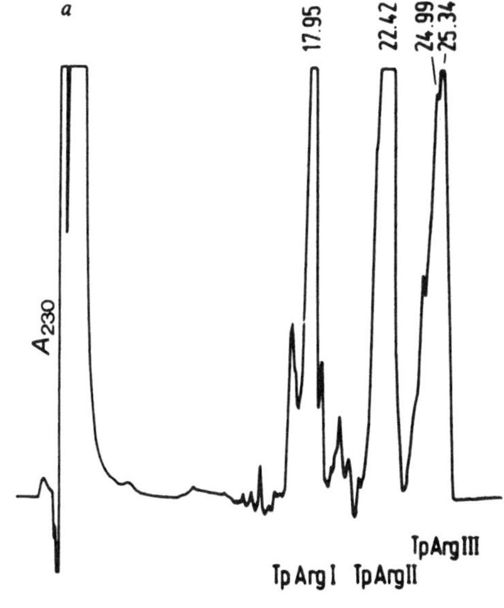

图 2. *a*，倭黑猩猩血红蛋白测序的重要肽段的氨基酸分析。Tp Arg，阻断赖氨酸侧链之后的胰蛋白酶肽段。Pro Pep.，天冬氨酸–脯氨酸键发生亲水解离后的 C–末端肽 [28]。大部分肽段的分离是通过高效液相色谱法进行的。通过高效液相色谱法将精氨酸肽段从 α– 链上分离下来（仪器装置：贝克曼仪器、反向二柱乙腈梯度仪和 pH 6.0 的醋酸铵缓冲液）。

b

Bonobo α–Chains

	Tp1 Arg	Tp2 Arg	Tp3 Arg	Pro Pep.	Chain Analysis
Pos.	1-31	32-92	93-141	65-141	
Lys	3.00	5.24	3.25	3.17	11.1
His	1.21	6.27	3.20	3.13	9.80
Arg	1.11	1.08	1.08	1.05	3.00
Asp	2.16	6.80	3.07	2.40	12.32
Thr	1.08	3.83	3.70	3.80	8.74
Ser	1.08	4.92	4.80	4.80	10.80
Glu	2.98	1.15	1.08	1.27	5.30
Pro	1.01	2.94	2.86	2.86	7.05
Gly	3.84	3.19			7.30
Ala	7.06	8.18	5.85	6.26	20.70
Cys			0.70	0.90	0.90
Val	3.05	4.15	5.90	4.80	12.70
Met		1.85			1.70
Leu	2.20	7.12	9.10	8.86	18.20
Tyr	1.01	1.08	0.95	1.05	2.90
Phe		4.22	3.12	3.13	7.05
Try	0.90				0.80
Total	31	61	49	47	141

Bonobo β–Chains

	Tp1 Arg	Tp2 Arg	Tp3 Arg	Tp4 Arg	CN1	CN2	Pro pep.	Chain Analysis
Pos.	1-30	31-40	41-74	75-146	1-55	56-146	70-146	
Lys	2.14		6.29	2.92	2.26	8.84	3.14	11.40
His	1.07		4.36	3.62	1.21	8.10	3.97	9.10
Arg	1.07	1.07	1.01		2.08	1.10	0.99	3.00
Asp	2.09		8.70	2.19	4.08	8.80	3.14	12.80
Thr	1.92	1.05	2.80	1.14	3.91	3.07	1.07	6.80
Ser	1.01		3.80		2.87	2.08		4.75
Glu	3.96	1.11	3.20	3.13	5.99	4.82	3.97	11.30
Pro	0.96	1.02	2.80	2.18	2.87	3.80	2.70	7.40
Gly	3.85		5.90	3.18	4.78	8.11	3.30	12.80
Ala	3.00		5.20	7.00	4.17	10.90	6.94	14.75
Cys			0.80	0.90		1.80	1.00	1.80
Val	4.80	1.75	4.30	6.70	7.04	10.70	7.10	17.80
Met			0.91					0.81
Leu	3.05	1.98	8.20	5.06	6.20	11.90	5.29	18.30
Tyr		1.03		1.98	1.04	1.96	1.90	3.10
Phe			5.89	2.14	3.21	5.10	2.97	8.10
Try	1.00	0.90			1.80			1.70
N-Ser					1.00			
Total	30	10	64	42	55	91	47	146

b

倭黑猩猩 α– 肽链

	Tp1 Arg	Tp2 Arg	Tp3 Arg	Pro Pep.	肽链分析
Pos.	1-31	32-92	93-141	65-141	
Lys	3.00	5.24	3.25	3.17	11.1
His	1.21	6.27	3.20	3.13	9.80
Arg	1.11	1.08	1.08	1.05	3.00
Asp	2.16	6.80	3.07	2.40	12.32
Thr	1.08	3.83	3.70	3.80	8.74
Ser	1.08	4.92	4.80	4.80	10.80
Glu	2.98	1.15	1.08	1.27	5.30
Pro	1.01	2.94	2.86	2.86	7.05
Gly	3.84	3.19			7.30
Ala	7.06	8.18	5.85	6.26	20.70
Cys			0.70	0.90	0.90
Val	3.05	4.15	5.90	4.80	12.70
Met		1.85			1.70
Leu	2.20	7.12	9.10	8.86	18.20
Tyr	1.01	1.08	0.95	1.05	2.90
Phe		4.22	3.12	3.13	7.05
Try	0.90				0.80
Total	31	61	49	47	141

倭黑猩猩 β– 肽链

	Tp1 Arg	Tp2 Arg	Tp3 Arg	Tp4 Arg	CN1	CN2	Pro pep.	肽链分析
Pos.	1-30	31-40	41-74	75-146	1-55	56-146	70-146	
Lys	2.14		6.29	2.92	2.26	8.84	3.14	11.40
His	1.07		4.36	3.62	1.21	8.10	3.97	9.10
Arg	1.07	1.07	1.01		2.08	1.10	0.99	3.00
Asp	2.09		8.70	2.19	4.08	8.80	3.14	12.80
Thr	1.92	1.05	2.80	1.14	3.91	3.07	1.07	6.80
Ser	1.01		3.80		2.87	2.08		4.75
Glu	3.96	1.11	3.20	3.13	5.99	4.82	3.97	11.30
Pro	0.96	1.02	2.80	2.18	2.87	3.80	2.70	7.40
Gly	3.85		5.90	3.18	4.78	8.11	3.30	12.80
Ala	3.00		5.20	7.00	4.17	10.90	6.94	14.75
Cys			0.80	0.90		1.80	1.00	1.80
Val	4.80	1.75	4.30	6.70	7.04	10.70	7.10	17.80
Me			0.91					0.81
Leu	3.05	1.98	8.20	5.06	6.20	11.90	5.29	18.30
Tyr		1.03		1.98	1.04	1.96	1.90	3.10
Phe			5.89	2.14	3.21	5.10	2.97	8.10
Try	1.00	0.90			1.80			1.70
N-Ser					1.00			
Total	30	10	64	42	55	91	47	146

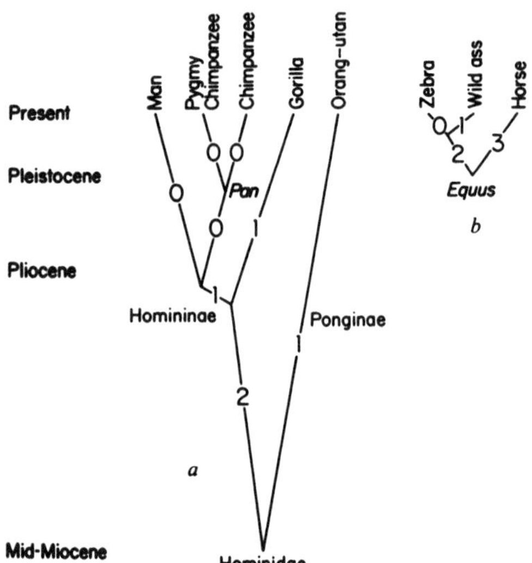

Fig. 3. Hominid (*a*) and equine (*b*) radiations as deduced from maximum parsimony trees of α- and β-haemoglobin sequences. Numbers of NRs are shown on the branches. They result from the following substitutions recorded in the single-letter code for amino acids[21]. *a*, Between Hominidae ancestral node and orangutan: α12A → T; between Hominidae and Homininae ancestral nodes: β87K → T, β125Q → P; between Homininae ancestral node and gorilla: β104R → K; between Homininae and *Pan-Homo* ancestral nodes: α23D → E. *b*, Between *Equus* ancestral node and horse: β52A → G, β87Q → A; between *Equus* and zebra–wild ass ancestral nodes: α20H → N, α131S → T; between zebra–wild ass ancestral node and wild ass: α23E → D. Immunological evidence for cladistically grouping pygmy chimpanzee with chimpanzee in the genus *Pan*[29] is now supported by amino acid sequence results on carbonic anhydrase I (D. Hewett-Emmett, unpublished data). The positioning of the divergence node for pygmy chimpanzee and chimpanzee reflects views[30] based on immunological and electrophoretic distance data.

Aside from supporting inclusion of *Pan* and *Gorilla* in Homininae, the parsimony evidence from amino acid sequence data hints at the possibility that *Pan* is more closely related to *Homo* than to *Gorilla.* One less NR is required with α-haemoglobin sequences by grouping *Pan and Homo* first before adding gorilla (Fig. 3*a*). This is due to aspartic acid at position α23 in New World monkey, gibbon, orang-utan and gorilla chains as compared with glutamic acid at this position in human, chimpanzee and pygmy chimpanzee chains. Amino acid sequences of γ-haemoglobin chains also point to a *Homo–Pan* clade distinct from gorilla. There are two non-allelic chains in the hominines, $^{G}\gamma$ characterized by glycine at position 136 and $^{A}\gamma$ with alanine at this position. The two sequences in chimpanzee as inferred from amino acid composition data are identical to their human counterparts[17]. In contrast, gorilla $^{A}\gamma$ (as deduced from nucleotide sequence data obtained on gorilla γ-haemoglobin genes)[18] has arginine at position 104 whereas human and chimpanzee $^{A}\gamma$ chains have lysine. The presence of arginine at position 104 in Old World monkey (*Macaca* and *Papio*) γ-haemoglobin chains[19] raises the possibility that human and chimpanzee $^{A}\gamma$ have the derived rather than primitive residue.

With mitochondrial DNA sequences there are two trees of lowest NR length grouping *Pan* either with *Gorilla* or with *Homo*[16]; phenetically, however, *Pan* shows more base matches

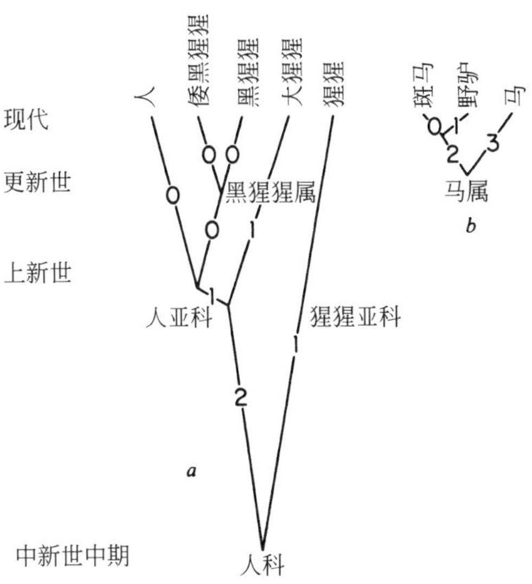

图 3. 基于 α– 血红蛋白和 β– 血红蛋白序列构建的最大简约进化树，推导出的人科动物（*a*）和马（*b*）的辐射进化关系。NR 数标在每个分支上。它们是由氨基酸密码子的一个字母替换产生的，如下[21]：*a*，人科的祖先节点与猩猩之间：α12A → T；人科与人亚科的祖先节点之间：β87K → T，β125Q → P；人亚科的祖先节点与大猩猩之间：β104R → K；人亚科与黑猩猩属 – 人属的祖先节点之间：α23D → E。*b*，马属的祖先节点与马之间：β52A → G，β87Q → A；马属与斑马 – 野驴的祖先节点之间：α20H → N，α131S → T；斑马 – 野驴的祖先节点与野驴之间：α23E → D。在进化支上将倭黑猩猩与黑猩猩划分到黑猩猩属[29]的免疫学证据现为碳酸酐酶 I 的氨基酸序列结果所支持（休伊特 – 埃米特，数据尚未发表）。倭黑猩猩和黑猩猩的分离节点的定位反映了基于免疫学和电泳距离数据的观点[30]。

除了支持黑猩猩属和大猩猩属属于人亚科之外，由氨基酸序列得到的简约性证据还提示黑猩猩属与人属的亲缘关系可能比它们与大猩猩属的亲缘关系更近。在加上大猩猩之前，如果想通过 α– 血红蛋白序列先将黑猩猩和人类进行聚类，那么需要减少一个 NR（图 3*a*）。这是因为处于人类、黑猩猩和倭黑猩猩这一进化链中的 α23 位点都是谷氨酸，与之相比，处于新大陆的猴、长臂猿、猩猩和大猩猩这一进化链中的相应位点却是天冬氨酸。γ– 血红蛋白链的氨基酸序列也表明人属 – 黑猩猩属进化支与大猩猩不同。人亚科的 γ– 血红蛋白有两条非等位链，这两条链的特征为：$^{G}\gamma$ 在位点 136 处为甘氨酸，$^{A}\gamma$ 的该位点为丙氨酸。从氨基酸组成数据得出黑猩猩的这两段序列与人类的相一致[17]。相比之下，大猩猩的 $^{A}\gamma$（从大猩猩的 γ– 血红蛋白基因的核苷酸序列推导出来的）[18] 在位点 104 处是精氨酸，而人类和黑猩猩的 $^{A}\gamma$ 链的该位点是赖氨酸。旧大陆猴（例如猕猴属和狒狒属）的 γ– 血红蛋白链的 104 位点处也是精氨酸[19]，这提出了一种可能性，即人类和黑猩猩的 $^{A}\gamma$ 的残基是派生而来的而不是原本就有的。

使用线粒体 DNA 序列得到了两棵具有最低 NR 长度的进化树，分别将黑猩猩属与大猩猩属聚到一类或者将黑猩猩属与人属聚到一类[16]；然而从表型上看，黑猩猩

with *Homo* than with *Gorilla*[13,16] as it also does with nuclear DNA sequences[20]. Clearly it will be difficult to resolve the *Homo–Pan–Gorilla* trichotomy into two branching events if, as may be suspected, the second branching occurred shortly after the first.

That proteins such as globins show a strong—although uneven—tendency to increasing sequence divergencies with elapsed time is evident in both amino acid difference matrices[21,22] and phylogenetic reconstructions[5]. Accepting empirical observations that this tendency operates in a majority of clades, our finding of almost no sequence differences among hominines in either their α- or β-haemoglobin chains argues for a relatively late ancestral separation of *Homo, Pan* and *Gorilla.* Nevertheless, even if common ancestry existed until 5–6 Myr BP (just before the appearance of *Australopithecus* fossils), the rate of hominine haemoglobin evolution was slow, averaging about only 6 NRs per 100 codons per 10^8 yr (that is, 6 NR%), in contrast to much faster rates averaging about 29 NR% for β-haemoglobin sequences and 25 NR% for α-sequences during eutherian phylogeny[5].

To highlight the slowness of haemoglobin evolution during the hominine radiation, we can compare hominine with equine rates, taking advantage of the fact that α- and β-haemoglobin sequences have been determined for horse (*Equus callabus*), zebra (*Equus zebra*) and wild ass (*Equus hemionus*)[23]. Preceded by *Pliohippus,* the genus *Equus* does not appear in the fossil record until the Pleistocene[24]. Yet, during the short span of time (2 Myr BP to present) of this equine radiation, 2–3 NRs accumulated per equine lineage for the two haemoglobin chain types (Fig. 3*b*), compared with no changes in *Homo* and *Pan.* Even by the conservative assumption that *Homo* and *Pan* had no older separation than zebra and horse, and given a Poisson distribution with a mean of 5 NRs based on *Equus*, the probability of obtaining zero NRs as for *Homo–Pan* is less that 0.007—thus eliminating the hypothesis of homogeneous rates.

The slow rate of globin sequence evolution in hominines indicates prior selection in the pre-hominines of haemoglobin molecules with finely tuned, perfected adaptations. As previously reported[5], we find that between eutherian and hominine ancestral nodes, the fastest evolving positions are $\alpha_1\beta_1$ contact sites, important functional positions which show few normal human variants[25]. Also mutations to proline were fixed at exterior helical positions αA2 and at βA2 and $\alpha_1\beta_1$ contact positions βD2 and βH3. These additional prolines, all at the beginning of helices, by producing more stably directed helices, presumably improved subtler functions in the haemoglobin molecule. Furthermore, almost none of the many seemingly normal haemoglobin variants[25] are at polymorphic frequencies, possibly because even these variants harm the finer adaptations of human haemoglobin. The slow rate of haemoglobin evolution in African apes and humans suggests that organismal evolution in our earlier ancestors may have depended on amino acid sequence changes in proteins as well as regulatory changes in gene expression.

We thank Drs A. Spiegel, R. Faust and H. Wiesner for the pygmy chimpanzee, chimpanzee

属与人属之间的碱基匹配比它们与大猩猩属之间匹配得更高[13,16]，这与核DNA序列分析的结果一致[20]。很明显，如果像我们猜想的那样，第二次分支是在第一次分支后不久发生的，那么要想将人属－黑猩猩属－大猩猩属这种三元关系分解为二分支事件将是很困难的。

诸如球蛋白之类的蛋白质有一个很强但不规则的趋势，即随着时间的流逝序列分支会增大，这一趋势在氨基酸的差异矩阵[21,22]及系统进化重建[5]中都很明显。如果认同该趋势对大多数进化支都起作用这一经验性观察，那么对于我们的发现——人亚科之间的α–血红蛋白或β–血红蛋白链序列几乎没有差异——支持人属、黑猩猩属和大猩猩属的祖先分离相对较晚。然而，即使距今500万到600万年前（就在南方古猿化石出现之前）共同祖先依旧存在，人亚科的血红蛋白进化速率还是很慢，大概平均每一亿年间每100个密码子只有6个位点发生核苷酸取代（记作6NR%）。与此相比，真兽亚纲的系统发育过程中，其血红蛋白的进化速率要快得多，其中β–血红蛋白序列的进化速率达到了29 NR%，而α–血红蛋白则达到了25 NR%[5]。

为了强调在人亚科的辐射进化中血红蛋白进化之慢，我们可以利用家马、斑马和野驴的α–血红蛋白和β–血红蛋白序列这一已经确定的事实[23]，将人亚科的血红蛋白进化速率与马的相比较。马属的化石记录最早可追溯至更新世[24]，首先就是上新马。然而，与人属和黑猩猩属的两条血红蛋白链序列毫无差异相比，马在其辐射进化的很短一段时间内（从距今200万年前到现在），这两种血红蛋白链在每种马家族中都积累了2~3个NR（图3*b*）。实际上根据最保守的假设——人属和黑猩猩间不比斑马和马分离的时间早，而且考虑到以5个NR作为马属的平均值而得到的泊松分布，得出人属－黑猩猩属间没有发生核苷酸取代的概率小于0.007，因此排除了均匀速率的假说。

人亚科的球蛋白序列进化速率之慢暗示：在人亚科形成之前，血红蛋白分子就通过细致调整、完善适应性进行了优先选择。正如之前报道的[5]，我们发现，在真兽亚纲和人亚科的祖先节点之间，进化最快的是重要的功能性位置，即$\alpha_1\beta_1$接触位点，以上几科未见过正常的人类变异体[25]。突变成的脯氨酸也被固定在外螺旋位置αA2、βA2以及$\alpha_1\beta_1$接触的位置βD2和βH3处。这些额外的脯氨酸都处于螺旋结构的开端，可能通过产生更稳定的定向螺旋结构来提高血红蛋白分子的精细功能。此外，几乎所有的表观正常的血红蛋白变异体[25]都不具有多态性频率分布，这可能是因为这些变异体会损害人类血红蛋白完善的适应性。非洲猿和人类血红蛋白进化速度之慢提示：我们祖先的生物进化过程可能依赖于蛋白质的氨基酸序列变化以及基因表达的调控变化。

感谢施皮格尔、福斯特和威斯纳博士为我们提供倭黑猩猩、黑猩猩和大猩猩的

and gorilla blood samples, also Dr Gerard Joswiak and John Czelusniak for valuable discussion of the statistical significance of our findings. Phylogenetic analysis of the sequence data was supported by NSF grant DEB 78-10717.

(**303**, 546-548; 1983)

Morris Goodman*, Gerhard Braunitzer†, Anton Stangl† and Barbara Schrank†

* Department of Anatomy, Wayne State University School of Medicine, Detroit, Michigan 48201, USA

† Max-Planck Institut für Biochemie, Abteilung Proteinchemie, D-8033 Martinsried bei Munchen, FRG

Received 30 December 1982; accepted 7 April 1983

References:

1. Goodman, M. *Ann. N. Y. Acad. Sci.* **102**, 219-234 (1962).
2. Andrews, P. & Cronin, J.E. *Nature* **297**, 541-546 (1982).
3. Sarich, V. M. & Wilson, A.C. *Science* **158**, 1200-1203 (1967).
4. Goodman, M. *Hum. Biol.* **54**, 247-264 (1982).
5. Goodman, M. *Prog. Biophys. Molec. Biol.* **38**, 105-164 (1981).
6. Sarich, V. M. & Cronin, J. E. in *Molecular Anthropology* (eds Goodman, M.& Tashian, R. E.) 141-170 (Plenum, New York, 1976).
7. Johanson, D. L. & White, T. D. *Science* **203**, 321-330 (1979).
8. Patterson, B., Behrensmeyer, A. K. & Sill, W. *Nature* **226**, 918-921 (1970).
9. Rifkin, D. B. & Konigsberg, W. *Biochim. Biophys. Acta* **104**, 457-461 (1965).
10. Zuckerkandl, E. & Schroeder, W. A. *Nature* **192**, 984-985 (1961).
11. Tashian, R. E., Hewett-Emmett, D. & Goodman, M. in *Isozymes* Vol. 7 (eds. Rattozzi, M. C., Scandalias, J. G., Siciliano, M. T.& Whitt, G.S.) 79-100 (Liss, New York, 1983).
12. Anderson, S. *et al. Nature* **290**, 457-465 (1981).
13. Brown, W. M., Prager, E. M., Wong, A. & Wilson, A. C. *J. Molec. Evol.* **18**, 225-239 (1982).
14. Anderson, S. *et al. J. Molec. Biol.* **156**, 683-717 (1982).
15. Bibb, M. H. *et al. Cell* **26**, 167-180 (1981).
16. Goodman, M., Olson, C. B., Beeber, J. E. & Czelusniak, J. *Acta Zool. Fenn.* **169**, 19-35 (1982).
17. DeJong, W. W. *Biochim. Biophys. Acta* **251**, 217-226 (1971).
18. Scott, A. F. *et al. Am. J. Hum. Genet.* **34**, 193A (1982).
19. Mahoney, W. C. & Nute, P. E. *Biochemistry* **19**, 4436-4442 (1980).
20. Hoyer, B. H., Van de Velde, N. W., Goodman, M. & Roberts, R. B. *J. Hum. Evol.* **1**, 645-649 (1972).
21. Dayhoff, M. O. *Atlas of Protein Sequence and Structure* Vol. 5 (National Biochemical Research Foundation, Silver Springs, Maryland, 1972).
22. Dickerson, E. R. & Geis, L. *Hemoglobin: Structure, Function, Evolution, and Pathology* (Benjamin/Cummings, Menlo Park, 1983).
23. Mazur, G. & Braunitzer, G. *Hoppe-Seyler's Z. Physiol. Chem.* **363**, 59-71 (1982).
24. Romer, A. S. *Vertebrate Paleontology* (University of Chicago Press, 1966).
25. Bunn, H.F., Forget, B. G. & Ranney, H. M. *Human Hemogloblins* (Saunders, Philadelphia, 1977).
26. Edman, P. & Begg, G. *Eur. J. Biochem.* **1**, 80-91 (1967).
27. Braunitzer, G. & Schrank, B. *Hoppe-Seyler's Z. Physiol. Chem.* **351**, 417-418 (1970).
28. Jauregui-Adell, J. & Marti, J. *J. Analyt. Biochem.* **69**, 468-473 (1975).
29. Goodman, M. *Hum. Biol.* **35**, 377-436 (1963).
30. Zihlman, A. L., Cronin, J. E., Cramer, D. L. & Sarich, V. M. *Nature* **275**, 744-746 (1978).

血样，同样感谢杰勒德·约斯韦克博士和约翰·泽鲁斯尼亚克对我们的发现的统计学意义进行的有价值的讨论。序列数据的系统进化分析由美国国家科学基金会资助的DEB 78-10717 项目所支持。

（刘皓芳 翻译；冯兴无 审稿）

A Conserved DNA Sequence in Homoeotic Genes of the *Drosophila* Antennapedia and Bithorax Complexes

W. McGinnis *et al.*

Editor's Note

The development of flies and other insects from their fertilized eggs is a complex matter. The successive developmental stages include the appearance of larvae, a period of apparent quiescence (as in chrysalis) and the emergence of an adult insect usually bearing no resemblance to the larval stage. The question for developmental biologists is how the adult stage is specified in the earliest stages of the embryo. The answer, inferred by classical biologists, is that even the larvae contain internal structures that develop into the wings, legs, antennae and other structures embodied in the adult. In 1984 Walter Gehring and his colleagues at the University of Basel in Switzerland found a way of identifying the genes in the fruitfly *Drosophila* that are responsible for the succession of changes. The homoeotic genes, as they are called, are localized on the third chromosome of the fruitfly and have a characteristic structure that makes it possible to recognize genes with similar function in other organisms.

A repetitive DNA sequence has been identified in the *Drosophila melanogaster* genome that appears to be localized specifically within genes of the bithorax and Antennapedia complexes that are required for correct segmental development. Initially identified in cloned copies of the genes *Antennapedia*, *Ultrabithorax* and *fushi tarazu*, the sequence is also contained within two other DNA clones that have characteristics strongly suggesting that they derive from other homoeotic genes.

MANY of the homoeotic genes of *Drosophila* seem to be involved in the specification of developmental pathways for the body segments of the fly, so that each segment acquires a unique identity. A mutation in such a homoeotic gene often results in a replacement of one body segment (or part of a segment) by another segment that is normally located elsewhere. Many of these homoeotic loci reside in two gene complexes, the bithorax complex and the Antennapedia (Antp) complex, both located on the right arm of chromosome 3 (3R).

The bithorax complex is located in the middle of 3R, and its resident genes impose specific segmental identities on the posterior thoracic and abdominal segments[1]. For example, inactivation of the *bithorax* gene of the complex causes a transformation of the anterior half of the third thoracic segment into the anterior half of the second thoracic segment, resulting in a fly having wing structures in a site normally occupied by haltere.

果蝇触角足和双胸基因复合体同源异型基因中的保守DNA序列

麦金尼斯等

编者按

苍蝇以及其他昆虫从它们的受精卵开始经历的发育是一个复杂的问题。连续的发育阶段包括幼虫期、一个看似静止的时期（例如：蛹）和通常与幼虫阶段没有相似处的成虫期。发育生物学家的问题是，成虫阶段是如何在胚胎的最早期阶段被特化的？根据经典生物学家的推断，回答是：即使是幼虫也含有可以发育成成虫翅、足、触角和其他结构的内部结构。1984 年瑞士巴塞尔大学的沃尔特·格林教授及其同事找到了一种鉴定负责果蝇连续变化的基因的方法。这个被他们称之为同源异型的基因位于果蝇的第 3 条染色体上，并且具有一个特征性结构，可用于识别其他生物中具有类似功能的基因。

在黑腹果蝇的基因组里发现了一段重复 DNA 序列，这段序列似乎特异地定位于正常体节发育所需的双胸和触角足基因复合体内部。起初这个序列是在果蝇触角足基因、超级双胸和*ftz* 基因的克隆拷贝中鉴定出来的，后来在其他两个 DNA 克隆中也发现了这种序列，它们所具有的特征强有力地说明了这些基因起源于其他的同源异型基因。

果蝇的许多同源异型基因似乎都参与了果蝇体节发育途径的特化，从而每个体节都有其独有的特征。此类同源异型基因中的一个突变常常导致一个体节（或一个体节的一部分）被通常位于其他位置的另一个体节所替代。许多同源异型基因座位于两种基因复合体——双胸基因复合体和触角足基因复合体中，二者都位于 3 号染色体右臂上。

双胸基因复合体位于 3 号染色体右臂的中间，其内的基因可以影响后胸部和腹部体节特异性的体节特征 [1]。例如，这个基因复合体的双胸基因失活会引起第 3 胸节的前半部转变成第 2 胸节的前半部，从而导致果蝇在通常平衡棒所在的位置长出翅的结构。该基因复合体的其他隐性突变可以引起类似的变化，导致后部身体结构

Other recessive mutations in the complex cause analogous transformations of posterior body structures into structures normally located in a more anterior position. Embryos having a deletion of the entire bithorax complex show a transformation of all the posterior body segments into reiterated segments with structures of the second thoracic segment. Based on the above results and others, Lewis has proposed a model in which segmental identity in the thorax and abdomen is controlled by a stepwise activation of additional bithorax complex genes in more posterior segments[1].

The Antp complex is localized nearer the centromere of 3R than the bithorax complex. The genes of the Antp complex appear to control segmental development in the posterior head and thorax, in a manner analogous to the way in which the bithorax complex operates in the more posterior segments[2–5]. A dominant mutation in the *Antp* locus, for example, can result in the transformation of the antenna of the fly into a second thoracic leg[6,7].

The homoeotic genes of both the bithorax and Antp complexes can be thought of as selector genes, using the nomenclature of Garcia-Bellido[8], that act by interpreting gradients of positional information. Based on their location in the gradient, a specific combination of selector genes are expressed, and thus different regions of the developing fly become selected to proceed down specific developmental pathways. Although the available evidence supports this model[1,9,10], the real situation appears to be more complex as there is also evidence that regulatory interactions between different homoeotic selector genes have a role in limiting their region of expression[10–12].

The physical proximity and similar but distinct functions of the bithorax complex genes led Lewis to propose that the genes of this cluster evolved by mutational diversification of tandemly repeated genes[1]. In the primitive millipede-like ancestors of *Drosophila*, an ancestral gene or genes would direct the development of repetitive segments having similar identities. The evolutionary transition to the Dipterans, with highly diverse segmental structures, might be achieved by duplication and divergence of ancestral genes. According to this model, null mutations in the present set of bithorax complex genes could result in a fly having a more primitive segmental array, that is, with legs on the abdominal segments, or with wings on the third thoracic segment, in addition to those on the second thoracic segment; both types of phenotype are known to result from reduction or loss of function of bithorax complex genes.

Although the bithorax and Antp complexes are widely separated on the third chromosome, their similar functions in specifying segmental identity suggests that both complexes might have evolved from a common ancestral gene or gene complex. A critical test for this hypothesis involves a test for conserved sequences in the genes of the two complexes. These conserved sequences could be relics of ancient gene duplications or regions specifically preserved by selection against mutational change. Here we show that there is DNA sequence homology between some genes of the bithorax complex and the Antp complex. We use this homology, which is imperfect and limited to small regions, to

变成通常位于更前部位的结构。整个双胸基因复合体缺失的胚胎，其所有后部体节转变成为具有第二胸节结构的重复体节。基于以上结果和其他信息，刘易斯提出了一个模型，即胸部和腹部的体节特征受控于位于更后部体节中其他双胸复合体基因的逐步激活 [1]。

触角足基因复合体位于第 3 号染色体右臂上，它比双胸基因复合体更接近着丝粒。这些触角足复合体中的基因对头后部和胸部的体节发育的控制，似乎是通过一种类似于双胸基因复合体操纵更后部体节发育的方式实现的 [2-5]。例如，触角足基因座的显性突变会导致果蝇的触角转变为第二个胸足 [6,7]。

根据加西亚 – 贝利多命名法 [8]，可以将双胸基因复合体和触角足基因复合体两者的同源异型基因命名为选择者基因，它们通过解读位置信息的梯度起作用。基于其梯度定位，表达一组特定组合的选择者基因，从而使正在发育的果蝇的不同区域选择进行特定的发育途径。虽然现有的证据支持这个模型 [1,9,10]，但真正的情况似乎更为复杂，因为也有证据表明，不同的同源异型选择者基因之间的相互调节作用也有限制它们自身表达区域的作用 [10-12]。

双胸复合体中的基因，它们在基因组上的位置邻近，功能相似但又有不同，这使得刘易斯提出，这个基因簇中的基因是通过串联重复基因的突变多样化进化而来 [1]。果蝇有一种原始的祖先，类似千足虫，其一个或多个祖先基因可指导具有相似特性的重复体节的发育。膜翅目昆虫具有高度多样的体节结构，其进化转换可能是通过祖先基因的复制和分化而实现的。根据这个模型，在现今的双胸基因复合体中的无效突变可能使果蝇具有一种更原始的体节排列，也就是，除了第二胸节上的足外，在腹节或第三胸节上也长有足。这两种表型都被认为由双胸复合体基因功能的下降或缺失所致。

虽然双胸基因复合体和触角足基因复合体在第 3 号染色体上相隔很远，但它们在决定体节特性时的相似功能表明，这两种复合物可能从一个共同的祖先基因，或基因复合物进化而来。对这种假说的一个关键性检验涉及对这两个复合物基因中保守序列的检测。这些保守序列可能是远古基因复制的遗迹，或是通过自然选择作用于突变变化而特异性地保留下来的区域。在此，我们指出双胸基因复合体和触角足基因复合体中的一些基因之间的 DNA 序列具有同源性。我们利用这种不严格的且只限于小区域的同源性，从果蝇基因组里分离出了其他的交叉杂交克隆。与这两个

isolate other cross-hybridizing clones from the *Drosophila* genome. The cytogenetic map locations and spatial and temporal patterns of expression for the genes homologous to two of the clones suggest that they represent other homoeotic genes.

Repeated Sequences

Genomic and cDNA clones from the *Antp* locus have been isolated and characterized by Garber *et al.*[13]. To test whether the *Antp* gene might be a member of a multigene family, we hybridized the 903 cDNA probe derived from the *Antp* locus to Southern blots of *Drosophila* genomic DNA. The 903 cDNA (see Fig. 1) is complementary to four non-contiguous chromosomal DNA regions spanning 100 kilobases (kb) at the *Antp* locus. Both normal- and reduced-stringency hybridization conditions were used with the 903 probe; in both types of hybridization conditions we detected many genomic fragments homologous to 903 that gave very strong signals, and many (>50) that were relatively weak. The weak signals were more prominent on the blot hybridized in reduced-stringency conditions (data not shown). A stringent wash of both blots (see Fig. 1 legend) removed the weak signals whereas the strong signals remained. The strongly hybridizing genomic fragments had the expected size for those portions of the genome represented in the 903 cDNA. The weakly hybridizing genomic fragments presumably possessed mismatched homology to one or more repeated sequences within the 903 cDNA.

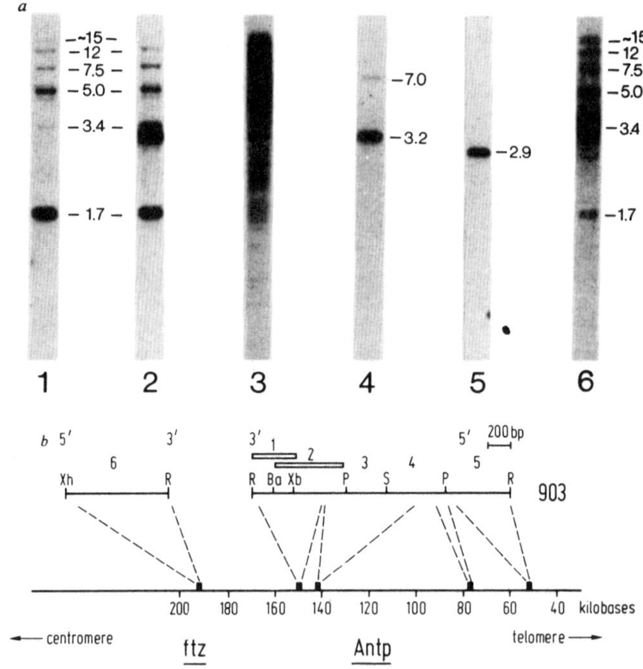

Fig. 1. Repeated sequences in the *Antp* and *ftz* genes. *a*, Individual lanes from *Drosophila* whole genomic Southern blots. The genomic DNA in each case was digested with *Eco*RI. The number below the lanes designates the fragment number used as a probe. The fragment number designations are shown above the respective fragments in *b*. The numbers alongside the lanes indicate the sizes in kilobases (kb) of the hybridizing genomic fragments. All the blots were hybridized and washed in the reduced-stringency conditions described below. Note that the two bands in lane 4 are due to the number 4 probe containing

克隆同源的基因的细胞遗传学图谱定位以及时空表达模式表明，它们代表了其他的同源异型基因。

重复序列

加伯等人从触角足基因座分离出了基因组和 cDNA 的克隆，并描述了其特征 [13]。为了检测触角足基因是否是多基因家族的成员，我们将来源于触角足基因座的 903 cDNA 探针与果蝇基因组 DNA 进行了 DNA 印迹法杂交。903 cDNA（见图 1）与触角足基因的 4 个非邻接的有 100 千碱基(kb)的染色体 DNA 区域互补。针对 903 探针，采用了正常型和低严紧型两种杂交条件；两种杂交条件下，我们检测到了许多与 903 探针具有同源性的基因组片段，它们有很多给出了非常强烈的信号，同时也有很多（>50）信号相对较弱。在低严紧型条件下的印迹杂交中，弱信号在数据中占优势（数据未列出）。对两种杂交印迹的严紧型洗涤(见图 1 注)可以去除弱信号而保留强信号。强杂交基因组片段显示了所预期的，由 903 cDNA 代表的基因组中部分序列的大小。弱杂交基因组片段大概具有与 903 cDNA 内的一个或多个重复序列错配的同源性。

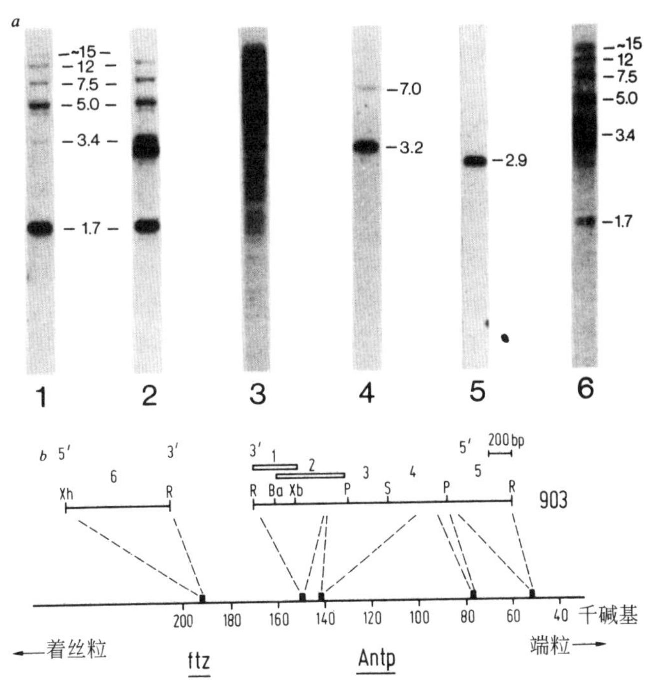

图 1. 触角足基因和 *ftz* 基因中的重复序列。*a*，每个泳道都是果蝇全基因组的 DNA 印迹法杂交片段。每个泳道的基因组 DNA 都用 *Eco*RI 消化。泳道下面的数字是用作探针的片段编号。片段编号列在 *b* 中各自片段的上面。靠泳道旁的数字是以千碱基对表示的杂交基因组片段的大小。所有的印迹都是在下述低严紧条件下进行杂交和洗涤的。注意：泳道 4 中的两个条带是由于 4 号探针含有两个触角足基因外显子。*b*，用作基因组印迹杂交探针的触角足基因和 *ftz* 基因片段的图谱。触角足基因 903 就是加伯等人 [13] 所述的一个 cDNA 克隆，其所含的基因片段来自图谱中以实心块标记的触角足基因座。虚线表示在每个基因

a sequence from each of two *Antp* exons. *b*, Map of the portions of the *Antp* and *ftz* genes used as hybridization probes for the genomic blots. *Antp* 903 is a cDNA clone described by Garber *et al.*[13] which contains the regions from the *Antp* locus marked by solid blocks. The broken lines indicate the approximate extent of the cDNA in each genomic location. The bottom line is a representation of the *Antp* region of chromosome 3, as taken from Garber *et al.*[13]; the numbers reflect the distance (in kb) from the *Humeral* chromosomal breakpoint. The 5′ and 3′ labels show the direction of transcription for the two loci (ref. 17 and A.K., unpublished results). Xh, *Xho*I; R, *Eco*RI; Ba, *Bam*HI; Xb, *Xba*I; P, *Pvu*II; S, *Sph*I.

Methods: Reduced-stringency hybridizations were done as follows. Southern blots[31] were prehybridized in 5×SSC, 0.1% bovine serum albumin, 0.1% Ficoll, 0.1% polyvinylpyrollidone, 250 μg ml^{-1} sonicated, boiled herring sperm DNA, 50 mM $NaPO_4$ *p*H 7, 0.1% SDS, 43% deionized formamide at 37 °C for 2–3 h. The prehybridization buffer was removed from the bag and replaced with the same buffer containing 10^6 c.p.m. ml^{-1} of hybridization probe. Blots were hybridized at 37 °C for 25–48 h, then washed twice in 2×SSC, 0.1% SDS for 5 min at room temperature, followed by two washes for 15 min each at 45–50 °C. Stringent hybridization and wash conditions differed only in the hybridization buffer, which contained 50% instead of 43% formamide, and in the final wash which was done in 0.2×SSC, 0.1% SDS at 65–70 °C.

To determine which region(s) within the 903 cDNA sequence were repetitive, the cDNA was subdivided into five restriction fragments of ~500 base pairs (bp) each, and these were individually hybridized to replica genomic Southern blots in reduced-stringency conditions (Fig. 1). The two left-most fragments, which overlap in the 3′ half of the 903 cDNA, detect the expected genomic fragments at *Antp*, and also cross-hybridize with seven other genomic fragments with less intensity. The next 903 fragment to the right (fragment 3) hybridizes to more than 50 genomic fragments. Finally, the two right-most 903 fragments (4 and 5) detect only their genomic homologues at the *Antp* locus.

Garber *et al.*[13] found weak homology between the 903 cDNA and a site to the left of the *Antp* locus, at position 190 on the map in Fig. 1. This site has subsequently been shown to be part of the transcription unit of the *fushi tarazu* (*ftz*) gene (A.K. and E.H., in preparation). The *ftz* gene is required for the determination of the correct number of segments in the *Drosophila* embryo[4,14]. Embryos that are homozygous for certain mutant alleles of *ftz* die early in development, and show deletions of alternate segment primordial. A 0.9-kb *Xho*I/*Eco*RI fragment (probe 6), containing a 3′ portion of the *ftz* transcription unit (A.K., unpublished results), was used as a probe of another Southern blot identical to those used for the five 903 fragments (Fig. 1). In addition to the strong signal contributed by the homologous genomic fragment from the *ftz* locus, eight other genomic fragments weakly cross-hybridized. Five of these weakly hybridizing genomic DNA fragments are identical in size to those detected by the two probes from the 3′ region of the 903 cDNA. Thus, the 3′ regions of the transcription units of the *Antp* and *ftz* genes share a common sequence, one that appears to be present at five or more locations in the *Drosophila* genome.

Subsequently, we will refer to this low-level repeat as the H repeat, and the high-level repeat in the middle of the 903 cDNA (fragment 3) as the M repeat. The M repeat is not detectable in the DNA of the *ftz* locus.

Presence of H Repeat in Bithorax Complex

Next, we performed experiments to test for the presence of the H repeat in other homoeotic genes. We hybridized fragments 2 and 6 (from *Antp* and *ftz* respectively; see

组位点的 cDNA 的大致范围。底线代表加伯等人所取的第 3 号染色体上的触角足基因区域[13]。数字代表与肱骨染色体断点的距离（以 kb 为单位）。5′和 3′标记两个基因座的转录方向（参考文献 17 和库若瓦未发表的结果）。Xh，*Xho*I；R，*Eco*RI；Ba，*Bam*H1；Xb，*Xba*I；P，*Pvu*II；S，*Sph*I。

方法：低严紧型杂交操作如下：用 5× 柠檬酸钠缓冲液、0.1% 牛血清蛋白、0.1% 聚蔗糖、0.1% 聚乙烯吡咯烷酮、250 μg · ml^{-1} 超声并煮沸的鲱精 DNA、50 mM $NaPO_4$ pH7、0.1% 十二烷基硫酸钠、43% 去离子甲酰胺，在 37℃下进行 DNA 印迹法[31]预杂交 2~3 小时。从杂交袋中去除预杂交缓冲液，加入含有 10^6 c.p.m. ml^{-1} 杂交探针的同样的缓冲液，在 37℃下印迹杂交 25~48 小时，然后在室温下用 2× 柠檬酸钠缓冲液、0.1% 十二烷基磺酸钠洗涤 2 次，每次 5 分钟。接着在 45~50℃下洗涤 2 次，每次 15 分钟。低严紧型杂交和洗涤与严紧型杂交和洗涤条件的差别仅在于所用的杂交缓冲液，后者用 50% 的甲酰胺代替 43% 的甲酰胺；另外，最后一次洗涤时，在 65~70℃下用 0.2× 柠檬酸钠缓冲液、0.1% 十二烷基硫酸钠进行。

为了确定 903 cDNA 序列内哪个（或哪些）区域存在重复，该 cDNA 被再分成 5 个各有约 500 碱基对（bp）的限制性酶切片段；在低严紧性条件下，将每个片段与复制基因组进行 DNA 印迹法杂交（图 1）。903 cDNA 两个最左端的片段在 903 cDNA 3′端相互重叠，以这两个序列为探针，检测到了所预期的触角足基因上的基因组片段，并且这两个片段也可以与另外 7 个基因组片段进行低强度地交叉杂交。向右数的下一个 903 片段（片段 3）可与超过 50 个基因组片段杂交。最后，两个最右端的 903 片段（4 和 5）只在触角足基因座检测到它们的基因组同源物。

加伯等人发现 903 cDNA 和触角足基因座左端某位点（在图 1 图谱的 190 位置）之间有弱的同源性[13]。随后，这个位点被证明是 *ftz* 基因的转录单位的一部分（库若瓦和哈芬，文章准备中）。*ftz* 基因是确保果蝇胚胎体节具有正常数目所必需的[4,14]。该等位基因的突变纯合子胚胎在发育的早期就会死亡，并且显示交替的体节原基的缺失。与其他 5 个 903 片段所用的探针相同，采用含有 *ftz* 转录单位 3′端部分（库若瓦，未发表的结果）的一段 0.9-kb 的 *Xho*I/*Eco*RI 片段（探针 6）作为另一个 DNA 印迹法杂交的探针（图 1）。除了 *ftz* 基因座的同源基因组片段有强杂交信号外，其他 8 个基因组片段有微弱的交叉杂交。其中 5 个微弱杂交的基因组 DNA 片段的大小与 903 cDNA 3′端区的 2 个探针检测到的片段相同。因此，触角足基因和 *ftz* 基因转录单位的 3′端区拥有一个共同的序列，在果蝇基因组中这样的序列存在 5 个以上。

随后，我们将这种低度重复称为 H 重复，将 903 cDNA 中间位置的高度重复片段（片段 3）称为 M 重复。在 *ftz* 基因座 DNA 中无法检测到 M 重复。

双胸基因复合体中存在的 H 重复

接下来，我们的实验是检验其他同源异型基因中 H 重复的存在。在低严紧条件下，我们用片段 2 和 6（分别来自触角足基因和 *ftz* 基因，见图 1）对双胸基因复合

Fig. 1) in reduced-stringency conditions to Southern blots of recombinant clones from the *Ultrabithorax* (*Ubx*) unit of the bithorax complex, specifically λ2229, λ2269, λ2288 and 2296 (ref. 15) (given by P. Spierer). A 3.2-kb *Bam*HI fragment common to λ2288 and λ2296 hybridized to both H repeat-containing probes (Fig. 2). This *Bam*HI fragment contains most or all of the 3′ exon of the *Ubx* transcription unit (refs 15, 16 and M. Goldschmidt-Clermont, personal communication). Cross-hybridization between the *Antp*, *ftz* and *Ubx* loci has been independently detected by M. Scott (personal communication). None of the *Ubx* region clones that we tested contained the M repeat.

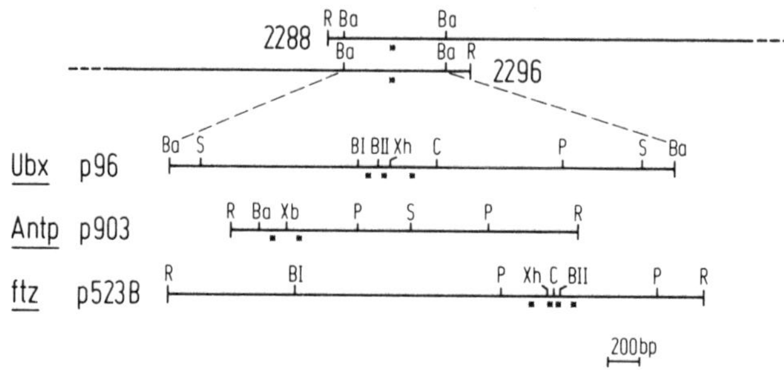

Fig. 2. Localization of the H repeat in clones from the *Ultrabithorax (Ubx)*, *Antennapedia (Antp) and fushi tarazu (ftz)* genes. The overlapping region from the *Ubx* locus between λ clones 2288 and 2296 (gifts from. P. Spierer) is shown at the top, as well as the *Bam*HI fragment common to both which hybridizes to fragments 2 and 6 (Fig. 1). Both these clones are described elsewhere[15], although λ2296 is incorrectly labelled as 2269. The indicated *Bam*HI fragment was subcloned into pAT153 and the resulting plasmid designated p96. A map of the p96 insert is shown. Restriction fragments containing the H repeat are marked by asterisks. *Antp* p903 is the same cDNA clone shown in Fig. 1. Restriction fragments to both sides of the *Xba*I site (Xb) show homology to the H repeat. *ftz* p523B is a *Drosophila* genomic clone from region 190 on the map in Fig. 1, an *Eco*RI fragment in pAT153. This fragment contains most of the *ftz* transcription unit (A.K., unpublished results). Again, the restriction fragments containing the H repeat are marked by asterisks.

Figure 2 shows more detailed restriction maps of the regions at each locus that contain the H repeat. The location of the H repeat within each of the *Antp*, *Ubx* and *ftz* clones was determined by *inter se* hybridizations to Southern blots of each clone digested with various restriction enzymes. The H repeat is the only detectable region of cross-homology between the three clones, and based on the intensity of signal the cross-homology is either very short (<100 bp), or larger but poorly matched. As the region of cross-homology overlaps both ends of the 90-bp *Xho*I/*Bgl*II fragments found in both *Ubx* and *ftz*, the latter possibility is more likely. The intensity of signal obtained from homologous hybridization compared with cross-hybridization due to the H repeat is shown in lane 1 of Fig. 1. Probe fragment 1, from the 3′ end of the *Antp* cDNA 903, hybridizes very strongly to its genomic homologue on a 1.7-kb *Eco*RI fragment, but with at least 10 times less intensity to the 3.4-kb and 7.5-kb genomic fragments in the same lane, which carry the *ftz* and *Ubx* H repeats respectively.

The cloned regions in Fig. 2 are shown with the same 3R chromosomal orientation: the centromere is to the left, and the telomere to the right. The transcriptional direction of the *Ubx* and *Antp* clones is from right to left, and for the *ftz* clone from left to right (refs

体的超双胸（*Ubx*）单位的重组克隆（具体即λ2229、λ2269、λ2288 和 λ2296 克隆（文献 15）（施皮雷尔提供））进行 DNA 印迹杂交。将 λ2288 和 λ2296 共有的 3.2-kb 的 *Bam*HI 片段与两个含有 H 重复的探针杂交 (图 2)。该 *Bam*HI 片段含有 *Ubx* 转录单位的大部分或全部的 3′ 端外显子（文献 15、16 和戈尔德施密特 – 克莱蒙，个人交流）。斯科特独立地进行了触角足、*ftz* 和超双胸基因座之间的交叉杂交（个人交流）。我们检测的 *Ubx* 区克隆都没有 M 重复。

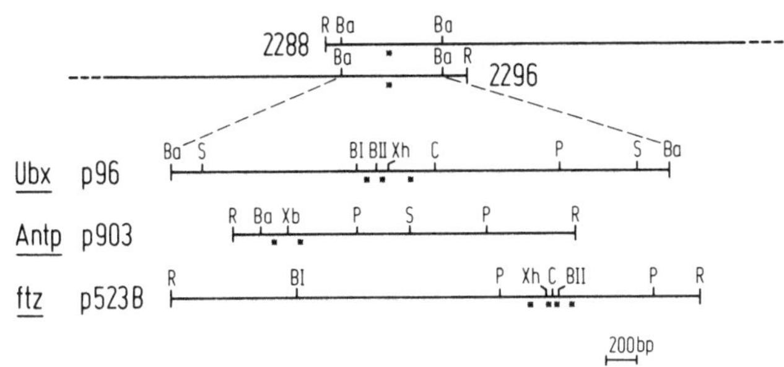

图 2. 超双胸、触角足和 *ftz* 基因克隆中的 H 重复的定位。λ2288 克隆和 λ2296 克隆（施皮雷尔赠送）之间的超双胸基因的重叠区以及可以与片段 2 和 6 杂交的两个基因共有的 *Bam*HI 片段都显示在图的顶端（图 1）。这两个克隆在其他文献中有描述[15]，尽管这些描述错误地将 λ2296 克隆标记为 2269。将图中标明的 *Bam*HI 片段亚克隆到 pAT153 质粒上，并将此质粒命名为 p96 质粒，其插入图谱如图所示。用星号标记有 H 重复的限制性酶切片段。触角足 p903 是与图 1 所示相同的 cDNA 克隆。限制性酶切片段上 *Xba*I 位点两侧序列显示与 H 重复有同源性。*ftz* 基因的 p523B 片段是图 1 图谱中 190 区的果蝇基因组克隆，是 pAT153 中的 *Eco*RI 片段，含有 *ftz* 转录单位的大部分（库若瓦，未发表的结果）。同样，其具有 H 重复的限制性酶切片段用星号标出。

图 2 标示了含有 H 重复的每个基因座区域内更详细的限制性酶切图谱。用多种限制性内切酶消化触角足、超双胸、*ftz* 基因克隆，每个克隆彼此之间进行 DNA 印迹法杂交，从而确定每个克隆的 H 重复位点。H 重复是这 3 个克隆之间唯一可检测的交叉同源区，而且基于信号强度，交叉同源物要么太短（<100 bp），要么较长但匹配很差。由于交叉同源区与超双胸和 *ftz* 都具有的两个 90-bp 长的 *Xho*I/*Bgl*II 片的末端重叠，所以后一种情况更有可能。图 1 的泳道 1 显示了基于 H 重复的交叉杂交和同源杂交信号强度间的对比。来自触角足基因 cDNA 903 的 3′末端的探针片段 1 与其基因组同源物在 1.7-kb 长的 *EcoRI* 片段上有非常强的杂交，但是与同一泳道的 3.4-kb 和 7.5-kb（分别携带着 *ftz* 和 *Ubx* 的 H 重复）的基因组片段杂交的强度相比，其强度至少弱 10 倍。

图 2 中的克隆区域显示的方向与 3R 染色体相同：左端是着丝粒，右端是端粒。超双胸和触角足基因克隆的转录方向从右到左，*ftz* 基因克隆的转录方向从左到右

16, 17 and A.K., unpublished results). The relative orientation of the *Xho*I and *Bgl*II sites within the cross-homologous regions of *Ubx* and *ftz,* is consistent with the polarity of the H repeat being the same with respect to the direction of transcription. Preliminary DNA sequencing results from these two regions (W.McG., unpublished results) and of the *Antp* 903 cDNA (R. Garber, unpublished results) also indicate that the H repeat is in the same orientation with respect to transcription at each locus.

Isolation of Clones Containing H Repeat

The observation that the H repeat was associated with three loci known to be crucial for proper segmental development in the fly suggested that it might be a common feature of many homoeotic loci in *Drosophila.* To test this, we first isolated *Drosophila* genomic clones that possessed homology to cloned H repeat sequences. Such clones were then subjected to the following three tests to implicate them as potential new homoeotic loci.

(1) Hybridization of the clones to *Drosophila* polytene chromosomes to determine whether their cytogenetic map locations corresponded to those of genetically characterized homoeotic loci.

(2) Hybridization of subclones containing the H repeat to Northern blots of RNA extracted from successive developmental stages to determine whether the regions were transcribed, and whether their transcription was developmentally regulated in a manner that might be expected for a homoeotic locus.

(3) Hybridization of subclones containing the H repeat to *Drosophila* embryonic tissue sections, to determine whether the transcripts homologous to the subcloned regions were distributed in a segmentally restricted manner, as has been shown for transcripts derived from the homoeotic loci *Antp* and *Ubx*[16,18,19].

Genomic Library Screen

Approximately 30,000 recombinant bacteriophages (three genome equivalents) from the Charon 4/*Drosophlia* library of Maniatis *et al.*[20] were screened using H repeat probes from *Antp* and *ftz* (fragments 2 and 6 in Fig. 1) as probes of duplicate filters. The hybridizations and washes were done in the reduced-stringency conditions described in Fig. 1 legend. A total of 74 plaques from the original plates hybridized to both probes, with a wide range of signal intensity. All were picked and re-screened with the same probes, and 24 were plaque-purified. DNA was extracted from each recombinant and digested with *Eco*RI, *Bam*HI and both enzymes together, then separated on an agarose gel, and transferred to nitrocellulose.

These blots were hybridized with probes 2 and 6 to determine which region of the insert contained the H repeat. In this way, we found that three of the clones were re-isolates of *ftz,* and two were re-isolates of *Antp.* The remaining 19 clones were divided into two classes on the basis of their extent of homology (as determined by signal intensity) with

（文献 16、17 和库若瓦未发表的结果）。超双胸和 *ftz* 基因的交叉同源区内的 *Xho*I 和 *Bgl*II 位点的相对方向及转录方向与 H 重复的极性是一致的。对这两个区域（麦金尼斯，未发表的结果）和触角足基因 903 cDNA（加伯，未发表的结果）的初步 DNA 测序结果也表明：H 重复的方向与每个基因座的转录方向是相同的。

含有 H 重复的克隆的分离

H 重复与已知对果蝇体节正常发育至关重要的 3 个基因座相关，这一观察结果说明，这种重复可能是果蝇中同源异型基因座的一个共同特征。为了检验这一点，我们首先分离了与克隆的 H 重复序列具有同源性的果蝇基因组克隆。然后将这些克隆进行下列 3 项试验，说明它们是潜在的新的同源异型基因座。

（1）将这些克隆与果蝇多线染色体杂交，以确定其细胞遗传学图谱位点是否与通过遗传学手段表征的同源异型基因座相对应。

（2）将含有 H 重复的亚克隆与从连续发育阶段提取的 RNA 进行印迹法杂交，确定这些区域是否被转录，以及其转录作用是否以预计的同源异型基因座的方式受到发育调控。

（3）将含有 H 重复的亚克隆与果蝇胚胎组织切片进行杂交，以确定与亚克隆区同源的转录物是否像同源异型基因座触角足和超双胸基因的转录物所显示的那样，以体节限定的方式分布 [16,18,19]。

基因组文库筛选

用触角足和 *ftz* 基因的 H 重复序列（图 1 中的片段 2 和 6）作为印影滤膜的探针，从曼尼阿蒂斯等人 [20] 的 Charon4/ 果蝇库中筛选到大约 30,000 重组噬菌体（三个基因组当量）。在图 1 注中所述的低严紧型条件下进行杂交和洗涤。原来的平板上总共有 74 个噬菌斑能与上述 2 个探针杂交，并表现出广范围的信号强度。挑出所有杂交的噬菌斑，用同样的探针进行再筛选，纯化出 24 株噬菌体。从每个重组子中提取 DNA，分别用 *Eco*RI、*Bam*HI 以及上述两种酶一起进行酶切，然后在琼脂糖凝胶上进行分离，并转移到硝酸纤维素膜上。

为确定哪个插入区域有 H 重复，我们用探针 2 和 6 进行了印迹杂交。通过这种方法发现，其中 3 个克隆是 *ftz* 基因的再分离物，2 个克隆是触角足基因的再分离物。基于与触角足和 *ftz* 基因的 H 重复序列的同源程度（由信号强度决定），将其余的 19 个克隆分成两类。有 7 个克隆交叉杂交相对较强，并且通过相互间杂交

the H repeats from *Antp* and *ftz*. Seven clones cross-hybridized relatively strongly and by *inter se* hybridizations we found that these were derived from two different genomic regions. Representative clones from each of the two regions were designated 93 and 99.

The H repeat lies on a 4.5-kb *Bam*HI/*Eco*RI fragment for clone 93; this fragment was subcloned into plasmid vector pAT153 and is designated p93. The p93 insert is a single-copy sequence in the *Drosophila* genome, when tested by genomic Southern blot analysis in stringent hybridization and wash criteria, and derives from the 15-kb genomic *Eco*RI fragment previously identified by genomic blotting (Fig. 1, lane 1). The clone 99 H repeat lies on a 5-kb *Eco*RI fragment which was cloned into pAT153 and designated p99. The p99 insert is also a single-copy *Drosophila* sequence (in stringent conditions), and corresponds to the 5-kb *Eco*RI genomic fragment detected in Fig. 1, lane 1. Although p99 and p93 were selected by hybridization with the H repeat they also hybridize with the probe 3 shown in Fig. 1, and therefore probably contain the M repeat also.

In situ Localization of p93 and p99

DNA from p93 and from p99 was labelled with biotinylated dUTP by nick-translation, and hybridized to squashes of salivary gland polytene chromosomes. The sites of hybridization were revealed by an immunoperoxidase detection protocol[21]. The stringency of hybridization was sufficient to allow only the detection of the genomic isologues of p93 and p99.

Clone p93 hybridizes to the 89E region on the right arm of chromosome 3 (Fig. 3), the cytogenetic location of the bithorax complex[1]. Bender *et al.*[15,22] have recently reported the isolation of the left half of the bithorax complex DNA, and Karch and Bender have subsequently isolated overlapping cloned genomic DNAs that include much of the right part of the complex (F. Karch and W. Bender, unpublished results); they provided us with a Southern blot containing cloned DNA from the entire cloned bithorax region. We find that the p93 insert is located within their cloned region, to the right of the *bithoraxoid/postbithorax* unit of the bithorax complex, near the *infra-abdominal-2* locus (data not shown). The same result has been obtained by S. Sakonju (personal communication) and M.S.L.

测试发现，它们来自两个不同的基因组区域。我们将这两个区域的代表性克隆定名为 93 和 99。

H 重复序列位于 93 号克隆的 4.5-kb 长的 *Bam*HI/*Eco*RI 片段上。将其亚克隆到质粒载体 pAT153 上，命名为 p93。经过在严紧的杂交和洗涤条件下进行的基因组 DNA 印迹法分析发现，p93 插入序列是果蝇基因组上的单拷贝序列，而且是由以前用基因组印迹鉴定（图 1 泳道 1）的 15-kb 长的基因组 *Eco*RI 片段衍生而来。99 号克隆的 H 重复序列位于克隆的载体质粒 pAT153 上的一段 5-kb 长的 *Eco*RI 片段上，该片段被克隆至 pTA153 质粒中并被命名为 p99。p99 插入序列也是单拷贝的果蝇序列（在严紧型条件下），与图 1 泳道 1 的 5-kb 长的 *Eco*RI 基因组片段相对应。虽然 p99 和 p93 是通过与 H 重复序列杂交选择出来的，但它们也能与图 1 所示的探针 3 杂交，因此，它们也可能有 M 重复。

p93 和 p99 的原位定位

通过切口平移，用生物素脱氧三磷酸尿苷标记 p93 和 p99 的 DNA，再将其与唾液腺多线染色体压片进行杂交。用免疫过氧化物酶检测方法显示杂交位点 [21]。杂交的严紧性足以保证只能检测到 p93 和 p99 基因组的同构体。

克隆 p93 与 3 号染色体右臂上的 89E 区（双胸基因复合体的细胞遗传位点 [1]）杂交(图 3)。本德尔等人 [15,22] 最近报道了双胸基因复合体 DNA 左半部的分离。随后，卡奇和本德尔又分离出了包括该复合物右端大部分的重叠克隆基因组 DNA（卡奇与本德尔，未发表的结果）。他们为我们提供的 DNA 印迹包含来自整个双胸克隆区的 DNA 克隆。我们发现 p93 插入序列定位于其克隆区内，在双胸复合体的双胸状的 / 后双胸基因单位的右面，靠近下腹 2 基因座（数据未出示）。左近允（个人交流）和莱文得到了同样的结果。

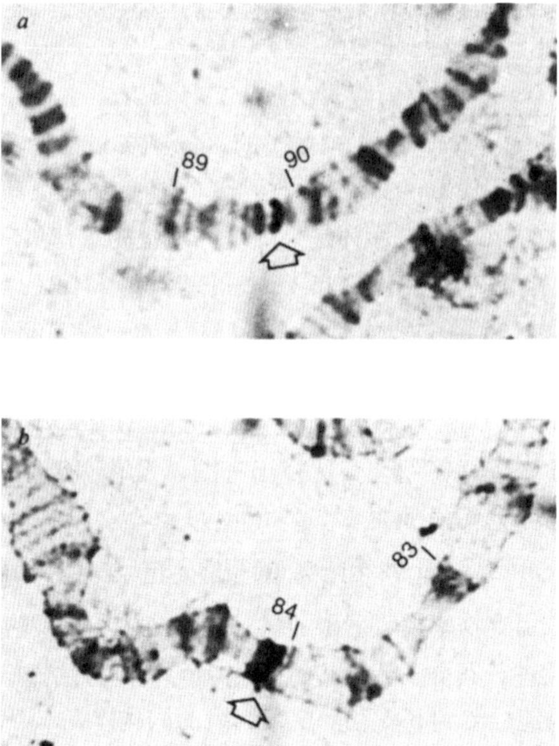

Fig. 3. *In situ* hybridization of H repeat clones p93 and p99 to polytene chromosomes. Clones p93 and p99 were nick-translated with biotinylated nucleotide, and hybridized to squashes of Ore-R chromosomes: p93 to *a*, and p99 to *b*. The hybridized probes were detected by an immunoperoxidase method[21] and the chromosomes stained with Giemsa and photographed. The chromosome 3 divisions are indicated and the sites of hybridization marked by arrowheads.

Clone p99 hybridizes to the 84A region on the right arm of chromosome 3; this is close to the cytogenetic locations of the *ftz* and *Antp* loci, both of which are in 84B1-2 (refs 2, 4, 5). The 84A region contains the Antp complex genes *proboscipedia, zerknüllt* and *Deformed,* all of which have been shown to affect the proper development of the posterior head segments of *Drosophila*[4,5,23].

Transcription

Genetic and developmental studies on the time of expression of homoeotic selector genes show that early to mid-embryo-genesis (0–12 h after oviposition) and pre- and early metamorphosis (late third instar larval and early pupal stages) are possible periods of high levels of expression[24–26]. *Antp* transcripts exhibit their highest levels during both these periods, and *ftz* transcripts are most abundant in early embryogenesis (A.K., unpublished results).

To test whether the regions homologous to p93 and p99 were transcribed, the two clones were hybridized to Northern blots containing *Drosophila* poly $(A)^+$ RNA from successive stages of development. The blots were hybridized and washed using the stringent conditions described in Fig. 1 legend.

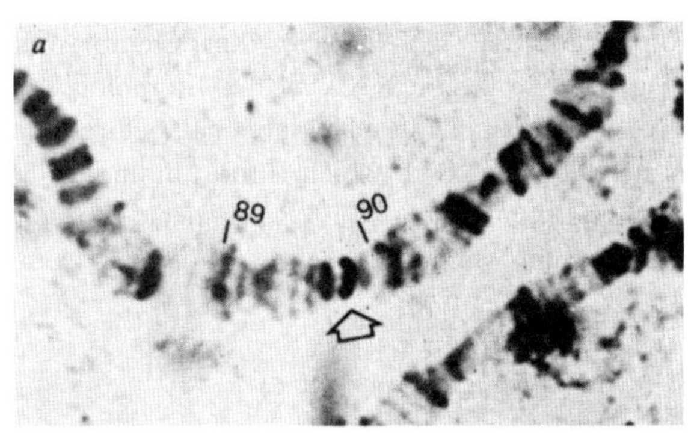

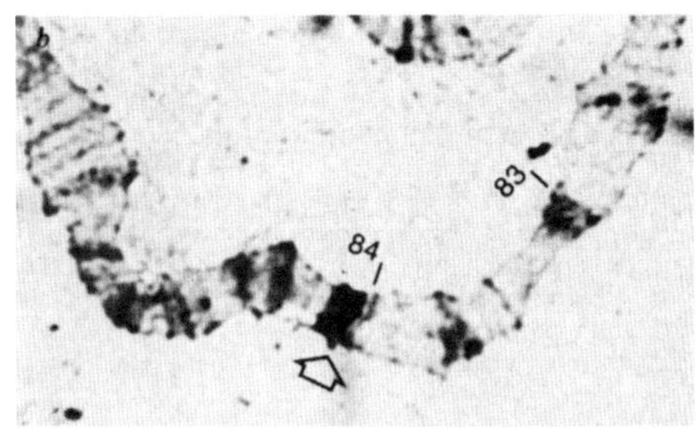

图 3. H 重复克隆 p93 和 p99 与多线染色体的原位杂交。p93 和 p99 克隆用生物素化的核苷酸作切口平移，并与 Ore-R 染色体压片杂交：p93 杂交示于 a，p99 杂交示于 b。用免疫过氧化物酶方法 [21] 检测杂交的探针，用吉姆萨染液给染色体染色并照相。图中显示染色体 3 分区，其中，箭头所示为杂交位点。

p99 克隆与 3 号染色体右臂上的 84A 区域进行杂交，该区域靠近位于 84B1-2 上的 *ftz* 和触角足基因座的细胞遗传学位点（文献 2，4，5）。84A 区有触角足复合体基因中的吻足基因、皱褶基因和变形基因，已知这些基因都能影响果蝇后头部体节的正常发育 [4,5,23]。

转　录

对同源异型选择者基因表达时序的遗传和发育研究显示，早期到中期的胚胎发生（产卵后 0~12 小时）、前期和早期的变态（三龄幼虫晚期和蛹期的早期）都是高表达的可能时期 [24-26]。触角足基因在这两个时期显示了最高的转录量，*ftz* 转录物在胚胎发生的早期最丰富（库若瓦，未发表的结果）。

为了检测与 p93 和 p99 同源的区域是否被转录，将这两个克隆与来自果蝇连续发育阶段的多聚腺嘌呤 RNA 进行印迹法杂交。用图 1 注所述的严紧型条件进行印迹杂交和洗涤。

Clone 93 is homologous to multiple RNA species during embryonic stages, especially in embryos 6–12 h old (Fig. 4). The largest RNA (5.4 kb) homologous to p93 at the 6–12-h stage is also abundant in late third instar larvae, just before pupation. Clone p99 is homologous to an RNA species of 2.8 kb which is most abundant at early embryogenesis (0–6 h) and in the early pupal stage (Fig. 4). The RNA species homologous to p93 and p99 are both present at approximately the same levels as transcripts from the *Antp* locus (A.K., unpublished results).

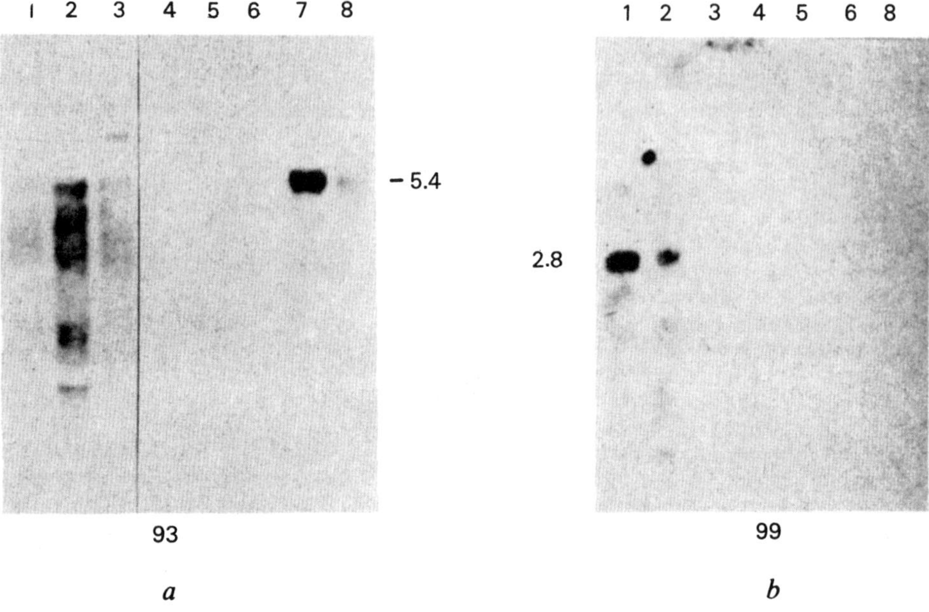

Fig. 4. Transcription from p93 and p99 during *Drosophila* development. Lanes 1, 2 and 3 in each panel contain embryonic RNA from 0–6, 6–12 and 12–18-h stages, respectively. Lanes 4 and 5 in each panel contain RNA from first and second instar larvae. Lanes 6 and 7 in each panel contain RNA from early and late third instar larvae, respectively. Lane 8 in each panel contains RNA from 1-day-old pupae. The numbers alongside the panels indicate the approximate size (in kb) of the largest hybridizing RNAs. On longer exposures a faint band of 2.8 kb was detected in the pupal lane (8) of the blot hybridized with p99 (not shown). Poly(A)$^+$ (10 μg) from successive stages of *Drosophila* development was run on formaldehyde agarose gels and blotted[32]. The blots were hybridized with nick-translated p93 and p99 in the stringent buffer described in Fig. 1 legend. After a stringent wash (also described in Fig. 1) the blots were used to expose X-ray film.

Localization of Transcripts

The most important experiment that we performed with the p93 and p99 clones was to test whether the transcripts homologous to these clones were spatially restricted during development. It has recently been shown that *Antp* and *Ubx* transcripts are restricted in a segmentally specific manner during embryonic development[16,18,19]. To a rough approximation, the embryonic segments and segmental anlagen that accumulate *Antp* and *Ubx* transcripts are those in which the function of these genes is believed to be required for proper development. Therefore, if the 93 and 99 clones represent other homoeotic selector loci, their transcripts should be restricted during embryonic stages to segments where 93 or 99 expression is required for proper development.

在胚胎时期，特别是6~12小时胚龄期间，克隆93与多种RNA同源（图4）。在6~12小时的阶段，与p93同源的最大的RNA（5.4 kb）在幼虫蛹化前的3龄期后期也很丰富。p99克隆与2.8 kb长的RNA同源，这一RNA在早期的胚胎发生（0~6小时）和早期蛹的阶段最丰富（图4）。与p93和p99同源的RNA都与触角足基因座的转录物以大约相同的量存在（库若瓦，未发表的结果）。

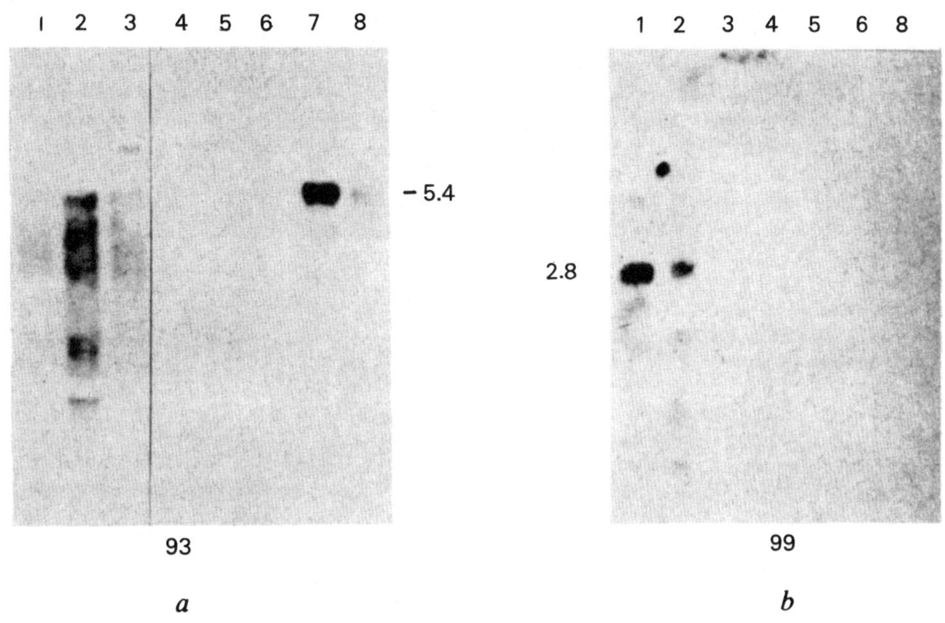

图 4. 果蝇发育期间p93和p99的转录情况。在每块胶板上，泳道1、2和3分别加0~6、6~12、12~18小时的胚胎RNA，泳道4和5分别加1龄和2龄幼虫的RNA，泳道6和7分别加早期和晚期3龄幼虫的RNA，泳道8加1天龄蛹的RNA。胶板侧面的数字为最大杂交RNA的近似大小（以kb为单位）。对p99板进行比较长时间的曝光，可检测到泳道(8)有一条2.8 kb的模糊条带(未显示)。将果蝇发育的连续阶段的多聚腺嘌呤（10 μg）进行甲醛琼脂糖凝胶电泳[32]。在图1注所述的严紧型缓冲液中与切口平移的p93和p99进行印迹杂交。严紧型洗涤后（如图1所述），印迹在X光片上曝光。

转录物的定位

我们用p93和p99克隆进行的最重要的实验是检测与这些克隆同源的转录物在发育期间是否受空间的限定。近来有研究表明，在胚胎发育期间，触角足基因和超双胸基因的转录产物在胚胎发育阶段以体节特异的方式受到限定[16,18,19]。据粗略估计，胚胎体节和体节原基中积累触角足基因和超双胸基因的转录物，在这些体节中，上述基因的功能是体节正常发育所必需的。因此，如果93和99号克隆代表其他同源异型选择者基因座，其转录物在胚胎阶段应该被限定在这些基因表达为正常发育所需的体节中。

Figure 5 shows that the transcripts homologous to both p93 and p99 show spatial restrictions during development. We have shown only one developmental stage, when transcript localization is striking, for each clone. More detailed studies on the expression of these cloned regions will be reported elsewhere.

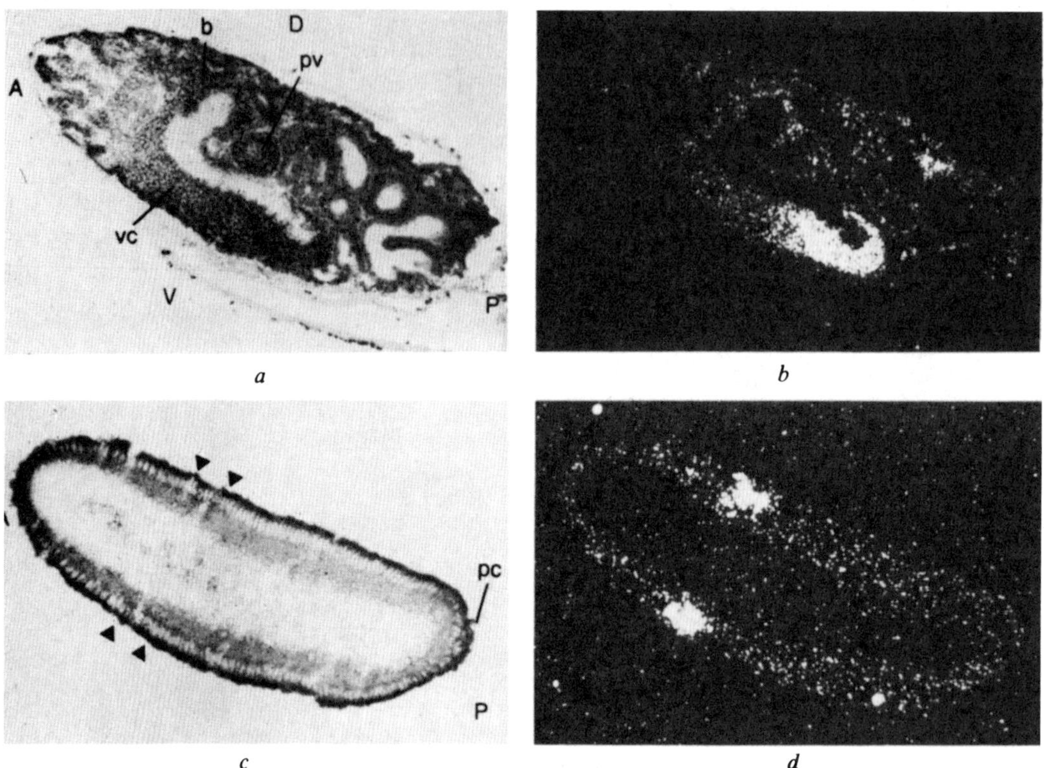

Fig. 5. Localization of transcripts homologous to p93 (*a*, *b*) and p99 (*c*, *d*) in embryonic tissue sections. The embryonic tissue sections were prepared and hybridized to ^{3}H-labelled probes as described by Hafen *et al.*[18]. Labelling of p93 is shown on a sagittal section of an 18–20-h embryo. Both brightfield (*a*) and darkfield (*b*) photographs of a 4-week autoradiographic exposure are shown. The p99 labelling is shown for an embryo 3 h old, just after the stage during which cell membranes envelop the nuclei at the periphery of the embryo. Both bright- and darkfield photomicrographs of a 4-week autoradiographic exposure are shown. The extent of hybridization of the brightfield view (*a*) is marked by arrowheads. A, anterior; P, posterior; vc, ventral nerve cord; b, brain; pv, proventriculus; pc, pole cells; D, dorsal; V, ventral.

The localization of transcripts homologous to p93 is shown on a sagittal section of an 18–20-h embryo. At this advanced stage of embryogenesis, the central nervous system includes the two brain hemispheres and the condensed ventral nerve cord (Fig. 5). Before condensation the ventral cord consists of 12 paired ganglia, the sub-oesophageal ganglion, three thoracic ganglia and eight abdominal ganglia. Each ganglion of the ventral cord innervates its corresponding body segment[27]. The entire central nervous system of the embryo section hybridized with p93 appeared to be labelled above background levels, but a striking and reproducible concentration of label was observed over the posterior region of the ventral nerve cord, encompassing at least posterior-most five or six abdominal neuromeres.

图 5 说明与 p93 和 p99 同源的转录产物在发育期间受到空间限定。在转录物的定位显著时，我们对每个克隆只展示出一个发育阶段。有关这些克隆区域表达更为详细的研究将另行发表。

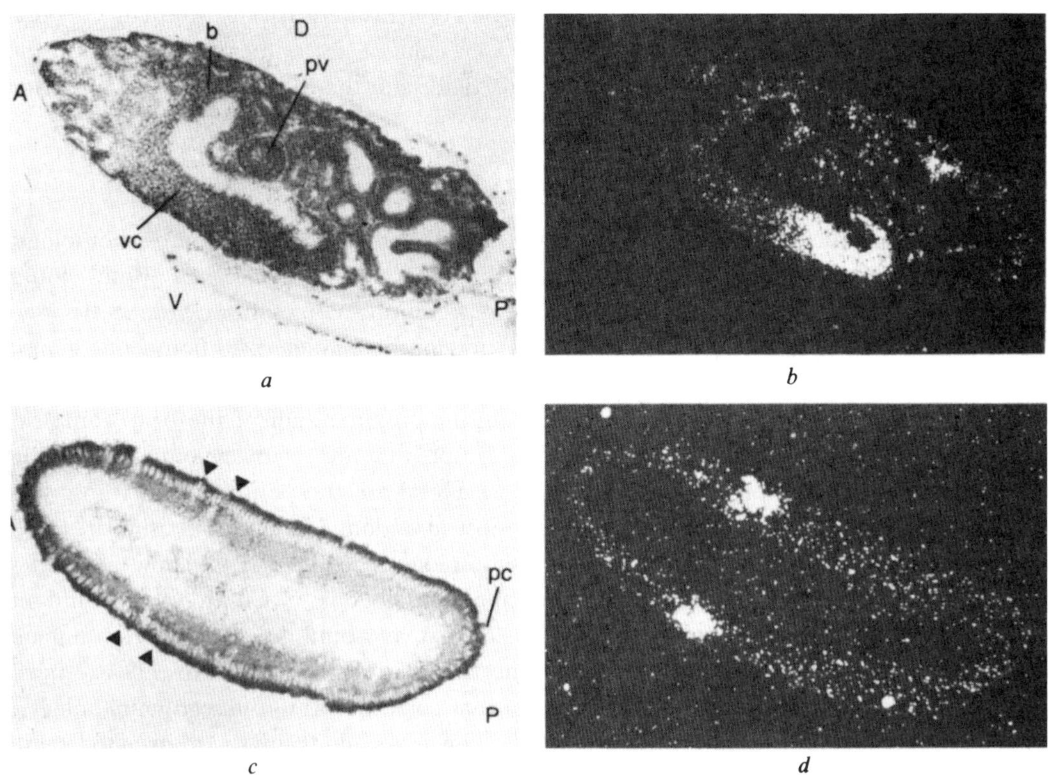

图 5. 胚胎组织切片中与 p93 (*a*, *b*) 和 p99 (*c*, *d*) 同源的转录产物的定位。胚胎组织切片的制备和与 ^{3}H 标记的探针杂交按哈芬等人的方法进行[18]。p93 的标记显示在一个 18~20 小时的胚胎矢状切面上。亮视野（*a*）和暗视野（*b*）照片是自动曝光四星期后所得。3 小时胚龄期的切片显示标记的 p99，这一时期正好在细胞膜包裹着胚边缘的细胞核的阶段之后。图中显示了自动曝光四星期的亮视野和暗视野的显微照片。亮视野（*a*）的杂交程度用箭头表示。A，前部；P，后部；vc，腹神经索；b，脑；pv，前胃；pc，极性细胞；D，背部；V，腹部。

18~20 小时的胚胎的矢状切面上显示了与 p93 同源的转录物的位置。在此胚胎发生的晚期阶段，中枢神经系统包括两个脑半球和凝缩的腹神经索（图 5）。在凝缩前，腹神经索由 12 对神经节组成，即食管下神经节、3 个胸神经节和 8 个腹神经节。腹神经索的每个神经节都支配着相应的体节[27]。与 p93 杂交的胚胎切面的整个中枢神经系统都显示了高于本底量的标记结果，但是在腹神经索的后区可观察到显著及可重现浓度的标记，环绕着至少最后部的 5 个或 6 个腹部神经管节。

The labelling pattern of p99 is shown at the cellular blastoderm stage, ~3 h after oviposition. The cellular blastoderm consists of a monolayer of morphologically identical cells. It is at this stage that cells first become restricted in their developmental potential, with different regions of the blastoderm acquiring separate determinative fates[28,29]. Transcripts homologous to p99 were found to be concentrated in cells in about 60–65% of the embryo length from the posterior pole. Cell ablation experiments in this region of the cellular blastoderm result in embryos having defects in the first thoracic and posterior head segments[30].

Conclusions

Our analyses of the 93 and 99 clones, both isolated with the H repeat cross-homology, strongly suggest that they represent other homoeotic loci of *Drosophila.* Both clones fulfilled all three criteria that we applied for representing clones from homoeotic loci. First, both hybridize to cytogenetic locations of previously characterized homoeotic genes; 93 to the right half of the bithorax complex in the chromosome region 89E, and 99 to chromosome region 84A, which contains genes in the proximal half of the Antp complex. Second, both 93 and 99 are homologous to transcripts that are relatively abundant during embryogenesis and just prior to metamorphosis. These are the periods when transcripts homologous to the homoeotic locus *Antp* are most abundant (A.K., unpublished results). Third, and most importantly, the transcripts homologous to 93 and 99 show a striking spatial restriction during development. Transcripts homologous to p93 are most abundant in the posterior abdominal neuromeres of the embryo, as would be expected from a gene in the right half of the bithorax complex. The transcripts homologous to p99 are most abundant in a region of the cellular blastoderm that corresponds to the segmental anlagen of the posterior head or first thoracic segments. This is also consistent with its cytogenetic location in 84A, which contains genes that affect the development of these segments.

The basis for the cross-homology is of great interest. The position of the H repeat in the 3′ region of the transcription units of *Antp*, *Ubx* and *ftz* is consistent with a conserved protein-coding sequence. The DNA sequence of the H repeats of *Antp*, *ftz* and *Ubx* leaves no doubt that the sequence conservation is due to a conserved protein-coding domain (W. McG. and R. Garber, unpublished results). Since faithful copies of the H repeat are strictly delimited and found only in homoeotic genes, we now call the H repeat the "homoeotic sequence". However, it seems clear that not all homoeotic genes carry the homoeobox, for example, we have been unable to detect it in the *bithoraxoid/postbithorax* unit of the bithorax complex (W.McG., F. Karch and W. Bender, unpublished results). It is possible, of course, that another subset of homoeotic genes contains another repeat.

On the basis of these results, we propose that a subset of the homoeotic genes are members of a multigene family, highly diverged but nonetheless detectable by DNA cross-homology. This suggests a common evolutionary origin for some genes of both the Antp and bithorax complexes, as proposed by Lewis[1] for the genes of the bithorax complex. The conspicuous evolutionary conservation of the homoeobox sequence in some homoeotic genes of *Drosophila* suggests that it might also be conserved in other animal

在产卵3小时后的细胞囊胚期就可显示p99标记模式。细胞囊胚由具有相同形态的单层细胞组成。正是在这个时期，细胞首次被限定了发育潜能，囊胚的不同区域获得了各自限定的命运[28,29]。与p99同源的转录物集中在后极胚胎长度60%~65%处的细胞内。该区域囊胚层的细胞消融实验，将导致胚胎在第一胸节和后部的头部体节方面有缺陷[30]。

结　论

我们对H重复交叉同源物分离到的两个克隆（93和99）的分析结果有力地表明，它们代表了果蝇的其他同源异型基因座。两个克隆都符合我们应用的同源基因位点代表性克隆的所有3项标准：首先，两个克隆都能与以前表现出同源异型基因特征的细胞遗传学位点杂交。93号克隆与位于染色体89E区的双胸基因复合体的右半部杂交，99号克隆与有触角足基因复合体近侧一半基因的染色体84A区杂交。其次，93和99号克隆两者都与那些胚胎发生期间，正好在变态前含量相当多的转录物同源。这个时期正是与同源异型基因座触角足基因同源的转录产物最丰富的时期（库若瓦，未发表的结果）。最后，也是最重要的，与93和99同源的转录物在发育期间显示出了明显的空间限定。正如根据双胸基因复合体右半部的一个基因所预期的那样，在胚胎后腹部神经管节中与p93同源的转录物最丰富。在与后头部的体节原基或第一胸节对应的细胞囊胚层中，与p99同源的转录物最丰富。这与其在84A上的细胞遗传学位点一致，该处含有影响这些体节发育的基因。

交叉同源物的基底很有趣。触角足、超双胸和*ftz*基因转录单位3′末端区的H重复的位置符合保守的蛋白质编码序列。毫无疑问，触角足、超双胸和*ftz*基因的H重复的DNA序列保守性归因于一个保守的蛋白质编码域（麦金尼斯和加伯，未发表的结果）。由于精确的H重复拷贝是有严格界限的，而且只在同源异型基因中发现，所以现在我们称H重复为“同源异型基因序列”。但是，很显然，并不是所有同源异型基因都携有同源异型框，例如，我们还没有在双胸基因复合体的双胸状的/后双胸基因单位中检测到这种框架（麦金尼斯、卡奇和本德尔，未发表的结果）。当然，有可能同源异型基因的另一个亚群具有另一种重复。

基于这些结果，我们提出，同源异型基因的亚群是多基因家族的成员，它们高度分化，但是可用DNA交叉同源进行检测。这表明，正如刘易斯[1]就双胸基因复合所提出的那样，触角足和双胸基因复合体两者的某些基因有共同的进化起源。果蝇的某些同源异型基因的同源框序列显著的进化保守性表明，其他动物种类中同源

species; preliminary experiments strongly support this view (W.McG., unpublished results). It is possible that a fundamental principle in development is to duplicate a gene specifying a segment identity, allowing one of the copies to diverge and acquire new functions, or new spatial restrictions in expression, or both; this might allow, within the limits of natural selection, a striking polymorphism in the different segments of an animal, and the acquisition of highly specialized functions in different segments.

We thank Nadine McGinnis for experimental assistance; Rick Garber for helpful comments during the early phases of this work; Pierre Spierer for the gifts of cloned DNA; and Welcome Bender and Francois Karch for the bithorax complex blot. We also thank Erika Wenger-Marquardt for preparation of the manuscript. M.S.L. and W.McG. were supported by Jane Coffin Childs fellowships. The work was made possible by a grant from the Swiss NSF and the Kanton Basel-Stadt.

(**308**, 428-433; 1984)

W. McGinnis, M. S. Levine, E. Hafen, A. Kuroiwa and W. J. Gehring
Department of Cell Biology, Biocenter, University of Basel, Klingelbergstrasse 70, CH-4056 Basel, Switzerland

Received 12 January; accepted 5 March 1984.

References:

1. Lewis, E. B. *Nature* **276**, 565-570 (1978).
2. Kaufman, T. C., Lewis, R. & Wakimoto, B. *Genetics* **94**, 115-133 (1980).
3. Lewis, R. A., Wakimoto, B. T., Denell, R. E. & Kaufman, T. C. *Genetics* **95**, 383-397 (1980).
4. Wakimoto, B. T. & Kaufman, T. C. *Devl. Biol.* **81**, 51-64 (1981).
5. Hazelrigg, T. & Kaufman, T. C. *Genetics* **105**, 581-600 (1983).
6. Le Calvez, J. *Bull. biol. Fr. Belg.* **82**, 97-113 (1948).
7. Hannah, A. & Strömnaes, O. *Drosoph. Inf. Serv.* **29**, 121-123 (1955).
8. Garcia-Bellido, A. *Am. Zool.* **17**, 613-629 (1977).
9. Duncan, I. & Lewis, E. B. *Developmental Order: Its Origin and Regulation,* 533-554 (Alan R. Liss, New York, 1982).
10. Struhl, G. *Proc. Natl. Acad. Sci. U.S.A.* **79**, 7380-7384 (1982).
11. Struhl, G. *J. Embryol. exp. Morph.* **76**, 297-331 (1983).
12. Hafen, E., Levine, M. & Gehring, W. J. *Nature* **307**, 287-289 (1984).
13. Garber, R. L., Kuroiwa, A. & Gehring, W. J. *EMBO. J.* **2**, 2027-2036 (1983).
14. Nüsslein-Volhard, C., Wieschaus, E. & Jurgens, G. *Verh. dt zool. Ges.* 91-104 (1982).
15. Bender, W. *et al. Science* **221**, 23-29 (1983).
16. Akam, M. *EMBO J.* **2**, 2075-2084 (1983).
17. Scott, M. P. *et al. Cell* (in the press).
18. Hafen, E., Levine, M., Garber, R. L. & Gehring, W. J. *EMBO J.* **2**, 617-623 (1983).
19. Levine, M., Hafen, E., Garber, R. L. & Gehring, W. J. *EMBO J.* **2**, 2037-2046 (1983).
20. Maniatis, T. *et al. Cell* **15**, 687-701 (1978).
21. Langer-Sofer, P. R., Levine, M. & Ward, D. C. *Proc. Natl. Acad. Sci. U.S.A.* **79**, 4381-4385 (1982).
22. Bender, W., Spierer, P. & Hogness, D. S. *J. Molec. Biol.* **168**, 17-33 (1983).
23. Kaufman, T. C. *Genetics* **90**, 579-596 (1978).
24. Morata, G. & Garcia-Bellido, A. *Wilhelm Roux's Arch. Dev. Biol.* **179**, 125-143 (1976).
25. Sanchez-Herrero, E. & Morata, G. *J. Embryol. exp. Morph.* **76**, 251-264 (1983).
26. Morata, G. & Kerridge, S. *Nature* **290**, 778-781 (1981).

序列进化也可能是保守的。初步的实验有力地支持这种观点（麦金尼斯，未发表的结果）。因此，发育的基本原理可能是复制一个指定体节同一性的基因，允许在拷贝中的某个拷贝出现分化从而获得新的功能或表达出新的空间限定，或者两者都有。这样，在自然选择的限度内，一个动物的不同体节就可以出现显著的多态性，以及不同体节可以获得高度特化的功能。

我们对娜丁·麦金尼斯进行的实验协助，里克·加伯对本工作早期的有益评论，皮埃尔·施皮雷尔赠给的克隆 DNA 以及韦尔科姆·本德尔与弗朗索瓦·卡奇做的双胸基因复合体印迹实验表示感谢。我们还要对埃丽卡·文格尔－马夸特准备的初稿表示感谢。莱文和麦金尼斯受到了简·科芬·蔡尔兹研究基金支持。本研究获得瑞士国家科学基金会及坎顿·贝泽尔－施塔特的资助。

（荆玉祥 翻译；沈杰 审稿）

27. Poulson, D. F. *Biology of Drosophila,* 168-274 (ed. Demereo, M.) (Wiley, New York, 1950).

28. Chan, L. N. & Gehring, W. J. *Proc. Natl. Acad. Sci. U.S.A.* **68**, 2217-2221 (1971).

29. Wieschaus, E. & Gehring, W. J. *Devl. Biol.* **50**, 249-263 (1976).

30. Underwood, E. M., Turner, F. R. & Mahowald, A. P. *Devl. Biol.* **74**, 286-301 (1980).

31. Southern, E. *J. Molec. Biol.* **98**, 503-517 (1975).

32. Goldberg, D. A. *Proc. Natl. Acad. Sci. U.S.A.* 77, 5794-5799 (1980).

Complete Nucleotide Sequence of the AIDS Virus, HTLV-III

L. Ratner *et al.*

Editor's Note

The early 1980s was a period of intense AIDS-related research and discovery. The AIDS-causing virus, then called human T-cell leukaemia virus III (HTLV-III), was identified and isolated, as were its modes of transmission. Here US biologist Robert Gallo and colleagues describe the complete nucleotide sequences of the AIDS virus, a milestone in AIDS research. Overall, the sequence resembles that of other RNA-encoded "retroviruses" containing the three hallmark genes *gag*, *pol* and *env*. But it also contains anomalies, not least the immense heterogeneity between clones. The genome has enabled researchers to define key viral genes and proteins, has shed light on the origins, nature and spread of HIV, and continues to influence diagnostic and drug development.

The complete nucleotide sequence of two human T-cell leukaemia type III (HTLV-III) proviral DNAs each have four long open reading frames, the first two corresponding to the *gag* and *pol* genes. The fourth open reading frame encodes two functional polypeptides, a large precursor of the major envelope glycoprotein and a smaller protein derived from the 3′-terminus long open reading frame analogous to the long open reading frame (*lor*) product of HTLV-I and -II.

HUMAN T-cell leukaemia (lymphotropic) viruses HTLV-I, -II and -III, a family of exogenous retroviruses, are associated with T-cell disorders including adult T-cell leukaemia lymphoma (ATLL) and the acquired immune deficiency syndrome (AIDS)[1-14]. These viruses share a number of biological and structural features: tropism for $OKT4^+$ lymphocytes[10,15-21], the ability to produce giant multinucleated cells in culture[10,17,22-24], immunological cross-reactivity of some virally encoded proteins[12,13], distant nucleic acid sequence similarities[8,9,25-28] and the preference for magnesium of the viral reverse transcriptase[29-31]. Moreover, the genome of HTLV-I and -II as well as the related bovine leukaemia virus (BLV) is somewhat longer than that of other retroviruses[8,9,25,32-34]. A sequence of 1,600–1,800 nucleotides is interposed between the 3′ end of the *env* gene and the long terminal repeat (LTR) sequences of these viruses[25,32-37], the 3′ portion of which is a long open reading frame (*lor*)[32,35,37]. Another feature that distinguishes HTLV-I, -II, -III and BLV from other retroviruses is the marked increase in the rate of transcription initiated within the viral LTR sequences in infected compared with uninfected cells[38,39]. This phenomenon, *trans*-acting transcriptional regulation, is not observed for other retroviruses[40] and is probably mediated by the *lor* gene product of these retroviruses[38].

艾滋病病毒HTLV-III的完整核苷酸序列

拉特纳等

编者按

20 世纪 80 年代早期是艾滋病相关研究和发现的密集时期。人们鉴定并分离出了引发艾滋病的病毒——后被称为 T 细胞白血病病毒 III（HTLV-III）并弄清了它的传播模式。在本文中，美国生物学家罗伯特 · 加洛及其同事描述了艾滋病病毒的完整核苷酸序列，这是艾滋病研究领域的一块里程碑。这个序列和其他 RNA 编码的“逆转录病毒”一样，都具有 *gag*、*pol* 和 *env* 这三个标志基因。但它也有一些不同，这些不同不仅仅局限于克隆间巨大的遗传异质性。基因组使得研究人员能够定义关键的病毒基因和蛋白，解释艾滋病的起源、性质和传播，进而影响诊断和药物的发展。

人类两种 T 细胞白血病 III（HTLV-III）的前病毒 DNA 的完整序列各有四个长的开放阅读框，前两个与 *gag* 和 *pol* 基因相关。第四个开放阅读框编码两个功能多肽，主要包膜糖蛋白的大前体和来源于 3′ 末端长开放阅读框的小蛋白，类似于 HTLV-I 和 -II 长开放阅读框（*lor*）的产物。

人类 T 细胞白血病（淋巴性）病毒 HTLV-I，-II 和 -III，属于外源逆转录病毒家族，它们与 T 细胞紊乱有关，包括成人 T 淋巴细胞白血病（ATLL）和获得性免疫综合征（AIDS）[1-14]。这些病毒具有许多共同的生物学和结构特征：对 OKT4⁺ 淋巴细胞的嗜性 [10,15-21]、在培养基中产生巨型多核细胞的能力 [10,17,22-24]、与一些病毒编码蛋白的免疫学交叉反应 [12,13]、远亲核苷酸序列之间的相似性 [8,9,25-28] 和病毒逆转录酶对镁的偏好 [29-31]。此外，HTLV-I 和 -II 与相关的牛白血病病毒（BLV）一样比其他的逆转录病毒的基因组都略长 [8,9,25,32-34]。在 *env* 基因的 3′ 末端和这些病毒的长末端重复序列（LTR）之间得到一段 1,600~1,800 个核苷酸的序列 [25,32-37]，该序列 3′ 部分是一个长的开放阅读框（*lor*）[32,35,37]。HTLV-I、-II、-III 和 BLV 区别于其他逆转录病毒的另一个特点是，被感染的细胞中 LTR 序列区域起始转录的速率比未被感染的细胞有显著增长 [38,39]。这种反式转录调控的现象在其他逆转录病毒中 [40] 没有观察到，可能是由这些逆转录病毒的 *lor* 基因的产物所介导的 [38]。

Despite the similarity between HTLV-III and the other members of the HTLV-BLV family of viruses, the biology and pathology of HTLV-III differ substantially. Infection with HTLV-III results often in profound immunosuppression (AIDS), consequent on the depletion of the OKT4+ cell population[10-14,41-43]. This effect is mirrored by a pronounced cytopathic rather than transforming effect of HTLV-III infection upon the OKT4+ cells in lymphocyte cultures *in vitro*[10,11,20]. In contrast, infection with HTLV-I results in a low incidence of T-cell leukaemia lymphoma (an OKT4+-cell malignancy)[1-6]. There is evidence also for some degree of immunodeficiency in HTLV-I patients[6,44]. Infection of primary lymphocytes in culture by HTLV-I and -II results in *in vitro* transformation of predominantly OKT4+ cells[45,46]. A cytopathic effect of HTLV-I infection on lymphocytes is apparent, but the effect is not as pronounced as that in HTLV-III[16,17,45-48]. HTLV-III differs also from HTLV-I and -II in the extent of infectious virion production *in vivo* and *in vitro*. High titres of cell-free infectious virions can be obtained from AIDS patient semen and saliva and from the supernatant of cultures infected with HTLV-III[10,49-51]. Very few, if any, cell-free infectious virions can be recovered from ATLL patients or from cultures infected with HTLV-I or -II[52].

To investigate the biological activity of these viruses *in vitro* and *in vivo* and to provide information useful for the development of diagnostic and therapeutic reagents for AIDS, we have determined the complete nucleotide sequence of the HTLV-III provirus.

Genomic Structure of HTLV-III

Several closely related clones of HTLV-III DNA were obtained from the H9 cell line infected with HTLV-III present in the blood pooled from several American AIDS patients[10]. The complete primary nucleotide sequence of three unintegrated viral clones of 8.9, 5.3 and 3.6 kilobases (kb) in length[27] was determined with the partial sequence from an integrated proviral clone[53] (Fig. 1).

The HTLV-III provirus is 9,749 base pairs (bp) long. The overall structure of the provirus resembles that of other retroviruses. The sequences that encode viral proteins are flanked by LTR sequences. The LTR itself is flanked by inverted repeated sequences two nucleotides long (Fig. 1). Four long open reading frames are identified in the viral DNA (Fig. 2).

Long Terminal Repeat

A detailed analysis of the HTLV-III LTR is presented elsewhere[54]; it is 634 nucleotides long with U3, R and U5 regions of 453, 98 and 83 nucleotides, respectively. The boundaries of these regions of the LTR were defined by localization of the 5′-cap site by S_1 nuclease mapping and by measurement of the length of the strong stop DNA transcript, as well as by determination of the 3′-terminus of the viral RNA by sequence analysis of cDNA clones. A TATAA sequence typical of eukaryotic promoters, as well as the consensus sequence for polyadenylation, are indicated in Fig. 1. A transfer RNA binding site complementary to the 3′ end of $tRNA^{Lys}$ is located 3′ to the 5′ LTR. DNA sequence homologies to the LTR of HTLV-I[32], -II[55] and BLV[56] are indicated in Fig. 3.

尽管 HTLV-III 与 HTLV-BLV 病毒家族其他成员之间有相似性，但是 HTLV-III 与它们的生物学和病理学特征有着根本上的不同。HTLV-III 的感染常常导致完全的免疫抑制（AIDS），随后就是 $OKT4^+$ 细胞数减少 [10-14,41-43]。这一效应是通过 HTLV-III 感染体外淋巴细胞培养物 $OKT4^+$ 造成的细胞病变反映出来的，而非转染效应 [10,11,20]。相反地，细胞感染 HTLV-I 后仅导致一种低发病率的 T 淋巴细胞白血病（一种 $OKT4^+$ 细胞恶性肿瘤）[1-6]，而且也有 HTLV-I 病人存在不同程度的免疫缺陷的例子 [6,44]。HTLV-I 和 -II 体外感染导致培养的原代淋巴细胞大部分变为 $OKT4^+$ 细胞 [45,46]。HTLV-I 感染淋巴细胞引起的细胞病变效应是明显的，但是这一效应不如 HTLV-III 引起的显著 [16,17,45-48]。HTLV-III 与 HTLV-I 和 -II 的不同之处还在于它们在体内和体外产生的具有感染能力的病毒颗粒的数量不同。从艾滋病患者的精液和唾液以及 HTLV-III 感染的细胞培养液上清中能够获得高效价的感染性病毒颗粒 [10,49-51]。从 ATLL 病人或者从 HTLV-I 和 -II 感染的细胞培养基中，很少能够获得游离的病毒粒子 [52]。

为了研究这些病毒在体内和体外的生物学活性并为艾滋病的诊断和治疗药物的开发提供有用的信息，我们完成了 HTLV-III 前病毒完整核苷酸序列的测序。

HTLV–III 的基因组结构

从几个美国艾滋病患者血液中提取 HTLV-III，然后感染 H9 细胞系，从中获得了几个密切相关的 HTLV-III DNA 克隆 [10]。利用一个完整前病毒克隆的部分序列 [53](图 1)，我们测定了长度 [27] 分别为 8,900、5,300 和 3,600 个碱基的三个不完整病毒克隆的全部初级序列。

HTLV-III 前病毒长 9,749 个碱基对。前病毒的总体结构类似于其他的逆转录病毒。编码病毒蛋白的序列两侧为 LTR 序列。LTR 自身的两侧是 2 个核苷酸长度的反向重复序列（图 1）。在病毒 DNA 中发现了四个长的开放阅读框（图 2）。

长末端重复序列

HTLV-III LTR 的详细分析已经被报道过 [54]；该序列长 634 个核苷酸，有 U3、R 和 U5 三个区域，长度分别是 453、98 和 83 个核苷酸。这些区域在 LTR 中的边界是通过 S_1 核酸酶谱和测量强制终止 DNA 转录产物的长度定位 5′端帽子位点，以及通过 cDNA 克隆的序列分析来确定病毒 RNA 的 3′末端，这些方法来确定的。在图 1 中标明了 TATAA 序列，它是真核启动子的典型序列，保守的多腺嘌呤序列也被标出。与 $tRNA^{Lys}$ 的 3′末端互补的转运 RNA 结合位点位于 5′LTR 的 3′末端。在图 3 中标出了与 HTLV-I[32]、-II[55] 和 BLV[56] 的 LTR 同源的序列。

```
CLONE                                                             核苷酸位置   氨基酸残基

                                          |— U3
                                          IR
BH10                                      TGGAAGGGCTAATTCACTCCCAACGAAGACAAGA   -420
BH8                                       ----------------------------------
                  (Bam HI)
BH10  TATCCTTGATCTGTGGATCTACCACACACAAGGCTACTTCCCTGATTAGCAGAACTACACACCAGGGCCAGGGAT   -345
BH8   ------------------C-------------------------G-----------------------AG-
BH10  CAGATATCCACTGACCTTTGGATGGTGCTACAAGCTAGTACCAGTTGAGCCAGAGAAGTTAGAAGAAGCCAACAA   -270
BH8   -----------------------------------------------------------A-----------T--
BH10  AGGAGAGAACACCAGCTTGTTACACCCTGTGAGCCTGCATGGAATGGATGACCCGGAGAGAGAAGTGTTAGAGTG   -195
BH8   ----------------------------------------------T----------------------------
BH10  GAGGTTTGACAGCCGCCTAGCATTTCATCACATGGCCCGAGAGCTGCATCCGGAGTACTTCAAGAACTGCTGACA   -120
BH8   -------------------------------------------------------------------------T-
BH10  TCGAGCTTGCTACAAGGGACTTTCCGCTGGGGACTTTCCAGGGAGGCGTGGCCTGGGCGGGACTGGGGAGTGGCG   -45
BH8   ---------------------------------------------------------------------------
                          TATA
                          BOX   Pvu II            U3—|
BH10  AGCCCTCAGATCCTGCATATAAGCAGCTGCTTTTTGCCTGTACT                                  -1
BH8   --------------------------------------------
      |— R               Bgl II         Sst I
BH10  GGGTCTCTCTGGTTAGACCAGATCTGAGCCTGGGAGCTC                                       39
BH8   ---------------------------------------

HXB2                                          TCTGGCTAACTAGGGAACCCACTGCTTAAGCCTCAA   75
        Hind III         R ——————|—— U5
HXB2  TAAAGCTTGCCTTGAGTGCTTCAAGTAGTGTGTGCCCGTCTGTTGTGTGACTCTGGTAACTAGAGATCCCTCAGA   150
                                 U5 —| — tRNA-lysine ———|—— Leader sequence
                                       IR
HXB2  CCCTTTTAGTCAGTGTGGAAAATCTCTAGCAGTGGCGCCCGAACAGGGACCTGAAAGCGAAAGGGAAACCA       221
      Sst I
BH10  GAGCTCTCTCGACGCAGGACTCGGCTTGCTGAAGCGCGCACGGCAAGAGGCGAGGGGCGGCGACTGGTGAGTACG   296
BH5   -------------------------- -- ---------------------------------------------
                          Leader sequence —|— GAG p17
BH10  CCAAAAATTTTGACTAGCGGAGGCTAGAAGGAGAGAGATGGGTGCGAGAGCGTCAGTATTAAGCGGGGGAGAATT   371
                                            MetGlyAlaArgAlaSerValLeuSerGlyGlyGluLeu           13
BH5   ---------------------------------------------------------------------------

BH10  AGATCGATGGGAAAAAATTCGGTTAAGGCCAGGGGGAAAGAAAAAATATAAATTAAAACATATAGTATGGGCAAG   446
       AspArgTrpGluLysIleArgLeuArgProGlyGlyLysLysLysTyrLysLeuLysHisIleValTrpAlaSer          38
BH5   ---------------------------------------------------------------------------

BH10  CAGGGAGCTAGAACGATTCGCAGTTAATCCTGGCCTGTTAGAAACATCAGAAGGCTGTAGACAAATACTGGGACA   521
       ArgGluLeuGluArgPheAlaValAsnProGlyLeuLeuGluThrSerGluGlyCysArgGlnIleLeuGlyGln          63
BH5   ---------------------------------------------------------------------------

BH10  GCTACAACCATCCCTTCAGACAGGATCAGAAGAACTTAGATCATTATATAATACAGTAGCAACCCTCTATTGTGT   596
       LeuGlnProSerLeuGlnThrGlySerGluGluLeuArgSerLeuTyrAsnThrValAlaThrLeuTyrCysVal          88
BH5   ---------------------------------------------------------------------------
                                      Hind III
BH10  GCATCAAAGGATAGAGATAAAAGACACCAAGGAAGCTTTAGACAAGATAGAGGAAGAGCAAAACAAAAGTAAGAA   671
       HisGlnArgIleGluIleLysAspThrLysGluAlaLeuAspLysIleGluGluGluGlnAsnLysSerLysLys          113
BH5   ---------------------------------------------------------------------------

                     Pvu II                          GAG p17 ——|- - GAG p24
BH10  AAAAGCACAGCAAGCAGCAGCTGACACAGGACACAGCAGTCAGGTCAGCCAAAATTACCCTATAGTGCAGAACAT   746
       LysAlaGlnGlnAlaAlaAlaAspThrGlyHisSerSerGlnValSerGlnAsnTyrProIleValGlnAsnIle          138
BH5   ---------------------------------------------------------------------------
                                          Aha III
BH10  CCAGGGGCAAATGGTACATCAGGCCATATCACCTAGAACTTTAAATGCATGGGTAAAAGTAGTAGAAGAGAAGGC   821
       GlnGlyGlnMetValHisGlnAlaIleSerProArgThrLeuAsnAlaTrpValLysValValGluGluLysAla          163
BH5   ---------------------------------------------------------------------------
                                                                    Aha III
BH10  TTTCAGCCCAGAAGTAATACCCATGTTTTCAGCATTATCAGAAGGAGCCACCCCACAAGATTTAAACACCATGCT   896
       PheSerProGluValIleProMetPheSerAlaLeuSerGluGlyAlaThrProGlnAspLeuAsnThrMetLeu          188
BH5   ---------------G-----------------------------------------------------------
                                                                      Pst I
BH10  AAACACAGTGGGGGGACATCAAGCAGCCATGCAAATGTTAAAAGAGACCATCAATGAGGAAGCTGCAGAATGGGA   971
       AsnThrValGlyGlyHisGlnAlaAlaMetGlnMetLeuLysGluThrIleAsnGluGluAlaAlaGluTrpAsp          213
BH5   ---------------------------------------------------------------------------

BH10  TAGAGTACATCCAGTGCATGCAGGGCCTATTGCACCAGGCCAGATGAGAGAACCAAGGGGAAGTGACATAGCAGG   1046
       ArgValHisProValHisAlaGlyProIleAlaProGlyGlnMetArgGluProArgGlySerAspIleAlaGly          238
BH5   ------G----------------------C---------------------------------------------

BH10  AACTACTAGTACCCTTCAGGAACAAATAGGATGGATGACAAATAATCCACCTATCCCAGTAGGAGAAATTTATAA   1121
       ThrThrSerThrLeuGlnGluGlnIleGlyTrpMetThrAsnAsnProProIleProValGlyGluIleTyrLys          263
BH5   ---------------------------------------------------------------------------

BH10  AAGATGGATAATCCTGGGATTAAATAAAATAGTAAGAATGTATAGCCCTACCAGCATTCTGGACATAAGACAAGG   1196
       ArgTrpIleIleLeuGlyLeuAsnLysIleValArgMetTyrSerProThrSerIleLeuAspIleArgGlnGly          288
BH5   -----------------------------------G--------T------------------------------
                                                                   Hind III
BH10  ACCAAAAGAACCTTTTAGAGACTATGTAGACCGGTTCTATAAAACTCTAAGAGCCGAGCAAGCTTCACAGGAGGT   1271
       ProLysGluProPheArgAspTyrValAspArgPheTyrLysThrLeuArgAlaGluGlnAlaSerGlnGluVal          313
BH5   ------G-----C----------------------------------------------------------A--
                                                                    Aha III
BH10  AAAAAATTGGATGACAGAAACCTTGTTGGTCCAAAATGCGAACCCAGATTGTAAGACTATTTTAAAAGCATTGGG   1346
       LysAsnTrpMetThrGluThrLeuLeuValGlnAsnAlaAsnProAspCysLysThrIleLeuLysAlaLeuGly          338
BH5   ---------------------------------------------------------------------------

BH10  ACCAGCGGCTACACTAGAAGAAATGATGACAGCATGTCAGGGAGTAGGAGGACCCGGCCATAAGGCAAGAGTTTT   1421
       ProAlaAlaThrLeuGluGluMetMetThrAlaCysGlnGlyValGlyGlyProGlyHisLysAlaArgValLeu          363
BH5   ---------------------------------------------------------------------------

BH10  GGCTGAAGCAATGAGCCAAGTAACAAATACAGCTACCATAATGATGCAGAGAGGCAATTTTAGGAACCAAAGAAA   1496
       AlaGluAlaMetSerGlnValThrAsnThrAlaThrIleMetMetGlnArgGlyAsnPheArgAsnGlnArgLys          388
BH5   ---------------------------T--A---------------A----------------------------
                                 SerThr
                         |————— Direct Repeat ——————————|
BH10  GATGGTTAAGTGTTTCAATTGTGGCAAAGAAGGGCACACAGCCAGAAATTGCAGGGCCCCTAGGAAAAAGGGCTG   1571
       MetValLysCysPheAsnCysGlyLysGluGlyHisThrAlaArgAsnCysArgAlaProArgLysLysGlyCys          413
A--T-------------------------------------T---A---------A---------A----GA-----
                                         Ile            Lys          Arg
              |——————— Direct Repeat ——————————|            |— POL    Bgl II
BH10  TTGGAAATGTGGAAAGGAAGGACACCAAATGAAAGATTGTACTGAGAGACAGGCTAATTTTTTAGGGAAGATCTG   1646
       TrpLysCysGlyLysGluGlyHisGlnMetLysAspCysThrGluArgGlnAlaAsnPheLeuGlyLysIleTrp          438
                                                             PhePheArgGluAspLeu             6
BH5   ---------------------------------------------------------------------------
                                |———————— Direct Repeat ——————————|—— Direct
BH10  GCCTTCCTACAAGGGAAGGCCAGGGAATTTTCTTCAGAGCAGACCAGAGCCAACAGCCCCACCATTTCTTCAGAG   1721
       ProSerTyrLysGlyArgProGlyAsnPheLeuGlnSerArgProGluProThrAlaProProPheLeuGlnSer          463
      AlaPheLeuGlnGlyLysAlaArgGluPheSerSerGluGlnThrArgAlaAsnSerProThrIleSerSerGlu            31
BH5   ---------------------------------------------------------------------------

          Repeat ——————————|
BH10  CAGACCAGAGCCAACAGCCCCACCAGAAGAGAGCTTCAGGTCTGGGGTAGAGACAACAACTCCCCCTCAGAAGCA   1796
       ArgProGluProThrAlaProProGluGluSerPheArgSerGlyValGluThrThrThrProProGlnLysGln          488
      GlnThrArgAlaAsnSerProThrArgArgGluLeuGlnValTrpGlyArgAspAsnAsnSerProSerGluAla            56
BH5   ------------------------------------------------------------------T---------
                                                                    Ser
                                                                   Leu
                                                               GAG p15 ———|
BH10  GGAGCCGATAGACAAGGAACTGTATCCTTTAACTTCCCTCAGATCACTCTTTGGCAACGACCCCTCGTCACAATA   1871
       GluProIleAspLysGluLeuTyrProLeuThrSerLeuArgSerLeuPheGlyAsnAspProSerSerGln            512
      GlyAlaAspArgGlnGlyThrValSerPheAsnPheProGlnIleThrLeuTrpGlnArgProLeuValThrIle            81
BH5   ---------------------------------------------------------------------------

BH10  AAGATAGGGGGGCAACTAAAGGAAGCTCTATTAGATACAGGAGCAGATGATACAGTATTAGAAGAAATGAGTTTG   1946
      LysIleGlyGlyGlnLeuLysGluAlaLeuLeuAspThrGlyAlaAspAspThrValLeuGluGluMetSerLeu           106
BH5   ---------------------------------------------------------------------------

BH10  CCAGGAAGATGGAAACCAAAAATGATAGGGGGAATTGGAGGTTTTATCAAAGTAAGACAGTATGATCAGATACTC   2021
      ProGlyArgTrpLysProLysMetIleGlyGlyIleGlyGlyPheIleLysValArgGlnTyrAspGlnIleLeu           131
BH5   ---------------------------------------------------------------------------

BH10  ATAGAAATCTGTGGACATAAAGCTATAGGTACAGTATTAGTAGGACCTACACCTGTCAACATAATTGGAAGAAAT   2096
      IleGluIleCysGlyHisLysAlaIleGlyThrValLeuValGlyProThrProValAsnIleIleGlyArgAsn           156
BH5   ---------------------------------------------------------------------------
                            Aha III
BH10  CTGTTGACTCAGATTGGTTGCACTTTAAATTTTCCCATTAGCCCTATTGAGACTGTACCAGTAAAATTAAAGCCA   2171
      LeuLeuThrGlnIleGlyCysThrLeuAsnPheProIleSerProIleGluThrValProValLysLeuLysPro           181
BH5   ---------------------------------------T--------A--------------------------

BH10  GGAATGGATGGCCCAAAAGTTAAACAATGGCCATTGACAGAAGAAAAAATAAAAGCATTAGTAGAAATTTGTACA   2246
      GlyMetAspGlyProLysValLysGlnTrpProLeuThrGluGluLysIleLysAlaLeuValGluIleCysThr           206
BH5   ---------------------------------------------------------------------------

BH10  GAAATGGAAAAGGAAGGGAAAATTTCAAAAATTGGGCCTGAGAATCCATACAATACTCCAGTATTTGCCATAAAG   2321
      GluMetGluLysGluGlyLysIleSerLysIleGlyProGluAsnProTyrAsnThrProValPheAlaIleLys           231
BH5   ---------------------------------------A-----------------------------------

BH10  AAAAAAGACAGTACTAAATGGAGAAAATTAGTAGATTTCAGAGAACTTAATAAGAGAACTCAAGACTTCTGGGAA   2396
      LysLysAspSerThrLysTrpArgLysLeuValAspPheArgGluLeuAsnLysArgThrGlnAspPheTrpGlu           256
BH5   -------------------------------------------------G-------------------------
                                                       Arg

BH10  GTTCAATTAGGAATACCACATCCCGCAGGGTTAAAAAAGAAAAAATCAGTAACAGTACTGGATGTGGGTGATGCA   2471
      ValGlnLeuGlyIleProHisProAlaGlyLeuLysLysLysLysSerValThrValLeuAspValGlyAspAla           281
BH5   --------G------------------------------------------------------------------

BH10  TATTTTTCAGTTCCCTTAGATGAAGACTTCAGGAAGTATACTGCATTTACCATACCTAGTATAAACAATGAGACA   2546
      TyrPheSerValProLeuAspGluAspPheArgLysTyrThrAlaPheThrIleProSerIleAsnAsnGluThr           306
BH5   ------------------------------------------------------------------T--------

BH10  CCAGGGATTAGATATCAGTACAATGTGCTTCCACAGGGATGGAAAGGATCACCAGCAATATTCCAAAGTAGCATG   2621
      ProGlyIleArgTyrGlnTyrAsnValLeuProGlnGlyTrpLysGlySerProAlaIlePheGlnSerSerMet           331
BH5   -------G-G-----------------------------------------------------------------
             SerGly
                          Aha III
BH10  ACAAAAATCTTAGAGCCTTTTAAAAAACAAAATCCAGACATAGTTATCTATCAATACATGGATGATTTGTATGTA   2696
      ThrLysIleLeuGluProPheLysLysGlnAsnProAspIleValIleTyrGlnTyrMetAspAspLeuTyrVal           356
BH5   -----------------G-------------------T-------------------------------------
                       Arg

BH10  GGATCTGACTTAGAAATAGGGCAGCATAGAACAAAAATAGAGGAGCTGAGACAACATCTGTTGAGGTGGGGACTT   2771
      GlySerAspLeuGluIleGlyGlnHisArgThrLysIleGluGluLeuArgGlnHisLeuLeuArgTrpGlyLeu           381
BH5   -------------------------------------------------------------------------T--
                                                                              Phe

BH10  ACCACACCAGACAAAAAACATCAGAAAGAACCTCCATTCCTTTGGATGGGTTATGAACTCCATCCTGATAAATGG   2846
      ThrThrProAspLysLysHisGlnLysGluProProPheLeuTrpMetGlyTyrGluLeuHisProAspLysTrp           406
BH5   ---------------------------------------------------------------------------
                                 Pvu II
BH10  ACAGTACAGCCTATAGTGCTGCCAGAAAAAGACAGCTGGACTGTCAATGACATACAGAAGTTAGTGGGGAAATTG   2921
      ThrValGlnProIleValLeuProGluLysAspSerTrpThrValAsnAspIleGlnLysLeuValGlyLysLeu           431
BH5   --GA-----------------------------------------------------------A------
         Ile

BH10  AATTGGGCAAGTCAGATTTACCCAGGGATTAAAGTAAGGCAATTATGTAAACTCCTTAGAGGAACCAAAGCACTA   2996
      AsnTrpAlaSerGlnIleTyrProGlyIleLysValArgGlnLeuCysLysLeuLeuArgGlyThrLysAlaLeu           456
BH5   ------------------T--------------------------------------------------------

BH10  ACAGAAGTAATACCACTAACAGAAGAAGCAGAGCTAGAACTGGCAGAAAACAGAGAGATTCTAAAAGAACCAGTA   3071
      ThrGluValIleProLeuThrGluGluAlaGluLeuGluLeuAlaGluAsnArgGluIleLeuLysGluProVal           481
BH5   ---------------------------------------------------------------------------

BH10  CATGGAGTGTATTATGACCCATCAAAAGACTTAATAGCAGAAATACAGAAGCAGGGGCAAGGCCAATGGACATAT   3146
      HisGlyValTyrTyrAspProSerLysAspLeuIleAlaGluIleGlnLysGlnGlyGlnGlyGlnTrpThrTyr           506
BH5   ---------------------------------------------------------------------------
                    Aha III
BH10  CAAATTTATCAAGAGCCATTTAAAAATCTGAAAACAGGAAAATATGCAAGAATGAGGGGTGCCCACACTAATGAT   3221
      GlnIleTyrGlnGluProPheLysAsnLeuLysThrGlyLysTyrAlaArgMetArgGlyAlaHisThrAsnAsp           531
BH5   ---------------------------------------------------------------------------
                                                                         Aha III
BH10  GTAAAACAATTAACAGAGGCAGTGCAAAAAATAACCACAGAAAGCATAGTAATATGGGGAAAGACTCCTAAATTT   3296
      ValLysGlnLeuThrGluAlaValGlnLysIleThrThrGluSerIleValIleTrpGlyLysThrProLysPhe           556
BH5   ---------------------------------------------------------------------------

BH10  AAACTACCCATACAAAAGGAAACATGGGAAACATGGTGGACAGAGTATTGGCAAGCCACCTGGATTCCTGAGTGG   3371
      LysLeuProIleGlnLysGluThrTrpGluThrTrpTrpThrGluTyrTrpGlnAlaThrTrpIleProGluTrp           581
BH5   ----------------A----------------------------------------------------------
                                     Kpn I
BH10  GAGTTTGTTAATACCCCTCCTTTAGTGAAATTATGGTACCAGTTAGAGAAAGAACCCATAGTAGGAGCAGAAACC   3446
      GluPheValAsnThrProProLeuValLysLeuTrpTyrGlnLeuGluLysGluProIleValGlyAlaGluThr           606
BH5   ---------------------------------------------------------------------------

BH10  TTCTATGTAGATGGGGCAGCTAACAGGGAGACTAAATTAGGAAAAGCAGGATATGTTACTAACAAAGGAAGACAA   3521
      PheTyrValAspGlyAlaAlaAsnArgGluThrLysLeuGlyLysAlaGlyTyrValThrAsnLysGlyArgGln           631
BH5   ---------------------G---------------------------------------T-G----------
                          Ser                                        Arg

BH10  AAGGTTGTCCCCCTAACTAACACAACAAATCAGAAAACTGAGTTACAAGCAATTTATCTAGCTTTGCAGGATTCA   3596
      LysValValProLeuThrAsnThrThrAsnGlnLysThrGluLeuGlnAlaIleTyrLeuAlaLeuGlnAspSer           656
BH5   --A------A--------C-----------------G-----A------------C-------------------G
               Thr      His                              Asn

BH10  GGATTAGAAGTAAACATAGTAACAGACTCACAATATGCATTAGGAATCATTCAAGCACAACCAGATAAAAGTGAA   3671
      GlyLeuGluValAsnIleValThrAspSerGlnTyrAlaLeuGlyIleIleGlnAlaGlnProAspLysSerGlu           681
BH5   --------------T------------------------------------------------------------
                                                                  Kpn I
BH10  TCAGAGTTAGTCAATCAAATAATAGAGCAGTTAATAAAAAAGGAAAAGGTCTATCTGGCATGGGTACCAGCACAC   3746
      SerGluLeuValAsnGlnIleIleGluGlnLeuIleLysLysGluLysValTyrLeuAlaTrpValProAlaHis           706
BH5   ---------------------------------------------------------------------------

BH10  AAAGGAATTGGAGGAAATGAACAAGTAGATAAATTAGTCAGTGCTGGAATCAGGAAAATACTATTTTTAGATGGA   3821
      LysGlyIleGlyGlyAsnGluGlnValAspLysLeuValSerAlaGlyIleArgLysIleLeuPheLeuAspGly           731
BH5   ---------------------------------------------------------------------------

BH10  ATAGATAAGGCCCAAGATGAACATGAGAAATATCACAGTAATTGGAGAGCAATGGCTAGTGATTTTAACCTGCCA   3896
      IleAspLysAlaGlnAspGluHisGluLysTyrHisSerAsnTrpArgAlaMetAlaSerAspPheAsnLeuPro           756
BH5   ----------------A----------------------------------------------------------
                              Pvu II
BH10  CCTGTAGTAGCAAAAGAAATAGTAGCCAGCTGTGATAAATGTCAGCTAAAAGGAGAAGCCATGCATGGACAAGTA   3971
      ProValValAlaLysGluIleValAlaSerCysAspLysCysGlnLeuLysGlyGluAlaMetHisGlyGlnVal           781
BH5   ---------------------------------------------------------------------------

BH10  GACTGTAGTCCAGGAATATGGCAACTAGATTGTACACATTTAGAAGGAAAAGTTATCCTGGTAGCAGTTCATGTA   4046
      AspCysSerProGlyIleTrpGlnLeuAspCysThrHisLeuGluGlyLysValIleLeuValAlaValHisVal           806
BH5   ---------------------------------------------------------------------------
                                                                    Aha III
BH10  GCCAGTGGATATATAGAAGCAGAAGTTATTCCAGCAGAAACAGGGCAGGAAACAGCATATTTTCTTTTAAAATTA   4121
      AlaSerGlyTyrIleGluAlaGluValIleProAlaGluThrGlyGlnGluThrAlaTyrPheLeuLeuLysLeu           831
BH5   ---------------------------------------------------------------------------

BH10  GCAGGAAGATGGCCAGTAAAAACAATACATACAGACAATGGCAGCAATTTCACCAGTGCTACGGTTAAGGCCGCC   4196
      AlaGlyArgTrpProValLysThrIleHisThrAspAsnGlySerAsnPheThrSerAlaThrValLysAlaAla           856
BH5   ---------------------------------------------------------------------------
                                Eco RI
BH10  TGTTGGTGGGCGGGAATCAAGCAGGAATTTGGAATTCCCTACAATCCCCAAAGTCAAGGAGTAGTAGAATCTATG   4271
      CysTrpTrpAlaGlyIleLysGlnGluPheGlyIleProTyrAsnProGlnSerGlnGlyValValGluSerMet           881
BH5   ---------------------------------------------------------------------------

BH10  AATAAAGAATTAAAGAAAATTATAGGACAGGTAAGAGATCAGGCTGAACATCTTAAGACAGCAGTACAAATGGCA   4346
      AsnLysGluLeuLysLysIleIleGlyGlnValArgAspGlnAlaGluHisLeuLysThrAlaValGlnMetAla           906
BH5   ---------------------------------------------------------------------------
                 Aha III
BH10  GTATTCATCCACAATTTTAAAAGAAAAGGGGGGATTGGGGGGTACAGTGCAGGGGAAAGAATAGTAGACATAATA   4421
      ValPheIleHisAsnPheLysArgLysGlyGlyIleGlyGlyTyrSerAlaGlyGluArgIleValAspIleIle           931
BH5   ---------------------------------------------------------------------------

BH10  GCAACAGACATACAAACTAAAGAATTACAAAAACAAATTACAAAAATTCAAAATTTTCGGGTTTATTACAGGGAC   4496
      AlaThrAspIleGlnThrLysGluLeuGlnLysGlnIleThrLysIleGlnAsnPheArgValTyrTyrArgAsp           956
BH5   ---------------------------------------------------------------------------

BH10  AGCAGAAATCCACTTTGGAAAGGACCAGCAAAGCTCCTCTGGAAAGGTGAAGGGGCAGTAGTAATACAAGATAAT   4571
      SerArgAsnProLeuTrpLysGlyProAlaLysLeuLeuTrpLysGlyGluGlyAlaValValIleGlnAspAsn           981
BH5   ---------------------------------------------------------------------------
                 |———— SOR
BH10  AGTGACATAAAAGTAGTGCCAAGAAGAAAAGCAAAGATCATTAGGGATTATGGAAAACAGATGGCAGGTGATGAT   4646
      SerAspIleLysValValProArgArgLysAlaLysIleIleArgAspTyrGlyLysGlnMetAlaGlyAspAsp           1006
                 CysGlnGluGluLysGlnArgSerLeuGlyIleMetGluAsnArgTrpGlnValMetIle               20
BH5   ---------------------------------------------------------------------------
                 POL ——|
BH10  TGTGTGGCAAGTAGACAGGATGAGGATTAGAACATGGAAAAGTTTAGTAAAACACCATATGTATGTTTCAGGGAA   4721
      CysValAlaSerArgGlnAspGluAsp                                                   1015
       ValTrpGlnValAspArgMetArgIleArgThrTrpLysSerLeuValLysHisHisMetTyrValSerGlyLys          45
BH5   -------------------------------------------------G-------------------------
                                                       Arg

BH10  AGCTAGGGGATGGTTTTATAGACATCACTATGAAAGCCCTCATCCAAGAATAAGTTCAGAAGTACACATCCCACT   4796
       AlaArgGlyTrpPheTyrArgHisHisTyrGluSerProHisProArgIleSerSerGluValHisIleProLeu          70
BH5   ---------------------------------------------------------------------------

BH10  AGGGGATGCTAGATTGGTAATAACAACATATTGGGGTCTGCATACAGGAGAAAGAGACTGGCATTTGGGTCAGGG   4871
       GlyAspAlaArgLeuValIleThrThrTyrTrpGlyLeuHisThrGlyGluArgAspTrpHisLeuGlyGlnGly          95
BH5   ---------------------------------------------------------------------------

BH10  AGTCTCCATAGAATGGAGGAAAAAGAGATATAGCACACAAGTAGACCCTGAACTAGCAGACCAACTAATTCATCT   4946
       ValSerIleGluTrpArgLysLysArgTyrSerThrGlnValAspProGluLeuAlaAspGlnLeuIleHisLeu          120
BH5   ----------------------G----------------------------------------------------
                            Arg

BH10  GTATTACTTTGACTGTTTTTCAGACTCTGCTATAAGAAAGGCCTTATTAGGACACATAGTTAGCCCTAGGTGTGA   5021
       TyrTyrPheAspCysPheSerAspSerAlaIleArgLysAlaLeuLeuGlyHisIleValSerProArgCysGlu          145
BH5   -C---------T---------------------------------------------------------------

BH10  ATATCAAGCAGGACATAACAAGGTAGGATCTCTACAATACTTGGCACTAGCAGCATTAATAACACCAAAAAAGAT   5096
       TyrGlnAlaGlyHisAsnLysValGlySerLeuGlnTyrLeuAlaLeuAlaAlaLeuIleThrProLysLysIle          170
BH5   -------------------------------------------------------------------------G-
                                                                           Val

BH10  AAAGCCACCTTTGCCTAGTGTTACGAAACTGACAGAGGATAGATGGAACAAGCCCCAGAAGACCAAGGGCCACAG   5171
       LysProProLeuProSerValThrLysLeuThrGluAspArgTrpAsnLysProGlnLysThrLysGlyHisArg          195
BH5   ---------------------------------------------------------------------------
                 SOR ————|
BH10  AGGGAGCCACACAATGAATGGACACTAGAGCTTTTAGAGGAGCTTAAGAATGAAGCTGTTAGACATTTTCCTAGG   5246
       GlySerHisThrMetAsnGlyHis                                                     203
BH5   ---A-----------------------------------------------------------------------

BH10  ATTTGGCTCCATGGCTTAGGGCAACATATCTATGAAACTTATGGGGATACTTGGGCAGGAGTGGAAGCCATAATA   5321
BH5   ---------------------------------------------------------------------------
       Eco RI                                      Sal I
BH10  AGAATTCTGCAACAACTGCTGTTTATCCATTTTCAGAATTGGGTGTCGACATAGCAGAATAGGCGTTACTCGACA   5396
BH5   -----------------------------------------------------------------------A---

BH10  GAGGAGAGCAAGAAATGGAGCCAGTAGATCCTAGACTAGAGCCCTGGAAGCATCCAGGAAGTCAGCCTAAAACTG   5471
BH5   ---------------------------------------------------------------------------

BH10  CTTGTACCAATTGCTATTGTAAAAAGTGTTGCTTTCATTGCCAAGTTTGTTTCATAACAAAAGCCTTAGGCATCT   5546
BH5   ---------C-----------------------------------------------------------------
                                       (Sst I)
BH10  CCTATGGCAGGAAGAAGCGGAGACAGCGACGAAGACCTCCTCAAGGCAGTCAGACTCATCAAGTTTCTCTATCAA   5621
BH5   -------------------------------------G---
BH8                                     --G---A--G-A-------------------------------------

BH10  AGCAGTAAGTAGTACATGTAATGCAACCTATACAAATAGCAATAGTAGCA[ ]TTAGTAGTAGCAATAATAATAGCAA   5696
BH8   -------------------C-----------C-----T--C-A---T-GCC----C-------------------

BH10  TAGTTGTGTGGTCCATAGTAATCATAGAATATAGGAAAATATTAAGACAAAGAAAAATAGACAGGTTAATTGATA   5771
BH8   ---------------------------------------------------------------------------
              |———————— ENV-LOR
BH10  GACTAATAGAAAGAGCAGAAGACAGTGGCAATGAGAGTGAAGGAGAAATATCAGCACTTGTGGAGATGGGGGTGG   5846
                LysGluGlnLysThrValAlaMetArgValLysGluLysTyrGlnHisLeuTrpArgTrpGlyTrp       22
BH8   ---------------------------------------------------------------------------

BH10  AGATGGGGCACCATGCTCCTTGGGATGTTGATGATCTGTAGTGCTACAGAAAAATTGTGGGTCACAGTCTATTAT   5921
      ArgTrpGlyThrMetLeuLeuGlyMetLeuMetIleCysSerAlaThrGluLysLeuTrpValThrValTyrTyr           47
BH8   -------------------------------------------------------------------------T-
                                                                             Phe
        Kpn I
BH10  GGGGTACCTGTGTGGAAGGAAGCAACCACCACTCTATTTTGTGCATCAGATGCTAAAGCATATGATACAGAGGTA   5996
      GlyValProValTrpLysGluAlaThrThrThrLeuPheCysAlaSerAspAlaLysAlaTyrAspThrGluVal           72
BH8   ---------------------------------------------------------------------------
```

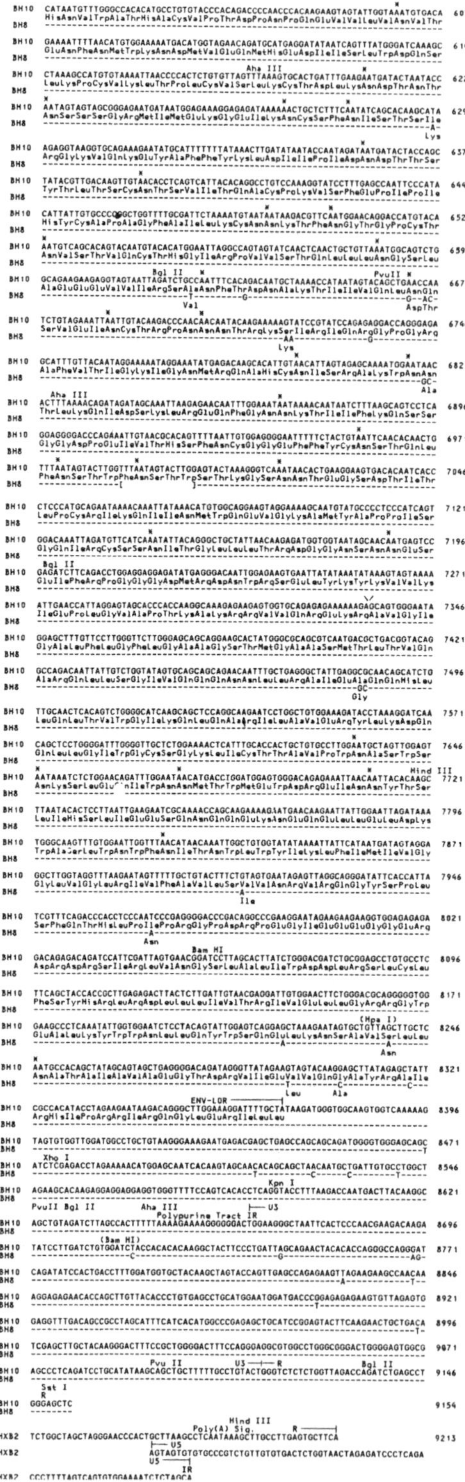

Fig. 1. Nucleotide sequence of HTLV-III. The complete nucleotide sequence of clone BH10 is shown together with the predicted amino acid sequence of the four largest open reading frames. The position of sequences encoding *gag* protein p17, the N-terminus of *gag* p24 and the C-terminus of *gag* p15 (which overlaps with the N-terminus of the *pol* protein) are indicated. The open reading frames for *pol*, *sor* and *env-lor* are indicated. The sequence of the remaining 182 bp of the HTLV-III provirus not present in clone BH10 (including a portion of R, U5, the tRNA primer binding site and a portion of the leader sequence) was derived from clone HXB2. The boundaries of R, U5 and U3, the positions of the polypurine tract, inverted repeated sequences (IR) and the transcriptional initiation (TATA) and termination (AATAA) signals are shown. The sequences of BH8 and BH5 are illustrated also; nucleotide and predicted amino acid differences compared with BH10 are listed and dashes are shown for identical sequences. Restriction enzyme sites are listed above the nucleotide sequence and sites present in clone BH8 but not BH10 are in parentheses. Deletions are also noted ([ ]) at nucleotides 251, 254, 5,671 and 6,987–7,001. The nucleotide positions (to the right of each line) start with the transcriptional initiation site and end with the viral RNA transcriptional termination site. The amino acid residues are numbered (to the right of each line) for the four largest open reading frames starting after the preceding termination codon in each case except *gag* which is enumerated from the first methionine codon. A proposed peptide cleavage site (v) and possible asparagine-linked glycosylation sites [77] are shown (*) for the *env-lor* open reading frame. The sequences in the LTR derived from clones BH8 and BH10 listed in the beginning of the figure are derived from the 3′ portion of each clone and are assumed to be identical to those present in the 5′ LTR of the integrated copies of these viral genomes. Clone HXB2 was derived from a recombinant phage library of *Xba*I-digested DNA from HTLV-III-infected H9 cell cloned in λJ1 (ref. 53). Clones BH10, BH8 and BH5 were derived from a library of *Sst*I-digested DNA from the Hirt supernatant fraction of HTLV-III-infected H9 cells cloned in λgtWes·λB (ref. 27). Both libraries were screened with a cDNA probe synthesized from virion RNA using oligo (dT) as a primer[26]. Clones BH8, BH5 and a portion of HXB2 were sequenced as described previously[91]. Clone BH10 was sequenced by the method of Sanger[92] modified by the use of oligonucleotides complementary to the M13 insert sequence as primers and using Klenow fragment of DNA polymerase I or reverse transcriptase as the polymerase.

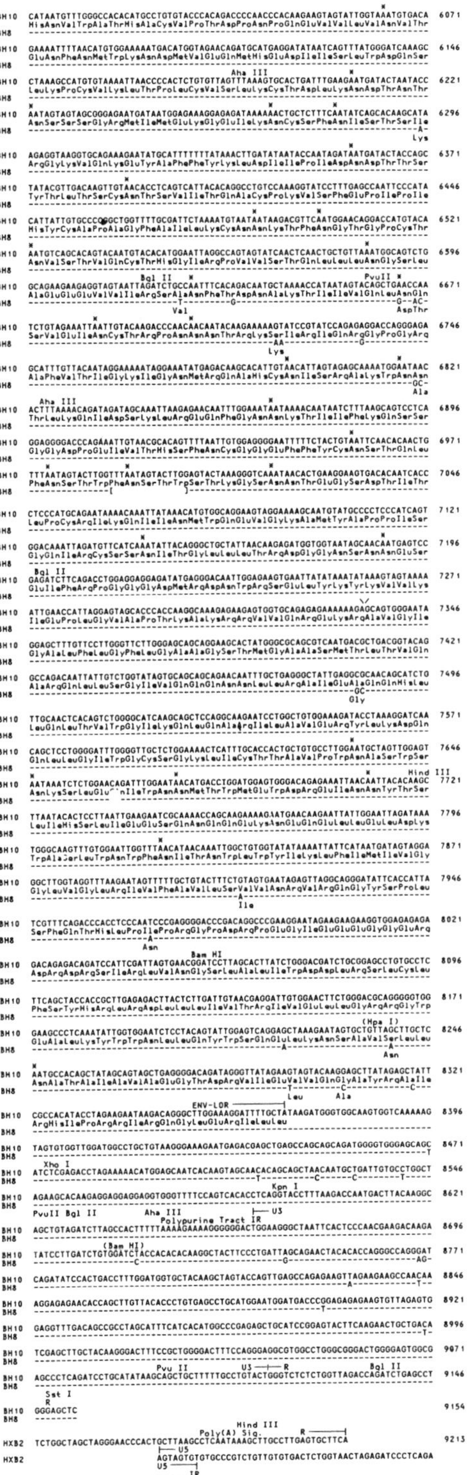

图 1. HTLV-III 的核苷酸序列。对应列出克隆 BH10 的完整核苷酸序列与预测的四个最大的开放阅读框的氨基酸序列。编码 *gag* 蛋白 p17 的序列位置，p24 的 N 端和 p15 的 C 端（它与 *pol* 蛋白的 N 端有重叠）序列的位置都已指出。*pol*，*sor* 和 *env-lor* 的开放阅读框都已标出。克隆 BH10 不含有的 182bp HTLV-III 前病毒序列（包括部分 R、U5、tRNA 引物结合位点和部分前导序列）来自克隆 HXB2。图中还标明了 R、U5 和 U3 的边界，多嘌呤序列、反向重复序列（IR）和转录起始（TATA）和终止信号（AATAA）的位置。BH8 和 BH5 的序列也做了说明；图中列出了与 BH10 相比存在差异的核苷酸和预测氨基酸序列，完全相同的序列用破折号来表示。在核苷酸序列上方还标出了限制性酶切位点，圆括号表示的是在 BH8 中出现而 BH10 中未出现的位点。在核苷酸 251、254、5,671 和 6,987~7,001 位中的缺失位点用（[ ]）来表示。核苷酸的位置的标注（每行的右边）由转录起始位点开始到病毒 RNA 转录终止位点结束。在四个大的开放阅读框中，对氨基酸残基进行编号（在每一行的右边），除了 *gag* 是从第一个甲硫氨酸密码子开始编号，其他开放读框都是从前一个终止密码子开始的。*env–lor* 开放阅读框中表示了一个已报道的肽剪切位点（v）和可能的天冬酰胺连接型糖基化位点[77]（*）。在图起始处列出的 LTR 中的序列来自克隆 BH8 和 BH10，起源于每个克隆的 3′端，被认为和这些病毒基因组整合拷贝的 5′LTR 含有的序列相同。克隆 HXB2 来源于一个重组噬菌体文库，此文库是 HTLV-III 感染的 H9 细胞中的 DNA 经 *XbaI* 消化后克隆在 λJ1 中获得的（参考文献 53）。克隆 BH10、BH8 和 BH5 所在的 DNA 文库，是 HTLV-III 感染的 H9 细胞中 Hirt 上清部分的 DNA 经 *Sst*I 消化后克隆在 λgtWes · λB 中的（参考文献 27）。以寡聚 dT 为引物利用病毒粒子的 RNA 合成 cDNA 探针[26]，对这两个文库进行筛选。克隆 BH8、BH5 和部分 HXB2 的序列用之前描述的方法进行了测序[91]。克隆 BH10 通过改良的桑格法[92]进行测序，利用与 M13 插入序列互补的寡聚核苷酸作为引物，并用 DNA 聚合酶 I Klenow 片段或逆转录酶作为聚合酶。

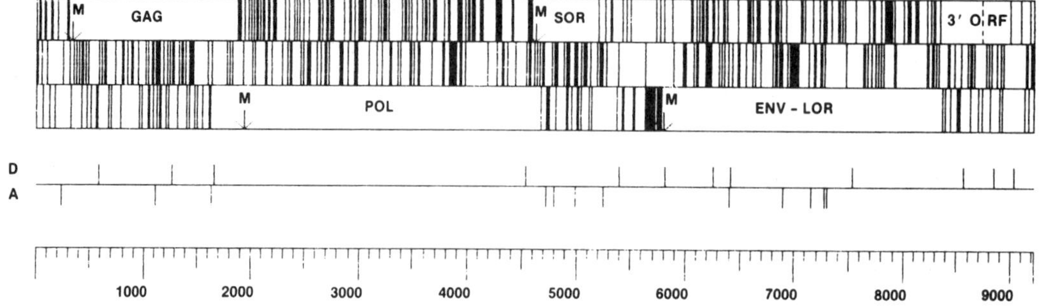

Fig. 2. Distribution of open reading frames in HTLV-III. The positions of termination codons are indicated for each of the three possible reading frames as a vertical line within each box. Thicker lines represent clusters of closely spaced termination codons. The positions of consensus splice donor (D) and acceptor (A) sites[93] are shown below. The nucleotide sequence positions are also indicated below, starting from the transcriptional initiation site. The positions of the four largest open reading frames are identified as *gag*, *pol*, *sor* and *env-lor*, and the position of the first ATG codon (M) in each reading frame is indicated. A fifth open reading frame (3′-*orf*) is found in clone BH8, but not BH10; the broken vertical line indicates the position of a termination codon in BH10.

```
                                                                                   PERCENT NUCLEOTIDE
                                                                                   IDENTITY TO HTLV-III

HTLV-III BH10 CLONE U3   8824   GAGAAGTTAGAAGAAGCCAACAAAGGAGAGA    8854
HTLV-III BH8  CLONE U3   8824   GAGAAGTAAGAAGAAGCCAATAAAGGAGAGA    8854
HTLV-I              U3    307   AATAAACTAGCAGGAGTCTATAAAAGCGTGG     337                    61
                                 * **  *** ** ** * ***** * * *
HTLV-II             U3    268   AATAAAAGATGCCGAGTCTATAAAGGCGCAA     298                    51
                                 * **  *     ** * ******* *  *

HTLV-III            U3   9067   TGGCGAGCCCTCAGATCCTGC  ATA   TAA    9093
BLV                 U3    152   TGCTGA  CCTCA   CCTGCTGATAAATTAA     178                   62
                                **  **  *****   *****  ***   ***

HTLV-III GAG p15         1886   ACTAAAGGAAGCTCTATTAGATACAGGAGCAGATGAT ACAGT ATT AGA AG   1935
HTLV-I BETWEEN GAG & POL 2223   ACTA TCGAAGCTTTACTAGATACAGGAGCAGA CATGACAGTCCTTCCGATAG   2274    76
                                ****   ****** ** ***************  ** *****  **   ** **

HTLV-III BETWEEN SOR & ENV-LOR 5615  CTATCAAAGCAGTAAGTAGTACATGTAATG   5644
HTLV-I ENV                     5831  CTA CAAAGCACTAATTA TACTTGCATTG   5858                     77
                                     *** ******* *** ** *** ** * **
```

Fig. 3. Nucleotide sequence homologies between HTLV-I, -II, -III and BLV. Asterisks indicate identity of nucleotides to that present at the same position in HTLV-III (clone BH10 except where indicated otherwise). The nucleotide position of each sequence is shown.

First Open Reading Frame (*gag*)

The structural proteins of the *gag* gene of retroviruses are derived by proteolytic cleavage of a polypeptide precursor encoded at the 5′ end of the genome[57]. Such a precursor, 512 amino acids long, could be encoded by the first long open reading frame of the HTLV-III provirus (nucleotides 310–1,869). The definitive assignment of amino-terminus of the *gag* precursor requires direct amino acid sequencing. A protein of relative molecular mass (M_r) 53,000 (53K) has been detected in immunoprecipitates of HTLV-III-infected cells using sera from an HTLV-III-infected person, as well as sera raised in animals by inoculation of HTLV-III virions[12,13,58]. This 53K protein probably corresponds to the *gag* precursor of HTLV-III.

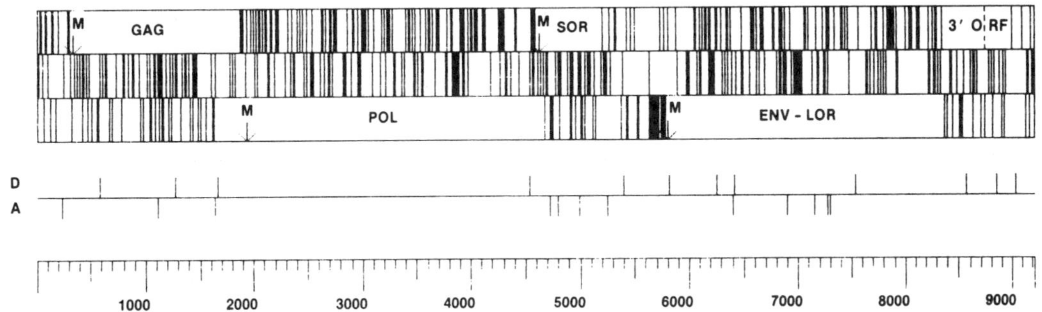

图 2. HTLV-III 中开放阅读框的分布。在每一个方框中，三个可能的阅读框的终止密码子的位置用垂线表示。粗线代表位置比较接近的终止密码子簇。保守的剪接供体 D 和受体 A[93] 的位置也在图中标明。核苷酸序列的位置也在图下方表示，位置编号从转录起始位点开始。四个最大的开放阅读框的位置标注为 *gag*、*pol*、*sor* 和 *env-lor*，每个开放阅读框的第一个 ATG 密码子（M）也被标出。第五个开放阅读框（*3′-orf*）在克隆 BH8 中发现，但是在 BH10 中没有；虚垂线表示在 BH10 中终止密码子的位置。

与HTLV-Ⅲ的相似度（%）

```
HTLV-III BH10 CLONE U3   8824   GAGAAGTTAGAAGAAGCCAACAAAGGAGAGA   8854
HTLV-III BH8  CLONE U3   8824   GAGAAGTAAGAAGAAGCCAATAAAGGAGAGA   8854
HTLV-I              U3    307   AATAAACTAGCAGGAGTCTATAAAAGCGTGG    337              61
                                 * **  *** ** ** * ***** * * *
HTLV-II             U3    268   AATAAAAGATGCCGAGTCTATAAAGGCGCAA    298              51
                                 * **  *     ** * ******* *  *

HTLV-III            U3   9067   TGGCGAGCCCTCAGATCCTGC  ATA   TAA   9093
BLV                 U3    152   TGCTGA  CCTCA   CCTGCTGATAAATTAA    178             62
                                **  **  *****   *****  ***   ***

HTLV-III GAG p15          1886   ACTAAAGGAAGCTCTATTAGATACAGGAGCAGATGAT ACAGT ATT AGA AG   1935
HTLV-I BETWEEN GAG & POL  2223   ACTA TCGAAGCTTTACTAGATACAGGAGCAGA CATGACAGTCCTTCCGATAG   2274      76
                                ****   ****** ** ***************  ** *****  **  ** **

HTLV-III BETWEEN SOR & ENV-LOR  5615   CTATCAAAGCAGTAAGTAGTACATGTAATG   5644
HTLV-I ENV                      5831   CTA CAAAGCACTAATTA TACTTGCATTG   5858          77
                                       *** ******* *** ** *** ** * **
```

图 3. HTLV-I，-II，-III 和 BLV 的核苷酸序列的同源性。星号表示在 HTLV-III 相同位置中出现的相同的核苷酸（克隆 BH10 另有注明）。每一条序列的核苷酸位置都已标出。

第一个开放阅读框（*gag*）

逆转录病毒 *gag* 基因的结构蛋白是基因组 5′ 末端编码的一段多肽前体经蛋白酶水解的产物[57]。这个前体有 512 个氨基酸，可能由 HTLV-III 前病毒的第一个长开放阅读框编码（核苷酸 310~1,869）。确认 *gag* 前体的氨基末端需要进行氨基酸测序。利用感染了 HTLV-III 的人的血清和接种了 HTLV-III 病毒粒子的动物的血清，与得到的感染了 HTLV-III 的细胞进行免疫共沉淀反应，得到了一种相对分子量为 53,000（53K）的蛋白[12,13,58]。这种 53K 的蛋白可能是 HTLV-III 的 *gag* 前体。

The identity of the first open reading frame as the *gag* gene is established by the correspondence of the sequence of the amino-terminal 15 amino acids of the p24 major capsid protein (our unpublished observations with S. Oroszolan) with the sequence predicted for the first open reading frame between nucleotides 730 and 774. Moreover, amino acid similarities between the *gag* gene products of HTLV-I and BLV were detected in regions corresponding to the p24 and p15 *gag* products[32,33,59] (Fig. 4).

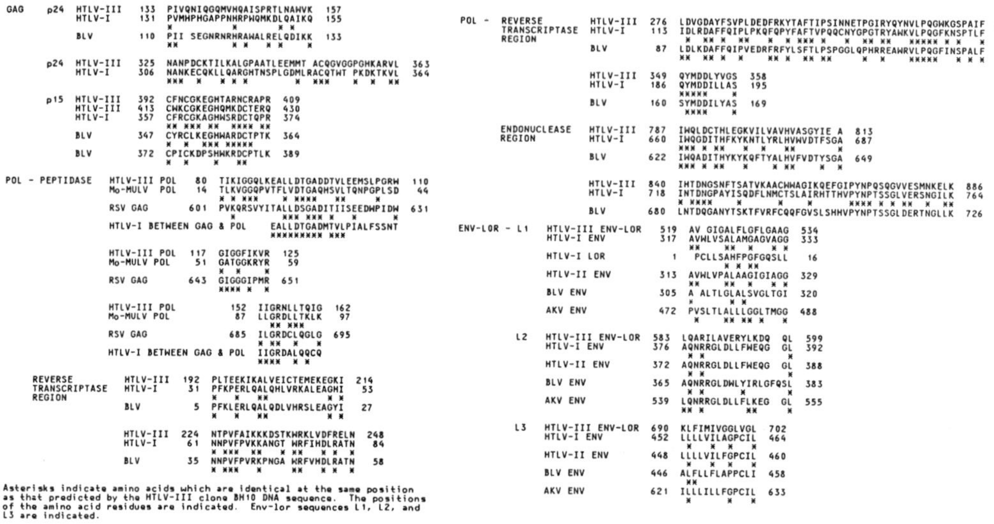

```
GAG   p24  HTLV-III  133  PIVQNIQGQMVHQAISPRTLNAWVK  157
           HTLV-I    131  PVMHPHGAPPNHRPWQMKDLQAIKQ  155
                          *          *       * *
           BLV       110  PII SEGNRNRHRAWALRELQDIKK  133
                          **         * *   * *    *

      p24  HTLV-III  325  NANPDCKTILKALGPAATLEEMMT ACQGVGGPGHKARVL  363
           HTLV-I    306  NANKECQKLLQARGHTNSPLGDMLRACQTWT PKDKTKVL  364
                          ***  *   * * *        *  ***    *  *  **

      p15  HTLV-III  392  CFNCGKEGHTARNCRAPR  409
           HTLV-III  413  CWKCGKEGHQMKDCTERQ  430
           HTLV-I    357  CFRCGKAGHWSRDCTQPR  374
                          ** *** **  **** **
           BLV       347  CYRCLKEGHWARDCTPTK  364
                          *  *  *** *****
           BLV       372  CPICKDPSHWKRDCPTLK  389
                          *  *    *   **

POL - PEPTIDASE  HTLV-III POL   80  TIKIGGQLKEALLDTGADDTVLEEMSLPGRW  110
                 Mo-MULV  POL   14  TLKVGGQPVTFLVDTGAQHSVLTQNPGPLSD   44
                                    * * ***    * ****   **      *
                 RSV GAG       601  PVKQRSVYITALLDSGADITIISEEDWPIDW  631
                                      *       **** *** *   *   *  *
                 HTLV-I BETWEEN GAG & POL    EALLDTGADMTVLPIALFSSNT
                                             ********* ***

                 HTLV-III POL  117  GIGGFIKVR  125
                 Mo-MULV POL    51  GATGGKRYR   59
                                    *  *    *
                 RSV GAG       643  GIGGGIPMR  651
                                    **** *  *

                 HTLV-III POL           152  IIGRNLLTQIG  162
                 Mo-MULV POL             87  LLGRDLLTKLK   97
                                               ** ***
                 RSV GAG                685  ILGRDCLQGLG  695
                                             * **  *   *
                 HTLV-I BETWEEN GAG & POL    IIGRDALQQCQ
                                             ****  * *

    REVERSE          HTLV-III  192  PLTEEKIKALVEICTEMEKEGKI  214
    TRANSCRIPTASE    HTLV-I     31  PFKPERLQALQHLVRKALEAGHI   53
    REGION                          *   *   **          * *
                     BLV         5  PFKLERLQALQDLVHRSLEAGYI   27
                                    *   *   **          * *

                     HTLV-III  224  NTPVFAIKKKDSTKWRKLVDFRELN  248
                     HTLV-I     61  NNPVFPVKKANGT WRFIHDLRATN   84
                                    * ***  **   * **   * *  *
                     BLV        35  NNPVFPVRKPNGA WRFVHDLRATN   58
                                    * ***   *     **   * *  *

Asterisks indicate amino acids which are identical at the same position
as that predicted by the HTLV-III clone BH10 DNA sequence.  The positions
of the amino acid residues are indicated.  Env-lor sequences L1, L2, and
L3 are indicated.

POL -  REVERSE        HTLV-III  276  LDVGDAYFSVPLDEDFRKYTAFTIPSINNETPGIRYQYNVLPQGWKGSPAIF  327
       TRANSCRIPTASE  HTLV-I    113  IDLRDAFFQIPLPKQFQPYFAFTVPQQCNYGPGTRYAWKVLPQGFKNSPTLF  164
       REGION                         *  ** *  **   *  * *** *   *  ** **   ***** * **  *
                      BLV        87  LDLKDAFFQIPVEDRFRFYLSFTLPSPGGLQPHRREAWRVLPQGFINSPALF  138
                                     **  ** *  *    ** *  ** **     *  *    *****   *** *

                      HTLV-III  349  QYMDDLYVGS  358
                      HTLV-I    186  QYMDDILLAS  195
                                     *****    *
                      BLV       160  SYMDDILYAS  169
                                      ****    *

       ENDONUCLEASE   HTLV-III  787  IWQLDCTHLEGKVILVAVHVASGYIE A  813
       REGION         HTLV-I    660  IWQGDITHFKYKNTLYRLHVWVDTFSGA  687
                                     *** * **   *  *   **       *
                      BLV       622  IWQADITHYKYKQFTYALHVFVDTYSGA  649
                                     *** * **   *    * **       *

                      HTLV-III  840  IHTDNGSNFTSATVKAACHWAGIKQEFGIPYNPQSQGVVESMNKELK  886
                      HTLV-I    718  INTDNGPAYISQDFLNMCTSLAIRHTTHVPYNPTSSGLVERSNGILK  764
                                     * ****    *      *    *      **** * * **  *  **
                      BLV       680  LNTDQGANYTSKTFVRFCQQFGVSLSHHVPYNPTSSGLDERTNGLLK  726

ENV-LOR - L1  HTLV-III ENV-LOR  519  AV GIGALFLGFLGAAG  534
              HTLV-I ENV        317  AVWLVSALAMGAGVAGG  333
                                     **    **  *   * *
              HTLV-I LOR          1   PCLLSAHFPGFGQSLL   16
                                           * * **
              HTLV-II ENV       313  AVWLVPALAAGIGIAGG  329
                                     **    **  *   * *
              BLV ENV           305  A ALTLGLALSVGLTGI  320
                                     *      * *
              AKV ENV           472  PVSLTLALLLGGLTMGG  488
                                      *    ** ** *   *

          L2  HTLV-III ENV-LOR  583  LQARILAVERYLKDQ  QL  599
              HTLV-I ENV        376  AQNRRGLDLLFWEQG  GL  392
                                      * *              *
              HTLV-II ENV       372  AQNRRGLDLLFWEQG  GL  388
                                      * *              *
              BLV ENV           365  AQNRRGLDWLYIRLGFQSL  383
                                      * *      *       *
              AKV ENV           539  LQNRRGLDLLFLKEG  GL  555
                                     ** *       **     *

          L3  HTLV-III ENV-LOR  690  KLFIMIVGGLVGL  702
              HTLV-I ENV        452  LLLLVILAGPCIL  464
                                      *   *  *   *
              HTLV-II ENV       448  LLLLVILFGPCIL  460
                                      *   *  *   *
              BLV ENV           446  ALFLLFLAPPCLI  458
                                      **
              AKV ENV           621  ILLLILLFGPCIL  633
                                      *      *   *
```

Fig. 4. Amino acid homologies between HTLV-III predicted proteins and those of other retroviruses. Asterisks indicate amino acids identical at the same position as that predicted by the HTLV-III clone BH10 DNA sequence. Positions of amino acid residues are indicated. Sequences L1, L2 and L3 in *env-lor* are noted in Fig. 5.

Cleavage of the putative *gag* gene product precursor between amino acids 132 and 133 to yield the observed amino-terminus of p24 would produce an amino-terminal polypeptide 132 amino acids long. This protein would be similar in length to the amino-terminus of the *gag* gene product of HTLV-I (126 amino acids)[32,34]. A 17K virion protein has been identified that probably corresponds to this product[12,20,58].

The immediate juxtaposition of the p24 protein with the amino-terminal p17 protein is like the structure of the *gag* gene products of HTLV-I[32] and BLV[33] which differs from murine, avian, feline and primate type-C and type-D retroviruses in this region, as HTLV-I and BLV seem to lack a short phosphoprotein encoded usually in other retroviruses by sequences between the amino-terminal *gag* protein and the major capsid protein coding sequences[34,57,60,61].

The third HTLV-III *gag* protein (p15) reveals significant amino acid homologies to the p15 proteins of HTLV-I[32,62] and BLV[33] (Fig. 4). There are direct repeats of the DNA sequence in this region of *gag* (Fig. 1), a feature common to both HTLV-I and BLV p15 proteins. The predicted p15 protein of HTLV-III is probably basic and binds nucleic acids, like

通过主要衣壳蛋白 p24（我们和欧罗斯兰未发表的观察结果）的氨基末端 15 个氨基酸序列与第一个开放阅读框 730 和 774 位核苷酸之间的预测序列的相关性，我们确定了 *gag* 基因就是第一个开放阅读框。此外，我们在与 p24 和 p15 *gag* 基因产物对应的区域中，发现了 HTLV -I 和 BLV 的 *gag* 基因产物的氨基酸相似性 [32,33,59]（图 4）。

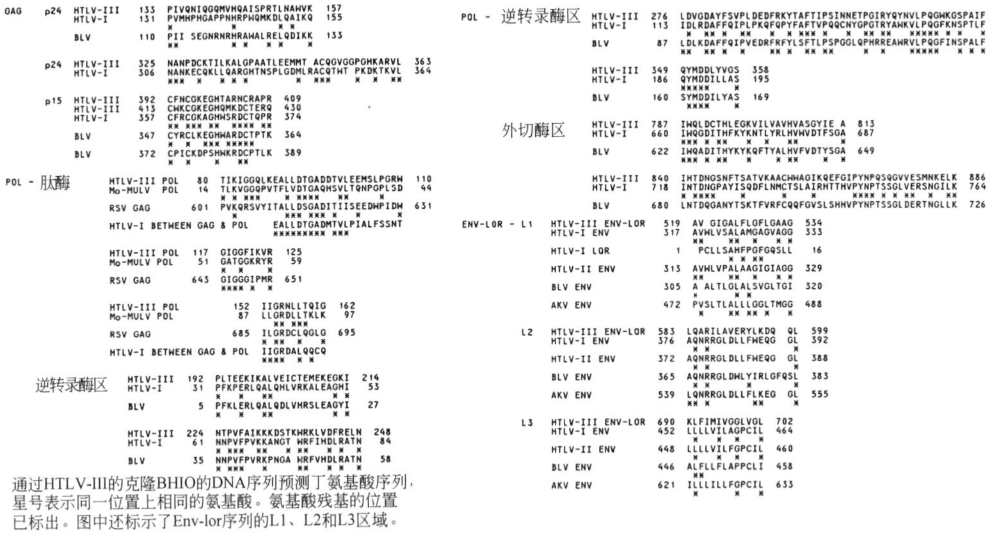

```
GAG   p24  HTLV-III  133  PIVQNIQGQMVHQAISPRTLNAWVK  157
           HTLV-I    131  PVMHPHGAPPNHRPWQMKDLQAIKQ  155
                          *          *     * *
           BLV       110  PII SEGNRNRHRAWALRELQDIKK  133
                          **         * *  * *   *

      p24  HTLV-III  325  NANPDCKTILKALGPAATLEEMMT ACQGVGGPGHKARVL  363
           HTLV-I    306  NANKECQKLLQARGHTNSPLGDMLRACQTWT PKDKTKVL  364
                          ***  *   * * *        *  ***    *  *  **

      p15  HTLV-III  392  CFNCGKEGHTARNCRAPR  409
           HTLV-III  413  CWKCGKEGHQMKDCTERQ  430
           HTLV-I    357  CFRCGKAGHWSRDCTQPR  374
                          ** *** **  **** **
           BLV       347  CYRCLKEGHWARDCTPTK  364
                          *  *  *** *****
           BLV       372  CPICKDPSHWKRDCPTLK  389
                          *  *    *   **

POL - 肽酶    HTLV-III POL   80  TIKIGGQLKEALLDTGADDTVLEEMSLPGRW  110
              Mo-MULV  POL   14  TLKVGGQPVTFLVDTGAQHSVLTQNPGPLSD   44
                                 * * ***   * ****   **    *
              RSV GAG       601  PVKQRSVYITALLDSGADITIISEEDWPIDW  631
                                   *       *** *** *   *     *
              HTLV-I BETWEEN GAG & POL     EALLDTGADMTVLPIALFSSNT
                                           ********* ***

              HTLV-III POL  117  GIGGFIKVR  125
              Mo-MULV POL    51  GATGGKRYR   59
                                 *  *    *
              RSV GAG       643  GIGGGIPMR  651
                                 **** *  *

              HTLV-III POL          152  IIGRNLLTQIG  162
              Mo-MULV POL            87  LLGRDLLTKLK   97
                                           ** ***
              RSV GAG               685  ILGRDCLQGLG  695
                                         * **  *   *
              HTLV-I BETWEEN GAG & POL   IIGRDALQQCQ
                                         ****  * *

    逆转录酶区  HTLV-III  192  PLTEEKIKALVEICTEMEKEGKI  214
               HTLV-I     31  PFKPERLQALQHLVRKALEAGHI   53
                              *  *  **            * *
               BLV         5  PFKLERLQALQDLVHRSLEAGYI   27
                              *  *  **            * *

               HTLV-III  224  NTPVFAIKKKDSTKWRKLVDFRELN  248
               HTLV-I     61  NNPVFPVKKANGT WRFIHDLRATN   84
                              * ***  **   * **  * *  *
               BLV        35  NNPVFPVRKPNGA WRFVHDLRATN   58
                              * ***   *     **  * *  *

POL - 逆转录酶区  HTLV-III  276  LDVGDAYFSVPLDEDFRKYTAFTIPSINNETPGIRYQYNVLPQGWKGSPAIF  327
                 HTLV-I    113  IDLRDAFFQIPLPKQFQPYFAFTVPQQCNYGPGTRYAWKVLPQGFKNSPTLF  164
                                 * ** *  **   *  * *** *   *  ** **   ***** * *   *
                 BLV        87  LDLKDAFFQIPVEDRFRFYLSFTLPSPGGLQPHRREAWRVLPQGFINSPALF  138
                                **  ** *  *    ** *  ** **    *  *    *****   * * *

                 HTLV-III  349  QYMDDLYVGS  358
                 HTLV-I    186  QYMDDILLAS  195
                                *****    *
                 BLV       160  SYMDDILYAS  169
                                 ****    *

    外切酶区      HTLV-III  787  IWQLDCTHLEGKVILVAVHVASGYIE A  813
                 HTLV-I    660  IWQGDITHFKYKNTLYRLHVWVDTFSGA  687
                                *** * **   *  *   **       *
                 BLV       622  IWQADITHYKYKQFTYALHVFVDTYSGA  649
                                *** * **   *    * **     *  *

                 HTLV-III  840  IHTDNGSNFTSATVKAACWWAGIKQEFGIPYNPQSQGVVESMNKELK  886
                 HTLV-I    718  INTDNGPAYISQDFLNMCTSLAIRHTTHVPYNPTSSGLVERSNGILK  764
                                * ****     *      *     *      **** * * **  *  **
                 BLV       680  LNTDQGANYTSKTFVRFCQQFGVSLSHHVPYNPTSSGLDERTNGLLK  726

ENV-LOR - L1  HTLV-III ENV-LOR   519  AV GIGALFLGFLGAAG  534
              HTLV-I ENV         317  AVWLVSALAMGAGVAGG  333
                                      **    **  *   * *
              HTLV-I LOR           1   PCLLSAHFPGFGQSLL   16
                                             * * **
              HTLV-II ENV        313  AVWLVPALAAGIGIAGG  329
                                      **    **  *   * *
              BLV ENV            305  A ALTLGLALSVGLTGI  320
                                      *       * *
              AKV ENV            472  PVSLTLALLLGGLTMGG  488
                                       *    ** ** *   *

          L2  HTLV-III ENV-LOR   583  LQARILAVERYLKDQ  QL  599
              HTLV-I ENV         376  AQNRRGLDLLFWEQG  GL  392
                                       * *               *
              HTLV-II ENV        372  AQNRRGLDLLFWEQG  GL  388
                                       * *               *
              BLV ENV            365  AQNRRGLDWLYIRLGFQSL  383
                                       * *      *        *
              AKV ENV            539  LQNRRGLDLLFLKEG  GL  555
                                      ** *        **     *

          L3  HTLV-III ENV-LOR   690  KLFIMIVGGLVGL  702
              HTLV-I ENV         452  LLLLVILAGPCIL  464
                                       *  *  *   *
              HTLV-II ENV        448  LLLLVILFGPCIL  460
                                       *  *  *   *
              BLV ENV            446  ALFLLFLAPPCLI  458
                                       **
              AKV ENV            621  ILLLILLFGPCIL  633
                                       *       *   *
```

通过HTLV-III的克隆BHIO的DNA序列预测丁氨基酸序列，星号表示同一位置上相同的氨基酸。氨基酸残基的位置已标出。图中还标示了Env-lor序列的L1、L2和L3区域。

图 4. HTLV-III 预测的蛋白与其他逆转录病毒蛋白的氨基酸同源性。星号表示某个位置的氨基酸与 HTLV-III 克隆 BH10 DNA 序列中预测的相同。氨基酸残基位置被标明。*env-lor* 的 L1、L2 和 L3 的序列见图 5。

蛋白酶在推测的 *gag* 基因前体产物的第 132 和 133 位氨基酸之间切割，能够产生一个 132 个氨基酸的多肽，即已观察到的 p24 的氨基末端。这个蛋白在长度上与 HTLV-I 的 *gag* 基因的氨基末端产物（126 个氨基酸）相似 [32,34]。已发现的一个 17k 的病毒颗粒蛋白可能就是这个产物 [12,20,58]。

p24 蛋白和 p17 蛋白的氨基末端是紧密相邻的，就像 HTLV-I[32] 和 BLV[33] 的 *gag* 基因产物的结构一样，它们不同于啮齿类、鸟类、猫科和灵长类的 C- 型和 D- 型逆转录病毒在该区域的产物，HTLV-I 和 BLV 似乎缺少一段短的磷蛋白，在其他逆转录病毒中通常由 *gag* 蛋白氨基末端和主要衣壳蛋白的编码序列之间的序列所编码 [34,57,60,61]。

第三个 HTLV-III *gag* 蛋白（p15）与 HTLV-I[32,62] 和 BLV[33] 的 p15 蛋白有显著的氨基酸同源性（图 4）。在 *gag* 的这个区域有正向的 DNA 重复序列（图 1），这是 HTLV-I 和 BLV p15 蛋白共有的一个特点。该预测的 HTLV-III 的 p15 蛋白可能是碱

the basic proteins of other retroviruses[34,63,64]. (There is also a proline-rich sequence at the carboxyl end of the *gag* precursor that may encode a small polypeptide similar to the progagtin described by S. Oroszolan and co-authors (personal communication).)

We conclude that the first long open reading frame of the HTLV-III genome encodes the *gag* protein precursor, based on the location of the open reading frame near the 5′ end of the viral genome and the correspondence between sequences predicted for the *gag* gene precursor and those observed for the major capsid p24 protein. The structure of the *gag* precursor resembles more closely that of HTLV-I and BLV than it does other retroviruses, as the amino-terminal *gag* protein is immediately juxtaposed to the major capsid protein.

Second Open Reading Frame (*pol*)

The predicted amino acid sequence of the second long open reading frame (nucleotides 1,629–4,673) that overlaps the first open reading frame by 80 amino acids, reveals numerous regions of amino acid similarity to the *pol* gene products of other retroviruses[34,65,66] (Fig. 4). These regions of conserved sequences are co-linear with those of other retroviruses, therefore we conclude that the second long open reading frame encodes the viral reverse transcriptase protein.

Similarities in amino acid sequences in the virally encoded protease products of other retroviruses are present at the amino-terminus of the HTLV-III *pol* protein precursor[57,60,61,67-70] (Fig. 4). Therefore, we suggest that the second open reading frame, in addition to encoding the reverse transcriptase, also encodes a protease.

The 3′ portion of the *pol* genes of Rous sarcoma virus (RSV) and Moloney murine leukaemia virus (Mo-MuLV) also encode an endonuclease of M_r~40K[67,71]. Regions of homology between the polymerase genes of RSV[61] and Mo-MuLV[60] with HTLV-III suggest that the *pol* gene of HTLV-III also encodes an endonuclease.

We note that the region of greatest nucleic acid sequence similarity of HTLV-III to HTLV-I[32] is near the 5′ end of the second open reading frame (nucleotides 1,886–1,935) (Fig. 3), but this region does not encode a protein product in HTLV-I[32].

We conclude that the second open reading frame encodes the *pol* gene, the reverse transcriptase and a protease at its 5′-terminus, plus an endonuclease at its 3′-terminus. We note that HTLV-III differs from HTLV-I in the amino-terminal region of the second open reading frame as the corresponding region of HTLV-I cannot encode a functional protease[32,34].

Third Open Reading Frame (*sor*)

The region between the second and fourth open reading frames (nucleotides 4,674–5,780) also includes an open reading frame (nucleotides 4,588–5,196), capable of encoding

性的且能与核酸结合，就像其他逆转录病毒的碱性蛋白那样[34,63,64]。（在*gag*蛋白前体的羧基末端也有一段富含脯氨酸的序列，它可能编码一个小的多肽，与欧罗斯兰及其合著者（个人交流）描述的*gag*蛋白前体相似）。

我们的结论是HTLV-III基因组的第一个长开放阅读框编码*gag*蛋白前体，因为这个开放阅读框的位置靠近病毒基因组的5′端，而且*gag*基因前体的预测序列和观察到的主要衣壳蛋白p24的序列相对应。与其他的逆转录病毒相比较，*gag*前体的结构和HTLV-I以及BLV的更为相似，因为*gag*蛋白的氨基端和主要衣壳蛋白在位置上是紧密相邻的。

第二个开放阅读框（*pol*）

由第二个长开放阅读框（核苷酸1,629~4,673）预测的氨基酸序列与第一个开放阅读框有80个氨基酸的重叠，揭示了该序列多个区域的氨基酸序列与其他逆转录病毒的*pol*基因产物[34,65,66]有相似性（图4）。这些保守区序列与其他逆转录病毒的保守区序列呈共线性，因此我们推测第二个长开放阅读框编码病毒逆转录酶。

HTLV-III *pol*蛋白前体的氨基末端与其他逆转录病毒编码的蛋白酶在氨基酸序列上具有相似性[57,60,61,67-70]（图4）。因此，我们认为第二个开放阅读框除了编码逆转录酶外，还编码蛋白酶。

劳氏肉瘤病毒（RSV）和莫洛尼小鼠白血病病毒（Mo-MuLV）的*pol*基因3′端也编码一个分子量约40K[67,71]的核酸内切酶。RSV[61]和Mo-MuLV[60]与HTLV-III聚合酶基因的同源区表明HTLV-III的*pol*基因也编码一个核酸内切酶。

我们发现，HTLV-III和HTLV-I[32]之间的核酸序列相似度最大的区域位于第二个开放阅读框的5′末端附近（核苷酸1,886~1,935）（图3），但是这个区域在HTLV-I中不编码蛋白[32]。

我们推测第二个阅读框编码*pol*基因，其5′端编码逆转录酶和蛋白酶，同时其3′端编码核酸内切酶。我们认为HTLV-III与HTLV-I在第二个开放阅读框的氨基端的不同之处在于HTLV-I的相应区域不能编码有功能的蛋白酶[32,34]。

第三个开放阅读框（*sor*）

第二和第四开放阅读框（核苷酸4,674~5,780）之间还有一个开放阅读框（核苷酸4,588~5,196）能编码一个203个氨基酸的蛋白，与*pol*基因的3′端有86个核苷

a protein 203 amino acids long, which overlaps with the 3′ end of the *pol* gene by 86 nucleotides. The 3′ portion of this region contains multiple termination codons in all three reading frames and therefore cannot encode a functional polypeptide (Fig. 2). Both the short open reading frame and non-coding region in this part of the HTLV-III genome have been identified by sequence analysis of two independent clonal isolates, BH5 and BH10. In addition, the restriction map of a biologically active HTLV-III proviral DNA clone is indistinguishable from that of the BH10 provirus throughout this region (our unpublished observations with A. Fisher and E. Collalti). For these reasons, we suggest that this unusual sequence is common to replication-competent HTLV-III viruses. We designate the 5′ portion of this sequence *sor* (short open reading frame).

The predicted product of the *sor* region would be a polypeptide of M_r 21K. This amino acid sequence shows no significant homology to any known viral or mammalian cellular genes, but there is a region of DNA sequence similarity in the non-coding region (nucleotides 5,615–5,644) with the sequences found in the *env* gene of HTLV-I[32] (Fig. 3). Also, the overlap in the open reading frame of the second and third open reading frames is like the structure of the envelope genes of HTLV-I[32] and of BLV[33]. Thus, it is possible that the *sor* region of HTLV-III and the flanking non-coding region represent a vestigial *env* gene.

Fourth Open Reading Frame (*env-lor*)

The fourth open reading frame (nucleotides 5,781–8,369) could encode a protein 863 amino acids long. There is a short tandem duplication (nucleotides 6,972–6,986 and 6,987-7,001) in the DNA sequence in this open reading frame in one of the two HTLV-III clones (BH10) sequenced (Fig. 1).

The predicted product of the fourth reading frame, *env-lor*, shares many features in common with the envelope gene precursors of other retroviruses, the most striking of which is a hydrophobic region near the middle of the protein (amino acids 519-534)[32-34,72-75] (Fig. 4). This region of amino acid conservation is preceded in the HTLV-III and other retroviral proteins by an arginine-rich hydrophilic region[37,76] which includes also the processing site for cleavage of the *env* protein precursors into exterior and transmembrane proteins (see Fig. 5).

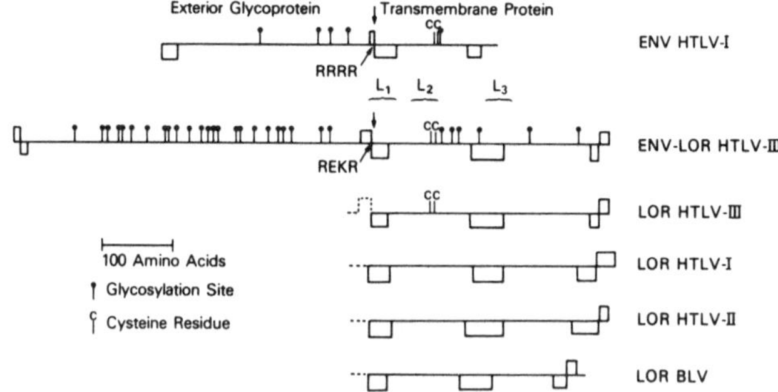

Fig. 5. Structure of *env-lor* product of HTLV-III. The structural features of the predicted *env-lor* product

酸重叠。在三种阅读框中，该区域的3′端均含有多个终止密码子，因此不能编码有功能的多肽（图2）。通过对两个独立的克隆BH5和BH10的序列进行分析，我们识别出HTLV-III基因组这一区域的短开放阅读框和非编码区。另外，在这部分区域上，有生物学活性的HTLV-III前病毒DNA克隆的限制性酶切图谱与BH10病毒前体的限制性酶切图谱非常相似（我们和费希尔以及科拉尔蒂未发表的观测结果）。由于这些原因，我们认为这个不寻常的序列为有复制能力的HTLV-III病毒所共有。我们将这条序列的5′部分命名为*sor*（意为短开放阅读框）。

*sor*区域的预测编码产物是一个相对分子量为21K的多肽。这条氨基酸序列与已知的任何病毒或者哺乳动物细胞基因都没有显著的同源性，但是其非编码区的DNA序列（核苷酸5,615~5,644）与HTLV-I的*env*基因中的序列有相似性[32]（图3）。第二和第三开放阅读框的重叠与HTLV-I[32]和BLV[33]的包膜基因的结构很相似。因此，HTLV-III的*sor*区和其相邻的非编码区可能是一个退化的*env*基因。

第四个开放阅读框（*env–lor*）

第四个开放阅读框（核苷酸5,781~8,369）编码一个863个氨基酸的蛋白。两个已测序的HTLV-III克隆中的一个（BH10）在这个开放阅读框的DNA序列中有一个短的串联倍增序列（核苷酸6,972~6,986和6,987~7,001）（图1）。

第四个开放阅读框*env-lor*的预测产物与其他逆转录病毒的包膜基因前体有许多共同特征，最显著的就是靠近蛋白中间部位的疏水区域（氨基酸519-534处）[32-34,72-75]（图4）。这段具有氨基酸保守性的区域在HTLV-III中通过富精氨酸疏水区得到延长，在其他逆转录病毒的蛋白中也是如此[37,76]，它还包括将*env*蛋白前体切割为膜外和跨膜蛋白的作用位点（见图5）。

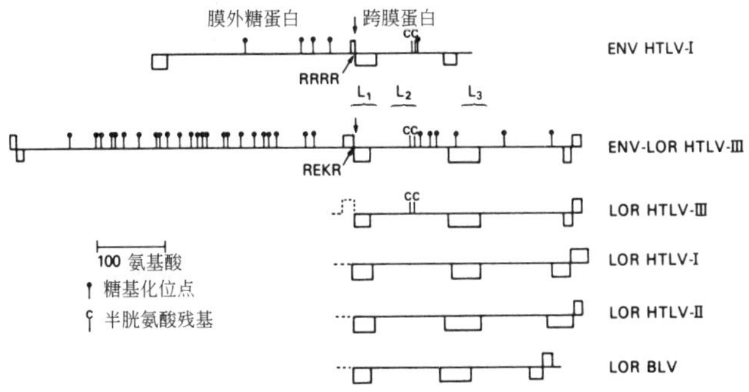

图5. HTLV-III的*env-lor*产物结构。将预测的HTLV-III的*env-lor*产物的结构特点与HTLV-I[32]、II[35]和

of HTLV-III is compared with those of HTLV-I[32], -II[35] and the BLV *lor* products[33,37] and HTLV-I *env* product[32]. Boxes above and below each line represent hydrophilic and hydrophobic regions, respectively[94,95]. Also shown are the positions of potential asparagine-linked glycosylation sites[77] and cysteine residues in the transmembrane proteins. Following asparagine-rich sequences (R, asparagine; E, glutamic acid; K, lysine), the proposed cleavage sites between the exterior glycoprotein and transmembrane protein are shown by an arrow. The scale for amino acid residues is shown at the left. L1, L2 and L3 refer to amino acid sequences which are similar to those of other retroviruses, shown in Fig. 4.

The amino-terminal domain of the translation product of the fourth open reading frame also resembles the *env* protein precursors of other retroviruses[34]. There is a short hydrophobic sequence at the amino-terminus (amino acids 17–37) which may correspond to the signal polypeptide cleaved from the *env* precursor in the maturation process. Moreover, the region of the protein between the putative signal peptide and transmembrane protein is hydrophilic and contains 24 potential asparagine-linked glycosylation sites[77].

The structure of the amino-terminal domain of the predicted transmembrane region resembles closely that of other retroviruses, with respect to both the relative distribution of hydrophobic and hydrophilic residues and the relative location of cysteine residues[34,37,73,78-80] (Fig. 5). The HTLV-III *env-lor* protein, however, differs from envelope proteins of other retroviruses in an additional 180 amino acids at the carboxy-terminus[32,34,37,60,61,73].

In summary, we believe that the fourth open reading frame encodes an *env* precursor, possibly 826 amino acids long, without the signal peptide sequence. In its mature form it is probably cleaved into a large heavily glycosylated exterior membrane protein about 481 amino acids long and a transmembrane protein 345 amino acids long which may be glycosylated (Figs 1,5). The size of these predicted products agrees with the detection of a large glycosylated protein of M_r 120–160K in HTLV-III-infected cells which is probably the glycosylated *env* gene precursor[58] and a smaller, virion-associated gp41 which is probably the transmembrane protein[12,13].

Does a *lor* Gene Exist?

The genomes of HTLV-I, -II and BLV differ from those of the non-acute retroviruses by the presence of a *lor* 3′ to the *env* gene[32,33,35,37,81]. This region of HTLV-I and -II is encoded by a 2-kb mRNA[81,82] which is translated into M_r 42K and 38K proteins, respectively[83,84]. There are several reasons to suspect that the 3′-portion of the fourth open reading frame of HTLV-III also encodes a protein with functions similar to those of the *lor* products of the HTLV-I and -II genomes. Most important is the presence of a 2-kb spliced mRNA species in HTLV-III-infected cells including the 3′, but not the 5′ portion, of the fourth open reading frame (our unpublished observations with S. K. Arya and G. Chan). This mRNA is long enough to include ~1,000 nucleotides of the 3′ portion of the fourth open reading frame.

The overall structure of the *lor* product thought to be synthesized from the carboxy-terminal 1,000 nucleotides of the fourth open reading frame of HTLV-III is similar in

BLV 的 *lor*[33,37] 产物，HTLV-I 的 *env* 产物[32] 相比较。每条线的上面和下面的方框分别代表亲水和疏水区域[94,95]。同时也标明了潜在的天冬酰胺连接型糖基化位点[77] 和跨膜蛋白中半胱氨酸残基的位置。富天冬酰胺序列（R, 天冬酰胺；E, 谷氨酸；K, 赖氨酸）膜外糖蛋白和跨膜蛋白之间可能的切割位点用箭头表示。氨基酸残基的比例尺在左边。L1、L2 和 L3 代表与其他逆转录病毒相似的氨基酸序列，如图 4。

第四开放阅读框翻译产物的氨基端区域也类似于其他逆转录病毒 *env* 蛋白前体[34]。在氨基末端有一个短的疏水区（氨基酸 17~37），它可能是 *env* 前体在成熟过程中切下的信号肽。另外，在预测的信号肽和跨膜蛋白之间的区域是亲水的，还包括 24 个潜在的天冬酰胺连接型糖基化位点[77]。

预测的跨膜区域的氨基末端结构域在结构上由于相关的疏水、亲水残基的分布以及半胱氨酸残基相对位置[34,37,73,78-80]，而与其他逆转录病毒的蛋白非常相似（图 5）。但是 HTLV-III 的 *env-lor* 蛋白比其他逆转录病毒包膜蛋白在羧基端多出 180 个氨基酸[32,34,37,60,61,73]。

总之，我们相信第四开放阅读框编码一个 *env* 前体蛋白，可能有 826 个氨基酸，没有信号肽序列。它可能被切割为一个约 481 个氨基酸的高度糖基化的膜外蛋白和一个 345 个氨基酸的可能被糖基化的跨膜蛋白，从而进入其成熟形式（图 1、图 5）。预测的蛋白大小与在感染的 HTLV-III 的细胞中检测到大的糖基化蛋白的分子量在 120~160K 之间的结果一致，它可能是糖基化的 *env* 前体[58] 和一个小的病毒颗粒相关蛋白 gp41，gp41 可能是跨膜蛋白[12,13]。

是否存在 *lor* 基因？

由于 *env* 基因的 3′端存在 *lor* 基因[32,33,35,37,81]，因此 HTLV-I、-II 和 BLV 的基因组与那些非急性逆转录病毒不同。HTLV-I、-II 的这部分区域由 2,000 个碱基的 mRNA 编码[81,82]，分别翻译成分子量为 42K 和 38K 的蛋白[83,84]。有许多理由怀疑，HTLV-III 第四开放阅读框的 3′端也编码与 HTLV-I 和 -II 基因组上 *lor* 编码产物有相似功能的蛋白。最重要的是，在感染 HTLV-III 的细胞中出现了 2,000 个碱基的 mRNA 剪接产物，它包含第四阅读框的 3′端而缺少 5′端（我们和阿里亚以及尚未发表的结果），这段 mRNA 的长度足够包含第四开放阅读框 3′端 1,000 个左右的核苷酸。

我们认为 *lor* 基因产物是由第四开放阅读框羧基端的 1,000 个核苷酸所合成的，其整体结构与通过 HTLV-I、-II 和 BLV 的 *lor* 区域的核苷酸序列预测的结构相似[32,35,37]。

structure to that predicted by the nucleic acid sequence of the *lor* regions of HTLV-I, -II and BLV[32,35,37]. This similarity includes the size of the *lor* polypeptide, the presence of three hydrophobic regions spaced approximately the same distance apart and the presence of a carboxy-terminal hydrophilic region (Fig. 5). Such a protein may mediate the observed *trans*-activation of transcription of the viral LTR in HTLV-III-infected cells[39], as suggested already for the *lor* product of HTLV-I, -II and BLV[38,85].

There is a sequence of 292 nucleotides between the end of the fourth open reading frame and the beginning of the LTR. The purine-rich sequence preceding the LTR is similar to that of the site of second-strand DNA initiation[86] (Fig. 1).

We propose that the fourth open reading frame encodes two functional polypeptides, one that serves as the precursor for the major envelope glycoprotein and a second derived from the 3′ end of the open reading frame, corresponding to the *lor* gene of other HTLV-BLV viruses.

Heterogeneity in HTLV-III Viruses

Sequences from different clones of HTLV-III allow an analysis of the level of sequence diversity of the virus. A comparison of clones BH8 and BH5 with BH10 (Fig. 1) demonstrates a 0.9% base pair polymorphism in the coding regions of the genome and a 1.8% base pair polymorphism in the non-coding regions. In addition, three 1–3-bp deletions are noted in the non-coding regions, as well as a 15-bp sequence present in one copy in clone BH8, but repeated tandemly in the *env-lor* region of clone BH10. In the coding regions, most nucleotide differences are in the third codon positions; of those that predict amino acid differences between clones, most are non-conservative substitutions. The resultant differences in viral protein function remain to be determined. Of note is the presence of a fifth open reading frame (nucleotides 8,344–8,991), designated 3′ *orf*, present in clone BH8 but truncated in BH10 (Fig. 2).

The heterogeneity among HTLV-III clones shown here could represent sequence divergence developing in culture in a given individual over a period of time, or polymorphic differences in viruses from different individuals. Diversity among different HTLV-III isolates seems to be greater than that between different HTLV-I isolates[6,53]. Different isolates of HTLV-III comprise a spectrum of closely to distantly related viruses. Heterogeneity of proviruses within a given individual, however, has not been noted. Furthermore, a cloned HTLV-III provirus from another individual differs in 18 of 31 restriction enzyme sites mapped compared with clone BH10; the restriction enzyme map of this isolate did not change over a period of several months in culture (our unpublished observations with B. Hahn and G. Shaw). Thus, it is likely that most of the divergence among the HTLV-III clones analysed here represents differences in strains in different individuals. The full extent of this diversity in the HTLV-III family of viruses, however, remains to be determined, and is probably necessary in estimating possible differences in antigenicity of viral proteins. These differences could be important factors in the design of agents useful for viral detection or for therapeutic agents.

相似性包括 *lor* 多肽的长度，三个间隔距离相似的疏水区和一个羧基端亲水区（图 5）。这个蛋白可以介导实验中观察到的 HTLV-III 感染细胞中病毒 LTR 转录的反式激活现象[39]，就像 HTLV-I、-II 和 BLV 的 *lor* 产物[38,85]一样。

在第四开放阅读框的末尾和 LTR 开始之间有一条 292 个核苷酸长度的序列。LTR 前端的富嘌呤序列与 DNA 第二链起始位点相似[86]（图 1）。

我们认为第四开放阅读框编码两个功能性多肽，一个是主要包膜糖蛋白的前体，另一个来自开放阅读框的 3′端，与其他 HTLV-BLV 病毒的 *lor* 基因相对应。

HTLV-III 病毒的异质性

不同克隆的 HTLV-III 序列可以用于分析病毒序列的多样性。比较克隆 BH8 和 BH5 与 BH10 的差异（图 1），发现在基因编码区有 0.9% 的碱基对多态性，在非编码区有 1.8% 的碱基对多态性。另外，在非编码区发现有三处 1~3 个碱基对的缺失现象，同时存在一个 15 碱基对序列，在克隆 BH8 中以单拷贝形式存在，但是在克隆 BH10 的 *env-lor* 区则为串联重复形式。在编码区，大多数核苷酸的差异出现在密码子第三位；不同克隆之间氨基酸序列的差异，大多数是非保守型的替换。由此导致的病毒蛋白功能的差异有待探究。值得注意的是，在 BH8 中出现了定义为 3′*orf* 的第五开放阅读框（核苷酸 8,344~8,991），但是它在 BH10 中被截短了（图 2）。

本文中 HTLV-III 克隆之间表现出来的异质性说明给定的个体培养一段时间后可能产生序列差异，或者来自不同个体的病毒存在多态性差异。不同 HTLV-III 病毒株之间的多样性似乎高于 HTLV-I 之间的多样性[6,53]。虽然尚未发现在一个给定个体内前病毒的异质性，但不同的 HTLV-III 分离株包含了相似度从小到大的病毒谱。此外，与克隆 BH10 相比，另一个体的 HTLV-III 克隆的前病毒 31 个限制性酶切位点中有 18 个存在差异；在几个月的时间里，培养的分离株的限制性酶切图谱都不会改变（我们和哈恩以及肖未发表的结果）。因此，这里分析的大部分 HTLV-III 克隆间的差异很可能代表了不同个体之间的差异。然而，HTLV-III 病毒家族之间差异性的范围仍旧需要鉴定，而且评价病毒蛋白抗原性可能的差异也是必要的。这些差异对于设计病毒检测方法和治疗试剂可能是重要的因素。

Discussion

HTLV-III demonstrates structural and functional similarities to other members of the HTLV-BLV family of retroviruses. Like HTLV-I[32] and BLV[33], the *gag* precursor of HTLV-III lacks a small phosphoprotein situated between the amino-terminal *gag* protein and the major capsid protein. Significant amino acid homologies exist between the major capsid and the basic *gag* proteins of HTLV-III and the corresponding proteins of HTLV-I and BLV. These homologies could account for the immunological cross-reactivity among these proteins[12,13]. Similarities of nucleic acid sequence in several areas of the genome are consistent also with previous hybridization data[26,27]. The presence of conserved amino acid sequences in the transmembrane region of the major envelope glycoprotein also might account for the cross-reactivity in the envelope gene products of HTLV-III and HTLV-I, -II and BLV[12,13,87,88]. The possibility that the 3′ end of the fourth open reading frame may encode a protein similar to the *lor* product of HTLV-I, -II and BLV[33,35,37,81,83,84] might explain the *trans*-acting transcriptional activation common to these viruses[38,39,85]. Alteration of the cellular transcriptional apparatus by an HTLV-III *lor* product may be responsible, at least in part, for the specific cytotoxic effect of HTLV-III on $OKT4^+$ cells.

Previous studies also have highlighted differences between HTLV-III and other retroviruses. Although detectable, nucleic acid homology between the genomes of HTLV-III and other members of the HTLV-BLV family of viruses is low, reflected in differences in primary nucleotide sequence. The structure of the envelope glycoprotein of HTLV-III is different also from that of HTLV-I, -II and BLV, consistent with recent reports of differences in the size of the major envelope glycoproteins[58] as well as the differences in the type of cell receptors recognized by these viruses (refs 89, 90 and our unpublished observations with M. Popovic). Another unusual feature of the HTLV-III genome is the presence of a short open reading frame, *sor*, between the *pol* and *env* genes. The location of the *sor* region suggests that it may be evolutionarily related to the *env* gene. Although the function of this region is unknown, it too may account for some of the unusual cytotoxic and immunological properties of the HTLV-III virus.

We thank Drs Craig Rosen, Joseph Sodroski and Takashi Okamoto, Augusto Cordova, Mr Miltiades C. Psallidopoulos, Dennis Perkins, Steven Untersee and Ms Debra Briggs for assistance with these experiments; Drs George Shaw, Beatrice Hahn and Suresh Arya in the Laboratory of Tumor Cell Biology for providing the HTLV-III clones; Dr Yogi Ikawa and co-workers for sharing unpublished BLV sequences and Ms Debra Lomb and Dr David Lipmann for assistance with the computer analysis. Part of this work was supported by NIH grant CA36974 and ACS grant RD-186.

(**313**, 277-283; 1985)

Lee Ratner[*], William Haseltine[†], Roberto Patarca[†], Kenneth J. Livak[‡], Bruno Starcich[*], Steven F. Josephs[*], Ellen R. Doran[‡], J. Antoni Rafalski[‡], Erik A. Whitehorn[‡], Kirk Baumeister[‡], Lucinda Ivanoff[‡], Stephen R. Petteway Jr[‡], Mark L. Pearson[‡], James A. Lautenberger[§], Takis S. Papas[§], John Ghrayeb[||], Nancy T. Chang[||], Robert C. Gallo[*] & Flossie Wong-Staal[*]

讨 论

HTLV-III 的结构和功能与 HTLV-BLV 家族的其他逆转录病毒具有相似性。就像 HTLV-I[32] 和 BLV[33] 那样，HTLV-III 的 *gag* 前体在 *gag* 蛋白的氨基末端和主要衣壳蛋白之间缺少一个小的磷蛋白。在 HTLV-III 主要衣壳蛋白和碱性 *gag* 蛋白之间存在的显著的氨基酸同源性，在 HTLV-I 与 BLV 的对应蛋白之间也存在。同源性可以解释这些蛋白之间免疫交叉反应 [12,13]。基因组中几个核酸区域的相似性与先前的杂交数据相一致 [26,27]。在主要包膜蛋白的跨膜区域出现保守氨基酸序列能够解释 HTLV-III 与 HTLV-I，-II 和 BLV 的包膜基因产物之间的交叉反应 [12,13,87,88]。第四开放阅读框的 3′ 末端可能编码一个与 HTLV-I，-II 和 BLV[33,35,37,81,83,84] 的 *lor* 基因产物相似的蛋白，这可以解释这些病毒之间普遍存在的反式转录激活 [38,39,85]。HTLV-III 的 *lor* 产物改变细胞的转录机器可能导致了 HTLV-III 对 $OKT4^+$ 的细胞毒性，至少部分如此。

前面的研究也发现 HTLV-III 与其他逆转录病毒有很大不同。虽然能够检测到同源性，但是 HTLV-III 基因组与 HTLV-BLV 病毒家族其他成员核酸序列同源性很低，仅反映在一级核酸序列上的差异。HTLV-III 包膜糖蛋白的结构与 HTLV-I、-II 和 BLV 也不同，这一结果与近来报道的主要包膜糖蛋白大小的 [58] 以及这些病毒识别的细胞受体类型的不同相一致（参考文献 89、90 和波波维奇未发表的结果）。HTLV-III 基因组另一个不寻常的特点是短开放阅读框 *sor* 的出现，它位于 *pol* 和 *env* 之间。*sor* 区域的位置暗示它可能在进化上与 *env* 基因相关。尽管这个区域的功能还未知，但是它足以说明 HTLV-III 病毒不同寻常的细胞毒性和免疫属性。

我们感谢克雷格 · 罗森博士、约瑟夫 · 索德劳斯基博士和冈本鹰司、奥古斯托 · 科多瓦、米尔蒂亚季斯先生、丹尼斯 · 珀金斯、史蒂文 · 翁特塞和德布拉 · 布里格斯女士对这些实验的帮助；感谢肿瘤细胞生物学实验室的乔治 · 肖博士，比阿特丽斯 · 哈恩博士和苏雷什 · 阿里亚博士提供 HTLV-III 克隆；感谢依仪井川博士和他的合作者们分享未发表的 BLV 序列；感谢德布拉 · 隆布女士和戴维 · 李普曼博士进行的计算机分析。这项工作的一部分是由 NIH 基金 CA36974 和 ACS 基金 RD-186 支持的。

（郑建全 翻译；孙军 审稿）

* Laboratory of Tumor Cell Biology, Developmental Therapeutics Program, Division of Cancer Treatment, National Cancer Institute, Bethesda, Maryland 20205, USA

† Dana-Farber Cancer Institute and Department of Pathology, Harvard Medical School, Department of Cancer Biology, Harvard School of Public Health, Boston, Massachusetts 02114, USA

‡ Central Research and Development Department, Experimental Station, E.I. duPont de Nemours, Wilmington, Delaware 19898, USA

§ Laboratory of Molecular Oncology, National Cancer Institute, Frederick, Maryland 21701, USA

‖ Centocor, 244 Great Valley Parkway, Malvern, Pennsylvania 19355, USA

Received 29 November; accepted 14 December 1984.

References:

1. Poiesz, B. J. *et al. Proc. Natl. Acad. Sci. U.S.A.* **77**, 7415-7419 (1980).
2. Kalyanaraman, V. S., Sarngadaharan, M. G., Bunn, P. A., Minna, J. D. & Gallo, R. C. *Nature* **294**, 271-273 (1981).
3. Robert-Guroff, M., Ruscetti, F. W., Posner, L. E., Poiesz, B. J. & Gallo, R. C. *J. exp. Med.* **154**, 1957-1964 (1981).
4. Yoshida, M., Miyoshi, I. & Hinuma, Y. *Proc. Natl. Acad. Sci. U.S.A.* **79**, 2031-2035 (1982).
5. Popovic, M. *et al. Nature* **300**, 63-65 (1982).
6. Gallo, R. C. in *Cancer Surveys* (eds Franks, L. M., Wyke, J. & Weiss, R. A.) 113-159 (Oxford University Press, 1984).
7. Kalyanaraman, V. S. *et al. Science* **218**, 571-573 (1982).
8. Chen, I. S. Y., McLaughlin, J., Gasson, J. C., Clark, S. C. & Golde, D. *Nature* **305**, 502-505 (1983).
9. Gelmann, E. P., Franchini, G., Manzari, V., Wong-Staal, F. & Gallo, R. C. *Proc. Natl. Acad. Sci. U.S.A.* **81**, 993-997 (1984).
10. Popovic, M., Sarngadharan, M. G., Read, E. & Gallo, R. C. *Science* **224**, 497-499 (1984).
11. Gallo, R. C. *et al. Science* **224**, 500-502 (1984).
12. Schupbach, J. *et al. Science* **224**, 503-505 (1984).
13. Sarngadharan, M., Popovic, M., Bruch, L., Schupbach, J. & Gallo, R. C. *Science* **224**, 506-508 (1984).
14. Barre-Sinoussi, F. *et al. Science* **220**, 868-871 (1983).
15. Gallo, R. C. *et al. Proc. Natl. Acad. Sci. U.S.A.* **79**, 4680-4683 (1983).
16. Popovic, M. *et al. Science* **219**, 856-859 (1983).
17. Popovic, M., Lange-Watzin, G., Sarin, P. S., Mann, D. & Gallo, R. C. *Proc. Natl. Acad. Sci. U.S.A.* **80**, 5402-5406 (1983).
18. Gallo, R. C. *et al. Cancer Res.* **43**, 3892-3899 (1983).
19. Wong-Staal, F. *et al. Nature* **302**, 626-628 (1983).
20. Klatzmann, D. *et al. Science* **225**, 59-63 (1984).
21. Chen, I. S. Y., Quan, S. G. & Golde, D. *Proc. Natl. Acad. Sci. U.S.A.* **80**, 7006-7009 (1983).
22. Hoshino, H. *et al. Proc. Natl. Acad. Sci. U.S.A.* **80**, 6061-6065 (1983).
23. Nagy, K., Clapham, P., Cheingong-Popov, R. & Weiss, R. A. *Int. J. Cancer* **32**, 321-328 (1983).
24. Clapham, P., Nagy, K. & Weiss, R. A. *Proc. Natl. Acad. Sci. U.S.A.* **81**, 3082-3086 (1984).
25. Shaw, G. M. *et al. Proc. Natl. Acad. Sci. U.S.A.* **81**, 4544-4548 (1984).
26. Arya, S. K. *et al. Science* **225**, 927-930 (1984).
27. Hahn, B. *et al. Nature* **312**, 166-169 (1984).
28. Shimotohno, K., Golde, D. W., Miwa, M., Sugimura, T. & Chen, I. S. Y. *Proc. Natl. Acad. Sci. U.S.A.* **81**, 1079-1083 (1984).
29. Rho, H. M., Poeisz, B., Ruscetti, F. W. & Gallo, R. C. *Virology* **112**, 355-360 (1981).
30. Yoshida, M., Miyoshi, I. & Hinuma, Y. *Proc. Natl. Acad. Sci. U.S.A.* **79**, 2031-2035 (1982).
31. Rey, M. A. *et al. Biochem. Biophys. Res. Commun.* **111**, 116-133 (1984).
32. Seiki, M., Hattori, S., Hirayama, Y. & Yoshida, M. *Proc. Natl. Acad. Sci. U.S.A.* **80**, 3618-3622 (1983).
33. Sagata, N. *et al. Proc. Natl. Acad. Sci. U.S.A.* (in the press).
34. Haseltine, W. A., Sodroski, J. G. & Patarca, R. *Curr. Topics Microbiol. Immun.* **115** (in the press).
35. Haseltine, W. A. *et al. Science* **225**, 419-421 (1984).
36. Shimotohno, K. *et al. Proc. Natl. Acad. Sci. U.S.A.* **81**, 6657-6661 (1984).
37. Rice, N. R. *et al. Virology* **138**, 82-93 (1984).
38. Sodroski, J., Rosen, C. & Haseltine, W. *Science* **225**, 381-385 (1984).
39. Sodroski, J. *et al. Science* (in the press).
40. Celander, D. & Haseltine, W. *Nature* **312**, 159-162 (1984).
41. Safai, B. *et al. Lancet* i, 1438-1440 (1984).

42. Laurence, J. *et al. New Engl. J. Med.* **311**, 1269-1273 (1984).
43. Gottlieb, M. S. *et al. New Engl. J. Med.* **305**, 1425-1431 (1981).
44. Essex, M. *et al.* in *Human T-Cell Leukemia-Lymphoma Virus* (eds Gallo, R. C., Essex, M. & Gross, L.) 355-362 (Cold Spring Harbor Laboratory, New York, 1984).
45. Miyoshi, I. *et al. Nature* **294**, 770-771 (1981).
46. Yamamoto, N., Okada, M., Koyangi, Y., Kanagi, M. & Hinuma, Y. *Science* **217**, 737-739 (1982).
47. Kinoshita, K. *et al.* in *Adult T-Cell Leukemia Related Diseases* (eds Hakaoka, M., Takatsuki, K. & Shimoyama, M.) 167-184 (Plenum, Tokyo, 1982).
48. Markham, P. D. *et al. Int. J. Cancer* **31**, 413-420 (1983).
49. Groopman, J. E. *et al. Science* **226**, 447-449 (1984).
50. Zagury, D. *et al. Science* **226**, 449-451 (1984).
51. Ho, D. D. *et al. Science* **226**, 451-453 (1984).
52. Gallo, R. C. & Wong-Staal, F. *Blood* **50**, 545-557 (1982).
53. Shaw, G. M. *et al. Science* **226**, 1165-1171 (1984).
54. Starcich, B. *et al. Science* (in the press).
55. Sodroski, J. *et al. Proc. Natl. Acad. Sci. U.S.A.* **81**, 4617-4621 (1984).
56. Sagata, N., Yasunaga, T., Ogawa, Y., Tsuzuka-Kawamura, J. & Ikawa, Y. *Proc. Natl. Acad. Sci. U.S.A.* **81**, 4751-4754 (1984).
57. Weiss, R., Teich, N., Varmus, H. & Coffin, J. (eds) *RNA Tumor Viruses* (Cold Spring Harbor Laboratory, New York, 1982).
58. Kitchen, L. *et al. Nature* **312**, 367-369 (1984).
59. Oroszlan, S. *et al. Proc. Natl. Acad. Sci. U.S.A.* **79**, 1291-1294 (1984).
60. Shinnick, T. M., Lerner, R. A. & Sutcliffe, J. G. *Nature* **293**, 543-548 (1981).
61. Schwartz, D. E., Tizard, R. & Gilbert, W. *Cell* **32**, 853-869 (1983).
62. Copeland, T. D., Oroszlan, S., Kalyanaraman, V. S., Sarngadaharan, M. G. & Gallo, R. C. *FEBS Lett.* **162**, 390-395 (1983).
63. Bolognesi, D. P., Luftig, R. & Shaper, J. H. *Virology* **56**, 549-564 (1973).
64. Smith, B. J. & Bailey, J. M. *Nucleic Acids. Res.* 7, 2055-2072 (1979).
65. Chiu, I-M., Tronick, S. R., Schlom, J. & Aaronson, S. A. *Science* **223**, 364-370 (1984).
66. Patarca, R. & Haseltine, W. A. *Nature* **309**, 728 (1984).
67. Levin, J. G., Hu, S. C., Rein, A., Messer, L. I. & Gerwin, B. I. *J. Virol.* **51**, 470-478 (1984).
68. Von der Helm, K. & Duesberg, P. H. *Proc. Natl. Acad. Sci. U.S.A.* **72**, 614-618 (1975).
69. Vogt, P., Wight, W. & Eisenman, R. *Virology* **98**, 154-167 (1979).
70. Dittmar, K. J. & Moelling, J. K. *J. Virol.* **28**, 106-118 (1978).
71. Schiff, R. D. & Grandgenett, D. P. *J. Virol.* **28**, 279-291 (1978).
72. Herr, W. *J. Virol.* **49**, 471-478 (1984).
73. Sodroski, J. *et al. Science* **225**, 378-381 (1984).
74. Cianciolo, G. J., Kipnis, R. J. & Snyderman, R. *Nature* **311**, 515 (1984).
75. Haseltine, W. A. & Patarca, R. *Nature* **309**, 728 (1984).
76. Lenz, J., Crowther, R., Straceski, A. & Haseltine, W. A. *J. Virol.* **42**, 519-529 (1982).
77. Bahl, O. P. & Shah, R. H. in *The Glycoconjugates* Vol. 1 (eds Horowitz, M. I. & Pigman, W.) 385-422 (Academic, New York, 1977).
78. Pinter, A. & Fleissner, E. *Virology* **83**, 417-422 (1983).
79. Segrest, J. P. & Feldman, R. J. *J. Molec. Biol.* **87**, 853-858 (1974).
80. Koch, W., Zimmerman, W., Oliff, A. & Friedrich, R. *J. Virol.* **49**, 828-840 (1984).
81. Wachsman, W., Shimotohno, K., Clark, S. C., Golde, D. W. & Chen, I. S. Y. *Science* **226**, 177-179 (1984).
82. Franchini, V., Wong-Staal, F. & Gallo, R. C. *Proc. Natl. Acad. Sci. U.S.A.* **81**, 6207-6211 (1984).
83. Lee, T. H. *et al. Science* **226**, 58-60 (1984).
84. Slamon, D. J., Shimotohno, K., Cline, M. J., Golde, D. W. & Chen, I. S. Y. *Science* **226**, 61-63 (1984).
85. Rosen, C., Sodroski, J., Kettman, R., Burney, A. & Haseltine, W. A. *Science* (in the press).
86. Czernilofsky, A. P. *et al. Nucleic Acids Res.* **8**, 2967-2984 (1980).
87. Essex, M. *et al. Science* **220**, 859-862 (1983).
88. Essex, M. E. *et al. Science* **221**, 1061-1064 (1983).
89. Dalgleish, A. *et al. Nature* **312**, 763-766 (1984).
90. Weiss, R. A. *et al.* in *Retroviruses in Human Lymphoma/ Leukemia* (ed. Miwa, M.) (Japan Sci. Soc. Press, Tokyo, in the press).
91. Maxam, A. M. & Gilbert, W. *Meth. Enzym.* **65**, 499-560 (1980).
92. Sanger, F., Nickelen, S. & Coulson, A. R. *Proc. Natl. Acad. Sci. U.S.A.* **74**, 5463-5467 (1977).
93. Mount, S. M. *Nucleic Acids Res.* **10**, 459-472 (1982).
94. Kyte, J. & Doolittle, R. F. *J. Molec. Biol.* **157**, 105-120 (1981).
95. Hopp, T. P. & Woods, K. R. *Proc. Natl. Acad. Sci. U.S.A.* **78**, 3824-3825 (1981).

2.5-Myr *Australopithecus boisei* from West of Lake Turkana, Kenya

A. Walker *et al.*

Editor's note

From the late 1970s, palaeoanthropologists began to realise that the story of human evolution was far from simple. The tale was shaped less like a ladder than a bush, with many evolutionary "experiments", some of which terminated without issue. This was certainly the case for the "robust" australopithecines, a group of primitive, small-brained, large-jawed hominids specialised for eating tough vegetable matter. Here Alan Walker and colleagues add a very robust-looking and extremely ancient "Black Skull" from Kenya to the well-known *Australopithecus robustus* form from South Africa and *A. boisei* from East Africa. Eventually named *Australopithecus aethiopicus*, this Black Skull illustrated the complexity of human evolution and how little we still knew about it.

Specimens of *Australopithecus boisei* have been found in 2.5-Myr-old sediments west of Lake Turkana, Kenya. The primitive morphology of these early *A. boisei* suggests that robust and hyper-robust *Australopithecus* developed many of their common features in parallel and further that *A. africanus* is unlikely to have been ancestral to *A. boisei*.

THE "hyper-robust" hominid *Australopithecus boisei* is well-known from several East African Plio-Pleistocene deposits dated between 2.2 and 1.2 Myr (refs 1, 2). It has been thought of variously, as: the northern vicar of the equally well-known *A. robustus*[3]; the extremely specialized end-member of the robust clade[4]; an already developed species which immigrated from another, unknown area[5]; and as representing individuals at the large end of a single *Australopithecus* species that also encompasses *A. robustus* and *A. africanus*[6].

There is a growing consensus that the east and south African samples are different enough to allow them to be placed in separate species[1-4]. The type specimen of *A. boisei* is Olduvai Hominid 5[7,8].

These two robust species are placed in the genus *Paranthropus* by some authors[9]. Although recognizing that the two known samples overlap in time, some have advocated an ancestor–descendant relationship with *A. robustus* giving rise to *A. boisei*. Perhaps the most compelling recent evidence for this last view is Rak's exemplary study of the structure and function of the australopithecine face[4]. He has followed an evolutionary scheme in which the origins of the robust clade are in *A. africanus*, which is thereby removed from consideration as a

肯尼亚图尔卡纳湖以西250万年前的南方古猿鲍氏种

沃克等

编者按

自20世纪70年代晚期起，古人类学家们开始意识到人类的演化历史远非人们想象的那么简单。与其说人类演化是按阶梯进行，不如说这段历史是由许多演化“实验”构成的繁芜庞杂的灌木丛，这些“实验”中许多还没有结果便终结了。南方古猿“粗壮种”显然是其中之一。这是一群原始、颅容量小、具有适应摄食粗糙植物的大颌骨的人科动物。本文中艾伦·沃克与其同事们将产自肯尼亚的外观十分粗壮且极为古老的“黑色颅骨”归入产自南非的著名南方古猿粗壮种和产自东非的南方古猿鲍氏种。最终，这件黑色颅骨被命名为南方古猿埃塞俄比亚种，它揭示了人类演化的复杂性以及我们对这段历史的浅见寡识。

在肯尼亚图尔卡纳湖西侧距今250万年的沉积物中发现了南方古猿鲍氏种标本。这些早期南方古猿鲍氏种原始的形态表明，粗壮的和超级粗壮的南方古猿发育了许多与南方古猿非洲种平行演化出的共同特征，并且进一步表明了南方古猿非洲种不可能是南方古猿鲍氏种的祖先。

众所周知，“超级粗壮的”人科动物——南方古猿鲍氏种产自距今220万年至120万年的几处东非上新世–更新世沉积物（参考文献1、2）。对于鲍氏种有多种解释：同等著名的南方古猿粗壮种的北方替代物种[3]；粗壮种分支极端特化的终端成员[4]；从另一未知地区迁徙而来的已经独立的物种[5]；包括南方古猿粗壮种和南方古猿非洲种在内的单一的南方古猿种的末期代表性个体[6]。

越来越多的人认同，东非和南非标本与其他地区标本的不同之处足以将他们划分到不同的物种中[1-4]。南方古猿鲍氏种的模式标本是奥杜威人科5号[7,8]。

有些学者将这两个粗壮种物种归入傍人属中[9]。尽管有些人认识到这两种已知的标本在时间上有重叠，但他们还是主张是南方古猿粗壮种衍生出了南方古猿鲍氏种，二者具有祖裔关系。可能最近对于后一个观点的最有说服力的证据就是拉克对南方古猿面部结构和功能的典型研究[4]。他所遵循的演化图解中，粗壮种分支起源于南方古猿非洲种，因此不能把粗壮种看作是人类祖先[10]。这种方案还没有得到普遍

human ancestor[10]. This scheme has not found universal acceptance[11,12].

Localities

Prospecting was carried out in 1985 in Pliocene sediments to the west of Lake Turkana, Kenya. It led to the recovery of two *A. boisei* specimens. A cranium and a partial mandible were discovered at two separate localities: sediments in the Lomekwi and Kangatukuseo drainages (see Fig. 1) at approximately 3°45′ N, 35°45′ E.

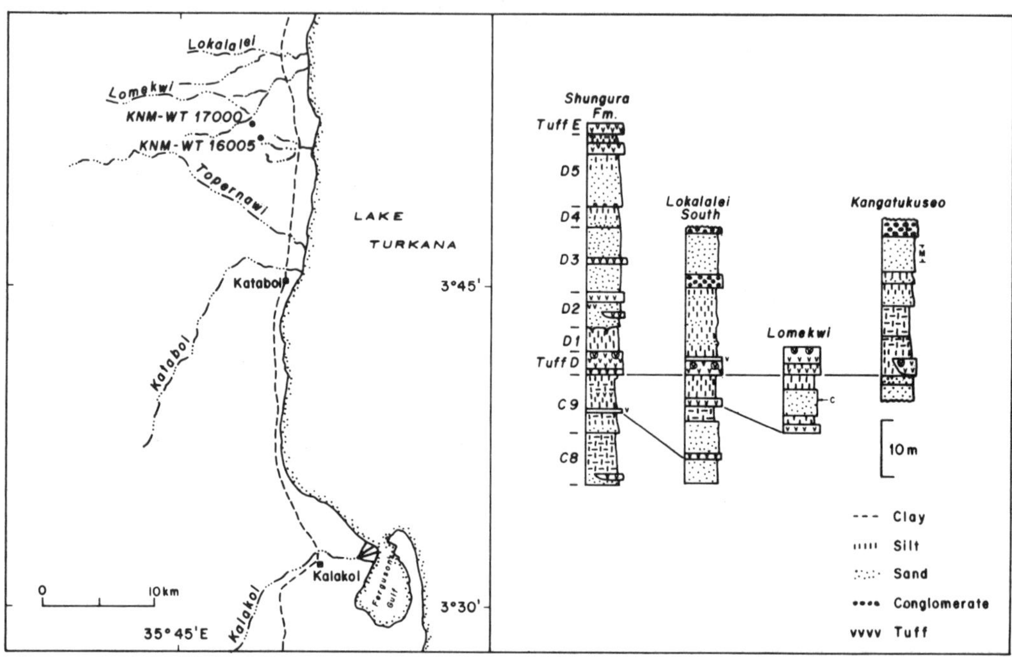

Fig. 1. Map and sections showing geographical and stratigraphic positions of KNM-WT 17000 and KNM-WT 16005. C, cranium; M, mandible.

The general dip of strata is to the east in the Lomekwi drainage, but the strata at Lomekwi I from which the cranium was derived are deformed into a syncline by drag along a fault that truncates the section about 50 m east of the site. Several small faults cut the section west of the site, but it has been possible to link the short sections together by analysis of volcanic ash layers in the sequence. The total thickness of section immediately surrounding the site is less than 15 m, and the volcanic ash layer which caps the section is compositionally indistinguishable from Tuff D of the Shungura Formation. Earlier[13], this tuff was referred to the informally designated Upper Burgi Tuff of the Koobi Fora Formation. With more numerous analyses it is now clear that the Upper Burgi Tuff also correlates with Tuff D, and an additional correlation datum is provided between the Shungura Formation and the Koobi Fora Formation. Tuff D has been dated at 2.52 ± 0.05 Myr[14]. This age is an average computed from samples from the Shungura Formation in Ethiopia, and from the correlative unit at Kangatukuseo. The cranium derives from a level 3.8 m below Tuff D. Around 10 m below Tuff D there is a second ash layer. In the Lokalalei drainage, about 4 km northwest of the site, there are three ash layers exposed in

认同[11,12]。

地　　点

1985年，研究人员在肯尼亚图尔卡纳湖西侧的上新世沉积物中进行了勘查工作。这次工作出土了两件南方古猿鲍氏种标本。在两处地点——大约位于北纬3°45′，东经35°45′处的罗埋奎流域和坎噶图库涩流域的沉积物中（见图1）分别发现了一件颅骨和一件不完整的下颌骨。

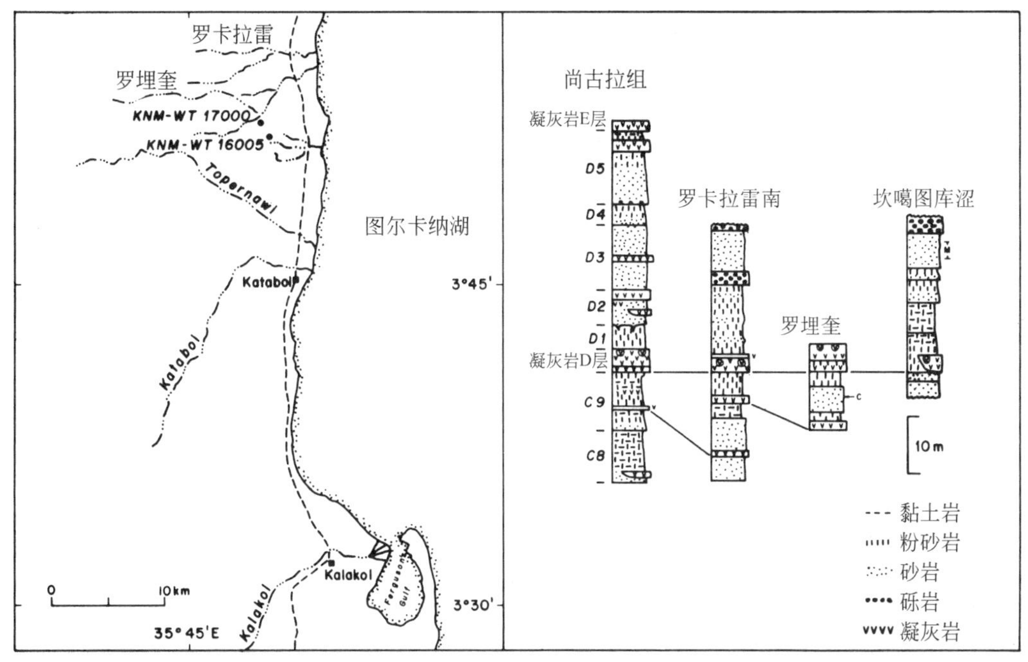

图1. KNM-WT 17000和KNM-WT 16005的地理位置和地层位置及柱状剖面。C，颅骨；M，下颌骨。

地层总体倾斜至罗埋奎流域的东部，但是发现颅骨的罗埋奎I号地点的地层由于沿断层的拖曳作用而形变成为向斜，该断层截去了这一遗址东面约50米的剖面。几处小断层则切去了该遗址西侧的剖面，但还是有可能通过分析地层层序中的火山灰层来将这些短的剖面联系起来。直接围绕该遗址的剖面总厚度不到15米，覆盖在该剖面之上的火山灰层在组分上与尚古拉组的凝灰岩D层几乎区分不出来。早先[13]，该凝灰岩层被认为是库比福勒组中非正式命名的上布基凝灰岩层。随着更多的分析，现在已经明确了上布基凝灰岩层也与凝灰岩D层层位相当，并且又提出了一个尚古拉组和库比福勒组之间的对比基准。经测定，凝灰岩D层的年龄为252万年 ±5万年[14]。这一年龄是根据在埃塞俄比亚的尚古拉组采集的样本和在坎噶图库涩研究地的相当层位采集的样本计算出来的平均值。颅骨是从凝灰岩D层之下3.8米的地层中得到的。凝灰岩D层之下约10米处还有一个灰层。在该遗址西北部约4千米处的罗

sequence. The lowest correlates with a tuff in submember C9 of the Shungura Formation, and the upper two correlate with the two ash layers exposed at the site of the cranium. On this basis, the cranium is shown to lie within strata correlative with submember C9 of the Shungura Formation. Based on the K/Ar chronology of the Shungura Formation and scaling on the basis of constant sedimentation rates there, the cranium is estimated to be 2.55 Myr old. Using palaeomagnetic polarity boundaries as the basis of chronological placement, the age of the cranium would be ~2.45 Myr, as there is a slight discordance between the two chronologies[15]. Therefore, we believe that the age of the cranium can be confidently stated as 2.50 ± 0.07 Myr, including all errors. The sediments from which the cranium derives are overbank deposits of a large perennial river, probably the ancestral Omo.

The mandible from Kangatukuseo III derives from a level about 19 m above Tuff D. Tuff E is not exposed in Kangatukuseo, but it is exposed in the northern part of the Lomekwi drainage, and in the southern Lokalalei drainage. There, sediments correlative with Member D of the Shungura Formation are at least 26 m thick. On this basis, the mandible is assigned to the central part of Member D of the Shungura Formation, the best age estimate for which is 2.45 ± 0.05 Myr. The section is faulted east of the mandible site, and older sediments are exposed along the drainage east of that point. In fact, the entire section exposed in Kangatukuseo lies within an interval from ~25 m above Tuff D to 5 m below that tuff. The mandible was collected from a sandstone layer ~6 m thick, deposited by a large river system. Thin basalt pebble conglomerates intercalated in this part of the section show that the site lay near the boundary between sediments deposited by this large river, and alluvial fan deposits derived from the west.

The interval of section from which the australopithecine specimens were collected has also yielded over 200 fossils representing more than 40 other mammalian species (Table 1), including the skeleton of a ground-dwelling colobine and a relatively complete camel mandible. Bovids are the most common elements of the fauna at this level and most represent species that are otherwise known from the lower portion of the Omo Shungura succession. But whereas alcelaphines and impalas predominate at lower horizons west of Lake Turkana, reduncine bovids are the commonest fossils at the australopithecine sites reported here and are taxonomically different from those recovered slightly higher in the sequence. *Elephas recki shungurensis* is the common elephant at the localities considered here while *Notochoerus scotti* and *Kolpochoerus limnetes* are the common suids. Although all three taxa have lengthy Pliocene distributions, the West Turkana specimens closely match samples of these species from Omo Shungura Members C and D—thus supporting the estimate of age derived from the tuff analyses. The age estimate is corroborated by the apparent absence of the elephantids *Loxodonta adaurora*, *Loxodonta exoptata* and *Elephas recki brumpti*, and of the suids *Nyanzachoerus kanamensis*, *Notochoerus euilus*, *Kolpochoerus afarensis* and *Potamochoerus* sp., all of which occur in older horizons in the upper reaches of the Laga Lomekwi. Also missing from the australopithecine-bearing assemblage are the reduncine bovid *Menelikia lyrocera* and equids of the genus *Equus*, both of which occur higher in the sequence.

卡拉雷流域，有三个灰层依次出露。最底层灰层与尚古拉组亚段 C9 中的一个凝灰岩层相当，上面两层与在发现颅骨的遗址处出露的两个灰层相当。基于这个对比，证明该颅骨的层位与尚古拉组亚段 C9 相当。根据尚古拉组的钾 / 氩年表以及对此处恒定沉积速率的测算，估计该颅骨的年龄为 255 万年。使用古地磁极性界限作为测年依据，得出该颅骨的年龄约为 245 万年，这种差异是由于两种测年方法之间存在微小的差异[15]。因此，考虑到所有的误差，我们可以确信该颅骨的年龄为 250 万年 ±7 万年。出土该颅骨的沉积物是一条大型常流河的河漫滩沉积物，这一常流河可能是古时的奥莫河。

在坎噶图库涩 III 发现的下颌骨是从位于凝灰岩 D 层之上约 19 米处的层位得到的。凝灰岩 E 层在坎噶图库涩没有出露，但是在罗埋奎流域的北部以及罗卡拉雷流域南部都有出露。在那里，与尚古拉组 D 段层位相当的沉积物至少有 26 米厚。根据这一点，该下颌骨被划分到了尚古拉组 D 段的中部，对其最佳的年代估计值是距今 245 万年 ±5 万年。该剖面在发现下颌骨的地点的东侧发生断层，沿该点东侧的流域有更为古老的沉积物出露。事实上，在坎噶图库涩出露的整个剖面都位于从凝灰岩 D 层之上约 25 米到凝灰岩 D 层之下 5 米之间。该下颌骨是从由一个大型水系沉积下来的厚约 6 米的砂岩层中发掘出来的。薄层玄武岩质砾岩嵌入到该剖面的这一部分中，这表明该处遗址位于这条大河沉积形成的沉积物与来自西部的冲积扇沉积物的分界线附近。

采集到南方古猿标本的剖面的间隔带中也发现了 200 多件分属 40 余种其他哺乳动物的化石（表 1），包括一副地表疣猴的骨架和一件相对完整的骆驼下颌骨。牛科动物是在该层位中最常见的动物群成员，也是已知的在奥莫尚古拉组下部发现的其他物种中最具代表性的种类。但是狷羚和黑斑羚在图尔卡纳湖西侧较低的层位占优势，在本文报道的南方古猿遗址处，小苇羚牛类是最常见的化石，与稍高层位上发掘的标本在分类学上存在差异。尚古拉巨象是研究地常见的一种象，而斯氏南方猪和沼泽裂尾猪是常见的猪科动物。尽管这三个类群在上新世时期都具有漫长的分布史，但是图尔卡纳西侧的标本与这些来自奥莫尚古拉组 C 段和 D 段的标本非常匹配——由此支持了根据凝灰岩分析得到的年代估计值。该年代估计值通过象科动物非洲祖象、扇羽非洲象和一种古菱齿象及猪科动物肿颊猪、南方猪、阿法裂尾猪和河猪属未定种等的明显缺失得到证实，所有这些物种在拉伽罗埋奎上游较古老的层位上都出现了。包含南方古猿的化石组合中还缺失了小苇羚牛类 *Menelikia lyrocera* 和（马科的）马属动物化石，这两者都是在较高的地层中出现。

Table 1. Fossil mammals from the new australopithecine localities west of Lake Turkana

Theropithecus brumpti	*Giraffa* sp.
T. cf. brumpti	*Aepyceros* sp.
Cercopithecidae large	*Connochaetes* sp.
Cercopithecidae medium	*Parmularius cf. braini*
Cercopithecidae small	Alcelaphini medium
Papionini medium/large	Alcelaphini small
Parapapio ado	*Menelikia* sp.
Paracolobus mutiwa	*Kobus sigmoidalis*
Australopithecus boisei	*Kobus* sp. A
Hyaena sp.	*Kobus* sp. B
Homotherium sp.	*Kobus* sp. C
Felidae	*Kobus* sp. D
Viverridae	*Kobus* sp. E
Carnivora indet.	*Kobus* sp. F
Deinotherium bozasi	Reduncini indet.
Elephas recki shungurensis	Reduncini large
Hipparion hasumensis	Reduncini medium
Ceratotherium sp.	Reduncini small
Diceros bicornis	*Tragelaphus nakuae*
Notochoerus scotti	*Tragelaphus* sp.
Kolpochoerus limnetes	Bovini
Hexaprotodon protamphibius	*Gazella aff. granti*
Hippopotamus imagunculus	*Antidorcas recki*
Camelus sp.	Antilopini
Sivatherium maurusium	

Specimens

KNM-WT 17000 is an adult cranium with the following parts missing: all of the tooth crowns except a half molar and the right P^3; some facial bone fragments which have spalled off the infilled maxillary sinus; most of the frontal processes and temporal plates of the zygomatics; the zygomatic arches themselves; a large part of the frontal and parietals superiorly (but a piece of the sagittal crest on the anterior part of the parietal is preserved); parts of both pterygoid regions inferiorly and the posterior part of the maxilla and palate on the right side; and the inferior part of the nuchal region of the occipital. There is no bilateral asymmetry and all bony contacts are sharp. There is no evidence of any plastic deformation and the brain case has retained its spheroidal shape (Fig. 2).

表 1. 图尔卡纳湖西侧南方古猿新遗址出土的哺乳动物化石

布兰普特兽猴	长颈鹿未定种
布兰普特兽猴相似种	高角羚未定种
大型长尾猴	角马未定种
中型长尾猴	*Parmularius cf. braini*
小型长尾猴	中型狷羚亚科
中 / 大型狒狒	小型狷羚亚科
阿拉伯狒狒	*Menelikia* 未定种
傍狨猴	S 型弯角水羚
南方古猿鲍氏种	水羚属未定种 A
缟鬣狗未定种	水羚属未定种 B
似剑虎未定种	水羚属未定种 C
猫科	水羚属未定种 D
灵猫科	水羚属未定种 E
食肉目科属种未定	水羚属未定种 F
恐象	小苇羚属种未定
尚古拉巨象	大型小苇羚
哈苏门三趾马	中型小苇羚
白犀未定种	小型小苇羚
双角犀	*Tragelaphus nakuae*
斯氏南方猪	林羚未定种
沼泽裂尾猪	牛科
原两栖六齿河马（中新古河马）	葛氏瞪羚近亲种
Hippopotamus imagunculus	*Antidorcas recki*
骆驼未定种	羚羊科
西瓦兽	

标　本

KNM-WT 17000 是一件成人颅骨，它的如下部分丢失了：除了半颗臼齿和右 P^3 外的所有齿冠；一些从已被填充的上颌窦剥落的面骨碎片；大部分额突和颧骨的颞板；颧弓自身；额骨的大部分和顶骨上部（但是在顶骨前部的一块矢状嵴保存了下来）；两侧翼状区下部与上颌骨后部的部分骨骼及右侧腭骨；枕骨颈区的下部。此颅骨双侧对称，所有骨质接触部位都很清晰。没有证据表明存在任何的塑性变形，因此该脑颅保留下了其球形形状（图 2）。

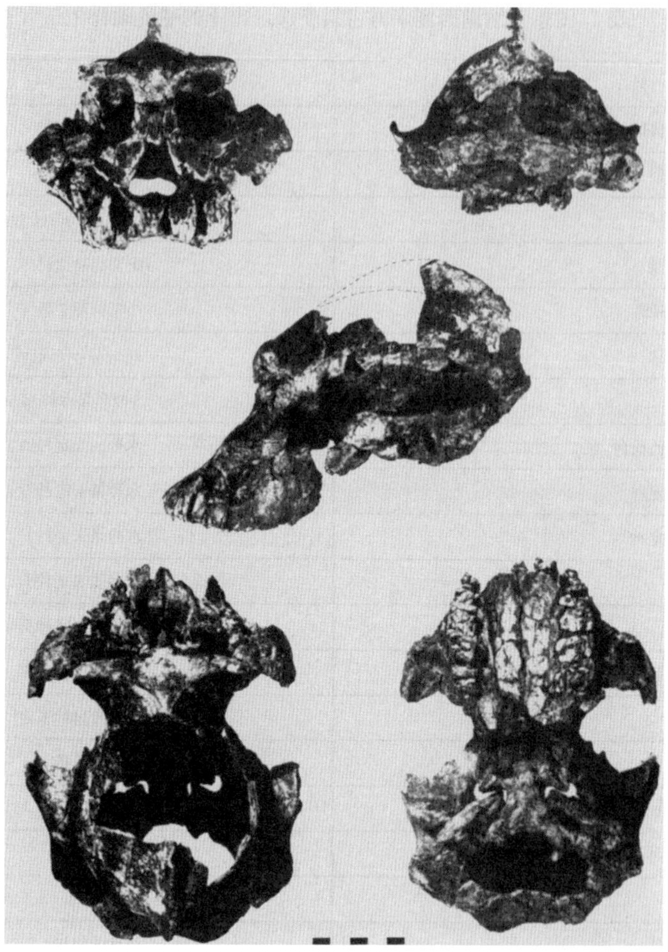

Fig. 2. Anterior, posterior, left lateral, superior and inferior views of cranium KNM-WT 17000. Scale in cm.

It is a massively-built cranium with a very large facial skeleton, palate and large cranial base, but with a small brain case. The palate and cranial base are roughly the same size as in Olduvai Hominid (O.H.) 5. The cranial base is about the same size as, but the palate is slightly larger than in, KNM-ER 406 (ref. 16). The cranial capacity is 410 ml (mean of five determinations by water displacement with a standard error of 4.32). This measurement is probably accurate since the orbital plates of the frontals, the cribiform plate region of the ethmoid, one anterior clinoid and both posterior clinoid processes are preserved together with the rest of the cranial base. The missing cranial vault fragments can be reconstructed with fair certainty by following the internal contours all around them. This is the smallest published cranial capacity for any adult fossil hominid, although A.L. 162-28 from Hadar[17] must have been smaller. Given the massive face and palate combined with a small brain case, it is not surprising that the sagittal crest is the largest ever in a hominid. Further, the sagittal crest joins completely to compound temporal-nuchal crests with no intervening bare area[18]. The foramen magnum position is far forward as in other robust *Australopithecus* specimens[19].

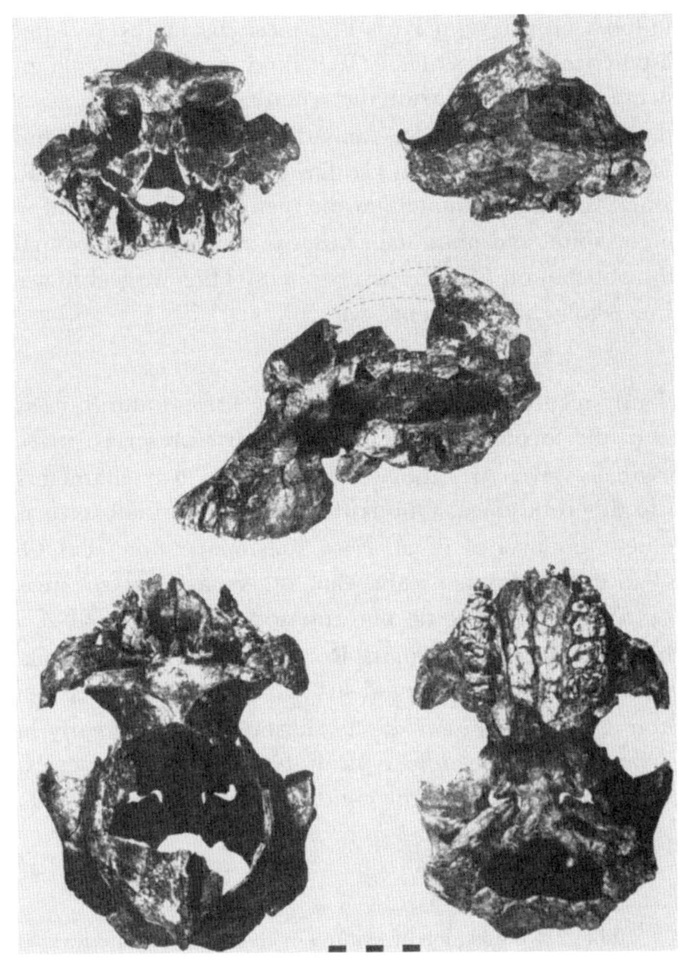

图 2. 颅骨 KNM-WT 17000 的前、后、左侧、上、下视图。标尺单位为厘米。

这是一件厚重的颅骨，具有很大的面部骨骼、腭骨和大的颅底，但是脑颅很小。腭骨和颅底与奥杜威人科（OH）5 号的尺寸大致相同。该颅底与 KNM-ER 406（参考文献 16）的大小几乎相同，但是腭骨稍大。其颅容量为 410 毫升（通过水置换法测得的五次测量值的平均值，标准误差为 4.32）。该测量值应该是准确的，因为额骨的眶板、筛骨的筛板区、一个前床突和两个后床突都与颅底的其余部分一起被保留了下来。丢失的颅骨碎片可以根据其周围所有的内部轮廓非常准确地予以重建。这是目前已公布的所有成年人科化石中颅容量最小的标本，尽管在哈达尔[17]发现的 A.L.162-28 肯定更小。鉴于该标本具有较大的面部和腭骨以及小型的脑颅，其矢状嵴在已发现的人科动物中最大就不足为奇了。此外，矢状嵴与颞－颈复合嵴完全接合在一起，中间没有裸区[18]。枕骨大孔的位置与其他粗壮的南方古猿属标本一样，位置都很靠前[19]。

The one complete tooth crown, right P^3, is 11.5 mesiodistally by 16.2 buccolingually. This is bigger mesiodistally (md) than O.H. 5 (10.9) and smaller buccolingually (bl) (17.0)[7,8]. These dimensions are completely outside the recorded range for *A. robustus* (9.2–10.7 md, 11.6–15.2 bl) and at the high end of the range for *A. boisei* (9.5–11.8 md, 13.8–17.0 bl)[20]. Only the largest *A. boisei* mandibles found so far (for example, KNM-ER 729 and 3230[21]) would fit this cranium. It is unfortunate that the region of the occipital which would show the grooves for the occipital and marginal sinuses is missing, but the small sigmoid sinuses appear to have no contribution from transverse ones. Thus we feel that enlarged occipital and a marginal sinuses may have been present.

Most of the previously recorded differences between *A. boisei* and *A. robustus* have involved greater robustness in the former. In fact for those parts preserved in KNM-WT 17000, the definitions originally given by Tobias[8] include only two characters that cannot be simply attributed to this robustness. One is that the supraorbital torus is "twisted" along its length. Subsequent discoveries of *A. boisei* specimens show that O.H. 5 is extreme in its supraorbital torus development and that others are not so "twisted". The other character is that *A. boisei* palates are deeper anteriorly than those of *A. robustus*, in which they tend to be shallow all along the length. Recently, Rak[4] has undertaken a study of the australopithecine face and has documented structural differences between the faces of all four species. Skelton *et al.*[22] have just made a cladistic analysis of early hominids; Table 2 lists some of the characteristics given as typical of *Australopithecus* species by these authors (see ref. 22 and refs therein) as well as the condition found in KNM-WT 17000. For most features the new specimen resembles *A. boisei*.

Table 2. Compilation of features of *Australopithecus* species

Feature	*A. afarensis*	*A. africanus*	*A. robustus*	*A. boisei*	KNM-WT 17000
Position of I^2 roots relative to nasal aperture margins	Lateral	Medial	Medial	Medial	Medial
Divergence of temporal lines relative to lambda	Below	Above	Above	Above	Below
Lateral concavity of nuchal plane	Present	Absent	Absent	Absent	Probably present
Depth of mandibular fossa	Shallow	Deep	Deep	Deep	Shallow
Temporal squama pneumatization	Extensive	Weak	Weak	Weak	Extensive
Flat, shallow palate	Present	Absent	Absent	Absent	Present
Subnasal prognathism	Pronounced	Intermediate	Reduced	Reduced	Pronounced
Orientation of tympanic plate	Less vertical	Intermediate	Vertical	Vertical	Intermediate
Flexion of cranial base	Weak	Moderate	Strong	Strong	Weak
Relative sizes of posterior to anterior temporalis	Large	Intermediate	Small	Small	Large
Position of postglenoid process relative to tympanic	Completely anterior	Variable	Merge superiorly	Merge superiorly	Completely anterior

那颗完整的齿冠右 P^3，近远中径（md）是 11.5 mm，颊舌径（bl）是 16.2 mm。其近远中径比 OH 5（10.9 mm）的要大，而颊舌径比 OH 5（17.0 mm）[7,8] 要小些。这些尺寸完全不在已记录的南方古猿粗壮种的范围（md 9.2~10.7 mm，bl 11.6 ~15.2 mm）之内，而处于南方古猿鲍氏种范围（md 9.5~11.8 mm，bl 13.8~17.0 mm）[20] 的上限。只有迄今发现的最大的南方古猿鲍氏种的下颌骨（例如，KNM-ER 729 和 3230[21]）与该颅骨尺寸对应。不幸的是，这件标本上能够显示出枕骨窦和边缘窦沟的枕骨区不见了，而从横窦处，小型的乙状窦似乎没有什么贡献。因此我们认为扩大的枕骨和边缘窦可能已经存在了。

之前所记录的南方古猿鲍氏种和南方古猿粗壮种之间的差异，大部分都提到前者更加粗壮。实际上，对于那些 KNM-WT 17000 中保存下来的部分，最初托拜厄斯 [8] 给出的描述仅包括两种特征，这两种特征不能简单地被归为粗壮性。一个特征是眶上圆枕沿其长轴是“扭曲的”。后来发现的南方古猿鲍氏种标本表明 OH 5 的眶上圆枕的发育是一种极端情况，并且其他标本的眶上圆枕并没有如此“扭曲”。另一个特征是南方古猿鲍氏种的腭在前部比南方古猿粗壮种的要深，后者似乎在其整个长轴上都较浅。最近，拉克 [4] 正在进行一项对南方古猿面部的研究，并且已经证明了所有四个物种的面部之间存在结构性差异。斯凯尔顿等 [22] 刚刚对早期人科进行了支序分析；表 2 列出了由这些作者给出的南方古猿物种的一些典型特征（见参考文献 22 及其中的参考文献）以及在 KNM-WT 17000 中发现的一些特征。新标本的大部分特征与南方古猿鲍氏种相似。

表 2. 南方古猿属物种特征汇总

特征	南方古猿阿法种	南方古猿非洲种	南方古猿粗壮种	南方古猿鲍氏种	KNM-WT 17000
I^2 根部相对于鼻孔边缘的位置	外侧	内侧	内侧	内侧	内侧
λ 形的颞线分歧	下方	上方	上方	上方	下方
项平面侧凹	存在	缺失	缺失	缺失	可能存在
下颌窝深度	浅	深	深	深	浅
鳞颞部气腔形成	广泛	弱	弱	弱	广泛
平、浅的腭	存在	缺失	缺失	缺失	存在
鼻下颌前突	显著	中等	减弱	减弱	显著
鼓板方向	略垂直	中度垂直	垂直	垂直	中度垂直
颅底屈曲	弱	中度	强	强	弱
颞骨从前至后的相对尺寸	大	中等	小	小	大
下颌窝后突相对于鼓骨的位置	完全在前部	位置多样	在上部融合	在上部融合	完全在前部

Continued

Feature	*A. afarensis*	*A. africanus*	*A. robustus*	*A. boisei*	KNM-WT 17000
Tubular tympanic	Present	Intermediate	No	No	Intermediate
Articular eminence	Weak	Intermediate	Strong	Strong	Weak
Foramen magnum relative to tympanic tips	Anterior	Intermediate	Anterior	Anterior	Anterior
Coronally placed petrous temporals	No	Variable	Yes	Yes	Yes
Distance between M^1 and temporomandibular joint	Long	Variable	Short	Short	Long
P^3 outline	Asymmetric	Intermediate	Oval	Oval	Oval
Relative size of C	Very large	Medium	Small	Small	Small
Anterior projection of zygomatic	Absent	Intermediate	Strong	Very Strong	Very Strong
Height of masseter origin	Lowest	Intermediate	High	High	High
Canine jugum separate from margin of pyriform aperture	Yes	Variable	No	No	No
Distinct subnasal and intranasal parts of clivus	Yes	Intermediate	No	No	No
Relative size of post-canine teeth	Moderate	Large	Very large	Very large	Very large
Robustness of zygomatic arches	Moderate	Strong	Very Strong	Very Strong	Very Strong
Common origin of zygomatic arch	M^1/P^4	P^4	P^4	P^3	P^3
C jugum	Prominent	Pronounced	Reduced	Lost	Lost
Inclination of nuchal plane	Steep	Less steep	—	Variable	Less steep
Compound temporonuchal crest	Present	Absent	Males only	Males only	Present
Asterionic notch	Present	Absent	Absent	Absent	Probably Present
Medial inflection of mastoids	Strong	Reduced	Reduced	Reduced	Reduced
Anterior facial pillars	Absent	Present	Present	Absent	Absent
Length of nuchal plane relative to occipital	Long	Intermediate	Long	Long	Long
Braincase relative to face	—	High	Low	Low	Low
Nasals wide above frontonasal suture	—	No	Yes	Yes	Yes
Nasoalveolar gutter present	No	No	Yes	Yes	Yes
Infraorbital foramen high	Yes	Yes	No	No	No
Maxillary fossula present	No	Yes	Yes	No	No
Inferior orbital margins soft laterally	—	No	Yes	No	Yes
Greatest orbital height	—	Middle	Middle	Medial	Middle
Foramen magnum heart-shaped	—	No	No	Yes	Yes

There are some features of KNM-WT 17000 that differ from all other "hyper-robust" specimens as well as from robust ones. The most obvious and important is the prognathic

续表

特征	南方古猿阿法种	南方古猿非洲种	南方古猿粗壮种	南方古猿鲍氏种	KNM-WT 17000
管状鼓板	存在	中等	无	无	中等
关节隆起	弱	中等	强	强	弱
枕大孔相对于鼓部尖端的位置	前部	中部	前部	前部	前部
颞岩骨冠状位排列	否	多样	是	是	是
M^1 与颞颌关节的距离	长	距离不等	短	短	长
P^3 轮廓	非对称	中度对称	卵形	卵形	卵形
C 的相对尺寸	非常大	中等	小	小	小
颧骨前突度	缺失	中度	强	非常强	非常强
咬肌起点高度	最低	中等	高	高	高
犬齿轭从梨状孔边缘脱离	是	形式多样	否	否	否
斜坡的鼻下与鼻腔部分独特	是	中等	否	否	否
颊齿的相对尺寸	中等	大	非常大	非常大	非常大
颧弓的粗壮性	中等	强	非常强	非常强	非常强
颧弓的共同起点	M^1/P^4	P^4	P^4	P^3	P^3
犬齿隆突	突出	显著	减弱	丢失	丢失
项平面的倾斜程度	陡	略陡	—	多样	略陡
颞 / 项复合脊	存在	缺失	仅雄性个体存在	仅雄性个体存在	存在
星穴	存在	缺失	缺失	缺失	可能存在
乳突内弯曲	强	减弱	减弱	减弱	减弱
前部面柱	缺失	存在	存在	缺失	缺失
项平面相对于枕骨的长度	长	中等	长	长	长
脑颅相对于面部的位置	—	高	低	低	低
额鼻缝上方的鼻部宽	—	否	是	是	是
存在鼻槽沟	否	否	是	是	是
眶下孔较高	是	是	否	否	否
存在上颌窝	否	是	是	否	否
下眶缘侧向柔软	—	否	是	否	是
最大眶高位置	—	位于中间	位于中间	位于内侧	位于中间
枕骨大孔心形	—	否	否	是	是

KNM-WT 17000的有些特征与所有其他的“超级粗壮”标本以及粗壮标本的都不同。最显著且最重要的是突颌的面中部和下部区域。从俯视图看，所有其他的粗

mid- and lower facial region. In superior view all other robust crania are so orthognathic that only a small part of the incisor region projects past the supraorbital tori. In KNM-WT 17000 the midface projects strongly past the tori and the anterior maxilla projects well forwards as a square muzzle. In summary, we regard this specimen as part of the *A. boisei* clade and view its differences from the younger sample as being either primitive, or part of normal intraspecific variation that has not been documented before, or both.

Mandible KNM-WT 16005 has the body preserved to the M_3 alveoli on the left and the M_2 alveoli on the right. The base is missing. The incisors and canines are, as judged from their roots, relatively very small and the post-canine teeth relatively very large. In its size, shape and proportions, KNM-WT 16005 is very similar to the Peninj mandible[23], except that the P_4 and M_1 of the latter are a little larger and the M_2 a little smaller than this specimen. KNM-WT 16005 is smaller than the mandible which KNM-WT 17000 possessed. Tooth measurements are given in Table 3, and the specimen is shown in Fig. 3.

Table 3. Tooth measurements of KNM-WT 16005 (mm)

	Mesiodistal	Buccolingual
Left P_3	10.7	13.8
P_4	(12.0)	(15.0)
M_1	15.7	14.3
M_2	(17.0)	16.7

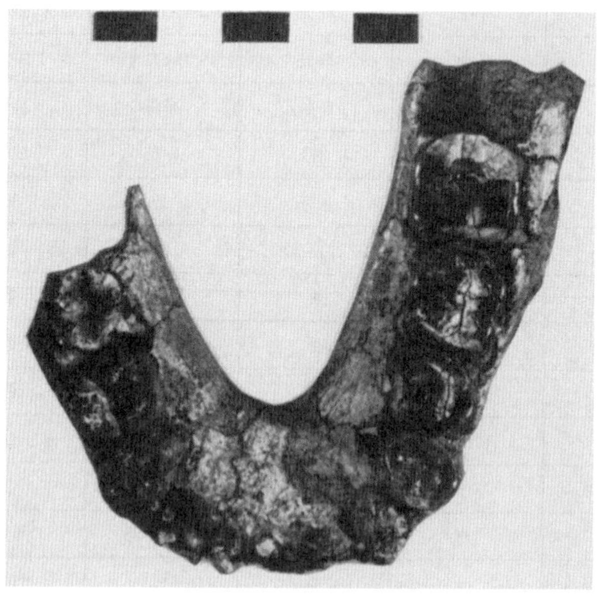

Fig. 3. Occlusal view of mandible KNM-WT 16005. Scale in cm.

壮颅骨都为直颌型，以至于只有一小部分门齿区突出越过眶上圆枕。在 KNM-WT 17000 中，面中部非常突出，越过圆枕，前面的上颌向前突出成一个正方形的吻突。总之，我们认为该标本属于南方古猿鲍氏种分支的一例，并且将其与年代较晚的标本比较后，认为其与其他标本的不同点较为原始，或者，该标本是尚无资料记载的正常种内差异的一部分，或许二者兼有。

下颌骨 KNM-WT 16005 的下颌体在左侧保存至 M_3 齿槽，在右侧保存至 M_2 齿槽。下颌骨的基底丢失了。正如依据齿根判断的那样，门齿和犬齿相对较小，而颊齿相对较大。KNM-WT 16005 与佩宁伊下颌骨 [23] 在大小、形状和比例上很相似，除了后者的 P_4 和 M_1 比该标本大些，以及 M_2 比该标本小些。KNM-WT 16005 比与 KNM-WT 17000 对应的下颌骨小。表 3 给出了牙齿尺寸，图 3 展示了该标本。

表 3. KNM-WT 16005 的牙齿测量数据（毫米）

	近远中径	颊舌径
左 P_3	10.7	13.8
左 P_4	(12.0)	(15.0)
左 M_1	15.7	14.3
左 M_2	(17.0)	16.7

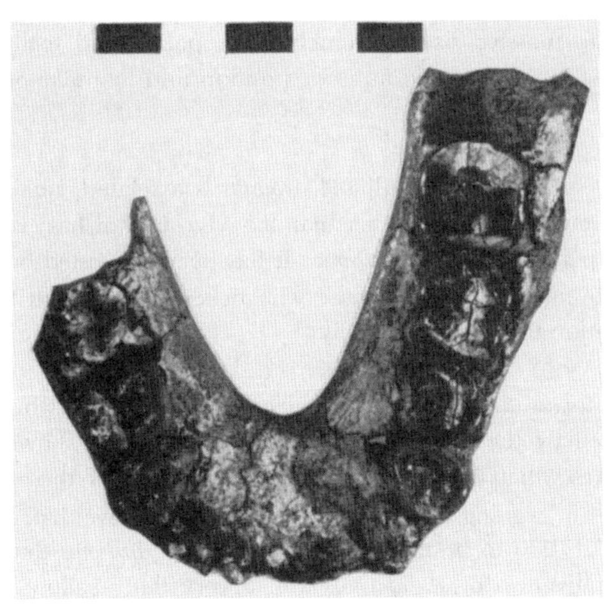

图 3. 下颌骨 KNM-WT 16005 嚼面视。标尺单位为厘米。

Although future finds may show that KNM-WT 17000 is well within the range of variation of *A. boisei*, it is also possible that the differences will prove sufficient to warrant specific distinction. If the latter proves to be the case we suggest that some specimens from the same time period and from the same sedimentary basin (for example, Omo 1967-18 from the Shungura Formation) will be included in the same species. Omo 1967-18 is the type specimen of *Paraustralopithecus aethiopicus* Arambourg and Coppens[24]. In our view, the appropriate name then would be *Australopithecus aethiopicus*.

Conclusions

The new specimens show that the *A. boisei* lineage was established at least 2.5 Myr ago and further that, in robustness and tooth size, at least some members of the early population were as large as any later ones. Although one authority suggested that the robust australopithecines became smaller in skull and tooth size with time[25], most have pointed out that the available sample showed the opposite, that within *A. boisei* there has been an increase in size and robustness of the skull and jaws. This was apparently an artefact of sampling and is no longer correct.

Although recognizing that at least some populations of *A. robustus* and *A. boisei* overlapped in their time ranges, Rak[4] hypothesized that the former was ancestral to *A. boisei*. This is no longer tenable. *A. robustus* shares with younger examples of *A. boisei* several features which are clearly derived from the condition seen in KNM-WT 17000. These include the cresting pattern—with the emphasis on the anterior and middle parts of the temporalis muscle—the orthognathism and the deep temporomandibular joint with strong eminence. At the same time KNM-WT 17000 is clearly a member of the *A. boisei* lineage, as demonstrated by the massive size, extremely large palate and teeth, the build of the infraorbital and nasal areas and the anterior position and low take-off of the zygomatic root.

Therefore, this new specimen shows that *A. robustus* is a related, smaller species that was either derived from ancestral forms earlier than 2.5 Myr and/or has evolved independently in southern Africa, perhaps from *A. africanus*. It has been suggested before that *A. robustus* was derived from *A. africanus*[4], but by those who believed *A. robustus* then gave rise to *A. boisei*—an interpretation that is now unlikely.

The idea that *A. africanus* was the earliest species of a lineage in which *A. robustus* led to *A. boisei* is challenged by the new evidence. KNM-WT 17000 shows that all known *A. africanus* share features which are derived relative to it. Many of these same features were cited by White *et al.*[20] in arguing that *A. afarensis* is more primitive than *A. africanus*. Features showing KNM-WT 17000 to be more primitive than *A. africanus* that were also used to distinguish the primitiveness of *A. afarensis* are: a very flat, shallow palate; pronounced subnasal prognathism; compound temporal/nuchal crests; sagittal crest with emphasis on posterior fibres of the temporalis muscle; an extensively pneumatized squamous temporal, which in KNM-WT 17000 is 11.5 thick just above the supraglenoid gutter; small occipital

尽管将来的发现可能会显示 KNM-WT 17000 正好处于南方古猿鲍氏种的变异范围之内，但是也有可能这些差异足以成为物种差异的根据。如果事实证明是后者的话，那么我们主张，有些来自同一时间段、同一沉积盆地的标本（例如，来自尚古拉组的奥莫 1967-18）可被归入同一物种中。奥莫 1967-18 是由阿朗堡和科庞[24]命名的埃塞俄比亚副猿的模式标本。那么在我们看来，其适当的名字应该是南方古猿埃塞俄比亚种。

结　论

这些新标本表明南方古猿鲍氏种谱系是在至少 250 万年前建立起来的，并且在粗壮性和牙齿大小方面，至少早期种群的一些成员与后来的成员是一样大的。尽管有位权威人士认为，粗壮种南方古猿的颅骨和牙齿会随着时间变小[25]，但是大部分学者指出，已得到的标本表明了相反的情况，即南方古猿鲍氏种的颅骨和颌骨的大小及其粗壮性都有所增加。这显然是由于取样造成的假象，不再是正确的。

尽管意识到至少有些南方古猿粗壮种和南方古猿鲍氏种群在其时间分布范围上有重叠，但拉克[4]还是假设前者是南方古猿鲍氏种的祖先。现在这一观点已经站不住脚。南方古猿粗壮种与年代较近的南方古猿鲍氏种有些特征是相同的，这些特征明显是从 KNM-WT 17000 中见到的特征衍生而来的。这些特征包括颅骨的嵴型——着重颞肌的前部和中部——直颌型和具有显著隆起的较深的颞颌关节。同时，KNM-WT 17000 明显是南方古猿鲍氏种谱系中的一员，通过较大的标本尺寸、极大的腭和牙齿、眶下和鼻区的构造以及颧骨根的前位和低的出发点可以证明。

因此，这一新发现的标本表明，南方古猿粗壮种是一个有亲缘关系的、较小的物种，该物种或者起源于早于 250 万年前的祖先，而且（或者）是在南非（可能是由南方古猿非洲种）独立进化而来的。以前就有人提出，南方古猿粗壮种起源于南方古猿非洲种[4]，但是现在对于那些相信南方古猿粗壮种后来演化成了南方古猿鲍氏种的人来说，这种解释是不太可能的。

南方古猿非洲种是谱系中最早的物种、在该谱系中南方古猿粗壮种产生了南方古猿鲍氏种，这一观点受到了新证据的挑战。KNM-WT 17000 表明所有已知的南方古猿非洲种都具有衍生而来的与其相关的特征。在关于南方古猿阿法种比南方古猿非洲种更加原始的证明中，怀特等[20]引用过很多这些相同的特征。表明 KNM-WT 17000 比南方古猿非洲种更加原始的特征也被用来区分南方古猿阿法种的原始性，这些特征包括：一个非常平坦的、浅的腭板；显著的鼻下突颌；颞 / 项复合嵴；以附着颞肌后纤维为主的矢状嵴；一个广泛布有气腔的鳞状的颞部，在 KNM-WT 17000 中有 11.5 mm 厚，其位置刚好在盂上沟之上；与项平面关联的小枕骨；侧向颅骨基底

relative to nuchal plane; pneumatization of lateral cranial base to produce strongly flared parietal mastoid angles; shallow and mediolaterally broad mandibular fossae; tympanics completely posterior to the postglenoid process. In KNM-WT 17000, the asterionic region is poorly preserved, but an asterionic notch was probably present, which is an additional feature also cited to demonstrate the primitiveness of *A. afarensis*.

Other primitive features found in KNM-WT 17000, but not known or much discussed for *A. afarensis*, are: very small cranial capacity; low posterior profile of the calvaria; nasals extended far above the frontomaxillary suture and well onto an uninflated glabella; low calvaria with receding frontal squama; and extremely convex inferolateral margins of the orbits such as found in some gorillas. Thus there are many features in which KNM-WT 17000 is more primitive than *A. africanus* and similar to *A. afarensis* yet KNM-WT 17000 is clearly a member of the *A. boisei* clade. Further, although the dating of the South African sites is admittedly still imprecise and populations of ancestral species may survive a speciation event, the time sequence of the fossils is becoming increasingly less supportive of the idea of an *africanus–robustus–boisei* lineage.

Finally, it is striking that many of the features of this cranium shared by *A. afarensis* are primitive and not found in *A. robustus* or later specimens of *A. boisei*. These primitive features shared by KNM-WT 17000 and *A. afarensis* are almost exclusively confined to the calvaria, despite the largely complete face of KNM-WT 17000 and the existence of several partial facial specimens at Hadar. However, not one individual adult specimen of *A. afarensis* preserves a facial skeleton attached to a calvaria. This observation raises two alternatives: first, that these features are primitive to the Hominidae and therefore not of great taxonomic value in determining relationships among hominids; second, that, as Olson[26] has suggested, the specimens identified as *A. afarensis* include two species, one of which gives rise directly to *A. boisei*. Whatever the final answer, these new specimens suggest that early hominid phylogeny has not yet been finally established and that it will prove to be more complex than has been stated.

We thank the Government of Kenya and the Governors of the National Museums of Kenya. This research is funded by the National Geographic Society, Washington, D.C., the Garland Foundation and the National Museums of Kenya. F.H.B. was funded for analyses by NSF grant BNS 8406737. We thank Bw. Kamoya Kimeu and his team for invaluable help. Many colleagues, but especially M. G. Leakey and P. Shipman, helped in various ways.

(**322**, 517-522; 1986)

A. Walker*, R. E. Leakey†, J. M. Harris‡ & F. H. Brown§

* Department of Cell Biology and Anatomy, Johns Hopkins University School of Medicine, 725 North Wolfe Street, Baltimore, Maryland 21205, USA

† National Museums of Kenya, PO Box 40658, Nairobi, Kenya

‡ Los Angeles County Museum of Natural History, 900 Exposition Boulevard, Los Angeles, California 90007, USA

§ Department of Geology and Geophysics, University of Utah, Salt Lake City, Utah 84112, USA

的气腔形成以产生强烈伸展的顶骨乳突角；浅的沿内外侧方向宽阔的下颌窝；完全在下颌后突之后的鼓室。在 KNM-WT 17000 中，星穴区保存状况差，但是星穴凹可能还存在，这是用以证实南方古猿阿法种原始性的另外一个特征。

已在 KNM-WT 17000 中发现、但在南方古猿阿法种中还未知或者探讨不多的其他原始特征包括：非常小的颅容量；低的颅盖后轮廓；鼻骨延伸远至额颌缝之上并恰好位于隆起的眉间之上；具有退化的额鳞的低的颅盖；极端凸出的眼眶下侧边缘，正如在某些大猩猩中所见到的。因此 KNM-WT 17000 有许多特征比南方古猿非洲种的更加原始，与南方古猿阿法种相似，然而 KNM-WT 17000 很显然是南方古猿鲍氏种分支的一员。而且，尽管不可否认南非遗址年代的测定仍不精确，祖先种的某些种群可能在一个物种形成事件后保留下来，然而这些化石的时序变得越来越无法为南非古猿非洲种 – 粗壮种 – 鲍氏种这一谱系提供支持。

最后，值得注意的是，该颅骨与南方古猿阿法种共有的许多特征都是原始的，这些原始特征在南方古猿粗壮种或者后来的南方古猿鲍氏种标本中都没有发现过。尽管我们有 KNM-WT 17000 基本完整的面部骨骼以及若干在哈达尔发现的部分面部骨骼标本，这些 KNM-WT 17000 和南方古猿阿法种共有的原始特征几乎只限于颅盖。然而，没有一个南方古猿阿法种的成年个体标本保存有与颅盖衔接的面部骨骼。这一观察结果产生了两种可能：首先，这些特征对人科来讲比较原始，因此在确定人科动物间的关系时没有太大的分类学价值；第二，正如奥尔森[26]提出的那样，这些被鉴定为南方古猿阿法种的标本包括两个物种，其中一个直接演化成南方古猿鲍氏种。无论最终的答案是什么，这些标本暗示着早期人科的系统发育尚未最终建立，并且系统发育将被证明其比已确定的关系更加复杂。

我们向肯尼亚政府和肯尼亚国家博物馆的管理者表示表示感谢。该研究得到了华盛顿哥伦比亚特区的国家地理学会、加兰基金会和肯尼亚国家博物馆的资助。布朗的研究受到了美国国家科学基金会 BNS 8406737 项拨款的资助。我们感谢卡莫亚 · 基梅乌及其团队提供的宝贵协助。许多同事，尤其是利基和希普曼都以不同方式提供了帮助。

（刘皓芳 翻译；董为 审稿）

Received 7 April; accepted 2 July 1986.

References:

1. Howell, F. C. in *Evolution of African Mammals* (eds Maglio, V. J. & Cooke, H. B. S.) 154-248 (Harvard, Cambridge, 1978).
2. Coppens, Y. in *Morphologie Evolutive—Morphogenese du Crane at Origine de l'Homme* (ed. Sakka, M.) 155-168 (CRNS, Paris, 1981); *Bull Mem. Soc. Anthrop. Paris* **3**, 273-284 (1983).
3. Tobias, P. V. *A. Rev. Anthrop.* **2**, 311-334 (1973).
4. Rak, Y. *The Australopithecine Face*, 1-169 (Academic, New York, 1983); *Am. J. Phys. Anthrop.* **66**, 281-288 (1985).
5. Boaz, N. T. in *New Interpretations of Ape and Human Ancestry* (eds Ciochon, R. L. & Corruccini, R. S.) 705-720 (Plenum, New York, 1983).
6. Wolpoff, M. H. *Palaeoanthropology* 131-157 (Knopf, New York, 1980).
7. Leakey, L. S. B. *Nature* **184**, 491-493 (1959).
8. Tobias, P. V. *Olduvai Gorge* Vol. 2 (Cambridge University Press, 1967).
9. Robinson, J. T. in *Evolution and Hominisation* (ed. Kurth, G.) 120 (Fischer, Stuttgart, 1962); in *Evolutionary Biology* (eds Dobzhansky, T., Hechi, M. K. & Steere, W.) (Appleton-Century Crofts, New York, 1967).
10. Johanson, D. C. & White, T. D. *Science* **203**, 321-330 (1979).
11. Tobias, P. V. *Palaeont. afr.* **23**, 1-17 (1980).
12. Olson, T. R. in *Aspects of Human Evolution* (ed. Stringer, C. B.) 99-128 (Taylor & Francis, London, 1981).
13. Harris, J. M. & Brown, F. H. *National Geographic Res.* **1**, 289-297 (1985).
14. Brown, F. H., McDougall, L., Davies, T. & Maier, R. in *Ancestors: the Hard Evidence* (ed. Delson, E.) 82-90 (Liss, New York, 1985).
15. Hillhouse, J. J., Cerling, T. E. & Brown F. H. *J. Geophys. Res.* (in the press).
16. Leakey, R. E. F., Mungai, J. M. & Walker, A. C. *Am. J. Phys. Anthrop.* **35**, 175-186 (1971).
17. Kimbel, W. H., Johanson, D. C. & Coppens, Y. *Am. J. Phys. Anthrop.* **57**, 453-499 (1982).
18. Dart, R. A. *Am. J. Phys. Anthrop.* **6**, 259-284 (1948).
19. Dean, M. C. & Wood, B. A. *Am. J. Phys. Anthrop.* **59**, 157-174 (1982).
20. White, T. D., Johanson, D. C. & Kimbel, W. H. *S. Afr. J. Sci.* 77, 445-470 (1981).
21. Leakey, M. G. & Leakey, R. E. *Koobi Fora Research Project*, 100, 169 (Clarendon, Oxford, 1978).
22. Skelton, R. R., McHenry, H. & Drawhorn, G. M. *Curr. Anthrop.* **27**, 21-43 (1986).
23. Leakey, L. S. B. & Leakey, M. D. *Nature* **202**, 5-7 (1964).
24. Arambourg, C. & Coppens, Y. *C. r. hebd. Séanc. Acad. Sci., Paris* **265**, 589-590 (1967).
25. Robinson, J. T. *Early Hominid Posture and Locomotion* (University of Chicago Press, 1972).
26. Olson, T. R. in *Ancestors: the Hard Evidence* (ed. Delson, E.) 102-119 (Liss, New York, 1985).

肯尼亚图尔卡纳湖以西250万年前的南方古猿鲍氏种

Forty Years of Genetic Recombination in Bacteria: A Fortieth Anniversary Reminiscence

Joshua Lederberg

Editor's Note

In 1944, Canadian-born researcher Oswald Avery and colleagues demonstrated that DNA was the molecule responsible for inheritance. Inspired by this discovery, US molecular biologist Joshua Lederberg began investigating his own hypothesis that, instead of reproducing asexually and passing down exact copies of genetic information, bacteria can sometimes enter a sexual phase in which genetic information is combined and merged. Here he reminisces on the experiments that were to later earn him and colleague Edward Tatum the 1958 Nobel Prize, which demonstrated that bacteria can mate and exchange genes. The results, which challenged dogma, prompted lively discussion when they were presented at the 1946 Cold Spring Harbor Symposium, but bacterial genetic recombination was soon incorporated into the textbooks and mainstream molecular biological research.

Between April and June 1946, Joshua Lederberg and Edward L. Tatum carried out a series of experiments that proved that bacteria can exchange their genes by sexual crossings. The experiments were reported in *Nature* just 40 years ago[1]. In the following pair of articles*, Joshua Lederberg first provides a personal reminiscence of the circumstances of the discovery and then, together with Harriet Zuckerman, considers it as a possible case of "postmature" scientific discovery.

Lederberg in 1945

A Fortieth Anniversary Reminiscence

IN September 1941, when I started as an undergraduate at Columbia University, the genetics of bacteria was still a no-man's-land between the disciplines of genetics

* Only the first of these two articles is reproduced here.

细菌遗传重组四十年——四十周年纪念

乔舒亚·莱德伯格

编者按

1944年加拿大裔研究者奥斯瓦尔德·埃弗里和他的同事证明DNA分子对遗传起作用。美国分子生物学家乔舒亚·莱德伯格受这一结论的启发，开始研究他自己提出的假说，即细菌有时可以进行有性生殖，在此过程中基因信息重组和融合，而不是通过无性生殖来传递精确拷贝的过程。在本文中，他回顾了证明细菌可交配并交换基因的实验，这项工作为莱德伯格和同事爱德华·塔特姆赢得了1958年的诺贝尔奖。当这些挑战传统法则的结果在1946年冷泉港学术研讨会上被提出时，引起了热烈讨论，而不久之后，细菌遗传重组理论就被纳入教科书和主流的分子生物学研究中。

从1946年4月到6月，乔舒亚·莱德伯格和爱德华·塔特姆开展的一系列实验证实了细菌可以通过有性杂交交换其基因。这些实验结果发表在40年前的《自然》杂志上[1]。在随后文章中*，乔舒亚·莱德伯格首次对这一发现的详细情况进行了个人回顾，在这之后他又与哈丽雅特·朱克曼共同对此进行了评价，将之称为一个“晚熟”的科学发现。

1945年的莱德伯格

四十周年纪念

1941年9月，当我开始在哥伦比亚大学读本科的时候，细菌遗传学仍然是介于遗传学和（医学的）细菌学之间的一个未知领域。关于“细菌是否像所有其他生物

* 本书只收录了两篇论文中的第一篇。

and (medical) bacteriology. The question whether "bacteria have genes, like all other organisms" was still unanswered, indeed rarely asked. My own thoughts at that moment lay elsewhere. I looked forward to a career in medical research applying chemical analysis to problems like cancer and the malfunctions of the brain. Cytotoxicology then appeared to be the most promising approach to cell biochemistry. It was Francis J. Ryan (d. 1963) who turned my attention to the sharper tools of genetics.

Ryan had spent 1941–42 as a postdoctoral fellow at Stanford University, where he had met G. W. Beadle and E. L. Tatum (d. 1975), and had become fascinated with their recent invention of nutritional mutations in *Neurospora* as a tool for biochemical genetic analysis[2]. Although working on a fungus like *Neurospora* did not go down smoothly in a Department of Zoology as at Columbia, where Ryan had accepted an instructorship, he established a laboratory to continue these studies. In January 1943 I was fortunate to get a job in his laboratory assisting in the preparation of media and handling of *Neurospora* cultures. Ryan's personal qualities as a teacher and the setting of serious research, discussion with him, other faculty members and graduate students in the department nourished my education as a scientist. On 1 July 1943, I was called to active duty in the United States Naval Reserve, and my further months at Columbia College alternated with spells of duty at the United States Naval Hospital, St Albans, Long Island. There, in the clinical parasitology laboratory, I had abundant opportunity to observe the life cycle of *Plasmodium vivax*. This experience dramatized the sexual stages of the malaria parasite, which undoubtedly sensitized me to the possibility of cryptic sexual stages in other microbes (perhaps even bacteria). In October 1944, I was reassigned to begin my studies at Columbia Medical School; but I continued working with Ryan at the Morningside Heights campus.

Discovery

The important biological discovery of that year, by Avery, MacLeod and McCarty, was the identification of DNA as the substance responsible for the *Pneumococcus* transformation[3]. This phenomenon could be viewed as the transmission of a gene from one bacterial cell to another; but such an interpretation was inevitably clouded by the obscure understanding of bacterial genetics at the time. Avery's work, at the Rockefeller Institute in New York, was promptly communicated to Columbia biologists by Theodosius Dobzhansky (who visited Rockefeller) and by Alfred Mirsky (of the Rockefeller faculty) who was a close collaborator of Arthur Pollister in the Zoology Department. The work was the focus of widespread and critical discussion among the faculty and students. Mirsky was a vocal critic of the purported chemical identification of the transforming agent, while applauding the central importance of the work. For my own part, the transcendent leap was simply the feasibility of knowing the chemistry of the gene. Whether this was DNA or protein would certainly be clarified quickly, provided the *Pneumococcus* transformation could be securely retained within the conceptual domain of gene transmission. I read the Avery, MacLeod and McCarty paper on 20 January 1945, prompted by Harriett Taylor (later Ephrussi-Taylor) a graduate student in Zoology who planned to pursue her postdoctoral studies with Avery. My excited response is recorded as..."unlimited in its im plications... Direct demonstration of the multiplication of transforming factor... Viruses are gene-type

一样含有基因”的问题一直没有答案，甚至都无人问津。当时，我对这方面也不感兴趣。我渴望从事将化学分析应用于癌症和大脑功能障碍等问题的医学研究方面的工作。于是，细胞毒理学似乎成为进入细胞生物化学研究的最有希望的途径。后来，我遇到了弗朗西斯·瑞安（于1963年去世），正是他使我的研究兴趣转移到这个更有效的研究工具——遗传学上。

1941年到1942年间瑞安在斯坦福大学开展博士后研究，他在那里遇到了比德尔和塔特姆（于1975年去世），并开始对将脉孢菌的营养缺陷型突变体作为工具进行生物化学遗传分析产生浓厚的兴趣[2]。尽管在动物系对诸如脉孢菌的真菌的研究工作并不如在哥伦比亚（他在那里获得了讲师职位）那般顺利，但是他建立了实验室并继续从事这些研究。1943年1月，我有幸在他的实验室得到了一份工作，帮助准备培养基和处理脉孢菌培养物。受到瑞安作为教师的个人品质、严谨的研究态度的影响以及与他、动物系的其他教员及研究生的讨论使我受益匪浅。1943年7月1日，我被征召进入美国海军预备队服役，本该在哥伦比亚学院度过的几个月，转而用于长岛圣奥尔本斯的美国海军医院工作。在那里的临床寄生虫学实验室，我有充足的机会观察间日疟原虫的生命周期。这一经历使我注意到了疟原虫的有性阶段，并促使我开始探索其他微生物（甚至是细菌）中是否也可能有隐蔽的有性阶段。1944年10月，我再次被分到了哥伦比亚医学院学习，但依然继续在晨边高地校区与瑞安共事。

发　现

那一年重要的生物学发现之一是埃弗里、麦克劳德和麦卡蒂发现DNA是肺炎球菌转化的物质基础[3]。这一现象可以被看作是一个细菌细胞的基因转移到另一个细菌细胞中；不过这种解释不可避免地因当时对细菌遗传学的模糊认识而变得疑云重重。艾弗里在纽约洛克菲勒研究所的工作被特奥多修斯·多布然斯基（曾访问过洛克菲勒）和阿尔弗雷德·米尔斯基（洛克菲勒的教员）迅速传达给了哥伦比亚的生物学家。阿尔弗雷德·米尔斯基与动物学系的阿瑟·波利斯特有密切的合作。当时这一研究成为倍受教员和学生关注的重要讨论焦点。米尔斯基毫无保留地批判所谓的转化物质的化学鉴定，但对艾弗里的核心工作非常推崇。在我看来，这一飞跃只不过是了解了基因的化学本质后的必然结果。如果说肺炎球菌的转化确实是基因传递，那么这一物质是DNA还是蛋白质肯定很快就被证明了。在动物系一个即将跟随埃弗里开展博士后研究的研究生——哈丽雅特·泰勒（后来改名为埃弗吕西·泰勒）的推荐下，我阅读了艾弗里、麦克劳德和麦卡蒂在1945年1月20日发表的论文。以下文字记录了我兴奋的反映：“……意义深远……直接证明了转化因子的增殖……病毒是基因类型的复合物”。

compounds."

At once, I thought of attempting similar transformations by DNA in *Neurospora.* This organism had a well understood lifecycle and genetic structure. The biochemical mutants opened up by Beadle and Tatum also allowed the efficient detection of nutritionally self-sufficient (prototrophic) forms, even if these were vanishingly rare. This would facilitate the assay of transformational events.

Between January and May, 1945, I shared this idea with Francis Ryan; in June, he invited me to work on the subject with him. To our dismay, we soon discovered that the leucine-minus *Neurospora* mutant would spontaneously revert to prototrophy[4], leaving us with no reliable assay for the effect of DNA in mediating genetic change in *Neurospora.* Questions about the biology of transformation would remain inaccessible to conventional genetic analysis if bacteria lacked a sexual stage. But was it true that bacteria were asexual? Rene Dubos' monograph, *The Bacterial Cell*[5], footnoted how inconclusive the claims were for or against any morphological exhibition of sexual union between bacterial cells.

My notes dated 8 July 1945 detail hypothetical experiments both to search for mating among *Monilia* (medically important yeast-like fungi) and to seek genetic recombination in bacteria (by the protocol that later proved to be successful). These notes coincide with the beginning of my course in medical bacteriology. They were provoked by the contrast of the traditional teaching that bacteria were *Schizomycetes*, asexual primitive plants, with an appreciation of sexuality in yeast[6], which was represented at Columbia by the graduate research work of Sol Spiegelman and Harriett Taylor.

Dubos[5] cited many unclear, and two clear-cut negative results[7,8] for sexuality in bacteria using genetic exchange methodology. But these two studies had no selective method for the detection of recombination and so would have overlooked the process had it occurred less often than perhaps once per thousand cells. With the use of a pair of nutritional mutants, say A^+B^- and A^-B^+, one could plate out innumerable cells in a selective medium and find a single A^+B^+ recombinant. In early July, I began experiments along these lines. In the first instance I used a set of biochemical mutants in *Escherichia coli*, which I began to accumulate in Ryan's laboratory. To avoid the difficulty that had arisen in our *Neurospora* experiments, a spontaneous reversion from A^-B^+ to A^+B^+, the strategy would be to use a pair of double mutants: $A^-B^-C^+D^+$ and $A^+B^+C^-D^-$. Sexual crossing should still generate $A^+B^+C^+D^+$ prototroph recombinants. These would be unlikely to arise by spontaneous reversions which, in theory, requires the coincidence of two rare events; $A^-\rightarrow A^+$ and $B^-\rightarrow B^+$. Much effort was devoted to control experiments to show that double reversions would follow this model, and so occur at a negligible frequency in the cultures handled separately. Thus the occurrence of prototrophs in the mixed cultures would be presumptive evidence of genetic recombination.

我立刻开始考虑在脉孢菌中使用 DNA 进行类似的转化实验，人们对于这种生物的生活史以及遗传结构有着比较清楚的认识。比德尔和塔特姆构建的生化突变体使营养自足型（原养型）的高效检测成为可能，即使其数量稀少。这也为转化事件的检测提供了便利。

1945 年 1 月到 5 月期间，我和弗朗西斯·瑞安讨论了这个想法；6 月份，他邀请我和他一起研究这个课题。让我们沮丧的是，不久我们发现脉胞菌的亮氨酸缺陷型突变体会自发回复成原养型[4]，这使得我们无法建立一个可靠的检测方法来验证 DNA 在介导脉胞菌的遗传改变方面的作用。如果细菌不存在有性生殖的话，那么有关转化的生物学过程的问题也就无法用传统的遗传分析方法来解释。但是，细菌真的是无性生殖么？勒内·杜博斯的专论——《细菌细胞》[5] 中明确指出，任何支持或否定细菌细胞间存在性连接的形态学表现都无明确的结论。

我在 1945 年 7 月 8 日详细记录了用于寻找念珠菌（一种医学上十分重要的类似酵母的真菌）中的交配以及细菌中的遗传重组（所采取的实验方案后来证明是成功的）的假设性实验。这些记录与我在医学细菌学课程一开始的内容一致。它们挑战了传统教科书认为细菌是裂殖类、属于无性生殖的原始植物的观点，并且赞同索尔·施皮格尔曼和哈丽雅特·泰勒主张酵母存在性别差异[6] 的观点，这是两人在哥伦比亚研究生时期的研究工作。

杜博斯[5] 引用了很多利用遗传交换方法研究细菌有性生殖的阴性结果[7,8]，其中很多结果尚不清楚，但有两个阴性结果很明确。不过这两个研究结果并没有建立用于检测重组的筛选方法，因此如果发生重组的概率低于千分之一，可能会忽略这个过程。通过使用一对营养突变体，分别命名为 A^+B^- 和 A^-B^+，就可以将无数的细胞涂布在选择性培养基上，而只有 A^+B^+ 型的重组细胞可以长出单克隆。在 7 月初，我开始按照这些方法开展实验。在第一个实验中，我使用了一系列大肠杆菌的生物化学突变体（来自瑞安的实验室）。为了避免在脉孢菌实验中出现的困境，即由 A^-B^+ 自发回复突变 A^+B^+，我们使用了下面一对双突变体：$A^-B^-C^+D^+$ 和 $A^+B^+C^-D^-$。有性杂交应该仍然会产生 $A^+B^+C^+D^+$ 型的原养型重组子。这些不太可能由自发性回复突变产生，因为在理论上，这需要从 A^- 到 A^+ 和 B^- 到 B^+ 同时发生两个小概率的突变。我们在对照实验上投入了很多精力，实验结果显示，双回复突变也按照这一模式进行，所以其发生的概率非常低以致可以忽略。因此，在混合培养物中出现的原养型突变体将证明有遗传重组发生。

Long Shot

Meanwhile at Stanford, Ed Tatum, whose doctoral training at Wisconsin had been in the biochemistry of bacteria, was returning to bacteria as experimental subjects, having published two papers on the production of biochemical mutants in *E. coli*[9], including double mutants like those described here. During the summer of 1945 Francis Ryan learned that Tatum was leaving Stanford to set up a new programme in microbiology at Yale. He suggested that, rather than merely ask Tatum to share these new strains, I apply to work with him and get the further benefit of his detailed experience and general wisdom. Tatum agreed and suggested that I arrive in New Haven in late March, to give him time to set up his laboratory. He hinted that he had some similar ideas of his own, but never elaborated them. The arrangement suited him by leaving him free to complete his work on the biochemistry of *Neurospora*, perform the heavy administrative duties of his new programme, and still participate in the long-shot gamble of looking for bacterial sex.

It took about six weeks, from the first serious efforts at crossing in mid-April 1946, to establish well-controlled, positive results. These experiments could be done overnight, so the month of June allowed over a dozen repetitions, and the recruitment of almost a dozen genetic markers in different crosses. Besides the appearance of $A^+B^+C^+D^+$ prototrophs, it was important to show that additional unselected markers in the parent stocks would segregate and recombine freely in the prototrophic progeny. This result left little doubt as to the interpretation of the experiments.

An immediate opportunity for public announcement presented itself at the international Cold Spring Harbor Symposium in July. This was dedicated to the genetics of microorganisms, signalling the postwar resumption of major research in a field that had been invigorated by the new discoveries with *Neurospora*, phage, and the role of DNA in the *Pneumococcus* transformation. Tatum was already scheduled to talk about his work on *Neurospora*. We were granted a last-minute improvisation in the schedule to permit a brief discussion of our new results.

The discussion was lively. The most principled criticism came from Andre Lwoff who worried about cross-feeding of nutrients between the two strains without their having in fact exchanged genetic information. Having taken great pains to control this possibility, I felt that the indirect genetic evidence was quite conclusive. Fortunately, Max Zelle mediated the debate, and generously offered to advise and assist me in the direct isolation of single cells under the microscope. These subsequent observations did quiet remaining concerns of the group that Lwoff had assembled at the Pasteur Institute, including Jacques Monod, Francois Jacob and Elie Wollman, who were to make the most extraordinary contributions to the further development of the field. The single cell methods were also useful in later investigations in several fields. A direct result of the Cold Spring Harbor meeting was the prompt ventilation of all the controversial issues. With a few understandable, but minor, points of resistance, genetic recombination in bacteria was

展 望

与此同时，毕业于威斯康星大学从事细菌生化研究的爱德华·塔特姆博士在斯坦福大学也重新开始使用细菌作为实验对象进行研究，并发表了两篇论文报道了他构建的大肠杆菌生物化学突变体[9]，其中也包括上文中提到的双突变体。在 1945 年的夏天，弗朗西斯·瑞安了解到塔特姆即将离开斯坦福大学并将在耶鲁大学启动一个新的微生物学研究项目。他建议我与其仅仅向塔特姆索取这些新的菌株来开展实验，还不如干脆申请和他一起工作从而获得更多这方面的经验和指导。塔特姆同意了我的申请，并建议我在三月下旬到纽黑文来，这期间他可以建立他的实验室。他向我提到，他自己也有一些类似的想法，不过并没有深入考虑。如此安排使得他可以在继续完成对脉孢菌的生物化学研究的同时，既能够有时间完成他新项目繁重的管理工作，又可以在细菌的性别方面开展一些探索性的远景研究。

从 1946 年 4 月中旬第一次尝试杂交实验开始，我花了大约 6 个星期的时间来得到一个拥有良好对照的阳性结果。这些实验可以过夜进行，因此我在 6 月份得以进行了超过 12 次重复实验，并且在不同的杂交实验中发现近 12 个遗传标记。很重要的是除了实验中出现的 $A^+B^+C^+D^+$ 原养型外，那些亲本中未用于筛选的其他遗传标记也会在原养型子代中自由地发生分离和重组。这些结果符合对实验的解释。

随后在 7 月举行的冷泉港国际论坛上，这一结果得以公布于众。这一成果被归到微生物遗传学，这也标志着在脉胞菌、噬菌体以及 DNA 在肺炎球菌转化中的作用等新发现的带动和鼓舞下，一个主流研究领域在战后的复苏。按照既定日程，塔特姆介绍了他在脉胞菌中开展的研究工作。我们则被允许在会议日程的最后进行一个即席演讲，来简短地讨论一下我们的新发现。

讨论开展得非常热烈，其中最主要的争议来自安德烈·利沃夫，他提出两个菌株也可能会在营养上交互共生而事实上它们之间并没有发生遗传信息的交换。基于之前已经花了大量的精力来排除这种可能，我认为间接的遗传证据已经非常具有说服力了。幸运的是，马克斯·泽尔调停了这场争论，他慷慨地给出建议并帮助我在显微镜下直接分离出单个细胞。这些后续的观察确实解决了在巴斯德研究所的利沃夫研究组所提出的质疑，这个小组中还有雅克·莫诺、弗朗西斯·雅各布和伊利·沃尔曼，他们对该领域的进一步发展做出了非凡的贡献。单细胞研究方法在后来的多个领域的研究中都有应用。冷泉港会议的直接结果是各种争议可以迅速得到沟通。虽然存在一些无可厚非的小阻力，细菌的遗传重组仍旧很快被纳入分子生物学的前沿研究的主流，并在大概十年后被写入细菌学的权威教科书里。人们仍然需要花上

soon incorporated into the mainstream of the burgeoning research in molecular biology, and after another decade or so into the standard texts of bacteriology. It still took some years to work out the intimate details of crossing in *E. coli*; some, including the crucial question of the physical mechanism of DNA transfer between mating cells, are still obscure.

The public image of the scientific fraternity today has seldom been so problematic and the system cannot avoid putting a high premium on competition and self-assertion. We can recall with gratification how the personalities of Ryan[10] and Tatum[11] exemplified norms of nurture, dignity, respect for others, and above all a regard for the advance of knowledge.

Experimental Luck

1. We have learned[12] that *E. coli* strain K-12 itself was a remarkably lucky choice of experimental material: only about one in twenty randomly chosen strains of *E. coli* would have given positive results in experiments designed according to our protocols. In particular, strain B, which has become the standard material for work on bacteriophage, would have been stubbornly unfruitful. Tatum had acquired K-12 from the routine stock culture collection in Stanford's microbiology department when he sought an *E. coli* strain to use as a source of tryptophanase in work on tryptophan synthesis in *Neurospora*[13]. The same strain was then in hand when he set out to make single, and then double mutants in *E. coli*[9]. In 1946, I was very much aware of strain specificities and was speculating about mating types (as in *Neurospora*). I have no way to say how many other strains would have been tried, or in how many combinations, had the June 1946 experiments not been successful.

2. An equally important piece of luck was that the selected markers Thr (threonine) and Leu (leucine) are found almost at the origin of the *E. coli* chromosome map[14]. The cognoscenti will recognize that in a cross $B^-M^-T^+L^+F^+ \times B^+M^+T^-L^-F^-$, the configuration used in June 1946, these chromosome localizations offer almost a maximum yield of selectible recombinants. We were therefore led stepwise into the complexities of mapping.

(**324**, 627-628; 1986)

Joshua Lederberg is at the Rockefeller University, New York, NY 10021. The research summarized in this article was supported in 1946 by a fellowship of the Jane Coffin Childs Fund for Medical Research.

References:

1. Lederberg, J. & Tatum, E. L. *Nature* **158**, 558 (1946).
2. Beadle, G. W. & Tatum, E. L. *Proc. Natl. Acad. Sci. U.S.A.* **27**, 499-506 (1941).
3. Avery, O. T., MacLeod, C. M. & McCarty, M. *J. exp. Med.* **79**, 137-158 (1944).
4. Ryan, F. J. & Lederberg, J. *Proc. Natl. Acad. Sci. U.S.A.* **32**, 163-173 (1946).
5. Dubos, R. *The Bacterial Cell* (Harvard, Cambridge, 1945).
6. Winge & Lausten, O. *C. R. Lab. Carlsberg, Ser. physiol.*, **22**, 99-119 (1937).
7. Sherman, J. M. & Wing, H. U. *J. Bact.* **33**, 315-321 (1937).

好几年的时间来阐明大肠杆菌有性杂交的内在细节，有些关键问题，如DNA在交配细胞间如何转移的物理机制等依然不清楚。

今天公众对科学界的质疑很少，并且这个体系不可避免采用高额奖金来鼓励竞争和独立思考。我们非常高兴地回忆起瑞安[10]和塔特姆[11]如此杰出的人格，他们为其他人在育人、尊严与尊重方面做出了示范，以及最重要的以知识为先的精神。

实验的运气

我们已经知道[12]选取大肠杆菌K-12菌株做实验材料本身就是一件非常幸运的事：根据我们的实验方案在随机挑选的20个大肠杆菌菌株中，只有大约1株可以在实验中得到阳性结果。尤其是B菌株，它成了噬菌体研究的标准实验材料，并且该菌株是高度不育的。塔特姆从斯坦福大学微生物系的常规培养物库存中得到了K-12，那时他打算寻找一个大肠杆菌菌株作为脉胞菌色氨酸合成研究[13]中色氨酸酶的来源。当他开始构建大肠杆菌的单突变体以及之后的双突变体[9]时，很快就得到了相同的菌株。1946年，我对菌株的特异性非常了解，并开始推断交配的方式（就像在脉胞菌中一样）。如果1946年6月的实验不成功的话，我不知道还要尝试多少其他的菌株，也不知道需要尝试多少种组合。

另一个同样重要的运气在于，实验所使用的筛选标记——苏氨酸和亮氨酸几乎位于大肠杆菌染色体图谱的起始部位[14]。内行人士都会知道，在1946年6月使用的$B^-M^-T^+L^+F^+ \times B^+M^+T^-L^-F^+$杂交组合中，这些染色体定位有利于产生几乎最高产量的重组子，因此我们被逐步引导进入定位的复杂性中。

（张锦彬 翻译；肖景发 审稿）

8. Gowen, J. W. & Lincoln, R. E. *J. Bact.* **44**, 551-554 (1942).
9. Gray, C. H. & Tatum, E. L. *Proc. Natl. Acad. Sci. U.S.A.* **30**, 404-410 (1944).
10. Lederberg, J. *in University on the Heights.* (ed. First, W.)105-109. (Doubleday, Garden City, New York, 1969).
11. Lederberg, J. *A. Rev. Genet.* **13**, 1-5 (1979).
12. Lederberg, J. *Science* **114**, 68-69 (1951).
13. Tatum, E. L. & Bonner, D. M. *Proc. Natl. Acad. Sci. U.S.A.* **30**, 30-37 (1944).
14. Bachmann, B. J. *Microb. Revs.* **47**, 180-230 (1983).

Mitochondrial DNA and Human Evolution

R. L. Cann *et al.*

Editor's Note

Before this paper, our understanding of human prehistory had come almost entirely from studying ancient bones and stones. But here Allan Wilson, a pioneer of "molecular evolution" techniques, and his colleagues use modern people as windows on the past. By comparing the mitochondrial DNA from 147 people of diverse origins, they show that modern humans shared a common, African ancestry some 200,000 years ago. This conclusion has since been validated by traditional palaeontological methods, but the paper caught the popular imagination by creating a virtual "ancestral celebrity". Because mitochondrial DNA is inherited maternally, Wilson and colleagues postulated the existence of a single woman from whom we all descended: "mitochondrial Eve".

Mitochondrial DNAs from 147 people, drawn from five geographic populations have been analysed by restriction mapping. All these mitochondrial DNAs stem from one woman who is postulated to have lived about 200,000 years ago, probably in Africa. All the populations examined except the African population have multiple origins, implying that each area was colonised repeatedly.

MOLECULAR biology is now a major source of quantitative and objective information about the evolutionary history of the human species. It has provided new insights into our genetic divergence from apes[1-8] and into the way in which humans are related to one another genetically[9-14]. Our picture of genetic evolution within the human species is clouded, however, because it is based mainly on comparisons of genes in the nucleus. Mutations accumulate slowly in nuclear genes. In addition, nuclear genes are inherited from both parents and mix in every generation. This mixing obscures the history of individuals and allows recombination to occur. Recombination makes it hard to trace the history of particular segments of DNA unless tightly linked sites within them are considered.

Our world-wide survey of mitochondrial DNA (mtDNA) adds to knowledge of the history of the human gene pool in three ways. First, mtDNA gives a magnified view of the diversity present in the human gene pool, because mutations accumulate in this DNA several times faster than in the nucleus[15]. Second, because mtDNA is inherited maternally and does not recombine[16], it is a tool for relating individuals to one another. Third, there are about 10^{16} mtDNA molecules within a typical human and they are usually identical to one another[17-19]. Typical mammalian females consequently behave as haploids, owing to a bottleneck in the genetically effective size of the population of mtDNA molecules within each oocyte[20]. This maternal and haploid inheritance means that mtDNA is more sensitive

线粒体DNA与人类进化

卡恩等

编者按

在这篇文章发表之前，我们对史前时代的了解全部来源于对古人骨骼和石器的研究。但是“分子进化”技术的先锋艾伦·威尔逊和他的同事们用现代人类作为了解过去的窗口。通过比较来源不同的147个人的线粒体DNA，他们发现现代人类共同拥有的20万年以前的非洲血统。这个结论已经通过传统的古生物学方法证实，但是这篇文章创造了一个虚拟的“祖先名人”从而吸引了大众的兴趣。因为线粒体DNA是母系遗传的，所以威尔逊和他的同事们假设了一个女人的存在，我们都是她的后裔，这就是“线粒体夏娃”。

我们分析了来自五个不同地域的群体、共147例人线粒体DNA的限制性酶切图谱。所有这些线粒体DNA都源于同一个我们假设的、生活在约20万年以前的非洲女性。除来自非洲的群体外，其他所有的群体都有多个起源，这暗示了每个地域的群体被入侵群体反复占据多次。

如今，分子生物学已是关于人类进化历史的量化和客观信息的主要来源。它为我们揭示人类自猿以来的遗传分化[1-8]和人与人之间遗传上相关联的方式[9-14]提供了新视角。然而人类遗传进化的研究被蒙上了一层迷雾，因为之前的研究主要基于细胞核基因的比较，而核基因突变的积累是很慢的。此外，核基因遗传自父母双方并在每一代中混合。这种混合模糊了个体的遗传背景，两亲本基因组之间可以发生遗传重组。如果不研究那些紧密连锁的位点，遗传重组会使得特定DNA片段的遗传历史难以追踪。

在全世界范围内对线粒体DNA（mtDNA）的调查研究在以下三个方面加强了我们对人类基因库历史的认识：第一，因为mtDNA的突变积累速度比核基因快好几倍[15]，这使我们更容易观察到人类基因库中的遗传多样性；第二，mtDNA是母系遗传，不会发生遗传重组[16]，从而能够将个体之间关联起来；第三，人体中约有10^{16}个mtDNA分子，它们的DNA序列通常是完全相同的[17-19]。由于每个卵母细胞中mtDNA分子遗传上有效群体大小的瓶颈效应[20]，所以典型的雌性哺乳动物是以单倍体遗传的。mtDNA的母系遗传和单倍体遗传意味着mtDNA对生物群体中个体数量

than nuclear DNA to severe reductions in the number of individuals in a population of organisms[15]. A pair of breeding individuals can transmit only one type of mtDNA but carry four haploid sets of nuclear genes, all of which are transmissible to offspring. The fast evolution and peculiar mode of inheritance of mtDNA provide new perspectives on how, where and when the human gene pool arose and grew.

Restriction Maps

MtDNA was highly purified from 145 placentas and two cell lines, HeLa and GM 3043, derived from a Black American and an aboriginal South African (!Kung), respectively. Most placentas (98) were obtained from US hospitals, the remainder coming from Australia and New Guinea. In the sample, there were representatives of 5 geographic regions: 20 Africans (representing the sub-Saharan region), 34 Asians (originating from China, Vietnam, Laos, the Philippines, Indonesia and Tonga), 46 Caucasians (originating from Europe, North Africa, and the Middle East), 21 aboriginal Australians, and 26 aboriginal New Guineans. Only two of the 20 Africans in our sample, those bearing mtDNA types 1 and 81 (see below) were born in sub-Saharan Africa. The other 18 people in this sample are Black Americans, who bear many non-African nuclear genes probably contributed mainly by Caucasian males. Those males would not be expected to have introduced any mtDNA to the Black American population. Consistent with our view that most of these 18 people are a reliable source of African mtDNA, we found that 12 of them bear restriction site markers known[21] to occur exclusively or predominantly in native sub-Saharan Africans (but not in Europeans, Asians or American Indians nor, indeed, in all such Africans). The mtDNA types in these 12 people are 2–7, 37–41 and 82 (see below). Methods used to purify mtDNA and more detailed ethnographic information on the first four groups are as described[17,22]; the New Guineans are mainly from the Eastern Highlands of Papua New Guinea[23].

Each purified mtDNA was subjected to high resolution mapping[22-24] with 12 restriction enzymes (*Hpa*I, *Ava*II, *Fnu*DII, *Hha*I, *Hpa*II, *Mbo*I, *Taq*I, *Rsa*I, *Hinf*I, *Hae*III, *Alu*I and *Dde*I). Restriction sites were mapped by comparing observed fragment patterns to those expected from the known human mtDNA sequence[25]. In this way, we identified 467 independent sites, of which 195 were polymorphic (that is, absent in at least one individual). An average of 370 restriction sites per individual were surveyed, representing about 9% of the 16,569 base-pair human mtDNA genome.

Map Comparisons

The 147 mtDNAs mapped were divisible into 133 distinct types. Seven of these types were found in more than one individual; no individual contained more than one type. None of the seven shared types occurred in more than one of the five geographic regions. One type, for example, was found in two Australians. Among Caucasians, another type occurred three times and two more types occured twice. In New Guinea, two additional types were found three times and the seventh case involved a type found in six individuals.

的锐减比核基因更敏感[15]。一对可育个体只遗传一种类型的mtDNA到后代中，但却有四套单倍体核基因遗传给后代。mtDNA的快速进化和独特的遗传模式将有助于我们重新理解人类基因库是何时何地以及怎样起源和演化的。

限制性酶切图谱

我们从145个胎盘和HeLa、GM3043两个细胞系中得到了高度纯化的mtDNA，这两个细胞系分别来源于一个美国黑人和一个南非土著居民（布须曼昆人）。大部分胎盘（98个）是从美国医院里获得的，其余则来自澳大利亚和新几内亚。样本中有5个地理区域的代表：20个非洲人（代表撒哈拉以南地区），34个亚洲人（来自中国、越南、老挝、菲律宾、印度尼西亚和汤加），46个高加索人（来自欧洲、北非和中东地区），21个澳大利亚土著居民和26个新几内亚原住民。我们样本中的20个非洲人中只有两个出生在撒哈拉沙漠以南的非洲地区，他们的mtDNA为1型和81型（见下文）。其他的18个人都是美国黑人，他们拥有许多非非洲人群核基因，这些基因可能主要来自高加索男性。这些高加索男性不可能将任何mtDNA引入美国黑人群体中。与我们的观点相一致，这18个美国黑人大多数是非洲人mtDNA的可靠来源，我们发现其中12个携带有撒哈拉以南非洲土著居民（而不是欧洲人、亚洲人或美洲第安人，当然也不是所有这些非洲人）所特有的或者主要的限制性酶切位点遗传标记[21]。这12例的mtDNA类型为2~7、37~41和82（见下文）。mtDNA的纯化方法和最初四组更详细的人群信息如前文献所述[17,22]，新几内亚人样本主要来自巴布亚新几内亚的东部高地[23]。

每一个纯化得到的mtDNA都用12个限制性酶（*Hpa*Ⅰ、*Ava*Ⅱ、*Fnu*DⅡ、*Hha*Ⅰ、*Hpa*Ⅱ、*Mbo*Ⅰ、*Taq*Ⅰ、*Rsa*Ⅰ、*Hinf*Ⅰ、*Hae*Ⅲ、*Alu*Ⅰ和*Dde*Ⅰ）进行酶切，然后绘制出高分辨率的图谱[22-24]。将限制性酶切后观察到的片段和已知的人mtDNA序列[25]中期望出现的片段相比较从而得到限制性酶切图谱。通过这种方法，我们鉴定出了467个独立酶切位点，其中195个呈现出多态性（即至少在一例个体中缺少这个位点）。平均每个个体中分析了370个限制性酶切位点，代表了人mtDNA全基因组16,569个碱基对中的9%。

图 谱 比 对

147个样品的mtDNA酶切图谱可分成133个不同的类别。其中有7类出现在了多个个体中，并且每个个体中只含一种类型的mtDNA。这7类中的每一类也只出现在五个地理区域中的一个，例如在两个澳大利亚人中只发现了一种类型的mtDNA；在高加索人样本中，有一类出现了三次，有两种类型出现了两次；在新几内亚人样本中，另两类出现了三次，第七种类型被发现存在于六个个体中。

A histogram showing the number of restriction site differences between pairs of individuals is given in Fig. 1; the average number of differences observed between any two humans is 9.5. The distribution is approximately normal, with an excess of pairwise comparisons involving large numbers of differences.

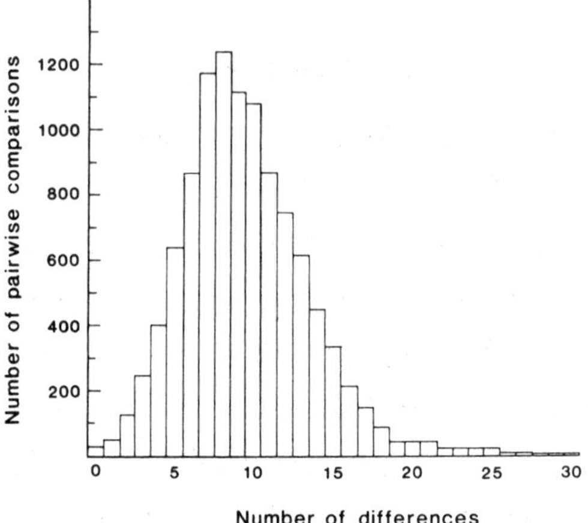

Fig. 1. Histogram showing the number of site differences between restriction maps of mtDNA for all possible pairs of 147 human beings.

From the number of restriction site differences, we estimated the extent of nucleotide sequence divergence[26] for each pair of individuals. These estimates ranged from zero to 1.3 substitutions per 100 base pairs, with an average sequence divergence of 0.32%, which agrees with that of Brown[17], who examined only 21 humans.

Table 1 gives three measures of sequence divergence within and between each of the five populations examined. These measures are related to one another by equation (1):

$$\delta = \delta_{xy} - 0.5(\delta_x + \delta_y) \quad (1)$$

where δ_x is the mean pairwise divergence (in percent) between individuals within a single population (X), δ_y is the corresponding value for another population (Y), δ_{xy} is the mean pairwise divergence between individuals belonging to two different populations (X and Y), and δ is a measure of the interpopulation divergence corrected for intrapopulation divergence. Africans as a group are more variable ($\delta_x = 0.47$) than other groups. Indeed, the variation within the African population is as great as that between Africans and any other group ($\delta_{xy} = 0.40$–0.45). The within-group variation of Asians ($\delta_x = 0.35$) is also comparable to that which exists between groups. For Australians, Caucasians, and New Guineans, who show nearly identical amounts of within-group variation ($\delta_x = 0.23$–0.25), the variation between groups slightly exceeds that within groups.

直方图图 1 显示了成对比较的个体间的限制性酶切位点差异的数量，观察任何两个人的限制性酶切位点的数量平均差异数量为 9.5。其分布近似服从正态分布，超出正态分布的两两比较包含了大量差异。

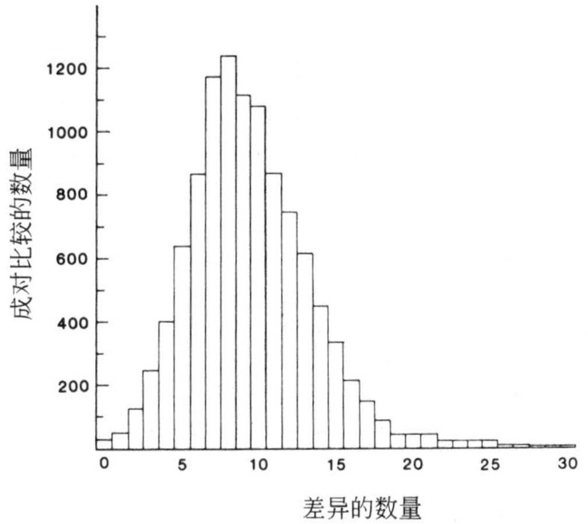

图 1. 直方图显示了 147 个人中所有可能成对的 mtDNA 限制性酶切图谱之间的位点差异的数量。

根据限制性酶切位点差异的数量，我们估算了每对个体间核苷酸序列分化的程度[26]。每 100 个碱基对中估计有 0~1.3 个碱基对发生了替换，平均序列分化程度为 0.32%，此结果与之前布朗得到的结果[17]一致，但是他们只研究了 21 例人的样本。

表 1 给出了对我们所研究的五个群体内或群体间序列分化程度的三种度量。所得到的值通过等式（1）计算群体之间的关联：

$$\delta = \delta_{xy} - 0.5\ (\delta_x + \delta_y) \qquad (1)$$

其中 δ_x 为单个群体（X）内个体差异平均值（百分比形式），δ_y 为另一个群体（Y）的对应值，δ_{xy} 为两个不同群体（X 和 Y）的个体之间分化的平均值，δ 是进行过群体内序列分化校正的群体间分化的度量值。非洲人群内变异（$\delta_x = 0.47$）比其他群体内大。实际上，非洲人群内部的变异程度与非洲人群和其他群体之间的分化程度（$\delta_{xy} = 0.40$~0.45）一样大。亚洲人群内分化程度（$\delta_x = 0.35$）也近似于群体之间的分化程度。澳大利亚人群、高加索人群和新几内亚人群的群体内的变异程度近似相等（$\delta_x = 0.23$~0.25），群体之间的分化程度稍稍大于群体内的变异程度。

Table 1. MtDNA divergence within and between 5 human populations

	% sequence divergence				
Population	1	2	3	4	5
1. African	0.47	0.04	0.04	0.05	0.06
2. Asian	0.45	0.35	0.01	0.02	0.04
3. Australian	0.40	0.31	0.25	0.03	0.04
4. Caucasian	0.40	0.31	0.27	0.23	0.05
5. New Guinean	0.42	0.34	0.29	0.29	0.25

The divergence is calculated by a published method[26]. Values of the mean pairwise divergence between individuals within populations (δ_x) appear on the diagonal. Values below the diagonal (δ_{xy}) are the mean pairwise divergences between individuals belonging to two different populations, X and Y. Values above the diagonal (δ) are interpopulation divergences, corrected for variation within those populations with equation (1).

When the interpopulational distances (δ_{xy}) are corrected for intrapopulation variation (Table 1), they become very small ($\delta = 0.01–0.06$). The mean value of the corrected distance among populations ($\delta = 0.04$) is less than one-seventh of the mean distance between individuals within a population (0.30). Most of the mtDNA variation in the human species is therefore shared between populations. A more detailed analysis supports this view[27].

Functional Constraints

Figure 2 shows the sequence divergence (δ_x) calculated for each population across seven functionally distinct regions of the mtDNA genome. As has been found before[24,27,28], the most variable region is the displacement loop ($\bar{\delta}_x = 1.3$), the major noncoding portion of the mtDNA molecule, and the least variable region is the 16S ribosomal RNA gene ($\bar{\delta}_x = 0.2$). In general, Africans are the most diverse and Asians the next most, across all functional regions.

表 1. 五个群体内和群体间的 mtDNA 分化程度

	序列分化程度（%）				
人群	1	2	3	4	5
1. 非洲人	0.47	0.04	0.04	0.05	0.06
2. 亚洲人	0.45	0.35	0.01	0.02	0.04
3. 澳大利亚人	0.40	0.31	0.25	0.03	0.04
4. 高加索人	0.40	0.31	0.27	0.23	0.05
5. 新几内亚人	0.42	0.34	0.29	0.29	0.25

通过已发表文献中的方法计算了分化程度[26]。对角线为群体内不同个体之间的平均成对分化值（δ_x）。对角线之下的为两个不同群体 X 和 Y 的个体之间的平均成对分化值（δ_{xy}）。对角线之上为群体之间的平均成对分化值（δ），该值已由等式（1）进行群体内变异的校正。.

群体间的分化程度（δ_{xy}）经群体内分化校正后（表 1）就非常小了（$\delta = 0.01 \sim 0.06$）。群体间校正后的平均值（$\delta = 0.04$）小于同一群体内不同个体间分化程度的七分之一（0.30）。因此大多数人群的 mtDNA 分化是发生在群体之间的。有一个更为详尽的分析能够支持此观点[27]。

线粒体基因组功能区的不同选择压力

图 2 显示了计算出的每一群体中 mtDNA 基因组全部七个不同功能区域的序列分化程度（δ_x）。正如同以前所发现的[24,27,28]，最容易发生变异的区域就是 mtDNA 控制区（D 环），是 mtDNA 分子中一段非编码区（$\bar{\delta}_x = 1.3$）；变异最小的区域是 16S 核糖体 RNA 基因（$\bar{\delta}_x = 0.2$）。通常来说，非洲人群在所有的功能区域中是最具多样性的，亚洲人群次之。

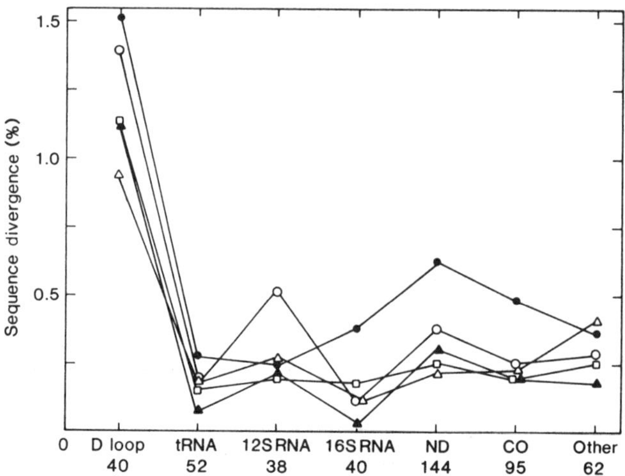

Fig. 2. Sequence divergence within 5 geographic areas for each of 7 functional regions in human mtDNA. Sequence divergence (δ_x) was estimated from comparisons of restriction maps[26]. Symbols for the 5 races are: ●, Africa; ○, Asia; △, Australia; □, Caucasian; ▲, New Guinea. Along the horizontal axis are the numbers of restriction sites in each functional region (D loop, transfer RNA genes, 12S and 16S ribosomal RNA genes, NADH dehydrogenase subunits 1-5, cytochrome oxidase subunits and other protein-coding regions).

Evolutionary Tree

A tree relating the 133 types of human mtDNA and the reference sequence (Fig. 3) was built by the parsimony method. To interpret this tree, we make two assumptions, both of which have extensive empirical support: (1) a strictly maternal mode of mtDNA transmission (so that any variant appearing in a group of lineages must be due to a mutation occurring in the ancestral lineage and not recombination between maternal and paternal genomes) and (2) each individual is homogeneous for its multiple mtDNA genomes. We can therefore view the tree as a genealogy linking maternal lineages in modern human populations to a common ancestral female (bearing mtDNA type a).

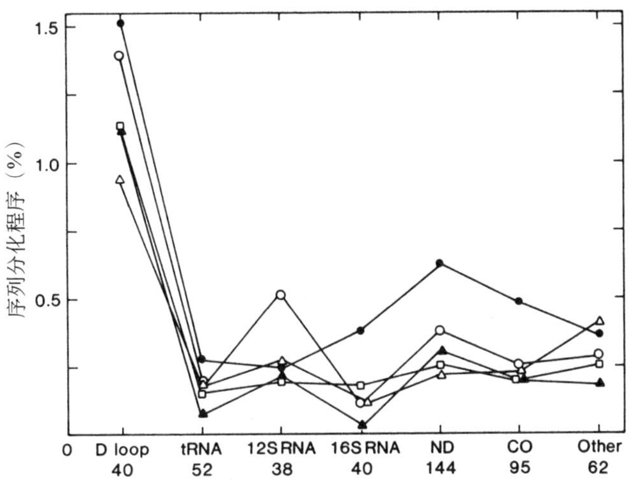

图 2. 五大地理区域人群 mtDNA 中的 7 个功能区域的序列分化程度。该序列分析程度（δ_x）是通过比较限制性酶切图谱来估算的[26]。五个族群的符号是：●，非洲；○，亚洲；△，澳大利亚；□，高加索；▲，新几内亚。横坐标为每一个功能区域（D 环、tRNA 基因、12S 核糖体 RNA 基因、16S 核糖体 RNA 基因、NADH 脱氢酶亚基 1-5、细胞色素氧化酶亚基和其他蛋白质编码区域）内限制性酶切位点的数目。

进 化 树

通过简约法构建了 133 种类型的人 mtDNA 和参照序列（图 3）的进化树。为了能够诠释此进化树，我们做了两个有大量实证支持的假设：(1) mtDNA 是严格遵守母系遗传模式的（即任何出现在直系亲属中的变异都是由直系亲属祖先的 mtDNA 发生变异造成的，并且母本和父本之间不会发生遗传重组）；(2) 每个个体在多个 mtDNA 基因组上是同质的。因此我们可以把此进化树看成是现代人群与一位共同的女性祖先（携带了 a 型 mtDNA）之间的母系谱系关系。

a

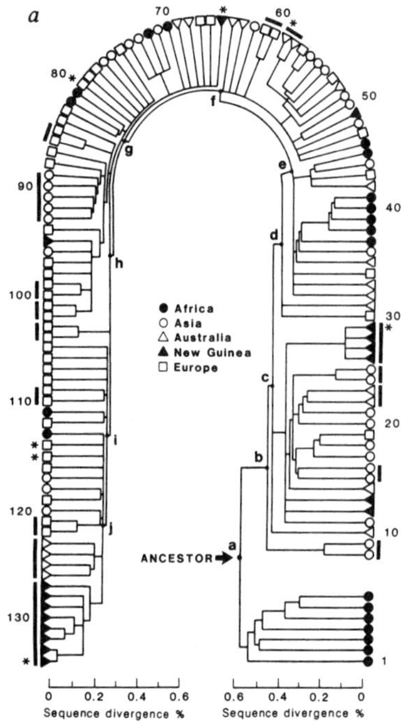

b **D LOOP:** *8*j 2 5 8 9 23 81-83 85 86 118; *64*i (*16494*) 26 ; *134*l 3; *207*h 128-134; *255*f 39; *259*a 106; *340*j 112 **12S rRNA:** *663*e 43 48 53; *712*i 90; 740j 16-19; *748*b 49; 1240a 7-9 37 110; *1403*a (*1448*) 11-18 20-29 34 49 50 55-57; 1463e 14; 1484e 2 4 5 8-11 14-17 19 23-25 30 32-35 41-43 45 52-58 61 107 110 120; *1536*f 31 **VAL tRNA:** 1610a 2 3 44 45 104; 1637c 47 102 108; 1667c 47 **16S rRNA:** 1715c 22 23 27 123; 1917a 70; 2208a 87; *2223*a (*2635*) 37; *2384*a (*2472*) 106; *2390*j 2 3 8 9 49 79 93; 2734a 18 22 54 73 101; 2758k 1-8; 2849k 112; 3123k 119 **ND 1:** 3315e 11; 3337k 94 118 133; *3391*e 41 105 122; 3537a 44; *3592*h 2-7 9 37-41; 3698f 130-132; *3842*e 44; 3849e 5 46; *3899*l 14 18 22 23 32 71 92 102 110 120; *3930*c 110; 3944 l 12; *4092*g 112 **MET tRNA:** 4411a 23 35 36 57; 4464k 129 **ND 2:** *4481*b (*10933*) 111 112; 4631a 2; *4643*k 87 100; *4732*k 115; *4769*a 110; *4793*e 76 85; 5176a 21 25 54-57; 5261e 118; 5269l 71; *5351*f 17-19 **TRP tRNA:** *5538*f 130; 5552c 10 **L. ORIGIN:** 5742i 15 44; *5754*i (*5755*) 81 **CO I:** 5978a 52; *5984*b/5983g 87; *5985*k/5983g 133 134; 5996a 87; 6022a 74; *6166*f (*6168*) 63; 6211g 37-41 82; 6260e 95; *6356*c 5 9; 6377c 68; *6409*l (*7854*) 68; *6501*i 78; *6610*g 12; *6699*b (*8719,8723*) 78; 6871g 18 19; *6915*k 5 49; 6931g 48; 6957e 44 45; 7025a 75-86 93 94 103-112; 7055a 3 6 7; *7241*k 72; 7335l 6 7 30 92 120; *7347*e 124 **SER tRNA:** 7461l 71; 7474a 121 **CO II:** *7617*f 55 73; 7750c 3-9 37 40 41 98; 7859j 120; *7970*g 10 24 71-74 77; 8074a 68; 8112i 1; 8150i 1 126 ; *8165*e 10; *8249* b (*8250*) /8250e 1 61 132 **LYS tRNA:** *8299*k 126 **ATP 8:** 8391e 31 120; 8515c 27 **ATP 6:** 8592j 21 23; 8783g 6 7; 8852f (8854, 8856) 109 110; 8994e60 61; *9009*a 15 16; 9053f 52 91-93 95-102 112; *9070*l 5-7; *9150*j 7 103 104 **CO III:** 9266e 28 49 67; 9294e 25; 9342e 27; 9380f 90; *9429*k 56 57; 9553e 37 117; *9714*e 98; 9746k 62; 9751l 48; *9859*g 86 **GLY tRNA:** *10028*a 126 **ND 3:** *10066*f 45; *10084*l 45; 10352a 110; 10364e 96; *10394*c 1-29 39-45 49 50 84-89 94-100 111 120-123 **ARG tRNA:** *10413*a (*10536*) 16 46 56 85 98 **ND 4L:** *10644*k 19 20; 10689e 96; *10694*a 104; *10725*e 22 46 96 **ND 4:** *10806*g 1-7 50 83 124; *10893*l 34; 11146c 61 110; *11161*i 6 7; *11329*e (*11690*) 2 4; *11350*a 15 37 110; *11806*a 65; 11922j 7; *12026*h 124 **ND 5:** *12345*k (*12350,12528*) 88 127-134; 12406h 91-93; 12560a 19 52 93; *12795*j (*12798,12806,13374*) 40 90 98 99; *12810*k 5-7; *12925*g 7; *12990*a (*13642*) 27; *13004*j (*13018,13182,13194*) 20; 13031g 1-6 9-16 18-23 26-35 37 39 40 42 43 45-64 82 83 92 93; 13051e 23 38-41; *13068*a 128; *13096*k 26; *13100*i 107; 13103g 7; 13208f 4 5 7 15 75 88; 13268g 65; 13367b 40 88 90 99; 13404l 8 9 79 ; *13635*l (*13641*) 39 115; 13702e 38-41; 14015a 14; *14050*l (*14366*) 125 **ND 6:** *14279*e 1; *14279*j 23; *14322*a 92; *14385*c 10; *14509*a 68 123-126; *14567*i 68; 14608c 5 **CYT b:** *14749*e 87; 14869j 86 126; 14956l 115; *15005*g 56 57; 15172e 10; *15195*j (*15221*) 44; 15234g 113; 15238c 84 100; 15250c 14-18 20 21 24 25 54 81 90 111; *15606*a 65 127-134; 15723g 43; *15790*j (*16373*) 82; 15883e 7 45 58 59 112 **THR tRNA:** *15897*k 97 129; *15907*k 46; *15912*i 81; 15925i 15 81 90 **PRO tRNA:** 15996c/16000g 48 **D LOOP:** 16049k 15 22 46 74 81; 16065g 121 122; *16089*k 21; 16096k 119; 16125k 3 9 74 78 88-91 93 105 119 120 122; *16178*l 26-29; 16208k 11 15 51 52 55; *16217*l 95; *16246*g 46; *16254*a 127 132; 16303k 36 58 59 82 89-93 111 112 131 132; 16310k 1-3 5 6 8-14 20 26-31 58 73 78 82 94-97 99-102 120 125 126 133 134; *16389*g/16390b 11 35-41 82 89; *16398*e 5 11 34 37 66 118 119; *16490*g 10; 16517e 2 11-13 15-24 26-29 51-55 58 59 82 97 105-134.

Fig. 3. *a*, Genealogical tree for 134 types of human mtDNA (133 restriction maps plus reference sequence); *b*, comprehensive list of polymorphic restriction sites used. The tree accounts for the site differences observed between restriction maps of these mtDNAs with 398 mutations. No other order of branching tested is more parsimonious than this one. This order of branching was obtained (using the computer program PAUP, designed by Dr. David Swofford) by ignoring every site present in only one type of mtDNA or absent in only one type and confining attention to the remaining 93 polymorphic sites. The computer program produces an unrooted network, which we converted to a tree by placing the root (arrow) at the midpoint of the longest path connecting the two lineages (see ref. 58). The numbers refer to mtDNA types, no. 1 being from the aboriginal South African (!Kung) cell line (GM 3043), no. 45 being from the HeLa cell line and no. 110 being the published human sequence[25]. Black bars, clusters of mtDNA types specific to a given geographic region; asterisks, mtDNA types found in more than 1 individual: type 134 was in six individuals, types 29, 65 and 80 each occurred thrice, and other types flagged with asterisks occurred twice. To place the nodes in the tree relative to the percent divergence scale, we took account of the differences observed at all 195 polymorphic sites. *b*, The numbering of sites is according to the published human sequence[25], with 12 restriction enzymes indicated by the following single letter code: a, *Alu*I; b, *Ava*II; c, *Dde*I; d, *Fnu*DII; e, *Hae*III; f, *Hha*I; g, *Hinf*I; h, *Hpa*I; i, *Hpa*II; j, *Mbo*I; k, *Rsa*I; l, *Taq*I. Italicized sites are present in the indicated mtDNA types and nonitalicized sites are absent in the indicated types; parentheses refer to alternative placements of inferred sites; sites separated by a slash are polymorphic for two different restriction enzymes caused by a single inferred nucleotide substitution; letters in bold face refer to noncoding regions and genes for transfer RNA, ribosomal RNA and proteins (ND, NADH dehydrogenase; CO, cytochrome oxidase; ATP, adenosine triphosphatase; CYT b, cytochrome *b*). For example, *8*j indicates a *Mbo*I site beginning at nucleotide position 8 in the D loop that was not found in mtDNA type 1 but was present in type 2, etc. Note that since this site is not present in the reference sequence (type 110), the sequence beginning at position 8 is actually a semisite, differing from the *Mbo*I recognition sequence at one position (see ref. 23 for a detailed description of the method of mapping such inferred sites). Not all sites were scored in all individuals: *8*j, 1484e and 7750c were not determined for types 1, 31, 59, 63, 68, and all of the New Guinea mtDNAs; mtDNAs 114 and 121 could not be typed with *Rsa*I. The locations of some sites differ from those reported before[23,24], as do the individuals in which some sites occur; these revisions are based on re-examination of previously-studied mtDNAs.

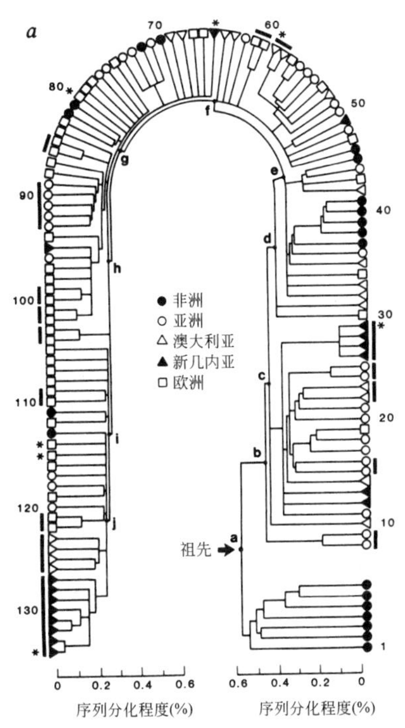

b **D LOOP:** *8*j 2 5 8 9 23 81-83 85 86 118; *64*i (*16494*) 26 ; *134*l 3; *207*h 128-134; *255*f 39; *259*a 106; *340*j 112 **12S rRNA:** *663*e 43 48 53; *712*i 90; 740j 16-19; *748*b 49; 1240a 7-9 37 110; *1403*a (*1448*) 11-18 20-29 34 49 50 55-57; 1463e 14; 1484e 2 4 5 8-11 14-17 19 23-25 30 32-35 41-43 45 52-58 61 107 110 120; *1536*f 31 **VAL tRNA:** 1610a 2 3 44 45 104; 1637c 47 102 108; 1667c 47 **16S rRNA:** 1715c 22 23 27 123; 1917a 70; 2208a 87; *2223*a (*2635*) 37; *2384*a (*2472*) 106; *2390*j 2 3 8 9 49 79 93; 2734a 18 22 54 73 101; 2758k 1-8; 2849k 112; 3123k 119 **ND 1:** 3315e 11; 3337k 94 118 133; *3391*e 41 105 122; 3537a 44; *3592*h 2-7 9 37-41; 3698f 130-132; *3842*e 44; 3849e 5 46; *3899*l 14 18 22 23 32 71 92 102 110 120; *3930*c 110; 3944 l 12; *4092*g 112 **MET tRNA:** 4411a 23 35 36 57; 4464k 129 **ND 2:** *4481*b (*10933*) 111 112; 4631a 2; *4643*k 87 100; *4732*k 115; *4769*a 110; *4793*e 76 85; 5176a 21 25 54-57; 5261e 118; 5269l 71; *5351*f 17-19 **TRP tRNA:** *5538*f 130; 5552c 10 **L. ORIGIN:** 5742i 15 44; *5754*i (*5755*) 81 **CO I:** 5978a 52; *5984*b/5983g 87; *5985*k/5983g 133 134; 5996a 87; 6022a 74; *6166*f (*6168*) 63; 6211g 37-41 82; 6260e 95; *6356*c 5 9; 6377c 68; *6409*l (*7854*) 68; *6501*i 78; *6610*g 12; *6699*b (*8719,8723*) 78; 6871g 18 19; *6915*k 5 49; 6931g 48; 6957e 44 45; 7025a 75-86 93 94 103-112; 7055a 3 6 7; *7241*k 72; 7335l 6 7 30 92 120; *7347*e 124 **SER tRNA:** 7461l 71; 7474a 121 **CO II:** *7617*l 55 73; 7750c 3-9 37 40 41 98; 7859j 120; *7970*g 10 24 71-74 77; 8074a 68; 8112i 1; 8150i 1 126 ; *8165*e 10; *8249* b (*8250*) /8250e 1 61 132 **LYS tRNA:** *8299*k 126 **ATP 8:** 8391e 31 120; 8515c 27 **ATP 6:** 8592j 21 23; 8783g 6 7; 8852f (8854, 8856) 109 110; 8994e60 61; *9009*a 15 16; 9053f 52 91-93 95-102 112; *9070*l 5-7; *9150*j 7 103 104 **CO III:** 9266e 28 49 67; 9294e 25; 9342e 27; 9380f 90; *9429*k 56 57; 9553e 37 117; *9714*e 98; 9746k 62; 9751l 48; *9859*g 86 **GLY tRNA:** *10028*a 126 **ND 3:** *10066*f 45; *10084*l 45; 10352a 110; 10364e 96; *10394*c 1-29 39-45 49 50 84-89 94-100 111 120-123 **ARG tRNA:** *10413*a (*10536*) 16 46 56 85 98 **ND 4L:** *10644*k 19 20; 10689e 96; *10694*a 104; *10725*e 22 46 96 **ND 4:** *10806*g 1-7 50 83 124; *10893*l 34; 11146c 61 110; *11161*i 6 7; *11329*e (*11690*) 2 4; *11350*a 15 37 110; *11806*a 65; 11922j 7; *12026*h 124 **ND 5:** *12345*k (*12350,12528*) 88 127-134; 12406h 91-93; 12560a 19 52 93; *12795*j (*12798,12806,13374*) 40 90 98 99; *12810*k 5-7; *12925*g 7; *12990*a (*13642*) 27; *13004*j (*13018,13182,13194*) 20; 13031g 1-6 9-16 18-23 26-35 37 39 40 42 43 45-64 82 83 92 93; 13051e 23 38-41; *13068*a 128; *13096*k 26; *13100*i 107; 13103g 7; 13208f 4 5 7 15 75 88; 13268g 65; 13367b 40 88 90 99; 13404l 8 9 79 ; *13635*l (*13641*) 39 115; 13702e 38-41; 14015a 14; *14050*l (*14366*) 125 **ND 6:** *14279*e 1; *14279*j 23; *14322*a 92; *14385*c 10; *14509*a 68 123-126; *14567*i 68; 14608c 5 **CYT b:** *14749*e 87; 14869j 86 126; 14956l 115; *15005*g 56 57; 15172e 10; *15195*j (*15221*) 44; 15234g 113; 15238c 84 100; 15250c 14-18 20 21 24 25 54 81 90 111; *15606*a 65 127-134; 15723g 43; *15790*j (*16373*) 82; 15883e 7 45 58 59 112 **THR tRNA:** *15897*k 97 129; *15907*k 46; *15912*i 81; 15925i 15 81 90 **PRO tRNA:** 15996c/16000g 48 **D LOOP:** 16049k 15 22 46 74 81; 16065g 121 122; *16089*k 21; 16096k 119; 16125k 3 9 74 78 88-91 93 105 119 120 122; *16178*l 26-29; 16208k 11 15 51 52 55; *16217*l 95; *16246*g 46; *16254*a 127 132; 16303k 36 58 59 82 89-93 111 112 131 132; 16310k 1-3 5 6 8-14 20 26-31 58 73 78 82 94-97 99-102 120 125 126 133 134; *16389*g/16390b 11 35-41 82 89; *16398*e 5 11 34 37 66 118 119; *16490*g 10; 16517e 2 11-13 15-24 26-29 51-55 58 59 82 97 105-134.

图 3. *a* 为根据 134 类人 mtDNA（133 个限制性酶切位点图谱和一个参照序列）构建的系统树；*b* 为多态性限制性酶切位点总表。系统树由包括 398 个突变的 mtDNA 的限制性酶切图谱之间的位点差异性构建而成。经测试其他任何的分支排布都不及此树简约。忽略每一个位点只在一种类型的 mtDNA 中出现或缺失的情况，将注意力集中到剩余的 93 个多态性位点，基于此而计算（使用了戴维·斯沃福德博士设计的计算机程序 PAUP）得到了系统树中各分支的排布顺序。计算机程序产生的是一个无根的网络，我们通过把最长一条路径的中点设置为树根（箭头所指）将该网络转化为一个系统树，这条路径将两个支系联系起来（见参考文献 58）。系统树中的数字代表了 mtDNA 的类型，1 型来自南非土著居民（布须曼昆人）的细胞系（GM3043），45 型来自 HeLa 细胞系，110 型是已发表的人的 mtDNA 序列[25]。黑色线条包含一个地域特异的 mtDNA 类型簇；星号表示在多个个体中发现的 mtDNA 类型：134 型出现在六个个体中，29、65 和 80 型 mtDNA 都出现了三次，其他由星号标记的 mtDNA 类型出现了两次。我们考虑到了所有 195 个多态性酶切位点之间的差异，以使系统树的节点与分化百分比相对应。*b* 图中，根据已发表的人类 mtDNA 序列[25]对酶切位点进行编号，分别用 12 个单字母对限制性酶编号：a，*Alu*Ⅰ；b，*Ava*Ⅱ；c，*Ded*Ⅰ；d，*Fnu*DⅡ；e，*Hae*Ⅲ；f，*Hha*Ⅰ；g，*Hinf*Ⅰ；h，*Hpa*Ⅰ；i，*Hpa*Ⅱ；j，*Mbo*Ⅰ；k，*Rsa*Ⅰ；l，*Taq*Ⅰ。斜体字标示的位点为出现在目的 mtDNA 类型中的位点，而非斜体字标示的位点为在目的 mtDNA 类型中没有的位点；括号内是推测位点的其他替代位置；由斜线分开的位点为由一个推断的核苷酸替换导致的两种不同限制性内切酶的多态性位点；黑体加粗的字母为非编码区和 tRNA、核糖体 RNA 及一些蛋白质（ND，NADH 脱氢酶；CO，细胞色素氧化酶；ATP，腺苷三磷酸酶；CYT b，细胞色素 b）的基因。例如，*8*j 就表示酶切位点 *Mbo*Ⅰ的起始位置是在 D 环 8 位核苷酸处，它存在于 2 型 mtDNA 中但不存在于 1 型 mtDNA 中。由于参照序列（110 型 mtDNA）中并没有此位点，8 位起始位置其实是一个半位点，不同于 1 位的 *Mbo*Ⅰ识别序列(参见参考文献 23 中确定这些位点位置的方法)。并不是所有的酶切位点在所有个体中都能找到，如在 1、31、59、63、68 型 mtDNA 和所有新几内亚人的 mtDNA 中没有找到 *8*j、1484e 和 7750c 的位点；114 和 121 型 mtDNA 中没有 *Rsa*Ⅰ位点。一些位点的位置与之前报道的结果[23,24]不同，一些位点出现的个体也与之前报道的结果不同；所做的这些修正是通过对所研究的 mtDNA 的复查得到的。

Many trees of minimal or near-minimal length can be made from the data; all trees that we have examined share the following features with Fig. 3. (1) two primary branches, one composed entirely of Africans, the other including all 5 of the populations studied; and (2) each population stems from multiple lineages connected to the tree at widely dispersed positions. Since submission of this manuscript, Horai *et al.*[29] built a tree for our samples of African and Caucasian populations and their sample of a Japanese population by another method; their tree shares these two features.

Among the trees investigated was one consisting of five primary branches with each branch leading exclusively to one of the five populations. This tree, which we call the population-specific tree, requires 51 more point mutations than does the tree of minimum length in Fig. 3. The minimum-length tree requires fewer changes at 22 of the 93 phylogenetically-informative restriction sites than does the population-specific tree, while the latter tree required fewer changes at four sites; both trees require the same number of changes at the remaining 67 sites. The minimum-length tree is thus favoured by a score of 22 to 4. The hypothesis that the two trees are equally compatible with the data is statistically rejected, since 22 : 4 is significantly different from the expected 13 : 13. The minimum-length tree is thus significantly more parsimonious than the population-specific tree.

African Origin

We infer from the tree of minimum length (Fig. 3) that Africa is a likely source of the human mitochondrial gene pool. This inference comes from the observation that one of the two primary branches leads exclusively to African mtDNAs (types 1–7, Fig. 3) while the second primary branch also leads to African mtDNAs (types 37–41, 45, 46, 70, 72, 81, 82, 111 and 113). By postulating that the common ancestral mtDNA (type a in Fig. 3) was African, we minimize the number of intercontinental migrations needed to account for the geographic distribution of mtDNA types. It follows that b is a likely common ancestor of all non-African and many African mtDNAs (types 8–134 in Fig. 3).

Multiple Lineages per Race

The second implication of the tree (Fig. 3)—that each non-African population has multiple origins—can be illustrated most simply with the New Guineans. Take, as an example, mtDNA type 49, a lineage whose nearest relative is not in New Guinea, but in Asia (type 50). Asian lineage 50 is closer genealogically to this New Guinea lineage than to other Asian mtDNA lineages. Six other lineages lead exclusively to New Guinean mtDNAs, each originating at a different place in the tree (types 12, 13, 26–29, 65, 95 and 127–134 in Fig. 3). This small region of New Guinea (mainly the Eastern Highlands Province) thus seems to have been colonised by at least seven maternal lineages (Tables 2 and 3).

从数据中可以得到很多最小长度或者接近最小长度的进化树，从图 3 中可以看出所有这些进化树有以下共同的特点：(1) 两个一级分支中，一个全部由非洲人群组成，而另外一个一级分支则包括了五个地域的所有人群；(2) 每个群体都源自在广泛分散的位置上与进化树相连接的多个谱系。就在我们提交本文后，宝来等人[29]将他们的日本人群样本结合我们的非洲人和高加索人样本，通过另外一种方法构建了进化树，他们的进化树也同样具有上述的两个特点。

在所研究的进化树中，其中一个进化树有五个一级分支，每个分别只代表了五大地域人群中的一个。此进化树我们称之为群体特异性进化树，与图 3 中最小长度的进化树相比它还需要 51 个点突变。在 93 个系统发育信息性限制性酶切位点中，与群体特异性进化树相比，最小长度的进化树需要在 22 个酶切位点发生更少数量的变异。然而群体特异性进化树需要在 4 个限制性酶切位点处发生更少的变异。在剩下的 67 个限制性酶切位点处两种进化树需要有相同数量的变异。因此最小长度进化树以 22∶4 占优势。22∶4 与预期的 13∶13 相差太大，因此可以否定关于两种进化树在同等程度上与所得到统计数据相符的猜测。因此最小长度进化树就远比群体特异性进化树简约了。

人群的非洲起源

我们从最小长度的进化树（图 3）推测非洲可能是人类线粒体基因库的起源地。这种推测是因为观察到进化树中的两个一级分支中的一个全部指向非洲人群 mtDNA（图 3 中的 1~7 型），第二个也指向非洲人群 mtDNA（37~41、45、46、70、72、81、82、111、113 型）。假设非洲人是所有人群 mtDNA（图 3 中的 a 型）的共同祖先，要满足不同类型 mtDNA 的地理分布需要在地域之间发生一定数量的迁移，我们将此数量进行了最小化。b 可能是所有非非洲人群和许多非洲人群 mtDNA 的共同祖先（图 3 中的 8~134 型）。

每个族群的多个亲缘系

进化树（图 3）给我们的第二个暗示就是每一个非非洲人群有多个起源，这可以用最简单的新几内亚人的例子来阐述。以 49 型 mtDNA 为例，与它亲缘关系最近的并不是新几内亚人，而是亚洲人群（50 型 mtDNA）。与其他亚洲人群的亲缘性相比，亚洲的 50 型谱系与新几内亚 49 型谱系的亲缘关系更近一些。指向新几内亚人群 mtDNA 的另外六个系，每一个都起源于进化树中不同的位置（图 3 中的 12、13、26~29、65、95、127~134 等型）。因此新几内亚的这个小区域（以东部高地省份为主）似乎已经经过了至少七个母系血统的繁衍（表 2 和 3）。

Table 2. Clusters of mtDNA types that are specific to one geographic region

Geographic region	Number of region-specific clusters	Mean pairwise distance within clusters*	Average age of clusters†
Africa	1‡	0.36	90–180
Asia	27	0.21	53–105
Australia	15	0.17	43–85
Europe	36	0.09	23–45
New Guinea	7	0.11	28–55

* For clusters represented by two or more individuals (and calculated for individuals, not for mtDNA types) in Fig. 3.
† Average age in thousands of years based on the assumption that mtDNA divergence occurs at the rate of 2–4% per million years[15,30].
‡ Assuming that Africa is the source, there is only one African cluster.

Table 3. Ancestors, lineages and extents of divergence in the genealogical tree for 134 types of human mtDNA

Ancestor	No. of descendant lineages or clusters specific to a region							Age*
	Total	Africa	Asia	Australia	Europe	N. Guinea	% divergence	
a	7	1	0	0	0	0	0.57	143–285
b	2	0	1	0	0	0	0.45	112–225
c	20	0	7	3	1	3	0.43	108–215
d	2	0	0	1	1	0	0.39	98–195
e	14	2	2	4	2	0	0.34	85–170
f	19	1	7	4	4	1	0.30	75–150
g	10	2	3	2	2	1	0.28	70–140
h	30	2	4	0	15	1	0.27	68–135
i	8	1	0	0	6	0	0.26	65–130
j	22	1	3	1	5	1	0.25	62–125
All	134	10	27	15	36	7	—	—

* Assuming that the mtDNA divergence rate is 2–4% per million years[15,30].

In the same way, we calculate the minimum numbers of female lineages that colonised Australia, Asia and Europe (Tables 2 and 3). Each estimate is based on the number of region-specific clusters in the tree (Fig. 3, Tables 2 and 3). These numbers, ranging from 15 to 36 (Tables 2 and 3), will probably rise as more types of human mtDNA are discovered.

Tentative Time Scale

A time scale can be affixed to the tree in Fig. 3 by assuming that mtDNA sequence divergence accumulates at a constant rate in humans. One way of estimating this rate is

表 2. 特定地理区域的 mtDNA 类型簇

地理区域	区域特异性簇的数量	簇内平均比对距离 *	簇的平均年代 †
非洲	1‡	0.36	90~180
亚洲	27	0.21	53~105
澳大利亚	15	0.17	43~85
欧洲	36	0.09	23~45
新几内亚	7	0.11	28~55

* 为图 3 中两个及两个以上个体（针对个体，而非 mtDNA 类型）中出现的簇。
† 根据 mtDNA 分化速率为每一百万年 2%~4% 的假设计算的平均年代（千年）[15,30]。
‡ 假设非洲是共同起源地，那么只有一个非洲人群 mtDNA 簇。

表 3. 人类 134 种 mtDNA 的系统树中的祖先、世系和分化程度

祖先	一个地区特定的后代世系或簇的数量							年代 *
	总数	非洲	亚洲	澳大利亚	欧洲	新几内亚	分化程度(%)	
a	7	1	0	0	0	0	0.57	143~285
b	2	0	1	0	0	0	0.45	112~225
c	20	0	7	3	1	3	0.43	108~215
d	2	0	0	1	1	0	0.39	98~195
e	14	2	2	4	2	0	0.34	85~170
f	19	1	7	4	4	1	0.30	75~150
g	10	2	3	2	2	1	0.28	70~140
h	30	2	4	0	15	1	0.27	68~135
i	8	1	0	0	6	0	0.26	65~130
j	22	1	3	1	5	1	0.25	62~125
总数	134	10	27	15	36	7	—	—

* 假设每一百万年 mtDNA 的分化速率是 2%~4%[15,30]。

用同样的方法，我们计算了在澳大利亚、亚洲和欧洲繁衍所需的母系血统的最小数量（表 2 和 3）。每个值都是基于系统树中的区域特异性簇的数量而估算出来的（图 3，表 2 和 3）。随着更多类型的人类 mtDNA 的发现，这些范围在 15 到 36 之间（表 2 和 3）的估算值可能也会增加。

尝试构建年代时间表

假定人类中 mtDNA 序列分化是以恒定的速率积累的，我们就能够在图 3 中的进化树上标示出进化的年代时间表。估计此积累速率的一种方法是考虑发生于新几

to consider the extent of differentiation within clusters specific to New Guinea (Table 2; see also refs 23 and 30), Australia[30] and the New World[31]. People colonised these regions relatively recently: a minimum of 30,000 years ago for New Guinea[32], 40,000 years ago for Australia[33], and 12,000 years ago for the New World[34]. These times enable us to calculate that the mean rate of mtDNA divergence within humans lies between two and four percent per million years; a detailed account of this calculation appears elsewhere[30]. This rate is similar to previous estimates from animals as disparate as apes, monkeys, horses, rhinoceroses, mice, rats, birds and fishes[15]. We therefore consider the above estimate of 2–4% to be reasonable for humans, although additional comparative work is needed to obtain a more exact calibration.

As Fig. 3 shows, the common ancestral mtDNA (type a) links mtDNA types that have diverged by an average of nearly 0.57%. Assuming a rate of 2–4% per million years, this implies that the common ancestor of all surviving mtDNA types existed 140,000–290,000 years ago. Similarly, ancestral types b–j may have existed 62,000–225,000 years ago (Table 3).

When did the migrations from Africa take place? The oldest of the clusters of mtDNA types to contain no African members stems from ancestor *c* and included types 11–29 (Fig. 3). The apparent age of this cluster (calculated in Table 3) is 90,000–180,000 years. Its founders may have left Africa at about that time. However, it is equally possible that the exodus occurred as recently as 23–105 thousand years ago (Table 2). The mtDNA results cannot tell us exactly when these migrations took place.

Other mtDNA Studies

Two previous studies of human mtDNA have included African individuals[21,28]; both support an African origin for the human mtDNA gene pool. Johnson *et al.*[21] surveyed ~40 restriction sites in each of 200 mtDNAs from Africa, Asia, Europe and the New World, and found 35 mtDNA types. This much smaller number of mtDNA types probably reflects the inability of their methods to distinguish between mtDNAs that differ by less than 0.3% and may account for the greater clustering of mtDNA types by geographic origin that they observed. (By contrast, our methods distinguish between mtDNAs that differ by 0.03%.) Although Johnson *et al.* favoured an Asian origin, they too found that Africans possess the greatest amount of mtDNA variability and that a midpoint rooting of their tree leads to an African origin.

Greenberg *et al.*[28] sequenced the large noncoding region, which includes the displacement loop (D loop), from four Caucasians and three Black Americans. A parsimony tree for these seven D loop sequences, rooted by the midpoint method, appears in Fig. 4. This tree indicates (1) a high evolutionary rate for the D loop (at least five times faster than other mtDNA regions), (2) a greater diversity among Black American D loop sequences, and (3) that the common ancestor was African.

内亚（表 2；参考文献 23 和 30）、澳大利亚[30]和新世界[31]各自簇内特异的分化程度。人类近期才在这些地区进行殖民：新几内亚、澳大利亚和新世界，开始的时间分别至少为 3 万年[32]、4 万年[33]和 1.2 万年[34]以前。通过这些时间我们可以计算出 mtDNA 分化的平均速率为每一百万年 2%~4%；关于此计算详细说明可参见其他文献[30]。此 mtDNA 分化速率与之前从猿、猴、马、犀牛、小鼠、大鼠、鸟类和鱼等不同的动物中得出的速率[15]相近。因此我们所估算的 2%~4% 的速率值对人类来说是一个相对合理的值，即使仍需要进一步的比较来做更准确的校正。

如图 3 所示，共同 mtDNA 祖先（a 型）与平均分化程度为 0.57% 的 mtDNA 类型相联系。假如进化速率为每一百万年 2%~4%，那么所有存留下来的 mtDNA 类型的共同祖先出现在 14 万年至 29 万年以前。同样推算，祖先类型 b~j 可能出现在 6.2 万年至 22.5 万年前（表 3）。

从非洲向外迁移是什么时候开始的呢？最古老的不含非洲人群 mtDNA 的 mtDNA 类型簇起源于祖先 *c*，包括了 11~29 类型（图 3）。此类型 mtDNA 的表观年龄达 9 万年至 18 万年（计算值见表 3），其祖先可能在这个时间已经离开了非洲。然而，大批迁移同样有可能发生在 2.3 万年至 10.5 万年前（表 2）。分析 mtDNA 的结果并不能准确地告诉我们迁移发生的时间。

其他 mtDNA 的研究

之前的两个关于人类 mtDNA 的研究所用的样本包括非洲的个体[21,28]，它们的结果都支持人类 mtDNA 基因库是从非洲人群起源的。约翰逊等人[21]在非洲、亚洲、欧洲和新世界的 200 例 mtDNA 样品中检测了约 40 个限制性酶切位点，发现了 35 种类型的 mtDNA。这么少的 mtDNA 类型的数目反映出他们的方法不能区分差别小于 0.3% 的 mtDNA，可能还解释了他们所观察到的由于地域起源而产生的更大的 mtDNA 类型簇（相比之下，我们的方法能够区分差异仅为 0.03% 的 mtDNA）。尽管约翰逊等人偏向于认为亚洲人群是人类 mtDNA 的起源，但他们也发现了非洲人群的 mtDNA 变异度最大，并且他们进化树中一个中点树根指向了非洲起源。

格林伯格等人[28]测定了四个高加索人和三个美国黑人 mtDNA 中包括替代环（D 环）在内的大段非编码区序列。图 4 是这 7 个 D 环序列的简约进化树，通过中点法得出树根。从进化树中可以看出：(1) D 环的进化速率很高（至少是其他 mtDNA 区域的五倍）；(2) D 环在美国黑人中的多样性要更大些；(3) 共同祖先是非洲人群。

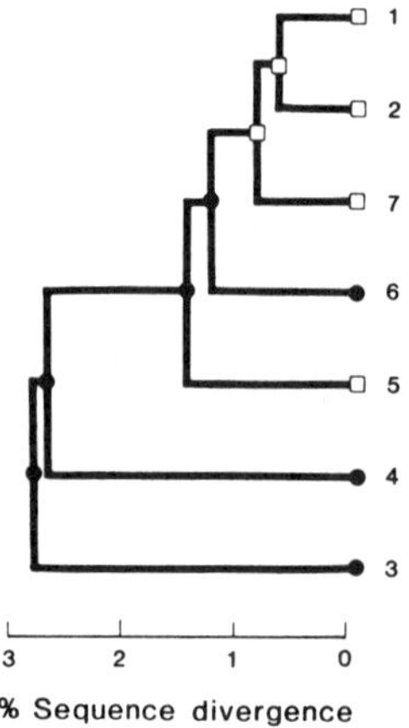

Fig. 4. Genealogical tree relating the nucleotide sequences of D loops from seven human mtDNAs. This tree, which requires fewer mutations than any other branching order, was constructed by the PAUP computer program from the 900-bp sequences determined by Greenberg *et al.*[28] and was rooted by the midpoint method. Symbols: ●, African origin (Black American); □, Caucasian.

Nuclear DNA Studies

Estimates of genetic distance based on comparative studies of nuclear genes and their products differ in kind from mtDNA estimates. The latter are based on the actual number of mutational differences between mtDNA genomes, while the former rely on differences in the frequencies of molecular variants measured between and within populations. Gene frequencies can be influenced by recombination, genetic drift, selection, and migration, so the direct relationship found between time and mutational distance for mtDNA would not be expected for genetic distances based on nuclear DNA. But studies based on polymorphic blood groups, red cell enzymes, and serum proteins show that (1) differences between racial groups are smaller than those within such groups and (2) the largest gene frequency differences are between Africans and other populations, suggesting an African origin for the human nuclear gene pool[11,12,35]. More recent studies of restriction site polymorphisms in nuclear DNA [14,36-42] support these conclusions.

Relation to Fossil Record

Our tentative interpretation of the tree (Fig. 3) and the associated time scale (Table 3) fits with one view of the fossil record: that the transformation of archaic to anatomically modern forms of *Homo sapiens* occurred first in Africa[43-45], about 100,000–140,000 years ago, and that all present-day humans are descendants of that African population. Archaeologists have observed that blades were in common use in Africa 80–90 thousand years ago, long before they replaced flake tools in Asia or Europe[46,47]. But the agreement between our molecular view and the evidence from palaeoanthropology and archaeology should be treated cautiously for two reasons. First, there is much uncertainty about the ages of these remains. Second, our placement of the common ancestor of all human mtDNA diversity in Africa 140,000–280,000 years ago need not imply that the transformation to anatomically modern *Homo sapiens* occurred in Africa at this time. The

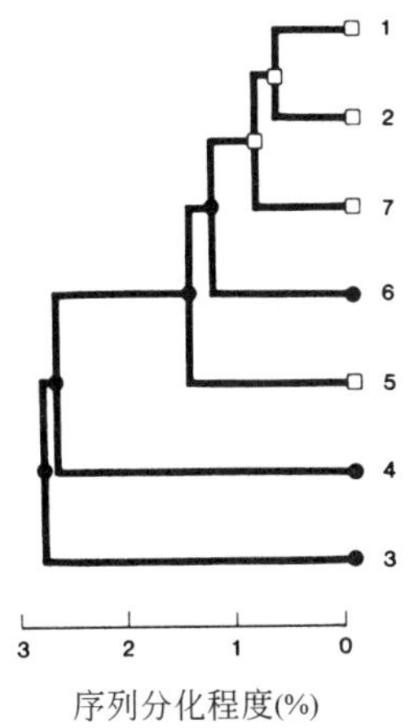

图 4. 根据来自七个人 mtDNA 分子的 D 环核苷酸序列而构建的系统树。此树是根据格林伯格等[28]测出的 900 个碱基对序列通过 PAUP 程序构建的；与其他任何分支顺序相比，此树需要更少数量的突变；中点法确定树根。符号：●，起源于非洲的人群（美国黑人）；□，高加索人。

对核 DNA 的研究

基于比较核基因以及它们的产物而估算出的遗传距离与通过 mtDNA 进化而估算出的遗传距离是不同的。后者基于 mtDNA 基因组之间突变差异的实际数目，而前者依赖于群体间或群体内分子变异的频率差异。而基因重组、遗传漂变、自然选择和迁移都能够影响基因频率，因此 mtDNA 中时间与突变距离之间的直接关系与基于核 DNA 的遗传距离是不能等同的。但是对于血型、红细胞酶和血清蛋白多态性的研究显示：(1) 族群之间的差异要小于族群内的差异，(2) 基因频率差异最大的是非洲人群和其他人群之间的差异，这也暗示了人类核基因库[11,12,35]起源于非洲人群。最近的一些关于核 DNA 限制性酶切位点多态性[14,36-42]的研究也支持这些结论。

与化石记录的联系

我们对系统树尝试性的阐述（图 3）和相关的年代时间表（表 3）与一定的化石记录相符合：化石记录显示解剖学上从远古人到现代人群智人的转化最初发生在 10 万年至 14 万年前的非洲[43-45]，所有现在的人都是那时候非洲人群体的后裔。考古学家发现 8 万年至 9 万年前刀片在非洲非常常用，很久之后在亚洲和欧洲所使用的石片工具才被这种刀片取代[46,47]。然而我们应该谨慎对待我们在分子水平上的观点与古人类学以及考古学证据的一致性，原因有两点。首先，这些遗物的年代还有很大的不确定性；其次我们提出的观点是所有人类 mtDNA 类型的共同祖先出现在 14 万年至 28 万年前的非洲，此观点不需要揭示解剖学上远古人到现代人的转化就

mtDNA data tell us nothing of the contributions to this transformation by the genetic and cultural traits of males and females whose mtDNA became extinct.

An alternative view of human evolution rests on evidence that *Homo* has been present in Asia as well as in Africa for at least one million years[48] and holds that the transformation of archaic to anatomically modern humans occurred in parallel in different parts of the Old World[33,49]. This hypothesis leads us to expect genetic differences of great antiquity within widely separated parts of the modern pool of mtDNAs. It is hard to reconcile the mtDNA results with this hypothesis. The greatest divergences within clusters specific to non-African parts of the World correspond to times of only 90,000–180,000 years. This might imply that the early Asian *Homo* (such as Java man and Peking man) contributed no surviving mtDNA lineages to the gene pool of our species. Consistent with this implication are features, found recently in the skeletons of the ancient Asian forms, that make it unlikely that Asian *erectus* was ancestral to *Homo sapiens*[50–52]. Perhaps the non-African *erectus* population was replaced by *sapiens* migrants from Africa; incomplete fossils indicating the possible presence of early modern humans in western Asia at Zuttiyeh (75,000–150,000 years ago) and Qafzeh (50,000–70,000 years ago) might reflect these first migrations[45,53].

If there was hybridization between the resident archaic forms in Asia and anatomically modern forms emerging from Africa, we should expect to find extremely divergent types of mtDNA in present-day Asians, more divergent than any mtDNA found in Africa. There is no evidence for these types of mtDNA among the Asians studied[21,54–56]. Although such archaic types of mtDNA could have been lost from the hybridizing population, the probability of mtDNA lineages becoming extinct in an expanding population is low[57]. Thus we propose that *Homo erectus* in Asia was replaced without much mixing with the invading *Homo sapiens* from Africa.

Conclusions and Prospects

Studies of mtDNA suggest a view of how, where and when modern humans arose that fits with one interpretation of evidence from ancient human bones and tools. More extensive molecular comparisons are needed to improve our rooting of the mtDNA tree and the calibration of the rate of mtDNA divergence within the human species. This may provide a more reliable time scale for the spread of human populations and better estimates of the number of maternal lineages involved in founding the non-African populations.

It is also important to obtain more quantitative estimates of the overall extent of nuclear DNA diversity in both human and African ape populations. By comparing the nuclear and mitochondrial DNA diversities, it may be possible to find out whether a transient or prolonged bottleneck in population size accompanied the origin of our species[15]. Then a fuller interaction between palaeoanthropology, archaeology and molecular biology will allow a deeper analysis of how our species arose.

We thank the Foundation for Research into the Origin of Man, the National Science

发生在此时的非洲。从已经灭绝的古人类到现代人在遗传和文化特征方面发生了什么样的转化，我们是无法从 mtDNA 的数据中找到答案的。

有证据表明人在亚洲和非洲都存在了至少一百万年 [48]，因此有另外一种关于人类进化的观点，即认为从远古人到解剖学意义上的现代人的转化并行发生在旧世界的不同区域 [33,49]。这个假设使得我们期望看到广泛分布有现代人类 mtDNA 库的地域在远古时候就有明显的基因差异。但是很难将 mtDNA 结果与该假设对应起来。世界上非非洲人群特异性簇内最大分化所对应的时间仅有 9 万年至 18 万年。这也许暗示了早期亚洲人属（比如爪哇猿人和北京猿人）没有存留下来的 mtDNA 血统贡献于当今人群的基因库。与此暗示相符的是，最近发现的一些古亚洲人骨骼的特征说明亚洲直立人不可能是智人的祖先 [50-52]。也许非洲以外的直立人被来自非洲的智人移民替换了，不完整的化石记录显示在西亚的祖提耶（7.5 万年至 15 万年以前）和卡夫泽（5 万年至 7 万年前）可能出现了最初的现代人群，这也许反映了这些最初的移民 [45,53]。

如果亚洲古代居民与非洲出现的解剖学意义上的现代人发生交配，我们应该能在当今亚洲人的 mtDNA 类型里找到一些极端分化的类型，其分化程度要比在非洲发现的 mtDNA 类型分化程度更高。但对亚洲人的研究中 [21,54-56] 没有任何这种 mtDNA 类型存在的证据。尽管这些古老的 mtDNA 类型可能在种群之间的杂交过程中会丢失，但是 mtDNA 血统在一个膨胀的种群中灭绝的可能性是很低的 [57]。因此我们推测亚洲的直立人被入侵的非洲智人取代了，而且他们之间基本没有遗传混合。

结论和展望

对人类 mtDNA 的研究为我们揭示了现代人是在何时何地以何种方式起源的，这些结果与古人类骨骼和所使用工具的证据相符。仍然需要更多的分子水平上的比较来改善我们构建的 mtDNA 进化树的树根和校正人类 mtDNA 的分化速率。这样能够提供更加可信的人群扩散时间表，也能够更加准确地估算在所发现的非非洲人群体里所包含的母系血统的数量。

获得更多人类和非洲猿的核 DNA 多样性整体程度的定量估计也很重要。通过比较核 DNA 和 mtDNA 的多样性，我们有可能发现伴随在人类起源历史中的 [15] 种群大小瓶颈效应究竟是短暂的还是长期的。古人类学、考古学和分子生物学的充分互动将有利于更深入地分析我们人类是怎样起源的。

在此我们要感谢人类起源研究基金、国家科学基金会和美国国立卫生研究院的

Foundation and the NIH for support. We also thank P. Andrews, K. Bhatia, F. C. Howell, W. W. Howells, R. L. Kirk, E. Mayr, E. M. Prager, V. M. Sarich, C. Stringer and T. White for discussion and help in obtaining placentas.

(**325**, 31-36; 1987)

Rebecca L. Cann*, Mark Stoneking & Allan C. Wilson
Department of Biochemistry, University of California, Berkeley, California 94720, USA.

Received 17 March; accepted 7 November 1986.

References:

1. Goodman, M. *Hum. Biol.* **35**, 377-424 (1963).
2. Sarich, V. M. & Wilson, A. C. *Science* **158**, 1200-1203 (1967).
3. King, M. C. & Wilson, A. C. *Science* **188**, 107-116 (1975).
4. Ferris, S. D., Wilson, A. C. & Brown, W. M. *Proc. Natl. Acad. Sci. U.S.A.* **78**, 2432-2436 (1981).
5. Brown, W. M., Prager, E. M., Wang, A. & Wilson, A. C. *J. Molec. Evol.* **18**, 225-239 (1982).
6. Sibley, C. G. & Ahlquist, J. E. *J. Molec. Evol.* **20**, 2-15 (1984).
7. Bianchi, N. O., Bianchi, M. S., Cleaver, J. E. & Wolff, S. *J. Molec. Evol.* **22**, 323-333 (1985).
8. O'Brien, S. J. *et al. Nature* **317**, 140-144 (1985).
9. Cavalli-Sforza, L. L. *Proc. R. Soc. Lond.* **B164**, 362-379 (1966).
10. Cavalli-Sforza, L. L. *Sci. Am.* **231(3)**, 81-89 (1974).
11. Nei, M. & Roychoudhury, A. K. *Evol. Biol.* **14**, 1-59 (1982).
12. Nei, M. in *Population Genetics and Molecular Evolution* (eds Ohta, T. & Aoki, K.) 41-64 (Japan Sci. Soc. Press, Tokyo, 1985).
13. Constans, J. *et al. Am. J. Phys. Anthrop.* **68**, 107-122 (1985).
14. Wainscoat, J. S. *et al. Nature* **319**, 491-493 (1986).
15. Wilson, A. C. *et al. Biol. J. Linn. Soc.* **26**, 375-400 (1985).
16. Olivo, P. D., Van de Walle, M. J., Laipis, P. J. & Hauswirth, W. W. *Nature* **306**, 400-402 (1983).
17. Brown, W. M. *Proc. Natl. Acad. Sci. U.S.A.* 77, 3605-3609 (1980).
18. Monnat, R. J. & Loeb, L. A. *Proc. Natl. Acad. Sci. U.S.A.* **82**, 2895-2899 (1985).
19. Monnat, R. J., Maxwell, C. L. & Loeb, L. A. *Cancer Res.* **45**, 1809-1814 (1985).
20. Hauswirth, W. W. & Laipis, P. J. in *Achievements and Perspectives in Mitochondrial Research*, Vol. 2 *Biogenesis* (eds E. Quagliariello, E. C. Slater, F. Palmieri, C. Saccone & A. M. Kroon) 49-60 (Elsevier, New York, 1986).
21. Johnson, M. J. *et al. J. Molec. Evol.* **19**, 255-271 (1983).
22. Cann, R. L. Ph. D. Thesis, Univ. California, Berkeley, California (1982).
23. Stoneking, M., Bhatia, K. & Wilson, A. C. in *Genetic Variation and Its Maintenance in Tropical Populations* (eds Roberts, C. F. & Destefano, G.) 87-100 (Cambridge University Press, Cambridge, 1986).
24. Cann, R. L., Brown, W. M. & Wilson, A. C. *Genetics* **106**, 479-499 (1984).
25. Anderson, S. *et al. Nature* **290**, 457-465 (1981).
26. Nei, M. & Tajima, F. *Genetics* **105**, 207-217 (1983).
27. Whittam, T. S., Clark, A. G., Stoneking, M., Cann, R. L. & Wilson, A. C. *Proc. Natl. Acad. Sci. U.S.A.* (in the press).
28. Greenberg, B. D., Newbold, J. E. & Sugino, A. *Gene* **21**, 33-49 (1983).
29. Horai, S., Gojobori, T. & Matsunaga, E. *Jap. J. Genet.* **61**, 271-275 (1986).
30. Stoneking, M., Bhatia, K. & Wilson, A. C. *Cold Spring Harb. Symp. Quant. Biol.* **51**, 433-439 (1986).
31. Wallace, D. C., Garrison, K. & Knowler, W. C. *Am. J. phys. Anthrop.* **68**, 149-155 (1985).
32. Jones, R. *A. Rev. Anthrop.* **8**, 445-466 (1979).
33. Wolpoff, M. H., Wu, X. Z. & Thorne, A. G. in *Origins of Modern Humans: A World Survey of the Fossil Evidence* (eds Smith, F. H. & Spencer, F.) 411-484 (Liss, New York, 1984).
34. Owen, R. C. in *Origins of Modern Humans: A World Survey of the Fossil Evidence* (eds Smith, F. H. & Spencer, F.) 517-564 (Liss, New York, 1984).
35. Mourant, A. E. *et al. The Distribution of the Human Blood Groups and Other Polymorphism* (Oxford University Press, Oxford, 1978).
36. Murray, J. C. *et al. Proc. Natl. Acad. Sci. U.S.A.* **81**, 3486-3490 (1984).

支持。还要感谢安德鲁斯、巴蒂亚、豪厄尔、豪厄尔斯、柯克、普拉格、迈尔、普拉格、萨里奇、斯特林格和怀特在研究结果讨论和获得胎盘样本方面的帮助。

（周晓明 姜薇 翻译；吕雪梅 审稿）

37. Cooper, D. N. & Schmidtke, J. *Hum. Genet.* **66**, 1-16 (1984).

38. Cooper, D. N. *et al. Hum. Genet.* **69**, 201-205 (1985).

39. Hill, A. V. S., Nicholls, R. D., Thein, S. L. & Higgs, D. R. *Cell* **42**, 809-819 (1985).

40. Chapman, B. S., Vincent, K. A. & Wilson, A. C. *Genetics* **112**, 79-92 (1986).

41. Chakravarti, A. *et al. Proc. Natl. Acad. Sci. U.S.A.* **81**, 6085-6089 (1984).

42. Chakravarti, A. *et al. Am. J. Hum. Genet.* **36**, 1239-1258 (1984).

43. Rightmire, G. P. in *Origins of Modern Humans: A World Survey of the Fossil Evidence* (eds Smith, F. H. & Spencer, F.) 295-326 (Liss, New York, 1984).

44. Bräuer, G. *Courier Forsch. Int. Senckenberg* **69**, 145-165 (1984).

45. Bräuer, G. in *Origins of Modern Humans: A World Survey of the Fossil Evidence* (eds Smith, F. H. & Spencer, F.) 327-410 (Liss, New York, 1984).

46. Isaac, G. *Phil. Trans. R. Soc. Lond.* B**292**, 177-188 (1981).

47. Clark, J. D. *Proc. Br. Acad. Lond.* **67**, 163-192 (1981).

48. Pope, G. G. *Proc. Natl. Acad. Sci. U.S.A.* **80**, 4988-4992 (1983).

49. Coon, C. S. *The Origin of Races* (Knopf, New York, 1962).

50. Stringer, C. B. *Courier Forsch. Inst. Senckenberg* **69**, 131-143 (1984).

51. Andrews, P. *Courier Forsch. Inst. Senckenberg* **69**, 167-175 (1984).

52. Andrew, P. *New Scient.* **102**, 24-26 (1984).

53. Stringer, C. B., Hublin, J. J. & Vandermeersch, B. in *Origins of Modern Humans: A World Survey of the Fossil Evidence* (eds Smith, F. H. & Spencer, F.) 51-136 (Liss, New York, 1984).

54. Horai, S., Goobori, T. & Matsunaga, E. *Hum. Genet.* **68**, 324-332 (1984).

55. Bonne-Tamir, B. *et al. Am. J. Hum. Genet.* **38**, 341-351 (1986).

56. Horai, S. & Matsunaga, E. *Hum. Genet.* **72**, 105-117 (1986).

57. Avise, J. C., Neigel, J. E. & Arnold, J. *J. Molec. Evol.* **20**, 99-105 (1984).

58. Farris, J. S. *Am. Nat.* **106**, 645-668 (1972).

Transmission Dynamics of HIV Infection

R. M. May and R. M. Anderson

Editor's Note

By 1987, it was clear that the disease called AIDS had become an international epidemic. There were many schemes for developing vaccines against the causative virus (known as HIV) and formal treatments of the epidemiology of the infective process, of which this is one of the most significant. Robert May (now Lord May of Oxford) was at the time at Princeton University in the United States; an Australian, he was trained as a physicist at the University of Sydney, before converting to biology and, after migrating to Oxford, England, became Chief Scientific Adviser to the British government and president of the Royal Society. Notice that, in 1987, the authors were unsure whether heterosexual transmission was significant.

Simple mathematical models of the transmission dynamics of human immunodeficiency virus help to clarify some of the essential relations between epidemiological factors, such as distributed incubation periods and heterogeneity in sexual activity, and the overall pattern of the AIDS epidemic. They also help to identify what kinds of epidemiological data are needed to make predictions of future trends.

DESPITE remarkable advances in understanding the basic biology of human immunodeficiency virus (HIV—the aetiological agent of AIDS, acquired immune deficiency syndrome)[1-5] public health planning continues to be hampered by uncertainties about epidemiological parameters[4-6]. Accurate information about the typical duration and intensity of infectiousness, or about the fraction of those infected who will go on to develop AIDS (and after how long), will emerge only from carefully designed studies on these same timescales, which is to say many years. In the absence of such information, mathematical models of the transmission dynamics of HIV cannot be used at present to make accurate predictions of future trends in the incidence of AIDS, but they can facilitate the indirect assessment of certain epidemiological parameters, clarify what data is required to predict future trends, make predictions under various specified assumptions about the course of infection in individuals and patterns of sexual activity within defined populations (or changes therein) and, more generally, provide a template to guide the interpretation of observed trends[7,8].

Whether an infection can establish itself and spread within a population is determined by the key parameter R_0, the basic reproductive rate of the infection[7]. R_0 is the average number of secondary infections produced by one infected individual in the early stages of an epidemic (when essentially all contacts are susceptible); clearly the infection can

HIV的传播动力学

梅，安德森

编者按

到 1987 年，很显然艾滋病已成为国际性传染病。已有许多开发疫苗的研究，都是致力于抵抗这种致病病毒（我们所知的 HIV）和其感染过程中流行病学方面的正规治疗，下面发表的文章就是其中之一。当时，澳大利亚人罗伯特·梅（即现在牛津的梅勋爵）在美国普林斯顿大学，在转到生物专业之前，他在悉尼大学学习物理学。移居到英国牛津以后，他担任英国政府首席科学顾问和英国皇家学会主席。值得注意的是，在 1987 年作者尚不确定异性传染是否为主要的传染方式。

人类免疫缺陷病毒传播动力学的简单数学模型帮助我们阐明了一些流行病学因素之间的重要关系，比如分布式潜伏期和异性之间的性活动以及艾滋病传染的所有形式。这些简单的数学模型也帮助我们确定了预测艾滋病的发展趋势都需要哪些类型的流行病学数据。

尽管在人类免疫缺陷病毒（HIV——艾滋病，即获得性免疫缺陷综合征的发病原因）[1-5] 的基本生物学特性上的认识已有显著进展，但是公共卫生计划仍然受到流行病学不确定因素的阻碍 [4-6]。要想获得关于感染性的典型持续时间和感染强度，或者被感染者发展为艾滋病的比例（以及多久之后发展为艾滋病）这些方面的准确信息，只有在这些事件发生的同时进行仔细设计的研究才可以，也就是说这需要很多年。缺少这些数据，目前 HIV 传染动力学的数学模型就不能用来准确预测艾滋病将来的发展趋势，但是它们有助于间接评价某些流行病学参数，弄清楚预测未来发展趋势需要哪些数据，在多种特定推测的基础上，对个人的传染过程和限定人群中性活动的模式（或者改变）做出预测，通俗地讲，也就是提供一个模板来指导对于已观察到的趋势进行解释 [7,8]。

感染是否可以自身发生并在人群中传播是由关键的参数 R_0，即感染的基本再生率决定的 [7]。R_0 是指在感染的早期阶段（这时基本上所有接触都是易感的）一个受感染的个体引起的继发性感染的平均数；很明显只要 R_0 超过 1[9,10]，病毒传染就可以

maintain itself within the population only if R_0 exceeds unity[9,10]. For a sexually transmitted disease (STD), R_0 depends on c, which is essentially the average rate at which new sexual partners are acquired, on β, the average probability that infection is transmitted from an infected individual to a susceptible partner (per partner contact) and on D, the average duration of infectiousness[7,11]. In what follows, we mainly restrict attention to the spread of HIV among homosexual males, now responsible for the bulk of AIDS cases (about 70–80% in the United States, and a similar proportion in European countries[6,12,13]).

Initial Stages of the Epidemic

The characteristics of most STDs cause their epidemiology to differ from that of common childhood viral infections[11,14,15]. Unlike infections caused by airborne transmission, the rate at which new infections are produced is not dependent on the density of the host. Second, the carrier phenomenon in which certain individuals harbour asymptomatic infection is often important (as in the spread of herpesvirus). Third, many STDs induce little or no acquired immunity on recovery (for example, gonorrhoea) and fourth, net transmission depends on the degree of heterogeneity in sexual activity prevailing in the population.

As Hethcote and Yorke[11] have shown in their studies of gonorrhoea, mathematical models for the dynamics of STDs need to take account of the substantial variations of sexual activity within the population at risk. A particular risk group (such as male homosexuals in San Francisco[12]) of total size N may be roughly partitioned into subgroups of size N_i, each of whom on the average acquire i new sexual partners per unit time (when $N = \Sigma_i N_i$). The probability that susceptible individuals in this ith group will become infected, per unit time, is thus $i\lambda$, where λ is the probability that infection is acquired from any one new partner. In turn, λ is equal to the product of the transmission probability β defined above and the probability that any one randomly-chosen partner is infected (with such partners being more likely to come from the sub-groups of individuals with high degrees of sexual activity).

Exponential Growth

When these assumptions are incorporated into a model for the transmission dynamics of HIV infection, the infected fraction of the population at risk (who are seropositive in tests for HIV) rises exponentially, as $\exp(\Lambda t)$, in the early stages of the epidemic. The exponential growth rate, Λ, is related to the basic epidemiological quantities defined above by:

$$\Lambda = \beta c - 1/D \tag{1}$$

The effective average over the distribution by degrees of sexual activity, c, is given explicitly as

$$c = \Sigma i^2 N_i / \Sigma_i N_i = m + \sigma^2/m \tag{2}$$

where m is the mean and σ^2 the variance of the distribution of the number of new sexual partners per unit of time[8]. Thus, c is not simply the mean but the mean plus the ratio

在人群中维系。对于性传播疾病（STD），R_0 受 c（获得新的性伙伴的平均速率）、β（从一个感染者传播到易感伙伴——每次与伙伴接触的平均概率）和 D（具有感染性的平均持续时间）的影响[7,11]。在下文中，我们的注意力主要集中在 HIV 在男性同性恋之间的传播，现在这在艾滋病病例中占有大部分的比例（在美国大概有 70%~80%，在欧洲国家比例也差不多[6,12,13]）。

流行病的最初阶段

大多数 STDs 的特点导致其与常见的儿童病毒感染的流行病学不同[11,14,15]。第一，不像空气传播引起的感染，新感染的概率不取决于宿主的密度。第二，某些个体感染后没有任何症状，这种携带者现象通常很重要（如疱疹病毒的传播）。第三，许多 STDs 在恢复期导致很少或没有获得性免疫力（如淋病）。第四，网状传播主要依赖于人群中性活动的异质性的程度。

就像赫思克特和约克[11]在淋病研究中发现的那样，STDs 动力学数学模型需要考虑危险人群中性活动的实质性变量。一个总数为 N 的特定的危险群体（比如旧金山的男性同性恋群体[12]）可以大致分成几个亚群 N_i，单位时间内平均每个人找到 i 个新的性伙伴（$N=\Sigma_i N_i$）。$i\lambda$ 是单位时间内，第 i 组易感个体被感染的可能性，λ 是来自任何一个新的性伙伴的感染概率。反过来，λ 等于上述定义的传染率 β 与任何一个随机选择的性伙伴被感染的概率的乘积（这些性伙伴更有可能来自具有高频率性活动的亚群）。

指 数 增 长

当把这些假设与 HIV 感染的传播动力学模型结合起来时，在流行病的早期阶段，危险人群（HIV 检测呈阳性）被感染的比例呈指数上升，记作 $\exp(\Lambda t)$。指数增长速率（Λ）与上面定义的基本流行病学数量相关。

$$\Lambda = \beta c - 1/D \tag{1}$$

性活动程度分布的有效平均数 c，如下面公式所描述

$$c = \Sigma i^2 N_i/\Sigma_i N_i = m + \sigma^2/m \tag{2}$$

其中 m 是平均数，σ^2 是单位时间内新的性伴侣的数量分布的方差[8]。因此，c 就不仅仅是平均值，而是平均值加上方差比例，通过高度活跃的个体（在性活动分布概

of variance to mean, which reflects the disproportionate role played by highly active individuals (in the tail of the probability distribution of sexual activity), who are both more likely to acquire infection and more likely to transmit it. The basic reproductive rate for HIV infection, R_0, is related to the parameters β, c and D, and hence to Λ by the formula

$$R_0 = \beta c\, D \tag{3}$$

In contrast with standard epidemiological models in homogeneous populations (where the exponential phase of rising incidence lasts until something like half the pool of susceptibles have been infected), the early exponential phase is of relatively short duration in our HIV models, giving way to a more nearly linear rise in the fraction infected (see Fig. 2).

This is because most susceptibles in the sexually highly active categories are infected in the early stages of the epidemic, producing saturation effects in these categories which decrease the exponential rise in incidence within them; although the incidence of infection continues to rise among individuals in less sexually active categories, the overall rate of increase is now slower than exponential.

Much less information is available about the rise in the number of individuals infected with HIV, as a function of time, than about the rise in the subsequent incidence of AIDS[46,12,13], largely because information about infection requires serological examination for antibodies to the HIV virus. Although the initial infection may produce symptoms[3,16,17] and, in some cases acute encephalopathy[18] and meningitis[16], it is not clear that such symptoms are always evoked: in any event, the symptoms are usually sufficiently mild to preclude systematic reporting. By contrast, the opportunistic infections cancers[4,5,12,19], and subsequent mortality characteristic of the destruction of the immune system in AIDS, leads to fairly reliable reporting[20]. There is, however, one study of hepatitis B virus (HBV) in a cohort of 6,875 homosexual and bisexual males in San Francisco, which resulted in serum samples being taken and preserved as early as 1978[12,21,22]; stored sera of a representative sample of 785 of these individuals gives the rise in the fraction seropositive for HIV, from 1978 to 1985, shown in Fig. 1.

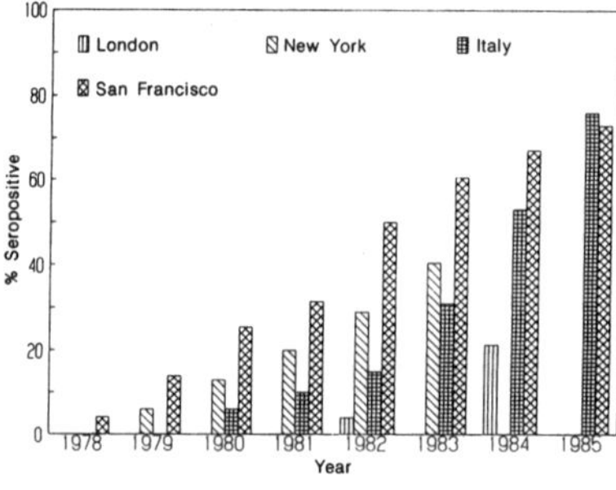

Fig. 1. The rise in seropositivity to HIV antigens in cohorts of patients over the period 1978–1985. The studies in San Francisco[12], London[23] and New York[35] were of homosexual/bisexual males. The study in Italy[24] is of drug addicts.

率的尾部）不成比例的作用，反映了这部分人群既更加有可能被传染也更有可能传播 HIV。HIV 传染的基本再生率 R_0 与 β，c 和 D 几个参数有关，并且根据 Λ 得到下面公式

$$R_0 = \beta c D \tag{3}$$

与同类人群中标准流行病学模型（发病率指数上升阶段持续到半数的易感人群已被感染）相比，在我们的 HIV 模型中，早期的指数阶段的持续时间相对较短，被感染的比例几乎呈直线上升（见图 2）。

这是因为性高度活跃人群中的大多数易感染者在疫情的早期阶段被感染，在这些人群中产生饱和效应，抑制了他们疾病发生率的指数增长。尽管在性活动较少的人群中感染的发生率持续增长，但是总的增长率仍低于指数增长速率。

随着时间的变化，与艾滋病并发率的增长相比，很少有关于感染 HIV 个体的数目增长的信息可以利用 [46,12,13]。主要是因为感染信息需要 HIV 病毒抗体的血清学检测。尽管感染初期可能会产生症状 [3,16,17]，如某些病例中的急性脑病 [18] 和脑膜炎 [16]，但是这些症状是否总是被诱发尚不清楚：通常情况下，这些症状十分温和不会引起预警系统报告。相比之下，机会性感染的癌症 [4,5,12,19] 和随后的艾滋病引起的免疫系统破坏的死亡特性产生清楚可靠的报告 [20]。然而，有一项对旧金山的 6,875 个男性同性恋和双性恋群体进行的乙肝病毒（HBV）的研究，使得早在 1978 年就开始采集和保存血清标本 [12,21,22]，储存的这些有代表性的血清标本中，从 1978 年到 1985 年，这些个体中有 785 个人的 HIV 血清阳性反应比例呈上升趋势，见图 1。

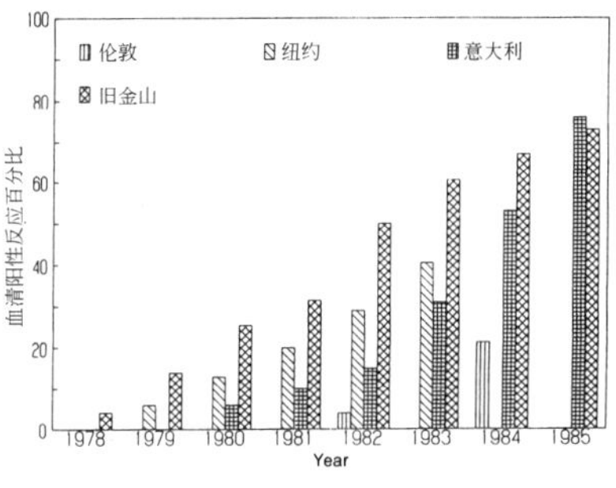

图 1. 1978~1985 年期间，被统计的病人中 HIV 抗原血清阳性反应呈增长趋势。在旧金山 [12]、伦敦 [23] 和纽约 [35] 研究的是男性同性恋和双性恋。在意大利 [24] 研究的是吸毒者。

The pattern of roughly linear rise shown in Fig. 1 is uncharacteristic of standard epidemics (in homogeneously mixed populations), but is suggested by our HIV models. In Britain and other countries in Europe, the virus seems first to have appeared several years later than in the United States (Fig. 2*a*, *b* and *c*), and the spread of infection is still in its early stages. As a result there are serological studies focused on HIV roughly from its initial appearance in Europe[23,24] (Fig. 1). The initially exponential rise in HIV infection may be characterized by a doubling time, t_d, related to the growth rate, Λ, of (1) by $t_d = (\ln 2)/\Lambda$.

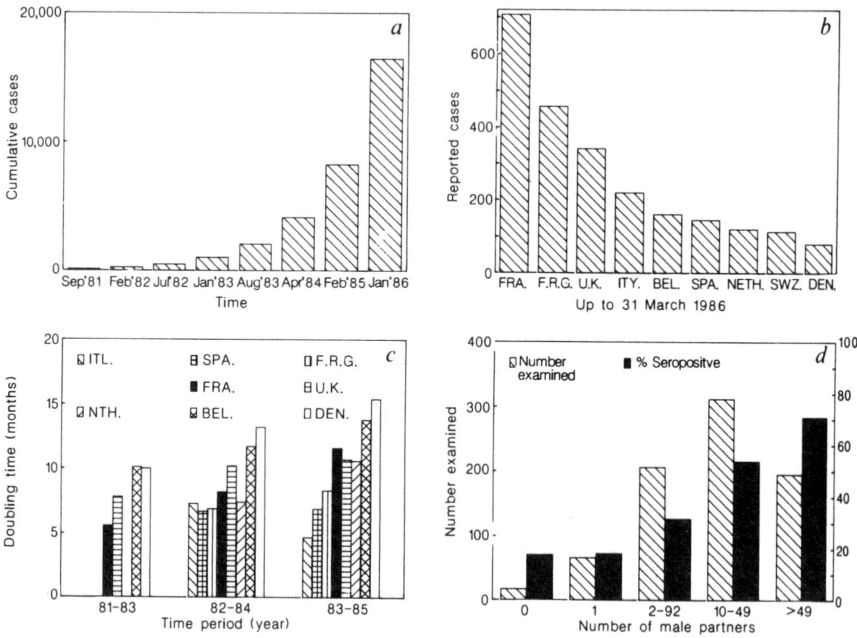

Fig. 2. *a*, The rise in the cumulative number of reported cases of AIDS in the USA over the interval September 1981–January 1986[13]. *b*, Reported cases of AIDS in 9 countries of the European Community up to 31 March 1986[40]. *c*, Doubling times in the cumulative incidence of AIDS (t_d) recorded in months for various European countries[40] over various time intervals (1981–83, 1982–84, 1983–85; DEN, Denmark; BEL, Belgium; NTH, Netherlands; FRA, France; E.C., European Community in total; F.R.G., Federal Republic of Germany; SPA, Spain; ITL, Italy; UK, United Kingdom). *d*, The relationship between sexual activity amongst a sample of homosexual/bisexual males (from San Francisco, USA) as measured by the number of male partners over a two-year period, and the percentage of each group (based on sex partners) who were seropositive for HIV antibodies (data from ref. 26).

Table 1 summarizes information about doubling times deduced from serological and case notification studies, which lead to a surprisingly consistent estimate of $t_d \sim$ 8–10 months in the early stages of the epidemic (Fig. 2*d*) giving an estimate of Λ of about 1.0 yr^{-1}.The characteristic duration of infection (and infectiousness), D, is probably not significantly less than the characteristic time from HIV infection to manifestation of AIDS.

我们的HIV模型表明，如图1中所示的大致呈直线增长的模式，并不是标准流行病（在同类的混合群体中）的特点。在英国和其他的欧洲国家，这种病毒首次出现的时间要比美国晚几年（图2*a*，*b*和*c*），传染病的蔓延仍旧处于早期阶段。结果大概从它最初在欧洲出现以后[23,24]（图1），才有集中于HIV的血清学研究。HIV感染的最初指数增长可以用倍增时间来描述，t_d，与生长速率Λ有关。公式（1）变为$t_d=(\ln 2)/\Lambda$。

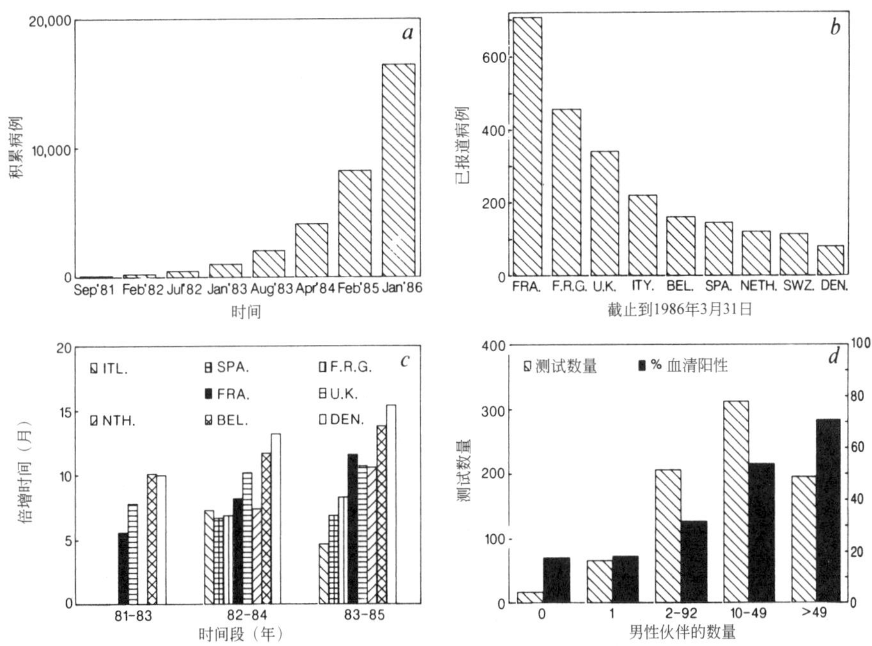

图2. *a*，在美国，1981年9月到1986年1月这段时间里[13]，所报道的艾滋病病例总数的增长。*b*，欧共体9个国家，截止到1986年3月31日报道的艾滋病病例的数量[40]。*c*，多个欧洲国家[40]在不同时间段内艾滋病累积发生的倍增时间(t_d)（1981~83、1982~84、1983~85；DEN，丹麦；BEL，比利时；NTH，荷兰；FRA，法国；E.C.，欧共体总计；F.R.G.，德国；SPA，西班牙；ITL，意大利；UK，英国）。*d*，取样男性同性恋/双性恋（美国旧金山），记录两年内男性伙伴的数量，比较性活动与每组（性伴侣为基础）HIV抗体反应阳性百分比（数据来自参考文献26）之间的关系。

表1，通过推断血清学和呈报病例的研究，总结了关于倍增时间的信息，给出了一个令人吃惊的一致性估计，在传染期的早期t_d约为8~10个月（图2*d*），得出Λ的估计值为大约每年1.0。传染（和传染性）的典型持续时间——*D*，可能不是显著少于从感染HIV到表现出AIDS的特征时间。

Table 1. Doubling time of the HIV epidemic (in the early stages)

Serological data		
Area	Period	Doubling time t_d (in months)
(a) Male homosexuals		
San Francisco, USA	1978–80	10–11
New York City, USA	1979–80	10–11
London, UK	1982–84	9–10
(b) Intravenous drug users		
Italy	1980–83	15–16
London, UK	1983–85	11–12
Switzerland	1983–84	8–9
Case notifications		
(a) All risk groups		
Australia	1982–85	4–5
Austria	1983–85	15–16
Belgium	1982–84	11–12
Canada	1981–85	9–10
Denmark	1982–84	13–14
Europe(EC)	1982–84	8–9
France	1982–84	8–9
Italy	1982–84	7–8
Netherlands	1982–84	7–8
Spain	1982–84	6–7
Sweden	1983–85	8–9
Switzerland	1983–85	9–10
United Kingdom	1982–84	10–11
United States	1982–83	5–6
West Germany	1982–84	6–7
(b) Heterosexuals		
United States	1982–84	9–10
Average		9–10

But D may be significantly longer if a substantial proportion of infected individuals remain asymptomatic carriers (with the epidemiology similar to hepatitis B virus[25]). On the other hand, recent studies observing that measurable HIV antigen (HIV-Ag, the

表 1. HIV 流行的倍增时间（在早期阶段）

血清学数据		
地区	时期	倍增时间t_d（月）
（a）男同性恋		
美国旧金山	1978~80	10~11
美国纽约	1979~80	10~11
英国伦敦	1982~84	9~10
（b）静脉注射吸毒者		
意大利	1980~83	15~16
英国伦敦	1983~85	11~12
瑞士	1983~84	8~9
呈报病例		
（a）所有危险群体		
澳大利亚	1982~85	4~5
奥地利	1983~85	15~16
比利时	1982~84	11~12
加拿大	1981~85	9~10
丹麦	1982~84	13~14
欧洲	1982~84	8~9
法国	1982~84	8~9
意大利	1982~84	7~8
荷兰	1982~84	7~8
西班牙	1982~84	6~7
瑞典	1983~85	8~9
瑞士	1983~85	9~10
英国	1982~84	10~11
美国	1982~83	5~6
西德	1982~84	6~7
（b）异性恋		
美国	1982~84	9~10
平均数		9~10

如果被感染个体表现为无症状携带者（流行病学与乙肝病毒相似[25]）的话，D 可能明显更长。另外，最近的研究表明可测量的 HIV 抗原（HIV-Ag，它的出现表明

presence of which indicates the presence of the virus) appears early and transiently in primary HIV infection, that antibody production follows (1–3 months after infection) and that HIV-Ag may then disappear could imply lower estimates of D, as would the apparent correlation of this persistence or reappearance of antigen with clinical, immunological and neurological deterioration[3].

In the absence of conclusive data on infectiousness during the incubation period, we shall assume that D is equal to the incubation period. Studies of cases of AIDS associated with transfusion suggest that the average incubation period is 4–5 years[31], but as such studies are extended, this estimate will rise (Fig. 3). The true average may be 8–10 years or more. Our estimate of Λ in conjunction with equation 1 then leads to the rough estimate.

$$\beta c \simeq 1 \text{ yr}^{-1} \qquad (4)$$

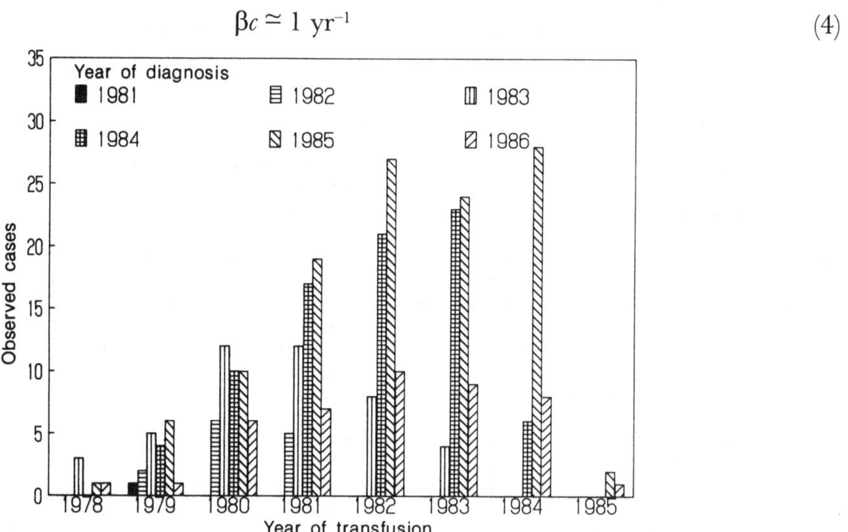

Fig. 3. Data on the distribution of the incubation period of AIDS derived from longitudinal studies of transfusion recipients (data from ref. 31). Observed cases of AIDS are recorded as a function of the year of transfusion (the assumed point of acquisition of infection) and the year of diagnosis. A Weibull distribution provides a good empirical description of this data with a mean incubation period of ~4–5 years.

Note that $\Lambda \simeq \beta c$ provided D is large (4–5 years plus). Thus data on changes in seropositivity over time have allowed us to infer the approximate magnitude of the combination of epidemiological parameters β and c, neither of which can easily be estimated directly.

Is this estimate consistent with what is known about β and c separately? Unfortunately, nearly all the information about degrees of sexual activity among male homosexuals has focused on average numbers of sexual partners, as distinct from average number of new partners per unit time[26,27] (Fig. 4). For less active individuals (say, 1–3 partners per 6-month interval), the rate of acquisition of new partners will be seriously overestimated by the average number of partners. On the other hand, the quantity c is disproportionately influenced by highly active individuals, most of whose partners are likely to be new, so that

病毒的出现）在 HIV 传染初期出现较早而且短暂，随后抗体产生（感染后 1~3 个月），然后 HIV-Ag 会消失，这意味着对 D 的估计值较低，这种抗原的持久性或再现与临床上、免疫学上和神经学上的恶化之间有明显的关系[3]。

在潜伏期，由于缺少关于传染性的确凿数据，我们假定 D 等于潜伏期。和输血相关的艾滋病病例研究表明平均潜伏期是 4~5 年[31]，但是随着研究的扩展，这个估计值将会上升（图 3）。真实的平均值可能是 8~10 年或者更高。Λ 的估值与方程 1 相关联，得到这个粗略估计。

$$\beta c \simeq 1\ \mathrm{yr}^{-1} \tag{4}$$

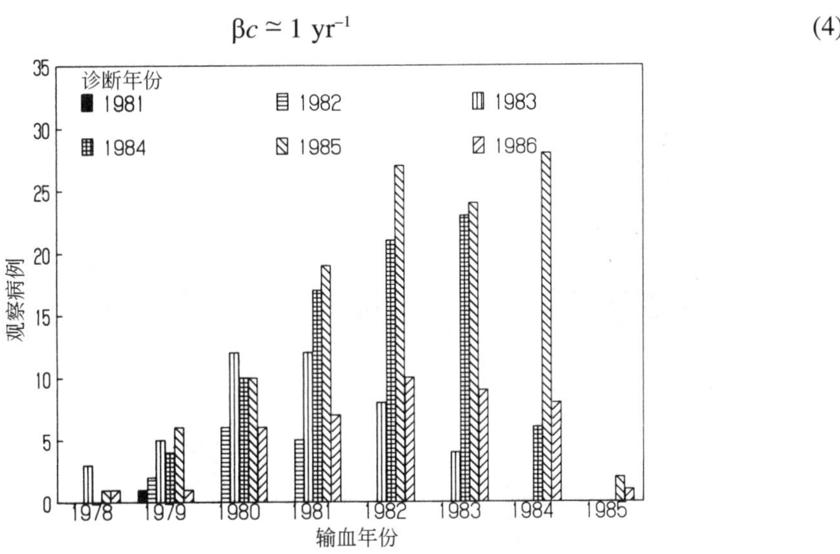

图 3. 来源于对输血接受者纵向研究的艾滋病潜伏期的分布数据（数据来自参考文献 31）。将观察到的艾滋病病例作为输液年份（假定的感染点）和诊断年份的函数。韦伯分布给出了数据的一个很好的实证描述，潜伏期的平均值大约为 4~5 年。

倘如 D 很大（4~5 年），那么 $\Lambda \simeq \beta c$。超过一定时间，血清反应阳性随时间改变的数据使得我们可以推断出流行病学参数 β 和 c 组合的大概量值，二者的值都不容易进行直接估计。

这个估计是否分别与已知的 β 和 c 相符合？不幸的是几乎所有的关于男性同性恋之间性活动程度的信息都集中在性伴侣的平均数量上，它与单位时间里获得新性伴侣的平均数是不同的[26,27]（图 4）。对于性活动少的个体来说（比如，每 6 个月 1~3 个伴侣），性伴侣的平均数大大超出他们获得新的性伴侣的概率。另外，c 的量受高度性活跃个体不成比例的影响，他们的性伴侣可能是新的，因此仅仅依据性伴

studies based simply on numbers of partners may give a rough guide to the magnitude of c (Fig. 4). Quantitative information on average values of β, whether for homosexuals or heterosexuals, is very limited at present. Estimates vary widely (from 0.05 to 0.5) although it appears that the average probability of transmission per partner contact is higher among male homosexuals than among heterosexuals, perhaps as a result of more frequent sexual activity that results in epithelial damage (for example anal intercourse)[26,29]. Our estimates of $\beta c \sim 1\ \text{yr}^{-1}$ together with the high estimates of c for homosexuals suggest that β may be small (~ 0.05). But estimates of c based on the reported number of partners per unit of time may significantly overestimate the number of new partners per unit of time, which, or that equating D to the incubation period, may overestimate the average duration of infectiousness. It may also be that the high values of c arise from sampling biased towards the high activity groups of homosexual communities.

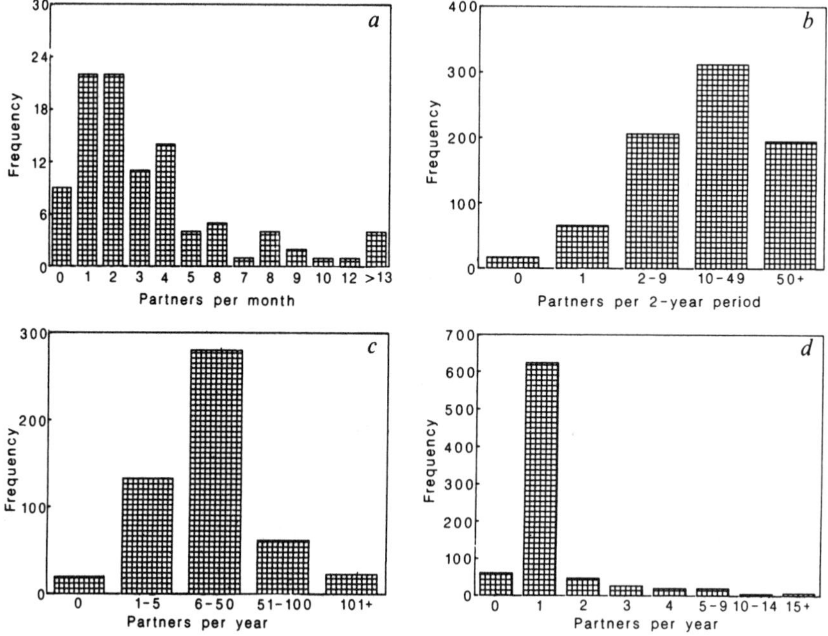

Fig. 4. Studies of sexual activity amongst *a–c*, male homosexual and *d* heterosexual communities. The graphs record frequency distributions of the number of sex partners per defined time period in samples of homosexual/bisexual males and heterosexuals. *a*, Homosexual/bisexual males resident in London surveyed in 1986 (unpublished data from C. A. Carne and I. V. Weller) (m = 4.7 per month, σ^2 = 56.7). Data denote male partners per month. *b*, Homosexual/bisexual males resident in San Francisco surveyed in 1984–85 (data from ref 26). Data denote male partners per 2-year period. *c*, Homosexual/bisexual males resident in London surveyed in 1984 (unpublished data from T. McManus). Data denote male partners per year. *d*, Heterosexuals between the ages of 18–44 years in England surveyed in November 1986 (unpublished data, Harris Research Organisation; R.M.A. and G. F. Medley). Data denote partners of the opposite sex per 1 year period (sample size = 823, m = 1.41, σ^2 = 4.36). A further survey of homosexual men (ref. 27) in San Francisco reveals a decline in the mean partners per month over the period November 1982 to November 1984 from 5.9 to 2.5.

侣数量进行的研究，可能对 c 值的大小给出一个粗略的指导（图 4）。不论是同性恋还是异性恋，β 的平均值的量化信息目前还很受限。尽管可能由于较频繁的性活动会导致上皮损伤（比如肛交）[26,29]，男性同性恋之间每次接触感染的平均可能性似乎大于异性恋之间的，但是估值范围很大（从 0.05 到 0.5）。结合同性恋的高 c 估值，我们对 βc 每年约等于 1 的估值表明 β 可能很小（约为 0.05）。以单位时间里得到的性伴侣的数目为基础的 c 的估值明显超过单位时间里新的性伴侣数量，如果将 D 等同于潜伏期，则会高估传染性的平均持续时间。高 c 值可能是由于取样偏向于同性恋群体中的高度活跃人群。

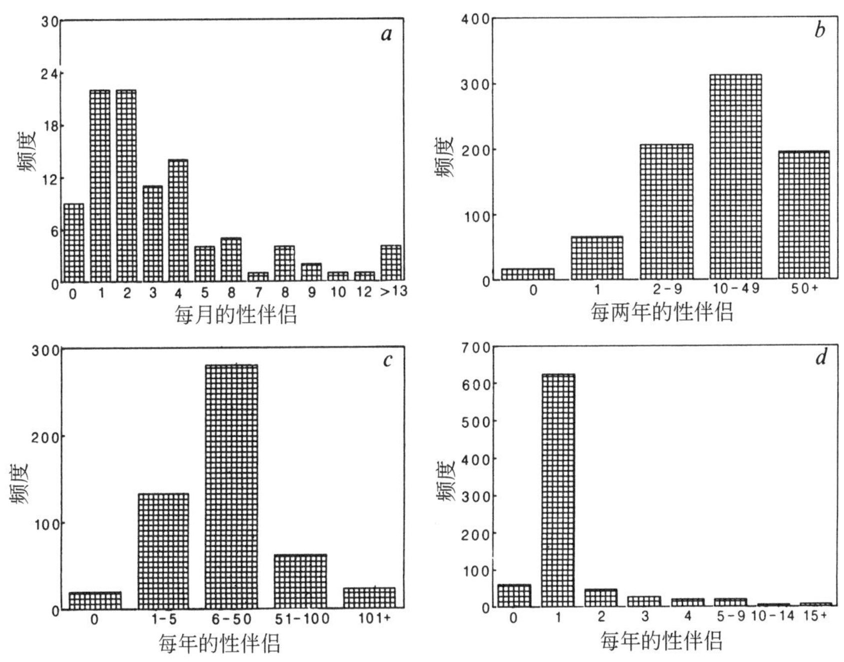

图 4. *a*~*c* 男同性恋和 *d* 异性恋群体内的性活动的研究。图表记录了男性同性恋或双性恋和异性恋者在一定时间里性伴侣数量的分布频率。*a*，1986 年调查的伦敦同性恋和双性恋男性居民（来自卡恩和韦勒的未发表数据）（m 的值为每月 4.7，σ^2=56.7）。数据表示每个月男性伴侣数量。*b*，1984~85 年，在旧金山调查到的男性同性恋和双性恋(数据来自参考文献 26)。数据表示每两年男性伴侣的数量。*c*，在 1984 年，伦敦调查到的同性恋和双性恋男性居民（来自麦克马纳斯未发表的数据）。数据表示每年的男性伴侣的数量。*d*，1986 年 11 月，在英格兰调查到的 18~44 岁之间的异性恋居民（哈里斯研究组织；安德森和梅德利未发表数据）。数据表示每年异性性伴侣的数量（样本大小 =823，m=1.41，σ^2=4.36）。进一步的调查发现（参考文献 27），从 1982 年 11 月到 1984 年 11 月，旧金山的男同性恋每月的伴侣平均数下降，从 5.9 降为 2.5。

As public awareness about AIDS has increased, there have been changes in patterns of sexual activity among male homosexuals in the United States (reflected, for example, in marked decreases in the incidence of rectal gonorrhoea[29,30]) which have presumably resulted in changes both in β and in *c*[27]. Our discussion, therefore, pertains mainly to the relatively early stages—1978 to the early 1980s—of the epidemic.

Incubation Period

Although much more information is available about the incidence of AIDS than of HIV infection (Table 1), it is harder to tease estimates of epidemiological parameters out of these. Incidence of AIDS depends not only on the transmission factors β and *c*, but also on the incubation period and on the fraction, *f*, of those infected who will eventually develop AIDS.

Significantly, estimates of both the incubation period and *f* have tended systematically to increase since the epidemic was first recognized[4,5,28,31,32]. Estimates of *f* range from 10% to 75% or more[19,28,33,37], with an incubation period of 4–5 years or more[31]. The progressive sequence of steps which eventually impair the ability of the immune system to respond to opportunistic infection seem not to be reversible. But whether all those infected with HIV are moving toward AIDS at different rates, or whether some will develop AIDS while others never will, remains unclear. Variability in the incubation period, and whether or not an infected person develops AIDS, could be accounted for by genetic heterogeneity within the host population (HLA-linked[38,39]), or could be associated with specific strains of the antigenically variable HIV virus[1,6].

Studies of the incubation period for those who develop AIDS suggest that the "hazard function", the probability of the disease manifesting itself as a function of the time since infection, increases with time (Fig. 3). Lui and co-workers[31] have assumed a Weibull distribution (a flexible two parameter probability distribution) for the incubation period with probability density function

$$h(t) = \gamma\nu^{-\gamma} t \exp[-(t/\gamma)^{\nu}] \qquad (5)$$

If indeed the probability per unit time, to develop AIDS (for that fraction *f* who do indeed develop it) increases linearly with time from infection as αt, the result is a Weibull distribution with $\gamma = 2$ and $\nu = \alpha$ for the hazard function[8]. This assumption differs from conventional epidemiological models, where infected individuals move through the incubation interval either at a fixed rate, or in a fixed time. But none of this resolves the question of what proportion of those infected will develop AIDS on what timescale: That issue will be resolved only by very long term (many decades) studies.

Fraction Eventually Infected

In a closed and homogeneously mixed population, the total fraction eventually infected depends only on the basic reproductive rate of the infection, R_0, defined above as shown[7]

由于公众对艾滋病的认识有所提高，美国男性同性恋之间的性活动模式已发生改变（比如，表现为直肠淋病的发病率明显下降[29,30]），可能导致β和c的改变[27]。因此，我们的讨论主要适合于流行病相对早期的阶段——1978年到20世纪80年代初。

潜 伏 期

尽管艾滋病发病率比HIV感染有更多可利用的信息（表1），但是从这些数据中更难找到流行病学参数的估值。艾滋病的发病率不仅与传播因素β和c有关，也与潜伏期和那些最终发展为艾滋病的感染者的比例（f）有关。

值得注意的是，自从第一次认识到这种流行病[4,5,28,31,32]，潜伏期和f的估计值都趋向于系统性的增加。f的估计值介于10%~75%或者75%以上之间[19,28,33,37]，潜伏期4~5年或者更长[31]。这种渐进性的过程最终损害了免疫系统，导致其不能对机会感染做出应答，而且这个过程似乎是不可逆的。但是否所有感染了HIV的人正在以不同速率发展成艾滋病，或者是否有些人会发展为艾滋病而其他的永远不会，这些尚不清楚。潜伏期的变化以及感染者是否会发展成艾滋病，可以通过宿主（连接有HLA[38,39]）基因异质性或者结合抗原变异的HIV病毒的特殊关系来确定[1,6]。

对发展为艾滋病的那些人的潜伏期的研究显示，风险函数即感染后显示出疾病的可能性，随时间而增加（图3）。卢伊和他的同事[31]结合概率密度函数推测出一个潜伏期的韦伯分布公式（一个灵活的双参数的概率分布）

$$h(t) = \gamma v^{-\gamma} t \exp[-(t/\gamma)^{v}] \tag{5}$$

如果从感染后定义为αt，单位时间里发展为艾滋病的概率随时间的变化呈线性增长，那么韦伯公式得出的风险率是 $\gamma = 2$ 和 $v = \alpha$[8]。这一假说与传统的流行病模型不同，在传统模型中受感染的个体以固定的概率或固定的时间度过潜伏期。但是这个解说也没有解决感染HIV的群体有多大比例及多长时间里会发展为艾滋病的问题：这个问题只有通过长期（数十年）的研究才能解决。

最终感染率

在一个封闭单一混合人群中，最终总体感染率仅仅与传染的基本再生率R_0有关，如图5中最上面的曲线图所示[7]。对性传染疾病比如HIV，这个结果可以引申

by the uppermost curve in Fig. 5. For sexually-transmitted infections such as HIV, the result can be extended to include the complications associated with a wide diversity in degrees of sexual activity.

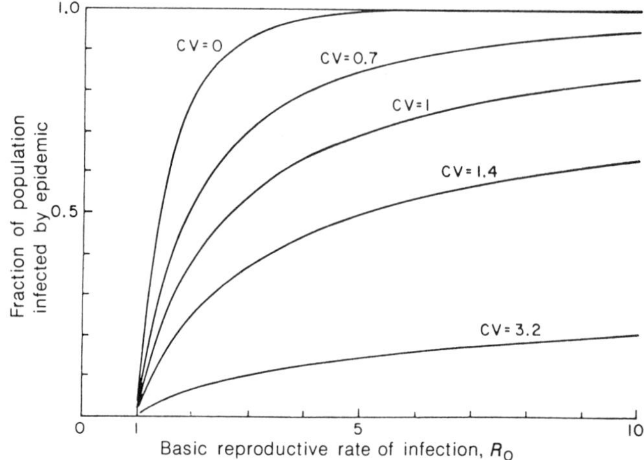

Fig. 5. The relationship between the eventual fraction infected (seropositive) in an epidemic of HIV and the basic reproductive rate R_0 (see text). The predictions are based on a model which assumes that sexual activity (defined as the number of new partners per unit of time) obeys a gamma distribution with varying coefficients of variation (CV) (mean m and variance σ^2; see text).

In a closed population, the eventual fraction seropositive will depend both on R_0 and on the actual distribution of rates of acquisition of sexual partners. Assuming a gamma distribution[8], we may characterize it by c and by its coefficient of variation ($CV = \sigma/m$). The resulting overall fraction infected is shown as a function of R_0, for a range of values of CV, in Fig. 5; for fixed R_0, the eventual seropositive fraction can be much lower than for $CV = 0$, if the variability in degrees of sexual activity (measured by CV) is high. This makes intuitive sense: the highly active individuals acquire infection, and eventually are removed, relatively early in the epidemic; transmission among the remaining, less active, individuals may be relatively weak.

Figure 5 may be used, in combination with two factual observations, to make a rough assessment of R_0 for HIV among male homosexuals. First, the studies indicate great variability in degrees of sexual activity among male homosexuals (with CV significantly in excess of unity[8,26,27]), thus confining attention to the lower curves. The second observation is that levels of seropositivity to HIV among male homosexuals in San Francisco in 1985 are variously reported as 70% or more[12] (in the HBV study which is probably biased towards more active individuals) and as around 50% (in a study carefully constructed to avoid bias[26]), providing a lower bound of 50–70% on the proportion ever seropositive. For CV noticeably in excess of unity, this can be achieved only if R_0 is in excess of 5.

Thus our early estimate of $\beta c \sim 1 \text{ yr}^{-1}$, in conjunction with the assessment that R_0 exceeds 5,

为与性活动程度多样化有关的综合征。

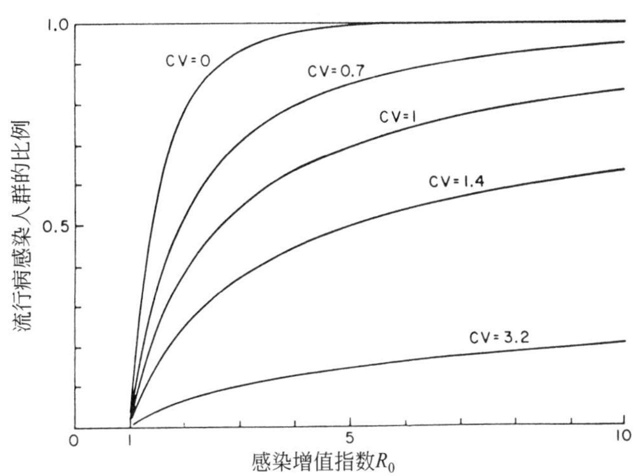

图 5. 在 HIV 流行病中，最终的感染比例（血清反应阳性）与基本再生数 R_0（见正文）之间的关系。推测是以模型为基础，该模型假定性活动（定义为单位时间内性伴侣的数量）遵从伽马分布，该分布的变异系数（CV）（平均值 m 及方差 σ；见正文）是变化的。

在一个封闭的人群里，最终血清反应阳性的比例与 R_0 以及获得性伴侣的比率的实际分布有关。假定一个伽马分布[8]，我们以 c 和它的变异系数（CV=σ/m）来表示。最终的总感染比例用 R_0 的函数来表示，范围为 CV 的值，如图 5；对于固定的 R_0 来说，如果性活动程度的变化（通过 CV 来测量）很大，那么最终血清反应阳性的比例可能更低，因为 CV=0。这会造成这样一种直觉：在流行病的早期，高度活跃的个体被感染，最后被排除；剩余的不太活跃的个体之间的传播可能相对较弱。

利用图 5，并结合两个实际观察结果，得出一个男性同性恋之间 HIV 的 R_0 粗略估值。首先，研究表明，男性同性恋之间的性活动程度有很大的可变性（CV 明显超过整体[8,26,27]），因此我们将注意力集中在较低的曲线上。第二个调查结果是，1985 年旧金山男同性恋的 HIV 血清反应阳性水平的报道分别为 70% 或高于 70%[12]（在 HBV 的研究中，很可能会偏向于更活跃的个体）和 50% 左右（一个为消除偏差而仔细构建的研究[26]），这给出了血清反应阳性比例的下界为 50%~70%。对于 CV 显著超过整体的，只有在 R_0 超过了 5 才能够实现。

因此结合 R_0 超过 5 的估值，我们早期 βc 每年约等于 1 的估值间接导致 D 的估

leads to an indirect estimate that D exceeds 5 years. Although R_0, like β and c changes with changing social and sexual habits, the data leading to our earlier estimate for βc come from the early stages of the rise in HIV infection, before such changes were significant. The estimate of R_0 depends importantly on observed levels of seropositivity, but these were also high before social changes became pronounced. Consequently, our estimate of D which depends only on the basic biology of HIV, is reasonably consistent. This independent estimate of D ~5 years accords with current estimates that the incubation period is 4–5 years or more.

An estimate of the value of R_0 in the early stages of the epidemic is also valuable in indicating the magnitude of the social changes needed to bring R_0 below unity. If R_0 is around 5–10 or more, then reductions by a factor of 5–10 or more in βc are needed. Because c depends disproportionately on those in the highly sexually active category, programmes aimed at getting them to change their habits—both to fewer partners and to "safe sex"—are most efficient. But if such individuals are less likely to respond to public health education, it will be harder to bring R_0 below unity.

Mortality

The frequent assumption that the severity of the epidemic, in terms of cumulative mortality, will be greatest if all those infected eventually develop AIDS and subsequently die is not necessarily true. Mortality depends critically on the duration of infectiousness of both those infected who develop AIDS and those infected who do not. If the latter have a similar life expectancy to those not infected, but remain infectious for life, they may contribute more to the net transmission of the virus, R_0, than those who die of AIDS. Much may be understood by recognizing that the overall net reproductive rate of the virus, R_0, is made up of two components, the reproductive rate of those who develop AIDS (R_{01}) and the equivalent rate of those who do not (R_{02}). If a fraction f develop AIDS

$$R_0 = fR_{01} + (1-f)R_{02} \tag{6}$$

where the two reproductive rates are defined by equation (3) with different parameters for the separate groups. Even if the asymptomatic carriers are less infectious than those who develop AIDS, if they remain infectious over, say, a 30-year span of sexual activity, R_{02} may be much larger than R_{01}, and, depending on f, the contribution of the asymptomatic carriers to R_0 may be dominant.

At present, it is not possible to tell whether the severity of the epidemic will be increased or decreased if a larger fraction of those infected develop AIDS, for the relative infectiousness of the two categories is unknown. For public health planning it is clearly important to attempt to acquire such data.

Dynamics of the Epidemic

The dynamics of an HIV epidemic within a homosexual community are represented by the results of our calculations given in Fig. 6, which shows the proportion seropositive

值超过了5年。尽管像β和c一样，R_0随着社会习性和性习惯的改变而改变，但是在这些改变变得很显著之前，我们用HIV感染早期阶段的增长水平来估计之前βc的值。R_0的估值主要受血清反应阳性水平的影响，但是在社会习性改变很明显之前，这些值也是很高的。因此，仅仅依据HIV的基本生物学所得D的估值是合理的。D的独立估计值大约是5年，与潜伏期的现行估计值（4~5年或更长）一致。

对在染病早期阶段R_0的估值是有用的，它可以表明使R_0低于整体值的社会习性变化幅度。如果R_0的值在5~10左右或者更高，那么βc需要减小5~10倍或更多。因为c不成比例地取决于那些使高度性活跃群体中的人改变自己习惯——较少的性伴侣和安全的性行为——的项目是最有效的。但是如果这些人不太可能响应公共健康教育，那么使R_0低于整体值是很困难的。

死亡率

如果感染者最终都发展为艾滋病并随后死亡，那么流行病的严重程度，也就是累计死亡率就会达到最大，这种推测是不准确的。死亡率主要取决于已经发展为AIDS的感染者和没有发展为AIDS的感染者二者的传染性的持续时间。如果后者与未被感染的人有相似的预期寿命，但是他们生活中仍保持传染性，那么他们对病毒的净繁殖率R_0比那些死于艾滋病的人起的作用更大。认识了病毒总的净繁殖率R_0是由那些发展为艾滋病的繁殖率（R_{01}）和那些没有发展成艾滋病的等同的繁殖率（R_{02}）两部分组成后，我们可以理解更多。如果发展为艾滋病的比例为f，那么

$$R_0 = fR_{01} + (1-f)R_{02} \tag{6}$$

公式（3）定义了两种繁殖率，两个单独的群体用不同的参数表示。即使无症状的携带者比发展为艾滋病的人有更小的传播性，如果他们在30年的性活动中保持传染性，那么R_{02}可能比R_{01}大很多，并且与f有关，无症状的携带者对R_0的影响可能比较明显。

目前，由于两类群体的相对传染力是未知的，如果较大比例感染者发展为艾滋病，那么就不可能预测到传染病的严重程度是增加还是降低。对于公共健康计划，显然努力获得这些数据是很重要的。

流行病的动力学

根据我们的计算结果建立的同性恋群体中HIV流行病的动力学如图6所示，它

and the incidence of cases of AIDS as a function of time since the start of the epidemic. It is assumed that 30% of those infected eventually manifest AIDS, with the incubation intervals obeying a Weibull distribution such that the average incubation period is 5 years[31]. Individuals who are incubating AIDS are assumed infectious throughout the incubation interval, and the 70% who remain asymptomatic are assumed to remain infectious for similar periods.

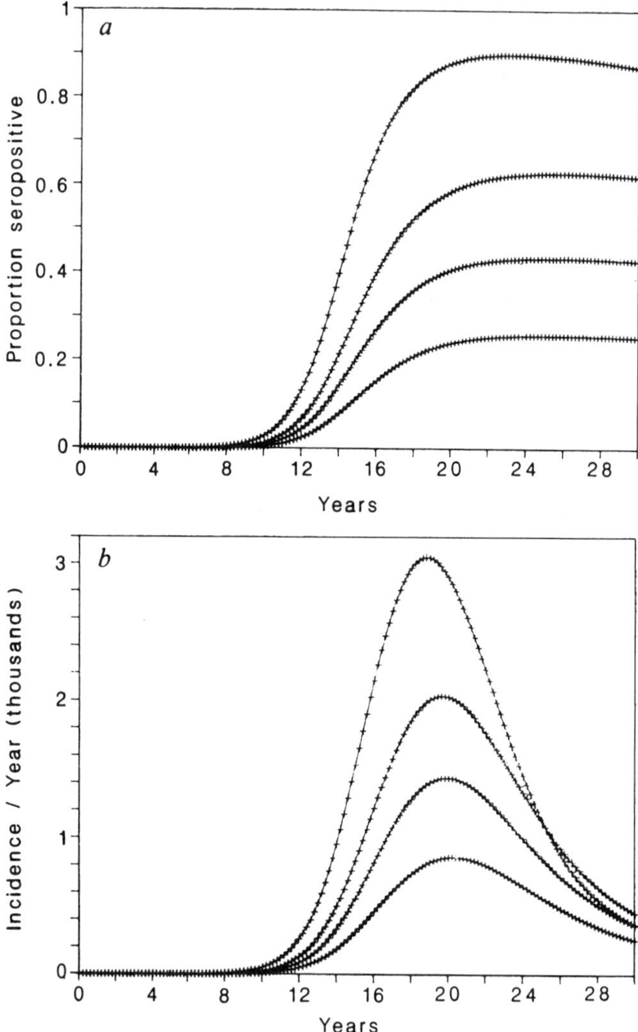

Fig. 6. The predictions of a model (see ref. 8) incorporating variable incubation periods, heterogeneity in sexual activity and recruitment of susceptibles. The two graphs record changes in seropositivity through time from the point of introduction of HIV into a community of 100,000 homosexual/bisexual males (graph *a*) and the incidence of AIDS yr^{-1} (graph *b*). Heterogeneity in sexual activity is described by a gamma distribution with a mean fixed at 5 partners yr^{-1} and variances 5, 25, 50 and 100 representing the predictions recorded by the four lines depicted in each graph. In *a* and *b* the smallest epidemic arises when the variance is largest and vice versa. Parameter values, $R_0 = 5$, $D = 5$ yr, $f = 0.3$ with the life expectancy of AIDS patient set at 1 yr from diagnosis and for the susceptible sexually active community at 32 yr from the point of joining the sexually active class. The 70% of infecteds who do not develop AIDS are assumed to be infectious for a period equal to D. The immigration of new susceptibles into the sexually active community was set at 100,000 per 32 yr.

以时间函数的形式描述了从流行病开始以来，血清反应阳性的比例和艾滋病的发病率。如果潜伏期遵从韦伯分布，即平均潜伏期为 5 年 [31]，据推测约 30% 的感染者最终会发展为艾滋病患者。我们推测处于艾滋病潜伏期的个体在整个潜伏期都有传染性，70% 的无症状者在同一时期也保持传染性。

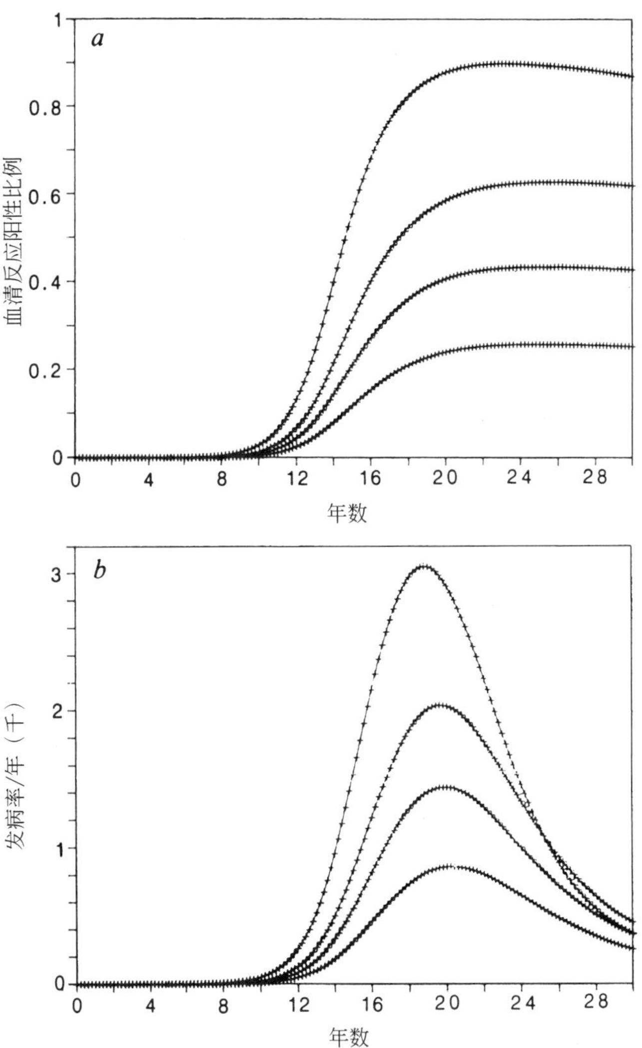

图 6. 一个结合可变的潜伏期、性活动不均一性和易感人群募集的模型（见文献 8）的预测。这两幅图记录了从 HIV 传入到 100,000 个男性同性恋和双性恋的群体的时间内，血清反应阳性的变化（表 *a*）和每年 AIDS 的发生率（表 *b*）。性活动的异质性以伽马分布描述，固定平均值为每年 5 个性伴侣，变量 5、25、50 和 100 代表了在表中描述的四条线的预测值。在表 *a* 和 *b* 中当变量最大时，传染增长最小，反之亦然。参数值：R_0=5，D=5 年，f=0.3，艾滋病病人的预期寿命设定为诊断后的一年，性活跃的易感人群的预期寿命设定为加入性活跃类别后的 32 年。假设 70% 没有发展成艾滋病的感染者在长度为 D 的时期内有传染性。性活跃群体中新的易感人数设定为每 32 年 100,000 人。

Many of the features presented in Fig. 6 show qualitative agreement with observation. The rise in incidence of infection (seropositivity) is initially exponential, but soon shows a more linear rise. And, the rise in incidence of AIDS lags that in the proportion infected, as seen.

It is easy to build epidemiological models of arbitrary complexity, which may appear beguilingly realistic, but we think there is little point in constructing them until more is known about the relevant epidemiological parameters. We distrust predictions made by using statistical procedures to fit polynomial or exponential curves to existing data on the incidence of AIDS, and then extrapolating[40,44]. The HIV epidemic is a dynamic process; to predict future trends, models must be based on the underlying epidemiological phenomena.

Heterosexual Transmission

In developed countries, the extent to which HIV infection can be transmitted by heterosexual contacts is uncertain[43,48]. HIV infections in females come from contact with bisexual males (the dominant sexually-transmitted route at present), transfusion recipients, haemophiliacs and intravenous drug users[45]. If such females are not themselves a significant source of infection back into the homosexual/bisexual community (through contacts with uninfected bisexuals), we would expect the incidence of HIV infections among the female partners of bisexuals initially to rise roughly in proportion to the incidence among homosexual males.

Specifically, we would expect the ratio of HIV infection among female partners of bisexuals to that among bisexual males to be $\sim \beta' c'/\beta c$, where β and c are as previously defined, β' is the transmission probability for male-to-female contact, and c' is the mean number of new female partners acquired by a bisexual male, per unit time. We expect this ratio to be significantly less than unity, because β' is less than β, and c' significantly less than c. The data (Table 1) suggest the doubling time for heterosexually-transmitted HIV infection is roughly equal to that among homosexual males, which is in accord with our simple expectation.

As the epidemic progresses, a high proportion of homosexual males become infected (Fig. 2*e*). On the other hand, the pool of female partners for bisexual males may be large, and we would expect incidence of HIV infection still to be rising roughly linearly among women at and beyond the point at which saturation effects limit the epidemic among homosexual/bisexual males. How long this rise continues, and how many females are eventually likely to be infected, depends on the average duration of infectiousness in the transmitting group of males (which could be long if a substantial fraction remain asymptomatic carriers).

The transmission of HIV infection to females by bisexual males is a process whose initial dynamics is essentially determined by R_0 for transmission among homosexual males, thereafter the question of its transmission and maintenance by purely heterosexual contact arises. The basic reproductive rate for such heterosexual transmission of HIV, R_0', is given by

图 6 中呈现出的许多特征与观察到的特征在性质上相一致。感染（血清反应阳性）最初呈指数增长，但是很快就表现为直线上升。如我们所见，艾滋病发生率增长滞后于感染比例增长。

建立任意复杂的流行病学模型是很容易的，但是它可能出现现实欺骗性，除非我们对相关流行病学参数有更多的了解，否则我们认为建立这个模型是没有意义的。利用统计学程序对已有的艾滋病发生率数据套用多项式和指数曲线然后进行推测[40,44]，对这样的预测结果我们深表怀疑。HIV 流行病是一个动力学过程；要预测未来的发展趋势，模型必须以根本的流行病学现象为基础。

异性性传播

在发达国家，通过异性性接触传播的 HIV 传染程度是不确定的[43,48]。女性感染 HIV 是由于与双性恋男性（目前最主要的性传播途径）、受血者、血友病患者和静脉吸毒者接触[45]。如果这样的女性自身不是一个再传染同性恋和双性恋群体的重要感染源（通过接触未被感染的双性恋），那么我们预测女性双性恋性伴侣之间的 HIV 传染率最初大致上与男性同性恋之间的概率成比例上升。

特别是，我们预期双性恋的女性伴侣的 HIV 感染率与双性恋男性感染率的比值大约为 $\beta'c'/\beta c$，β 和 c 在前面已经定义过，β' 是男性与女性接触传染的可能性，c' 是单位时间内双性恋男性获得新女性性伴侣的平均数。我们预期这个比率显著小于整体，因为 β' 小于 β，c' 显著小于 c。表 1 的数据显示，异性传染 HIV 的倍增时间约等于男性同性恋之间的倍增时间，与我们的预期值相同。

随着流行病的进一步发展，男性同性恋感染的比率很高（图 2*e*）。另一方面，双性恋男性的女性性伴侣范围可能比较广，我们预计女性之间的 HIV 感染率仍旧大致以线性方式增长，大于或等于同性恋或异性恋男性流行病的饱和效应限定点。这种增长持续多久，以及多少女性最终会被感染与男性传染群体具有传染性的平均持续时间有关（如果很大一部分能保持无症状携带者的状态，那么这段时间可能会很长）。

双性恋男性传染 HIV 给女性只是一个过程，它最初的动力学主要是由同性恋男性之间的传播 R_0 决定的，其后出现了纯粹异性恋接触的传播和持续时间的问题。这种异性恋之间 HIV 传播的感染增值指数 R_0' 的计算公式是

$$R_0' = (\beta_1\beta_2c_1c_2)^{1/2}D \tag{7}$$

Here β_1 and β_2 are the transmission parameters for contacts between infected females and susceptible males and between infected males and susceptible females, respectively; c_1 and c_2 are as before given by (2), for the distribution in rates of acquiring new partners of the other sex by females and males, respectively.

Data are very limited on the transmission and sexual activity parameters, but the data in Fig. 4 suggest that c_1 and c_2 are significantly smaller than c among homosexual males. Further, it seems likely that $\beta_1 < \beta_2$ and that both are less than the MM for homosexual males. Thus overall, the factor $(\beta_1\beta_2c_1c_2)^{1/2}$ seems likely to be much smaller than c for homosexual males, which suggests that in developed countries, R_0 for purely heterosexual transmission is probably significantly smaller than R_0 for purely male homosexual transmission. Whether R_0 is greater than unity, such that HIV infection can maintain itself and spread by purely heterosexual transmission, is at present unclear. There is an urgent need for studies to measure c_1 and c_2 in different communities (stratified by age and social status) and to assess how these parameters change as a consequence of educational programmes and publicity campaigns on AIDS. The use of professional opinion poll organizations to gather quantitative data on rates of partner change over a series of specified time intervals by interview and questionnaire (Fig. 4*d*) could help to fill this gap in our knowledge, but estimates of β_1 and β_2 will come only from long term studies of the heterosexual partners of infected patients.

If R_0' does exceed unity, the incidence of HIV infection in the heterosexual community will initially grow exponentially, at a rate given by the analogue of equation (1):

$$\Lambda = (\beta_1\beta_2c_1c_2)^{1/2} - 1/D = (R_0' - 1)D \tag{8}$$

The estimates above indicate that initial doubling times will be significantly longer than the 9 months or so for HIV among homosexual males; the slow initial growth will be difficult to discern against a background of homosexual transmission among males and bisexual transmission to females.

These observations are not necessarily inconsistent with the epidemiological situation for HIV in sub-Sahara Africa[6,49,56,57]. In contrast to the United States and the United Kingdom, where male/female ratios of AIDS cases have been of the order of 14:1 to 20:1, in certain parts of central Africa including areas in Zaire, Rwanda and Uganda, sex ratios approaching unity have been reported[45,50-53,56,57]. Very high prevalences of HIV antibodies have been found in males and females from surveys in urban and rural areas[52,56,57]. These points suggest that heterosexual transmission has been frequent in both directions and horizontal studies have shown that infection is associated with the age-related degree of sexual activity amongst heterosexuals[27,47,48,56,57]. We note, however, that in the early and approximately exponential phase of the epidemic, the ratio of the number of seropositive males to seropositive females is not unity, but is roughly $(\beta_1c_1/\beta_2c_2)^{1/2}$.

$$R_0' = (\beta_1\beta_2 c_1 c_2)^{1/2} D \tag{7}$$

β_1 和 β_2 分别是被感染的女性和易受感染的男性之间接触以及被感染的男性和易受感染的女性之间接触的传播参数；c_1 和 c_2 在前面（2）中已经给出，分别是女性和男性获得新的异性性伴侣的比率分布。

虽然数据受传播和性活动的参数的限制，但是图 4 的数据显示 c_1 和 c_2 明显地比男性同性恋之间的 c 小。此外，似乎 $\beta_1<\beta_2$，它们两个都小于男性同性恋之间的 MM。因此，总的来说，$(\beta_1\beta_2 c_1 c_2)^{1/2}$ 因素似乎可能比男性同性恋的 c 小得多，表明在发达国家，纯粹异性传染的 R_0 可能明显的小于纯粹男性同性恋传染的 R_0。R_0 是否大于整体的值，譬如 HIV 感染可以通过单纯的异性传播来维持自身和传播，目前还不清楚。在不同群体（按照年龄和社会地位分类）中测量 c_1 和 c_2 的值以及评估这些参数如何随着关于艾滋病的教育项目和宣传活动而改变的研究急待开展。利用专业的民意测验系统，通过访谈和调查问卷收集一系列特定时间间隔内性伴侣改变速率的定量数据（图 4*d*），能够填补我们在这方面认识的空缺。但是只能通过对被感染病人的异性性伴侣进行长期研究才能获得 β_1 和 β_2 的估值。

如果 R_0' 确实超过了整体，那么在异性恋群体中，HIV 的传染发生率最初会以指数形式增长，通过模拟公式（1），给出了这个增长速率：

$$\Lambda = (\beta_1\beta_2 c_1 c_2)^{1/2} - 1/D = (R_0' - 1)D \tag{8}$$

上面的估值显示，在男性同性恋之间，HIV 最初的倍增时间显著长于 9 个月左右；最初的慢速增长在男性同性恋之间的传播和双性恋传染给女性的背景下很难识别。

这些观察结果不一定与非洲撒哈拉以南地区的 HIV 流行状况不一致 [6,49,56,57]。与美国和英国不同，那里的男性和女性感染艾滋病的比例分别是 14：1 到 20：1，在中非的某些区域包括扎伊尔（刚果民主共和国的旧称）、卢旺达和乌干达，报道称他们的性别比例接近整体的性别比例 [45,50-53,56,57]。在城市和农村地区的调查中发现男性和女性中普遍具有 HIV 的抗体 [52,56,57]。这些观点表明异性传染很频繁，纵向和横向的研究表明传染与异性恋年龄相关的性活动程度有关 [27,47,48,56,57]。然而，我们发现在流行病学的早期和近似指数期，血清反应阳性的男性数量与女性数量的比例不是统一的，但是近似等于 $(\beta_1 c_1/\beta_2 c_2)^{1/2}$。

It is generally thought that β_1 is less than β_2 for HIV, although the facts are uncertain (for gonorrhea, for instance, male-to-female transmission, β_2 is roughly twice β_1). Obviously the average number of heterosexual partners of females and males, m_1 and m_2, are equal, but c_1 could significantly exceed c_2 if the variance of the distribution of rate of acquiring new sexual partners by females (associated with the concentrated activities of female prostitutes) is greater than that for males. This effect could partly offset β_1 being smaller than β_2. Although there is no *a priori* reason to expect the ratio $\beta_1 c_1/\beta_2 c_2$ to be exactly unity, its square root could easily be close to unity, which would explain the roughly equal proportions of seropositive males and females. Alternatively, the roughly equal proportions could be explained if homosexual transmission among males had coincidentally raised the seropositive proportion among males to around the level among females, or by transmission by contaminated needles in public and private medical services[6]. In any event, the rough equality of the seropositive proportions among males and females is a puzzle to be explained, and is not by itself evidence for purely heterosexual transmission.

Discussion

The ideas presented above are based on relatively simple mathematical models, with the aim of making clear some of the essential relations between epidemiological parameters and the overall course of HIV infection within various populations. Such models help to clarify what kinds of epidemiological data are needed to make predictions. As such data become available, the models can be made more detailed and realistic.

For public health planning, the dominant unknown is f, the fraction infected who will eventually develop AIDS. Estimates of this parameter have been increasing in recent years, but on present evidence the possibility cannot be ruled out that it is as low as 20% or as high as virtually 100%. Thus any current predictions about the number of homosexuals likely to acquire AIDS are uncertain by at least a factor 5 or so. Better understanding of the mechanisms of interactions between virus and host may help to determine f, but it is possible that only epidemiological data gathered on a decade-long timescale, as cases accumulate, will resolve this question.

The duration of infectiousness, and the way this duration is distributed among different infectives, is also relevant to estimates of R_0 and thence of the eventual number infected; more studies directed towards eliciting this information, including looking for virus in the blood, excretions and secretions of infected individuals over time, together with longitudinal studies of the partners of infected patients, are needed[3].

More generally, there is need for more studies that combine information about the epidemiological history of individuals with information about their sexual habits, such as the important study by Winkelstein *et al.*[26] of an unbiased sample of homosexuals in San Francisco, which demonstrated the association between the number of sexual partners and probability of acquiring infection. We emphasize that what is epidemiologically important is the average rate of acquiring new sexual partners, not necessarily the same as the

尽管还不确定，但是一般认为对 HIV 来说 β_1 小于 β_2（以淋病为例，男性传染给女性，β_2 大概是 β_1 的两倍）。很明显，男性和女性的异性性伴侣的平均值 m_1 和 m_2 相等，但是如果女性获得性的性伴侣的概率分布的方差高于男性（与妓女的集中活动有关），那么 c_1 就会明显的超过 c_2。这个效应会部分抵消 β_1 比 β_2 小的部分。没有先验的理由认为比率 $\beta_1 c_1/\beta_2 c_2$ 必须精确地等于 1，但是它的平方根很容易接近 1，这可以解释男性和女性血清反应阳性的比例大致相等。或者说，如果男性同性恋之间传播或者通过公共和私人医疗服务污染的针头传播碰巧将男性血清反应阳性比例增长到与女性相同的水平，那么也可以解释大略相等的比例[6]。不管怎样，男性和女性血清反应阳性比例的大致相等都很难解释，而且它本身不是一个纯粹异性传播的证据。

讨　论

为了弄清楚流行病学参数和 HIV 在多种人群中传染的全过程之间的重要联系，上述观点是以较简单的数学模型为依据的。这些模型帮助我们阐明做出预测都需要什么类型的流行病学数据。当可以获得这样的数据时，模型可以做得更加详细和实际。

对于公共健康计划，关键的未知数是 f，也就是最终会发展为艾滋病感染者的比例。最近几年这个参数的估值已经增加了，但是在当前的证据下，低至 20% 或几乎高达 100% 的可能性都不能够排除。因此，当前与可能得艾滋病的同性恋数目有关的预测都是不确切的，至少在 5 倍左右范围内。更好地理解病毒和宿主之间的相互作用机制可以帮助我们确定 f，但是只有以长达十年的时间为标度来，随着病例的积累收集流行病学数据，才有可能解决这个问题。

感染的持续时间和不同感染阶段持续的方式，都与 R_0 的估值和从那以后终被传染的人数有关；更多的研究直接指向引出这些信息，包括这段时间里的血液病毒检查、被感染个体的排泄物和分泌物检查以及对被感染病人性伴侣进行纵向研究也是必需的[3]。

更为普遍的是，我们需要将个体的流行病史和他们的性习惯方面的信息结合起来进行更多的研究，如温克尔斯坦等[26]在旧金山所做的一个同性恋无偏样本的研究，揭示了性伴侣的数量和被感染的概率之间的联系。我们强调，流行病学上最重要的是获得新的性伴侣的平均概率不一定要和单位时间里性伴侣的平均数相同。一些作者认识到，高度性活跃的个体在传染动力学中起到了不成比例的重要作用；公式（2）

average number of partners per unit time. Some authors have recognized that sexually highly active individuals play a disproportionate role in the transmission dynamics; equation (2) quantifies this observation, making it clear that the epidemiologically relevant quantity is not the mean number of new partners but, rather, the mean-square divided by the mean.

In developed countries, at present and into the near future, it is probable that sexually-transmitted HIV infections among females are likely to come mainly from bisexual males. Whether subsequent spread of infection from such females to heterosexual male partners is likely to reach significant levels, and more importantly whether purely heterosexual transmission of HIV infection may be selfsustaining ($R_0' > 1$), depends on estimates of the transmission parameters β_1, β_2, c_1 and c_2.

We have shown how c for transmission among homosexual males can be estimated indirectly from data on initial doubling times, but corresponding estimates of $\beta_1 c_1$ and $\beta_2 c_2$ are much harder, partly because the corresponding doubling rates are likely to be longer and partly because these infections are likely to be masked by homosexual/bisexual transmission among males, and by bisexual-to-female transmission among females (both of which processes depend simply on c). Attempts to estimate these quantities directly, and thence to estimate R_0, are urgently needed.

From present knowledge, it is not possible to assess whether R_0' is greater or less than unity in developed countries, and it is thus not possible to say whether HIV infections could spread epidemically by purely heterosexual transmission. The evidence from Africa, however, clearly argues that the sexually active population as a whole should be regarded as at risk[6,47,56,57].

We have greatly benefited from discussions with Anne Johnson, Mike Adler, John Pickering, Graham Medley, Stephen Blythe and Jenny Crombie. Financial support from the MRC and the NSF is gratefully acknowledged. We thank C. A. Carne, I. V. Weller and T. McManus for permission to quote unpublished data (Fig. 4).

(**326**, 137-142; 1987)

Robert M. May is in the Biology Department, Princeton University, Princeton, New Jersey 08544, USA. Roy M. Anderson is in the Parasite Epidemiology Research Group, Department of Pure and Applied Biology, Imperial College, University of London, London SW72BB, UK.

References:

1. Hahn B. H. *et al. Science* **232**, 1548-1552 (1986).

2. Goedert, J. J. *et al. Lancet* **ii** 711-715 (1984).

3. Goudsmit, J. *et al. Lancet* **ii** 177-180 (1986).

4. Peterman, T. A., Drotman, D. P. & Curran, J. W. *Epidem. Rev.* 7, 1-21 (1985).

5. Curran, J. W. *Ann. Intern. Med.* **103**, 657-662 (1985).

量化了这个观察结果，明确了流行病相关的数量不是新伴侣的平均值，而是均方除以平均值之后的结果。

在现在和不久的将来，在发达国家女性中通过性传播 HIV 可能主要来自男性双性恋。从这样的女性到异性恋的男性伴侣随后的传播是否有可能达到显著的水平和最重要的是纯粹的异性恋之间的 HIV 传播是否可以自我维持（$R_0'>1$），都取决于传播参数 β_1，β_2，c_1 和 c_2 的估计值。

我们已经阐明如何间接地从原始的倍增时间估算出男性同性恋之间的传染 c 值，但是 β_1c_1 和 β_2c_2 相应的估算较困难，一部分是因为相应的倍增概率可能比较长，一部分是这些传染可能会被男性同性恋、双性恋和双性恋男性传染给女性（这两个过程只与 c 有关）所遮盖。尽量直接估计这些数据然后再估计 R_0 是急切需要的。

就现在的知识，我们很难估计出发达国家的 R_0 是否大于或小于 1，因此我们不可能确定 HIV 传染能否纯粹依靠异性传播。尽管来自非洲的证据明确指出性活跃群体作为一个整体应该被认为是高危人群[6,47,56,57]。

我们从和安妮 · 约翰逊、迈克 · 阿德勒、约翰 · 皮克林、格雷厄姆 · 梅德利、斯蒂芬 · 布莱思和珍妮 · 克龙比的讨论中获益匪浅，资金支持来自医学研究理事会和国家科学基金会。我们感谢卡恩、韦勒和麦克马纳斯准许我们使用他们尚未发表的数据（图 4）。

（郑建全 翻译；孙军 审稿）

6. Acheson, E. D. *Lancet* i 662-676 (1986).
7. Anderson, R. M. & May, R. M. *Nature* **318**, 323-329 (1985).
8. Anderson, R. M., May, R. M., Medley, G. F. & Johnson, A. *IMAJ. Math. Med. Biol.* (in the press).
9. Dietz, K. *Lect. Notes. Biomaths.* **11**, 1-15 (1976).
10. Anderson, R. M. in *Theoretical Ecology* (ed. May, R. M.) 318-355 (Blackwell, Oxford, 1981).
11. Hethcote, H. W. & Yorke, J. A. *Lect. Notes. Biomaths.* **56**, 1-105 (1984).
12. Centers for Disease Control MMWR **34**, 573-589 (1985).
13. Centers for Disease Control MMWR **35**, 17-20 (1986).
14. Anderson, R. M. & May, R. M. *Nature* **280**, 361-367 (1979).
15. May, R. M. & Anderson, R. M. *Nature* **280**, 455-461 (1979).
16. Ho, D. D. *et al. New Engl. J. Med.* **313**, 1606 (1985).
17. Ho, D. D. *et al. Ann. Intern. Med.* **103**, 880-883 (1985).
18. Carne, C. A. *et al. Lancet* ii 1206-1208 (1985).
19. Wong-Staal, F. & Gallo, R. C. *Nature* **317**, 395-403 (1985).
20. Centers for Disease Control MMWR **34**, 373-375 (1985).
21. Schreeder, M. T. *et al. J. Infect. Dis.* **146**, 7-15 (1982).
22. Jaffe, H. W. *et al. Ann. Intern. Med.* **103**, 210-214 (1985).
23. Carne, C. A. *et al. Lancet* i, 1261-1262 (1985).
24. Angarano, G. *et al. Lancet* ii 1302 (1985).
25. Francis, D. P. *Rev. Infect. Dis.* **5**, 322-329 (1983).
26. Winkelstein, W. *et al. J. Am. Med. Assoc.* **257**, 321-325 (1987).
27. McKusick, L. *et al. Pub. Hlth. Repts.* **100**, 622-628 (1985).
28. Curran, J. W., Morgan, W. M. & Hardy, A. M. *Science* **229**, 1352-1357 (1985).
29. Weller, I. V. D., Hindley, D. J. & Adler, M. W. *Brit. med. J.* **289**, 1041 (1984).
30. Centers for Disease Control MMWR **34**, 613-615 (1985).
31. Lui, K. J. *et al. Proc. Natl. Acad. Sci. U.S.A.* **83**, 3051-3055 (1986).
32. Peterman, T. A. *et al. J. Am. med. Ass.* **254**, 2913-2917 (1985).
33. Moss, A. R. *et al. J. infect. Dis.* **152**, 152-161 (1985).
34. Weber, J. N. *et al. Lancet* i 1179-1182 (1986).
35. Stevens, C. E. *et al. J. Am. med. Assoc.* **255**, 2167-2171 (1985).
36. Goedert, J. J. *et al. Science* **231**, 992-995 (1986).
37. Brodt, H. R. *et al. Dtsch. med. Wschr.* **111**, 1175-1180 (1986).
38. Weiss, R. A. in *Virus Resistance* (eds Mahy, B. W. J., Mison, A. C. & Dorby, G. K.) 267-288 (Cambridge Univ., London, 1982).
39. Scorza Smeraldi, R. *et al. Lancet* ii 1187-1189 (1986).
40. McEvoy, M. & Tillett, H. *Lancet* ii, 541-542 (1985).
41. Artalego, F. R. *et al. Lancet* i 378 (1986).
42. Downs, A. M., Ancelle, R. & Brunet, J. B. (in the press).
43. Barnes, D. M. *Science* **232**, 1589-1590 (1986).
44. Gonzalez, J. J. & Keoch, M. G. *W. H. O. Report on Euro meeting on AIDS containment* (WHO, Geneva, 1986).
45. Vogt, M. W. *et al. Lancet* i, 525-527 (1986).
46. Wofsy, C. *et al. Lancet* i, 527-529 (1986).
47. Centers for Disease Control MMWR, **34**, 561-563 (1985).
48. de Perre, P. V. *et al. Lancet* ii 524.
49. Harris, C. *et al. New Engl. J. Med.* **308**, 1181-1184 (1985).
50. Centers for Disease Control MMWR **31**, 697-698 (1984).
51. Newmark, P. *Nature* **322**, 6 (1986).
52. Biggar, R. J. *Lancet* i 79-82 (1986).
53. Kreiss, J. K. *et al. New Engl. J. Med.* **314**, 414-417 (1985).
54. Papervangelou, G., Roumeliotou-Karayannis, A., Kallinikos, G. & Papoutsakis, G. *Lancet* ii 1018 (1985).
55. Tirelli, U. *et al. Lancet* ii 1424 (1985).
56. Quinn, T. C., Mann, J. M., Curran, J. W. & Piot, P. *Science* **234**, 955-963 (1986).
57. Melbye, M. *et al. Lancet* ii 1113-1117 (1986).

Structure of the Repressor–operator Complex of Bacteriophage 434

J. E. Anderson *et al.*

Editor's Note

In 1967, molecular biologist Mark Ptashne demonstrated negative gene regulation in the form of a repressor protein binding directly to specific DNA sequences. Two decades later, with the help of colleagues John E. Anderson and Stephen C. Harrison, Ptashne describes the crystal structure of a specific complex between the DNA-binding domain of a bacteriophage repressor protein and a synthetic operator DNA. The structure reveals how the repressor recognizes both a particular conformation of DNA and an array of base-pair contacts, a phenomenon driving repressor specificity. Structures such as this have enabled the controlled switching of gene expression to become a reality and provided a framework for understanding the mechanisms governing gene regulation.

The crystal structure of a specific complex between the DNA-binding domain of phage 434 repressor and a synthetic 434 operator DNA shows interactions that determine sequence-dependent affinity. The repressor recognizes its operators by its complementarity to a particular DNA conformation as well as by direct interaction with base pairs in the major groove.

WE describe here the structure of a specific complex between the DNA-binding domain of a bacteriophage repressor protein and a synthetic operator DNA. The description is based on an X-ray crystallographic analysis of crystals that diffract to 3.2 Å resolution in some directions and to about 4.5 Å in others. The repressor is encoded by coliphage 434, a close relative of phage λ. It binds to a set of six similar but non-identical 14-base-pair (bp) operators in the phage genome. Its differential affinity for these sites, coupled with the distinct differential affinity of the homologous cro protein, creates a regulatory switch determining the choice between lysogeny and lytic growth[1]. The structure of the repressor–operator complex, taken together with experiments reported in an accompanying paper and elsewhere[2,3], reveals the basis of specific binding and its modulation by the exact DNA sequence of the operator.

In an account of this structure at 7 Å resolution[4], we described the general features of the complex. The repressor DNA-binding domain (R1-69), contains the first 69 residues of the complete polypeptide. Its conformation is predominantly α-helical, similar to the first four helices of the λ repressor[5]. The second and third alpha helices (α2 and α3) form a

噬菌体434阻遏蛋白–操纵基因复合物的结构

安德森等

编者按

1967年，分子生物学家马克·普塔什尼证明了基因负调控以阻遏蛋白直接结合于特异DNA片段的形式实现。二十年后，在其同事约翰·安德森和斯蒂芬·哈里森的协助下，普塔什尼描述了一种噬菌体阻遏蛋白的DNA结合域与人工合成的操纵基因DNA所形成的特异复合物的晶体结构。这一结构揭示了阻遏蛋白如何同时识别DNA的特殊构象并与一排碱基对相接触的，该现象使阻遏蛋白具有了特异性。这样的结构使基因表达的可控开关成为可能，并为了解基因调控机制提供了框架。

噬菌体434阻遏蛋白的DNA结合域与人工合成的434操纵基因DNA所形成的特异复合物的晶体结构显示了决定序列依赖性亲和力的相互作用方式。阻遏蛋白通过和特异DNA构象的互补作用及与DNA大沟中的碱基对直接作用的方式识别其操纵基因。

本文所描述的是噬菌体阻遏蛋白的DNA结合域与人工合成的操纵基因DNA所形成的特异复合物的结构。所有描述都是基于对晶体的X射线晶体学衍射图谱的分析，该晶体的衍射分辨率在某些方向上达到3.2 Å，在其余方向约为4.5 Å。阻遏蛋白是由与λ噬菌体非常相似的大肠杆菌噬菌体434基因编码的。它结合于噬菌体基因组中六个长度为14个碱基对、序列相似但不完全相同的操纵基因上。它对六个位点的不同亲和力，加上和同源的cro蛋白有着明显的亲和性差异，形成了一个调控开关来决定噬菌体的生长处于溶源还是裂解周期[1]。这个阻遏蛋白–操纵基因复合物的结构，加上本期杂志中另一篇相关文章及其他文献报道的实验结果[2,3]，揭示了特异结合的原理以及操纵基因中特定DNA序列对它的调节。

在过去该结构的分辨率为7 Å时[4]，我们描述了这一复合物的大致特征。阻遏蛋白的DNA结合域（R1-69）包含该蛋白全长多肽起始的69个氨基酸残基。它的构象主要是α-螺旋，类似于λ噬菌体阻遏蛋白起始的四个螺旋[5]。第二和第三个α螺旋（α2和α3）形成一个螺旋–转角–螺旋基序，这在其他某些DNA结合蛋白的

helix–turn–helix motif, found in crystal structures of certain other DNA binding proteins[6-8] and believed on the basis of amino-acid sequence similarity to occur in many more[9]. The conformation of the 14-bp synthetic operator (14-mer) is B-DNA-like, and individual 14-mers are stacked end-to-end to form pseudocontinuous double helices running through the crystals. Two R1-69 subunits are bound to each 14-mer, and α3 of each monomer rests in the major groove of a half-site.

The higher resolution structure presented here shows the interaction in molecular detail. Hydrogen bonds between amide nitrogens of the peptide backbone and phosphate groups on DNA strongly constrain the fit of DNA to protein, and three glutamine side chains on α3 are in position to form a pattern of van der Waals and hydrogen-bond contacts with the five outer base pairs. Thus, the repressor recognizes in DNA both a particular conformation and an array of base-pair contacts.

Structure Determination

Crystals of R1-69 and the 14mer were prepared as described previously[10]. Data-collection procedures and statistics are presented in Table 1. Initial single isomorphous replacement (SIR) phases were obtained using crystals with DNA containing 5-bromodeoxyuracil instead of thymine at position 7. Phases between 15 Å and 7 Å were refined using real space noncrystallographic symmetry averaging[11-13] about a local 3_1 axis parallel to $[1\bar{1}1]$ and intersecting [110] 33.5 Å from the origin[4]. The small magnitude of the heavy-atom differences and weak intensities beyond 7 Å made conventional refinement of heavy-atom positions impractical. We therefore used a two-step strategy to extend phases, first from 7 Å to 5 Å and then from 5 Å to 3.2 Å. In the first step, heavy-atom positions were refined at 7 Å resolution by averaging a 15–7 Å heavy-atom difference map about a series of local 3_1 axes, parallel to $[1\bar{1}1]$ and intersecting [110] in the neighbourhood of 33.5 Å from the origin. The position of the 3_1 axis that gave the strongest average heavy-atom peaks was selected, and new heavy-atom positions were determined by inspection of the corresponding map. SIR phases between 7 and 5 Å resolution were computed using these positions, combined with the previously refined 15 to 7 Å phases, and used to generate a 5 Å map. Eight cycles of noncrystallographic symmetry averaging[13] about the new local 3_1 (34.0 Å from the origin along [110]) gave an averaging R-factor, $R_{av} = 0.28$. In the second step, refined phases to 5 Å resolution were used in an analogous way to adjust the origin of the threefold screw axis and to improve the heavy-atom coordinates. The noncrystallographic symmetry phase refinement was unstable if all phases from 15 to 3.2 Å were allowed to change, but fixing phases from 15 to 5 Å at their best values from the 5 Å refinement allowed the computation to converge with $R_{av} = 0.30$ for all data from 15 to 3.2 Å (Table 2, upper part). This way of refining heavy-atom coordinates is similar to the procedure introduced by Hogle for work on poliovirus[14].

晶体结构中也有发现[6-8]，并且基于氨基酸序列的相似性可以相信在更多的 DNA 结合蛋白中均有该基序的存在[9]。14 个碱基对的人工合成操纵基因 (14 碱基对) 具有类 B 型 DNA 构象，单独的 14 碱基对以末端相接的方式堆积成伪连续的双螺旋贯穿晶体。每个 14 碱基对结合两个 R1-69 亚基，每个亚基中的 α3 螺旋位于一个半位点的大沟中。

本文所描述的更高分辨率的结构显示了分子细节上的相互作用。蛋白质多肽骨架中氨基氮和 DNA 的磷酸基团之间的氢键极大地限制了 DNA 对蛋白质的匹配，而 α3 螺旋中三个谷氨酰胺侧链则与五个外缘碱基对形成范德华力和氢键模式的接触。这样，阻遏蛋白对 DNA 的识别既依赖 DNA 的特殊构象，又需要与一串碱基对的接触。

结 构 测 定

R1-69 和 14 碱基对 DNA 复合物的晶体制备参见前述[10]。数据收集步骤和统计学结果如表 1 所示。初始的单对同晶置换（SIR）相位来自在 DNA 第七位上用 5– 溴代尿苷取代了胸腺嘧啶的晶体。15 Å 到 7 Å 之间的相位修正采用的是实空间非晶体学对称平均[11-13]的方法，围绕一个平行于 $[1\bar{1}1]$ 并在距离原点 33.5 Å 处与 [110] 交叉的局部 3_1 轴[4]进行。重原子差异很小，分辨率超过 7 Å 的信号强度也很弱，因此不能使用常规的重原子位置修正。为此我们采用两步法来拓展相位，先从 7 Å 到 5 Å，再由 5 Å 到 3.2 Å。第一步，通过将围绕着一系列局部 3_1 轴的 15~7 Å 分辨率的重原子差值图进行平均，使重原子的位置在 7 Å 的分辨率上得到修正，这些局部 3_1 轴距原点 33.5 Å，平行于 $[1\bar{1}1]$ 并交叉于 [110]。选择平均重原子峰值最高的 3_1 轴的位置，通过观察相应图谱确定新的重原子位置。利用这些位置信息，再结合先前修正的 15 Å 到 7 Å 的相位信息，计算出 7 Å 到 5 Å 间单对同晶置换的相位，并用来生成 5 Å 分辨率的密度图。围绕新的局部 3_1 轴（沿着 [110] 的方向距原点 34 Å）进行了八轮非晶体学对称平均[13]，得到 R 因子平均值，R_{av}=0.28。第二步，利用 5 Å 分辨率的修正相位通过类似的方式来调整三次螺旋轴的起点，并修正重原子的坐标。如果 15 Å 到 3.2 Å 的相位都设置为可变，则非晶体学对称相位的修正就不稳定。但是将 15 Å 到 5 Å 的相位固定在 5 Å 修正的最佳值，便使得从 15 Å 到 3.2 Å 的所有数据中计算出的 R 因子收敛于 0.30，即 R_{av}=0.30（表 2，上部）。这种修正重原子坐标的方法与霍格尔在脊髓灰质炎病毒研究中使用的方法类似[14]。

Table 1. Statistics for data between 45 and 3.2 Å resolution

Native				
d_{min} (Å)	<*F*>	R_{sym} (N_{mult})	*N* (% $F>2\sigma$)	N_{poss} (% $F>2\sigma$)
9.90	7,568	0.080 (148)	368 (98.4)	622 (58.2)
7.08	6,559	0.096 (585)	929 (98.5)	622 (58.2)
5.81	3,476	0.107 (721)	1,099 (97.0)	1,260 (84.6)
5.04	2,916	0.138 (815)	1,283 (95.9)	1,466 (83.9)
4.51	2,955	0.157 (920)	1,399 (94.1)	1,651 (79.8)
4.12	3,022	0.170 (951)	1,535 (94.2)	1,817 (79.6)
3.82	2,695	0.212 (982)	1,675 (92.0)	1,961 (78.6)
3.58	2,491	0.242 (941)	1,705 (88.7)	2,090 (72.3)
3.37	2,161	0.284 (921)	1,756 (82.8)	2,237 (65.0)
3.20	2,535	0.212 (834)	1,708 (79.7)	2,328 (58.5)
Overall	3,129	0.161 (7,818)	13,457(90.7)	16,444(74.2)
Derivative				
d_{min} (Å)	<*F*>	R_{sym} (N_{mult})	*N* (% $F>2\sigma$)	N_{poss} (% $F>2\sigma$)
9.90	7,194	0.041 (59)	229 (97.4)	622 (35.9)
7.08	6,225	0.047 (240)	664 (96.8)	1,013 (63.5)
5.81	3,341	0.093 (387)	866 (88.6)	1,260 (60.9)
5.04	2,879	0.135 (442)	970 (84.7)	1,466 (56.1)
4.51	2,902	0.179 (471)	1,079 (85.1)	1,651 (55.6)
4.12	2,905	0.212 (424)	1,132 (83.7)	1,817 (52.2)
3.82	2,790	0.282 (366)	1,244 (79.7)	1,961 (50.6)
3.58	2,630	0.313 (274)	1,200 (75.5)	2,090 (43.3)
3.37	2,430	0.274 (244)	1,141 (70.9)	2,237 (36.2)
3.20	3,405	0.194 (261)	1,089 (75.0)	2,328 (35.1)
Overall	3,224	0.169 (3,168)	9,614 (81.6)	16,444(47.7)

Mean isomorphous difference = $\sum|F_{nat}-F_{der}|/\sum|F_{nat}|$ = 0.22, calculated after exclusion of data (between 45 and 3.2 Å) for which $|F_{nat}-F_{der}| > 5\sigma(\Delta F)$.

$R_{sym}=\sum_i\sum_j|I_{ij}-\bar{I}_i|/\sum_i\sum_j I_{ij}$

N_{mult} = number of reflections measured more than once; only these contribute to R_{sym}.

N = number of reflections measured (after rejections) including those measured only once.

N_{poss} = maximum number of reflections possible in corresponding resolution range.

The percentage of observed and possible data with $F > 2\sigma(F)$ is indicated. Data from native crystals were collected to 3.2 Å using CEA-25 X-ray film and oscillation photography with Elliot GX-6 and GX-13 X-ray generators. The films were scanned with an Optronics P1000 film scanner on a 50 μm raster. Integrated intensities were obtained with the film-scanning program SCANFILM, a derivative of SCAN12 (ref. 34). Data to 7 Å were also collected on a prototype Xentronics area detector with the Harvard software package[35]. These measurements provided intensities for native reflections that were too bright to be measured on film. Derivative data to 3.2 Å were collected on the area detector from isomorphous crystals prepared with DNA containing 5-bromodeoxyuracil instead of thymine at position 7 of the 14mer[4]. The native data from detector and film were processed and merged, then scaled to the processed derivative data. All programs for data processing and scaling were from P. Evans (MRC Laboratory of Molecular Biology).

表 1. 分辨率 45 Å 到 3.2 Å 间的数据统计

母体				
d_{min}(Å)	<F>	$R_{sym}(N_{mult})$	$N(\% F>2\sigma)$	$N_{poss}(\% F>2\sigma)$
9.90	7,568	0.080（148）	368（98.4）	622（58.2）
7.08	6,599	0.096（585）	929（98.5）	622（58.2）
5.81	3,476	0.107（721）	1,099（97.0）	1,260（84.6）
5.04	2,916	0.138（815）	1,283（95.9）	1,446（83.9）
4.51	2,955	0.157（920）	1,399（94.1）	1,651（79.8）
4.12	3,022	0.170（951）	1,535（94.2）	1,817（79.6）
3.82	2,695	0.212（982）	1,675（92.0）	1,961（78.6）
3.58	2,491	0.242（941）	1,705（88.7）	2,090（72.3）
3.37	2,161	0.284（921）	1,756（82.8）	2,237（65.0）
3.20	2,535	0.212（834）	1,708（79.7）	2,328（58.5）
总体	3,129	0.161（7,818）	13,457（90.7）	16,444（74.2）
衍生物				
d_{min}(Å)	<F>	$R_{sym}(N_{mult})$	$N(\% F>2\sigma)$	$N_{poss}(\% F>2\sigma)$
9.90	7,194	0.041（59）	229（97.4）	622（35.9）
7.08	6,225	0.047（240）	664（96.8）	1,013（63.5）
5.81	3,341	0.093（387）	866（88.6）	1,260（60.9）
5.04	2,879	0.135（442）	970（84.7）	1,466（56.1）
4.51	2,902	0.179（471）	1,079（85.1）	1,651（55.6）
4.12	2,905	0.212（424）	1,132（83.7）	1,817（52.2）
3.82	2,790	0.282（366）	1,244（79.7）	1,961（50.6）
3.58	2,630	0.313（274）	1,200（75.5）	2,090（43.3）
3.37	2,430	0.274（244）	1,141（70.9）	2,237（36.2）
3.20	3,405	0.194（261）	1,089（75.0）	2,328（35.1）
总体	3,224	0.169（3,168）	9,614（81.6）	16,444（47.7）

平均同型差异 $= \sum |F_{nat} - F_{der}| / \sum |F_{nat}| = 0.22$，是将 45 Å 到 3.2 Å 的数据中 $|F_{nat} - F_{der}| > 5\sigma(\Delta F)$ 的数据排除后计算得到的结果。

$R_{sym} = \sum_i \sum_j |I_{ij} - \bar{I}_i| / \sum_i \sum_j I_{ij}$

N_{mult}= 测到多次的反射数；仅限于对 R_{sym} 有贡献的反射。

N= 测到的反射数（除掉了不符合条件的数据），包括只测到一次的反射。

N_{poss}= 在相应分辨率范围内可能的最大反射数。

观察到的与可能的数据中符合 $F > 2\sigma$ (F) 的百分比在表中列出。母体晶体收集到了分辨率达 3.2 Å 的数据，使用的是 CEA-25 型 X 射线胶片和振荡摄影术以及埃利奥特 GX-6、GX-13 型 X 射线发射机。胶片用 Optronics P1000 型胶片扫描仪以 50 μm 光栅产生的光进行扫描。积分强度则由程序 SCAN12（参考文献 34）衍生出的一个胶片扫描程序 SCANFILM 获得。分辨率达 7 Å 的数据也是在使用哈佛软件包的 Xentronics 面探测器原型机上收集的 [35]。这些测量提供了因光太强而不能在胶片上进行测量的母体反射强度。分辨率达 3.2 Å 的衍生物数据由同晶置换的晶体在面探测器上收集，该同晶置换中将 14 碱基 DNA 中第七位的胸腺嘧啶改变为 5 溴代脱氧核糖 [4]。由面探测器和胶片得到的母体数据经过处理和合并后，缩放到与处理过的衍生物数据一样的大小。数据处理和缩放的所有软件来自埃文斯实验室（医学研究协会，分子生物学实验室）。

Table 2. Phase refinement by noncrystallographic symmetry averaging

Refinement of SIR phases*						
	Phases refined 15→3.2 Å			Phases refined 5→3.2 Å		
Cycle	R_{av}	r	Mean phase change	R_{av}	r	Mean phase change
1	0.58	0.537	53.3	0.55	0.346	75.3
2	0.45	0.734	40.1	0.41	0.600	34.7
3	0.39	0.785	21.9	0.35	0.678	17.1
4	0.38	0.791	14.5	0.33	0.718	10.2
5	0.38	0.780	11.4	0.31	0.737	7.3
6	0.39	0.780	10.2	0.31	0.750	5.7
7	0.40	0.781	10.5	0.30	0.758	4.8
8	0.43	0.747	11.0	0.30	0.785	4.1
Overall			65.5			82.0

Refinement of phases determined from initial model†			
Cycle	R_{av}	r	Mean phase change
1	0.37	0.792	23.2
2	0.32	0.837	9.2
3	0.30	0.858	6.3
4	0.29	0.866	4.7
5	0.28	0.873	3.8
6	0.28	0.878	3.3
7	0.27	0.881	2.7
Overall			43.4

* SIR phases were refined by symmetry averaging maps calculated with coefficients $(2F_o-F_c)\exp(i\alpha_c)$. The molecular envelopes required for symmetry averaging[13] were generated from model coordinates by enclosing each atom in a sphere of 7 Å radius. The initial model for the R1-69 : 14mer complex (Anderson *et al.* 1985), used to produce the envelope for 5 Å symmetry averaging (see text), contained idealized straight B-DNA. Inspection of the 5 Å map revealed that a 14mer with a slight bend would fit the density better. We therefore adjusted the positions of the two half-sites as rigid bodies using the real space refinement option of FRODO, and carried out the eight cycles of 3.2 Å refinement using an envelope from this model. Mean phase changes $(\sum|\Phi_c-\Phi_o|/N)$, for cycles subsequent to the first, are for all data present in the native data set, whether or not present in the initial SIR data set.

† Phases calculated from coordinates fit to the averaged maps from *a* were refined by averaging $F_o\exp(i\alpha_c)$ maps. The envelope was constructed from the coordinates on which the initial phases were based. $R_{av}=\sum|F_o-F_c|/\sum|F_o|$

$$r=\frac{\sum|F_o||F_c|-(\sum|F_o|\sum|F_c|)}{[\sum|F_o|^2-(\sum|F_o|)^2]\cdot[\sum|F_o|^2-(\sum|F_c|)^2]}$$

表 2. 经过非晶体学对称平均的相位修正

单对同晶置换（SIR）相位修正 *						
	相位修正 15 → 3.2Å			相位修正 5 → 3.2Å		
轮数	R_{av}	r	平均相位改变	R_{av}	r	平均相位改变
1	0.58	0.537	53.3	0.55	0.346	75.3
2	0.45	0.734	40.1	0.41	0.600	34.7
3	0.39	0.785	21.9	0.35	0.678	17.1
4	0.38	0.791	14.5	0.33	0.718	10.2
5	0.38	0.780	11.4	0.31	0.737	7.3
6	0.39	0.780	10.2	0.31	0.750	5.7
7	0.40	0.781	10.5	0.30	0.758	4.8
8	0.43	0.747	11	0.30	0.785	4.1
总体			65.5			82.0

初始模型的相位修正 †			
轮数	R_{av}	r	平均相位改变
1	0.37	0.792	23.2
2	0.32	0.837	9.2
3	0.30	0.858	6.3
4	0.29	0.866	4.7
5	0.28	0.873	3.8
6	0.28	0.878	3.3
7	0.27	0.881	2.7
总体			43.4

* 通过用系数 $(2F_o - F_c)$ exp $(i\alpha_c)$ 计算的对称平均图来进行 SIR 相位修正。用于对称平均[13]的分子边界来自囊括半径为 7 Å 的球体内每个原子的模型坐标。R1-69 和 14 碱基对 DNA 复合物的初始模型（安德森等，1985）用来产生 5 Å 平均对称的分子边界（见正文），其包含的是理想的、直链 B 型 DNA。对 5 Å 分辨率图谱的观察提示 14 碱基对的 DNA 微弱弯曲将使密度匹配的更好。于是我们使用软件 FRODO 的实空间修正选项将两个半位点作为刚体调整其位置，然后使用这个模型的分子边界进行八轮 3.2 Å 的修正。随后的每一轮相对第一轮的平均相位改变 $(\sum|\Phi_c - \Phi_o|/N)$ 是针对所有的母体数据集合，不论其是否存在于初始的 SIR 数据集合中。

† 与 a 的平均密度图相匹配的坐标系计算得到的相位通过平均 $F_o\exp(i\alpha_c)$ 图来修正。分子边界是基于初始相位的坐标而建立的。$R_{av} = \sum|F_o - F_c|/\sum|F_o|$

$$r = \frac{\sum|F_o||F_c| - (\sum|F_o|\sum|F_c|)}{[\sum|F_o|^2 - (\sum|F_o|)^2]\cdot[\sum|F_o|^2 - (\sum|F_c|)^2]}$$

The map thus computed showed clear double-helical DNA density, each deoxyribose-phosphate backbone having a staircase like shape, with sugars corresponding to the risers. Base pairs were poorly defined by density, but an unambiguous model could be built using backbone density and hydrogen-bonding restraints as a guide. Four α-helices of the protein appeared as twisted rods. A number of side chains were well-defined; others appeared as truncated bulges. For model-building, we used the program FRODO[15] with separate files for DNA and protein coordinates. The 14mer was broken into single stranded mono-, di- and trinucleotide segments, the phosphates and sugars were fitted to the DNA backbone density, and the torsion angles were adjusted to restore approximately correct base-pair geometry. The Hendrickson-Konnert restrained least squares refinement programs[16], modified by G. Quigley (MIT) to accept nucleic acid coordinates, were used with base-pair hydrogen bonding restraints to restore idealized geometry. Further manual adjustment followed, with periodic regularization. Residues 1–58 of the protein were built into the map and adjusted with the regularization routines in FRODO. Finally, protein and DNA coordinates were regularized with the constrained-restrained least squares refinement program CORELS[17], with base-pair hydrogen bonding restraints imposed on the DNA.

The initial model, constructed as just described, was used to obtain starting phases for a second round of noncrystallographic symmetry phase refinement. The order in the crystals is anisotropic, with intensities measurable to spacings of 3.2 Å in the directions of noncrystallographic symmetry axes formed by stacked protein–DNA complexes, but only to about 4.5 Å in other directions[10]. The map can therefore be regarded as density at 3.2 Å resolution, selectively "smeared" in directions perpendicular to the non-crystallographic symmetry axis. When building a model, however, stereochemical constraints couple positions along the local symmetry axis with those normal to it. Therefore, the effective resolution is intermediate between the two limits. To 5 Å, the intensity fall-off is relatively isotropic and computed structure factors at this resolution were used in a translation search program (A. Aggarwal, personal communication) to carry out a final adjustment of the position of the 3_1 axis. The anisotropy and non-crystallographic symmetry were incorporated into FFT (fast Fourier transform) structure factor calculations[18] by using the Bricogne density reconstitution procedure[13] to generate a full crystallographic asymmetric unit from a half-complex "map" that is computed in a coordinate frame with the noncrystallographic symmetry axis along *Z*. This choice of frame simplified introduction of a uniaxial, anisotropic "temperature factor". Various ratios of Bx = By to Bz were examined, by comparing calculated and observed amplitudes. The optimum occurred at 1.75, as judged both by *R* factor minimum and by relative lack of noise in corresponding $2F_o$–F_c and F_o–F_c maps.

Phases from the initial model were used to initiate eight further cycles of noncrystallographic symmetry refinement of phases between 15 and 3.2 Å. The computation converged stably to R_{av} = 0.27, with an overall phase change of 43° (Table 2, lower part). Significant adjustments were indicated by this map, and we extended the model from residue 58 to residue 63 (Fig. 1). Two further rounds of smaller adjustments

这样计算得到的密度图可以清晰地看到 DNA 双螺旋的密度，每个脱氧核糖–磷酸骨架的形状就像是楼梯，而糖对应着楼梯的侧板。碱基对在密度图中并不清晰，但是利用骨架的密度和氢键的限制作为指导，可以得到一个很清楚的模型。蛋白质的四个 α 螺旋就像扭缠在一起的棒。一些侧链可以很清晰地看到，另一些则如同被截去一部分的凸起。我们利用 FRODO 程序[15]以及 DNA 和蛋白质各自的坐标文件构建了模型。含 14 碱基对的片段被打碎成单链的单核苷酸、双核苷酸和三核苷酸片段的形式，磷酸和糖与 DNA 骨架的密度相匹配，扭转角则经调整恢复至大致正确的碱基对几何学结构。亨德里克森–康纳特的限制最小二乘修正程序[16]，经由奎格利（麻省理工学院）改进后可以接受核酸的坐标，用于加入碱基对的氢键限制，从而重构理想的几何图形。接下来进行更深入的手动调整，进行周期性正则化。使用 FRODO 程序，蛋白质 1~58 位的氨基酸残基被构建到密度图中，并进行正则化调整。最后，利用约束–限制最小二乘法修正软件 CORELS[17]，将碱基对的氢键限制加到 DNA 上，使蛋白质和 DNA 坐标正则化。

按如上所述构建的初始模型，可用来获得第二轮非晶体学对称相位修正的初始相位。该晶体存在各向异性，堆积的蛋白质–DNA 复合物在非晶体学对称轴的方向上可以在 3.2 Å 的分辨率处测得强度，而在其他方向则只能在 4.5 Å 的分辨率处测得[10]。因此，可以将密度图看作是 3.2 Å 分辨率处获得的，但在垂直于非晶体学对称轴的方向上有些选择性“模糊”。然而，在搭建模型时，沿着常规对称轴的位置伴随着立体化学的限制。因此有效的分辨率介于两个极限之间。就 5 Å 的分辨率而言，衍射强度的衰减相对来说是各向相同的，在此分辨率下计算得到的结构因子被用于一个转化搜索程序（阿加沃尔，个人交流）来进行 3_1 轴位置的最终调整。通过布里科涅密度重建程序[13] 将各向异性和非晶体学对称性结合到快速傅里叶变换（FFT）结构因子计算当中[18]，从以非晶体学对称轴为 Z 轴的坐标系中计算出的半复合物“密度图”得到了完整的晶体学非对称单元。这种坐标系的选择简化了单轴向、各向异性的“温度因子”的引入。通过对振幅计算值和观察值的比较，测定了 Bx=By 与 Bz 的不同比值。结合 R 因子最小值和相应 $2F_o–F_c$ 和 $F_o–F_c$ 差值图噪音的相对缺乏判定最佳比值为 1.75。

利用初始模型的相位对 15 Å 到 3.2 Å 之间的非晶体学对称相位进行进一步的八轮修正。计算值稳定地收敛于 R_{av}=0.27，整体的相位角改变为 43°（表 2，下部）。重要的调整在密度图中做了标示，并且我们将模型从第 58 位的残基拓展到 63 位（图 1）。根据 $2F_o–F_c$ 和 $F_o–F_c$ 差值密度图又进一步进行了两轮更小的调整。不同阶段的密度

were based on $2F_o–F_c$ and $F_o–F_c$ maps. Maps and coordinates at various stages are shown in Fig. 2. It is evident that the second round of non-crystallographic-symmetry refinement substantially improved phase determination. Base-pair and side-chain densities became far clearer and peptide-backbone density more continuous. Poor placement of some side chains of the initial model did not prevent appearance of density in correct positions in the final map, showing that errors in that model did not unduly bias the outcome of the phase refinement (Fig. 2*b* and *c*). The use of an intermediate model to initiate a second round of phase refinement was used in determining the turnip crinkle virus structure, where similar corrections to an actual structure were generated[19]. The power of threefold non-crystallographic redundancy in phase refinement has been amply demonstrated in previous work—for example, in the determination of the influenza virus haemagglutinin structure, with data of comparable accuracy to ours[20]. Reciprocal-space refinement of the model is in progress. The *R*-factor of the model presented here is 0.44. (H. Holley, J.E.A. and S.C.H. unpublished data).

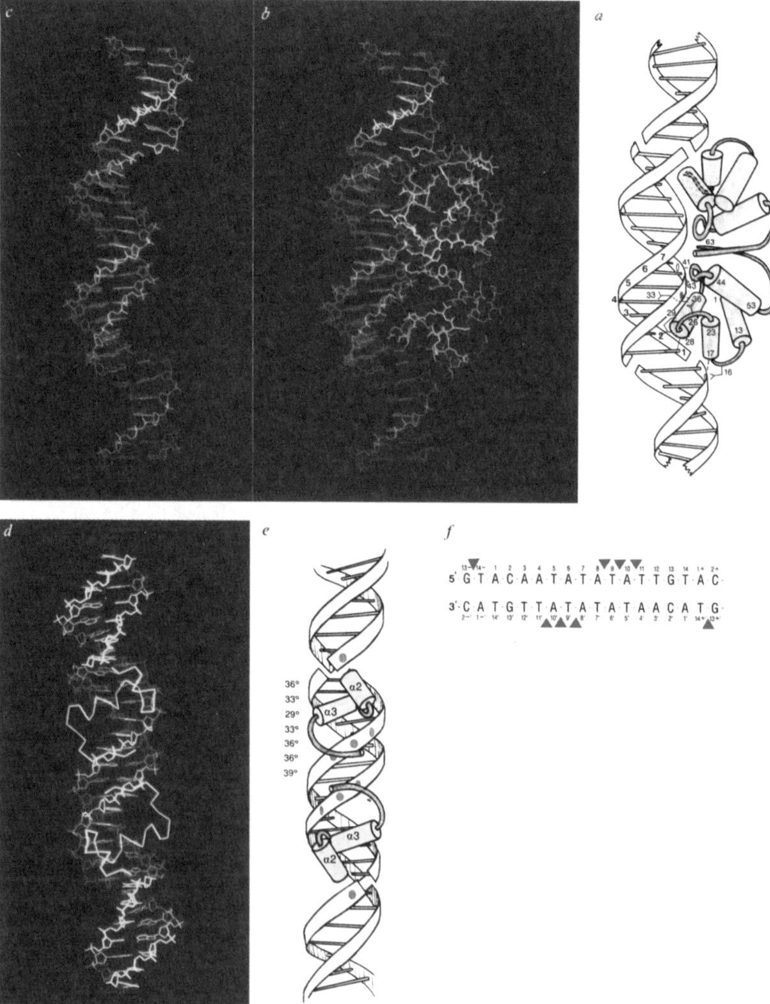

Fig. 1. Diagram summarizing the elements of the 434 repressor–operator complex, in an orientation

图和坐标系如图 2 所示。很明显，非晶体学对称性的第二轮修正从根本上改善了相位的确定。碱基对和侧链的密度变得更加清晰，多肽骨架的密度也更加连续。起始模型中某些侧链较差的定位并没有影响其在最终密度图中的正确位置，这说明模型中的错误没有使相位修正后的结果偏移（图 2*b* 和 *c*）。在芜菁皱缩病毒结构的测定过程中曾利用中间模型来起始第二轮的相位修正，也生成了对实际结构的类似修正 [19]。三重非晶体学冗余度在相位修正中的作用已在先前的工作中得到充分展示——例如，在测定流感病毒血细胞凝集素的结构时，数据的准确度与我们的相近 [20]。模型的倒易空间修正正在进行中。本文中模型的 *R* 因子为 0.44（霍利、安德森和哈里森，未发表数据）。

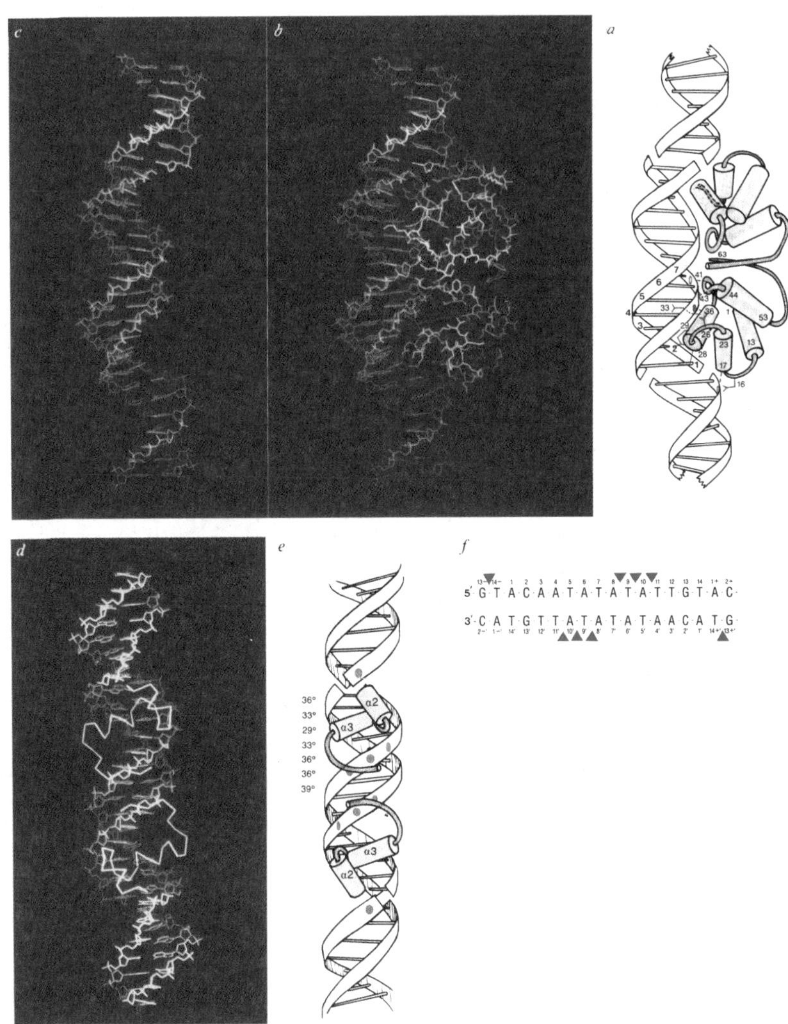

图 1. 434 阻遏蛋白–操纵基因复合物中各要素的总结性图示，其方向与图 *b* 所示的计算机模拟图对应。

corresponding to the computer-graphics display in *b*. R1-69 α-helices are represented as cylinders and non-helical polypeptide chain as tubes. In the lower R1-69 monomer, key residues are shown and the numbers of the first and last residues in each α-helix are also entered. Numbers along ribbon indicate DNA base pairs (See *f*). *b*, Overall view of the 434 repressor–operator complex. The central 14mer (red) is bound to two R1-69 domains (light blue and dark blue). Adjacent 14mers, stacked as in the crystal, are also shown (violet). The view is perpendicular to the operator dyad, which lies in the centre of the diagram. *c*, DNA alone, viewed as in *b*, to show bend. Note variation of minor groove width. *d*, DNA and part of protein, seen as if viewed from the right in *a* or *b*. Residues 16–44 are represented by Cα backbone. *e*, Diagram summarizing DNA conformation and backbone contacts, drawn to correspond to the view in *d*. The numbers to the left of the diagram indicate twist in degrees between base pairs, calculated using the method of Kabsch[36] after obtaining the best visual superposition of each base pair on the next. *f*, Sequence of 14mer, together with numbering scheme used in the text. Symbols + and – refer to adjacent 14mers as packed in the crystal. Two base pairs of each adjacent 14mer are shown. Arrows refer to phosphates at which ethylation interferes with binding[22].

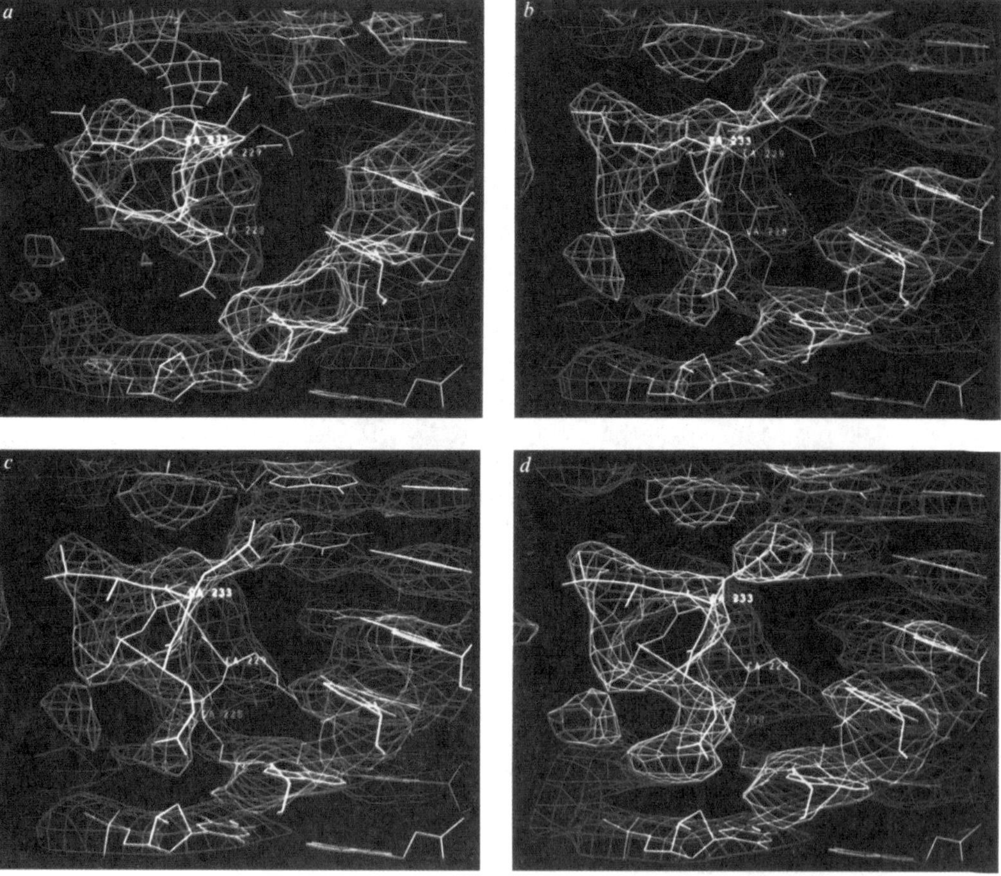

Fig. 2. Part of the electron density maps at different stages of the structure determination, showing the N-terminal half of α3 and portions of base pairs 1–4. Models are superimposed on density. The helix α3 is viewed from the C-terminal end, with residues 28–34 (QQSIEQL) shown and α-carbons of residues 28, 29 and 33 labelled (228, 229 and 233, respectively). *a*, Map after refinement of SIR phases, with initial model superimposed. *b*, Map after refinement of phases determined from initial model, with initial model superimposed. Note clear indications for repositioning of side chains, especially Gln 28 and Gln 29 (background) and Gln 32 (foreground). *c*, Same map as in *b*, but with rebuilt model. *d*, Final ($2F_o$–F_c) map, with rebuilt model.

R1-69 的 α 螺旋显示为柱状图形，非螺旋的多肽链则显示为管状。在位于下方的 R1-69 单体中显示了关键残基，也标注了每个螺旋起始和结束的残基序号。在 DNA 带上的数字则是 DNA 的碱基对编号（见图 *f*）。*b*，434 阻遏蛋白 – 操纵基因复合物的整体视图。中间的 14 碱基对（红色）DNA 序列与两个 R1-69 结构域（浅蓝色和深蓝色）结合。在晶体中堆叠的相邻 14 碱基对 DNA 在图中也有显示（紫色）。观察方向与反向重复的操纵基因是垂直的，它位于图的中央。*c*，将 DNA 单独画出以显示其弯曲，观察方向与 *b* 一样。请注意小沟宽度的变化。*d*，DNA 和部分蛋白质，观察方向是从 *a* 或 *b* 的右侧看过来。残基 16 ~ 44 用 Cα 骨架标出。*e*，DNA 构象和骨架接触的总结性图示，观察方向与 *d* 对应。图形左侧的数字为碱基对之间的扭转角，在获得了每个碱基对相对于下一个的最佳视觉重合效果之后使用卡卜什法 [36] 计算得到。*f*，14 碱基对的序列以及正文中使用的编号体系。符号 + 和 – 表示在晶体中相邻的 14 碱基对。图中标示了两头相邻的 14 碱基对中的两个碱基对。箭头指示乙酰化会干扰其结合能力的磷酸基团 [22]。

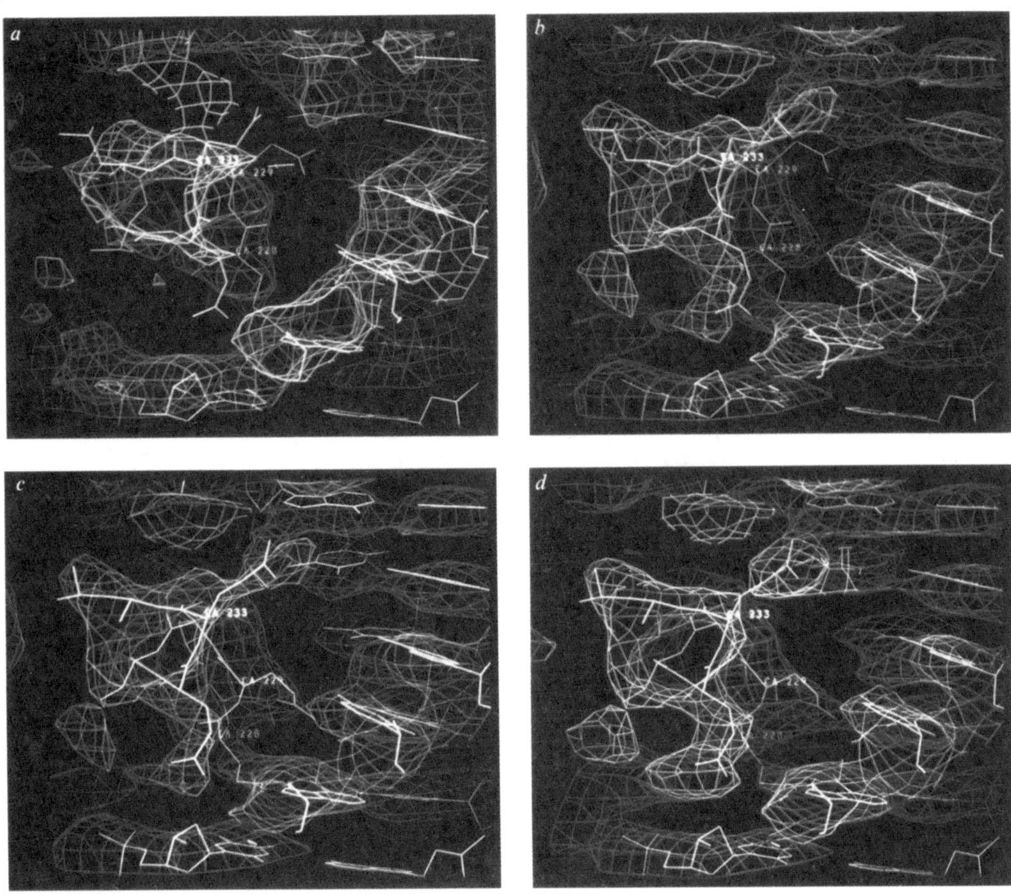

图 2. 结构测定过程中不同阶段的部分电子密度图，显示的是 α3 的 N 末端和碱基对 1~4 的部分。模型叠加在密度图上。观察方向是从螺旋 α3 的 C 端看该螺旋，显示了残基 28~34（QQSIEQL）和标记的残基 28、29 和 33 的 α 碳原子（分别对应 228、229 和 233）。*a*，单对同晶置换相位修正后的密度图，初始模型叠加其上。*b*，起始模型相位修正后的密度图，初始模型叠加其上。注意侧链位置变清晰了，特别是第 28 位谷氨酰胺（Gln 28）和 Gln 29（背景）以及 Gln 32（前景）。*c*，和 *b* 中相同的密度图，但叠加了重建的模型。*d*，最后的 $(2F_o-F_c)$ 图，叠加了重建的模型。

As in any structure determination at this resolution, interactions such as hydrogen bonds must be inferred from the position and orientation of the participating groups. We believe our present structure to be sufficiently well determined that the protein–DNA interactions can be described correctly and that likely hydrogen bonds can be assigned. More precise conformational details, accurate hydrogen-bond geometry, contact distances and features such as the positions of water molecules—all of which may be significant for complete understanding of specificity—will only be visible at the higher resolution afforded by a different R1-69–operator crystal currently being studied (A. Aggarwal and S.C.H. unpublished data).

DNA Conformation

The DNA of the 14mer forms a B-type helix throughout its length. The 14mers are stacked accurately on adjacent DNA segments to form a pseudocontinuous helix. The local twist varies from about 39° to 29° per base pair—that is, from 9.2 to 12.3 base pairs per turn (Fig. 1*e*). A similar range of variation in twist has been found in crystalline B-DNA[21]. With respect to the average of 10.5 base pairs per turn, the 14mer is overwound at its centre and underwound at its ends. The helix axis bends somewhat, causing the DNA to curl slightly around helix 3 of each monomer. The bend is sharpest between base pairs 4 and 5 (and 10 and 11), with relatively more gentle bending between base pairs 1 and 4, between base pairs 5 and 10, and between base pairs 11 and 14. Overwinding and bending in the centre of the DNA fragment narrows the minor groove. Underwinding and bending widens it at the ends. This variation is evident in Figs 1 and 3.

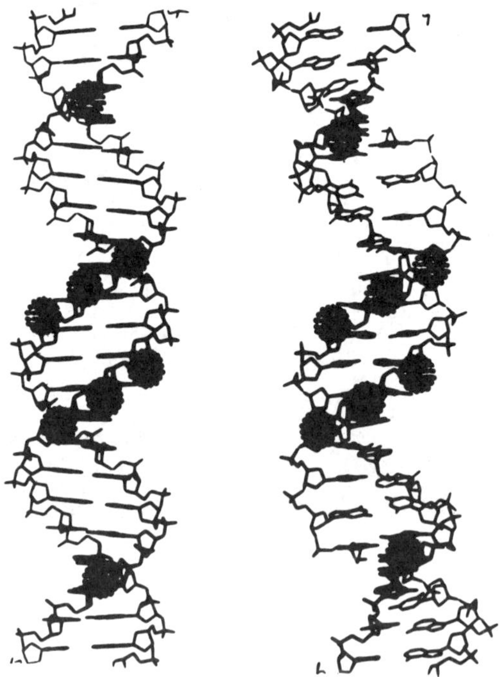

Fig. 3. Conformation of DNA in the crystalline complex (right) compared with idealized, 10.5-bp-per-turn B-DNA (left). Contacted phosphates are highlighted. It is clear from this figure, and from Fig. 1, that model building with idealized B-DNA could not correctly predict the protein–DNA contacts.

如同在这个分辨率下的任何结构测定，像氢键这样的相互作用必须从参与基团的位置和取向来推断。我们相信现在的结构已经得到了非常好的确定，可以正确的显示蛋白质–DNA 的相互作用以及可能的氢键。更多精确的构象细节、准确的氢键几何图形、作用距离及其他特征，如水分子的位置等——所有这些都可能对完全理解 DNA 与蛋白质结合的特异性具有重要的意义——将会在一个不同的 R1-69 阻遏蛋白晶体所提供的更高分辨率结构中得以展示，该晶体目前还在研究中（阿加沃尔和哈里森，未发表数据）。

DNA 的构象

14 碱基对的 DNA 在其整个长度上形成了一个 B 型螺旋。相邻的 14 碱基对 DNA 精确地堆叠成伪连续螺旋。每个碱基对形成的局部扭转从 39°到 29°不等——这样每一圈就有 9.2 到 12.3 个碱基对（如图 1*e*）。类似的扭转变化范围在结晶的 B 型 DNA 中已经发现[21]。就平均每圈 10.5 个碱基对而言，14 碱基对的 DNA 在中心部分卷得太紧，而在末端则太松。DNA 螺旋轴轻微弯曲，使得它围着每个蛋白单体中的螺旋 3 微小卷曲。这种弯曲在碱基对 4 和 5（及 10 和 11）之间最剧烈，而在碱基对 1 到 4、5 到 10 及 11 到 14 之间相对缓和。DNA 片段在中心的过度扭转和弯曲使得小沟变窄，扭转不足和弯曲则使得末端变宽。在图 1 和图 3 中这种变化非常明显。

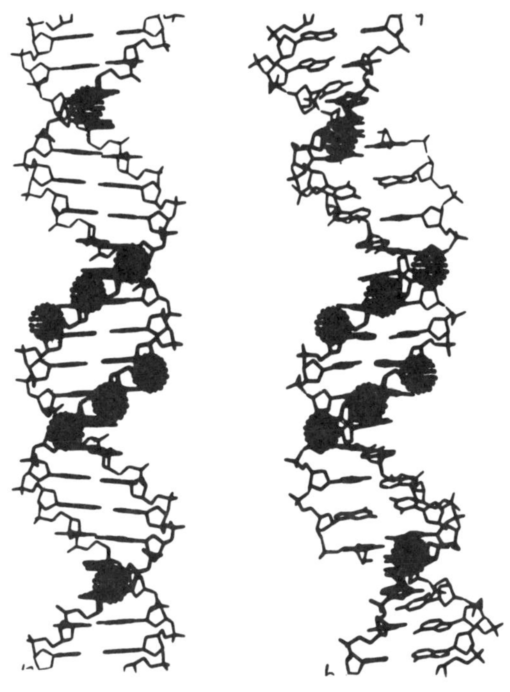

图 3. 结晶复合物中 DNA 的构象（右）与理想状态下每圈 10.5 个碱基对的 B 型 DNA（左）对比。参与接触的磷酸加亮。从这幅图和图 1 中可以很清楚地看到，依据理想的 B 型 DNA 构建的模型不能正确预测蛋白质与 DNA 的接触。

Protein Conformation

R1-69 is a cluster of four α-helices, with a C-terminal extension. Two of the helices, α2 and α3, form a helix–turn–helix motif. The residues found in each of the four helices are shown in Table 3. For comparison, the corresponding residues in the N-terminal domain of λ repressor[5], determined by comparison of the two models, are also listed. As observed in our earlier paper[4], if the 434 and λ α2–α3 structures are superposed, the axes of α1 and α4 in the two repressors coincide to within about 2–3 Å. The α-carbon positions suggest that 434 repressor has "deletion" with respect to λ repressor of two residues between α1 and α2 and of one residue at the beginning of α4. The two models diverge at residue 58 of 434 (corresponding to 75 of λ), precisely the point at which the two amino-acid sequences cease to be similar. Various hydrophobic side-chains form the interior of the four-helix cluster, which is also stabilized by several polar linkages—notably, Gln 17–Glu 32 and Arg 5–Glu 35. The last six residues of R1-69 are poorly defined in our map. They appear to extend toward DNA backbone near P11.

Table 3. Amino-acid residues in α-helices

	434 repressor	λ repressor
α1	1–13	9–23
α2	17–23	33–39
α3	28–36	44–52
α4	44–53	61–69

The Cα backbone of λ repressor N-terminal domain[5] was superimposed on the R1-69 backbone, optimizing coincidence of α2 and α3 (ref. 4). Assignments of helical residues were made by visual inspection of backbone conformation. Residue 15 of λ repressor closely corresponds in three-dimensional position to residue 1 of 434 repressor; 434 repressor lacks the N-terminal arm of λ repressor and the first few residues of α1. The first helix of λ repressor, as built by Pabo and Lewis, ends "earlier" than α1 of 434 repressor (residue 23 of λ corresponds to 9 of 434). The region at the end of α1 was not well-defined in the original λ repressor map, however, and the actual Cα coincidence in α1 may be even greater than suggested by the assignments above. In the present 434 repressor model, residue 27 might also be assigned to helix 3, since its carbonyl appears to hydrogen bond to N31.

Interactions occur between the two R1-69 subunits bound to one 14mer. The two monomers are in contact across the dyad near residues 57–58. The aliphatic chain of Arg 41 and a residue near the C-terminus in one subunit lie against the aromatic ring of Phe 44 in the other subunit. These contacts may be important in determining the relative orientations of the DNA-binding surfaces of the two monomers, thereby defining the required spatial orientation of one operator half-site with respect to the other.

蛋白质的构象

R1-69 是一个由 4 个 α 螺旋构成的簇，并且在 C 端存在延伸。其中螺旋 α2 和 α3 形成一个螺旋 – 转角 – 螺旋基序。4 个螺旋中每个螺旋的氨基酸残基如表 3 所示。为了方便比较，在表中还列出了 λ 阻遏蛋白[5] N 端结构域中相应的残基，这些位置是通过比较两个模型而确定的。如我们以前发表的文章所述[4]，若将噬菌体 434 阻遏蛋白和 λ 阻遏蛋白的螺旋 α2~α3 重叠，可以发现这两个阻遏蛋白的 α1 和 α4 的轴几乎重合，仅有 2~3Å 的微小差别。α 碳的位置显示，相比于 λ 阻遏蛋白，434 阻遏蛋白缺失了螺旋 α1 和 α2 间的两个残基以及 α4 起始处的一个残基。两个模型在 434 的 58 位残基(对应于 λ 的 75 位残基）出现分歧，也正是从这个位置开始两个蛋白的氨基酸序列不再相似。各种疏水侧链组成四螺旋簇的内部结构，该结构还因几个极性键而得以稳定——尤其是 Gln 17–Glu 32 间和 Arg 5–Glu 35 间的联接。R1-69 的最后六个残基在我们的电子密度图上不是很明确。它们似乎朝着 P11 附近的 DNA 骨架延伸。

表 3. α 螺旋中的氨基酸残基

	434 阻遏蛋白	λ 阻遏蛋白
α1	1~13	9~23
α2	17~23	33~39
α3	28~36	44~52
α4	44~53	61~69

λ 阻遏蛋白 N 端结构域的 Cα 骨架[5] 叠加在 R1-69 骨架上，使 α2 和 α3 的相合性最佳（参考文献 4)。α 螺旋残基的分配是根据骨架构象的视图而定的。三维位置中 λ 阻遏蛋白的 15 位残基与 434 阻遏蛋白的第 1 位残基相对应；434 阻遏蛋白缺少 λ 阻遏蛋白的 N 末端臂和 α1 起始位置的一些残基。λ 阻遏蛋白的第一个螺旋是由帕博和刘易斯构建的，比 434 阻遏蛋白的 α1 结束的“更早”（λ 的残基 23 与 434 的残基 9 相对应)。然而，在原先的 λ 阻遏蛋白密度图中 α1 的末端区域并不清晰，实际的 α1 的 Cα 相合性可能要比上述残基分配所得到的更好些。在现有的 434 阻遏蛋白的模型中，27 位残基也可能属于螺旋 3，因为它的羰基似乎与 N31 形成氢键。

与一个 14 碱基对的 DNA 结合的两个 R1-69 亚基间会发生相互作用。两个单体在残基 57~58 处相接触。一个亚基中的 Arg 41 的脂肪族侧链和邻近 C 端的一个残基依靠在另一个亚基的 Phe 44 的芳香族环上。这些接触可能在确定两个单体的 DNA 结合表面的相对取向上很重要，从而决定了操纵基因的半位点相对于另一个半位点所需的空间取向。

Protein–DNA Contacts

The polypeptide chain is so folded that in the complex the N-termini of α-helices 2, 3 and 4 all point toward the DNA backbone, α3 lies in the major groove, and the loop joining α3 and α4 runs along the DNA backbone (Fig. 1*d*). Contacts to DNA backbone from one R1-69 subunit occur on both sides of the major groove that is occupied by α3 and possibly across the minor groove that contains the operator dyad. Major-groove interactions with base pairs 1–5 and 10–14 are made by three glutamine side chains projecting from α3. These relationships are summarized in Figs 1*d* and *e*, and described in detail below.

Backbone contacts. Four phosphate groups per half-site, corresponding precisely to the phosphates implicated by ethylation interference experiments[22], appear to interact with protein. Considering half of the complex, the contact phosphate groups lie 5′ to bases 14-, 9′, 10′ and 11′ (see Fig. 1*f* for numbering scheme). The P14- contact occurs between different complexes in the crystal. Because the stacking of 14mers mimics continuous DNA, we believe that the interaction we observe at P14- closely reflects the one actually made in continuous DNA.

The three tightest contacts (as judged both from the structure and by the strength of ethylation interference) all appear to involve peptide backbone NH groups. At P14-, the NH of Gln 17 approaches the phosphate oxygens closely enough to form a hydrogen bond. Residue 17 is at the N-terminus of α2, and its peptide NH group is expected to make a particularly strong hydrogen bond to a negatively charged ion, as a result of oriented peptide dipoles in the α-helix[23]. Analogous interactions occur in the *Salmonella typhimurium* sulphate binding protein, where peptide NH groups at the N-termini of α helices are ligands of bound SO_4^{2-} ions[24]. The side chain of Asn 16 also appears to be involved in the contact to P14-. At P9′, there is close approach of the main chain NH groups of Lys 40 and Arg 41; the side-chain conformation of Arg 41 permits a salt link with the same phosphate (Fig. 4). At P10′, there is close approach of the peptide NH groups of both Arg 43 and Phe 44 (Fig. 4). Phe 44 is the amino-terminal residue of α4. The use of peptide NH groups to contact phosphates may be a common feature in proteins that bind DNA and nucleotides. Dreusicke and Schulz[25] have described a "giant anion hole" at the site in adenylate kinase where a nucleotide phosphoryl group is believed to be located. The polypeptide chain loops in such a way that five peptide NH groups coordinate a bound sulphate in their crystal structure.

蛋白质–DNA 的相互作用

复合物中的多肽链高度折叠，以至于 α 螺旋 2、3 和 4 的 N 端都指向 DNA 骨架，α3 位于 DNA 大沟内，且连接 α3 和 α4 的环也沿着 DNA 的骨架行走（如图 1*d*）。一个 R1-69 亚基与 DNA 骨架的接触发生在被 α3 占据的大沟的两侧，并有可能穿过包含有反向重复操纵基因的小沟。与碱基对 1~5 以及 10~14 发生的大沟相互作用是通过从螺旋 α3 中伸出的三个谷氨酰胺侧链实现的。这些作用关系在图 1*d* 和 *e* 中进行了总结，详情如下所述。

骨架的接触。每个半位点有四个磷酸基团可能与蛋白质相互作用，这些磷酸基团精确对应于乙酰化干扰实验所指向的磷酸基团 [22]。就复合物的一半而言，参与接触的磷酸基团位于碱基 14-、9′、10′ 和 11′ 的 5′ 端（见图 1*f* 的编号体系）。P14- 的接触发生在晶体中不同的复合物之间。因为 14 碱基对的堆叠模拟了连续的 DNA，所以我们认为我们观察到的 P14- 的接触作用近似地体现了连续 DNA 实际发生作用时的情况。

三个最紧密的接触（根据结构和乙酰化干扰强度来判断）看来都涉及多肽骨架的 NH 基团。在 P14- 位，氨基酸 Gln 17 的 NH 基团足够接近磷酸基团中的氧，形成氢键。残基 17 位于螺旋 α2 的 N 末端，由于 α 螺旋中定向的肽偶极 [23]，使得肽的 NH 基可以与负离子产生非常强烈的氢键作用。相似的相互作用见于鼠伤寒沙门氏杆菌硫酸结合蛋白中，α 螺旋 N 末端的肽 NH 基团是其上结合的硫酸根离子的配体 [24]。Asn 16 的侧链也可能与 P14- 相接触。在 P9′ 位点上，主链上 NH 基团与 Lys 40 和 Arg 41 很接近；Arg 41 侧链的构象可以使其与同一个磷酸基团形成盐键（图 4）。在 P10′ 位点上，肽链 NH 基团与 Arg 43 和 Phe 44 都非常接近（图 4）。Phe 44 是螺旋 α4 的氨基末端的残基。利用肽 NH 基团与磷酸基团作用可能是结合 DNA 和核苷酸的蛋白质的普遍特征。德勒西克和舒尔茨 [25] 已经描述了在腺苷酸激酶中有一个“巨大的阴离子洞”，该位置被认为结合了一个核苷酸磷基团。它们的晶体结构中多肽链以一定的方式环化，使五个肽 NH 基团与一个硫酸根相协调。

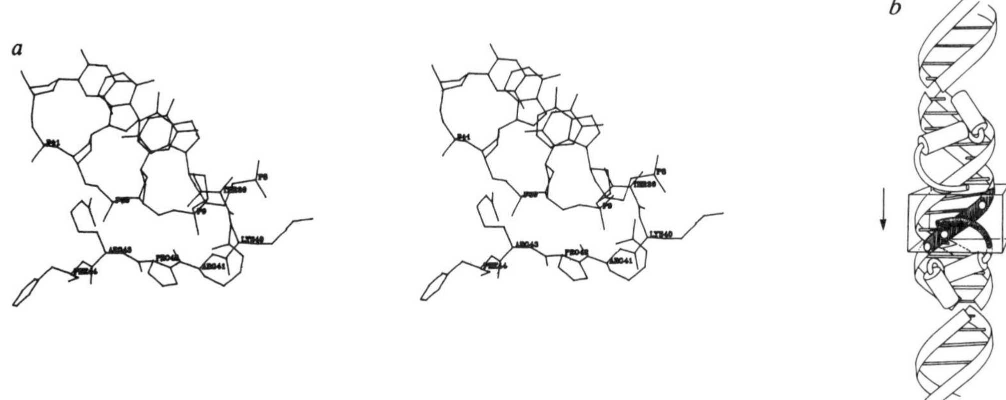

Fig. 4. Stereo view of contacts between protein and DNA phosphates P9′ and P10′. *b*, Schematic diagram as a key to *a*. The view is of the region indicated in the (shaded structures), looking approximately along the DNA axis (arrow). Compare key with Fig. 1*e*. Residues 39–44 of the protein are shown, together with nucleotides 8′–11′. (Primes are omitted in the labels.)

The side chain of Arg 43 projects into the minor groove near base pair 7. The groove is somewhat narrower than average at this position, and the guanidinium group could probably have significant coulombic interaction with phosphates 9 and 10′ on either side of the groove (Fig. 4). The possibility of direct interaction with base pair 7 is discussed in the next section and in the accompanying paper.

The ethylation of P11 interferes more weakly with binding than ethylation of P14-, P9′ or P10′ (ref. 22). Thr 26 appears to contact P11′, and the C-terminal residues of the other monomer may also lie in this vicinity. Because the C-terminal residues are not well defined in our map, we cannot at present determine whether any actually engage in backbone contacts. If they do, the interaction may help establish the favoured relative orientation of the two operator half-sites, because an R1-69 domain bound in one half-site would make contact by its C terminus with DNA backbone in the other half-site.

Base-pair contacts. Three glutamines (residues 28, 29 and 33) project from α3 into the major groove (Fig. 5). Gln 28 approaches A1 (base pair 1) in such a way that hydrogen bonds can readily form between N7 of the base and Nε of the glutamine and between N6 and Oε. "Bidentate" H-bonding of this type was suggested as a mode of adenine-specific interaction by Seeman *et al.*[26]. Gln 29 lies along α3. Its Cβ and Cγ are in van der Waals contact with the 5-methyl group of T12′ (base pair 3), and its Nε can hydrogen bond to O6 (and perhaps also N7) of G13′ (base pair 2). The Oε of Gln 29 may hydrogen bond to its main chain NH, an interaction occasionally seen with glutamine at amino termini of α-helices[27]. This hydrogen bond would fix the orientation of the glutamine and point the Nε donor group directly at G13′. Gln 33 projects toward T11′ and A10′ (base pairs 4 and 5). There could be a hydrogen bond between Nε of Gln 33 and O4 of T11′ (base pair 4), and perhaps between Oε and N6 of A10′ (base pair 5); a refined model will be required for confident assignment.

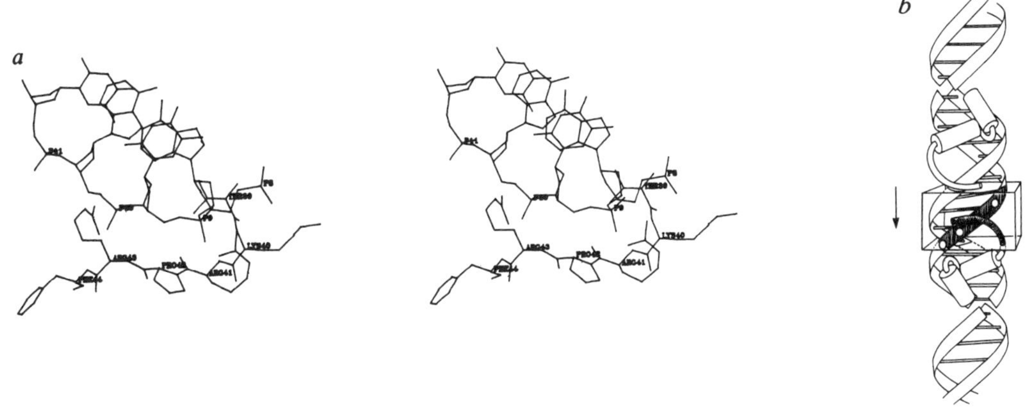

图 4. 蛋白质和 DNA 磷酸基团 P9′ 和 P10′ 接触的立体视图。*b*，作为 *a* 图索引的概要图。图中标出了 *a* 图所描述的具体位置（阴影结构），观察方向是沿着 DNA 的轴向（箭头所指）。参考图 1*e* 的标示。显示的是蛋白质 39～44 的残基和核酸的 8′～11′ 的碱基对。(标注省略了)

Arg 43 的侧链伸入小沟中第 7 个碱基对附近。此位置的小沟要比平均宽度更窄些，胍基很可能与位于小沟任一侧的磷酸 9 和 10′ 有显著的库仑相互作用（图 4）。与碱基对 7 直接作用的可能性将在下一节和本期另一篇相关文章里讨论。

P11 的乙酰化对结合的干扰作用比 P14-、P9′ 或 P10′ 弱（参见参考文献 22）。Thr 26 看来与 P11′ 相接触，而另一个单体的 C 端残基可能也位于这一区域附近。由于在我们的密度图上 C 端残基不清晰，因而现在不能确定其是否参与了骨架的接触。如果它们参与了接触，这个相互作用可能会帮助两个操纵基因半位点建立最惠的相对取向，因为结合在一个半位点上的 R1-69 结构域会使其 C 末端与另一个半位点的 DNA 骨架相接触。

碱基对的接触。螺旋 α3 中的三个谷氨酰胺（残基 28、29 和 33）从大沟中伸出（如图 5）。Gln 28 接近 A1（碱基对 1），从而使碱基的 N7 与 Gln 的 N*ε*、碱基的 N6 与 Gln 的 O*ε* 之间形成氢键。这种“双配位”氢键类型被西曼等人描述为腺嘌呤特异性的相互作用模式 [26]。Gln 29 位于螺旋 α3。它的 C*β* 和 C*γ* 与 T12′（碱基对 3）的 5–甲基基团通过范德华作用相接触，其 N*ε* 可以与 G13′（碱基对 2）的 O6（也可能是 N7）形成氢键。Gln 29 的 O*ε* 可能与其主链上的 NH 基团形成氢键，这种相互作用偶尔存在于 α 螺旋氨基末端的 Gln[27]。这个氢键可以固定 Gln 的取向，将 N*ε* 供体基团直接指向 G13′。Gln 33 指向 T11′ 和 A10′（碱基对 4 和 5）。在 Gln 33 的 N*ε* 和 T11′（碱基对 4）的 O4 之间可能有氢键，在 O*ε* 和 A10′（碱基对 5）的 N6 之间也可能有；这需要一个修正的模型来准确地推断。

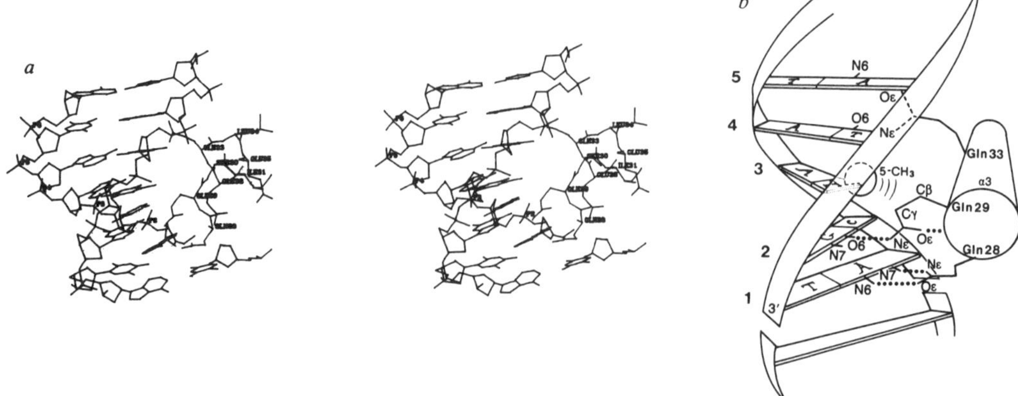

Fig. 5. *a*, Stereo view of helix α3 and base pairs 1–5, viewed from the N-terminus of α3. One base pair of the adjacent 14-mer is also present. Included are residues 28–35 of R1-69. *b*, Schematic diagram as a key to *a*. Note the similarity to view of α3 and surrounding structure in the lower monomer of Fig. 1*c*. Gln 28, Gln 29 and Gln 33 are shown, with hydrogen bonds suggested by the current model drawn in dotted lines. Functional groups on bases and on Gln 33 that might participate in additional hydrogen bonds are also shown. The proposed non-polar contact between the 5-CH_3 group of T12′ (base pair 3) and Cβ and Cγ of Gln is indicated.

The present model shows no direct interactions with base-pairs 6 and 7. Although residues 64–69 are not well defined in the map they are clearly excluded from the major groove on the near side of the operator by α3 and from the minor groove by residues 40–44. Moreover, model building indicates that there is no reasonable possibility of base-pair contact from these residues in the major groove on the far side of the DNA.

Contributions to Specificity

Three different features of the DNA structure appear to be significant for specificity: (1) the groups accessible in the major groove on base pairs 1–5, (2) the conformational details of the half-sites, and (3) the overwinding of the central base pair steps with a narrowing of the minor groove. These features are relevant to different classes of interactions made by the protein. (1) Side chains from α3 contact base pairs in the major groove. (2) Peptide amino groups, as well as side chains from various residues, contact backbone phosphates in ways that constrain the conformation of the operator half-sites. (3) Dimer interactions between repressor monomers favour a particular spatial relationship between half-sites.

Base-pair contacts. The interactions of Gln 28 with base pair 1 and of Gln 29 with base-pairs 2 and 3 are consistent with the conservation of base pairs 1–3 in all 12 operator half-sites, because changes in any of these base pairs would destroy one or more of the hydrogen bonds or disturb van der Waals contacts with the thymine methyl group. (For a list of all naturally-occurring operator sequences, see accompanying paper, ref. 2.) Wharton[3] used a random mutagenesis scheme to generate all possible changes in amino acids 28 and 29. No viable mutants were detected, and indeed no likely substitutions in the model can be imagined. Wharton and Ptashne[28] have also shown that a mutant operator

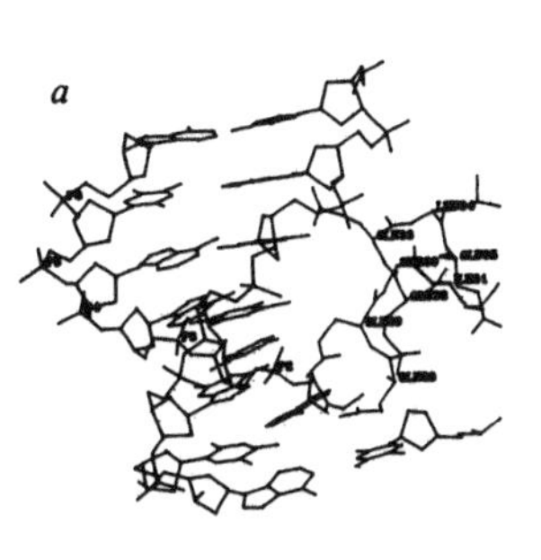

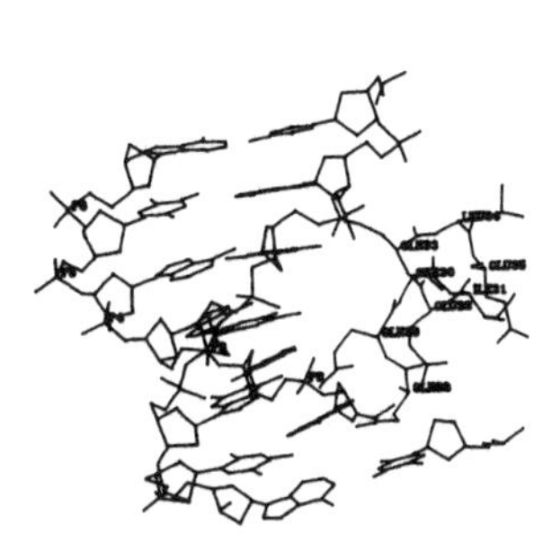

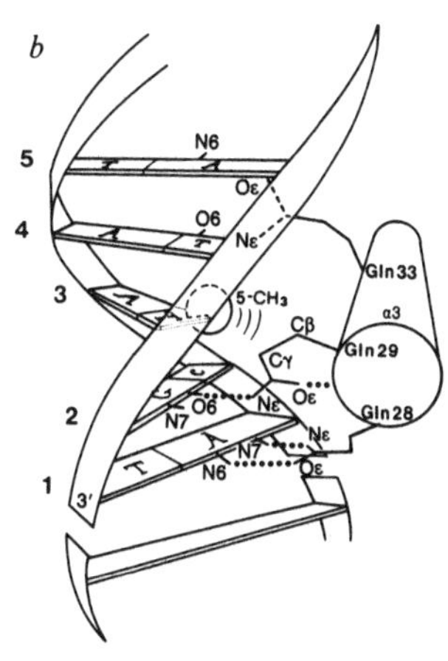

图 5. *a*，螺旋 α3 和碱基对 1~5 的立体视图，从 α3 的 N 末端观察。并显示了与 14 碱基对相邻的一个碱基对。包括的残基从 R1-69 的 28 至 35 位。*b*，作为 *a* 图索引的概要图。注意本图与图 1*c* 下方单体中 α3 及周围的结构相似。图中显示了 Gln 28、Gln 29 和 Gln 33，当前模型中的氢键用虚线表示。碱基 Gln 33 中可能参与附加氢键作用的功能基团也有所显示。我们提出的 T12′（碱基对 3）的 5–甲基基团与 Gln 的 Cβ 和 Cγ 之间的非极性接触也标示了。

目前的模型显示碱基对 6 和 7 之间没有直接作用。虽然残基 64~69 在密度图上不能很清晰地看到，然而它们清楚地被螺旋 α3 排除在操纵基因近侧大沟外并被残基 40~44 排除在小沟外。而且，模型构建显示位于 DNA 大沟较远一侧的残基没有合理的可能性去参与碱基对接触。

对特异性的贡献

DNA 结构的三个不同特征似乎对于特异性有重要意义：(1) 大沟中碱基对 1~5 的基团是容易接近的，(2) 半位点的构象细节，(3) 中间碱基对的过旋使得小沟变窄。这些特征是和蛋白质不同类型的相互作用相关的。(1) α3 的侧链与大沟中的碱基对接触。(2) 肽链上的氨基基团以及来自不同残基的侧链与磷酸骨架接触，从而限制操纵基因半位点的构象。(3) 阻遏蛋白单体之间形成二聚体的相互作用促成半位点之间形成的特殊空间关系。

碱基对的接触。Gln 28 与碱基对 1 以及 Gln 29 与碱基对 2 和 3 的相互作用与所有 12 个操纵基因半位点中碱基对 1~3 的保守性是一致的，因为这些碱基对中任意一个的改变都会破坏一个或者多个氢键，或者打乱与胸腺嘧啶甲基基团的范德华接触。(所有天然操纵基因序列的清单请参见本期另一篇相关文章，参考文献 2) 沃顿[3] 使用了一种随机突变策略在氨基酸 28 和 29 位置上产生所有可能的变化。我们没有检测到可行的突变，事实上，在模型中不可能存在合适的取代。沃顿和普塔什尼[28] 已

with T instead of A at position 1 is recognized specifically by a mutant repressor with alanine at position 28. These coordinated changes can be built into the model, generating a van der Waals contact between the —CH_3 group of T1 and the —CH_3 side chain of Ala 28.

The side chain of Gln 33 is in position to contact base pairs 4 and 5. The model shows that this residue could hydrogen bond to either base 11′ or 10′, or to both (Fig. 5). Wharton[3] has studied the effects of substituting Ala for Gln 33 on the binding of repressor to various operators. His results, taken with others (R. Wharton and M.P. unpublished data), are consistent with the idea that residue 33 indeed contacts positions 4 and 5.

Backbone contacts. Within a half-site, contacts to the DNA backbone position a subunit of the protein against the operator. The relationship is likely to be quite exact, because main-chain peptide amino groups participate in the contacts and the main chain of a folded protein is generally less flexible than the side chains. The backbone contacts, especially the one at P14-, require a gently bent outer operator[22]. This curvature is essential for at least one of the major groove contacts: without it, the interaction between Gln 28 and A1 would not have the correct geometry. It is not yet clear whether the bend is present in unbound operator DNA, or whether protein binding induces it.

Central base pairs. The configuration of the operator at its central base pairs is consistent with sequence preferences at positions 6–9. A more detailed discussion appears in the accompanying paper, but we summarize the arguments here. The twist is about 36° between base pairs 6 and 7 and base pairs 8 and 9 and 39° between base pairs 7 and 8 (Fig. 1*e*). The minor groove is consequently 2.5 Å narrower than in idealized B-DNA. It is believed that GC or CG base pairs are not well accommodated in a narrow minor groove[29], and repressor indeed discriminates against GC or CG at positions 6–9 (ref. 2). The sequence preferences at central base pairs persist when alanine is substituted for Arg 43, which projects into the minor groove. The position of the guanidinium group in the present model does not indicate hydrogen bonding to bases, and the binding affinities of the Ala 43 mutant show that any minor-groove contacts that are made do not contribute significantly to sequence discrimination at positions 6–9. We therefore believe that it is the DNA conformation, rather than contacts to the bases themselves, that influences binding.

How does the structure in the centre of the operator influence affinity? One explanation is that a particular conformation is required for correct relative orientation of the two half-sites. Though monomeric at most concentrations in solution, the two R1-69 domains in the complex with DNA have significant interactions, including complementary packing of parts of each subunit that are important for DNA backbone contacts (especially residues 41–44: see Fig. 4). Thus, the relative orientation of monomer DNA-binding surfaces is influenced by interactions between individual R1-69 subunits. The tightly structured repressor domain binds to a half-site in a precise way. Any variation in twist or bend at the central base pairs changes the spatial relationship of the two half-sites and requires compensatory variation in the protein dimer contacts, with a corresponding loss

经证实第一位的 A 突变为 T 的操纵基因可以被 28 位突变为丙氨酸的阻遏蛋白特异性识别。这些相互协调的改变可以构建进模型中，在 T1 的甲基基团和 Ala 28 的甲基侧链之间产生范德华接触。

Gln 33 的侧链所处的位置使其能与碱基对 4 和 5 相接触。模型显示这个残基可以与碱基 11′或 10′或二者形成氢键（图 5）。沃顿[3]已经研究了将阻遏蛋白的 Gln 33 突变为丙氨酸后与不同操纵基因的结合能力。他的结果与其他人的结果（沃顿和普塔什尼的未发表数据）一致，即残基 33 的确与碱基对 4 和 5 位点相接触。

骨架的接触。在一个半位点内，与 DNA 骨架的接触促使阻遏蛋白的一个亚基处于与操纵基因相对的位置。这种关系可能是非常精确的，因为主链的多肽氨基基团参与到了接触中，且已经折叠好的蛋白质的主链通常不如侧链柔软。骨架的接触，特别是在 P14- 处的接触，需要一个微微向外弯曲的操纵基因[22]。这种弯曲对至少一个数量的大沟接触是必要的：如果没有它，Gln 28 与 A1 间的相互作用将不会有正确的几何结构。目前还不清楚这种弯曲是否存在于未被结合的操纵基因 DNA 中，或者是否由蛋白的结合所诱导。

中央的碱基对。操纵基因中央碱基对的构型与其在 6~9 的位置上的序列偏好性一致。在本期另一篇文章中有更加详尽的讨论，这里我们只进行观点的总结。在碱基对 6 和 7 之间、8 和 9 之间的扭转角都为 36°，在碱基对 7 和 8 之间的扭转角为 39°（图 1*e*）。结果，小沟比理想状态的 B 型 DNA 窄了 2.5 Å。据信碱基对 GC 或 CG 碱基对不能很好地安置在狭窄的小沟中[29]，事实上阻遏蛋白不识别位于 6~9 的 GC 和 CG（参考文献 2）。当 Arg 43 突变为丙氨酸后，伸入小沟中，中间碱基对序列的偏好性依旧存在。在本模型中胍基并没有与碱基形成氢键，Ala 43 突变体的结合亲和力实验显示任何与小沟的接触没有明显影响其对 6~9 序列的辨别力。因此，我们认为是 DNA 的构象而不是与碱基对自身的接触影响了结合。

位于操纵基因中央的结构是怎样影响亲和性的呢？一种解释是两个半位点之间正确的相对取向需要特定的构象。尽管在多种浓度的溶液中该蛋白均以单体形式存在，然而在与 DNA 形成的复合物中两个 R1-69 的结构域有显著的相互作用，包括每个亚基中对 DNA 骨架的接触非常重要的那些部分互补堆积（特别是残基 41~44：见图 4）。因此与 DNA 结合的单体表面的相对取向受到单个 R1-69 亚基间相互作用的影响。结构紧密的阻遏蛋白结构域精确地结合到半位点上。中央碱基扭转或弯曲的任何变化都会改变两个半位点的空间关系，要求在蛋白二聚体的接触上做出补偿性的改变，导致复合物的整体能量得到相应的损失。我们估计完整的阻遏蛋白二聚

in the overall energy of the complex. We expect dimer contacts to be more extensive and less deformable in the intact repressor, and indeed alterations in central base-pair sequence have greater effects on the binding of the intact molecule than on R1-69 (ref. 2). Analogous mechanisms have been suggested on the basis of indirect evidence as a contribution to the specificity of SV40 T antigen[30].

The conformation of operator bound to a dimer of R1-69 is likely to differ from the equilibrium structure of the DNA in solution, but the base sequence will in any case affect flexibility and hence its propensity to reconfigure on binding. An even greater perturbation has been seen in the complex of *Eco*RI endonuclease with its DNA binding site, where an effective unwinding of about 25° occurs at the centre of the symmetrical sequence[31]. The unwinding induced by the endonuclease is important for appropriate alignment of protein–DNA interactions on either side.

The general notion that sequence-dependent aspects of DNA conformation can influence affinity of specific binding proteins has been discussed by Dickerson[32]. The characteristics that we see here to be necessary for a good fit, both within a half-site and between half-sites, appear to be examples of such conformational effects, but the specific "rules" developed by Dickerson[32] may not be straightforwardly applicable[2].

Operator Recognition

The tight fit between 434 repressor and DNA could not be achieved with a random 14-bp sequence. We suppose that in a non-specific complex, the repressor would be displaced by 2–3 Å outward from the DNA axis, to accommodate non-optimal major-groove interactions, and that the peptide backbone–DNA backbone hydrogen bonds would be lost. Residues such as Arg 41 (which must curl back toward a phosphate in the specific complex) and Arg 43 (which lies in the minor groove) could extend to provide compensating electrostatic stabilization. Such non-specific complexes may be intermediates in operator binding, because proteins like 434 repressor are believed to locate their sites by initial nonspecific interaction with distant sequences, followed by diffusion along the backbone. Record *et al.*[33] have postulated a similar distinction between complexes of *lac* repressor with operator and non-operator sequences.

We do not yet have a complete description of how the 434 repressor distinguishes among its six naturally-occurring operator sites. To summarize our current picture of specific repressor–operator interactions, it is useful to distinguish three zones in a 7-bp half-site. (1) The first three base pairs (ACA...), invariant in all operators (see ref. 2), interact directly with Gln 28 and Gln 29 on α3. Changes at any one of these positions decrease repressor affinity over 150-fold (R. Wharton, personal communication), and indeed no other DNA sequence can be properly complementary to the contact surface presented by the glutamine side chains. (2) The fourth and fifth base pairs both interact with Gln 33. We therefore expect effects of altered base sequence at these positions to be correlated. One of the naturally-occurring operators (O_R3) differs from the others at position 4 in

体的接触范围更广且更不易改变，事实上中央碱基对序列的改变对完整分子的结合能力的影响比对 R1-69 单体更大（参考文献 2）。已有间接证据提示类似的机制对 SV40 T 抗原的特异性有贡献[30]。

结合了 R1-69 二聚体的操纵基因的构象很可能与在溶液中 DNA 的平衡结构不同，但不论在何种状态，碱基序列都会影响蛋白质的柔性，进而影响其在结合时发生重新装配的倾向。在 *Eco*RI 限制性内切酶和 DNA 结合位点形成的复合物中发现了一个更大的扰动，在该复合物中回文序列中心产生了大约 25° 的有效解旋[31]。这种由限制性内切酶诱导产生的解旋对在回文序列任一侧的蛋白质–DNA 相互作用的适当对齐是重要的。

关于 DNA 构象的序列依赖性可以对特异性结合蛋白质的亲和力产生影响的一般观点，迪克森已经讨论过[32]。我们这里看到的对于无论是在单个半位点内还是半位点间形成好的匹配所必需的构象特征，看来可以作为这种构象效应的一个范例，而迪克森[32]推导的特殊“规则”可能并不能直接应用[2]。

操纵基因的识别

434 阻遏蛋白和 DNA 的紧密匹配并不能用随机的 14 个碱基对序列实现。我们假设在非特异性的复合物中，阻遏蛋白将从 DNA 轴外移 2~3Å，以适应非最佳的大沟相互作用，则多肽骨架和 DNA 骨架之间的氢键作用也将消失。像 Arg 41（在特异性复合物中它必定向磷酸基团后回旋）和 Arg 43（它位于小沟中）这样的残基将可能伸展以补偿静电稳定。这种非特异性的复合物可能是与操纵基因结合的中间体，因为像 434 阻遏物这样的蛋白被认为首先通过非特异性的相互作用与距离较远的序列结合，然后再沿着骨架扩散来定位。雷科德等人[33]对 *lac* 阻遏蛋白与操纵基因和与非操纵基因形成的复合物的区别提出过类似的假说。

我们现在还不能完整地解释 434 阻遏蛋白是如何区分其天然存在的六个操纵基因的位点的。为了总结我们现有的对特异的阻遏蛋白 – 操纵基因相互作用的理论框架，在 7 个碱基对的半位点中区分三个区域是很有帮助的。(1) 前三位的碱基对（ACA···），在所有的操纵基因中是不变的（参见参考文献 2），直接与螺旋 α3 上的 Gln 28 和 Gln 29 作用。改变这些位置中的任何一个碱基都会使阻遏蛋白的亲和力下降超过 150 倍（沃顿，个人交流），事实上，其他 DNA 序列都不能与谷氨酰胺侧链的接触表面正确地互补。(2) 第四和第五位碱基对，都与 Gln 33 作用。因此我们预计在这些位置替换碱基序列所造成的影响将是相关的。有一个天然操纵基因（O_R3）

one half-site, and there is considerable variation among sites at position 5. (3) The sixth and seventh base pairs form no direct contact with protein, and modulation of affinity by sequence variation at these positions occurs through effects on DNA conformation. The corresponding free energy changes are modest (1–2 kcal mol^{-1}), but natural operator sites differ in binding free energy by just such small amounts. The sequences of the six sites vary widely in this zone.

The distinction between these zones is only heuristic and it is clear that a perturbation at one position will affect the fit elsewhere. For example, the entire operator half-site must be configured correctly against the repressor monomer for any of the interactions with base pairs to be meaningful. Given such a configuration, the present description appears to explain many of the observed effects of operator mutations.

We thank Marie Drottar for laboratory assistance and preparation of crystals; R. Crouse for maintaining X-ray apparatus; J. Katz for artwork; A. Aggarwal, A. Mondragon, J. Moulai and G. Quigley for programs and advice; J. Koudelka and R. Wharton for discussion of unpublished results; C. Wolberger and D. C. Wiley for comments and suggestions. J. E. Anderson was a Burroughs Wellcome Fund Fellow of the Life Sciences Research Foundation. The work was supported by a grant from the NIH to S.C.H. and M. P.

(**326**, 846-852, 1987)

J. E. Anderson*, **M. Ptashne & S. C. Harrison**
Department of Biochemistry and Molecular Biology, Harvard University, Cambridge, Massachusetts 02138, USA
* Present address: Cold Spring Harbor Laboratory, Cold Spring Harbor, Long Island, New York 11724, USA

Received 8 January; accepted 1 April 1987.

References:

1. Ptashne, M. *A Genetic Switch* (Cell Press, Cambridge, Massachusetts, 1986).
2. Koudelka, J., Harrison, S. C. & Ptashne, M. *Nature* **326**, 886-888 (1987).
3. Wharton, R. thesis, Harvard Univ. (1985).
4. Anderson, J., Ptashne, M. & Harrison, S. C. *Nature* **316**, 596-601 (1985).
5. Pabo, C. O. & Lewis, M. *Nature* **298**, 443-447 (1982).
6. Anderson, W. F., Ohlendorf, D. H., Takeda, Y. & Matthews, B. W. *Nature* **290**, 754-758 (1981).
7. McKay, D. B. & Steitz, T. A. *Nature* **290**, 744-749 (1981).
8. Schevitz, R. W., Otwinowski, Z., Joachimiak, A., Lawson, C. L. & Sigler, P. B. *Nature* **317**, 782-786 (1985).
9. Pabo, C. O. & Sauer, R. *Ann. Rev. Biochem.* **53**, 293-321 (1984).
10. Anderson, J., Ptashne, M. & Harrison, S. C. *Proc. Natl. Acad. Sci. U.S.A.* **181**, 1307-1311 (1984).
11. Rossmann, M. G. & Blow, D. M. *Acta crystallogr.* **16**, 39-45 (1973).
12. Bricogne, G. *Acta crystallogr.* **A130**, 395-405 (1974).
13. Bricogne, G. *Acta crystallogr.* **A132**, 832-847 (1976).
14. Hogle, J., Chow, M. & Filman, D. *Science* **229**, 1359 (1985).
15. Jones, T. A. *J. appl. Crystallogr.* **11**, 268-272 (1978).
16. Hendrickson, W. A. & Konnert, J. H. *Biomolecular Structure, Conformation, Function and Evolution* Vol. I (ed. Srinivasan, R.) 43-57 (Pergamon, Oxford, 1981).
17. Sussmann, J., Holbrook, S. R., Church, G. M. & Kim, S.-H. *Acta crystallogr.* **33**, 800-804 (1977).
18. Ten Eyck, L. *Acta. crystallogr.* **A33**, 486-492 (1977).
19. Hogle, J., Maeda, A. & Harrison, S. C. *J. Molec. Biol.* **191**, 625-638 (1986).

在一个半位点中的位置 4 与其他操纵基因不同，而且在位置 5 的位点中有相当大的改变。(3) 第六和第七个碱基，与蛋白质没有直接相互作用，这些位置的序列变化可以通过影响 DNA 的构象来调节与蛋白质的亲和力。相应的自由能变化是不大的 (1~2 kcal · mol^{-1})，然而天然操纵基因位点在结合自由能上的区别就是通过这样微小的量来体现的。这个区域内六个位点的序列变化是很广泛的。

这些区域间的区别仅仅是启发式的，很显然一个位置的扰动将影响其他位置的匹配。例如，完整的操纵基因半位点必须有对于阻遏蛋白正确的构型从而使得其与碱基对的任何相互作用有意义。有了这样的构型，目前的描述看来能够解释观察到的许多操纵基因突变所产生的影响。

感谢玛丽 · 德罗塔尔协助实验室工作并制备晶体；感谢克劳斯维护 X 射线装置；感谢卡茨为本文作插图；感谢阿加沃尔、蒙德拉贡、穆莱和奎格利提供程序和建议；感谢考德尔卡和沃顿对未发表结果的讨论；感谢沃尔贝格和威利的评论和建议。安德森是生命科学研究基金会伯勒斯 · 韦尔科姆基金研究员。这项研究受到国立卫生研究院给予斯蒂芬 · 哈里森和马克 · 普塔什尼的基金支持。

（侯彦婕 翻译；吴琳 审稿）

20. Wilson, I. A., Wiley, D. C. & Skehel, J. J. *Nature* **289**, 373-378 (1981).
21. Drew, H. R. *et al. Proc. Natl. Acad. Sci. U.S.A.* **78**, 2179-2183 (1981).
22. Bushman, R., Anderson, J. E., Harrison, S. C. & Ptashne, M. *Nature* **316**, 651-653 (1985).
23. Hol, W. G. J., van Duijnen, P. T. & Berendsen, H. J. C. *Nature* **273**, 443-446 (1978).
24. Pflugrath, J. W. & Quiocho, F. A. *Nature* **314**, 257-260 (1985).
25. Dreusicke, D. & Schulz, G. *FEBS Lett.* **208**, 301-304 (1986).
26. Seeman, N. C., Rosenberg, J. N. & Rich, A. *Proc. Natl. Acad. Sci. U.S.A.* **173**, 804-808 (1976).
27. Baker, E. N. & Hubbard, R. E. *Progr. Biophys. molec. Biol.* **144**, 97-179 (1984).
28. Wharton, R. & Ptashne, M. *Nature* **326**, 888-891 (1987).
29. Drew, H. R. & Travers, A. A. *Cell* **37**, 491-502 (1984).
30. Ryder, K., Silver, S., DeLucia, A. L., Fanning, E. & Tegtmeyer, P. *Cell* **44**, 719-725 (1986).
31. McClarin, J. A. *et al. Science* **234**, 1526-1541 (1986).
32. Dickerson, R. E. *J. Molec. Biol.* **166**, 419-441 (1983).
33. Record, M. T., deHaseth, P. L. & Lohman, T. M. *Biochemistry* **16**, 4791-4796 (1977).
34. Crawford, J. thesis, Harvard Univ. (1977).
35. Durbin, R. *et al. Science* **232**, 1127-1132 (1986).
36. Kabsch, W. *Acta crystallogr.* A **33**, 922 (1976).

GAL4 Activates Transcription in *Drosophila*

J. A. Fischer *et al.*

Editor's Note

The GAL4-UAS system is used to study gene expression and function. It has two parts: the *gal4* gene, which encodes the transcription-activating protein GAL4, and the upstream activation sequence (UAS), which binds GAL4 to activate transcription. Here molecular biologist Mark Ptashne and colleagues demonstrate that GAL4 activates transcription in the fruit fly *Drosophila*. GAL4-activated transcription can be seen in a tissue-specific manner thanks to the inclusion of the *lacz* reporter gene, and the technique offered researchers a way to alter gene expression in specific tissue types. Although GAL4 is a yeast protein not normally found in other organisms, it can promote transcription in many other species, including humans and the model organisms *Xenopus* and zebrafish.

GAL4 is a yeast regulatory protein that binds to specific sites within a DNA sequence called UAS_G (galactose upstream activating sequence) and activates transcription of linked genes[1-6]. This activation requires two functions of the protein[7,8]: a DNA binding domain located near the amino terminus[8] , and one or more "activating regions"[7-11]. The "activating regions" are highly acidic[9-11] (see also ref. 12) and can be replaced, for example, by a short peptide designed to form a negatively charged, amphipathic α-helix[13] . GAL4, as well as deletion derivatives bearing one or more "activating regions" attached to the DNA binding domain, activates transcription in cultured mammalian cells from mammalian promoters linked to a UAS_G (refs 14, 15). Here we show that GAL4, when expressed in particular tissues of *Drosophila* larvae, stimulates tissue-specific transcription of a *Drosophila* promoter linked to GAL4 binding sites.

WE constructed an effector gene that expresses GAL4 in a tissue specific manner in *Drosophila*, and a reporter gene that allows the visualization of GAL4-activated transcription by a histochemical stain for β-galactosidase activity (Fig. 1). In the effector gene, expression of GAL4 is driven by a *Drosophila Adh* promoter. This promoter, when fused to the *lacz* gene of *Escherichia coli*, expresses β-galactosidase mainly in four larval tissues—fat body (fb), Malpighian tubules (mt), anterior midgut (amg) and middle midgut (mmg): expression in the hindgut (hg) and tracheae (tr) is variable (Figs 3, 4; refs 16, 34). In the reporter gene, the TATA-box and 5′-untranslated region of the *Drosophila* heat shock gene *hsp70* are fused to *lacz* (Fig. 1). This fusion gene is transcriptionally inactive (see below): the *hsp70* promoter is normally heat inducible, but sequences upstream of −43 from the *hsp70* transcription start site were deleted, thus eliminating the regulatory elements required for that induction[17-20]. We inserted four copies of a 17 base pair (bp)

在果蝇中GAL4激活转录

费希尔等

编者按

GAL4-UAS 系统用于基因表达与功能的研究。它由两部分组成：编码转录激活蛋白 GAL4 的 *gal4* 基因和上游激活序列（UAS），该序列通过结合 GAL4 可激活转录。在本文中，分子生物学家马克·普塔什尼和他的同事们证明了在果蝇中 GAL4 可激活转录。*lazc* 报告基因的引入使得研究者可以通过组织特异性的方式观察到 GAL4 激活的转录，这项技术给他们提供了在特定组织类型中改变基因表达的方法。尽管 GAL4 是一种仅在酵母内发现的蛋白，但它能够在包括人类和模式生物如非洲爪蟾和斑马鱼在内的许多其他的物种中启动转录。

GAL4 是一种酵母调节蛋白，它可与 UAS_G（半乳糖上游活化序列）上的特异性位点结合并激活连锁基因的转录[1-6]。激活需要这种蛋白的两个功能[7,8]：一个临近氨基末端的 DNA 结合域[8]，以及一个或多个“激活域”[7-11]。激活域高度酸化[9-11]（也见参考文献 12）并且可以被替换，例如它可被一段带负电的双性 α 螺旋短肽所取代[13]。GAL4 及其包含一个或多个与 DNA 结合域相连的“激活域”的缺失衍生物，在人工培养的哺乳动物细胞中激活与 UAS_G 相连的哺乳动物启动子的转录（参考文献 14、15）。我们在本文展示的是，当 GAL4 在果蝇幼虫特定组织中表达时，可激活一个与 GAL4 结合位点相连的果蝇启动子的组织特异性转录。

我们构建了一个在果蝇体内以组织特异性的方式表达 GAL4 的效应基因和一个可以通过组织化学染色反映 β– 半乳糖苷酶活性从而观察到 GAL4 激活转录的报告基因（图 1）。在效应基因中由果蝇的 *Adh* 启动子驱动 GAL4 的表达。该启动子与大肠杆菌的 *lacz* 基因融合后，主要在幼虫的以下四种组织中表达 β－半乳糖苷酶：脂肪体（fb）、马氏管（mt）、前部中肠（amg）和中部中肠（mmg）中；GAL4 在后肠（hg）和气管（tr）中的表达是变化的（图 3、4；参考文献 16、34）。在报告基因中果蝇热激基因 *hsp70* 的 TATA 盒和 5′ 非翻译区被融合到 *lacz* 中（图 1）。这个融合基因是转录失活的（见下文）：*hsp70* 启动子通常可被热激诱导，但将 *hsp70* 转录起始位点到上游 −43 位点的序列缺失后，会造成诱导所需的调节元件被删除[17-20]。我们插入了具有 17 个碱基对（bp）的四个拷贝，即 17-mers 的 *hsp70* TATA 盒上游序列：这些

sequence called the 17-mer upstream of the *hsp70* TATA-box: these 17-mers are closely related to the GAL4 binding sites in UAS_G (ref. 8) and are recognized by GAL4 (ref. 15) (see legend to Fig. 1).

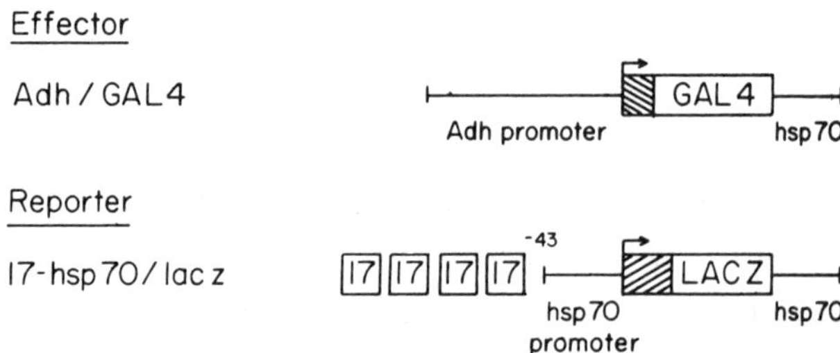

Fig. 1. Effector and reporter genes introduced into the *D. melanogaster* genome by P element transformation. The effector gene (*Adh/GAL4*) is a hybrid gene in which a *Drosophila Adh* promoter drives transcription of the yeast GAL4 coding sequences. The *Adh* promoter in the effector gene is that of the *D. mulleri Adh* −1 gene[24,25], with additional copies of its natural enhancer, BOX B[16,34] inserted upstream to increase its level of expression. The reporter gene (*17-hsp70/lacz*) is a hybrid gene in which the TATA-box of the *Drosophila hsp70* gene, with four GAL4 17-mers immediately upstream, drives transcription of the *E. coli* gene encoding β-galactosidase (*lacz*). The 17-mers are high affinity GAL4-binding sites (see below). Each gene contains transcriptional termination sequences from *hsp70*. The effector and reporter genes were separately introduced into the *D. melanogaster* genome by P element transformation[21,22].

Methods. The *Adh* promoter in the effector gene contains a *Bgl*II/*Bal*I fragment from the 5′-flanking region of the *D. mulleri Adh*–1 gene which extends from −1.5 kilobase (kb) to +58 bp from the transcription start site, immediately downstream of the ATG start codon. A *Bam*HI linker was inserted at the *Bal*I site, and three copies of an 100 bp *Sac*I-linked DNA fragment (−282 to −181 from the *Adh*–1 transcription start site) containing BOX B were cloned into the *Sna*BI site at −1.45 kb. GAL4 coding sequences were fused to the *Adh* promoter at the *Bam*HI site inserted downstream of the ATG of *Adh*–1 via a *Hin*dIII site within the untranslated leader sequence of GAL4 in pLPK-C15[26]. Site-directed mutagenesis with a mismatched primer[27] generated a "perfect" fusion of the *Adh*–1 leader sequences to the ATG start codon of GAL4. The reporter gene was constructed from an *hsp70/lacz* gene (a gift of John Lis and colleagues) containing *Drosophila hsp70* sequences from −43, 10 bp upstream of the TATA-box, to +265 (thus including the promoter and the first seven codons), fused in-frame to an *E. coli lacz* gene. Four copies of a 29 bp *Kpn*I/*Pst*I fragment of pMH100[15], containing a high affinity (Melvyn Hollis and M. Ptashne., unpublished data; see ref. 15) 17-mer GAL4 recognition site (the *Sca*I 17-mer: 5′-CGGAGTACTGTCCTCCG-3′) were cloned into the *Kpn*I site just upstream of the *hsp70/lacz* gene in pSP73lac2 (a gift of Dean Falb), which contains the *hsp70/lacz* gene cloned as a *Sal*I fragment into the *Sal*I site of pSP73 (ref. 28). A 4.8 kb *Xba*I fragment containing the effector gene and a 3.6 kb *Xba*I fragment containing the reporter gene were each cloned into the P element transformation vector C70TX (a gift of Dean Falb), in the same transcriptional orientation as the *rosy* gene. C70TX was constructed by inserting an *Xba*I linker into the *Sal*I site of C70T (a gift of John Lis and colleagues), which is a derivative of Carnegie 20 (ref. 29) containing, in the *Sal*I site, a 250 bp *Sal*I/*Xho*I fragment of *hsp70* gene terminator sequences. The P element plasmids were purified by banding in CsCl-EtBr gradients and coinjected at a concentration of 300 μg ml^{-1} with the helper plasmid pπ25.7wc[30] (70 μg ml^{-1}) into embryos of *D. melanogaster* strain ry^{506} as described[21,22]. The effector plasmid was also injected at a concentration of 100 μg ml^{-1}, because very few embryos injected even with the lower concentration survived past the first instar larval stage. Flies from embryos that survived microinjection were individually backcrossed to ry^{506}. Among the backcross progeny, germ-line transformants were distinguished by their wild type (ry^{+}) eye colour. The backcross progeny, heterozygous for the P element, were used to perform the effector × reporter crosses. All enzymatic reactions and DNA manipulations were carried out using standard conditions and techniques[31]. Flies were grown at 25 °C on standard cornmeal food.

17-mer 上游序列与 UAS_G 上的 GAL4 的结合位点非常接近（参考文献 8）并且可以被 GAL4 识别（参考文献 15）（见图 1 注）。

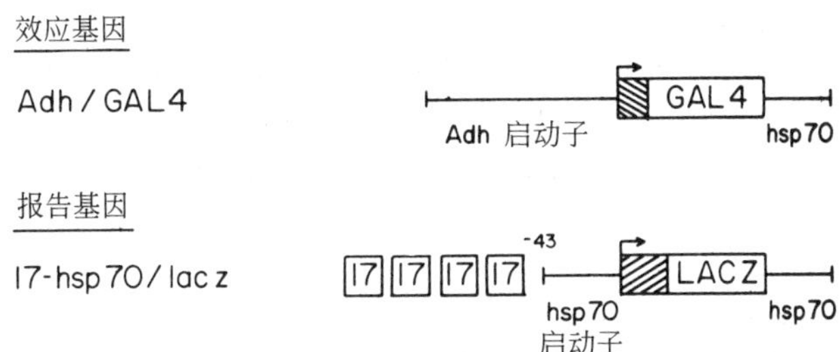

图 1. 效应基因和报告基因通过 P 因子转化转入黑腹果蝇基因组内。效应基因（*Adh/GAL4*）是一个复合基因，由果蝇的 *Adh* 启动子启动酵母 GAL4 编码序列的转录。效应基因的 *Adh* 启动子是果蝇（*D. mulleri*）*Adh*–1 基因的启动子[24,25]，并具有其自身增强子的额外拷贝，BOX B[16,34] 插入到上游以提高其表达水平。报告基因（*17-hsp70/lacz*）也是复合基因，在果蝇 *hsp70* 基因上的 TATA 盒，携带紧邻上游的四个 GAL4 17-mer 序列，这些序列将启动大肠杆菌 β– 半乳糖苷酶编码基因（*lacz*）的转录。17-mer 序列是 GAL4 高亲和性的结合位点（见下文）。每个基因都含有来自 *hsp70* 的转录终止序列。效应基因和报告基因分别通过 P 因子转入黑腹果蝇体基因组内[21,22]。

方法。效应基因的 *Adh* 启动子包含一个来自果蝇 *Adh*–1 基因 5′ 侧翼区的 *Bgl*II/*Bal*I 片段，侧翼区包括自转录起始位点至紧邻 ATG 起始密码子下游的 –1.5 kb 到 +58 bp 之间的区域。在 *Bal*I 位点插入一个 *Bam*HI 接头，包含 BOX B 的长度为 100 bp 的 *Sac*I 连接的 DNA 片段（*Adh*–1 转录起始位点 –282~–181 的区域）的三个拷贝被克隆到 –1.45 kb 处的 *Sna*BI 位点。GAL4 编码序列在 *Bam*HI 位点与 *Adh* 启动子融合，该 *Bam*HI 位点通过 pLPK-C15[26] 上 GAL4 非翻译引导序列内的一个 *Hind*III 位点插入在 *Adh*–1 基因的 ATG 起始密码子下游。利用错配引物造成的定点突变[27] 产生 *Adh*–1 引导序列与 GAL4 ATG 起始密码子的“完美”融合。报告基因是由 *hsp70/lacz* 基因（约翰 · 利斯及同事惠赠）开始构建，该基因包含果蝇 *hsp70* 中从 –43 即 TATA 盒上游 10 bp 到 +265 的一段序列（因此包含启动子和开始的七个密码子），并以符合读码框架的形式与大肠杆菌的 *lacz* 基因融合。pMH100[15] 的 29 bp *Kpn*I/*Pst*I 片段的四个拷贝被克隆到 pSP731ac2（迪安 · 法尔布惠赠）中 *hsp70/lacz* 基因上游的 *Kpn*I 位点，上述片段包含高亲和性的（梅尔文 · 霍利斯和马克 · 普塔什尼，未发表数据；参考文献 15）17-mer GAL4 识别位点（*Sca*I 17-mer：5′-CGGAGTACTGTCCTCCG-3′），而 pSP731ac2 则是将 *hsp70/lacz* 基因作为一个 *Sal*I 片段克隆到 pSP73（参考文献 28）的 *Sal*I 位点。一个包含效应基因的 4.8 kb 的 *Xba*I 片段和一个包含报告基因的 3.6 kb 的 *Xba*I 片段分别以与红眼基因（*rosy*）相同的转录方向克隆到 P 因子转化载体 C70TX（迪安 · 法尔布惠赠）中。C70TX 是通过将一个 *Xba*I 接头插入在 C70T（约翰 · 利斯及同事惠赠）的 *Sal*I 位点上构建出来的，C70T 是 Carnegie 20（参考文献 29）的一个衍生物，在 *Sal*I 位点处包含 *hsp70* 基因终止序列的一个 250 bp *Sal*I/*Xho*I 片段。P 因子质粒通过氯化铯–溴化乙锭梯度分带纯化，并以 300 $\mu g \cdot ml^{-1}$ 的浓度与辅助质粒 pπ25.7wc[30]（70 $\mu g \cdot ml^{-1}$）共同注射到上文说过的黑腹果蝇 ry^{506} 株系的胚胎中[21,22]。效应质粒的注射浓度为 100 $\mu g \cdot ml^{-1}$，因为即使注射浓度再低一些也只有很少的胚胎能活过一龄幼虫期。胚胎显微注射后成活的成虫与 ry^{506} 株系果蝇回交。在回交后代中，通过野生型（ry^{+}）眼睛颜色分辨出种系转化体。含有杂合 P 因子的回交后代被用来进行效应基因与报告基因的杂交。所有酶促反应和 DNA 操作均采用标准条件和技术进行。果蝇在 25 ℃用标准燕麦培养。

The effector and reporter genes were separately introduced into the *D. melanogaster* genome by P element transformation[21,22] (Fig. 1 legend). Three independent effector transformant lines were separately mated to nine independent reporter transformant lines. Twelve larval progeny of each cross were tested for expression of the reporter gene by a histochemical staining assay for β-galactosidase activity (Fig. 4 legend). The flies transformed with the effector and reporter genes were each heterozygous for the respective P elements (Fig. 1 legend), and so we expected one fourth of their larval progeny to carry one copy of each gene.

In each of the 27 crosses, we detected β-galactosidase activity in~one quarter of the larval progeny. In all of these larvae, enzyme activity was observed at high levels in the fat body and anterior midgut (Fig. 2), two of the tissues in which we expect the *Adh* promoter of the effector gene to be active. Also as expected, we detected variable levels of β-galactosidase activity in the hindgut and tracheae (Figs 2 and 3). In the middle midgut and Malpighian tubules, two other tissues in which we expect the *Adh* promoter of the effector gene to be active, we detected enzyme activity in only a small number of larvae, and then only in a few cells (data not shown); perhaps the *Adh/GAL4* transcript or GAL4 itself is particularly unstable in these two tissues. β-galactosidase activity was never detected in tissues other than those in which we expect the *Adh* promoter of the effector gene to be active (Figs 3 and 4), and no β-galactosidase activity (data not shown) nor reporter transcripts (Fig. 2) were detected in larvae transformed with the reporter gene alone. Furthermore, neither β-galactosidase activity (data not shown; see Fig. 3) nor transcripts (Fig. 2) were detected from a reporter gene lacking the 17-mers. GAL4-activated transcription of the reporter gene was initiated at the same start point as was transcription of the reporter gene activated by a *Drosophila Adh* enhancer (Fig. 2). GAL4 bound to the 17-mers stimulated transcription as efficiently as the *Drosophila* enhancer (Fig. 2 legend).

通过P因子转化将效应基因及报告基因分别转入到黑腹果蝇基因组内[21,22](图1注)。三个独立的效应基因转化系分别与九个独立的报告基因转化系交配。用组织化学染色法分别对每个杂交组合12个幼虫后代中报告基因表达的β–半乳糖苷酶活性进行检测(图4注)。转化了效应基因和报告基因的果蝇各自所含的P因子都是杂合的(图1注),因此我们预计其幼虫后代中有四分之一各含有每个基因的一个拷贝。

对27个杂交组合,我们检测了约四分之一幼虫后代的β–半乳糖苷酶活性。在所有幼虫的脂肪体和前部中肠都检测到了高水平的酶活性(图2),我们预计这两个组织中效应基因的*Adh*启动子应处于激活状态。而在后肠和气管中我们观察到β–半乳糖苷酶活性水平是变化的,这也与我们的预期相符(图2和3)。在我们期望的效应基因的*Adh*启动子呈激活状态的另外两个组织——中部中肠和马氏管中,我们仅在少量幼虫的少数细胞中观察到了酶活性(数据未显示),这可能是*Adh/GAL4*转录本或GAL4自身在这两个组织中极不稳定。除了我们预计的效应基因的*Adh*启动子呈激活状态的组织外,未在其他组织中检测到β–半乳糖苷酶活性(图3和4),并且在报告基因单独转化的幼虫中既未检测到β–半乳糖苷酶活性(数据未显示)也未检测到报告基因转录本(图2)。此外,缺少17-mer的报告基因转化幼虫中也均未检测到β–半乳糖苷酶活性(数据未显示,见图3)和转录本(图2)。GAL4与果蝇*Adh*增强子激活的报告基因的转录起始点相同(图2)。GAL4结合到17-mer上激活转录的效率与果蝇增强子相同(图2注)。

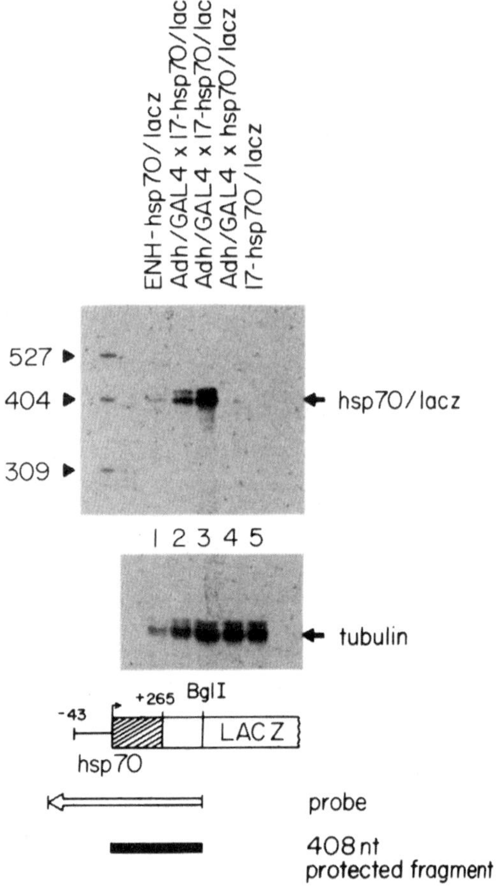

Fig. 2. Reporter gene transcripts in transformed larvae. The transcription start site of the reporter gene (*17-hsp70/lacz*) in the transformed larvae or larval progeny of the crosses indicated was assayed by quantitative RNAse protection[32] using the uniformly ^{32}P-labelled RNA probe shown, complementary to the 5′ end of the reporter transcripts, and also an RNA probe complementary to the endogenous α1-*tubulin* gene as an internal control for mRNA levels. ENH-*hsp70/lacz* (lane 1) are transformants (a gift of Dean Falb) carrying a gene like the reporter, except the enhancer of the *D. melanogaster Adh* gene's distal promoter (−660 to −128 from the distal transcription start site; Dean Falb and T.M., unpublished data) was installed upstream of the truncated *hsp70* promoter instead of the GAL4 17-mers. Two independent effector (*Adh/GAL4*) transformant lines were crossed with the same reporter (*17-hsp70/lacz*) line in lanes 2 and 3. Identical fragments of the reporter transcripts were protected by the probe in lanes 1–3. The appearance of two bands is probably an artefact of the RNAse digestion. Consistent with the β-galactosidase activity assay, no protected fragments were detected in larval progeny of a cross of the effector and a reporter with no 17-mers (*hsp70/lacz*; lane 4), or in larvae transformed with the reporter gene alone (lane 5). (Flies transformed with the *hsp70/lacz* gene, which lacks the 17-mers, were a gift of Dean Falb.) An approximation of the relative strengths of the GAL4 enhancer and the particular *Drosophila Adh* enhancer in ENH-*hsp70/lacz* was obtained by densitometry of appropriate autoradiographic exposures of the *hsp70/lacz* and tubulin protected fragments in lanes 1–3. The ratios of *hsp70/lacz* to tubulin signals were corrected for gene copy number and then normalized to the ratio obtained for lane 1, which was arbitrarily assigned the value 1.0: the results were 0.3 (lane 2) and 1.2 (lane 3). Thus, GAL4 bound to the 17-mers is comparable with the *Adh* enhancer in its ability to activate transcription of the reporter gene.

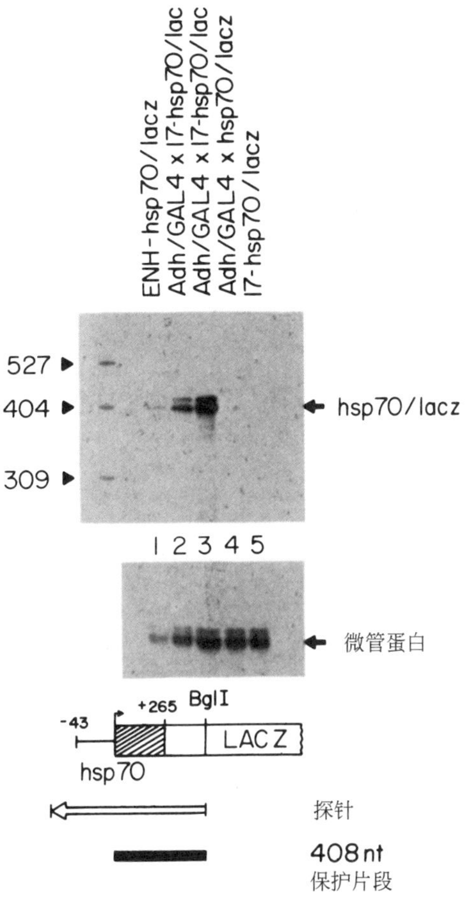

图 2. 转化幼虫中报告基因的转录产物。转化幼虫或杂交的幼虫后代中报告基因（*17-hsp70/lacz*）转录起始点通过定量 RNA 酶保护分析，用一段与报告基因转录本 5′ 末端互补的 ^{32}P 均一标记的 RNA 探针显示，还有一个与内源 α1 微管蛋白基因互补的 RNA 探针作为 mRNA 水平的内参。ENH-*hsp70/lacz*（泳道 1）是一个携带与报告基因相似的基因的转化子（迪安 · 法尔布赠），二者的唯一区别是将黑腹果蝇 *Adh* 基因远端启动子的增强子（远端转录起始位点 –660 到 –128 的区域；迪安 · 法尔布和汤姆 · 马尼阿蒂斯，未发表数据）而不是 GAL4 17-mer 序列安装在截短的 *hsp70* 启动子上游。泳道 2 和 3 分别显示两个独立的效应基因（*Adh/GAL4*）转化系与同一个报告基因（*17-hsp70/lacz*）转化系杂交的结果。在 1~3 泳道中，相同的报告基因转录物片段被探针保护。泳道上显示的两条带可能是 RNA 酶消化的人为效应。与 β– 半乳糖苷酶活性实验结果相一致，在效应基因和缺少 17-mer 序列的报告基因（*hsp70/lacz*，泳道 4）的杂交幼虫后代或报告基因独自转化的幼虫（泳道 5）中未检测到保护片段。（缺少 17-mer 序列的 *hsp70/lacz* 基因转化的果蝇成虫为迪安 · 法尔布惠赠。）通过对泳道 1~3 中的 *hsp70/lacz* 和微管蛋白保护片段适当曝光的放射自显影照片的密度计量，显示 GAL4 增强子和 ENH-*hsp70/lacz* 中果蝇的 *Adh* 增强子的相对强度接近。*hsp70/lacz* 与微管蛋白的信号比率通过基因拷贝数修正，并对泳道 1 的数据进行归一化，设定其值为 1.0 后：实验结果是 0.3（泳道 2）和 1.2（泳道 3）。因此，结合于 17-mer 序列的 GAL4 与 *Adh* 增强子在激活报告基因转录的能力上是相当的。

Methods. RNA was prepared as previously described[24,25] from actively feeding third instar larvae, or from ENH-*hsp70/lacz* adults. Two uniformly ^{32}P-labelled RNA probes were transcribed with SP6 RNA polymerase as described[33], from the plasmids SP6-αtub[25] (a gift of Vicki Corbin) and SP6-hslac. SP6-hslac was constructed by ligating a ~450 bp *Bgl*II/*Bgl*I fragment (the *Bgl*I end was treated with T4 DNA polymerase) of pSP73lac2 (see Fig. 1 legend) into pSP72 (ref. 28) digested with *Bgl*II and *Sma*I. Before transcription, SP6-hslac was digested with *Bgl*II, resulting in a ~450 nucleotide (nt) probe. The SP6-αtub probe[25] protected a 90 nt fragment of the endogenous α1-tubulin gene transcripts. As shown, the SP6-hslac probe protected a 408 nt fragment of the *hsp70/lacz* fusion gene transcripts. Five-forty μg of each RNA preparation was hybridized simultaneously with the ^{32}P-labelled RNA probes as previously described[25]. RNAse digestion conditions were also as described[25]. Samples were electrophoresed on a 6% denaturing acrylamide gel. The top portion of the gel is a longer autoradiographic exposure than the bottom portion.

Summary of β-galactosidase Activity in Larval Tissues

	fb	mt	amg	mmg	hg	tr	gc	pro	pmg	di	br	sg	ov/ts
Adh/lacz (6)	+	+	+	+	var	var	–	–	–	–	–	–	–
Adh/GAL4 (3) ; 17-hsp70/lacz (9)	+	+/-	+	+/-	var	var	–	–	–	–	–	–	–

Fig. 3. Tissue specificity of β-galactosidase activity in transformed larvae. The tissue specificity of β-galactosidase activity, detected by histochemical staining, in *Adh/lacz* transformant larvae or in larvae carrying both the effector (*Adh/GAL4*) and reporter (*17-hsp70/lacz*) genes is summarized. The *Adh/lacz* gene is described in the legend to Fig. 4, and the effector and reporter genes are described in Fig. 1. The numbers in parentheses indicate the number of independent transformant lines assayed. The tissues indicated are: fat body (fb), Malpighian tubules (mt), anterior midgut (amg), middle midgut (mmg), hindgut (hg), tracheae (tr), gastric caecae (gc), proventriculus (pro), posterior midgut (pmg), imaginal discs (di), brain (br), salivary glands (sg), ovaries and testes (ov/ts). +, High staining intensities in all lines. –, Staining was never seen, +/–, Staining was observed in some larvae, and only at a low level in a few cells. Variable staining is indicated as var: by variable we mean that the staining intensity varied from high to none within or between different lines or crosses. No β-galactosidase activity was detected with a reporter gene lacking 17-mers, nor in larvae transformed with the reporter gene alone (data not shown).

a

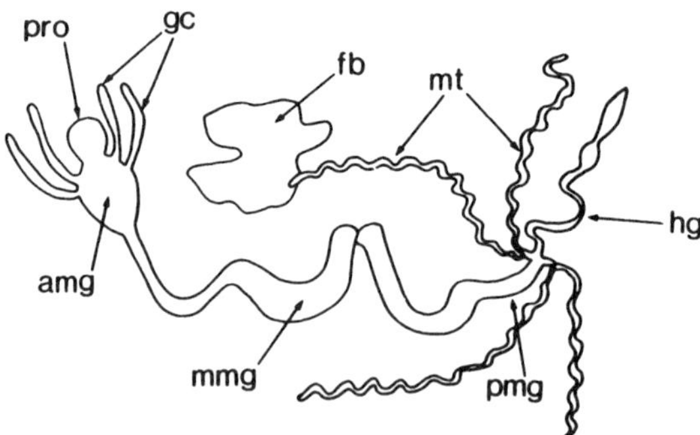

方法。RNA 来自活跃进食的三龄幼虫或 ENH-*hsp70*/*lacz* 成虫，制备方法按前人描述[24,25]。如前人描述[33]，用 SP6 RNA 聚合酶从质粒 SP6-αtub[25]（维基 · 科尔宾惠赠）和 SP6-hslac 转录出两个均一 ^{32}P 标记的 RNA 探针。SP6-hslac 的构建是通过将 pSP73*1ac2*(见图 1 注)的一个约 450 bp 的 *Bgl*II/*Bgl*I 片段(*Bgl*I 末端用 T4 DNA 聚合酶处理）连接到用 *Bgl*II 和 *Sma*I 消化的 pSP72（见参考文献 28）上完成的。转录前 SP6-hslac 用 *Bgl*II 消化，产生一段约 450 核苷酸（nt）的探针。SP6-αtub 探针[25] 保护内源 α1- 微管蛋白基因转录本的一个 90 nt 片段。如图所示，SP6-hslac 探针保护 *hsp70*/*lacz* 融合基因转录本的一个 408 nt 片段。如前所述，5～40 μg 的 RNA 样品与 ^{32}P 标记的 RNA 探针同时进行杂交[25]。RNA 酶消化条件同样参照文献描述[25]。样品经 6%的变性丙烯酰胺凝胶电泳。凝胶上部曝光时间比底部略长。

在幼虫组织中β–半乳苷酶活性总结

	fb	mt	amg	mmg	hg	tr	gc	pro	pmg	di	br	sg	ov/ts
Adh/lacz (6)	+	+	+	+	var	var	–	–	–	–	–	–	–
Adh/GAL4 (3) ; 17-hsp70/lacz (9)	+	+/-	+	+/-	var	var	–	–	–	–	–	–	–

图 3. 在转化幼虫中 β– 半乳糖苷酶活性的组织特异性。通过组织化学染色观察，在 *Adh*/*lacz* 转化幼虫或者同时携带效应基因（*Adh*/*GAL*4）和报告基因（*17-hsp70*/*lacz*）的幼虫中 β– 半乳糖苷酶活性组织特异性总结如图 3。*Adh*/*lacz* 基因在图 4 注解中有描述，效应基因和报告基因在图 1 注解中有描述。括号中的数字表示检测的独立转化系数目。组织表示符号如下：脂肪体（fb）、马氏管（mt）、前部中肠（amg）、中部中肠（mmg）、后肠（hg）、气管（tr）、胃盲囊 (gc)、前胃 (pro)、后部中肠 (pmg)、成虫盘 (di)、脑 (br)、唾液腺 (sg)、卵巢和睾丸 (ov/ts)。+，所有种系中染色强度都很高；–，从未观察到染色；+/–，只在部分幼虫中的少数细胞中观察到低强度的染色。染色变化用 var 表示：所谓染色变化是指在不同株系或杂交后代之间染色强度从高到无。在缺少 17-mer 的报告基因或报告基因单独转化的幼虫中未检测到 β– 半乳糖苷酶活性（数据未显示）。

a

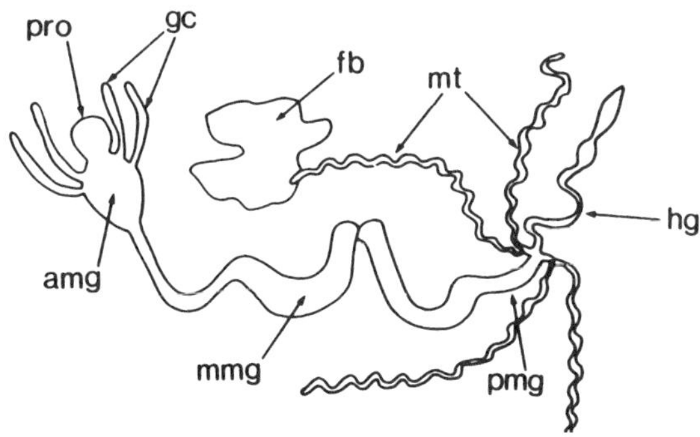

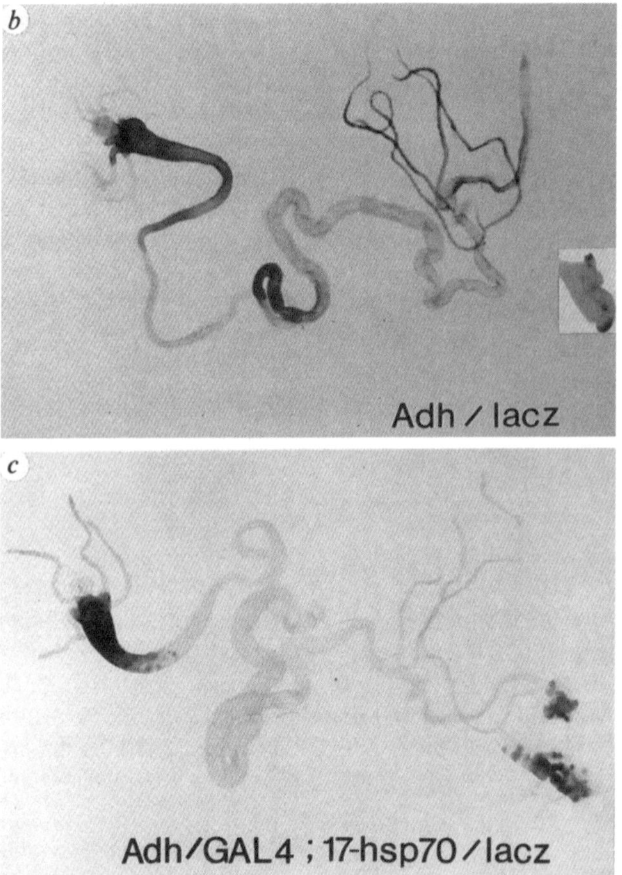

Fig. 4. β-galactosidase activity in larval tissues of P element transformants. *a*, Diagram of the larval tissues in the photographs beneath: gastric caecae (gc), proventriculus (pro), anterior midgut (amg), middle midgut (mmg), posterior midgut (pmg), Malpighian tubules (mt), fat body (fb), and hindgut (hg) are indicated. *b*, β-galactosidase activity in third instar larvae of six independent lines transformed with an *Adh/lacz* hybrid gene was visualized by a histochemical stain. The hybrid gene contains the same *Adh* promoter as in the effector gene, but here, fused in-frame to the *E. coli lacz* gene rather than to GAL4[16,34]. Blue staining is observed only in the fat body, Malpighian tubules, anterior midgut, middle midgut, and in this particular larva, also in the hindgut. Expression of β-galactosidase in the hindgut, and also in the tracheae, was variable (see refs 16, 34 and Fig. 3). *c*, A third instar larva carrying both the effector (*Adh/GAL4*) and the reporter (*17-hsp70/lacz*) genes expressed β-galactosidase in the fat body and anterior midgut. Other larvae expressed β-galactosidase in the hindgut and/or tracheae, and low levels of β-galactosidase activity were sometimes detected in a few cells of the middle midgut or Malpighian tubules (see Fig. 3).

Methods. Transformant larvae were dissected in a solution of 1% glutaraldehyde in 0.1 M $NaPO_4$ *p*H 7.0, 1 mM $MgCl_2$. After 15 min, the tissues were transferred to a stain solution consisting of 10 mM $NaPO_4$ *p*H 7.0, 150 mM NaCl, 1 mM $MgCl_2$, 3.3 mM $K_4Fe(CN)_6$ $3H_20$, 3.3 mM $K_4Fe(CN)_4$ $3H_2O$, and 0.2% X-gal dissolved as a 2% solution in dimethylformamide (Pieter Wensink, personal communication). After staining for 15 min (*a*) or 1 h (*b*), the tissues were placed in a drop of 70% glycerol on a slide, covered with a coverslip and photographed.

We have shown that a single protein, GAL4, when expressed in a tissue specific manner, generates tissue specific gene expression in *Drosophila*. Typical tissue specific enhancers (the

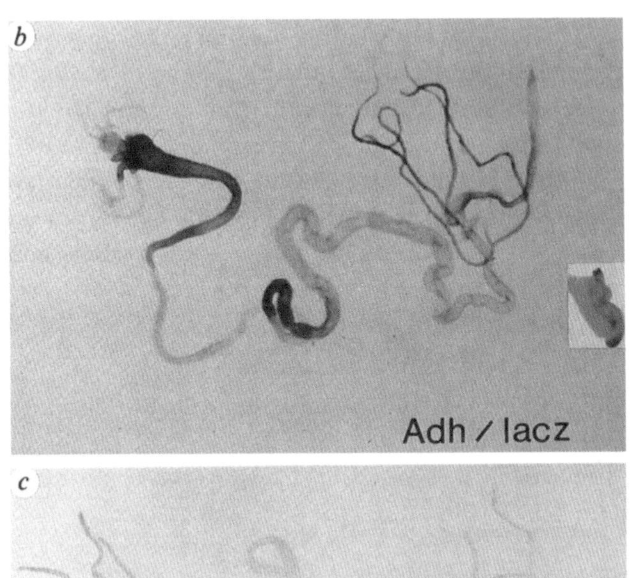

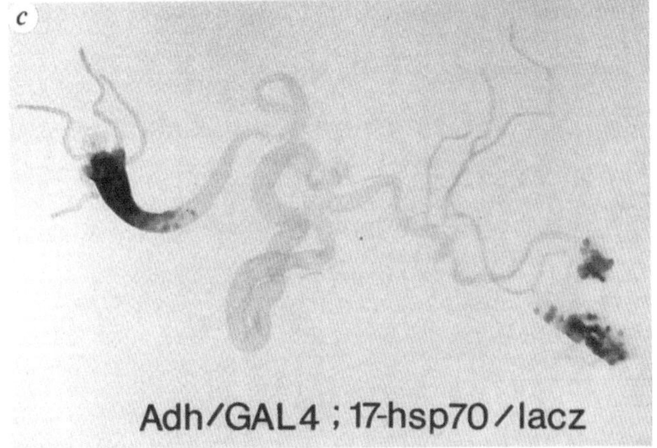

图4. P因子转化体幼虫组织的β–半乳糖苷酶活性。*a*，下面照片中幼虫组织的示意图：胃肠盲囊(gc)、前胃(pro)、前部中肠（amg)、中部中肠（mmg)、后部中肠(pmg)、马氏管（mt)、脂肪体（fb）和后肠（hg)。*b*，通过组织染色法观察到的*Adh*/*lacz*杂合基因转化的六个独立体系的三龄幼虫中β–半乳糖苷酶活性。杂合基因包含与效应基因相同的*Adh*启动子，但它以符合读码框架的形式与大肠杆菌*lacz*基因而不是GAL4融合[16, 34]。蓝色染料仅在脂肪体（ft)、马氏管（mt)、前部中肠（amg)、中部中肠(mmg）观察到，在这只特定幼虫的后肠（hg）中也观察到。β–半乳糖苷酶在后肠（hg）和气管（tr）中的表达是变化的（见参考文献16、34和图3)。*c*，同时携带效应基因（*Adh*/*GAL4*）和报告基因（*17hsp70*/*lacz*）的三龄幼虫在脂肪体（ft）和前部中肠（amg）中表达β–半乳糖苷酶。其他幼虫在后肠（hg）和（或）气管（tr）中表达β–半乳糖苷酶，有时在中肠或马氏管的少数细胞中可观察到低水平的β–半乳糖苷酶活性（见图3)。

方法。在含0.1 M $NaPO_4$ pH 7.0，1 mM $MgCl_2$的1%的戊二醛溶液中解剖转化体幼虫。15分钟后，将组织转移到染液中，染液包含10 mM $NaPO_4$ pH 7.0，150 mM NaCl，1 mM $MgCl_2$，3.3 mM $K_4Fe(CN)_6 \cdot 3H_2O$, 3.3 mM $K_4Fe(CN)_4 \cdot 3H_2O$及0.2%的半乳糖苷，半乳糖苷是在二甲基甲酰胺中溶解成2%的溶液（彼得·文辛克，个人交流)。分别在染色15分钟（*a*）或1小时后（*b*）将组织放在载玻片上的70%的甘油液滴中，盖上盖玻片并拍照。

我们展示了当一个GAL4蛋白在果蝇中以组织特异性方式表达时，可造成组织特异的基因表达。典型的组织特异性增强子（研究最清楚的是哺乳动物基因的增强

best characterized are those of mammalian genes) interact with several different proteins (ref. 23 and refs therein). Our results suggest that the apparent complexity of tissue specific enhancers is not necessary for their function as transcriptional activators. Perhaps that complexity is exploited to obtain intricate patterns of gene control from a relatively small number of regulatory proteins. The ability of GAL4 to activate mammalian and *Drosophila* promoters suggests that the protein with which DNA-bound GAL4 interacts to stimulate transcription in yeast—perhaps RNA polymerase—has a close homologue in higher eukaryotes. It seems likely that the mechanism of action of at least one class of gene activators is conserved between yeast, mammalian cells and *Drosophila*.

We thank Dean Falb for generous gifts of unpublished plasmids and transformant lines, and Gerald Rubin, Melvyn Hollis, Liam Keegan and John Lis for plasmids. This work was supported by grants from the American Cancer Society (to M.P.) and the National Institutes of Health (to T.M.).

(**332**, 853-856; 1988)

Janice A. Fischer, Edward Giniger, Tom Maniatis & Mark Ptashne
Department of Biochemistry and Molecular Biology, Harvard University, 7 Divinity Avenue, Cambridge, Massachusetts 02138, USA

Received 29 February; accepted 14 March 1988.

References:

1. Guarente, L., Yocum, R. R. & Gifford, P. *Proc. Natl. Acad. Sci. U.S. A.* **79**, 7410-7414 (1982).
2. Yocum, R. R., Hanley, S., West, R. W., Jr. & Ptashne, M. *Molec. Cell. Biol.* **4**, 1985-1998 (1984).
3. West, R. W., Jr., Yocum, R. & Ptashne, M. *Molec. Cell. Biol.* **4**, 2467-2478 (1984).
4. Johnston, M. & Davis, R. W. *Molec. Cell. Biol.* **4**, 1440-1448 (1984).
5. Bram, R. & Kornberg, R. D. *Proc. Natl. Acad. Sci. U.S.A.* **82**, 43-47 (1985).
6. Giniger, E., Varnum, S. M. & Ptashne, M. *Cell* **40**, 767-774 (1985).
7. Brent, R. & Ptashne, M. *Cell* **43**, 729-736 (1985).
8. Keegan, L., Gill, G. & Ptashne, M. *Science* **231**, 699-704 (1986).
9. Ma, J. & Ptashne, M. *Cell* **48**, 847-853 (1987).
10. Ma, J. & Ptashne, M. *Cell* **51**, 113-119 (1987).
11. Gill, G. & Ptashne, M. *Cell* **51**, 121-126 (1987).
12. Hope, I. A. & Struhl, K. *Cell* **46**, 885-894 (1986).
13. Giniger, E. & Ptashne, M. *Nature* **330**, 670-672 (1987).
14. Kakidani, H. & Ptashne, M. *Cell* **52**, 161-167 (1987).
15. Webster, N., Jin, J. R., Green, S., Hollis, M. & Chambon, P. *Cell* **52**, 169-178 (1987).
16. Fischer, J. A. thesis Harvard Univ. (1987).
17. Pelham, H. R. B. *Cell* **30**, 517-538 (1982).
18. Lis, J. T., Simon, J. A. & Sutton, C. A. *Cell* **35**, 403-410 (1983).
19. Dudler, R. & Travers, A. A. *Cell* **38**, 391-398 (1984).
20. Cohen, R. S. & Meselson, M. *Proc. Natl. Acad. Sci. U.S.A.* **81**, 5509-5513 (1984).
21. Spradling, A. C. & Rubin, G. M. *Science* **218**, 341-347 (1982).
22. Rubin, G. M. & Spradling, A. C. *Science* **218**, 348-353 (1982).
23. Maniatis, T., Goodbourn, S. & Fischer, J. A. *Science* **236**, 1237-1245 (1987).
24. Fischer, J. A. & Maniatis, T. *Nucleic Acids Res.* **13**, 6899-6917 (1985).

子）可以与几个不同的蛋白相互作用（参考文献 23 及其中的文献）。我们的实验结果表明组织特异性增强子的复杂性并不是其转录活化功能所必需的。或许这种复杂性是通过一小部分的调节蛋白的调控来达到精细的基因控制模式。GAL4 具有激活哺乳动物和果蝇启动子的能力，这提示我们，在酵母中一个与结合了 DNA 的 GAL4 相互作用的蛋白（或许是 RNA 聚合酶）可激活转录，它与高等真核生物中的 RNA 聚合酶高度同源。这表明至少有一类基因活化因子的作用机制很可能在酵母、哺乳动物细胞和果蝇之间是保守的。

我们感谢迪安 · 法尔布慷慨地赠予未发表的质粒和转化体系，感谢杰拉尔德 · 鲁宾、梅尔文 · 霍利斯、利亚姆 · 基根和约翰 · 利斯赠予质粒。本研究受到美国癌症协会(由马克·普塔什尼承担)和美国国家卫生研究院(由汤姆·马尼阿蒂斯承担)的资助。

（李梅 翻译；胡松年 审稿）

25. Fischer, J. A. & Maniatis, T. *EMBO J.* **5**, 1275-1289 (1986).

26. Silver, P. A., Keegan, L. P. & Ptashne, M. *Proc. Natl. Acad. Sci. U.S.A.* **81**, 5951-5955(1984).

27. Zoller, M. J. & Smith, M. *Meth. Enzym.* **100**, 469-500 (1983).

28. Krieg, P. A. & Melton, D. A. *Meth. Enzym. Recomb. DNA.* **155**, 397-415 (1987).

29. Rubin, G. M. & Spradling, A. C. *Nucleic Acids Res.* **11**, 6341-6351 (1983).

30. Karess, R. E. & Rubin, G. M. *Cell* **38**, 135-146 (1984).

31. Maniatis, T., Fritsch, E. F., Sambrook, J. *Molecular Cloning: A Laboratory Manual* (Cold Spring Harbor Laboratory, New York, 1982).

32. Zinn, K., DiMaio, D. & Maniatis, T. *Cell* **34**, 865-879 (1983).

33. Melton, D. A., Krieg, P. A., Rebagliati, M. R., Maniatis, T., Zinn, K. & Green, M. *Nucleic Acids Res.* **12**, 7035-7056.

34. Fischer, J. A. & Maniatis, T. *Cell* **53** (in the press).

Mysteries of HIV: Challenges for Therapy and Prevention

J. A. Levy

Editor's Note

Here virologist Jay Levy, who co-discovered HIV five years earlier, describes the unsolved problems in HIV research. Four key areas deserve special attention, namely the viral and cellular determinants of infection, the mechanisms generating different strains, the factors influencing progression to disease and an understanding of how HIV causes disease. Subsequent research into these areas has, as Levy correctly speculated, greatly influenced therapeutic and prophylactic design. In particular, the development and refinement of anti-retroviral drugs paved the way for combination therapy, which continues to reduce morbidity and increase quality of life in HIV-positive people. Although no cure or vaccine currently exists, research into these and other areas should prove crucial in the fight against HIV.

A number of problems still surround infection by the human immunodeficiency virus and the pathogenesis of AIDS. Solutions to the problems would provide valuable information for the development of antiviral therapy and a vaccine.

DESPITE substantial progress in our understanding of the human immunodeficiency virus (HIV), a number of mysteries remain concerning the virus, its target cells and the responses of the infected host. If a rational approach to treatment and therapy of AIDS is to be successful, solutions to the mysteries must be found. From my perspective, the four major questions are: what are the viral and cellular determinants of infection, what mechanisms generate the different HIV strains, how does HIV cause disease, and what determines the course from infection to disease?

What Governs Viral Tropism?

Recognition that the CD4 antigen on the surface of human helper T lymphocytes is a major cellular receptor for the HIV envelope protein gp120[1] led to the inference that endocytosis of the virus by cells expressing the CD4 molecule is an initial step in infection. But although mouse cells expressing human CD4 (as a result of experimental transfection) can bind HIV to the cell surface, they do not produce virus[2]. Moreover, HIV can infect cells lacking CD4, such as brain astrocyte cell lines[3] and human fibroblasts[4], and the virus is detected in endothelial[5] and epithelial[6] cells of seropositive individuals. Together, these observations strongly suggest that there are also mechanisms that do not involve CD4 by which HIV initially interacts with some cells.

Fusion of the target cell membrane with the HIV transmembrane envelope protein gp41

神秘的HIV：它的治疗和预防面临的挑战

利维

编者按

五年前共同发现HIV的病毒学家杰伊·利维描述了HIV研究中未解决的问题。四个关键的领域值得给予特殊关注：病毒与细胞感染的决定因素，不同病株的产生机制，影响疾病进展的因素和对于HIV引发疾病的理解。正如利维所说，随后对于这些领域的研究，极大地影响了治疗和预防药物的开发。特别是抗逆转录病毒药物的开发和完善为联合治疗铺平了道路，联合治疗可以持续减少发病率和提高HIV阳性患者的生活质量。尽管现在没有治愈的方法和疫苗，但是在这些领域和其他领域内的研究对抵抗HIV是至关重要的。

围绕人类免疫缺陷病毒感染和艾滋病的发病机理仍旧存在许多问题。这些问题的解决将会为抗病毒治疗和疫苗研发提供有价值的信息。

尽管我们在人类免疫缺陷病毒（HIV）的理解上已经取得了实质性的进展，但是关于 HIV 病毒还有很多未解之谜，如它的靶细胞和感染后宿主的反应。如果要寻找成功治疗 AIDS 的合理方法，那么我们必须解决这些难题。在我看来，这四个主要问题是：病毒和细胞感染的决定性因素是什么，不同 HIV 株的产生机制是什么，HIV 怎样引发疾病以及从感染到发病的过程中的决定因素是什么？

支配病毒亲嗜性的因素

我们认识到，人辅助 T 淋巴细胞表面的 CD4 抗原是 HIV 病毒包膜蛋白 gp120[1] 主要的细胞受体，细胞表达 CD4 分子导致病毒通过胞吞作用进入细胞，这是感染的第一步。尽管表达人 CD4 的小鼠细胞（转染的实验结果）可以在细胞表面结合 HIV，但是它们不能产生 HIV 病毒颗粒 [2]。另外，HIV 可以感染缺乏 CD4 的细胞，比如脑星形胶质细胞系 [3] 和人成纤维细胞 [4]，血清反应阳性的个体在内皮 [5] 和上皮 [6] 细胞里也发现 HIV 病毒。总之，这些观察结果有力地表明还存在另外一些机制，即在 HIV 病毒与某些细胞早期的相互作用过程中并不涉及 CD4。

靶细胞膜和 HIV 跨膜的包膜蛋白 gp41 融合可能是病毒进入细胞的一种方法。

is one possible alternative means of virus entry. It may operate alone or in concert with gp120. In support of a fusion process, infection by HIV appears to be *p*H-independent[7]. Moreover, part of the sequence of gp41 is similar to a sequence coding for the fusogenic glycoproteins of paramyxoviruses[8], and antibodies to the gp41 protein neutralize HIV[9]. In addition, antibody-dependent enhancement of virus infection can mediate HIV infection of cells[10]. As previously described for dengue and other viruses[11], virus-antibody complexes permit HIV infection of macrophages and T cells, most likely via the complement and/or Fc receptor.

These findings should prompt re-evaluation of the likelihood that soluble CD4 will be of major value in the prevention of HIV infection of cells and encourage the further search for methods to control the first steps in virus infection. It could be worth directing attention to the second conserved region of the viral gp120, which is not involved in CD4 attachment but is essential for early stages of infection[12].

The establishment of HIV infection requires several processes that are influenced by both viral and cellular factors. Strain variations in specific viral genes may, for example, account for the fact that, compared with blood isolates of HIV, isolates from the central nervous system replicate better in macrophages—the main HIV-expressing cell type of the brain[5]—than in T lymphocytes[13]. The properties of brain isolates may be reflected in the clinical observation of some infected individuals who have neurological defects without signs of immune deficiency[14]. Strain variations in specific viral genes such as the heterogenous *orf-B* region (see below) may also account for the lack of replication of some HIV isolates despite their efficient attachment to human $CD4^+$ T cells[15].

In other studies, cellular factors appear to determine virus replication. For instance, a single HIV isolate displays different levels of productive infection in peripheral blood mononuclear cells from various individuals[15]. Cultured mouse cells, in contrast to human cells, do not produce a substantial number of virus progeny after transfection with a biologically active DNA clone of HIV[16]. Cellular factors, such as the NF-κB protein, have been shown to interact with regulatory regions of HIV and enhance virus replication[17].

Defining the viral and cellular genes governing the host-range specificity of HIV is a major avenue of study. Experiments using recombinant HIVs, in which portions of macrophage-tropic and lymphocyte-tropic strains are combined, or using site-directed mutagenesis of biologically active DNA clones of these viruses should help clarify the viral genes involved. Assays of proteins found in selected cells or the use of somatic cell hybrids (mouse-human, for example) could uncover cellular factors influencing viral tropism. Clearly, the wide host range of HIV and the factors affecting virus replication are important variables to be defined for therapeutic strategies against viral infection of all cell types.

What Causes HIV Heterogeneity?

Biological heterogeneity of HIV is reflected by differences in host range, replicative

它可能单独起作用或者与 gp120 共同起作用。由于需要一个融合的过程，所以 HIV 的感染表现为非 pH 依赖性 [7]。此外，gp41 的部分序列与副粘病毒膜融糖蛋白的编码序列相似 [8]，且 gp41 蛋白的抗体可中和 HIV[9]。此外，病毒感染的抗体依赖性增强可介导 HIV 对细胞的感染 [10]。就像以前报道的登革热病毒和其他病毒 [11] 一样，病毒 – 抗体复合物最有可能通过补体(和)或 Fc 受体允许 HIV 感染巨噬细胞和 T 细胞。

这些发现促进我们重新评估可溶的 CD4 扮演预防 HIV 感染细胞的主要角色的可能性，并激励我们深入寻找控制病毒感染第一步的方法。我们值得将注意力转向病毒 gp120 的第二个保守区，它没有参与 CD4 附着，但是在感染的早期是非常重要的 [12]。

HIV 病毒感染的完成需要几个过程，这几个过程都受病毒和细胞因子的影响。某些病毒基因发生株变异可以说明这个事实，与从血液中分离出的 HIV 相比，从中枢神经系统分离的 HIV 病毒在巨噬细胞中——脑部主要表达 HIV 的细胞类型 [5]——比在 T 淋巴细胞 [13] 中有更好的复制能力。这种脑部分离株的特性可以在一些有神经缺陷但是没有免疫缺陷迹象的感染者的临床观察中反映出来 [14]。某些病毒基因的株变异，如 *orf-B* 区的异质性（如下），也可以解释某些 HIV 分离株虽然能有效地黏附人 $CD4^+$ T 细胞，但是无法复制的现象 [15]。

在其他的研究中发现细胞因子对病毒的复制起到决定性作用。比如单个 HIV 分离株在不同个体的外周血单核细胞中表现出不同水平的生产性感染 [15]。培养的小鼠细胞与人源细胞相比，经过具有生物活性的 HIV 的 DNA 克隆转染后，没有产生大量的病毒后代 [16]。细胞因子，如 NF-κB 蛋白，已被证明可与 HIV 的调控区域相互作用，并增强病毒的复制 [17]。

明确病毒和细胞基因调控 HIV 宿主范围的特异性是研究的主要途径。利用嗜巨噬细胞和嗜淋巴细胞病毒株的部分基因形成的重组 HIV 或者利用定点诱变的有生物学活性的病毒 DNA 克隆进行实验，有助于阐明所涉及的病毒基因。分析在选定的细胞或者杂交的体细胞（如小鼠和人）中发现的蛋白，可以揭示影响病毒亲嗜性的细胞因子。显然，对于所有类型细胞抗病毒感染的治疗策略来说，HIV 广泛的宿主范围和影响病毒复制的细胞因子都是需要定义的重要变量。

是什么引起了 HIV 的异质性?

HIV 的生物异质性反映在宿主范围、复制特性和感染细胞中的细胞病变作用的

properties and cytopathic effects in infected cells[18]. Restriction enzyme[19] and sequence analyses[20] of HIV isolates, as well as the patterns of neutralization of different isolates by antibodies,[21] also demonstrate genetic diversity, particularly in the envelope glycoprotein. (This is true not only of HIV-1 but also of the more recently identified HIV-2 subtype[22]). The mechanism responsible for generating these varying strains of virions is puzzling. One theoretical possibility is that the unintegrated proviral copies of HIV that accumulate during acute replicative infection[23] can undergo efficient genomic recombination leading to the evolution of infectious variants. The lack of genetic changes in HIV during long-term passage of the integrated virus in persistently infected cells[24] is consistent with this hypothesis.

Three explanations may account for the occurrence of envelope variants. First, the immunological reaction of the host selects specific spontaneous mutants which differ in their external envelope region. Thus far, this mechanism has only been observed with certain animal lentiviruses[25]. Second, mutations in the envelope, in contrast to other structural genes, can be tolerated during HIV replication. Third—and most speculatively—the HIV reverse transcriptase may be most error-prone when dealing with sequences for glycosylated proteins.

What Determines Pathogenesis?

An understanding of how HIV causes disease is a major research objective. As with any infectious agent, both viral and host determinants are involved.

Neuropathy. The disease in the central nervous system has been attributed by some investigators to the production by infected macrophages of cytokines that affect normal brain function[5,26]. HIV-infected brain capillary endothelial cells could compromise the maintenance of the blood–brain barrier, so permitting entry of toxic materials from the blood[5,26]. HIV also infects oligodendrocytes and astrocytes[3,5,14]. Compared with brain macrophages, only a small number of these glial cells produce viral RNA[4,5] but persistent or low replication of HIV in glial cells might affect their function. Since astrocytes maintain the integrity of the blood–brain barrier[27] and oligodendrocytes produce myelin, which is required for nerve conduction, the participation of these infected cells in AIDS neuropathy should be further evaluated. Finally, whether the neurologic disease results from long-term effects of HIV infection or specific pathogenic properties of a neurotropic strain[13,14] remains to be elucidated.

Enteropathy. HIV infection of enterochromaffin cells in the intestinal mucosa[6], perhaps by a process similar to that in the brain, could explain the chronic diarrhoea and malabsorption (particularly with duodenal involvement) observed in AIDS patients in the absence of any known bowel pathogens. These neuroendocrine cells migrate during embryogenesis from the neural crest and help regulate motility and digestive functions of the intestine. How their infection specifically accounts for the pathology is not known and the possible toxic effects of cytokines secreted by infected macrophages in the lamina propria needs to be considered.

不同[18]，HIV 分离株的限制性内切酶[19]和序列分析[20]，以及不同分离株中和抗体的模式[21]，也证明遗传的多样性，尤其是包膜糖蛋白。（这是事实，不仅是 HIV-1，还有最近发现的 HIV-2 亚型也是如此[22]）。产生这些不同病毒株的机制令人费解。理论上的可能性是，未整合的 HIV 前病毒的拷贝在急性复制感染期过程中累积[23]，能发生高效的基因重组，导致有感染性的变异体的演变。在持续感染的细胞中的整合病毒在长期传代过程中，HIV 缺乏遗传变化[24]，与这个假说一致。

三个解释可以说明包膜变异体的发生。首先，宿主免疫反应筛选特定的自发突变体，它与其他突变体的外部包膜区域不同。迄今为止，这个机制只在某些动物的慢病毒中观察到[25]。第二，包膜的突变与其他结构基因相比，在 HIV 复制的过程中可被耐受。第三，最大胆的猜想是，处理糖基化蛋白的序列时，HIV 逆转录酶是最容易出错。

是什么决定了发病机理?

对 HIV 是如何导致疾病的理解是一个主要的研究目标。与任一感染原相同，是由病毒和宿主共同参与的。

神经病变 一些研究者将中枢神经系统疾病归因于被感染的巨噬细胞产生的细胞因子对脑正常功能的影响[5,26]。被 HIV 感染的脑部毛细血管内皮细胞可能损害血脑屏障,使血液中的有毒物质能够进入[5,26]。HIV 还感染少突胶质细胞和星形胶质细胞[3,5,14]。与脑巨噬细胞相比，仅有少量的神经胶质细胞能够产生病毒 RNA[4,5]，但是 HIV 在神经胶质细胞中持续或者低水平的复制可能影响它们的功能。由于星形胶质细胞维持血脑屏障[27]的完整性和少突胶质细胞产生神经传导所需要的髓磷脂，因此感染细胞在艾滋病神经病变中的作用应进一步评估。最终，神经疾病是由于 HIV 感染的长期效应所引起的还是由一个嗜神经变异株的特殊病理症状所导致[13,14]，仍有待阐明。

肠病变 HIV 感染肠黏膜中的肠嗜铬细胞[6]，也许是通过与感染脑部相似的过程，这就可以解释在缺少任何已知的肠病原体的情况下，艾滋病人中观察到的慢性腹泻和吸收不良（尤其是十二指肠的参与）。在胚胎发生期，这些神经内分泌细胞从神经嵴中迁移出来，帮助调节肠道的蠕动和消化功能。感染的机制如何通过病理学来解释还是未知的，并且需要考虑固有层中被感染的巨噬细胞分泌的细胞因子可能的毒性作用。

Immunopathology. How HIV causes immune suppression presents the greatest enigma. The immune dysfunctions identified in HIV infection were first linked to the cytopathic effects of HIV on helper T lymphocytes that play a vital role in regulating immune function. These cells infected with HIV *in vitro* undergo syncytial formation by recruiting many uninfected cells before proceeding to cell death[1].

Several observations, however, have challenged this suggested explanation for HIV-induced immunopathology. First, despite extensive histopathological studies of infected individuals, little evidence exists for syncytial cell formation *in vivo*[28]. Second, the number of HIV-infected cells in the blood (10^2—10^3 per ml)[29] does not account for the quantity of cells lost over time in the infected host; moreover, $CD4^+$ cells infected by HIV can survive several weeks in culture[30]. Third, many HIV isolates obtained from immunologically suppressed individuals are not highly cytopathic *in vitro*[18,31] and non-cytopathic isolates of HIV have sometimes been associated with disease[22]. Finally, abnormalities of immune function are observed not only in HIV-infected helper T cells and macrophages but also in uninfected cells of the haematopoietic system (Table 1)[32].

Table 1. Immune function abnormalities in AIDS

T lymphocytes	**B lymphocytes**
(1) Decreased proliferative responses to mitogens, soluble antigens, and allogeneic cells. (2) Decreased lymphokine production (IL-2, gamma interferon) in response to antigen. (3) Decreased cytotoxic T lymphocyte activity against virus-infected cells.	(1) Polyclonal activation with hypergammaglobulinaemia and spontaneous plaque forming cells. (2) Decreased humoral response to immunization. (3) Production of autoantibodies.
Monocytes	**NK cells**
(1) Decreased chemotaxis. (2) Decreased IL-1 production (or production of an inhibitor of IL-1). (3) Decreased microbiocidal activity.	(1) Decreased cytotoxic activity.

These abnormalities appear to begin with acute depletion of T helper cells and proliferation of B cells; other defects, many revealed by *in vitro* studies, accumulate over time. Several of the immune abnormalities may result from the decrease in $CD4^+$ lymphocytes and cytokine production[32].

What other mechanisms could explain the immunological features of HIV pathogenesis (Table 2)? HIV infection is known to cause epiphenomena including decreased expression of immunological recognition sites, such as CD4 and the interleukin-2 (IL-2) receptor, and reduced production of cytokines such as IL-1, IL-2, and gamma interferon (Table 2)[32]. Loss of these molecules could have far reaching effects on other cells of the immune system. Moreover, HIV infection of progenitor cells could reduce the replenishment of circulating lymphocytes and macrophages. The disarray in immune function could generate T suppressor cells or factors[33] that further compromise the immune response. HIV envelope proteins have been found to be toxic or inhibitory to immune cells[34]. By circulating in the blood, these proteins may block the ability of lymphocytes to recognize or respond to foreign antigens and may interfere with the function of antigen-presenting cells.

免疫病理学 HIV 如何引起免疫抑制是最大的谜题。HIV 感染引起的免疫功能紊乱，首先与 HIV 对辅助 T 淋巴细胞的病变效应联系起来，该细胞在免疫功能调节中发挥重要作用。体外感染 HIV 的这些细胞在死亡之前募集未感染细胞形成合胞体 [1]。

然而数个观察结果对 HIV 诱导的免疫病理学的解释提出质疑。首先，尽管有大量受感染的个体的组织病理学研究，但是很少有证据证明在体内可形成合胞体细胞 [28]。第二，血液中感染 HIV 的细胞数量（每毫升 10^2~10^3 个）[29] 不能说明被感染宿主一段时间里丧失的细胞数量，而且被 HIV 感染的 $CD4^+$ 细胞在培养基中可以存活几周 [30]。第三，在体外实验中许多从免疫抑制个体中获得的 HIV 分离株没有引起严重的细胞病变 [18,31]，非细胞病变 HIV 分离株有时与疾病有关 [22]。最后，免疫功能失常不仅仅在 HIV 感染的辅助 T 细胞和巨噬细胞中被观察到，在非感染的造血系统细胞中也有（表 1）[32]。

表 1. 艾滋病引起的免疫功能异常

T 淋巴细胞	B 淋巴细胞
(1) 减少有丝分裂原、可溶性抗原和异源细胞的增殖反应。 (2) 减少抗原反应产生的淋巴因子（IL-2和γ-干扰素）。 (3) 减少对病毒感染的细胞毒性T淋巴细胞的活性。	(1) 多克隆活化高γ球蛋白血症和自发空斑形成细胞。 (2) 减少体液免疫反应。 (3) 产生自身抗体。
单核细胞	**NK 细胞**
(1) 趋化性下降。 (2) 减少IL-1的生成（或是生成IL-1抑制剂）。 (3) 杀菌活性下降。	(1) 细胞毒素活性下降。

这些异常似乎伴随着 T 辅助细胞的急速消耗和 B 细胞的增殖；体外实验证明其他的缺陷随着时间而积累。几个免疫异常可能由于 $CD4^+$ 淋巴细胞的减少和细胞因子产生的减少导致的 [32]。

还有什么其他的机制可以解释 HIV 发病机理的免疫学特征呢（表 2）？我们已知 HIV 感染引起的附带现象包括免疫识别位点的表达的减少，如 CD4 和白细胞介素 –2（IL-2）受体，和细胞因子产生的减少，如 IL-1、-2 和 γ 干扰素（表 2）[32]。这些分子的丧失对免疫系统其他细胞影响深远。另外，HIV 感染祖细胞能够减少外周血淋巴细胞和巨噬细胞的补充。免疫功能的紊乱能够产生 T 抑制细胞或因子 [33]，它们能够进一步损害免疫应答。已发现 HIV 包膜蛋白对免疫细胞有毒害或抑制作用 [34]。通过血液循环，这些蛋白可能会阻断淋巴细胞识别或应答外源抗原的能力，并且可能会干扰抗原提呈细胞的功能。

Table 2. Mechanisms of immune suppression by HIV infection

Direct mechanisms	Indirect mechanisms
(1) HIV cytocidal effect on $CD4^+$ lymphocytes. (2) Functional defects in infected $CD4^+$ cells: (*a*) decreased expression of cell surface proteins (for example, IL-2 receptor, CD4); (*b*) impaired production of lymphokines such as IL-2 or gamma interferon. (3) Impaired antigen presentation and/or monokine production by infected macrophages; cell death.	(1) Generation of suppressor T cells and/or factors. (2) Toxic or inhibitory effects of viral protein. (3) Immune complex formation. (4) Induction of autoimmune phenomena: (*a*) autoantibodies resulting from polyclonal B cell activation or antigen mimicry; (*b*) virus mediated, enhanced immunogenicity of normal cellular proteins. (5) Cytotoxic cell activity against viral or self proteins.

In addition, HIV is associated, often early in the infection, with a polyclonal activation of B cells resulting in hypergammaglobulinaemia[32]. Some of these antibodies form immune complexes that can be detrimental to the immune system; some react with self-proteins, leading to autoimmunity[35]. The destruction of activated helper T cells by lymphocytotoxic autoantibodies directed against a normal cellular protein (p18) has been suggested as one mechanism for the substantial loss of $CD4^+$ cells in advanced disease[35]. Antibodies and/or cytotoxic cells directed against other shared antigens on immune cells can induce further abnormalities in immune function or haematopoiesis[36,37]. Finally, antibody-dependent cellular cytotoxicity reacting against envelope proteins bound to ligands on uninfected $CD4^+$ cells could be a factor in the pathogenic process[38]. Clearly, multiple effects of HIV infection on the immune system, both direct and indirect, need to be fully evaluated to appreciate their potential role in the immunological abnormalities observed.

Cytopathology. A critical question surrounding HIV pathogenesis is how the virus kills the cell. In general, helper T lymphocytes are most susceptible to the cytopathic effects of HIV; in culture, they often undergo fusion before cell death. Macrophage killing by some strains of the virus has also been observed[1]. Syncytial formation by T cells, however, does not always proceed cell death; infected cells can die without undergoing fusion[39]. The viral envelope gp120, alone or through its interaction with CD4, could be responsible for cell death[34,40,43]. Direct inoculation of inactivated virus or the envelope gp120 onto peripheral blood mononuclear cells can also produce cytopathic effects[34,41]. Cell death may result from direct membrane disruption involving calcium channels[42] and/or phospholipid synthesis[43]. The accumulation of unintegrated proviral copies of HIV DNA is an attractive explanation for cytopathology since it is associated with cell death in other retrovirus systems[44]. Finally, whether the cytopathology of HIV in cell culture mirrors pathogenesis in the host awaits development of an appropriate animal model.

What Influences Progression to Disease?

Long-term follow-up studies of HIV seropositive individuals indicate that about a third will remain free of symptoms for at least seven years[45]. Moreover, some healthy seropositive individuals lose a large proportion of their $CD4^+$ lymphocytes, yet do not

表 2. HIV 感染引起免疫抑制的机制

直接机制	间接机制
(1) HIV对 CD4+ 淋巴细胞的杀细胞效应。 (2) 被感染的 CD4+ 细胞的功能缺失： (*a*) 细胞表面蛋白表达减少(比如 IL-2受体，CD4)； (*b*) 减少淋巴因子的产生，如IL-2 或 γ 干扰素。 (3) 减少被感染的巨噬细胞的抗原提呈和/或者单核因子的产生；细胞死亡。	(1) 产生抑制性 T 细胞和/或者因子。 (2) 病毒蛋白的毒性或抑制效应。 (3) 免疫复合物形成。 (4) 诱导自身免疫现象： (*a*) 多克隆 B 细胞激活或抗原模拟导致自身抗体的产生； (*b*) 病毒介导、增强正常细胞蛋白的免疫原性。 (5) 细胞毒性细胞对抗病毒或自身蛋白的活性。

另外，通常在感染初期，HIV 与多克隆 B 细胞活化导致血丙种球蛋白过多有关[32]。有些抗体形成的免疫复合物可损害免疫系统；有些抗体与自身蛋白反应导致自身免疫性[35]。针对正常细胞蛋白 p18 的淋巴细胞毒性自身抗体导致活化的辅助性 T 细胞的损伤，被认为是疾病发展中 CD4+ 细胞大量丧失的一个机制[35]。针对免疫细胞上其他共享抗原的抗体或（和）细胞毒性细胞，可进一步导致免疫功能或造血作用失常[36,37]。最后，针对结合在未感染的 CD4+ 细胞配体上的包膜蛋白应答的抗体依赖性细胞的细胞毒性，可能是致病过程中的一个因素[38]。显然，HIV 感染对免疫系统的多种效应，无论直接的或是间接的都需要我们进行全面的评估，从而理解它们在观察到的免疫异常中的潜在作用。

细胞病理学 一个关于 HIV 发病机理的关键问题是病毒怎样杀死细胞。一般情况下，辅助 T 淋巴细胞最容易受 HIV 导致的细胞病变的影响；在培养基中，它们常常在死亡之前发生细胞融合。某些病毒株也被观测到正在杀死巨噬细胞[1]。然而，T 细胞合胞体的形成并不总是导致细胞死亡；感染的细胞可以不经历融合而死亡[39]。病毒包膜蛋白 gp120，可以独自或通过与 CD4 相作用，导致细胞死亡[34,40,43]。直接接种灭活的病毒或者包膜蛋白 gp120 到外周血单核细胞也能够产生细胞病变效应[34,41]。细胞死亡可能是由于细胞膜直接被破坏，涉及钙通道[42]（和）或者磷脂合成[43]。对于细胞病理学来说，未整合的 HIV 前病毒 DNA 拷贝的累积是一个有吸引力的解释，因为它与其他逆转录病毒系统中的细胞死亡有关[44]。最后，在培养细胞中 HIV 的细胞病理学是否反映了在宿主体内的发病机理，要等待建立合适的动物模型才可以证明。

是什么影响了疾病的发展？

对 HIV 血清反应阳性者进行长期随访的研究表明，大概有三分之一的人在至少七年的时间里没有症状[45]。此外，一些健康的血清反应阳性的个体，在丧失了大量 CD4+ 淋巴细胞后，仍旧没有出现症状[46]。一些艾滋病人（多数为患有波氏肉瘤）长

develop symptoms[46]. Several AIDS patients (many with Kaposi's sarcoma) have remained clinically stable for up to six years. Two patients followed at our medical center have had only 5% of their normal $CD4^+$ cell number for over a year without any new symptoms. Thus, predictions on the development of opportunistic infections or cancers based solely on a decrease in $CD4^+$ cells could be misleading; other presently unrecognized functions of the immune system may be fundamental in warding off disease.

$CD8^+$ cell activity. What determines the resistant state? Cytotoxic T lymphocytes that react with cells expressing HIV proteins[47,48] have been noted in infected individuals, but the clinical importance of this antiviral response is unknown. In our laboratory, studies of asymptomatic individuals have revealed that their peripheral blood mononuclear cells do not readily release HIV when placed in culture. Nearly all, however, yield virus when a subset of their $CD8^+$ lymphocytes is removed from the blood sample[49]. Similar observations have recently been made with the primate immune deficiency virus[50]. These $CD8^+$ cells apparently prevent HIV replication not by killing infected cells, but by producing a diffusible suppressor factor or factors. Among seropositive individuals, the level of this $CD8^+$ cell activity varies, and can be reflected in clinical status. Peripheral blood mononuclear cells cultured from many patients with disease readily produce virus and their $CD8^+$ cells show very little antiviral activity.

This variation in HIV replication in cultured cells probably mirrors the increase of viral p25 antigen in the plasma of individuals as they advance in disease[51]. Whether this observation reflects enhanced virus production or a decrease in antibodies to p25 is still not clear. Nevertheless, the information does suggest that the resistant state in infected individuals is mediated by cellular immune responses operating soon after infection; once these are reduced, renewed production of HIV can occur. The resumption of HIV replication could enhance progression to disease because of the emergence, by mutation or selection, of new HIV variants that replicate rapidly to high titre in a variety of cell types, and that are highly cytopathic[18]. This observation correlates clinically with the increased loss of $CD4^+$ cells[46] and the high levels of p25 antigenaemia[51] associated with development of disease. Nevertheless, whether the progression to disease results first from a reduced immune response or from the eventual emergence of a more pathogenic HIV strain, or from both, needs to be clarified.

Taken together, the data suggest steps in HIV infection that might explain variations in the course of the disease (Fig. 1). Levels of antiviral immune response, the inherent sensitivity of the host cell to virus replication and the relative virulence of the virus strain are major variables to be defined. What ends the resistant state (that is, triggers progression to disease) is not known, but could be a variety of factors including antilymphocyte antibodies, enhanced production of virus by activating events, progressive destruction of $CD4^+$ lymphocytes by intermittent periods of HIV replication and a decreased production by $CD4^+$ lymphocytes of cytokines required for the growth and function of antiviral $CD8^+$ cells.

达六年临床表现稳定。在我们医疗中心，有两个病人正常 $CD4^+$ 细胞数目只有 5%，一年多没有任何新的症状。因此，仅仅以 $CD4^+$ 细胞的减少为依据预测机会性感染或癌症的发展可能产生误导；目前其他还没有被发觉的免疫系统的功能也许是预防疾病的基础。

$CD8^+$ 细胞活性 是什么决定了抵抗状态？人们已经注意到，在被感染个体中，细胞毒性 T 淋巴细胞可与表达 HIV 蛋白的细胞发生作用 [47,48]，但是这种抗病毒反应的临床意义还是未知数。在我们实验室，对无症状个体的研究显示，当进行培养时，其外周血单核细胞不容易释放 HIV 病毒。然而从他们血液样本中分离出的 $CD8^+$ 淋巴细胞从血液样本中分离后，几乎都产生病毒 [49]。最近对灵长类免疫缺陷病毒也观察到相似的结果 [50]。很显然，这些 $CD8^+$ 细胞不是通过杀死感染的细胞来阻止 HIV 复制，而是通过产生一种或多种可扩散的抑制因子。血清反应阳性的不同个体中，$CD8^+$ 细胞的活性水平是有差异的，并且可以反映其临床状态。提取许多患者的外周血单核细胞进行培养，很容易产生病毒，而他们的 $CD8^+$ 细胞表现出很低的抗病毒活性。

在培养的细胞中，HIV 复制发生变异，可能反映出在感染个体血浆中随着疾病的发展 [51] 病毒 p25 抗原的增加。仍不清楚这个观察结果反映的是病毒复制增强还是 p25 的抗体减少。然而，这个信息确实显示，在被感染个体中，细胞免疫应答在感染以后很快介导了抵抗反应；一旦这些反应减弱，HIV 便开始重新产生了。HIV 病毒复制的重新开始可能加快了疾病的发展，因为通过突变或者选择，出现了新的 HIV 变异体，这些变异体在许多类型细胞中，都是迅速复制达到很高的滴度，并且导致高度的细胞病变 [18]。这个观察结果与临床上随着疾病的发展 $CD4^+$ 细胞 [46] 的缺失增加以及高水平的 p25 抗原血症 [51] 有关。然而，疾病的发展首先是由于免疫应答减弱还是最终高致病性 HIV 株的出现或是两者都有，仍需要进一步研究。

总之，有数据表明 HIV 感染的阶段可以解释病程的变化（图 1）。抗病毒免疫应答的水平，宿主细胞对病毒复制的内在敏感性和病毒株的相对毒性是需要明确的主要变量。现在仍不知道是什么终止了抵抗状态（即触发了疾病进展），但可能是多因素的，包括抗淋巴细胞抗体，激活事件导致病毒复制率增加，HIV 复制间歇期 $CD4^+$ 淋巴细胞进行性的破坏以及抗病毒的 $CD8^+$ 细胞生长和发挥作用所需的 $CD4^+$ 淋巴细胞的细胞因子产生的减少。

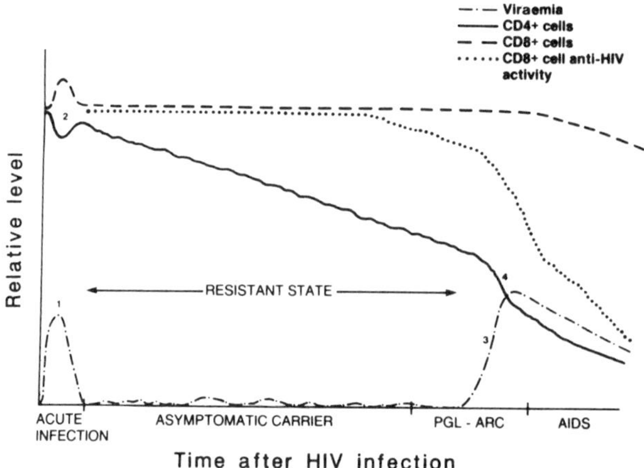

Fig. 1. A model of possible stages in the course of HIV infection. (1) Acute infection is characterized by the presence of free-virus (or its antigen) in the blood in the absence of antibodies to HIV. (2) A $CD8^+$ cells increase rapidly with a concomitant decrease in $CD4^+$ cells; thus the helper/suppressor ratio is dramatically reduced. After a short period (2–12 weeks) a resistant state develops in which $CD8^+$ cells return to above normal levels, $CD4^+$ cells decrease at a slow rate, and free virus is not readily detected in the blood, probably because it is produced episodically. Subsequently, after a variable time lag, the resistant state wanes for reasons that are not known. (3) The virus enters multiple replicative cycles during which a variant strain that can be resistant to the immune response and highly cytopathic can emerge and further compromise the antiviral state. (4) Viraemia is accompanied by enhanced destruction of $CD4^+$ cells and progression of disease.

Cofactors in pathogenesis. A major unresolved issue is whether other infectious agents can affect this asymptomatic period. Concomitant viral or parasitic infections may activate immune cells so they produce more virus, or become more sensitive to HIV infection[1]. Such events might lead to the suggested intermittent periods of HIV production. As demonstrated *in vitro*, agents such as the herpes viruses may act on the regulatory regions of HIV to enhance virus replication within the cell[52]. Continued surveillance of individuals with long asymptomatic periods, as well as studies of infected chimpanzees that have not yet shown any clinical abnormalities, should provide insight into factors influencing resistance to HIV.

Latency. A long incubation period might also be explained by a latent infection. During the latent period of retrovirus infection very little viral protein or RNA is made and no infectious progeny are produced by the infected cell[53]. When cells that have been latently infected with HIV in culture are treated with activating agents, such as halogenated pyrimidines and cytokines, virus replication begins and is then either maintained or reverts to latency[14,54].

Studies of the *orf-B* (or *3′ orf*) gene of HIV suggest it could be responsible for latency. Deletion of *orf-B* leads to a 5–10 fold increase in virus replication compared with the wild-type virus[55]. Conceivably, the *orf-B* gene product (p27 protein), which has GTPase and phosphorylating properties[56], interacts with cellular factors to down-regulate virus replication (see above) in a continuum that can proceed to latency.

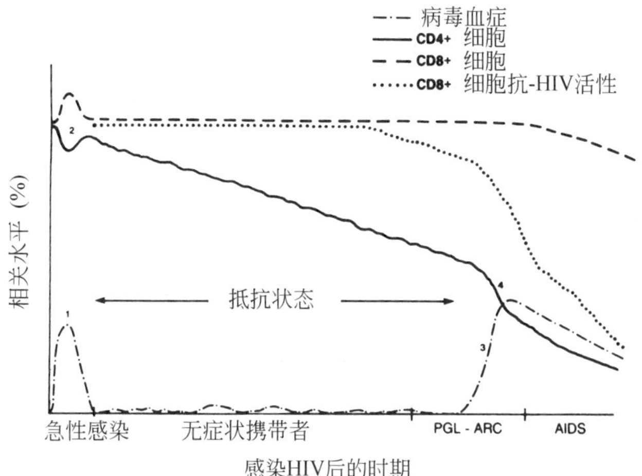

图 1. HIV 感染的过程中可能的阶段模型。（1）在缺少 HIV 抗体时，血液中出现游离病毒（或者它的抗原）是急性感染的特征。（2）伴随着 CD4$^+$ 细胞的减少 CD8$^+$ 细胞快速增加；从而辅助 T 细胞 / 抑制 T 细胞的比率明显下降。短期后（2~12 周），形成抵抗状态，此时 CD8$^+$ 细胞恢复到正常水平以上，CD4$^+$ 细胞也以较慢的速度减少，游离病毒在血清中不容易检测到，可能因为它的产生是偶发的。随后，经过一段不定的滞后时间，不知何种原因，抵抗状态衰退。（3）当对免疫应答有抵抗力的变异株和高度细胞病变出现后，病毒进入多复制周期，进一步危害抗病毒状态。（4）病毒血症的产生伴随着 CD4$^+$ 细胞的损坏增强和疾病的发展。

发病机理的辅助因素 一个重大的悬而未决的问题是其他的感染原是否影响无症状期。伴随病毒或寄生虫感染可以激活免疫细胞，使他们产生更多病毒，或对 HIV 更加敏感[1]。这些事情可以导致暗示的 HIV 复制间歇期。体外研究证明，其他传染病源如疱疹病毒可以在 HIV 的调控区域发生作用，以提高病毒细胞内的复制[52]。持续观察长期处于无症状时期的个体以及研究没有任何临床异常的被感染的猩猩，能够深入了解影响 HIV 抵抗力的因素。

潜伏期 潜伏期的感染也可以解释长时间的潜伏期。在逆转录病毒感染的潜伏期，产生很少病毒蛋白或者 RNA，感染的细胞不产生感染性的后代[53]。当培养被 HIV 病毒感染而处于潜伏期的细胞时，加入激活剂，如卤代嘧啶和细胞因子，病毒开始复制，然后或继续复制或回到潜伏期[14,54]。

HIV 的 *orf-B*（或者 *3′ orf*）基因的研究表明它可能与潜伏感染有关。*orf-B* 的缺失可导致病毒复制速率比野生型病毒增加 5~10 倍[55]。可以想象，*orf-B* 基因的产物（p27 蛋白）具有 GTP 酶和磷酸化特性[56]，通过与细胞因子相互作用持续下调病毒复制水平（见上文），从而进入潜伏期。

In rare cases, individuals who have been HIV seropositive become seronegative. In some of these individuals the presence of a latent HIV infection in peripheral mononuclear cells can be detected by means of the polymerase chain reaction; in others no HIV can be detected[57]. This potentially encouraging observation suggests that HIV infection in some individuals might be eliminated completely, but most likely the virus remains latent at other sites. Defining the factors governing latency should provide valuable information for the development of antiviral strategies. Moreover, the importance of latency to viral transmission must be assessed in "false negative" serological states.

Conclusions

Answers to many questions about the viral and host determinants of HIV pathogenesis could assist the prospects for therapy and prevention of AIDS. On current information, there are several features of HIV that need to be taken into account (Table 3).We should target antiviral drug strategies at the vulnerable sites of HIV replication, both before and after integration. In this regard, understanding how an HIV strain evolves into a more cytopathic (and potentially pathogenic) agent[18] would be valuable. We must find methods for inducing strong intracellular and cellular host responses against the virus: intracellular production of the *orf-B* protein or stimulation of the $CD8^+$ cell population responsible for suppressing HIV replication might produce long asymptomatic periods. Preventing the formation of antibodies to lymphocytes would be another promising direction. For vaccine development we need novel approaches that will define both specific epitopes of HIV and the appropriate adjuvant to elicit strong cross-reacting immune responses not generally observed in natural infection. Toward this objective, elimination of those epitopes responsible for antibody-dependent enhancement[10] would appear important. The immunized host must respond not only against free virus, but most importantly against productively and latently infected cells that can be major sources of HIV transmission[58]. Concentration on these areas of research should provide valuable information to help in the attack against HIV. In the process, we will learn a great deal more about viruses and the function of the immune system.

Table 3. Features of HIV of relevance to antiviral therapy

(1) Virus infection involves integration of the viral genome into the chromosome of the infected cell. This cell is a protective environment for the virus and a reservoir for persistent virus production.
(2) The infected cell is a major source of virus transmission and can pass HIV by cell-to-cell contact.
(3) The infected cells can remain "latent" and express very few viral antigens. Can these cells be recognized and eliminated by an antiviral response?
(4) HIV transmission occurs at specific sites in the host (such as the rectum). Prevention requires immune response at these local sites.
(5) Several independent serotypes and subtypes of HIV can be identified. Can they all be controlled by one strategy?
(6) Portions of HIV proteins resemble normal cellular proteins. Immunization may induce autoantibodies.
(7) Vaccination may induce antibodies that enhance HIV infection.

Studies conducted by the author were supported by the California State Universitywide Task Force on AIDS, the American Foundation for AIDS Research, and the National Institutes of Health. I would like to thank Drs S. Levy and J. Ziegler as well as Drs C. Cheng-Mayer, L. Evans, J. Homsy, J. Hoxie, J. Leong, M. McGrath and C. Walker for

在极少的病例中，HIV 血清反应阳性的个体会变为血清反应阴性。在一些这样的个体中，外周血单核细胞中存在的 HIV 潜伏感染可通过聚合酶链式反应检测，另外一些则检测不到 HIV 病毒 [57]。这个可能令人鼓舞的观察结果表明，HIV 在一些感染的个体中可能完全被消除，但是最有可能病毒潜伏在其他位置。明确潜伏期的调节因素为抗病毒策略的发展提供了有价值的信息。此外，潜伏期对于病毒传播的重要性，须考虑假阳性这一血清学参数。

结　论

解答在 HIV 发病机理中病毒和宿主的决定因素的问题，会为艾滋病的治疗和预防提供帮助。就目前的资料而言，我们需要考虑 HIV 的几个特点（表 3）。整合前后，我们都应该将抗病毒药物的目标定位在 HIV 复制时容易受攻击的部位。就这一点而言，了解一个 HIV 株怎样发展为一个能引发细胞病变（及潜在病变的）的病毒株 [18] 会有很大的意义。我们必须寻找能诱导强的细胞内和细胞宿主抵抗病毒的方法：细胞内产生 *orf-B* 蛋白或者依靠刺激 $CD8^+$ 细胞分裂来抑制 HIV 的复制可能产生长时间的无症状期。阻止淋巴细胞抗体的形成可能是另一种有前途的方向。对于疫苗的研制，我们需要新的方法来找到 HIV 的特定抗原表位和适当的佐剂引起免疫系统强烈的交叉反应，这在自然感染中一般是观察不到的。为了实现这个目标，消除这些抗体依赖性增强的抗原表位显得更重要 [10]。免疫的宿主不仅要抵抗游离的病毒，最重要的是抵抗复制期和潜伏期的感染细胞，它们可能是主要的 HIV 传播源 [58]。集中精力在这些领域进行研究，可以为攻击 HIV 提供有价值的信息。在这个过程中，我们将了解大量的关于病毒和免疫系统功能的知识。

表 3. HIV 与抗病毒治疗的相关特征

(1) 病毒感染包括病毒的基因组整合到被感染细胞的染色体中。细胞为病毒提供了一个安全的环境并成为病毒持续复制的一个储存器。
(2) 被感染细胞是病毒传播的主要源泉，病毒能够通过细胞之间的接触进行传播。
(3) 感染细胞处于潜伏状态，表达少量的病毒抗原，这些细胞能够被抗病毒应答所识别和消除吗？
(4) HIV传播发生在宿主的特定部位(如直肠)。这些部位的免疫应答可对病毒起到预防作用。
(5) 几个独立的HIV血清型和亚型能被鉴定。能通过一种方法全部控制吗？
(6) 部分HIV蛋白类似于正常的细胞蛋白，免疫自身诱导自身抗体。
(7) 疫苗可能会产生加强HIV感染的抗体。

作者进行的研究得到了加利福尼亚州立大学艾滋病研究小组、美国艾滋病研究基金会和美国国家卫生研究院的支持。我还要感谢利维、齐格勒、郑迈耶、埃文斯、霍姆西、霍克西、梁、麦格拉思和沃克博士对于这篇文章有价值的评论和贝格林

their helpful comments on this article, and C. Beglinger for its preparation.

(**333**, 519-522; 1988)

Jay A. Levy
Department of Medicine and Cancer Research Institute, University of California School of Medicine, San Francisco, California 94143-0128, USA

References:

1. Fauci, A. S. *Science* **239**, 617-622 (1988).
2. Maddon, P. J. *et al. Cell* **47**, 333-348 (1986).
3. Cheng-Mayer, C. *et al. Proc. Natl. Acad. Sci. U.S.A.* **84**, 3526-3530 (1987).
4. Tateno, M. & Levy, J. A. IV Int. Conf. AIDS, Stockholm (abstr.) (1988).
5. Wiley, C. A. *et al. Proc. Natl. Acad. Sci. U.S.A.* **88**, 7089-7093 (1986).
6. Nelson, J. A. *et al. Lancet* **i**, 259-262 (1988).
7. Stein, B. S. *et al. Cell* **49**, 659-668 (1987).
8. Gallaher, W. R. *Cell* **50**, 327-328 (1987).
9. Chanh, T. C. *et al. EMBO J.* **5**, 3065-3071 (1986).
10. Robinson, W. E. *et al. Lancet* **i**, 790-795 (1988).
11. Halstead, S. B. & O' Rourke, E. J. *J. exp. Med.* **146**, 201-217 (1977).
12. Ho, D. D. *et al. Science* **239**, 1021-1023 (1988).
13. Gartner, S *et al. Science* **233**, 215-219 (1986).
14. Levy, J. A. *et al. Ann. Inst. Pasteur* **138**, 101-111 (1987).
15. Evans, L. A. *et al. J. Immun.* **138**, 3415-3418 (1987).
16. Levy J. A. *et al. Science* **232**, 998-1001 (1986).
17. Nabel, G & Baltimore, D. *Nature* **326**, 711-713 (1987).
18. Cheng-Mayer, C. *et al. Science* **240**, 80-82 (1988).
19. Hahn, B. H. *et al. Science* **232**, 1548-1553 (1986).
20. Starcich, B. R. *et al. Cell* **45**, 637-648 (1986).
21. Weiss, R. A. *et al. Nature* **324**, 572-575 (1986).
22. Evans, L. A. *et al. Science* **240**, 1522-1525 (1988).
23. Luciw, P. A. *et al. Nature* **312**, 760-763 (1984).
24. Robert-Guroff, M. *et al. J. Immun.* **137**, 3306-3309 (1986).
25. Carpenter, S. *et al. J. Virol.* **61**, 3783-3789 (1987).
26. Price, R. W. *et al. Science* **239**, 586-592 (1988).
27. Fontana, A. *et al. Nature* **307**, 273-276 (1984).
28. Cohen, M. B. & Beckstead, J. in *AIDS: Pathogenesis and Treatment* (ed. Levy, J. A.) (Dekker, New York, in the press).
29. Harper, M. E. *et al. Proc. Natl. Acad. Sci. U.S.A.* **83**, 772-776 (1986).
30. Hoxie, J. A. *et al. Science* **229**, 1400-1402 (1985).
31. Asjo, B *et al. Lancet* **2**, 660-662 (1986).
32. Koenig, S. & Fauci, A. S. in *AIDS: Etiology, Diagnosis, Treatment and Prevention*, 2nd edn. (eds DeVita, V., Hellman, S. & Rosenberg, S.) (Lippincott, Philadelphia, in the press).
33. Laurence, J. *et al. J. clin. Invest.* **72**, 2072-2081 (1983).
34. Shalaby, M. R. *et al. Cell Immun.* **110**, 140-148 (1987).
35. Stricker, R. B. *et al. Nature* **327**, 710-713 (1987).
36. Ziegler, J. & Stites, D. P. *Clin. Immunol. Immunopathol.* **41**, 305-313 (1986).
37. Donahue, R. E. *et al. Nature* **326**, 200-203 (1987).
38. Weinhold, K. J. *et al. Lancet* i, 902-904 (1988).
39. Somasundaran, M. & Robinson, H. L. *J. Virol.* **61**, 3114-3119 (1987).
40. Hoxie, J. *et al. Science* **234**, 1123-1127 (1986).
41. Rasheed, S. *et al. Virology* **154**, 395-400 (1986).
42. Gupta, S. & Vayuvegula, B. *J. clin. Immun.* 7, 486 (1987).
43. Lynn, W. S. *et al. Virology* **163**, 43-51 (1988).
44. Keshet, E. & Temin, H. M. *J. Virol.* **31**, 376-388 (1979).

格所做的准备工作。

（郑建全 翻译；孙军 审稿）

45. Rutherford, G. W. & Werdegar, D. in *AIDS: Pathogenesis and Treatment* (ed. Levy, J. A.) (Dekker, New York, in the press).
46. Lang, W. *et al.* Abstr. Int. on Conf. AIDS, Stockholm (abstr.) (1988).
47. Walker, B. *et al. Nature* **328**, 345-348 (1987).
48. Plata, F. *et al. Nature* **328**, 348-351 (1987).
49. Walker, C. M. *et al. Science* **234**, 1563-1566 (1986).
50. Kannagi, M. *et al. J. Immun.* **140**, 2237-2242 (1988).
51. Lange, J. M. A. *et al. Br. med. J.* **293**, 1459-1462 (1986).
52. Mosca, J. D. *et al. Proc. Natl. Acad. Sci. U.S.A.* **84**, 7408-7412 (1987).
53. Rojko, J. L. *et al. Nature* **298**, 385-388 (1982).
54. Folks, T. M. *et al. Science* **231**, 600-602 (1986).
55. Luciw, P. A. *et al. Proc. Natl. Acad. Sci. U.S.A.* **84**, 1434-1438 (1987).
56. Guy, B. *et al. Nature* **330**, 266-269 (1987).
57. Farzadegan, H. e*t al. Ann. int. Med.* **108**, 785-790 (1988).
58. Levy, J. A. *J. Am. med. Ass.* **259**, 3037-3038 (1988).

Human Basophil Degranulation Triggered by Very Dilute Antiserum against IgE

E. Davenas *et al.*

Editor's Note

This is one of the most controversial papers *Nature* has published, reflected in the fact that its category ("Scientific paper") was never used before or since. Jacques Benveniste and his co-workers of the French medical research organization INSERM claimed to have found that antibodies remain able to trigger an immune response from a class of white blood cells called basophils even when the antibodies are diluted until no molecules should still be present in solution. This activity, they claimed, recurs in a periodic manner with increasing dilution. This appeared to offer some basis for homeopathy, which uses such ultra-dilute solutions. The study was never clearly replicated subsequently, but it gave rise to the notion that water has a "memory" of what it has dissolved.

When human polymorphonuclear basophils, a type of white blood cell with antibodies of the immunoglobulin E (IgE) type on its surface, are exposed to anti-IgE antibodies, they release histamine from their intracellular granules and change their staining properties. The latter can be demonstrated at dilutions of anti-IgE that range from 1×10^2 to 1×10^{120}; over that range, there are successive peaks of degranulation from 40 to 60% of the basophils, despite the calculated absence of any anti-IgE molecules at the highest dilutions. Since dilutions need to be accompanied by vigorous shaking for the effects to be observed, transmission of the biological information could be related to the molecular organization of water.

THE antibodies responsible for human immediate hypersensitivity belong to the IgE isotype[1]. The most salient feature of IgE is its capacity to bind to mast cell and polymorphonuclear basophil membranes through receptors with high affinity[2]. Human basophils are specifically challenged by immunological stimuli such as allergens or anti-IgE antiserum that can bridge IgE molecules in membrane[3]. This process triggers transmembrane and intracellular signals followed by granule exocytosis with the release of histamine and loss of metachromatic staining of basophil granules by a basic dye such as toluidine blue. Optical basophil degranulation is well correlated with other *in vitro* and *in vivo* procedures for the diagnosis of allergy[4-7].

In preliminary experiments, degranulation of human basophils contained in leukocyte suspensions was induced not only by the usual concentration of anti-IgE antibody (1×

高度稀释的抗IgE抗血清引发人嗜碱性粒细胞脱颗粒

达弗纳等

编者按

本文为《自然》发表的最有争议的文章之一，事实上，《自然》此前和此后再没有发表类似的“科学论文”。法国国家健康与医学研究院的雅克·邦弗尼斯特和他的同事们声称他们发现即使在抗体稀释到其分子不存在于溶液中的情况下，仍然可以引发一类被称为嗜碱性粒细胞的白细胞作出免疫应答。他们称，这种活性随着逐步稀释呈周期性的反复出现。这似乎为使用超稀溶液的顺势疗法提供了支持。后来该研究从来没有被清晰地重复出来，但是它给人们带来一种观念，即水对已溶解物质有“记忆”。

当人的多形核嗜碱性粒细胞——一种表面表达 E 型免疫球蛋白（IgE）抗体的白细胞，遇到抗 IgE 抗体时，它们会释放细胞内颗粒中的组胺，并改变自身的染色特性。后者可以通过使用不同稀释倍数的抗 IgE 抗体（1×10^{2} ~ 1×10^{120} 倍稀释）来证实；在这个范围里，即使最大稀释度时已经无法测算抗 IgE 分子，也可以观察到 40% ~60%嗜碱性粒细胞脱颗粒的一系列峰。由于只有在伴随剧烈摇动的稀释中才能观察到这些现象，因此生物信息的传递可能与水的分子结构有关。

造成人速发型超敏反应的抗体属于 IgE 同种型[1]。IgE 最突出的特点是它能够通过高亲和力的受体结合到肥大细胞和嗜碱性多形核粒细胞的膜上[2]。在过敏原或抗 IgE 抗血清（可以与细胞膜上 IgE 分子结合）等免疫原的刺激下，人嗜碱性粒细胞的反应最为明显[3]。这一过程会触发跨膜和细胞内信号，随即引起颗粒胞吐及组胺释放并导致嗜碱性颗粒对甲苯胺蓝等碱性染料的异染性消失。视检嗜碱性粒细胞脱颗粒现象与其他体内和体外的过敏诊断程序之间具有很大的关联[4-7]。

在初步的实验中我们发现，不仅常规浓度的抗 IgE 抗血清（1×10^{3} 倍稀释抗 IgE 抗血清，相当于该法中有 2.2×10^{-9} M 抗 IgE 抗体）可以引起白细胞悬液中的人

10^3 dilution of anti-IgE antiserum, corresponding to 2.2×10^{-9} M anti-IgE antibody in the assay), but also by very low concentrations of this antibody ($2.2\times10^{-16/18}$ M), where the number of IgG anti-IgE molecules in the assay is supposedly too low to trigger the process. We then further explored this phenomenon.

Serial tenfold dilutions of goat anti-human IgE (Fc) antiserum (1 mg specific antibody per ml) were prepared in HEPES-buffered Tyrode's solution containing human serum albumin (HSA) down to 1×10^{60} dilution, corresponding to a 2.2×10^{-66} M theoretical concentration (th) in the assay (see Fig.1 legend for methods). The expected basophil degranulation, which was assessed by counting cells with metachromatical properties, was observed after exposure of leukocyte preparations to low antiserum dilutions with a maximum at $\sim1\times10^3$ dilution. Successive peaks of degranulation varying between 40 and 60% were then found down to 1×10^{60} dilution, with periods of 6 to 9 tenfold dilutions (Fig. 1*a*). In other experiments, the antiserum was serially diluted a hundred-fold down to 1×10^{120} (to give 2.2×10^{-126} M th in the assay) and similar results were obtained (Fig. 1*b*). Degranulation induced by high dilutions of anti-IgE antiserum was observed in ten experiments on the full range of dilutions down to 1×10^{60}, when at least 70 similar results were obtained at one or the other part of the high dilution scale in the participating laboratories (Toronto, preliminary results). As controls, goat anti-human IgG (Fc) antiserum (Fig.1*b*, n = 4) or Tyrode's solution containing HSA (n = 5) were diluted down to 1×10^{120} and 1×10^{30}, respectively. Cells incubated in conditions identical to those with anti-IgE anti-serum gave no significant degranulation. The repetitive waves of anti-IgE-induced degranulation were reproducible, but the peaks of degranulation could shift by one or two dilutions with every fresh sequential dilution of anti-IgE and depended on the blood sample. The waves of basophil degranulation were also seen with substances other than anti-IgE anti-serum at high and low dilutions, such as monoclonal anti-human IgE antibodies, specific antigen in allergic patients or in peroxidase-immunized rabbits, phospholipase A_2 from bee venom or porcine pancreas, the Na^+ ionophore monensin (up to 90% degranulation at 1×10^{-30} M th) and the Ca^{2+} ionophores A23187 and ionomycin (1×10^{-38} M th). The specificity of the observed effects at high dilutions (already noted when comparing antiserum against IgE with antiserum against IgG) was further strikingly illustrated in the ionophore experiments, because removing the corresponding ion from the cellular environment blunted basophil degranulation.

嗜碱性粒细胞脱颗粒，而且极低浓度的抗体（$2.2 \times 10^{-16/18}$ M）也可以引起这一过程，通常人们认为 IgG 抗 IgE 分子的数量已经太少，不足以引发此过程。于是我们对这一现象进行了深入研究。

我们用含有人血清白蛋白（HSA）的 HEPES 缓冲蒂罗德液对羊抗人 IgE（Fc 段）抗血清（每毫升含 1mg 特异性抗体）进行 10 倍梯度稀释，一直稀释到 1×10^{60} 倍，该法中理论浓度（th）相当于 2.2×10^{-66}M（方法参见图 1 注）。将低浓度的抗血清稀释液加入白细胞后可以发生预期的脱颗粒现象，通过对异染性细胞计数可以计算嗜碱性粒细胞的脱颗粒程度，观察发现不超过 1×10^{3} 倍稀释液能刺激产生脱颗粒。而随着抗血清稀释倍数逐渐增大到 10^{60}，脱颗粒的连续峰在 40%～60%之间波动，并且以每 10^{6}~10^{9} 倍稀释为一个周期出现一个峰值（图 1*a*）。在另一个实验中，我们采用 100 倍梯度稀释抗血清直到 1×10^{120} 倍（相当于 2.2×10^{-126} M th）也可以得到类似的结果（图 1*b*）。在稀释倍数降到 1×10^{60} 倍的全范围稀释的 10 次实验中观察到通过抗 IgE 抗血清引起的脱颗粒现象，同时在参与实验的其他实验室中（多伦多，初步结果）也观察到在高度稀释的范围中至少存在 70 个类似的实验结果。作为对照，我们将羊抗人 IgG（Fc 段）抗血清（图 1*b*，$n = 4$）或含 HSA 的蒂罗德液（$n = 5$）分别梯度稀释到 1×10^{120} 和 1×10^{30} 倍。发现培养的细胞在与抗 IgE 抗血清同样的条件下，并不能引发嗜碱性粒细胞显著的脱颗粒现象。由抗 IgE 引起的脱颗粒曲线的峰谷交错是可重复的，不过曲线的峰值可能会由于每次新制的抗 IgE 系列稀释液及血样的不同而偏移 1 或 2 个稀释度。除了高、低稀释度的抗 IgE 抗血清外，其他物质如单克隆抗人 IgE 抗体、过敏患者体内或过氧化物酶免疫过的兔体内的特异性抗原、蜜蜂毒液或猪胰腺中的磷脂酶 A_2、钠离子载体莫能菌素（1×10^{-30} M th 可引起超过 90%的细胞脱颗粒）以及钙离子载体 A23187 和离子霉素（1×10^{-38} M th）都可以引起这样的嗜碱性粒细胞脱颗粒曲线。因为细胞环境中相关离子的去除钝化了嗜碱性粒细胞脱颗粒，所以高倍稀释时其作用的特异性（在比较抗 IgE 抗血清和抗 IgG 抗血清时已注明）将在离子载体实验中做进一步深入阐明。

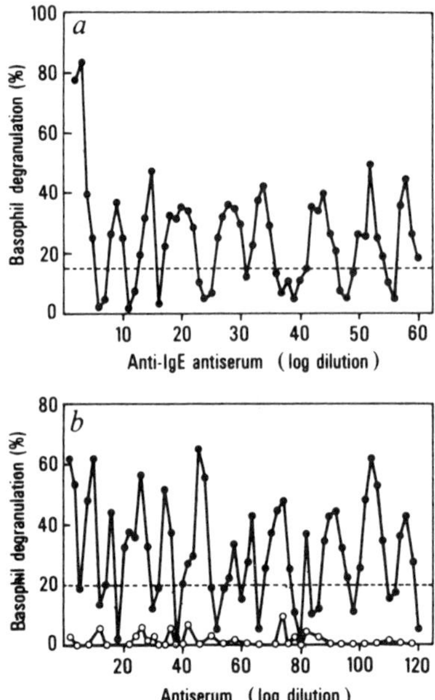

Fig. 1. Human basophil degranulation induced either by anti-IgE antiserum (●) diluted tenfold from 1×10^{2} down to 1×10^{60} (*a*) or hundredfold down to 1×10^{120} (*b*) or by anti-IgG antiserum (○) diluted hundredfold from 1×10^{2} down to 1×10^{120} (representatives of at least 10 experiments for anti-IgE and 4 experiments for anti-IgG). The significant ($P<0.05$) percentage of degranulation was 15% (*a*) and 20% (*b*). (....) relation to the number of counted basophils from control wells[15].

Methods. Goat anti-human IgE(Fc) antiserum or as a control, goat anti-human IgG (Fc) antiserum (Nordic Immunology, The Netherlands) was serially diluted as indicated above in HEPES-buffered Tyrode's solution (in g l^{-1}: NaCl, 8; KCl, 0.195; HEPES, 2.6; EDTA-Na_4, 1.040; glucose, 1 human serum albumin (HSA), 1.0; heparin, 5,000 U per 1; *p*H 7.4). Between each dilution, the solution was thoroughly mixed for 10 s using a Vortex. Given the molecular weight of IgG molecules (150,000), the 1×10^{60} and 1×10^{120} dilutions correspond in the assay to 2.2×10^{-66} M (th) and 2.2×10^{-126} M (th) respectively. Venous blood (20 ml) from healthy donors was collected using heparin (1 U per ml) and a mixture of 2.5 mM EDTA-Na_4/2.5 mM EDTA-Na_2(final concentrations) as anticoagulants and allowed to sediment. The leukocyte-rich plasma was recovered, twice washed by centrifugation (400*g*, 10 min) and finally resuspended in an aliquot of HEPES-buffered Tyrode's solution. The cell suspension (10 μl) was deposited on the bottom of each well of a microtitre plate containing 10 μl $CaCl_2$ (5 mM final) and 10 μl of either of anti-IgE or anti-IgG antiserum dilutions. To a control well were added 10 μl $CaCl_2$ and 10 μl Tyrode's but no anti-IgE or anti-IgG antiserum. Plates were then incubated at 37°C for 30 min. Staining solution (90 ml; 100 mg toluidine blue and 280 μl glacial acetic acid in 100 ml 25% ethanol, *p*H 3.2–3.4) was added to each well and the suspension thoroughly mixed. Specifically redstained basophils (non-degranulated basophils) were counted under a microscope using a Fuchs-Rosenthal haemocytometer. The percentage of basophil degranulation was calculated using the following formula: Basophil no. in control–basophil no. in sample/basophil no. in control × 100. Between 60 and 120 basophils were counted in cell suspensions from control wells after incubation either in the absence of anti-IgE antiserum, or in the presence of anti-IgG antiserum.

To confirm these surprising findings, four blind experiments were carried out (Table 1). In all cases the results were clear-cut, with typical bell-shaped degranulations at anti-IgE dilutions from 1×10^{32} to 1×10^{37}. The replicates were usually very close and of high significance (ANOVA test). In a fifth experiment, 7 control tubes and 3 tubes containing a dilution previously determined as active (1×10^{34}) were counted blind: basophil

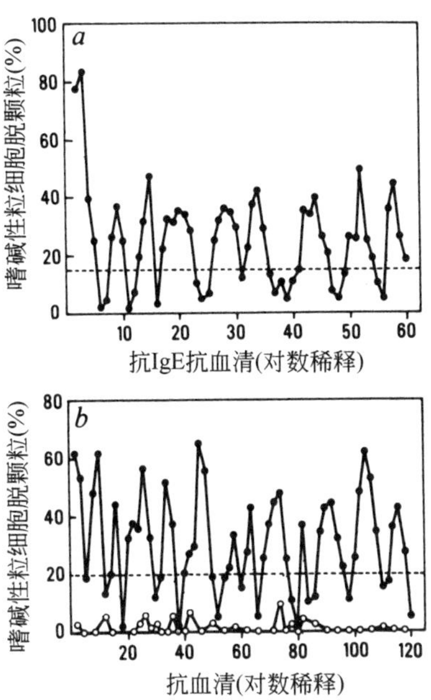

图 1. 抗 IgE 抗血清（●）和抗 IgG 抗血清（○）引起的人嗜碱性粒细胞脱颗粒，抗 IgE 抗血清从 1×10^2 倍稀释液十倍梯度稀释到 1×10^{60} 倍（*a*）或百倍梯度稀释到 1×10^{120}（*b*）；抗 IgG 抗血清从 1×10^2 倍稀释液百倍梯度稀释到 1×10^{120} 倍（图中的数据至少反映了使用抗 IgE 抗血清进行的 10 次实验和使用抗 IgG 抗血清进行的 4 次实验）。脱颗粒的显著（$P<0.05$）比例分别为 15%（*a*）和 20%（*b*）。（····）代表对照孔中嗜碱性粒细胞的计数 [15]。

方法。将羊抗人 IgE（Fc）抗血清或作为对照的羊抗人 IgG（Fc）抗血清（荷兰诺迪克免疫公司）分别用上文所述的 HEPES 缓冲蒂罗德液（含 8 g/l NaCl；0.195 g/l KCl；2.6 g/l HEPES；1.040 g/l EDTA–Na_4；1 g/l 葡萄糖 1 g/l 人血清白蛋白（HSA）；5,000 U/l 肝素；pH 7.4）梯度稀释。在每次稀释后，使用涡旋混合器振荡 10 秒以使溶液充分混匀。假设 IgG 分子的分子量为 150,000，那么 1×10^{60} 和 1×10^{120} 倍稀释液中分别为 2.2×10^{-66} M（th）和 2.2×10^{-126} M（th）。用肝素（1 U/ml）和 2.5 mM EDTA–Na_4/2.5 mM EDTA–Na_2 混合物（终浓度）作为抗凝血剂，从健康受试者身上采集 20 ml 静脉血，静置沉淀。两次离心收集富含白细胞的血浆（400*g*，10 分钟），最后用 HEPES 缓冲蒂罗德液重悬。在已加入 10 μl $CaCl_2$（终浓度为 5 mM）和 10 μl IgE 或 IgG 抗血清稀释液的微量滴定板的小孔底部，加入细胞悬液（10 μl）。对照孔中加入 10 μl $CaCl_2$ 和 10 μl 蒂罗德液，但不加抗 IgE 或抗 IgE 抗血清稀释液。将整个滴定板放置在 37℃孵育 30 分钟。向各孔中加入染色液（90 ml；将 100 mg 甲苯胺蓝和 280 μl 冰醋酸溶于 100 ml 25%乙醇，pH 3.2~3.4）并充分混匀。使用富 – 罗式血细胞计数器在显微镜下对特异性红染的嗜碱性粒细胞（未脱颗粒的嗜碱性粒细胞）计数。采用下列公式计算嗜碱性粒细胞脱颗粒的比例：（对照组中嗜碱性粒细胞的数量 – 实验组中嗜碱性粒细胞的数量）/ 对照组中嗜碱性粒细胞的数量 ×100。在对照孔中，不管是加入抗 IgG 抗血清的还是不加入抗 IgE 抗血清，都数到 60~120 个嗜碱性粒细胞。

为了进一步确认这些惊人的发现，我们又进行了 4 组双盲实验（表 1）。所有情况下结果都是明确的，在 1×10^{32} 倍到 1×10^{37} 倍的抗 IgE 稀释液间脱颗粒曲线均为典型的钟形曲线。重复实验的结果也非常相近，并具有很高的显著性（方差分析）。在第 5 次实验中，7 个对照样品和 3 个事先验证过具有活性的（1×10^{34} 倍

degranulation was 7.7±1.4% for the controls, and 44.8, 42.8 and 45.7% for the tubes containing diluted anti-IgE. The random chance in all these experiments was 2% and therefore the cumulative results statistically confirm the measured effect.

Table 1. Basophil counts after exposure to anti-IgE antiserum at low and high dilutions

Samples	Experiment 1	Experiment 2	Experiment 3	Experiment 4
Tyrode's-HSA*	81.3±1.2†	89.0±3.1	81.7±2.2	106.7±1.8
Tyrode's-HSA	81.6±1.4	87.7±1.4	83.0±1.0	105.0±1.2
Tyrode's-HSA	80.0±1.5	88.0±2.3	81.7±1.8	105.7±0.9
aIgE 1×10^3*	35.5±1.8(56)‡	42.3±4.8(53)	27.7±0.7(66)	40.0±1.5(62)
aIgE 2×10^{32}	77.6±0.8(4)	87.3±1.2(3)	66.3±2.3(18)	93.7±1.9(12)
aIgE 1×10^{33}	76.0±1.1(6)	88.7±1.8(1)	77.7±1.8(4)	74.7±2.8(30)
aIgE 1×10^{34}	53.6±1.4(33)	52.7±1.4(41)	38.0±0.6(53)	48.3±2.4(55)
aIgE 1×10^{35}	45.0±0.5(44)	35.0±1.0(61)	41.3±1.8(49)	49.3±1.2(54)
aIgE 1×10^{36}	49.0±1.7(40)	50.3±0.7(44)	55.0±2.1(32)	74.3±2.3(31)
aIgE 1×10^{37}	79.0±2.3(2)	85.3±0.7(5)	73.3±1.7(10)	105.3±0.7(0)

Blind experiments: test tubes were randomly coded twice by two independent pairs of observers and assayed. The codes were simultaneously broken at the end of all experiments. Dilutions of anti-IgE antiserum were performed as described in legend to Fig. 1.

* Uncoded additional tubes for negative (Tyrode's-HSA) or positive (aIgE 1×10^{-3}) controls.

† Data represent the mean ± s.e. of basophil number actually counted in triplicate (see legend to Fig. 1 for methods).

‡ Number in parenthesis indicates percentage degranulation compared with Tyrode's-HSA.

Two further blind experiments were performed using the usual dilution procedure: of the 12 tubes used in the first experiment (Table 2), 2 tubes contained goat anti-human antiserum IgE at 1×10^2 and 1×10^3 dilutions, 6 tubes contained dilutions from 1×10^{32} to 1×10^{37}, and 4 tubes buffer-HSA alone. The tubes were then randomly coded twice by three parties, one of which kept the two codes. The 12 tubes were each divided into 4. Three batches of 12 tubes were lyophilized, one of which was used for gel electrophoresis, one for assay of monoclonal antibodies, and the last (with the unlyophilized sample) for gel electrophoresis and basophil degranulation. By comparing the results of the different tests it was easy to identify the tubes containing IgE at normal concentrations compared with the tubes containing highly diluted IgE and the control tubes. When the codes were broken, the actual results exactly fitted those predicted, but HSA and its aggregates were present in all solutions and complicated interpretation of the gel electrophoresis. So we performed another almost identical experiment, using 6 tubes containing unlyophilized samples and buffer without HSA. Four tubes contained antibody at 1×10^2, 1×10^3, 1×10^{35} and 1×10^{36} dilutions, and 2 contained buffer alone. These tubes were coded and assayed according to the above protocol. The decoded results were clear-cut, high basophil degranulation being obtained with 1×10^2, 10^3, 10^{35} and 10^{36} dilutions, but no anti-IgE activity or immunoglobulins were detected either in the control tubes or in assays containing the 1×10^{35} and 10^{36} dilutions (Tables 2 and 3 and Fig.2). Thus there is no doubt that there was basophil degranulation in the absence of any detectable anti-IgE molecule.

稀释)样品采取了双盲检测：对照组嗜碱性粒细胞脱颗粒的百分比为 7.7 ± 1.4%，而含有稀释抗 IgE 的样品分别为 44.8%、42.8%和 45.7%。在所有这些实验中，随机因素的影响为 2%，因此这些结果再次从统计学上证实了上述检测结果。

表 1. 不同稀释度抗 IgE 抗血清刺激后的嗜碱性粒细胞计数

样品	实验一	实验二	实验三	实验四
蒂罗德液–HSA*	81.3 ± 1.2†	89.0 ± 3.1	81.7 ± 2.2	106.7 ± 1.8
蒂罗德液–HSA	81.6 ± 1.4	87.7 ± 1.4	83.0 ± 1.0	105.0 ± 1.2
蒂罗德液–HSA	80.0 ± 1.5	88.0 ± 2.3	81.7 ± 1.8	105.7 ± 0.9
抗IgE 1×10^{3}*	35.5 ± 1.8(56)‡	42.3 ± 4.8(53)	27.7 ± 0.7(66)	40.0 ± 1.5(62)
抗IgE 2×10^{32}	77.6 ± 0.8(4)	87.3 ± 1.2(3)	66.3 ± 2.3(18)	93.7 ± 1.9(12)
抗IgE 1×10^{33}	76.0 ± 1.1(6)	88.7 ± 1.8(1)	77.7 ± 1.8(4)	74.7 ± 2.8(30)
抗IgE 1×10^{34}	53.6 ± 1.4(33)	52.7 ± 1.4(41)	38.0 ± 0.6(53)	48.3 ± 2.4(55)
抗IgE 1×10^{35}	45.0 ± 0.5(44)	35.0 ± 1.0(61)	41.3 ± 1.8(49)	49.3 ± 1.2(54)
抗IgE 1×10^{36}	49.0 ± 1.7(40)	50.3 ± 0.7(44)	55.0 ± 2.1(32)	74.3 ± 1.2(31)
抗IgE 1×10^{37}	79.0 ± 2.3(2)	85.3 ± 0.7(5)	73.3 ± 1.7(10)	105.3 ± 0.7(0)

双盲实验：待测样品分别由两组独立的观察者进行两次随机编号和实验，在全部实验结束后，同时公布所有的编号。抗 IgE 抗血清的稀释方法参见图 1 注。

* 未编号的额外样品，分别为阴性对照（蒂罗德液 – HSA）和阳性（aIgE 1×10^{-3}）对照。

† 数据为 3 次重复实验对嗜碱性粒细胞计数的平均值 ±s.e.（方法参见图 1 注）。

‡ 括号中的数据为与蒂罗德液 –HSA 处理组相比脱颗粒细胞的百分比。

我们进一步使用普通的稀释程序进行了两次双盲实验：在第 1 次实验的 12 个样品中（表 2），2 个分别为羊抗人 IgE 抗血清的 1×10^{2} 倍和 1×10^{3} 倍稀释液、6 个为从 1×10^{32} 倍到 1×10^{37} 倍稀释液、4 个为只含 HSA 的缓冲液。这些样品被 3 组人员随机编号 2 次，其中一组保存 2 个编号。然后，把这 12 个样品分成 4 份，3 份冻干后，1 份用于凝胶电泳实验，1 份用于单克隆抗体实验，最后 1 份（和未冻干样品）用于凝胶电泳与嗜碱性粒细胞脱颗粒实验。通过比较这些不同实验的结果，我们很容易鉴定出正常浓度和高倍稀释的 IgE 抗血清以及对照。最后我们公布样品编号，发现实验结果与预期完全相符，不过由于 HSA 和其聚合物在所有样品中都存在，导致电泳实验的结果解释起来比较复杂。因此我们又进行了另外一个几乎完全相同的实验，用 6 个含未冻干样品和缓冲液但不含人血清白蛋白的样品进行实验。其中 4 个样品分别为抗体的 1×10^{2}、1×10^{3}、1×10^{35} 和 1×10^{36} 倍稀释液，2 个只含缓冲液。我们采用与上述实验相同的方法对这些样品进行了双盲编号和分析。结果是明确的，即 1×10^{2}、1×10^{3}、1×10^{35} 和 1×10^{36} 倍稀释液均能引发嗜碱性粒细胞高度脱颗粒，但在对照组以及 1×10^{35} 和 1×10^{36} 倍稀释液处理组中检测不到任何抗 IgE 活性和免疫球蛋白（表 2、表 3 和图 2）。因此，毫无疑问地，即使抗 IgE 抗体的浓度低到无法检测，也可以引发嗜碱性粒细胞脱颗粒。

Table 2. Comparison of basophil degranulation with the presence of immunoglobulins and anti-IgE activity in dilutions performed in HSA-containing Tyrode's

Samples	Basophil degranulation (%)*			Gel electrophoresis†		Anti-IgE activity
	I	II	III	*A*	*B*	μ ml^{-1}
Tyrode's-HSA	0	0	0	–	–	$<1\times10^{-3}$
Tyrode's-HSA	0	0	0	–	–	$<1\times10^{-3}$
Tyrode's-HSA	0	0	0	–	–	$<1\times10^{-3}$
Tyrode's-HSA	0	0	0	–	–	$<1\times10^{-3}$
aIgE 1×10^{-2}‡	53	50	33	++§	++	ND
aIgE 1×10^{-2}	51	44	37	++	++	10.6
aIgE 1×10^{-3}	65	38	45	+?	–	1.1
aIgE 1×10^{-32}	7	26	22	–	–	$<1\times10^{-3}$
aIgE 1×10^{-33}	37	0	13	–	–	$<1\times10^{-3}$
aIgE 1×10^{-34}	45	37	20	–	–	$<1\times10^{-3}$
aIgE 1×10^{-35}	39	41	34	–	–	$<1\times10^{-3}$
aIgE 1×10^{-36}	31	29	39	–	–	$<1\times10^{-3}$
aIgE 1×10^{-37}	23	12	29	–	–	$<1\times10^{-3}$

Blind experiments and dilution protocols as in Table 1. –, Lack of strained bands. ND, not determined. A faint band corresponding to IgG appeared after reduction by 2-mercaptoethanol.

* Basophil degranulation tests I, II, III were performed using 3 different blood samples (see Fig. 1). Percentage basophil degranulation induced by aIgE, as compared to Tyrode's HSA, was calculated from duplicates.

† Electrophoresis (polyacrylamide 7–15%, revealed by silver staining) was carried out in Rehovot (*A*) and at INSERM U 200 (*B*).

‡ Uncoded additional tube for positive control.

§ ++,+ Bands correspond to IgG present in large or small amounts.

Table 3. Comparison of basophil degranulation with the presence of immunoglobulins and anti-IgE activity in dilutions performed in Tyrode's without HSA.

Samples	Basophil degranulation (%)		Gel electrophoresis		Anti-IgE activity
	I	II	*A*	*B*	(μ ml^{-1})
Tyrode's	0	0	–	–	$<1\times10^{-3}$
Tyrode's	0	0	–	–	$<1\times10^{-3}$
aIgE 1×10^{-2}*	85	48	++	++	ND
aIgE 1×10^{-2}	81	47	++	++	32.6
aIgE 1×10^{-3}*	ND	ND	+	+	ND
aIgE 1×10^{-3}	75	53	+	+	ND
aIgE 1×10^{-35}	35	31	–	–	$<1\times10^{-3}$
aIgE 1×10^{-36}	40	35	–	–	$<1\times10^{-3}$

* Uncoded tubes for positive control of basophil degranulation and/or gel electrophoresis.

ND, not determined.

表 2. 在含 HSA 的蒂罗德液中比较免疫球蛋白引起的嗜碱性粒细胞脱颗粒与稀释液的抗 IgE 活性

样品	嗜碱性粒细胞脱颗粒(%)*			凝胶电泳†		抗IgE活性
	Ⅰ	Ⅱ	Ⅲ	*A*	*B*	μ ml^{-1}
蒂罗德液–HSA	0	0	0	–	–	<1 × 10^{-3}
蒂罗德液–HSA	0	0	0	–	–	<1 × 10^{-3}
蒂罗德液–HSA	0	0	0	–	–	<1 × 10^{-3}
蒂罗德液–HSA	0	0	0	–	–	<1 × 10^{-3}
抗IgE 1 × 10^{-2}‡	53	50	33	++ §	++	ND
抗IgE 1 × 10^{-2}	51	44	37	++	++	10.6
抗IgE 1 × 10^{-3}	65	38	45	+?	–	1.1
抗IgE 1 × 10^{-32}	7	26	22	–	–	<1 × 10^{-3}
抗IgE 1 × 10^{-33}	37	0	13	–	–	<1 × 10^{-3}
抗IgE 1 × 10^{-34}	45	37	20	–	–	<1 × 10^{-3}
抗IgE 1 × 10^{-35}	39	41	34	–	–	<1 × 10^{-3}
抗IgE 1 × 10^{-36}	31	29	39	–	–	<1 × 10^{-3}
抗IgE 1 × 10^{-37}	23	12	29	–	–	<1 × 10^{-3}

双盲实验和稀释方法与表 1 同。 – ，表示无染色带。ND，表示未检出。使用 2– 巯基乙醇还原后可以看见一条模糊的 IgG 的带。

* 嗜碱性粒细胞脱颗粒实验 Ⅰ 、Ⅱ 、Ⅲ分别是采用 3 个不同的血样进行的（见图 1）。与蒂罗德液 – HSA 处理相比，抗 IgE 引起的嗜碱性粒细胞脱颗粒的百分比是两次计算的平均值。

† 凝胶电泳实验（7%~15% 聚丙烯酰胺，银染）分别在雷霍沃特（*A*）和法国国家健康与医学研究院 U 200（*B*）实验室进行。

‡ 阳性对照，未进行双盲编码。

§ ++、+ 代表 IgG 条带含量的多少。

表 3. 在不含 HSA 的蒂罗德液中比较免疫球蛋白引起的嗜碱性粒细胞脱颗粒与稀释液的抗 IgE 活性

样品	嗜碱性粒细胞脱颗粒(%)		凝胶电泳		抗IgE活性
	Ⅰ	Ⅱ	*A*	*B*	(μ ml^{-1})
蒂罗德液	0	0	–	–	<1 × 10^{-3}
蒂罗德液	0	0	–	–	<1 × 10^{-3}
抗IgE 1 × 10^{-2}*	85	48	++	++	ND
抗IgE 1 × 10^{-2}	81	47	++	++	32.6
抗IgE 1 × 10^{-3}*	ND	ND	+	+	ND
抗IgE 1 × 10^{-3}	75	53	+	+	ND
抗IgE 1 × 10^{-35}	35	31	–	–	<1 × 10^{-3}
抗IgE 1 × 10^{-36}	40	35	–	–	<1 × 10^{-3}

* 未进行双盲编码的阳性对照的嗜碱性粒细胞脱颗粒和（或）凝胶电泳实验。

ND，表示未检出。

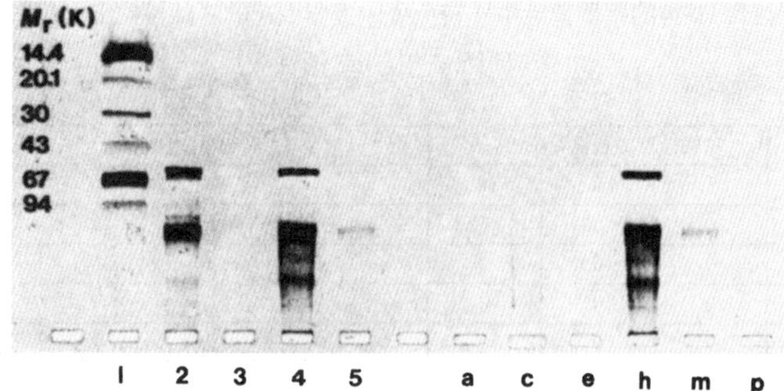

Fig. 2. Electrophoresis (polyacrylamide 7–15%, bands revealed by silver staining): samples numbered 1 to 5 are standards for the blind experiments *a, c, e, h, m, p*. Lane 1, Molecular weight standards for electrophoresis; lane 2, monoclonal IgG added with human serum albumin; lane 3, Tyrode's buffer without human serum albumin; lane 4, 1×10^{2} anti-IgE dilution; lane 5, 1×10^{3} dilution. Samples tested blind: *a* and *c*, buffer; *e*, 1×10^{36} anti-IgE dilution; *h*, 1×10^{2} anti-IgE dilution; *m*, 1×10^{3} anti-IgE dilution; *p*, 1×10^{35} anti-IgE dilution.

These results may be related to the recent double-blind clinical study of Reilly *et al.*[8] which showed a significant reduction of symptoms in hay-fever patients treated with a high dilution (1×10^{60}) of grass pollen versus placebo, and to our *ex vivo* experiments in the mouse[9]. We have extended these experiments to other biological systems: using the fluorescent probe fura-2, we recently demonstrated changes in intracellular Ca^{2+} levels in human platelets in the presence of the Ca^{2+} ionophore ionomycin diluted down to 1×10^{-39} M th (F.B. *et al.*, unpublished results).

Using the molecular weight of immunoglobulins and Avogadro's number, we calculate that less than one molecule of antibody is present in the assay when anti-IgE antiserum is diluted to 1×10^{14} (corresponding to 2.2×10^{-20} M). But in the experiments reported here we have detected significant basophil degranulation down to the 1×10^{120} dilution. Specific effects have also been triggered by highly diluted agents in other *in vitro* and *in vivo* biological systems[8-11], but still remain unexplained. The valid use of Avogadro's number could be questioned, but we are dealing with dilutions far below the Avogadro limit (1×10^{100} and below). It could be argued that our serial dilution procedure is subject to experimental error, but this is ruled out because: (1) pipette tips and glass micropipettes were discarded between each dilution (performed under laminar flow hood). (2) The c.p.m. in tubes containing serially diluted radioactive compounds decreased in proportion to the degree of dilution down to the background (data not shown). (3) Contamination would not explain the successive peaks of activity that evoke a periodic phenomenon and not a monotonous dose–effect curve, as usually observed when concentration of an agonist decreases. (4) To eliminate the possibility of contaminating molecules present in the highly diluted solutions, we carried out two series of experiments which can be summarized as follows. An Amicon membrane with molecular weight cut-off 10K retained the basophil degranulating IgG (150K) present at low dilutions (1×10^{2}, 1×10^{3}) in anti-IgE antiserum. By contrast, the activity present at high dilutions (1×10^{27}, 1×10^{32}) was totally recovered in the 10K Amicon filtrate. Anion or cation exchange chromatography, according to the

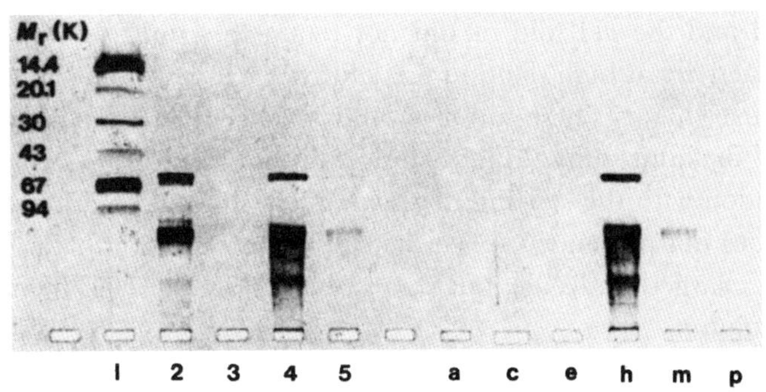

图 2. 凝胶电泳（7% ~15%聚丙烯酰胺，银染）：1~5 为双盲实验中 *a*、*c*、*e*、*h*、*m*、*p* 的比对标准。1 道为凝胶电泳的标准分子量；2 道为加了人血清白蛋白的单克隆抗体 IgG；3 道为不含人血清白蛋白的蒂罗德液；4 道为抗 IgE 的 1×10^{2} 倍稀释液；5 道为 1×10^{3} 倍稀释液。双盲测定样品：*a* 和 *c* 是蒂罗德液；*e* 为抗 IgE 的 1×10^{36} 倍稀释液；*h* 为抗 IgE 的 1×10^{2} 倍稀释液；*m* 为抗 IgE 的 1×10^{3} 倍稀释液；*p* 为抗 IgE 的 1×10^{35} 倍稀释液。

这些实验结果可能与赖利等人最近进行的临床双盲研究[8]存在一定关系，他发现与安慰剂相比，使用高倍稀释（1×10^{60}）的草花粉治疗枯草热患者可以显著地缓解症状，与我们进行的小鼠离体实验的结果[9]也相关。我们在其他生物学系统中进一步扩展了这些实验：使用 fura-2 作为荧光探针，我们证明了稀释到 1×10^{-39} M 的钙离子载体离子霉素可以引起人血小板中细胞内钙离子水平的改变（博韦等，未发表的结果）。

利用免疫球蛋白的分子量和阿伏伽德罗常量，我们可以计算出当抗 IgE 抗血清稀释 1×10^{14} 倍时（相当于 2.2×10^{-20} M），在前面所述的每个实验中平均只含有不到 1 个抗体分子。但是，我们的结果却表明即使稀释 1×10^{120} 倍，抗血清也可以引发显著的嗜碱性粒细胞脱颗粒。在其他体内、体外的生物系统中，人们也观察到了高倍稀释的试剂具有一些特殊的效应[8-11]，但尚不清楚其中的奥妙所在。也许在此处使用阿伏伽德罗常量是否合适尚存疑问，不过我们的稀释倍数已经远远低于阿伏伽德罗界限（1×10^{100} 甚至更低）。也许有人会怀疑我们的系列稀释方法会受到实验误差的影响，不过这个可能性已经被排除了，因为：(1) 稀释用的移液器吸头和玻璃微量移液器都是一次性的（在超净台中进行）；(2) 在含有放射性化合物的样品中，其放射性强度随着稀释过程相应成比例降低，直到与背景值一样（结果未显示）；(3) 杂质污染无法解释脱颗粒曲线呈现周期性，而不是单一的剂量－效应曲线（通常随激动剂浓度的降低而改变）；(4) 为了排除在高倍稀释的样品中含有杂质分子的可能，我们进行了下述两个系列的实验。使用截留分子量为 10K 的 Amicon 膜可以将抗 IgE 抗血清中的 IgG（分子量 150K）保留下来，从而抑制低倍数稀释液（1×10^{2}、1×10^{3}）引发嗜碱性细胞脱颗粒的能力。相反，对于高稀释倍数液（1×10^{27}、1×10^{32}）来说，使用分子量为 10K 的 Amicon 膜过滤处理并不能抑制该抗

type of resin used and the *p*H, did or did not retain the anti-IgE IgG at low dilutions, whereas the same activity at high dilution was always excluded from the columns and fully recovered in the first eluate. These filtration and ionexchange experiments demonstrated that the activity of the antiserum at high dilution cannot result from contamination of the highly diluted solution with the starting material. They showed, in addition, that the high-dilution activity does not present in space the steric conformation of an IgG molecule as it acts like a 150K charged molecule, but is not retained by the 10K filter or by a charged chromatography column.

We then investigated the physical chemical nature of the entity active at high dilution. Our results can be summarized as follows. (1) The importance of agitation in the transmission of information was explored by pipetting dilutions up and down ten times and comparing with the usual 10-s vortexing. Although the two processes resulted in the same dilution (degranulations at 1×10^2 and 1×10^3 were superimposable whatever the dilution process), degranulation did not occur at high dilution after pipetting. Ten-second vortexing was the minimum time required, but vortexing for longer (30 or 60 s) did not increase high-dilution activity. So transmission of the information depended on vigorous agitation, possibly inducing a submolecular organization of water or closely related liquids. (2) The latter is possible as ethanol and propanol could also support the phenomenon. In contrast, dilutions in dimethylsulphoxide did not transmit the information from one dilution to the other, but increasing the proportion of water in dimethylsulphoxide resulted in the appearance and increment of the activity at high dilutions. (3) Heating, freeze-thawing or ultrasonication suppressed the activity of highly diluted solutions, but not the activity of several active compounds at high concentrations. A striking feature was that molecules reacted to heat according to their distinctive heat sensitivity, whereas all highly diluted solutions ceased to be active between 70 and 80 °C. This result suggests a common mechanism operating at high dilution, independent of the nature of the starting molecule.

Therefore we propose that none of the starting molecules is present in the dilutions beyond the Avogadro limit and that specific information must have been transmitted during the dilution/shaking process. Water could act as a "template" for the molecule, for example by an infinite hydrogen-bonded network[12], or electric and magnetic fields[13,14]. At present we can only speculate on the nature of the specific activity present in the highly diluted solutions. We can affirm that (1) this activity was established under stringent experimental conditions, such as blind double-coded procedures involving six laboratories from four countries; (2) it is specific for the ligand first introduced, as illustrated when goat antiserum (IgG) anti-human IgE, but not goat IgG anti-human IgG supported this phenomenon. The link between high and low anti-IgE dilutions is shown as we could not detect basophil degranulation at high dilutions if it did not occur within the classical range. High dilutions of histamine, but not of its carboxylated precursor histidine, inhibited IgE-dependent basophil degranulation. Finally, ionophores at high dilution did not work when the specific ion was removed from the cell suspension (F.B., unpublished results). (3) Using six biochemical and physical probes, we demonstrated that what supports the activity at high dilutions is not a molecule. (4) Whatever its nature, it is capable of "reproducing" subtle

血清引发脱颗粒的能力。此外，不论根据树脂型号和 pH 值进行的阴离子或阳离子交换层析能或不能保留低倍稀释液中的抗 IgE 抗血清中的 IgG，但高倍稀释液在柱层析之前及其第一次洗脱之后却始终具有相同的脱颗粒活性。这些过滤和离子交换实验的结果证明高倍稀释液的血清的活性不可能是由于高倍稀释的原始样品被污染而造成的。此外，高倍稀释液的活性也与 IgG 分子空间构象的改变无关，因为它虽然表现得像 150K 带电分子，却不会被截流分子量为 10K 的滤膜和离子交换柱拦截。

我们进一步对高倍稀释液中活性实体物质的物理化学性质进行了分析。其结果如下：(1) 我们分别对样品采取了抽吸 10 次或涡旋振荡 10 秒的处理，以分析搅动方式对于信息传递的重要性。尽管这两种方法都能达到相同的稀释度（两种处理方法得到的 1×10^2 倍和 1×10^3 倍稀释液分别发生相同的脱颗粒现象），但枪头抽吸处理得到的高倍稀释液却不能产生脱颗粒现象。 涡旋振荡 10 秒是混匀所需的最少时间,不过进一步延长涡旋的时间(30 秒或 60 秒)并不能增强高倍稀释液脱颗粒的活性。据此我们认为某种信息的传递依赖于剧烈摇晃，这种摇晃可能诱导水或密切相关的液体中亚分子的组装；(2) 相关液体可以为乙醇和丙醇，以它们为溶剂可以支持这一现象。相反，二甲基亚砜则不能在稀释过程中传递信息，但逐渐增加二甲基亚砜中水的含量可以使高倍稀释液的活性出现并逐渐增加；(3) 加热、冻融或超声都会抑制高倍稀释液的活性，但不会影响几种高浓度活性化合物的活性。令人惊奇的是，基于不同的热敏感性，分子可对热作出反应，而高倍稀释的样品在 70℃ ~ 80℃活性逐渐消失。这表明在高倍稀释液中存在一种共同的机制，其与初始样品的性质无关。

据此，我们认为当稀释倍数超出阿伏伽德罗界限后，已经没有初始的分子残留在溶液中，而在稀释 – 摇晃过程中则一定有某种特定信息在传递。水分子在这个过程中可能起到分子“模板”作用，比如它可以形成一个无限的氢键网络 [12] 或电场和磁场 [13,14]。目前我们只能基于这种高倍稀释液的特殊活性进行推断。我们确认：(1) 这种活性是在严格的实验条件下被发现的，例如由 4 个国家的 6 个实验室参与的双盲实验；(2) 正如此前证明的只有羊抗血清（IgG）抗人 IgE 而不是羊抗血清（IgG）抗人 IgG 才能够引起这种现象，说明这种活性对首次引入的配体是特异性的。如果传统浓度范围内不能引发嗜碱性粒细胞脱颗粒，那么高倍稀释液也不行，所以高低倍抗 IgE 稀释液的活性是存在联系的。高倍稀释的组胺可以抑制 IgE 依赖的嗜碱性粒细胞脱颗粒，而其羧基化的前体组氨酸则不行。最后，当细胞悬液中的特定离子被去除后，高倍稀释液的离子载体将不能发挥作用（博韦，未发表的结果）；(3) 我们使用了 6 种生物化学的和物理的探针证明高倍稀释液的活性并不是由某种

molecular variations, such as the rearrangement of the variable region of an IgG (anti-ϵ versus anti-γ) molecule.

The precise nature of this phenomenon remains unexplained. It was critical that we should first establish the reality of biological effects in the physical absence of molecules. The entities supporting this "metamolecular" biology can only be explored by physical investigation of agitation causing interaction between the original molecules and water, thus yielding activity capable of specifically imitating the native molecules, though any such hypothesis is unsubstantiated at present.

We thank Professor Z. Bentwich from Ruth Ben Ari Institute for supervision of experiments conducted in Rehovot. The participation of J. Geen (Univ. Toronto), B. Descours and C. Hieblot (INSERM U 200) in experiments and of V. Besso in editing is gratefully acknowledged. This work is dedicated to the late Michel Aubin, who played a decisive role in initiating it.

Editorial Reservation

Readers of this article may share the incredulity of the many referees who have commented on several versions of it during the past several months. The essence of the result is that an aqueous solution of an antibody retains its ability to evoke a biological response even when diluted to such an extent that there is a negligible chance of there being a single molecule in any sample. There is no physical basis for such an activity. With the kind collaboration of Professor Benveniste, *Nature* has therefore arranged for independent investigators to observe repetitions of the experiments. A report of this investigation will appear shortly.

(**333**, 816-818; 1988)

E. Davenas, F. Beauvais, J. Amara*, M. Oberbaum*, B. Robinzon†, A. Miadonna‡, A.Tedeschi‡, B. Pomeranz§, P. Fortner§, P. Belon, J. Sainte-Laudy, B. Poitevin & J. Benveniste||

INSERM U 200, Université Paris-Sud, 32 rue des Carnets, 92140 Clamart, France

* Ruth Ben Ari Institute of Clinical Immunology, Kaplan Hospital, Rehovot 76100, Israel

† Department of Animal Sciences, Faculty of Agriculture, PO Box 12, The Hebrew University of Jerusalem, Rehovot 76100, Israel

‡ Department of Internal Medicine, Infectious Diseases and Immunopathology, University of Milano, Ospedale Maggiore Policlinico, Milano, Italy

§ Departments of Zoology and Physiology, Ramsay Wright Zoological Laboratories, University of Toronto, 25 Harbord Street, Toronto, Ontario M5S 1A1, Canada

|| To whom correspondence should be addressed

Received 24 August 1987; accepted 13 June 1988.

References:

1. Ishizaka, K., Ishizaka, T. & Hornbrook M. M. *J. Immun.* **97**, 75-85 (1966).

2. Metzger, H. *et al. A. Rev. Immun.* **4**, 419-470 (1986).

3. Ishizaka, T., Ishizaka, K., Conrad, D. H. & Froese, A. *J. Allergy clin. Immun.* **61**, 320-330 (1978).

分子来实现的；(4) 不管它的本质是什么，它具有对分子进行“再造”引入一些精细改变的能力，例如对 IgG（抗 ε 不抗 γ）分子可变区的重排。

上述现象的精确本质仍有待探索。重要的是，我们在物理上尚未发现分子存在的情况下，首先确定了这种生物功能的真实存在。支持这种“超分子”生物学的本质，只能通过物理上的探索，研究晃动引起原始分子（如抗 IgE，编者注。）和水产生的相互作用，是否使液体特异的模仿本来分子的能力而产生活性，不过对于这些假设目前尚无法证实。

我们感谢露丝本阿里学院的本特威奇教授对在雷霍沃特进行的实验的指导。衷心感谢多伦多大学的吉恩、法国国家健康与医学研究院 U 200 的德库尔和耶布洛参与实验，感谢贝索的编辑工作。把这个工作献给米歇尔·奥宾先生，他在这些工作的启动中发挥了决定性的作用。

编辑的保留

本文的读者可能会分享在过去的几个月中对几种版本做出评论的评论者所做出的多种质疑。本文结果的实质为，即使把抗体稀释到样品中不可能含有一个分子的浓度时，其水溶液仍然保持了激发生物反应的能力。这种活性不具备物理基础。在与邦弗尼斯特教授友好沟通之后，《自然》已经安排独立的调查员前去观察实验重复。调查报告将于近期公布。

（张锦彬 翻译；胡卓伟 审稿）

4. Benveniste, J. *Clin. Allergy* **11**, 1-11 (1981).

5. Camussi, G., Tetta, C., Coda, R. & Benveniste, J. *Lab. Invest.* **44**, 241-251 (1981).

6. Pirotzky, E. *et al. Lancet* **i**, 358-361 (1982).

7. Yeung-Laiwah, A. C., Patel, K. R., Seenan, A. K., Galloway, E. & McCulloch, W. *Clin. Allergy* **14**. 571-579 (1984).

8. Reilly, D. T., Taylor, M. A., McSharry, C. & Aitchison, T. *Lancet* **ii**, 881-886 (1986).

9. Davenas, E., Poitevin, B. & Benveniste, J. *Eur. J. Pharmac.* **135**, 313-319 (1987).

10. Bastide, M., Doucet-Jacoeuf, M. & Daurat, V. *Immun. Today* **6**, 234 (1985).

11. Poitevin, B., Davenas, E. & Benveniste, J. *Br. J. clin. Pharmac.* **25**, 439-444 (1988).

12. Stanley, H. E., Teixeira, J., Geiger, A. & Blumberg, R. L. *Physics* **106A**, 260-277 (1981).

13. Fröhlich, H. *Adv. Electron. & Electron. Phys*, **53**, 85-152 (1980).

14. Smith, C.W. & Aarholt, E. *Hlth Phys.* **43**, 929-930 (1988).

15. Petiot, J. F., Sainte-Laudy, J. & Benveniste, J. *Ann. Biol. Clin.* **39**, 355-359 (1981).

"High-dilution" Experiments a Delusion

John Maddox *et al.*

Editor's Note

The controversy stemming from the publication in *Nature* of a paper alleging biological activity in solutions of biomolecules at "homeopathic" high dilutions was only heightened when this investigation of the research appeared a month later. *Nature*'s editor John Maddox went with "fraud debunkers" James Randi (a professional magician) and Walter Stewart to the laboratory of French immunologist Jacques Benveniste, who had made the claims, and asked him to repeat the experiments. The French scientists were unable to reproduce their results, and Maddox and colleagues made various accusations of poor technique (but not fraud). Benveniste denounced the investigation as a stunt, and continued to work on high-dilution effects until his death in 2004. But his claims were dismissed by other scientists.

The now celebrated report by Dr J. Benveniste and colleagues elsewhere is found, by a visiting *Nature* team, to be an insubstantial basis for the claims made for them.

THE remarkable claims made in *Nature* (**333**, 816; 1988) by Dr Jacques Benveniste and his associates are based chiefly on an extensive series of experiments which are statistically ill-controlled, from which no substantial effort has been made to exclude systematic error, including observer bias, and whose interpretation has been clouded by the exclusion of measurements in conflict with the claim that anti-IgE at "high dilution" will degranulate basophils. The phenomenon described is not reproducible in the ordinary meaning of that word.

We conclude that there is no substantial basis for the claim that anti-IgE at high dilution (by factors as great as 10^{120}) retains its biological effectiveness, and that the hypothesis that water can be imprinted with the memory of past solutes is as unnecessary as it is fanciful.

We use the term "high dilution" reluctantly; these solutions contain no molecules of anti-IgE, and so are not solutions in the ordinary sense. "Solute-free solution" would similarly be illogical.

Our conclusion is based on a week-long visit to Dr Benveniste's laboratory, the INSERM unit for immunopharmacology and allergy (otherwise INSERM 200) at Clamart, in the western suburbs of Paris, during the week beginning 4 July. Among other things, we were

高倍稀释实验的错觉

约翰·马多克斯等

编者按

发表在《自然》杂志上的一篇论文宣称"顺势疗法"似的超高倍稀释的生物分子溶液仍有生物学活性。这一论文引起的争议被一个月后的研究调查激化。《自然》杂志的编辑约翰·马多克斯与"骗局揭露者"詹姆斯·兰迪（一位职业魔术师）及沃尔特·斯图尔特去了得出此结论的法国免疫学家雅克·邦弗尼斯特的实验室，并请他重复这一实验。法国科学家不能重复他们的结果，而马多克斯和他的同事们对粗劣的实验技术（但并非造假）提出了多项指控。邦弗尼斯特指责这次调查是一场闹剧，并继续高倍稀释效应的工作，直至其 2004 年去世。但是他的观点并未得到其他科学家的理会。

一个《自然》杂志调查小组认为邦弗尼斯特博士和他的同事们在他们的著名论文中提出的观点缺乏可靠证据。

在《自然》杂志（**333**，816；1988）中，雅克·邦弗尼斯特博士和他的助手们提出了一个著名的观点，然而他们用以得出结论的大规模系列实验缺乏统计学对照，研究者也没有进行实质性的努力来排除包括实验观测者的偏好在内的系统误差。并且，他们将与"高倍稀释的抗 IgE 抗体会引发嗜碱性粒细胞脱颗粒"这一结论相矛盾的测试都予以排除，这使得他们所阐述的观点变得不太可信。此外，他们所描述的实验现象在通常情况下没有可重复性。

我们认为"高倍稀释（稀释倍数高达 10^{120}）的抗 IgE 抗体仍具有生物学效应"的说法是缺乏可靠依据的，并且关于"水可以对之前的溶质形成记忆"的假设也太过离奇。

我们勉强地使用"高倍稀释溶液"这个名词，因为在所谓的高倍稀释的溶液中，已经没有抗 IgE 抗体，因此也不再是通常意义上的溶液了。"无溶质的溶液"在逻辑上也是讲不通的。

我们的结论是基于对邦弗尼斯特博士所在的实验室——位于法国巴黎西部郊区克拉马尔的法国国家健康和医学研究院免疫药理学与过敏反应研究组（也称 INSERM 200）进行的为期一周的调查访问（始于 7 月 4 日）后作出的。我们了解到

dismayed to learn that the salaries of two of Dr Benveniste's coauthors of the published article are paid for under a contract between INSERM 200 and the French company Boiron et Cie., a supplier of pharmaceuticals and homoeopathic medicines, as were our hotel bills.

Benveniste's results are being widely interpreted as support for homoeopathic medicine. In the light of our investigation, we believe that such use amounts to misuse.

Our visit and investigation were preconditions for the publication of the original article. We acknowledge that we are an oddly constituted group. One of us (J.R.) is a professional magician (and also a MacArthur Foundation fellow) whose presence was originally thought desirable in case the remarkable results reported had been produced by trickery. Another of us (W.W.S.) has been chiefly concerned, during the past decade, in studies of errors and inconsistencies in the scientific literature and with the subject of misconduct in science. The third (J.M.) is a journalist with a background in theoretical physics. None of us has first-hand experience in the field of work at INSERM 200.

We acknowledge that we might well have found ourselves unable to get to grips with the work of the laboratory. But, on the basis of our experience, we are confident that the design of the experiments reported by INSERM 200 is inadequate as a basis for the claims made last month and that the defects we shall catalogue are a sufficient explanation of the remarkable results then reported.

We believe that experimental data have been uncritically assessed and their imperfections inadequately reported. We believe that the laboratory has fostered and then cherished a delusion about the interpretation of its data.

We are grateful to Dr Jacques Benveniste for his openness in discussing most of the questions we raised with him. He allowed us to borrow and to photocopy the relevant laboratory notebooks, which were invaluable for our investigation. We have every reason to believe that Dr Benveniste was (and, perhaps, still is) convinced of the reality of the phenomena reported in his article. We are also in the debt of several of Dr Benveniste's colleagues, especially to Dr Elisabeth Davenas. On her fell most of the burden of demonstrating the standard dilution experiments and of repeating them in a blinded protocol under our scrutiny. We know that our report will be a disappointment to the laboratory. We are sorry.

What follows is a narrative account of our visit and a summary of our conclusions.

Our investigations concentrated exclusively on the experimental system on which the publication was based. During our week in Paris, we resisted several proffered opportunities to examine other systems in which high dilution is claimed not to diminish the biological effectiveness of a molecule.

和邦弗尼斯特博士所发表论文的两名共同作者的薪水是按 INSERM 200 和一家法国公司 Boiron et Cie. 所签署的一份合同来支付的。Boiron et Cie. 公司是顺势疗法药物的供应商，它也负担了我们此次访问的住宿费用。我们对上述情况感到很不安。

邦弗尼斯特的实验结果被广泛地宣传以支持顺势疗法医学理论。经过调查，我们认为这是一种滥用。

此次访问和调查源于这篇文章的发表。我们承认这个调查小组是一个奇特的队伍。小组中的一员（詹姆斯·兰迪）是一名专业的魔术师（也是麦克阿瑟基金会的成员），邀请他的初衷是希望他能够发现这个著名的实验结果中是否含有某种欺骗性的成分。另一个成员（沃尔特·斯图尔特）在过去十年中参与研究科学论文中的错误和矛盾，以及科学界的学术不端问题。第三个成员（约翰·马多克斯）是一位具有理论物理学背景的记者。所有成员都没有 INSERM 200 相关领域的直接工作经历。

我们承认我们可能无法快速了解这个实验室的工作。但是基于经验，我们充分相信 INSERM 200 设计的实验方案不足以支持他们上个月所得出的结论，并且我们所罗列的这些实验设计的缺陷足以解释他们所观察到的实验现象。

我们相信他们对实验数据的评估不够严格，而实验数据的缺陷也没有被充分地显示。我们相信这个实验室对这些数据进行解释时形成了错觉，并一直维持着这一错觉。

我们非常感谢雅克·邦弗尼斯特在与我们讨论多数问题时的坦诚。他允许我们借出并复印相关的实验记录本，这对于我们的调查来说弥足珍贵。我们完全相信邦弗尼斯特博士确信（也许现在仍是）他的论文中所报道的实验现象的真实性。我们也对邦弗尼斯特博士的一些同事，尤其是伊丽莎白·达弗纳博士深表感谢。在我们的详细审查下，验证标准稀释实验和按双盲原则重复它们的压力都落在她的身上。我们知道我们的报告会让这个实验室失望，对此我们很遗憾。

下面是我们叙述性的访问报告以及我们结论的总结。

我们的调查完全集中在邦弗尼斯特博士的论文所依赖的实验系统上。在我们访问巴黎的一周时间内，我们多次拒绝了对方提供的检查其他系统的机会，据说在那些系统中经过高倍稀释后分子的生物学效应仍未消失。

The experimental system has evolved from a test for assessing the susceptibility of people to specific allergens. The guiding principle is that blood-borne allergens have the specific effect of interacting with the leukocytes known as basophils, causing them to degranulate—that is, to release the contents of cytoplasmic granules carrying histamine and other active substances provoking the symptoms of asthma and hay-fever.

These allergic reactions are apparently mediated at least in part by IgE molecules attached to the surfaces of basophils (in the blood) or mast cells (in tissues). Normally, degranulation is triggered by the interaction of anchored IgE molecules with an antigen, but the same effect can be brought about by the use of anti-IgE—antibody prepared by injecting human IgE into an animal of another species. (INSERM 200 uses goat anti-IgE at a concentration of 1 mg cm^{-3} sold by the Dutch company Nordic.)

The laboratory notebooks provide ample evidence that this expected degranulation is a maximum between log (dilution) 2 and 4.

Benveniste described the published procedure as a "simple experiment". A buffered solution of anti-IgE is serially diluted by a factor of 10 by transferring measured volumes from one test-tube to another. Pipette tips are discarded after each transfer. Measured volumes of resuspended white cells derived from human blood are transferred to wells in a polystyrene plate. To each of these is added a measured volume of serially diluted anti-IgE or buffer as a control. The wells are incubated for 30 minutes at 37°C. An acidic solution of toluidine blue, which stains intact but not degranulated basophils red, is added and the numbers of recognizable basophils counted on a haemocytometer slide. Anti-IgG, which does not degranulate basophils, is used as a control.

We were surprised to learn that the experiments do not always "work". There have been periods of several months at a time during which solutions at high dilution have not degranulated basophils. Indeed, the laboratory had just emerged from such a period. (Speculation at the laboratory is that the distilled water may have been contaminated, or otherwise made unsuitable.) It also appears that bloods that "do not degranulate" are often encountered; we were informed that, in this event, data are recorded but not included in analyses prepared for publication. Even so, the source of blood for the experiments is not controlled, except that an attempt is made not to use blood from people with an allergy.

We witnessed a total of seven runs of this experiment, of which three were routine repetitions of the standard procedure. For the fourth experiment, samples of diluted IgE were transferred by one of us (W.W.S.) to wells in a plastic plate in a random sequence and then read blind by Dr Davenas. All four of these experiments, the last after decoding, gave results described as positive by Benveniste, but three further sequences of counts of stained basophils in three further strictly blind experiments gave negative results (see below).

这个实验系统是从一个检测人们对特定过敏原的易感性的实验演化而来的。其指导原则是血源性过敏原可以与嗜碱性粒细胞等白细胞相互作用并发挥效应，引起这些细胞脱颗粒，释放细胞质颗粒物中的组胺和其他活性物质，从而引起哮喘和枯草热的症状。

这些过敏反应（至少部分）是由附着在嗜碱性粒细胞（血液中）或肥大细胞（组织中）表面的 IgE 分子介导的。一般来讲，脱颗粒过程由锚定于细胞表面的 IgE 分子与抗原相互作用而触发，但是抗 IgE 抗体也可以引发相同的效应。将人 IgE 分子注射到其他物种的动物中便可以得到抗 IgE 抗体。（INSERM 200 使用了荷兰 Nordic 公司出售的山羊抗人 IgE 抗体，浓度为 $1\,mg\cdot cm^{-3}$）

实验记录本提供了充分的证据表明这一脱颗粒过程在抗 IgE 抗体稀释倍数为 10^2 倍与 10^4 倍之间达到最大值。

邦弗尼斯特将发表的实验程序描述为一个“简单的实验”。将含抗 IgE 抗体的缓冲溶液分步稀释，每次把一定体积的溶液从一个试管移入另一个试管并稀释 10 倍。枪头在每次移液后都会被扔掉。一定体积的来自人血液中的白细胞重悬液被加入聚苯乙烯板的培养孔中。实验人员将一定体积的经系列稀释的抗 IgE 抗体或作为对照的缓冲液加入这些培养孔中。在 37℃孵育 30 分钟后，向孔中加入甲苯胺蓝的酸性溶液，它可以将完整但没有脱颗粒的嗜碱性粒细胞染成红色。利用血球计数板可以对可见的嗜碱性粒细胞进行计数。此外，不能够引发嗜碱性粒细胞脱颗粒的抗 IgG 抗体被用作阴性对照。

我们很惊奇地发现，这个实验并不是总能“奏效”。曾经有几个月的时间那些高倍稀释的抗血清并不能够引发嗜碱性粒细胞脱颗粒。事实上，这个实验室也刚经历过这样的时期。（对此该实验室推测可能是由于蒸馏水被污染或者其他制备方法不合适。）并且，似乎他们也经常遇到那些“不能够脱颗粒”的血液样品。我们获悉，在遇到这种情况时虽然实验数据被记录下来，但是并没有包含在用于发表的分析中。即便如此，除了尽量避免使用有过敏倾向的人的血液之外，实验中所用的血液样品的来源并没有受到人为的控制。

我们总共见证了七轮实验，其中三次是按照标准步骤按部就班的重复。在第四次实验中，稀释的抗 IgE 抗体样品由调查组的一个成员（沃尔特·斯图尔特）随机加到含有嗜碱性粒细胞的培养孔中，然后由达弗纳博士在双盲的情况下观察细胞脱颗粒的情况。所有这四次实验，其中最后一次是在揭盲后，邦弗尼斯特得出了支持自己结论的阳性结果。但是在接下来的三个严格按照双盲规定进行的实验中，染色的嗜碱性粒细胞的计数给出了阴性结果（如下）。

Figure 1 shows results gathered in the first group of experiments. The ordinate is the decrease (compared with the control) of the numbers of stained basophils at dilutions ranging from 10^{-2} to 10^{-30}. In each case, the left-most peak is that expected from the interaction between anti-IgE and IgE bound to basophils. The number of stained basophils increases to near its control value at log (dilution) of between 5 and 7 (0 percent degranulation, called "achromasie"); the unexpected phenomenon is that the graphs then reach a series of three or four further peaks with increasing dilution.

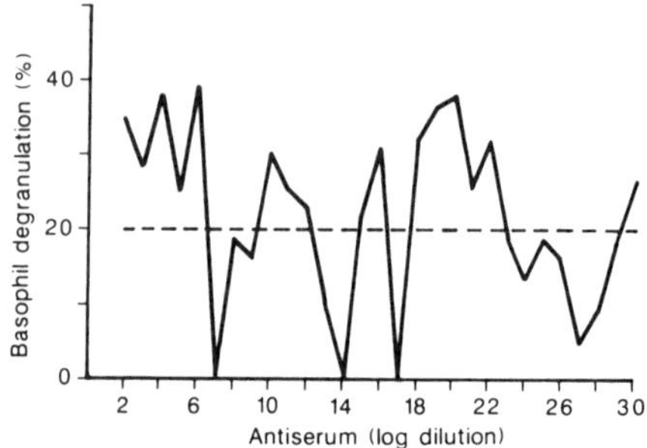

Fig. 1. A demonstration degranulation, the first of the three open experiments.

These are the successive peaks of activity said in the original article to occur in a periodic fashion, and whose position was said to be reproducible. It is clear from the four graphs that this claim is not obviously supported by this data. The laboratory notebooks confirm that the position of the peaks varies from one experiment to another.

The data in the fourth experiment appear different from those recorded earlier in the laboratory. Indeed, Benveniste volunteered that "we've not seen one like this before". The odd feature of the curve is that the activity of the diluted anti-IgE is, at its peak, identical with that of anti-IgE at log (dilution) 3—presumably the point at which the natural degranulating effect of anti-IgE is a maximum.

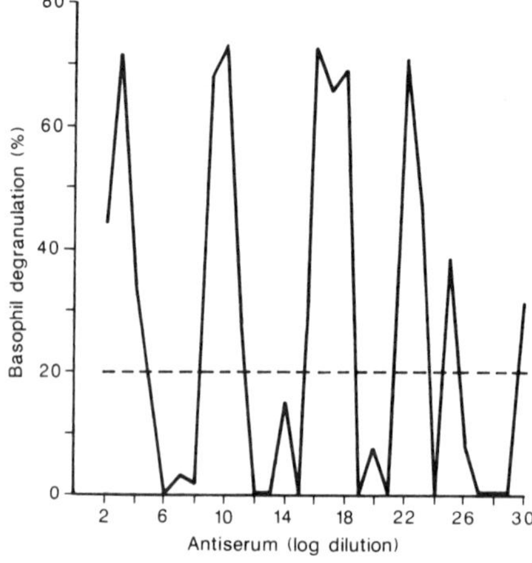

Fig. 2. The fourth demonstration experiment (read "blind") with unexpectedly high peaks (see text).

图 1 所示为第一组实验得到的结果。纵坐标是在稀释度 10^{-2}~10^{-30} 倍之间的实验组中染色的嗜碱性粒细胞减少的百分数（与对照组比较后）。最左边的峰是抗 IgE 抗体与嗜碱性粒细胞表面的 IgE 分子结合后所引起的。当稀释倍数增大到 10^5~10^7 时，实验组中染色的嗜碱性粒细胞的数量与对照组的相当（0%脱颗粒，称为“色盲”）；出乎意料的是，在此之后，随着稀释倍数的增大，又出现了 3~4 次脱颗粒峰值。

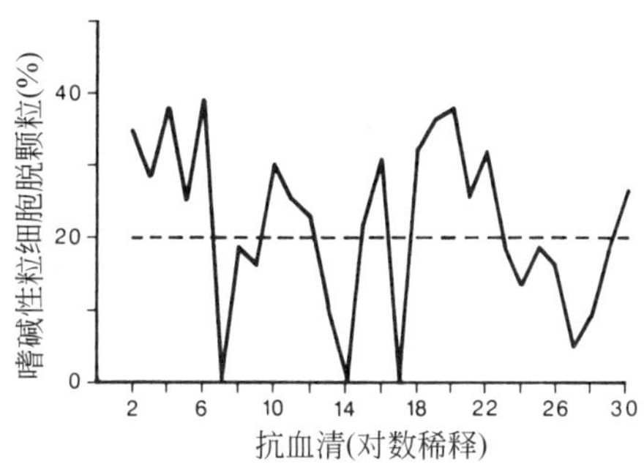

图 1. 脱颗粒实验的图示，三个公开实验中的第一次实验结果

上图即为邦弗尼斯特论文中所提到的呈周期性的连续的峰，并且这些峰值所处的位置被认为是可重复的。从本文 4 张图中我们可以清楚地看出，这些数据并不能够充分地支持这一论点。他们的实验记录也证明这些峰的位置在不同的实验中是不同的。

第四次实验的数据似乎与这个实验室之前所记录的结果有所不同。确实，邦弗尼斯特自己也说“我们以前没有看见过这样的实验结果”。这个曲线的奇怪之处在于随着抗 IgE 抗体的稀释，后续出现的各峰值与稀释 10^3 倍时的峰值一样——而我们知道，10^3 倍可能是抗 IgE 抗体引发的嗜碱性粒细胞脱颗粒的最大值的稀释倍数。

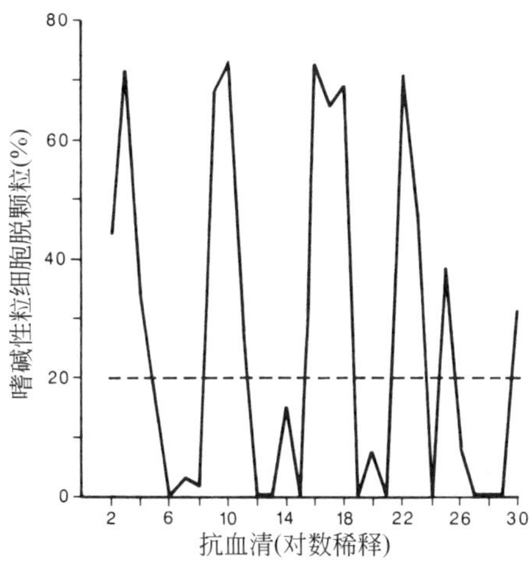

图 2. 在第四次实验中（读取实验结果时采取双盲的方式）观察到的出乎意料的高峰值（见正文）。

We raised with Dr Benveniste and his colleagues the obviously relevant question of the sampling error. We were astonished to learn, in the discussion of our conclusions at the end of our visit, that neither Dr Benveniste nor his colleagues seemed to be aware of what sampling errors are. We provided a simple explanation, complete with an account of what happens when one pulls a handful of differently coloured balls from a bag, to argue that the sampling error of any counting measurement must be of the order of the square root of the number to be counted. On several occasions, Benveniste called these "theoretical objections".

Ironically, he is himself one of the three authors of a paper published in 1981, in which just this issue had been addressed in a superficially similar situation (Petoit, J. F., Sainte-Laudy, J. & Benveniste, J. *Ann. Biol. clin.* **39**, 355; 1981), and which appears to be the justification of the dotted line drawn at about 20 percent (corresponding to two standard deviations) on the percent degranulations of intact basophils after axis.

That brief paper deals exclusively with the effect of sampling errors (not other kinds of errors) on the interpretation of measurements of intact basophils after white-cell suspensions had been allowed to react with allergens via their attached IgE molecules. Even now, at the Clamart laboratory, provision is made for the measurement of two control samples. Among other things, the paper provides a statistical test for telling when the difference between the two control values is statistically significant at the 5 percent level, in which case people using the procedure as a diagnostic test of allergy are advised to start their experiment all over again.

At INSERM 200, there seems to have grown up a less formal way of dealing with problems of this kind; when the reading of a diluted sample is greater than the control counts, the experimenter often counts the control sample again, on the grounds that the first reading "must have been wrong". This happened when Dr Davenas was counting the first of the first group of experiments.

This procedure exaggerates to some extent the amount of basophil degranulation measured with reagents at high dilution. The practice makes the control values unreliable, and is a significant pointer to the laboratory's disregard of statistical principles.

In these circumstances, it is natural that we should eagerly have accepted Benveniste's invitation to devise a blind experiment. We set out to devise a procedure that would be watertight. We asked that three samples of blood should be run. The serial dilutions would be prepared by Dr Davenas, secretly coded by us before being transferred to wells for incubation and staining by her.

In a small laboratory, procedures like this are inevitably and understandably disruptive. At INSERM 200, the sense of melodrama was further heightened by the general recognition of the importance of the trial, and by the precautions necessary to ensure that the code would not be known to others than ourselves as well as by the need that the one of us with

对此，我们向邦弗尼斯特博士和他的同事们提出了关于采样误差的疑问。在我们此次访问结束前的总结讨论中，我们很惊讶地发现，邦弗尼斯特博士和他的同事们似乎对采样误差毫不了解。我们对此进行了简单的解释，并以从一个装有各种颜色彩球的袋子中掏出一把不同颜色的球为例，排除计数方法的抽样误差，必须考虑样本量的平方根。对此，邦弗尼斯特称之为“理论上的反对意见”。

具有讽刺意味的是，他自己曾是 1981 年发表的一篇涉及相似问题的论文的三个作者之一（珀图瓦特、圣洛迪、邦弗尼斯特，《生物临床年报》第 39 卷，第 355 期，1981 年）。在这篇文章中，他们认为应在完整嗜碱性粒细胞的脱颗粒百分比的 20% 处（对应两个标准差）划一条点线作为判据，点线之下为可信的读数。

那篇文章专门研究了白细胞悬浮液通过表面 IgE 分子与抗原相互作用后，采样误差（不是别的误差）对检测脱颗粒后完整嗜碱性粒细胞的影响。即使现在，在克拉马尔实验室，也仍然规定要测量两个对照样品。此外，这篇文章还给出了用于验证两个对照值之间的差异是否具有统计学显著性（$p<0.05$）的检验方法。这使得那些通过观察嗜碱性粒细胞脱颗粒的情况检测过敏反应的研究人员开始重新设计他们的实验。

在 INSERM 200，对于此类问题似乎存在一个不那么正式的处理方法。当实验组的读值比对照组还要高的时候，实验员通常会以第一次读值“一定有问题”为理由而重新计算对照组的数值。这一情况在达弗纳博士统计第一组实验的第一个实验的结果时便发生了。

这种实验方法在某种程度上夸大了高倍稀释试剂中检测到的嗜碱性粒细胞脱颗粒的数量。这一行为使得对照组的数值变得不可信，这是这个实验室无视统计学原理的重要体现。

在这种情况下，我们自然热切地接受了邦弗尼斯特请我们设计一个双盲实验的邀请。我们着手设计了一个完善的实验流程。我们要求测试 3 个不同的血液样品，并且由我们对达弗纳博士系列稀释后的抗 IgE 抗体进行秘密编号，然后由她将这些样品加入培养板中与嗜碱性粒细胞共孵育和染色。

在一个小型实验室，这样的实验流程的混乱是不可避免的，也是可以理解的。不过在 INSERM 200，由于大家都认为此次实验非常重要，同时由于我们想尽办法防止他人获知试管编号，还由于我们不让以变戏法著称的詹姆斯·兰迪接触到盛有

a reputation for sleight of hand (J.R.) could be shown to have been kept away from the test-tubes containing the serial dilutions.

This was done by arranging that Davenas should carry the diluted anti-IgE solutions in stoppered test-tubes to a separate room, where their contents would have been transferred to previously labelled tubes as determined by counters drawn at random from a bag. The coding procedure was monitored by a video camera operated by Randi, who was thereby prevented from touching anything else. (We have a record of the proceedings on an unbroken reel of tape.)

We made two last-minute changes in the planned procedure. First, we included 5 control tubes containing only buffer. Second, having been warned that homoeopathists might regard the data as invalid if solutions were decanted from one set of tubes to another, we removed the numbers written with a felt pen on the original tubes, replacing them with numbered labels which Randi assured us were tamper-proof. The code itself was eventually folded in aluminium foil, enclosed in an envelope specially sealed by Randi and then taped to the laboratory ceiling for the duration of the experiment.

We also arranged a second step of coding just before the slides were counted. One of us (W.W.S.) took responsibility for pipetting, after securing the agreement of Davenas and Benveniste that his technique was satisfactory. Both the laboratory procedures and the codes themselves were recorded on video tape. The plates containing the stained cell suspensions were stored in a box (sealed by Randi) in a cold room until read, in random order. The second plate took longer to read, partly because each well was read in duplicate by each observer (Dr Davenas and her colleague, Dr Francis Beauvais), partly because the cells of the second plate were only faintly stained and were thus difficult to read.

Whatever the three runs would provide, we were especially anxious to derive some objective estimate of the intra- and inter-observer measurement errors. We had been told at the outset, by Benveniste, that Dr Davenas was not merely exceptionally devoted to her work but the one in whose hands the experiment most often "works". He said that she usually "counts more cells" than other people. Dr Beauvais, who was also said to be exceptionally skilled, read the slides separately from Dr Davenas, but at the same time. On this occasion, the sampling errors missing from most of the laboratory records did indeed appear.

The duplicate measurements in our strictly blinded experiments were especially important. First, they show that sampling errors do indeed exist, and are not "theoretical objections". Second, they show that the two observers were counting as accurately as could be expected, which gives the lie to the later complaint that the results of the double-blind experiments might be unreliable because the observers had been exhausted by our demands.

系列稀释试剂的试管，一切变得更富戏剧性。

整个实验过程是这样的：达弗纳将稀释好的抗 IgE 抗体溶液装到一个带塞子的试管中，并拿到另一个房间；在这个房间里，我们将这些溶液转移到事先做好标记的试管中，这些试管是由计数者从一个袋子里随机取出的。整个编号的过程由兰迪使用摄像机进行记录，从而避免他触碰任何其他东西。（我们有整个编号过程的完整的录影带。）

对于上述实验流程，我们还做了两个“最后一分钟的改变”。第一，我们在样品中混入了五个只含有空白缓冲液的试管作为阴性对照。第二，为了避免“顺势疗法”论者以“溶液被从一个试管中倒入另一个试管中”为理由称此次实验数据不能说明问题，我们将原始试管上用毡笔写的数字擦去，并贴上标有数字的标签，兰迪向我们确保它们不会被做手脚。最后用铝箔把这些编号包起来，并装入信封中由兰迪进行特殊密封，用胶带贴在实验室的天花板上并在实验过程保持这种状态。

在对脱颗粒情况进行计数前，我们还安排了第二次编号的步骤。我们中的一员（沃尔特·斯图尔特）负责移取液体，他的技术已经事先得到了达弗纳和邦弗尼斯特的认可。这些实验室操作和样品编码都被摄像记录下来。在染色后，含有细胞悬液的培养板被随机放到一个盒子中（由兰迪密封）并置于冷藏室，直到读数前才取出来。第二块板读数花的时间比较长，一部分原因是每个孔都分别由达弗纳博士和她的同事弗朗西斯·博韦重复观察。之所以这样做，部分是由于第二块板染色比较模糊，不易分辨结果。

无论这三个实验会产生什么样的结果，我们非常渴望对源自观测者自身和不同观测者之间的实验误差进行客观的评价。在最开始我们便从邦弗尼斯特博士那里获知达弗纳博士不仅对她自己的工作异常投入，而且她经手的实验通常多数是“奏效”的。邦弗尼斯特博士说达弗纳博士通常会比别人“数出更多的细胞”。博韦博士也是一个技术很精湛的研究人员，她与达弗纳博士同时分别读数。在这种情况下，大多数实验记录中缺乏的采样误差确实地出现了。

在我们严格的双盲实验中，双重测量是非常重要的。第一，它们证明采样误差是确实存在的，并且不是“理论反对意见”。第二，它们显示两名观测者都在尽可能准确地计数，这也反驳了后来有些人的抱怨，他们认为实验人员被我们的各种要求弄得筋疲力尽，从而整个实验的结果是不可信的。

Others working in this field recognize the difficulty of counting basophils (roughly 1 in 100 among leukocytes), preferring instead to measure the histamine released on degranulation. This practice is not followed at INSERM 200 because, we were told, of previous failure to record histamine release (as distinct from the disappearance of stained basophils) at high dilution (whence the term "achromasie").

We began to break the codes by lunchtime on our last day, the Friday. When the slides had been matched to the wells from which their samples had derived, but before the appropriate dilutions had been assigned to them, there was a great sense of light-heartedness in the laboratory, no doubt at the prospect that the ordeal would soon be at an end. Benveniste, glancing at the half-decoded data, even offered to predict where the peaks and troughs would fall in the data. His offer was accepted. But his predictions proved to be entirely wrong.

We asked at this stage for criticisms of the conduct of the trials, but were given none. To the question what would be said if the two observers had recorded degranulation peaks, but at different high-dilution values, Benveniste said that would still constitute success.

Opening sealed envelopes is Randi's expertise. He found that the sealed flap of the envelope had detached itself at a surprisingly straight angle when the scotch tape attaching the code to the ceiling was pulled away, but inspection of the aluminium foil allowed him to pronounce himself satisfied that the code had not been read. Then came the decoding—one person singing out numbers to another.

So do the numbers make sense? Six numbers into the record of the first plate to be read, Benveniste said "that patient isn't degranulating, try another". So we did—first the parallel readings by Dr Beauvais, then the remaining two experiments. In the event, the results of all three experiments were similar. The anti-IgE at conventional dilutions caused degranulation, but at "high dilution" there was no effect. Blood from three sources in a row degranulated at ordinary dilutions but not at homoeopathic dilutions. Each of the three experiments was a failure.

Conclusions

We conclude that the claims made by Davenas *et al.* are not to be believed. Our conclusion, not based solely on the circumstance that the only strictly double-blind experiments we had witnessed proved to be failures, may be summarized as follows:

■ **The care with which the experiments reported have been carried out does not match the extraordinary character of the claims made in their interpretation.** What we found, at Clamart, was a laboratory procedure possibly suitable for the application of a well-tested

相关领域的其他人员认识到嗜碱性粒细胞的计数是一个很难的事情（100 个白细胞中约有 1 个），因此，他们更倾向于测定在脱颗粒的过程中释放出的组胺的含量。不过，在 INSERM 200 并没有采用这一方法，是因为我们被告知在先前的实验中他们在高倍稀释下（发生“色盲”）检测不到组胺的释放（显著区别于染色的嗜碱性粒细胞的消失）。

在我们访问的最后一天，也就是星期五的午餐时间，我们公布了样品的编号情况。在将稀释样品与最终脱颗粒情况对应起来之前，我们先公布了玻片上的样品与样品孔的对应关系。此时，实验室的气氛是轻松的，大家认为这个痛苦的过程总算要结束了。邦弗尼斯特瞟了一眼这个解密了一半的数据，甚至提出要预测哪里是脱颗粒曲线的波峰、哪里是波谷。大家接受了他的提议。不过事实证明他的预测是完全错误的。

这个时候我们问大家对此次实验的执行是否有疑问，不过没有得到回应。当我们问到如何看待“两名观测人员虽然都记录到了脱颗粒的峰值，但是对应的是不同的稀释倍数”这个问题时，邦弗尼斯特认为这仍证明脱颗粒实验是成功的。

打开密封的信封是兰迪的拿手好戏。在把粘在天花板上的编号的透明胶带除去时，他发现信封的封口已经自己展成了令人惊讶的直角。通过检查包在外面的铝箔，他确认编号并没有被人偷看。然后我们开始解码——一个人把编号唱给另一个人。

那么结果与预期相符合么？读取了第一块板测的 6 个样品的脱颗粒情况之后，邦弗尼斯特说“这个病人的血样不能发生脱颗粒，试试另一个”。于是我们照做了——首先是博韦博士对第一块板的平行读数，然后继续分析剩下的两个实验结果。最终大家发现，所有三个实验的结果都是相似的：正常稀释倍数的抗 IgE 抗体可以引起脱颗粒，而“高倍稀释”的则无效；三个不同来源的血液样品都可以被正常稀释的抗 IgE 抗体引发脱颗粒，但不能被所谓的“顺势疗法”的高倍稀释抗体引发。从这个意义上说，这三个实验都是“失败”的。

结　论

我们认为达弗纳等人所得出的结论是不可信的。我们所作出的结论不仅仅是基于我们所见证的那次严格的双盲实验，还包括以下方面的内容：

■ **他们所进行的实验的细致程度达不到能证明他们在论文中所阐述的那些非同寻常的结论的要求**。在克拉马尔，我们所看到的这些实验流程可能只适合用于已经被广泛认可的生物实验，而不适合在这里用于证明稀释 10^{120} 倍的抗 IgE 抗体仍能引

bioassay, but unsuitable as a basis for claiming that anti-IgE retains its biological activity even at a log (dilution) of 120. In circumstances in which the avoidance of contamination would seem crucial, no thought seemed to have been given to the possibility of contamination by misplaced test-tube stoppers, the contamination of unintended wells during the pipetting process and general laboratory contamination (the experiments we witnessed were carried out at an open bench). We have no idea what would be the effect on basophil degranulation of the organic solvents and adhesives backing the scotch tape used to seal the polystyrene wells overnight, but neither does the laboratory.

The design of the experiments hardly matches the nature of their interpretation. For example, one would have thought that counting wells at least in duplicate would have been an elementary precaution against gross errors. The second of our strictly blinded experiments seems to be one of the few in which something of this kind had been attempted.

The laboratory seems to have been curiously uncritical of the reasons why its experiments do not, on many occasions, "work". For example, we were told that the best results were obtained when cells were left in the cold room overnight before counting, but there has been no investigation of that phenomenon, or of the reports that taking a second sample from a single well gives odd results (an effect not apparent in our double-blind experiments).

■ **The phenomena described are not reproducible, but there has been no serious investigation of the reasons.** We have referred to the fact that some blood yields negative results, and that there are periods of time when no experiments work. But the laboratory notebooks show great variability in the positions at which peaks occur.

■ **The data lack errors of the magnitude that would be expected, and which are unavoidable.** This is best illustrated by Fig. 3, whose two graphs have been constructed from data recorded by Dr Davenas from samples supposedly identical with each other, usually measurements of control samples but also including some duplicate runs. The recorded values have been normalized by subtracting the mean and dividing by the square root of the mean (the expected sampling error). If the only source of error were sampling error, the standard deviation of the plotted curve should be unity (1). Other sources of error, for example, experimental variability, could only increase the standard deviation. But Fig. 3 shows that repeat observations agree more closely than would be expected from the underlying distribution. This is a well-known effect that sometimes affects duplicate readings by the same individual, but the magnitude of the effect in this case calls into question the validity of the readings. This artefact is nevertheless not apparent in the blinded duplicated readings.

发嗜碱性粒细胞脱颗粒这样的结论。很显然，排除样品污染对于证明这个结论来说非常重要。不过，他们似乎并没有考虑试管塞的错误放置，移液过程中无意造成的样品孔污染和普通的实验室污染（我们看到实验是在开放的实验台上进行的）的可能性。我们不知道有机溶剂以及用于聚丙烯微量培养板密封过夜的透明胶带背面的黏合剂会对嗜碱性粒细胞脱颗粒的影响，不过这个实验室的人员似乎也没有考虑过这个问题。

实验设计难以揭示本质现象。例如在进行细胞计数时，至少应该采取两次读数取平均值的方式以减少粗差，这是人们通常的想法。然而，似乎只有在我们进行的第二个严格的双盲实验中，他们才对这种方法进行了尝试。

这个实验室似乎对于为什么有时候他们的实验不能“奏效”这个问题缺乏思考。例如，他们告诉我们，当把细胞放在冷藏室过夜后可以获得最好的脱颗粒实验的结果，但他们对这个现象并没有进行研究，也没有去解释为什么从同一个孔中再取一个样进行测试会得到奇怪的结果（在我们的双盲实验中这种效应并不明显）。

■ **他们所描述的现象是不可重复的，但他们并没有对其中的原因进行严肃地调查**。我们在前面提到过一些血液样品总是产生阴性的结果，并且曾经有一段时期这个实验都做不出预期的结果。此外，实验记录显示脱颗粒峰值的位置在不同的实验中波动很大。

■ **实验数据缺少在合理范围内不可避免会存在误差**。这在图 3 中得到了很好的体现。这两个图都来自达弗纳博士的数据，这些数据被认为是完全相同的样品的测试结果，通常是对照实验的结果，也包括一些重复实验的结果。我们把实验数据减去该组实验的平均值然后再除以平均值的平方根（预期采样误差），即可将测量数据标准化。如果整个实验系统中误差的唯一来源是采样误差，那么图中曲线的标准差则应该是统一的（1）。其他的误差来源，如实验可变性，只会导致标准偏差的增大。图 3 的数据说明重复测量的结果比根据潜在分布做出的预期更加吻合。这是一种众所周知的效应，它会影响同一个人的重复读数。不过在这个实验中，这种效应的影响程度已经导致了读数可靠性的问题。在双盲实验中，这种人为因素造成的假象并不显著。

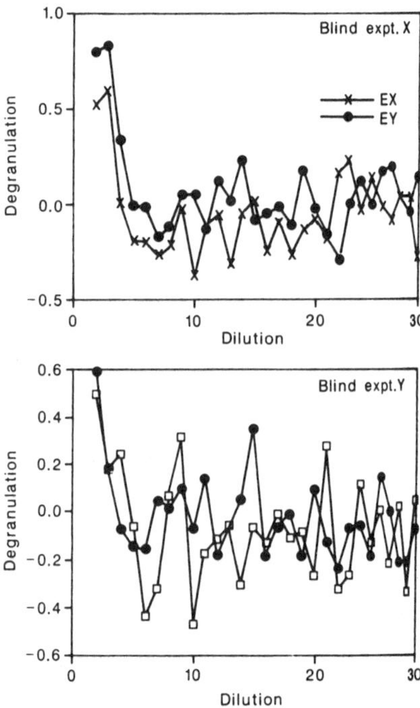

Fig. 3. Records for the first two blind experiments (5–7 inclusive). showing sampling noise only below the expected decline of degranulation with increasing dilution. Note that the ordinate extends below zero on the degranulation scale (to accommodate sampling errors above as well as below the control values).

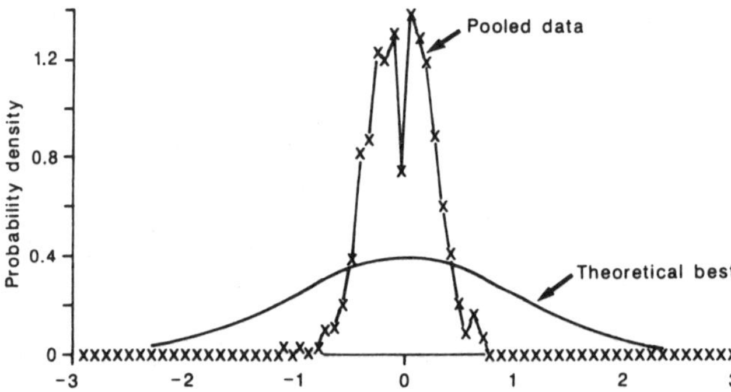

Fig. 4 . Comparison of measured departures of duplicate normalized readings from their means with the gaussian distribution expected.

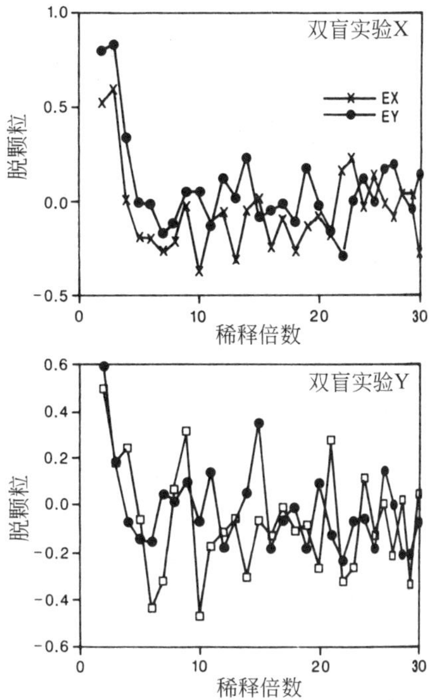

图 3. 前两个双盲实验的结果（包括 5~7）显示随着抗血清稀释倍数的增大，由采样误差导致的脱颗粒的数值仅比预期由抗血清所引发的数值略低。值得注意的是，纵坐标中细胞脱颗粒的百分比已经低于零（这是为了让这些由采样误差引起的值能够得以体现出来）。

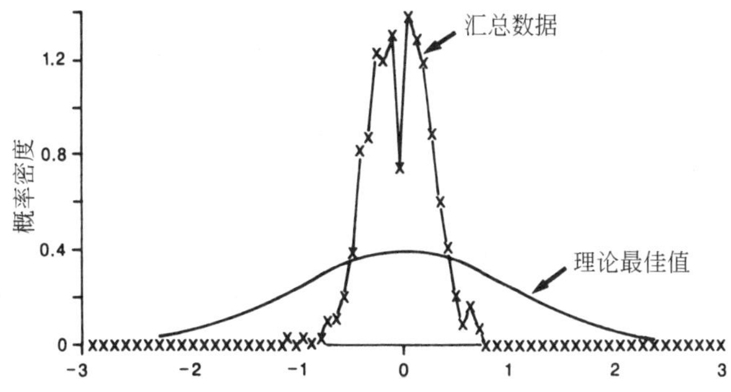

图 4. 将脱颗粒实验结果的两次读数的平均值标准化之后与预期的高斯分布进行测量偏离的比较。

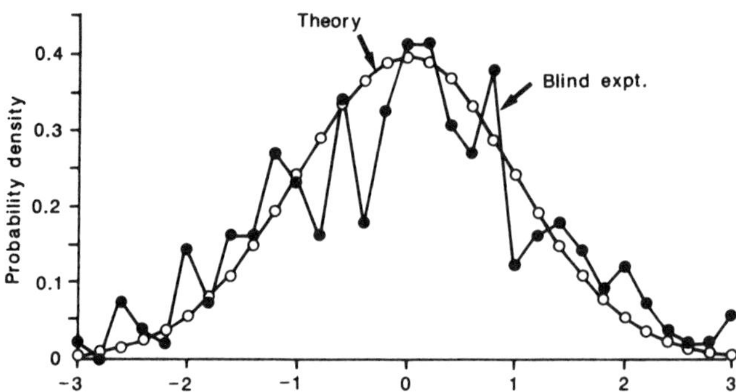

Fig. 5. Same as Fig. 4 except that data derive from duplicated readings within the blind experiments only.

■ **No serious attempt has been made to eliminate systematic errors, including observer bias.** It is true that the laboratory notebooks record experiments in which anti-IgG has been used as a control; we were surprised to find that the IgG control run reported by Davenas *et al.* (their Fig. 1*b*) was carried out at a different time from the run with IgE published in the same figure.

Most of the data recorded in the laboratory notebooks derive from experiments in which the same person has been responsible for the sequential dilution, plating out and counting. Given the shared belief at Clamart in the reality of the phenomenon reported last month, and its potential importance, it is mystifying that duplicate and blind counting is not routine.

■ **The climate of the laboratory is inimical to an objective evaluation of the exceptional data.** So much is readily apparent from the way in which experiments are described as successes and failures, by the use of the word "working" to describe experiments yielding a positive result, and by the several speculations we were offered, without experimental evidence in their support, to explain the several failures the laboratory has experienced. The folklore of high-dilution work pervades the laboratory, as epitomized by the suggestion that decanting diluted solution from one tube to another might spoil the effect and the report that the repeated serial dilution by factors of three and seven (rather than ten) always yields negative results.

Collaborations

We have not been able to pay as much attention as we would have wished to the data collected at other laboratories and cited in Davenas *et al.*, but we have examined documentary evidence available.

Supporting data were said to have come from Rehovot (Israel), Milan and Toronto. Dr Benveniste told us we could not see the Toronto data, described as preliminary, without the consent of the authors, who could not be telephoned.

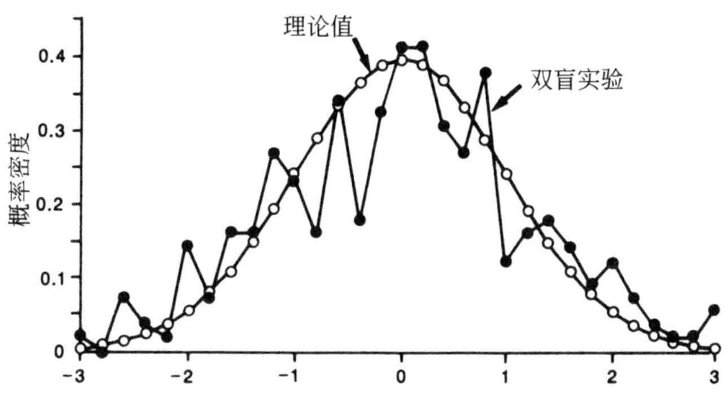

图 5. 同图 4，只是本图数据仅来自双盲实验中重复读数。

■ **他们没有认真尝试采取相关措施排除包括观测者偏好在内的系统误差。**实验室记录中确实显示研究人员使用了抗 IgG 抗血清作为实验的阴性对照，不过我们却惊奇地发现由达弗纳等人进行的 IgG 对照实验（图 1*b*）与发表在同一图表的 IgE 实验是在不同时间进行的。

实验记录显示，大多数实验都是由同一个人全程完成系列稀释、铺板、计数等一系列操作步骤的。考虑到克拉马尔的研究人员对“高倍稀释的抗血清可以引发嗜碱性粒细胞脱颗粒”这个实验现象真实性以及潜在的重要性都很有信心，我们对他们缺乏重复实验和双盲实验的意识而感到迷惑。

■ **这个实验室不能客观地评价实验中不支持自己假说的数据。**从实验通常被描述为“成功”的和“失败”的诸多事实已经证明了这一点。当得到阳性的结果时，人们会说这个实验“奏效”了；而对于好几次被认为“阴性”的结果，用我们提出的若干推测而非实验证据来解释。这个实验室还很流行高倍稀释效应这样的民间传说，并且还认为把稀释后的溶液从一个试管中倒入另一个试管会破坏其疗效，以及报告称先后稀释 10^3 倍和 10^7 倍（而不是 10^{10} 倍）的抗血清总是会产生阴性的结果。

合　作

虽然我们很希望能够对由其他实验室获得并被达弗纳等人引用的数据进行同样的分析，但受时间和精力所限，我们只能对其中可以获得的一些文件进行检查。

雷霍沃特（以色列）、米兰、多伦多等多个地方的实验室都参与了这项工作。邦弗尼斯特博士告诉我们：不能在没有得到作者的同意的情况下看多伦多的数据，该数据被视为最起始的数据，而作者电话打不通。

The data gathered in Israel and Milan are, apparently, significant. Figure 6 is typical of the data from Milan. Though there are no duplicate measurements and therefore no direct evidence of sampling error, there is also some evidence of degranulation at high dilution. Without knowing more about the circumstances, we are unable to comment.

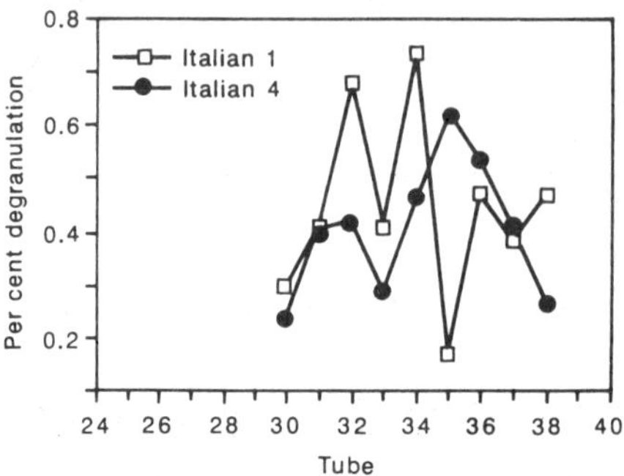

Fig. 6. Two duplicate Italian runs showing high degranulation, but discordantly.

The Israeli data are more extensive. The first trials were in March 1987, during a visit to Rehovot by Dr Davenas. The most remarkable of several successful trials was her correct identification of seven high-dilution tubes out of ten presented to her blind. Even so, the report (to Benveniste) of the trials was cautious. Later, analysis of the tubes which had tested positive in this trial revealed not merely immunoglobins but other protein contaminants apparently identical with materials in the original IgE vial. One of the participants (Professor Meir Shinitsky of the Weizmann Institute) then withdrew as a putative co-author.

Since then, there have been two developments in Israel— a series of experiments carried out independently of Benveniste's laboratory and a further blinded experiment. Data from the latter are unfortunately not available. Maitre Simart, a legal official at Clamart who held the codes, is said not to have had time to decode them.

These measurements are nevertheless, to judge from the documents we have seen, stronger evidence than any we found at Dr Benveniste's laboratory to support his claims. But we do not have the information to evaluate them.

Postscript

We presented the substance of these conclusions to Dr Benveniste and his colleagues immediately after the strictly blinded experiments were decoded. The discussion that followed was inevitably tense. Benveniste acknowledged that his experimental design may not have been "perfect", but insisted (not for the first time) that the quality of his data was no worse than that of many papers published in *Nature* and other such journals.

以色列和米兰的数据显然是很重要的。图 6 是一个具有代表性的米兰的数据。尽管他们没有对结果进行重复测量，也因此无法计算采样误差，但仍有一些证据表明在他们的实验中，高倍稀释的抗血清可以引发脱颗粒。由于无法获知更多的实验相关的信息，我们无法对此结果作出评论。

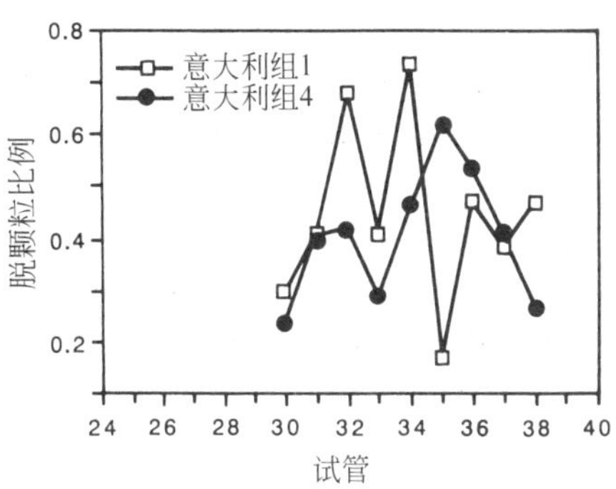

图 6. 两个来自意大利的重复实验结果，它们都显示存在很高的脱颗粒比例，但两个实验的数据并不一致。

来自以色列的结果相对更具有普遍意义。第一次实验是在 1987 年 3 月达弗纳博士到雷霍沃特访问时进行的。在几个成功的实验中，最著名的是她在单盲实验中正确地从 10 个样品中识别出了 7 个高倍稀释样品。虽然如此，她给邦弗尼斯特的报告中对这次实验的表述也是很谨慎的。后来，对该实验中盛阳性样品的试管进行分析发现，其中不仅含有免疫球蛋白，还含有与原始 IgE 样品中同样的蛋白质污染物。于是这个实验的参与者之一——魏茨曼科学研究所的梅厄 · 希尼特斯克伊教授放弃成为共同作者。

之后，我们又调查了在以色列进行的取得进展的另两个实验：一个是独立于邦弗尼斯特实验室的一系列实验，另外一个是进一步的双盲实验。不幸的是，我们没有拿到双盲实验的数据。克拉马尔的一位法律官员迈特尔 · 西马尔掌握着这些数据的密码，但据称他没有时间将它们解密。

从文件判断，以色列的这些结果比我们在邦弗尼斯特博士的实验室看见的更能支持其结论。但由于缺乏相关的信息，我们也无法对此进行评价。

后　记

在严格的双盲实验解码结束后，我们即向邦弗尼斯特和他的同事们通报了上述结果。不可避免的，接下来的讨论是非常紧张的。邦弗尼斯特承认他的实验设计或许不够“完美”，但仍然一再坚持他的实验数据的质量不比在《自然》和其他类似杂志上发表的论文差。

One of us (J.M.) said it would be best if Benveniste would withdraw the published article, or at least write to *Nature* to qualify his findings and their interpretation, in which case we would not publish this report. It was mutually agreed that nothing would be said publicly until 28 July. But Benveniste said that the laboratory would work through the weekend "and all next week" to prove the reality of the phenomenon.

Our greatest surprise (and disappointment) is that INSERM 200 seems not to have appreciated that its sensational claims could be sustained only by data of exceptional quality. Randi put the point best, during our Friday discussion, by saying: "Look, if I told you that I keep a goat in the backyard of my house in Florida, and if you happened to have a man nearby, you might ask him to look over my garden fence, when he'd say 'That man keeps a goat'. But what would you do if I said, 'I keep a unicorn in my backyard'?" We have no way of knowing whether the point was taken.

Eventually, there was no more to say. We shook hands all round, sped past the common-room filled with champagne bottles destined now not to be opened and into the lens of a news agency photographer summoned for the happier event.

(**334**, 287-290;1988)

John Maddox, James Randi & Walter W. Stewart

我们小组中的约翰·马多克斯说如果邦弗尼斯特愿意撤销已发表的论文或者至少向《自然》杂志写信对他的发现和阐述进行修改,那么我们就不发表此次调查报告。双方都同意在 7 月 28 日之前不就此事发表公开声明。但邦弗尼斯特也表示将利用这个周末以及下周的时间验证他们所发现的现象的真实性。

最令我们惊讶同时也让我们失望的是，INSERM 200 似乎并没有意识到他们宣称的轰动性结论仅仅被一些非正常的数据支持。在我们周五的讨论中，兰迪指出，“如果我告诉你我在佛罗里达州的房子的后院里面养了一只山羊，而且你恰巧有一个朋友在那里附近，你就可以请他从篱笆外面看一下，他会说‘那人养了一只山羊’。但如果我说‘我在后院养了一只独角兽’时，你会怎么做呢？”我们不知道他们是否理解了兰迪的意思。

最后，大家都没什么要说的了。我们和所有人握手告别，从摆满了现在无法开启的香槟酒瓶子的公共休息室中穿过，旁边是期待着捕捉圆满结局画面的新闻摄影师。

（张锦彬 翻译；秦志海 审稿）

Resolution of Quantitative Traits into Mendelian Factors by Using a Complete Linkage Map of Restriction Fragment Length Polymorphisms

A. H. Paterson *et al.*

Editor's Note

Genetic mapping has proved a powerful approach for identifying genes and biological processes underlying inheritance-influenced traits. Its simplest form, linkage analysis, uses the meiotic recombination of "linked" genes to work out the location of a gene relative to a known sequence or "marker". But the genetic regions underlying complex traits were at this time hard to map, because the inheritance of an entire genome could not be studied with genetic markers. Here geneticist Eric Lander and colleagues show that complex quantitative traits can be resolved into discrete inheritable (Mendelian) factors. They achieve this by using a complete linkage map of restriction fragment length polymorphisms (RFLPs), DNA variations that become detectable when enzymatically cleaved fragments are analysed on a gel.

The conflict between the Mendelian theory of particulate inheritance[1] and the observation of continuous variation for most traits in nature was resolved in the early 1900s by the concept that quantitative traits can result from segregation of multiple genes, modified by environmental effects[2-5]. Although pioneering experiments[6-9] showed that linkage could occasionally be detected to such quantitative trait loci (QTLs), accurate and systematic mapping of QTLs has not been possible because the inheritance of an entire genome could not be studied with genetic markers[7]. The use of restriction fragment length polymorphisms[10] (RFLPs) has made such investigations possible, at least in principle. Here, we report the first use of a complete RFLP linkage map to resolve quantitative traits into discrete Mendelian factors, in an interspecific back-cross of tomato. Applying new analytical methods, we mapped at least six QTLs controlling fruit mass, four QTLs for the concentration of soluble solids and five QTLs for fruit *p*H. This approach is broadly applicable to the genetic dissection of quantitative inheritance of physiological, morphological and behavioural traits in any higher plant or animal.

The parents for the back-cross were the domestic tomato *Lycopersicon esculentum* cv. UC82B (denoted E) and a wild South American green-fruited tomato *L. chmielewskii*[11] accession LA1028 (denoted CL). These strains have very different fruit masses (E~65 g; CL~5 g) and concentrations of soluble solids[12] (E~5%; CL~10%)—traits of agricultural importance, because they jointly determine the yield of tomato paste. In addition, the

通过限制性酶切片段长度多态性的完整连锁图对孟德尔因子数量性状的解析

佩特森等

编者按

遗传作图已被证明是用于识别基因和遗传性状背后的生物学过程的有力工具。连锁分析，是遗传作图最简单的形式，它通过利用“连锁”基因的减数分裂重组来定位一个基因相对于一段已知序列或“标记”的位置。但由于仅通过遗传标记无法研究整个基因组的遗传，故对决定复杂性状的遗传区域的定位成为一大难题。本文中遗传学家埃里克·兰德及其同事的工作表明，复杂的数量性状可被分解为独立的遗传因子（孟德尔因子）。通过对限制性酶切片段长度多态性完整连锁图的分析可得出以上结论，DNA 变异可以通过酶切片段的凝胶电泳分析来检测。

孟德尔的颗粒遗传理论和自然界生物的大部分性状中所观察到的连续变异之间存在矛盾，这一矛盾在二十世纪初期通过一个关键观点的提出而得到解决，即经环境因素修饰的多基因的分离产生了数量性状 [2-5]。尽管早期一些开拓性的实验 [6-9] 已经表明偶尔能够测到这些数量性状基因座（QTL）的连锁，但是对其更加准确和系统地作图却还不能实现，因为只通过一些基因遗传标记无法对整个基因组的遗传进行研究 [7]。限制性酶切片段长度多态性（RFLP）[10] 的应用使得这一研究成为可能，至少在理论上此方法是可行的。在此，我们首次在番茄的种间回交实验中用完整的 RFLP 连锁图谱将数量性状分解为孟德尔遗传因子来研究。采用这种新的分析方法，我们鉴定出了六个控制果实质量的 QTL、四个控制可溶性固形物含量的 QTL 以及五个控制果实 pH 的 QTL。此方法也可广泛地应用于其他高等动植物的生理学特征、形态学特征和行为学特征数量性状的遗传学研究中。

回交品种的亲本是驯养品种番茄 *Lycopersicon esculentum* cv. UC82B(标注为 E) 和野生的南美绿果番茄 *L. chmielewskii*[11] accession LA1028(标注为 CL)。这些品种在果实质量（E 约为 65 克，CL 约为 5 克）和可溶性固形物含量 [12]（E 约为 5%；CL 约为 10%）方面有很大差异，而这两个性状体现在作物的农业价值上是很重要的，

strains are known to be polymorphic for genes affecting fruit *p*H, which is important for the optimal preservation of tomato products[13]; the difference in *p*H between the parental strains is, however, small.

A total of 237 back-cross plants, with E as the recurrent parent, were grown in the field at Davis, California. Between five and 20 fruit from each plant were assayed[13] for fruit mass, soluble-solids concentration (°Brix; see Fig. 1 legend for definition) and *p*H, each of which showed continuous variation (Fig. 1). Soluble-solids concentration correlated negatively with fruit mass ($r = -0.42$) and positively with *p*H ($r = +0.33$).

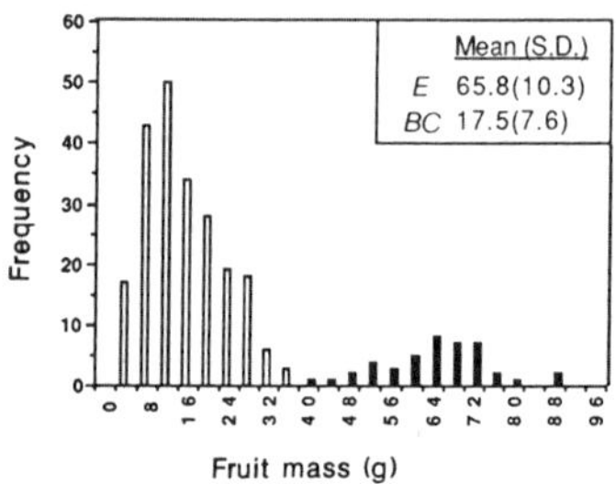

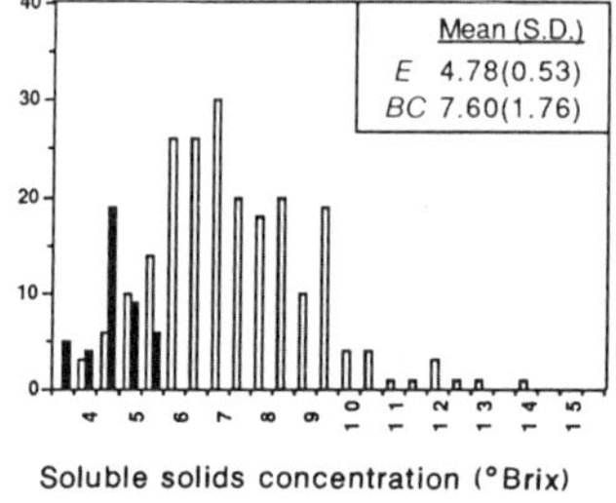

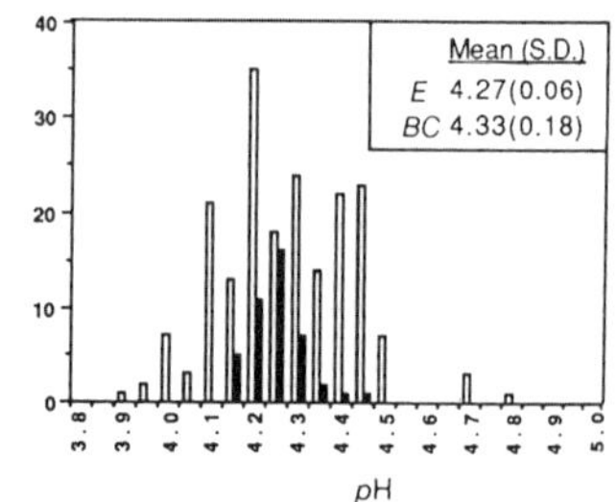

Fig. 1. Frequency distribution for fruit mass, soluble-solids concentration (°Brix, a standard refractometric measure primarily detecting reducing sugars, but also affected by other soluble constituents; 1°Brix is approximately 1% w/w) and *p*H in the E parental strain and in the back-cross (BC) progeny. The tomatoes were grown in the field at Davis, California, in a completely randomized design including 237 BC plants (with E as the recurrent pistillate parent), as well as E, CL and the F_1 as controls. Neither CL nor the F_1 progeny matured completely, as is typical in the central valley of California. Among the BC plants, six failed to mature and 12 produced too few fruit to assay reliably for quantitative traits. The absence of quantitative trait data for these few progeny should yield at most a slight bias in our analyses. Means and standard deviations for the distributions of the E parental strain (*E* filled bars) and the BC progeny (*BC* open bars) appear in the upper right of each histogram. The distributions for soluble-solids concentration and *p*H are approximately normal. The distribution of the BC progeny for fruit weight is clearly skewed; $\log_{10}$ (fruit mass) was studied throughout to achieve approximate normality (see ref. 5; $E = 1.81\pm0.07$; $BC = 1.20\pm0.19$). The proportion of variance due to environment was estimated to be the square of the ratio of the standard deviations (*E*/*BC*), for log-mass, solids and *p*H.

We had previously constructed a genetic linkage map of tomato[14] with over 300 RFLPs and 20 isozyme markers, by analysing 46 F_2 individuals derived from *L. esculentum* cv. VF36×*L. pennellii* accession LA716 (E×P). The map is essentially complete: it has linkage groups covering all 12 tomato chromosomes with an average spacing of 5 cM between markers (1 cM is the distance along the chromosome which gives a recombination frequency of one percent). For QTL mapping, we selected a subset of markers spaced at approximately 20 cM intervals and displaying polymorphism between the E and CL strains. These included 63 RFLPs and five isozyme markers. In addition, the E and CL strains differ in two easily-scored, simply-inherited morphological traits: determinacy (described below) and uniform ripening, controlled by the *sp* and *u* genes, respectively. Although a few distal regions did not contain appropriate markers, we estimate that about 95% of the tomato genome was detectably linked to the markers used.

因为它们共同决定了番茄酱的产量。并且影响果实 pH 值的基因在这两个品种中呈现多态性，果实 pH 值对于番茄制品的优化保存是很重要的[13]；而亲本品种的 pH 值差异很小。

以 E 作为轮回亲本，在加利福尼亚州戴维斯的试验田中一共种植了 237 株回交品种植株，然后分析了每个植株上 5~20 个不等的果实[13]，得到了这些植株的果实质量、可溶性固形物含量 (单位为白利度，见图 1 注中对其的定义) 和 pH 值性状指标，每一个性状都表现出连续变异的特点（见图 1）。可溶性固形物含量与果实质量（$r=-0.42$）呈负相关，而与果实的 pH（$r=+0.33$）呈正相关。

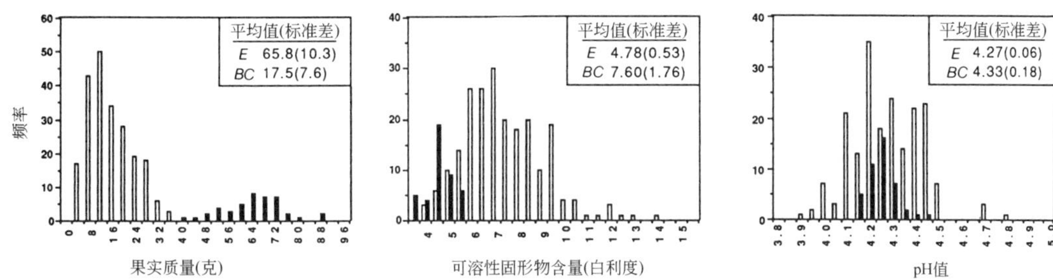

图 1. E 亲本和回交 (BC) 后代中果实质量、可溶性固形物含量（白利度是主要用于测定还原糖的标准折射率测定法，但也会受到其他可溶性成分的影响。1 白利度约等于 1% 质量浓度）和 pH 的频率分布。番茄以完全随机的方式种植于加利福尼亚州戴维斯的试验田里，包括 237 株 BC 植株，（以 E 作为雌性轮回亲本），以 E、CL 和 F_1 作为对照组。由于是生长在加利福尼亚的中央谷中，CL 和 F_1 子代没有完全成熟。在所采集的 BC 植株中，有 6 个未成熟，有 12 个所产果实由于量过少以至于不能准确分析其数量性状。这些少数样本的数量性状数据的缺失对我们的数据分析影响并不大。每个柱状图的右上方显示了 E 亲本（实心柱所示）和 BC 子代（空心柱所示）频率分布的平均值和标准差。可溶性固形物含量和 pH 的频率分布近似正态分布，但 BC 子代的果实质量的频率分布发生了明显的偏移。将果实质量进行对数作图以求其近似正态分布性（详见参考文献 5，E=1.81±0.07，BC=1.20±0.19），通过计算标准差（E/BC）的平方评估了环境所引起的果实质量对数、可溶性固形物含量和 pH 值这些性状比例的变化。

在此之前，我们已经通过分析 46 株 *L. esculentum* cv. VF36×*L. pennellii* accession LA716（E×P）的 F_2 代植株，绘制出了番茄的基因连锁图谱，其中包括超过 300 个 RFLP 和 20 个同工酶标记。所得遗传图谱基本完整，包含的连锁群覆盖了所有 12 条番茄染色体，其上遗传标记之间的平均遗传距离为 5 厘摩（1 厘摩为同一染色体上重组率为百分之一的两位点的距离）。在定位 QTL 时，我们选择了遗传距离在 20 厘摩左右、并且在 E 和 CL 品种中呈现多态性的一组遗传标记。所选的这些遗传标记包括了 63 个 RFLP 和 5 个同工酶标记。另外，E 和 CL 这两个品种在生长习性（见下文解释）和成熟度上存在明显差异，这两个分别由 *sp* 和 *u* 基因控制的形态性状比较容易评价且遗传行为简单。尽管一些基因组远端区域没有合适的遗传标记，但我们推测番茄基因组中约 95% 的基因可利用我们设计的生物遗传标记检测到。

These 70 genetic markers were scored for each of the 237 E×CL back-cross progeny (as described in ref. 13), and a linkage map was constructed *de novo* using MAPMAKER[15]. The map covers all 12 chromosomes with an average spacing of 14.3 cM. Although the linear order of markers inferred from the E×CL cross essentially agreed with that inferred from the E×P cross (but see Fig. 3 legend), genetic distances differed markedly in certain intervals (for example, 51 cM in E×P and 11 cM in E×CL, for the distance between the 45S ribosomal repeat and *TG1B* on chromosome 2). In total, the markers scored in both crosses span 852 cM in the E×CL map versus 1,103 cM in the E×P map, a highly significant ($P<0.01$) difference. Skewed segregation ($P<0.05$) was detected for 48 of the 70 markers, comprising 21 distinct regions distributed over all 12 chromosomes. The heterozygote (E/CL) was overabundant in 12 cases, whereas in nine cases the homozygote (E/E) was favoured. Overall, the effects of skewing approximately cancelled each other out: on average, the back-cross contained the expected 75% E genome (Fig. 2).

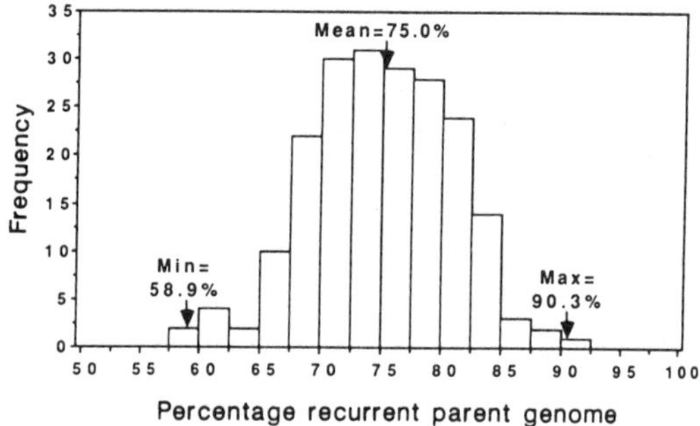

Fig. 2. Distribution of percentage of recurrent parent (E) genotype in the 237 back-cross progeny, estimated on the basis of the marker genotypes and their relative distances. Determination of marker genotypes was as previously described[13]. Estimates of the percentage of recurrent parent genome were produced by the recently developed computer program HyperGene™ (N. D. Young, A.H.P. and S.D.T., unpublished results). Although the average agreed closely with the Mendelian expectation of 75% for a back-cross, values for individual plants ranged from 59% to over 90%. The distribution of the proportion of recurrent-parent genome agrees with the mathematical expectation[35,36]. The individual with >90% E appears to carry only five fragments from CL (ranging from 9 to 47 map units in length) and could be returned to essentially 100% *E* with two additional back-crosses of fewer than 100 plants each, or one additional back-cross of about 550 plants. This is far more rapid than the 6–8 back-crosses routinely used to eliminate donor genome in the absence of markers.

a

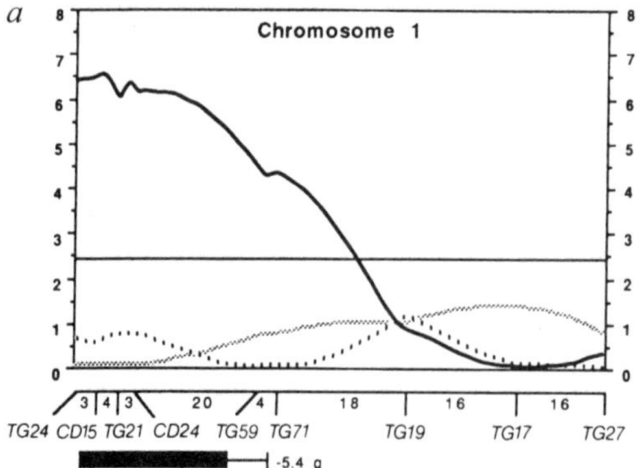

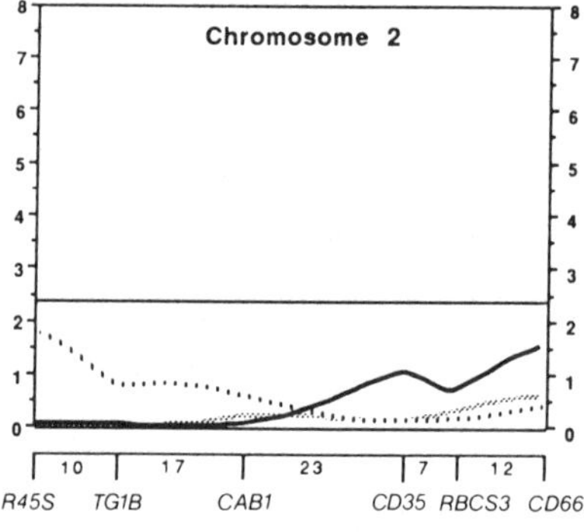

分别对 237 个 E×CL 回交子代（如参考文献 13 所述）的 70 个遗传标记进行遗传距离测定，并用 MAPMAKER[15] 从头构建连锁图。此连锁图覆盖了番茄的全部 12 条染色体，平均遗传距离为 14.3 厘摩。尽管从 E×CL 杂交种中推测出来的所有遗传标记的线性顺序与 E×P 杂交种中的顺序基本一致（见图 3 注），但是某些区间的遗传距离却有很大的差别（例如，45S 核糖体重复序列与 2 号染色体上 *TG1B* 之间的遗传距离在 E×P 和 E×CL 中分别为 51 厘摩和 11 厘摩）。总之，在两次杂交实验中，这些遗传标记在 E×CL 图谱中跨越了 852 厘摩，而在 E×P 图谱中跨越了 1,103 厘摩，统计检验表明这一差异高度显著（$P<0.01$）。经卡方检验，70 个遗传标记中，48 个发生了偏分离（$P<0.05$），这 48 个标记分布于所有 12 条染色体，包含 21 个不同的区域。杂合体（E/CL）在 12 例中丰度较高，而在 9 例中纯合体（E/E）较占优势。回交种中平均携带了预期的 75% 的 E 基因组，总体上，偏移的影响基本上可以被相互抵消（图 2）。

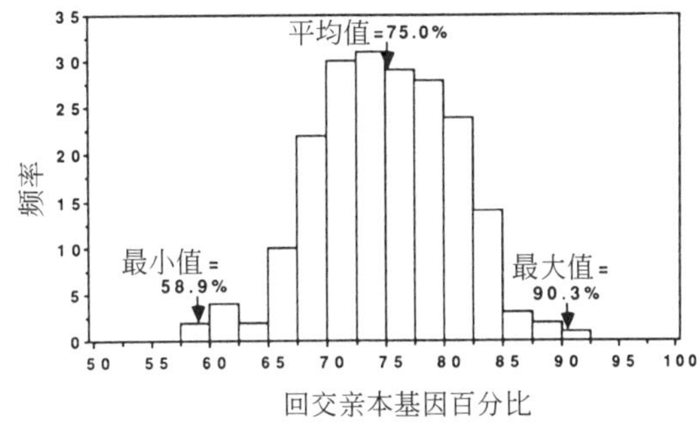

图 2. 根据遗传标记的基因型和它们之间的相对距离，评估轮回亲本 E 基因型在 237 株回交子代中所占百分比的分布情况。对于检测遗传标记基因型的方法如前所述[13]。在计算轮回亲本基因组所占比例时采用了近期新研发的计算机程序—HyperGene™（扬、佩特森和坦克斯利未发表结果）。尽管对整个回交子代来讲所计算百分比的平均值与孟德尔遗传规律所期望的 75% 很接近，但是对个体的植株来说此百分比则介于 59% 至 90% 以上。轮回亲本基因组在子代中所占百分比的分布情况与数学期望一致[35,36]。携带有超过 90% 亲本 E 基因组的个体似乎只携带了五个来自 CL 的片段（在长度上介于 9 到 47 染色体图距单位），这些个体通过两代（每代不超过 100 株植株）的回交或一次约 550 株植株的回交就可以基本恢复到携带 100% E 亲本基因组。在缺少标记的情况下，这与通常需要 6 至 8 代回交才能消除供体基因组相比快了很多。

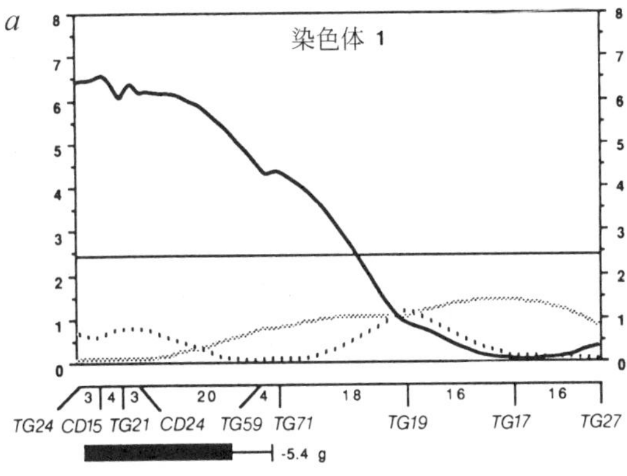

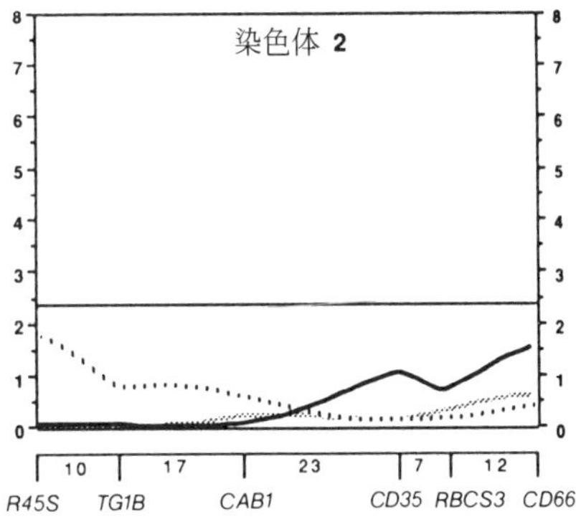

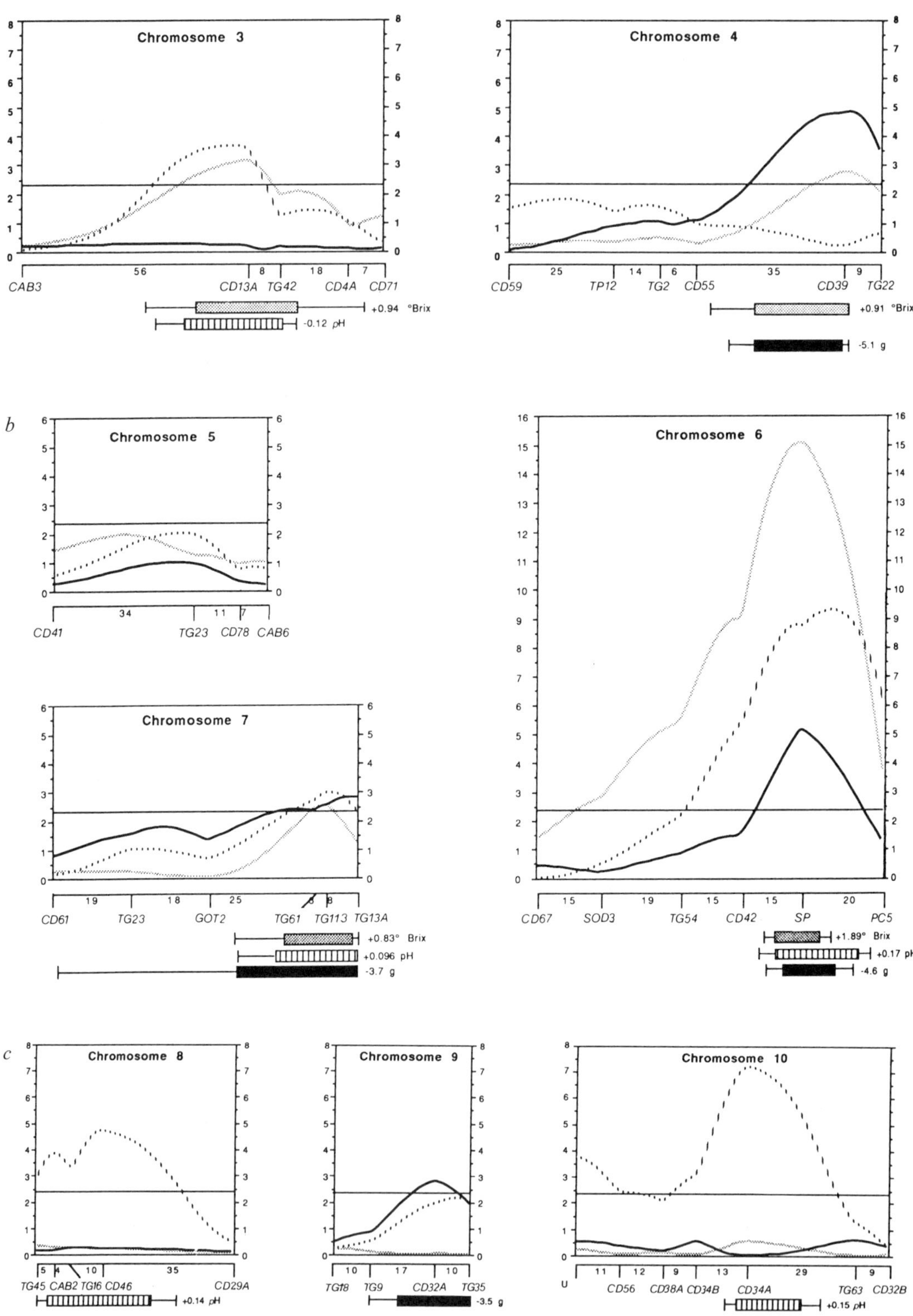
Chromosome 3
CAB3
CD13A
TG42
CD4A
CD71
+0.94 °Brix
-0.12 pH
Chromosome 4
CD59
TP12
TG2
CD55
CD39
TG22
+0.91 °Brix
-5.1 g
b
Chromosome 5
CD41
TG23
CD78
CAB6
Chromosome 6
CD67
SOD3
TG54
CD42
SP
PC5
+1.89° Brix
+0.17 pH
-4.6 g
Chromosome 7
CD61
TG23
GOT2
TG61
TG113
TG13A
+0.83° Brix
+0.096 pH
-3.7 g
c
Chromosome 8
TG45
CAB2
TG16
CD46
CD29A
+0.14 pH
Chromosome 9
TG18
TG9
CD32A
TG35
-3.5 g
Chromosome 10
CD56
CD38A
CD34B
CD34A
TG63
CD32B
+0.15 pH

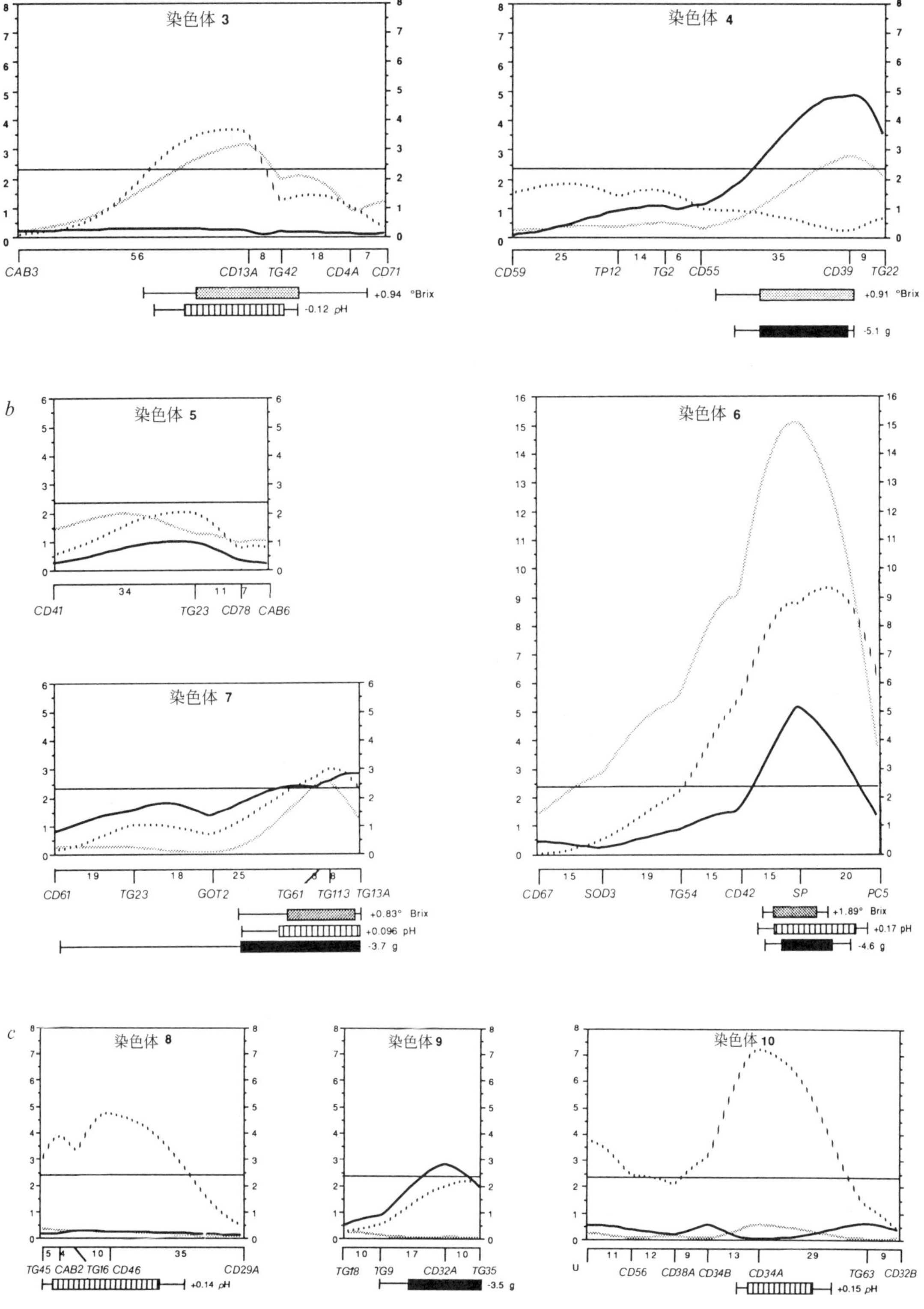

染色体 3
CAB3
CD13A
TG42
CD4A
CD71
+0.94 °Brix
-0.12 pH
染色体 4
CD59
TP12
TG2
CD55
CD39
TG22
+0.91 °Brix
-5.1 g
b
染色体 5
CD41
TG23
CD78
CAB6
染色体 6
CD67
SOD3
TG54
CD42
SP
PC5
+1.89° Brix
+0.17 pH
-4.6 g
染色体 7
CD61
TG23
GOT2
TG61
TG113
TG13A
+0.83° Brix
+0.096 pH
-3.7 g
c
染色体 8
TG45
CAB2
TG16
CD46
CD29A
+0.14 pH
染色体 9
TG18
TG9
CD32A
TG35
-3.5 g
染色体 10
CD56
CD38A
CD34B
CD34A
TG63
CD32B
+0.15 pH

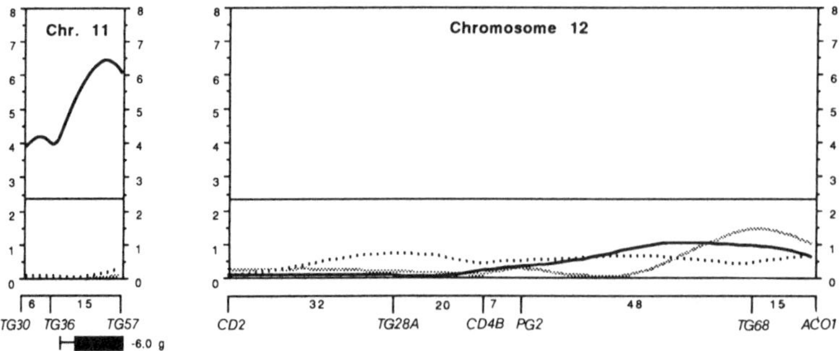

Fig. 3. QTL likelihood maps indicating lod scores for fruit mass (solid lines and bars), soluble-solids concentration (dotted lines and bars) and *p*H (hatched lines and bars), throughout the 862 cM spanned by the 70 genetic markers. The RFLP linkage map used in the analysis is presented along the abscissa, in Kosambi[37] cM. (The order of the markers agrees with the previously published map[14] of the E×P cross, except for three inversions of adjacent markers: (*TG24-CD15*), (*TG63-CD32B*) and (*TG30-TG36*). In the first case, re-analysis of the E×P data with MAPMAKER[15] indicates that the order shown here is the more likely in both E×P and E×CL. For the other two, the orders shown here are more likely in E×CL by odds of 10^4:1 and 10^7:1, but the inverse is more likely in E×P by 11 : 1 and 8 : 1 odds. These differences will be investigated in a larger E×P population.) Soluble-solids concentration and *p*H were analysed in °Brix and *p*H units, respectively; allele effects on fruit mass are presented in g; log-transformation of fruit mass was used in all analyses to achieve approximate normality. The maximum likelihood effect of a putative QTL, as well as the lod score in favour of the existence of such a QTL, have been determined at points spaced every 1 cM throughout the genome, according to Lander and Botstein[16] and a smooth curve plotted through the points. The height of the curve indicates the strength of the evidence ($\log_{10}$ of the odds ratio) for the presence of a QTL at each location—not the magnitude of the inferred allelic effect. The horizontal line at a height of 2.4 indicates the stringent threshold that the lod score must cross to allow the presence of a QTL to be inferred (see text). Information about the likely position of the QTL can also be inferred from the curve. The maximum likelihood position of the QTL is the highest point on the curve. Bars below each graph indicate a 10 : 1 likelihood support interval[16] for the position of the QTL (the range outside which the likelihood falls by a lod score of 1.0), whereas the lines extending out from the bars indicate a 100 : 1 support interval. Phenotypic effects indicated beside the bars are the inferred effect of substituting a single CL allele for one of the two E alleles at the QTL. Several regions show sub-threshold effects on one or more traits (chromosome one near *TG19*, chromosome five near *TG32* and chromosome 12 near *TG68*) which may represent QTLs but this requires additional testing. The region near *TG68* may be particularly interesting, as it is the only instance found where the CL allele seems to decrease soluble-solids concentration (by about 0.7 °Brix). In the case of chromosome 10, the lod score for *p*H crosses the significance threshold in two places. Controlling for the presence of a QTL near *CD34A*, we tested for the presence of a second QTL near *u* (by comparing the maximum lod scores assuming the presence of only the first QTL to the maximum lod score assuming the presence of two QTLs). Allowing for a QTL in the region of *CD34A*, the residual lod score near *u* falls below the required threshold. Thus, the evidence is not yet sufficient to support the presence of a QTL near *u*.

Methods. The lod score and the maximum likelihood estimate (MLE) of the phenotypic effect at any point in the genome is computed assuming that the distribution of phenotypes in the BC progeny represents a mixture of two normal distributions (of equal variance) with means depending on the genotype at a putative QTL at the given position. (Note that QTLs are considered individually and thus we did not assume that different QTL effects can be added—except in studying the possibility of two QTLs on chromosome 10 affecting *p*H.) Specifically, at a given position in the genome, the likelihood function for individual *i* with quantitative phenotype Φ_i is given by $L_i(\alpha, \sigma) = (2\pi\sigma^2)^{-1/2}\{p_1 \exp(-\Phi_i)^2 / 2\sigma^2 + p_2 \exp(-(\Phi_i) - \alpha)^2 / 2\sigma^2\}$, where α is the effect of substituting a CL allele for an E allele at a putative QTL in the given position, σ^2 is the phenotypic variance not attributable to the QTL and p_1 and p_2 are the probabilities that individual *i* has genotype E/E and E/CL, respectively, at the QTL (which can be computed on the basis of the genotypes

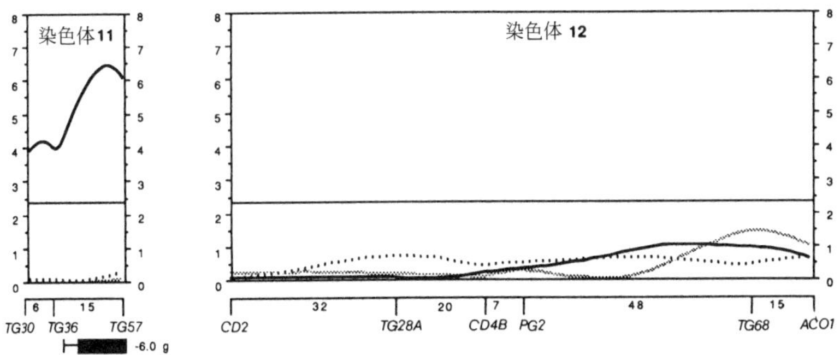

图 3. QTL 似然图谱显示了跨越 862 厘摩遗传距离的 70 个遗传标记在果实质量（实线和实线柱）、可溶性固形物含量（点线和点线柱）和 pH 值（阴影线和阴影柱）上的对数优势比。横坐标为 RFLP 连锁图谱，单位为 Kosambi cM[37]。(遗传标记的顺序与之前发表的 E×P 杂交品种的图谱[14]一致，但例外的是 (*TG24-CD15*)、(*TG63-CD32B*) 和 (*TG30-TG36*) 这三个相邻的倒位遗传标记。在第一种情况下，用 MAPMARKER[15] 重新分析 E×P 的数据，结果表明这里显示的顺序更可能同时存在于 E×P 和 E×CL 两种情况中。而对于另外两种情况，其顺序则在 E×CL 中的概率为 10^4：1 和 10^7：1，但是其反向顺序在 E×P 中的概率则为 11：1 和 8：1。这些差异仍需研究更大的 E×P 种群。) 可溶性固形物含量、pH 值分别以白利度和 pH 为单位进行分析；等位基因对果实质量的影响以克表示；为获得近似的正态分布，在所有分析中将果实质量进行对数转换。假定的 QTL 的最大似然效应和这一 QTL 存在的对数优势比已在整个基因组上每 1 厘摩距离的位点中确定，此计算是基于兰德和博特斯坦的方法[16]和这些点拟合后的平滑曲线完成的。曲线的高度代表了每个位置存在 QTL 的证据的强度（优势比以 10 为底的对数值），而不是推测的等位基因效应强度。图中 2.4 位置处的水平线是所设定的严格阈值，只有高于这个阈值时，才认定存在 QTL（详见正文）。根据曲线也可以推测出 QTL 可能的位置信息，曲线上的最高点代表了 QTL 的最大似然位置。每个图下方的横柱显示了 QTL 所在位置的 10：1 似然支持区间[16]（范围以外的似然值下降了 1.0 个对数优势比），而从横柱延伸出的线段则代表了 100：1 的支持区间。除了根据横柱估计，还推测出由单个 CL 等位基因替代 QTL 处的两个 E 等位基因中的一个后引起的表型效应。若干区域在一个或多个性状上表现出了亚阈值效应（1 号染色体靠近 *TG19* 的区域、5 号染色体上靠近 *TG32* 的区域和 12 号染色体上靠近 *TG68* 的区域），这些性状可能说明了 QTL 的存在，但需要进一步检验。靠近 *TG68* 的区域尤其值得关注，因为它是仅有的 CL 等位基因可能减少可溶性固形物含量的例子（减幅约为 0.7 白利度）。10 号染色体中，pH 性状的对数优势比曲线有两处都高于显著性阈值。控制 *CD34A* 附近存在的 QTL，我们检验了 *u* 附近存在另一个 QTL 的可能性（通过比较只存在一个 QTL 时的最大优势比与两个 QTL 都存在的最大优势比）。若在 *CD34A* 区域存在一个 QTL 时，*u* 附近区域其余的优势比则低于要求阈值。因此，目前的证据还不能充分证明 *u* 附近区域存在 QTL。

方法。假设回交子代中表型效应的分布代表了两个混合的正态分布（具有相等方差），其平均值与给定位置处所假定的 QTL 的基因型相关。然后计算基因组中任何位点上表型影响的对数优势比和最大似然估计（MLE）。(注意：在此 QTL 是作为单独个体来研究的，因此，除研究 10 号染色体上两个 QTL 对 pH 值的影响外，我们并没有假设不同 QTL 引起的效应是可以相加。) 尤其是在基因组的给定位点，对于有数量性状表型 Φ_i 的个体 i，其似然函数为 $L_i(\alpha,\sigma)=(2\pi\sigma^2)^{-1/2}\{p_1\exp(-\Phi_i)^2/2\sigma^2+p_2\exp(-(\Phi_i)-\alpha)^2/2\sigma^2\}$，其中 α 是在给定位置的假定 QTL 上由一个 CL 等位基因代替 E 等位基因后的效应。σ^2 是非数量性状位点引起的表型差异。p_1 和 p_2 分别是个体 i 在 QTL 处（QTL 位置可以通过侧翼遗传标记的基因型和 QTL 与标记的距离计算得到）拥有 E/E 和 E/CL 基因型的概率。整个种群的似然函数为 $L=\prod L_i$。同样，α* 和 σ* 表示 MLE 所允许出现 QTL 的可能性（最大值为 L），σ** 表示 σ 的 MLE，即限定无连锁的 QTL（α=0）。对数优势比为 $\log_{10}\{L(\alpha^*, \sigma^*)/L(0, \sigma^{**})\}$。此方法在数量性状位点图谱绘制上的发展在参考文献 16 中有更充分的展示。

at the flanking markers and the distance to the flanking markers). The likelihood function for the entire population is $L = \Pi L_i$. Also α^* and σ^* denote the MLEs allowing the possibility of a QTL at the location (the values which maximize L) and σ^{**} denotes the MLE of σ, subject to the constraint that no QTL is linked ($\alpha = 0$). The lod score is then given by $\log_{10}\{L(\alpha^*, \sigma^*) / L(0, \sigma^{**})\}$. This method for QTL mapping is developed more fully in ref. 16.

We then turned to the question of mapping the Mendelian factors that underly continuous variation in fruit mass, soluble-solids concentration and *p*H. The method of maximum likelihood and lod scores, commonly used in human linkage analysis[17], has recently been adapted[16] to allow interval mapping of QTLs. At each position in the genome, one computes the "most likely" phenotypic effect of a putative QTL affecting a trait (the effect which maximizes the likelihood of the observed data arising) and the odds ratio (the chance that the data would arise from a QTL with this effect divided by the chance that it would arise given no linked QTL). The lod score, defined as the $\log_{10}$ of the odds ratio, summarizes the strength of evidence in favour of the existence of a QTL with this effect at this position; if the lod score exceeds a pre-determined threshold, the presence of a QTL is inferred. The traditional approach[8,9] to mapping QTLs involves standard linear regression, which accurately measures the effect of QTLs falling at marker loci only, underestimating the effects of other loci in proportion to the amount of recombination between marker and QTL. In contrast, interval mapping allows inference about points throughout the entire genome and avoids confounding phenotypic effects with recombination, by using information from flanking genetic markers. In the special case when a QTL falls exactly at a marker locus, interval mapping reduces to linear regression. A computer program, MAPMAKER-QTL, was written (S.E.L. and E.S.L., unpublished) to implement interval mapping.

Due to the large number of markers tested, an extremely high lod score threshold must be adopted to avoid false positives. Given the genetic length of the tomato genome and the density of markers used, a threshold of 2.4 gives a probability of less than 5% that even a single false positive will occur anywhere in the genome[16]. This is approximately equivalent to requiring the significance level for any single test to be 0.001.

QTL likelihood maps, showing how lod scores for fruit mass, soluble-solids concentration and *p*H change as one moves along the genome, reveal multiple QTLs for each trait and estimate their location to within 20–30 cM (Fig. 3).

(1) Factors for fruit mass were found on six chromosomes (1, 4, 6, 7, 9 and 11). In each case, CL alleles decrease fruit mass (by 3.5 to 6.0 g), adding to a total reduction of 28.1 g inferred for back-cross progeny carrying a CL allele at all six loci. This accounts for about half of the approximately 60 g difference between E and CL.

(2) Factors for soluble-solids concentration were found on four chromosomes (3, 4, 6 and 7). In each case, CL alleles elevate soluble-solids concentration (by 0.83 to 1.89 °Brix), adding to a total of 4.57 °Brix (versus a difference of ~5 °Brix between the parental strains). This

我们再回到最初的问题上，如何定位果实质量、可溶性固形物含量和pH值这三个连续变异背后的孟德尔遗传因子呢？最大似然法和对数优势比通常被用于人的遗传连锁分析[17]，最近这些方法也被用于数量性状位点（QTL）的区间作图[16]。在基因组的每一个位置计算出QTL影响性状的“最大可能”表型效应（此效应将所观测到数据的似然最大化）和优势比（数据来自与此效应相关的QTL的概率除以与此效应无关的QTL概率的比值）。对数优势比定义为以10为底优势比的对数值，它代表了在某一位置存在与表型效应相关的QTL的证据的强度，假如对数优势比超过了预设的阈值，则认为存在QTL。作为定位QTL的传统方法[8,9]之一，标准线性回归精确地测量了仅位于遗传标记位点的QTL所产生的效应，却低估了与基因标记和QTL重组相关的其他位点所产生的效应。但区间作图运用了侧翼遗传标记的信息，从而可以推测出贯穿整个基因组的位点，并且能够避免混淆重组造成的表型效应。在特殊情况下，当QTL严格地位于遗传标记位点时，间距图谱可以归纳为线性回归。MAPMAKER-QTL是一个专门为区间作图而编写的计算机程序（兰德和林肯，尚未发表）。

由于检验了较大数量的遗传标记，所以需要设定一个极高的对数优势比的阈值以避免得到假阳性结果。鉴于番茄基因组的遗传学长度和使用的遗传标记的密度，当预设的阈值为2.4时，出现假阳性的概率不到5%，即使这样也意味着单个假阳性会出现在基因组的任何位置[16]。这相当于在任何单次检验中其显著性水平都需要为0.001。

QTL的似然图谱显示了每一性状的多个QTL及其在20~30厘摩遗传距离范围内的定位（图3），能够让我们看出在整个基因组水平上针对果实质量、可溶性固形物含量和pH值的对数优势比是怎样变化的。

（1）我们在六条染色体（1、4、6、7、9和11）上发现了能够影响果实质量的因子。在每一个位点，CL的等位基因都能降低果实质量（减少量从3.5~6克不等），推测在携带所有6个CL等位基因的回交子代中果实质量总减少量达28.1克。这也许能解释为什么E和CL的果实质量存在的约30克的差距。

（2）在四条染色体（3、4、6和7）上发现了能够影响可溶性固形物含量的因子。每一个位点上，CL等位基因都能够提高可溶性固形物的含量（从0.83~1.89白利度不等），增加量总和为4.57白利度（与亲本品种之间存在的约5白利度的差异相对应）。

large effect in the back-cross is consistent with previous reports that high soluble-solids concentration exhibits dominance[12] and overdominance[13]. The QTL alleles for both fruit mass and soluble-solids concentration all produce effects in the direction predicted by the difference between the parental strains.

(3) Factors for *p*H were found on five chromosomes (3, 6, 7, 8 and 10). In addition, the lod score for a putative QTL on chromosome 9 fell just below our threshold. Because the parental strains do not differ greatly in *p*H, we suspected that CL alleles might not all produce effects in the same direction. In fact, *p*H was increased by four QTLs and decreased by two, including the likely QTL on chromosome 9. This provides a genetic explanation for the observation that many back-cross progeny exhibited more extreme phenotypes than the parental strains (Fig. 1), a phenomenon known as transgression[18].

Together, the QTLs identified for fruit mass, soluble solids and *p*H account for 58%, 44% and 48%, respectively, of the phenotypic variance among the back-cross progeny, with another 13%, 9% and 11% attributable to environment.

The numbers of QTLs reported for each trait must be considered a minimum estimate. Because an extremely stringent threshold was used to avoid any false positives, some sub-threshold effects probably represent real QTLs. For example, the regions near *TG19* on chromosome 1, *CD41* on chromosome 5 and *TG68* on chromosome 12 may affect soluble-solids concentration and merit further attention in larger populations. Similarly, the region near the *u* locus on chromosome 10 may contain an additional QTL affecting *p*H (see Fig. 3 legend). Moreover, we cannot rule out the presence of many additional QTLs with tiny phenotypic effects—postulated in evolutionary theory[19] and supported by some experimental evidence[20]. Also, it is conceivable that some of our apparent QTLs actually represent several closely-linked QTLs, each with small phenotypic effects in the same direction—a phenomenon that might arise particularly in regions of genetic map compression. Finally, we should emphasize that the QTL mapping here strictly applies only to the specific environment tested and to heterozygosity for CL alleles. In principle, homozygosity for CL alleles could have been studied by using an F_2 self between E and CL, but in practice too many of the progeny are sterile.

Some regions of the genome clearly exert effects on more than one trait (for example, chromosome 6; Fig. 3), providing a genetic explanation for at least some of the correlation between the traits. Although the present data are insufficient to distinguish between pleiotropic effects of a single gene and independent effects of tightly-linked loci, the frequent coincidence of QTL locations for different traits makes it likely that at least some of the effects are due to pleiotropy.

The region near *sp* on chromosome 6 has the largest effects on soluble solids and *p*H, as well as a substantial effect on fruit mass. The *sp* gene affects plant-growth habit: the dominant CL allele causes continuous apical growth (indeterminate habit), whereas the recessive E allele causes termination in an inflorescence ("determinate" or "self-pruning"

在回交品种中产生如此巨大的影响，这与之前关于可溶性固形物的高含量呈显性[12]或超显性[13]的报道相一致。控制果实质量和可溶性固形物含量的QTL等位基因产生的影响都与根据亲本品系之间差异所预期的影响一致。

(3) 发现控制pH的位点存在于五条染色体（3、6、7、8和10）上，此外9号染色体上一个假定的QTL对数优势比恰好低于我们所设定的阈值。因为在亲本中pH差异就很小，所以我们推测CL等位基因并没有都朝同一个方向起作用。事实上，包括存在于9号染色体上疑似的QTL在内，有四个QTL引起pH值的升高，两个位点造成pH值的降低。这在遗传学角度上解释了为什么许多回交子代与亲代相比具有更加极端的表型（图1），也就是所谓的越亲现象[18]。

总体上，在鉴定出的回交子代的果实质量、可溶性固形物含量和pH值表型变异中，各有58%、44%和48%受到QTL的控制，而其他各13%、9%和11%的变异则归因于环境因素。

对控制每个性状的QTL数目必须有个最小估算。为了避免任何假阳性，设置了极端严格的对数优势比阈值，但一些真正的QTL产生的效应也可能是亚阈值。例如，1号染色体上靠近*TG19*的区域、5号染色体*CD41*附近的区域和12号染色体*TG68*附近的区域可能影响了可溶性固形物含量，值得在更大的种群范围内更深入地关注。相似的情况还有10号染色体上靠近*u*基因座区域可能存在一个影响pH值的额外QTL（见图3注）。我们不能否认，那些仅产生细微表型变化的数量性状位点的存在是进化论[19]的推论，而且可以得到实验证据[20]的支持。可想而知，一些我们认为显著的QTL事实上代表了若干紧密连锁的位点，每一个位点都产生细微的同一趋势的表型效应，特别是在遗传图谱位点基因密集区会出现这样的现象。最后还要特别强调的是这里的QTL遗传图谱只能严格应用于特定环境下的检测和CL等位基因杂合子的情形。理论上，CL等位基因纯合子能够通过E和CL的F_2自交后代来研究，但实际上非常多的子二代是不育的。

我们已经很清楚，基因组上的一些区域能够影响不止一种性状（如图3所示的6号染色体），这给出了性状之间至少存在某种关联的遗传学解释。尽管现有的数据还不足以区分单个基因的多效性和紧密连锁的多个基因座各自独立的效应，但是同一位置的QTL时常影响着不同的遗传性状，说明可能至少某些效应是基因多效性造成的。

6号染色体上接近*sp*基因的区域对可溶性固形物和pH值能够造成最显著的影响，对果实质量也能产生巨大的影响。*sp*基因能够影响植物的生长习性，其显性CL等位基因能够导致连续性的顶端生长（无限生长习性），而隐性E等位基因会终

habit)[21]. Although indeterminacy has been reported previously[22] to elevate both fruit mass and soluble-solids concentration within *L. esculentum*, we associated it with reduced fruit mass in both E×CL and another interspecific cross (E×*L. cheesmanii*; A.H.P., S. Damon, J.D.H. and S.D.T., unpublished data). These differing results might be due to a second, tightly-linked locus or to unlinked modifier genes.

Overall, pairwise epistatic interactions between intervals were not common (about 5% of two-way analysis-of-variance tests were significant at 0.05). An interesting exception was the region near *TG16* on chromosome 8, at which the CL allele significantly enhanced the effect of three of the four QTLs for soluble-solids concentration. *TG16* also showed the most extreme segregation distortion of any marker scored (about 4 : 1 in favour of the E/E homozygote) and is in a region known to exhibit skewed segregation in back-crosses to other green-fruited tomato species[23,24]. The unusual properties of this region of CL clearly merit further study.

The QTLs identified here may well differ from those that would be fixed by repeated back-crossing with continuing selection for a trait, a classical method for introgressing quantitative traits. Work on LA1563, a strain with increased soluble solids produced[12] through back-crossing a different strain of E to CL has provided some suggestive evidence. By surveying RFLPs, Tanksley and Hewitt[13] recently found that LA1563 has maintained three separate regions from CL: near *CD56* on chromosome 10, near *Got2* on chromosome 7 and near *TG13* on chromosome 7. Here, we detected above-threshold effects in the last of these three regions only (which, interestingly, failed to show effects on soluble solids in a single-environment test by Tanksley and Hewitt[13]). Moreover, we detected QTLs affecting soluble-solids concentration in regions that did *not* seem to be retained. Unfortunately, the results of the two experiments are not directly comparable due to the use of a different E strain by Rick, possible environmental differences between the experiments, the possibility that small CL fragments containing QTLs went undetected in LA1563, the possibility that the region near *TG13* retained in LA1563 may not contain the QTL we detected here and the possibility that some of our sub-threshold effects are real. Although more detailed studies are clearly needed, it is interesting to speculate about why repeated back-crossing may fix a narrower class of QTLs than found by QTL mapping. Because such breeding programs[12] demand horticultural acceptability, they are likely to select against otherwise-desirable QTLs which are closely linked to undesirable effects from the wild parent. If such QTLs can first be identified by mapping, it may be feasible to remove linked deleterious effects by recombination.

Having mapped several QTLs with relatively large effects, we are now making crosses to isolate them in near-isogenic lines. These lines will be used to characterize the QTLs in various dosages, genetic backgrounds, environments and combinations. By re-assembling selected CL alleles in an otherwise E genotype, we hope to engineer an agriculturally-

止花序的顶端生长（有限生长或自修剪习性）[21]。尽管之前的研究认为无限生长能够提高 *L. esculentum* 品种中果实的质量和可溶性固形物的含量，但是我们发现它与 E×CL 杂交品种和其他一些种间杂交品种（E×*L. cheesmanii* 杂交品种；佩特森、戴蒙、约翰和坦克斯利，未发表的数据）的果实质量的降低有关。此差异可能是由另一个与基因座紧密连锁的基因或者不连锁的修饰基因造成的。

总的来说，位点之间成对的上位相互作用并不是很普遍（约 5% 的双向方差分析检验的显著度是 0.05）。一个有趣的特例是在 8 号染色体上靠近 *TG16* 的区域，CL 等位基因能够显著地增强四个影响可溶性固形物含量的 QTL 中三个的效应。*TG16* 也显示出与任何遗传标记的极端偏分离（E/E 纯合子中约 4∶1），*TG16* 所在的区域在与其他绿果番茄品种回交时显示出偏分离 [23,24]。所以 CL 基因组中这个非同寻常的区域是值得进一步研究的。

我们所鉴定的 QTL 与传统经典的方法获得的 QTL 有很大不同，因为传统的鉴定基因渗入的数量性状是通过反复回交连续筛选同一性状。LA1563 是一个可溶性固形物含量增加了的品种，它是通过 E 的另一个种系回交 CL 得到的，对 LA1563 的研究为我们提供了一些提示性证据。坦克斯利和休伊特最近通过观察 LA1563 的 RFLP 发现 LA1563 保留了三个来自 CL 的独立区域，分别是 10 号染色体上靠近 *CD56* 的区域、7 号染色体上靠近 *Got2* 的区域、7 号染色体上靠近 *TG13* 的区域。在此我们仅在这三个区域中的最后一个中检测到超出临界阈值的效应（有意思的是坦克斯利和休伊特并未在单一环境检验中检测到这一区域对可溶性固形物浓度的影响）。并且我们又检测了那些看似不能保留下来的影响可溶性固形物含量的 QTL。然而这两个实验的结果并不能直接进行比较。因为以下几方面原因：第一，里克用了 E 的另外一个不同的种系；第二，两实验的实验环境也可能不一样；第三，一些含有 QTL 的 CL 基因组的小片段在 LA1563 中可能是检测不到的；第四，也许我们所检测的保留在 LA1563 中的 *TG13* 附近的区域是不含 QTL 的；第五，一些亚阈值效应的 QTL 也许是存在的。尽管需要更多的详细研究，但有趣的是，仍可以推测为什么反复回交鉴定的 QTL 的类别要少于绘制 QTL 图谱所鉴定出来的位点的原因。因为这种反复回交的育种方案要求有园艺学上的可接受性，需要筛选掉其他一些野生亲本中与非期望效应紧密连锁的 QTL。假如这种遗传作图的方法能在最初就鉴定出这些 QTL，它将有望通过重组消除不利的连锁效应。

由于已经定位了若干引起相对显著效应的 QTL，我们正在通过杂交从近等基因系中分离出这些位点。这些近等基因系可用于在不同的剂量水平、遗传背景、环境因素以及上述各种影响因素组合的背景下鉴定 QTL。我们希望通过从其他 E 基因型中重组

useful tomato with a higher yield of soluble solids.

The general approach of QTL mapping is broadly applicable to a wide range of biological endeavours. In agriculture, it might be desirable to transfer to domestic strains many quantitative traits harboured in wild species, including resistance to diseases and pests, tolerance to drought, heat, cold and other adverse conditions, efficient use of resources and high nutritional quality[25,26]. In mammalian physiology, selective breeding has generated rodent strains which differ greatly in quantitative traits such as hypertension, atherosclerosis, diabetes, predispositions to cancer, drug sensitivities and various behavioural patterns; information on the number, location and nature of these QTLs would be of value in medicine[16,27]. In evolutionary biology, the process of speciation can be investigated by studying the number and nature of genes underlying reproductive isolation[28].

The availability of detailed RFLP linkage maps[14,29-34] makes it possible to dissect quantitative traits into discrete genetic factors (QTLs): all regions of a genome can be assayed and accurate estimates of phenotypic effects and genetic position derived from interval analysis[16]. Once QTLs are mapped, RFLP markers permit genetic manipulations such as rapid construction of near-isogenic lines: flanking markers may be used to retain the QTL and the study of the remaining markers may be used to speed progress by identifying individuals with a fortuitously high proportion of the desired genetic background[38] (Fig. 2 legend). Using isogenic lines, the fundamental tools of genetics and molecular biology may be brought to bear on the study of QTLs—including testing of complementation, dominance and epistasis; characterization of physiological and biochemical differences between isogenic lines; isolation of additional alleles by mutagenesis (at least in favourable systems); and, eventually, physical mapping and molecular cloning of genetic factors underlying quantitative traits.

We thank Janice Chen, Mark Daly, Gerald Dickinson and Mitzi Aguirre for technical assistance. We thank our colleagues and the referees for comments on the manuscript. This work was supported in part by grants from the NSF (to E.S.L. and to S.D.T.), from the System Development Foundation (to E. S. L.), and from the US Department of Agriculture (to S.D.T.).

(**335**,721-726;1988)

Andrew H. Paterson*, Eric S. Lander†‡, John D. Hewitt§, Susan Peterson*, Stephen E. Lincoln† & Steven D. Tanksley*
* Department of Plant Breeding and Biometry, Cornell University, Ithaca, New York 14853, USA
† Whitehead Institute for Biomedical Research, 9 Cambridge Center, Cambridge, Massachusetts 02142, USA
‡ Harvard University, Cambridge, Massachusetts 02138, USA
§ Department of Vegetable Crops, University of California, Davis, California 95616, USA

Received 8 July; accepted 9 September 1988.

筛选 CL 等位基因从而构建出具有农用价值的可溶性固形物产量更高的番茄品种。

QTL 定位的一般方法可以在诸多生物领域有广泛的应用。在农业上，可以将一些野生品种中特有的 QTL 转移到驯化品种中来，包括对疫病和害虫的抗性、对干旱、热、冷等逆境条件的耐受能力，以及对资源的高效利用能力和高营养价值[25,26]。在哺乳动物生理学中，选择育种已经能够培育出数量性状差异很大的啮齿类动物品种。这些数量性状包括高血压、动脉粥样硬化、糖尿病、癌症易感体质、药物敏感性和各种行为模式等方面的差异。这些 QTL 的数量、位置以及性质方面的信息具有一定的医学价值。在进化生物学中，可以通过研究造成生殖隔离的基因的数目和性质来研究物种形成的过程[28]。

精细的 RFLP 遗传连锁图谱[14,29-34]的获得使我们能够将数量性状分解成很多离散的遗传因子（QTL）来研究。这样就可以分析基因组中所有的区域，通过区间分析[16]准确地估算表型效应和基因位置。一旦 QTL 被定位，就能利用 RFLP 遗传标记进行一些遗传操作，例如快速构建近等基因系：侧翼遗传标记可用于保持 QTL，对保留下来的遗传标记的研究可以用来加速对具有罕见高比例的期望遗传背景的个体识别进程[38]（图 2 注）。作为遗传学和分子生物学的基本工具，等基因系可用于 QTL 的研究，包括基因互补作用、显性和上位效应、等基因系之间生理生化特性差异的鉴定、利用基因突变对其他等位基因的分离（至少是在良好的体系中的突变）以及最终对这些决定数量性状的遗传因子进行物理定位和分子克隆。

在此，我们要感谢贾尼丝·陈、马克·戴利、杰拉尔德·迪金森和米蒂兹·阿吉雷在技术上的帮助。还要感谢我们的同事和审稿人对稿件提出的宝贵意见。本工作得到了美国国家科学基金会、系统发育基金和农业部的部分资助。

（周晓明 翻译；方向东 审稿）

References:

1. Mendel, G. *Verh. naturf. Ver. in Brunn.* **4**, (1868).
2. Johannsen, W. *Elemente der exakten Erblichkeitsllehre* (Fischer, Jena, 1909).
3. Nilsson-Ehle, H. *Kreuzunguntersuchungen an Hafer und Weizen* (Lund, 1909).
4. East, E. M. *Genetics* **1**, 164-176 (1915).
5. Wright, S. *Evolution and the Genetics of Populations* (University of Chicago Press, 1968).
6. Sax, K. *Genetics* **8**, 552-556 (1923).
7. Thoday, J. M. *Nature* **191**, 368-369 (1961).
8. Tanksley, S. D., Medina-Filho, H. & Rick, C. *Heredity* **49**, 11-25 (1982).
9. Edwards, M. D., Stuber, C. W. & Wendel, J. F. *Genetics* **116**, 113-125 (1987).
10. Botstein, D., White, R. L., Skolnick, M. & Davis, R. W. *Am. J. Hum. Genet.* **32**, 314-331 (1980).
11. Chmielewski, T. *Genet. pol.* **9**, 97-124 (1968).
12. Rick, C. M. *Hilgardia* **42**, 493-510 (1974).
13. Tanksley, S. D. & Hewitt, J. *Theor. appl. Genet.* **75**, 811-823 (1988).
14. Tanksley, S. D., Miller, J., Paterson, A. & Bernatzky, R. *Proc. 18th Stadler Genet. Symp.* (in the press).
15. Lander, E. S. *et al. Genomics* **1**, 174-181 (1987).
16. Lander, E. S. & Botstein, D. *Genetics* (in the press).
17. Ott, J. *Analysis of Human Genetic Linkage* (Johns Hopkins, Baltimore, 1985).
18. Simmonds, N. W. *Principles of Crop Improvement*, 82-85 (Longman, New York, 1981).
19. Lande, R. *Heredity* **50**, 47-65 (1983).
20. Shrimpton, A. E. & Robertson, A. *Genetics* **11**, 445-459 (1988).
21. Yeager, A. F. *J. Hered.* **18**, 263-265 (1927).
22. Emery, G. C. & Munger, H. M. *J. Am. Soc. Hort. Sci.* **95**, 410-412 (1966).
23. Zamir, D. & Tadmor, Y. *Bot. Gaz.* **147**, 355-358 (1986).
24. Tanksley, S. D. in *Isozymes in Plant Genetics and Breeding* (eds Tanksley, S. D. & Orton, T. J.) 331-338 (Elsevier, Amsterdam, 1983).
25. Rick, C. M. in *Genes, Enzymes and Populations* (ed. A. M. Srb) 255-268 (Plenum, New York, 1973).
26. Harlan, J. R. *Crop Sci.* **16**, 329-333 (1976).
27. Festing, M. F. W. *Inbred Strains in Biomedical Research* (Oxford, New York, 1979).
28. Coyne, J. A. & Charlesworth, B. *Heredity* **57**, 243-246 (1986).
29. Helentjaris, T. *Trends Genet.* **3**, 217-221 (1987).
30. Landry, B. S., Kessell, R., Leung, H. & Michelmore, R. W. *Theor. appl. Genet.* **74**, 646-653 (1987).
31. Burr, B., Burr, F. A., Thompson, K. H., Albertsen, M. C. & Stuber, C. W. *Genetics* **118**, 519-526 (1988).
32. Chang, C., Bowman, J. L., DeJohn, A. W., Lander, E. S. & Meyerowitz, E. M. *Proc. Natl. Acad. Sci. U.S.A.* **68**, 6856-6860.
33. McCouch, S. R. *et al. Theor. appl. Genet.* (in the press).
34. Bonierbale, M. W., Plaisted, R. L. & Tanksley, S. D. *Genetics* (in the press).
35. Franklin, I. A. *Theor. Populat. Biol.* **11**, 60-80 (1977).
36. Stam, P. *Genet. Res.* **25**, 131-155 (1980).
37. Kosambi, D. D. *Ann. Eugen.* **12**, 172-175 (1944).
38. Tanksley, S. D. & Rick, C. M. *Theor. appl. Genet.* **57**, 161-170 (1980).

DNA Fingerprinting on Trial

E. S. Lander

Editor's Note

DNA fingerprinting, the forensic technique used to analyse DNA from crime scenes, was developed by British geneticist Alec Jeffreys in the mid 1980s. It quickly caught on, and by 1989 had already been used in more than 80 criminal trials in the United States. But here US geneticist Eric Lander cautions against its hasty introduction to the courts. Lander, who had acted as an advisor to the defence on recent murder trials, is concerned that appropriate controls are sometimes lacking and that interpretation of results is not standardized. Here he calls for the scientific community to agree on clear guidelines for procedures and standards in order that the true power of DNA fingerprinting for forensic identification be realized.

In the rush to use the tremendous power of DNA fingerprinting as a forensic tool, the need for standards has been overlooked.

WITH the exception of identical twins, no human beings have identical DNA sequences. Of the 3,000 million nucleotides which we inherit from each parent, about 1 in 1,000 is a site of variation, or polymorphism, in the population. These DNA polymorphisms are most conveniently detected when they alter the length of the DNA fragments produced by the action of restriction enzymes, giving rise to restriction fragment length polymorphisms (RFLPs). In standard practice, the length of the fragments is measured by the rate at which they move in an electrophoresis gel.

More than 3,000 RFLPs have been identified to date, including some 100 highly polymorphic loci at which dozens of variant alleles are present in the population. By using RFLPs to trace the inheritance of chromosomal regions in families afflicted with genetic disorders, human geneticists have been able to pinpoint the location of the genes causing diseases such as Huntington's disease, cystic fibrosis and others—in the process spawning the field of DNA diagnostics.

Forensic science has more recently latched onto RFLPs, but with a different purpose: to identify the individual origin of blood or semen samples found in criminal investigations based on their distinctive RFLP patterns. In the United States, forensic RFLP testing has been pioneered by two private laboratories—Lifecodes Corporation of Valhalla, New York and Cellmark Diagnostics of Germantown, Maryland—and is also used by the Federal Bureau of Investigation, which began testing earlier this year.

DNA指纹图谱技术在庭审中的应用

兰德

编者按

DNA 指纹图谱是一种用于分析来自犯罪现场 DNA 的法医学技术，于 20 世纪 80 年代中期由英国遗传学家亚历克·杰弗里斯建立。这种技术迅速流行，截至 1989 年在美国已经有超过 80 起刑事案件的审理应用了此项技术。但是在本文中，美国遗传学家埃里克·兰德警告说这项技术引入庭审有些草率。作为近期几起谋杀案庭审的辩方顾问，兰德注意到该技术有时缺少合适的对照并且对检测结果的解释也没有标准化。他呼吁科学界就操作方法和标准建立清晰的指导原则，以发挥 DNA 指纹图谱技术在法医学鉴定中的真正威力。

人们忙于将 DNA 指纹图谱技术作为强有力的法医学工具进行使用，却忽视了建立相关标准的需求。

除了同卵双胞胎以外，没有人和其他人拥有一模一样的 DNA 序列。就整个人群而言，在我们从父母那里遗传而来的三十亿核苷酸中，大约每一千个中就会有一个变异位点，也称为多态性。当 DNA 被限制性酶切割时，由于某些位点发生变异使酶切产生的片段长度发生变化，从而形成限制性酶切片段长度多态性（RFLP），这种 DNA 的多态性最容易被检测出来。标准的检测手段是通过电泳凝胶中 DNA 片段的迁移速率来区分它们的长度。

迄今为止，人们已经鉴定出 3000 多种 RFLP，其中包括 100 个高度多态性的位点。这些位点分布在人群中已经发现的几十个变异的等位基因上。随着 DNA 诊断学的蓬勃发展，通过 RFLP 对患有遗传疾病的家族进行染色体区域的遗传追踪，人类遗传学家已经可以精确定位导致亨廷顿病以及囊性纤维化等遗传疾病的基因。

出于另外一个目的，即基于每个人所独有的 RFLP 图谱，可以鉴定在刑事侦查中发现的血液或精液样本来自何人，法医学如今也与 RFLP 更加紧密联系起来。在美国，法医学 RFLP 检测由两家私人实验室（位于纽约瓦尔哈拉的生命密码公司和位于马里兰州日耳曼敦的细胞标记诊断公司）率先开创。联邦调查局也于今年早些时候开始了此项检验。

Since the first use of DNA "fingerprinting" in a trial in Florida in 1988, DNA "fingerprint" evidence has already been used in more than 80 criminal trials in the United States. Applying the legal standard for the admissibility of novel scientific evidence defined in *United States v. Frye* in 1923, trial judges have raced to admit DNA fingerprinting as evidence on the grounds that the methods are "generally accepted in the scientific community", citing the application of RFLPs in DNA diagnostics and accepting claims that false positives are virtually impossible.

With due respect, the courts have been too hasty. Although DNA fingerprinting clearly offers tremendous potential as a forensic tool, the rush to court has obscured two critical points: first, DNA fingerprinting is far more technically demanding than DNA diagnostics; and second, the scientific community has not yet agreed on standards that ensure the reliability of the evidence.

DNA diagnostics requires simply identifying whether each parent has passed to a child the RFLP pattern inherited from his or her mother or father. Because the four discrete patterns are known in advance, these investigations have built-in consistency checks which guard against many errors and artefacts.

DNA fingerprinting, by contrast, is more like analytical biochemistry: one must determine whether two completely unknown samples are identical. Because hypervariable RFLP loci often involve 50–100 alleles yielding restriction fragments of very similar lengths, reliably recognizing a match is technically demanding. At one commonly used locus, for example, most alleles lie within a mere 2 percent of the length of the gel.

Few molecular geneticists, in such circumstances, would declare a match without performing a mixing experiment, in which a 50:50 mixture of the two samples is shown to yield the same pattern as each sample separately. In the rare event that a mixing experiment could not be carried out, most molecular geneticists would at least insist on using internal controls—probes which detect non-polymorphic DNA fragments within each lane—to verify that the lanes have run at equal speeds and to provide standards against which fragment sizes can be measured precisely.

Yet the DNA fingerprinting results now being introduced into the US criminal courts are often based on much flimsier evidence. Not only are mixing experiments and internal controls often omitted, but some laboratories use no objective standards whatsoever for declaring a match.

Unlike DNA diagnostics, DNA fingerprinting also depends on inferences about the frequency with which matching RFLP patterns will be found by chance, which in turn rest on simplifying assumptions about population genetics whose accuracy has not yet been rigorously tested for highly polymorphic RFLP loci. For example, it is assumed without convincing proof that Caucasians, Blacks and Hispanics can each be regarded as homogeneously mixed populations, without significant subgroups, even when considering

自从1988年佛罗里达州第一次将DNA“指纹图谱”用于案件审理以来，在美国已经有超过80起刑事案件的审理中用到DNA“指纹图谱”证据。根据1923年弗赖伊案中定义的允许采纳新科学证据的法律标准，法官们争先恐后地采用DNA指纹图谱作为证据。因为该技术是“在科学界被普遍接受的”，并以RFLP在DNA诊断学上的应用为引证，接受了假阳性几乎是不可能的这一观点。

冒昧地说，法院的这个做法太草率了。虽然DNA指纹图谱技术作为法医学工具有着巨大的潜力，但是法院的仓促决定忽视了非常关键的两点：第一，DNA指纹图谱的技术要求比DNA诊断要高得多；第二，科学界尚未就如何确保该证据的可靠性达成统一的标准。

DNA诊断只需要简单的鉴定父母双亲是否将他们从自己父母那里继承的RFLP图谱遗传给他们的子女。由于四个独立的样本图谱已经预先得知，这些检查可以通过一致性检验来防止许多可能的错误和假象。

相反地，DNA指纹图谱技术更像是分析生物化学。它必须要确定两个完全未知的样品是否一模一样。由于高度变异的RFLP位点通常与50~100个等位基因相关，这些等位基因产生的限制性酶切片段长度非常相似，准确地识别一对匹配样品对技术的要求很高。举例来说，针对一个经常用到的位点，大部分等位基因都分布在仅占凝胶总长度2%的范围内。

在这样的情况下几乎不会有分子遗传学家不做混合实验就宣布样品是匹配的。混合实验是指将两种样品按50∶50的比例混合，查看混合样品是否能获得与两种样品分别单独电泳相同的图谱。在很罕见的情况下，混合实验不能进行，这时大部分分子遗传学家至少会坚持用内参——即在每条泳道中用来检测非多态性DNA片段的探针——来确定每条泳道中样品的迁移速率一致，并提供精确测量片段大小的标准。

然而，目前美国刑事法庭所采纳的DNA指纹图谱结果都基于非常薄弱的实验证据。不仅这些证据常常省略了混合实验和内参，甚至有些实验室没有定义判断匹配与否的客观标准。

与DNA诊断不同，DNA指纹图谱技术还依赖于对RFLP图谱随机匹配发生频率的推测。相反地，这一推测则基于简化的群体遗传学假定，而这些假设在高多态性的RFLP位点的精确性并未得到严格检验。例如，在没有令人信服的证据的情况下，即使考虑那些从群体遗传学的角度看来形成相对较晚的大部分等位基因位点，

loci at which most alleles are relatively young from the perspectives of population genetics.

Yet despite such fundamental uncertainties, forensic laboratories blithely cite breathtaking frequencies: a recent report based on the study of only four RFLPs announced that the chance of an alleged match occurring at random was 1 in 738,000,000,000,000.

It is my belief that we, the scientific community, have failed to set rigorous standards to which courts, attorneys and forensic-testing laboratories can look for guidance—with the result that some of the conclusions presented to courts are quite unreliable.

My concern is not merely academic: during the past five months, I have been an advisor for the defence and given six days of testimony in what has turned out to be the longest and most searching pretrial Frye hearing in the United States on the admissibility of DNA evidence—a murder case in the Bronx, one of the boroughs of New York City. I have also had occasion to investigate several other DNA fingerprinting cases in the course of preparing a report at the request of the US Office of Technology Assessment, to be delivered later this summer. (My own field is medical genetics: both projects arose as unintended consequences of accepting an invitation to a conference on DNA forensics at Cold Spring Harbor's Banbury Center in late 1988.)

It is my contention that DNA forensics sorely lacks adequate guidelines for the *interpretation* of results—both in molecular biology and in population genetics. To illustrate this, I will draw examples from cases with which I am personally familiar. I should emphasize that the focus on specific cases is not intended to criticize particular testing laboratories: with scientific consensus lacking, similar disagreements about interpretation would surely arise in many other cases involving other laboratories.

The Castro Case

On 5 February 1987, Vilma Ponce and her 2-year-old daughter were stabbed to death in their Bronx apartment. Acting on a tip, police interrogated a neighbourhood handyman, Jose Castro. Detectives noticed a small bloodstain on Castro's watch, which was sent for analysis to Lifecodes. Company scientists extracted about 0.5 μg of DNA from the bloodstain, which they compared with DNA from the two victims.

The DNA was digested with the restriction enzyme *Pst*1, size-fractionated on an agarose gel, and transferred onto a Southern blot. The blot was then hybridized with probes for three RFLP loci; DXYS14, D2S44 and D17S79, as well as a probe for a Y-chromosome locus to identify sex. (Human loci detected by random DNA probes are named according to chromosome and order of discovery.) On 22 July 1987, Lifecodes issued a formal report[1] to the district attorney (Table 1) stating that the DNA patterns on the watch and the mother matched, and reporting the frequency of the pattern to be about 1 in 100,000,000 in the Hispanic population. The report indicated no difficulties or ambiguities. Yet there are several fundamental difficulties, as follows:

高加索人、黑人和西班牙裔仍分别被看作是没有显著亚群体特征的同种混合的种群。

然而很多法医学实验室无视这些基本的不确定因素，满怀喜悦地引用着激动人心的数据：最近一项仅仅基于四种 RFLP 的研究宣称，那些被宣布匹配的样本是随机匹配的可能性仅为七百三十八万亿分之一。

我相信科学界还没有建立起可供法庭、律师和法医鉴定实验室寻求指导的严格标准，因此一些呈送给法庭的结论是非常不可信的。

我的担心并不仅仅是学术上的：在过去的五个月里，我在弗赖伊预审听证会上作为辩方顾问用六天时间向法庭提供了证词，这是美国历史上最长也是最深入的关于 DNA 证据可采纳性的预审听证会——关于一起发生在纽约市布朗克斯区的谋杀案。应美国技术评估办公室之邀，我也有机会参与了其他几个 DNA 指纹图谱相关案件的调查，调查报告预计于今年夏天交付。（我个人的研究领域是医学遗传学：上述两个项目都是我接受 1988 年底在冷泉港的班伯里中心举行的 DNA 法医学会议邀请的意外结果。）

我的观点是，从分子生物学和群体遗传学的角度来讲，DNA 法医学都极度缺乏适宜的准则来对测试结果进行解释。为了阐述这一观点，我将用几个我熟悉的案件来举例说明。我需要强调的是，聚焦于特定的案件并不是要批评某些参与测试的实验室：因为缺少科学性的共识，其他实验室对其他案件的解释也会引发相似的分歧。

卡斯特罗案

1987 年 2 月 5 日，比尔马·庞塞与其两岁的女儿被刺死在她们位于布朗克斯的公寓里。根据线报，警察审问了隔壁的勤杂工——乔斯·卡斯特罗。侦探们注意到卡斯特罗的手表上有滴很小的血迹，就将其送往生命密码公司进行分析。公司的科学家从血迹中提取出了 0.5 微克的 DNA，并用这些 DNA 与两名受害者的 DNA 进行了比较。

科学家用限制性酶 *Pst*1 对 DNA 进行酶切，用琼脂糖凝胶进行 DNA 片段分离，然后转膜进行 DNA 印迹。之后将膜上印迹的 DNA 与三个 RFLP 位点的探针进行杂交。这三个位点分别是：DXYS14、D2S44 和 D17S79，以及用于检测性别的 Y 染色体位点的探针。（被随机 DNA 探针检测到的人类 RFLP 位点是根据其所在的染色体以及发现顺序命名的。）1987 年 7 月 22 日，生命密码公司向地方检察官出具了一份正式报告（表 1），指出手表上发现的 DNA 与受害母亲的 DNA 图谱匹配，并称该种图谱在西班牙裔中出现的频率约为一亿分之一。该报告没有指出任何困难或者不清楚的地方。然而，这份报告的确存在着以下几个基本的难点：

Table 1. Reported fragment sizes (in kilobases) at three RFLPs in *New York v. Castro,* as given in Lifecodes' formal report to the district attorney

	D2S44	D17S79		DXYS14		
Blood from watch	10.25	3.87	3.50	4.83	3.00	1.94
Deceased mother	10.25	3.87	3.50	4.83	3.00	1.94
Deceased daughter	ND	3.87	3.50	4.83	—	1.94

DXYS14: Identifying bands. Contrary to the forensic report (Table 1), the only autoradiogram involving DXYS14 shows five bands in the watch lane and only three bands in the mother's (Fig. 1). In his testimony, Michael Baird, Lifecodes' director of paternity and forensics, agreed that the watch lane showed two additional non-matching bands, but he asserted that these bands could be discounted as being contaminants "of a non-human origin that we have not been able to identify".

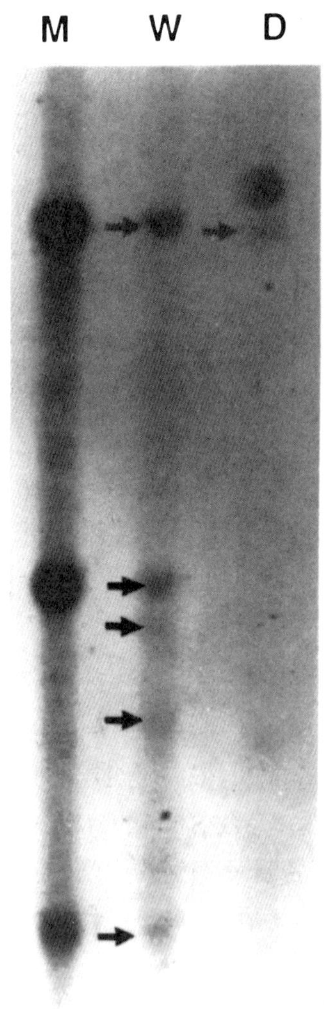

Fig. 1. Hybridization of probe 29C1 for the locus DXYS14 to *Pst*1-digested DNA from deceased mother (M), blood speck from defendant's watch (W) and deceased daughter (D), performed by Lifecodes in *New*

表1. 纽约卡斯特罗案中报告的三个RFLP的片段大小(以千碱基对计)。数据来自生命密码公司向地方检察官提交的正式报告。

	D2S44	D17S79		DXYS14		
手表上的血迹	10.25	3.87	3.50	4.83	3.00	1.94
已故的母亲	10.25	3.87	3.50	4.83	3.00	1.94
已故的女儿	未检出	3.87	3.50	4.83	-	1.94

DXYS14：条带的鉴定。与法医报告（表1）相矛盾的是，和DXYS14有关的唯一一张放射性自显影照片显示在手表上的血迹样品的泳道里有五条条带，而母亲样品的泳道中却只有三条（图1）。在生命密码公司亲子关系与法医学部门负责人迈克尔·贝尔德的证词中，他承认手表上的血迹样品的泳道中多出了两条没有匹配的条带，但他坚称这些多出的条带是“一种我们无法鉴定的非人类来源的污染物”因而不予考虑。

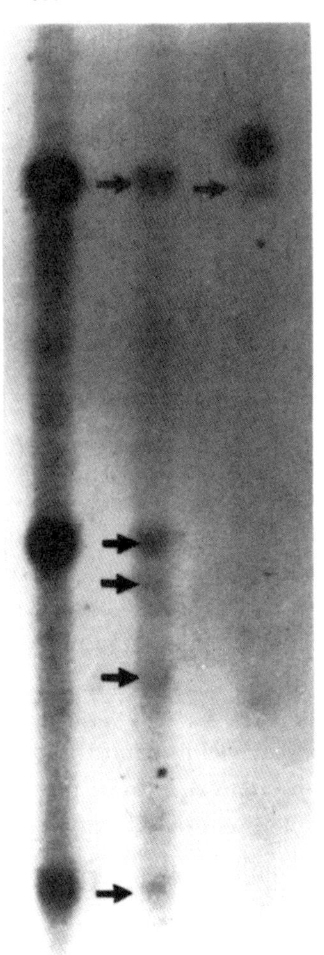

图1. 纽约卡斯特罗案中已故的母亲（M），被告手表上的血迹（W）和已故的女儿（D）的DNA在*Pst*1消化后所得片段与检测DXYS14位点的探针29C1杂交的结果。实验由生命密码公司完成。尽管放射性

York v. Castro. Although the autoradiogram shows three bands in lane M, five bands in lane W and one band in lane D, Lifecodes recorded three bands in identical positions in all three lanes. The autoradiogram is an overnight exposure; no further exposures or hybridizations involving this probe were performed.

In my opinion, it is impossible to know whether the bands are non-human without demonstrating that they are absent when the experiment is repeated with an uncontaminated probe. How then did Lifecodes reach its judgment? Baird stated that, in his experience, DXYS14 should exhibit a pattern of fragments whose intensities decrease in proportion to their length: the extra bands could be ignored because their intensities were "not in the proportion I would expect to see".

In fact, the published scientific literature shows that DXYS14 actually yields patterns (ranging from one to more than six polymorphic fragments) whose intensities obey no ironclad rule. Ironically, the prosecution itself had put into evidence the very article which proved the point: the original paper[2] defining DXYS14, by Howard Cooke of the Medical Research Council in Edinburgh, containing photographs of *Pst*l Southern blots showing lanes in which hybridization intensity and fragment length are clearly uncorrelated (Figs 6*c* and 7*a* in ref. 2).

David Page of the Whitehead Institute later introduced *Pst*l blots from his laboratory showing arbitary patterns of intensities at DXYS14. Finally, Cooke himself, who had provided the probe to Lifecodes, testified that the DXYS14 autoradiogram had to be considered to exclude the defendant in the absence of any experiments to explain away the non-matching bands.

Lifecodes' discounting of the two non-matching bands in the watch lane suggests that its identification of bands may have been influenced by making direct comparisons between lanes containing different DNA samples, rather than by considering each lane in its own right. Additional support for this hypothesis is provided by two further examples:

(1) In the dead daughter's pattern at DXYS14, all other expert witnesses for the prosecution and the defence identified only a single band (Fig. 1). However, Lifecodes' laboratory records show that it recorded three bands in this lane—in precisely the same positions as those recorded for the mother and the watch. (For reasons I do not know, the forensic report listed only two of these bands.)

(2) Although D17S79 is expected to yield at most two bands, the dead daughter's lane exhibited four bands in the appropriate size range (not shown) in the only hybridization with the probe completed before the forensic report was issued. The report listed only two of these four bands—the two in the same position as in the mother and the watch.

The tendency to use lane-to-lane comparison to distinguish between bands and artefacts is perfectly natural; such comparison can be quite helpful in certain experiments. However, in my opinion, it is inappropriate in DNA fingerprinting analysis of unknown samples—as

自显影照片显示 M 泳道有 3 条条带，W 泳道有 5 条条带，D 泳道有 1 条条带，生命密码公司却记录说在三条泳道的同一位置都有三条条带。放射性自显影照片曝光过夜；后续没有用该探针进行其他杂交或曝光实验。

我认为，除非证明这些多余的条带在使用未被污染的探针进行重复实验后消失，否则就无法得知这些条带是否为非人类来源。生命密码公司是如何做出他们的判断的呢？贝尔德声称在他的经验中，DXYS14 应该呈现一种随片段长度减少而信号强度逐渐递减的模式，而那些多余的条带之所以被忽略，是因为它们的强度“没有呈现出我所预计的比例”。

事实上，已发表的科学文献显示，DXYS14 图谱（可以是从 1 条到 6 条以上的多态性片段）的信号强度模式并没有一定之规。讽刺的是，检方自己将证明如下观点的文章列为了证据：在爱丁堡医学研究理事会的霍华德·库克定义 DXYS14 的原始文献中，有用 *Pst*1 酶切后做的 DNA 印迹的图片，上面显示杂交的信号强度与片段的长度明显不具有相关性（见参考文献 2 中的图 6*c* 和 7*a*）。

怀特黑德研究所的戴维·佩奇从库克的实验室引进了 *Pst*1 印迹法，并用此法显示了 DXYS14 的强度模式是无规可循的。最终，向生命密码公司提供探针的库克自己作证说 DXYS14 的放射性自显影照片在缺乏实验来解释不匹配条带的情况下，只能认为被告的嫌疑被排除。

生命密码公司对手表上的血迹样品中两条不匹配的条带不予考虑，暗示了影响条带鉴定的因素可能是对载有不同 DNA 样本的泳道进行直接比较，而不是从每条泳道自身考虑。为了进一步支持这个假说，我们另外再举两个例子：

(1) 在死去的女儿的 DXYS14 图谱中，所有控方和辩方的其他专家都确认只有一条条带（图 1）。然而，生命密码公司的实验室记录显示这条泳道中有三条条带——恰好精准的位于与母亲和手表上的血迹样品中同样的位置上。（出于我所不知道的原因，法医学报告中仅列出了其中的两条条带。）

(2) 尽管 D17S79 被预计为最多出现两条条带，但死去的女儿的泳道中却在大小合适的位置上出现了四条条带（未显示），这一结果出现在法医报告出具以前完成的唯一一次该探针的杂交实验中。报告最终只列出了四条条带中的两条——两条与母亲和手表上的血迹样品在同一位置的条带。

用泳道之间的对比来区分条带与假象是非常自然的事情，这种方法在特定的实验中非常有用。然而我认为，这种方法并不适用于对未知样品的 DNA 指纹图谱进

one runs the risk of discounting precisely those differences that would exonerate an innocent defendant. Forensic laboratories should be required to use objective criteria for identifying the bands in each lane, and to use experiments to rule out proposed artefacts.

When a result is reported to have an error rate of 1 in 100,000,000, it seems essential that the underlying data are not left as a matter of subjective opinion.

D2S44 and D17S79: declaring a match. To obtain objective measurements of a band's position, Lifecodes uses a computer-digitizing apparatus[3-5]. The approach is reported to be highly accurate: when identical fragments are electrophoresed in different lanes, the difference between their positions is reported to show a standard deviation (s.d.) equal to 0.6 percent of molecular weight[3,5]. Based on these experiments, Lifecodes defined a formal matching rule: two fragments are said to match when their positions differ by less than 3 s.d.s. The matching rule was explicitly stated in a recent population study[5] ("two DNA fragments were considered to be of different size if their values differed by more than 3 s.d.s") and in the formal forensic reports[1] in the Castro case ("fragments with measurements that are within [3 s.d.s] of each other … are considered indistinguishable and their average size reported".)

Because the fragments at D2S44 and D17S79 did not appear to match perfectly, the defence examined Lifecodes' computer measurements. In fact, the bands fell outside the declared matching rule; as shown in Table 2, the bands at D2S44 differ by 3.06 s.d.s and the lower bands at D17S79 differ by 3.66 s.d.s. Under the objective matching rule, the bands were non-matches.

Table 2. Measured fragment sizes (in base pairs) at three RFLPs in *New York v. Castro,* as shown in Lifecodes' records from its computer-digitizing apparatus produced in response to subpoena

	DS244	D17S79		DXYS14		
Blood from watch	10,350	3,877	3,541	4,858	2,995	1,957
Deceased mother	10,162	3,869	3,464	4,855	2,999	1,946
Difference (percent of average size)	1.83	0.21	2.20	0.06	0.13	0.56
Difference in number of s.d.s	3.06*	0.34	3.66*	0.10	0.22	0.94

According to Lifecodes' published papers[3,5] and its formal reports to district attorney[1], the difference between two identical bands in different lanes shows an s.d. equal to 0.6 percent of fragment size. Asterisks, bands differing by > 3 s.d.s.

Why then was a match declared? Lifecodes stated that it did not actually use the objective threshold of 3 s.d.s. for declaring a forensic match: its decisions were based on subjective visual comparison. Agreeing that the explicit statements in the forensic report implied that the objective criterion had been used, Baird allowed that the statement "may not be the best explanation" of the company's actual procedures. As far as I can see, there is also no mention of the use of visual matching in the company's scientific papers or forensic reports. Clearly, there has been a significant misunderstanding about the matching rule which Lifecodes has been using.

行分析——它会带来无视某些差异的风险，而那些差异本可以让无辜的被告无罪开释。法医学实验室应该使用客观的标准来鉴定泳道里的每一条条带，并用实验来排除那些预测出的假象。

作为一个重要的前提，当一个结果被报告为有着一亿分之一的错误率时，一些潜在的数据可能不会由于主观原因未被记录下来。

D2S44 和 D17S79：匹配的认定。为了获得条带的客观测量位置，生命密码公司使用了计算机数字化设备[3-5]。据报告，这项技术是高度精确的：当相同的片段在不同的泳道进行电泳时，它们之间位置差异的标准差（s.d.）相当于分子量的 0.6%[3,5]。基于上述实验，生命密码公司定义了正式的匹配规则：当两个片段的位置差异小于 3 个标准差时，就认为这两个片段是匹配的。该匹配原则在一份最近的种群研究[5]（“当两个 DNA 片段之间的值差异超过 3 个标准差时就认为这两个片段大小不同”）以及卡斯特罗案正式的法医报告中[1]（“片段之间测量值差异小于 [3 个标准差]……被认为无法区分并按它们的平均大小进行报告”）都有明确的陈述。

由于 D2S44 和 D17S79 的片段看上去并不完全匹配，辩护方检查了生命密码公司的计算机测量结果。事实上，这些条带不符合他们声明的匹配规则：如表 2 中所示，D2S44 的条带之间差了 3.06 个标准差，D17S79 中较低的条带相差 3.66 个标准差。按客观的匹配规则来看，这些条带是不匹配的。

表 2. 纽约卡斯特罗案中测量的三个 RFLP 的片段大小（以碱基对计）。数据来自生命密码应法庭传讯所出示的计算机数字化设备记录。

	D2S44	D17S79		DXYS14		
手表上的血迹	10,350	3,877	3,541	4,858	2,995	1,957
已故的母亲	10,162	3,869	3,464	4,855	2,999	1,946
差异（占平均片段大小的百分比）	1.83	0.21	2.20	0.06	0.13	0.56
差异与标准差的倍数比	3.06*	0.34	3.66*	0.10	0.22	0.94

根据生命密码公司发表的文章[3,5]及其向地方检察官提供的正式报告[1]，在不同泳道中的相同条带之间差异的标准差（s.d.）相当于片段大小的0.6%。星号表示条带差异大于3个标准差。

那为什么还要宣称这些条带是匹配的呢？生命密码公司陈述说他们实际上并不是用 3 个标准差这一客观阈值来确定条带是否在法医学上匹配：他们的结果是基于主观的肉眼比较来决定的。法医学报告中明确的陈述暗示客观标准已被使用，贝尔德同意这一观点，同时他承认那份陈述“可能不是”公司实际操作的“最好解释”。就我看来，不管是在公司的科学文献还是法医学报告中，都没有提到过用肉眼识别匹配。很明显，人们对生命密码公司使用的匹配原则存在着非常明显的误解。

In my opinion, visual matching is inappropriate in DNA fingerprinting, inasmuch as (1) many alleles have very similar sizes; (2) the accuracy of the measurement process is reported to be known; and (3) without an objective definition of a match, there is no meaningful way to determine the probability that a declared match might have arisen by chance (see below).

DYZ1: Use of controls. The DYZ1 locus provides a convenient method of identifying the sex of a sample: the locus is repeated about 2,000 times on the distal long arm of the Y chromosome, giving rise to an intense 3.7-kilobase (kb) *Pst*1 band in males and no such band in females. Based on a hybridization with a probe for DYZ1, the blood on the watch was said to have come from a female.

Indeed, the mother, daughter and watch DNAs showed no male-specific band. But neither did the lane marked control. Who was the control?

(1) Initially, Baird testified that the control DNA came from the female-derived HeLa cell line.

(2) Two weeks later, however, the technician who actually performed the experiments testified that the control DNA came not from HeLa cells but rather, he recalled, from a male scientist.

(3) Baird then explained the absence of a positive signal in the now-male control lane by telling the court that the male scientist has a "short" Y chromosome which "does not react with this repeat sequence", a condition which is "fairly rare, but it does happen". In conventional genetic terminology, the individual was a genetic mutant deleted for the region.

(4) I then testified that the population frequency of such complete deletions is about 1 in 10^3–10^4, with most causing phenotypic abnormalities usually including complete sterility (D. Page, H. Cooke and K. Smith, personal communication). A normal male with such a deletion would be so rare as to be publishable.

(5) Baird then reported that the control DNA came not from the male scientist, but from a female technician. Although no precise record had been kept of which DNA preparation had been used, he said that he had managed to identify the source of the control lane by studying its RFLP pattern—an unforeseen use of DNA fingerprinting.

The confusion had probably resulted from faulty recollections (by Baird and the technician) and faulty inferences (about the male scientist), but it underscored the need for meticulous record-keeping in DNA forensics, which may not originally have been as clear.

Leaving aside the identity of the control DNA, there is a more important question: should

鉴于以下原因，我认为肉眼识别匹配对DNA指纹图谱技术是不适合的：(1) 许多等位基因的大小非常相似；(2) 测量过程的精确性是已知的；(3) 没有对匹配的客观定义，就没有有意义的方法来计算所谓的匹配偶然发生的概率（见下文）。

DYZ1：对照组的使用。DYZ1位点提供了鉴定样本性别的简便方法：在男性中，这个位点在Y染色体长臂远端重复了约2,000次，在*Pst*1酶切后产生了一条3.7千碱基对的强信号条带，女性没有该条带。基于与DYZ1探针的杂交结果，手表上的血液被认为是来自女性。

母亲、女儿和手表上的DNA都确定没有男性特异性的条带。但是没有一条泳道被标记为对照组。谁是对照组呢？

(1) 最初，贝尔德作证说对照组DNA来自女性来源的海拉细胞系。

(2) 然而两周后，实际操作这个实验的实验员却作证说，根据他的回忆，对照组DNA来源于一位男性科学家，而不是海拉细胞。

(3) 贝尔德随后向法庭解释了这个现在是男性的对照组泳道里为什么没有阳性信号。他说这位男性科学家有一个“短的”Y染色体，因而“不能对该重复序列产生反应”，这个情况“虽然罕见，但是的确发生了”。用传统遗传学的术语来说，该个体是该区域缺失的遗传突变体。

(4) 我随即证实，人群中发生这种完全缺失的频率大约是1/1,000~1/10,000。并且大多数情况下该突变会导致异常表型，通常包括完全不育（与戴维·佩奇，霍华德·库克及史密斯进行的个人交流）。一个正常的男性有这样的缺失是非常罕见的，如果有的话一定会被发表出来。

(5) 之后贝尔德又报告说对照组DNA并非是来源于那位男性科学家，而是来源于一名女性实验员。尽管没有准确的记录表明使用了哪份DNA样品，贝尔德却说他通过研究该样品的RFLP图谱设法鉴定了对照组的来源——这真是DNA指纹图谱的意外用法。

这些困惑可能是源于错误的回忆（来自贝尔德和实验员）和错误的推断（关于那名男性科学家），但是它们却强调了DNA法医学对严谨的实验记录的要求，而这种要求可能并不是一开始就很清楚。

撇开对照组DNA的来源不谈，我们还有一个更重要的问题：放射性自显影照片中

a sex test be considered reliable without seeing a positive control on the autoradiogram to prove that the experiment had worked correctly? Baird testified that such a result could be considered "reliable". I would vigorously disagree.

D2S44: Analysis of degraded DNA. Based on seeing only a single band in the watch DNA, the sample was reported to be homozygous for a 10.25-kilobase (kb) band at the D2S44 locus. Although the conclusion may seem reasonable at face value, there is a serious problem: the small quantity of DNA on the watch was clearly degraded, and nearly 90 percent of alleles in the Hispanic population lie above 10.25 kb.

How can one be sure that the sample was not a heterozygote, with a higher band undetected due to degradation? Estimating the extent of degradation by eye, Baird stated that the photographs of the ethidium bromide stain of the gel "gives you some indication that there is enough material present to be able to get a signal" in the 12–15-kb range", but that "I['d] hate to bet the ranch" on it.

Wouldn't a probe detecting a non-polymorphic single-copy band at about 15 kb have provided a definitive positive control? He replied as follows:

"If you are making a decision based only on a single locus, whether or not a pattern you saw was homozygous or whether or not you are missing a band, you'd have to have some way to absolutely be sure that you were seeing everything that was there and the control that you mentioned would be very helpful to do that.

"Now, in addition to the D2S44 locus, we also looked at two additional loci, D17S79 and DXYS14. By looking at the combination of loci, and seeing whether or not there was a pattern that matched or not, allowed us to determine or help[ed] us to determine that the pattern that we saw with the D2S44 locus is [a] homozygote."

Personally, I do not understand how the presence of matches at D17S79 and DXYS14 has any bearing on the determination of a match at D2S44: each test must be evaluated independently, especially as the individual probabilities of a match for each locus are multiplied together at the end (see below).

Probe contamination. To explain various artefacts, Lifecodes invoked four separate instances of probe contamination: human probes were said to be contaminated with bacterial sequences, and bacterial and plasmid probes were said to be contaminated with human sequences. Moreover, Baird testified that the company continued to use probes even after learning that they were contaminated, while apparently keeping no precise record of when such probes had been used.

Although the use of contaminated probes may be permissible in some types of experiment, it is, in my opinion, inappropriate in DNA forensics. Because the samples are

没有出现一个可以证明实验正确运行的阳性对照组。在这种情况下，性别测试还能被认为是可信的吗？贝尔德作证说这样的结果可以被认为是“可信的”。我对此表示强烈反对。

D2S44：对降解的DNA的分析。由于在手表上的血迹样品的DNA中只看到了单一条带，因此报告称该样品在D2S44位点上是一个含有10.25千碱基对的纯合子。尽管表面上看这个结论是合理的，但实际上存在一个很严重的问题：手表上的少量DNA很明显被降解了，因为在西班牙裔人群中近90%的这种等位基因的酶切片段都大于10.25千碱基对。

如果在更高处有条带却因为降解而未被检测到，又怎能确定这个样品不是杂合子呢？通过肉眼观察对降解的程度进行估测，贝尔德声称在溴化乙锭染色的胶图中“有迹象表明在12~15千碱基对范围内有足够的DNA信号”，但是“我并不打包票”。

在15千碱基对处被探针检测到的非多态性单拷贝条带不是正好提供了明确的阳性对照吗？对此他是这样回应的：

“如果你仅基于一个单一位点来决定你所见的图谱是否是纯合子，或者你是否遗失了一条条带，你必须采取一些手段来确保你看见了图谱上所有的东西，在这种情况下你所提到的对照组会非常有用。

“现在，除了D2S44位点，我们还查看了其他两个位点：D17S79和DXYS14。综合观察这些位点，并且查看图谱是否存在匹配，可以让我们判定或者帮助我们判定在D2S44位点看见的图谱是纯合子。”

从个人角度来说，我并不理解在D17S79和DXYS14位点出现的匹配与判定D2S44位点上匹配与否之间有任何关系：每一个测试都必须独立地进行评估，尤其在每个位点独立匹配的概率最终要相乘的情况下（见下文）。

探针污染。为了解释各种假象，生命密码公司列举了四个不同的探针污染的例子：据称被细菌序列污染的人类探针，以及据称被人类序列污染的细菌和质粒探针。此外，贝尔德作证说公司在得知这些探针被污染后依然继续使用，同时显然对这些探针何时被使用也没有准确的记录。

尽管在一些类型的实验中允许使用被污染的探针，然而在我看来，这样做在DNA法医学中是不合适的。因为样品的来源未知并且经常被污染，（使用污染过的

of unknown origin and often contaminated, false matches can result but may be hard to recognize.

Population genetics. After declaring a forensic match, testing laboratories apply a three-step procedure to calculate the probability that the match might have arisen by chance in the population: (1) for each allele, one counts the frequency with which matching bands occur in a previously-drawn population sample (Lifecodes used a database[5] of US Hispanics in the Castro case); (2) for each locus, one then computes the probability of observing a matching genotype by applying the classical Hardy–Weinberg equations[5,6], under the assumption that population is freely intermixing and thus contains no heterogeneous subgroups; and (3) for the complete RFLP pattern, one then multiplies the three single-locus probabilities, under the assumption that the genotypes at the loci are in linkage equilibrium (are uncorrelated). In fact, none of these steps stood up to scientific scrutiny in the Castro case.

Probability of a match: inconsistent matching rules. Whatever matching rule is used to declare a forensic match, it is axiomatic that the same matching rule must then be used for counting the matches occurring in the population database. In fact, Lifecodes' calculations did not use the same matching rule.

■ To declare a forensic match with a given band, Lifecodes' published matching rule calls for examining a range of ±3 s.d.s around the band. Obviously, the chance of a match arising at random is just the proportion of bands in this range.

■ To calculate the probability of a matching band occurring by chance, however, Lifecodes uses a completely different approach: the reported probability is essentially the frequency of bands occurring in a window of size ±2/3 s.d.s. More exactly, as described in an as-yet-unpublished paper[8], it is a weighted-average of frequencies corresponding to such intervals, with the terms weighted according to a gaussian distribution centred on the estimated allele size.

Because the method involves averaging frequencies corresponding to intervals that are 4.5-fold smaller than those allowed for declaring a forensic match, the probability reported for each allele will typically be too small by a factor of about 4.5. Because each RFLP involves two alleles, the probability may thus be understated by a factor of about $(4.5)^2$, or about 20, for each locus. For a three-locus genotype, the error may thus be about 8,000-fold. (Of course, these calculations assume that one uses the 3 s.d. matching threshold. If the less stringent standard of visual matching is used, the discrepancy may be even greater.) Such a statistical procedure is like catching a match with a 10-foot-wide butterfly net, but then attempting to prove the difficulty of the feat by showing how hard it is to catch matches with a 6-inch-wide butterfly net.

探针）很可能导致假匹配并且很难被识别出来。

群体遗传学。在宣布一个法医学上的匹配以后，测试的实验室会计算该匹配在人群中偶然发生的概率，该程序分三个步骤：(1）对每对检测的等位基因，都要计算匹配条带在已统计的人口样本（在卡斯特罗案中，生命密码公司使用了一个美国西班牙裔数据库[5]）中的发生频率；(2）对每个位点，都通过经典的哈迪－温伯格方程计算观测到匹配的基因型的概率[5,6]。该方程基于人群是自由混合并且没有异质性亚群这一假设；(3）根据位点基因型的连锁平衡（无关联）假设，对于完整的RFLP图谱，要将三个单基因位点各自的概率相乘。事实上，在卡斯特罗案中，上述的任何一个步骤都禁不起严格的科学检验。

匹配的概率：矛盾的匹配原则。无论使用什么样的匹配规则来宣布一个法医学上的匹配，都必须使用相同的规则对群体数据库中的匹配进行计数，这一点无疑是不言自明的。而实际上，生命密码公司的计算并没有使用同一个匹配规则。

■ 为了报告给定条带的法医学匹配结果，生命密码公司公布了一种匹配规则。该规则需要检测条带附近3个标准差的范围。显然，随机匹配发生的概率仅和该范围内条带所占的比例有关。

■ 然而，为了计算条带匹配偶然发生的概率，生命密码公司采用了完全不同的方法：报告中的概率基本上是条带在 ±2/3 个标准差范围内出现的频率。更准确地说，正如一篇尚未发表的论文中所描述的那样[8]，它是以某一区间内频率的加权平均值来计算的。这个区间通过以等位基因大小预估值为中心的高斯分布来加权获得。

因为该方法用来计算平均频率的区间比用于宣布法医学匹配的区间小了4.5倍，所以每个等位基因在报告中的概率也都小了4.5倍。因为每个RFLP都包含两个等位基因，所以每个位点的概率可能都因此被低估了4.5的平方倍，或者说大概20倍左右。对于一个包含三个位点的基因型，这个误差可能有大约8,000倍。(当然，这些计算都假设他们用了3个标准差作为匹配阈值。如果他们用到了肉眼判断匹配这样更不严格的标准，差异可能会更大。）这样的统计法就好比用10英尺宽的捕蝶网来抓住一只匹配的蝴蝶，却又试图通过展示用6英寸宽的捕蝶网抓住匹配的蝴蝶的困难程度来证明前一项技艺的困难。

How does Lifecodes justify its approach? Once a forensic match is declared between two bands, Lifecodes apparently considers the average fragment size to represent the allele present in the sample. It then estimates the population frequency of this allele, essentially by counting the bands within a range so narrow that it may not even include either of the two actual measurements. This approach substantially underestimates the true chance of a forensic match occurring at random, as it takes no account of the actual threshold used for declaring matches.

Heterogeneity within the Hispanic population. To justify applying the classical formulas of population genetics in the Castro case, the Hispanic population must be in Hardy–Weinberg equilibrium. In fact, Lifecodes' own data show that it is not. The classical test for Hardy–Weinberg equilibrium is based on Wahlund's principle[7,9] that the rate of homozygosity in a population containing distinct subgroups will be higher than would be expected under the assumption of random mating. Applying this test to the Hispanic sample, one finds spectacular deviations from Hardy–Weinberg equilibrium: 17 percent observed homozygotes at D2S44 and 13 percent observed homozygotes at D17S79 compared with only 4 percent expected at each locus, indicating, perhaps not surprisingly, the presence of genetically distinct subgroups within the Hispanic sample. (The expected 4 percent frequency of homozygotes is based on the empirical probability of randomly drawing two alleles from the population sample that are either identical or so close together as to be scored as a single band; the minimum size difference needed to discriminate between one versus two bands in Lifecodes' experiments was stated explicitly in testimony and in a paper[8].)

Once a population is known to be heterogeneous, one also cannot assume linkage equilibrium even for loci on different chromosomes: if an individual possesses an allele common among Puerto Ricans at one locus, it is more likely that he will do so at a second locus as well.

In fact, Lifecodes' population study[5] is scientifically valuable. From an evolutionary point of view, highly polymorphic loci contain many young alleles which may not be uniformly distributed within the Caucasian, Black or Hispanic populations. Studies of RFLP frequencies can reveal the detailed substructure of the human population, shedding light on migrations and trading patterns.

Such complexities, however, can undermine the use of simplified calculations in DNA forensics. Without the assumption of Hardy–Weinberg equilibrium and linkage equilibrium, there is no reliable way to convert allele frequencies into overall genotype frequencies: applying the classical equations can lead to spuriously low probabilities of a match. Possible solutions include empirical studies to identify ethnic subgroups that are in Hardy–Weinberg equilibrium and theoretical studies to derive appropriate correction factors for heterogeneity.

生命密码公司怎样证明这一方法是合理的呢？一旦两条条带被宣布为法医学匹配，生命密码公司很明显就会考虑用这些片段的平均大小来代表样品中出现的等位基因。然后他们就通过对这些条带计数来估计等位基因在人群中出现的频率，而这些条带值的范围是如此之窄以至于它们很可能不包括一对等位基因的任何实际测量值。因为没有考虑到宣布匹配实际用到的阈值，这样的方法本质上低估了法医学匹配随机发生的真实概率。

西班牙裔群体中的异质性。在卡斯特罗案中，为了证明使用群体遗传学经典方程的合理性，西班牙裔群体必须符合哈迪－温伯格平衡。实际上，生命密码公司自己的数据显示事实并非如此。对哈迪－温伯格平衡的经典检验基于瓦隆德原理[7,9]，即一个含有不同的亚群体的种群，在随机交配的假设下，其纯合子的比率要高于预期值。将这一检验应用于西班牙裔样本中，我们发现这一样本与哈迪－温伯格平衡有着惊人的差异：17% 的人在 D2S44 上是纯合子，13% 的人在 D17S79 上是纯合子。而相比较之下，这两个位点都是纯合子的预期值仅有 4%。这一结果提示我们一个也许并不奇怪的结论，那就是在西班牙裔样本中存在着在遗传学上不同的亚群体。(纯合子为 4% 这一预期值是基于经验概率，即从种群样本中随机抽取两个等位基因，它们完全一致或者相似到不能区分的概率；在生命密码公司的实验中，区分一条条带还是两条条带的最小差异在其证词和一篇文章中已经被明确阐述[8]。)

一旦一个群体被确定是异质性的，那么即使是对不同染色体上的位点，也不能假设它们是连锁平衡的：如果一个人在某个位点上拥有一个在波多黎各人中常见的等位基因，那么他很可能在另一个位点上也是如此。

事实上，生命密码公司的种群研究[5]在科学上是很有价值的。从进化的角度来看，那些高度多态性的位点包含许多可能并没有在高加索人、黑人或西班牙裔中均匀分布的新基因。研究 RFLP 的频率可以揭示人类种群微观结构，从而有助于人们阐明人类迁徙和流动的模式。

然而这样的复杂性会损害简化计算在 DNA 法医学上的应用。除了基于哈迪－温伯格平衡和连锁平衡的假设，并没有一个可靠的方法将等位基因频率转换为所有基因型的频率。但是应用这些经典方程又会大大降低匹配的概率。可能的解决方法包括通过经验研究的方法鉴定符合哈迪－温伯格平衡的亚种群，以及用理论研究的方法得到适用于异质群体的关联因子。

The Experts' Statement

After stepping down from the witness stand in late April, I attended the next day a Cold Spring Harbor conference on genome mapping co-organized by Richard Roberts, who had been the prosecution's lead witness when the Castro case had begun in mid-February. As Roberts explained, his testimony had been intended simply to provide the court with a primer on DNA analysis. Concerned about the issues that had come to light, Roberts conceived a novel plan: a joint scientific meeting of the experts who had testified for either side to review the evidence.

At the meeting, held on 11 May, the experts agreed upon a consensus statement declaring that "the DNA data in this case are not scientifically reliable enough to support the assertion that the samples … do or do not match. If these data were submitted to a peer-reviewed journal in support of a conclusion, they would not be accepted. Further experimentation would be required". In particular, the statement cited the inappropriateness of (1) discounting the extra bands at DXYS14; (2) declaring a match between bands at D2S44 and D17S79 whose measured positions differed by more than Lifecodes' announced threshold of 3 s.d.s; and (3) using a less-strict matching rule when declaring forensic matches than when searching the population database. (Baird was unable to attend the experts' meeting because of a prior engagement.)

At first, the prosecution indicated that it would withdraw the DNA evidence based on the advice of its own scientific experts. Eventually, however, the district attorney decided to press ahead.

The hearing then resumed, with former prosecution witnesses testifying now for the defence. The prosecution's efforts to mount a rebuttal case fizzled:

■ To rebut the non-matching bands at D2S44 and D17S79, the prosecution stated that the use of a new measurement system now reportedly showed that the bands actually lay within 3 s.d.s. The judge ruled this evidence inadmissible, however, calling a last-minute switch of methodology "highly unscientific".

■ To rebut the problem with degradation above 10 kb, Lifecodes probed the Southern blot with the human *Alu* repeat sequence and determined that it showed hybridization up to the 23-kb molecular mass marker. In my opinion, the experiment itself was meaningless (because the ability to detect a sequence repeated 300,000 times in the genome has no bearing on the ability to detect single-copy sequences), but it was unnecessary to explain this to the court. Defence attorney Peter Neufeld, by now a veteran reader of autoradiograms, noticed that someone had accidentally misread the size markers: the *Alu* hybridization actually extended only to the 9.8-kb marker.

■ To rebut the population-genetic issues, the prosecution made eleventh hour phone calls to various scientists, but they refused to testify for the state.

专家声明

四月底出庭作证后的第二天，我参加了在冷泉港举行的关于基因组作图的会议。会议的组织者之一——理查德·罗伯茨曾在二月中旬卡斯特罗案开始时出任检方的首席证人。罗伯茨解释说，他出庭作证只是想向法庭提供一种用于 DNA 分析的引物。考虑到已经暴露出来的问题，罗伯茨提出了一个新的计划：召开一个由控辩双方专家共同参加的联合科学会议来审查证据。

在 5 月 11 日举行的会议上，专家们达成共识并发表声明，宣布“本案中 DNA 数据在科学上的可信度不足以支持这些样品匹配与否的论断。如果这些数据被提交至一份同行评审的刊物用以支持某个结论，它们是不会被接受的，而是会被要求进一步实验”。声明中特别引出了如下的不恰当之处：(1) 未将 DXYS14 中多出的条带计算在内；(2) 在位置差异大于生命密码公司公布的 3 个标准差阈值的情况下，仍然宣称 D2S44 和 D17S79 的条带是匹配的；(3) 用于证明法医学匹配的原则比搜索种群数据库时所用原则的严格度低。(贝尔德由于之前另有安排，未能参加此次专家会议。)

最初检方表明，基于他们自己的科学家给出的建议，他们将收回这些 DNA 证据。然而最终，地方检察官却决定继续使用。

听证会继续召开，之前的检方证人现在开始为辩方作证。检方发起反驳的努力失败了：

■ 为了反驳 D2S44 和 D17S79 条带不匹配的说法，检方声明使用一种新的检测系统后，两条条带落在了 3 个标准差的范围内。然而法官判定该证据不被采纳，原因是最后关头换用方法是“非常不科学的”。

■ 为了反驳 10 千碱基对以上降解的问题，生命密码公司用探针对人类 *Alu* 重复序列进行 DNA 印迹，并判定在高达 23 千碱基对分子量标记处 DNA 仍有杂交。在我看来，这个实验本身就没有意义（因为检测一个在基因组中重复 30 万次的序列的能力和检测一个单拷贝序列的能力根本就没有关系），不过没有必要向法庭解释这些。辩方律师彼得·诺伊费尔德现在已经是个阅读放射自显影照片的老手了，他发现有人偶然地误读了分子量大小标记：*Alu* 杂交事实上只延伸到了 9.8 千碱基对标记处。

■ 为了反驳群体遗传学上的问题，检方最后向多位科学家打电话，但他们都拒绝在这一问题上作证。

In the end, the prosecution's rebuttal case consisted only of the contention that the DYZ1 sex-test hybridization had actually worked correctly—based on the fact that Lifecodes had located another experiment done on the same day which showed a male-specific band in a rape suspect. (This would be an adequate, if unorthodox, control if it could be proved that both hybridizations had been carried out in the same plastic bag.)

The hearing concluded on 26 May, about 15 weeks after it began. Lawyers for both sides are now preparing briefs. Justice Gerald Sheindlin is expected to issue a decision in July. Sometime thereafter, the murder trial itself will commence—with or without the DNA evidence.

Other Illustrative Cases

The general issues of interpretation are unique neither to the Castro case nor to Lifecodes, as the following examples show.

Georgia v. Caldwell. In a death-penalty case currently in progress in Atlanta, James Caldwell stands accused of raping and killing his daughter Sarah. According to the forensic report, Caldwell's blood matches semen samples from the crime with 67,500,000,000:1 odds against the match arising at random. Testifying at the Frye hearing, Lifecodes scientist Kevin McElfresh described the process of declaring a match as a "very simple straight-forward operation", asserting that "there are no objective standards about making a visual match. Either it matches or it doesn't. It's like if you walk into the parking lot and see two blue Fords parked next to each other. That's the situation here."

In fact, the patterns clearly do not match by eye (Fig. 2). McElfresh agreed, but asserted that "There is, however, a consistent non-alignment of the bands throughout the test, telling us there's a match." In other words, McElfresh contended that the differences were due to one lane having run faster than the other, although Lifecodes presented no internal controls to support this explanation.

D S

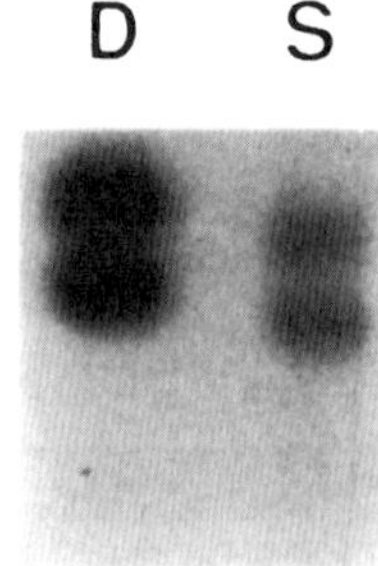

Fig. 2. Hybridization of probe pAC256 for the locus D17S79 to *Pst*1-digested DNA from defendant (D) and semen sample (S), performed by *Lifecodes in Georgia v. Caldwell*.

Although the hearing had been expected to last only a week, prosecutors asked for a month-long delay to prepare their rebuttal. Observers speculated that Lifecodes might use

最终，检方的反驳理由只剩下了 DYZ1 的性别检验杂交事实上是正确的这一个观点。这一观点的根据是生命密码公司在同一天做的另一个实验，在一个强奸嫌疑人的样品中显示了一条男性特异性的条带。（如果能证明两个杂交是在同一个塑料袋中进行的，那么这就是一个不那么正统但是差强人意的对照组。）

在开始了 15 周后，听证会于 5 月 26 日结束。双方律师正在准备辩护状，法官杰拉尔德·沙因德林预计于 7 月作出判决。在这之后，将会开始这件谋杀案的审理——使用或者不使用 DNA 证据。

其他案例

大部分案件的解读与卡斯特罗案以及生命密码公司没有什么不同，如下例所示。

佐治亚州考德威尔案。正在亚特兰大审理的一桩死刑案中，詹姆斯·考德威尔被控强奸并杀害了他的女儿萨拉。根据法医学报告，考德威尔的血液与犯罪现场提取的精液样品匹配，该匹配随机发生的概率为 675 亿分之一。在弗赖伊听证会的证言中，生命密码公司的科学家凯文·麦克尔弗雷休描述判断匹配的过程是一个“非常简单直接的操作”，并断言“没有关于视觉判断匹配的客观标准。条带要么匹配要么不匹配，这就像你走进一个停车场看见两辆蓝色的福特汽车并排停在一起一样，情况就是这样的。”

事实上，凭肉眼就能看出这两条条带的模式明显不匹配（图 2）。麦克尔弗莱希同意这种说法，但是他断言“然而整个实验过程中，有很一致的条带没有对齐的现象，这告诉我们这是匹配的。”换言之，麦克尔弗莱希主张这些差异是由于一条泳道比另一条泳道跑得快，尽管生命密码公司没有提供任何内参来支持这一解释。

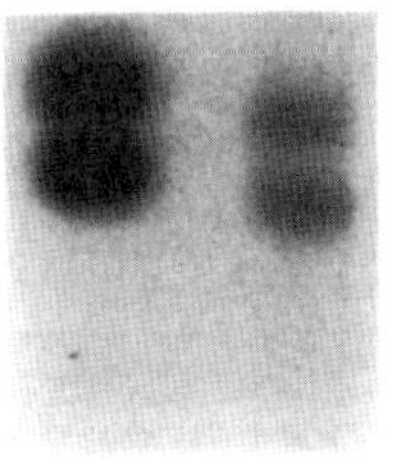

图 2. 佐治亚州考德威尔案中被告（D）与精液样品（S）中的 DNA 经 *Pst*I 消化后，与检测 D17S79 位点的 pAC256 探针杂交的结果。该实验由生命密码公司完成。

尽管听证会预期只进行一个星期，检方还是申请了一个月的延期来准备他们的反证。观察员推测说生命密码公司可能利用这段时间来测试他们的内参。如果是这

the time to test an internal control. If so, it would raise the question of whether Frye hearings are becoming a substitute for having generally accepted laboratory protocols in the first place.

New York v. Neysmith. In a Bronx case in 1987, Hamilton Neysmith was charged with rape based on the victim's identification. Asserting his innocence, Neysmith hired Lifecodes to compare his blood with semen samples from the assailant: the laboratory declared an exclusion. Protesting that the defendant may not have sent his own blood, prosecutors obtained a court order to compel a second blood sample: Lifecodes reported in August 1988 that the two blood samples came from different people. Based on this evidence, Bronx assistant district attorney Karen Yaremko asked the court to revoke Neysmith's bail, planning to charge him with obstruction of justice. The judge declined to revoke bail and the defendant, rather than leaving town, maintained his innocence, and demanded a third blood sample. After Yaremko pressed Lifecodes, she said, the company determined that an error had indeed occurred. Having come close to losing his liberty over inaccurate DNA results, Neysmith was finally exonerated after blood and semen samples were sent to Cellmark Diagnostics which confirmed the original exclusion. (Lifecodes declined last week to comment about the incident, which may have been nothing more than the sort of sample mix-up that can occur in any clinical laboratory. In view of the infallibility with which many jurors regard DNA fingerprinting, however, it may be that even stricter sample-handling procedures should be required.)

In *New York v. McNamara* in November 1988, another Bronx defendant sought to prove his innocence with DNA fingerprinting. Assistant district attorney Renee Myatt opposed the request, telling the court, "the office policy in dealing with a particular agency that does testing with respect to DNA [is] that their testing has been inaccurate, and therefore, unreliable." Notwithstanding this policy, the same District Attorney's office sought to introduce DNA evidence three months later in *New York v. Castro.*

Texas v. Hicks. In this rape–murder, the odds of the declared match occurring at random were reported to be 1 in 96,000,000. Apart from the issue of the matching rule used for searching the database, the population-genetic analysis took no account of the fact that the crime occurred in a small, inbred Texas town founded by a handful of families. The defendant was convicted and sentenced to death.

The case of the abandoned baby. When the president of an insurance agency in Ocean City, Maryland left her car to be towed for repairs to a garage in February 1988, the mechanic claimed that he had discovered a dead infant in the back seat. Although the woman insisted and a detailed medical examination confirmed that she had not been pregnant, Cellmark Diagnostics reported that its DNA analysis showed that she was the mother. (The evidence consisted of one hybridization with a mixture of four RFLP probes in which the woman shared four of eight bands with the child, as well as two hybridizations using probes detecting certain multi-locus repeats.) Local papers reported the sensational news. No murder charges were filed, however, after the state medical examiner determined that the baby had been stillborn on about 4 February.

样，那么就引出了这样一个问题：弗赖伊听证会是否正在取代被普遍接受的实验流程成为首要规则？

纽约内史密斯案。在1987年布朗克斯区的一桩案件中，基于受害者的指证，汉密尔顿·内史密斯被控犯有强奸罪。内史密斯坚称自己是清白的，并雇佣生命密码公司将他的血液与袭击者的精液样品进行比较：结果实验室宣布内史密斯的嫌疑被排除。为防止被告没有提供其本人的血液样本，检方得到了法庭的命令，强制获得了第二份血样。生命密码在1988年8月报告说这两份血样来自不同的人。基于这个证据，布朗克斯区助理地方检察官卡伦·亚列姆科请求法庭撤销内史密斯的保释，并准备起诉他妨碍司法公正。法官拒绝撤销保释，被告在不离开本市的情况下依然无罪，并要求获取第三份血样。亚列姆科在向生命密码公司施压后说，该公司承认的确犯了个错误。内史密斯由于错误的DNA检测结果而险些失去自由，在其血液和精液样品被送至细胞标记诊断公司，证实了最初内史密斯的嫌疑被排除的结论后，他最终被判定无罪。（生命密码公司上周拒绝就此次事件做出评论。即使评论，也很可能就是些将样品搞混了之类的每个临床实验室都可能发生的事情。然而，鉴于许多陪审员都认为DNA指纹图谱是绝对可靠的，样品处理流程应该需要更严格的规定。）

在1988年11月的纽约麦克纳马拉案中，另一位布朗克斯区的被告试图通过DNA指纹图谱技术来证明自己的清白。地区助理检察官勒妮·米亚特反对该请求，并告诉法庭："负责处理该案的有关部门的确检测了关于DNA的一些项目，但是他们的检测结果是不准确的，因而是不可信的。"尽管如此，同一个地方检察院在三个月后的纽约卡斯特罗案中还是引入了DNA证据。

得克萨斯州希克斯案。在这个奸杀案中，匹配来源于随机匹配的概率被宣称为9,600万分之一。且不论用于搜索数据库的匹配原则，在群体遗传分析中根本没有考虑到这个案件发生在得克萨斯州一个在几户人家基础上建立起来的近亲繁殖的小镇上。被告被宣布有罪并判处死刑。

弃婴案。1988年2月，马里兰州大洋城一家保险代理公司的董事长的车被拖去车库修理。机械师声称他在车的后座发现了一个死婴。尽管这个女人坚称自己没有怀孕，细致的医学检查也证实了这一点，但细胞标记诊断公司仍报告说DNA分析显示她就是婴儿的母亲。（证据包括一次用四个RFLP混合探针做的杂交，显示在小孩的八条条带中，该女子有四条相同。另外还包括两次与用于检测某多位点重复序列的探针进行的杂交。）当地报纸报道了这则轰动性的新闻。然而在州验尸官确认这名婴儿在2月4日出生时已经夭折后，未提出谋杀指控。

As it happens, the woman later gave birth to a full-term baby girl on October 24—conceived on about January 29, according to sonograms carried out by her obstetrician. In view of the apparent contradiction, Cellmark last week invited a group of outside scientists to reanalyse remaining DNA samples from the baby.

Setting Standards

Readers should not conclude from this article that DNA fingerprinting is not a powerful tool for forensic identification or that current testing labs are not competent: as in the early stages of any new technology, some difficulties are to be expected. Rather, there is an urgent need for the scientific community to agree on clear guidelines for the procedures and standards needed to ensure reliable DNA fingerprinting. Legislators should also consider whether licensing and proficiency testing should be required in forensics. At present, forensic science is virtually unregulated—with the paradoxical result that clinical laboratories must meet higher standards to be allowed to diagnose strep throat than forensic labs must meet to put a defendant on death row.

An appropriate start would be a US National Academy of Sciences committee, charged with preparing a report on guidelines for DNA fingerprinting. There is ample precedent: when voice-print evidence began to be introduced in the 1970s, the academy convened such a group to examine the technology. An academy study on DNA fingerprinting had been planned for last year, but was postponed indefinitely when the National Institute of Justice would not finance it. As one justice official told me, the study was unwelcome: scientists had done their part by discovering DNA; it was not their job to tell forensic labs how to use it.

(**339**, 501-505; 1989)

Eric S. Lander is in the Whitehead Institute for Biomedical Research, Nine Cambridge Center, Cambridge, Massachusetts 02142, USA.

References:

1. *Lifecodes' formal reports to District Attorney* (22 July 1987, 12 August 1988 and 6 February 1989).
2. Cooke, H. *Nature* **317**, 687-692 (1985).
3. Baird, M. *et al. Am. J. Hum. Genet.* **39**, 489-501 (1986).
4. Baird, M. *et al. Adv. forens. Haemogenet.* **2**, 396-402 (1988).
5. Balasz, I., Baird, M., Clyne, M. & Meade, E. *Am. J. Hum. Genet.* **44**, 182-190 (1989).
6. Cavalli-Sforza, L. L. & Bodmer, W. F. *The Genetics of Human Populations* (Sinauer, Sunderland, Mass., 1989).
7. Hartl, D. L. & Clark, A. G. *Principles of Population Genetics* (Sinauer, Sunderland, Massachusetts, 1989).
8. Morris, J. W., Sanda, A. I. & Glassberg, J. *J. forens. Sci.* (in the press).
9. Wahlund, S. *Hereditas* **11**, 65-106 (1928).

Acknowledgements. The Castro defence was a joint effort with my colleagues Howard Cooke, Lorraine Flaherty, Conrad Gilliam, Philip Green and David Page, together with attorneys Peter Neufeld and Barry Scheck. I also thank Richard Roberts for his efforts to help resolve this hearing and Carl Dobkin for participating in the experts' meeting. I am grateful to Susan Lee and Stephen Lincoln for technical assistance.

碰巧的是，这名女子后来在10月24日生下了一名足月的女婴——根据她的产科医生扫描的超声波图，她在1月29日就已怀孕。考虑到明显的前后不一致，细胞标记诊断公司上周邀请了一组第三方的科学家对婴儿剩余的DNA样品进行重新检测。

建立标准

读者们不应从这篇文章得出结论说DNA指纹图谱技术不是法医学鉴定的有力工具，或者现在的测试实验室不能胜任：任何新技术发展的最初阶段，出现困难都是预料之中的。更确切地说，现在迫切需要科学界就所需的操作步骤和实验标准达成明确一致的准则，以确保DNA指纹图谱的可靠性。立法者也应考虑在法医中是否需要颁发执照和进行能力测试。目前，法医科学实际上是不规范的——矛盾的是，临床实验室诊断脓毒性咽喉炎所需达到的标准都比法医学实验室把被告扔进死囚牢里所需的标准高。

由美国国家科学委员会负责起草关于DNA指纹图谱的指导原则是一个良好的开端。有大量这样的先例：当声纹证据最初在20世纪70年代被引入时，科学委员会就召集了这样一个小组来检测该技术。一项关于DNA指纹图谱技术的学术研究本来计划在去年进行，但是由于美国国家司法研究所不予资助而无限期延后。一位司法官员告诉我，这项研究不受欢迎：科学家做好自己发现DNA的本职工作就好了，告诉法医学实验室怎么使用它不属于他们的职责范畴。

（周晓明 翻译；肖景发 审稿）

A Membrane-targeting Signal in the Amino Terminus of the Neuronal Protein GAP-43

M. X. Zuber *et al.*

Editor's Note

Growth-associated protein 43 (GAP-43) is a neuronal protein expressed in high levels in growth cones, specialized structures found at the end of growing nerve fibres that help guide them to their destination. Research had suggested that the protein's amino terminus might be important for binding the otherwise cytoplasmic protein to the growth cone membrane. Here Mauricio Zuber and colleagues use mutational analysis and confocal microscopy to confirm the hypothesis. They show that the first ten amino acids of the molecule act as membrane-targeting signals. GAP-43, also called neuromodulin, is now known to guide developing axons during development —mice lacking the protein die shortly after birth due to axon path-finding defects. GAP-43 is also involved in regeneration and plasticity.

Neurons and other cells, such as those of epithelia, accumulate particular proteins in spatially discrete domains of the plasma membrane. This enrichment is probably important for localization of function, but it is not clear how it is accomplished. One proposal for epithelial cells is that proteins contain targeting signals which guide preferential accumulation in basal or apical membranes[1]. The growth-cone membrane of a neuron serves as a specialized transduction system, which helps to convert cues from its environment into regulated growth. Because it can be physically separated from the cell soma, it has been possible to show that the growth-cone membrane contains a restricted set of total cellular proteins[2], although, to our knowledge, no proteins are limited to that structure. One of the most prominent proteins in the growth-cone membrane is GAP-43 (refs 3–5; for reviews see refs 6 and 7). Basi *et al.* have suggested that the N-terminus of GAP-43 might be important for the binding of GAP-43 to the growth-cone membrane[8]. Skene and Virag[9] recently found that the cysteines in the N-terminus are fatty-acylated and that this post-translational modification correlates with membrane-binding ability. We investigated the binding of GAP-43 to the growth-cone membrane by mutational analysis and by laser-scanning confocal microscopy of fusion proteins that included regions of GAP-43 and chloram-phenicol acetyltransferase (CAT). We found that a short stretch of the GAP-43 N-terminus suffices to direct accumulation in growth-cone membranes, especially in the filopodia. This supports a previous proposal[8,9] for the importance of this region of GAP-43 in determining the membrane distribution of GAP-43.

神经元蛋白GAP-43氨基末端中的膜导向信号

朱伯等

编者按

生长相关蛋白43(GAP-43)是一种神经元生长锥中高表达的蛋白，生长锥是在正在生长的神经纤维末端发现的特殊结构，有助于指引这些纤维到达它们的目的地。研究揭示，此蛋白的氨基末端可能对于使本来分布于细胞质的蛋白质分子结合到生长锥膜是重要的。在这里毛里西奥·朱伯和他的同事运用突变分析和共聚焦显微技术来证实上述假说。他们的结果显示，此分子的前10个氨基酸起到了膜导向信号的作用。GAP-43，也被称为神经调制蛋白，现在已知它在发育过程中指引正在发育的轴突——缺乏此蛋白的小鼠在出生后很快死亡，因为轴突的路径寻找功能存在缺陷。GAP-43在神经再生和可塑性中也有作用。

神经元和其他细胞，例如上皮细胞，在原生质膜上不同的区域中积累特定的蛋白。这一蛋白富集可能对于功能定位来说是重要的，但并不清楚此过程是如何实现的。一个关于上皮细胞的设想是：蛋白质分子具有在细胞膜的基部或顶端优先积累的导向信号[1]。神经元的生长锥膜作为特化的信号转导系统，有助于将来自周围环境的信号转化为受调控的生长。生长锥与细胞体空间上分隔，已经发现生长锥膜含有细胞总蛋白有限的一部分[2]，尽管据我们所知，没有蛋白仅仅分布于这一结构内。在生长锥膜上，最引人注目的蛋白之一就是 GAP-43（参考文献 3~5；综述见参考文献 6 和 7)。巴锡等人指出，GAP-43 的 N 端可能对于使它与生长锥膜结合来说是重要的[8]。斯基恩和维拉格[9]最近发现，N 端中的半胱氨酸是脂肪酸酰基化的，并且这一翻译后修饰与膜结合活性相关。通过突变分析和激光扫描共聚焦显微镜对 GAP-43 部分区域与氯霉素乙酰转移酶（CAT）的融合蛋白的分析，我们研究了 GAP-43 与生长锥膜的结合。我们发现，GAP-43 N 端的一小段序列足够指导该分子在生长锥膜，特别是丝状伪足中的积累。这支持以前的设想[8,9]，即 GAP-43 的这一区域在决定 GAP-43 膜分布中具有重要意义。

ALTHOUGH GAP-43 lacks the hydrophobic amino-acid sequences that are characteristic of integral membrane proteins[9,10], it resists dissociation by agents that usually remove peripheral membrane proteins[9,11-14]. Hence, the mechanism of membrane binding is not clear. We have found that transfected GAP-43 binds to the membrane in COS cells and other non-neuronal cells[15]. The high level of GAP-43 expression in COS cells permitted preliminary screening to determine which regions were important in membrane association. Whereas membrane binding persists after deletion of a large internal segment (residues 41–189 out of 226 residues), or after deletion of the C-terminal eight amino acids and substitution of an unrelated dodecapeptide, the deletion of amino acids 2–5 markedly diminishes membrane binding (data not shown). Contained in the N-terminus are the only two cysteines in GAP-43, at positions 3 and 4, and it seemed reasonable that this unusual structure of adjacent sulphydryl groups might contribute to membrane binding. As shown in Fig. 1, in which cytosolic and membrane cell-fractions are compared, GAP-43 binds predominantly to the membranes, whereas replacement of Cys 3 or Cys 4, or both, by threonine, causes most of the protein to remain cytosolic. These two cysteines are palmitylated and this correlates with membrane association in brain[14]. Thus, the N-terminus of GAP-43, especially its N-terminal cysteine residues, seem to be necessary for the binding of GAP-43 to membranes.

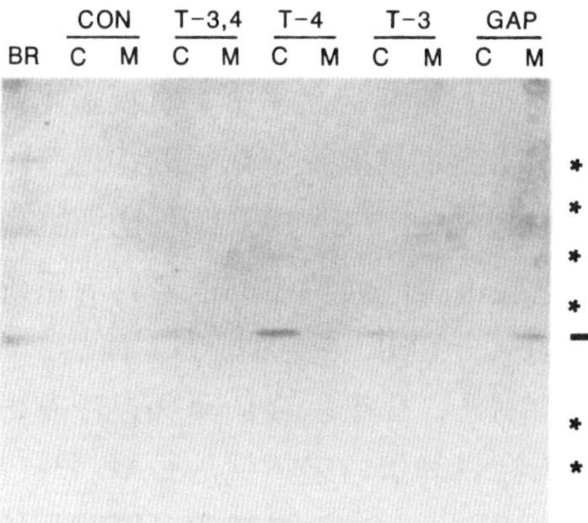

Fig. 1. GAP-43 N-terminal mutations prevent membrane association. COS cells were transfected with the expression vector pGAP (GAP) or with plasmids containing point mutations (Cys to Thr) at position 3 (T-3), position 4 (T-4) or position 3 and 4 (T-3, 4). Immunoblots of membrane (M)- and cystosolic (C) fractions stained with anti-GAP antibody show that GAP-43 immunoreactivity co-migrated with purified GAP-43 protein (bar on right) and was greatly enriched in the membrane fraction. By contrast, the mutant GAP proteins were all more concentrated in the cytosol. Non-specific staining by the antibody was negligible in control fractions from CDM8-transfected COS cells (CON). A crude membrane fraction from rat brain (BR) shows immunostaining at a band corresponding to the same relative molecular mass (M_r) as the band for the GAP-transfected COS cells. The migration positions of protein standards with M_rs of 116,000 (116K), 84K, 58K, 48.5K, 36.5K and 26.6K are indicated by the asterisks.

Methods. Point mutations in GAP-43 were generated from single-stranded DNA made from pGAP and used to alter the cysteines to threonines. The oligonucleotides GGCATGCTGACCTGTATG, ATGCTGTGCACTATGAGA, and GCAGGCATGCTGACCACTATGAGAAGAACC were used to mutate Cys 3 to Thr, Cys 4 to Thr, and Cys 3, 4 to Thr, respectively. After mutagenesis, a 200-base pair (bp) fragment containing the mutation was subcloned into a non-mutated plasmid. The sequence was confirmed after mutagenesis by dideoxy chain-termination sequencing. Other procedures are as described in Fig. 2.

虽然 GAP-43 缺乏作为整合型膜蛋白 [9,10] 特征的疏水氨基酸序列，但是它可以耐受由通常能移除外周膜蛋白 [9,11-14] 的试剂造成的质膜解离。因此其膜结合的机制尚不清楚。我们已经发现，在 COS 细胞和其他非神经元细胞 [15] 中，转染的 GAP-43 与膜结合。GAP-43 在 COS 细胞中的高度表达使我们得以初步筛选决定哪一区域对膜结合是重要的。在一个大片段（226 个残基中的 41~189 位残基）被删除后，或在 C 端 8 个氨基酸被删除，并被某一无关的十二肽所取代后，膜结合依然存在。而第 2~5 位的氨基酸残基删除则显著减少了膜结合（数据未展示）。GAP-43 仅有的 2 个半胱氨酸在 N 端第 3 位和第 4 位，而这种相邻巯基团的不寻常结构可能有助于膜结合，这看上去是合理的。如图 1 所示，将胞浆和膜组分做比较，GAP-43 主要与膜结合，然而用苏氨酸替换第 3 位或第 4 位半胱氨酸，或者二者都被替换为苏氨酸，使此蛋白多数保留在胞浆中。这两个半胱氨酸是棕榈酰化的，而且这与大脑 [14] 中的膜结合是对应的。因此，GAP-43 的 N 端，特别是它的 N 端半胱氨酸残基，看上去对于 GAP-43 与膜的结合来说是必要的。

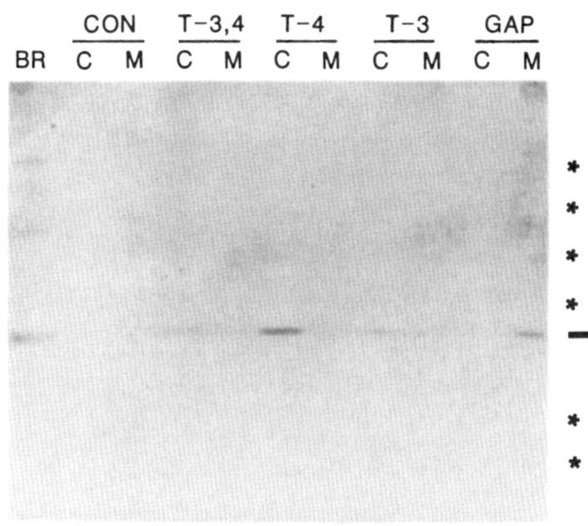

图 1. GAP-43 N 端突变阻断膜结合。用 pGAP（GAP）表达载体或含有点突变（半胱氨酸突变为苏氨酸）的质粒转染 COS 细胞，质粒点突变的位置在第 3 位（T-3）或第 4 位（T-4）或第 3 位和第 4 位（T-3,4）。用抗 GAP 抗体染细胞膜（M）和细胞质（C）组分的免疫印迹实验显示，GAP-43 的免疫反应性与纯化的 GAP-43 蛋白共迁移（由右侧横线显示），并且在膜组分中极大富集。作为对比，突变 GAP 蛋白在胞质更集中。在来自 CDM8 转染的 COS 细胞（CON）的对比组分中，抗体的非特异染色可以忽略。来自大鼠大脑（BR）的细胞膜粗提取物显示，有一条带的免疫染色与 GAP 转染 COS 细胞的条带具有相同的相对分子量（M_r）。星号指明了蛋白分子量标记的迁移位置，分子量分别为 116K、84K、58K、48.5K、36.5K 和 26.6K。

方法。GAP-43 中的点突变产生于 pGAP 制造的单链 DNA，并用来将半胱氨酸转化为苏氨酸。寡核苷酸链 GGCATGCTGACCTGTATG、ATGCTGTGCACTATGAGA 和 GCAGGCATGCTGACCACTATG-AGAAGAACC 被分别用来使第 3 位半胱氨酸突变为苏氨酸，使第 4 位半胱氨酸突变为苏氨酸，以及使第 3 位和第 4 位半胱氨酸突变为苏氨酸。突变形成后，一个含有突变的长 200 碱基对（bp）的片段被亚克隆加入一个非突变质粒。突变发生后通过双脱氧链终止测序确定序列。其他过程如图 2 中所述。

To determine whether the N-terminus is sufficient to confer on GAP-43 the ability to bind to membranes, we made constructs that encoded differing amounts of the GAP-43 N-terminus fused to a reporter peptide, and expressed them in COS and PC12 cells. We chose chloramphenicol acetyltransferase (CAT) as the reporter peptide because it is cytosolic when expressed in eukaryotic cells and is very stable. We constructed plasmids that encode fusion proteins containing either the first 10 amino acids of GAP-43, Met-Leu-Cys-Cys-Met-Arg-Arg-Thr-Lys-Gln, fused to the N-terminus of the complete CAT protein ($GAP_{10}CAT$), or the first 40 amino acids of GAP-43 fused to CAT ($GAP_{40}CAT$). As shown by immunoblotting, the CAT that was expressed in COS cells (Fig. 2*a*) or PC12 cells (Fig. 2*b*) was present only in the cytosolic fraction. By contrast, the chimaeric proteins $GAP_{10}CAT$ and $GAP_{40}CAT$ were membrane-associated. We were able to extract the fusion protein with detergent, but not with sodium chloride, calcium chloride or EGTA (data not shown). Thus, the nature of this membrane binding was similar to that of native GAP-43 in rat brain[11-14].

a

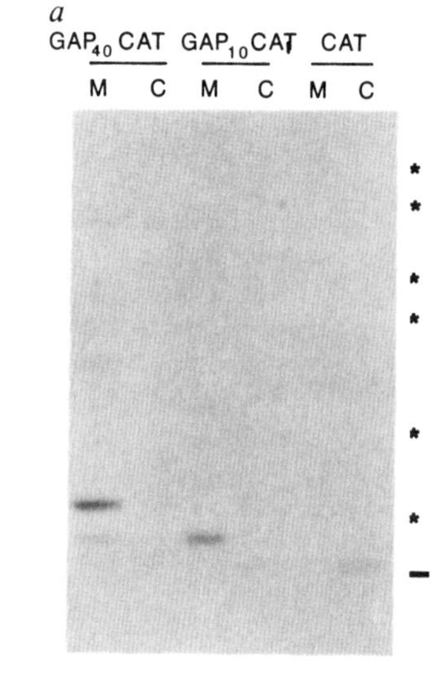

b

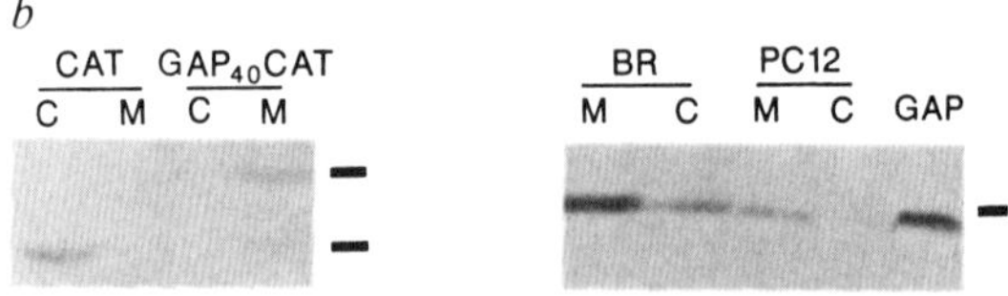

Fig. 2. Membrane association of GAP-43 and GAP-CAT fusion proteins. *a*. Chimaeric proteins with the N-terminus of GAP-43 fused to CAT associate with COS cell membranes. The constructs CAT, $GAP_{10}CAT$ and $GAP_{40}CAT$ were transiently expressed in COS cells. Immunoblots of membrane (M)- and cytosolic (C) fractions from each transfection were prepared using anti-CAT antibody. Note that in the CAT-transfected cells, immunoreactivity was found only in the cytosolic fraction and co-migrated with purified CAT protein (indicated by the bar). In the $GAP_{40}CAT$- and $GAP_{10}CAT$-transfected cells, nearly all of the immunoreactivity was membrane-associated and migrated more slowly than that for CAT-transfected cells, as expected for fusion proteins with an M_r of 4K or 10K greater than CAT. Protein

为了确定N端是否足够赋予GAP-43与膜结合的能力，我们构建了不同长度的GAP-43 N端与报告多肽融合蛋白的表达载体，并使它们在COS和PC12细胞中表达。我们选择氯霉素乙酰基转移酶（CAT）作为报告多肽，因为它在真核细胞胞浆中表达，并且非常稳定。我们构建编码融合蛋白的质粒，该融合蛋白或者将GAP-43的前10个氨基酸，甲硫氨酸–亮氨酸–半胱氨酸–半胱氨酸–甲硫氨酸–精氨酸–精氨酸–苏氨酸–赖氨酸–谷氨酰胺，融合到完整CAT蛋白的N端（$GAP_{10}CAT$），或者是GAP-43的前40个氨基酸融合到CAT（$GAP_{40}CAT$）。如免疫印迹所示，在COS细胞中（图2*a*）或PC12细胞中（图2*b*）表达的CAT只存在于可溶性胞质组分中。与此相对照，嵌合蛋白$GAP_{10}CAT$和$GAP_{40}CAT$与膜相关。我们能够用去垢剂，而不是用氯化钠、氯化钙或EGTA（数据未展示）提取融合蛋白。因此，此蛋白的膜结合属性与大鼠脑[11-14]中的天然GAP-43类似。

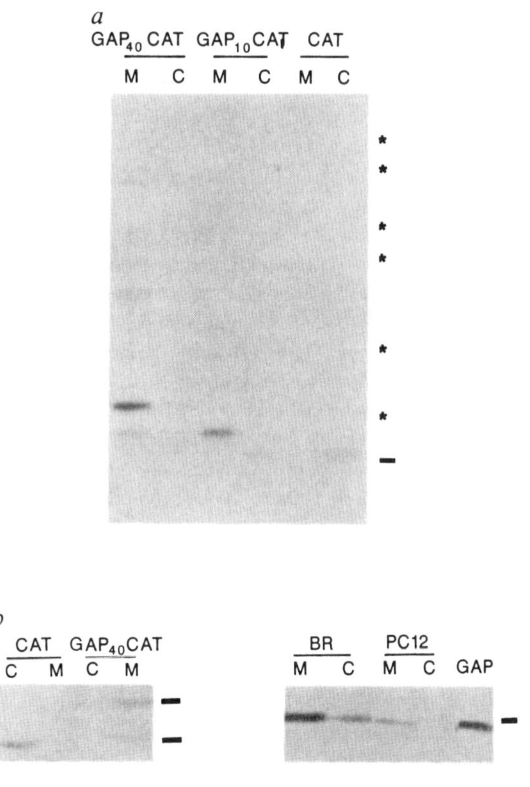

图2. GAP-43及GAP-CAT融合蛋白的膜结合。*a*. GAP-43 N端与CAT融合形成的嵌合蛋白与COS细胞膜结合。CAT、$GAP_{10}CAT$和$GAP_{40}CAT$载体都在COS细胞中瞬时表达。使用免疫印迹的方法通过抗CAT抗体检测来自每一转染的膜（M）和胞质（C）组分。在CAT转染细胞中，免疫反应性只存在于胞质组分中，并且与纯化的CAT蛋白共迁移（由横线显示）。在$GAP_{40}CAT$和$GAP_{10}CAT$转染细胞中，几乎所有的免疫反应性都在膜组分中，并且比CAT转染的细胞迁移得更慢。正如预期的那样，融合蛋白的分子量比CAT大4K或10K。星号所示蛋白标准的分子量为116K、84K、58K、48.5K、36.5K和

standards with M_rs of 116K, 84K, 58K, 48.5K, 36.5K and 26.6K migrated as indicated by the asterisks. *b*. Membrane association of GAP and $GAP_{40}CAT$ in PC12 cells. Stably transfected PC12 cells expressing CAT, $GAP_{40}CAT$ or GAP were selected as described below. Immunoblots of membrane (M)-and cytosolic (C) fractions were stained with anti-CAT antibodies (left panel) or anti-GAP (right panel) antibodies. CAT-transfected cells (CAT) contained immunoreactivity in the cytosolic, but not in the membrane fraction, and this immunoreactive CAT co-migrated with purified CAT (lower bar to right of left panel). In contrast, $GAP_{40}CAT$ transfected cells contained membrane-associated CAT immunoreactivity, which migrated more slowly (upper bar). Fractions from rat brain (BR) demonstrate that most, but not all, endogenous GAP-43 immunoreactivity was membrane-associated. In transfected PC12 cells over-expressing GAP-43 (PC12), nearly all of the GAP-immunoreactivity was membrane-associated and co-migrated with purified GAP-43 (GAP; indicated by far-right bar).

Methods. In the GAP-43-expression plasmid, pGAP, the GAP-43-coding sequence replaced the stuffer region at the *Xba*I sites of the CDM8 plasmid described by Seed[25]. The inserted GAP-43 sequence included the entire coding sequence of rat GAP-43, from the *Nla*III site at the start of translation to the *Sau*3AI site 68 bp downstream from the termination codon[8]. For the CAT expression plasmid, pCAT, the *Hin*dIII-*Bam*HI fragment containing the CAT-coding sequence and polyadenylation site from pSV2CAT[26] replaced the *Hin*dIII-*Bam*HI fragment of CDM8 containing the stuffer region and polyadenylation site. Plasmids $pGAP_{40}CAT$ and $pGAP_{10}CAT$ included sequences encoding the first 40 or 10 amino acids of GAP-43, respectively, fused in-frame with *CAT* in pCAT by the use of polylinkers. For transient transfection of COS cells we used DEAE-dextran and chloroquine as described[27]. For stable transfection of PC12 cells we used a neomycin-resistance plasmid co-transfected with the plasmid of interest on a 1:10 ratio as described[15]. During selection of PC12 cells, 400 μg ml^{-1} of active Geneticin (GIBCO) were used. Transient transfection of PC12 cells was by electroporation with the BioRad electroporation system using 300 V and 960 μF. After 8 h the medium was changed. Twenty-four hours after electroporation the cells were plated on poly-D-lysine-coated coverslips in the presence of 50 ng ml^{-1} NGF and analysed 24 h later. For immunochemical assays, rabbit anti-GAP-43 antibodies were made by immunizing rabbits against four peptides including amino-acid residues 1–24, 35–53, 53–69, and 212–228 of rat GAP-43. Anti-GAP-43 antibody was affinity-purified on GAP peptide agarose. Anti-GAP antibody was bound to a resin which contained 10 mg ml^{-1} of each peptide coupled to agarose by the cyanogen bromide method, and the antibody was eluted at pH 3.5. Rabbit anti-CAT antibodies were obtained from 5 Prime-3 Prime Inc. Secondary antibodies were obtained from Organon Teknika, Jackson Immunologicals and Vector Labs. For cell fractionation, COS or PC12 cells were scraped from 100-mm confluent Petri dishes and pelleted at 2,000*g* for 10 min. The pelleted cells were homogenized by Polytron in 10 mM Tris-HCl, 1 mM EDTA, pH 7.6 (300 μl per dish) and centrifuged at 250,000*g* for 30 min at 4°C. The supernatant was collected as the cytosolic fraction. The pellet was washed by homogenization and centrifugation in the same buffer, and then resuspended to the same volume as the cytosol fraction. Rat brain was obtained from 3-day-old rats and homogenized by Polytron in 10 mM Tris-HCl, 1 mM EDTA, pH 7.6 (10 ml per gram of wet-weight tissue). The cytosolic and washed-membrane fractions were prepared by centrifugation as described for the cell extracts. GAP-43 protein was purified from rat brain by a modification of the method of Andreasen *et al.*[28] and used as a positive control for immunostaining. The same volume of cytosolic or membrane fraction (usually 100 μl) was electrophoresed on polyacrylamide gels[29]. Proteins were electrophoretically transferred to nitrocellulose and excess sites were blocked with 4% BSA. Membranes were then incubated for 24 h at 4°C with 40 μg ml^{-1} affinity-purified anti-GAP, or a 1:1,000 dilution of anti-CAT antibodies. Bound antibody was detected using anti-rabbit horseradish-peroxidase method (Vectastain) according to the manufacturer's instructions. Tetramethyl benzidine (Kirkegaard and Perry, Gaithersburg, Maryland) was used as peroxidase substrate.

We investigated the cellular distribution of GAP-43 and the GAP-CAT chimaeric proteins in nerve growth factor (NGF)-treated transfectants of PC12 cells by confocal microscopy to determine whether the N-terminus accounts for the growth-cone enrichment of GAP-43 in neuronal cells. In this assay, CAT remained cytosolic (Fig. 3*a*), whereas GAP-43 was distributed in a punctate pattern with notable enrichment in many growth cones (Fig. 3*b*), a pattern similar to that of native GAP-43 in neurons. The N-terminus of GAP-43, which

26.6K。*b*. PC12 细胞中 GAP 和 GAP_{40}CAT 与膜结合，按如下所述选择 CAT、GAP_{40}CAT 或 GAP 表达稳定的转染 PC12 细胞。使用免疫印迹的方法通过抗 CAT 抗体（左图）或抗 GAP 抗体（右图）检测膜（M）和胞质（C）组分。CAT 转染细胞（CAT）含有的免疫反应性在胞质中而非膜中，而且这一具免疫反应性的 CAT 与纯化的 CAT（左图右下横线）共迁移。与此形成对比的是 GAP_{40}CAT 转染细胞在膜组分中具有 CAT 免疫反应性，并且迁移得更慢（上部横线）。来自大鼠脑（BR）的组分显示大部分但非所有内源 GAP-43 的免疫反应性在膜组分中。在转染的 PC12 细胞中过表达的 GAP-43（PC12），几乎所有免疫反应性都在膜中，并且与纯化的 GAP-43（GAP；如最右侧横线所示）共迁移。

方法。在 GAP-43 表达质粒 pGAP 中，GAP-43 编码序列取代了位于锡德 [25] 所述 CDM8 质粒的 *Xba*I 位点的填充区域。插入的 GAP-43 序列包括大鼠 GAP-43 的全编码序列，从位于翻译起点的 *Nla*Ⅲ位点直到距离下游终止密码子 [8]68 个碱基对的 *Sau*3AI 位点。CAT 表达质粒 pCAT 用含有 CAT 编码序列的 *Hind*Ⅲ-*Bam*HI 片段和来自 pSV2CAT[26] 的聚腺苷酸化位点取代含有填充区域和聚腺苷酸化位点的 CDM8 的 *Hind* Ⅲ -*Bam*HI 片段。质粒 $pGAP_{40}$CAT 和 $pGAP_{10}$CAT 分别包括编码 GAP-43 的前 40 或前 10 个氨基酸的序列，通过含多克隆位点的接头在同一阅读框中与 pCAT 中的 *CAT* 融合。如同以前描述 [27] 的那样，我们使用了 DEAE 葡聚糖和氯喹对 COS 细胞进行瞬时转染。至于 PC12 细胞的稳定转染，如同以前描述 [15] 的那样，我们使用按 1：10 比例混合的具有新霉素抗性质粒和感兴趣的质粒共同转染细胞。在 PC12 细胞的筛选过程中，使用了 400 μg · ml^{-1} 的活性遗传霉素（GLBCO）。用 BioRad 电穿孔系统，通过电穿孔（300 V，960 μF）转染 PC12 细胞。8 小时后换培养基。电穿孔 24 小时后，将细胞置于多聚赖氨酸包被的盖玻片上，培养液中 NGF 浓度为 50 ng · ml^{-1}，并在 24 小时后分析。为了进行免疫组化实验，用包含大鼠 GAP-43 的氨基酸残基 1~24、35~53、53~69 以及 212~228 的 4 个肽段免疫兔子，以制备兔源抗 GAP-43 抗体。在 GAP 肽琼脂糖上亲和纯化抗 GAP-43 抗体。通过溴化氰方法使含有 10 mg · ml^{-1} 的每一多肽的树脂与琼脂糖耦联，再使抗 GAP 抗体结合到这一树脂，并在 pH 3.5 的条件下洗脱抗体。从 5 Prime-3 Prime 公司获得兔源抗 CAT 抗体。从欧加农泰尼克，杰克逊免疫学和载体实验室获得二级抗体。为了提取细胞各种组分，从长满细胞的直径 100 mm 培养皿上刮下 COS 或 PC12 细胞，2,000*g* 离心 10 分钟。在包含 10 mM Tris-HCl 和 1 mM EDTA 且 pH 7.6（每个培养皿 300 μl）缓冲液中用 Polytron 对离心获得的细胞进行匀浆，并于 4℃ 250,000*g* 离心 30 分钟。收集上清作为胞质可溶性组分。剩下的沉淀在同一缓冲溶液中匀浆和离心，然后重新悬浮达到与胞质组分相同体积。取 3 日龄大鼠的大脑，并在包含 10 mM Tris-HCl 和 1 mM EDTA 且 pH 7.6 溶液中（每克湿重组织用 10 ml 溶液）用 Polytron 匀浆。像描述的那样，通过离心制备可溶性胞质和洗脱过的膜组分。使用安德烈亚森等所提出的修订后的方法 [28] 来纯化大鼠脑的 GAP-43 蛋白。并以此作为免疫染色的阳性对照。在聚丙烯酰胺胶 [29] 上电泳分离相同体积的可溶性胞质或膜组分（通常是 100 μl）。电转蛋白到硝酸纤维膜上，并用 4%BSA 封闭过量位点。用 40 μg ml^{-1} 亲和纯化的抗 GAP 抗体或 1：1000 稀释的抗 CAT 抗体在 4℃条件下孵育膜 24 小时。使用抗兔辣根过氧化物酶方法（Vectastain 试剂盒），遵照厂商的使用说明检测结合抗体。使用四甲基联苯胺（马里兰州盖瑟斯堡的柯克加德和佩里实验室）作为过氧化物酶的底物。

我们用共聚焦显微技术研究了 GAP-43 和 GAP-CAT 融合蛋白在神经生长因子（NGF）处理过的转染 PC12 细胞中的分布，以确定 N 端是否在 GAP-43 在神经细胞生长锥富集过程中起作用。在这一实验中，CAT 保留在可溶性胞质中（图 3*a*），而 GAP-43 呈点状分布并明显富集在很多生长锥中（图 3*b*），与神经元中的天然

was fused to CAT, caused the resulting fusion protein to acquire a distribution that closely resembled that of GAP-43 itself (Fig. 3*c*, *d*). Perinuclear labelling for both GAP-43 and the chimaeric protein was detected at a low level, and may have been due to localization to the Golgi, as has been observed for native GAP-43[16]. Glutaraldehyde fixation provided better histological preservation of the finer processes of the growth cones, and revealed that the chimaeric protein accumulated especially within filopodia (Fig. 4).

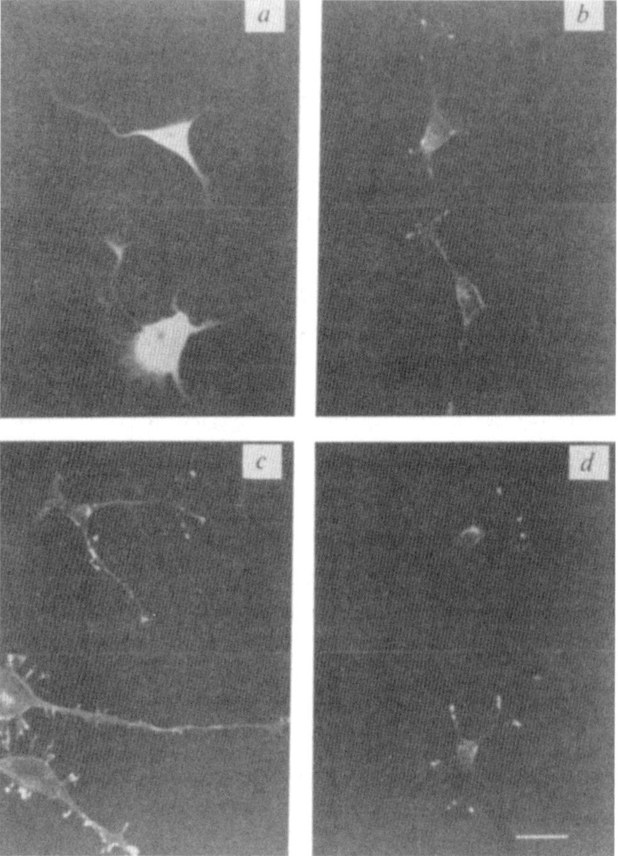

Fig. 3. Subcellular localization of CAT, GAP-43 and fusion proteins in transfected PC12 cells. Confocal immunofluorescence of CAT (*a*), GAP-43 (*b*), $GAP_{40}CAT$ (*c*), and $GAP_{10}CAT$ (*d*) in PC12 cells. Two representative fields of cells transfected with each plasmid are shown. CAT labelling is diffuse and cytosolic, whereas GAP-43 is localized to the membrane in a punctate fashion with some enrichment in the growth cones. When either the N-terminal 40 amino acids ($GAP_{40}CAT$) or 10 amino acids ($GAP_{10}CAT$) were fused to CAT, the immunofluorescent distribution resembled that for GAP-43, including enrichment in growth cones. All cells were treated with NGF for 24 h before fixation. Anti-CAT antibody was used for *a*, *c*, and *d*, whereas anti-GAP-43 antibody was used for *b*. Control PC12 cells of this variant expressed undetectable levels of GAP-43 and CAT immunoreactivity. Scale bar, 25 μm.

Methods. PC12 cells were transferred to poly-D-lysine-coated coverslips 24 h before immunofluorescence in the presence of 50 μg ml^{-1} NGF, fixed with 3.7% formaldehyde for 7 min, and made permeable with 0.1% Triton X-100 for 3 min. The samples were blocked with 4% BSA in PBS for 1 h, incubated for 1 h in primary antibody, rinsed with PBS, incubated in 0.3% H_2O_2 in PBS for 15 min (to reduce background), rinsed again and incubated for 1 h in secondary antibody. After washing with PBS several times, coverslips were rinsed with water and mounted with Gelvatol containing 0.4% n-propyl gallate to decrease bleaching. Immunofluorescence was not detectable above background when cells did not contain specific antigens or when the primary or secondary antibodies were omitted.

GAP-43 类似。GAP-43 N 端与 CAT 融合，使融合蛋白获得了非常类似 GAP-43 自身的分布（图 3*c*、*d*）。检测到 GAP-43 和融合蛋白在细胞核周围标记水平都很低，这可能因为它们和观察到的天然 GAP-43[16] 一样定位到高尔基体。戊二醛固定对生长锥的精细突起形态保存更好，并揭示了融合蛋白的积累，尤其是在丝状伪足中（图 4）。

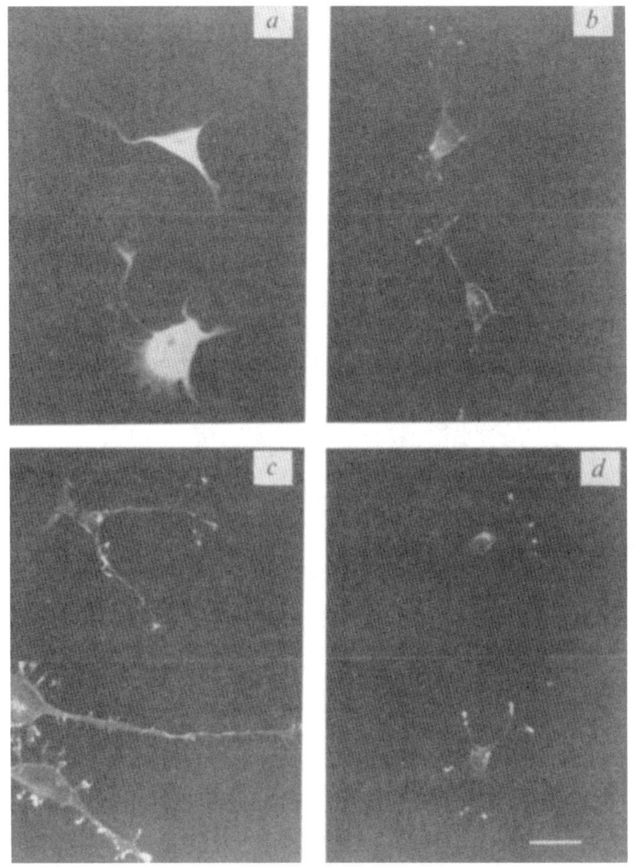

图 3. CAT、GAP-43 和融合蛋白在被转染的 PC12 细胞中的亚细胞定位。在 PC12 细胞中的 CAT（*a*）、GAP-43（*b*）、$GAP_{40}CAT$（*c*）以及 $GAP_{10}CAT$（*d*）的共聚焦免疫荧光结果。图片显示了用每一质粒转染的两个典型视野。CAT 标记是弥散的，并且分布在可溶性胞质中，而 GAP-43 位于膜上，呈点状分布，在生长锥中有一些富集。当 N 端的 40 个氨基酸（$GAP_{40}CAT$）或 10 个氨基酸（$GAP_{10}CAT$）与 CAT 融合，免疫荧光的分布类似 GAP-43 的结果，包括在生长锥中富集。所有细胞在固定前用 NGF 处理 24 小时。对 *a*、*c* 以及 *d* 使用抗 CAT 抗体，而对 *b* 使用抗 GAP-43 抗体。作为对照的 PC12 细胞表达的 GAP-43 和 CAT 的免疫反应性在可检测水平以下。比例尺为 25 μm。

方法。将 PC12 细胞转移到多聚赖氨酸覆盖的盖玻片上，用 50 $\mu g \cdot ml^{-1}$ NGF 培养液培养 24 小时后进行免疫荧光实验，用 3.7% 甲醛溶液固定 7 分钟，0.1% Triton X-100 处理 3 分钟，使细胞变得通透。用含 4%BSA 的 PBS 封闭 1 小时，加一抗孵育 1 小时，用 PBS 冲洗，在含 0.3%H_2O_2 的 PBS 溶液中孵育 15 分钟（以降低背景），再次冲洗，然后加二抗孵育 1 小时。在用 PBS 洗涤数次后，用水冲洗盖玻片，然后加含 0.4% 3,4,5- 三羟基苯甲酸正丙酯的 Gelvatol 以减少褪色。当细胞不含有特异性抗原或忽略一抗二抗时，背景上的免疫荧光是检测不到的。

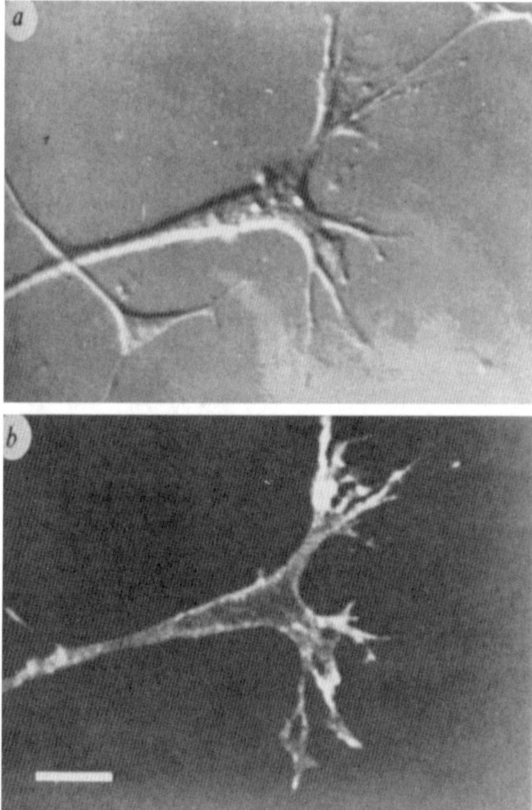

Fig. 4. Localization of $GAP_{40}CAT$ in the growth cone of a PC12 cell. A high power comparison of PC12 cells expressing $GAP_{40}CAT$ viewed with Nomarski optics (top) and scanning confocal immunofluorescence, labelled with anti-CAT antibodies (bottom). Cells had been treated with NGF for 7 days. One growth cone is brightly labelled, but the smaller one is not. Unequal labelling of different growth cones, even of the same cells, occurs for native GAP-43 in neurons[30] and did for the cells here. Comparison of the two images shows that filopodia are especially labelled. Similar results were seen for $GAP_{10}CAT$ (data not shown). Scale bar, 5 μm.

Methods. For high resolution confocal microscopy, the cells were fixed with freshly made 4% paraformaldehyde and 0.5% glutaraldehyde, which was essential to preserve the fine structure of the filopodia, and then with 0.1% Triton X-100 for 3 min and 2 mg ml^{-1} sodium borohydrate in PBS for 10 min. Confocal analysis used a BioRad MRC-500 scanning confocal imaging system and a Zeiss Axioplan microscope.

Thus, the first 10 amino acids of GAP-43 suffice to direct the cellular distribution of GAP-43. We are not aware of other proteins that have a sequence closely related to the GAP-43 N-terminus, although at least one other non-integral membrane protein that accumulates in growth-cone membranes, SCG 10 (ref. 17) has two cysteines in close proximity (at positions 22 and 24). In polarized epithelial cells, different proteins accumulate in the apical and basolateral plasma membranes[1,18,19], a process believed to depend on sorting signals in the protein, similar to the signals that direct traffic of membrane- and secreted proteins to their particular destinations[20-22]. For epithelial cells, such signals would also

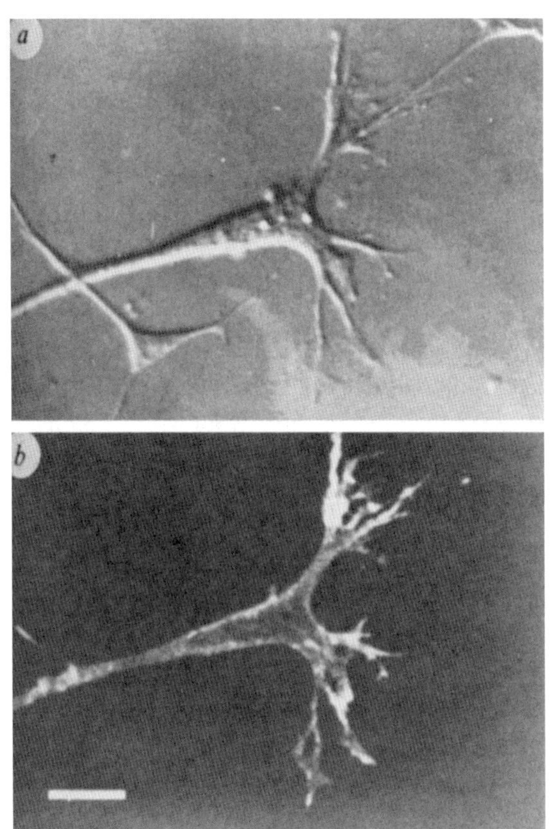

图 4. $GAP_{40}CAT$ 在 PC12 细胞生长锥中的定位。表达 $GAP_{40}CAT$ 的 PC12 细胞在高倍下诺马斯基光学图像（顶部）与用抗 CAT 抗体染色的免疫荧光共聚焦扫描结果（底部）比较。细胞用 NGF 处理了 7 天。一个生长锥被染得很亮，但是稍小一点的生长锥不是这样。不同生长锥标记程度不同，甚至同一细胞的生长锥染色也不相同，这种情况发生于神经元[30]中的天然 GAP-43，而这里的细胞也是如此。两幅图的比较显示，特别是丝状伪足被染色。类似的结果也见于 $GAP_{10}CAT$（数据未展示）。比例尺为 5 μm。**方法。**为进行高分辨率共聚焦显微分析，用新鲜配制的 4% 多聚甲醛和 0.5% 戊二醛固定细胞，后者对于保存丝状伪足的精细结构来说是必需的，然后用 0.1% Triton X-100 处理 3 分钟，并用含 2 mg · ml^{-1} 硼氢化钠的 PBS 溶液处理 10 分钟。使用 BioRad MRC-500 扫描共聚焦成像系统和蔡司 Axioplan 显微镜做共聚焦分析。

因此，GAP-43 的前 10 个氨基酸足以指导它在细胞中的分布。虽然至少有一个在生长锥膜中积累的非整合蛋白，SCG10（参考文献 17）有两个邻近的半胱氨酸（在位置 22 和 24），但我们没有发现含有 GAP-43 N 端序列相似序列的其他蛋白。在极化的上皮细胞中，不同的蛋白在细胞膜顶部和底部[1,18,19]积累，此过程被认为依赖于蛋白中的分选信号，类似于指导膜蛋白以及分泌型蛋白运输到它们特定目的地的信号[20-22]。对于上皮细胞，这样的信号也会识别细胞膜的顶部和底部。在神经元中，

recognize different regions of the plasma membrane as apical or basolateral. In neurons, the growth-cone membrane is also distinctive, although not unique, in its protein make-up. One interesting possibility is that the growth-cone membrane has binding sites that recognize and enhance the binding of the palmitylated N-terminus of GAP-43. Although possible, it seems less likely that the palmitylated residues interact with the lipid bilayer directly, because that would probably cause a more uniform membrane distribution for GAP-43. Along these lines, the fatty acid moiety of another acylated protein, N-myristylated VP4 of the poliovirus, has been shown by X-ray diffraction to interact with specific amino-acid residues of other viral proteins and not with the lipid bilayer[23, 24]. Because GAP-43 and GAP-CAT fusion proteins bind to the membrane of non-neuronal cells, similar or identical binding sites must be present in other cells types. However, because GAP-43 is neuron-specific, these sites would presumably be targets for different proteins in non-neuronal cells. Sorting and selective transport, however, are as likely to account for growth-cone accumulation[31] as is the existence of specialized membrane binding-sites.

It is notable that the sorting domain of GAP-43 caused enrichment especially in many filopodia. This is the normal location of GAP-43 in these cells, as shown by electron microscopy[16]. Given the previous observation that transfected GAP-43 can enhance the propensity of non-neuronal cells to extend filopodia[15], it will be of interest to correlate GAP-43 location with the motile activity of particular filopodia. Finally, the sequence described here may be useful in the delivery of histological markers or other agents to the growth-cone membrane.

(**341**, 345-348; 1989)

Mauricio X. Zuber, Stephen M. Strittmatter & Mark C. Fishman

Developmental Biology Laboratory of the Massachusetts General Hospital Cancer Center, Departments of Medicine and Neurology of Harvard Medical School, and the Howard Hughes Medical Institute, Wellman 4, Massachusetts General Hospital, Boston, Massachusetts 02114, USA

Received 12 June; accepted 14 August 1989.

References:

1. Matlin, K. S. *J. Cell Biol.* **103**, 2565-2568 (1986).
2. Ellis, L., Wallace, I., Abreu, E. & Pfeninger K. H. *J. Cell Biol.* **101**, 1977-1989 (1985).
3. Meiri, K. F., Pfenninger, K. H. & Willard, M. B. *Proc. Natl. Acad. Sci. U.S.A.* **83**, 3537-3541 (1986).
4. Skene, J. H. P. *et al. Science* **233**, 783-786 (1986).
5. De Graan, P. N. E. *et al. Neurosci. Lett.* **61**, 235-241 (1985).
6. Benowitz, L. I. & Routtenberg, A. *Trends Neurosci.* **10**, 527-532 (1987).
7. Skene, J. H. P. A. *Rev. Neurosci.* **12**, 127-156 (1989).
8. Basi, G. S., Jacobson, R. D., Virag, I., Schilling, J. & Skene, J. H. P. *Cell* **49**, 785-791 (1987).
9. Skene, J. H. P. & Virag, I. *J. Cell Biol.* **108**, 613-624 (1989).
10. Karns, L. R., Ng, S.-C., Freeman, J. A. & Fishman, M. C. *Science* **236**, 597-600 (1987).
11. Cimler, B. M., Andreasen, T. J., Andreasen, K. I. & Storm, D. R. *J. Biol. Chem.* **260**, 10784-10788 (1985).
12. Perrone-Bizzozero, N. I., Weiner, D., Hauser, G. & Benowitz, L. I. *J. Neurosci. Res.* **20**, 346-350 (1988).
13. Oestreicher, A. B., Van Dongen, C. J., Zwiers, H. & Gispen, W. H. *J. Neurochem.* **41**, 331-340 (1983).
14. Chan, S. Y., Murakami, K. & Routtenberg, A. *J. Neurosci.* **6**, 3618-3627 (1986).

生长锥膜的蛋白组成也是有特点的，虽然并非独一无二。一个有趣的可能性是，生长锥膜具有能够识别并增强与 GAP-43 的棕榈酰化 N 端结合的结合位点。虽然可能，但棕榈酰化的残基直接与脂质双分子层相互作用的可能性不大，因为可能会引起 GAP-43 更均匀的膜分布。同理，X 射线衍射显示另一乙酰化蛋白，脊髓灰质炎病毒的 N- 十四烷基化的 VP4 蛋白质分子的脂肪酸部分与其他病毒蛋白的特异氨基酸残基相互作用，而并不与脂质双分子层相互作用[23,24]。由于 GAP-43 和 GAP-CAT 融合蛋白与非神经元细胞的细胞膜结合，类似或相同结合位点必然存在于其他细胞类型中。尽管如此，因为 GAP-43 是神经元特异性的，所以在非神经元细胞中，这些位点大概是其他蛋白的结合靶点。虽然如此，分选和选择性转运就像特化的膜结合位点一样，可能对其生长锥中的积累[31]起作用。

很明显，GAP-43 的分选结构域造成了蛋白质的富集，特别是在很多丝状伪足中。如电子显微镜结果所示，在这类细胞中，这是 GAP-43 的正常定位[16]。根据此前的观察结果，转染的 GAP-43 能够增强非神经元细胞延伸丝状伪足的倾向[15]，这揭示 GAP-43 的定位与某些丝状伪足的运动活性之间的关系值得研究。最后，这里描述的氨基酸序列可能有助于将组织学标记或其他蛋白质分子定向转运到生长锥膜。

（周平博 翻译；刘佳佳 审稿）

15. Zuber, M. X., Goodman, D. W., Karns, L. R. & Fishman, M. C. *Science* **244**, 1193-1195 (1989).
16. Van Hooff, C. O. M. *et al. J. Cell Biol.* **108**, 1115-1125 (1989).
17. Stein, R., Mori, N., Matthews, K., Lo. L.-C. & Anderson, D. J. *Neuron* **1**, 463-476 (1988).
18. Rodriguez-Boulan, E. J. & Sabatini, D.D. *Proc. Natl. Acad. Sci. U.S.A.* 75, 5071-5075 (1978).
19. Simmons, K. & Fuller, S. D. *Ann. Rev. Cell Biol.* **1**, 243-288 (1985).
20. Wickner, W. T. & Lodish, H. F. *Science* **230**, 400-407 (1985).
21. Verner, K. & Schatz, G. *Science* **241**, 1307-1313 (1988).
22. Pfeffer, S. R. & Rothman, J. E. *Ann. Rev. Biochem.* **56**, 829-852 (1987).
23. Schultz, A. M., Henderson, L. E. & Oroszlan, S. *A. Rev. Cell Biol.* **4**, 611-647 (1988).
24. Chow, M. *et al. Nature* **327**, 482-486 (1987).
25. Seed, B. *Nature* **329**, 840-846 (1987).
26. Gorman, C. M., Moffat, L. F. & Howard, B. H. *Molec. Cell. Biol.* **2**, 1044-1051 (1982).
27. Zuber, M. X., Simpson, E. R. & Waterman, M. R. *Science* **234**, 1258-1261 (1986).
28. Andreasen, T. J., Leutje, C. W., Heideman, W. & Storm, D. R. *Biochemistry* **22**, 4615-4618 (1983).
29. Laemmli, U. K. *Nature* **227**, 680-685 (1970).
30. Goslin, K., Schreyer, D. J., Skene, J. H. P. & Banker, G. *Nature* **336**, 672-674 (1988).
31. Sargent, P. B. *Trends Neurosci.* **12**, 203-205 (1989).

Acknowledgements. We thank D. Goodman and L. Karns for help with plasmid construction, D. Rosenzweig and A. Pack for technical assistance, J. Garriga, R. Horvitz, P. Matsudaira and M. Shafel from MIT and J. White from the MRC for assistance with and access to confocal microscopy , and J. Jackson for preparation of the manuscript. M.X.Z. is the recipient of a postdoctoral fellowship from the American Cancer Society.

Rescue of *bicoid* Mutant *Drosophila* Embryos by Bicoid Fusion Proteins Containing Heterologous Activating Sequences

Wolfgang Driever *et al.*

Editor's Note

The patterning and fate of cells in developing embryos is influenced by so-called "morphogen gradients" of diffusible molecules. German biologists Wolfgang Driever and Christiane Nüsslein-Volhard had previously shown that the transcription factor Bicoid (Bcd) exists along a concentration gradient in *Drosophila* embryos, which in turn influences the expression pattern of other body-patterning genes. Here they team up with molecular biologists Mark Ptashne and Jun Ma to investigate the morphogenetic properties of Bcd. Mutants lacking the *bicoid* (*bcd*) gene develop duplicate posterior structures but lack heads and thoraces. But the embryos can be rescued by injecting them with mRNA encoding fusion proteins consisting of acidic transcriptional activating sequences—peptides that trigger transcription in many eukaryotes—and the *bicoid* DNA-binding domain.

The maternal gene *bicoid* (*bcd*) determines pattern in the anterior half of the *Drosophila* embryo. It is reported here that the injection of *bcd*$^-$ mutant embryos with messenger RNAs that encode proteins consisting of heterologous acidic transcriptional activating sequences fused to the DNA-binding portion of the *bcd* gene product, can completely restore the anterior pattern of the embryo.

THE *bcd* gene organizes development in the anterior half (head and thorax) of the *Drosophila* embryo[1]. The molecular basis for this morphogen function is a concentration gradient of Bcd protein spanning the anterior two thirds of the syncytial blastoderm stage embryo[2]. The Bcd protein is synthesized from *bcd* mRNA that is localized at the anterior tip of the embryo[3,4], and the concentration of Bcd protein determines positional values along the anterio-posterior axis[5]. Genetic analyses indicate that the zygotic expression of *hunchback* (*hb*), a gap gene that is expressed early in a distinct domain in the anterior 45% of the embryo, is regulated by the Bcd protein[6-9]. Recent experiments demonstrate that Bcd protein binds specifically to the *hb* promoter region and induces the expression of *hb* in both *Drosophila* Schneider cells and the embryo[10]. The binding of Bcd protein to multiple copies of Bcd-binding sites in the upstream region of fusion genes is sufficient to generate an *hb*-like expression pattern in the early embryo[11].

通过含异源激活序列的Bicoid融合蛋白拯救*bicoid*突变果蝇胚胎

沃尔夫冈·德里费尔等

编者按

胚胎发育过程中细胞的模式建成与命运受到可扩散分子形成的所谓“形态发生素梯度”的影响。德国生物学家沃尔夫冈·德里费尔和克里斯蒂安娜·福尔哈德早前已经揭示转录因子 Bicoid（Bcd）在果蝇胚胎中以浓度梯度的形式存在，此梯度反过来影响其他身体形态建成相关基因的表达模式。本文中他们与分子生物学家马克·普塔什尼和马俊合作研究 Bcd 在形态发生方面的特性。缺失 *bicoid*（*bcd*）基因的突变体发育出重复的后部结构但是缺少头部和胸部。但是这些胚胎可以通过注射某种信使 RNA 获得拯救，这种信使 RNA 编码一种融合蛋白，这种蛋白包含酸性转录激活序列（在许多真核生物中启动转录的多肽）和 *bicoid* 的 DNA 结合域。

母体基因 *bicoid*（*bcd*）决定果蝇胚胎前半部的发育模式。本文报道，将融合蛋白的信使 RNA（mRNA）注入 bcd^- 突变胚胎中，该信使 RNA 编码一个蛋白，该蛋白由异源酸性转录激活序列与 *bcd* 基因产物的 DNA 结合域融合而成，可以完全恢复胚胎前部的构造。

bcd 基因决定果蝇胚胎前半部（头和胸）的发育[1]。这种形成发生素功能的分子基础是横跨合胞体胚盘期胚胎前三分之二区域的 Bcd 蛋白浓度梯度[2]。Bcd 蛋白由定位于胚胎前端的 *bcd* mRNA 合成[3,4]，Bcd 蛋白的浓度决定胚胎前后轴的位置信息[5]。遗传分析表明一个合子表达的驼背基因（*hb*），是一种早期在胚胎前部 45% 的特定区域内表达的裂隙基因，受 Bcd 蛋白的调控[6-9]。最近的实验表明 Bcd 蛋白与 *hb* 启动子区特异性结合，并诱导 *hb* 在果蝇的施耐德细胞和胚胎中表达[10]。早期胚胎中，Bcd 蛋白与融合基因上游多拷贝的 Bcd 结合位点结合足以产生类似 *hb* 的表达模式[11]。

The N-terminal portion of the 489-amino-acid Bcd protein[4] contains a homoeodomain (residues 92–151), a protein sequence which is conserved among many gene products participating in embryonic pattern formation[12-14], and which is believed to direct specific DNA binding[15-18]. Recent experiments indicate that at least several of the homoeodomain proteins involved in *Drosophila* embryonic pattern formation are transcriptional regulatory proteins[19-21]. Near the N-terminal end of the Bcd protein there is a repeat that is found in the *paired* gene product[3]; this repeat is a sequence of ~30 amino acids, rich in histidine and proline, which is also present in other gene products in *Drosophila*. The function of this repeat, however, is unknown. The C-terminal part of the Bcd protein contains an acidic region (residues 347–414, with 16 acidic and 5 basic residues) as well as a glutamine-rich region (M- or opa-repeat; residues 256–289)[12]. The Bcd protein is also rich in hydroxyl amino acids (37 serines, 27 threonines and 19 tyrosines). In *Drosophila* Schneider cells, and most probably in the embryo, the Bcd protein is phosphorylated at several sites[10]. Wild-type Bcd protein, as well as certain other homoeodomain-containing proteins in *Drosophila*, activate transcription in yeast[18,22], and the protein product of several mutant *bcd* alleles have been analysed for their transcriptional function in yeast[22].

Recent studies on yeast transcriptional activators indicate that a typical activator contains two separable functions: one that directs the activator to specific DNA sequences and another that interacts with some component of the transcriptional machinery[23-26]. For example, the yeast transcriptional activators GAL4 and GCN4, which are involved in galactose metabolism and amino-acid synthesis, respectively, contain acidic activating sequences that interact with the transcriptional machinery[27,28]. The activating sequences of GAL4 can be functionally replaced either by acidic peptides encoded by *Escherichia coli* DNA, or by an acidic peptide designed to form an amphiphilic α-helix[29,30]. Not all acidic peptides function as activating sequences; some aspect of structure, perhaps an α-helix, is required[30]. GAL4, or its derivatives that contain one or more acidic activating sequences fused to the DNA-binding portion of GAL4, also stimulate gene expression in mammalian, insect and plant cells, provided that the sequence to which GAL4 binds is inserted near the promoter of the test gene[31-34]. Other classes of activating sequences have been described[35,36]; none of these, to our knowledge, has been shown to function in many different eukaryotic cells as do the acidic sequences, and some of them have been shown to work only in restricted cell types[36].

Here we describe experiments that show that the Bcd protein activates transcription from the Bcd-binding sites in three different systems—yeast cells, *Drosophila* Schneider cells, and *Drosophila* embryos. We also report that the N-terminal half of the Bcd protein (residues 1–246), which contains the DNA-binding function, has little transcriptional activation function in all three systems. When tested in *Drosophila* embryos by injecting the corresponding mRNA, residues 1–246 of Bcd only partially rescues the mutant bcd^- phenotype, whereas fusion proteins consisting of residues 1–246 of Bcd attached to heterologous acidic activating sequences fully rescue the bcd^- phenotype.

含489个氨基酸的Bcd蛋白N末端[4]包含一个同源异型结构域（第92~151位残基），这个同源域在许多参与胚胎构造形成的基因产物中都是保守的[12-14]，可以认为指导特异性的DNA结合[15-18]。最近的实验表明参与果蝇胚胎构造形成的同源异型结构域蛋白中至少有几个是转录调控蛋白[19-21]。Bcd蛋白N末端附近有一段在*prd*基因产物中发现的重复序列[3]，这个重复序列约含30个氨基酸，富含组氨酸和脯氨酸，该重复序列在果蝇其他基因产物也存在。然而这个重复序列的功能尚未知。Bcd蛋白的C末端含一个酸性区（第347~414位残基，有16个酸性残基和5个碱性残基）和一个富含谷氨酰胺的区域（甲硫氨酸或opa重复；第256~289位残基）[12]。Bcd蛋白也富含羟基氨基酸（37个丝氨酸、27个苏氨酸和19个酪氨酸）。在果蝇的施耐德细胞中并且最可能在胚胎中，Bcd蛋白几个位点是磷酸化的[10]。研究人员已经分析了若干*bcd*突变等位基因的蛋白产物在酵母中的转录功能[22]，发现果蝇的野生型Bcd蛋白及其他某些含同源域的蛋白，可在酵母中激活转录[18, 22]。

最近对酵母转录激活子的研究表明一个典型的激活子包含两个独立的功能：一个是介导激活子结合到特异DNA序列上，另一个是与转录机器的某些组件相互作用[23-26]。例如，酵母转录激活子GAL4和GCN4，分别参与半乳糖代谢和氨基酸合成，它们包含与转录机器相互作用的酸性激活序列[27,28]。GAL4激活序列功能可被大肠杆菌DNA编码的酸性多肽或可形成两性α螺旋结构的酸性多肽所取代。并不是所有酸性多肽都具有激活序列的功能；某种特定的结构，可能是一个α螺旋，是必需的[30]。GAL4或它的衍生物包含一个或多个与其DNA结合域融合的酸性活性序列，也能在哺乳动物、昆虫和植物细胞中激活基因表达，只要与GAL4结合的序列插入在检测基因的启动子附近[31-34]。其他类型的激活序列也有过报道[35. 36]；据我们所知，它们都不能像酸性序列一样在许多不同的真核细胞中起作用，其中有一些仅在有限的细胞类型中起作用[36]。

本文描述的实验证明在三种不同体系——酵母细胞、果蝇施耐德细胞和果蝇胚胎中，Bcd蛋白通过Bcd结合位点激活转录。我们也报道了一半的有DNA结合功能的Bcd蛋白N末端（第1~246位残基）在上述三种体系中很少有转录激活功能。通过向果蝇胚胎注射相应的mRNA做检测时，Bcd的1~246位残基只能部分拯救突变体*bcd*$^-$的突变表型，而Bcd 1~246位残基与异源酸性激活序列的融合蛋白则能完全拯救突变体*bcd*$^-$的表型。

Transcriptional Activation by Bcd Derivatives

Transcriptional activity in yeast was measured in cells bearing an integrated *GAL1–lacZ* fusion gene[37] containing three strong Bcd-binding sites[10,11]. These yeast cells also contained plasmids expressing, from the yeast *ADH1* promoter[38], genes encoding the *Drosophila* Bcd protein or one of its C-terminal deletion derivatives. The activities of Bcd derivatives in *Drosophila* Schneider cells were measured by co-transfecting the cells with a reporter plasmid containing the gene for chloramphenicol acetyl transferase (CAT) fused to the *hb* gene, and effector plasmids expressing the Bcd protein or one of its derivatives. For the assay in embryos, mRNAs encoding the Bcd protein or one of its derivatives were synthesized *in vitro*. These were then co-microinjected with a plasmid containing the reporter gene *hb-298CAT* into the anterior half of pre-polecell stage embryos from females homozygous for bcd^{E1}, a strong *bcd* allele bearing a large deletion[1,4,22]. The injected mRNA is efficiently translated, as revealed by immunohistological stainings of the injected embryos for Bcd protein (data not shown). The reporter DNA might enter the nuclei during the rapid syncytial nuclear divisions and the *hb-CAT* fusion gene is transcribed under the control of the Bcd protein[10]. About 3 h after injection, at the onset of gastrulation, CAT enzyme activity was measured in extracts prepared from the injected embryos (Table 1).

Table 1. Activities of Bcd protein and its deletion derivatives

Effector	Yeast	*Drosophila* Schneider cells	*Drosophila* embryos	
	β-gal activities (hb-GAL1–lacZ)	CAT activities (hb–298CAT)	CAT activities (hb–298CAT)	Rescue of bcd^- mutant phenotype
Wt Bcd (1–489)	125	45	40	complete
Bcd (1–396)	12	36	35	complete
Bcd (1–348)	12	9*	19	complete
Bcd (1–246)	13	8	8	partial
Bcd (1–111)	<1	2	2	none
None	<1	2	2	—

The stimulatory effect of Bcd protein on *hb* transcription was assayed in yeast, *Drosophila* Schneider cells and *Drosophila* embryos. Yeast—plasmids expressing Bcd or its deletion derivatives were transformed into a yeast strain (GGY1:: MA630R) containing an integrated *GAL1–lacZ* fusion gene; the *hb* promoter element bearing the Bcd protein binding sites[10] was located upstream of the *GAL1* gene. *Drosophila* Schneider cells—the effector plasmids, which express from the *Drosophila* metallothionein promoter various Bcd deletion derivatives, were transfected into Schneider cells together with the reporter plasmid hbCAT-289 which expresses the *E. coli CAT* gene from the *hb* promoter bearing the Bcd-protein-binding sites. CAT activities (percent acetylated forms of total chloramphenicol) are shown.* In Schneider cells, instead of Bcd (1–348), the derivative Bcd (1–338) was used. *Drosophila* embryos—*in vitro* transcribed mRNAs encoding Bcd protein and its deletion derivatives were co-injected with the reporter plasmid hbCAT-298 into the anterior half of pre-pole cell stage embryos derived from females homozygous for bcd^{E1}. CAT activities were determined at the onset of gastrulation. The rescue of the bcd^- mutant phenotype is described in detail in Fig. 2 and Table 3*a*.
The yeast strain used in these experiments is derived from GGY1 (*ura3*, *leu2*, *his3*, *Δgal4*, *Δgal80*) (ref. 46) by integrating at the *URA3* locus a modified *GAL1–lacZ* fusion gene. The plasmid pMA630R, which was used to generate GGY1:: MA630R, was constructed by inserting the *Hin*dIII–*Mlu*I fragment from the plasmid hbCAT-298 (ref. 10) into the *Xho*I site of LR1Δ1Δ2μ(ref. 37). The plasmids expressing Bcd deletion proteins were constructed in two steps. First, the *Xba*I linker (5'-CTAGTCTAGACTAG-3'), which contains translational stop codons in all three reading

由 Bcd 衍生物引起的转录激活

在含有整合的 *GAL1-lacZ* 融合基因 [37] 的细胞中检测酵母中的转录活性，该基因含三个强 Bcd 结合位点 [10,11]。这些酵母细胞也包含质粒，这些质粒通过酵母 *ADH1* 启动子 [38] 表达编码果蝇 Bcd 蛋白或它的一个 C 端缺失衍生物的基因。果蝇施耐德细胞中 Bcd 衍生物的活性是通过将细胞与一个报告基因质粒和一个效应基因质粒共同转染测定的，报告基因质粒中氯霉素乙酰转移酶（CAT）与 *hb* 基因融合，而效应基因质粒表达 Bcd 蛋白或它的一个衍生物。在胚胎实验中编码 Bcd 蛋白或其衍生物的 mRNA 在体外合成。然后与含报告基因 *hb-298CAT* 的质粒共同显微注射到 bcd^{E1} 雌性纯合子的前极细胞期胚胎的前半部，bcd^{E1} 是 *bcd* 一个大片段缺失的强等位基因 [1,4,22]。在注射过的胚胎中用胚胎免疫组织学染色检测 Bcd 蛋白（数据未显示），结果显示注射的 mRNA 被有效地翻译了。报告基因的 DNA 可能在快速合胞体核分裂期间进入核中，并且 *hb-CAT* 融合基因在 Bcd 蛋白控制下转录 [10]。注射约 3 小时后，在原肠胚形成初期，注射胚胎的提取物中检测到 CAT 酶活性（表 1）。

表 1. Bcd 蛋白及其缺失衍生物的活性

效应基因	酵母	果蝇施耐德细胞	果蝇胚胎	
	β-半乳糖苷酶活性（hb-GAL1-lacZ）	CAT活性（hb-298CAT）	CAT活性（hb-298CAT）	bcd^- 突变表型的拯救
野生型Bcd（1~489）	125	45	40	完全
Bcd（1~396）	12	36	35	完全
Bcd（1~348）	12	9 *	19	完全
Bcd（1~246）	13	8	8	部分
Bcd（1~111）	<1	2	2	无
无	<1	2	2	—

在酵母、果蝇施耐德细胞和果蝇胚胎中检验 Bcd 蛋白对 *hb* 转录的激活效应。酵母——将表达 Bcd 或其缺失衍生物的质粒转化到一个酵母菌株（GGY1∷MA630R）中，这个菌株包含整合的 *GAL1-lacZ* 融合基因；含有 Bcd 蛋白结合位点 [10] 的 *hb* 启动子元件位于 *GAL1* 基因的上游。果蝇施耐德细胞——效应基因质粒通过果蝇金属硫蛋白启动子表达各种 Bcd 缺失衍生物，与报告基因质粒 hbCAT-289 一起转入施耐德细胞，这种报告基因质粒通过具有 Bcd 蛋白结合位点的 *hb* 启动子表达大肠杆菌 *CAT* 基因。CAT 活性（总氯霉素乙酰化形式的百分比）如表所示。* 在施耐德细胞中用的不是 Bcd（1~348），而是衍生 Bcd（1~338）。果蝇胚胎—体外转录的编码 Bcd 蛋白及其缺失衍生物与报告基因质粒 hbCAT-298 一起注射到 bcd^{E1} 雌性纯合子产生的前极细胞期胚胎前半部分。在原肠胚形成初期测定 CAT 活性。对 bcd^- 突变表型拯救的详细描述见图 2 和表 3*a*。

这些实验中用的酵母菌株是通过在 GGY1（*ura3*、*leu2*、*his3*、*Δgal4*、*Δgal80*）（参考文献 46）的 *URA3* 位点整合一个修饰过的 *GAL1-lacZ* 融合基因得到的。用于产生 GGY1∷MA630R 的质粒 pMA630R 是通过将质粒 hbCAT-298（参考文献 10）的 *Hin*dIII-*Mlu*I 片段插入到 LR1Δ1Δ2μ（参考文献 37）的 *Xho*I 位点构建起来的。构建表达 Bcd 缺失蛋白的质粒分两步。第一，将在所有三个阅读框中包含转录终止密码子的 *Xba*I 接头

frames, was inserted at different positions in the *bcd* gene in the plasmid pTN3bcd, an expression vector (Sp6 promoter) for the Bcd-encoding region with the frog globin mRNA leader[39]. Then, either the intact or modified Bcd encoding regions were inserted into the *Hin*dIII site of the yeast plasmid AAH5 containing the yeast *ADH1* promoter[38]. The *Xba*I linker was inserted at the following positions: *Pst*I site (at residue 111), *Sal*I site (at residue 246), *Acc*I site (at residue 348) and *Bgl*II site (at residues 396). The β-galactosidase (β-gal) activities were measured in yeast cells grown in synthetic minimal medium lacking leucine with glycerol and ethanol as carbon sources[47]. For Schneider cells, modified Bcd encoding regions containing stop codons at various positions were inserted into plasmid pRmHa-3, a *Drosophila* expression vector bearing the metallothionein promoter[48]. Unlike the Bcd (1–348) protein, which was generated by inserting the stop codon linker at the *Acc*I site, we used Bcd (1–338) in Schneider cells, the stop linker being inserted at the *Ava*I site. The procedures for Schneider-cell transfection, the reporter plasmid hbCAT-298 and CAT assay were as described previously[10,49]. Twelve hours after transfection, the culture medium was replaced by the medium containing 1 mM $CuCl_2$ to induce the metallothionein promoter, and the CAT activities were assayed 24 h later. The transfection efficiency was normalized by the β-gal activity expressed from the third transfected plasmid pPAclacZ (ref. 10). For the assays in yeast and Schneider cells, synthesis of the various Bcd proteins was tested by immunoblotting analysis (Fig. 1). For *Drosophila* embryos, mRNAs coding for Bcd protein or its deletion derivatives were transcribed *in vitro* by Sp6 RNA polymerase[50] from plasmid pTN3bcd and its derivatives (see above). The mRNAs (0.5 μg μl^{-1}) were co-injected with the reporter plasmid phbCAT-298(1 μg μl^{-1}) into the anterior half of pre-pole cell stage embryos from females homozygous for bcd^{E1}. At the onset of gastrulation, extracts were prepared and CAT activities were determined as described previously[10]. All assays described above were done in quadruplicate. For assays in yeast and Schneider cells, the standard error of the mean (s. e. m.) was normally <20%; for assays in *Drosophila* embryos the s. e. m. was 20% (for wild-type (wt) Bcd), 26% (for Bcd1–396) and 50–60% (for the shorter truncations).

Table 1 describes transcriptional activation by the Bcd protein and various deletion derivatives in the three assays, as well as the results of experiments measuring rescue of bcd^- embryos (see later). Wild-type Bcd protein activates transcription in all three systems. Deletion of 83 C-terminal residues from the Bcd protein creates a derivative containing residues 1–396, Bcd (1–396), which activates transcription only 10% as efficiently in yeast as does the wild-type Bcd protein, but 80% as efficiently as does the wild-type protein in *Drosophila* cells and embryos. Derivative Bcd (1–338) has lost most of its ability to activate transcription in Schneider cells, but activates ~50% as efficiently as wild-type Bcd protein in *Drosophila* embryos. Derivative Bcd (1–246) activates transcription only slightly in all three systems, and Bcd (1–111), which contains only a third of the homoeodomain, is inactive. Figure 1 shows that proteins of the expected size are synthesized in yeast and Schneider cells. However, the results are obscured somewhat by our lack of knowledge of protein levels in the three test systems but, when taken with the results of the following section, indicate a correlation between the ability of a Bcd derivative to activate transcription in the embryo and its biological function as a morphogen. The fragment Bcd (1–396) has lost part of the acidic sequence of the Bcd protein, and Bcd (1–348) has lost all of it; thus this acidic sequence is not required for all of the transcriptional activation function of the Bcd protein in the embryo.

(5'-CTAGTCTAGACTAG-3')，插入到质粒 pTN3bcd 中 *bcd* 基因的不同位置，pTN3bcd 是一个表达 Bcd 编码区的载体（Sp6 启动子），它带有蛙珠蛋白 mRNA 前导序列 [39]。然后，将完整的或修饰过的 Bcd 编码区插入到含酵母 *ADH1* 启动子的酵母质粒 AAH5 的 *Hin*dIII 位点 [38]。将 *Xba*I 接头插入到以下位置：*Pst*I 位点（位于 111 位残基）、*Sal*I 位点（位于 246 位残基）、*Acc*I 位点（位于 348 位残基）、*Bgl*II 位点（位于 396 位残基）。β– 半乳糖苷酶活性以在丙三醇和乙醇作碳源且缺少亮氨酸的合成基本培养基上培养的酵母细胞中测定 [47]。对施耐德细胞，将修饰过的含有位于不同位置的终止子 Bcd 编码区插入到质粒 pRmHa-3 中，pRmHa-3 是具有金属硫蛋白启动子的果蝇表达载体 [48]。不像将终止密码子接头插入 *Acc*I 位点得到 Bcd（1~348）蛋白，在施耐德细胞中我们用 Bcd（1~338），是将终止密码子接头插在 *Ava*I 位点。施耐德细胞的转染策略，报告基因质粒 hbCAT-298 和 CAT 检测已有描述 [10,49]。转染后 12 小时，将培养基换为含 1mM 氯化铜的培养基，以诱导金属硫蛋白启动子，并在 24 小时后检测 CAT 活性。转染效率通过第三个转化质粒 pPAclacZ（参考文献 10）表达的 β– 半乳糖苷酶的活性校正。在酵母和施耐德细胞实验中，各种 Bcd 蛋白的合成通过免疫印迹分析检测（图 1）。果蝇胚胎实验中，Bcd 蛋白或其缺失衍生物的 mRNA 在体外通过 Sp6 RNA 聚合酶由 pTN3bcd 和它的衍生物转录而来（见上文）。mRNA（0.5 μg · μl^{-1}）与报告基因质粒 phbCAT-298（1 μg · μl^{-1}）一起注射到 *bcd*^{E1} 雌性纯合子的极细胞前期胚胎的前半部。在原肠胚形成初期，按以前的描述制备提取物和测定 CAT 活性 [10]。上述所有实验都做了四次。酵母和施耐德细胞实验的平均标准误差（s.e.m.）通常 < 20%；果蝇胚胎实验的平均标准误差是 20%（野生型（wt）Bcd）、26%（Bcd1~396）及 50%~60%（更短片段）。

表 1 描述了在三个实验中 Bcd 蛋白及其各种缺失衍生物的转录激活，及拯救 *bcd*$^{-}$ 胚胎的实验检测结果（见后文）。野生型 Bcd 蛋白在三种实验体系中都激活转录。Bcd 蛋白缺失 C 末端 83 个残基产生一个 1~396 个残基的衍生物，Bcd（1~396），它在酵母中激活转录的效率只有野生型 Bcd 蛋白的 10%，但在施耐德细胞和果蝇胚胎中的效率为野生型蛋白的 80%。衍生物 Bcd（1~338）在施耐德细胞中丢失了大部分转录激活功能，但在果蝇胚胎中的转录激活效率为野生型 Bcd 的 50%。衍生物（1~246）在三种体系中只能微弱地激活转录，Bcd（1~111）只包含三分之一的同源异型结构域，无活性。图 1 显示了在酵母和施耐德细胞中预期大小的蛋白的合成。然而，由于我们缺乏这三个检测系统的蛋白质水平的知识，这些结果有些不明朗，但当此结果与随后的部分结合，则表明一种 Bcd 衍生物在胚胎中激活转录的能力与它作为形态发生素的生物学功能之间的相互关系。片段 Bcd（1~396）缺失 Bcd 蛋白酸性序列的一部分，Bcd（1~348）缺失了所有酸性序列；因此这个酸性序列不是胚胎中 Bcd 蛋白所有转录激活功能所必需的。

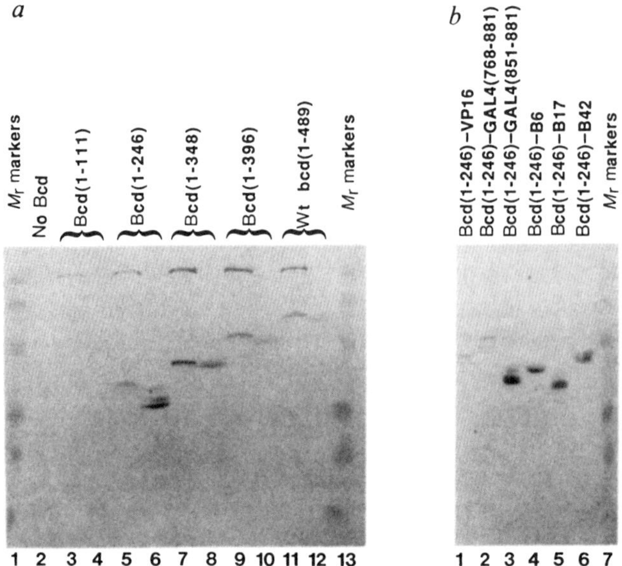

Fig. 1. Immunoblotting assays. *a*, The intact Bcd protein and deletion derivatives synthesized in yeast (lanes 4, 6, 8, 10 and 12; lane 2, yeast extracts without Bcd protein) and *Drosophila* Schneider cells (lanes 3, 5, 7, 9 and 11). The predicted relative molecular masses (M_r) are ($M_r \times 10^{-3}$): wild-type Bcd, 54 (ref. 4); Bcd (1–396), 44; Bcd (1–348), 39; Bcd (1–246), 28; and Bcd (1–111), 13. The monoclonal antibody does not react with Bcd (1–111). The apparent M_r of the proteins including wild-type Bcd protein are higher than predicted. The electrophoretic mobilities of the proteins synthesized in yeast are relatively higher than those of the proteins from Schneider cells; the possible effect of different post-translational modifications in yeast and Schneider cells is discussed in the text. *b*, The Bcd fusion proteins synthesized in yeast. The predicted M_r of the fusion proteins are ($M_r \times 10^{-3}$): Bcd (1–246) –VP16 (326 amino acids), 36; Bcd (1–246) –GAL4 (768–881) (361 amino acids), 40; Bcd (1–246) –GAL4 (851–881) (278 amino acids), 31; Bcd (1–246) –B6 (334 amino acids), 37; Bcd (1–246) –B17 (278 amino acids), 31; and Bcd (1–246)–B42 (337 amino acids), 37. Apparently, the fusion protein Bcd (1–246) –B6 synthesized in yeast migrates faster than expected, and it is possible that most of this protein is degraded; this result may be related to the observation that Bcd (1–246) –B6 is not very active in yeast cells (see Table 2). The sizes of M_r markers (from BioRad) are ($M_r \times 10^{-3}$): 130, 75, 50, 39, 27 and 17.

Methods. Procedures of preparing proteins from yeast and Schneider cells were as previously described[10,51]. For the yeast proteins in *a*, the yeast cells were grown in 10 ml synthetic minimal medium (lacking leucine) with glycerol and ethanol (optical density at 600 nm (OD_{600}) at 0.8–1.1), and for Bcd fusion proteins in *b*, the yeast cells were grown in 5 ml synthetic minimal medium with glucose (OD_{600} at 1.0–1.7). The amount of protein samples loaded on the protein gel was normalized by the density of the yeast cultures (OD_{600}). Proteins from Schneider cells were expressed from the *Drosophila* metallothionein promoter as described in the legend to Table 1. The derivative Bcd (1–348) in Schneider cells should be Bcd (1–338) (see legend to Table 1 for details). A 12% SDS gel was used to separate the proteins. The first antibody (mouse anti-Bcd)[2] was diluted 1:1,000 and the second antibody (alkaline phosphatase conjugated rabbit anti-mouse IgG) was diluted 1 : 1,500. Staining procedure was according to the manufacturer's instructions (BRL)

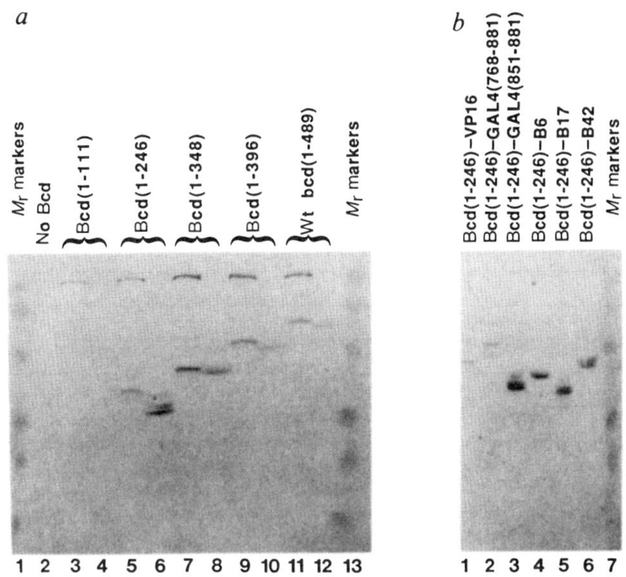

图 1. 免疫印迹实验。*a*，酵母（泳道 4、6、8、10 和 12；泳道 2 是不含 Bcd 蛋白的酵母提取物）和果蝇施耐德细胞（泳道 3、5、7、9 和 11）中合成的完整 Bcd 蛋白和缺失衍生物。预测相对分子量（M_r）为（$M_r \times 10^{-3}$）：野生型 Bcd，54（参考文献 4）；Bcd（1~396），44；Bcd（1~348），39；Bcd（1~246），28 和 Bcd（1~111），13。单克隆抗体不与 Bcd（1~111）作用。包括野生型 Bcd 蛋白，这些蛋白表观相对分子量比预测的大。酵母中合成的蛋白电泳迁移率与施耐德细胞合成的相比较高；可能是酵母细胞和施耐德细胞中不同的翻译后修饰引起的，下文对此进行了讨论。*b*，酵母中合成的 Bcd 融合蛋白。融合蛋白的预测相对分子量为（$M_r \times 10^{-3}$）：Bcd（1~246）–VP16（326 个氨基酸），36；Bcd（1~246）–GAL4（768~881）（361 个氨基酸），40；Bcd（1~246）–GAL4（851~881）（278 个氨基酸）31；Bcd（1~246）–B6（334 个氨基酸），37；Bcd（1~246）–B17（278 个氨基酸），31；Bcd（1~246）–B42（337 个氨基酸），37。显然地，酵母中合成的 Bcd（1~246）–B6 迁移速度比预测快，可能是这个蛋白的大部分降解了，这个结果可能与观察到的 Bcd（1~246）–B6 在酵母细胞中不很活跃有关（见表 2）。相对分子量标记（购自伯乐公司）的大小为（$M_r \times 10^{-3}$）：130、75、50、39、27 和 17。

方法。从酵母和施耐德细胞制备蛋白的步骤参照以前的描述[10,51]。对 *a* 中的酵母蛋白，酵母细胞在 10 ml 含丙三醇和乙醇的人工基本培养基（缺亮氨酸）中培养（600 nm 光密度（OD_{600}）0.8~1.1），对于 *b* 中的 Bcd 融合蛋白，酵母细胞在 5 ml 含葡萄糖的人工基本培养基中培养（OD_{600} 为 1.0~1.7）。蛋白凝胶的蛋白上样量通过酵母培养物 600 nm 处光密度（OD_{600}）归一化。施耐德细胞的蛋白由果蝇金属硫蛋白启动子表达，在表 1 注中描述过了。衍生物 Bcd（1~348）在施耐德细胞中应为 Bcd（1~338）（详细说明见表 1 注）。用 12% 的 SDS 凝胶分离蛋白质。一抗（鼠抗 Bcd）[2] 按 1∶1,000 稀释，二抗（碱性磷酸酶结合的兔抗鼠球蛋白 IgG）按 1∶1,500 稀释。染色步骤按产品操作说明（BRL 公司）。图中出现的单词 M_r Marker：分子量标记；No Bcd：无 Bcd 蛋白。

Table 2. Activities of Bcd fusion proteins

Effector	Yeast β-gal activities (hb–GAL1–lacZ)	*Drosophila* embryos Rescue of *bcd*⁻ mutant phenotype
Wt Bcd	125	complete
Bcd (1–246)	13	partial
Bcd (1–246) –VP16	1,307	lethal
Bcd (1–246) –GAL4 (768–881)	953	lethal
Bcd (1–246) –GAL4 (851–881)	342	complete
Bcd (1–246) –B6	43	complete
Bcd (1–246) –B17	105	complete
Bcd (1–246) –B42	367	lethal
None	<1	—

Yeast—plasmids expressing Bcd fusion proteins were transformed into a yeast strain, and β-gal activities measured. See legend to Table 1 for further details. *Drosophila* embryos—the rescue of the *bcd*⁻ mutant phenotype by injection of *in vitro* transcribed mRNAs coding for Bcd protein derivatives is described in detail in Fig. 2 and Table 3*b*. Various activating sequences were fused to the *bcd* gene in the plasmid pTN3bcd; these plasmids were also used to synthesize the mRNAs for the assays in *Drosophila* embryos. The activating sequences were obtained from the *Hin*dIII-*Sal*I fragment of the plasmid β58dTX for VP16, and the *Eco*RI-*Sal*I fragments for all the others[28,29]. Then, these *bcd* fusion genes were inserted into the *Hin*dIII site of the yeast expression vector AAH5 (ref. 38). The yeast strain used was GGY1:: MA630R. The assays were done in quadruplicate and the standard error was normally <20%.

Morphogenetic Activities

We tested the morphogenetic activity of the Bcd deletion derivatives by injecting small mounts of corresponding *in vitro*-synthesized mRNAs into the anterior tip of embryos obtained from bcd^{E1} mutant females and analysing the cuticle phenotype after one day of development (Tables 1 and 3*a*; Fig. 2). Embryos not injected do not develop heads and thoraces, but instead form duplications of posterior structures anteriorly (Fig. 2*a*). Depending on the concentration, injection of mRNA coding for wild-type Bcd protein can completely rescue the mutant phenotype, and larvae frequently hatch, giving rise to adult flies[39]. Of the deletion derivatives, both Bcd (1–396) and Bcd (1–348) are also able to rescue the mutant phenotype completely. By contrast, Bcd (1–111) does not rescue any aspect of the bcd^- mutant phenotype. The derivative Bcd (1–246) only partially restores the wild-type pattern: it suppresses the formation of posterior structures at the anterior, and induces with high frequency thoracic structures, and more rarely, structures of the segmented region of the head, the labial, maxillary or mandibular segment. But even when the mRNA coding for the derivative Bcd (1–246) is injected at high concentration (5 $\mu g\ \mu l^{-1}$; Table 3*a*), the embryos never hatch (see Fig. 2*d*). Embryos injected with mRNA encoding the fragment Bcd (1–246) resemble those from females that are homozygous for two of the weak *bcd* alleles, bcd^{E5} and bcd^{111} (ref.1), which are caused by amber mutations at amino acids 259 and 257, respectively[22]. Thus our data indicate that the C-terminal third of the protein, containing the acidic region (residues 347–414), is not essential for the ability of the protein to rescue the mutant phenotype. Our results also indicate that only those forms of Bcd that activate transcription in the embryo—wild type, Bcd (1–396) and Bcd (1–348)—can completely rescue bcd^- embryos. We now show that Bcd (1–246), a

表 2. Bcd 融合蛋白活性

效应基因	酵母 β-半乳糖苷酶活性(hb–GAL1–lacZ)	果蝇胚胎拯救*bcd*⁻突变表型
野生型Bcd	125	完全
Bcd(1~246)	13	部分
Bcd(1~246)–VP16	1,307	致死
Bcd(1~246)–GAL4(768~881)	953	致死
Bcd(1~246)–GAL4(851~881)	342	完全
Bcd(1~246)–B6	43	完全
Bcd(1~246)–B17	105	完全
Bcd(1~246)–B42	367	致死
无	<1	—

酵母——将表达 Bcd 融合蛋白的质粒转入酵母菌株，并测定 β- 半乳糖苷酶活性。其他细节见表 1 注。果蝇胚胎——注射体外转录的 Bcd 蛋白衍生物 mRNA 拯救 *bcd*⁻突变表型，详细描述见图 2 和表 3*b*。在质粒 pNT3bcd 中各种激活序列被融合到 *bcd* 基因上；这些质粒用来合成果蝇胚胎实验所用的 mRNA。VP16 的激活序列从质粒 β58dTX 的 *Hin*dIII-*Sal*I 片段而来，其他的激活序列从 *Eco*RI-*Sal*I 片段而来[28,29]。然后，将这些 *bcd* 融合基因插入到酵母表达载体 AAH5 的 *Hin*dIII 位点（参考文献 38）。所用酵母菌株为 GGY1∶∶MA630R。所有实验重复四次，标准误差通常小于 20%。

形态发生活性

我们检测了 Bcd 缺失衍生物的形态发生活性，通过向 bcd^{E1} 突变母体产生的胚胎前端注射少量相应的体外合成的 mRNA，发育一天后分析表皮表型（表 1 和 3*a*；图 2）。未注射胚胎没有发育出头部和胸部，而是形成了后部结构向前的重复（图 2*a*）。依赖于 mRNA 浓度，注射编码野生型 Bcd 蛋白的 mRNA 可以完全拯救突变表型，大部分幼虫孵化，并发育为成虫[39]。在缺失衍生物中 Bcd（1~396）和 Bcd（1~348）也可以完全拯救突变表型。相反，Bcd（1~111)不能在任何方面拯救 *bcd*⁻ 突变表型。衍生物 Bcd（1~246）只能部分修复野生型模式：它抑制后部结构在前部的形成，高频率地诱导胸部结构发生，但很少出现头、唇、上颌或下颌的体节结构。但即使以高浓度注射（$5\ \mu g\cdot\mu l^{-1}$；表 3*a*）衍生物 Bcd（1~246）的 mRNA，胚胎也从不孵化（见图 2*d*）。用片段 Bcd（1~246）mRNA 注射的胚胎类似于两个含纯合的弱 *bcd* 等位基因 bcd^{E5} 和 bcd^{111}（参考文献 1）的雌性产生的胚胎，它们分别由 259 位氨基酸和 257 位氨基酸的琥珀型突变引起[22]。因此，我们的数据表明这个蛋白靠近 C 末端的三分之一，包括酸性区域（347~414 位残基）不是它拯救突变表型的能力所必需的。我们的结果还表明只有在胚胎中激活转录的几种 Bcd——野生型、Bcd（1~396）和 Bcd（1~348）——可以完全拯救 *bcd*⁻ 胚胎。我们现在展示的 Bcd（1~246），是一个在三个体系中都能微弱地激活转录的衍生物，当它与异源酸性激活序列融合

derivative that activates transcription poorly in all three systems, can rescue *bcd*⁻ mutant embryos when fused to heterologous acidic activating sequences.

Table 3*a*. Rescue of *bcd*⁻ embryos by Bcd and its deletion derivatives

	Bcd(1–111)				Bcd(1–246)					Bcd(1–348)				Bcd(1–396)				Wild-type Bcd			
RNA concentration ($\mu g\ \mu l^{-1}$)	1.20	0.40	0.10	0.02	5.00	1.20	0.40	0.10	0.02	1.20	0.40	0.10	0.02	1.20	0.40	0.10	0.02	1.20	0.40	0.10	0.02
Phenotype n	31	16	34	21	33	56	62	60	63	65	47	56	39	84	61	59	55	58	71	54	65
1. Anterior not developed	3	6	0	0	48	38	15	5	0	25	15	14	8	25	18	5	4	26	7	13	2
2. *bcd*⁻ mutant	97	94	100	100	3	9	58	93	100	0	4	27	85	1	18	88	96	2	9	41	97
3. Partial rescue	0	0	0	0	9	5	18	2	0	28	38	52	8	35	36	7	0	21	32	37	2
4. Partial rescue (anterior open)	0	0	0	0	40	48	7	0	0	29	19	7	0	20	21	0	0	29	18	9	0
5. Head and thorax completely rescued	0	0	0	0	0	0	0	0	0	18	23	0	0	19	7	0	0	22	35	0	0

Table 3*b*. Rescue of *bcd*⁻ embryos by Bcd and Bcd fusion proteins

	Bcd(1–246)–VP16				Bcd(1–246)–GAL4 (768–881)				Bcd(1–246)–GAL4 (851–881)				Bcd(1–246)–B6				Bcd(1–246)–B17				Bcd(1–246)–B42			
RNA concentration ($\mu g\ \mu l^{-1}$)	1.20	0.40	0.10	0.02	1.20	0.40	0.10	0.02	1.20	0.40	0.10	0.02	1.20	0.40	0.10	0.02	1.20	0.40	0.10	0.02	1.20	0.40	0.10	0.02
Phenotype n	19	59	52	33	24	73	64	43	46	34	73	51	29	57	64	38	25	45	72	54	40	35	56	67
1. Anterior not developed	95	92	96	55	58	18	28	14	13	3	21	2	10	5	23	5	12	0	18	30	63	12	60	28
2. *bcd*⁻ mutant	5	7	4	45	0	5	13	65	2	26	10	98	7	2	17	95	0	2	11	20	5	14	5	34
3. Partial rescue	0	0	0	0	0	60	56	21	6	29	38	0	7	32	30	0	20	40	50	50	5	43	20	37
4. Partial rescue (anterior open)	0	0	0	0	42	15	3	0	59	29	22	0	60	25	28	0	64	35	11	0	27	31	14	0
5. Head and thorax completely rescued	0	0	0	0	0	0	0	0	19	12	10	0	17	37	3	0	4	20	11	0	0	0	0	0

Shown are the results of micro-injections of mRNAs encoding different Bcd derivatives into embryos that were obtained from homozygous bcd^{E1} females[1]. The cuticles of the developed embryos were analysed and the phenotypes classified (on the left of each Table) as described in Fig. 2. The mRNAs encoding various Bcd derivatives are shown at the top of each Table; concentrations as indicated. *n*, Number of cuticles analysed that developed from 100 injected embryos. The phenotypic classes are specified as a percentage of total embryos analysed. See legend to Fig. 2 for details of experiments. To test the statistical significance of the data obtained, several series of injections were repeated and the average difference between two injection series were <50%; different phenotypic classes were never obtained.

时能拯救 bcd^- 突变胚胎。

表 3*a*. Bcd 及其缺失衍生物拯救 bcd^- 胚胎

	Bcd(1~111)				Bcd(1~246)					Bcd(1~348)				Bcd(1~396)				野生型Bcd			
RNA浓度 (μg·μl^{-1})	1.20	0.40	0.10	0.02	5.00	1.20	0.40	0.10	0.02	1.20	0.40	0.10	0.02	1.20	0.40	0.10	0.02	1.20	0.40	0.10	0.02
表型 *n*	31	16	34	21	33	56	62	60	63	65	47	56	39	84	61	59	55	58	71	54	65
1. 前部未发育	3	6	0	0	48	38	15	5	0	25	15	14	8	25	18	5	4	26	7	13	2
2. bcd^-突变	97	94	100	100	3	9	58	93	100	0	4	27	85	1	18	88	96	2	9	41	97
3. 部分拯救	0	0	0	0	9	5	18	2	0	28	38	52	8	35	36	7	0	21	32	37	2
4. 部分拯救(前部开放)	0	0	0	0	40	48	7	0	0	29	19	7	0	20	21	0	0	29	18	9	0
5. 头和胸完全拯救	0	0	0	0	0	0	0	0	0	18	23	0	0	19	7	0	0	22	35	0	0

表 3*b*. Bcd 及 Bcd 融合蛋白拯救 bcd^- 胚胎

	Bcd(1~246)–VP16				Bcd(1~246)–GAL4 (768~881)				Bcd(1~246)–GAL4 (851~881)				Bcd(1~246)–B6				Bcd(1~246)–B17				Bcd(1~246)–B42			
RNA浓度 (μg·μl^{-1})	1.20	0.40	0.10	0.02	1.20	0.40	0.10	0.02	1.20	0.40	0.10	0.02	1.20	0.40	0.10	0.02	1.20	0.40	0.10	0.02	1.20	0.40	0.10	0.02
表型 *n*	19	59	52	33	24	73	64	43	46	34	73	51	29	57	64	38	25	45	72	54	40	35	56	67
1. 前部未发育	95	92	96	55	58	18	28	14	13	3	21	2	10	5	23	5	12	0	18	30	63	12	60	28
2. bcd^-突变体	5	7	4	45	0	5	13	65	2	26	10	98	7	2	17	95	0	2	11	20	5	14	5	34
3. 部分拯救	0	0	0	0	0	60	56	21	6	29	38	0	7	32	30	0	20	40	50	50	5	43	20	37
4. 部分拯救(前部开放)	0	0	0	0	42	15	3	0	59	29	22	0	60	25	28	0	64	35	11	0	27	31	14	0
5. 头和胸完全拯救	0	0	0	0	0	0	0	0	19	12	10	0	17	37	3	0	4	20	11	0	0	0	0	0

表中为向 bcd^{E1} 雌性纯合子产生的胚胎进行不同 Bcd 衍生物 mRNA 的显微注射的结果 [1]。对发育后胚胎表皮的分析和表型分类（在每个表的左侧）在图 2 中有描述。编码各种 Bcd 衍生物的 mRNA 标注在每个表的顶部，浓度也在表内标明。*n*，从 100 个注射过的胚胎发育来的已分析的表皮数目。每种表型标明了其占已分析的全部胚胎的百分数。实验细节见图 2 附注。为检验所获数据的统计显著性，重复了几个系列的注射，两个注射系列之间的平均差异 < 50%；未得到不同类型的表型。

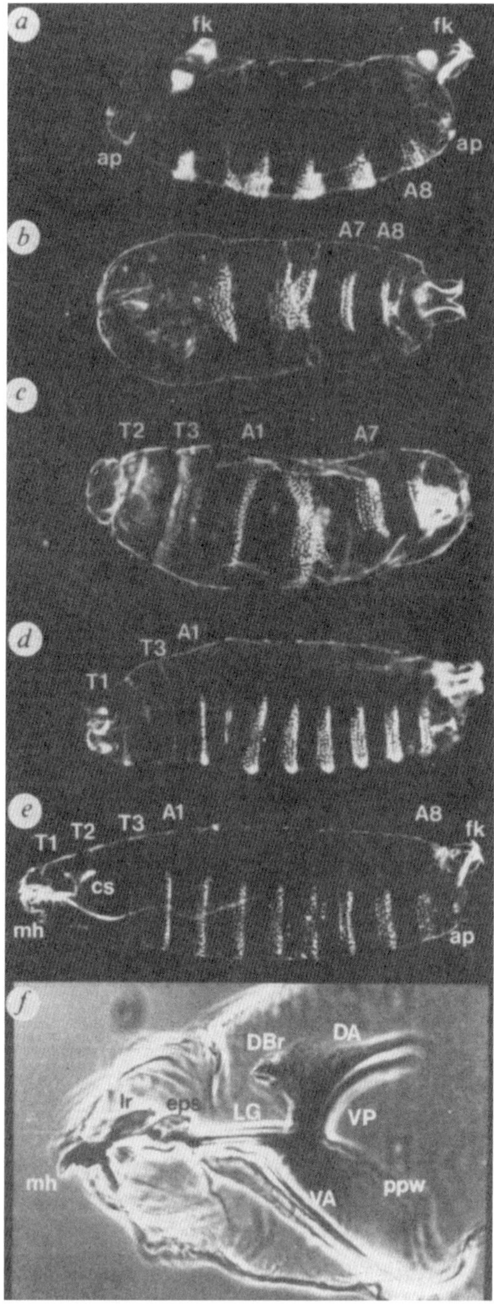

Fig. 2. Rescue of *bcd*⁻ mutant *Drosophila* embryos by Bcd-acid-sequence fusion proteins. Shown are cuticular preparations of *bcd*⁻ embryos injected with *in vitro* synthesized mRNAs encoding Bcd derivatives. The phenotypes of these embryos are grouped into five classes (also see legend to Table 3*a*, *b*). *a*, Strong *bcd*⁻ phenotype (class 2 in Table 3*a*, *b*). Heads and thoraces of the embryos are replaced by a duplication of posterior telson structures (for example, filzkörper, and anal plates)[1]. All embryos containing filzkörper structures anteriorly are classified as *bcd*⁻ mutant, including a few embryos that form thoracic structures as well as filzkörper anteriorly. *b*, Anterior structures not developed (class 1). Neither anterior nor posterior structures develop at the anterior of the embryo, and the anterior-most identifiable structures are the

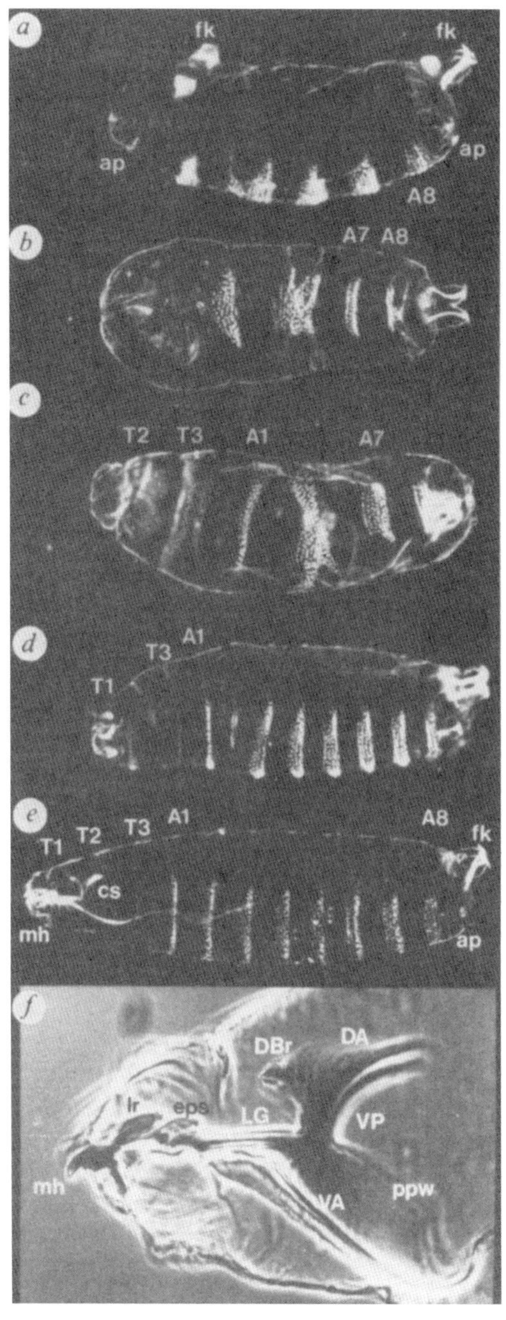

图 2. Bcd 酸性序列融合蛋白拯救 *bcd*⁻突变胚胎。图中是注射体外合成 Bcd 衍生物 mRNA 的 *bcd*⁻胚胎的表皮标本。将这些胚胎的表型分为五类（也见表 3*a*，*b* 注）。*a*，强 *bcd*⁻表型（表 3*a*，*b* 第 2 类）。头和胸被重复的尾节结构取代（例如，filzkörper 和肛板）[1]。所有前部包含 filzkörper 结构的都归为 *bcd*⁻突变体，包括少量前部形成胸和 filzkörper 的胚胎。*b*，前部结构未发育（第 1 类）。胚胎前部既不发育前部结构也不发育后部结构，近前部可识别结构是第三或第四腹齿带。虽然有些个体前部是封闭的，但通常表皮的前端都有一个大洞。*c*，带有开放前端的部分拯救（第 4 类）。胚胎发育具有高的可变性，发

third or fourth abdominal denticle belt. Usually the cuticles have a large hole at the anterior end, though in some cases (as shown) the anterior end is closed. *c*, Partial rescue with open anterior ends (class 4).The embryos develop, with a high variability, one to three thoracic denticle bands, an antennal sense organ, a maxillary sense organ, and/or mouth hooks, but they never develop a complete head skeleton. The cuticles have a hole of varying size at the anterior. Frequently some of the abdominal segments are fused or deleted. Embryos of this class form headfolds shifted towards the posterior of the embryos. Thus, the abdominal anlagen are obviously compressed and the segmentation distorted, whereas the head region is vastly expanded and can no longer be involuted—a possible cause for the hole at the anterior end. *d*, Partial rescue (class 3), The phenotype of this class resembles those of weak *bcd*$^{-}$ mutants (for example, *bcd*^{E5})[1]. Thoracic and gnathal structures but no structures derived from the anterio-most 20% of the blastoderm fate map, for example, the labrum, are formed. *e*, Complete rescue (class 5). Such embryos develop complete head and thorax including the anterior-most structure, the labrum. In a few cases the proportioning of the head skeleton deviates slightly from the wild-type form, and distortions of the abdominal segmentation appear frequently. After rescue by either wild-type Bcd protein or Bcd fusion proteins, 30–50% of the larvae in this class hatch, and some develop into adult flies. *f*, Higher magnification of the head region of a rescued *bcd*$^{-}$ mutant embryo. All the structures present in wild-type embryos can be identified here.

Methods. See legends to Table 1 and Table 2 for a description of the expression vectors used in our study. The mRNAs, synthesized *in vitro* by Sp6 RNA polymerase[50], were extracted with phenol-chloroform and chloroform, precipitated with ethanol, washed with 70% ethanol, and diluted in diethyl-pyrocarbonate-treated water to different concentrations. The mRNAs obtained were tested in an *in vitro* wheat-germ translation extract (Amersham) in the presence of [^{35}S] methionine. Measurement of ^{35}S incorporation into the proteins showed that all the mRNAs were of similar quality, and SDS-PAGE revealed proteins of the expected M_r (data not shown). The mRNAs were then injected into the anterior tips of the pre-pole-cell stage embryos obtained from homozygous mutant *bcd*^{E1} females according to standard procedures[1]. Embryos developed for 40 h at 18°C and the cuticles were prepared as previously described[1]. The anterior is always at the left and dorsal at top when lateral views are shown (*a*, *d*–*f*); *b* and c show ventral views. The mRNAs injected were (in μg μl^{-1}): Bcd (1–246) –GAL4 (851–881), 0.1 (*a*); Bcd (1–246) –VP16, 1.2 (*b*); Bcd (1–246) –GAL4 (768–881), 1.2 (*c*) ; Bcd (1–396), 0.8 (*d*); Bcd (1–246) –B17, 0.4 (*e*); Bcd (1–246) –B6, 0.4 (*f*). A1–A8, Ventral abdominal dentical bands; ap, anal plates; cs, cephalopharyngeal skeleton; DBr, dorsal bridge; DA, dorsal arm; eps, epistomal sclerite; fk, filzkörper; LG, lateralgräte; lr, labrum; mh, mouth hook; ppw, wall of pharynx posterior to ventral arm; T1–T3, thoracic denticle bands; VA, ventral arm; VP, vertical plate[52].

Bcd Fusion Proteins

We attached Bcd (1–246) fragments to (1) the acidic activating sequences of the herpes vital activator VP16 (residues 411–491) [40,41]; (2) the acidic activating region II of GAL4 (residues 768–881) [28]; (3) part of region II of GAL4 (residues 851–881) [28]; or (4) the *E. coli*-derived acidic activating sequences B6, B17, and B42 (ref. 29) (Fig. 1). As expected, each of these fusion proteins activates transcription in yeast cells more efficiently than does the N-terminal fragment of Bcd alone (Table 2).

We tested the morphogenetic activity of the Bcd fusion proteins by injecting the respective mRNAs synthesized *in vitro* into the anterior tip of embryos obtained from mutant females homozygous for *bcd*^{E1}. Three of these mRNAs, at appropriate concentrations, can completely rescue the *bcd*$^{-}$ mutant phenotype (Table 3*b*). The phenotypes obtained are described in detail in the legend to Fig. 2 and are discussed below. A complete rescue of head and thorax was obtained when the embryos were injected with mRNAs encoding fusion proteins of fragment Bcd (1–246) attached to either GAL4 (851–881) or the *E. coli* DNA-derived activating sequences B6 and B17. Up to 19%, 37% and 20% of the

育出一至三个胸齿带、一个触角感觉器官、一个上颌感觉器官和（或）口钩，但从没发育出完整的头部骨骼。在前部表皮上有一个尺寸不一的洞。一些腹部体节经常融合或缺失。这类胚胎形成头褶并向胚胎后部移动。因此腹部原基显著压缩且体节扭曲，而头部极度延伸并不向内卷—可能是前端形成洞的一个原因。*d*，部分拯救（第 3 类）。这类表型类似弱 bcd^- 突变体（如，bcd^{E5}）[1]。形成胸和颚结构，但胚盘囊胚发育图前 20% 的结构都没形成，例如上唇。*e*，完全拯救（第 5 类）。这些胚胎发育出完整的头和胸，包括近前部结构，上唇。在一些个体中头部骨骼比例与野生型相比发生细微地偏离，腹部体节扭曲频繁出现。经野生 Bcd 蛋白或融合 Bcd 蛋白拯救后，这类幼虫中有 30%~50% 孵化出来，有一些发育为成蝇。*f*，一个拯救的 bcd^-突变胚胎头部的高倍放大。野生型胚胎中出现的所有结构在这里都能识别。

方法。我们研究用的表达载体的详细描述见表 1 注和表 2 注。通过 Sp6 RNA 聚合酶 [50] 在体外合成的 mRNA 用酚 – 氯仿和氯仿抽提，乙醇沉淀，70% 的乙醇洗涤，用焦炭酸二乙酯处理的水稀释到不同浓度。得到的 mRNA 在体外含 [S^{35}] 甲硫氨酸的麦胚翻译提取物（安玛西亚公司）中检测。^{35}S 结合蛋白的测量显示所有 mRNA 质量相似，SDS 聚丙烯酰胺凝胶电泳反映蛋白的预测分子量（数据未显示）。然后根据标准步骤，将 mRNA 注射到突变 bcd^{E1} 雌性纯合子产生的前极细胞期胚胎的前端 [1]。胚胎在 18℃ 发育 40 小时，按以前的方法制备表皮 [1]。前部总是在左，侧视展示时背部在上（*a*, *d~f*）；*b* 和 *c* 展示腹面观。注射的 mRNA（单位为 μg · μl^{-1}）：Bcd(1~246)–GAL4(851~881)，0.1（*a*）；Bcd(1~246)–VP16，1.2（*b*）；Bcd(1~246)–GAL4(768~881)，1.2（*c*）；Bcd(1~396)，0.8（*d*）；Bcd(1~246)–B17，0.4（*e*）；Bcd(1~246)–B6，0.4（*f*）。A1–A8，腹侧腹齿带；ap，肛板；cs，头咽骨；DBr，背桥；DA，背腕；eps，额唇骨片；fk，filzkörper；LG，lateralgräte；lr，上唇；mh，口钩；ppw，后咽壁到腹腕；T1–T3，胸齿带；VA，腹腕；VP，垂直面 [52]。

Bcd 融合蛋白

我们将 Bcd 片段（1~246）结合到（1）疱疹病毒重要的激活子 VP16 的酸性激活序列（411~491 位残基）[40,41]；(2）GAL4 的酸性激活域 II（768~881 位残基）[28]；(3）GAL4 激活域 II 的一部分（851~881 位残基）[28]；(4）源于大肠杆菌的酸性激活序列 B6、B17 和 B42（参考文献 29）（图 1）。与预期相同，在酵母细胞中这些融合蛋白激活转录的效率比单独的 Bcd N 末端片段高（表 2）。

通过将体外合成的 mRNA 分别注射到 bcd^{E1} 突变雌性纯合子产生的胚胎前端，检测 Bcd 融合蛋白的形态发生活性。其中的三种 mRNA，在浓度适当时可完全拯救 bcd^- 的突变表型（表 3*b*）。获得的表型在图 2 注中有详细的描述，并在下文进行了讨论。当用片段 Bcd（1~246）与 GAL4（851~881）或源于大肠杆菌的酸性活性序列 B6、B17 的融合蛋白的 mRNA 注射胚胎时，头和胸得到完全恢复。通过这些激活子，拯救胚胎发育的比率分别达到 19%、37%和 20%。在每个案例中，就像注射

developing embryos were rescued, respectively, by each of these activators. In each case, as is the case when wild-type *bcd* mRNA was injected, the frequency of hatching larvae was ~30% of the number of embryos with morphologically normal heads.

With the exception of the three potent activators described below, the phenotypes obtained by injecting mRNAs encoding the various fusion proteins and deletion derivatives can, depending on the concentration, result in embryos with phenotypes mimicking those of various *bcd* alleles. These phenotypes include that of wild-type flies (complete rescue; class 5 in Table 3, and Fig. 2*e*, *f*), the phenotype of weak alleles (partial rescue; class 3 and Fig. 2*d*), and the phenotype of strong alleles (no rescue; class 2 and Fig. 2*a*). In addition, a novel phenotype is observed in embryos that are injected with these mRNAs at high concentrations. In some cases, abdominal segmentation is distorted (seen in embryo of classes 3, 4 and 5), whereas in severe cases, in addition to defective abdominal development, the head anlagen are apparently shifted so far towards the posterior that head involution cannot occur, and embryos are generated bearing a hole in the cuticle anterior to the thorax (class 4). The latter phenotype can be explained by a Bcd protein gradient with a much higher overall concentration throughout the embryo than that of the wild-type Bcd protein gradient. As Bcd protein determines positions along the anterio–posterior axis in a concentration-dependent manner[5], overall higher concentrations of Bcd would result in a shift of the blastoderm anlagen towards the posterior.

The injection of mRNAs that encode potent Bcd fusion proteins bearing VP16, GAL4 (768–881) or B42 cause another novel phenotype (class 1, see Fig. 2*b*). In these embryos, neither anterior nor posterior structures form anteriorly, but development is suppressed, resulting in embryos with a partial abdominal pattern and an open anterior end. By injecting the latter two mRNAs at lower concentrations, at least a partial rescue is obtained, whereas the most potent activator, the Bcd (1–246) –VP16 fusion protein, has a deleterious effect on the anterior development even when its mRNA is injected at a very low concentration (Table 3*b*). These three potent Bcd fusion proteins could cause the toxic effects by: (1) activating the Bcd-protein target gene(s) to some intolerable level; (2) activating some otherwise non-target genes because of single Bcd-binding sites that may be randomly distributed in the *Drosophila* genome; or (3) having the general inhibitory effect on gene expression called "squelching"(refs 42, 43).

Discussion

Here we have described experiments in which we assayed the morphogenetic properties of Bcd and various Bcd derivatives by injecting mRNAs into embryos derived from *bcd*$^-$ mutant females. We have also described the transcriptional activating properties of Bcd and various Bcd derivatives in yeast, *Drosophila* Schneider cells and *Drosophila* embryos. Our results with the Bcd (1–246) derivative are particularly indicative. This fragment, which contains the DNA-binding homoeodomain, has a weak transcriptional activation function as assayed in all systems, compared with wild-type Bcd protein, and can only partially rescue *bcd*$^-$ mutant embryos. However, when fused to certain acidic activating sequences—

野生型 *bcd* 的 mRNA 一样，头部形态正常的胚胎孵化成幼虫的频率约为 30%。

除下面描述的三种有效的激活子外，以一定浓度注射各种融合蛋白和缺失衍生物的 mRNA 可以产生与各种 *bcd* 等位基因表型相似的胚胎。这些表型包括野生型果蝇（完全恢复；图 3 第 5 类和图 2*e*，*f*）、弱等位基因表型（部分恢复；第 3 类和图 2*d*）和强等位基因表型（未恢复；第 2 类和图 2*a*）。此外，在高浓度注射这些 mRNA 时，在胚胎中观察到一种新表型。在一些情况下，腹部体节扭曲（见第 3、4、5 类胚胎），然而在一些极端的案例中，除腹部发育缺陷外，头部原基显著后移以致头部不能发育，产生的胚胎在胸前部的表皮上有一个洞（第 4 类）。整个胚胎中，Bcd 蛋白梯度的总体浓度比野生型 Bcd 蛋白梯度的浓度高可解释后者表型。由于 Bcd 蛋白以浓度依赖方式决定果蝇前后轴的位置信息 [5]，整体较高的 Bcd 浓度会导致腹部原基向后移。

注射含 VP16、GAL4（768~881）或 B42 的有效的 Bcd 融合蛋白的 mRNA 可产生另一种新表型（第 1 类，见图 2*b*）。在这些胚胎前部中既没形成前部结构也没形成后部结构，但是发育受抑制，产生的胚胎有部分腹部模式并且前端开放。低浓度注射后两种 mRNA，至少得到部分拯救，而最有效的激活子 Bcd（1~246）–VP16 融合蛋白即使以很低浓度 mRNA 注射也对前部发育有害（表 3*b*）。这三种有效的 Bcd 融合蛋白可产生毒性效果：(1) 过度激活 Bcd 蛋白靶基因；(2) 因为单个 Bcd 结合位点可能随机分布在果蝇基因组中，激活其他一些非靶标基因；(3) 对基因表达有普遍的抑制效果，称为“噪音抑制”（参考文献 42 和 43）。

讨　论

通过向 bcd^- 突变雌性产生的胚胎注射 mRNA，我们分析了 Bcd 和各种 Bcd 衍生物的形态发生特性。我们也描述了在酵母、果蝇施耐德细胞和果蝇胚胎中 Bcd 和各种 Bcd 衍生物的转录激活特性。这对 Bcd（1~246）衍生物的结果尤其有指示意义。这个片段包含 DNA 结合同源域，与野生型相比在三种实验体系中只有微弱的转录激活功能，只能部分拯救 bcd^- 突变胚胎。然而，当与特定的酸性激活序列——已知在很多真核生物中激活转录的多肽——融合时，Bcd（1~246）可完全拯救 bcd^- 突变胚胎。

peptides known to confer transcriptional activation in many eukaryotes—Bcd (1–246) fully rescues *bcd*$^-$ mutant embryos.

Our results with Bcd deletion derivatives also indicate a correlation between the ability of a given *bcd* allele to activate transcription as assayed in the embryo, and its ability to rescue the *bcd*$^-$ mutant phenotype. Thus, for example, of the derivatives tested, three (wild type, Bcd (1–396) and Bcd (1–348)) rescued the wild-type phenotype, and these derivatives stimulated transcription most efficiently in the embryo. We note that two of these proteins (Bcd (1–396) and Bcd (1–348)) are severely impaired in the activation function as assayed in yeast. These results are consistent with the idea that the fragments Bcd (1–396) and Bcd (1–348) are modified in the embryo, perhaps by phosphorylation, to potentiate or to acquire an activating function. It is also possible that, for example, the sequence of residues 247–348, which is rich in Gln, functions as an activating sequence in the embryo, but not in Schneider cells or in yeast.

The results of the mRNA injection experiments also allowed us to begin to analyse the effects of activating strengths and concentrations of the Bcd protein derivatives on embryonic development.

First, the concentration of the activators can vary within a certain range to allow normal embryonic development. For example, the Bcd fusion proteins bearing GAL4 (851–881), B6 or B17 fragments can rescue the *bcd*$^-$ mutant phenotype when their mRNAs are injected at concentrations differing by a factor of > 10 (Table 3*b*), although injection of mRNAs at high concentration tends to distort abdominal development. This observation is consistent with previous genetic data showing that the embryos from females bearing as many as six copies of the *bcd* gene show normal development, but when there are eight copies of the *bcd* gene in the female, the abdominal segmentation frequently is abnormal and the embryos may die (refs 4 and 5; T. Berleth and W.D., unpublished results). The "posteriorward" shift of position on the fate map with increased concentrations seems to be largely independent of the activator strength, because the weak activator Bcd (1–246) also shows this effect (data not shown).

Second, the activities of the Bcd derivatives can also vary within a certain range, and still allow the development of normal larvae. For example, both Bcd (1–348) and wild-type Bcd protein can completely rescue the *bcd*$^-$ mutant phenotype, although their transcriptional stimulatory activities, as determined in *Drosophila* embryos, differ by a factor of two (Table 1).

Our experiments demonstrating that Bcd protein activates transcription in yeast cells from the *hb* upstream element further support the idea that one function of Bcd protein during development is to stimulate the expression of *hb* gene by directly binding to the sites located upstream of the *hb* gene promoter. In addition to activating the target gene(s), the maternal Bcd protein could have other functions during early development of *Drosophila* embryos. For example, Bod protein could inhibit the translation of the maternal *caudal* gene mRNA (refs 2 and 44), and Bcd protein would be required to define the domain of

我们关于 Bcd 缺失衍生物的实验结果也表明在胚胎中检测到的给定 *bcd* 等位基因激活转录的能力和拯救 *bcd*⁻ 突变表型的能力之间有相关性。因此，举例来讲，在检测过的衍生物中，有三个（野生型、Bcd（1~396）和 Bcd（1~348））修复了野生型表型，这些衍生物可在胚胎中最有效地激活转录。在酵母的实验中我们注意到其中两个蛋白 Bcd（1~396）和 Bcd（1~348）的激活功能严重受损。这些结果与片段 Bcd（1~396）和 Bcd（1~348）在胚胎中可能被磷酸化修饰而增强或获得激活功能的观点一致。也可能，例如，富含谷氨酰胺的 247~348 位残基的序列，在胚胎中起到激活序列的作用，而在施耐德细胞或酵母中不能。

mRNA 注射实验的结果也使得我们能够分析激活强度和 Bcd 蛋白衍生物的浓度对胚胎发育的影响。

首先，激活子的浓度可以在一个区间内变化，使胚胎正常发育。例如，当含 GAL4（851~881）、B6 或 B17 片段的 Bcd 融合蛋白的 mRNA 以大于 10 倍的浓度变化（表 3*b*）注射时可以拯救 *bcd*⁻ 突变表型，尽管高浓度注射容易扭曲腹部发育。这些观察结果与以前的遗传学数据相符：母体含 6 拷贝 *bcd* 基因时产生的胚胎发育正常，但母体含 8 拷贝 *bcd* 基因时其胚胎腹部体节往往不正常并且胚胎会死亡。（参考文献 4 和 5；T. Berleth 和沃尔夫冈 · 德里费尔，未发表结果）。浓度升高导致囊胚发育图上的位置后移似乎与激活子强度没多大关系，因为弱激活子 Bcd（1~246）也产生这个效果（数据未显示）。

第二，Bcd 衍生物的活性也可在一个特定范围内变化，并使胚胎发育为正常幼虫。例如，Bcd（1~348）和野生型 Bcd 蛋白，虽然它们的转录激活活性有 2 倍的差异（表 1），如在果蝇胚胎中确定的，但它们都可完全拯救 *bcd*⁻ 突变表型。

我们的研究证明了 Bcd 蛋白在酵母细胞中通过 *hb* 上游元件激活转录，进一步支持了这样的观点，即 Bcd 蛋白在发育过程中的一个功能是通过直接结合到位于 *hb* 基因启动子上游的位点激活 *hb* 基因的表达。除激活靶基因外，母性 Bcd 蛋白在果蝇胚胎早期发育中还有其他功能。例如，Bcd 蛋白可抑制母源尾部基因 mRNA 的翻译（参考文献 2 和 44），并且 Bcd 蛋白可能是决定与 *hb* 无关的另一个间隔基因（跛

expression of another gap gene (*Krüppel*) independent of *hb* (ref. 45). Our results indicate that if these effects of Bcd protein are direct and if they are essential to the development of *Drosophila* embryos, they must be specified by the N-terminal portion (residues 1–246) of Bcd which is present in the Bcd fusion proteins that are able to rescue the *bcd*$^-$ mutant phenotype.

(**342**,149-154;1989)

Wolfgang Driever*, **Jun Ma**†, **Christiane Nüsslein-Volhard*** **& Mark Ptashne**†

* Max-Planck-Institut für Entwicklungsbiologie, Abteilung Genetik, Spemannstrasse 35/III, D-7400 Tübingen, FRG

† Department of Biochemistry and Molecular Biology, Harvard University, 7 Divinity Avenue, Cambridge, Massachusetts 02138, USA

Received 3 May; accepted 5 September 1989.

References:

1. Frohnhöfer, H. G. & Nüsslein-Volhard, C. *Nature* **324**, 120-125 (1986).
2. Driever, W. & Nüsslein-Volhard, C. *Cell* **54**, 83-93 (1988).
3. Frigerio, G., Burri, M., Bopp, D., Baumgartner, S. & Noll, M. *Cell* **47**, 735-746 (1986).
4. Berleth, T. *et al. EMBO J.* **7**, 1749-1756 (1988).
5. Driever, W. & Nüsslein-Volhard, C. *Cell* **54**, 95-104 (1988).
6. Tautz, D. *Nature* **332**, 281-284 (1988).
7. Schröder, C. *et al. EMBO J.* **7**, 2881-2888 (1988).
8. Lehmann, R. & Nüsslein-Volhard, C. *Devl. Biol.* **119**, 402-417 (1987).
9. Tautz, D. *et al. Nature* **327**, 383-389 (1987).
10. Driever, W. & Nüsslein-Volhard, C. *Nature* **337**, 138-143 (1989).
11. Driever, W., Thoma, G. & Nüsslein-Volhard, C. *Nature* **340**, 363-367 (1989).
12. McGinnis, W. *et al. Nature* **308**, 428-433 (1984).
13. Laughon, A. & Scott, M. P. *Nature* **310**, 25-31 (1984).
14. Gehring, W. J. *Science* **236**, 1245-1252 (1987).
15. Desplan, C., Theis, J. & O'Farrell, P. *Nature* **318**, 630-635 (1985).
16. Hoey, T. & Levine, M. *Nature* **332**, 858-861 (1988).
17. Desplan, C., Theis, J. & O'Farrell, P. *Cell* **54**, 1081-1090 (1988).
18. Hanes, S. & Brent, R. *Cell* **57**, 1275-1283 (1989).
19. Jaynes, J. & O'Farrell, P. *Nature* **336**, 744-749 (1988).
20. Fitzpatrick, D. & Ingles, J. *Nature* **337**, 666-668 (1989).
21. Han, K., Levine, M. & Manley, J. *Cell* **56**, 573-583 (1989).
22. Struhl, G., Struhl, K. & Macdonald, P. *Cell* **57**, 1259-1273 (1989).
23. Brent, R. & Ptashne, M. *Cell* **43**, 729-736 (1985).
24. Keegan, L., Gill, G. & Ptashne, M. *Science* **231**, 699-704 (1986).
25. Ptashne, M. *Nature* **322**, 697-701 (1986).
26. Struhl, K. *Cell* **49**, 295-297 (1987).
27. Hope, I. & Struhl, K. *Cell* **46**,885-894 (1986).
28. Ma, J. & Ptashne, M. *Cell* **48**, 847-853 (1987).
29. Ma, J. & Ptashne, M. *Cell* **51**, 113-119 (1987).
30. Giniger, E. & Ptashne, M. *Nature* **330**, 670-672 (1987).
31. Kakidani, H. & Ptashne, M. *Cell* **52**, 161-167 (1988).
32. Webster, N., Jin, J. R., Green, S., Hollis, M. & Chambon, P. *Cell* **52**, 169-178 (1988).
33. Fisher, J., Giniger, E., Ptashne, M. & Maniatis, T. *Nature* **332**, 853-856 (1988).
34. Ma, J., Przibilla, E., Hu, J., Bogorad, L. & Ptashne, M. *Nature* **334**, 631-633 (1988).
35. Courey, A. & Tjian, R. *Cell* **55**, 887-898 (1988).
36. Tora, L. *et al. Cell* (in the press).

子）表达区域所必需的（参考文献 45）。我们的结果暗示，如果 Bcd 蛋白的这些效应是直接的，而且如果它们是果蝇胚胎发育必需的，那么很可能它们的 N 末端部分（1~246 位残基）起决定作用，它们存在于 Bcd 融合蛋白中，能完全拯救 bcd^- 突变表型。

（李梅 翻译；沈杰 审稿）

37. West, R., Yocum, R. & Ptashne, M. *Molec. Cell. Biol.* **4**, 2467-2478 (1984).

38. Ammerer, G. *Meth. Enzymol.* **101**, 192-201 (1983).

39. Driever, W., Siegel, V. & Nüsslein-Volhard, C. *Science* (in the press).

40. Triezenberg, S., Kingsbury, R. & McKnight, S. *Genes Dev.* **2**, 718-729 (1988).

41. Sadowski, I., Ma, J., Triezenberg, S. & Ptashne, M. *Nature* **335**, 563-564 (1988).

42. Gill, G. & Ptashne, M. *Nature* **334**, 721-724 (1988).

43. Ptashne, M. *Nature* **335**, 683-689 (1988).

44. Mlodzik, M. & Gehring, W. *Development* **101**, 421-435 (1987).

45. Gaul, U. & Jäckle, H. *Cell* **51**, 549-555 (1987).

46. Gill, G. & Ptashne. M. *Cell* **51**, 121-126 (1987).

47. Yocum, R., West, R. & Ptashne, M. *Molec. Cell. Biol.* **4**, 1985-1998 (1984).

48. Bunch, T. A., Grinblat, Y. & Goldstein, L. S. B. *Nucleic Acids Res.* **16**, 1043-1061 (1988).

49. Rio, D. C. & Rubin, G. M. *Molec. Cell. Biol.* **5**, 1833-1838 (1985).

50. Melton, D. A. *et al. Nucleic Acids Res.* **12**, 7035-7056 (1984).

51. Silver, P., Chiang, A. & Sadler, J. *Genes Dev.* **2**, 707-717 (1988).

52. Jürgens, G. *et al. Wilhelm Roux's Arc. dev. Biol.* **196**, 141-157 (1986)

Acknowledgements. We thank our colleagues in both MPI and Harvard, especially V. Siegel, S. Roth, M. Klingler, L. Stevens, T. Goto, L. Keegan and H. Himmelfarb, as well as P. O'Farrell at UCSF, for discussions and comments on the manuscript; G. Thoma for technical assistance; T. Berleth and D. Ruden for communicating unpublished results; and V. Siegel, T. Bunch, D. Melton and S. McKnight for plasmids. This work was supported by the DFG (Leipnitz Programm) and the NIH.

The *C. elegans* Genome Sequencing Project: a Beginning

J. Sulston *et al.*

Editor's Note

Here British biologist John Sulston and colleagues describe developments during the first year of the collaborative *Caenorhabditis elegans* genome sequencing project. Sulston believed the small genome would prove a useful pilot for sequencing the human genome. And given the worm's popularity as a lab model for studying gene function and development, the researchers hoped its DNA sequence would also shed light on human biology. In this first phase, the team developed methods amenable to large-scale sequencing and report the sequence of three gene-rich DNA sections. The finished sequence was published six years later, making *C. elegans* the first multicellular organism to have its genome completely sequenced. The complete human genome, which benefited from the techniques developed here, was published in 2003.

The long-term goal of this project is the elucidation of the complete sequence of the *Caenorhabditis elegans* genome. During the first year, methods have been developed and a strategy implemented that is amenable to large-scale sequencing. The three cosmids sequenced in this initial phase are surprisingly rich in genes, many of which have mammalian homologues.

THE realization that the human genome can be sequenced in its entirety has stimulated great interest in genome analysis[1,2] and efforts have already begun to construct genetic and physical maps[3] and to improve DNA sequencing methods[4]. In pursuit of this great enterprise, it will be necessary to sequence and analyse the smaller genomes of experimentally tractable organisms which will serve as pilot systems for evaluating technology and also provide information essential for interpreting the human sequence. The genomes of single-celled organisms such as *Escherichia coli* and *Saccharomyces cerevesiae* will reveal features peculiar to the basic functions shared by all living things. Analysis of simple animal genomes intermediate in size and complexity between those of yeast and man will help us to understand the more complex features of mammals.

The genome of the small nematode *C. elegans* is a good candidate for complete sequence analysis[5]. This organism has been used to investigate animal development and behaviour[6,7]. Its small size and short generation time facilitate genetic analysis, and more than 900 loci have now been identified through mutations[8]. Each animal develops with essentially

秀丽隐杆线虫基因组测序计划：一个开端

萨尔斯顿等

编者按

在本文中，英国生物学家约翰·萨尔斯顿和他的同事们描述了秀丽隐杆线虫基因组计划第一年的进展。萨尔斯顿相信这个小基因组将被证明是人类基因组计划的有效先导。线虫作为实验室模式生物，在研究基因功能与发育方面有很高的普及性，研究人员希望它的DNA序列也能为人类生物学提供线索。在第一阶段，研究团队建立了适用于大规模测序的实验方法，并报道了3个基因密集区的序列。全部基因组序列于6年后发表，使得秀丽隐杆线虫成为第一个完成全基因组测序的多细胞生物。在2003年发表的人类全基因组序列从本项目建立的技术中获益良多。

本项目的长期目标是阐明秀丽隐杆线虫全基因组序列。第一年，研究者们建立了测序方法，并制定了一个适用于大规模测序的策略。在初始阶段所测序的三个黏粒（cosmid）中，基因之丰富令人惊讶，其中许多都有哺乳类的同源基因。

由于认识到人类基因组可以被完全测序，激发出人们对基因组分析的巨大兴趣[1,2]，人们已经开始努力构建遗传和物理图谱[3]以及改进DNA测序的方法[4]。为追求这一伟大目标，有必要对基因组较小且易于实验操作的模式生物进行测序和分析。这些模式生物将作为前期实验系统用于技术评估，并为解析人类基因序列提供基础信息。大肠杆菌和酿酒酵母等单细胞生物的基因组，尤其能够揭示所有生命体共同拥有的基本功能特征。通过对大小和复杂性介于酵母和人类之间的简单动物基因组进行分析，将帮助我们理解哺乳类更复杂的特征。

一种小型线虫——秀丽隐杆线虫是开展全基因组序列分析得很好选择[5]。这种生物已被广泛用于研究动物的发育和行为[6,7]。其个体小、世代短，便于进行遗传分析，目前已经通过分析突变体鉴定了900多个基因位点[8]。每个个体发育时的细胞分化模式基本相同，而这整个模式是已知的[9-11]。这种线虫的解剖结构简单，只有

the same pattern of cell divisions, and this entire pattern is known[9-11]. The anatomy is simple, with only 959 somatic cells, and the ultrastructure established — for example, the complete connectivity of its 302 neurons has been determined[12].

Many genes required for normal development and behaviour are being studied. With the aid of an active transposon system[13-15] and a physical map of the genome[16-18], they can be readily cloned. Methods for transformation have been developed which allow reintroduction of engineered genes[19-21]. The similarity of many of these genes to those in mammals is often extensive, supporting the contention that information obtained in the nematode will be relevant to understanding the biology of man.

The *C. elegans* genome contains an estimated 100 megabases (10^8 bases), less than the size of an average human chromosome. Generally genes in *C. elegans* have smaller and fewer introns than their mammalian counterparts[22] and the gene density is high (see below). A clonal physical map of the nematode genome is nearly complete[18]. A combination of cosmid and yeast artificial chromosomal (YAC) clones has been used to reconstruct more than 95 megabases (Mb) of the genome. More than 90 Mb have been positioned along the chromosomes using genetically mapped sequences and through *in situ* hybridization of cloned sequences. Fewer than 40 gaps now remain in the map and progress towards closure is proceeding steadily (A.C. *et al.*, unpublished results).

We have embarked on a project to determine the entire sequence of the nematode genome. The region where we began is the centre of chromosome III (Fig. 1). There is good cosmid coverage over most of several megabases, and the few areas lacking cosmid coverage are spanned by YACs. The region lies in the central gene-rich cluster of chromosome III, where several genes of interest have been mapped.

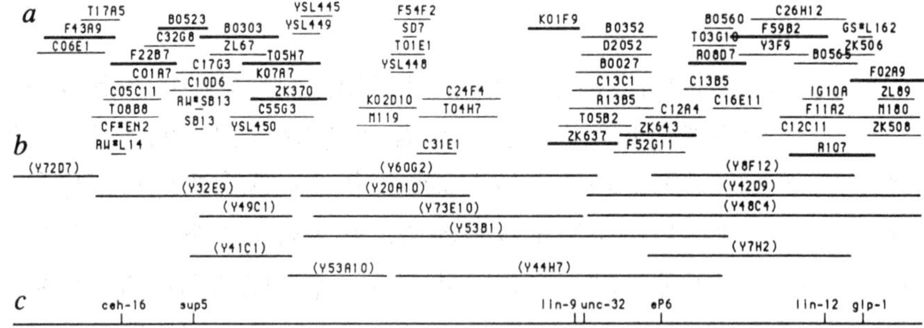

Fig. 1. Physical map of the region where the sequencing project has begun. *a*, Selected overlapping cosmid and lambda clones are represented by the horizontal lines. The cosmids which have been sequenced and reported here are underlined in bold, as well as the additional cosmids underway at present. *b*, The overlapping YAC clones from the region bridge the segments not represented in cosmid clones above. Methods must still be developed to capture sequence efficiently from the spans presently cloned in YACs only. *c*, Six genes and markers known to lie in the regions from genetic and other studies[50]. The genes *unc-32* and *sup-5* are less than 0.1 centimorgans apart on the genetic map of chromosome III. Because of the way the physical map was constructed[16,17], the physical distance separating these two genes can only be estimated to be more than 150 kb, yielding a ratio of more than 1.5 Mb per centimorgan. This metric is typical of the central chromosomal clusters in *C. elegans*.

959 个体细胞，并已建立超微结构——例如，其 302 个神经元全部连通的情况 [12] 已经被确定。

我们正在对正常发育和行为所需的许多基因进行研究。成功克隆这些基因需要一个活跃的转座子系统 [13-15] 和一张基因组物理图谱 [16-18] 的辅助。我们还建立了允许再引入工程基因 [19-21] 的转化方法。这些基因与哺乳动物的基因十分相似，支持了关于从该线虫研究中获得的信息将有助于理解人类生物学的观点。

秀丽隐杆线虫的基因组包含大约 100 兆碱基对（10^8 碱基），小于一条人类染色体的平均大小。通常来说，秀丽隐杆线虫基因的内含子比哺乳动物的内含子 [22] 更小更少，而基因的密度较高（如下）。该线虫基因组的物理图谱克隆已基本完成 [18]。使用黏粒和酵母人工染色体（YAC）克隆相结合的方法重构了该基因组中超过 95 Mb 基（Mb）的区域；采用遗传定位序列和克隆序列原位杂交法，已经沿染色体定位了超过 90 Mb。目前该图谱上仅剩不到 40 个空缺，填补缺口工作正在稳步进行中（库尔松等，未发表结果）。

我们已经着手开展一个项目，以确定该线虫基因组的全序列。我们从第Ⅲ染色体的中部开始（图 1）。几个兆碱基的序列大多与黏粒重叠较好，而少数缺少黏粒覆盖的区域则采用酵母人工染色体进行扩充。该区域位于第Ⅲ染色体中部基因丰富的簇中，在这里我们已经定位了一些人们感兴趣的基因。

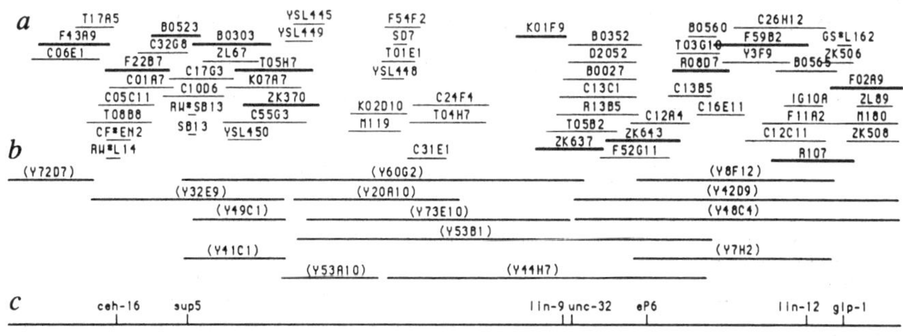

图 1. 测序项目起始区域的物理图谱。*a*，水平线代表选定的重叠黏粒和 λ 克隆。这里已经测序并报道的黏粒下面用粗线标记，其余正在测序的黏粒也是如此。*b*，该区域重叠的酵母人工染色体克隆连接了上述黏粒克隆没有体现的片段。仍需建立仅从克隆在酵母人工染色体中的扩展区有效捕获序列的方法。*c*，从遗传学和其他研究 [50] 中获知位于该区域的六个基因和遗传标记。在第 III 染色体的遗传学图谱上，基因 *unc-32* 和 *sup-5* 的遗传距离小于 0.1 厘摩。由于构建该物理图谱所采用的方法 [16,17]，只能估计分隔这两个基因的物理距离超过 150 kb，达到了每厘摩大于 1.5 Mb 的比例。这一尺度在秀丽隐杆线虫的中央染色体簇上很典型。

Strategy and Methods

The physical map of *C. elegans* was constructed using cosmid and YAC clones, but directed sequencing of even cosmid clones proved impractical because of the presence of repeat sequences and problems with obtaining sufficient quantities of template DNA. Thus we began by generating random subclones from sheared, sized DNA[23,24]. Libraries of small inserts (1–3 kilobases (kb)) were convenient and larger insert libraries (6–9 kb) were useful for establishing continuity in gap closure.

To collect the sequence data, we relied largely on two fluorescent-based sequence-gel readers, the Applied Biosystems ABI 373A and the Pharmacia ALF, which provide data directly in machine-readable form. All data were transferred to Unix-based Sun workstations. A display editor was developed to allow rapid clipping of the vector from the 5′ end and unreliable sequence from the 3′ end[25].

In the first phase of data generation, single reads of about 400 base pairs (bp) were taken from one end of the random clones using the ABI 373 instrument and *Taq* polymerase with a cycle sequencing protocol that reduced the amount of template necessary[26,27]. After 100–350 reads were obtained (the optimal number will ultimately depend on relative costs and has not been established; Table 1) and assembled into contigs using Staden's assembly program[28], the project switched to a directed phase for closure and finishing. As a preliminary directed step, a reverse read was sometimes obtained from the opposite end of selected inserts to help establish linkage, double stranding and gap closure. Further directed sequencing required custom oligonucleotide primers, whose selection was aided by OSP (for oligonucleotide selection program)[29]. For technical reasons, the Pharmacia ALF was more convenient for reads from custom primers[30]. After gap closure and double stranding, in some cases further reads had to be taken, often with different chemistries[31,32], to resolve ambiguities. Final editing, and indeed editing throughout the project, was assisted by the ability to recall the original trace data for any region of concern from the editor program. Further details of our sequencing strategy will be published elsewhere[33,34].

Table 1. Sequencing strategies and statistics

Cosmid clone	B0303	ZK637	ZK643
Insert DNA size (bp)	41,071	40,699	39,528
Random subclone libraries size range:	5–6 kb	9–14 kb	1–2 kb, 6–9 kb, 9–14 kb
cloning vector:	pUC118[47]	pBS[48]	M13mp18[49], pEMBL9, pBS
Random sequencing method	ds, ABI	ds, ABI; ss, ^{32}P	ss, ABI
Number of random subclones	197	102	360
Number of reverse primer readings	119	70	0
Closure method	ss/ds, ALF	ds, ^{32}P	ss, ALF

策略和方法

采用黏粒和酵母人工染色体克隆来构建秀丽隐杆线虫的物理图谱时，由于存在重复序列，以及无法获得足够数量的模板 DNA，对黏粒克隆进行直接测序还是不现实的。因而我们首先使用剪切成一定大小的 DNA 生成随机亚克隆 [23,24]。小的插入片段（1~3 kb）文库使用十分方便，而较大的插入片段（6~9 kb）文库可用于在弥合空缺时建立连续性。

我们主要依靠美国应用生物系统公司的 ABI 373A 和法玛西亚公司的 ALF 两部荧光测序凝胶阅读仪来收集测序数据，这些设备可以直接以机器可读格式提供数据。所有数据都转入基于 Unix 的 Sun 工作站。我们还开发了一个演示编辑器，以便迅速从 5′端去掉载体序列，从 3′端删除 [25] 不可靠序列。

在生成数据的第一阶段，采用可降低模板需求量的环形测序法 [26,27]，使用 ABI 373 设备和 *Taq* 聚合酶，从随机克隆的一端得到了大约 400 个碱基对（bp）的单个测序片段。在获得 100~350 个测序片段（理想数据量最终取决于相关测序费用，该统计结果尚未建立；表 1），并采用施塔登氏（Staden's）拼接程序 [28] 拼接成重叠群（contigs）之后，该项目转入了定向的消除缺口与完成阶段。作为一个初步定向步骤，有时会从选定插入片段的另一端获取一个反向测序片段，以帮助建立关联、形成双链和消除缺口。进一步的定向测序需要定制的寡聚核苷酸引物，用 OSP（寡聚核苷酸选择程序）[29] 帮助选择引物。由于技术原因，法玛西亚公司 ALF 对来自定制引物的测序结果更为方便 [30]。在消除缺口和形成双链之后，有时必须获取更多的测序片段来解决不确定的序列，通常采用不同的化学方法 [31,32]。就任何所关注区域，通过编辑程序回溯原始序列峰图数据的能力，将会有助于最后的编辑，乃至贯穿整个项目的编辑工作。关于我们测序策略的更多细节将发表在其他地方 [33,34]。

表 1. 测序策略和统计结果

黏粒克隆	B0303	ZK637	ZK643
插入的 DNA 大小（bp）	41,071	40,699	39,528
随机亚克隆文库的大小范围	5~6 kb	9~14 kb	1~2 kb, 6~9 kb, 9~14 kb
克隆载体	pUC118[47]	pBS[48]	M13mp18[49], pEMBL9, pBS
随机测序方法	ds, ABI	ds, ABI; ss, ^{32}P	ss, ABI
随机亚克隆的数量	197	102	360
反向引物测序的数量	119	70	0
消除缺口的方法	ss/ds, ALF	ds, ^{32}P	ss, ALF

Continued

Cosmid clone	B0303	ZK637	ZK643
Oligonucleotide primers required	102	417	100
Total number of readings	440	589	496
Average bp per read (vector sequence removed)	415	339	390
Total bp read (for assembly)	171,375	184,000	186,437
Final sequence redundancy	3.6	3.4	4.3

Abbreviations: ss, single-stranded template; ds, double-stranded template; AB1, Applied Biosystems Inc. 373A sequence-gel reader; ALF, Pharmacia ALF sequence-gel reader.

Once the final sequence was obtained, the databases were searched for similarities using the algorithm BLAST[35]. In addition, we have started to interpret the sequence directly. The program GENEFINDER (P.G. and L.H., unpublished) uses a statistically rigorous treatment of likelihoods to find possible genes. Other features, such as repeated sequences, are also being examined. The annotated sequences have been submitted to Genbank and EMBL databases. To present the sequences in the context of our knowledge about the worm, we developed a *C. elegans* database, ACEDB, which holds not only the available sequences for the nematode, but also physical and genetic map information, along with reference lists and strain information (R.D. and J.T.-M., unpublished).

Sequences

Three cosmids have so far been completed (sequences submitted to the EMBL and Genbank databases) during the development of our strategy, each with method variations to improve efficiency (Table 1). The first cosmid, ZK637, was sequenced using a minimum of random clones and radioactively labelled primers for walking. The number of primers required proved costly, and the use of films to collect data made editing difficult. ZK643 was partly sequenced using restriction enzyme partial digestion for the random clone production: although subclone recovery was high, these clones were biased in their representation. B0303 DNA was sheared to produce the subclones and, as for ZK637, reverse primer sequencing was used to establish linkage before beginning directed sequencing. Closure for both ZK643 and B0303 was done using directly labelled primers on the Pharmacia ALF.

The overlap of the manual, radiolabelled sequence with the fluorescent shotgun reads in ZK637 amply confirms the fidelity of the fluorescent method. The accuracy of the base calling was tested by comparing a sampling of individual sequence reads with the final edited sequence (Fig. 2). Generally the only errors found in the first 300 bases of a read could be attributed to compressions or stops due to gel or enzyme limitations. Above 300, software limitations predominate, with increasing numbers of sites at which either no base call can be made or errors in the base calling occur. The failure to call a base is not a serious problem as no false information enters the database. By far the most common error is overcalling the number of bases in a run, but undercalls and miscalls also occur; these errors are easily corrected, once attention has been drawn to them, by reference to the traces.

续表

黏粒克隆	B0303	ZK637	ZK643
所需寡聚核苷酸引物数量	102	417	100
总测序片段的数量	440	589	496
每个测序片段的平均碱基对数（去除载体序列）	415	339	390
测序碱基对总数（用于拼接）	171,375	184,000	186,437
最终序列冗余	3.6	3.4	4.3

缩写：ss，单链模板；ds，双链模板；ABI，应用生物系统公司 373A 测序凝胶测序仪；ALF，法玛西亚公司 ALF 测序凝胶测序仪。

在获得最终序列之后，采用 BLAST 算法[35] 在数据库中查找相似序列。此外，我们还开始直接解释序列。GENEFINDER 程序（格林和伊利耶，未发表）用统计上严格的似然性处理寻找可能的基因。对重复序列等其他特征序列，也进行了分析。解释过的序列已经提交到 Genbank 和 EMBL 数据库。为结合我们关于该线虫的知识来介绍这些序列，我们建立了一个秀丽隐杆线虫数据库 ACEDB，该数据库不仅包含这种线虫已有的序列，而且包括其物理和遗传图谱资料，以及相关文献列表和品系信息（德宾和蒂里－米格，未发表）。

序　列

在我们制定策略的过程中，已经完成了三个黏粒的测序（序列已提交到 EMBL 和 Genbank 数据库），每个都有提高效率的改进方案（表 1）。第一个黏粒，ZK637，是采用最小随机克隆和放射性标记引物进行步移来完成测序。所需引物数量很大，且采用胶片收集数据使得编辑非常困难。ZK643 的部分测序是采用限制性酶部分消化随机克隆产物来完成的：尽管亚克隆的回收率很高，但是这些克隆在代表性上有偏好。B0303 DNA 被剪切以生成亚克隆，并按照对 ZK637 使用的方法，在开始定向测序之前用反向引物测序建立片段连锁。ZK643 和 B0303 消除缺口都是在法玛西亚公司 ALF 上使用直接标记引物完成的。

人工放射性标记序列与 ZK637 中的荧光鸟枪法测序片段的重叠，充分证实了荧光法的可靠性。通过比较单个序列测序片段样本和最终的编辑后序列，检验了碱基读取的精确性（图 2）。一般来说，在某一测序片段的前 300 个碱基中发现的错误，可能是由于凝胶或酶的限制而导致密集或停顿。超过 300 个碱基，则主要是软件的限制，导致碱基无法读取或出现读取错误的位点数量上升。无法读取碱基不是一个严重问题，因为并没有错误信息进入数据库。目前最常见的错误是在跑胶时重复读取碱基，然而少读或误读也会发生；一旦注意到这些错误，则很容易通过查找原始数据来纠正。

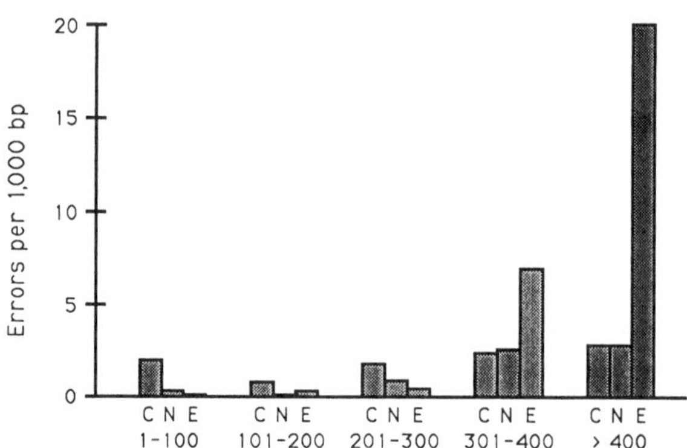

Fig. 2. The types of errors found when comparing the initial machine reads with the final edited sequence. The errors are broken down into the following categories: errors due to either the enzyme or the gel, resulting in compressions or stops (C) and errors due to the software, where there was either no call made (N, nocall) or an error made in base calling (E). In the last case the software has either inserted an extra base (an overcall), failed to recognize a base (an undercall) or simply called the wrong base (a miscall). Nocalls are less troublesome than an error in base calling, and nocalls accounted for most of the software failures in the first 300 bases of the reads. Software errors, particularly base calling errors, predominated above 300 bases. The majority of these errors were overcalls (87), as opposed to undercalls (20) or miscalls (16). The data are taken from 89 reads of single-stranded templates, with average length of 427 bases (Fig.1), done on cosmid F59B2, for which changes in our data-handling software made the analysis easier. The data quality, however, should be similar in the cosmid sequences presented here.

The accuracy of the final sequence is difficult to estimate without extensive independent tests, but several factors give us confidence that the fidelity to the true genomic sequence is high. All sequences were determined at least once on both strands, with the exception of certain long tandem repeat sequences. Regions with compressions or other unresolved conflicts were resequenced. Possible errors indicated by the similarity searches and GENEFINDER analysis were checked. To detect major cloning artefacts, we compared the polymerase chain reaction products derived from the cosmids with both the predicted lengths and the products from genomic DNA across the sequenced regions . The walking primers often proved useful for this purpose. For one cosmid, part of the vector sequence was analysed which was derived from the random phase alone so that not every region had been sequenced on both strands; here only one base in 2,000 was at variance with the available sequence.

Using similarity searches and GENEFINDER to analyse these sequences, a high density of likely genes was revealed (Fig.3). One of the most striking similarities was found between the 22–26-kb region of ZK637 and the 116K subunit (relative molecular mass 116,000) of the rat vacuolar proton pump (Fig.4). The similarity falls into several blocks which GENEFINDER suggests are probably exons, and spans from residues 14 to 824 of the 838-amino-acid protein. The amino-terminal half is most highly conserved between the two sequences; in one stretch 99 of 139 residues (71%) are identical and in a second, 223 of 314 (71%) are identical, with no gaps introduced into the alignment. The carboxy-terminal half has less similarity but contains sequences similar to all eight

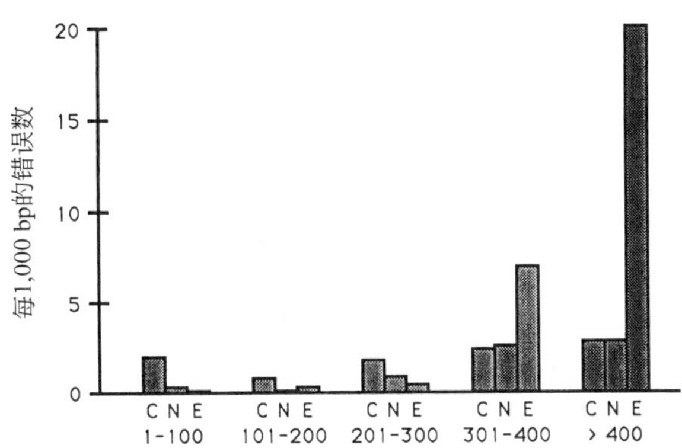

图 2. 与最终编辑后的序列比较，从最初机器所读取序列中发现的错误类型。这些错误被分为以下几类：由于酶或凝胶导致的密集或停顿而产生的错误（C），和因软件导致的错误，无信号（N）或是碱基读取错误（E）。在最后一种情况下，该软件或是插入了一个额外碱基（多读），未能读取一个碱基（少读）或是错误读取一个碱基（误读）。无信号比错误读取碱基带来的麻烦要少，而且在测序片段的前 300 个碱基，无信号占软件错误的大部分。超过 300 个碱基，软件错误，特别是碱基读取错误变成了主要问题。与少读（20）或误读（16）相比，这些错误中主要是多读（87 个）。这些数据是用黏粒 F59B2 进行实验获得的，涉及其单链模板的 89 个测序片段，平均长度为 427 个碱基（图 1），数据处理软件做了些改变，以易于分析。黏粒 F59B2 的数据质量应当与本文显示的黏粒序列相似。

没有广泛的独立检验，最终测序的精确性很难估计，但若干因素使我们确信，反映真实基因组序列的准确性是很高的。除一些较长的串联重复序列之外，所有序列至少在 DNA 的两条链上各确认了一次。那些密集区域或有其他不确定的区域被再次测序。对在相似性查找和 GENEFINDER 分析中产生的可能的错误进行了检验。为检测主要的人工克隆假象，我们比较了黏粒的聚合酶链式反应产物与基因组 DNA 相应区域的长度和产物。事实证明引物步移对实现这一目标非常有用。就其中一个黏粒来说，我们对载体序列的一部分进行了分析，因其仅来自随机阶段，因而并非两条 DNA 链的每个区域都进行了测序；在 2,000 个碱基中，只有一个与已有的序列不同。

采用相似性搜索和 GENEFINDER 来分析这些序列，发现了高密度的疑似基因（图 3）。最令人惊讶的是，ZK637 的 22~26 kb 区域和大鼠细胞液泡质子泵的 116K 亚基（相对分子量为 116,000）具有高度相似性（图 4）。两者的相似处分布于若干区段，GENEFINDER 将其读取为外显子，从这种有 838 个氨基酸的蛋白质的第 14 残基延伸至第 824 残基处。这两段序列的氨基端的一半高度保守；在其中一段，139 个残基中有 99 个是相同的（71%），而在第二段，314 个残基中有 223 个是相同的（71%），且在序列比对中没有出现空缺。而羧基端的一半相似性较低，但含有与我们推测的 8 个跨膜结构域相似的序列。其他具有广泛相似性的基因包括乙酰辅酶 A 乙酰基转

postulated transmembrane domains. Other extensive similarities were found to acetyl-CoA acetyltransferase, glutathione reductase, the arsenical pumpdriving ATPase, the hypothetical transposase TcA of the nematode transposon Tcl, and the host protective factor of the parasitic nematode *Trichostrongylus colubriformis*. Less extensive but still highly significant similarities (BLAST scores of >100 over one or two exons) were found with the 50S ribosomal protein L11 and the neutrophil oxidase factor. This last similarity includes a motif shared between the oxidase factor, yeast actinbinding protein ABP-1, acanthamoeba myosin IC and *src*-related kinase[36]. The hypothetical nematode protein has three copies of this motif. Finally, other more limited but still significant similarities of likely *C. elegans* coding regions were found with phenylethanolamine-*N*-methyltransferase, adenylyl cyclase, giant secretory protein, the yeast *CDC25* cell-cycle gene (and the *Drosophila string* homologue), glucose transporter and immediate early protein IE110 of herpes simplex virus. For adenylyl cyclase, several adjacent segments of B0303 showed similarity with a repeated motif of the cyclase sequence.

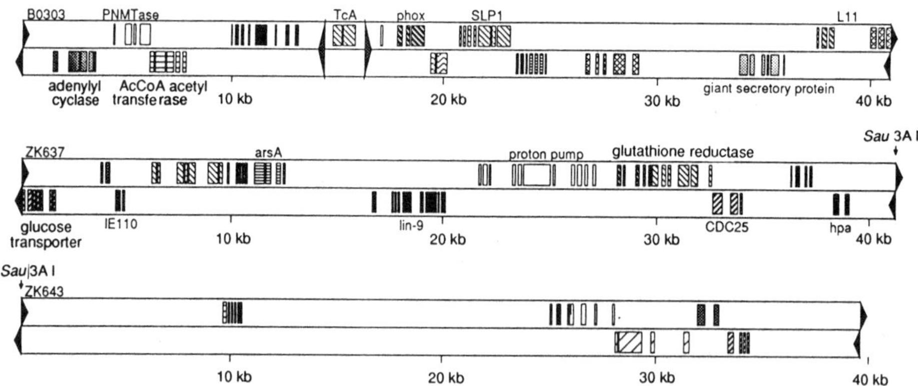

Fig. 3. The genes in the cosmids sequenced, as determined through homology searches, GENEFINDER analysis, and available cDNA sequences. The exons of predicted genes are indicated by shaded blocks, with different shading patterns indicating distinct genes. Those shown above centre for each cosmid are encoded by the top strand and those below by the bottom strand. The genes with significant similarities to genes in the databases are indicated by the name of the most similar sequence (see Fig. 4 for details). Also shown is the position of the *lin-9* gene, as determined from cDNA sequence (G. Beitel and R. Horvitz, personal communication). ZK637 and ZK643 overlap by the single *Sau*3a site indicated. The inverted repeats flanking the TcA homologue are indicated by shaded triangles. The region at 25.1 to 28.1 kb of ZK643 contains two predicted genes that overlap by 52 bases but use distinct reading frames in the overlapped segment. This is indicated in the diagram by different shading within the same block, but has not been confirmed by experiment. PNMTase, phenylethanolamine-*N*-methyltransferase; phox: neutrophil oxidase factor.

移酶、谷胱甘肽还原酶、砷驱动泵的三磷酸腺苷酶、线虫转座子 Tc1 的假想转座酶 TcA 以及寄生性线虫蛇形毛圆线虫的宿主保护因子。50S 的核糖体蛋白 L11 和嗜中性粒氧化酶因子，虽然序列跨度较短但仍在其中发现了高度相似性（一或两个外显子的 BLAST 值 >100），最后一个相似性包括了一个氧化酶因子、酵母肌动蛋白结合蛋白 ABP-1、棘阿米巴肌球蛋白 IC 和 *src* 相关激酶所共有的功能域[36]。预测的线虫蛋白中有三个该功能域的拷贝。最后，其他长度更短但仍然与秀丽隐杆线虫预测编码区很相似的，包括苯基乙醇胺 –*N*– 甲基转移酶、腺苷酸环化酶、巨分泌蛋白、酵母 *CDC25* 细胞周期基因（以及果蝇 *string* 基因的同源基因）、葡萄糖转运蛋白和单纯疱疹病毒的即时早期蛋白 IE110。关于腺苷酸环化酶，B0303 的若干相邻片段与该环化酶序列的一个重复功能域显示出相似性。

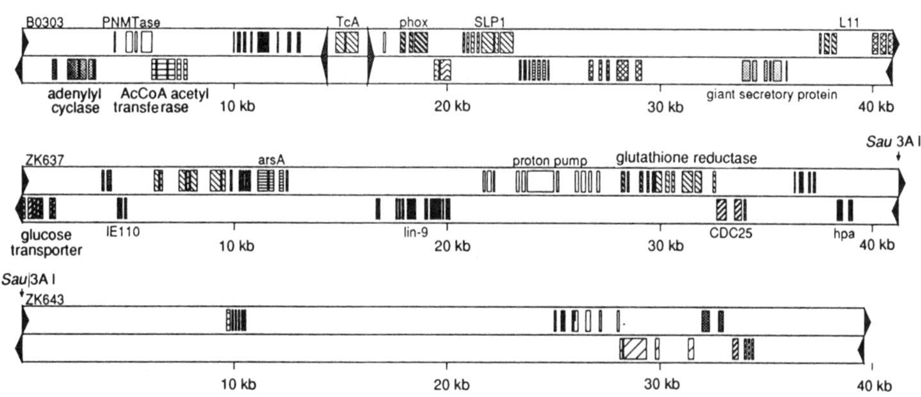

图 3. 在所测的黏粒中，采用同源搜索、GENEFINDER 分析和已知的 cDNA 序列确认的基因。阴影块表示所预测基因中的外显子，不同的阴影涂布方式表示不同的基因。标注在每个黏粒上部的是用上链编码的，而标注在下部的是用底链编码的。那些与数据库中基因显著相似的基因，用与其最相似的序列的名称标注（细节见图 4）。同时显示的还有用 cDNA 序列（比特尔 (Beitel) 和霍维茨，个人通信）确定的 *lin-9* 基因的位置。ZK637 和 ZK643 在单一 *Sau*3a 位点的重叠如图所示。在 TcA 同源基因侧翼的反向重复序列用阴影三角形表示。ZK643 的 25.1 到 28.1 kb 区域包含两个预测到的基因，它们有 52 个碱基的重叠，但在重叠片段使用各自的阅读框架。这一点在图上用同一方块的不同阴影来表示，但尚未经实验证明。PNMTase：苯基乙醇胺 –*N*– 甲基转移酶；phox：嗜中性粒氧化酶因子。

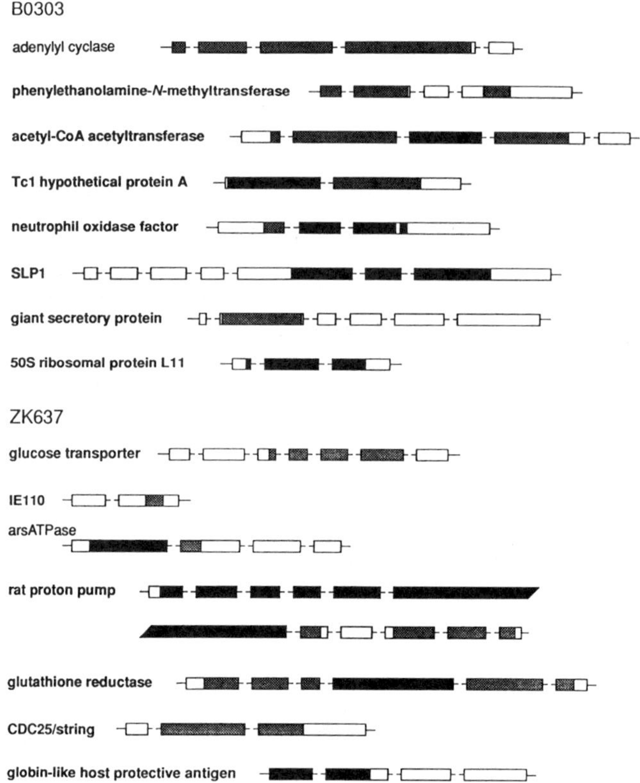

Fig. 4. The similarities of the predicted genes to previously known genes are shown in detail. The gene with the highest similarity to the proposed nematode gene is given on the left. On the right the exons of the predicted gene are indicated as blocks; homologous regions are shaded. Intervening sequences are indicated by lines and have been truncated. Hatched shading indicates regions with BLAST scores of 70 to 100, and solid shading indicates BLAST scores of greater than 100. Once similarities were detected with BLAST, the amino acids predicted from the gene were aligned with the matched gene using the FASTA algorithm[51]. The extent of the shading reflects the region with significant similarity determined from the FASTA output. For adenylyl cyclase, the locus CYAA$YEAST was used[52]; for phenylethanolamine, -*N*-methyl transferase, PNMT$HUMAN[53,54]; for acetylCoA acetyltransferase, THIL$RAT[55]; for Tc1 hypothetical protein A, CELTCB2[56]; for neutrophil oxidase factor, NOF$HUMAN[57]; for SLP1, SLP1$YEAST[58]; for giant secretory protein, BARG$CHITH[59]; for 50S ribosomal protein L11, RL11$ECOLI[60-62]; for glucose transporter, ATHSTP1[63]; for IE110, IE11$HSV11[64,65]; for arsenic ATPase ARSA$ECOLI[66]; for rat proton pump, RATPRPU[67]; for glutathione reductase, GSHR$ECOLI[68]; for CDC25/string, CC25$SCHPO[69]; for globin-like host protective antigen, TCSHGPR[70].

Altogether, GENEFINDER predicts a total of 33,573 bases (27%) of coding sequence in the total of 121,298 bases sequenced. For the individual cosmids the percentage of coding sequences are 37% for B0303, 33% for ZK637, and 13% for ZK643. The number of different genes represented is difficult to estimate, as at present there are no clear rules by which to distinguish ends of genes from introns. Taking into account factors such as the distinct homologies and the spacing and strandedness of exons, we would estimate B0303 contains 14 genes, ZK637 12 genes and ZK643 6 genes. Transcript analysis will be required to test these predictions. Several studies have already centred on genes in the

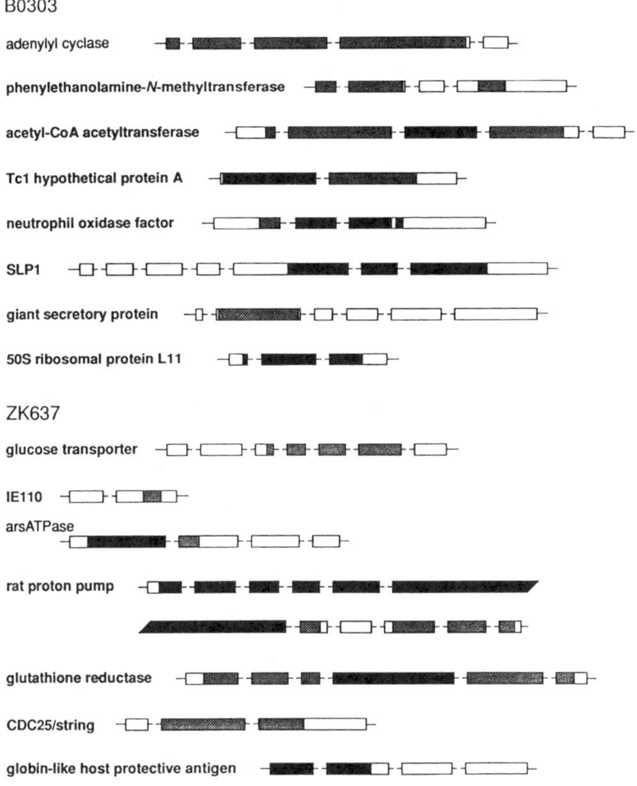

图 4. 详细展示了预测基因与此前已知基因的相似性。左侧是和该线虫基因具有最高相似性的基因。右侧用方块表示预测基因的外显子；同源区域用阴影表示。外显子之间的序列用虚线表示。浅色阴影表示 BLAST 值在 70 到 100 之间的区域，实阴影则表示 BLAST 值超过 100 的区域。一旦 BLAST 检测到了相似性，就采用 FASTA 算法，把从该基因中预测到的氨基酸与相应基因匹配[51]。阴影的长度反映了根据 FASTA 结果认为有显著相似性区域。对于腺苷酸环化酶，采用了 CYAA$YEAST 基因位点[52]；而对于苯基乙醇胺 –*N*– 甲基转移酶，采用 PNMT$HUMAN[53,54]；乙酰辅酶 A 乙酰基转移酶，THIL$RAT[55]；Tc1 假想蛋白 A，CELTCB2[56]；嗜中性粒氧化酶因子，NOF$HUMAN[57]；SLP1，SLP1$YEAST[58]；巨分泌蛋白，BARG$CHITH[59]；50S 的核糖体蛋白 L11，RL11$ECOLI[60-62]；葡萄糖转运蛋白，ATHSTP1[63]；IE110，IE11$HSV11[64,65]；砷三磷酸腺苷酶，ARSA$ECOLI[66]；大鼠质子泵，RATPRPU[67]；谷胱甘肽还原酶，GSHR$ECOLI[68]；CDC25/ 链，CC25$SCHPO[69]；类球蛋白宿主保护抗原，TCSHGPR[70]。

总之，在被测序的 121,298 个碱基序列中，GENEFINDER 总计预测了 33,573 个碱基（27%）的编码序列。就各黏粒来说，B0303 的编码序列占序列全长的 37%，ZK637 占 33%，而 ZK643 占 13%。关于所代表的不同基因数量则很难估计，因为目前没有明确的规则把基因的末尾从内含子中辨别出来。考虑到各计算因素如外显子的有区别的同源性、间隔和链型等，我们估计 B0303 包含 14 个基因，ZK637 含 12 个，而 ZK643 含 6 个。转录分析对检验上述预测是必需的。若干研究已经将该区域的基因作为重点。基因 *sup-5*（文献 37）已经被定位在 B0303 的末端附近。实

region. The gene *sup-5* (ref. 37) had been mapped close to the end of B0303; indeed, the B0303 sequence overlaps with the 4.2-kb region previously sequenced around the transfer RNA gene *sup-5* (K. Kondo and R.W., unpublished results). The genes *lin-9* (ref. 38) and *unc-32* (ref. 6) were known by transformation rescue to lie within ZK637 before sequencing began. The gene *lin-9* has been shown to correspond to the predicted gene on the bottom strand of ZK637 in the region 20.2–16.6, and the complementary DNA sequence used to refine the predictions of the GENEFINDER analysis (G. Beitel and R. Horvitz, personal communication). From its genetic position, *unc-32* is likely to be one of the genes immediately to the right of *lin-9*.

The sequences have also been scanned for repeats. Large, almost perfect inverted repeats (465/468) flank the region of B0303 near the 15-kb point (Fig. 3) that shares similarity with the TcA transposase; subsequent analysis revealed this region to represent a copy of the Tcl-related transposon Tc3 (D. Schneider, J. Collins and P. Anderson, personal communication). Examples of short tandemly repeated sequences include 21 copies of a 59-bp motif at 35 kb in ZK643 and two segments (11 and 13 copies) of an 11-bp motif inverted with respect to one another at 13.3 and 13.7 kb in ZK637. A larger duplicated segment is present spanning the join of ZK637 and ZK643, where blocks of 99,305 and 789 bp spread over 1.6 kb are almost exactly duplicated 4 kb away. A 500-bp region near 7 kb in ZK643 shares several fragments with strong homology (>90%) to an 800-bp region near 32 kb in B0303, the order and orientation of the fragments being different in the two cases. A 1-kb region near 38-kb in ZK643 contains large stretches of the nucleotide motif NGG tandemly repeated; the same motif is found at the fragile X site[39-41]. Some 20 copies of a 94-bp consensus sequence were found dispersed throughout the three cosmids, many in inverted pairs separated by up to 130 bp. Some copies showed a good match to the consensus (80–90% identity), whereas others diverged strongly or were incomplete. A search of Genbank showed the sequence to be specific to *C. elegans* and present in four copies in other entries. Although they are relatively few compared with those in mammalian DNA and are confined to local regions, these repeats would make impossible a walking strategy that relied primarily on cosmid templates.

Discussion

The first three cosmids of the nematode sequencing project are complete. They represent only 0.1% of the total genome, but the methods developed have the potential to be scaled up effectively to production level. We have used two commercially available sequencing machines, but because we are proceeding on a cosmid-by-cosmid basis we have the flexibility to change as new methods and machines are developed.

The essence of our approach is to follow a conventional "shotgun" phase, in which an initial read is taken from each subclone using a universal primer, with a "walking" phase, in which additional reads are taken from selected subclones by the use of selected custom primers. Each subclone is large enough to allow several walking steps to be taken. Thus the shotgun phase serves not only to provide much of the final sequence but also to map the subclones for the walking phase, and the traditional problem of closure is solved

际上，B0303 序列与先前测序的，在转运 RNA 基因 *sup-5* 附近约 4.2 kb 的区域重叠 (孔多和威尔逊，未发表结果)。在测序开始前，通过转化恢复试验（transformation rescue），已知基因 *lin-9*（文献 38）和 *unc-32*（文献 6）位于 ZK637 中。已经表明，基因 *lin-9* 与 ZK637 的 20.2–16.6 区域底链的预测基因相对应，并使用互补 DNA 序列来优化 GNENFINDER 分析得到的预测（比特尔和霍维茨，个人通信）。从遗传学位置得知，*unc-32* 可能是紧临 *lin-9* 基因右侧的基因之一。

我们也对序列中的重复序列进行了分析。大片段的几乎完美的反向重复序列（465/468）位于 B0303 的 15 kb 位点附近的侧翼（图 3），与 TcA 转座酶有相似性。随后的分析表明，这一区域代表了与 Tc1 相关的转座子 Tc3 的一个拷贝（施奈德、柯林斯、安德森，个人通信）。短串联重复序列的例子，包括在 ZK643 的 35 kb 位置的 59 bp 保守序列的 21 份拷贝；分别位于 ZK637 的 13.3 和 13.7 kb 区域，11 bp 保守序列的两个互为反向的片段（分别为 11 和 13 个拷贝）。在 ZK637 和 ZK643 连接的区域出现了一个较大的重复片段，覆盖了 1.6 kb 长度的 99,305 bp 和 789 bp 区段在间隔 4 kb 对处精确复制。在 ZK643 靠近 7 kb 处的 500 bp 的区域，与 B0303 靠近 32 kb 处的 800 bp 的区域相比，若干片段有很高的同源性（>90%），但二者的顺序和方向不同。在 ZK643 靠近 38 kb 处的一个 1 kb 区域，包含大片段核苷酸保守序列 NGG 的串联重复；在具有脆性的 X 位点 [39-41] 也发现了同样的保守序列。在这三个黏粒上，发现散布着一个 94 bp 保守序列的约 20 份拷贝，许多是被多达 130 bp 分隔的反向重复对。某些拷贝显示出与保守序列的良好匹配（80%~90% 相同），而其他则变异较大或是不完整。在 Genbank 中检索，发现该序列是秀丽隐杆线虫特有的，在其他线虫记录中有四个拷贝。尽管它们与哺乳类 DNA 相比较少，并限定在局部区域，这些重复序列将会使主要依赖于黏粒模板的步移策略变得不可能。

讨　论

该线虫测序计划前三个黏粒已经完成。它们仅代表整个基因组的 0.1%，但所建立的方法有可能被有效规模化至生产水平。我们使用了两种商业测序设备，但由于我们是以逐个黏粒测序为基础进行的，当新的方法和设备产生以后，我们可以灵活调整。

我们策略的本质是传统的“鸟枪法”，使用一个通用引物，可以从每个亚克隆获得一个初始测序片段；下一阶段采取了“步移”法，即使用筛选出的特异引物，从选定的亚克隆中获得更多的测序片段。每个亚克隆大到足以允许采取若干步移步骤。因而鸟枪阶段不但提供了许多最终序列，而且可以对步移阶段的亚克隆进行定位。

efficiently without the requirement for high redundancy. Exactly when the switch is made from random clones to walking will be dictated by the costs and convenience of the two approaches.

The number of predicted genes in these cosmids is higher than would have been expected on the basis of genetic estimates of essential gene number[42] and on a previous transcript analysis around the vitellogenin genes[43]. The high gene density (one every 3–4 kb) may arise in part because the sample comes from the gene-rich cluster on chromosome III. But each chromosome contains a central gene-rich cluster, and the physical map shows that about half the genome is contained in such clusters. Extrapolating then from our (admittedly small) sample, *C. elegans* will have about 15,000 genes in the clusters alone. The other half of the DNA is not devoid of genes, but we do not yet have an accurate estimate of the gene density there. The high gene density makes genomic DNA sequencing an efficient approach to discovering complete coding sequences in *C. elegans*.

The discrepancy between genetic estimates of essential genes and our estimates of total genes based on sequence is similar to that found in other organisms. In yeast, where studies are most advanced, it is estimated that no more than one in six genes yields scorable phenotypes when function is eliminated[44]. In *Drosophila*, in the region between *rosy* and *Bithorax*, more than three times as many transcripts have been found as there are mutationally defined genes, even though this is a region held to be saturated genetically[45,46].

More accurate prediction of genes will be aided by the accumulation of more data, providing a better basis for statistical analysis. The GENEFINDER predictions are of limited use in this regard as they necessarily reflect the biases on which they are based. Nonetheless, the success of GENEFINDER at this early stage encourages our confidence that genomic sequence is an efficient means for finding not only genes but also the other information stored in the genome. In turn, the fraction of genes that yield similarities, already about a third, will only increase and these provide immediate insights on which to design experiments to test function.

Each cosmid sequence is submitted, as it is completed, to the databases. In addition, the sequence is entered in the *C. elegans* database being developed as part of this project. Here the sequence will be available in conjunction with the physical map, the genetic map, references and strains. The ability to view the sequence in the context of much of the available knowledge about the worm should speed the assignment of function to each sequence.

The test of our strategy will come in the next two years as we scale up our efforts: we will sequence 800 kb in the first year and 2,000 kb in the second. We anticipate the cost, currently estimated at $1 per base with current methods applied on production, will come down as we move to a mock production mode. Operator time should decrease with experience, new methods and increased automation. For example, we are experimenting with ways of preparing templates more rapidly and cheaply. Improved throughput for the

同时，无需大量的冗余测序，就把消除缺口的传统问题有效地解决了。从随机克隆转换到步移的具体时机是由这两种方法的成本和方便性来决定的。

与基于遗传学估计所预测到的主要基因数量[42]以及以前对卵黄蛋白原基因转录分析所得基因数量[43]相比，在这些黏粒中预测到的基因数量要多一些。较高基因密度（每 3~4 kb 一个基因）部分是由于样品来自第 III 染色体基因富集簇。不过每个染色体都包含一个重要的基因富集簇，物理图谱表明大约一半的基因组包含在这种簇里。根据我们的样品（应承认样本较小）推测，秀丽隐杆线虫仅在这些簇就有大约 15,000 个基因。基因组 DNA 的另一半并非没有基因，只是我们对该处的基因密度还没有精确估计。较高的基因密度使得基因组 DNA 测序成为揭示秀丽隐杆线虫完整编码序列的有效途径。

对重要基因的遗传学估计和我们根据序列对总基因数的估计存在差异，在其他生物中也存在这种类似差异。酵母的研究最为成熟，据计算当某种功能缺失时，仅有不超过六分之一的基因会产生可观测的表型[44]。在果蝇中，在基因 *rosy* 和 *Bithorax* 之间的区域，尽管这是一个遗传饱和区[45,46]，但所发现的转录物数量是以突变手段鉴定的基因数量的三倍以上。

积累更多的数据，提供更好的统计分析基础，将有助于更精确的基因预测。在这方面 GENEFINDER 预测的作用有限，并必然反映出了所依赖的方法学上的偏好。无论如何，GENEFINDER 在这一初期阶段的成功，促使我们相信，基因组测序是找到基因和其他储存在基因组中的信息的有效手段。反过来，存在相似性的基因片段目前已经达到 1/3，这个数目一定会增加，这为我们设计检验各种功能的实验提供了直接的线索。

每个黏粒的测序一旦完成，结果立即提交到数据库。此外，作为本项目的内容之一，该序列还会被录入秀丽隐杆线虫数据库。在这里可获得与物理图谱、遗传学图谱、文献及品系相关联的序列信息。根据线虫的大部分已知信息来考察序列的能力，将加速与序列对应的功能解释。

随着投入不断增加，今后两年我们将对该策略进行检验：我们将在第一年测序 800 kb，在第二年测序 2,000 kb。采用目前方法进行生产，估计成本为每碱基 1 美元，随着我们转向类似于生产的模式，预计成本将下降。随着经验增多、新方法出现和自动化程度提升，操作时间将缩短。例如，我们正在采用更迅速而便宜的模板

sequencing machines, either by increasing the number of samples per run or by reducing the run time, is also under development by the manufacturers. Oligonucleotide costs are steadily declining with both reduced reagent usage and reduced reagent costs. Local preparation may make savings possible for some reagents in a large-scale operation. With these improvements, a reduction in costs to $0.50 per base seems realistic.

(**356**, 37-41;1992)

J. Sulston*, Z. Du†, K. Thomas*, R. Wilson†, L. Hillier†, R. Staden*, N. Halloran†, P. Green†, J. Thierry-Mieg‡, L. Qiu†, S. Dear*, A. Coulson*, M. Craxton*, R. Durbin*, M. Berks*, M. Metzstein*, T. Hawkins*, R. Ainscough* & R. Waterston†

* MRC Laboratory of Molecular Biology, Hills Road, Cambridge CB2 2QH, UK

† Department of Genetics, Box 8232, Washington University School of Medicine, 4566 Scott Avenue, St Louis, Missouri 63110, USA

‡ CNRS-CRBM et Physique-Mathématique, PO Box 5051, Montpellier 34044, France

Received 3 September 1991; accepted 4 February 1992.

References:

1. Watson, J. D. *Science* **248**, 44-49 (1990).
2. Cantor, C. R. *Science* **248**, 49-51 (1990).
3. Roberts, L. *Science* **249**, 1497 (1990).
4. Hunkapiller, T., Kaiser, R. J., Koop, B. F. & Hood, L. *Science* **254**, 59-67 (1991).
5. Roberts, L. *Science* **248**, 1310-1313 (1990).
6. Brenner, S. *Genetics* 77, 71-94 (1974).
7. Wood, W. B. *et al. The Nematode Caenorhabditis elegans* (Cold Spring Harbor Laboratory. New York, 1988).
8. Edgley, M. L. & Riddle, D. L. *Genetic Maps* **5**, 3 (1990).
9. Sulston, J. E. & Horvitz, H. R. *Devl. Biol.* **56**, 110-156 (1977).
10. Kimble, J. E. & Hirsh, D. I. *Devl. Biol.* **70**, 396-417 (1979).
11. Sulston, J. E., Schierenberg, E., White, J. G. & Thomson, J. N. *Devl. Biol.* **100**, 64-119 (1983).
12. White, J. G., Southgate, E., Thomson, J. N. & Brenner, S. *Phil. Trans. R. Soc.* **314**, 1-340 (1986).
13. Emmons, S. W., Yesner, L., Ruan, K. S. & Katzenberg, D. *Cell* **32**, 55-65 (1983).
14. Eide, D. J. & Anderson, P. *Proc. Natl. Acad. Sci. U.S.A.* **82**, 1756-1760 (1985).
15. Moerman, D. G., Benian, G. M. & Waterston, R. H. *Proc. natn. Acad. Sci. U.S.A.* **83**, 2579-2583 (1986).
16. Coulson, A. R., Sulston, J. E., Brenner, S. & Karn, J. *Proc. Natl. Acad. Sci. U.S.A.* **83**, 7821-7825 (1986).
17. Coulson, A. R., Waterston, R. H., Kiff, J. E., Sulston, J. E. & Kohara, Y. *Nature* **335**, 184-186 (1988).
18. Coulson, A. *et al. BioEssays* **13**, 413-417 (1991).
19. Stinchcomb, D. T., Shaw, J. E., Carr, S. H. & Hirsh, D. I. *Molec. Cell. Biol.* **5**, 3484-3496 (1985).
20. Fire, A. *EMBO J.* **5**, 2673-2680 (1986).
21. Fire, A. & Waterston, R. H. *EMBO J.* **8**, 3419-3428 (1989).
22. Blumenthal, T. & Thomas, J. H. *Trends Genet.* **4**, 305-308 (1988).
23. Schriefer, L. A., Gebauer, B. K., Qiu, L. Q. Q., Waterston, R. H. & Wilson, R. K. *Nucleic Acids Res.* **18**, 7455-7456 (1990).
24. Deininger, P. L. *Analyt. Biochem.* **129**, 216-223 (1983).
25. Gleeson, T. & Hillier, L. *Nucleic Acids Res.* **19**, 6481-6483 (1991).
26. Smith, L. M. *et al. Nature* **321**, 674-679 (1986).
27. Craxton, M. *Methods: A Companion to Methods in Enzymology* **3**, 20-26 (1991).
28. Dear, S. & Staden, R. *Nucleic Acids Res.* **19**, 3907-3911 (1991).
29. Hillier, L. & Green, P. *PCR Meth. Appls* **1**, 124-128 (1991).
30. Ansorge, W., Sproat, B. S., Stegemann, J. & Schwager, C. *J. biophys. Biphys. Meth.* **13**, 315-323 (1986).
31. Mizusawa, S., Nishimura, S. & Seela, F. *Nucleic Acids Res.* **14**, 1319-1324 (1986).
32. Hawkins, T. L. & Sulston, J. E. *Nucleic Acids Res.* **19**, 2784 (1991).

制备方法。制造商也正在开发更高通量的测序设备，通过增加每次运行的样品数量，或是缩短运行时间。通过减少试剂使用量和降低试剂成本，寡聚核苷酸的成本正在稳步下降。在规模化的运作模式下，某些试剂本地生产也将节约成本。伴随着这些改进，把测序成本降低到每碱基 0.50 美元看来是可实现的。

（周志华 翻译；吕雪梅 审稿）

33. Hawkins, T. L., Du, Z., Halloran, N. D. & Wilson, R. K. *Electrophoresis* (manuscript submitted).
34. Craxton, M. in *DNA Sequencing: Laboratory Protacols* (eds Griffin, H. G. & Griffin, A. M.) (Humana, NJ, 1992).
35. Altschul, S. F., Gish, W., Miller, W., Myers, E. W. & Lipman, D. J. *J. Molec. Biol.* **215**, 403-410 (1990).
36. Drubin, D. G., Mulholland, J., Zhu, Z. & Botstein, D. *Nature* **343**, 288-290 (1990).
37. Wills, N. *et al. Cell* **33**, 575-583 (1983).
38. Ferguson, E. L. & Horvitz, H. R. *Genetics* **110**, 17-72 (1985).
39. Oberle, I. *et al. Science* **252**, 1097-1102 (1991).
40. Yu, S. *et al. Science* **252**, 1179-1181 (1991).
41. Verkerk, A. J. M. H. *et al. Cell* **65**, 905-914 (1991).
42. Herman, R. K. in *The Nematode* Caenorhabditis elegans (eds Wood, W. B. *et al.*) 17-45 (Cold Spring Harbor Laboratory, New York, 1988).
43. Heine, U. & Blumenthal, T. *J. Molec. Biol.* **188**, 301-312 (1986).
44. Olson, M. in *Genome Dynamics, Protein Synthesis and Energetics* (eds Broach, J. R., Pringle, J. R. & Jones, E. W.) 1-41 (Cold Spring Harbor, NY. 1991).
45. Hall, L. M. C., Mason, P. J. & Spierer, P. *J. Molec. Biol.* **169**, 83-96 (1983).
46. Bossy, B., Hall, L. M. C. & Spierer, P. *EMBO J.* **3**, 2537-2541 (1984).
47. Vieira, J. & Messing, J. *Meth. Enzym.* **153**, 3-11 (1987).
48. Short, J. M., Fernandez, J. M., Sorge, J. A. & Huse, W. D. *Nucleic Acids Res.* **16**, 7583-7600 (1988).
49. Yanisch-Perron, C., Vieira, J. & Messing, J. *Gene* **33**, 103-119 (1985).
50. Burglin, T. R., Finney, M., Coulson, A. & Ruvkin, G. *Nature* **341**, 239-243 (1989).
51. Pearson, W. R. & Lipman, D. J. *Proc. Natl. Acad. Sci. U.S.A.* **85**, 2444-2448 (1988).
52. Kataoka, T., Broek, D. & Wigler, M. *Cell* **43**, 493-505 (1985).
53. Kaneda, N. *et al. J. boil. Chem.* 263, 7672-7677 (1988).
54. Sasaoka, T., Kaneda, N., Kurosawa, Y.,Fujita, K. & Nagatsu, T. *Neurochem. Int.* **15**, 555-565 (1989).
55. Fukao, T. *et al. J. Biochem., Tokyo* **106**, 197-204 (1989).
56. Prasad, S. S., Harris, L. J., Baillie, D. L. & Rose, A. M. *Genome* **34**, 6-12 (1991).
57. Leto, T. L. *et al. Science* **248**, 727-730 (1990).
58. Wada, Y., Kitamoto, K., Kanbe, T., Tanaka, K. & Anraku, Y. *Molec. Cell. Biol.* **10**, 2214-2223 (1990).
59. Lendahl, U. & Wieslander, L. *Cell* **36**, 1027-1034 (1984).
60. Dognin, M. J. & Wittman-Liebold, B. *Eur. J. Biochem.* **112**, 131-151 (1980).
61. Post, L. E., Strycharz, G. D., Nomura, M., Lewis, H. & Dennis, P. P. *Proc. Natl. Acad. Sci. U.S.A.* **76**, 1697-1701 (1979).
62. Downing, W. L., Sullivan, S. L., Gottesman, M. E. & Dennis, P. P. *J. Bact.* **172**, 1621-1627 (1990).
63. Sauer, N., Friedl, K. & Wicke, U. Genbank Accession Number X55350 (1991).
64. McGeoch, D. J. *et al. J. gen. Virol.* **69**, 1531-1574 (1988).
65. Perry, L. J., Rixon, F. J., Everett, R. D., Frame, M. C. & McGeoch, D. J. *J. gen. Virol.* **67**, 2365-2380 (1986).
66. Chen, C.-M., Misra, T. K., Silver, S. & Rosen, B. P. *J. biol. Chem.* **261**, 15030-15038 (1986).
67. Perin, M. S., Fried, V. A., Stone, D. K., Xie, X.–S. & Sudhof, T. C. *J. biol. Chem.* **266**, 3877-3881 (1991).
68. Greer, S. & Perham, R. N. *Biochemistry* **25**, 2736-2742 (1986).
69. Russell, P. & Nurse, P. *Cell* **45**, 145-153 (1986).
70. Frenkel, M. J., Dopheide, T. A., Wagland, B. M. & Ward, C. W. Genbank Accession Number M63263 (1991).

Acknowledgements. We thank G. Beitel and R. Horvitz for information on *lin-9*; P. Anderson for sequence of Tc3; M. Jier and R. Shownkeen for synthesis of oligonucleotides; and P. Kassos for preparing the manuscript. The work was supported by grants from the NIH Human Genome Center and the MRC HGMP, as well as our respective institutions.

SNAP Receptors Implicated in Vesicle Targeting and Fusion

T. Söllner *et al.*

Editor's Note

Vesicle merging—the fusion of tiny sac-like structures to a membrane—has been implicated in numerous processes, including nerve cell communication and various secretory pathways. Here cell biologist (and future Nobel laureate) James E. Rothman and colleagues propose a mechanism to explain how different vesicles reach and merge with their specific target in the brain. They show that vesicles and target membranes contain different types of synaptic receptors called SNARE proteins, and suggest that specificity occurs when compartment-specific SNAREs in the vesicle and target membrane pair up. The classification of SNAREs into vesicle-associated (v-SNAREs) and target membrane-associated (t-SNAREs) remains today, although the proteins can also be characterised according to their structure.

The *N*-ethylmaleimide-sensitive fusion protein (NSF) and the soluble NSF attachment proteins (SNAPs) appear to be essential components of the intracellular membrane fusion apparatus. An affinity purification procedure based on the natural binding of these proteins to their targets was used to isolate SNAP receptors (SNAREs) from bovine brain. Remarkably, the four principal proteins isolated were all proteins associated with the synapse, with one type located in the synaptic vesicle and another in the plasma membrane, suggesting a simple mechanism for vesicle docking. The existence of numerous SNARE-related proteins, each apparently specific for a single kind of vesicle or target membrane, indicates that NSF and SNAPs may be universal components of a vesicle fusion apparatus common to both constitutive and regulated fusion (including neurotransmitter release), in which the SNAREs may help to ensure vesicle-to-target specificity.

TRANSACTIONS among the membrane-bound compartments in the cytoplasm of eukaryotic cells are generally executed by transport vesicles which bud from one membrane and fuse selectively with another[1]. The mechanism by which each vesicle chooses its target is currently unknown, but must embody the essence of compartmental specificity. Choice of target is implicit in all vesicular fusion events, from the ubiquitous steps in the constitutive secretory and endocytotic pathways (such as the fusion of a vesicle from the endoplasmic reticulum with the Golgi) to specialized and tightly regulated forms of exocytosis (such as the triggered fusion of a synaptic vesicle containing neurotransmitter with the axonal membrane). Here we describe a fundamental relationship between these processes, suggesting that a general apparatus is used to bring about fusion, and that specificity is established by the pairing of compartment-specific proteins in the transport vesicle and target membrane, respectively.

囊泡靶向及融合过程中 SNAP 受体的功能

泽尔纳等

编者按

囊泡融合（微小的囊状结构与膜的融合）涉及神经细胞交流和各种分泌途径等众多过程。在这篇文章中，细胞生物学家、之后的诺贝尔奖获得者詹姆斯·罗思曼及其同事提出了一种机制，来说明大脑中不同的囊泡如何到达它们的特异性靶位点并与之融合。他们指出囊泡和靶细胞膜包含不同类型的突触受体（称为 SNARE 蛋白），并提出当囊泡和靶细胞膜上的区室特异性 SNARE 蛋白配对时会发生特异性融合。虽然现在可以根据 SNAREs 的结构特点分类，但是按照定位将 SNARE 分为囊泡相关蛋白（v-SNARE）和靶细胞膜相关蛋白（t-SNARE）两类的分类方法到现在仍在沿用。

N–乙基马来酰亚胺敏感融合蛋白（NSF）和可溶性 NSF 附着蛋白（SNAP）似乎是细胞内膜融合装置的必需组分。根据这些蛋白与它们靶分子的天然结合特性建立了一种亲和纯化方法，利用该方法从牛的脑组织中分离 SNAP 受体（SNARE）。值得注意的是，分离出来的四种主要蛋白质都是与神经突触有关的蛋白。其中一种分布于突触小泡上，另一种则定位在细胞膜上，这提示它们可能参与囊泡停靠过程中的某种简单机制。目前已经发现了大量的 SNARE 相关蛋白，并且每种蛋白都特异地对应某种囊泡或者靶细胞膜，这表明 NSF 和 SNAP 可能是组成型和调控型囊泡融合（包括神经递质的释放）装置的通用组分，而 SNARE 则可能负责保证囊泡与靶位点结合的特异性。

真核细胞的细胞质中，由膜包裹的区室间的运输通常都是通过运输小泡来实现的。这些囊泡从一个区室的膜表面出芽形成，之后选择性地与另一个膜发生融合[1]。目前对于囊泡选择靶位点的机制尚不了解，但其肯定与细胞内的区室特异性有关。从组成型分泌和内吞途径中普遍存在的步骤（例如来源于内质网的囊泡与高尔基体融合）到特异性的、高度调控的胞吐过程（例如含有神经递质的突触囊泡与神经轴突细胞膜的融合），所有的囊泡融合事件都涉及靶位点的选择。在这篇文章中，我们描述了这些过程之间的基本关系，并提出细胞内融合过程具有一个通用装置，而融合的特异性则是分别通过运输小泡和靶细胞膜上的区室特异性蛋白之间的配对实现的。

The *N*-ethylmaleimide-sensitive fusion protein is a soluble tetramer of 76K subunits which was purified[2,3] on the basis of its ability to restore intercisternal Golgi transport in a cell-free system[4,5]. It is required for transport vesicle fusion, as vesicles accumulate at the acceptor membrane in its absence[6,7]. In yeast, NSF is encoded by the *SEC18* gene[8], originally shown to be necessary for transport from the endoplasmic reticulum (ER) to the Golgi[9]. The SEC18 protein can replace NSF in a mammalian system for cell-free Golgi transport[8] and is required at every discernible step of the secretory pathway *in vivo*[10]. NSF thus appears to participate in a variety of intracellular fusion processes.

NSF requires additional cytoplasmic factors to attach to Golgi membranes[11]. Three species of monomeric soluble NSF attachment proteins, termed α-, β- and γ-SNAP (M_rs of 35, 36 and 39K, respectively), have been purified from brain[12,13] and SNAP activity is necessary for vesicle fusion *in vitro*[13]. α-SNAP (but not β- or γ-SNAP) can restore animal cell Golgi transport activity to cytosol prepared from *sec17* mutant yeast, implying[13] that the *SEC17* gene encodes α-SNAP in yeast. SEC17 is now known to be functionally equivalent to α-SNAP[14], and the two proteins are clearly related[15]. In the absence of functional SEC17 (or SEC18), ER to Golgi transport stops in yeast, and transport vesicles accumulate[16]. SNAPs, like NSF, thus appear to be components of a general intracellular membrane fusion apparatus common to all eukaryotic cells.

SNAPs bind to distinct sites in membranes which up until now have only been operationally defined, and NSF will only interact with SNAPs that are already attached to these sites[17]. α-SNAP and β-SNAP compete for binding to the same receptor site with low nM affinity. γ-SNAP binds to a noncompetitive site in the same complex[17], and, although not essential for NSF binding, increases the complexes' affinity for NSF[18]. Crosslinking studies suggest that the SNAP receptor contains an α-SNAP-binding subunit of 30–40K (ref. 17). When membrane-bound NSF–SNAP–SNAP receptor complexes are solubilized with detergent, they sediment as a distinct multisubunit particle at 20S (ref. 18), which may form the core of a generalized apparatus catalysing bilayer fusion at its point of assembly[6,13]. The 20S particles are also formed when detergent extracts of Golgi membranes containing SNAP receptors are mixed with SNAPs and NSF. When NSF and SNAPs are added in excess, all SNAP receptor activity in the membrane extract is incorporated into 20S particles[18].

NSF is an ATPase containing two ATP-binding sites in separate domains[19], and the binding and hydrolysis of ATP are critical in determining the stability of NSF[3] and its attachment to membranes[2,18]. Stable 20S particles can be formed in the presence of either Mg–ATP-γS (a non-hydrolysable analogue of ATP) or ATP without magnesium. In the presence of Mg–ATP, however, particles rapidly dissociate even at 0 °C, liberating NSF in a process that requires ATP hydrolysis[18]. This disassembly may be an intrinsic step in the fusion mechanism[18].

NSF 是一个可溶的四聚体，亚基分子量为 76K，是基于其可以在一个无细胞系统中恢复高尔基体各片层间转运的性质[4,5]而分离纯化[2,3]出来的。NSF 是运输小泡融合所必需的，如果没有 NSF 就会导致囊泡在受体膜附近积累[6,7]。NSF 在酵母中由 *SEC18* 基因编码[8]，最初发现它在内质网(ER)到高尔基体的转运中起着必不可少的作用[9]。在无细胞高尔基体转运实验中[8]，SEC18 蛋白可以替代哺乳动物系统中 NSF 的功能，并且对于体内分泌途径的每个可辨别的步骤来说都是必需的[10]。因此，NSF 似乎参与多种细胞内融合过程。

NSF 需要额外的细胞质因子才能结合到高尔基体膜上[11]。研究人员从脑组织中纯化出了三种可以与 NSF 结合的单体可溶性蛋白，分别命名为 α-，β- 和 γ-SNAP (分子量分别为 35K，36K 和 39K)[12,13]。SNAP 的活性是体外囊泡融合所必需的[13]。α-SNAP(而非 β- 和 γ-SNAP) 可以恢复利用 *sec17* 酵母突变体制备的胞质溶胶辅助的动物细胞高尔基体转运的活性，说明[13]在酵母中 *SEC17* 基因编码 α-SNAP。现在已知 SEC17 具有与 α-SNAP 相同的功能[14]，并且这两种蛋白是明确相关的[15]。当缺失功能性 SEC17 (或 SEC18)时，酵母中 ER 到高尔基体的转运过程便停止了，运输小泡则会出现积累[16]。因此，与 NSF 一样，SNAP 也是所有真核细胞中膜融合装置的一个通用组分。

SNAP 可以与膜上的特异性位点结合。到现在为止，对特异性位点的定义仍停留在操作性定义的层面上。NSF 只能与已经结合到膜上的 SNAP 相互作用[17]。α-SNAP 和 β-SNAP 以纳摩尔级别的亲和力竞争结合同一个受体位点，而 γ-SNAP 则以非竞争的形式与相同的复合物结合[17]。尽管 γ-SNAP 并不是 NSF 的结合所必需的，但是其可以提高 NSF 与复合物相互作用的亲和力[18]。交联实验表明，SNAP 受体包含一个可以与 α-SNAP 结合的 30K~40K 的亚基(参考文献 17)。当使用去垢剂溶解结合在膜上的 NSF–SNAP–SNAP 受体复合物时，这个复合物会以一个多亚基颗粒的形式沉淀下来，沉降系数为 20S(参考文献 18)。这一复合物在组装时可以形成某种通用装置的核心，用以催化脂双层膜的融合[6,13]。用去垢剂抽提高尔基体膜得到的膜抽提物中含有 SNAP 受体，当将其与 SNAP 和 NSF 混合时，也会形成 20S 的颗粒。当加入过量的 NSF 和 SNAP，膜抽提物中所有 SNAP 受体的活性都会融合到 20S 颗粒中去[18]。

NSF 是一种 ATP 酶，它的两个分开的结构域中分别含有一个 ATP 结合位点[19]。ATP 的结合以及水解对于 NSF 的稳定性[3]以及其与膜的结合过程[2,18]具有重要的影响。当存在 Mg–ATP-γS(一种不能被水解的 ATP 类似物)或者存在 ATP 但无 Mg 时，都可以形成稳定的 20S 颗粒。然而当 Mg–ATP 存在时，这些颗粒即使在 0 ℃也会迅速地解离，并随着 ATP 的水解而释放出 NSF[18]。这一解离的过程或许是膜融合过程中固有的一个步骤[18]。

Purification of SNAP Receptors

To purify SNAP receptors, we used the specificity inherent in the assembly and disassembly of 20S particles as the basis of an affinity purification technique (schematized in Fig. 1): 20S particles were formed and attached to a solid matrix, allowing purified SNAP receptor to be released by particle disassembly when ATP was subsequently hydrolysed.

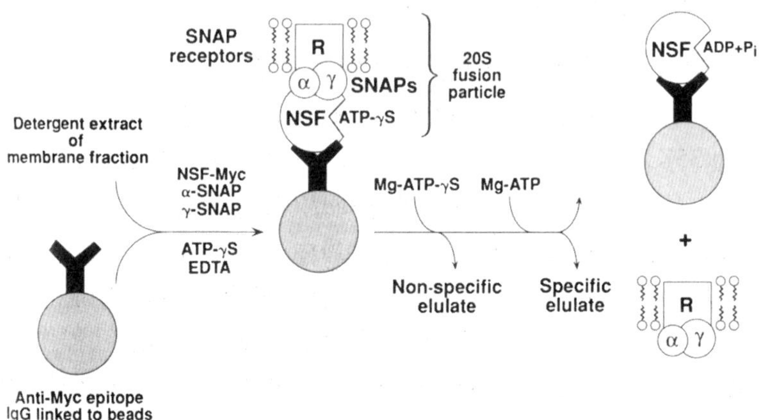

Fig. 1. Procedure used to purify SNAP receptors (SNAREs). Recombinant NSF, α-SNAP, and γ-SNAP are assembled into 20S particles by SNAREs present in a crude detergent extract of membranes. The SNAREs are incorporated stoichiometrically into the particles, which are then bound to beads by means of NSF. For this purpose, the NSF is epitope-tagged with Myc, and an anti-Myc monoclonal IgG is linked to the beads. The beads are washed and then eluted first with Mg–ATP-γS (nonspecific eluate) and then with Mg–ATP (specific eluate). The bound 20S particles disassemble in the presence of Mg–ATP (but not Mg–ATP-γS), releasing stoichiometric amounts of SNAPs and SNAREs. NSF remains bound to the beads. Experimental details are given in Fig. 2 legend.

A recombinant form of NSF, epitope-tagged with a Myc peptide[20] (EQKLISEEDL) at its carboxy terminus, was expressed in *E. coli* and shown to be functional[21]. The 20S particles were formed in solution by mixing a Triton X-100 extract of membranes with pure, recombinant α- and γ-SNAPs expressed in *E. coli* and NSF–Myc at 0 °C (ref. 15) in the presence of ATP-γS and EDTA (to chelate any magnesium). A monoclonal anti-Myc IgG (9E10; ref. 22) attached to protein G beads was then added to bind the 20S particles through their NSF–Myc subunits. The anti-Myc IgG does not inhibit NSF–Myc function in cell-free transport assays (our unpublished results).

The beads were formed into a column and washed extensively before a first (nonspecific) elution with Mg–ATP-γS to elute any proteins attached in a Mg^{2+}-sensitive fashion, or by Mg–ATP-γS *per se.* A second (specific) elution was then done with Mg–ATP (replacing Mg–ATP-γS). Only proteins released as the immediate consequence of ATP hydrolysis on the column will be recovered in this specific eluate. This scheme is based on that described

SNAP 受体的纯化

为了纯化 SNAP 受体，我们使用了一种亲和纯化技术。这一技术基于 20S 颗粒在组装和解离方面所固有的特性（如图 1 所示）：20S 颗粒形成后附着到固体基质上，然后再通过 ATP 水解，使颗粒解离，从而得到纯化的 SNAP 受体。

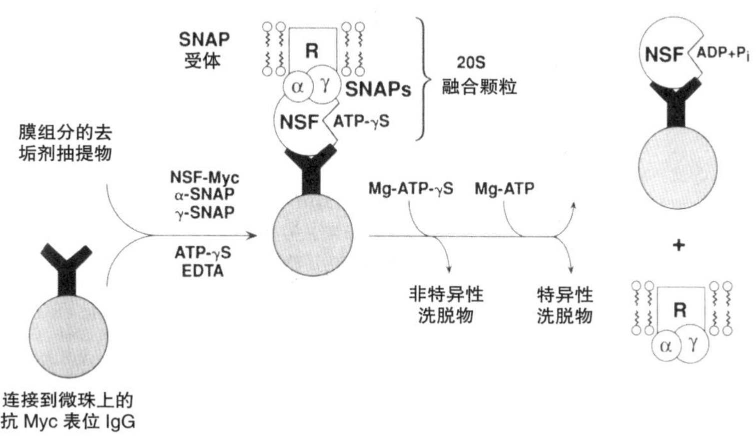

图 1. SNAP 受体（SNARE）的纯化步骤。重组 NSF、α-SNAP 和 γ-SNAP 被加入使用去垢剂抽提得到的细胞膜粗提物中，并与其中的 SNARE 结合，形成 20S 颗粒。SNARE 会按化学计量比包含在 20S 颗粒中，并进一步通过其中的 NSF 与微珠结合。为了实现这个目的，NSF 的 C 端加上了一个 Myc 表位标签，同时微珠的表面偶联了抗 Myc 单克隆 IgG。微珠用缓冲液冲洗后，首先用 Mg–ATP-γS 溶液洗脱一次（非特异性洗脱），然后再用 Mg–ATP 进行洗脱（特异性洗脱）。在 Mg–ATP（而不是 Mg–ATP-γS）存在的情况下，20S 颗粒会发生解离，从而将一定化学计量数的 SNARE 和 SNAP 释放出来，而 NSF 则仍然结合在微珠上。实验细节请见图 2 注。

我们在大肠杆菌中表达了在 C 端加一个 Myc 表位标签 [20]（EQKLISEEDL）的重组 NSF 蛋白，并且实验证明该蛋白是有活性的 [21]。利用 Triton X-100 抽提膜组分，然后与纯化的、在大肠杆菌中表达的重组 α-SNAP、γ-SNAP 和纯化的 NSF–Myc 在 0 ℃混合（参考文献 15），并向体系中加入 ATP-γS 和 EDTA（以螯合溶液中的镁离子），促进 20S 颗粒的形成。另一方面，将抗 Myc 单克隆 IgG（9E10；参考文献 22）连接到表面包被了蛋白 G 的微珠上，将它们加入含有 20S 颗粒的溶液中，从而可以通过抗 Myc IgG 与 NSF–Myc 的特异结合将 20S 颗粒分离出来。在无细胞转运系统中，抗 Myc IgG 并不会抑制 NSF–Myc 的功能（未发表数据）。

将分离出来的微珠填充到一个柱子中，用大量缓冲液冲洗，然后首先（非特异性）使用含有 Mg–ATP-γS 的溶液洗脱附着在上面的所有对镁离子或 Mg–ATP-γS 敏感的蛋白，再用 Mg–ATP 溶液（替代 Mg–ATP-γS）进行第二次（特异性）洗脱。在特异性洗脱过程中只有那些在 ATP 水解后立即被释放出来的蛋白质才会被收集到特异性洗

previously[18] but includes a number of key modifications detailed in the figure legends. As NSF remains on the beads, the specific eluate should consist of a fraction of the added SNAPs, together with additional polypeptides representing SNAP receptors, ideally in stoichiometric amounts.

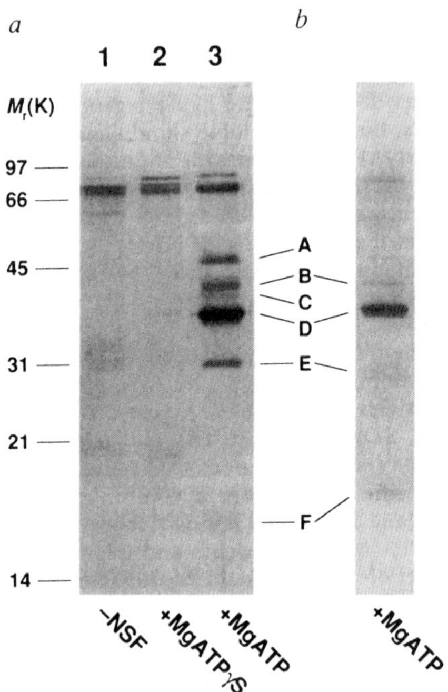

Fig. 2. Identification of proteins released from NSF after ATP hydrolysis. *a*, Polyacrylamide gel stained with Coomassie blue. Lane 1, control, Mg–ATP eluate of control binding reaction in the absence of NSF; lane 2, "nonspecific" eluate from complete binding reaction with NSF–Myc and Mg–ATP-γS; lane 3, "specific" eluate of the same column as for lane 2 following the exchange of ATP for ATP-γS (in the presence of EDTA) and addition of Mg^{2+} to allow ATP hydrolysis (Fig. 1). *b*, Silver-stained Laemmli gel of the specific (Mg–ATP) eluate.

Methods. All manipulations were performed at 0–4 °C. Bovine brain tissue was initially stripped of meninges, and all but the grey matter removed and discarded. The resulting tissue (30 g) was homogenized with 30 strokes of a Dounce homogenizer in buffer A (20 mM Tris–HCl, pH 8.0, 1 M KCl, 250 mM sucrose, 2 mM $MgCl_2$, 1 mM DTT, 1 mM phenylmethylsulphonyl fluoride (PMSF). A total particulate fraction was isolated by centrifugation in a Ti-45 rotor (Beckman) for 60 min at 33,000 r.p.m. The pellet was then resuspended in buffer A by Dounce homogenization and the membranes collected by centrifugation. The resulting pellet was washed once in buffer B (10 mM HEPES/KOH, pH 7.8, 100 mM KCl, 2 mM $MgCl_2$, 1 mM DTT) and then resuspended in 100 ml buffer B. Triton X-100 was added slowly with mixing to a final concentration of 4% (v/v), and the suspension incubated on ice with frequent mixing. After 45 min, the suspension was clarified by centrifugation (Ti-45 rotor) for 60 min at 33,000 r.p.m. and the supernatant dialysed overnight against 100 vol 25 mM Tris–HCl, pH 7.8, 50 mM KCl, 1 mM DTT, 1% (v/v) Triton X-100. After dialysis, the material was clarified by centrifugation (Ti-45 rotor for 60 min at 33,000 r.p.m.), aliquoted, and stored at −80 °C. Protein concentration was measured using the BCA reaction (Pierce) with ovalbumin as standard. This bovine brain extract (2 mg protein) was preincubated in the presence of His_6-α-SNAP (12 μg protein), His_6-γ-SNAP (4 μg protein) and NSF–Myc (12 μg protein) in buffer C (25 mM HEPES/KOH, pH 7.0, 0.75% (w/v) Triton X-100, 75 mM KCl, 1 mM DTT, 2 mM

脱物中。这一实验流程是基于以前的一篇报道[18]形成的，但对一些关键的步骤进行了修改(如图注中所示)。由于 NSF 会继续与免疫微珠结合，因此特异性的洗脱液中会含有一部分 SNAP 和额外的代表 SNAP 受体的多肽，并且这两种蛋白的分子数理论上应该是符合一定化学计量比例的。

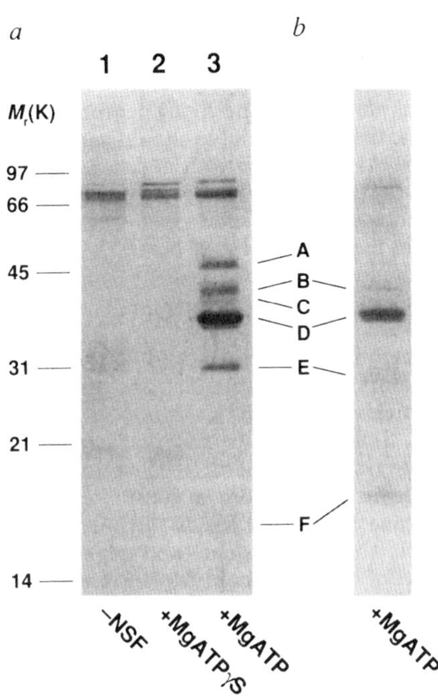

图 2. ATP 水解后，从 NSF 上解离下来的蛋白质的鉴定。*a*，使用考马斯蓝染色的聚丙烯酰胺凝胶。1 道为没有加 NSF 的对照，也使用 Mg–ATP 洗脱；2 道为使用 Mg–ATP-γS 洗脱的、与 NSF–Myc 非特异性结合的样品；3 道为使用 Mg–ATP(含 EDTA) 洗脱的、与 NSF–Myc 特异性结合的样品(图 1)。*b*，Mg–ATP 特异性洗脱下来的样品，使用银染显色的莱氏凝胶电泳。

方法。所有的操作都在 0~4 ℃进行。剥去牛脑组织的脑膜，然后将除灰质以外的部分都丢弃。将约 30 g 的灰质在杜恩斯匀浆器中使用缓冲液 A(20 mM Tris–HCl，pH 8.0，1 M KCl，250 mM 蔗糖，2 mM $MgCl_2$，1 mM DTT，1 mM 苯甲基磺酰氟(PMSF)) 匀浆 30 次。使用 Ti-45 转子(贝克曼公司)33,000 rpm，离心 60 分钟。用溶液 A 重悬沉淀并用杜恩斯匀浆器再次匀浆，离心后收集膜组分。离心后的沉淀使用缓冲液 B(10 mM HEPES/KOH，pH 7.8，100 mM KCl，2 mM $MgCl_2$，1 mM DTT) 冲洗一次，然后在 100 ml 缓冲液 B 中重悬。缓慢加入 Triton X-100 至终浓度为 4%(体积比)。将悬浮液在冰上孵育并不时混匀。45 分钟后，使用 Ti-45 转子 33,000 rpm，离心 60 分钟，将上清液使用 100 倍体积的 25 mM Tris–HCl，pH 7.8，50 mM KCl，1 mM DTT，1%(体积比)Triton X-100 混合溶液透析过夜。第二天，使用 Ti-45 转子 33,000 rpm，离心 60 分钟，将上清液分装并储存于 −80 ℃冰箱中。使用 BCA 法(皮尔斯公司)测定溶液中的蛋白浓度，以卵清蛋白为标准。在 4 ℃条件下，将牛脑组织抽提物(2 mg 蛋白质)与 His_6-α-SNAP(12 μg 蛋白质)、His_6-γ-SNAP(4 μg 蛋白质)和 NSF–Myc(12 μg 蛋白质)在缓冲液 C(25 mM HEPES/KOH，pH 7.0，0.75%(质量体积比)Triton X-100，75 mM KCl，1 mM DTT，2 mM EDTA，0.5 mM ATP(1 道)或 0.5 mM ATP-γS(2 道和 3 道)，1%(质量体积比)聚乙二醇(PEG)4000，0.4 mM PMSF) 中

EDTA, 0.5 mM ATP (for lane 1) or 0.5 mM ATP-γS (for lanes 2 and 3), 1% (w/v) polyethylene-glycol (PEG) 4000, 0.4 mM PMSF) for 30 min at 4 °C in a final volume of 2 ml. Mouse anti-Myc monoclonal antibody (200 µg protein, termed 9E10; ref. 22) covalently coupled to protein G–Sepharose 4 fast-flow (Pharmacia) was added and the incubation continued for 2 h with constant agitation. (The anti-Myc IgG was covalently coupled to the protein G–Sepharose using 20 mM dimethyl suberimidate as described[47].) The beads were packed into a column, washed with 10 vol buffer D (20 mM HEPES/KOH, pH 7.0, 0.5% (w/v) Triton X-100, 100 mM KCl, 1 mM DTT, 2 mM EDTA, 0.5 mM ATP (lane 1) or 0.5 mM ATP-γS (lanes 2 and 3), 0.4 mM PMSF) and the proteins eluted with 6 column-volumes of buffer D containing 8 mM $MgCl_2$ (resulting in a concentration of free Mg^{2+} of 4 mM). The column containing ATP-γS (lanes 2 and 3) was washed with an additional 2 column-volumes of buffer D containing 4 mM EDTA to complex the free Mg^{2+}, 3 column-volumes of buffer D containing ATP to exchange for bound ATP-γS, and then eluted with 6 column-volumes of buffer D containing 8 mM $MgCl_2$ to allow ATP hydrolysis. The appropriate fractions containing the eluate were pooled, precipitated with trichloroacetic acid, boiled in sample buffer[48] and analysed by electrophoresis on Tris–urea/SDS–polyacrylamide (18% acrylamide, 6 M urea, 750 mM Tris–HCl, pH 8.85, 50 mM NaCl, 0.1% SDS[49]) and stained with Coomassie blue R-250. For sequencing, the reaction was scaled up 25-fold.

A detergent extract of a crude, salt-washed total particulate fraction from the grey matter of bovine brain was used as the source of potential SNAP receptors. Membranes were washed with 1 M KCl before use to remove most of their endogenous SNAP supply[12]. As recombinant SNAPs were added in great excess for the binding reaction (Fig. 1), these should compete out the remaining endogenous SNAPs, preventing them from forming particles on the beads. Figure 2*a* shows a Coomassie blue-stained SDS–urea–high Tris–polyacrylamide gel of the specific eluate (lane 3), the nonspecific eluate fraction (lane 2), and the specific eluate from a control experiment omitting NSF–Myc (lane 1). Several bands (labelled A–F) appear only in the specific (Mg–ATP) eluate and depend on the presence of NSF. A set of bands at about M_r 70K are present in all three lanes, but apart from these, bands A–F are substantially pure in the specific eluate fraction. Band F runs as a sharper band in a standard Laemmli gel (Fig. 2*b*; specific eluate, stained with silver) than in the high Tris–urea gel (Fig. 2*a*). Bands A and C virtually co-electrophorese with the abundant band D in a Laemmli gel (Fig. 2*b*; see below).

Identification of SNAP Receptors

To identify specific bands, they were excised from blots and digested with trypsin, giving peptides that were then separated by high-pressure liquid chromatography and microsequenced (see Fig. 3 for details). Bands B and D co-electrophoresed with recombinant γ-SNAP and α-SNAP, respectively, both of which were His_6-tagged[15] and thus slightly larger than their endogenous counterparts (not shown). These identifications were confirmed by microsequencing peptides from bands B and D (not shown).

预孵育 30 分钟，最后总体积为 2 ml。加入共价偶联到蛋白 G–Sepharose 4 fast-flow（法玛西亚公司）上的小鼠抗 Myc 单克隆抗体（200 μg，称为 9E10；参考文献 22），继续孵育 2 小时，并持续振荡。（使用 20 mM 的辛二亚氨酸二甲酯作为交联剂可以将抗 Myc 抗体共价偶联到蛋白 G–Sepharose 上[47]。）然后将微珠装入层析柱中，使用 10 倍体积的缓冲液 D（20 mM HEPES/KOH，pH 7.0，0.5%（质量体积比）Triton X-100，100 mM KCl，1 mM DTT，2 mM EDTA，0.5 mM ATP（1 道）或 0.5 mM ATP-γS（2 道和 3 道），0.4 mM PMSF）冲洗，使用 6 倍柱体积的含有 8 mM $MgCl_2$（游离的 Mg^{2+} 浓度为 4 mM）的缓冲液 D 洗脱结合在柱子上的蛋白。使用额外的 2 倍柱体积的缓冲液 D（含有 4 mM EDTA）冲洗含有 ATP-γS 的柱子（2 和 3 道）以螯合游离的 Mg^{2+}，再使用 3 倍柱体积的缓冲液 D（含有 ATP）置换 ATP-γS，然后再用 6 倍柱体积的缓冲液 D（含有 8 mM $MgCl_2$）洗脱，使得 ATP 水解。收集洗脱下来的蛋白组分用三氯乙酸沉淀后，用样品缓冲液[48]煮沸，然后使用 Tris–尿素 / SDS–聚丙烯酰胺（18% 丙烯酰胺，6 M 尿素，750 mM Tris–HCl，pH 8.85，50 mM NaCl，0.1% SDS[49]）凝胶电泳分离，考马斯蓝 R-250 染色。如果所得样品用于测序，则将反应体系扩大 25 倍。

使用去垢剂抽提并经盐冲洗后的牛脑灰质粗提物作为潜在的 SNAP 受体来源。首先使用 1 M KCl 溶液冲洗膜组分以去除大部分内源性的 SNAP[12]。当过量的重组 SNAP 被加入样品中之后（图 1），它们应该可以竞争过残余的内源性 SNAP，阻止其在微珠上形成微粒。图 2*a* 显示了经 SDS–尿素–高 Tris–聚丙烯酰胺凝胶分离的特异洗脱样品（3 道）、非特异洗脱的样品（2 道）和对照组（未加入重组 NSF–Myc 蛋白）的特异洗脱样品（1 道）用考马斯蓝染色后的结果。只有在实验组的特异洗脱（使用 Mg–ATP）且存在 NSF 的样品中才可以看到多个条带（分别标记为 A~F），而 1 道和 2 道都没有。三个道中都有一条分子量约为 70K 的条带。除此以外，在特异洗脱部分中，条带 A~F 基本上是纯的。使用标准的莱氏凝胶（图 2*b*；特异性洗脱，银染）可以得到比高浓度 Tris–尿素凝胶（图 2*a*）边缘更清晰的条带 F，在莱氏凝胶中条带 A 和 C 会与大量的条带 D 共电泳（图 2*b*；见下文）。

SNAP 受体的鉴定

为了鉴定这些特异的条带，我们将这些条带从印迹中切下来并用胰蛋白酶消化得到多肽，然后使用高压液相层析法（HPLC）将其分离并进行微量测序（详细信息见图 3）。条带 B 和 D 分别与重组的 γ-SNAP 和 α-SNAP 共电泳。由于这两个重组的 SNAP 都带有一个 His_6 标签[15]，因此比细胞内的 SNAP 要稍大一些（数据未展示）。对从条带 B 和 D 中得到的多肽进行微量测序进一步证实了这一结果（数据未展示）。

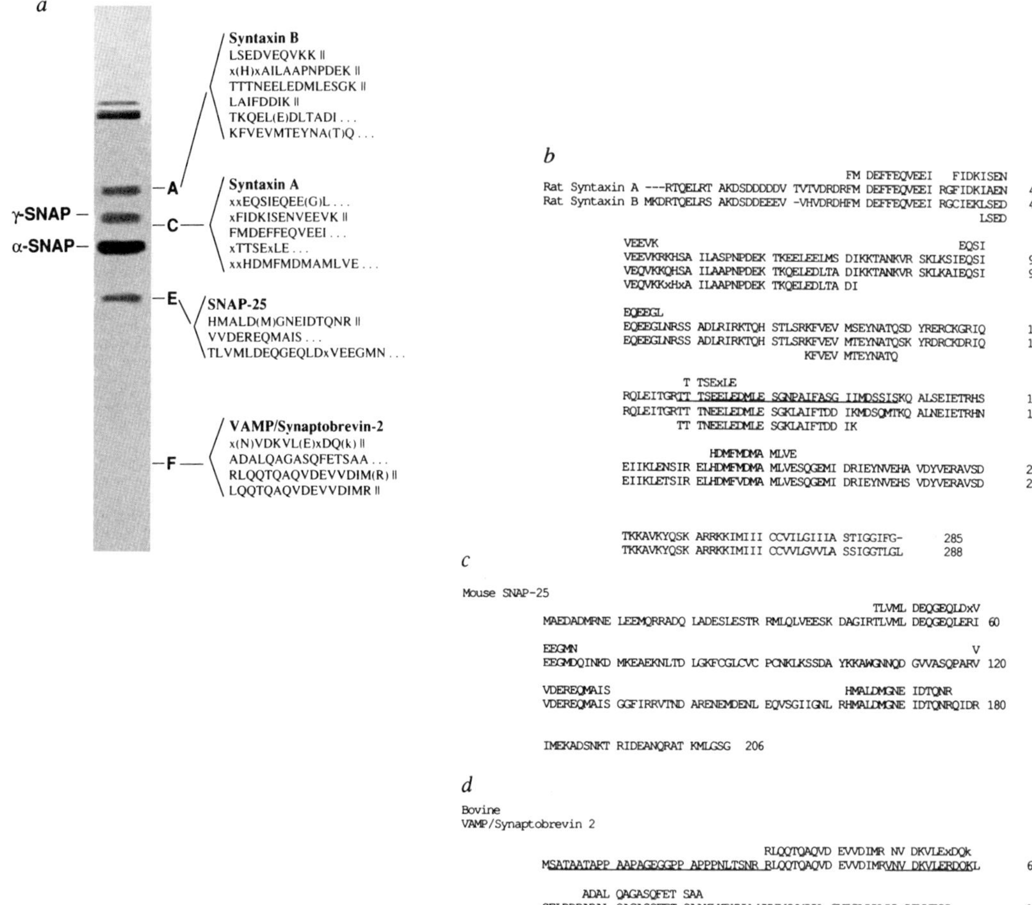

Fig. 3. Identification of SNAP receptors by amino-acid sequencing. Each of bands A–F was excised from an electroblot onto a nitrocellulose membrane after staining with Ponceau S, digested with trypsin, and the fragments separated by reverse-phase HPLC. *a*, Peptide sequences from each of the indicated bands are shown on the right (one-letter code). An "X" means that no residue could be identified at this position. A capital letter in parentheses means that the residue was identified with a lower degree of confidence; a lower-case letter in parentheses means that the residue was present but in very small amounts. The symbol ‖ indicates that the C terminus of the peptide was known. A dotted line means no more interpretable signals were obtained, but that this was unlikely to be the C terminus. The identify of bands B and D (Fig. 2) as γ-SNAP and α-SNAP, respectively, was confirmed by peptide sequencing. In the case of the syntaxin A peptide XTTSExLE... (from band C), only this limited sequence was obtained because of instrument failure. However, mass analysis of this peptide (on 1/25th of the sample) gave a value of $m/z = 3{,}274.8$; this is in good agreement with the value ($MH^+ = 3{,}276.67$) calculated for the predicted tryptic peptide (residues 156–186 of syntaxin A) which contains the limited sequence, so confirming the identify of the entire peptide. Mass analysis was also critical in the identification of band F as VAMP/synaptobrevin-2, confirming the presence of an *N*-acetylated peptide corresponding to residues 2–31 in the digest of band F (see text for details). *b*, Amino-acid sequences of syntaxins A (top) and B (below) from rat brain[23]. Shown above syntaxin A are the peptide sequences obtained from band C, and below syntaxin B are the peptide sequences obtained from band A. *c*, Complete amino-acid sequence of synaptosome-associated protein 25 (SNAP-25) from mouse brain[24]. Above are the peptide sequences obtained from band E. Underlined sequences in *d* and *b* are peptides identified by mass analysis.

Methods. Tryptic digestions and peptide separation. Ponceau S staining bands were excised, digested

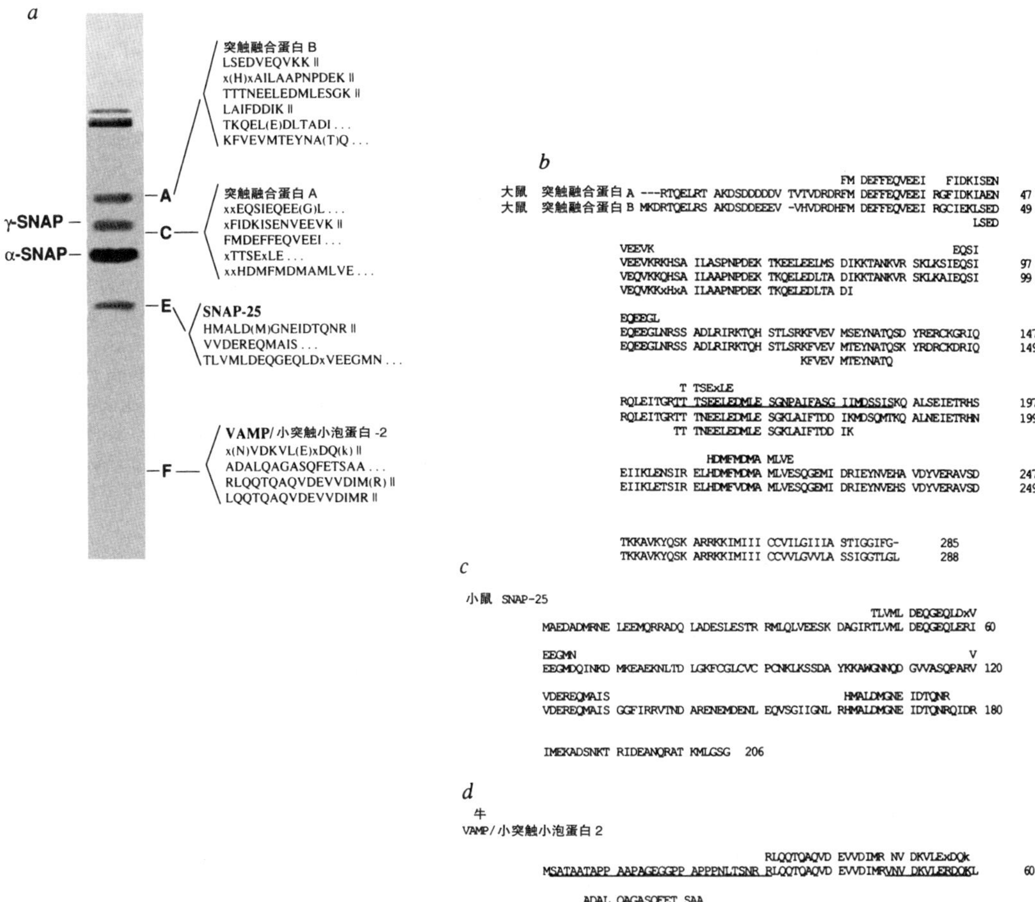

图 3. 通过氨基酸测序鉴定 SNAP 受体。蛋白转印到硝酸纤维素膜上后，用丽春红 S 染色，然后将 A~F 的各个条带切下来并使用胰蛋白酶消化，最后使用反相 HPLC 对消化得到的肽段进行分离。*a*，各个条带分析得到的多肽序列标记在各个带的右侧(使用氨基酸单字母代码)。其中“X”表示该位点的多肽残基未测出来；括号中的大写字母表示该位点鉴定的残基可信度较低；括号中的小写字母表示存在该残基，但量非常少；“‖”表示该多肽的 C 端已知；虚线表示虽然没有可以识别的信号，但这不太可能是 C 端。在图 2 中，已经鉴定出条带 B 和 D 分别是 γ-SNAP 和 α-SNAP，多肽测序的结果证实了这一点。由于仪器错误，在突触融合蛋白 A 的一个肽段中只测出了有限的序列 XTTSExLE…(从条带 C 中)。然而，这一多肽的质谱分析(1/25 的样品量)结果显示，其质荷比为 3,274.8，与理论预计的包含有限序列的胰蛋白酶消化肽段(突触融合蛋白 A 的残基 156~186)的质荷比(3,276.67)非常吻合，因此可以确定这一多肽就是突触融合蛋白 A 的片段。在鉴定条带 F 就是 VAMP / 小突触小泡蛋白 -2 的过程中，质谱分析也提供了有力的证据。其证明在 2~31 位的多肽中存在 *N*- 乙酰化修饰(详情请见正文)。*b*，大鼠脑组织中突触融合蛋白 A(上面)和 B(下面)的氨基酸序列 [23]。图中所示的突触融合蛋白 A 的序列来自条带 C，而突触融合蛋白 B 的序列来自条带 A。*c*，小鼠脑组织中突触小体相关蛋白 25 (SNAP-25)的氨基酸全序列 [24]。其多肽序列来自条带 E。在 *b* 和 *d* 中带下划线的序列是质谱中鉴定出的片段。

方法。胰蛋白酶酶解与多肽的分离。经丽春红 S 染色的条带剪切后使用胰蛋白酶原位消化(1 μg 测序级的胰蛋白酶，伯林格 – 曼海姆公司)，所产生的多肽使用 RP-HPLC(参考文献 50)进行分离。HPLC 系统如文献所述 [50]，采用的是 Separations Group 公司(希斯皮里亚市，美国加利福尼亚州)的

in situ with trypsin (1 μg sequencing grade trypsin; Boehringer–Mannheim), and the resulting fragments separated by RP-HPLC (ref. 50). Generally, HPLC systems were as described[50] and equipped with a 2.1 × 250 mm Vydac C4 (214TP54) column from the Separations Group (Hesperia, CA). Peptides resulting from digests of band F were fractionated on a 1 × 100 mm Inertsil 100GL-I-ODS-110/5 C18 column from SGE (Ringwood, Australia). The 1-mm column was operated in a system consisting of a model 140B syringe pump (ABI, Foster City, CA) and an ABI model 783 detector, fitted with a LC-Packings Kratos-compatible capillary flow cell which was directly connected to the column outlet (system assembly by C. Elicone); gradient slope was 1% B per min at a flow of 30 μl min^{-1}. Fractions were stored at −70 °C before sequence or mass analysis. Peptide sequencing: Peptides were sequenced with the aid of an ABI model 477A automated sequencer, optimized for femtomol phenylthiohydantoin analysis as described[51,52]. Fractions were always acidified (20% TFA final concentration) before application on the sequencer disc. Mass analysis: Aliquots (typically 1/50–1/25) of selected peak fractions were analysed by matrix-assisted laser desorption time-of-flight mass spectrometry using a Vestec (Houston, TX) LaserTec instrument with a 337 nm output nitrogen laser operated according to ref. 53. The matrix was α-cyano-4-hydroxy cinnamic acid and a 25 kV ion acceleration and 3 kV multiplier voltage were used. Laser power was adjusted manually as judged from optimal deflections of specific maxima, using a Tektronix TDA 520 digitizing oscilloscope.

Sequences from six major peptides containing a total of 70 residues were obtained from band A, each precisely matching the sequence of syntaxin B (p35B) from rat brain[23] (Fig. 3*b*). The sequences of these and peptides from the other specific bands are shown in association with the band from which they were obtained (Fig. 3*a*). No sequences or fragments attributable to any other protein were found, indicating that the bulk of material in band A is syntaxin B.

Five major peptides containing a total of 53 residues were sequenced from band C (Fig. 3*a*), and proved to differ at only one position from the published sequence of syntaxin A from rat brain[23]; the difference presumably reflects the species involved. Expected sequence differences between the 84% identical syntaxins A and B were found in the peptides from bands C and A (Fig. 3*b*). The syntaxins run more slowly than expected from their relative molecular masses (about 35K) in the high percentage Tris–SDS–urea–polyacrylamide gels were used, although they virtually co-electrophorese with α-SNAP (35K) in standard Laemmli gels (Fig. 2*b*, and data not shown).

Three major peptides (Fig. 3*a*) from band E gave sequences (Fig. 3*c*) matching a protein predicted from a mouse brain complementary DNA clone which by coincidence has been termed SNAP-25 (for synaptosome-associated protein of 25K; ref. 24). Of 45 residues obtained, 42 were identical to the published sequence of mouse SNAP-25. The discrepancies, two of which are conservative, are probably due to species differences. SNAP-25 migrates at M_r 31K in the SDS–urea gel.

All five peptides (containing 43 identified residues) sequenced from band F (Fig. 3*a*) exactly matched the sequence of VAMP/synaptobrevin-2 from bovine brain[25] (Fig. 3*d*). VAMP-2, the major isoform of VAMP/synaptobrevin in brain, differs slightly from VAMP-1 (refs 26, 27); all the sequences obtained from band F correspond to VAMP-2. All HPLC peptide peaks from band F, other than those attributable to trypsin autodigestion, were shown to be derived from VAMP/synaptobrevin. As a control, the region corresponding to band F

2.1 mm × 250 mm 的 Vydac C4（214TP54）柱子。条带 F 中消化得到的多肽则使用 SGE 公司（灵伍德，澳大利亚）的 1 mm × 100 mm 的 Inertsil 100GL-I-ODS-110/5 C18 柱进行分离。1 mm 的柱子是在含有 140B 型注射泵（ABI 公司，福斯特城，美国加利福尼亚州）和 ABI 783 型检测器的系统下进行操作，并通过一个 LC-Packings Kratos-compatible 毛细管流动吸收池与柱子流出管直接相连（该系统由 C. Elicone 组装）。梯度为每分钟 1% B，流速为 30 μl · min^{-1}。各组分在进行测序或质谱分析前在 −70 ℃中保存。多肽测序：使用 ABI 477A 型自动序列分析仪进行多肽序列分析。针对飞摩尔级乙内酰苯硫脲分析进行了优化，实验方法如文献所述[51,52]。在上样前，各组分通常都会先进行酸化处理（使用终浓度为 20% 的三氟乙酸）。质谱分析：将等份的（一般是 1/50~1/25）收集的峰组分使用"基质辅助激光解吸时间飞行质谱"进行分析，激光器是 Vestec 公司（休斯敦，美国得克萨斯州）的 LaserTec 仪器，氮气激光，波长为 337 nm，操作方法如文献 53 所述。使用的基质是 α-氰基-4-羟基桂皮酸，离子加速电压 25 kV，倍增器电压 3 kV。使用 Tektronix TDA 520 数字示波器，激光的功率根据特定峰值的最佳偏转来手动调节。

对从条带 A 中得到的 6 个主要多肽（总共含有 70 个氨基酸残基）进行测序，发现每个肽段都可以精确地与大鼠脑组织中的突触融合蛋白 B（p35B）的序列吻合[23]（图 3*b*）。条带 A 和其他特异条带的多肽序列分别显示在其对应的条带旁（图 3*a*）。从条带 A 中消化得到的序列或片段没有一个与其他蛋白吻合，说明条带 A 中的主要物质就是突触融合蛋白 B。

从条带 C 中得到的 5 个主要的多肽（总共含有 53 个氨基酸残基）同样经测序后发现（图 3*a*）：除了一个位点不同外，其与已公布的来源于大鼠脑组织的突触融合蛋白 A[23] 均相同。这一差别可能反映了物种的差异。突触融合蛋白 A 和 B 的序列相似性为 84%，这一序列差异也在条带 C 和 A 的测序结果中得到体现（图 3*b*）。尽管在标准的莱氏凝胶中突触融合蛋白几乎与分子量为 35K 的 α-SNAP 共电泳，但是在高比例 Tris-SDS-尿素-聚丙烯酰胺凝胶中它们的迁移速度远低于其相对分子质量（约 35K）对应的预期速度（图 2*b*，数据未显示）。

从条带 E 中得到的 3 个主要多肽（图 3*a*）的序列（图 3*c*）可以与小鼠脑组织中 cDNA 克隆编码的一个名为 SNAP-25（名字源于 25K 的突触小体相关蛋白；参考文献 24）的蛋白相匹配。45 个已获得的残基中，有 42 个与已公布的小鼠 SNAP-25 的序列吻合。另外 3 个序列中有 2 个是保守的，这种差异可能是由于物种不同引起的。在 SDS-尿素凝胶中，SNAP-25 迁移速度与分子量为 31K 的蛋白相同。

从条带 F 中得到的 5 个多肽（含有 43 个已鉴定的残基）的序列（图 3*a*）与牛脑组织中的 VAMP（突触小泡相关膜蛋白）/ 小突触小泡蛋白-2 [25] 完全吻合（图 3*d*）。VAMP-2 是 VAMP / 小突触小泡蛋白在大脑中的主要亚型，其与 VAMP-1 有细微的差别（参考文献 26，参考文献 27）；从条带 F 中得到的所有序列都与 VAMP-2 匹配。在 HPLC 分析中，除了胰蛋白酶自身消化的片段外，所有的肽段峰都来自 VAMP /

from the blot of the gel of the nonspecific (Mg–ATP-γS) eluate was subject to tryptic digestion and HPLC analysis in parallel with band F from the Mg–ATP eluate in two separate experiments. None of the VAMP/synaptobrevin-2 peptide peaks from band F were present in the HPLC profile of the control; only the autolytic tryptic peptides were detected.

Two peptides from the digest of band F failed to give any sequence, suggesting that they had blocked N termini. Mass spectroscopy indicated that these components had m/z values of 2,680.90 and 2,838.36 respectively, differing by 157.46, or approximately the molecular mass of Arg (156.19 average isotopic mass). The tryptic cleavage site closest to the N terminus in bovine VAMP/synaptobrevin-2 consists of an Arg-Arg sequence (at positions 30–31), so the expected partial cleavage would yield two N-terminal-derived peptides. If it is assumed that the initiator Met is removed, and that the newly generated N-terminal Ser is acetylated, adding 42.04 to the M_r, the two masses correspond almost perfectly with the predicted M_rs [MH^+] of the bovine VAMP/synaptobrevin-2 peptides spanning residues 2–30 and 2–31 (2,681.90 and 2,838.09, respectively). The N terminus of VAMP/synaptobrevin-2 thus appears to be processed and acetylated.

The material in the 70K region of all three eluates (specific, nonspecific and control) (Fig. 2) was microsequenced (from the specific eluate) in an attempt to determine if any synaptotagmin/p65 was present, as this protein can be immunoprecipitated together with syntaxins[23]. No evidence for its presence was found, although it cannot be excluded. Of three peptides sequenced, two were from Hsp70, an ATP-binding protein[28], and one was from a protein not found in the Genbank database.

Scanning and integrating the Coomassie blue stain in the gel shown in Fig. 2*a* gives the following ratios when staining is corrected for molecular weight, expressed as a fraction of α-SNAP; syntaxin B, 0.18; SNAP-25, 0.15; VAMP/synaptobrevin-2, 0.23. The syntaxin A band (band C) is weak compared with syntaxin B and could not readily be quantitated. The sum of all of the SNAP receptor species is 0.56 mol per mol α-SNAP, and so is approximately stoichiometric, given that proteins vary in staining by Coomassie blue, and the stoichiometry of SNAP and receptor in complexes does not need to be 1 : 1. The relative amounts of the different SNAP receptors in such a co-purified mixture may not reflect their relative abundance in the starting material, as their affinities for SNAPs may differ. The ratio of γ-SNAP to α-SNAP was ~0.2.

The role of syntaxins, SNAP-25, and synaptobrevin as SNAP receptors is supported by (1) their absence when NSF was omitted from the purification; (2) the need for ATP hydrolysis, rather than the mere presence of ATP, to elute them from the beads; (3) the approximately stoichiometric amounts purified relative to SNAP itself; and (4) the substantial purity of the proteins obtained. The fact that all of these proteins originate from the synapse further strengthens this conclusion.

As further confirmation, we attempted to follow the incorporation of α-SNAP receptor into

小突触小泡蛋白。作为对照，我们将非特异性洗脱（Mg–ATP-γS）样品的凝胶中与条带 F 位置相同的区域切下来后，和 Mg–ATP 洗脱中的条带 F 分别在两个独立实验中平行进行胰蛋白酶消化和 HPLC 分析。结果发现除了胰蛋白酶自溶肽段外，对照组的 HPLC 色谱图中不含任何条带 F 中的 VAMP / 小突触小泡蛋白 –2 的多肽峰。

条带 F 的消化产物中有两个多肽的序列不能够被测定，说明他们具有封闭的 N 端。质谱分析显示它们的质荷比分别为 2,680.90 和 2,838.36，相差 157.46，大约为一个精氨酸（平均同位素量 156.19）的分子量。胰蛋白酶在牛 VAMP / 小突触小泡蛋白 –2 上最靠近 N 端的切割位点是一个 Arg-Arg 序列（位于多肽残基的第 30~31 位）。因此，预期的不完全切割可能会产生两种 N 端来源的肽段。假使起始的甲硫氨酸被去除，而新的 N 端——丝氨酸发生乙酰化，那么该 N 端肽段的分子量将增加约 42.04。这样计算的话，牛 VAMP / 小突触小泡蛋白 –2 第 2~30 位和第 2~31 位肽段的分子量分别为 2,681.90 和 2,838.09，与上面质谱的结果相吻合。因此，VAMP / 小突触小泡蛋白 –2 的 N 端似乎发生了乙酰化。

对三种洗脱情况（特异、非特异及对照组）（图 2）中，分子量为 70K 的物质也进行了微量测序，以确定是否有突触结合蛋白 / p65 存在。这是由于该蛋白被证明可以与突触融合蛋白免疫共沉淀 [23]。虽然无法完全排除其存在的可能性，但没有证据表明 70K 条带含有该蛋白。对所有三个多肽的测序发现，其中两个来自 Hsp70（一种 ATP 结合蛋白 [28]），另一个在 Genbank 数据库中找不到与之匹配的蛋白。

当染色经分子量校正后，扫描并分析考马斯蓝染色的胶（图 2*a*）发现，各蛋白以占 α-SNAP 的比例表示如下：突触融合蛋白 B，0.18；SNAP-25，0.15；VAMP / 小突触小泡蛋白 –2，0.23。与突触融合蛋白 B 相比，突触融合蛋白 A（条带 C）的含量太低以至于无法定量。所有的 SNAP 受体与 α-SNAP 的摩尔比为 0.56∶1。由于不同蛋白经考马斯蓝染色深度不一样，且 SNAP 和受体结合时并不一定要 1∶1 结合，因此这是一个近似的化学计量比例。在共纯化的混合物中，不同 SNAP 受体之间的相对含量并不一定能够反映出它们在起始物质中的相对丰度，这是因为不同受体与 SNAP 的亲和力可能不一样。γ-SNAP 和 α-SNAP 的比率大约为 0.2。

以下证据支持突触融合蛋白、SNAP-25 和小突触小泡蛋白是 SNAP 受体：(1) 不加入 NSF 纯化时，不存在这些蛋白；(2) 只有当 ATP 存在且发生水解的情况下，它们才能够从微珠上洗脱下来；(3) 纯化下来的这些蛋白与 SNAP 具有近似的化学计量数；(4) 获得的这些蛋白纯度都很高。此外，这些蛋白都来自突触的事实也进一步巩固了这一结论。

为了进一步验证上述结论，我们制备了一个抗 SNAP-25 的一个 C 端多肽的抗

20S particles formed from crude extracts of bovine brain membranes using an antibody raised against a C-terminal peptide from SNAP-25. Detergent extract from salt-washed membranes was incubated with excess recombinant α- and γ-SNAPs with (Fig. 4*c*) or without (Fig. 4*b*) excess recombinant NSF, and the products were separated by velocity centrifugation in a linear glycerol gradient. In the absence of NSF, SNAP-25 sedimented at about 5S, but in its presence, almost all SNAP-25 cosedimented with NSF, α-SNAP, and γ-SNAP in the 20S particle (Fig. 4*d*). Assembly of NSF and SNAPs into the 20S particle required bovine brain membrane extract (Fig. 4*a*), as expected from earlier studies with Golgi membranes[18].

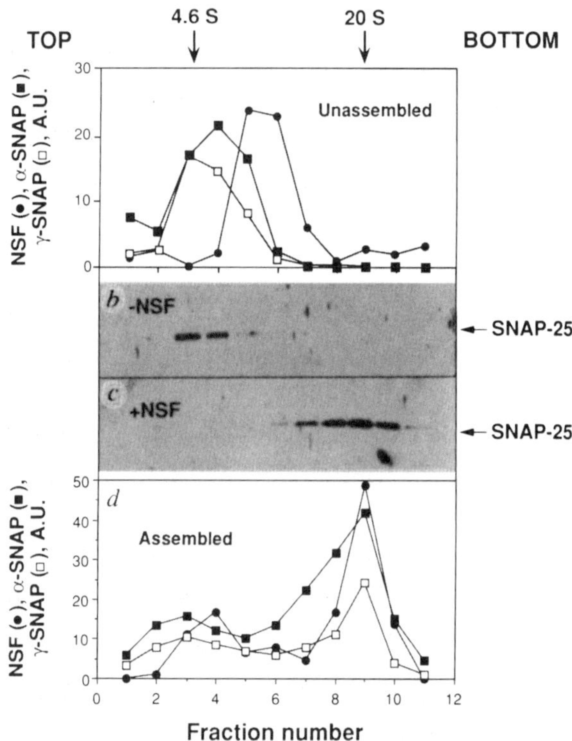

Fig. 4. Incorporation of SNAP-25 and other components into 20S fusion particles. *a*, *b*, Conditions in which 20S particles do not form (for example, controls); *c*, *d*, conditions in which 20S particles assemble. Components of the 20S fusion particle were incubated and then fractionated by glycerol gradient centrifugation, and the fractions analysed by SDS–PAGE to determine the location of α-SNAP and γ-SNAP, and to determine the locations of NSF and SNAP-25. Bovine serum albumin (4.6S) and α_2-macroglobulin (20S) were used as standards.

Methods. Reaction conditions were arranged so that the protein of interest would be maximally incorporated into 20S particles; thus, results in *a* and *d* are composites of several experiments. In the case in which the SNAPs were examined, *in vitro*-translated ^{35}S-labelled α-SNAP (■) and γ-SNAP (□) (ref. 17) (~2.1×10^6 c.p.m. for α- and 1.4×10^5 c.p.m. for γ-SNAP) were incubated on ice with bovine brain extract (1.2 mg protein), in the absence (*a*) or presence (*d*) of His_6-NSF (100 μg protein) in 20 mM HEPES–KOH, pH 7.4, 100 mM KCl, 2 mM DTT, 2 mM EDTA, 0.5 mM ATP, 0.5% (v/v) Triton X-100 in a final volume of 0.5 ml. After 15 min, reactions were loaded onto a 10–35% (w/v) glycerol gradient[18] and centrifuged in an SW41 rotor (Beckman) for 18 h at 40,000 r.p.m. Gradients were fractionated from the bottom at 1 ml per min, and an aliquot of each fraction analysed by SDS–PAGE and autofluorography. Recovery of both radiolabelled SNAPs

体，以此来追踪α-SNAP 受体整合到 20S 颗粒（来自牛脑组织细胞膜粗提物）的过程。我们将盐冲洗过的细胞膜抽提物在加（图 4*c*）或不加（图 4*b*）过量的重组 NSF 蛋白的情况下与过量的重组α-SNAP 和γ-SNAP 蛋白孵育，然后使用线性甘油梯度离心法分离产物。当不加 NSF 时，SNAP-25 的沉降系数为 5S；而加入 NSF 后，几乎所有的 SNAP-25 都与 NSF、α-SNAP 和γ-SNAP 一起沉淀于 20S 的颗粒中（图 4*d*）。此外，与早期使用高尔基体膜的实验得到的预期[18]一致的是，由 NSF 和 SNAP 组成的 20S 颗粒的形成需要牛脑组织细胞膜抽提物（图 4*a*）。

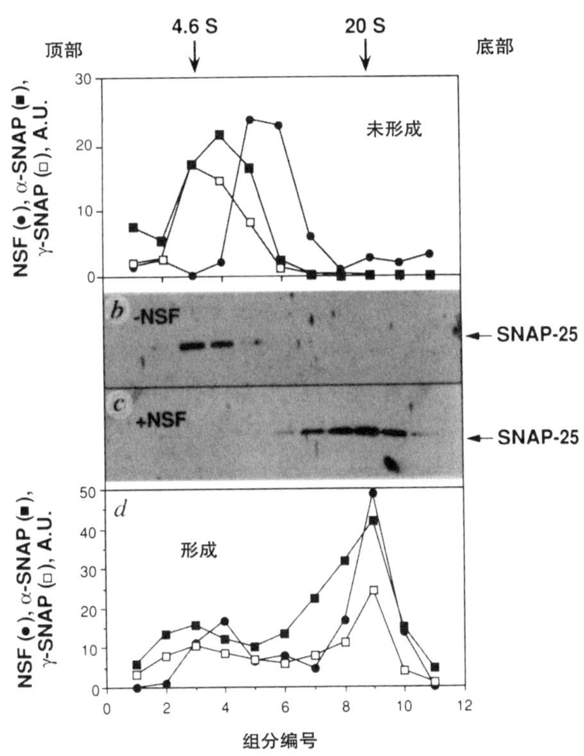

图 4. SNAP-25 与其他组分组装成 20S 融合颗粒的过程。*a*, *b*，没有形成 20S 颗粒的情况（例如，对照组）；*c*, *d*，形成 20S 颗粒的情况。20S 颗粒的组分在孵育后使用甘油梯度离心的方法进行分离，然后不同的组分再使用 SDS–PAGE 进行分析，来鉴别出哪些是α-SNAP 和γ-SNAP，哪些是 NSF 和 SNAP-25。以牛血清白蛋白（4.6S）和α_2–巨球蛋白（20S）为标准。

方法。我们对反应条件进行了优化，使得我们感兴趣的蛋白能够最大限度地组装成 20S 颗粒。因此，*a* 和 *d* 所示的结果是由多次实验产生的。在检验 SNAP 的实验中，我们使用了体外翻译的 ^{35}S 标记的α-SNAP（■）和γ-SNAP（□）（参考文献 17）（浓度分别约为 2.1×10^6 cpm 和 1.4×10^5 cpm），分别在 His_6-NSF（100 μg 蛋白）存在（*d*）和不存在（*a*）的情况下，在冰上与牛脑组织抽提物（1.2 mg 蛋白）共同孵育，溶液中含有 20 mM HEPES–KOH，pH 7.4，100 mM KCl，2 mM DTT，2 mM EDTA，0.5 mM ATP，0.5%（体积比）Triton X-100，总体积为 0.5 ml。15 分钟后，将反应产物加到 10%~35%（质量体积比）甘油梯度[18]中，使用 SW41 转子（贝克曼公司）40,000 rpm，离心 18 小时。离心结束后，将各梯度的组分从离心管底部以 1 ml · min^{-1} 的速度放出，分别收集，然后用 SDS–PAGE 和放射自显影法分析各组分。结果表明，放

exceeded 90%. Autofluorographs were scanned with a ScanJet Plus (Hewlett Packard) and the images integrated using Scan Analysis software (BioSoft, Cambridge). To examine NSF (●), His_6-NSF (ref. 15) (10 μg protein) was incubated with each of the His_6-SNAPs (20 μg each protein), either in the absence (*a*) or presence (*d*) of bovine brain extract (1.5 mg protein). Samples were incubated and fractionated as described. Aliquots were analysed by western blotting using an anti-His_6 antibody and blots were scanned as described[17]. To determine the extent to which SNAP-25 is incorporated into the 20S particle, bovine brain extract was added in limiting amounts (300 μg protein) and incubated either in the absence (*b*) or presence (*c*) of both His_6-SNAPs and His_6-NSF (30 μg each protein). Samples were processed and aliquots of each fraction analysed by western blotting using an antibody generated against the C-terminal peptide of SNAP-25. For this purpose, a peptide of the sequence used in ref. 24 was synthesized but with an additional cysteine at the N terminus, and affinity-purified antibodies generated. Blots were visualized using the ECL System (Amersham). AU, arbitrary unit.

Discussion

In this study, we have purified SNAP receptors from a crude brain membrane fraction on the basis of their ability to assemble 20S particles from SNAPs and NSF, and their release from NSF upon ATP hydrolysis (studies using Golgi membranes as source material will be described elsewhere). The simplest explanation for the fact that multiple polypeptides are obtained is that each is a distinct SNAP receptor. In particular, there is no evidence that any of these polypeptides associate, for example, as heterodimers[54]. All of them must thus have either the capacity to assemble a 20S particle, or a very high affinity for one that has been assembled. As our method is biased to isolating receptors of the highest abundance and affinity, we have probably isolated only some of the receptor types that are present.

By exploiting the same enzyme cycle used in the vesicle fusion process to offer two layers of biological specificity, we have purified SNAP receptors on the basis of their function, giving a largely pure preparation of four proteins from the crudest possible extract of brain, all of which originate from synapses. This striking selection for synaptic components probably reflects the degree to which the brain is specialized for synaptic vesicle fusion. The fact that each SNAP receptor purified corresponds to a previously cloned and sequenced gene no doubt reflects the exhaustive structural characterization of synaptic components by molecular biologists who have focused on the synapse because of its central importance in neuronal development and function[29,30,54]. Despite this effort, however, the roles of these proteins are unknown because of the lack of functional assays.

Of the SNAP receptors we have isolated, SNAP-25 is found in the presynaptic terminals of neurons[24] but has not been more precisely localized. Although its sequence suggests it is hydrophilic, it behaves as an integral membrane protein[24] and is palmitylated at one or more of the four cysteine residues between positions 85 and 92 (ref. 55). Fatty acyl-CoA is required for NSF-dependent fusion[56], and SNAP-25 or related proteins may serve as the relevant acyl acceptors. Multiple acylations of such proteins could create a hydrophobic surface as a trigger for fusion.

VAMP/synaptobrevin is inserted into synaptic vesicles by a single hydrophobic C-terminal

射性标记的 SNAP 的回收率超过 90%。放射自显影的图像使用 ScanJet Plus 扫描仪（惠普公司）扫描，并使用扫描分析软件（BioSoft 公司，剑桥）对图像进行整合。为了检测 NSF（●），我们分别在牛脑组织抽提物（1.5 mg 蛋白）存在（*d*）和不存在（*a*）的情况下，将 His_6-NSF（10 μg 蛋白）（参考文献 15）与 His_6-SNAP（每种蛋白 20 μg）一起孵育。样品采用相同的方法进行孵育和分离，然后使用抗 His_6 的抗体进行蛋白质印迹检测，并采用与上述相同的方法对那些杂交出来的点进行扫描和分析 [17]。为了确定 SNAP-25 整合在 20S 颗粒中的比例，我们向一定量（300 μg 蛋白）的牛脑组织抽提物中分别加入（*c*）或不加（*b*）His_6-SNAP 蛋白和 His_6-NSF 蛋白（每种蛋白均 30 μg）并进行孵育。然后再采用与上述相同的方法分离样品，并使用抗 SNAP-25 C 端多肽位点的抗体进行蛋白质印迹检测。为了达到这个目的，我们合成了文献 24 中所述的多肽序列并在 N 端加上一个额外的半胱氨酸，以此作为抗原，制备亲和纯化的抗体。蛋白质印迹杂交的结果使用电化学发光（ECL）系统（安马西亚公司）进行显色。AU，任意单位。

讨　论

本研究中，我们利用 SNAP 受体可以与 SNAP 和 NSF 结合形成 20S 颗粒并可以随着 ATP 的水解而从 NSF 上解离下来的性质（利用高尔基体膜作为材料源进行的研究发表在别的杂志上），从脑组织细胞膜组分粗提物中纯化出了 SNAP 受体。对于这些纯化出来的多肽，最简单的解释就是他们每一个都是 SNAP 受体。尤其值得注意的是，没有任何证据表明这些多肽之间会相互结合，例如形成异源二聚体等 [54]。因此，所有这些多肽要么具有组装成 20S 颗粒的能力，要么具有与那些已经形成的颗粒高亲和力结合的能力。由于我们的实验方法偏好于分离出具有最高丰度和亲和力的受体，因此我们可能只分离出了存在的受体中的一部分。

通过探索在囊泡融合过程中提供两层生物特异性的同一种酶循环，我们利用 SNAP 受体的功能将其纯化出来，最终从脑组织的粗提物中分离出了四种很纯的蛋白质。这些蛋白质都来自突触，这一惊人发现很可能反映出大脑为突触囊泡融合而特化的程度。此次纯化出的各个 SNAP 受体分别与之前克隆或基因序列分析的结果一致的事实也毫无疑问地体现了分子生物学家对突触结构鉴定的巨大努力。这些分子生物学家之所以会关注突触，是因为突触在神经元发育和功能中具有核心作用 [29,30,54]。虽然如此，由于缺乏相应的功能性实验，目前对于这些蛋白的功能仍缺乏了解。

在分离得到的 SNAP 受体中，我们发现 SNAP-25 分布于神经元的突触前末端 [24]，但更精确的定位尚不清楚。尽管其序列表明它是一个亲水性的蛋白，但是其种种表现却更像是一个整合膜蛋白 [24]。另外，在 SNAP-25 的 85 到 92 位的四个半胱氨酸残基中，有一个或多个发生了棕榈酰基化修饰（参考文献 55）。由于依赖 NSF 的融合过程需要脂酰辅酶 A 的参与 [56]，SNAP-25 或其他相关蛋白可能扮演着相关酰基受体的角色。这一类蛋白质的多重酰化可以形成一个疏水的界面从而引发膜融合过程。

VAMP / 小突触小泡蛋白通过其 C 端一个疏水性的片段插入到突触囊泡上，而

segment, and the remainder of the protein is cytoplasmic. Both tetanus and botulinum B toxins, potent inhibitors of neurotransmitter release, are proteases specific for VAMP/synaptobrevin-2 (ref. 31), suggesting that this SNAP receptor is essential for synaptic vesicle fusion *in vivo*. Syntaxin[23] has the same topography as VAMP/synaptobrevin, but is concentrated in "active zones" of the presynaptic membrane[23] at which a subpopulation of synaptic vesicles is docked[32] to allow fusion within 200 μs of the calcium influx induced by an action potential[33]. Components of the fusion machinery may have to be largely preassembled at these sites to allow such a rapid response.

In this light, the fact that 20S particles can contain VAMP/synaptobrevin (from the synaptic vesicle) and syntaxin (from the plasma membrane) is of interest, because these complexes might form part of the fusion apparatus docking the vesicle to its target in the active zone. It is not yet possible to say whether syntaxin and VAMP/synaptobrevin exist together in a single 20S particle (Fig. 5*a*), or whether separately assembled particles join together to form an attachment site (Fig. 5*b*), perhaps with the aid of additional proteins.

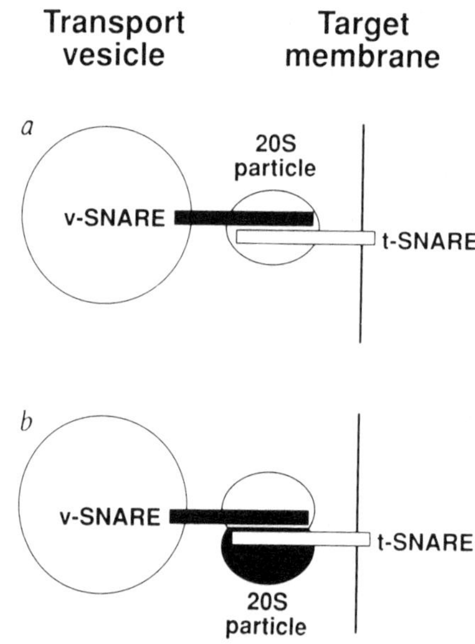

Fig. 5. Models to explain vesicle targeting based on the finding that SNAREs isolated in 20S fusion particles can originate from either the transport vesicle (v-SNAREs) or from the target membrane (t-SNAREs). A 20S particle (containing NSF and SNAPs) that simultaneously binds a v-SNARE and a t-SNARE (*a*) would attach a vesicle to its target. Alternatively (*b*), 20S particles, each capable of binding only one SNARE at a time, could interact to attach vesicle to target, a process that perhaps requires other proteins to assemble together.

The fact that molecules previously implicated in regulated exocytosis at the synapse are SNAP receptors implies that the same NSF- and SNAP-dependent machinery necessary for many constitutive fusion events also underlies triggered release of neutotransmitters at synapses. By extension, we would expect that NSF and SNAP are similarly employed in other forms of regulated exocytosis[34].

其余的部分则仍位于细胞质中。破伤风毒素和肉毒杆菌 B 毒素(有效的神经递质释放抑制剂)都是 VAMP / 小突触小泡蛋白 –2 的特异性蛋白酶(参考文献 31)。这表明在体内 SNAP 受体对于突触囊泡的融合来说至关重要。突触融合蛋白[23]具有与 VAMP / 小突触小泡蛋白相同的拓扑结构，但是却集中分布于突触前膜的“活性区”[23]——在这些区域有一个亚群的突触囊泡停靠[32]，当有动作电位引发钙离子内流时，在 200 μs 内便可以发生囊泡的融合[33]。融合装置的各个组分需要预先在这些位点大量组装好，从而可以在很短的时间内迅速作出反应。

这样看来，20S 颗粒中含有 VAMP / 小突触小泡蛋白(来自突触囊泡)和突触融合蛋白(来自细胞膜)就显得很有意思了，这意味着这些复合物也许可以形成将囊泡停靠在特定活性区域所需要的融合装置的一部分。不过目前仍无法肯定在单个 20S 颗粒中是否同时含有突触融合蛋白和 VAMP / 小突触小泡蛋白(图 5*a*)，也不确定是否由各个独立的颗粒来联合组装成一个附着位点(图 5*b*)。也许另外还有别的蛋白对这一过程有辅助作用。

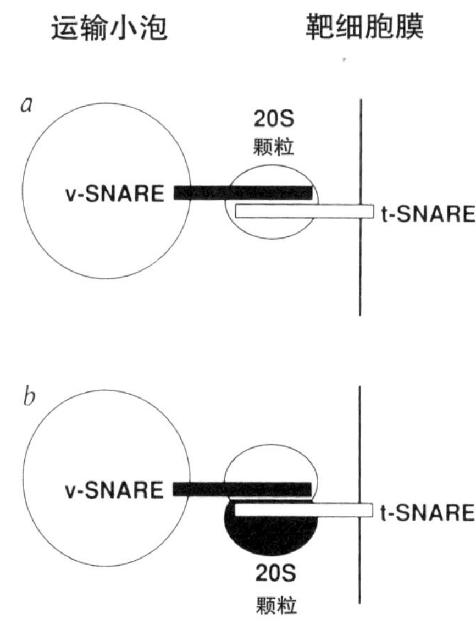

图 5. 囊泡靶向机理模式图。20S 融合颗粒中的 SNARE 有的来自运输小泡(v-SNARE)，有的来自靶细胞膜(t-SNARE)。当一个 20S 颗粒(含有 NSF 和 SNAP)同时与 v-SNARE 和 t-SNARE(*a*)结合时，便会将囊泡与其靶位点结合在一起。或者如(*b*)所示，如果每个 20S 颗粒一次只能结合一个 SNARE，则可以通过颗粒间的相互作用将囊泡与靶位点结合在一起，这一过程可能需要借助其他蛋白来完成组装。

突触的调控型胞吐过程有 SNAP 受体的参与，说明参与组成型融合的 NSF 及 SNAP 机器也同样参与了诸如突触神经递质释放这样的调控型融合过程。引申开来，我们认为 NSF 和 SNAP 在其他形式的调控型胞吐过程中也具有类似的功能[34]。

How can constitutive fusion machinery also be used in triggered exocytosis? An inhibitory component(s) must apparently prevent the 20S fusion particle from engaging, or from completely assembling in the first place. Such a fusion clamp could either cover an active site in NSF, SNAP or the SNAP receptors, or could prevent any of the additional cytoplasmic subunits of the fusion machinery[35] from binding. In the case of the synapse, a calcium trigger would remove the clamp. A strong candidate for such a clamp is the synaptic vesicle membrane protein synaptotagmin[36,37]. Synaptotagmin co-immunoprecipitates with syntaxin[23], and the recombinant proteins can bind each other[23]. It also undergoes a calcium-dependent conformational change[38]. Although synaptotagmin is not one of the major components of the 20S particles characterized here, this is not surprising, as the interaction between syntaxin and synaptotagmin is disrupted by 0.5 M salt[23], and we used a 1 M salt wash.

Rab3A, a Ras-related GTP-binding protein, is another candidate for a clamp, as this protein dissociates from synaptic vesicles upon stimulation of fusion[39]. Fusion clamps may respond to a variety of second messengers such as cyclic AMP, activated G proteins, protein phosphorylation, and so on, in addition to calcium, depending on cell type and physiological context.

SNAREs may Encode Specificity

Both syntaxin and VAMP/synaptobrevin have homologues in yeast[40,42]. The yeast protein SED5 is related to syntaxin[40], and its absence blocks ER-to-Golgi transport, leading to the accumulation of transport vesicles; it appears to reside in the Golgi[40]. Another homologue, PEP12, is found primarily on the vacuole membrane, and is required for Golgi-to-vacuole transport[41]. The yeast *SEC22*, *BET1* and *BOS1* gene products are related to VAMP/ synaptobrevin, and are required for fusion with the Golgi *in vivo.* BOS1 at least is a component of ER-derived transport vesicles[16,42-45]. Other recent examples are discussed in ref. 46. SNAP-25 is distantly related to SED5, PEP12, and the syntaxins.

The synaptic SNAP receptors we have identified are thus members of a family of proteins which appear to be distributed in a compartmentally specific fashion, with one set attached to the transport vesicles (v-SNAREs) and another set attached to target membranes (t-SNAREs). This suggests a model in which each transport vesicle contains one or more members of the v-SNARE superfamily obtained on budding from a corresponding donor compartment, and every target compartment in a cell contains one or more members of the t-SNARE superfamily. Specificity in membrane transactions would be assured by the unique and non-overlapping distribution of v-SNAREs and t-SNAREs among the different vesicles and target compartments. In the simplest view, that is, if there were no other source of specificity, only when complementary v-SNARE and t-SNARE pairs engage would a productive fusion event be initiated. Thus, a v-SNARE from the ER (possibly SEC22)

组成型融合装置在引发胞吐的过程中如何发挥作用呢？应该有某种(或多种)抑制性组分可以显著抑制 20S 融合颗粒的啮合或全部组装。这种融合“钳子”要么会掩盖 NSF、SNAP 或 SNAP 受体的活性位点，要么会阻止融合装置中其他胞质亚基[35]的结合。对于突触而言，钙触发剂可以移除这种“钳子”。突触囊泡膜蛋白——突触结合蛋白[36,37]是这种“钳子”强有力的候选蛋白。突触结合蛋白可以与突触融合蛋白免疫共沉淀[23]，并且这两种重组蛋白可以相互结合[23]。此外，突触结合蛋白还会受钙离子的调控发生构象变化[38]。虽然在 20S 颗粒中，突触结合蛋白并没有作为一个主要成分被鉴定出来，但我们对此并不感到奇怪，因为突触融合蛋白和突触结合蛋白的相互作用会被 0.5 M 的盐破坏[23]，而我们在实验中却使用了 1 M 的盐进行冲洗。

Rab3A 是一个 Ras 相关的 GTP 结合蛋白，它是“钳子”的另一个候选蛋白。当融合过程被激活时，这个蛋白会从突触囊泡上解离下来[39]。除了钙离子之外，融合“钳子”还可能会随着细胞种类和生理环境的不同而对多种第二信使发生应答，比如环腺苷酸(cAMP)、激活的 G 蛋白、蛋白质磷酸化等等。

SNARE 也许是特异编码的

突触融合蛋白和 VAMP / 小突触小泡蛋白在酵母中都有同源蛋白[40,42]。酵母中的 SED5 蛋白与突触融合蛋白同源[40]，其缺失会抑制从 ER 到高尔基体的转运，导致运输小泡的积累。SED5 似乎分布于高尔基体中[40]。另一个同源分子——PEP12 则主要分布于液泡膜上，是从高尔基体到空泡转运所必需的[41]。酵母 *SEC22*，*BET1* 和 *BOS1* 基因的产物是 VAMP / 小突触小泡蛋白的同源分子，对于体内高尔基体的融合来说是必需的。BOS1 至少是来自 ER 运输小泡的一种成分[16,42-45]。其他最近的例子在参考文献 46 中进行了讨论。SNAP-25 与 SED5、PEP12 和突触融合蛋白的关系比较远。

因此，我们已经鉴定出来的突触 SNAP 受体属于一个蛋白家族，他们的分布具有区室特异性，其中一部分与运输小泡(v-SNARE)结合，另一部分与靶细胞膜(t-SNARE)结合。这提示了一种囊泡靶向机制——每个运输小泡中含有一种或多种 v-SNARE 超家族的成员，这些 v-SNARE 来自囊泡出芽的相应供体区；细胞中每个靶向区域则含有一个或多个 t-SNARE 超家族的成员。通过 v-SNARE 和 t-SNARE 在不同囊泡和靶位点上独特的、无重叠的分布，可以实现特异的膜融合。简而言之，如果没有其他决定特异性的因素的话，那么只有当 v-SNARE 和 t-SNARE 相匹配的时候，囊泡的融合过程才能够被启动。因此，来自 ER 的 v-SNARE(可能是 SEC22)可以与来自高尔基体的 t-SNARE(可能是 SED5)结合，而不与来自溶酶体(可能是

could engage a t-SNARE from the Golgi (possibly SED5), but not one from the lysosome (possibly PEP12). For the proposed mechanism to work (again, assuming no other source of specificity), non-complementary SNAREs would have to be excluded from the same 20S particle (Fig. 5*a*) or multi-particle assemblies (Fig. 5*b*), although any one SNARE might be able to form a 20S particle by itself while waiting for a partner to arrive. If this is correct, the 20S fusion particle assembly reaction might be used as a cell-free "read-out" system to test whether candidate proteins are SNAREs (as shown here for SNAP-25), and to establish which SNAREs, if any, form specific cognate pairs. Localization of the SNARE proteins *in situ* would then establish which are v-SNAREs and which are t-SNAREs and which compartments are involved in each pairing.

A related mechanistic question is whether the SNAPs (and NSF) are associated primarily with a v-SNARE or with a t-SNARE, or whether an essentially symmetrical structure is formed at the junction of vesicle and target. NSF and SNAPs can associate with Golgi membranes to form 20S particles in the absence of transport vesicles[18], implying that 20S particles can be formed with only one SNARE partner. As native NSF is a tetramer of subunits with two ATPase domains each, it is easy to imagine several ways in which a 20S particle containing one NSF could form a symmetrical vesicle–target junction by simultaneously binding a v-SNARE and a t-SNARE (Fig. 5*a*).

Although many important details need to be established, our findings imply a general role for NSF and SNAP in regulated and constitutive intracellular membrane fusion processes, and in synaptic transmission in particular. The SNAREs we have identified appear to be members of compartment-specific membrane protein multigene families which may form attachment sites between specific vesicles and their correct target membranes together with NSF and SNAPs.

(**362**, 318-324; 1993)

Thomas Söllner, Sidney W. Whiteheart, Michael Brunner, Hediye Erdjument-Bromage, Scott Geromanos, Paul Tempst & James E. Rothman
Rockefeller Research Laboratory, Memorial Sloan-Kettering Cancer Center, 1275 York Avenue, New York, New York 10021, USA

Received 25 February; accepted 9 March 1993.

References:

1. Palade, G. E. *Science* **189**, 347-358 (1975).
2. Glick, B. S. & Rothman, J. E. *Nature* **326**, 309-312 (1987).
3. Block, M. R., Click, B. S., Wilcox, C. A., Wieland, F. T. & Rothman, J. E. *Proc. Natl. Acad Sci. U.S.A.* **85**, 7852-7856 (1988).
4. Fries, E. & Rothman, J. E. *Proc. Natl. Acad. Sci. U.S.A.* 77, 3870-3874 (1980).
5. Balch, W. E., Dunphy, D. W., Braell, W. A. & Rothman, J. E. *Cell* **39**, 405-416 (1984).
6. Malhotra, V., Orci, L., Glick, B. S., Block, M. R. & Rothman, J. E. *Cell* **54**, 221-227 (1988).
7. Orci, L., Malhotra, V., Amherdt, M., Serafini, T. & Rothman, J. E. *Cell* **56**, 357-368 (1989).
8. Wilson, D. W. *et al. Nature* **399**, 355-359 (1989).

PEP12)的某种蛋白结合。为使所假设的机制成立(再一次假设没有其他决定特异性的因素存在)，尽管任何一个 SNARE 也许都能够形成 20S 颗粒，但非互补性的 SNARE 将不会存在于同一个 20S 颗粒(图 5*a*)或多颗粒的复合体(图 5*b*)中。如果这个推论正确的话，那么 20S 融合颗粒的组合实验也许可以被作为一个无细胞体系，用于检验候选蛋白是否为 SNARE(例如这里说的 SNAP-25)，还可以用于研究哪些 SNARE 可以形成特异的配对。SNARE 蛋白的原位定位实验则可以用于鉴别哪些是 v-SNARE，哪些是 t-SNARE，以及哪些细胞内的区室参与配对。

对于这个机制来说，仍存在一个问题，即是否 SNAP(以及 NSF)主要与某种 v-SNARE 或 t-SNARE 结合，或者在囊泡和靶位点结合的时候是否会形成一种基本对称的结构。在没有运输小泡存在的情况下，NSF 和 SNAP 也可以与高尔基体膜结合形成 20S 颗粒[18]，这表明即使只有一个 SNARE 配体也可以形成 20S 颗粒。由于天然 NSF 由 4 个亚基组成，每个亚基有两个 ATP 酶结构域，因此很容易想象含有一个 NSF 的 20S 颗粒通过同时结合 v-SNARE 和 t-SNARE 形成一个对称的囊泡–靶位点汇合处的多种方法(图 5*a*)。

尽管许多重要的细节仍有待研究，但我们的发现揭示了 NSF 和 SNAP 在调控型和组成型细胞内膜融合过程中，尤其是在突触信号的传递中的普遍性作用。我们所鉴定的多种 SNARE 蛋白似乎是某种区室特异性膜蛋白家族的成员，它们可以与 NSF 和 SNAP 一起介导囊泡与其靶位点形成特异的结合。

(张锦彬 翻译；石磊 审稿)

9. Novick, P., Ferro, S. & Schekman, R. *Cell* **25**, 461-469 (1981).
10. Graham, T. R. & Emr, S. D. *J. Cell Biol.* **114**, 207-218 (1991).
11. Weidman, P. J., Melancon, P., Block, M. R. & Rothman, J. E. *J. Cell Biol.* **108**, 1589-1596 (1989).
12. Clary, D. O. & Rothman, J. E. *J. Biol. Chem.* **265**, 10109-10117 (1990).
13. Clary, D. O., Griff, I. C. & Rothman, J. E. *Cell* **61**, 709-721 (1990).
14. Griff, I. C., Schekman, R., Rothman, J. E. & Kaiser, C. A. *J. Biol. Chem.* **267**, 12106-12115 (1992).
15. Whiteheart, S. W. *et al. Nature* **362**, 353-355 (1993).
16. Kaiser, C. A. & Schekman, R. *Cell* **61**, 723-733 (1990).
17. Whiteheart, S. W., Brunner. M., Wilson, D. W., Wiedmann, M. & Rothman, J. E. *J. Biol. Chem.* **267**, 12239-12243 (1992).
18. Wilson, O. W., Whiteheart, S. W., Wiedman, M., Brunner, M. & Rothman, J. E. *J. Cell Biol.* **117**, 531-538 (1992).
19. Tagaya, M., Wilson, D. W., Brunner, M., Arango, N. & Rothman, J. E. *J. Biol. Chem.* **269**, 2662-2666 (1993).
20. Munro, S. & Pelham, H. R. B. *Cell* **48**, 899-907 (1987).
21. Wilson, D. W. & Rothman, J. E. *Meth. Enzym.* **219**, 309-318 (1992).
22. Evan, G. I., Lewis, G. K., Ramsey, G. & Bishop, M. J. *Molec. Cell. Biol.* **5**, 3610-3616 (1985).
23. Bennett, M. K., Calakos, N. & Schelller, R. H. *Science* **257**, 255-259 (1992).
24. Oyler, G. A. *et al. J. Cell Biol.* **109**, 3030-3052 (1989).
25. Sudhof, T. C. *et al. Neuron* **2**, 1475-1481 (1989).
26. Elferink, L. A., Trimble, W. S. & Scheller, R. H. *J. Biol. Chem.* **264**, 11061-11064 (1989).
27. Archer, B. T., Ozcelik, T., Jahn, R., Franke, U. & Sudhof, T. C. *J. Biol. Chem.* **265**, 17267-17273 (1990).
28. Chappell, T. G. *Cell* **45**, 3-13 (1986).
29. Greengard, P., Valturta, F., Czernik, A. J. & Benfenati, F. *Science* **259**, 780-785 (1993).
30. Trimble, W. S., Linial, M. & Sheller, R. H. *A. Rev. Neurosci.* **14**, 93-122 (1991).
31. Schiavo, G. *et al. Nature* **359**, 832-835 (1992).
32. Landis, D. M. D., Hall, A. K., Weinstein, L. A. & Reese, T. S. *Neuron* **1**, 201-209 (1988).
33. Llinas, R., Steinberg, I. Z. & Walton, K. *Biophys. J.* **33**, 323-352 (1981).
34. Kelley, R. B., *Science* **230**, 25-32 (1985).
35. Waters, R. G., Clary, D. O. & Rothman, J. E. *J. Cell Biol.* **118**, 1015-1026 (1992).
36. Perin, M. S., Fried, V. A., Mignery, G. A., Jahn. R. & Sudhof, T. C. *Nature* **34**, 260-263 (1990).
37. Wendland, B., Miller. K. G., Schilling, J. & Scheller, R. H. *Neuron* **6**, 993-1007 (1991).
38. Brose, N., Petrenko, A. G., Sudhof, T. C. & Jahn, R. *Science* **256**, 1021-1025 (1992).
39. Fischer von Mollard, G., Sudhof, T. C. & Jahn, R. *Nature* **349**, 79-81 (1991).
40. Hardwick, K. G. & Pelham, H. R. B. *J. Cell Biol.* **119**, 513-521 (1992).
41. Preston, R. A. *et al. Molec. Cell. Biol.* **11**, 5801-5812 (1991).
42. Dascher, C., Ossig, R., Gallwitz, D. & Schmitt, H. D. *Molec. Cell. Biol.* **11**, 872-885 (1991).
43. Newman, A. P., Groesch, M. E. & Ferro-Novick, S. *EMBO J.* **11**, 3609-3617 (1992).
44. Newman, A. P. *et al. Molec. Cell. Biol.* **12**, 3663-3664 (1992).
45. Shim, J., Newman, A. P. & Ferro-Novick, S. *J. Cell Biol.* **113**, 55-64 (1991).
46. Bennett, M. K. & Scheller, R. H. *Proc. Natl. Acad. Sci. U.S.A.* (in the press).
47. Harlow, E. & Lane, O. *Antibodies: A Laboratory Manual* (Cold Spring Harbor Laboratory Press, New York, 1988).
48. Laemmli, U. K, *Nature* **227**, 680-685 (1970).
49. Schlenstedt, G., Gudmundsson, G. H., Boman, H. G. & Zimmermann, R. *J. Biol. Chem.* **265**, 13960-13968 (1990).
50. Tempst, P., Link, A. J., Riviere, L. R., Fleming, M. & Elicone, C. *Electrophoresis* **11**, 537-553 (1990).
51. Tempst, P. & Riviere, L. *Analyt. Biochem.* **183**, 290-300 (1989).
52. Erdjument-Bromage, H., Geromanos, S., Chodera, A. & Tempst, P. *Techniques in Protein Chemistry IV* (Academic, San Diego, in the press).
53. Beavis, R. C. & Chait, B. T. *Rapid Commun. Mass Spect.* **3**, 432-435 (1989).
54. Sudhof, T. C. & Jahn, R. *Neuron* **6**, 665-677 (1991).
55. Hess, D. T., Slater, T. M., Wilson, M. C. & Skene, J. H. P. *J. Neurosci.* **12**, 4634-4641 (1992).
56. Pfanner, N., Glick, B. S., Arden, S. R. & Rothman, J. E. *J. Cell Biol.* **110**, 955-961 (1990).

Acknowledgements. We thank M. Hum, L. Lacomis and M. Lui for help with protein digestions, HPLC separations and peptide sequencing; C. Elicone for custom assembly of the microbore HPLC system; M. Wiedmann for discussion; and W. Patton for gel scans and analysis. This work was supported by the Mathers Charitable Foundation, and by an NIH grant (to J.E.R.), a Fellowship of the Deutsche Forschungsgemeinschaft (to T.S.), a Fellowship from the Jane Coffin Childs Memorial Fund for Medical Research (to S.W.W.), and by an EMBO postdoctoral fellowship (to M.B.). The Sloan-Kettering microchemistry core facility is supported, in part, by an NCI core grant.

B-cell Apoptosis Induced by Antigen Receptor Crosslinking is Blocked by a T-cell Signal Through CD40

T. Tsubata *et al.*

Editor's Note

The controlled elimination of self-reactive lymphocytes by programmed cell death (apoptosis) is an essential feature of the immune system that helps prevent autoimmunity. Here Takeshi Tsubata and colleagues use a mouse B cell line, which can readily be induced to apoptose, to study anti-apoptotic mechanisms. The team show that co-culture with T helper cells prevents B-cell apoptosis, and that the effect may be due to the interaction of the B cell CD40 protein binding to its ligand on the T helper cell. They conclude that the presence or absence of T-cell help through CD40 may be crucial in determining whether B cells are activated or killed upon interaction with antigens, with implications for understanding and treating autoimmune disease.

In mice transgenic for an autoantibody, self-reactive B cells have been shown to be eliminated upon interaction with membrane-bound self-antigens in the periphery[1,2] as well as in the bone marrow[3-5], suggesting that both immature and mature B cells are eliminated by multimerization of surface immunoglobulins (sIg). Activation of mature B cells by antigens may thus require a second signal that inhibits sIg-mediated apoptosis. Such a second signal is likely to be provided by T helper cells, because B-cell tolerance is more easily induced in the absence of T helper cells[6-9]. To assess the molecular nature of the signal that inhibits sIg-mediated apoptosis, we used anti-IgM-induced apoptotic death of WEHI-231 B lymphoma cells[10,11] as a model system. Here we report that the signal for abrogating sIg-mediated apoptosis is generated by association of the CD40L molecule on T cells with the CD40 molecule on WEHI-231 cells. T-cell help through CD40 may thus determine whether B cells are eliminated or activated upon interaction with antigens.

WEHI-231 cells undergo apoptotic cell death when the sIg is crosslinked by anti-IgM antibody[10,11]. To test whether T helper cells are able to block sIg-mediated apoptosis of B cells, we treated WEHI-231 cells with anti-IgM in the presence or absence of A3.4C6 helper T hybridoma cells[12]. After 2 days, cells were collected and stained with phycoerythrin-labelled anti-IgM and fluorescein-labelled anti-B220. A known number of $sIgM^-/sIgG^+$ B lymphoma K46 cells were added to the collected cells before staining so that we could calculate the number of WEHI-231 cells in the culture. Flow cytometry analysis separated and quantified WEHI-231 cells ($sIgM^+$, $B220^+$), K46 cells ($sIgM^-$, $B220^+$) and T cells ($sIgM^-$, $B220^-$) in the mixture (Fig. 1*a*). Treatment with anti-IgM alone markedly

T 细胞通过 CD40 抑制抗原受体交联引起的 B 细胞凋亡

锷田武志等

编者按

免疫系统的一个基本特征是能够通过程序性细胞死亡（凋亡）有计划地清除自身反应性淋巴细胞，从而防止自身免疫病发生。在这篇文章中锷田武志及其同事利用容易发生诱导性凋亡的小鼠细胞系研究了 B 细胞的抗凋亡机制。结果显示，B 细胞凋亡在有辅助 T 细胞存在时被显著抑制，而 B 细胞的 CD40 蛋白与其在辅助 T 细胞上配体的结合可能介导了这种抑制作用。该研究证明，B 细胞能否通过 CD40 获得 T 细胞的帮助可能决定了抗原结合后 B 细胞的命运（活化或死亡），为理解和治疗自身免疫病提供了启示。

在自身抗体转基因小鼠的外周血[1,2]和骨髓[3-5]中，自身反应性 B 细胞与膜表面结合的自身抗原结合后被清除，这说明非成熟或者成熟 B 细胞在表面免疫球蛋白（sIg）的多聚化后均会死亡。因此，成熟 B 细胞在被抗原活化时也许需要通过第二信使来抑制 sIg 介导的细胞凋亡。在缺失辅助 T 细胞时[6-9]，B 细胞更容易产生抗原耐受。因而，辅助 T 细胞极有可能提供了抑制 B 细胞凋亡的第二信使。我们利用抗 IgM 抗体诱导 B 淋巴瘤细胞系 WEHI-231 凋亡模型[10,11]来研究了第二信使抑制 sIg 介导凋亡的分子机理。研究发现，表达于 T 细胞的 CD40 配体与 WEHI-231 细胞上 CD40 结合后能够产生抑制信号，影响 sIg 介导的 B 细胞凋亡。因此，通过 CD40 获得 T 细胞的帮助决定了 B 细胞在抗原结合后是被活化还是被清除。

抗 IgM 抗体可以引起 sIg 的交联，并进一步诱发 WEHI-231 细胞的凋亡性死亡[10,11]。为了验证辅助 T 细胞是否可以阻断 sIg 介导的 B 细胞凋亡，我们分别在 A3.4C6 辅助 T 杂交瘤细胞[12]存在或不存在时检验了抗 IgM 抗体对 WEHI-231 细胞的作用。反应 2 天后，将这些细胞收集起来并使用藻红蛋白标记的抗 IgM 抗体和荧光素标记的抗 B220 抗体染色。并且在染色前，我们向收集的细胞中加入了一定数量的 $sIgM^-/sIgG^+$ K46 B 淋巴瘤细胞，以便计算 WEHI-231 细胞的数量。我们使用流式细胞术分析将细胞混合液中的 WEHI-231 细胞（$sIgM^+$，$B220^+$）、K46 细胞（$sIgM^-$，

reduced the number of WEHI-231 cells, whereas the growth of anti-IgM-treated WEHI-231 cells was restored in the presence of A3.4C6 cells. A similar result was obtained when a T helper clone DB14 (ref. 13) was used (data not shown). On the other hand, other T-helper hybridoma lines 2B4 (ref. 14) and N3-6-71 (ref. 15), as well as a cytotoxic T-cell clone CTLL-2, failed to rescue anti-Ig-treated WEHI-231 cells from apoptotic death (Fig. 1*b*). These results indicate that some T-helper lines are able to block sIg-mediated apoptotic death of B cells.

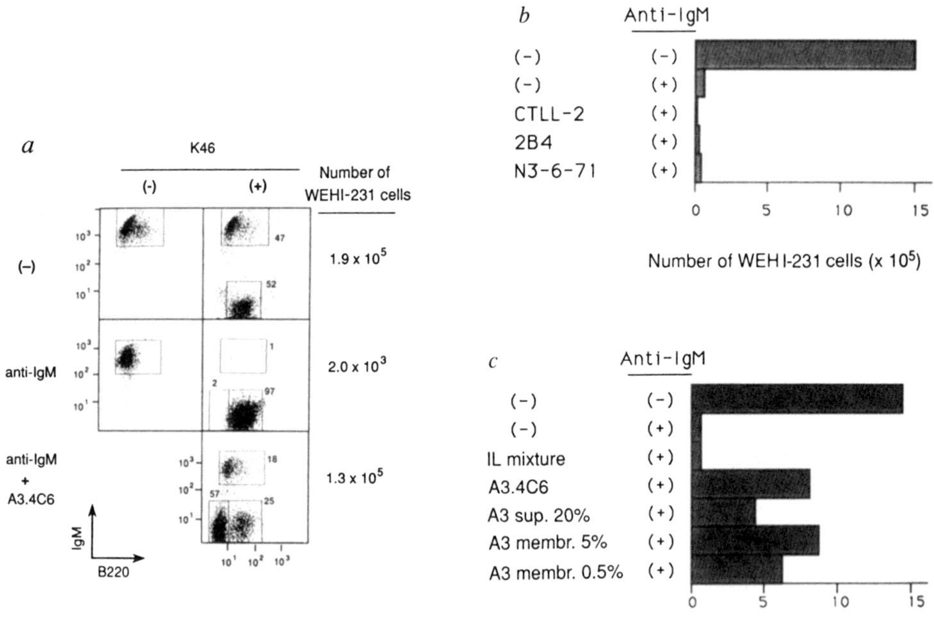

Fig. 1. Abrogation of sIg-mediated cell death of WEHI-231 cells by T helper cells. WEHI-231 cells were cultured with 10 μg ml^{-1} of goat anti-mouse IgM antibody (Cappel) in the presence or absence of various cells or reagents in 1 ml RPMI-1640 medium supplemented with 10% fetal calf serum, 50 μM 2-mercapto-ethanol and antibiotics for 2 days. Cells were subsequently collected and, after adding 2×10^5 K46 cells into each sample, stained by fluorescein-conjugated anti-B220 (RA3-6B2) and phycoerythrin-conjugated goat anti-mouse IgM antibody (Southern Biotechnology). Cells were analysed on a FACScan (Becton–Dickinson), and the fraction of WEHI-231 cells and K46 cells in the culture determined. The number of WEHI-231 cells was calculated as (the fraction of WEHI-231/fraction of K46) × (2×10^5). *a*, Two-colour flow cytometry analysis for the growth of WEHI-231 cells. WEHI-231 cells (2×10^4) were cultured with anti-IgM in the presence or absence of 1×10^5 A3.4C6 cells for 2 d. Percentages of WEHI-231 (IgM^+B220^+), K46 (IgM^-B220^+) and A3.4C6 (IgM^-B220^-), and calculated numbers of WEHI-231 cells recovered from each culture are indicated. As controls for staining, untreated or anti-Ig-treated WEHI-231 cells were stained in parallel without adding K46 cells. Note that the condition of flow cytometry for untreated WEHI-231 cells is slightly different from that for the others and that sIgM expression of anti-Ig treated WEHI-231 cells is significantly higher than that of K46 cells, although anti-Ig treatment reduced sIgM expression of WEHI-231 cells 20-fold. *b*, Growth of anti-IgM-stimulated WEHI-231 cells cultured with T-cell lines. WEHI-231 (1×10^5) were cultured with 10 μg ml^{-1} of anti-IgM in the presence of 1×10^5 cells of the T-cell lines CTLL-2, 2B4 or N3-6-71 for 2 days. *c*, Growth of anti-IgM-stimulated WEHI-231 cells cultured with reagents to replace helper T cells. A3.4C6 cells (1×10^5), the culture supernatant

B220⁺）和 T 细胞（sIgM⁻，B220⁻）进行了分离并计算了数量（图 1*a*）。单独使用抗 IgM 抗体处理会导致 WEHI-231 细胞数量显著减少，而在 A3.4C6 细胞存在时，抗 IgM 抗体处理过的 WEHI-231 细胞恢复生长。使用 DB14 辅助 T 细胞（参考文献 13）也可以获得类似的结果（结果未展示）。然而另一些辅助 T 杂交瘤细胞系——2B4（参考文献 14）、N3-6-71（参考文献 15）和杀伤性 T 细胞 CTLL-2 则不能阻断抗 IgM 抗体所引起的 WEHI-231 细胞凋亡（图 1*b*）。这些结果表明，某些辅助 T 细胞系可以阻断 sIg 介导的 B 细胞凋亡。

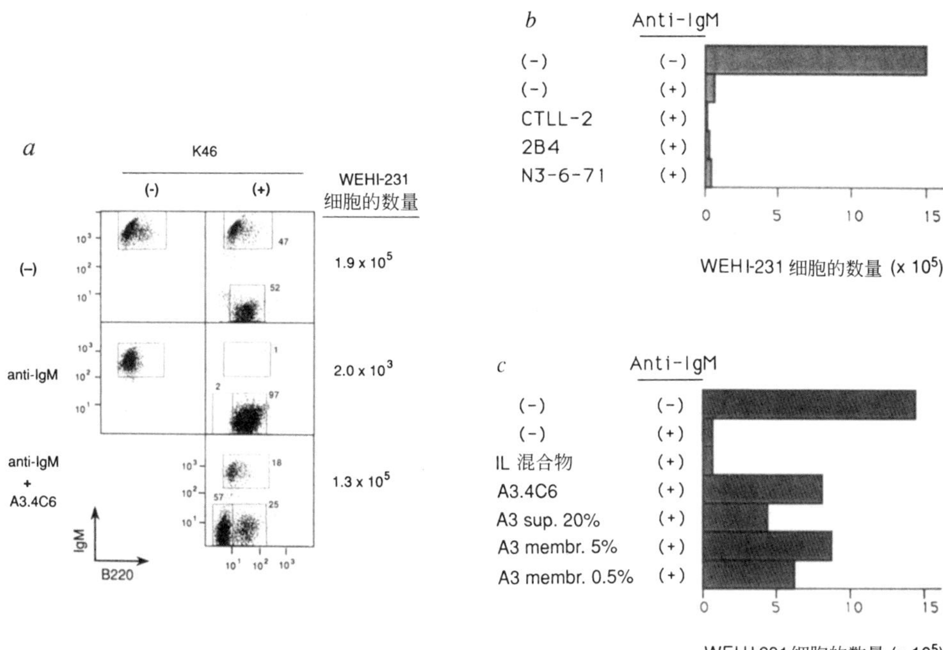

图 1. 辅助 T 细胞阻断 sIg 介导的 WEHI-231 细胞死亡。在有或没有各种细胞或因子条件下，WEHI-231 细胞在含 10% 胎牛血清、50 μM 2–巯基乙醇和抗生素的 1 ml RPMI-1640 培养基中与 10 μg · ml⁻¹ 的羊抗鼠 IgM 抗体（Cappel 公司）共同培养 2 天。随后收集细胞，在每个样品中加入 2×10^5 个 K46 细胞后，用荧光素标记的抗 B220 抗体（RA3-6B2）和藻红蛋白标记的羊抗鼠 IgM 抗体（Southern Biotechnology 公司）染色并使用流式细胞仪（美国贝克顿–迪金森公司）进行分析，确定 WEHI-231 与 K46 的比例。WEHI-231 细胞的数量等于 WEHI-231 与 K46 荧光强度之比乘以 2×10^5。*a*，WEHI-231 细胞的双染流式细胞术分析。在 1×10^5 个 A3.4C6 存在和不存在的情况下，将 2×10^4 个 WEHI-231 细胞与抗 IgM 抗体共培养 2 天。每个培养基中 WEHI-231 细胞（IgM⁺B220⁺）、K46 细胞（IgM⁻B220⁺）和 A3.4C6 细胞（IgM⁻B220⁻）的百分比以及 WEHI-231 细胞的数量都标注在图中。在没加 K46 细胞的情况下，未处理或抗 IgM 抗体共培育处理的 WEHI-231 细胞平行染色作为对照。需要注意的是未处理的 WEHI-231 细胞进行流式分析的条件与其他的细胞有轻微的不同，并且尽管使用抗 IgM 抗体处理的 WEHI-231 细胞的 sIgM 的表达量降低为 1/20，但仍然显著高于 K46 细胞。*b*，抗 IgM 抗体刺激的 WEHI-231 细胞与 T 细胞系共培养。1×10^5 个 WEHI-231 细胞和 10 μg · ml⁻¹ 的抗 IgM 抗体一起分别与 1×10^5 个 T 细胞系 CTLL-2、2B4、N3-6-71 共培养 2 天。*c*，抗 IgM 抗体刺激的 WEHI-231 细胞与各种物质共培养。1×10^5 个 WEHI-231

of A3.4C6 (A3 sup.), the crude membrane fraction of A3.4C6 (A3 membr.), or a mixture of interleukins (IL) were cultured with 1×10^5 WEHI-231 cells in the presence of 10 μg ml^{-1} anti-IgM for 2 days. Culture supernatants of A3.4C6 were added at concentrations of 5 or 20%. The crude membrane fraction was prepared from 5×10^7 A3.4C6 cells as described[31] and suspended in 500 μl phosphate-buffered saline. The membrane fraction was subsequently added to the culture at concentrations of 0.5 or 5%. Culture with the interleukin mixture contained 25 U recombinant IL-1, 1 nM recombinant IL-2, 100 U recombinant IL-5, 1,300 U recombinant IL-6, and 5% culture supernatant of X63Ag8.653 cells transfected with an expression vector carrying mouse IL-4 cDNA[32].

To identify a molecule responsible for the inhibition of sIg-mediated B-cell death, we tested whether the membrane fraction and the culture supernatant of A3.4C6 cells could replace intact T cells for the rescuing activity. The membrane fraction of A3.4C6 completely substituted for the intact A3.4C6 cells, and the culture supernatant from A3.4C6 cells partially rescued WEHI-231 cells from sIg-mediated death (Fig. 1*c*). But none of interleukins 1, 2, 4, 5 and 6, either alone or in combination, were able to rescue WEHI-231 cells. These results suggest that sIg-mediated death of WEHI-231 cells may be prevented by interaction with membrane-bound molecules on A3.4C6, a fraction of which are also released into the supernatant, presumably by proteolysis.

T cells express the membrane-bound ligands for CD40 (ref. 16) and CD72 (ref. 17), which are expressed on the B-cell surface and are capable of transducing a signal for activation of B cells[18,23]. Furthermore, signalling through CD40 blocks spontaneous apoptosis of human germinal centre B cells cultured *in vitro*[24] and may thus play a part in the selection of B cells undergoing somatic hypermutation in germinal centres. To test whether either of the ligands for CD72 or CD40, that is, Ly1 or CD40L, respectively, is responsible for rescuing WEHI-231 cells, we tested A3.4C6 cells for expression of these molecules. As controls we examined 2B4 and N3-6-71 cells, which are inactive in rescuing anti-Ig-treated WEHI-231. Ly1 is expressed on all the T cell hybridomas tested (data not shown). In contrast, CD40L messenger RNA is expressed in A3.4C6 cells but not in 2B4 nor N3-6-71 cells (Fig. 2)

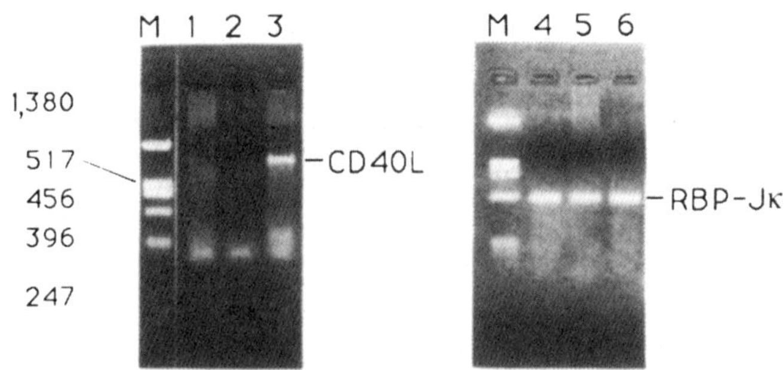

Fig. 2. Assay of CD40L by reverse transcriptase–polymerase chain reaction (RT-PCR). Total RNA was prepared from 2B4 (lanes 1 and 4), N3-6-71 (lanes 2, 5) and A3.4C6 (lanes 3, 6), followed by cDNA synthesis as described[33]. One μg of each cDNA was amplified using the pair of primers for CD40L (lanes 1–3): 5′-CG GAATTCAGTCAGCATGATAGAAAC-3′ (sense) and 5′-AAGTCGACAGCGCACTGTTCAGAGT-3′ (antisense). After 30 cycles, PCR products were resolved by agarose gel electrophoresis, followed by staining with ethidium bromide. As controls, the same cDNA samples were analysed for expression of RBP-Jκ[34]

细胞和 10 μg · ml^{-1} 的抗 IgM 抗体一起分别与 1×10^5 个 A3.4C6 细胞、A3.4C6 细胞培养上清（A3 sup.）、A3.4C6 细胞膜粗提物（A3 membr.）或白介素（IL）的混合物共培育 2 天。加入的 A3.4C6 细胞培养上清浓度为 5% 或 20%。细胞膜粗提物组分是参照文献描述从 5×10^7 个 A3.4C6 细胞中制备的[31]，并最终悬于 500 μl 磷酸盐缓冲液。加入的该粗提物浓度为 0.5% 或 5%。白介素混合物包括 25 U 重组 IL-1、1 nM 重组 IL-2、100 U 重组 IL-5、1,300 U 重组 IL-6 和 5%转染了小鼠 IL-4 cDNA 的 X63Ag8.653 细胞的培养上清[32]。

为了进一步鉴定抑制 sIg 介导的 B 细胞凋亡的分子，我们分别测试了 A3.4C6 细胞的细胞膜成分和培养上清是否能够代替完整的 T 细胞发挥阻断作用。结果发现 A3.4C6 的细胞膜成分可以完全替代完整细胞的功能，而细胞培养上清则可以部分抑制 sIg 介导的 WEHI-231 细胞的凋亡（图 1*c*）。我们也测试了 IL-1、IL-2、IL-4、IL-5、IL-6，发现它们单独或联合使用都不具有上述功能。这些结果表明 WEHI-231 细胞通过与 A3.4C6 膜表面结合分子相互作用从而抑制 sIg 介导的凋亡作用，而小部分结合分子据推测经由蛋白质水解作用释放到细胞培养上清中。

此前的研究表明 T 细胞可以表达 CD40（参考文献 16）和 CD72（参考文献 17）的膜结合型配体，而 CD40 和 CD72 在 B 细胞表面有表达，并介导信号传递促进 B 细胞的激活[18,23]。此外，CD40 介导的信号还可以阻断体外培养的人生发中心 B 细胞的自然凋亡[24]，并且其也许能将在生发中心进行体细胞高频突变的 B 细胞筛选出来。鉴于此，为了验证 CD72 或 CD40 的配体（即 Ly1 或 CD40L）是否与抑制 WEHI-231 细胞的凋亡有关，我们首先检测了 A3.4C6 细胞中这两个分子的表达情况，并使用 2B4 和 N3-6-71 这两个不具有抑制 B 细胞凋亡功能的细胞作为对照。结果表明，Ly1 在所有检测的 T 杂交瘤细胞中都有表达（结果没显示），而 CD40L mRNA 只表达于 A3.4C6 细胞中，而在 2B4 和 N3-6-71 细胞中没有表达（图 2）。

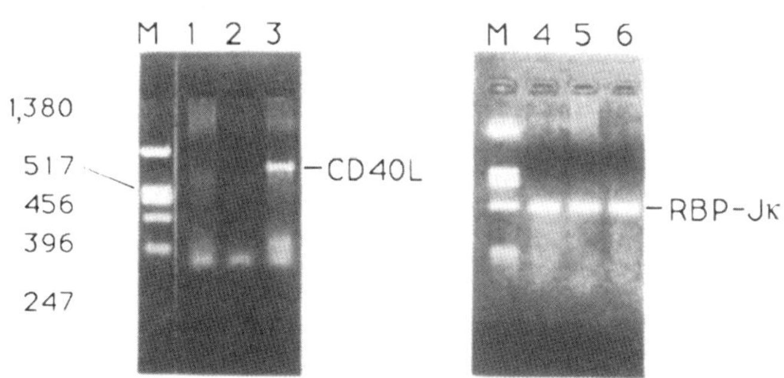

图 2. 使用逆转录聚合酶链式反应（RT-PCR）检测 CD40L 的表达。我们首先提取了 2B4（1 道，4 道）、N3-6-71（2 道，5 道）和 A3.4C6（3 道，6 道）细胞的所有 RNA，然后合成了 cDNA[33]。取 1 μg cDNA，使用下列引物扩增 CD40L（1~3 道）：5′-CGGAATTCAGTCAGCATGATAGAAAC-3′（正义链），5′-AAGTCGACAGCGCACTGTTCAGAGT-3′（反义链）。经过 30 个循环后，用琼脂糖凝胶电泳分离 PCR 产物，并用溴化乙锭（EB）染色。另外，使用下列引物扩增 RBP-Jκ[34] 作为对照（4~6 道）：

by RT-PCR using the pair of primers (lanes 4–6): 5′-TGGACGACGACGAGTCGGAA-3′ (sense) and 5′-CTTGAGAAAGGCAGAAGTAC-3′ (antisense). M, molecular size marker (*Hin*fI-digested BS-KS plasmid (Stratagene)); sizes are shown on the left in base pairs.

To test whether CD40L on the T-cell surface is responsible for rescuing B cells from apoptosis, we stimulated WEHI-231 cells with anti-CD40, together with anti-IgM. As antibodies against mouse CD40 are not available, we introduced the expression plasmid pMIK-hCD40 encoding human CD40 into WEHI-231 cells and treated the transfectant WEHI-hCD40 with anti-IgM and anti-human CD40 antibodies. Growth of WEHI-hCD40 cells was measured by the number of cells recovered after treatment (Fig. 3*a*) and by tritiated thymidine (^{3}H-TdR) uptake (Fig. 3*b*). Anti-Ig treatment markedly inhibited cell growth. This growth inhibition was mostly abrogated in the presence of anti-CD40 antibody, indicating that sIg-mediated apoptosis is inhibited by signalling through CD40. This conclusion is supported by a quantitative assay for apoptosis using propidium iodide[25] (Fig. 3*c*). WEHI-hCD40 cells were treated with anti-IgM and/or anti-CD40, and nuclei were prepared by treating cells with hypotonic solution. DNA contents of nuclei were measured by staining with propidium iodide, followed by flow cytometry. Almost all nuclei of the WEHI-hCD40 cells treated with anti-IgM contained reduced amounts of DNA, indicating extensive fragmentation of nuclei, presumably as a result of apoptosis. In contrast, the DNA contents of almost all nuclei were intact in cells treated with both anti-Ig and anti-CD40 antibodies. Moreover, growth inhibition of anti-IgM-treated WEHI-231 cells is blocked by co-culture with X63Ag8.653 myeloma cells transfected with an expression vector for CD40L (X63CD40L), but not with parent X63Ag8.653 cells (Fig. 3*d*). Taken together, signal transduction through CD40 that is induced by either anti-CD40 antibody or the ligand for CD40 inhibits sIg-mediated apoptotic cell death of WEHI-231 cells. It is, however, unlikely that signalling through CD40 blocks all the pathways of signal transduction generated by sIg multimerization. Indeed, treatment of WEHI-hCD40 cells with anti-Ig and anti-CD40 antibodies gives enhanced Ia expression, one of the markers for B-cell activation[26], whereas the increase in Ia expression by stimulation with anti-CD40 alone is marginal (Fig. 4). Moreover, Ia expression of the few WEHI-hCD40 cells that survived anti-IgM treatment was enhanced, but was less than that of WEHI-hCD40 cells treated with both anti-IgM and anti-CD40 antibodies. This result indicates that B cells are activated by a combination of signals through CD40 and sIg.

5′-TGGACGACGACGAGTCGGAA-3′(正义链)，5′-CTTGAGAAAGGCAGAAGTAC-3′(反义链)。M 道为分子大小标尺(使用 *Hin*fI 消化的 BS-KS 质粒，Stratagene 公司)。标尺的大小以碱基对为单位标注在左侧。

接下来，为了验证 T 细胞表面的 CD40L 是否与抑制 B 细胞凋亡有关，我们同时使用了抗 CD40 抗体和抗 IgM 抗体刺激 WEHI-231 细胞。由于目前还没有抗鼠 CD40 的抗体，于是我们向 WEHI-231 细胞中转染了可以表达人 CD40 的 pMIK-hCD40 质粒(所得转染细胞记为 WEHI-hCD40)，然后再用抗 IgM 抗体和抗人 CD40 抗体来处理这个细胞。我们分别使用了细胞计数(图 3*a*)和 ^{3}H 胸腺嘧啶(^{3}H-TdR)示踪(图 3*b*)的方法来测量 WEHI-hCD40 细胞的生长。结果显示，抗 Ig 抗体处理后可以显著地抑制细胞的生长；如果同时使用抗人 CD40 抗体来处理细胞，则可以消除细胞的生长抑制。这表明 CD40 介导的信号可以抑制 sIg 介导的细胞凋亡。使用碘化丙啶 [25] 定量检测凋亡的实验结果进一步支持了上述结论(图 3*c*)。我们首先使用抗 IgM 抗体和(或)抗 CD40 抗体刺激 WEHI-hCD40 细胞，然后使用低渗溶液处理细胞从而分离出细胞核；接下来再用碘化丙啶给细胞核中的 DNA 染色并用流式细胞仪检测。结果发现几乎所有经抗 IgM 抗体处理过的细胞的细胞核中的 DNA 含量都减少了，表明细胞核的碎片化较严重，而这可能是细胞凋亡引起的结果。相反，同时使用抗 IgM 抗体和抗 CD40 抗体处理的细胞其细胞核中的 DNA 含量几乎不变，相应地 DNA 的含量也较高。此外，使用转染了 CD40L 基因的 X63Ag8.653 骨髓瘤细胞(X63CD40L) 与抗 IgM 抗体处理的 WEHI-231 细胞共培养，也可以阻断抗 IgM 抗体引起的 WEHI-231 细胞的生长抑制，而亲代 X63Ag8.653 骨髓瘤细胞则不具有这种作用(图 3*d*)。综合考虑上述结果，我们可以得出结论——使用抗 CD40 抗体或者 CD40 配体都可以引发 CD40 产生信号转导从而抑制 sIg 介导的 WEHI-231 细胞的凋亡。不过，上述 CD40 介导的信号似乎不能阻断所有的 sIg 多聚化引发的信号转导通路。事实上，同时使用抗 IgM 抗体和抗 CD40 抗体处理 WEHI-hCD40 细胞可以引起该细胞中 Ia(B 细胞激活的标志分子之一 [26]) 表达水平的升高，而单独使用抗 CD40 抗体则不会产生显著的效果(图 4)。此外，在那些经抗 IgM 抗体处理后仍存活下来的细胞中，Ia 的表达水平也是升高的，只不过比同时使用抗 IgM 抗体和抗 CD40 抗体处理的细胞要低一些。这表明 B 细胞的激活是 CD40 介导的信号和 sIg 共同作用的结果。

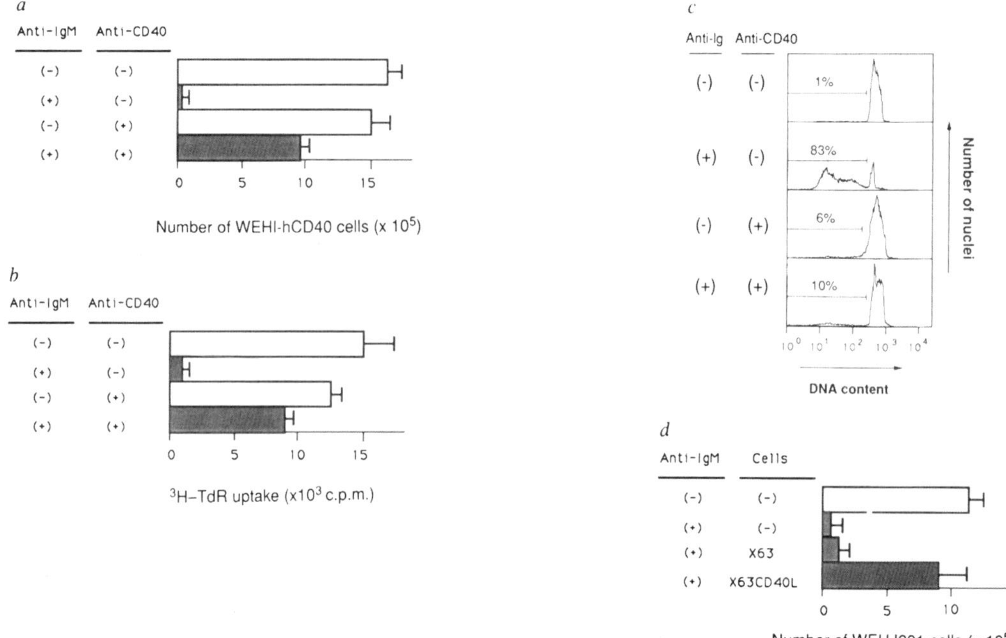

Fig. 3. Signalling through CD40 abrogates sIg-mediated cell death of WEHI-231 cells. Human CD40 transfectants of WEHI-231 (WEHI-hCD40) were cultured with 10 μg ml^{-1} anti-IgM, 10 μg ml^{-1} anti-human CD40 (5C3; a gift from H. Kikutani), or both, for 2 days. Alternatively, WEHI-231 cells were cultured either with X63Ag8.653 (X63) cells or X63 cells transfected with a CD40L vector (X63CD40L) for 2 days. *a*, Growth of WEHI-hCD40 cells. WEHI-hCD40 cells (1×10^5) were cultured in 1 ml culture medium with antibodies as indicated. All cells in the culture were collected and viable cells excluding trypan blue counted. Data represent mean cell number ± s.d. *b*, Proliferation of WEHI-hCD40 cells. WEHI-231 cells (4×10^4) were cultured in triplicate in 200 μl culture medium with the antibodies indicated for 2 days. Cultures were pulsed with 0.2 μCi ^{3}H-TdR (Amersham) for the last 16 h of culture. Cells were collected and ^{3}H-TdR uptake measured by liquid scintillation counting (Packard). Data represent the mean ^{3}H-TdR uptake (c.p.m.) ± s.d. *c*, Apoptosis of WEHI-hCD40 cells. WEHI-hCD40 cells (1×10^5) were cultured in 1 ml culture medium with the antibodies indicated. Cells were lysed in a hypotonic solution (0.1% sodium citrate, 0.1% Triton-X100) containing 10 μg ml^{-1} of propidium iodide, and the DNA content of the nuclei analysed on a FACScan as described[25]. The percentage of fragmented nuclei is indicated. *d*, Growth of WEHI-231 cells. WEHI-231 cells (1×10^5) were cultured with either 1×10^5 X63 cells or the same number of X63CD40L cells. All cells were collected and viable cells excluding trypan blue were counted. The number of WEHI-231 cells was determined by using flow cytometry analysis as in Fig. 1. Data represent mean cell number ± s.d.

Methods. The expression plasmid pMIK-hCD40 coding for human CD40 was constructed by cloning the *Xho*I fragment of CDM8/CDw40[35] containing human CD40 cDNA into *Xho*I-digested pMIKNeo (−) plasmid (gift from K. Maruyama). The pMIK-hCD40 plasmid was introduced into WEHI-231 cells by electroporation and the transfectant WEHI-hCD40 obtained by G418 selection. For generating X63CD40L cells, murine CD40L cDNA was amplified from A3.4C6 cells by RT-PCR (Fig. 2 legend) using the following primers: 5′-CGGAATTCAGTCAGCATGATAGAAAC-3′ (sense) and 5′-AAGTCGACAGCGCACTGTTCAGAGT-3′ (antisense). The amplified cDNA was digested with *Eco*RI and *Sal*I, and cloned into the pMIKneo plasmid digested with both *Eco*RI and *Xho*I, resulting in the pMIK-CD40L vector. X63Ag8.653 cells were transfected with the pMIK-CD40L vector and transfectants obtained by G418 selection.

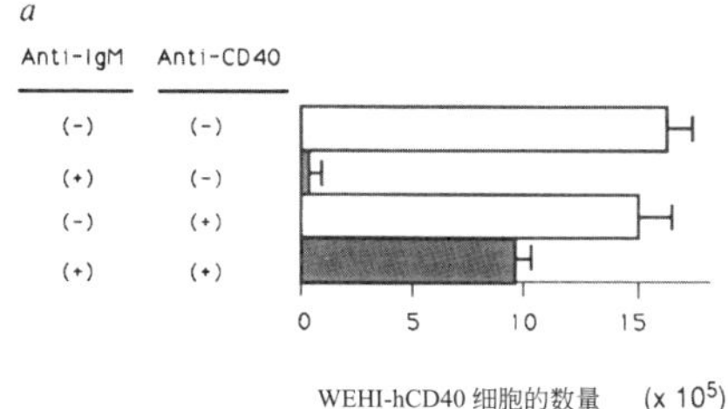

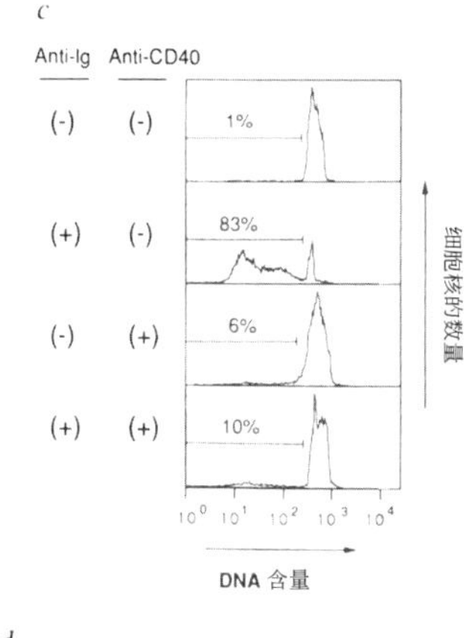

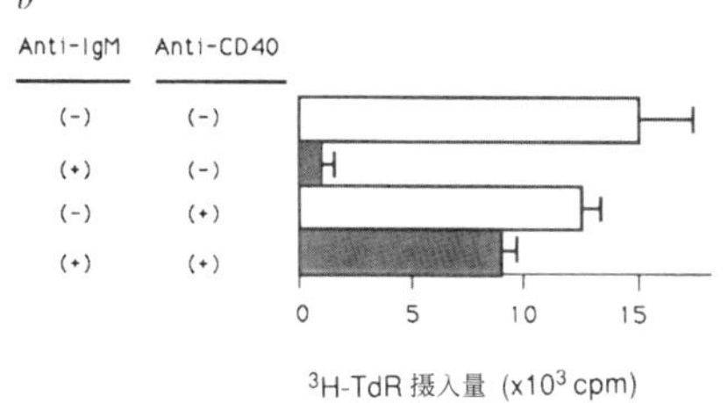

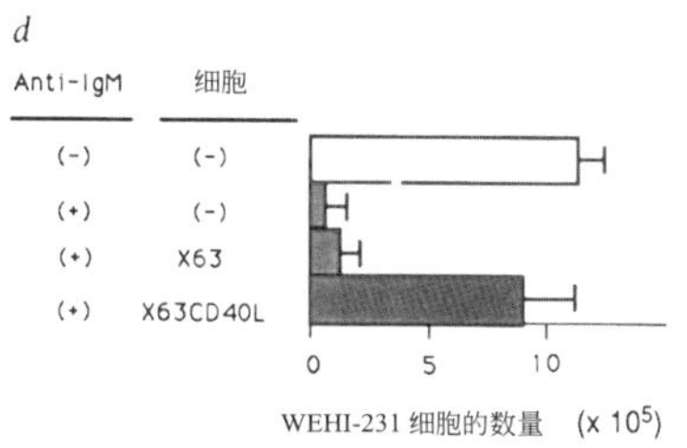

图 3. CD40 介导的信号能抑制 sIg 介导的 WEHI-231 细胞的凋亡。人 CD40 基因转染的 WEHI-231 细胞 (WEHI-hCD40) 与 10 μg · ml^{-1} 的抗 IgM 抗体和(或)10 μg · ml^{-1} 的抗人 CD40 抗体(5C3；菊谷惠赠)培养 2 天。或者，WEHI-231 细胞与 X63Ag8.653 (X63) 骨髓瘤细胞或转染了 CD40L 基因的 X63 细胞 (X63CD40L) 共培养 2 天。*a*, WEHI-hCD40 细胞的生长情况。1×10^5 个 WEHI-hCD40 细胞在 1 ml 培养基中与抗体共培养后，收集细胞，用台盼蓝染色后计算活细胞数量。数据以细胞平均数 ± 标准差的形式表示。*b*, WEHI-hCD40 细胞的增殖。4×10^4 个 WEHI-hCD40 细胞在 200 μl 培养基中与抗体共培养 2 天。每个都做了 3 个平行实验。在培养的最后 16 小时使用 0.2 μCi ^{3}H-TdR(安马西亚公司) 冲击，收集细胞，使用液体闪烁计数器计算 ^{3}H-TdR 被细胞摄入的量。数据以 ^{3}H-TdR 摄入平均值 ± 标准差(cpm)的形式表示。*c*, WEHI-hCD40 细胞的凋亡。1×10^5 个 WEHI-hCD40 细胞在 1 ml 培养基中与抗体共培养。然后使用含 10 μg · ml^{-1} 碘化丙啶的低渗溶液(0.1% 柠檬酸钠，0.1% Triton-X100) 裂解细胞，使用流式细胞仪分析细胞核中 DNA 含量 [25]。图中显示了细胞核碎片化的百分比。*d*, WEHI-231 细胞的生长情况。1×10^5 个 WEHI-231 细胞与 1×10^5 个 X63 细胞或同样数量的 X63CD40L 细胞共培养。收集细胞后使用台盼蓝染色计算活细胞数量。WEHI-231 细胞的数量采用与图 1 相同的流式细胞术分析。数据以细胞平均数 ± 标准差的形式表示。

方法。通过克隆含人 CD40 cDNA 的 CDM8/CDw40[35] *Xho*I 片段，并将其插入有 *Xho*I 酶切位点的 pMIKNeo(−) 质粒(丸山惠赠)，构建出编码人 CD40 的 pMIK-hCD40 质粒。通过电穿孔将 pMIK-hCD40 质粒转入 WEHI-231 细胞，使用 G418 筛选出 WEHI-hCD40 细胞。为了获得 X63CD40L 细胞，使用如下引物从 A3.4C6 细胞中通过 RT-PCR 扩增鼠 CD40L cDNA(图 2 图注)：5′-CGGAATTCAGTCAGCATGATAGAAAC-3′(正义链)，5′-AAGTCGACAGCGCACTGTTCAGAGT-3′(反义链)。使用 *Eco*RI 和 *Sal*I 酶切扩增出的 cDNA，插入到 *Eco*RI 和 *Xho*I 酶切后的 pMIKneo 质粒中，构建得到 pMIK-CD40L 载体。用 pMIK-CD40L 载体转染 X63Ag8.653 细胞并使用 G418 筛选得到 X63CD40L 细胞。

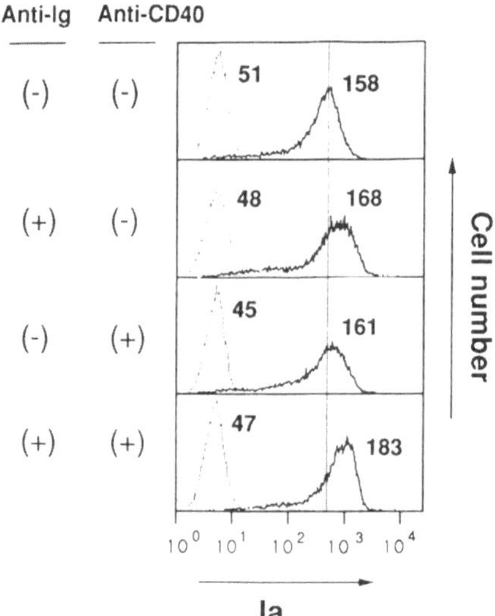

Fig. 4. Expression of the Ia antigen on the surface of WEHI-hCD40 cells. WEHI-hCD40 cells were cultured with anti-CD40 alone, anti-IgM alone, or both anti-IgM and anti-CD40, as described for Fig. 3*a*. After 2 days, cells were collected and stained for Ia expression (thick traces) by incubating with rat anti-mouse Ia (M5/114.15.2), followed by reaction with fluorescein-conjugated MARK1 (mouse anti-rat κ). Cells were subsequently analysed on a FACScan. As a control, cells were stained with fluorescein-conjugated MARK1 alone (thin traces). Mean fluorescence intensities are indicated. The vertical line in the flow cytometry histograms shows the peak channel for Ia expression of unstimulated WEHI-231 cells.

We have shown here that CD40L-producing T hybridoma cells A3.4C6, anti-CD40 antibody, or myeloma transfectants producing CD40L, can all rescue WEHI-231 cells from apoptosis induced by sIg multimerization. These results indicate that signalling through CD40 alone is sufficient for abrogating sIg-mediated apoptosis of WEHI-231 cells. Signalling through CD40 not only inhibits sIg-mediated apoptosis but also, together with signalling through sIg, activates WEHI-231 cells. T-helper cell lines not only express CD40L on the surface[16,23] but also secrete CD40L (ref. 27). The rescuing ability for WEHI-231 cells detected in the culture supernatant of A3.4C6 cells may thus be attributed to secreted CD40L.

Mature B cells undergo apoptosis by signalling through sIg *in vivo* (ref. 1; and T.T., unpublished observation). Activation of mature resting B cells *in vivo* probably requires an additional signal that abrogates sIg-mediated apoptosis. Our results indicate that such a signal is likely to be generated through the CD40 molecule, which is expressed on almost all B cells. The presence or absence of T-cell help through CD40 may, therefore, be crucial in determining whether antigen-stimulated B cells are activated or killed. As we have detected significant enhancement of bcl-2 protein production in WEHI-hCD40 cells after anti-CD40 treatment (data not shown), bcl-2 may be involved in CD40-mediated abrogation of B-cell apoptosis. But constitutive expression of bcl-2 does not prevent the death of WEHI-231 cells treated with anti-Ig (ref. 28; and A. Strasser and S. Cory, personal communication), nor does it abrogate clonal deletion of self-reactive B cells in autoantibody transgenic mice[29], suggesting that bcl-2 alone is not sufficient to block apoptosis of B cells. As CD40L

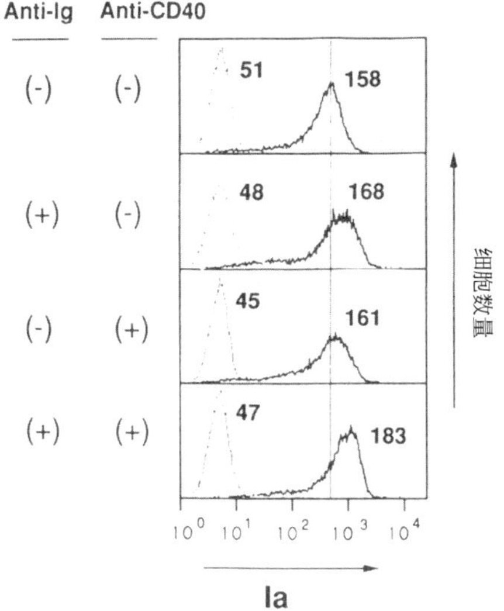

图 4. WEHI-hCD40 细胞表面 Ia 抗原的表达。如图 3*a* 所述，分别使用抗 CD40 抗体、抗 IgM 抗体或二者一起与 WEHI-hCD40 细胞共同培养。2 天后收集细胞，将该细胞与大鼠抗小鼠 Ia 抗体（M5/114.15.2）共培养，然后与荧光素标记的 MARK1（小鼠抗大鼠 κ 链）反应，检测 Ia 抗原的表达情况（图中的粗线）。利用流式细胞仪分析细胞数量。仅使用荧光素标记的 MARK1 直接染色的细胞作为对照（图中的细线）。图中标出了平均荧光强度。图中的垂直线为未刺激的 WEHI-hCD40 细胞中 Ia 抗原表达的峰值。

在这篇文章中，我们证明了表达 CD40L 的 T 杂交瘤细胞 A3.4C6、抗 CD40 抗体或转了 CD40L 的骨髓瘤细胞都可以抑制 sIg 多聚化所引起的 WEHI-231 细胞的凋亡。这些结果表明 CD40 介导的信号途径完全可以抑制 sIg 介导的 WEHI-231 细胞的凋亡。不仅如此，CD40 介导的信号还可以与 sIg 信号一起激活 WEHI-231 细胞。有研究表明辅助 T 细胞系不仅在细胞表面表达 CD40L[16,23]，而且可以分泌 CD40L 到细胞外（参考文献 27）。这也解释了 A3.4C6 细胞培养上清为什么也可以部分抑制 WEHI-231 细胞的凋亡——可能归功于分泌的 CD40L。

在体内实验中，sIg 信号会引起成熟 B 细胞的凋亡（参考文献 1 以及锷田武志尚未发表的观察结果）。在体内，处于静息期的成熟 B 细胞的激活可能需要额外的信号来阻断 sIg 介导的凋亡。我们的实验结果证明这个信号很可能是通过 CD40 分子来产生的。这一分子几乎在所有的 B 细胞中都有表达。因此，是否有辅助 T 细胞激活 B 细胞上的 CD40 信号成为那些被抗原激活的 B 细胞是活化还是死亡的决定性因素。我们在实验中观察到经抗 CD40 抗体处理过的 WEHI-hCD40 细胞中 bcl-2 蛋白的表达水平也有显著的提高（结果未展示），这表明 bcl-2 也可能参与了 CD40 介导的阻断 B 细胞凋亡过程。不过组成型表达的 bcl-2 并不能抑制抗 IgM 抗体处理的 WEHI-231 的凋亡（参考文献 28；以及与斯特拉瑟和科里的个人交流），也不能消除自身抗体转

is expressed on activated T helper cells but not on resting T cells[16,23,30], inhibition of sIg-mediated apoptosis and activation of B cells probably take place in an antigen-specific manner only in the presence of antigen-presenting cells, antigens, and helper T cells specific for the antigens. Self-reactive B cells may thus be deleted, presumably as a result of the absence of self-reactive T cells. But if T-cell tolerance breaks down, autoreactive T cells may initiate the pathogenesis of antibody-mediated autoimmunity by antagonizing clonal deletion of self-reactive mature B cells.

(**364**, 645-648; 1993)

Takeshi Tsubata, Jing Wu & Tasuku Honjo
Department of Medical Chemistry, Kyoto University Faculty of Medicine, Kyoto 606, Japan

Received 31 December 1992; accepted 4 June 1993.

References:

1. Murakami, M. *et al. Nature* **357**, 77-80 (1992).
2. Russell, D. M. *et al. Nature* **354**, 308-311 (1991).
3. Nemazee, D. A. & Bürki, K. *Nature* **337**, 562-566 (1989).
4 Hartley, S. B. *et al. Nature* **353**, 765-769 (1991).
5. Okamoto, M. *et al. J. Exp. Med.* **175**, 71-79 (1992).
6. Nossal, G. J. V. *A. Rev. Immun.* **1**, 33-62 (1983).
7. Goodnow, C. C. *A. Rev. Immun.* **10**, 489-518 (1992).
8. Metcalf, E. S. & Klinman, N. R. *J. Exp. Med.* **143**, 1327-1340 (1976).
9. Metcalf, E. S. & Klinman, N. R. *J. Immun.* **118**, 2111-2116 (1977).
10. Benhamou, L. E., Cazenave, P. A. & Sarthou, P. *Eur. J. Immun.* **20**, 1405-1407 (1990).
11. Hasbold, J. & Klaus, G. G. B. *Eur. J. Immun.* **20**, 1685-1690 (1990).
12. Ozaki, S., Durum, S. K., Muegge, K., York-Jolley, J. & Berzofsky, J. A. *J. Immun.* **141**, 71- 78 (1988).
13. Ogasawara, K., Maloy, W. L., Beverly, B. & Schwartz, R. H. *J. Immun.* **142**, 1448-1456 (1989).
14. Ashwell, J. D., Cunningham, R. E., Noguchi, P. D. & Hernandez, D. *J. Exp. Med.* **165**, 173- 194 (1987).
15. Odaka, C., Kizaki, H. & Tadakuma, T. *J. Immun.* **144**, 2096-2101 (1990).
16. Armitage, R. J. *et al. Nature* **357**, 80-82 (1992).
17. Van de Velde, H., von Hoegen, I., Luo, W., Parnes, J. R. & Thielemans, K. *Nature* **351**, 662-664 (1991).
18. Clark, E. A. & Ledbetter, J. A. *Proc. Natl. Acad. Sci. U.S.A.* **83**, 4494-4498 (1986).
19. Paulie, S. *et al. J. Immun.* **142**, 590-595 (1989).
20. Rousset, F., Garcia, E. & Banchereau, J. *J. Exp. Med.* **173**, 705-710 (1991).
21. Subbarao, B. & Mosier, D. E. *J. Immun.* **130**, 2033-2037 (1983).
22. Subbarao, B. & Mosier, D. E. *J. Exp. Med.* **159**, 1796-1801 (1984).
23. Noelle, R. J. *et al. Proc. Natl. Acad. Sci. U.S.A.* **89**, 6550-6554 (1992).
24. Liu, Y.-J. *et al. Nature* **342**, 929-931 (1989).
25. Nicoletti, I., Migliorati, M. C., Grignani, F. & Riccardi, C. *J. Immun. Meth.* **139**, 271-279 (1991).
26. Mond, J. J., Seghal, E., Kung, J. & Finkelman, F. D. *J. Immun.* **127**, 881-888 (1981).
27. Armitage, R. J. *et al. Eur. J. Immun.* **22**, 2071-2176 (1992).
28. Cuende, E. *et al. EMBO J.* **12**, 1555-1560 (1993).
29. Hartley, S. B. *et al. Cell* **72**, 325-335 (1993).
30. Noell, R. J., Ledbetter, J. A. & Aruffo, A. *Immun. Today* **13**, 431-433 (1992).
31. Robb, R. J. *Proc. Natl. Acad. Sci. U.S.A.* **83**, 3992-3996 (1986).
32. Karasuyama, H. & Melchers, F. *Eur. J. Immun.* **18**, 97-104 (1988).

基因小鼠中自身反应性 B 细胞的克隆排除[29]。这表明只有 bcl-2 是不足以阻断 B 细胞的凋亡。由于 CD40L 只在激活的辅助 T 细胞上表达而不在静息的 T 细胞中表达[16,23,30]，因此抑制 sIg 介导的凋亡和促进 B 细胞的激活可能是抗原特异性的。也就是说，只有当抗原呈递细胞、抗原以及针对该抗原特异性的辅助 T 细胞一起存在时才能激发 B 细胞的第二信使。如果自身反应性 B 细胞发生凋亡，那么极有可能是没有自身反应性 T 细胞导致的。但是如果 T 细胞耐受被打破，自身反应性 T 细胞就会通过对抗自身反应性成熟 B 细胞的克隆排除，从而引发自身抗体介导的自身免疫疾病。

（张锦彬 翻译；秦志海 审稿）

33. Shimizu, A., Nussenzweig, M. C., Mizuta, T.-R., Leder, P. & Honjo, T. *Proc. Natl. Acad. Sci. U.S.A.* **86**, 8020-8023 (1989).
34. Matsunami, N. *et al. Nature* **342**, 934-937 (1989).
35. Stamenkovic, I., Clark, E. A. & Seed, B. *EMBO J.* **8**, 1403-1410 (1989).

Acknowledgements. We thank S. Ozaki, J. A. Berzofsky, K. Ogasawara, E. A. Clark, H. Kikutani, K. Maruyama, A. Tominaga and J. Reed for reagents, M. Paumen for critically reading the manuscript, R. Sakai for technical help, and K. Hirano for help in preparing the manuscript. This work was supported in part by a grant from the Ministry of Education, Science and Culture of Japan.

Music and Spatial Task Performance

F. H. Rauscher *et al.*

Editor's Note

In appearing to verify the long-standing suspicion that music enhances intelligence, this report presented by Frances Rauscher and colleagues in California attracted immense interest. They reported that college students show slightly better results in spatial reasoning tests (boosting IQ by 8–9 points) after listening to ten minutes of Mozart's music than after either listening to a "relaxation tape" or sitting in silence. Follow-up studies drew diverse conclusions: some apparently supported the claim, others found no effect, or results not specific to Mozart. It now seems that such effects are minor and transient, and are caused by mood and arousal rather than anything specific to music. All the same, "brain music" for children has become a small marketing industry.

THERE are correlational[1], historical[2] and anecdotal[3] relationships between music cognition and other "higher brain functions", but no causal relationship has been demonstrated between music cognition and cognitions pertaining to abstract operations such as mathematical or spatial reasoning. We performed an experiment in which students were each given three sets of standard IQ spatial reasoning tasks; each task was preceded by 10 minutes of (1) listening to Mozart's sonata for two pianos in D major, K488; (2) listening to a relaxation tape; or (3) silence. Performance was improved for those tasks immediately following the first condition compared to the second two.

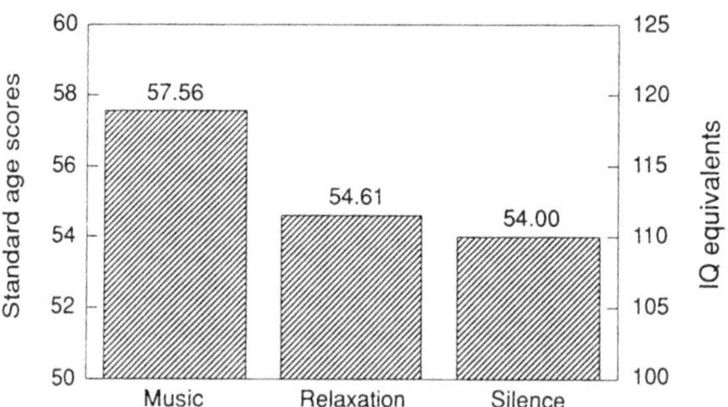

Standard age scores for each of the three listening conditions.

Testing procedure. In the music condition, the subject listened to 10 min of the Mozart piece. The relaxation condition required the subject to listen to 10 min of relaxation instructions designed to lower blood pressure. The silence condition required the subject to sit in silence for 10 min. One of three abstract reasoning tests taken from the Stanford–Binet intelligence scale[4] was given after each of the listening conditions. The abstract/spatial reasoning tasks consisted of a pattern analysis test, a multiple-choice matrices test and a multiple-choice paper-folding and cutting test. For our sample, these three tasks correlated at the 0.01 level of significance. We were thus able to treat them as equal measures of abstract reasoning ability.

Scoring. Raw scores were calculated by subtracting the number of items failed from the highest item number administered. These were then converted to SAS using the Stanford–Binet's SAS conversion table of normalized standard scores with a mean set at 50 and a standard deviation of 8. IQ equivalents

音乐和空间任务表现

劳舍尔等

编者按

这篇由加利福尼亚大学的弗朗西丝·劳舍尔和她的同事所写的报告似乎证实了人们长期以来的怀疑，即音乐可以提高智力。这引起了人们极大的兴趣。在他们的报告中，大学生在听了十分钟莫扎特的音乐之后在空间推理测试中的成绩比听放松磁带或静坐的情况下稍微好些(IQ 提高了 8~9 分)。然而,后续研究却得到了不同的结论：一些显然支持这种说法，一些没有发现相同的效果或是效果不特定于莫扎特的音乐。现在看来，这种效果较小且短暂，更可能是由情绪和唤醒状态而不是由音乐特定引起的。尽管如此，儿童的“大脑音乐”已经成为一个小的营销行业。

音乐认知和其他“高级脑功能”在相关性上[1]、历史上[2]和民间传闻中[3]都有关联。但音乐认知和与抽象操作(数学或者空间推理等)有关的认知之间的因果关系至今还没有被证实。我们进行了一项实验，在实验中每个学生都被要求完成三套标准 IQ 空间推理测试，学生在每份测试之前先进行 10 分钟的活动：(1)听莫扎特的 D 大调双钢琴奏鸣曲 K488,(2)听一盘放松的磁带，或者(3)静坐。相比第二和第三项活动，学生在紧跟着第一项活动后的测试中表现有所提高。

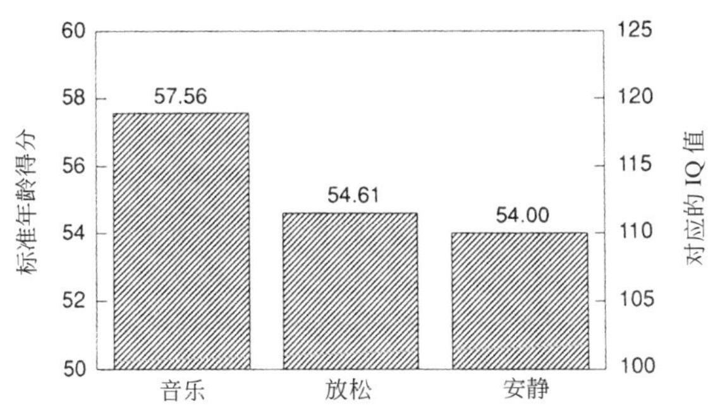

在三种听觉情境下各自的标准年龄分数

测试步骤。在音乐情境下，测试者听 10 分钟莫扎特的作品。在放松情境下，测试者听 10 分钟可降低血压的放松磁带。在安静情境下，测试者静坐 10 分钟。在每一种听力情境结束后，测试者完成斯坦福–比奈智力量表[4]中三份抽象推理测验中的一份。抽象(空间)推理任务包括一个图形分析测试、一个多项选择矩阵测试和一个多项选择的折纸剪纸测试。对于我们的样本，这三项测试在 0.01 水平下显著相关，因此我们把它们作为对抽象推理能力的等效测试。

得分。原始分数由测试所给最高次数减去失误总次数得到。该分数再通过斯坦福–比奈标准年龄得分(SAS)转化表转化成平均值为 50，标准差为 8 的 SAS 值。对应 IQ 值的计算方法为先将每个 SAS 值乘

were calculated by first multiplying each SAS by 3 (the number of subtests required by the Stanford–Binet for calculating IQs). We then used their area score conversion table, designed to have a mean of 100 and a standard deviation of 16, to obtain SAS IQ equivalents.

Thirty-six college students participated in all three listening conditions. Immediately following each listening condition, the student's spatial reasoning skills were tested using the Stanford–Binet intelligence scale[4]. The mean standard age scores (SAS) for the three listening conditions are shown in the figure. The music condition yielded a mean SAS of 57.56; the mean SAS for the relaxation condition was 54.61 and the mean score for the silent condition was 54.00. To assess the impact of these scores, we "translated" them to spatial IQ scores of 119, 111 and 110, respectively. Thus, the IQs of subjects participating in the music condition were 8–9 points above their IQ scores in the other two conditions. A one-factor (listening condition) repeated measures analysis of variance (ANOVA) performed on SAS revealed that subjects performed better on the abstract/spatial reasoning tests after listening to Mozart than after listening to either the relaxation tape or to nothing ($F_{2,35} = 7.08$; $P = 0.002$). The music condition differed significantly from both the relaxation and the silence conditions (Scheffe's $t = 3.41$, $P = 0.002$; $t = 3.67$, $P = 0.0008$, two-tailed, respectively). The relaxation and silence conditions did not differ ($t = 0.795$; $P = 0.432$, two-tailed). Pulse rates were taken before and after each listening condition. A two-factor (listening condition and time of pulse measure) repeated measures ANOVA revealed no interaction or main effects for pulse, thereby excluding arousal as an obvious cause. We found no order effects for either condition presentation or task, nor any experimenter effect.

The enhancing effect of the music condition is temporal, and does not extend beyond the 10–15-minute period during which subjects were engaged in each spatial task. Inclusion of a delay period (as a variable) between the music listening condition and the testing period would allow us quantitatively to determine the presence of a decay constant. It would also be interesting to vary the listening time to optimize the enhancing effect, and to examine whether other measures of general intelligence (verbal reasoning, quantitative reasoning and short-term memory) would be similarly facilitated. Because we used only one musical sample of one composer, various other compositions and musical styles should also be examined. We predict that music lacking complexity or which is repetitive may interfere with, rather than enhance, abstract reasoning. Also, as musicians may process music in a different way from non-musicians, it would be interesting to compare these two groups.

(**365**, 611; 1993)

Frances H. Rauscher , Gordon L. Shaw*, Katherine N. Ky
Center for the Neurobiology of Learning and Memory, University of California, Irvine, California 92717, USA
* Also at Department of Physics.

References:

1. Hassler, M., Birbaumer, N. & Feil, A. *Psychol. Music* 13, 99-113(1985).
2. Allman, G. J. *Greek Geometry from Thales to Euclid* p. 23 (Arno, New York, 1976).
3. Cranberg, L. D. & Albert, M. L. in *The Exceptional Brain* (eds Obler, L. K. & Fein, D.) 156 (Guilford, New York, 1988).
4. Thorndike, R. L., Hagen, E. P. & Sattler, J. M. *The Stanford–Binet Scale of Intelligence* (Riverside, Chicago, 1986).

以 3（斯坦福–比奈量表测试 IQ 值需要的子测验数目）。然后我们使用各项对应的分数换算表得到平均值为 100，标准差为 16 的 SAS 对应 IQ 值。

36 名大学生参与了所有三种情景下的测试。每种听觉情景结束后，我们立刻通过斯坦福–比奈智力量表[4]测试他们的空间推理能力。三种情景下测试所得的平均标准年龄得分（SAS）如图所示。听音乐情景下平均 SAS 为 57.56；听放松磁带情景下平均 SAS 为 54.61；安静情景下平均 SAS 为 54.00。为了评估这些分数的影响，我们将它们“翻译”为空间 IQ 分数，该分数分别为 119，111 和 110。因此，测试对象在音乐情境下的 IQ 分数要比另外两种情境下的 IQ 分数高 8~9 分。标准年龄得分的单因素（听觉情境）重复测量方差分析（ANOVA）显示，测试对象听完莫扎特音乐后在抽象（空间）推理能力测试中的表现好于听放松磁带或者什么都不听（$F_{2,35} = 7.08$；$P = 0.002$）。听音乐后的表现与听放松磁带后或者安静情况下的表现有显著区别（Scheffe 差别检验法，$t = 3.41$，$P = 0.002$；$t = 3.67$，$P = 0.0008$，双尾检验）。而听放松磁带后的表现和安静情况下的表现无差异（$t = 0.795$；$P = 0.432$，双尾检验）。我们在三种听觉活动前后分别测量了测试者的脉搏次数。对脉搏次数的双因素（听觉情境和脉搏测量时间）重复测量的 ANOVA 表明听觉情境和测量时间对脉搏次数没有交互作用或主效应，因此排除了唤醒状态这一明显的导致因素。我们没有发现听觉情境或任务的顺序效应，也未发现实验者效应。

音乐情境的增强效应是暂时的，没有超过测试对象完成空间测试任务所需的 10~15 分钟。在听音乐和测试阶段之间加入时间延迟（作为一个变量）可以让我们定量地测定音乐效应的衰减常数。改变听音乐的时间长短来优化音乐的增强效应，并且检测听音乐对其他一般智力（语言推理、计量推理、短时记忆）是否有相似的提升作用，将会很有趣。因为我们只用了一位作曲家的一首乐曲，其他的曲目和音乐风格也应该被研究。我们预测缺少复杂性的或者重复的音乐将会妨碍而不是提高抽象推理能力。另外，因为音乐家加工音乐的方式可能与非音乐家不同，比较这两个组的表现也将会很有趣。

（苏怡汀 翻译；杜忆 审稿）

A Cell Initiating Human Acute Myeloid Leukaemia after Transplantation into SCID Mice

T. Lapidot *et al.*

Editor's Note

This paper by Canadian cell biologist John E. Dick and colleagues marks a significant shift in cancer biology. Dick isolated putative stem cells from blood samples of patients with acute myeloid leukaemia (AML) and injected them into mice with deficient immune systems, causing the animals to develop leukaemia. The experiment suggested that these rare cancer stem cells cause AML, and offered a potential explanation for why chemotherapy, which targets fast-growing cancer cells rather than slow-growing stem cells, sometimes fails. Reaction was guarded: many researchers thought that leukaemia, a cancer of the blood, might be an exceptional case. But opinion changed in 2003 when researchers found cancer stem cells in two different solid tumours—human breast cancer and brain cancer.

Most human acute myeloid leukaemia (AML) cells have limited proliferative capacity, suggesting that the leukaemic clone may be maintained by a rare population of stem cells[1-5]. This putative leukaemic stem cell has not been characterized because the available *in vitro* assays can only detect progenitors with limited proliferative and replating potential[4-7]. We have now identified an AML-initiating cell by transplantation into severe combined immune-deficient (SCID) mice. These cells homed to the bone marrow and proliferated extensively in response to *in vivo* cytokine treatment, resulting in a pattern of dissemination and leukaemic cell morphology similar to that seen in the original patients. Limiting dilution analysis showed that the frequency of these leukaemia-initiating cells in the peripheral blood of AML patients was one engraftment unit in 250,000 cells. We fractionated AML cells on the basis of cell-surface-marker expression and found that the leukaemia-initiating cells that could engraft SCID mice to produce large numbers of colony-forming progenitors were $CD34^{+}CD38^{-}$; however, the $CD34^{+}CD38^{+}$ and $CD34^{-}$ fractions contained no cells with these properties. This *in vivo* model replicates many aspects of human AML and defines a new leukaemia-initiating cell which is less mature than colony-forming cells.

THE success in transplanting normal human haematopoietic cells[8,9] and acute lymphoid leukaemia cells[10,11] into immuno-deficient SCID mice[12] suggested that this system might be useful for studying human myeloid leukaemias. But previous experiments indicated that primary AML cells could not be grafted into SCID mice after intravenous injection[13-15], and only a few samples grew locally after implantation into the peritoneum or

一种植入SCID小鼠中引发人急性髓性白血病的细胞

拉皮多特等

编者按

加拿大细胞生物学家约翰·迪克和同事发表的这篇文章标志着肿瘤生物学研究的一个重大转变。迪克从急性髓性白血病患者的血液样品中分离得到了假定的干细胞，并将它们注入免疫系统缺陷的小鼠，使后者患上白血病。实验表明少数的癌症干细胞即能引起急性髓性白血病，这也解释了为什么以快速生长的肿瘤细胞而不是生长缓慢的干细胞为靶点的化学疗法有时候会失败。一开始人们的反应比较谨慎，许多研究者认为白血病作为一种血液肿瘤发现干细胞可能是一个例外。但是在2003年当研究人员在实体瘤——人乳腺肿瘤和脑肿瘤中也发现了癌症干细胞，人们对它的看法才发生了变化。

大多数人急性髓性白血病(AML)细胞的增殖能力有限，这暗示可能存在着数量非常稀少的肿瘤干细胞来维系白血病细胞克隆的存在[1-5]。由于现有的体外实验只能检测出增殖和替换能力有限的祖细胞，因此人们一直未能鉴定出这种白血病干细胞[4-7]。本文中，我们通过对重度联合免疫缺陷(SCID)小鼠进行细胞移植实验，发现了一种可以引发AML的细胞。这些细胞会归巢于骨髓中，并且在细胞因子的刺激下在体内会快速增殖，产生与初始白血病患者体内形态和模式均相似的白血病细胞。有限稀释实验显示，这些能够引发白血病的细胞在AML患者的外周血中的频率为1/250,000，即每25万个外周血淋巴细胞可以通过移植引发一例新的白血病。我们根据细胞表面标志分子的表达对AML细胞进行了分选，发现那些能够通过细胞移植引发SCID小鼠产生大量集落形成前体细胞的白血病起始细胞的表型是$CD34^+CD38^-$，而表型为$CD34^+CD38^+$以及$CD34^-$的细胞则不具备这种能力。这个体内实验模型再现了人AML的许多特征。利用这一模型，我们发现了一类新的白血病起始细胞。这类细胞在发育上较那些集落形成细胞来说还不够成熟。

在过去的研究中，人们成功地在免疫缺陷的SCID小鼠[12]体内植入正常的人造血细胞[8,9]和急性淋巴细胞白血病细胞[10,11]，这提示我们可以利用这一模型开展人髓性白血病的研究。不过，以往的实验表明通过静脉注射方式无法将原代AML细胞植入SCID小鼠中[13-15]。如果采用腹膜内植入或肾小囊下植入的方式，也只有少部分

under the renal capsule[13,14]. As AML cells have stringent cytokine requirements *in vitro*[7], we tested the effect of treating SCID mice, transplanted with peripheral blood cells (PBL) from AML patients newly diagnosed according to the French–American–British classification (FAB) as Ml, with cytokine PIXY321 (Fig. 1 legend) and human mast-cell growth factor (MGF). DNA analysis indicated that the bone marrow from mice treated for 30–60 days contained 10–100-fold more human cells than those from untreated control mice (data not shown). We have now examined a large number of samples ($n = 17$) from patients with newly diagnosed AML of different FAB subtypes (AML Ml, M2, M4) for their ability to proliferate in SCID mice. The cell source was either fresh bone marrow, fresh PBL, or banked frozen samples. All transplanted mice were treated with growth factors for the duration of the experiment (30–45 days). DNA and cytological analysis indicated that the bone marrow from 60 of 70 mice (86%) contained 10–100% human cells (Fig. 1), leading to almost complete replacement of murine haematopoiesis. Moreover, AML cells from all of the FAB subtypes (16 of 17 patients) engrafted SCID mice to high levels, indicating high reproducibility of the transplant system.

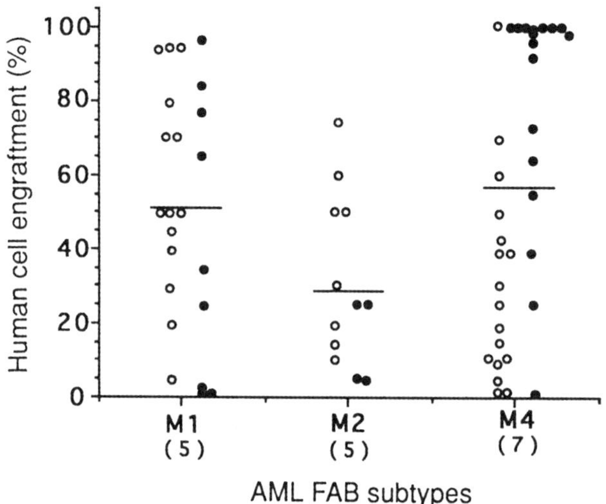

Fig. 1. Summary of human cell engraftment in the bone marrow of SCID mice transplanted with cells from AML patients, FAB subtypes M1, M2 or M4, where: M1 is AML without maturation; M2 is AML with granulocytic maturation; M4, acute myelomonocytic leukaemia; M5, acute monocytic leukaemia[22]. Number of donors is indicated. Engraftment was quantified by DNA analysis (open circles) or by cytology (filled circles) of Wright-stained bone marrow touch preparations 17–45 days post-transplantation. Lines represent mean level of engraftment for each FAB group.

Methods. Bone marrow and peripheral blood were obtained after informed consent according to procedures approved by the Human Experimentation Committee. Fresh or thawed AML cells were enriched by Ficoll-density gradient centrifugation and washed in IMDM medium containing 10% FCS. For transplantation of human cells, cells (1×10^7–4×10^7) were injected into the tail vein of sublethally irradiated (400 cGy, using a ^{137}Cs irradiator) SCID mice according to our established protocols[9]. PIXY321 (a fusion protein of human granulocyte–macrophage colony-stimulating factor with human IL-3) (7 μg) and human mast-cell growth factor (hMGF; c-*kit*-ligand; 10 μg) were administered on alternate days by intraperitoneal injection. Mice were bred and maintained in a defined flora colony (Ontario Cancer Institute). Grafting of human cells was quantified as described[9]. Briefly, 5 μg phenol-extracted DNA was digested with *Eco*RI, blotted onto a nylon membrane (Amersham) and probed with p17H8, a human

细胞可以在植入部位存活[13,14]。由于人 AML 细胞在体外的生长依赖于细胞因子[7]，我们对 PIXY321（参见图 1 图注）和人肥大细胞生长因子（MGF）在治疗 SCID 小鼠方面的效果进行了测试。其中，实验用的 SCID 小鼠均植入新诊断的 AML 患者（按照法国–美国–英国（FAB）分类标准定义为 M1）的外周血细胞（PBL）。DNA 分析显示，经细胞因子处理 30~60 天的小鼠骨髓中，人淋巴细胞的数量是未处理过的小鼠中的数量的 10~100 倍（数据未展示）。我们现已检测了 17 例不同 FAB 分型的 AML 患者（AML M1、M2 和 M4）的淋巴细胞在 SCID 小鼠中的增殖能力。这些细胞有的来源于新鲜骨髓，有的来源于新鲜 PBL，还有的是冷冻的库存样品。实验期间（30~45 天），所有移植后小鼠都使用生长因子处理。DNA 和细胞学分析结果表明，70 只小鼠中的 60 只，即 86%的骨髓中含有 10%~100%的人淋巴细胞（图 1），这导致小鼠造血细胞几乎完全被替代了。此外，所有 FAB 亚型的 AML 细胞（17 例中有 16 例）移植到 SCID 小鼠中都会出现高水平扩增，表明这一实验模型具有很高的可重复性。

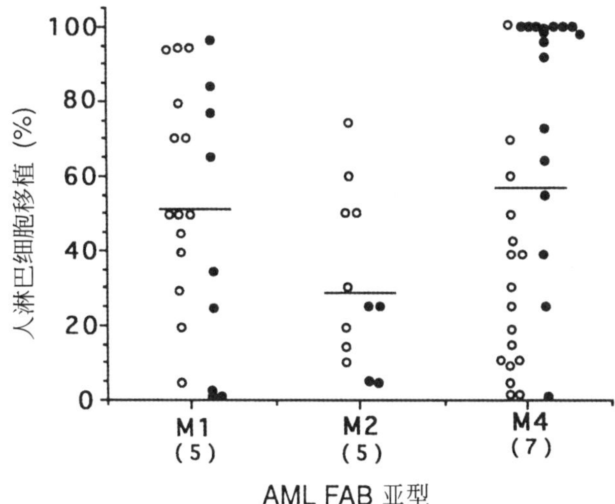

图 1. 不同 FAB 亚型（M1、M2、M4）AML 患者细胞植入 SCID 小鼠后小鼠骨髓中人类细胞的移植概况。M1 为 AML 未分化型；M2 为 AML 部分分化型；M4 为急性粒 – 单核细胞白血病；M5 为急性单核细胞白血病[22]。供体的数量如图所示。在植入细胞后的 17~45 天中，使用 DNA 分析（图中的空心圆）或使用细胞学（图中的实心圆）分析经瑞氏染色的骨髓印片标本，从而对移植进行量化研究。图中的横线代表各组的平均值。

方法。按照人体试验委员会批准的程序，在获得 AML 患者知情同意后，采集患者的骨髓和外周血样品。使用聚蔗糖密度梯度离心法富集新鲜或解冻的 AML 细胞，然后用含 10%胎牛血清的 IMDM 培养基洗涤细胞。按照我们先前建立的方法进行人类细胞的移植，将 1×10^7~4×10^7 个细胞通过尾静脉注射到经亚致死剂量（400 cGy，使用 ^{137}Cs 作为放射源）照射的 SCID 小鼠体内[9]。隔日腹腔注射 7 μg PIXY321（人粒细胞 – 巨噬细胞集落刺激因子和白介素 3 的融合蛋白）和 10 μg 人肥大细胞生长因子（hMGF，c-*kit* 配体）。小鼠在安大略肿瘤研究所饲养。扩散到骨髓中的人淋巴细胞的数量按文献中描述的方法计算[9]。简单地说，5 μg 使用苯酚抽提的 DNA 经 *Eco*RI 消化后转印到尼龙膜上（安马西亚公司）。然后使用特异性识别人 17 号染色体序列的 α 卫星探针——p17H8 作为探针进行检测[23]。通过比较样品与人 – 小鼠

α-satellite probe specific for human chromosome-17 sequences[23]. The percentage of human cells present in the mouse tissues was estimated by comparison of the intensity of the characteristic 2.7-kb band with standard human/mouse DNA mixtures (0, 0.1, 1.0, 10 and 50% human DNA).

Many morphological and dissemination features characteristic of the donor's disease were reproduced in the SCID-leukaemia mice. The bone marrow of mice transplanted with AML Ml cells was extensively infiltrated with undifferentiated blast cells (Fig. 2*a*); Auer rods were also seen in some leukaemic cells (Fig. 2*b*). SCID mice transplanted with AML M4 cells containing an inversion of chromosome 16 had many characteristically abnormal eosinophils with large basophilic granules (Fig. 2*c*). Flow cytometric analysis, using CD33 and CD13, of leukaemic cells from the engrafted murine bone marrow indicated that they had an immunophenotype identical to the donor leukaemic cells (data not shown). In addition, leukaemic cells were also present in the peripheral blood of engrafted SCID mice (Fig. 2*d*). In contrast to mice transplanted with AML M1 and M2, some with AML M4 cells became sick or died as early as 10–20 days post-transplant, with dissemination of leukaemic blasts to the liver (Fig. 2*e*), lungs, spleen and kidney (data not shown). Clinically, leukaemic blasts from patients with the monocytic subtypes AML M4 and M5 disseminate more extensively to extra-medullary sites than those from patients with AML M1/M2, suggesting that the SCID-leukaemia model accurately reflects biological differences between different AML subtypes.

To determine whether immature leukaemic blast colony-forming units (AML-CFU) were present in the bone marrow of highly engrafted mice, single-cell suspensions of marrow were plated in methylcellulose cultures. AML-CFU were present in mice transplanted with 11 of 11 donor samples, regardless of the FAB classification (Fig. 3*a*), and no progenitors of normal lineages were detected. Leukaemic cells, before and after transplantation into SCID mice, were plated at limiting dilution to compare the frequencies of AML-CFU. The assay was linear, and similar frequencies were obtained from the patient sample and the mouse bone marrow, 0.9 versus 0.3% respectively (Fig. 3*b*). Interestingly, the response in culture to interleukin-3 (IL-3) and human MGF of AML-CFU from the patient and the transplanted mouse was identical, indicating that neither the murine environment nor exogenous cytokine treatment selected for clones with altered responses to growth factors (Fig. 3*b*). Kinetic experiments were done to measure the number of human cells and AML-CFU present in the murine bone marrow over 28 days following transplantation of either 10^6 or 1.3×10^7 cells. At both cell doses, AML-CFU increased by > 100-fold over 14–28 days relative to the number detected in bone marrow one day after transplantation (Fig. 3*c*). The total number of human cells increased by 1,000-fold over the same period (data not shown). The presence of large numbers of growth-factor-responsive AML-CFU and their extensive expansion in the bone marrow of cytokine-treated mice implied that a leukaemic stem cell more immature than AML-CFU was maintaining the progenitor pool.

DNA 混合标准品（0、0.1%、1.0%、10%以及 50%人 DNA）中特征性的 2.7 kb 条带的浓度即可估算出人细胞在小鼠组织中的含量。

在 SCID 白血病小鼠中，供体的许多形态学和扩散性疾病特征得以再现。如植入 AML M1 型细胞的小鼠骨髓被未分化的母细胞高度浸润（图 2*a*）；在某些白血病细胞中可以观察到 Auer 小体的存在（图 2*b*）。植入 16 号染色体倒置的 AML M4 型细胞的 SCID 小鼠体内有明显异常的嗜酸性粒细胞，这些细胞中含有巨大的嗜碱性颗粒（图 2*c*）。使用流式细胞仪分析这些小鼠骨髓中白血病细胞表面 CD33 和 CD13 的表达情况时发现，它们具有和供体白血病细胞相同的免疫表型（数据未展示）。此外，在 SCID 白血病小鼠的外周血中也可以检测到白血病细胞（图 2*d*）。与植入 AML M1 型和 M2 型细胞的小鼠相比，一些植入 AML M4 型细胞的小鼠在植入后 10~20 天会由于白血病母细胞扩散到肝脏（图 2*e*）、肺部、脾脏和肾脏（数据未展示）而开始生病或死去。从临床上来讲，AML M4 和 M5 单核细胞亚型患者体内的白血病母细胞在骨髓外的扩散范围要比 AML M1 和 M2 型患者的更广。这表明 SCID– 白血病小鼠模型可以精确地反映出不同 AML 亚型的生物学差异。

为了明确植入细胞的小鼠骨髓中是否存在不成熟的白血病母细胞集落形成单位（AML-CFU），我们用甲基纤维素培养基制备了骨髓单细胞悬液。在全部 11 个供体样本中（不考虑 FAB 分型）都可以检测到 AML-CFU（图 3*a*），而检测不到正常细胞系的祖细胞。为了比较 AML-CFU 的频率，我们对移植前后的白血病细胞进行了有限稀释培养。分析结果表明两者呈线性关系，并且在患者样品和小鼠骨髓中 AML-CFU 出现的频率相近，分别为 0.9%和 0.3%（图 3*b*）。有意思的是，在培养基中患者和移植小鼠的 AML-CFU 对白介素 3 和人 MGF 的反应是一致的，表明小鼠体内环境和外源细胞因子的处理都与白血病细胞针对不同生长因素的克隆选择无关（图 3*b*）。在 10^6 或 1.3×10^7 个细胞植入小鼠后的 28 天内，我们对小鼠骨髓中人类细胞和 AML-CFU 的数量进行了动力学研究。结果显示，在两种细胞剂量下，骨髓中检测到的 AML-CFU 的数量在细胞植入后 14~28 天内相较于第一天都增加至原来的 100 倍以上（图 3*c*），而人类细胞的数量在相同时期内则增加至原来的 1,000 倍（数据未展示）。大量生长因子应答型 AML-CFU 的存在及其在细胞因子处理后的小鼠骨髓中广泛的扩增表明，很可能有一种比 AML-CFU 分化程度更低的白血病干细胞存在，从而使得祖细胞的数量得以维持。

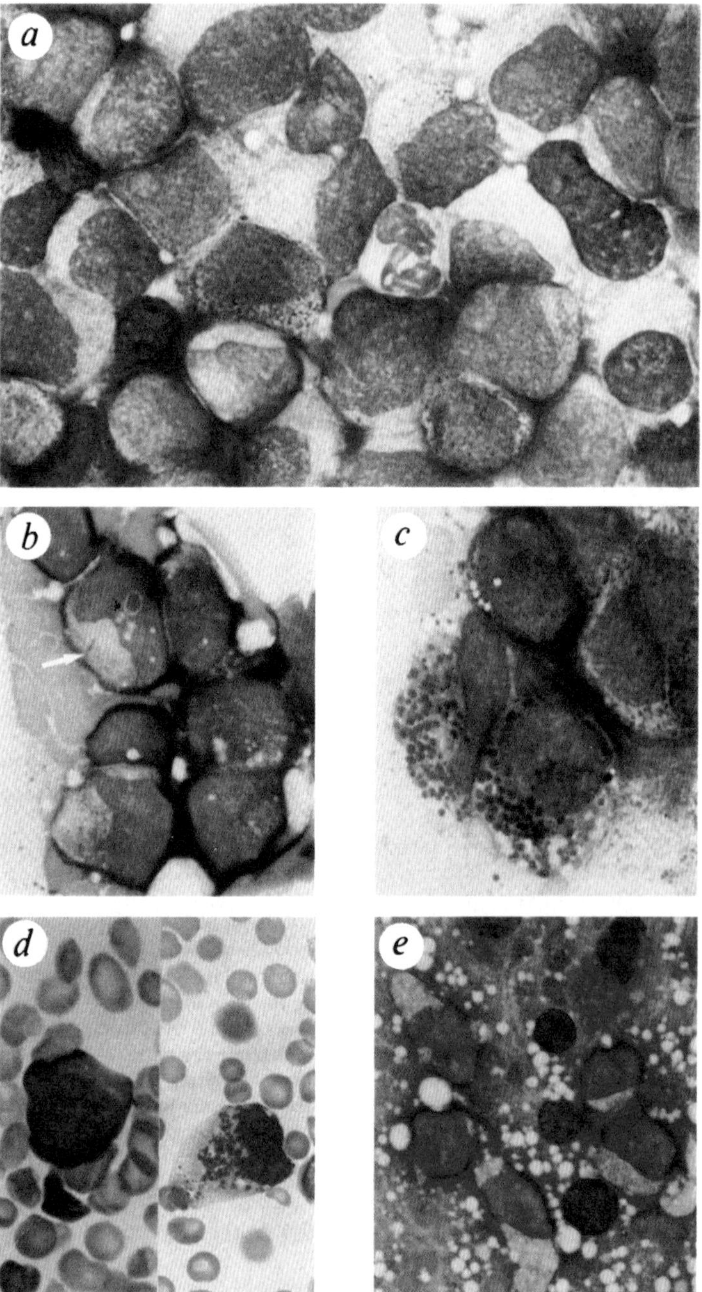

Fig. 2. Histology of cytokine-treated SCID mice injected with AML cells 3–5 weeks post-transplantation. *a*, Bone marrow touch preparation of a mouse highly infiltrated with AML Ml cells. Only one mouse neutrophil with a ring-shaped nucleus is present among the characteristic AML Ml blast cells. *b*, Mouse marrow repopulated with cells from a different AML Ml patient containing an Auer rod (arrow). *c*, Mouse marrow repopulated with cells from an AML M4 (with eosinophils and inversion of chromosome 16) patient containing characteristic eosinophils with large basophilic granules. *d*, Peripheral blood smear of a mouse repopulated with AML M4 cells. Human blast cells (left) and eosinophils (right) were present in the circulation of the mouse. *e*, Liver touch preparation of a mouse engrafted with AML M4 cells. Infiltrating leukaemic blast cells can be seen among mouse hepatocytes.

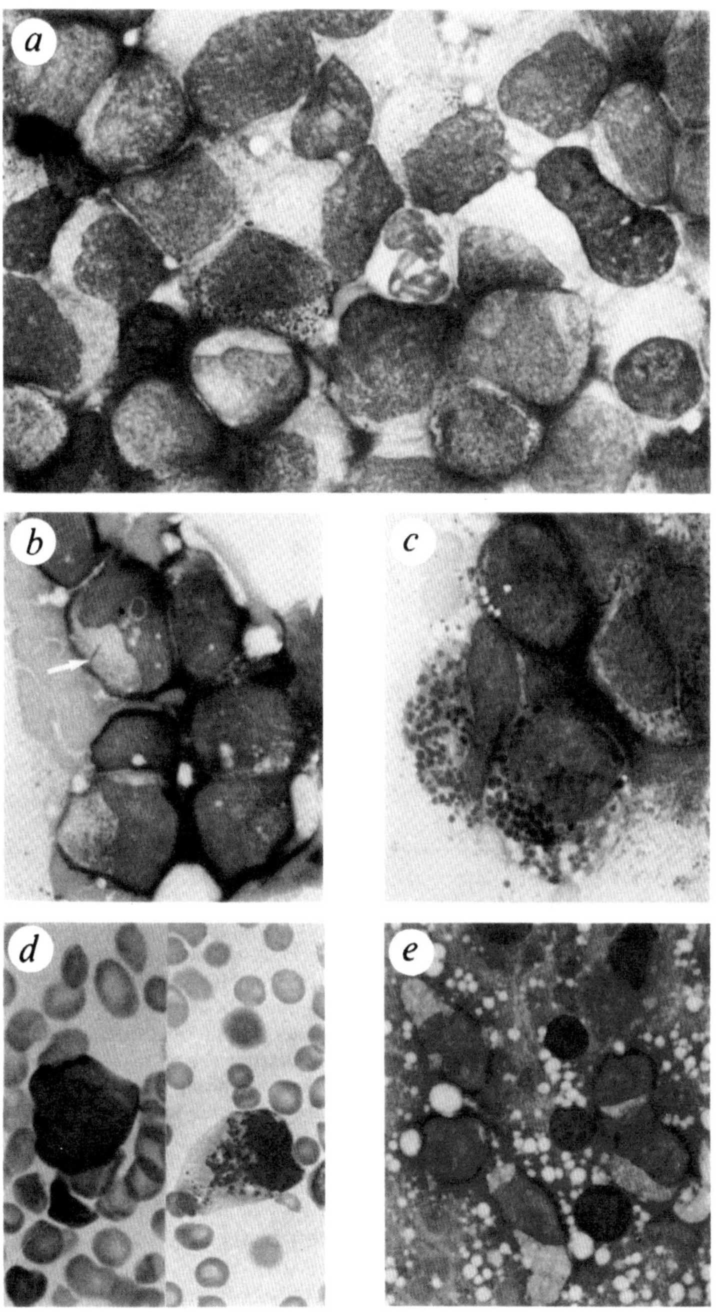

图2. 经细胞因子处理后的SCID小鼠植入AML细胞3~5周后的组织学图片。*a*，受人AML M1细胞高度浸润的小鼠骨髓印片。在AML M1母细胞中只有一个小鼠中性粒细胞（环状核）。*b*，另一个M1型AML患者的细胞在小鼠骨髓中的扩散情况。箭头所指为Auer小体。*c*，将M4型AML患者（含有嗜酸性粒细胞，并且16号染色体倒置）的细胞植入小鼠，在骨髓中可以看见含有巨大的嗜碱性颗粒的嗜酸性粒细胞。*d*，植入M4型AML患者细胞的小鼠外周血涂片。在小鼠的血液循环系统中可以看见人母细胞（左）和嗜酸性粒细胞（右）。*e*，植入M4型AML患者细胞的小鼠肝脏印片。在小鼠肝细胞周围可以看见渗入的白血病母细胞。

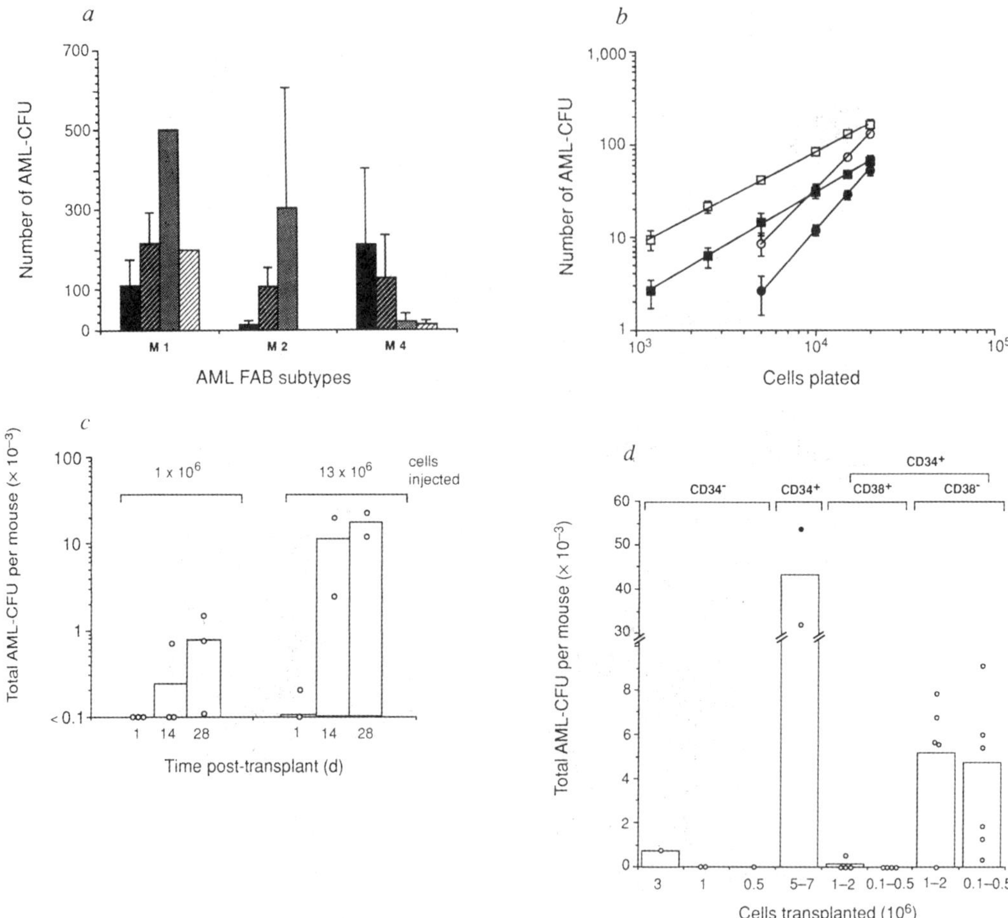

Fig. 3. Recovery of leukaemic colony-forming units (AML-CFU) from the bone marrow of transplanted SCID mice. *a*, Summary of the number of AML-CFU from the bone marrow of SCID mice transplanted with cells from AML M1, M2 or M4 patients and treated with hMGF and PIXY321 for 21 to 45 days. Each bar represents a unique donor sample showing the mean ± s.d. Error bars are shown only for groups containing 3 mice or more. The AML-CFU assay was performed in duplicate by plating 2×10^5 bone marrow cells in 0.9% methylcellulose-containing fetal bovine serum (15%), human plasma (15%), hMGF (50 ng ml^{-1}), PIXY 321 (5 ng ml^{-1}), hGM–CSF (1 U ml^{-1}), hIL-3 (10 U ml^{-1}) and human erythropoietin (2 U ml^{-1}). At day 7, leukaemic blast colonies were scored and their leukaemic identity confirmed by cytology and chromosomal analysis where applicable. No murine colonies were obtained under these selective culture conditions[8,9]. *b*, Frequency of AML-CFU before and after transplantation. Limiting dilution analysis of AML-CFU present in the PBL of an AML M1 patient before (open symbols) and after (filled symbols) transplantation. Cell numbers indicated were plated in methylcellulose culture containing either hIL-3 (1 nM) alone (circles) or in combination with hMGF (323 pM) (squares). Bone marrow from the transplanted mouse was analysed 35 days after transplantation and treatment with hMGF and PIXY321. *c*, Expansion of AML-CFU in the bone marrow of transplanted SCID mice. SCID mice were transplanted with AML Ml cells at the cell numbers indicated and killed at 1, 14 and 28 days post-transplant. The total number of AML-CFU per mouse was determined by multiplying the AML-CFU per 2×10^5 cells plated by the total number of bone marrow cells present in the mouse. The limit of detection was 100 AML-CFU per mouse. *d*, Measurement of AML-CFU from SCID mice transplanted with AML cells fractionated according to CD34 and CD38 expression. Three independent cell-sorting experiments were done on thawed cells containing 75% CD34⁺ cells from an AML Ml patient; of these 40% were

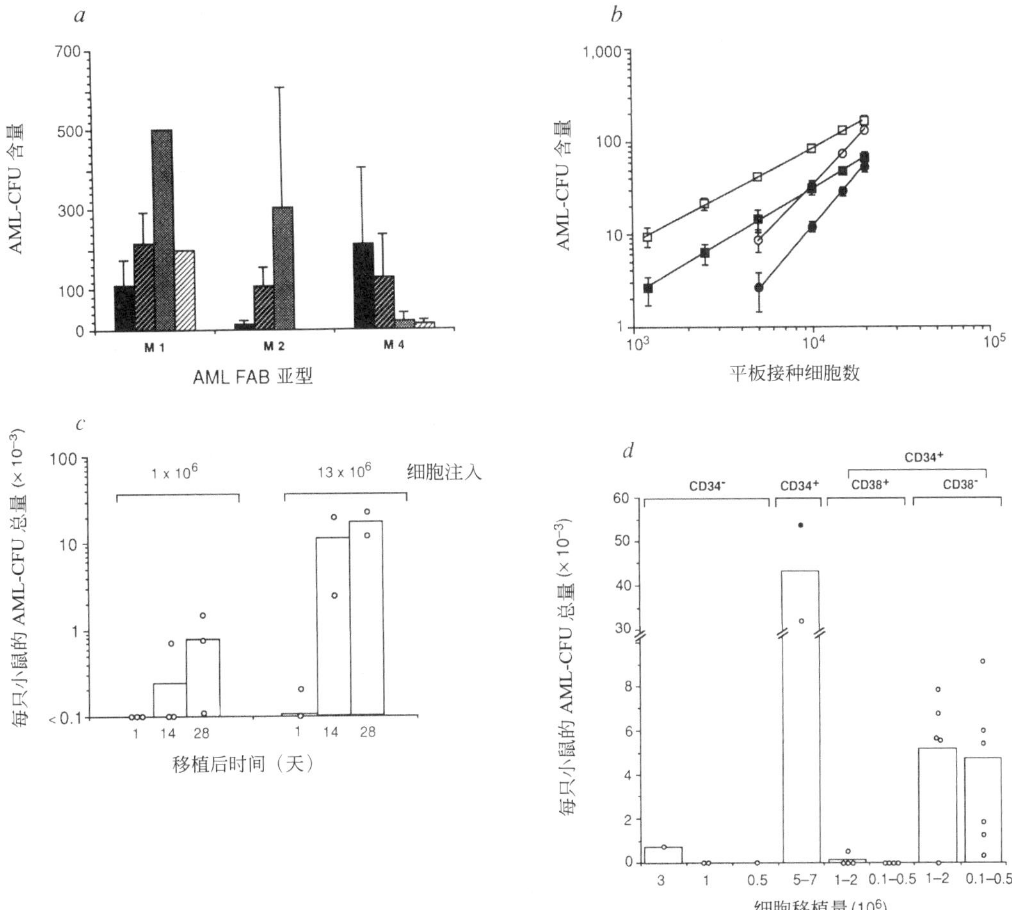

图 3. 从移植后小鼠骨髓中分离的白血病母细胞集落形成单位（AML-CFU）。*a*，植入 AML M1、M2 和 M4 细胞并且使用 hMGF 和 PIXY321 处理 21~45 天的小鼠骨髓中 AML-CFU 的含量。每个柱代表同一个植入样本在不同小鼠中 AML-CFU 含量的平均值 ± 标准差。只有每组小鼠数量大于等于 3 只时才计算误差。AML-CFU 测定按如下方法重复进行：将 2×10^5 个骨髓细胞接种在含有 15%胎牛血清、15%人血浆、50 ng · ml^{-1} 人肥大细胞生长因子、5 ng · ml^{-1} PIXY321、1 U · ml^{-1} 人粒细胞 – 巨噬细胞集落刺激因子、10 U · ml^{-1} 人白介素 3 和 2 U · ml^{-1} 人促红细胞生成素的 0.9% 甲基纤维素中。在第七天的时候计算白血病母细胞集落数。此时通过细胞学和染色体分析的方法可以鉴别他们的白血病特征。在这种选择培养条件下，没有出现鼠源细胞集落 [8,9]。*b*，在细胞植入前后，AML-CFU 出现的频率。使用有限稀释法分析 M1 型 AML 患者外周血淋巴细胞中 AML-CFU 的含量。空心符号为植入前，实心符号为植入后。圆圈表示甲基纤维素中含有人白介素 3（1 nM）的实验组，方框为同时含有人白介素 3（1 nM）和人肥大细胞生长因子（323 pM）的实验组。骨髓样品取自植入细胞 35 天后并使用人肥大细胞生长因子和 PIXY321 处理的小鼠。*c*，AML-CFU 在 SCID 小鼠骨髓中的扩增。向 SCID 小鼠中植入了如图所示数量的 AML M1 细胞，并分别在植入后 1 天、14 天和 28 天处死。每只小鼠的 AML-CFU 总量等于每 2×10^5 个细胞中 AML-CFU 的数量乘以小鼠中总骨髓细胞数。检测的下限是每只小鼠 100 个 AML-CFU。*d*，植入了不同 CD34 和 CD38 表达分型的 AML 细胞的小鼠中，AML-CFU 数量的测量。我们使用 3 个相互独立的细胞分选实验对来自 M1 型 AML 患者解冻的细胞进行检测，其中 75%为 $CD34^+$，40%为 $CD38^+$。分别使用荧光激活细胞分选法（FACStar^{PLUS}，BD 公司）或使用 CD34 亲和层析柱（Cell–Pro 公司，图中为

CD38$^+$. Cells were purified by fluorescence-activated cell sorting (FACStar^{PLUS}; Becton–Dickinson) or using a CD34 affinity column (Cell-Pro) (filled circle). In the first experiment, cells were separated on the basis of CD34 expression; the CD34$^+$ and CD34$^-$ populations were each 98% pure; in experiments 2 and 3, respectively, the CD34$^+$/CD38$^-$ cells were 88 and 97% pure, and the CD34$^+$/CD38$^+$ cells were 80 and 73% pure, being contaminated with 10 or 15% CD34$^+$/CD38$^-$ cells. The CD34$^+$/CD38$^-$ and the CD34$^+$/CD38$^+$ fractions from both experiments 2 and 3 contained an average of 6,525 and 8,736 AML-CFU per 2×10^5 cells, respectively. Mice were transplanted with the indicated number of purified cells and treated with cytokines. After 45 days (experiment 1) or 30 days (experiment 2 and 3), the total number of AML-CFU was determined.

Flow-sorting was used to characterize and purify the leukaemia-initiating cells. CD34 is a cell-surface marker normally expressed on a small population of bone marrow cells, including progenitor cells and pluripotent stem cells[16]. Expression of CD38 on CD34$^+$ cells is an important marker for lineage commitment and therefore the CD34$^+$CD38$^-$ phenotype defines an immature human cell in normal bone marrow. Although the expression of CD34 and CD38 on AML cells is very heterogeneous[17], the CD34$^+$CD38$^-$ phenotype is present on immature AML-CFU[18]. Peripheral blood cells from an AML M1 patient were separated into CD34-positive and -negative fractions and transplanted into SCID mice. Leukaemic cell proliferation and high levels of AML-CFU were observed in the bone marrow of mice transplanted with CD34$^+$ cells, as in mice transplanted with unsorted populations (Fig. 3*d*). By contrast, four mice transplanted with CD34$^-$ cells were poorly engrafted (0–0.1%) and contained no AML-CFU, except for one mouse that had been transplanted with a high cell dose and contained only a few colonies (Fig. 3*d*). In two further experiments mice were transplanted with CD34$^+$CD38$^-$ or CD34$^+$CD38$^+$ cells purified from the same donor. Leukaemia cell proliferation and high levels of AML-CFU were only seen in mice transplanted with CD34$^+$CD38$^-$ cells (Fig. 3*d*). Interestingly, both populations contained comparably high numbers of AML-CFU before injection into SCID mice, providing evidence that at least CD34$^+$CD38$^+$ AML-CFU could not engraft SCID mice. CD34$^+$CD38$^-$ cells transplanted into SCID mice did not produce any detectable normal human progenitors; all of the colony-forming cells tested (40/40) contained the same t(2;4) chromosomal translocation found in the patient's leukaemic clone.

We next investigated whether there was a linear relationship between the number of cells injected and leukaemic engraftment, in order to develop a quantitative assay for AML-initiating cells. Peripheral blood cells from AML patients were diluted in a tenfold series from 2×10^7 cells to 2×10^4 cells before transplanting into SCID mice. In a representative experiment, DNA analysis indicated that as few as 2×10^5 cells were sufficient to initiate leukaemic proliferation (Fig. 4). Statistical analysis of the proportion of mice that had leukaemic proliferation at each cell dose, using data from four different donors transplanted into 40 mice, indicated that the frequency of the leukaemia-initiating cell in the PBL of AML-M1 patients was 1 engraftment unit per 2.5×10^5 cells (range: 1 in 1.2×10^5 to 1 in 5.3×10^5); the engraftment followed single-order kinetics as measured by the χ^2 test (95% confidence limit).

实心圆）对细胞进行纯化。在第一个实验中，我们根据 CD34 的表达情况对细胞进行了分离。$CD34^+$ 细胞和 $CD34^-$ 细胞的纯度都达到了 98%。在实验二和实验三中，$CD34^+/CD38^-$ 细胞的纯度分别为 88% 和 97%，$CD34^+/CD38^+$ 细胞的纯度分别为 80%和 73%，其中可能污染了 10%或 15%的 $CD34^+/CD38^-$ 细胞。实验二和实验三中，每 2×10^5 个 $CD34^+/CD38^-$ 细胞以及 $CD34^+/CD38^+$ 细胞中分别含有 6,525 和 8,736 个 AML-CFU。向小鼠植入图中所示数量的纯化细胞并使用细胞因子处理，并在 45 天后（实验一）或 30 天后（实验二和实验三）确定其 AML-CFU 总量。

我们使用流式分选技术对这种白血病起始细胞进行分离纯化。CD34 是一个细胞表面标记，正常情况下仅表达于小部分骨髓细胞中，包括祖细胞和多能干细胞[16]。$CD34^+$ 细胞中 CD38 分子的表达是一个重要的细胞系定型标志。$CD34^+CD38^-$ 表型是正常骨髓中未成熟细胞的标志。尽管在 AML 细胞中，CD34 和 CD38 的表达差别很大[17]，但 $CD34^+CD38^-$ 却是未成熟 AML-CFU 细胞的一个特征[18]。我们将 AML M1 型白血病患者的外周血细胞分成 $CD34^+$ 和 $CD34^-$ 两部分，并分别把它们植入 SCID 小鼠体内。结果发现，在那些植入 $CD34^+$ 细胞和未分选细胞的小鼠骨髓中，可观察到白血病细胞的增殖和高水平的 AML-CFU（图 3*d*）。相反，四只植入 $CD34^-$ 细胞的 SCID 小鼠中植入成功率很低（0~0.1%），并且不含有 AML-CFU 细胞（除一只植入大剂量细胞且含有少量集落的小鼠外）（图 3*d*）。在两个后续实验中，我们将来自同一供体纯化的 $CD34^+CD38^-$ 或 $CD34^+CD38^+$ 细胞移植到小鼠体内。结果显示，只有在植入 $CD34^+CD38^-$ 细胞的小鼠体内能够看到白血病细胞的增殖和高水平的 AML-CFU（图 3*d*）。有趣的是，这两群细胞在植入 SCID 小鼠之前都包含同等高水平的 AML-CFU。这证明至少 $CD34^+CD38^+$ 的 AML-CFU 细胞不能植入 SCID 小鼠。那些 $CD34^+CD38^-$ 细胞在植入 SCID 小鼠后不能产生可检测的正常人祖细胞，在检测的 40 个集落形成细胞中都含有与患者白血病细胞中一样的染色体易位 t(2; 4)。

接下来，我们研究了接种细胞数与白血病植入数之间是否存在线型关系，以便建立一个 AML 起始细胞的定量分析模型。在向 SCID 小鼠植入细胞前，我们将 AML 患者的外周血细胞进行了 10 倍梯度稀释，从 2×10^7 一直稀释到 2×10^4。在一个代表性实验中，DNA 分析结果显示，最少 2×10^5 个细胞就足以引起白血病细胞的增殖（图 4）。我们利用四个不同患者的细胞植入 40 只小鼠后的数据，针对不同细胞接种剂量中出现白血病细胞增殖的小鼠比例进行了统计学分析，结果发现 AML-M1 患者的外周血中，白血病起始细胞的频率为每 2.5×10^5 个细胞中 1 个植入单位，这一细胞数的波动范围为 1.2×10^5~5.3×10^5。使用卡方检验（χ^2，95%置信限）发现白血病小鼠的移植变化趋势呈一级动力学曲线。

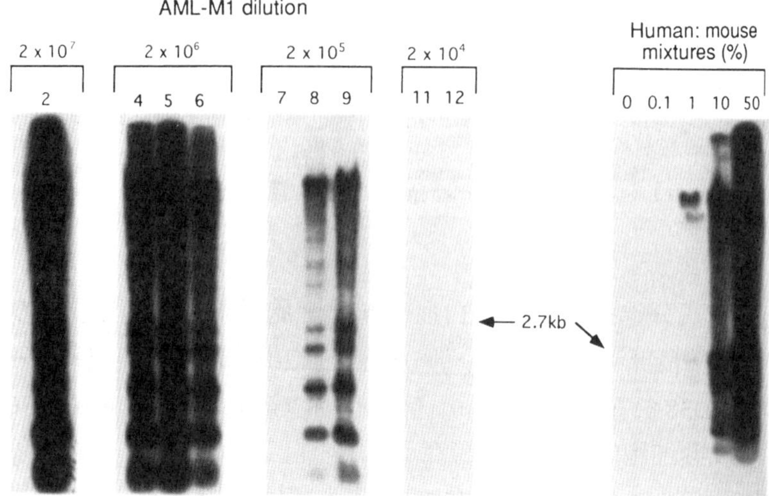

Fig. 4. Determination of the frequency of the SCID leukaemia-initiating cell (SL-IC) engraftment unit by limiting dilution analysis. PBL cells from an AML M1 patient were thawed and different cell doses (2×10^4, 2×10^5, 2×10^6 and 2×10^7 cells) were transplanted into groups of 3 or 4 mice. Mice were treated with hMGF and PIXY321 for 1 month, after which DNA from the bone marrow was analysed for human cells as described for Fig. 1. The Southern blot is representative of four experiments with four different donors. SCID mice containing > 5% leukaemic cells in the bone marrow were considered positive for the statistical analysis used to determine the frequency of SL-IC by the method of Porter and Barry[24]. The negative mice contained from 0.1% to undetectable human cells. Ethidium bromide staining indicated that equal amounts of DNA were loaded in each gel lane.

We have identified an AML-initiating cell on the basis of its ability to establish human leukaemia in SCID mice (the SCID leukaemia-initiating cell, or SL-IC). Three pieces of evidence suggest that there may be a hierarchy of leukaemic stem cells in human AML, where SL-IC are more immature than AML-CFU. First, the frequency of SL-IC in the PBL of AML Ml patients is at least 1,000-fold lower than the frequency of AML-CFU. Second, only mice transplanted with $CD34^+CD38^-$ cells developed leukaemia whereas $CD34^+CD38^+$ mice did not, despite the fact that similar numbers of AML-CFU were present in both populations before transplantation. Third, based on the low proliferative capacity of AML-CFU in liquid or long-term cultures[4,19,20] even with maximal growth-factor stimulation, it is likely that their large expansion in SCID mice for > 45 days post-transplantation is due to the proliferation and differentiation of SL-IC. In future, autologous transplantation with purged cells may address the relationship between SL-IC and the leukaemic stem cell that maintains the disease in patients. But the fact that SL-IC shares a $CD34^+CD38^-$ expression pattern similar to normal stem cells indicates that purging strategies may be difficult to develop. It will also be possible to create complementary DNA libraries[21] from single cells to characterize genes that are expressed in SL-IC and compare them with those from normal stem cells and the more differentiated AML-CFU. Finally, a SCID-leukaemia model that reproduces many features of human AML should help us to understand the processes governing the transformation and progression of leukaemic stem cells and to test new therapeutic strategies.

(**367**, 645-648; 1994)

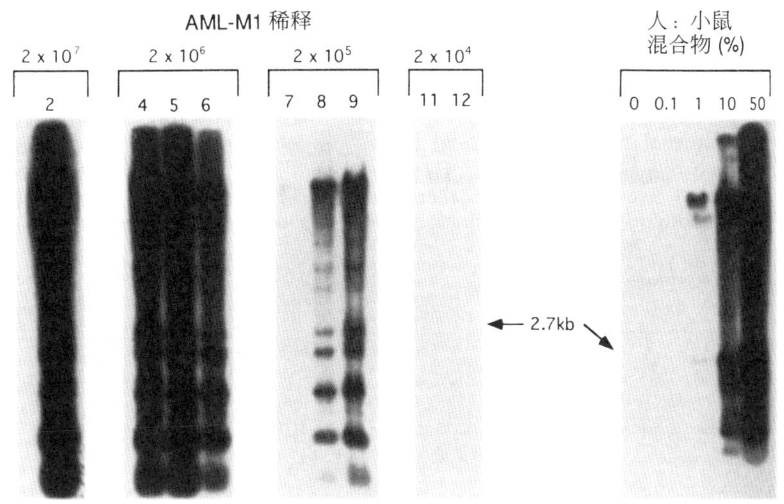

图 4. 利用有限稀释法计算 SCID 白血病起始细胞（SL-IC）扩散的频率。将解冻后不同细胞剂量的 M1 型 AML 患者的外周血淋巴细胞（2×10^4、2×10^5、2×10^6 和 2×10^7）分别植入各组小鼠中（每组含 3 或 4 只小鼠），使用人肥大细胞生长因子和 PIXY321 处理一个月，然后从小鼠骨髓中提取 DNA 并采用图 1 中提及的方法计算人细胞的含量。图中所示的是四个不同供体的 DNA 印迹结果。采用波特和巴里的方法 [24] 确定 SL-IC 的频率时，从统计学分析来讲骨髓中白血病细胞含量超过 5%即视为阳性。而人细胞含量在 0.1% 以下或无法检测则视为阴性。使用溴化乙锭染色可以看到每个孔中总 DNA 的上样量是一致的。

通过衡量细胞在 SCID 小鼠中引发人白血病的能力，我们鉴定了一类 AML 起始细胞（称为 SCID 白血病起始细胞或 SL-IC）。有三方面的证据表明在人 AML 中可能存在白血病干细胞群体，并且在这个群体中 SL-IC 比 AML-CFU 分化程度更低。第一，在 AML M1 型患者的外周血中 AML-CFU 的频率至少是 SL-IC 的 1,000 倍。第二，虽然在移植前 $CD34^+CD38^+$ 和 $CD34^+CD38^-$ 两种细胞中都含有大致相等的 AML-CFU，但是只有 $CD34^+CD38^-$ 细胞可以在 SCID 小鼠中引发白血病。第三，即使使用最大剂量的生长因子刺激，AML-CFU 在液体或长期培养中的增殖能力都很低 [4,19,20]。在植入细胞超过 45 天后，SCID 小鼠中 AML-CFU 的大幅扩增看起来更像是由于 SL-IC 细胞的增殖和分化。在将来，使用净化的细胞进行自体移植实验或许可以揭示 SL-IC 和维持病人生病状态的白血病干细胞之间的关系。不过，由于 SL-IC 和大多数干细胞一样也是 $CD34^+CD38^-$ 型的，这使得将其与干细胞区分开来变得很困难。人们有可能通过从单个细胞中建立 cDNA 文库 [21] 的方法来比较 SL-IC 与正常干细胞和 AML-CFU 中基因表达的差异。最后，这样一个再现了许多人 AML 特点的 SCID 白血病小鼠模型对于我们理解白血病干细胞分化的过程和试验新的治疗方案具有很大的帮助。

（张锦彬 翻译；秦志海 审稿）

Tsvee Lapidot, Christian Sirard, Josef Vormoor, Barbara Murdoch, Trang Hoang*, Julio Caceres-Cortes*, Mark Minden†, Bruce Paterson‡, Michael A. Caligiuri§ & John E. Dick‖
Department of Genetics, Research Institute, Hospital for Sick Children and Department of Molecular and Medical Genetics, University of Toronto, 555 University Avenue, Toronto, Ontario M5G 1X8, Canada
* Clinical Research Institute, Montreal, Quebec H2W 1R7, Canada
† Department of Medicine and ‡Department of Oncologic Pathology, Princess Margaret Hospital, Toronto, Ontario M4X 1K9, Canada
§ Department of Medicine, Roswell Park Cancer Institute, Buffalo, New York 14263-0001, USA

Received 30 September; accepted 29 November 1993.

References:

1. Sawyers, C., Denny, C. & Witte, O. *Cell* **64**, 337-350 (1991).
2. Fearon, E., Burke, P., Schiffer, C., Zehnbauer, B. & Vogelstein, B. *New Engl. J. Med.* **315**, 15-24 (1986).
3. Keinänen, M., Griffin, J., Bloomfield, C., Machnicki, J. & de la Chapelle, A. *New Engl. J. Med.* **318**,1153-1158 (1988).
4. Griffin, J. & Löwenberg, B. *Blood* **68**, 1185-1195 (1986).
5. Grier, H. & Civin, C. in *Hematology of Infancy and Childhood* (eds Nathan, D. G. & Oski, F. A.) 1288-1318 (Saunders, Philadelphia, 1993).
6. McCulloch, E., Izaguirre, C., Chang, L. & Smith, L. *J. Cell. Physiol. Suppl.* **1**,103-111 (1982).
7. Löwenberg, B. & Touw, I. *Blood* **81**, 281-292 (1993).
8. Kamel-Reid, S. & Dick, J. E. *Science* **242**, 1706-1709 (1988).
9. Lapidot, T. *et al. Science* **255**, 1137-1141 (1992).
10. Kamel-Reid, S. *et al. Science* **246**, 1597-1600 (1991).
11. Kamel-Reid, S. *et al. Blood* **78**, 2973-2981 (1991).
12. Dick, J., Lapidot, T. & Pflumio, F. *Immun. Rev.* **124**, 25-43 (1991).
13. Cesano, A. *et al. Oncogene* **7**, 827-836 (1992).
14. Sawyers, C., Gishizky, M., Quan, S. Golde, D. & Witte, O. *Blood* **79**, 2089-2098 (1992).
15. De Lord, C. *et al. Expl Hemat.* **19**, 991-993 (1991).
16. Civin, C. *et al. J. Immun.* **133**, 157-165 (1984).
17. Terstappen, L. *et al. Leukemia* **6**, 993-1000 (1992).
18. Terstappen, L., Huang, S., Safford, M., Lansdorp, P. & Loken, M. *Blood* **77**, 1218-1227 (1991).
19. Coulombel, L., Eaves, C., Kalousek, D., Gupta, C. & Eaves, A. *J. Clin. Invest.* **75**, 961-969 (1985).
20. Schiró, R. *et al. Blut* **61**, 267-270 (1990).
21. Brady, G., Barbara, M. & Iscove, N. *Meth. Molec. Cell. Biol.* **2**, 17-25 (1990).
22. Bennett, J. *et al. Br. J. Haemat.* **33**, 451-458 (1976).
23. Waye, S. & Willard, H. *Molec. Cell. Biol.* **6**, 3156-3165 (1986).
24. Porter, E. & Berry, R. *Brit. J. Cancer* **17**, 583-595 (1964).

Acknowledgments. We thank F. Pflumio, R. A. Phillips, A. Bernstein, N. Iscove and M. Buchwald for reviewing the manuscript; D. Williams (Immunex) for growth factors; and P. Laraya, D. Brown and L. Harton for assistance. Supported by grants from the MRC of Canada, the National Cancer Institute of Canada (NCIC), with funds from the Canadian Cancer Society, the NIH (M.A.C.), Coleman Leukemia Research Fund (M.A.C.), a studentship award (Hospital for Sick Children) (C.S.), postdoctoral fellowships from the NCIC (T.L.) and the Deutsche Forschungsgemeinschaft (J.V.), and a Research Scientist award from the NCIC (J.E.D.). T.L. and C.S. contributed equally to this work.

2.2 Mb of Contiguous Nucleotide Sequence from Chromosome III of *C. elegans*

R. Wilson *et al.*

Editor's Note

The collaborative effort to sequence the genome of the nematode worm *Caenorhabditis elegans* having been underway for three years, British biologist John Sulston and colleagues provide a progress update. Nearly 2.2 million base pairs out of the total 100 million base pairs had been sequenced, confirming the previously reported high gene density and yielding several intriguing features. Putative gene duplications and other repeats raised potential evolutionary implications, and the complete *C. elegans* gene catalogue was already valuable to those using the model organism to study development, behaviour and gene function. The project also suggested that megabase-scale DNA sequencing at a reasonable cost was possible. An essentially complete version of the genome was published, on schedule, in 1998.

As part of our effort to sequence the 100-megabase (Mb) genome of the nematode *Caenorhabditis elegans*, we have completed the nucleotide sequence of a contiguous 2,181,032 base pairs in the central gene cluster of chromosome III. Analysis of the finished sequence has indicated an average density of about one gene per five kilobases; comparison with the public sequence databases reveals similarities to previously known genes for about one gene in three. In addition, the genomic sequence contains several intriguing features, including putative gene duplications and a variety of other repeats with potential evolutionary implications.

THE free-living nematode *Caenorhabditis elegans* is an excellent model organism for the study of development and behaviour, and its small size and short life cycle greatly facilitate genetic analysis[1,2]. Because a nearly complete clonal physical reconstruction of the genome is available[3-5], and 1,200 genetic loci have been identified, the nematode is also a good candidate for complete DNA sequence analysis. The haploid genome of *C. elegans* contains approximately 100 Mb (megabases) distributed on six chromosomes[2]. Many of the genes required for normal development and behaviour in the nematode have extensive similarity to their mammalian counterparts. However, compared with mammalian genes, genes in *C. elegans* typically have smaller and fewer introns[2,6], thus simplifying the identification of previously uncharacterized genes. Many of these newly identified genes may in turn be used to probe for as-yet unidentified mammalian genes.

As previously reported[6], we have embarked on a collaborative project to sequence the

秀丽隐杆线虫 3 号染色体上的 2.2 Mb 连续核苷酸序列

威尔逊等

编者按

秀丽隐杆线虫基因组序列的测定工作已经进行了三年，英国生物学家约翰·萨尔斯顿和他的同事们不断更新进展。一亿个碱基对中有近 220 万碱基对已完成测序，证实了前期关于其高基因密度的报道并且发现了几个非常有趣的特征。推断的基因重复和其他序列重复提高了其潜在的进化意义，并且完整的秀丽隐杆线虫基因目录对于用模式生物研究发育、行为和基因功能发挥了重要作用。这个项目也表明以合理的花费进行兆碱基规模 DNA 测序是可行的。一个基本完整的基因组在 1998 年如期发表。

对线虫纲秀丽隐杆线虫 100 兆碱基（Mb）的基因组进行测序是我们研究的一部分，我们已经完成了秀丽隐杆线虫 3 号染色体的重要基因簇上连续 2,181,032 个碱基对的测序。对完成测序的序列的分析表明基因的平均密度为每五千个碱基含一个基因；与公开序列数据库的比较显示，这与已知基因的密度（每三千个碱基含一个基因）相似。此外，基因组序列包含几个有趣的特征，包括具有潜在进化意义的推断的基因重复和各种其他重复。

自由生活的线虫纲秀丽隐杆线虫是研究发育和行为极好的模式生物。它个体小，生命周期短，大大方便了遗传分析 [1,2]。因为基因组克隆序列重建几乎已经完成 [3-5]，并已鉴定了 1,200 个遗传基因座，所以线虫也是很好的分析完整 DNA 序列的候选者。秀丽隐杆线虫单倍体基因组约为 100 Mb，分布在六条染色体上 [2]。线虫正常发育和行为所必需的许多基因与哺乳动物对应基因有广泛的相似性。然而，与哺乳动物基因相比，秀丽隐杆线虫基因的内含子较小且较少 [2,6]，因此简化了之前未知基因的鉴定。许多新鉴定的基因反过来可作为探针寻找目前还未鉴定的哺乳动物基因。

正如以前报道的那样 [6]，我们早就着手秀丽隐杆线虫全基因组测序的合作项目。

entire genome of *C. elegans.* Here we report some of the results from the first three years of the pilot phase, a point at which each laboratory has completed one contiguous megabase of genomic sequence. The combined data represent a sequence spanning more than 2.1 Mb. Most of the sequence derives from cosmid clones that were mapped to chromosome III by restriction fingerprinting[3]. At this physical map location, there are two large cosmid contigs, each more than 1 Mb long and bridged by a yeast artificial chromosome (YAC) clone. A small cosmid contig of 92 kilobases (kb) lies near the centre of the YAC bridge. Our genomic sequence analysis of this large region of chromosome III has confirmed the high gene density that we found in the first three sequenced cosmids[6], and has resulted in the identification of many more genes and other interesting sequence features.

Sequencing Strategy

At the start of the pilot phase, we experimented with a primer-directed or "walking" strategy in which site-specific oligonucleotide primers were used to extend sequences sequentially from a limited number of starting points[7]. This was initially done using cosmid DNA as template for the sequencing reactions. However, because cosmid DNA proved difficult to purify in sufficient quantities and was troublesome because of the presence of repeated sequences, we used random phagemid and M13 subclones as templates for the walking strategy[8,9]. During this work, we found that a combination of small insert (1–2 kb) and large insert (6–9 kb) subclones provided representative coverage for a typical cosmid. Even with subclones, primer-directed sequencing was occasionally problematic because of the presence of repeated sequences. Thus, our strategy has evolved to the point where most sequence data come from the initial readings of 600–800 random subclones. As the use of automated fluorescent DNA sequencing machines for data collection provides high-throughput sample processing[10,11], this approach is efficient and cost-effective. This random or "shotgun" phase not only provides much of the final sequence, but also maps the subclones needed for closing gaps and completing the complementary strand. At this point, a walking strategy can be successfully exploited for completion, as most repeated sequences in the cosmid insert can be identified and selectively avoided. Many of the details of our sequencing strategy have been reported elsewhere[6,12-23].

Sequences were considered to be finished when every base had been determined on both strands and all ambiguities had been resolved. At this point, the finished sequence was compared with the public sequence databases using the BLASTX program for protein similarities and BLASTN for nucleotide similarities[24]. When similarity searches were complete, likely genes were identified using GENEFINDER (P.G. and L.H., unpublished), and in some cases interactively edited using ACEDB (R.D. and J.T.-M., unpublished). Finished sequences were annotated with regard to likely genes, homologies, *trans*-spliced leaders[26], complementary DNA matches[27,28] and other features, such as structural RNA genes and transposons, and submitted to the GenBank and EMBL databases. Also, to present genomic sequence data in the context of the physical and genetic maps of the nematode, all sequences are deposited in the *C. elegans* database ACEDB, which is available to the research community.

在这里，我们公布试验期第一个三年的一些结果，至此各个实验室已经完成了连续的一兆碱基的基因组序列。汇总的数据代表了长度大于 2.1 Mb 的序列。大部分序列来自利用限制性指纹识别的定位于 3 号染色体上的黏粒克隆 [3]。在这个物理图谱上有两个大的黏粒重叠群，每个长度都大于 1 Mb，并由一个酵母人工染色体（YAC）克隆连接。有一个 92 kb 的小黏粒重叠群邻近 YAC 桥中心。对 3 号染色体上这个大区域的基因组序列的分析证实了我们在前三个完成测序的黏粒中发现的高基因密度 [6]，并鉴定了更多基因和其他有趣的序列特征。

测序策略

试验阶段开始时，我们使用引物引导或“步移”策略，用位置特异的寡核苷酸引物从有限的起始点顺序扩展序列 [7]。最初用黏粒 DNA 做模板进行测序反应。然而，因为黏粒 DNA 很难纯化到足够的量，并且由于存在重复序列而非常棘手，所以我们用随机噬菌粒和 M13 亚克隆做模板进行步移 [8,9]。在此工作期间，我们发现一个小插入（1~2 kb）亚克隆和大插入（6~9 kb）亚克隆的组合代表性地覆盖了一个典型的黏粒。因为重复序列的存在，即使采用亚克隆，引物引导的测序偶尔也会出现问题。因此，我们的策略已经发展到从最初解读的 600~800 个随机亚克隆中获得多数序列数据的程度。由于使用自动荧光 DNA 测序仪收集数据可以高通量地处理样品 [10,11]，所以这个方法经济有效。这个随机或“鸟枪”时期不仅提供了大量最终的序列，也为封闭缺口和完成互补链所需的亚克隆进行了作图。至此，步移策略得以成功运用，黏粒插入的大多数重复序列可被鉴定和选择性避免。我们测序策略的很多细节也已在别处报道过 [6,12-23]。

当两条链上的所有碱基都已确定，所有不明确都已解决时就认为序列已经完成。这时，用 BLASTX 软件比较最终序列与公开序列数据库的蛋白相似性，用 BLASTN 比较核酸相似性 [24]。完成相似性查找后，用 GENEFINDER（格林和希利尔，未发表）鉴定可能的基因，有些情况下交互使用 ACEDB（德宾和蒂埃里 – 米格，未发表）进行编辑。对最终序列上的可能基因、同源物、反式剪接前导序列 [26]、互补 DNA 匹配物 [27,28] 和其他特征（如结构 RNA 基因和转座子）进行注释，并提交到 GenBank 和 EMBL 数据库。而且，将基因组序列数据绘制到线虫基因组物理和遗传图谱中，所有数据都放入线虫数据库 ACEDB 中，这些数据可供研究团体使用。

Sequences

Table 1 gives the cosmid clones that were sequenced, along with their database accession numbers and lengths. Most of the redundant overlapping data have been omitted from the database entries, although neighbouring sequences typically contain a small number of overlapping bases to facilitate joining or to keep a gene intact. The total non-overlapping sequence assembled from all of the clones in Table 1 spans 2.181 Mb. A map of this region showing the sites of predicted genes and previously known loci is presented in Fig. 1. Several additional cosmids which extend the sequence more than 2,000 kb to the left and 500 kb to the right are in various stages of completion.

Table 1. Sequences submitted to the GenBank and EMBL databases

Cosmid name	Locus name	Acc. no.	Length (bp)	Cosmid name	Locus name	Acc. no.	Length (bp)
ZK112	CELZK112	L14324	38,269	K02D10	CELK02D10	L14710	18,683
ZC97	CELZC97	L14714	4,166	F54F2	CELF54F2	L23645	39,573
ZK686	CELZK686	L17337	11,435	F44E2	CELF44E2	L23646	33,651
C08C3	CELC08C3	L15201	44,025	pAR3	CELPAR3	U00026	4,590
C27D11	CELC27D11	L23650	9,973	K01F9	CEK01F9	Z22175	19,834
ZK652	CELZK652	L14429	36,052	ZK637	CE1	Z11115	40,699
C02C2	CELC02C2	L23649	20,495	ZK638	CEZK638	Z12018	1,762
ZK688	CELZK688	L16621	36,977	ZK643	CEZK643	Z11126	39,534
C29E4	CELC29E4	L23651	40,050	R08D7	CER08D7	Z12017	27,368
F54H12	CELF54H12	L25599	19,168	F59B2	CEF59B2	Z11505	43,782
C06G4	CELC06G4	L25598	29,122	R107	CER107	Z14092	40,970
F44B9	CELF44B9	L23648	36,327	F02A9	CEF02A9	Z19555	26,242
K12H4	CELK12H4	L14331	38,582	ZK507	CEZK507	Z29116	13,501
K06H7	CELK06H7	L15314	22,073	ZK512	CEZK512	Z22177	36,997
C14B9	CELC14B9	L15188	43,492	F54G8	CEF54G8	Z19155	31,613
D2007	CELD2007	L16560	13,651	ZC84	CEZC84	Z19157	38,955
C50C3	CELC50C3	L14433	44,733	T23G5	CET23G5	Z19158	26,926
C30A5	CELC30A5	L10990	27,743	T02C1	CET02C1	Z19156	10,308
C02F5	CELC02F5	L14745	22,333	M01A8	CEM01A8	Z27081	19,001
F09G8	CELF09G8	L11247	41,449	K01B6	CEK01B6	Z22174	34,002
F10E9	CELF10E9	L10986	32,733	C40H1	CEC40H1	Z19154	27,271
ZC262	CELZC262	L23647	4,166	K04H4	CEK04H4	Z27078	33,930
R05D3	CELR05D3	L07144	38,810	C38C10	CEC38C10	Z19153	34,193
ZK353	CELZK353	L15313	24,916	T26G10	CET26G10	Z29115	30,251
ZK1236	CELZK1236	L13200	28,878	F54C8	CEF54C8	Z22178	23,000

序　列

表 1 给出已经测序的黏粒克隆及它们的数据库编号和长度。虽然相邻序列间经常包含促进连接处结合或保持基因完整的少量重叠碱基，但大部分冗余重叠数据已从数据库条目中去除。从表 1 的所有克隆中收集的非重叠序列横跨 2.181 Mb。图 1 是显示这个区域预测基因位置和已知基因座的图。使序列向左延伸 2,000 kb、向右延伸 500 kb 的多个黏粒处在测序的不同阶段。

表 1. 提交到 GenBank 和 EMBL 数据库的序列

黏粒名称	基因座名称	编号	长度（bp）	黏粒名称	基因座名称	编号	长度（bp）
ZK112	CELZK112	L14324	38,269	K02D10	CELK02D10	L14710	18,683
ZC97	CELZC97	L14714	4,166	F54F2	CELF54F2	L23645	39,573
ZK686	CELZK686	L17337	11,435	F44E2	CELF44E2	L23646	33,651
C08C3	CELC08C3	L15201	44,025	pAR3	CELPAR3	U00026	4,590
C27D11	CELC27D11	L23650	9,973	K01F9	CEK01F9	Z22175	19,834
ZK652	CELZK652	L14429	36,052	ZK637	CE1	Z11115	40,699
C02C2	CELC02C2	L23649	20,495	ZK638	CEZK638	Z12018	1,762
ZK688	CELZK688	L16621	36,977	ZK643	CEZK643	Z11126	39,534
C29E4	CELC29E4	L23651	40,050	R08D7	CER08D7	Z12017	27,368
F54H12	CELF54H12	L25599	19,168	F59B2	CEF59B2	Z11505	43,782
C06G4	CELC06G4	L25598	29,122	R107	CER107	Z14092	40,970
F44B9	CELF44B9	L23648	36,327	F02A9	CEF02A9	Z19555	26,242
K12H4	CELK12H4	L14331	38,582	ZK507	CEZK507	Z29116	13,501
K06H7	CELK06H7	L15314	22,073	ZK512	CEZK512	Z22177	36,997
C14B9	CELC14B9	L15188	43,492	F54G8	CEF54G8	Z19155	31,613
D2007	CELD2007	L16560	13,651	ZC84	CEZC84	Z19157	38,955
C50C3	CELC50C3	L14433	44,733	T23G5	CET23G5	Z19158	26,926
C30A5	CELC30A5	L10990	27,743	T02C1	CET02C1	Z19156	10,308
C02F5	CELC02F5	L14745	22,333	M01A8	CEM01A8	Z27081	19,001
F09G8	CELF09G8	L11247	41,449	K01B6	CEK01B6	Z22174	34,002
F10E9	CELF10E9	L10986	32,733	C40H1	CEC40H1	Z19154	27,271
ZC262	CELZC262	L23647	4,166	K04H4	CEK04H4	Z27078	33,930
R05D3	CELR05D3	L07144	38,810	C38C10	CEC38C10	Z19153	34,193
ZK353	CELZK353	L15313	24,916	T26G10	CET26G10	Z29115	30,251
ZK1236	CELZK1236	L13200	28,878	F54C8	CEF54C8	Z22178	23,000

Continued

Cosmid name	Locus name	Acc. no.	Length (bp)	Cosmid name	Locus name	Acc. no.	Length (bp)
C30C11	CELC30C11	L09634	30,865	B0464	CEB0464	Z19152	40,090
F42H10	CELF42H10	L08403	28,687	F55H2	CEF55H2	Z27080	22,950
C04D8	CELC04D8	L16687	10,433	ZK1098	CEZK1098	Z22176	37,310
ZC21	CELZC21	L16685	36,087	C48B4	CEC48B4	Z29117	35,000
K10C7 (C02D5)	CELC02D5	L16622	28,735	F58A4	CEF58A4	Z22179	38,000
C06E1/F43A9	CELC06E1	L16560	40,216	C15H7	CEC15H7	Z22173	28,000
C13G5	CELC13G5	L14730	10,883	C07A9/C40D10	CEC07A9	Z29094	66,004
F22B7	CELF22B7	L12018	40,222	T05G5	CET05G5	Z27079	36,180
B0523	CELB0523	L07143	14,334	R10E11	CER10E11	Z29095	32,254
B0303	CELB0303	M77697	41,071	ZK632	CEZK632	Z22181	36,000
ZK370	CELZK370	M98552	37,675	K11H3	CEK11H3	Z22180	33,000
pAR2	CELPAR2	U00025	12,721	ZK757	CEZK757	Z29121	31,000

The most striking result from genomic DNA sequencing is the continued high number of predicted genes in the region. In the 2.181 Mb of genomic sequence reported here, 483 putative genes were identified by similarity searches and GENEFINDER analysis. Start points for these candidate genes are indicated for both strands of the genomic sequence in Fig. 1. Generally, genes seem to be dispersed evenly throughout the region, with only a few examples of apparent clustering. One of the most gene-dense regions is contained in cosmids C30A5, C02F5 and F09G8 (Fig. 1, 492–553 kb). Here a 61-kb region contains 141 exons (in 24 putative genes), which account for 47% of the sequence. When the introns are also considered, more than 80% of the bases in this region are within predicted genes. Also in a 69.5-kb stretch of genomic DNA (F55H2, ZK1098; Fig. 1, 1,770–1,840 kb), 109 exons (in 19 putative genes) are predicted, with 40% of this region representing coding sequence and a total of 65% contained in genes. The longest stretch of genomic sequence that does not contain a likely gene is ~25 kb (pAR3, K01F9; Fig. 1, 1,145–1,170 kb). By comparison, the entire 2.181-Mb region is 29% coding sequence, with a total of 48% representing putative exons and introns. In our analysis of the first 0.1% of the genome, genes were found every 3–4 kb on average[6]. With more than 2% of the genomic DNA sequence now completed, the density is one gene every 4.5 kb. Because the region is expected to be relatively gene-rich[2], we cannot use this density to extrapolate the total gene number. However, an estimate independent of gene density can be made using the number of tagged cDNAs which hit candidate genes in the sequence[27]. Because 125 of the 4,615 *C. elegans* cDNA tags match predicted genes in the sequence, we can now estimate that the genome contains about 17,800 genes ($483 \times 4615/125$) for an average density of one gene every 5.6 kb.

There are no clear examples of genes that overlap, either on the same or on complementary strands. Several cases of orphan open reading frames that have high GENEFINDER scores and that overlap or are contained within candidate genes were observed, although

续表

黏粒名称	基因座名称	编号	长度(bp)	黏粒名称	基因座名称	编号	长度(bp)
C30C11	CELC30C11	L09634	30,865	B0464	CEB0464	Z19152	40,090
F42H10	CELF42H10	L08403	28,687	F55H2	CEF55H2	Z27080	22,950
C04D8	CELC04D8	L16687	10,433	ZK1098	CEZK1098	Z22176	37,310
ZC21	CELZC21	L16685	36,087	C48B4	CEC48B4	Z29117	35,000
K10C7 (C02D5)	CELC02D5	L16622	28,735	F58A4	CEF58A4	Z22179	38,000
C06E1/F43A9	CELC06E1	L16560	40,216	C15H7	CEC15H7	Z22173	28,000
C13G5	CELC13G5	L14730	10,883	C07A9/C40D10	CEC07A9	Z29094	66,004
F22B7	CELF22B7	L12018	40,222	T05G5	CET05G5	Z27079	36,180
B0523	CELB0523	L07143	14,334	R10E11	CER10E11	Z29095	32,254
B0303	CELB0303	M77697	41,071	ZK632	CEZK632	Z22181	36,000
ZK370	CELZK370	M98552	37,675	K11H3	CEK11H3	Z22180	33,000
pAR2	CELPAR2	U00025	12,721	ZK757	CEZK757	Z29121	31,000

基因组 DNA 测序最引人注目的结果是这个区域内有连续的大量预测基因。这里报道的 2.181 Mb 基因组序列中，通过相似性查找和 GENEFINDER 分析鉴定了 483 个假定基因。图 1 表明了这些候选基因在基因组序列两条链中的起始点。总体上，基因好像均匀地分散在整个区域，只有少数出现明显的簇集。基因密度最大的一个区域包含在黏粒 C30A5、C02F5 和 F09G8 中（图 1，492~553 kb）。有一个 61 kb 的区域包含 141 个外显子（在 24 个假定基因中），占这个序列的 47%。当考虑内含子时，这个区域 80% 以上的碱基都在预测基因上。在一段 69.5 kb 的基因组 DNA 内（F55H2、ZK1098；图 1，1,770~1,840 kb），预测了 109 个外显子（在 19 个假定基因中），这个区域 40% 的序列代表编码序列，总共 65% 包含在基因中。不包含可能基因的最长基因组序列片段约为 25 kb（pAR3、K01F9；图 1，1,145~1,170 kb）。经过比较，整个 2.181 Mb 区域中 29% 是编码序列，48% 代表预测的外显子和内含子。在我们分析的前 0.1% 基因组中，平均每 3~4 kb 发现一个基因 [6]。现在完成了多于 2% 的基因组 DNA 序列测序，基因的密度为每 4.5 kb 含一个基因。因为预测这个区域相对富含基因 [2]，所以我们不能用这个密度推断总的基因数目。然而，可以不依赖基因密度而是用击中序列中候选基因的标记 cDNA 数目估算总的基因数目 [27]。因为 4,615 个秀丽隐杆线虫 cDNA 标记中有 125 个与这段序列中的预测基因匹配，所以我们估算整个基因组包含 17,800 个基因（483 × 4615/125），平均 5.6 kb 含一个基因。

在相同或互补链上都没有基因重叠的明显例子。我们观察到几个有高 GENEFINDER 值的单独开放阅读框，它们重叠或包含在候选基因中，但没有数据表

there are no data to indicate that any of these is expressed. When the translated sequence is used in all six reading frames to search the public databases using the program BLAST, approximately one-third of the genes find significant matches to proteins from organisms other than *C. elegans*, with several very highly conserved genes indicated (Table 2). As can be seen in Table 2, a wide variety of different functions are represented, and with the exception of a homeobox cluster (see below), there is no obvious clustering of genes with related function. Most of the matches represent cross-phylum matches, for very little sequence from other nematodes is present in the databases. The fraction of genes with cross-phylum matches will increase as more genes from other organisms are entered into the database, but our previous analysis of these ancient conserved regions indicates that it is unlikely to rise above 40%, as most ancient conserved regions are already represented in the databases[29]. Although the data are inconclusive, preliminary evidence suggests that the fraction of matches to vertebrate proteins will be similar to this for any non-vertebrate genome.

Table 2. Gene candidates showing significant similarity to existing protein sequences

Gene name	Closest DB hit, Acc. no.	Closest block	Description
ZK112.2	TVMSBF, A40951	_	Kinase-related transforming protein
ZK112.7	A41087	BL00232	Tumour suppressor
ZK686.2	DB73_DROME, P26802	BL00039	ATP-dependent RNA helicase
ZK686.8	A24148	_	*N*-acetyllactosamine synthetase
C08C3.1	HM11_CAEEL, P17486	BL00027	*C. elegans* Hox protein egl-5 (ceh-11) gene
C08C3.3	HMMA_CAEEL, P10038	BL00027	*C. elegans* Hox protein mab-5
ZK652.4	R5RT35, A34571	BL00579	60S ribosomal protein L35
ZK652.5	A34218	BL00027	Distal-less homeotic protein
ZK652.6	S29962	_	ref (2) P protein, Zn-finger region
ZK652.8	C35815	_	Myosin heavy chain-3
C02C2.1	TYRO_STRGA, P06845	BL00497	Tyrosinase
C02C2.3	ACHG_RAT, P18916	BL00236	Acetylcholine receptor
C02C2.4	S27951	_	Sodium/phosphate transport protein
ZK688.8	A24148	_	*N*-acetyllactosamine synthase
C29E4.3	RNA1_YEAST, P11745	_	RNA production/processing
C29E4.7	S16267	_	Auxin-induced protein
C29E4.8	JS0422, JS0422	BL00113	Adenylate kinase
F54H12.1	ACON_YEAST, P19414	BL00450	Aconitate hydratase
F54H12.6	EF1B_BOMMO, P29522	_	Elongation factor 1β
C06G4.2	CAP3_RAT, P16259	_	Calpain
C06G4.5	SSR3_MOUSE, P30935	BL00237	Somatostatin receptor
F44B9.1	ACPH_RAT, P13676	BL00708	Acylamino-acid-releasing enzyme

明它们表达。用所有六个阅读框的翻译序列通过 BLAST 程序搜索公开数据库，发现约三分之一的基因与秀丽隐杆线虫以外的生物蛋白显著匹配，并发现了几个保守性很高的基因（表 2）。如表 2 中看到的，预测基因代表了种类广泛的不同的功能，并且除了同源框基因簇以外（见下文）没有相关功能基因的明显集合。因为只有很少其他线虫序列出现在数据库中，所以大多数匹配代表跨门匹配。随着其他生物的更多基因录入到数据库中，跨门匹配的基因比例会增加，但我们以前对这些古老保守区域的分析表明基因跨门匹配的比例不可能上升到 40% 以上，因为大多数古老的保守区域已经存在于数据库中了 [29]。虽然数据还不确定，但初步证据表明对于任何非脊椎动物而言，与脊椎动物蛋白匹配的比例都是相近的。

表 2. 候选基因与已有蛋白序列间有明显的相似性

基因名称	最接近的数据库检索编号	最接近的序列块	描述
ZK112.2	TVMSBF, A40951	—	激酶相关转化蛋白
ZK112.7	A41087	BL00232	肿瘤抑制蛋白
ZK686.2	DB73_DROME, P26802	BL00039	依赖 ATP 的 RNA 解旋酶
ZK686.8	A24148	—	*N*–乙酰氨基乳糖合成酶
C08C3.1	HM11_CAEEL, P17486	BL00027	秀丽隐杆线虫同源框蛋白 egl-5（ceh-11）基因
C08C3.3	HMMA_CAEEL, P10038	BL00027	秀丽隐杆线虫同源框蛋白 mab-5
ZK652.4	R5RT35, A34571	BL00579	60S 核糖体蛋白 L35
ZK652.5	A34218	BL00027	Distal-less 同源异型蛋白
ZK652.6	S29962	—	参考文献（2）P 蛋白，锌指域
ZK652.8	C35815	—	肌球蛋白重链 3
C02C2.1	TYRO_STRGA, P06845	BL00497	酪氨酸酶
C02C2.3	ACHG_RAT, P18916	BL00236	乙酰胆碱受体
C02C2.4	S27951	—	钠 / 磷酸盐转运蛋白
ZK688.8	A24148	—	*N*–乙酰氨基乳糖合成酶
C29E4.3	RNA1_YEAST, P11745	—	RNA 产物 / 加工
C29E4.7	S16267	—	植物生长素诱导蛋白
C29E4.8	JS0422,JS0422	BL00113	腺苷酸激酶
F54H12.1	ACON_YEAST, P19414	BL00450	乌头酸水合酶
F54H12.6	EF1B_BOMMO, P29522	—	延伸因子 1β
C06G4.2	CAP3_RAT, P16259	—	钙蛋白酶
C06G4.5	SSR3_MOUSE, P30935	BL00237	生长抑素受体
F44B9.1	ACPH_RAT, P13676	BL00708	酰基氨基酸释放酶

Continued

Gene name	Closest DB hit, Acc. no.	Closest block	Description
F44B9.8	A45253	_	Replication factor C
F44B9.9	PARB12, S11060	BL00125	Protein phosphatase
K12H4.1	JQ1397	_	*Drosophila melanogaster* Prospero
K12H4.4	SPC2_CHICK, P28687	_	Signal peptidase
K12H4.8	DEAD_ECOLI, P23304	BL00039	ATP-dependent RNA helicase dead
K06H7.1	S22127	BL00107	Protein kinase
K06H7.3	IPPI_YEAST, P15496	_	Isopentenyl-diphosphate δ-isomerase
K06H7.4	S24168	_	Sec7
K06H7.8	YCK1_YEAST, P23291	_	Casein kinase I
C14B9.1	CRAB_HUMAN, P02511	_	α-B-crystallin
C14B9.2	ER72_MOUSE, P08003	BL00194	Deoxycytidine kinase
C14B9.4	S22127, S22127	BL00098	Protein kinase
C14B9.7	R5RT21, A33295	_	Ribosomal protein L21
C14B9.8	S24109	_	Phosphorylase kinase
D2007.5	KERB_AVIER, P00535	_	ERB-B tyrosine kinase
C50C3.3	SPCN_CHICK, P07751	BL00545	Spectrin α-chain
C50C3.5	TPC1_BALNU, P21797	BL00018	Calmodulin
C50C3.7	OCRL_HUMAN, Q01986	_	Inositol polyphosphate-5-phosphatase
C50C3.11	JH0565	_	Calcium channel α-2b chain
C30A5.1	GRR1_YEAST, P24814	_	GRR1 protein (same as C02F5.7)
C30A5.3	S30854	BL00125	Phosphoprotein phosphatase
C30A5.4	SYB_DROME, P18489	BL00417	Synaptobrevin
C30A5.6	UN86_CAEEL, P13528	BL00035	Unc-86 alternate protein
C30A5.7	UN86_CAEEL, P13528	BL00035	Unc-86 protein
C02F5.3	JC1349	_	GTP-binding protein
C02F5.7	GRR1_YEAST, P24814	_	Glucose-induced repressor (GRR1)
C02F5.9	PRC5_HUMAN,P20618	BL00631	Proteasome component C5
F09G8.3	RS9_BACST, P07842	BL00360	Ribosomal protein S9
F09G8.6	A37122	_	Cuticle collagen
ZC262.3	A43425	_	N-CAM Ig domain
ZC262.5	ATPE_ BOVIN, P05632	_	ATP synthase ε-chain
R05D3.1	TOPB_HUMAN, Q02880	BL00177	DNA topoisomerase II homologue
R05D3.3	ZG44_XENLA, P18721	_	Gastrula zinc-finger protein
R05D3.6	ATPE_BOVIN, P05632	_	ATP synthase ε-chain
R05D3.7	KINH_LOLPE, P21613	BL00411	Kinesin heavy chain
ZK353.6	AMPA_RICPR, P27888	BL00631	Aminopeptidase

续表

基因名称	最接近的数据库检索编号	最接近的序列块	描述
F44B9.8	A45253	—	复制因子 C
F44B9.9	PARB12, S11060	BL00125	蛋白磷酸酶
K12H4.1	JQ1397	—	黑腹果蝇 Prospero 蛋白
K12H4.4	SPC2_CHICK, P28687	—	信号肽酶
K12H4.8	DEAD_ECOLI, P23304	BL00039	依赖 ATP 的 RNA 解旋酶 dead
K06H7.1	S22127	BL00107	蛋白激酶
K06H7.3	IPPI_YEAST, P15496	—	异戊烯二磷酸 δ 异构酶
K06H7.4	S24168	—	Sec7
K06H7.8	YCK1_YEAST, P23291	—	酪蛋白激酶 I
C14B9.1	CRAB_HUMAN, P02511	—	α-B 晶体蛋白
C14B9.2	ER72_MOUSE, P08003	BL00194	脱氧胞苷激酶
C14B9.4	S22127, S22127	BL00098	蛋白激酶
C14B9.7	R5RT21, A33295	—	核糖体蛋白 L21
C14B9.8	S24109	—	磷酸化酶激酶
D2007.5	KERB_AVIER, P00535	—	ERB-B 酪氨酸激酶
C50C3.3	SPCN_CHICK, P07751	BL00545	血影蛋白 α 链
C50C3.5	TPC1_BALNU, P21797	BL00018	钙调蛋白
C50C3.7	OCRL_HUMAN, Q01986	—	肌醇多磷酸−5−磷酸酶
C50C3.11	JH0565	—	钙通道 α-2b 链
C30A5.1	GRR1_YEAST, P24814	—	GRR1 蛋白(同 C02F5.7)
C30A5.3	S30854	BL00125	磷蛋白磷酸酶
C30A5.4	SYB_DROME, P18489	BL00417	小突触小泡蛋白
C30A5.6	UN86_CAEEL, P13528	BL00035	Unc-86 替代蛋白
C30A5.7	UN86_CAEEL, P13528	BL00035	Unc-86 蛋白
C02F5.3	JC1349	—	GTP 结合蛋白
C02F5.7	GRR1_YEAST, P24814	—	葡萄糖诱导的阻遏物(GRR1)
C02F5.9	PRC5_HUMAN,P20618	BL00631	蛋白酶体组分 C5
F09G8.3	RS9_BACST, P07842	BL00360	核糖体蛋白 S9
F09G8.6	A37122	—	表皮胶原蛋白
ZC262.3	A43425	—	N-CAM Ig 结构域
ZC262.5	ATPE_ BOVIN, P05632	—	ATP 合酶 ε 链
R05D3.1	TOPB_HUMAN, Q02880	BL00177	DNA 拓扑异构酶 II 同源物
R05D3.3	ZG44_XENLA, P18721	—	原肠胚锌指蛋白
R05D3.6	ATPE_BOVIN, P05632	—	ATP 合酶 ε 链
R05D3.7	KINH_LOLPE, P21613	BL00411	驱动蛋白重链
ZK353.6	AMPA_RICPR, P27888	BL00631	氨肽酶

Continued

Gene name	Closest DB hit, Acc. no.	Closest block	Description
ZK1236.1	LEPA_ECOLI, P07682	BL00301	LepA
ZK1236.2	NUCL_RAT, P13383	_	Nucleolin
C30C11.2	DXA2_MOUSE, P14685	_	Diphenol oxidase
C30C11.4	S30788	BL00297	HSP Msi3p
F42H10.4	GYRTI, A03270	_	Cysteine-rich intestinal protein
C04D8.1	SPCB_DROME, Q00963	_	Spectrin β-chain
ZC21.2	JH0588	_	Trp protein
ZC21.3	S14548	_	Dual bar protein
ZC21.4	S29956	_	Breakpoint cluster region (Bcr) protein
C02D5.1	ACDL_RAT, P15650	BL00072	Acyl-CoA dehydrogenase
C06E1.10	S22609, S22609	BL00690	Hypothetical protein
C06E1.4	B40171	_	Glutamate receptor
C06E1.8	B60191	_	Zn-finger
C06E1.9	A31922	_	ATP-dependent RNA helicase
C13G5.1	F34510	BL00027	Engrailed homeotic protein
F22B7.4	S18345	_	Environmental stress protein
F22B7.5	DNAJ_ECOLI, P08622	BL00636	DnaJ
F22B7.6	MUCB_ECOLI, P07375	_	Mucb protein
F22B7.7	S09048	_	Potassium channel protein Hak-6
B0523.1	S00904	BL00239	Tyrosine kinase
B0523.5	GELS_MOUSE, P13020	_	Gelsolin (flightless-1)
B0303.1	CYYA_YEAST, P08678	_	Adenylate cyclase
B0303.2	PNMT_BOVIN, P10938	_	Phenylethanolamine-*N*-methyltransferase
B0303.3	THIL_RAT, P17764	BL00098	Acetyl-CoA acetyltransferase
B0303.5	YT31_CAEEL, P03934	_	Tc3
B0303.7	NCF2_HUMAN, P19878	_	SH3 domain
B0303.9	SLP1_YEAST, P20795	_	SLP-1
ZK370.3	TALI_MOUSE, P26039	_	Talin
ZK370.4	KAPR_DICDI, P05987	_	Cyclic AMP-dependent protein kinase
ZK370.5	BCKD_RAT, Q00972	_	3-Methyl-2-oxobutanoate dehydrogenase
K02D10.1	S16088, S16088	_	4-Nitrophenylphosphatase
K02D10.5	S07258, S07258	_	*Escherichia coli* plasmid *RK2* gene for P116
F54F2.1	ITAP_DROME, P12080	BL00242	Vitronectin receptor α-subunit
F54F2.2	A44067	_	109K basic protein H
F44E2.1	S08405	_	Protease
F44E2.3	S28589	_	DnaJ

续表

基因名称	最接近的数据库检索编号	最接近的序列块	描述
ZK1236.1	LEPA_ECOLI, P07682	BL00301	LepA
ZK1236.2	NUCL_RAT, P13383	—	核仁蛋白
C30C11.2	DXA2_MOUSE, P14685	—	二酚氧化酶
C30C11.4	S30788	BL00297	HSP Msi3p
F42H10.4	GYRTI, A03270	—	富半胱氨酸肠蛋白
C04D8.1	SPCB_DROME, Q00963	—	血影蛋白 β 链
ZC21.2	JH0588	—	Trp 蛋白
ZC21.3	S14548	—	双杆蛋白
ZC21.4	S29956	—	裂点簇区（Bcr）蛋白
C02D5.1	ACDL_RAT, P15650	BL00072	脂酰辅酶 A 脱氢酶
C06E1.10	S22609, S22609	BL00690	假定蛋白
C06E1.4	B40171	—	谷氨酸受体
C06E1.8	B60191	—	锌指
C06E1.9	A31922	—	依赖 ATP 的 RNA 解旋酶
C13G5.1	F34510	BL00027	Engrailed 同源异型蛋白
F22B7.4	S18345	—	环境应激蛋白
F22B7.5	DNAJ_ECOLI, P08622	BL00636	DnaJ
F22B7.6	MUCB_ECOLI, P07375	—	Mucb 蛋白
F22B7.7	S09048	—	钾通道蛋白 Hak-6
B0523.1	S00904	BL00239	酪氨酸激酶
B0523.5	GELS_MOUSE, P13020	—	凝溶胶蛋白（flightless-1）
B0303.1	CYYA_YEAST, P08678	—	腺苷酸环化酶
B0303.2	PNMT_BOVIN, P10938	—	苯基乙醇胺 –*N*– 甲基转移酶
B0303.3	THIL_RAT, P17764	BL00098	乙酰辅酶 A 乙酰转移酶
B0303.5	YT31_CAEEL, P03934	—	Tc3
B0303.7	NCF2_HUMAN, P19878	—	SH3 结构域
B0303.9	SLP1_YEAST, P20795	—	SLP-1
ZK370.3	TALI_MOUSE, P26039	—	踝蛋白
ZK370.4	KAPR_DICDI, P05987	—	依赖 cAMP 的蛋白激酶
ZK370.5	BCKD_RAT, Q00972	—	3– 甲基 –2– 氧桥丁酸脱氢酶
K02D10.1	S16088, S16088	—	4– 硝基苯磷酸酶
K02D10.5	S07258, S07258	—	P116 的大肠杆菌质粒 *RK2* 基因
F54F2.1	ITAP_DROME, P12080	BL00242	玻连蛋白受体 α 亚基
F54F2.2	A44067	—	109K 碱性蛋白质 H
F44E2.1	S08405	—	蛋白酶
F44E2.3	S28589	—	DnaJ

Continued

Gene name	Closest DB hit, Acc. no.	Closest block	Description
F44E2.4	S03430	_	LDL receptor
F44E2.6	PILB_NEIGO, P14930	_	PILB protein
F44E2.7	CALD_CHICK, P12957	_	Caldesmon
ZK637.1	STP1_ARATH, P23586	_	Sugar transporter
ZK637.10	GSHR_HUMAN, P00390	BL00076	Glutathione reductase
ZK637.11	CC25_SCHPO, P06652	_	CDC25
ZK637.13	GLBH_TRICO, P27613	_	Globin
ZK637.14	PICO_HSV11, P08393	BL00518	Transactivator ICPO (motif 1)
ZK637.5	ARSA_ECOLI, P08690	_	ArsA
ZK637.8	VPP1_RAT, P25286	_	Proton pump
ZK643.2	DCTD_BPT2, P00814	_	DCMP deaminase
ZK643.3	CALR_PIG, P25117	BL00649	G-protein-coupled receptor
R08D7.5	CATR_CHLRE, P05434	BL00018	Calcium-binding protein
R08D7.6	CNAG_BOVIN, P14099	BL00126	Cyclic GMP phosphodiesterase
F59B2.3	NAGA_ECOLI, P15300	_	*N*-acetyl-glucosamine-6-phosphate deacetylase
F59B2.7	RAB6_HUMAN, P20340	_	Rab6 (Ras protein)
R107.7	GTP_CAEEL, P10299	_	Glutathione S-transferase P subunit
LIN12A.cds	LI12_CAEEL, P14585	BL00022	Lin-12/Notch, EGF and ankyrin repeats
F02A9.5	PCCB_RAT, P07633	_	Propionyl-CoA carboxylase
GLPIA.cds	GLP1_CAEEL, P13508	BL00022	Lin-12/Notch, EGF and ankyrin repeats
ZK507.1	HR25_YEAST, P29295	BL00107	HRR25 protein kinase
ZK507.6	CG2A_PATVU, P24861	BL00292	G2/M cyclin A
ZK512.2	SPB4_YEAST, P25808	BL00039	RNA helicase
ZK512.4	SRP9_CANFA, P21262	_	Signal recognition particle 9K protein
F54G8.2	KDGL_PIG, P20192	BL00479	Diacylglycerol kinase
F54G8.3	ITA3_CRISP, P17852	BL00242	Integrin α-chain
F54G8.4	KRET_HUMAN, P07949	BL00518	Ret zinc-finger region
ZC84.1	LACI_RABIT, P19761	BL00280	Serine protease inhibitor
ZC84.2	CNGC_RAT, Q00195	_	Cyclic nucleotide gated olfactory channel
ZC84.4	GASR _RAT, P30553	BL00237	G-protein-coupled receptor
T23G5.1	RIR1_HUMAN, P23921	BL0089	Ribonucleoside-disphosphate reductase Ig chain
T23G5.2	SC14_KLULA, P24859	_	SEC14 (yeast)
T23G5.5	NTTN_HUMAN, P23975	BL00610	Neurotransmitter transporter
T02C1.1	RA18_YEAST, P10862	BL00518	RAD-18 DNA-binding protein
M01A8.4	BIK1_YEAST, P11709	_	Nuclear fusion protein
C40H1.1	S24577	_	Ovarian protein (*D. melanogaster*)

续表

基因名称	最接近的数据库检索编号	最接近的序列块	描述
F44E2.4	S03430	—	LDL 受体
F44E2.6	PILB_NEIGO, P14930	—	PILB 蛋白
F44E2.7	CALD_CHICK, P12957	—	钙调蛋白结合蛋白
ZK637.1	STP1_ARATH, P23586	—	糖转运蛋白
ZK637.10	GSHR_HUMAN, P00390	BL00076	谷胱甘肽还原酶
ZK637.11	CC25_SCHPO, P06652	—	CDC25
ZK637.13	GLBH_TRICO, P27613	—	珠蛋白
ZK637.14	PICO_HSV11, P08393	BL00518	反式激活蛋白 ICPO(motif 1)
ZK637.5	ARSA_ECOLI, P08690	—	ArsA
ZK637.8	VPP1_RAT, P25286	—	质子泵
ZK643.2	DCTD_BPT2, P00814	—	DCMP 脱氨酶
ZK643.3	CALR_PIG, P25117	BL00649	G 蛋白偶联受体
R08D7.5	CATR_CHLRE, P05434	BL00018	钙结合蛋白
R08D7.6	CNAG_BOVIN, P14099	BL00126	cGMP 磷酸二酯酶
F59B2.3	NAGA_ECOLI, P15300	—	*N*–乙酰葡糖胺–6–磷酸脱乙酰酶
F59B2.7	RAB6_HUMAN, P20340	—	Rab6 (Ras 蛋白)
R107.7	GTP_CAEEL, P10299	—	谷胱甘肽 S–转移酶 P 亚基
LIN12A.cds	LI12_CAEEL, P14585	BL00022	Lin-12/Notch,EGF 和锚蛋白重复序列
F02A9.5	PCCB_RAT, P07633	—	丙酰辅酶 A 羧化酶
GLPlA.cds	GLP1_CAEEL, P13508	BL00022	Lin-12/Notch,EGF 和锚蛋白重复序列
ZK507.1	HR25_YEAST, P29295	BL00107	HRR25 蛋白激酶
ZK507.6	CG2A_PATVU, P24861	BL00292	G2/M 细胞周期蛋白 A
ZK512.2	SPB4_YEAST, P25808	BL00039	RNA 解旋酶
ZK512.4	SRP9_CANFA, P21262	—	信号识别颗粒 9K 蛋白
F54G8.2	KDGL_PIG, P20192	BL00479	二酰甘油激酶
F54G8.3	ITA3_CRISP, P17852	BL00242	整联蛋白 α 链
F54G8.4	KRET_HUMAN, P07949	BL00518	Ret 锌指区
ZC84.1	LACI_RABIT, P19761	BL00280	丝氨酸蛋白酶抑制蛋白
ZC84.2	CNGC_RAT, Q00195	—	环核苷酸控制的嗅觉通道
ZC84.4	GASR _RAT, P30553	BL00237	G 蛋白偶联受体
T23G5.1	RIR1_HUMAN, P23921	BL0089	核苷二磷酸还原酶 Ig 链
T23G5.2	SC14_KLULA, P24859	—	SEC14 (酵母)
T23G5.5	NTTN_HUMAN, P23975	BL00610	神经递质转运体
T02C1.1	RA18_YEAST, P10862	BL00518	RAD-18 DNA 结合蛋白
M01A8.4	BIK1_YEAST, P11709	—	核融合蛋白
C40H1.1	S24577	—	卵巢蛋白(黑腹果蝇)

Continued

Gene name	Closest DB hit, Acc. no.	Closest block	Description
C40H1.4	YCS4_YEAST, P25358	BL00030	Yeast hypothetical protein
K04H4.1	CA14_CAEEL, P17139	_	Collagen
C38C10.1	NK3R_RAT, P16177	BL00237	G-protein-coupled receptor
C38C10.5	RGR1_YEAST, P19263	_	Glucose repression regulatory protein RGR1
T26G10.1	DEAD_ECOLI, P23304	BL00039	RNA helicase
T26G10.3	RS24_HUMAN, P16632	BL00529	Ribosomal protein S24
F54C8.1	HCDH_PIG, P00348	BL00067	3-hydroxyacyl-CoA dehydrogenase
F54C8.2	H31_SCHPO, P09988	BL00322	Histone H3
F54C8.4	Y19K_NPVAC, P24656	_	ACMNPV hypothetical protein
F54C8.5	RAS_LENED, P28775	_	Ras family
B0464.1	SYD2_HUMAN,P14868	BL00179	Aspartyl--tRNA synthetase
B0464.5	KCLK_MOUSE, P22518	BL00107	Serine/threonine kinase
F55H2.1	SODC_BOVIN, P00442	BL00087	Superoxide dismutase
F55H2.2	MTPG_SULAC, P22721	_	Membrane-associated ATPase γ-chain
F55H2.5	C561_BOVIN, P10897	_	Cytochrome b_{561}
ZK1098.10	TPMX_RAT, P18342	_	Coiled-coil protein
ZK1098.4	GCN3_YEAST, P14741	_	GCN3 (yeast transcription activator)
C48B4.1	CA01_RAT, P07872	BL00072	Acyl-CoA oxidase I
C48B4.2	RHOM_DROME, P20350	_	Rhomboid (*D. melanogaster*)
C48B4.4	NODI_RHILO, P23703	BL00211	ATP-binding transport protein
C48B4.5	LIVG_SALTY, P30293	BL00211	ATP-binding transport protein
F58A4.10	UBC7_WHEAT, P25868	BL00183	Ubiquitin-conjugating enzyme
F58A4.3	H3_VOLCA, P08437	BL00322	Histone H3
F58A4.4	PRI1_MOUSE, P20664	_	DNA primase 49K subunit
F58A4.5	RTJK_DROME, P21328	_	Mobile element Jockey-rev. transcriptase
F58A4.7	A36394	BL00038	Transcription factor AP-4
F58A4.8	TBG_XENLA, P23330	BL00227	γ-Tubulin
F58A4.9	RPC9_YEAST, P28000	_	RNA Pol I/III 16K polypeptide
C15H7.2	KFPS_DROME, P18106	BL00790	Tyrosine kinase
C15H7.3	TCPT_HUMAN, P17706	_	Protein tyrosine phosphatase
C07A9.2	G10_XENLA, P12805	_	G10 protein
C07A9.3	KRAC_HUMAN, P31749	BL00107	Ser/Thr kinase
C07A9.4	S20969	BL00470	Na/Ca, K antiporter
C07A9.5	SPCA_DROME, P13395	BL00018	Spectrin α-chain
C07A9.6	UDP2_RAT, P09875	BL00375	UDP-glucuronosyltransferase
T05G5.3	CC2_HUMAN, P06493	BL00107	CDC2 kinase

续表

基因名称	最接近的数据库检索编号	最接近的序列块	描述
C40H1.4	YCS4_YEAST, P25358	BL00030	酵母假定蛋白
K04H4.1	CA14_CAEEL, P17139	—	胶原蛋白
C38C10.1	NK3R_RAT, P16177	BL00237	G 蛋白偶联受体
C38C10.5	RGR1_YEAST, P19263	—	葡糖阻遏调节蛋白 RGR1
T26G10.1	DEAD_ECOLI, P23304	BL00039	RNA 解旋酶
T26G10.3	RS24_HUMAN, P16632	BL00529	核糖体蛋白 S24
F54C8.1	HCDH_PIG, P00348	BL00067	3-羟脂酰辅酶 A 脱氢酶
F54C8.2	H31_SCHPO, P09988	BL00322	组蛋白 H3
F54C8.4	Y19K_NPVAC, P24656	—	ACMNPV 假定蛋白
F54C8.5	RAS_LENED, P28775	—	Ras 家族
B0464.1	SYD2_HUMAN,P14868	BL00179	天冬氨酰 tRNA 合成酶
B0464.5	KCLK_MOUSE, P22518	BL00107	丝氨酸 / 苏氨酸激酶
F55H2.1	SODC_BOVIN, P00442	BL00087	超氧化物歧化酶
F55H2.2	MTPG_SULAC, P22721	—	膜结合 ATP 酶 γ 链
F55H2.5	C561_BOVIN, P10897	—	细胞色素 b_{561}
ZK1098.10	TPMX_RAT, P18342	—	卷曲螺旋蛋白质
ZK1098.4	GCN3_YEAST, P14741	—	GCN3（酵母转录激活因子）
C48B4.1	CA01_RAT, P07872	BL00072	脂酰辅酶 A 氧化酶 I
C48B4.2	RHOM_DROME, P20350	—	Rhombiod（黑腹果蝇）
C48B4.4	NODI_RHILO, P23703	BL00211	ATP 结合转运蛋白
C48B4.5	LIVG_SALTY, P30293	BL00211	ATP 结合转运蛋白
F58A4.10	UBC7_WHEAT, P25868	BL00183	泛素缀合酶
F58A4.3	H3_VOLCA, P08437	BL00322	组蛋白 H3
F58A4.4	PRI1_MOUSE, P20664	—	DNA 引发酶 49K 亚基
F58A4.5	RTJK_DROME, P21328	—	可动元件 Jockey-rev. 转录酶
F58A4.7	A36394	BL00038	转录因子 AP-4
F58A4.8	TBG_XENLA, P23330	BL00227	γ 微管蛋白
F58A4.9	RPC9_YEAST, P28000	—	RNA 聚合酶 I/III 16K 多肽
C15H7.2	KFPS_DROME, P18106	BL00790	酪氨酸激酶
C15H7.3	TCPT_HUMAN, P17706	—	蛋白酪氨酸磷酸酶
C07A9.2	G10_XENLA, P12805	—	G10 蛋白
C07A9.3	KRAC_HUMAN, P31749	BL00107	丝氨酸 / 苏氨酸激酶
C07A9.4	S20969	BL00470	Na/Ca，K 反向转运体
C07A9.5	SPCA_DROME, P13395	BL00018	血影蛋白 α 链
C07A9.6	UDP2_RAT, P09875	BL00375	UDP 葡糖醛酸转移酶
T05G5.3	CC2_HUMAN, P06493	BL00107	CDC2 激酶

Continued

Gene name	Closest DB hit, Acc. no.	Closest block	Description
T05G5.5	S27735	–	Hypothetical protein A (*Thermus aquaticus*)
T05G5.6	ECHM_RAT, P14604	BL00166	Enoyl-CoA hydratase
T05G5.10	IF5A_HUMAN, P10159	–	Initiation factor 5A
R10E11.1	FSH_DROME, P13709	BL00633	Bromodomain
R10E11.2	VATL_DROME, P23380	BL00605	Vacuolar ATP synthase subunit
R10E11.3	UBP2_YEAST, Q01476	–	Ubiquitin-specific processing protease
R10E11.4	NALS_MOUSE, P15535	–	Galactosyltransferase
ZK632.1	MCM3_YEAST, P24279	–	Mcm 2/3
ZK632.3	S26727	–	Hypothetical protein 186 (*Thermoplasma acidophilum*)
ZK632.4	MANA_ECOLI, P00946	–	Mannose 6-phosphate isomerase
ZK632.6	CALX_HUMAN, P27824	BL00803	Calnexin
ZK632.8	ARF2_YEAST, P19146	–	ADP-ribosylation factor
K11H3.1	GPDA_DROVI, P07735	–	Glycerol-3-phosphate dehydrogenase
K11H3.3	UCP_HUMAN, P25874	BL00215	Mitochondrial carrier family
ZK757.2	TPCL_HUMAN, P28562	–	Protein tyrosine phosphatase

Known Genes

Several interesting genes had been positioned in this region before our sequence analysis. The first 120 kb of the sequence reported here is part of the HOM homeobox gene complex which extends an additional 150 kb to the left[30-33]. The region is centred around the *Antennapedia*-like gene *mab-5*, which is responsible for pattern formation in a posterior body region[31]. GENEFINDER predicted a gene candidate with sequence identity to the *mab-5* cDNA sequence (GenBank M22751) on the complementary strand of the cosmid clone C08C3 (annotated as C08C3.3). Approximately 30 kb to the right of *mab-5*, the *abdominal-B*-like gene *egl-5(ceh-11)*[31,32], which is required for normal development of several cell types in the tail region, was located (C08C3.1). Interestingly, the *mab-5* and *egl-5* genes are encoded in opposite orientation. A third homeobox gene, *ceh-23*, lies 23 kb to the right of *egl-5* (ZK652.5). The *ceh-23* gene, which is similar to the *Drosophila* genes *Distal-less* and *empty spiracles*, was located by identity to a cDNA clone[32]. Two additional homeobox genes, *lin-39(ceh-15)* and *ceh-13*, lie approximately 200 kb to the left of *mab-5*. Preliminary sequence data indicate that the order of these two genes relative to *mab-5* is *lin-39*, *ceh-13*, and not *ceh-13*, *lin-39* as originally reported[32,33]. An unrelated gene, *egl-45*, with no similarity to any known *Drosophila* genes, maps between the *egl-5* and *ceh-23* genes (tentatively correlated to gene candidate *C27D11.1*), and several other putative genes are contained within the HOM region.

Other genes previously mapped to the region were correlated to distinct loci (shown in

续表

基因名称	最接近的数据库检索编号	最接近的序列块	描述
T05G5.5	S27735	—	假定蛋白 A(栖热水生菌)
T05G5.6	ECHM_RAT, P14604	BL00166	烯酰辅酶 A 水合酶
T05G5.10	IF5_HAUMAN, P10159	—	起始因子 5A
R10E11.1	FSH_DROME, P13709	BL00633	布罗莫结构域
R10E11.2	VATL_DROME, P23380	BL00605	液泡 ATP 合酶亚基
R10E11.3	UBP2_YEAST, Q01476	—	泛素特异性加工蛋白酶
R10E11.4	NALS_MOUSE, P15535	—	半乳糖基转移酶
ZK632.1	MCM3_YEAST, P24279	—	Mcm2/3
ZK632.3	S26727	—	假定蛋白 186 (嗜酸热原体)
ZK632.4	MANA_ECOLI, P00946	—	甘露糖 6-磷酸异构酶
ZK632.6	CALX_HUMAN, P27824	BL00803	钙连蛋白
ZK632.8	ARF2_YEAST, P19146	—	ADP 核糖基化因子
K11H3.1	GPDA_DROVI, P07735	—	甘油-3-磷酸脱氢酶
K11H3.3	UCP_HUMAN, P25874	BL00215	线粒体载体家族
ZK757.2	TPCL_HUMAN, P28562	—	蛋白酪氨酸磷酸酶

已知基因

在我们进行序列分析前，已经有几个有趣的基因定位在这个区域了。这里报道的第一个 120 kb 序列是 HOM 同源框基因复合体的一部分，HOM 同源框基因复合体额外往左延伸 150 kb[30-33]。这个区域以 *Antennapedia* 类似基因 *mab-5* 附近为中心，*mab-5* 控制身体后部区域模式的形成 [31]。GENEFINDER 在黏粒克隆 C08C3 (注释为 C08C3.3) 的互补链上预测到一个与 *mab-5* cDNA 序列 (GenBank M22751) 具有序列一致性的候选基因。*abdominal-B* 类似基因 *egl-5* (*ceh-11*)[31,32] 定位于 *mab-5* 右侧约 30 kb 处 (C08C3.1)，它是尾部几个细胞类型正常发育所必需的。有趣的是，*mab-5* 和 *egl-5* 两个基因的编码方向相反。第三个同源框基因 *ceh-23* 基因，位于 *egl-5* (ZK652.5) 右侧 23 kb 处。*ceh-23* 基因与果蝇基因 *Distal-less* 和 *empty spiracles* 相似，通过一致性定位到一个 cDNA 克隆上 [32]。另外两个同源框基因，*lin-39* (*ceh-15*) 和 *ceh-13* 大约定位于 *mab-5* 左侧 200 kb 处。初始序列数据表明这两个基因相对 *mab-5* 的顺序依次是 *lin-39*、*ceh-13*，而不是最初报道的 *ceh-13*、*lin-39*[32,33]。一个与任何已知果蝇基因都无相似性的无关基因——*egl-45*，定位在 *egl-5* 和 *ceh-23* 基因 (暂时与候选基因 *C27D11.1* 相关) 之间，HOM 区域还包含其他几个假定基因。

通过基因组 DNA 测序，我们将之前定位在这个区域的其他基因与特定的基因

parentheses) by genomic DNA sequencing. These include *egl-45* (tentatively C27D11.1), *lin-36* (F44B9.6), *unc-36* (C50C3.11), *unc-86* (C30A5.7), *mig-10* (tentatively F10E9.6), *unc-116* (R05D3.7), *ceh-16* (C13G5.1), *dpy-19* (F22B7.10), *sup-5* (B0523.5), *unc-32* (ZK637.10), *lin-9* (ZK637.7), *gst-1* (R107.7), *lin-12* (LIN12A in cosmid R107), *glp-1* (GLP1A in cosmid F02A9), *emb-9* (K04H4.1), *tbg-1* (F58A4.8) and *ncc-1* (T05G5.3). The *unc-86* and *emb-9* genes and part of the *ceh-16* gene has been sequenced previously[34-36]. In the cases of *egl-45* (M. Basson and H. R. Horvitz, personal communication), *lin-36* (J. Thomas and H. R. Horvitz, personal communication), *unc-36* (L. Loebel and H. R. Horvitz, personal communication), *unc-116* (ref. 37), *lin-9* (G. Beitel and H. R. Horvitz, personal communication), *gst-1* (ref. 38), *lin-12* (ref. 39) and *glp-1* (ref. 40), cDNA sequences provided by researchers within the *C. elegans* community enabled gene assignments to be made. For *tbg-1* and *ncc-1*, cDNAs from the consortium tag-sequencing project[27] allowed assignment to genomic loci. For *mig-10, dpy-19* and *unc-32*, transgenic rescue experiments with mutant phenotypes localized the genes to a particular restriction fragment (J. Manser; S. Colloms; D. Thierry-Mieg, personal communications). In some cases, availability of the genomic sequence facilitated this type of analysis.

A few kilobases of genomic DNA sequence which included *sup-5* had been reported previously[41]. However, additional genomic sequencing revealed that the *sup-5* locus, a gene for transfer RNA^{Trp}, lies within an intron in the same transcriptional orientation as a homologue of *Drosophila melanogaster flightless I* (ref. 42) (B0303.1). As the *sup-5* mutation has been shown to suppress specific alleles of many unrelated genes[43], it is known to be a functional tRNA gene. Further, the *C. elegans fl I* homologue is expressed and spliced as predicted (ref. 42, and R. Wilson, unpublished data).

tRNA Genes

The haploid genome of *C. elegans* has been estimated to contain about 300 tRNA genes[44]. Thus, we would expect to find an average of three tRNA genes per megabase of genomic sequence. However, using tRNAscan[45], in the 2.181 Mb of the sequence reported here, at least 14 tRNA genes were identified. These are indicated in Fig. 1. Strikingly, two cosmid clones, C14B9 and F22B7, contain seven of these tRNA genes. Like the *sup-5* tRNA gene, a $tRNA^{Ser}$ and two $tRNA^{Phe}$ genes lie within introns of likely genes.

Repeats

Several types of repeated sequence are present in the 2.181-Mb region. Figure 1 indicates the locations of the larger and more complex repeats. Detailed analysis of three major types of repeated sequences (inverted, tandem and interspersed) reveals several interesting features.

Inverted repeats, in which a segment of genomic sequence lies within a few to several

座(括号中显示)一一对应。这些基因包括 *egl-45* (暂定 C27D11.1)、*lin-36* (F44B9.6)、*unc-36* (C50C3.11)、*unc-86* (C30A5.7)、*mig-10* (暂定 F10E9.6)、*unc-116* (R05D3.7)、*ceh-16* (C13G5.1)、*dpy-19* (F22B7.10)、*sup-5* (B0523.5)、*unc-32* (ZK637.10)、*lin-9* (ZK637.7)、*gst-1* (R107.7)、*lin-12* (黏粒 R107 中的 LIN12A)、*glp-1* (黏粒 F02A9 中的 GLP1A)、*emb-9* (K04H4.1)、*tbg-1* (F58A4.8) 和 *ncc-1* (T05G5.3)。*unc-86* 和 *emb-9* 基因及部分 *ceh-16* 基因以前已测过序[34-36]。秀丽隐杆线虫群落的研究者提供 cDNA 序列帮助我们确定基因 *egl-45* (巴松和霍维茨，个人交流)、*lin-36* (托马斯和霍维茨，个人交流)、*unc-36* (勒贝尔和霍维茨，个人交流)、*unc-116* (参考文献 37)、*lin-9* (比特尔和霍维茨，个人交流)、*gst-1* (参考文献 38)、*lin-12* (参考文献 39)、*glp-1* (参考文献 40) 的排布。进行标签测序项目[27]的团队提供的 cDNA 允许我们将 *tbg-1* 和 *ncc-1* 定位到基因组基因座上。通过突变表型的转基因拯救实验将基因 *mig-10*、*dpy-19* 和 *unc-32* 定位到一个特定的限制性片段上(曼瑟、科洛姆斯、蒂埃里 – 米格，个人交流)。在一些情况下，利用基因组序列能方便这类分析。

包含 *sup-5* 在内的几千个碱基的基因组 DNA 序列以前已经报道过[41]。然而，额外的基因组测序反映了 *sup-5* 基因座(一个 $tRNA^{Trp}$ 基因) 位于一个内含子内，并且作为黑腹果蝇 *flightless I*(参考文献 42) (B0303.1) 的同源物转录方向与其相同。因为 *sup-5* 突变表现出对许多无关基因的特异等位基因的抑制[43]，所以它被认为是有功能的 tRNA 基因。此外秀丽隐杆线虫 *fl I* 同源物和预测一样进行表达和剪接(参考文献 42 和威尔逊的未发表数据)。

tRNA 基 因

经估计秀丽隐杆线虫的单倍体基因组约包含 300 个 tRNA 基因[44]。因此，我们期望基因组序列中平均每兆碱基发现 3 个 tRNA 基因。然而，在这报道的 2.181 Mb 序列中，用 tRNAscan[45] 鉴定了至少 14 个 tRNA 基因。这些 tRNA 基因表示在图 1 中。令人惊奇的是，C14B9 和 F22B7 这两个黏粒克隆，包含了其中的 7 个 tRNA 基因。像 *sup-5* tRNA 基因一样，一个 $tRNA^{Ser}$ 和两个 $tRNA^{Phe}$ 基因位于可能基因的内含子中。

重 复 序 列

这 2.181 Mb 区域中出现了几类重复序列。图 1 表明较大和较复杂的重复序列的定位。对三种主要重复序列类型的详细分析(反向的，串联的和散在的)反映了几个有趣的特征。

在反向重复序列中，一段基因组序列位于它反向拷贝的几个到几百个碱基之内。

hundred bases of an inverted copy of itself, are the most common type of repeat that we have found. Considering only those inverted repeats of up to 1 kb end to end, with at least 70% identity, on average an inverted repeat is found every 5.5 kb. Most of these are quite small, with an average segment length of 70 base pairs (bp) and an average loop size of 164 bp. A relatively high proportion of these repeats (43%) occur in introns, which represent only 20% of the genome. Most inverted repeats fall into families and may be remnants of mobile elements. In particular, there were examples of inverted repeat elements from known transposons Tc*3* (ref. 46) and Tc*4* (ref. 47).

Tandem repeats, in which a segment of genomic sequence lies adjacent to one or more copies of itself, occur on average every 10 kb. As with inverted repeats, most of these are small, with an average segment length of 17 bp and an average copy number of 14. Interestingly, only 17% of tandem repeats were found in introns, whereas 63% occurred between genes. The most common category of tandem repeats were triplets, some of which were found in predicted exons. One of the most complex tandem repeats found was a 95-bp sequence which was repeated more than 30 times in the clone pAR3 (Fig. 1; 1,144–1,150 kb). This region, which was missing from the cosmid map, had to be recovered from the YAC clone Y53B1 by targeted gap rescue in yeast. Also a large (7.9 kb) tandem repeat, flanked by more complex short repeats, was found in cosmid ZC262 (Fig. 1; 595–633 kb). Sequence assembly for these repeat regions was accomplished with very stringent parameters and, in some cases, aided by map information.

There are several examples of short repetitive sequences that are scattered throughout the genome, including the 94-bp consensus of the repeat element from most common inverted repeat families, previously observed in *lin-12* and *glp-1* introns[39,40]. These elements are widely dispersed, with an average of 14 copies every 100 kb at varying levels of conservation. Often, the repeat elements of inverted repeat families are also found in singleton or tandem arrangements. For example, a tandem pair of degenerate elements (69% identical over 77 bp) flanks part of the predicted gene *F58A4.2*. As discussed below, this seems to have been duplicated from a region 200 kb away.

Gene Duplications

In addition to the short interspersed repeats discussed above, there are sequences that are repeated tens or hundreds of kilobases apart, with up to 98% similarity. In some cases, these apparent duplications have a complex structure wherein several segments from one region are repeated in a second location, but with different spacings and orientations. Many of these long-range repeats involve coding regions. In particular, there seems to be a recent gene duplication involving *F22B7.5* and *C38C10.4*, which are approximately 750 kb apart but more than 95% similar. The predicted genes *C38C10.3* and *F58A4.2,* mentioned above, are more than 200 kb apart but share a 1.4-kb region that is 98% similar. Two predicted exons are contained within the repeat. If the GENEFINDER predictions are correct, this would be an example of exon shuffling. Perhaps more likely, the *F58A4.2*

反向重复是我们发现的最普遍的重复序列类型。只考虑那些端到端达到 1 kb 且至少有 70% 相似性的反向重复序列，平均每 5.5 kb 发现一个反向重复序列。重复序列大多数都很小，平均片段长度 70 bp，平均环大小为 164 bp。占基因组 20% 的内含子中出现这些重复的比率相对较高（43%）。大多数反向重复序列属于家族，可能是可动元件的残迹。尤其是，有例子表明有的反向重复元件来自已知的转座子 Tc*3*（参考文献 46）和 Tc*4*（参考文献 47）。

在串联重复中，一段基因组序列与它自身的一个或多个拷贝相连。平均每 10 kb 含一个串联重复。与反向重复一样，大多数串联重复很小，平均片段长度为 17 bp，平均拷贝数为 14。有趣的是，只有 17% 的串联重复存在于内含子中，而 63% 的串联重复存在于基因间。最普遍的串联重复是三拷贝串联，其中一些在预测的外显子中发现。发现的最复杂的串联重复之一是一个 95 bp 的序列，它在 pAR3 克隆中重复了三十多次（图 1；1,144~1,150 kb）。这个区域从黏粒图中缺失了，必须通过目标间隔拯救从酵母 YAC 克隆 Y53B1 中恢复。在黏粒 ZC262 中也发现了一个大的（7.9 kb）串联重复，它两侧有更复杂的短重复（图 1；595~633 kb）。借助严格的参数，有时辅以图谱信息，完成了这些重复区域的序列组装。

有几个短重复序列分散在基因组中，其中包括一个以前在 *lin-12* 和 *glp-1* 内含子中发现的来自最常见反向重复家族的重复元件的 94 bp 共有序列[39,40]。这些元件分散广泛，保守程度不同，平均每 100 kb 含 14 个拷贝。反向重复家族的重复元件也经常以单独或串联的形式被发现。例如，一个简并元件的串联对（69% 相似性超过 77 bp）在部分预测基因 *F58A4.2* 的两侧。如下文讨论的那样，这好像是从 200 kb 以外的区域复制过来的。

基因重复

除了上面讨论的短散在重复，还有长度达几万或几十万碱基的重复序列，相似性高达 98%。有些例子中，这些明显的重复有一个复杂结构，其中一个区域的几个片段在第二个位置重复，但重复的间隔和方向都不同。许多大范围重复涉及编码区。特别是 *F22B7.5* 和 *C38C10.4* 好像发生了较新的基因重复，二者相距约 750 kb 但相似性高于 95%。上面提到的预测基因 *C38C10.3* 和 *F58A4.2* 之间距离大于 200 kb，但共享一个相似性为 98% 的 1.4 kb 的区域。这个重复中包含两个预测的外显子。如果 GENEFINDER 预测正确的话，这可能是一个外显子混编的例子。*F58A4.2* 中那重复的部分或许更可能是 *C38C10.3* 中一个功能基因的一部分的非转录拷贝。上述黏粒

version is a non-transcribed copy of part of a functional gene in *C38C10.3*. The 7.9-kb repeats in cosmid ZC262 described above contain segments of three different gene candidates, including a kinesin heavy-chain locus which was identified as *unc-116* (ref. 37). The *unc-116* gene begins outside the second copy of the repeat, with the last two exons present in the 7.9-kb. The same two exons in the first copy of the repeat are predicted to splice to a different exon in the second copy of the repeat. In addition to these recent gene duplications, we have identified similarities between other genes in the region. However, the degree of similarity suggests that these are more ancient in origin and may be examples of new gene families in *C. elegans*.

Prospects

The sequence reported here is already proving useful. In the very narrow sense, several genes previously under study have been sequenced, speeding the analysis and further study of these genes. The full sequence of the homeobox region will clarify its relationship to the *Drosophila Antennapedia* complex. More importantly, the 483 genes identified through genomic sequencing, together with the 1,194 genes discovered in the cDNA tagging project[27], are providing new and fruitful avenues for *C. elegans* research.

Furthermore, our experience in the pilot phase indicates that megabase-scale DNA sequencing at a reasonable cost is feasible with current methods and technology[6,48]. At the same time we feel that significant improvements are possible at almost every step. For example, we have developed an automated DNA template preparation capable of producing 400 M13 templates daily[18]. Initial plaque picking can also be done robotically[49], and instruments are under development that can perform large numbers of small-volume sequencing reactions automatically. Longer read lengths and greater sample capacity for the present generation of fluorescent gel readers are being developed, and more powerful instruments are being designed. Further improvements in the software will soon eliminate much of the sequence editing, which is currently a tedious task. Software tools are already available which simplify the selection of templates and oligonucleotides for directed sequencing (R.S., L.H. and S.D., unpublished). With these improvements and some increase in the scale of effort, production of more than 10 megabases of finished sequence per site per annum seems feasible. With this capacity in both halves of the consortium, the *C. elegans* genome sequence should be essentially completed before the end of 1998. In addition, both laboratories are contributing resources to speed the completion of the *S. cerevisiae* genome sequence. The complete genome sequences of these two organisms will provide insight into the genes that are likely to be common to all eukaryotes, and those specific to metazoans.

ZC262 中 7.9 kb 的重复序列包含三个不同候选基因的片段，包括一个鉴定为 *unc-116*（参考文献 37）的驱动蛋白重链基因座。*unc-116* 基因起始于这个重复第二个拷贝的外侧，且最后两个外显子在这个 7.9 kb 重复序列内。在这个重复的第一个拷贝中同样的两个外显子预测会剪接到该重复第二个拷贝的另一个外显子上。除这些最近的基因重复外，我们也鉴定了这个区域其他基因间的相似性。然而，相似性的水平表明它们的起源更古老，而且有可能是秀丽隐杆线虫的新基因家族的例子。

前　景

实践已经证明本文报道的序列是有用的。从狭义上讲，完成了几个以前研究的基因的测序，加快了对这些基因的分析和进一步研究。同源框区域的所有序列会阐明它与果蝇 *Antennapedia* 复合体的相互关系。更重要的是，通过基因组测序鉴定的 483 个基因与 cDNA 标记项目 [27] 中发现的 1,194 个基因，一同为秀丽隐杆线虫的研究提供了新的富有成效的途径。

此外，我们在试验阶段的经验表明，用目前的方法和技术，在合理的花费下进行兆碱基规模的 DNA 测序是可行的 [6,48]。同时，我们感到几乎每一步都可能有明显的改进。例如，我们已经发展了一种自动的 DNA 模板制备方法，每天可制备 400 个 M13 模板 [18]。最初的噬菌斑挑取现在也可以用机器完成 [49]，并且可自动进行大量小体积测序反应的仪器也在开发中。有更长的读取长度和更大的样品容量的荧光凝胶显示仪正在开发，更有效的机器也正在设计。序列编辑目前是一个乏味的工作，进一步地改善软件可以消除大量的序列编辑工作。软件工具的使用简化了定向测序的模板和寡聚核苷酸的选择（施塔登、希利尔和迪尔，未发表）。随着这些进步和更大范围的努力，每个单位每年产生 10 多兆碱基的最终序列似乎是可行的。以这样的速度，团队双方在 1998 年年底之前一定可以完成秀丽隐杆线虫基因组的测序。此外，双方实验室正为加快完成酿酒酵母基因组的测序贡献资源。这两个生物完整的基因组序列可以帮助大家了解可能所有真核生物中都普遍存在的基因，及后生动物特异的基因。

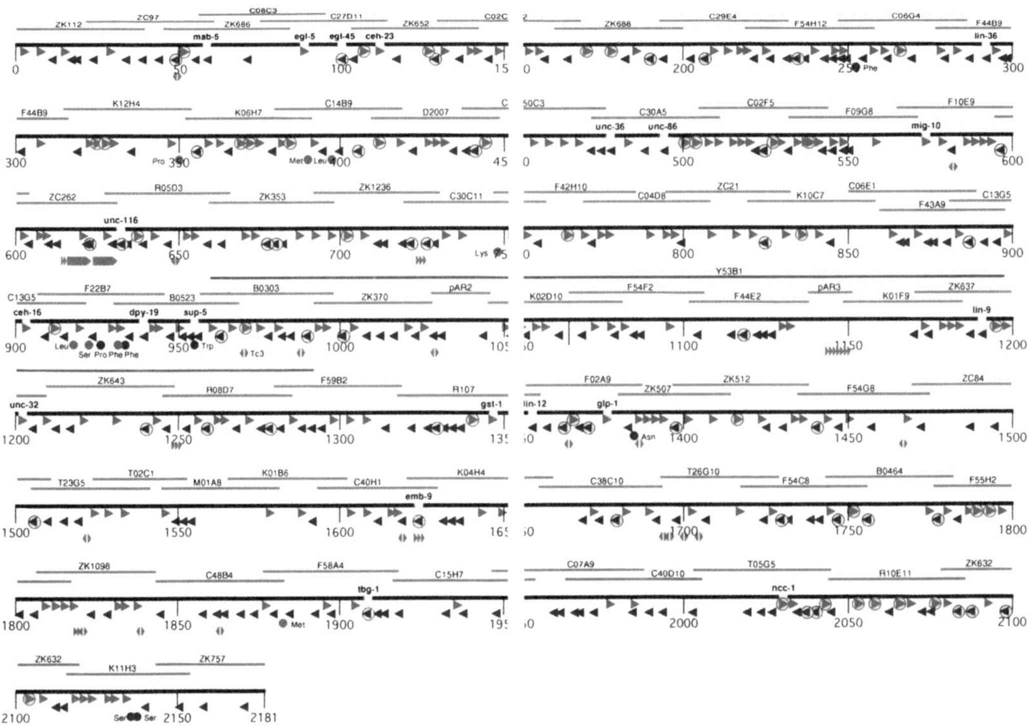

Fig. 1. The region of chromosome III described here. Each line represents 300 kb, with cosmid clones indicated by blue bars; a magenta bar indicates the YAC clone Y53B1. Red and blue arrowheads indicate the approximate starting points of gene candidates for both strands. Circled arrowheads indicate cDNA hits. Red or blue circles indicate the positions of tRNA genes. Green arrowheads indicate the position and type of major repeats. Genetic markers previously mapped to the region and assigned to genomic loci after DNA sequencing are indicated by vertical yellow bars.

Methods. Subclone libraries were prepared and sequencing reactions performed as previously described[6,12,14]. For data collection, automated fluorescent DNA sequencers were used. Both the Applied Biosystems 373A and Pharmacia ALF instruments were initially tested, although the 373A was preferred for the shotgun phase because of its greater sample capacity. Both laboratories currently complete two daily runs of each sequencer with 36 lane gels. Thus, with only four of these sequencing instruments, typically more than 1,700 samples are processed per week. After subtracting for failures and contaminating vector sequences, this is sufficient for yearly production of more than 20 Mb of raw sequence data. Two additional sequencing instruments are available for directed sequencing to close gaps and complete the complementary strand. At the conclusion of each sequencer run, the ABI trace files were transferred from the host Macintosh to UNIX workstations for all further work. During the shotgun phase all processing on the UNIX systems is fully automated using a single script. This includes reformating the trace files using MAKESCF (S.D., unpublished), selection of the quality data from each reading using AUTOTED (L.H., unpublished), clipping off cosmid and sequencing vector sequences using VEP (R.S., unpublished), and assembly using BAP[13]. After the shotgun phase, XBAP[13], which includes the oligo selection engine of OSP[22], is used to edit and finish each project. GENEFINDER and COP (S.D., unpublished) were used to check the finished sequence for errors. GENEFINDER is useful for identifying insertions and deletions in regions containing predicted genes, and COP, which compares the final sequence back to the raw data from which it was produced, is useful for identifying editing errors. We and others have previously described an evaluation of the accuracy of raw data from automated fluorescent DNA sequencers[6,25].

(**368**, 32-38; 1994)

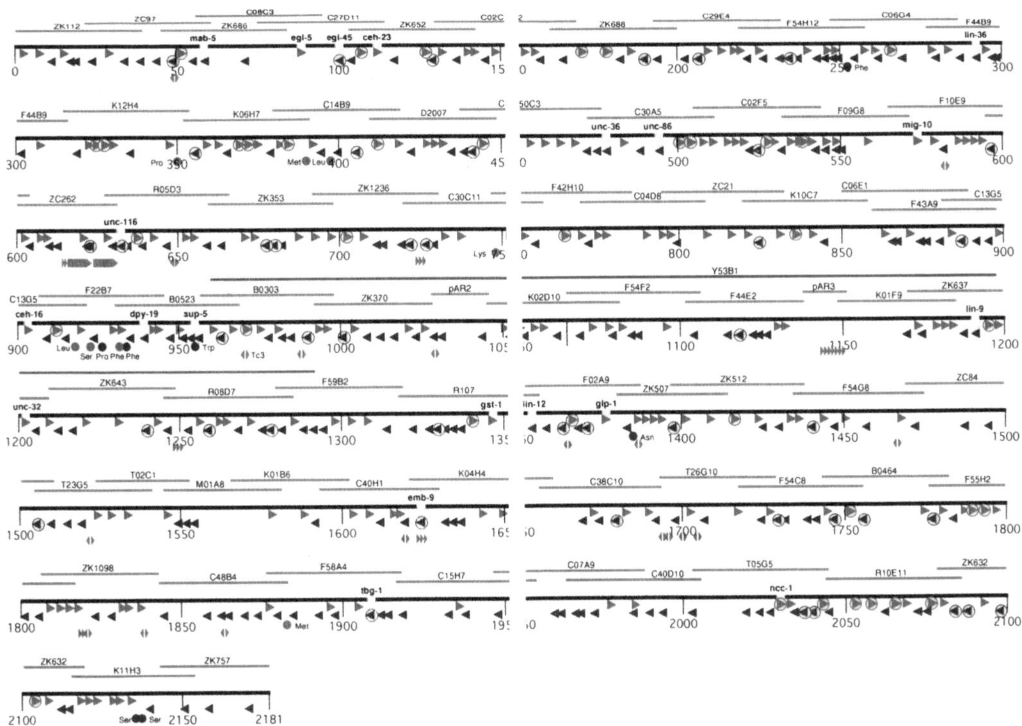

图 1. 本文描述的 3 号染色体的区域。每条线代表 300 kb，蓝色线段代表黏粒克隆；紫红色线段表示 YAC 克隆 Y53B1。红色和蓝色箭头指示候选基因在两条链的大致起点。带圈箭头指示 cDNA 位点。红色或蓝色的圈指示 tRNA 基因的位置。绿色箭头代表主要重复的位置和类型。用垂直的黄色线条表示之前定位到这个区域的遗传标记和 DNA 测序后确定到基因组基因座的遗传标记。

方法。按以前描述的方法制备亚克隆文库，进行测序反应 [6,12,14] 。用自动荧光 DNA 测序仪收集数据。美国应用生物系统公司的 373A 和法玛西亚的 ALF 仪器最初都进行过测试，但 373A 因为其样品容量大所以是鸟枪时期更倾向使用的仪器。现在双方实验室每台测序仪每天可运行 36 泳道凝胶两次。因此只用四个测序设备每周可处理 1,700 多个样品。减去失败和污染的载体序列后，每年足以产生 20 多兆碱基的原始序列数据。另外两个测序仪器可用来定向测序以闭合间隙及完成互补链序列。每台测序仪完成测序后，从 Macintosh 主机将 ABI 跟踪文件转到 UNIX 工作站进行下一步处理。鸟枪时期 UNIX 系统所做的所有处理是完全自动化的，并使用单一脚本。这包括用 MAKESCF 改变追踪文件格式（迪尔，未发表），用 AUTOTED 从每个读取数据中选择质量好的数据（希利尔，未发表），用 VEP 剪掉黏粒和测序载体序列（施塔登，未发表），以及用 BAP 组装序列 [13]。鸟枪时期以后，用包含 OSP[22] 寡核苷酸选择机器的 XBAP[13] 进行编辑并完成每个项目。用 GENEFINDER 和 COP（迪尔，未发表）检查完成序列的错误。GENEFINDER 对鉴定含有预测基因的区域中的插入和缺失很有用；COP 用来比较最终数据和原始数据，对鉴定编辑错误很有用。我们和其他人都已经评价过自动荧光 DNA 测序仪所得原始数据的准确性 [6,25]。

（李梅 翻译；杨崇林 审稿）

2.2 Mb of Contiguous Nucleotide Sequence from Chromosome III of *C. elegans*

R. Wilson[*], R. Ainscough[†], K. Anderson[*], C. Baynes[†], M. Berks[†], J. Bonfield[†], J. Burton[†], M. Connell[*], T. Copsey[†], J. Cooper[*], A. Coulson[†], M. Craxton[†], S. Dear[†], Z. Du[*], R. Durbin[†], A. Favello[*], A. Fraser[†], L. Fulton[*], A. Gardner[†], P. Green[*], T. Hawkins[†], L. Hillier[*], M. Jier[*], L. Johnston[*], M. Jones[†], J. Kershaw[†], J. Kirsten[*], N. Laisster[†], P. Latreille[†], J. Lightning[†], C. Lloyd[†], B. Mortimore[†], M. O'Callaghan[†], J. Parsons[*], C. Percy[†], L. Rifken[*], A. Roopra[*], D. Saunders[†], R. Shownkeen[†], M. Sims[†], N. Smaldon[†], A. Smith[†], M. Smith[†], E. Sonnhammer[†], R. Staden[†], J. Sulston[*], J. Thierry-Mieg[‡], K. Thomas[†], M. Vaudin[*], K. Vaughan[*], R. Waterston[*], A. Watson[†], L. Weinstock[*], J. Wilkinson-Sproat[†] & P. Wohldman[*]

[*] Department of Genetics and Genome Sequencing Center, Washington University School of Medicine, St Louis, Missouri 63110, USA

[†] MRC Laboratory of Molecular Biology and Sanger Center, Hinxton Hall, Cambridge CB10 IRQ, UK

[‡] CNRS-CRBM et Physique-Mathematique, Montpellier 34033, France

Received 15 November 1993; accepted 5 January 1994.

References:

1. Brenner, S. *Genetics* 77, 71-94 (1974).
2. Wood, W. B. *et al. The Nematode* Caenorhabditis elegans (Cold Spring Harbor Laboratory Press, New York, 1988).
3. Coulson, A. R., Sulston, J. E., Brenner, S. & Karn, J. *Proc. Natl. Acad. Sci. U.S.A.* **83**, 7821-7825 (1986).
4. Coulson, A., Waterston, R., Kliff, J., Sulston, J. & Kohara, Y. *Nature* **335**, 184-186 (1988).
5. Coulson, A. *et al. Bioessays* **13**, 413-417 (1991).
6. Sulston, J. *et al. Nature* **356**, 37-41 (1992).
7. Strauss, E. C., Kobori, J. A., Siu, G. & Hood, L. E. *Analyt. Biochem.* **154**, 353-360 (1986).
8. Deininger, P. *Analyt. Biochem.* **129**, 216-223 (1983).
9. Bankier, A. T. & Barrell, B. G. *Tech. Nucleic Acid Biochem.* **B5**, 1-34 (1983).
10. Smith, L. M. *et al. Nature* **321**, 674-679 (1986).
11. Connell, C. R. *et al. BioTechniques* **5**, 342-348 (1987).
12. Craxton, M. *Methods: A Comparison to Methods in Enzymology* Vol. 3 (ed. Roe, B.) 20-26 (Academic, San Diego, 1991).
13. Dear, S. & Staden, R. *Nucleic Acids Res.* **19**, 3907-3911 (1991).
14. Halloran, N., Du, Z. & Wilson, R. K. in *Methods in Molecular Biology* Vol. 10: *DNA Sequencing: Laboratory Protocols,* (eds Griffin, H. G. & Griffin, A. M.) 297-316 (Humana, Clifton, New Jersey, 1992).
15. Schriefer, L., Gebauer, B. K., Qiu, L. Q. Q., Waterson, R. H. & Wilson, R. K. *Nucleic Acids Res.* **18**, 7455-7456 (1990).
16. Lee, L. *et al. Nucleic Acids Res.* **20**, 2471-2483 (1992).
17. Hawkins, T. L., Du, Z., Halloran, N. D. & Wilson, R. K. *Electrophoresis* **13**, 552-559 (1992).
18. Watson, A., Smaldon, N., Lucke, R. & Hawkins, T. *Nature* **362**, 569-570 (1993).
19. Du, Z., Hood, L. & Wilson, R. K. *Meth. Enzym.* **218**, 104-121 (1993).
20. Gleeson, T. & Hillier, L. *Nucleic Acids Res.* **19**, 6481-6483 (1991).
21. Gleeson, T. J. & Staden, R. *Comput. Appl. Biosci.* 7, 398 (1991).
22. Hillier, L. & Green, P. *PCR Meth. Appl.* **1**, 124-128 (1991).
23. Dear, S. & Staden, R. *DNA Sequence* **3**, 107-110 (1992).
24. Alschul, S. F., Gish, W., Miller, W., Myers, E. W. & Lipman, D. J. *J. Molec. Biol.* **215**, 403-410 (1990).
25. Koop, B. F., Rowan, L., Chen, W.-Q., Deshpande, P., Lee, H. & Hood, L. *BioTechniques* **14**, 442-447 (1993).
26. Krause, M. & Hirsh, D. *Cell* **49**, 753-761 (1987).
27. Waterston. R. *et al. Nature Genet.* **1**, 114-123 (1992).
28. McCombie, W. R. *et al. Nature Genet.* **1**, 124-131 (1992).
29. Green, P. *et al. Science* **259**, 1711-1716 (1993).
30. Burglin, T. R. *et al. Nature* **351**, 703 (1991).
31. Chisholm, A. *Development* **111**, 921-932 (1991).
32. Wang, B. B. *et al. Cell* 74, 29-42 (1993).
33. Clark, S.,Chisholm, A. & Horvitz, H. R. *Cell* 74, 43-55 (1993).
34. Finney, M., Ruvkun, G. & Horvitz, H. R. *Cell* **55**, 757-769 (1988).
35. Guo, X., Johnson, J. J. & Kramer, J. M. *Nature* **349**, 707-709 (1991).
36. Naito, M., Kohara, Y. & Kurosawa, Y. *Nucleic Acids Res.* **20**, 2967-2969 (1992).
37. Patel, N., Thierry-Mieg, D. & Mancillas, J. R. *Proc. Natl. Acad. Sci. U.S.A.* **90**, 9181-9185 (1993).
38. Weston, K., Yochem, J. & Greenwald, I. *Nucleic Acids Res.* **17**, 2138 (1989).

39. Yochem, J., Weston, K. & Greenwald, I. *Nature* **335**, 547-550 (1988).

40. Yochem, J. & Greenwaid, I. *Cell* **58**, 553-563 (1989).

41. Waterston, R. H. GenBank locus CESUP5 (Acc. no. X54122) (1990).

42. Campbell, H. D. *et al. Proc. Natl. Acad. Sci. U.S.A.* **90**, 11386-11390 (1993).

43. Waterston, R. H. & Brenner, S. *Nature* **275**, 715-719 (1978).

44. Sulston, J. E. & Brenner, S. *Genetics* 77, 95-104 (1974).

45. Fichant, G. & Burks, C. *J. Molec. Biol.* **220**, 659-671 (1991).

46. Collins, J., Forbes, E. & Anderson, P. *Genetics* **121**, 47-55 (1989).

47. Yuan, J., Finney, M., Tsung, N. & Horvitz, H. R. *Proc. Natl. Acad. Sci. U.SA.* **88**, 3334- 3338 (1991).

48. Sulston, J. *Nature* **357**, 106 (1992).

49. Uber, D. C., Jaklevic, J. M., Theil, E. H., Lishanskaya, A. & McNeely, M. R. *BioTechniques* **11**, 642-647 (1991).

Acknowledgements. We thank T. Schedl for critical reading of the manuscript, P. Kassos and J. Rogers for administrative support, and other members of the *C. elegans* Genome Consortium for technical support. Database searches with St Louis sequences were done remotely using the NCBI BLAST server. ACEDB is available by anonymous ftp from cele.mrc-lmb.cam.ac.uk and from ncbi.nlm.nih.gov. This work was supported by the NIH National Center for Human Genome Research and the MRC Human Genome Mapping Project.

The First Skull and Other New Discoveries of *Australopithecus afarensis* at Hadar, Ethiopia

W. H. Kimbel *et al.*

Editor's Note

Almost 300 specimens of the primitive hominid *Australopithecus afarensis* had been found at Hadar in Ethiopia, the most famous being the skeleton known as Lucy. A frustrating absence was a skull, without which many questions remained unanswered—such as whether *A. afarensis* constituted a single, sexually dimorphic species, or two distinct species. This report, of a skull from Hadar, solved many riddles. Although the dentition marked it unequivocally as *afarensis,* it was bigger than any *Australopithecus* skull ever reported: evidence for sexual dimorphism in a single species. At three million years old, it was also one of the latest-known *afarensis* specimens. With the oldest at 3.9 million, Kimbel and colleagues claimed stasis as the rule in this species' 900,000-year tenure.

The Hadar Formation in Ethiopia is a prolific source of Pliocene Hominidae attributed to the species *Australopithecus afarensis*[1]. Since 1990, three seasons of field work have contributed 53 new specimens to the hominid inventory from Hadar, including the first fairly complete adult skull. Ranging from 3.0 to 3.4 million years in age (Fig. 1)[2-4], the new specimens bear on key debates in hominid palaeontology, including the taxonomic implications of sample variation and the reconstruction of locomotor behaviour. They confirm the taxonomic unity of *A. afarensis* and constitute the largest body of evidence for about 0.9 million years of stasis in the earliest known hominid species.

THE Hadar site yielded nearly 250 hominid fossils during the 1970s, including 40% of a female skeleton, A.L. 288-1 ("Lucy", ~3.18 million years), and the partial remains of at least 13 individuals from A.L. 333 (~3.20 Myr). Subsequent study of this collection and the sample from Laetoli, Tanzania, led to the recognition of the craniodentally primitive, bipedal species *Australopithecus afarensis*[5]. Morphological and metric variation in the *A. afarensis* hypodigm were initially explained as intraspecific, partly because of sexual dimorphism[6]. An alternative view attributed the differences to species-level taxonomic variation[7]. Debate was fuelled by the absence of a complete skull[8-10]. Similarly, although some interpreted the postcranial anatomy as indicating a complete adaptation to terrestrial bipedality[11,12], others saw evidence of a significant arboreal component in the locomotor repertoire of the species[13,14]. The new Hadar discoveries address both of these issues.

在埃塞俄比亚的哈达尔发现的第一例南方古猿阿法种的头骨和其他新发现

金贝尔等

编者按

在埃塞俄比亚的哈达尔地点发现了大约300块原始人类——南方古猿阿法种的标本，其中最著名的化石标本被称为露西骨架。令人遗憾的是这具标本缺乏头盖骨化石，而这令许多的问题仍然没有答案——例如，南方古猿阿法种是否是单一的、具有性二形性的物种，或是两个不同的物种。这份哈达尔头盖骨的报告解决了许多此前未能解释的谜题。虽然本文发现的头盖骨齿系明确地将之标记为南猿阿法种，但是它的头盖骨大于曾经发现报道过的所有南猿属：是同一物种性二形性的证据。在近300万年的历史中，这是最新发现的南猿阿法种标本之一。最古老的是在390万年前，金贝尔和他的同事断言此物种存在90万年的停滞期。

埃塞俄比亚的哈达尔组出土了丰富的上新世人科化石，这些化石标本都属于南方古猿阿法种[1]。自1990年以来，对哈达尔进行的三个季度的野外挖掘工作已经为原始人类化石记录贡献了53个新标本，包括第一例非常完整的成年头骨。新标本的年代在距今300万年到340万年之间（图1）[2-4]，这些新标本导致了古人类学中的一些重要争议，包括样本变异的分类学意义及其行动方式的重建等。他们确认了南方古猿阿法种可作为单独的分类学单元存在，并构成了证明已知的最早原始人类物种存在约90万年停滞期的最大证据。

哈达尔地点在20世纪70年代出土了将近250件原始人类化石，包括一具雌性骨架的40%、AL 288-1（“露西”，约318万年前）和来自AL 333（约320万年前）的至少13个个体的部分残骸。对这些化石和来自坦桑尼亚莱托里地点的样本进行的后续研究使得在颅骨和牙齿上表现出原始性的两足直立行走物种——南方古猿阿法种得以辨别出来[5]。南方古猿阿法种种型群的形态和测量变异最初被解释成种间差异，部分是因为性二形性[6]。另一种观点将这种差异归结为物种水平的分类学差异[7]。由于缺少完整的头骨，所以这方面的争议较大[8-10]。相似地，尽管有些人将颅后骨骼特征解释为对陆地直立行走的完全适应[11,12]，但是其他人将其视为该物种的行动方式中树栖行为占重要地位的证据[13,14]。在哈达尔地点新发现的标本可以解决这两个问题。

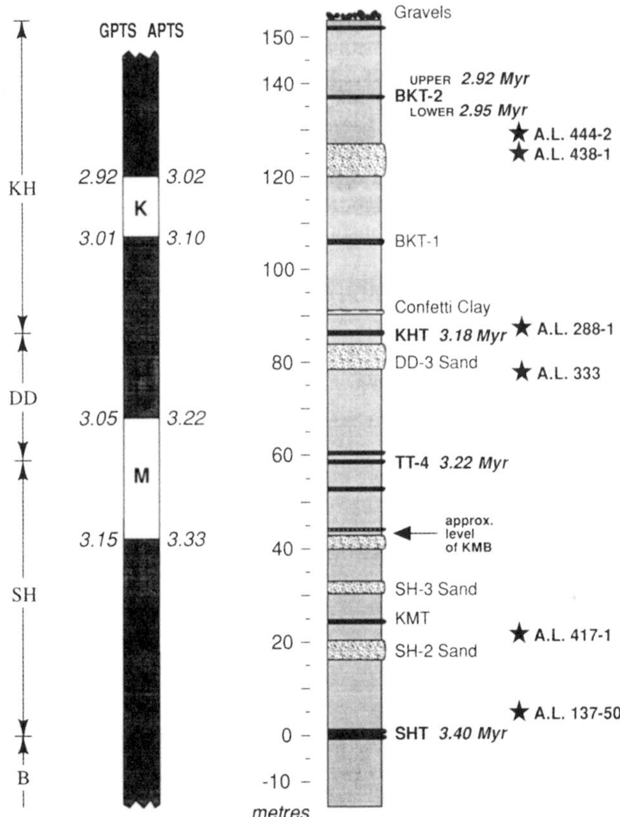

Fig. 1. Composite stratigraphic section of the Hadar Formation, based on revisions by R. C. Walter and J. L. Aronson through the 1993 field season, with positions of hominid specimens discussed in the text indicated. Single-crystal laser fusion (SCLF) $^{40}Ar/^{39}Ar$ ages for the SHT (3.40 ± 0.03 Myr), TT-4 (3.22 ± 0.01 Myr), KHT (3.18 ± 0.01 Myr) and BKT-2 (2.92–2.95 Myr) confirm that no Hadar hominid is older than 3.40 Myr, and establish ages of ~3.18 Myr for the A.L. 288-1 ("Lucy") skeleton and ~3.20 Myr for the A.L. 333 hominid assemblage[2,3]. The Kadada Moumou basalt (KMB) does not occur in the type-area, but its relative stratigraphic position has been correlated into the type-section in the field. A whole-rock $^{40}Ar/^{39}Ar$ plateau age of 3.28 ± 0.04 Myr has been obtained for the KMB (ref. 4). New specimens from A.L. 438 and 444 represent the first Hadar hominid discoveries stratigraphically above A.L. 288-1 ("Lucy"), extending the temporal range of *A. afarensis* at Hadar from ~3.4 to ~3.0 Myr. At left are shown the stratigraphic ranges of Basal (B), Sidi Hakoma (SH), Denen Dora (DD) and Kada Hadar (KH) Members of the Hadar Formation. Also depicted is the composite magnetostratigraphic section with the global polarity time scale (GPTS) and the astronomical polarity time scale (APTS); the new $^{40}Ar/^{39}Ar$ and magnetostratigraphic data support the APTS calibration of the Mammoth (M) and Kaena (K) reversed subchrons[3,4].

The new skull A.L. 444-2, from the middle Kada Hadar (KH) Mb. (~3.0 Myr), is clearly that of an adult male (Fig. 2). It has large canines and, with a biasterionic breadth of 106 mm, it is the largest *Australopithecus* cranium on record. Although the maxillary postcanine teeth are larger than any of those in the limited Hadar + Laetoli sample from the 1970s, relative to canine size they are not unusual[15,16]. The mandibular corpus is quite large, but its robusticity index falls below the Hadar mean (Table 1). The specimen's attribution to *A. afarensis* is warranted by its primitive constellation of mandibular, facial and calvarial

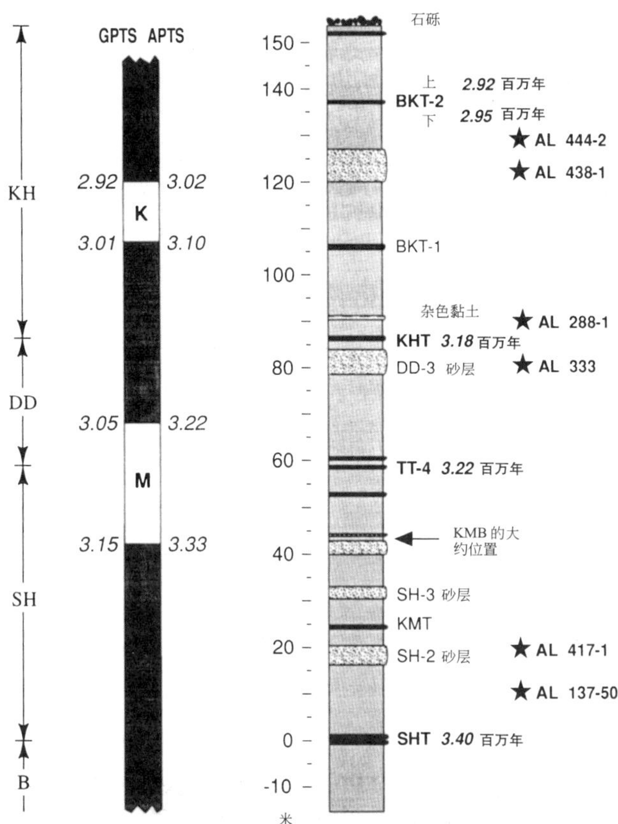

图 1. 哈达尔组剖面合成图，基于沃尔特和阿伦森 1993 野外工作进行的修正，图中标出了本文所指并讨论的原始人类标本的位置。SHT（340 万 ± 3 万年）、TT-4（322 万 ± 1 万年）、KHT(318 万 ± 1 万年）和 BKT-2（292 万至 295 万年）的单晶体激光核聚变（SCLF）$^{40}Ar/^{39}Ar$ 年代证实了哈达尔原始人类的年代早于 340 万年前，确立了 AL 288-1（"露西"）骨架的年代约为 318 万年前，而 AL 333 原始人类系列的年代约为 320 万年前 [2,3]。在该类型区域没有发现卡达达穆穆玄武岩（KMB），但是其相对地层位置与该野外典型剖面相关联。一块 KMB 整岩的 $^{40}Ar/^{39}Ar$ 坪年龄为 328 万 ± 4 万年（参考文献 4）。来自 AL 438 和 AL 444 的新标本代表了在地层上位于 AL 288-1（"露西"）之上发现的第一例哈达尔原始人类，将哈达尔南方古猿阿法种的时间范围从约 340 万年前扩展到了约 300 万年前。左侧给出了哈达尔组的底部（B）、西迪哈克玛（SH）、德嫩多拉（DD）和卡达哈达尔（KH）段的地层学范围。图中也描述了考虑了地球极性时间表（GPTS）和天文极性时间表（APTS）的复合地磁地层学剖面；新的 $^{40}Ar/^{39}Ar$ 和地磁地层学数据支持猛犸（M）和凯纳（K）反向亚时事件的 APTS 校正 [3,4]。

在卡达哈达尔（KH）段中部新发现的 AL 444-2 新头骨（约 300 万年前）明显属于一个成年雄性个体（图 2）。此例标本具有大的犬齿，星点间宽为 106 毫米，是可考的最大的南方古猿头盖骨。尽管该上颌前臼齿比发现于 20 世纪 70 年代的有限的哈达尔和莱托里样本的都要大，但是相对于犬齿尺寸而言，它们的大小还算正常 [15,16]。该样本的下颌体相当大，但是其粗壮指数低于哈达尔样本的均值（表 1）。通过与其他

characters relative to that of other *Australopithecus* species[15-20]. Its morphology is consistent with that of Hadar specimens from A.L. 333 that comprise the major components of the composite reconstruction of a male *A. afarensis* skull based on the 1970s collection[18,21,22]. The new Hadar skull refutes suggestions that the reconstruction incorporated facial and calvarial remains from two contemporaneous hominid species[9,10], and supports the interpretation of Hadar hominid cranial variation as intraspecific.

Table 1. Mandible corpus measurements

	Breadth (mm)	Height (mm)	Robusticity (%)
A.L. 417-1a (left)	18.0	36.0	50.0
A.L. 444-2b (right)	22.0	43.2	50.9
1970s Hadar sample	$\bar{x}$ = 18.5	$\bar{x}$ = 33.6	$\bar{x}$ = 55.1
	n = 12	n = 10	n = 10
	σ = 2.1	σ = 4.1	σ = 6.4
	range = 15.6–22.4	range = 28.0–40.5	range = 45.5–63.8

Mandible corpus measurements of A.L. 417-la, compared with those of A.L. 444-2b and the 1970s Hadar sample[26]. All measurements recorded here are for P4/M1 position, since measurements at M1 on A.L. 444-2b are not possible without further reconstruction of the corpus.

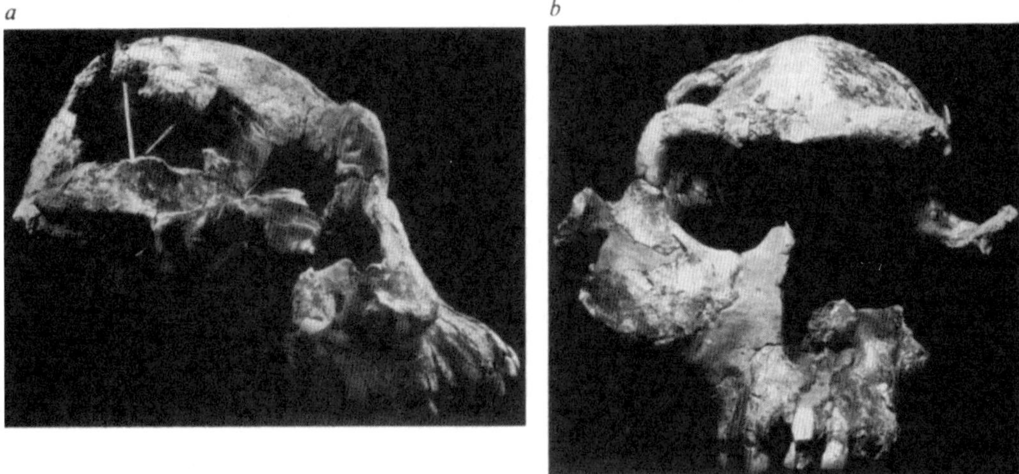

Fig. 2. Lateral (*a*) and anterior (*b*) views of Hadar skull A.L. 444-2, preliminary reconstruction, 47.5% natural size. Thirteen major pieces and many small fragments were clustered together in a small gully on a steep surface of Kada Hadar Member silts. Excavation failed to reveal additional hominid remains, but *in situ* mammal bone fragments with patina and preservation identical to that of the hominid were recovered upslope from the skull in a stratigraphic level 10.5 m below marker tuff BKT-2 (Lower). We believe this level to be the source horizon of the skull (Fig.1). It comprises the right mandible and symphysis (not shown) with fragmentary left and right incisors, partial right C and damaged P_4–M_1, the maxilla with right I^1, C, P^4–M^3 and left I^1, C, P^3–M^3, the right zygomatic, most of the nasal bones (not shown), most of the frontal with attached anteromedial segments of the parietals, the left parietal, the occipital with attached posterior right parietal, and both temporals. The anterior cranial base is missing, as are most of both temporal squames. Although there are no contacts between the frontal,

南方古猿物种的下颌骨、面部和头盖骨等特征的对比，将该标本划分为南方古猿[15-20]。其形态学特征与来自 AL 333 的哈达尔标本一致，研究人员根据这些 20 世纪 70 年代出土的标本综合重建出了一个雄性南方古猿阿法种头骨的主要部分[18,21,22]。新发现的哈达尔头骨驳斥了重建结果是两个同时代的原始人类物种的面部和头盖骨特征的拼接[9,10]的看法，而支持哈达尔原始人类头盖骨变异是种内变异的观点。

表 1. 下颌体的测量值

	宽度（毫米）	高度（毫米）	粗壮指数（%）
AL 417-1a（左）	18.0	36.0	50.0
AL 444-2b（右）	22.0	43.2	50.9
20 世纪 70 年代的哈达尔样本	$\bar{x}=18.5$	$\bar{x}=33.6$	$\bar{x}=55.1$
	$n=12$	$n=10$	$n=10$
	$\sigma=2.1$	$\sigma=4.1$	$\sigma=6.4$
	范围 ＝15.6~22.4	范围 ＝28.0~40.5	范围 ＝45.5~63.8

AL 417-1a 的下颌体测量值与 AL 444-2b 的那些样本和 20 世纪 70 年代的哈达尔样本的测量值进行比较[26]。这里记录的所有测量值都是在 P4/M1 位置测得的，因为 AL 444-2b 的下颌体没有得到进一步的重建，所以不可能测量其 M1 位置处的尺寸。

a

b

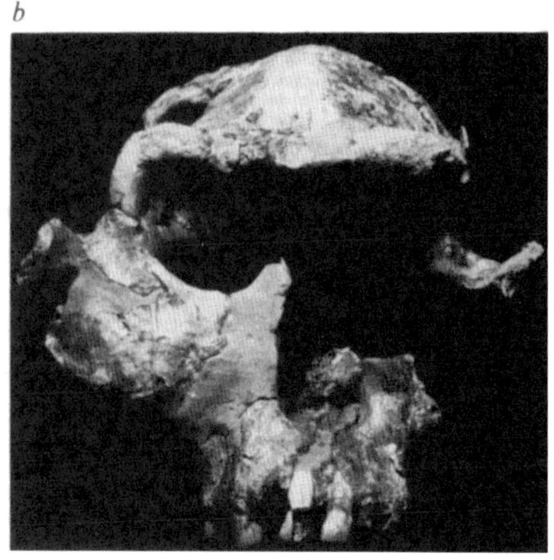

图 2. 哈达尔头骨 AL 444-2 初步重建结果的侧面视图（*a*）和前面视图（*b*），相当于自然尺寸的 47.5%。13 个大碎片和许多小碎片都聚集在卡达哈达尔地层淤泥陡面上的一个小峡谷中。挖掘工作中没有发现其他的原始人类遗迹，但是在发现该头骨的标志性凝灰岩 BKT-2（下层）之下 10.5 米的地层学水平向上，挖掘出了一些原位埋藏的保存状况与原始人类相同的生有绿锈的哺乳动物骨骼碎片。我们相信该水平是该头骨的来源地层（图 1）。该头骨由右侧下颌骨和正中联合部位（没有展示）、上颌骨、右颧骨、大部分鼻骨（没有展示）、大部分额骨（上面连有附着的顶骨的前正中部分）、左顶骨、附着右后顶骨的枕骨和两块颞骨构成。下颌骨上有成碎片了的左、右门齿，部分右 C 和损坏的 P_4~M_1；上颌骨上带有右 I^1、C、P^4~M^3 和左 I^1、C、P^3~M^3。颅底前部和两块颞骨鳞部的大部分都没有保存下来。尽管在额骨、上颌骨和颧骨之间没有任何衔接，但是保存下来的部分足以提供他们彼此之间和相对于脑颅位置的初步信息。

the maxilla and the zygomatic, there is sufficient preserved to provide a preliminary idea of their positions relative to one another and to the calvaria. Post-depositional deformation has compressed the palate along the midline, uplifted the right side of the frontoparietal fragment and supraorbital, twisted the left and elevated the right zygomatic arch, and pushed the nuchal plane of the occipital into the cranial cavity. The *A. afarensis* character complex present in A.L. 444-2 comprises a higher number of hominid symplesiomorphies than in any other species of the genus (including those attributed by some to *Paranthropus*). These primitive characters include strong prognathism relative to both the calvaria and the P^4/M^1 take-off of the zygomatic process of the maxilla, marked projection of sagittally and transversely convex premaxilla anterior to bicanine plane, horizontally inclined nasoalveolar clivus, procumbent maxillary incisors, extensive intranasal component of premaxilla with step-down to nasal cavity floor, sharp lateral nasal aperture margins, canine fossae, curved zygomaticoalveolar crests, anteriorly flat palate, $I^2/\underline{C}+\overline{C}/P_3$ diastemata, posteriorly convergent temporal lines with low, probably bifid, posterior sagittal crest, compound temporal/nuchal crest, asterionic notch sutural pattern, massive mastoid processes inflected beneath the cranial base with tips independent of occipitomastoid crests, shallow mandibular fossae, tubular tympanic plates that sit completely posterior to large postglenoid processes, weak petrous crest of tympanics, and coexistent transverse-sigmoid and occipital-marginal venous drainage systems (the polarity of this last character is indeterminate). Estimation of cranial capacity should be possible, but awaits final reconstruction of the calvaria. Measurements of the better-preserved maxillary teeth are (in mm, buccolingual breadth × mesiodistal length, the latter corrected for interproximal wear): Right C: 11.5 × 10.4; Left P4: 14.5 × 10.6, LM1: 15.0 × 13.7, LM2: 15.8 × 14.2, LM3: 16.2 × 14.9.

The frontal is a taxonomically important cranial region under-represented in the previous sample of *A. afarensis.* In A.L. 444-2 the supraorbital torus is vertically thick laterally and the very low squama has neither a chimpanzee-like supratoral sulcus nor a frontal trigone such as is found in "robust" *Australopithecus* crania. The distance across the postorbital constriction in A.L. 444-2 is large compared to that of other *Australopithecus* species, both absolutely (77 mm) and relative to facial breadth (frontobiorbital breadth index ~80%)[18]. These features are shared with the ~3.9 Myr Belohdelie frontal BEL-VP-1/1 (ref. 23), confirming the previously tentative attribution of the latter specimen to *A. afarensis*[24].

Maxilla A.L. 417-ld from the middle Sidi Hakoma (SH) Mb. (~3.25 Myr), associated with a partial mandible and cranial base fragments, constitutes the first relatively intact face of an *A. afarensis* female (Fig. 3*a–d*). It features small canines and less prognathism than A.L. 444-2, consistent with the great ape model of intraspecific sexual dimorphism[18,25], and otherwise demonstrates the same morphological pattern as do larger *A. afarensis* faces[17]. The mandible conforms to the *A. afarensis* plan[26]. Its low corpus robusticity index is identical to that of the A.L. 444-2 male mandible (Table 1). The P_3 lacks a lingual cusp, as in other *A. afarensis* females (A.L. 128-23, 288-1i) and some males (A.L. 277-1), but another new female jaw, A.L. 315-22, has a bicuspid P_3, confirming that P_3 polymorphism cuts across the size range of Hadar mandibles.

The A.L. 438-1 partial upper limb skeleton from middle KH Mb. (~3.0 Myr) includes the most complete *Australopithecus* ulna known (Fig. 3*e*, *f*); based on the associated mandible (corpus height at P_4/M_1 = 42.5 mm) and frontal fragment, it is almost certainly a male. Maximum ulnar length (excluding styloid process) of 268 mm is 22% greater than our length estimate for the A.L. 288-1 female ulna (220 mm). The size disparity between

沉积后的变形将上颚沿中线压缩了，将顶骨碎片和眶上部分的右侧提高了，左颧弓扭曲，右颧弓抬高，而枕骨的顶平面则被推进了颅腔。AL 444-2 中存在众多南方古猿阿法种的复合特征，其原始人类共同祖征（包括被某些人归属到傍人属中的那些特征）比该属其他任何物种都要多。这些原始特征包括相对于顶骨和上颌骨颧突（P^4/M^1 上方）而言强烈的凸颌、位于犬齿平面前的颌骨在矢状和横向方向明显突出、水平倾斜的鼻槽斜坡、平伏的上颌门齿、较大的下降到鼻腔底的前上颌骨的鼻内部分、锐利的梨状孔边缘、犬齿窝、弯曲的颧骨齿槽脊、前面平坦的上颚、$I^2/\underline{C}+\overline{C}/P_3$ 间隙裂、低的可能分叉的后矢状脊、颞线后方收敛、复合的颞骨 / 项脊、星点切迹缝形状、弯曲的位于颅底下的具有独立于枕乳脊的尖端的巨大乳突、浅的下颌窝、完全位于大型下颌窝后突之后的管状鼓板、鼓室的微弱岩脊以及同时存在乙状窦沟和枕边缘窦沟（最后这一特征的极性还不确定）。对颅容量的估计应该是可能的，但是还需等待对头盖骨的最终重建。保存较好的上颌牙齿的测量尺寸如下（单位：毫米，颊舌宽度 × 近远中长度，后者根据牙间磨损进行了修正）：右 C：11. 5 × 10.4；左 P4：14.5 × 10.6，LM1：15.0 × 13.7，LM2：15.8 × 14.2，LM3：16.2 × 14.9。

额骨在分类学上是一个很重要的颅骨区域，此前的南方古猿阿法种样本中并没有相应的代表。在 AL 444-2 中，眶上圆枕两侧在垂直方向厚，鳞部很低，既没有像黑猩猩一样的圆枕上沟，也没有在“粗壮”南方古猿的头盖骨中观察到的额三角。与其他南方古猿物种比起来，AL 444-2 的眶后缩窄绝对值（77 毫米）和相对于面部宽度的相对值（额眶宽指数约为 80%）[18] 都较大。这些特征在大约 390 万年前的贝洛德利样本的额区 BEL-VP-1/1（参考文献 23）中也存在，证实了以前将后者归为南方古猿阿法种这一暂时决定的正确性 [24]。

来自西迪哈克玛（SH）段中部的上颌骨 AL 417-1d（约 325 万年前），与部分下颌骨和颅底碎片连在一起，构成了第一例相对完整的雌性南方古猿阿法种的面部（图 3*a*~*d*）。其特征是犬齿较小，凸颌程度比 AL 444-2 低，这与大型猿类的种内性二形模式是一致的 [18,25]，另外，证实了其与较大型南方古猿阿法种的面部具有同样的形态学模式 [17]。下颌骨与南方古猿阿法种相符 [26]。其下颌体的粗壮指数较低，这与 AL 444-2 雄性下颌骨一致（表 1）。P_3 缺少舌侧牙尖，这与其他南方古猿阿法种雌性（AL 128-23，AL 288-li）和一些雄性（AL 277-1）是一样的，但是另一个新发现的雌性下颌骨 AL 315-22 的 P_3 有两个牙尖，证实了 P_3 在哈达尔下颌骨中就有不一致性。

来自 KH 段中部的 AL 438-1 的部分上肢骨骼（约 300 万年前）包括已知的最完整的南方古猿尺骨（图 3*e* 和 *f*）；参照与之相关联的下颌骨（下颌体高度 P_4/M_1 = 42.5 毫米）和额部碎片，几乎可以肯定该样本是一个雄性。尺骨最大长（不算茎状突）是 268 毫米，比 AL 288-1 雌性尺骨的长度估计值（220 毫米）大 22%。这两个哈达尔尺骨之间的尺寸差异与 AL 333x-5 和 AL 333w-36 的尺骨差异一样大。可以看出其前后的

these two Hadar ulnae is equalled by proximal ulnae A.L. 333x-5 and 333w-36. With anteroposterior shaft curvature visibly less than in chimpanzees, a tall olecranon process, and an anteriorly oriented trochlear notch, the A.L. 438-1 ulna diverges from the modern African ape condition and matches the human morphology of other Hadar hominid ulnae[27,28].

The large, presumably male, humerus A.L. 137-50 from the lower SH Mb. (Fig. 3*e*, *f*) is a remarkably close match for the recently reported Maka humerus MAK-VP-1/3, with which it is penecontemporaneous (~3.4 Myr)[23], as well as for Hadar proximal humerus A.L. 333-107 (~3.2 Myr)[28]. The new Hadar specimen has a retroflexed shaft and appears to be very robust relative to length, with thick cortical bone, pronounced deltoid and humeral adductor insertion sites, and a strong lateral supracondylar crest.

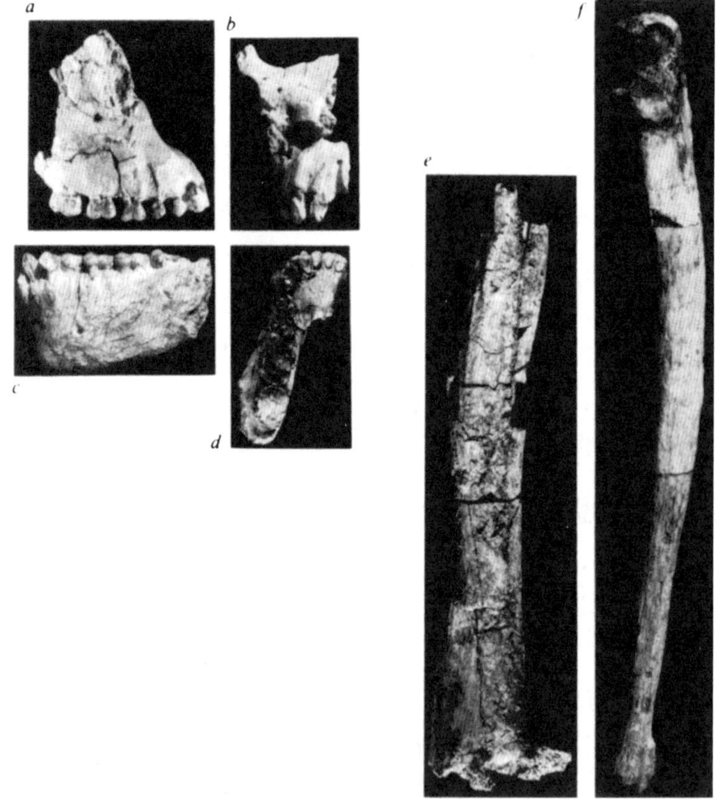

Fig. 3. Lateral (*a*) and anterior (*b*) views of A.L. 417-1d maxilla; lateral (*c*) and occlusal (*d*) views of A.L. 417-1a mandible, 50% natural size. Diagnostic *A.afarensis* morphology combined in the specimen is primitive relative to that of other *Australopithecus* species and includes: mandible corpus hollowed laterally, mental foramen positioned below midcorpus and opens anterosuperiorly, extramolar sulcus high and narrow, postcanine tooth row slightly concave buccally, low corpus robusticity, flat, vertical infraorbital bone plate, sagittally and transversely convex premaxilla that protrudes in front of the bicanine plane, and sharp lateral nasal aperture margin. Measurements of the dentition are (recorded as described in Fig. 2 legend): *Maxilla*, RC: 9.9 × 9.4, RP3: 11.7 × 8.3, RP4: 12.2 × 8.9, RM1: 13.4 × 12.3, RM2: 14.7 × 13.2, RM3: 14.8 × 13.0; *Mandible*, LP3: 10.8 × 8.9, LP4: 11.2 × 8.6, LM1: 11.6 × 12.4, LM2: 13.1 × 13.0, RM3: 13.3 × 14.9. *e*, Anterior view of right humerus A.L. 137-50; *f*, lateral view of left ulna A.L. 438-1a,

骨干曲率比黑猩猩的小，AL 438-1 中高的鹰嘴突出和向前的滑车切迹都与现代非洲猿类的情况不同，而与其他的哈达尔原始人类尺骨的人类形态学相一致[27,28]。

来自下 SH 段的较大的、推测是男性的肱骨 AL 137-50（图 3*e* 和 *f*）与近期报道的马卡肱骨 MAK-VP-1/3 具有非常密切的一致性，是几乎同时期的（约 340 万年前）[23]，哈达尔近端肱骨 AL 333-107（约 320 万年前）[28] 也是这样。新发现的哈达尔标本骨干呈翻转状，其相对长度似乎很粗壮，具有厚的皮质、突出的三角肌和肱骨内收肌附着处以及强壮的外侧髁上脊。

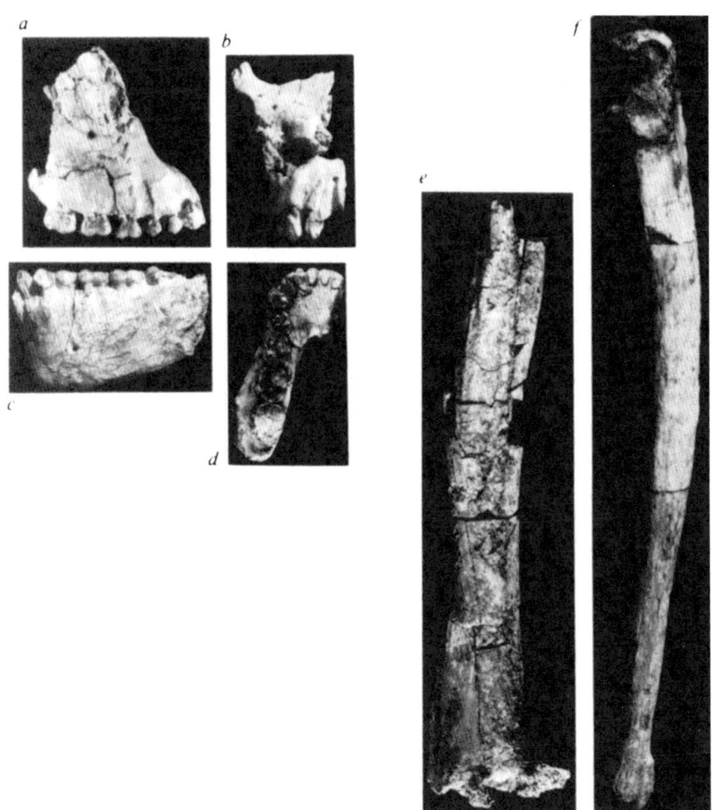

图 3. AL 417-1d 上颌骨的侧面（*a*）和前面（*b*）视图；AL 417-1a 下颌骨的侧面（*c*）和咬合面（*d*）视图，相当于自然尺寸的 50%。该标本中整合的南方古猿阿法种决定性的形态学特征相对于其他南方古猿物种来说比较原始，包括：侧面凹陷的下颌体、位于下颌体中部下向前上方开口的颏孔、高而窄的臼齿后沟、颊侧稍微凹陷的前臼齿、下颌骨体粗壮度较低、平而垂直的眶下骨板、在双侧犬齿平面沿纵向和横向突出的前颌骨以及锐利的梨状孔边缘。齿系的尺寸如下（按图 2 图注的描述记录）：上颌骨，RC：9.9 × 9.4，RP3：11.7 × 8.3，RP4：12.2 × 8.9，RM1：13.4 × 12.3，RM2：14.7 × 13.2，RM3：14.8 × 13.0；下颌骨，LP3：10.8 × 8.9，LP4：11.2 × 8.6，LM1：11.6 × 12.4，LM2：13.1 × 13.0，RM3：13.3 × 14.9。*e*，右肱骨 AL 137-50 的前面视图；*f*，左尺骨 AL 438-1a 的侧面视图，都相当于自然尺寸的 50%。AL 438 尺骨与

both 50% natural size. The A.L. 438 ulna is associated with a partial mandible, a frontal fragment, three metacarpals, as well as shaft portions of the right ulna, left and right radii and right humerus. Two fragments of the complete ulna were recovered *in situ* in a KH Member sand unit, 12 m below marker tuff BKT-2 (Lower) (Fig. 1).

Our preliminary estimate of humeral length for A.L. 137-50 is 295 mm, ~24% greater than that of A.L. 288-1 (238 mm, ref. 27). Using the A.L. 438-1 ulna as a counterpart to A.L. 137-50 yields an ulna/humerus length index of ~91%. The index for the A.L. 288-1 female is ~92.5%. If it is true that in *A. afarensis* humerus length is roughly that expected in a modern human of comparable body size[29], then these figures indicate relatively long ulnae (and, presumably, forearms) for the Hadar hominids, distinctly closer to the relative ulnar length of chimpanzees ($\bar{x}$ = 95%, σ = 3%, r = 88–101%, n = 20) than to that of modern humans ($\bar{x}$ = 80%, σ = 3%, r = 74–88%, n = 20). The combination of a relatively short, but robust, humerus and a long forearm is unlikely to resolve the debate about locomotion in *A. afarensis*—which has been concerned as much with incompatible evolutionary models for the interpretation of functional morphology as with divergent interpretations of the fossils themselves[11,12,30].

The new Hadar fossils support the taxonomic unity of *A. afarensis.* A single hominid species is now well documented in all three fossiliferous members of the Hadar Formation, spanning 0.4 Myr. The new discoveries for the first time extend the *A. afarensis* stratigraphic range well up into the Kada Hadar Member, by which time in the Hadar Fm. a variety of other macromammal clades had gone through taxonomic turnover[31]. Furthermore, with ages of ~3.0 Myr for the A.L. 444-2 skull and ~3.9 Myr for the Belohdelie frontal, the 0.9 Myr temporal range of *A. afarensis* is bracketed by specimens of high diagnostic value. The Hadar sample forms nearly all of the post-3.4 Myr portion of the *A. afarensis* hypodigm and, at close to 300 specimens, it is the most persuasive body of evidence for prolonged stasis within the oldest known hominid species[6,23].

(**368**, 449-451; 1994)

William H. Kimbel*, Donald C. Johanson* & Yoel Rak*†

* Institute of Human Origins, 2453 Ridge Road, Berkeley, California 94709, USA

† Department of Anatomy, Sackler School of Medicine, Tel Aviv University, Tel Aviv, Israel

Received 10 January; accepted 22 February 1994.

References:

1. Johanson, D. C., Taieb, M. & Coppens, Y. *Am. J. Phys. Anthrop.* 57, 373-402 (1982).
2. Walter, R. C. & Aronson, J. L *J. Hum. Evol.* 25, 229-240 (1993).
3. Walter, R. C. *Geology* 22, 6-10 (1994).
4. Renne, P., Walter, R., Verosub, K., Sweitzer, M. & Aronson, J. *Geophys. Res. Lett.* 20, 1067-1070 (1993).
5. Johanson, D. C., White, T. D. & Coppens, Y. *Kirtlandia* 28, 1-14 (1978).
6. Johanson, D. C. & White, T. D. *Science* 203, 321-329 (1979).
7. Olson, T. R. in *Ancestors: The Hard Evidence* (ed. Delson, E.) 102-119 (Liss, New York, 1985).
8. Leakey, R. E. & Walker, A.C. *Science* 207, 1103 (1980).

部分下颌骨、一个额骨碎片、三块掌骨以及右尺骨的骨干部分、左右桡骨和右肱骨连在一起。完整尺骨的两块碎片是在位于标志性凝灰岩 BKT-2（下层）之下 12 米的 KH 段的沙质单元原位出土的（图 1）。

我们对 AL 137-50 的肱骨长度的初步估计是 295 毫米，比 AL 288-1 的肱骨（238 毫米，参考文献 27）约长 24%。如果将 AL 438-1 作为 AL 137-50 的对照，可以得出尺骨 / 肱骨长度指数约为 91%。AL 288-1 雌性样本的该指数约为 92.5%。如果南方古猿阿法种的肱骨长度大体上与身体大小相当的现代人一样[29]，那么这些数字就说明哈达尔原始人类具有相对较长的尺骨（前臂可能也相应较长）；与现代人的（$\bar{x}=80\%$，$\sigma=3\%$，$r=74\%\sim88\%$，$n=20$）相比，它与黑猩猩的尺骨相对长度（$\bar{x}=95\%$，$\sigma=3\%$，$r=88\%\sim101\%$，$n=20$）更接近。一个相对较短但粗壮的肱骨和一个较长前臂的组合不太可能解决关于南方古猿阿法种行动方式的争议——这个问题不但与对功能形态学解释（引出的）不相容的进化模式有关，还与化石本身的不同解释有关[11,12,30]。

新发现的哈达尔化石支持南方古猿阿法种这一分类单元。一个单一的原始人类物种得到了很好的保存，分散在哈达尔组的三个化石层中，前后共跨越了 40 万年。这些新发现第一次将南方古猿阿法种的地层学范围扩展到了卡达哈达尔段，哈达尔组的许多其他的巨型哺乳动物进化枝都经历了分类学的颠覆[31]。另外，AL 444-2 头骨的年代距今约为 300 万年，贝洛德利额骨的年代距今约为 390 万年，可靠的标本将南方古猿阿法种 90 万年的时间范围框定起来。哈达尔样本组成了过去的 340 万年中几乎所有的南方古猿阿法种种型群，有将近 300 个标本，这是证明在已知的最古老的原始人类物种中存在长期停滞的最具有说服力的证据[6,23]。

（刘皓芳 翻译；崔娅铭 审稿）

9. Shipman, P. *Discover* 7, 87-93 (1986).
10. McKee, J. K. *Am. J. Phys. Anthrop.* **80**, 1-9 (1989).
11. Lovejoy, C. O. *Sci. Am.* **256**, 118-125 (1988).
12. Latimer, B. M. in *Origine(s) de la Bipédie chez les Hominidés* (eds Coppens, Y. & Senut, B.) 169-176 (CNRS, Paris, 1991).
13. Susman, R. L., Stern, J. T. & Jungers, W. L. *Folia Primatol.* **43**, 113-156 (1984).
14. Stern, J. T. & Susman, R. L. *Am. J. Phys. Anthrop.* **60**, 279-318 (1983).
15. White, T. D., Johanson, D. C. & Kimbel, W. H. S. *Afr. J. Sci.* 77, 445-470 (1981).
16. Kimbel, W. H., White, T. D. & Johanson, D. C. in *Ancestors: The Hard Evidence* (ed. Delson, E.) 122-137 (Liss, New York, 1985).
17. Rak, Y. *The Australopithecine Face* 1-169 (Academic, New York, 1983).
18. Kimbel, W. H., White, T. D. & Johanson, D. C. *Am. J. Phys. Anthrop.* **64**, 337-388 (1984).
19. Kimbel, W. H. & Rak, Y. *Am. J. Phys. Anthrop.* **66**, 31-54 (1985).
20. Ward, S. C. & Kimbel, W. H. *Am. J. Phys. Anthrop.* **61**, 157-171 (1983).
21. Kimbel, W. H., Johanson, D. C. & Coppens, Y. *Am. J. Phys. Anthrop.* 57, 453-499 (1982).
22. Kimbel, W. H. & White, T. D. *J. Hum. Evol.* 17, 545-550 (1988).
23. White, T. D. *et al. Nature* **366**, 261-265 (1993).
24. Asfaw, B. *J. Hum. Evol.* **16**, 611-624 (1987).
25. Wood, B. A., Yu, L. & Willoughby, C. *J. Anat.* **174**, 185-205 (1991).
26. White, T. D. & Johanson, D. C. *Am. J. Phys. Anthrop.* **57**, 501-544 (1982).
27. Johanson, D. C. *et al. Am. J. Phys. Anthrop.* **57**, 403-451 (1982).
28. Lovejoy, C. O., Johanson, D. C. & Coppens, Y. *Am. J. Phys. Anthrop.* 57, 637-650 (1982).
29. Jungers, W. L. & Stern, J. T. *J. Hum. Evol.* **12**, 673-684 (1983).
30. Stern, J. T. & Susman, R. L. in *Origine(s) de la Bipédie chez les Hominidés* (eds Coppens, Y. & Senut, B.) 99-111 (CNRS, Paris, 1991).
31. White, T. D., Moore, R. V. & Suwa, G. *J. Vert. Paleont.* **4**, 575-583 (1984).

Acknowledgements. We thank the Center for Research and Conservation of Cultural Heritage (Ethiopian Ministry of Culture and Sports Affairs) and the National Museum of Ethiopia for permission to conduct field work and for logistical support. Field work was funded by grants from the National Science Foundation, the National Geographic Society, the L. S. B. Leakey Foundation and Lufthansa Airlines. We thank R. Walter, chief geologist and project co-leader, J. Aronson, R. Bernor, M. Black, G. Eck, J. Harris, E. Hovers, N. Kahn, P. Renne, S. Semaw, C. Vondra and T. Yemane for scientific contributions to the project, T. White for the discovery of A.L. 137-50 during his visit to Hadar in 1990, B. Latimer and L. Jellema for human and chimpanzee data from the Hamann-Todd Osteological Collection, Cleveland Museum of Natural History and our Ethiopian colleagues T. Asebwork, A. Asfaw, Z. Assefa, H. Bolku, M. Fesseha, T. Hagos, M. Kassa, M. Tesfaye, S. Teshome, T. Wodajo and the Afar people of Eloha village without whose assistance our field work would not have been possible.

A Diverse New Primate Fauna from Middle Eocene Fissure-fillings in Southeastern China

K. C. Beard *et al.*

Editor's Note

The Eocene epoch (56–34 million years ago) was a golden age for mammalian evolution, particularly primates. However, until the mid-1990s, knowledge of Eocene primates was biased towards Europe and North America, and most consisted of two extinct groups—the adapids and omomyids—whose relationships with later forms was obscure. Barely a handful of fragmentary Eocene primates were known from Asia. This paper changed all that. Eocene deposits from Shanghuang in Jiangsu, China, offered a wealth of adapids and omomyids, a tarsier, and that most elusive of Eocene primates—a simian, *Eosimias sinensis*, then the earliest-known representative of the lineage which would eventually lead to monkeys, apes and man.

We report the discovery of a fauna of primates from Eocene (~45 Myr) deposits in China having a diversity greater than in European and North American localities of similar antiquity. From the many forms that will illuminate questions of primate phylogeny comes evidence for a basal radiation of primitive simians.

AT present, the fossil record of early primate evolution is strongly biased geographically, with the stratigraphically dense and intensively studied samples of Eocene primates from North America and Europe standing in sharp contrast to the paucity of similar data for Africa and Asia[1,2]. For example, only eight species of Eocene primates have been described from Asia[3-10], and most of these are represented by fragmentary or unique fossils. As a result, the phylogenetic affinities of Eocene Asian primates are controversial[1,11-14], and little is known about how this Asian record relates to those of other continents[15].

Since January 1992 we have worked on richly fossiliferous fissure-fillings located near the village of Shanghuang, in southern Jiangsu Province in the People's Republic of China (PRC) (Fig. 1 has been omitted in this edited version). These are the first such fillings of Eocene age to be discovered in Asia. The fissures have yielded a large and varied mammalian fauna[16], of which the diversity of Eocene primates is greater than that of the rest of Asia combined.

中国东南部中始新世裂隙堆积中新发现的一个多样化灵长类动物群

比尔德等

编者按

始新世(5,600 万 ~ 3,400 万年前)是哺乳动物演化的一个黄金时代，特别是灵长类动物。然而，直到 20 世纪 90 年代中期，对始新世灵长类动物的认知都来自欧洲和北美洲，大部分由两个灭绝的类群——兔猴类和鼩猴类组成，他们与后来的类型之间的关系是模糊不清的。在亚洲已知的仅仅是少数零星破碎的始新世灵长类化石。本文改写了这一切。中国江苏省上黄村的始新世堆积物中发现了丰富的兔猴类和鼩猴类、一种跗猴以及最难捉摸的始新世灵长类——中华曙猿(一种真猴类)。在当时，中华曙猿是最终分化出猴、猿和人这一谱系的已知最早代表。

我们报道了从中国的始新世(约 4,500 万年前)堆积物中发现的灵长类动物群，该动物群的多样性比欧洲和北美洲年代相仿地点的更为丰富。这些为灵长类系统发育关系问题带来光明的众多类型，也为原始真猴类的基群辐射带来了证据。

现在，早期灵长类演化的化石记录在地理位置上存在严重的偏向，地层中分布密集且研究比较集中的北美洲和欧洲的始新世灵长类化石样本与非洲和亚洲相似化石材料的缺乏形成鲜明对比 [1,2]。例如，亚洲到目前为止只描述过 8 种始新世时期的灵长类 [3-10]，其中大部分只有破碎的或唯一的化石材料。因此，始新世时期亚洲灵长类的系统发育亲缘关系就存在争议 [1,11-14]，而且关于这一亚洲记录与其他大陆发现的灵长类的关系如何还知之甚少 [15]。

自 1992 年 1 月以来，我们一直致力于几处含有丰富化石的裂隙堆积物的野外发掘工作，它们位于中国江苏省南部的上黄村附近(图 1；该图显示上黄裂隙堆积物的地理位置，此版本中省略)。这是亚洲首次发现始新世时期的这种堆积物。从这些裂隙中得到了一个多样的大型哺乳动物群 [16]，其中始新世灵长类的多样性比亚洲其余地方发现的总和还要丰富。

Like many Eocene localities in North America and Europe, the primate fauna from the Shanghuang fissures includes both adapiforms and omomyids. But the fillings have also yielded distinctive fossil primates that are not found elsewhere: these include the first Eocene tarsier and fossils that we interpret as basal simians (anthropoids or "higher primates"). This co-occurrence of adapiforms, omomyids, tarsiids and basal simians is unique, and provides fresh insight into primate phylogeny in general and the biogeographic role of Asia during early phases of primate evolution in particular.

Age and Geological Setting

The Shanghuang fissure-fillings were formed as middle Eocene karstic infillings into surrounding Triassic carbonates. So far, four fossiliferous fissure-fillings (IVPP locality 93006 A–D) of Eocene age have been sampled. All are exposed in an active commercial quarry near Shanghuang, where the Triassic carbonates are being mined for cement production. Available evidence indicates that the fissure-fillings span only a short interval (probably 1–2 Myr) of middle Eocene time, with the fauna from fissure D being slightly older than the others.

Biostratigraphically, the mammalian faunas from the fissures appear to represent the Irdinmanhan and early Sharamurunian Land Mammal Ages (LMAs), on the basis of taxa such as the lagomorph *Lushilagus lohoensis*[16], the carnivore *Miacis lushiensis*[16], the brontothere *Microtitan,* and the cricetid rodent *Pappocricetodon.* Unfortunately, direct radiometric dating of middle Eocene LMAs in Asia has not yet proven possible. However, the Irdinmanhan LMA can be correlated with the Bridgerian and early Uintan LMAs of North America on the basis of a major episode of intercontinental mammalian dispersal[17]. In North America the Bridgerian/Uintan boundary occurs at about 46 Myr[18]. As a rough appraisal based on these considerations, we estimate the age of the Shanghuang fissures to be about 45 Myr.

Shanghuang Adapiforms

Adapiforms[1] are lemur-like early Cenozoic primates that seem to be closely related to living strepsirhines[19]. At least two species are known from the Shanghuang fissures (Fig. 2), where they are most abundant in fissure D.

像北美洲和欧洲的许多始新世时期的化石地点一样，来自上黄裂隙的灵长类动物群既包括兔猴型类，也有鼩猴类(已经灭绝的鼩猴科的成员)。该堆积物也出产了没有在其他地方发现过的独特的灵长类化石，这些化石包括第一种始新世的跗猴化石和我们认为是基底真猴类(类人猿或“高等灵长类”)的化石。兔猴型类、鼩猴类、跗猴类(或称眼镜猴类)和基底真猴类的同时出现是独一无二的。这提供了对灵长类的总体系统发育关系，尤其是对灵长类早期演化过程中亚洲的生物地理作用的一些新见解。

年代和地质概况

上黄裂隙堆积物是由中始新世时期的喀斯特填充物堆积到三叠纪碳酸盐围岩裂隙中形成的。到现在为止，已经发掘了四处始新世时期的含化石的裂隙堆积(IVPP 地点编号 93006 A~D)。所有裂隙堆积都暴露在上黄附近的一处活跃的商业采石场里，那里的三叠纪碳酸盐岩被开采出来进行水泥生产。现在得到的证据显示这些裂隙堆积的形成只横跨了中始新世一个短暂的年代区间(约 100 万年到 200 万年)。从裂隙 D 得到的动物群要比其余裂隙的稍微古老一些。

从生物地层方面来看，根据兔形类——洛河卢氏兔[16]、食肉类——卢氏小古猫[16]、王雷兽类——小雷兽和仓鼠类——祖仓鼠等类群判断，从这些裂隙发掘出来的哺乳动物群应该代表了陆地哺乳动物期(LMA)中的伊尔丁曼哈期和沙拉木伦早期。不幸的是，现在想对亚洲中始新世时期的 LMA 进行直接的同位素年代测定还是不可能的。然而，基于各大陆间哺乳动物扩散的主要时期，伊尔丁曼哈期 LMA 可以与北美洲的布里杰期和早尤因塔期的 LMA 相对比[17]。在北美洲，布里杰期 / 尤因塔期的界线大约是在 4,600 万年前[18]。根据这些考虑粗略地评估了一下，我们估计上黄裂隙的年代约为 4,500 万年前。

上黄兔猴型类

兔猴型类[1]是类似狐猴的早新生代时期的灵长类动物，它们似乎与现生的曲鼻猴类具有亲密的关系[19]。来自上黄裂隙的兔猴型类现在已知的至少有两种(图 2)，所有上黄裂隙中，它们在裂隙 D 的化石最为丰富。

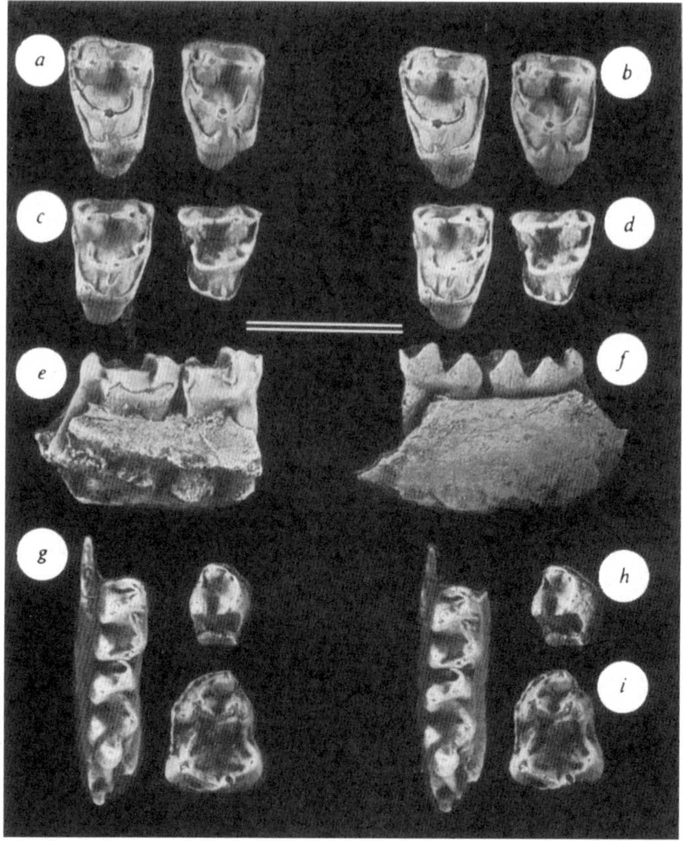

Fig. 2. Adapiform and omomyid primates from the Shanghuang fissure-fillings. *a*, *b*, *Europolemur*-like adapiform: IVPP V11019 (*a*), and IVPP V11020 (*b*), each consisting of an isolated left M^1 or M^2 from fissure D, occlusal view. *c–g*, *Adapoides troglodytes*: IVPP V11021 (*c*), isolated right M^1 or M^2 from fissure D, occlusal view; IVPP V11022 (*d*), isolated right deciduous P^4 from fissure D, occlusal view; *e–g*, IVPP V11023, right mandible fragment with M_{2-3} from fissure B, holotype of *Adapoides troglodytes*, in buccal (*e*), lingual (*f*) and occlusal (*g*) views. *h*, *i*, *Macrotarsius macrorhysis:* IVPP V11025 (*h*), isolated right P_4 from fissure D, holotype of *Macrotarsius macrorhysis*, occlusal view; IVPP V11024 (*i*), isolated left M_1 from fissure D, occlusal view. Occlusal views are stereopairs. Scale bar, 5 mm.

Aside from several isolated teeth from the Kuldana Formation of Pakistan[9], the Shanghuang adapiforms comprise the only unequivocal evidence for adapiforms in Asia during the Eocene.

A relatively primitive adapiform is represented by isolated teeth from fissure D. Cheek teeth of this species resemble those of *Europolemur*, an adapiform from the middle Eocene of Europe[1,20]. In particular, upper molars from Shanghuang resemble those of *E. koenigswaldi* from Messel, Germany[20], in that they lack a hypocone. However, absence of a hypocone in these two species is a shared primitive feature, and thus does not reflect a special relationship between them. The Shanghuang species is smaller than any previously named species of *Europolemur*, but further specimens will be required to ascertain whether it represents the first Asian record of this genus.

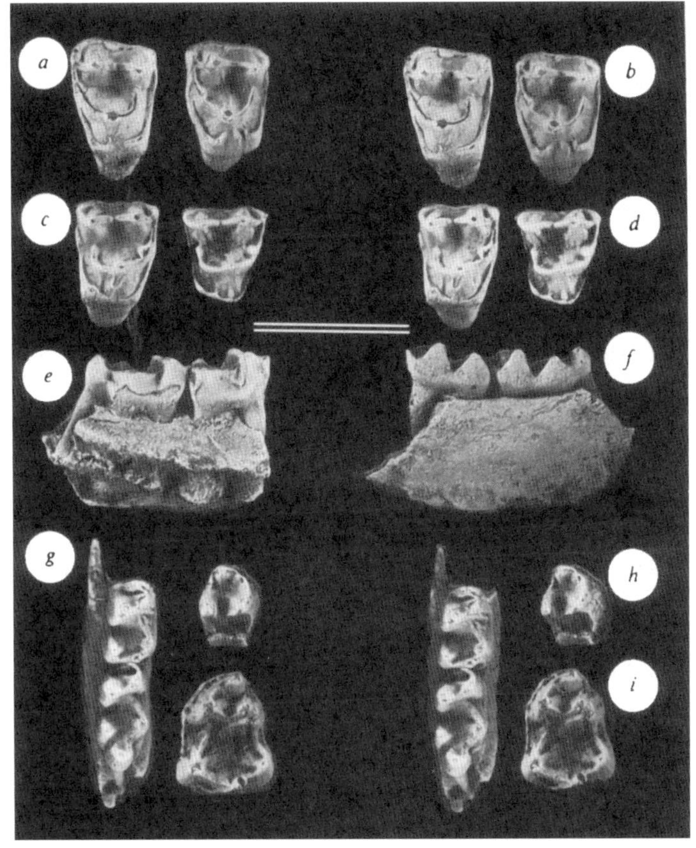

图 2. 来自上黄裂隙堆积的兔猴型类和鼩猴类。*a*，*b*，似欧狐猴兔猴型类：IVPP V11019 (*a*) 和 IVPP V11020 (*b*)，每个都包括一颗来自裂隙 D 的游离左 M^1 或 M^2，咬合面视图。*c*~*g*，穴居似兔猴：IVPP V11021 (*c*)，来自裂隙 D 的游离右 M^1 或 M^2，咬合面视图；IVPP V11022 (*d*)，来自裂隙 D 的游离右 dP^4，咬合面视图；*e*~*g*，IVPP V11023，来自裂隙 B 的带有 $M_{2\text{-}3}$ 的右下颌体残段，穴居似兔猴正模标本，颊面 (*e*)、舌面 (*f*) 和咬合面 (*g*) 视图。*h*，*i*，扬子大跗猴：IVPP V11025 (*h*)，来自裂隙 D 的游离右 P_4，扬子大跗猴的正模标本，咬合面视图；IVPP V11024 (*i*)，来自裂隙 D 的游离左 M_1，咬合面视图。咬合面视图是立体像对。比例尺，5 毫米。

除了巴基斯坦的库尔达纳组地层 [9] 中的几颗游离牙齿以外，上黄兔猴型类提供了始新世时期唯一一个确凿的亚洲兔猴型类的证据。

来自裂隙 D 的几颗游离牙齿代表了一种相对原始的兔猴型类。这一种类的颊齿与欧狐猴属的相似，后者是一种欧洲中始新世时期的兔猴型类 [1,20]。尤其是，上黄的上臼齿与在德国梅瑟尔发现的孔氏欧狐猴的很相似 [20]，因为它们缺少一个次尖。然而，这两个种类中次尖的缺失是一个共有的原始特征，因此并不能反映出它们之间具有特别的关系。上黄的种类比其他之前已命名的欧狐猴属内的种类都要小，但是还需要更多的标本才能确定它是否代表了该属在亚洲的首次记录。

A second adapiform is represented by more nearly complete fossils, which permit its description here.

Order Primates Linnaeus, 1758
Suborder Strepsirhini Geoffroy, 1812
Family Adapidae Trouessart, 1879
Subtribe Adapina Trouessart, 1879
Adapoides troglodytes, new genus and species

Holotype. IVPP V11023, right mandibular fragment preserving M_{2-3} (Fig. 2).

Type locality. IVPP 93006, fissure B.

Known distribution. Middle Eocene of Jiangsu Province, PRC.

Diagnosis. Smaller than *Adapis*, *Cryptadapis* and *Leptadapis.* Lower molars relatively longer and narrower than in *Leptadapis.* Lower molars without metastylids, in contrast to *Cryptadapis*, *A. parisiensis* and *L. magnus.* Lower molars with high, continuous crest between protoconid and metaconid and deep talonid notch, in contrast to *Microadapis* and *Leptadapis.*

Etymology. From the generic name *Adapis*+Greek suffix *-oides* (like). Trivial name from Greek *troglodytes* (inhabitant of caves).

Description. *Adapoides troglodytes* is a small adapinan, with lower molars (M_2L: 2.65 mm, M_2W: 1.90 mm; M_3L: 3.55 mm, M_3W: 1.80 mm) similar in size to those of *Microadapis sciureus.* The lower molars are primitive in lacking metastylids but derived in having a deep talonid notch and a strong crest uniting protoconid and metaconid. Referred upper molars show most of the derived features typical of adapinans (buccal crests form a single mesiodistally oriented shearing surface, metaconule weak or absent, occlusal outline of crown mesiodistally longer and transversely narrower than in *Europolemur* and its close relatives), but primitively lack well-developed hypocones.

Discovery of the Shanghuang adapiforms illuminates two abiding questions about adapiform evolution and biogeography. First, the controversial adapiform *Mahgarita stevensi*, known only from the Duchesnean (late Eocene) of western Texas[21-23], has been suggested to be an immigrant from Asia[24]. Previously, however, there were no Asian adapiforms of suitable morphology to represent a close relative of *Mahgarita,* the affinities of which lie near *Europolemur*[1,21]. Discovery of a *Europolemur*-like species at Shanghuang provides the first direct evidence that adapiforms of suitable morphology to represent the ancestral stock for *Mahgarita* inhabited Asia during the Eocene, thus corroborating the hypothesis of an Asian origin for this enigmatic genus.

另一种兔猴型类由更加接近完整的化石材料所代表，我们在此对其进行描述。

灵长目 Primates Linnaeus, 1758
曲鼻猴亚目 Strepsirhini Geoffroy, 1812
兔猴科 Adapidae Trouessart, 1879
兔猴亚族 Adapina Trouessart, 1879

穴居似兔猴（新属、新种）*Adapoides troglodytes*, n. gen. et sp.

正模标本。IVPP V11023，保存有 M_{2-3} 的右下颌体残段（图 2）。

产地。IVPP 93006，裂隙 B。

分布。中华人民共和国江苏省，中始新世。

特征。比兔猴属、隐兔猴属和丽兔猴属小。下臼齿与丽兔猴属相比，相对较长且狭窄。与隐兔猴属、巴黎兔猴和硕丽兔猴不同，下臼齿没有下后附尖。与倭兔猴和丽兔猴不同，下臼齿在下原尖和下后尖之间有高而连续的脊，并且有很深的下跟座缺。

词源。来源于属名 *Adapis*（兔猴）+ 希腊语后缀 *-oides*（类似）。种名取自希腊语 *troglodytes*（穴居者）。

描述。穴居似兔猴是一种小型兔猴亚族成员，其下臼齿（M_2L：2.65 毫米，M_2W：1.90 毫米；M_3L：3.55 毫米，M_3W：1.80 毫米）的尺寸与松鼠倭兔猴的相似。下臼齿比较原始，缺少下后附尖，但是具有深的下跟座缺和非常发育的将下原尖和下后尖联结在一起的脊。归入的上臼齿显示出大多数兔猴亚族成员的典型衍生特征（颊脊形成单一的近远中方向的剪切面，后小尖发育弱或缺失，齿冠咬合面轮廓在近远中方向上比欧狐猴属及其近缘属更长，横向则更窄），但是比较原始的特征是缺少发达的次尖。

上黄兔猴型类的发现为探讨兔猴型类演化和生物地理这两个长久以来一直存在的问题带来了光明。首先，存在争议的兔猴型类玛氏兔猴仅发现于得克萨斯州西部的杜申阶（晚始新世）地层[21-23]，有人认为其是从亚洲迁移过来的[24]。但是，之前并没有具有适当形态学特征的亚洲兔猴型类可以代表玛氏兔猴属的亲缘属，因此一直认为它们的亲缘关系与欧狐猴属接近[1,21]。在上黄发现的这种似欧狐猴属样的种类首次提供了直接证据，表明代表玛氏兔猴的祖先类群并具有相应形态学特征的兔猴型类于始新世期间生活在亚洲，因此证实了这一神秘属的亚洲起源假说。

Second, Shanghuang adapiforms also shed light on the phylogeny and biogeography of the subtribe Adapina (including *Adapis*, *Leptadapis*, *Cryptadapis*[1,25] and *Adapoides*). Anatomically, the Adapina are among the best known Eocene primates, being represented by well-preserved skulls, jaws and numerous postcranial elements[1,26-28]. However, the phylogenetic and geographic origin of the group has been obscure because the fossil record of these animals was previously limited to Europe, where they appeared suddenly at the beginning of the Robiacian LMA (late Eocene)[20,29]. Discovery of the small, anatomically primitive adapinan *Adapoides troglodytes* in the middle Eocene of China suggests that adapinans immigrated into Europe from Asia rather than Africa, as had been thought[20].

The Shanghuang adapiforms show no special resemblances to the enigmatic Sivaladapinae from the Miocene of India, Pakistan and Yunnan Province, PRC[30,31].

Shanghuang Tarsiiforms

Both omomyid and tarsiid primates have been recovered from Shanghuang (Figs 2, 3). Omomyids are usually considered to be primitive tarsiiforms[1,15,32], although this has been disputed[2]. Remains of omomyids are the rarest primate fossils recovered to date from Shanghuang, where only two isolated teeth have been found in fissure D. Despite their rarity, they are of great interest because they represent a new species of *Macrotarsius*, a genus otherwise known only from western North America[1,33,34].

Infraorder Tarsiiformes Gregory, 1915
Family Omomyidae Trouessart, 1879
Macrotarsius macrorhysis, new species

Holotype. IVPP V11025, isolated right P_4 (P_4L, 2.65 mm; P_4W, 2.05 mm) (Fig. 2).

Type locality. IVPP 93006, fissure D.

Known distribution. Middle Eocene of Jiangsu Province, PRC.

Diagnosis. P_4 smaller, relatively narrower and with simpler talonid than in other species of *Macrotarsius*. M_1 smaller than in *M. montanus* and *M. siegerti* and without complete ectocingulid, in contrast to *M. roederi*.

Etymology. Trivial name from Greek *makros* (long) + Greek *rhysis* (river), in allusion to the Yangtze River.

Description. *Macrotarsius macrorhysis* closely resembles North American species of *Macrotarsius* in comparable aspects of anatomy: P_4 is mesiodistally short and bears a strongly molarized trigonid, and a well developed crest unites the metaconid and entoconid

其次，上黄兔猴型类也对兔猴亚族（包括兔猴属、丽兔猴属、隐兔猴属[1,25]和似兔猴属）的系统发育关系和生物地理有所启示。从解剖学上说，兔猴亚族属于了解得最为清楚的始新世灵长类，已经发现保存很好的头骨、下颌和大量头后骨骼[1,26-28]。但是，该类群的系统发育和地理上的起源一直都很模糊，因为这些动物的化石记录之前都只局限在欧洲，在那里他们是在 LMA 罗比亚克期（晚始新世）的初期突然出现的[20,29]。在中国的中始新世时期发现这种小型的、解剖学特征原始的兔猴亚族成员——穴居似兔猴，表明兔猴亚族是从亚洲迁移到欧洲的，而非以前所认为的是从非洲迁移过去的[20]。

上黄兔猴型类没有显示出与印度、巴基斯坦和中国云南省中新世的神秘的西瓦兔猴亚科有特别的相似之处[30,31]。

上黄跗猴型类

鼩猴类和跗猴类这两类灵长类在上黄都有发现（图 2，图 3）。鼩猴类通常被认为是原始的跗猴型类[1,15,32]，尽管这一点一直都争议不断[2]。鼩猴类的化石是到现在为止从上黄发掘到的灵长类化石中最为稀少的，只从裂隙 D 中发现了两颗游离的牙齿。尽管它们的化石很稀少，但是它们仍具有重大意义，因为它们代表了一个大跗猴属的新种类，而大跗猴属之前仅在北美洲西部被发现过[1,33,34]。

跗猴型下目 Tarsiiformes Gregory, 1915

始镜猴科 Omomyidae Trouessart, 1879

扬子大跗猴（新种）*Macrotarsius macrorhysis*，n. sp.

正模标本。IVPP V11025，游离右 P_4（P_4L：2.65 毫米；P_4W：2.05 毫米）（图 2）。

产地。IVPP 93006，裂隙 D。

分布。中华人民共和国江苏省，中始新世。

特征。与大跗猴属的其他种相比，P_4 较小，相对较窄，并且下跟座较简单。M_1 比蒙大拿长跗猴和西氏长跗猴的小，没有完整的下外齿带，与罗氏长跗猴正好相反。

词源。种名取自希腊语 *makros*（长）+ 希腊语 *rhysis*（河），暗指长江。

描述。扬子大跗猴在相应的解剖学特征方面，与大跗猴属中北美洲的种类很像：P_4 近远中方向短，具有臼齿化强烈的下齿座，一条发达的脊将 M_1 的下后尖和下外

on M_1. In terms of size and crown structure, M_1 in *M. macrorhysis* is practically indistinguishable from that in *M. jepseni*, known from the early Uintan (middle Eocene), Uinta Basin, Utah[33].

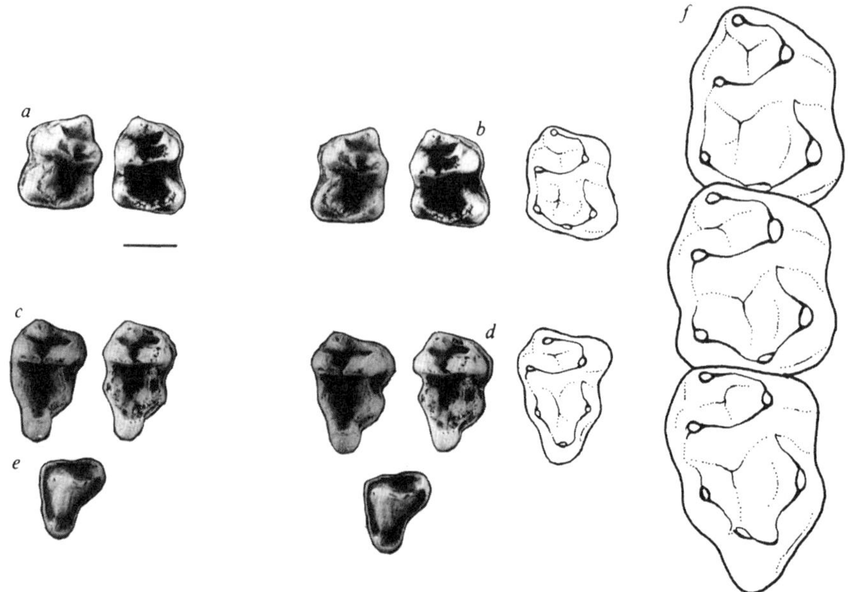

Fig. 3. *a–e*, *Tarsius eocaenus* from the Shanghuang fissure-fillings: IVPP V11026, isolated left M_1 from fissure A, occlusal view (*a*); IVPP V11030, isolated right M_1 from fissure C, holotype of *Tarsius eocaenus*, occlusal view (*b*); IVPP V11027, isolated right M_3 from fissure A, occlusal view (*c*); IVPP V11031, isolated right M_3 from fissure C, occlusal view (*d*); IVPP V11029, isolated left P^3 from fissure C, occlusal view (*e*). All views are stereopairs; scale equals 1 mm. (*f*), Dental morphology of *Tarsius eocaenus* (IVPP V11030 and IVPP V11027) compared with that of living *Tarsius bancanus* (USNM 300917), drawn to same scale.

Family Tarsiidae Gray, 1825
Tarsius eocaenus, new species

Holotype. IVPP V11030, isolated right M_1 (M_1L, 1.65 mm; M_1W, 1.55 mm) (Fig. 3).

Type locality. IVPP 93006, fissure C.

Known distribution. Middle Eocene of Jiangsu Province, PRC.

Diagnosis. Smaller than other species of *Tarsius.*

Etymology. Trivial name recognizes the Eocene age of this species.

Description. Isolated cheek teeth from fissures A and C attributed to *T. eocaenus* are virtually identical to those of living *Tarsius* in terms of crown structure (Fig. 3). Several aspects of dental anatomy distinguish *T. eocaenus* from omomyids and *Afrotarsius*[35]. For

尖联结在一起。从尺寸和牙冠结构方面看，扬子大跗猴的 M_1 与犹他州尤因塔盆地早尤因塔期（中始新世）的杰氏长跗猴的几乎没有区别[33]。

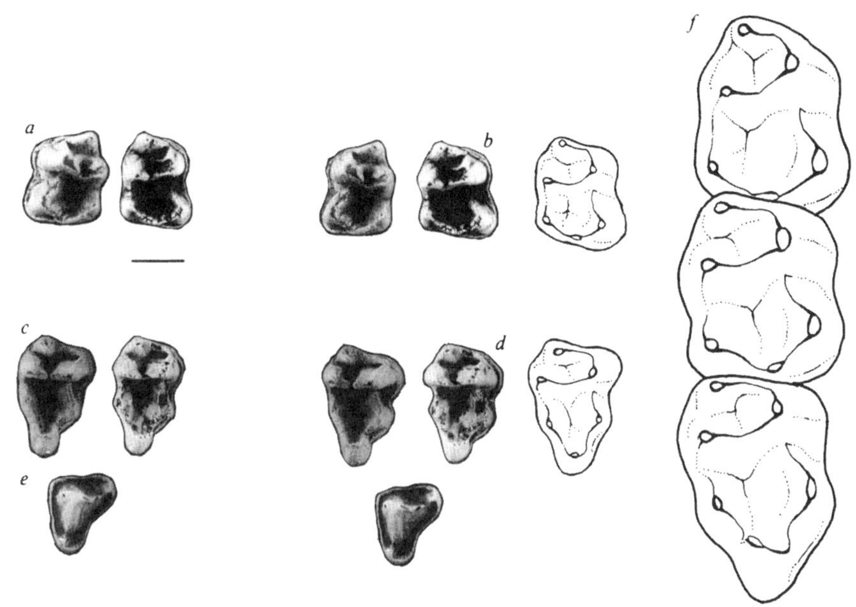

图 3. *a~e*，来自上黄裂隙堆积的始新跗猴：IVPP V11026，来自裂隙 A 的游离左 M_1，咬合面视图（*a*）；IVPP V11030，来自裂隙 C 的游离右 M_1，始新跗猴的正模标本，咬合面视图（*b*）；IVPP V11027，来自裂隙 A 的游离右 M_3，咬合面视图（*c*）；IVPP V11031，来自裂隙 C 的游离右 M_3，咬合面视图（*d*）；IVPP V11029，来自裂隙 C 的游离左 P^3，咬合面视图（*e*）。所有视图都是立体像对，比例尺是 1 毫米。（*f*），始新跗猴（IVPP V11030 和 IVPP V11027）与现生马来跗猴（USNM 300917）的牙齿形态学对比，绘图的比例相同。

跗猴科 Tarsiidae Gray, 1825

始新跗猴（新种）*Tarsius eocaenus*，n. sp.

正模标本。IVPP V11030，游离右 M_1（M_1L：1.65 毫米；M_1W：1.55 毫米）（图 3）。

产地。IVPP 93006，裂隙 C。

分布。中华人民共和国江苏省，中始新世。

特征。比跗猴属的其他种小。

词源。其种名指示该种年代是始新世。

描述。从裂隙 A 和 C 发掘出来的游离颊齿虽然被归入始新跗猴，但是这些颊齿在齿冠结构上与现生的跗猴属实际上是一样的（图 3）。牙齿的解剖学特征可从几个

example, the lower molars of *T. eocaenus* possess shelf-like paraconids situated near the lingual margin of the trigonid; a well developed entocristid uniting the entoconid and metaconid; entoconid positioned near the base of the trigonid, so that the talonid is short on M_1 and the entoconid is mesial to the level of the hypoconid on M_3 (M_2 is unknown); an angular, robust hypoconid; M_3 with a simple, unicuspid, distally projecting hypoconulid lobe. Cheek teeth of *T. eocaenus* are uniformly smaller than those of living species of *Tarsius*, including *T. pumilus*[36].

The tarsiiform primates from Shanghuang demonstrate for the first time that tarsiids and omomyids were virtually, if not certainly, sympatric during the middle Eocene in southeastern China. Discovery of *Macrotarsius* in Asia confirms previous predictions[15] regarding the feasibility of omomyid dispersal between North America and Asia during the middle Eocene. More dramatically, *Tarsius eocaenus* nearly triples the palaeontologically documented antiquity of tarsiers via a stratigraphic range extension from the Miocene[37] down to the middle Eocene. This brings the fossil record for Tarsiidae into much greater concordance with predictions based on cladistic reconstructions of the phylogenetic position of this group[2,32,38].

Basal Simians

The most diverse group of primates from the Shanghuang fissures, represented by at least four species, belongs to a further radiation of Eocene primates, interpreted here as basal simians. This radiation has been poorly sampled in the fossil record until now. Although the retention of several primitive features distinguishes these animals from more advanced simian taxa, such a combination of primitive and derived traits is not unexpected in basal simians[39-42]. Available evidence does not allow us to determine whether the basal simian taxa sampled to date from Shanghuang constitute a monophyletic or paraphyletic assemblage. Here we describe the basal simian from Shanghuang that is best represented in our current sample.

Suborder Anthropoidea Mivart, 1864
Eosimiidae, new family

Type genus. *Eosimias*, n. gen.

Diagnosis. Differs from Strepsirhini and Tarsiiformes in having the following combination of characters: lower dental formula 2-1-3-3; I_{1-2} small and vertically implanted; C_1 larger than I_{1-2} and P_2; P_2 single-rooted; P_{3-4} mesiodistally short and obliquely oriented with respect to tooth row (mesial root more labial than distal root); M_{1-2} with cuspidate paraconid, metaconid widely separated from enlarged protoconid, entoconid situated mesial to level of hypoconid, and distally-projecting hypoconulid.

方面将始新跗猴与鼩猴类和非洲跗猴区分开来[35]。例如，始新跗猴的下臼齿具有搁板状的位于下齿座舌侧边缘附近的下前尖；具有发达的将下内尖和下后尖联结在一起的下内脊；其下内尖位于下齿座基部附近，所以 M_1 的下跟座很短，而下内尖在 M_3 上的位置较下次尖更靠近中侧（M_2 的情况尚不知道）；具有一个尖锐的、粗壮的下次尖；其 M_3 有一个简单的、单尖的、远端突出的下次小尖叶。始新跗猴的颊齿比现生的包括小跗猴在内的跗猴属内的种类都要小[36]。

来自上黄的跗猴型类第一次从实际上（即使不确切）呈现了跗猴类和鼩猴类在中始新世的中国东南部的分布区是重叠的。在亚洲发现大跗猴属化石证实了之前对于中始新世期间鼩猴类在北美洲和亚洲之间扩散的可能性的预测[15]。更引人注目的是，始新跗猴将跗猴的地层分布下限从中新世[37]向下扩展至中始新世，使跗猴在古生物上记录到的年代范围向前延伸了两倍。这使得跗猴科的化石记录与建立在该类群系统位置的支序系统重建基础上的预测更加一致了[2,32,38]。

基底真猴类

来自上黄裂隙的最具多样性的灵长类类群至少代表了四个种类，它们属于始新世灵长类一次进一步辐射的产物，这里将其解释为基底真猴类。直到现在，化石记录中对这次辐射的记载都很不完善。尽管几种原始特征的保留将这些动物与更进步的真猴类类群区别了开来，但是这种原始特征与衍生特征的结合在基底真猴类中并不意外[39-42]。我们还无法根据现有证据确定目前为止从上黄采集的基底真猴类是构成了单系还是并系类群。这里我们描述了来自上黄的现有标本中最具有代表性的基底真猴类。

类人猿亚目 Anthropoidea Mivart, 1864

曙猿科（新科）Eosimiidae，n. fam.

模式属。*Eosimias*，n. gen.

特征。在如下几个特征方面不同于曲鼻猴亚目和跗猴型下目：下齿式为 2-1-3-3；I_{1-2} 小，垂直嵌入；C_1 比 I_{1-2} 和 P_2 大；P_2 单根；P_{3-4} 近远中方向短，就齿列而言是斜向的（近中根比远中根更靠近唇侧）；M_{1-2} 具有多尖的下前尖，下后尖与膨大的下原尖完全分离开了，下内尖位于下次尖近中侧，下次尖向远中突出。

Differs from other simians (except possibly *Catopithecus*[41], *Proteopithecus*[41], *Serapia*[42], *Arsinoea*[42] and *Afrotarsius*[35,37,40]) in retaining an unfused mandibular symphysis and prominent paraconids on M_{1-2}. Differs from simians except Parapithecidae, *Arsinoea* and possibly *Afrotarsius* in having P_4 metaconid situated well distal and inferior to protoconid, without strong transverse crest uniting these cusps. Differs from Platyrrhini and *Afrotarsius* in having prominent, distally-projecting hypoconulids on M_{1-2}. Differs from Old World simians except *Afrotarsius*, *Amphipithecus* and *Pondaungia* in lacking twinned entoconid and hypoconulid cusps on M_{1-2}.

Eosimias sinensis, new genus and species

Holotype. IVPP V10591, right mandible preserving P_4–M_2 and roots or alveoli for C_1, P_{2-3}, and M_3 (Fig. 4).

Type locality. IVPP 93006, fissure B.

Known distribution. Middle Eocene of Jiangsu Province, PRC.

Diagnosis. As for the family (currently monotypic).

Etymology. Greek *eos* (dawn, early) and Latin *simia* (ape). Trivial name refers to geographic provenance of this species.

Description. Aspects of the anterior lower dentition can be inferred from IVPP V10592, a referred specimen from fissure A (Fig. 4). Alveoli for I_{1-2} preserved in this specimen are diminutive. In contrast, the C_1 alveolus is relatively enormous, being much larger than those for I_{1-2} and P_2. The relative proportions of I_1–P_2 in eosimiids are plausibly interpreted as primitive for primates, since they resemble the conditions in other Paleogene simians, adapiforms[24] and certain omomyids (e.g., *Washakius*)[43]. Both IVPP V10591 and V10592 possess an unfused mandibular symphysis (more completely preserved in IVPP V10592). In contrast to adapiforms and omomyids, the symphyseal region of the mandible is dorsoventrally deep rather than strongly procumbent. In this derived feature, *Eosimias* resembles simians.

The crown of P_3 is unknown. Its alveoli, like those of P_4, are obliquely oriented with respect to the tooth row, the mesial alveolus being situated labial to the distal alveolus. This derived condition does not occur among Eocene adapiforms and omomyids, but is common among Fayum simians (e.g., *Qatrania*)[44]. A similar condition exists in *Amphipithecus* from the Eocene of Burma, a genus widely[5,12] but not universally[1] considered to be a primitive simian.

在保留未融合的下颌联合部以及 M_{1-2} 上的突出的下前尖方面，与其他真猴类(可能除了下猿属[41]、锋猿属[41]、塞拉皮斯猿属[42]、阿尔西诺伊猿属[42]和非洲跗猴属[35,37,40])不同。与除副猿科、阿尔西诺伊猿属、可能还有非洲跗猴属之外的真猴类也都有所不同，其 P_4 上下后尖位置比下原尖更靠近远中并比下原尖矮，没有发达的横脊将这些齿尖联结在一起。在 M_{1-2} 上具有显著的、向远中方向突出的下次小尖方面，与阔鼻猴小目和非洲跗猴属不同。在 M_{1-2} 缺乏成对下内尖和下次小尖方面，与除非洲跗猴、半猿、蓬当猿以外的旧世界真猴类不同。

中华曙猿(新种)*Eosimias sinensis*，n. sp.

正模标本。IVPP V10591，保存了 P_4~M_2 以及 C_1、P_{2-3} 和 M_3 的齿根或齿槽的右下颌骨(图 4)。

产地。IVPP 93006，裂隙 B。

分布。中华人民共和国江苏省，中始新世。

特征。与该科的特征一致(目前是单型属)。

词源。希腊语 *eos*(黎明,早的)和拉丁文 *simia*(猿)。其种名是指该种的地理产地。

描述。前下齿列的特征可以在 IVPP V10592 上观察到，这是一个来自裂隙 A 的标本(图 4)。该标本保存下来的 I_{1-2} 的齿槽都是小型的。相比之下，C_1 的齿槽相对庞大，比 I_{1-2} 和 P_2 的要大得多。曙猿中 I_1~P_2 的相对比例似乎可以理解为对灵长类而言是原始的，因为它们与其他古近纪真猴类、兔猴型类[24]和某些跗猴类(例如，华谢基猴属)的形态相似[43]。IVPP V10591 和 V10592 都拥有一个未融合的下颌联合部(IVPP V10592 中保存得更完整)。与兔猴型类和跗猴类不同，该下颌骨的联合部在背腹方向上很深，并不是非常平伏。对于该衍生特征，曙猿属与真猴类相像。

P_3 的齿冠形态还是未知的。其齿槽与 P_4 的一样，相对于齿列来说是斜向的，近中侧的齿槽位于远中侧齿槽的唇侧。这种衍生状态在始新世兔猴型类和跗猴类中并没出现，但是在法尤姆的真猴类中很常见(例如，夸特拉尼猿属)[44]。来自缅甸始新世的半猿中也存在类似的情况，这是一个广泛[5,12]但不是普遍[1]被认为是原始真猴类的属。

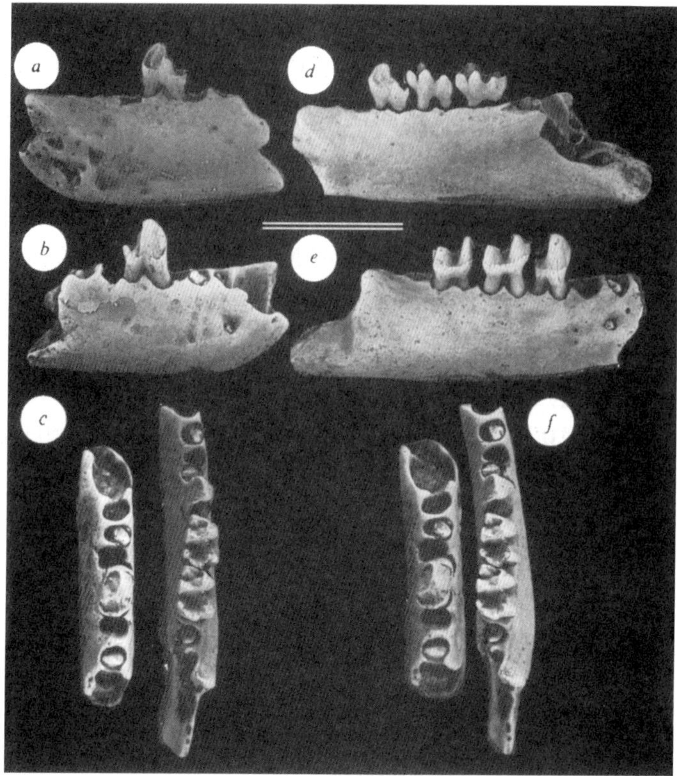

Fig. 4. Eosimiid partial mandibles from the Shanghuang fissure-fillings. *a–c*, IVPP 10592, right mandible fragment with P_4 and alveoli for anterior teeth from fissure A, in *a*, lingual; *b*, labial; and *c*, occlusal views. *d–f*, IVPP 10591, right mandible fragment with P_4–M_2 and alveoli for C_1–P_3 and M_3 from fissure B, holotype of *Eosimias sinensis*, in *d*, lingual; *e*, labial; and *f*, occlusal views. Occlusal views are stereopairs. Scale bar, 5 mm.

P_4 in the holotype is mesiodistally short (1.45 mm) and buccolingually wide (1.25 mm), a derived condition by comparison with such primitive, early Eocene primates as *Cantius* and *Steinius*[45] (Fig. 5). The moderately exodaenodont (labially bulging) crown is dominated by the protoconid. A weak metaconid occurs distal and well inferior to the protoconid. This position resembles that in parapithecids and *Arsinoea* but contrasts with the condition in other simians, in which the protoconid and metaconid are transversely aligned and united by a strong crest[40]. Neither a distinct paraconid nor an entoconid is present. The short talonid is open distolingually but is bounded labially by the hypoconid and cristid obliqua. A weak cingulid is discernible labially.

M_1 (length, 1.85 mm; width, 1.40 mm) bears a cuspidate paraconid. Despite this primitive retention, the M_1 trigonid shares derived characters with simians that are not found among Eocene adapiforms and omomyids (Fig. 5). In *Eosimias* the protoconid is enlarged relative to the metaconid and these two cusps are widely separated, as is typical for simians. Among Eocene adapiforms and omomyids these cusps are similar in size and more closely spaced[1,45]. A premetacristid, also typically absent among Eocene adapiforms and omomyids[1,45], unites

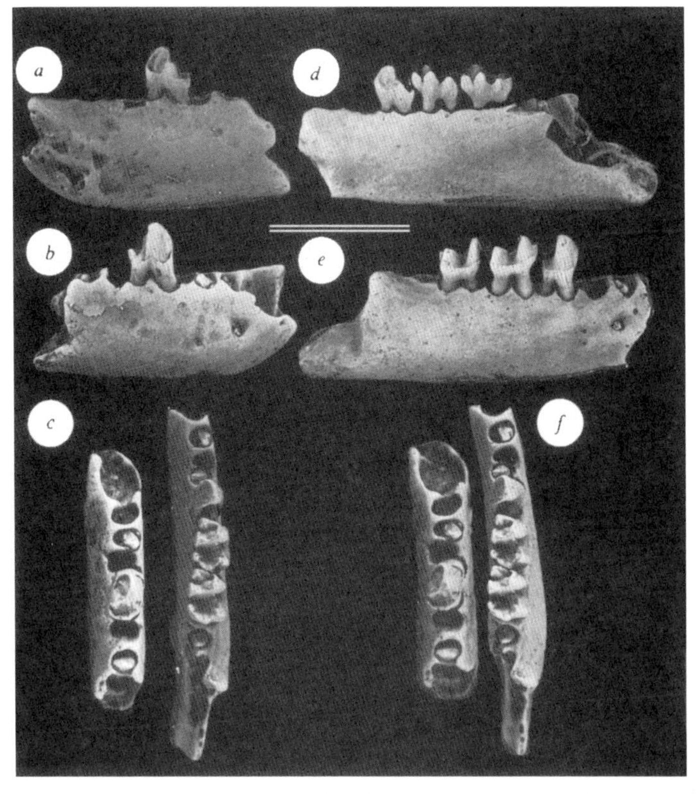

图 4. 来自上黄裂隙堆积的曙猿的残破下颌骨。*a*~*c*，IVPP 10592，来自裂隙 A 的带有 P_4 和前齿齿槽的右下颌骨残段。*a*，舌面视图；*b*，唇面视图；*c*，咬合面视图。*d*~*f*，IVPP 10591，来自裂隙 B 的带有 P_4~M_2 以及 C_1~P_3 齿槽和 M_3 齿槽的右下颌骨残段，中华曙猿的正模标本。*d*，舌面视图；*e*，唇面视图；*f*，咬合面视图。咬合面视图是立体像对。比例尺，5 毫米。

正模标本的 P_4 近远中方向上很短(1.45 毫米)，颊舌方向很宽(1.25 毫米)，与肯特猴和斯氏猴 [45] 等原始的早始新世灵长类相比，其为衍生特征(图 5)。胖边形齿(唇侧隆起)齿冠发育程度中等，并主要由下原尖占据。在下原尖远中侧有一个不发达并且明显比其矮的下后尖。该位置与副猿类和阿尔西诺伊猿属的情况相像，但与其他真猴类的情况相反，后者的下原尖和下后尖横向排列，由一个强脊将它们联结起来 [40]。不存在明显的下前尖和下内尖。短小的下跟座是向远中舌侧开放的，但在唇侧以下次尖和下斜脊为界。可以辨别出唇侧有一个不发达的下齿带。

M_1 (长度为 1.85 毫米，宽度为 1.40 毫米)具有一个多尖的下前尖。尽管这是一种原始特征的保留，但是 M_1 的下齿座与真猴类共有某些衍生特征，这些特征在始新世兔猴型类和鼩猴类中都没有观察到过(图 5)。曙猿属中，下原尖相对于下后尖来说扩大了，而且这两个齿尖被远远隔开，这是真猴类的典型特征。始新世兔猴型类和鼩猴类中，这些齿尖在大小上接近，彼此间距更加紧密 [1,45]。始新世兔猴型类和

the metaconid with the base of the paraconid. The talonid, which is wider than the trigonid, is unusual in having the entoconid situated farther mesial than the hypoconid and in possessing a prominent, distally projecting hypoconulid. The buccal cingulid is weak.

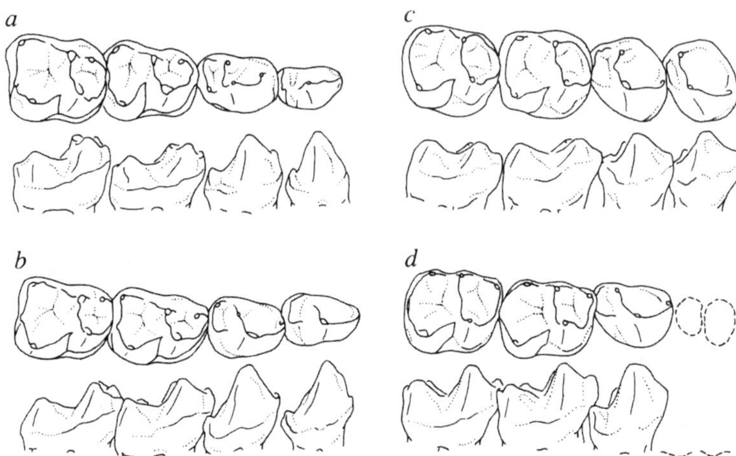

Fig. 5. Comparative schematic drawings of P_3–M_2 in selected fossil primates (not to same scale). In each pair of drawings, the upper view is occlusal and the lower view is buccal. *a,* The early Eocene adapiform *Cantius frugivorus* (CM 37448); *b,* the early Eocene omomyid *Steinius vespertinus* (USGS 25027); *c,* the middle Miocene simian *Neosaimiri fieldsi* (UCMP 39205); *d,* the middle Eocene simian *Eosimias sinensis* (IVPP V10591; crowns of P_4–M_2 only). Note obliquely oriented and mesiodistally short $P_{3\text{-}4}$ and similar molar morphology in the simians, in contrast to conditions in *Cantius* and *Steinius.* See text for further discussion.

M_2 (length, 1.85 mm; width 1.55 mm) is very similar to M_1, from which it differs in being relatively wider and having a less distally projecting hypoconulid. The paraconid is conspicuous and widely separated mesially from the metaconid. Among Eocene adapiforms and omomyids that retain paraconids on M_2, this cusp is typically much more connate with the metaconid[1,45].

Despite its retention of such primitive characters as diminutive body size (mean estimates based on regressions of body size against M_1 area in living primates[46] range from 67–137 g), an unfused mandibular symphysis and prominent paraconids on $M_{1\text{-}2}$, *Eosimias* appears to be a primitive simian based on its possession of the following dental synapomorphies: lower dental formula 2-1-3-3; P_2 single-rooted; mesiodistally short, obliquely oriented, and moderately exodaenodont $P_{3\text{-}4}$; $M_{1\text{-}2}$ with enlarged protoconid widely separated from metaconid, premetacristid present, and mesially situated entoconid.

Recent palaeontological discoveries[41,42] have revealed that the dentitions of early simians retained many primitive features, including lower molar paraconids. These symplesiomorphies complicate attempts to reconstruct phylogenetic relationships among early simians and their relatives, particularly on the basis of dental and jaw anatomy alone. Nevertheless, the combination of characters listed above for *Eosimias* and many early simians is absent in adapiforms and omomyids, suggesting that they are best

鼩猴类[1,45]通常所不具备的下后尖前脊，将下后尖与下前尖基部联结起来。下跟座比下齿座更宽，具有位置比下次尖更靠近近中侧的下内尖以及显著的向远中突出的下次小尖，这些特征都是不同寻常的。颊侧齿带微弱。

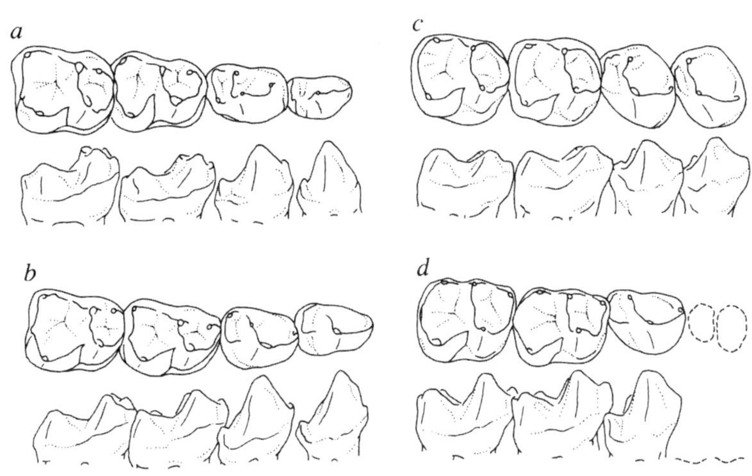

图 5. 相关灵长类化石的 P_3~M_2 的比较示意图（比例尺不同）。每对图片中，上面一排是咬合面视图，下面一排是颊面视图。*a*，早始新世兔猴型类——食果肯特猴（CM 37448）；*b*，早始新世鼩猴类——夜斯氏猴（USGS 25027）；*c*，中新世中期的真猴类——菲氏新松鼠猴（UCMP 39205）；*d*，中始新世的真猴类——中华曙猿（IVPP V10591；只有 P_4~M_2 的齿冠）。请注意，真猴类中斜向的、近远中方向短的 P_{3-4} 和相似的臼齿形态学特征与肯特猴属和斯氏猴属的情况正好相反。见正文中的进一步讨论。

M_2（长度为 1.85 毫米，宽度为 1.55 毫米）与 M_1 非常相似，但前者相对较宽并且下次小尖向远中侧突出程度弱于后者。下前尖明显，在近中部与下后尖远远隔开。M_2 上保留了下前尖的始新世兔猴型类和鼩猴类中，该齿尖通常与下后尖融合在一起[1,45]。

尽管曙猿属保留了一些原始特征，例如体型极小（根据现生灵长类[46]的 M_1 面积对身体大小进行的回归分析估计出来的平均值范围在 67~137 g 之间）、未融合的下颌联合部以及 M_{1-2} 显著的下前尖等，但从曙猿属拥有如下牙齿的共有裔征来看，其应是一种原始真猴类：下齿式为 2-1-3-3；P_2 单根；近远中方向短、斜向、中度胖边形齿的 P_{3-4}；M_{1-2} 具有膨大的下原尖并与下后尖远远隔开，下后尖前脊存在以及下内尖位置靠近近中侧。

最近的古生物学发现[41,42]揭示出早期真猴类的齿列保留了许多原始特征，包括下臼齿的下前尖。这些共有祖征使得尝试重建早期真猴类与其亲缘类群的系统发育关系变得更加复杂，尤其是只根据牙齿和下颌骨的解剖学特征来重建的时候。然而，上面列出的曙猿属和许多早期真猴类的特征组合在兔猴型类和鼩猴类中都不存在，表明对它们最好的解释就是真猴类共有裔征。曙猿属的衍生特征一致地指明其与真

interpreted as simian synapomorphies. Derived features of *Eosimias* consistently point toward simian affinities, whereas we know of no derived characters for *Eosimias* that suggest an alternative phylogenetic position. Moreover, the specifically simian-like mode of dental reduction and premolar compaction in *Eosimias* differs from that found in certain omomyids (in which this usually occurs in conjunction with hypertrophy of I_1) and adapiforms (none of which evolved obliquely oriented lower premolars as a means of compressing the antemolar dentition)[1]. Within Anthropoidea, *Eosimias* appears to occupy a very basal phylogenetic position, certainly before the diversification of Parapithecidae, Oligopithecinae, Platyrrhini and Catarrhini (Fig. 6).

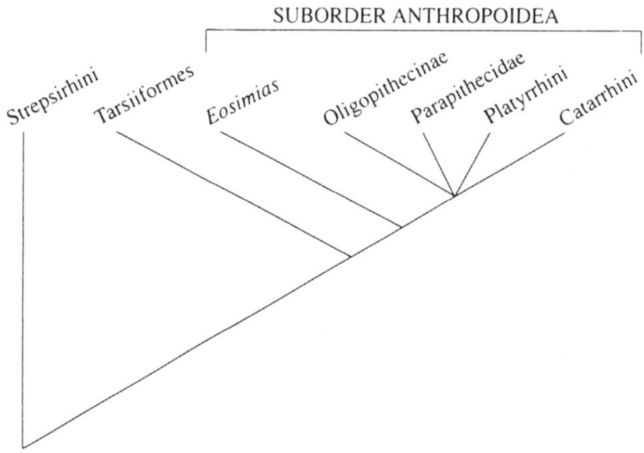

Fig. 6. Cladogram illustrating basal position of *Eosimias* with respect to the evolutionary radiation of simians. Relationships among more advanced simian clades are left unresolved, reflecting the lack of consensus regarding the phylogenetic resolution of this node[40,41,50].

Given this position, *Eosimias* yields new insight into the phylogenetic relationships of simians with respect to other primates. One hypothesis holds that early simians evolved from Eocene adapiforms[23,24,47,48]. The age and anatomy of *Eosimias* cast doubt on this idea. *Eosimias* is smaller than all known adapiforms except the specialized and temporally late *Anchomomys.* Hence, one must postulate an episode of phyletic dwarfing, which is rare among mammals, to derive *Eosimias* from adapiforms other than *Anchomomys.* Moreover, derivation of *Eosimias* from adapiforms such as *Mahgarita* (recently cited[23] as a possible close relative of early simians) would entail at least two implausible character reversals: loss of mandibular symphyseal fusion and neomorphic reacquisition of lower molar paraconids. Rather, *Eosimias* and other recent finds[32,41,42,48-50] suggest that Anthropoidea is a very ancient clade that did not evolve from either Eocene adapiforms or omomyids. This is consistent with the monophyly of either Haplorhini (Anthropoidea+Tarsiiformes) or Prosimii (Strepsirhini+Tarsiiformes), but runs counter to a monophyletic Simiolemuriformes (Strepsirhini+Anthropoidea). A wide body of evidence supports the monophyly of Haplorhini among living primates[1,2], and we prefer this phylogenetic hypothesis at present (Fig. 6).

猴类的亲缘关系，但是我们知道曙猿属没有那些指示其具有其他系统发育位置的衍生特征。此外，曙猿属中与真猴类相似的牙齿减少和前臼齿紧缩的模式与某些鼩猴类（鼩猴类中这种情况通常是与 I_1 的过度增大一起出现的）和兔猴型类（该类群中没有任何种类进化出斜向的下前臼齿以压缩臼前齿齿列）中观察到的情况不同[1]。类人猿亚目中，曙猿属肯定在副猿科、渐新猿亚科、阔鼻猴小目和狭鼻猴小目发生分化之前，就已经占据了一个非常基底的系统发育位置（图 6）。

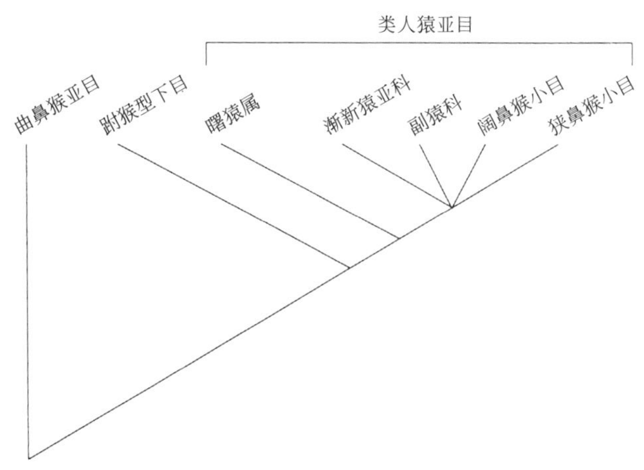

图 6. 从真猴类的演化辐射角度描述了曙猿属基底位置的支序图。更高级的真猴类演化分支间的相互关系还没有得到解决，这反映了对于该结点的系统发育关系问题的解决还缺乏一致意见[40,41,50]。

鉴于曙猿属的这个系统发育位置，其提供了真猴类系统发育关系的新见解。其中一个假说认为早期真猴类是从始新世兔猴型类进化而来的[23,24,47,48]。曙猿属的年代和解剖学特征使人们对这种假说产生了怀疑。除了特化的、在年代上稍晚的近鼩猴属以外，曙猿属比所有已知的兔猴型类都要小。因此，人们肯定会假定存在一个系统发育过程中的矮小化时期，这一现象在哺乳动物中很罕见，这一过程使曙猿属从兔猴型类衍生出来而非从近鼩猴属。此外，果真曙猿从诸如玛氏兔猴属（最近被引用[23]为早期真猴类的一种可能的关系密切的亲属）之类的兔猴型类衍生出来的话，将必须假定至少两个难以置信的特征逆向演化：下颌联合部的融合的丢弃以及下臼齿下前尖这一新生性状的重新获得。但更确切地说，曙猿属和其他最近的发现[32,41,42,48-50]显示类人猿亚目是一个非常古老的演化分支，它们既不是从始新世兔猴型类进化而来的，也不是从鼩猴类进化而来的。这与直鼻猴类（类人猿亚目 + 跗猴型下目）或原猴类（曲鼻猴亚目 + 跗猴型下目）的单系性是一致的，但是与类人猿狐猴型亚目（曲鼻猴亚目 + 类人猿亚目）的单系性却是背道而驰的。大量证据支持现生灵长类中直鼻猴类具有单系性[1,2]，我们现在更倾向于接受这种系统发育假说（图 6）。

The discovery of *Eosimias* also bears on hypotheses concerning the biogeographic centre of origin for higher primates. Recent discoveries of Eocene simians in Africa[41,42,48-50] have emphasized the antiquity of higher primates there, leading some to suggest that simians may have originated on that continent[40,49,50]. The primitive nature and great age of *Eosimias* suggest, at least, that southeastern Asia was also an important theatre of early simian evolution. The possibility that simians originated in Asia rather than Africa[11,12] cannot be rejected without further palaeontological evidence from both continents.

If *Eosimias* and related taxa from Shanghuang are basal simians as we suggest, they may help clarify a longstanding controversy concerning the younger Asian primates *Amphipithecus*, *Pondaungia* and *Hoanghonius*. All of these taxa have been compared favourably with early simians[1,2,47], and several workers have suggested that some or all of them represent early Asian higher primates[3,5,11,12,39]. The affinities of these taxa cannot be resolved here, but we note substantive similarities between *Eosimias* and *Amphipithecus*, particularly in the morphology of the lower premolars. These suggest that *Amphipithecus*, and possibly *Pondaungia* and/or *Hoanghonius*, may be younger, more derived members of the basal simian radiation sampled at Shanghuang. If so, the simian affinities of the entire radiation may be better documented by the Shanghuang primates, because of the acquisition of uniquely specialized features through time in the younger *Amphipithecus*, *Pondaungia* and *Hoanghonius*.

Discussion

The Shanghuang primates show surprisingly varied biogeographic affinities, reflecting the complex role played by Asia during early primate evolution. Both Shanghuang adapiforms exhibit clear affinities with European taxa, although the *Europolemur*-like species may be related to North American *Mahgarita* as well. *Macrotarsius* from Shanghuang, like *Asiomomys changbaicus* from Jilin Province, PRC[15], is closely related to North American omomyids. The tarsiid is endemic. The basal simian radiation sampled at Shanghuang, although probably related to younger Asian taxa such as *Amphipithecus* and *Hoanghonius*, may also indicate an ancient (?Palaeocene) biogeographic link with Africa, where early Cenozoic simians are becoming increasingly well known. Although these findings demonstrate that the Eocene primate fauna of Asia interacted with those of other continents through dispersal, many factors underlying the different dispersal patterns of individual taxa remain unknown.

(**368**, 604-609; 1994)

K. Christopher Beard*, Tao Qi†, Mary R. Dawson*, Banyue Wang† & Chuankuei Li†

* Section of Vertebrate Paleontology, Carnegie Museum of Natural History, 4400 Forbes Avenue, Pittsburgh, Pennsylvania 15213, USA

† Institute of Vertebrate Paleontology and Paleoanthropology (IVPP), Academia Sinica, Beijing 100044, People's Republic of China

Received 29 December 1993; accepted 8 February 1994.

曙猿属的发现也对高等灵长类起源的生物地理中心方面的假说产生了影响。最近在非洲发现的始新世真猴类[41,42,48-50]强调了那里的高等灵长类的古老性，这使得有些人认为真猴类可能是在非洲大陆上起源的[40,49,50]。至少曙猿属的原始特征和古老年代显示了亚洲东南部也是早期真猴类演化的一个重要舞台。如果亚洲和非洲这两块大陆不能找到进一步的古生物学证据的话，那么真猴类是从亚洲而非非洲起源的[11,12]这种可能性就不能排除。

如果正如我们所提出的，来自上黄的曙猿属及相关类群是基底真猴类的话，那么它们可能有助于澄清长久以来一直存在争议的关于比较年轻的亚洲灵长类——半猿属、蓬当猿属和黄河猴属的问题。所有这些类群都被倾向性地认为可与早期真猴类对比[1,2,47]，有些学者提出这些灵长类中的部分或者全部都代表了早期的亚洲高等灵长类[3,5,11,12,39]。这些类群的亲缘关系在这里还不能得以解决，但是我们注意到，曙猿属和半猿属之间存在着实质性的相似性，尤其是下前臼齿的形态特征方面。这些意味着半猿属，还可能包括蓬当猿属和（或）黄河猴属，可能是上黄基底真猴类演化辐射的更年轻的、更进步的成员。如果是这样的话，那么整个真猴类辐射的亲缘关系就可能通过上黄灵长类得到了更好的记录，因为更年轻的半猿属、蓬当猿属和黄河猴属经过岁月的长河获得了独特且特化的特征。

讨　论

上黄灵长类显示出了令人惊奇的多样的生物地理亲缘关系，反映了亚洲在早期灵长类演化过程中所扮演的复杂角色。尽管似欧狐猴的种类可能也与北美洲玛氏兔猴属有关，但是两种上黄兔猴型类都展示出与欧洲类群清晰的亲缘关系。来自上黄的大跗猴属就像来自中国吉林省的长白亚洲鼩猴[15]一样，与北美洲鼩猴类具有密切的关系。跗猴类具有地方性。在上黄记录到的基底真猴类辐射，尽管可能与诸如半猿属和黄河猴属等年轻亚洲类群有关，但是也可能暗示着一种与非洲古老的（？古新世）生物地理联系，我们对那里的早新生代的真猴类的了解正在变得越来越清楚。尽管这些发现展现了亚洲的始新世灵长类动物群通过扩散与其他大陆的那些灵长类相互影响，但是关于各个类群的不同扩散模式的许多潜在因素，我们仍然不清楚。

（刘皓芳 翻译；张颖奇 审稿）

References:

1. Szalay, F. S. & Delson, E. *Evolutionary History of the Primates* (Academic, New York, 1979).
2. Martin, R. D. *Nature* **363**, 223-234 (1993).
3. Pilgrim, G. E. *Mem. Geol. Surv. India* **14**, 1-26 (1927).
4. Zdansky, O. *Palaeontol. Sinica* **6**, 1-87 (1930).
5. Colbert, E. H. *Am. Mus. Novitates* **951**, 1-18 (1937).
6. Chow, M. *Vert. PalAsiatica* **5**, 1-5 (1961).
7. Dashzeveg, D. & McKenna, M. C. *Acta Palaeontol. Polonica* **22**, 119-137 (1977).
8. Russell, D. E. & Gingerich, P. D. *C. R. Acad. Sci., Paris* (Sér. D) **291**, 621-624 (1980).
9. Russell, D. E. & Gingerich, P. D. *C. R. Acad. Sci., Paris* (Sér. II) **304**, 209-214 (1987).
10. Wang, B. & Li, C. *Vert. PalAsiatica* **28**, 165-205 (1990).
11. Maw, B., Ciochon, R. L. & Savage, D. E. *Nature* **282**, 65-67 (1979).
12. Ciochon, R. L., Savage, D. E., Tint, T. & Maw, B. *Science* **229**, 756-759 (1985).
13. Rose, K. D. & Krause, D. W. *J. Mamm.* **65**, 721-726 (1984).
14. Gingerich, P. D., Dashzeveg, D. & Russell, D. E. *Geobios* **24**, 637-646 (1991).
15. Beard, K. C. & Wang, B. *Am. J. Phys. Anthrop.* **85**, 159-166 (1991).
16. Qi, T., Zong, G.-F. & Wang, Y.-Q. *Vert. PalAsiatica* **29**, 59-63 (1991).
17. Russell, D. E. & Zhai, R.-J. *Mém. Mus. Natl. Hist. Nat.* (Sér. C) **52**, 1-488 (1987).
18. Prothero, D. R. & Swisher, C. C. in *Eocene-Oligocene Climatic and Biotic Evolution* (eds Prothero, D. R. & Berggren, W. A.) 46-73 (Princeton Univ. Press, Princeton, 1992).
19. Beard, K. C., Dagosto, M., Gebo, D. L. & Godinot, M. *Nature* **331**, 712-714 (1988).
20. Franzen, J. L. *Cour. Forsch.-Inst. Senckenberg* **91**, 151-187 (1987).
21. Wilson, J. A. & Szalay, F. S. *Folia Primatol.* **25**, 294-312 (1976).
22. Wilson, J. A. & Stevens, M. S. *Univ. Wyoming Contrib. Geol. Spec. Pap.* **3**, 221-235 (1986).
23. Rasmussen, D. T. *Int. J. Primatol.* **11**, 439-469 (1990).
24. Gingerich, P. D. in *Evolutionary Biology of the New World Monkeys and Continental Drift* (eds Ciochon, R. L. & Chiarelli, A. B.) 123-138 (Plenum, New York, 1980).
25. Godinot, M. *C. R. Acad. Sci., Paris* (Sér. II) **299**, 1291-1296 (1984).
26. Gingerich, P. D. & Martin, R. D. *Am. J. Phys. Anthrop.* **56**, 235-257 (1981).
27. Dagosto, M. *Folia Primatol.* **41**, 49-101 (1983).
28. Godinot, M. & Jouffroy, F. K. in *Actes du Symposium Paléontologique Georges Cuvier* (eds Buffetaut, E., Mazin, J. M. & Salmon, E.) 221-242 (Le Serpentaire, Montbeliard, 1984).
29. Franzen, J. L. & Haubold, H. *Mod. Geol.* **10**, 159-170 (1986).
30. Gingerich, P. D. & Sahni, A. *Int. J. Primatol.* **5**, 63-79 (1984).
31. Pan, Y. *J. Hum. Evol.* **17**, 359-366 (1988).
32. Beard, K. C., Krishtalka, L. & Stucky, R. K. *Nature* **349**, 64-67 (1991).
33. Krishtalka, L. *Ann. Carnegie Mus.* **47**, 335-360 (1978).
34. Kelly, T. S. *Nat. Hist. Mus. Los Angeles County, Contrib. Sci.* **419**, 1-42 (1990).
35. Simons, E. L. & Bown, T. M. *Nature* **313**, 475-477 (1985).
36. Musser, G. G. & Dagosto, M. *Am. Mus. Novitates* **2867**, 1-53 (1987).
37. Ginsburg, L. & Mein, P. *C. R. Acad. Sci., Paris* (Sér. II) **304**, 1213-1215 (1987).
38. Norell, M. A. & Novacek, M. J. *Science* **255**, 1690-1693 (1992).
39. Deison, E. & Rosenberger, A. L. in *Evolutionary Biology of the New World Monkeys and Continental Drift* (eds Ciochon, R. L. & Chiarelli, A. B.) 445-458 (Plenum, New York, 1980).
40. Fleagle, J. G. & Kay, R. F. *J. Hum. Evol.* **16**, 483-532 (1987).
41. Simons, E. L. *Proc. Natl. Acad. Sci. U.S.A.* **86**, 9956-9960 (1989).
42. Simons, E. L. *Proc. Natl. Acad. Sci. U.S.A.* **89**, 10743-10747 (1992).
43. Covert, H. H. & Williams, B. A. *J. Hum. Evol.* **21**, 463-467 (1991).
44. Simons, E. L. & Kay, R. F. *Am. J. Primatol.* **15**, 337-347 (1988).
45. Rose, K. D. & Bown, T. M. *Proc. Natl. Acad. Sci. U.S.A.* **88**, 98-101 (1991).
46. Conroy, G. C. *Int. J. Primatol.* **8**, 115-137 (1987).
47. Rasmussen, D. T. & Simons, E. L. *Folia Primatol.* **51**, 182-208 (1988).

48. Simons, E. L. *Science* **247**, 1567-1569 (1990).
49. de Bonis, L., Jaeger, J.-J., Coiffat, B. & Coiffat, P.-E. *C. R. Acad. Sci., Paris* (Sér. II) **306**, 929-934 (1988).
50. Godinot, M. & Mahboubi, M. *Nature* **357**, 324-326 (1992).

Acknowledgements. We thank J. L. Carter for help with photography and A. D. Redline for artwork. Invaluable help in the field was provided by Wang Qingqing and Di Fubao of the Liyang Cultural Museum. This research was supported by the US NSF, the Chinese NSF, and M. Graham Netting and O'Neil Research Funds (Carnegie Museum of Natural History).

Australopithecus ramidus, a New Species of Early Hominid from Aramis, Ethiopia

T. D. White *et al.*

Editor's Note

The further one travels back in time, the scarcer the fossil evidence of human antiquity becomes. It is sparse enough beyond four million years; beyond five, it almost disappears completely (or did so until recent discoveries such as *Sahelanthropus* and *Orrorin*). Hence the significance of this announcement of a collection of extremely primitive fragments from Ethiopia, dated at 4.4 million years. With only a small number of dental features marking it out as hominid, *Australopithecus ramidus* had some claim to be close to the root of the divergence between the chimp and human lineages. In fact, White and colleagues later considered the form to be too primitive even for *Australopithecus*, and re-named it *Ardipithecus*.

Seventeen hominoid fossils recovered from Pliocene strata at Aramis, Middle Awash, Ethiopia make up a series comprising dental, cranial and postcranial specimens dated to around 4.4 million years ago. When compared with *Australopithecus afarensis* and with modern and fossil apes the Aramis fossil hominids are recognized as a new species of *Australopithecus*—*A. ramidus* sp. nov. The antiquity and primitive morphology of *A. ramidus* suggests that it represents a long-sought potential root species for the Hominidae.

WORK in southern Africa established *Australopithecus* as a human ancestor and revealed specific diversity within that genus. Subsequent work in eastern Africa extended the known geographical and temporal distribution of the genus. Until now, the earliest hominid species known was *Australopithecus afarensis*, dated to between 3 and 4 Myr. *A. afarensis* narrowed the temporal and morphological gap between Miocene hominoids and other early hominids[1]. Its primitive craniodental anatomy offered some support for molecular work[2] which had suggested a late Miocene or early Pliocene age for the common ancestor of hominids and African apes. Because details of the ape and human divergence are poorly understood[3-9], taxonomically diagnostic hominoid fossil evidence antedating the existing record of *A. afarensis* has been eagerly anticipated.

在埃塞俄比亚的阿拉米斯发现的一种新型的早期原始人类物种——南方古猿始祖种

怀特等

编者按

当我们向过去回溯的时间越古老，古人类的化石证据就越稀少。当超过四百万年时，这类化石是非常稀少的；超过五百万年后，这类化石几乎完全消失了（直到最近发现的托麦人属（乍得撒海尔人）和千禧人属）。因此本篇文章的意义在于在埃塞俄比亚收集到的一批非常原始的化石碎片，年代在440万年前。由于只有少量的牙齿特征标记出化石属于原始人类，有些人声称南方古猿始祖种接近于黑猩猩和人类谱系之间的分歧的根部。实际上，怀特和他的同事们后来经过考虑认为这一标本即使是相对于南方古猿属来说也过于原始，于是将其重命名为地猿属。

从埃塞俄比亚中阿瓦什阿拉米斯的上新世地层中挖掘到了17个人科动物化石，包括牙齿、颅骨和颅后骨骼的系列标本，年代可以追溯到约440万年前。通过与南方古猿阿法种及现代猿和化石猿相比，阿拉米斯原始人类化石被认为是南方古猿的一个新种——南方古猿始祖种（*A. ramidus* sp. nov.）（“sp. nov.”是“species nova”的缩写，即“新种”之意）。南方古猿始祖种的古老型和原始形态特征都提示它可能是人们长久以来寻找的人科祖根物种。

南部非洲的工作将南方古猿确立为人类的祖先之一，并且揭示了南方古猿属内的多样性。接下来在东非的工作扩展了该属已知的地理和时间分布。迄今为止，已知最早的人类是南方古猿阿法种，其可追溯到距今300万到400万年前。南方古猿阿法种将中新世人猿超科与其他早期人类之间[1]时间和形态学缺口缩窄了。其原始的颅骨牙齿解剖学特征为分子学研究提供了支持[2]，分子工作提示原始人类和非洲猿的共同祖先生活的年代是晚中新世或早上新世。由于对猿类和人类的分化细节理解得还不够[3-9]，所以人们热切渴望能够找到具有分类学鉴别意义的比现存的南方古猿阿法种记录更早的人科动物化石证据。

Description of *A. ramidus*

Order Primates Linnaeus 1758
Suborder Anthropoidea Mivart 1864
Superfamily Hominoidea Gray 1825

Australopithecus DART 1925

Australopithecus ramidus sp. nov.

Etymology. In recognition of the Afar people who occupy the Middle Awash study area and contribute to fieldwork there. The name is from the Afar language. "Ramid" means "root" and it applies to both people and plants.

Holotype. ARA-VP-6/1 (Fig. 1*a*) is an associated set of teeth from one individual that includes: upper left I^1, $\underline{C}$, P^3, P^4, right I^1, $\underline{C}$ (broken), P^4, M^2; and lower right P_3 and P_4. Found by Gada Hamed on Wednesday, 29 December 1993. Holotype and paratype series housed at the National Museum of Ethiopia, Addis Ababa.

Paratypes. Table 1 lists the holotype and paratype series, all from Aramis. Included are associated postcranial elements, two partial cranial bases, a child's mandible, associated and isolated teeth.

Locality. Aramis localities 1–7 are in the headwaters of the Aramis and Adgantoli drainages, west of the Awash river in the Middle Awash palaeoanthropological study area, Afar depression, Ethiopia[10]. Aramis VP Locality 6 is at 10° 28.74′ north latitude; 40° 26.26′ east longitude; ~625 m elevation.

Horizon and associations. All hominid specimens were surface finds located in the section within 3 m of the Daam Aatu Basaltic Tuff. The immediately underlying Gàala Vitric Tuff Complex is dated at 4.39 ± 0.03 Myr (ref. 10).

Diagnosis. *A. ramidus* is a species of *Australopithecus* distinguished from other hominid species, including *A. afarensis*, by the following: upper and lower canines larger relative to the postcanine teeth; lower first deciduous molar narrow and obliquely elongate, with large protoconid, small and distally placed metaconid, no anterior fovea, and small, low talonid with minimal cuspule development; temporomandibular joint without definable articular eminence; absolutely and relatively thinner canine and molar enamel; lower third premolar more strongly asymmetrical, with dominant, tall buccal cusp and steep, posterolingually directed transverse crest; upper third premolar more strongly asymmetric, with relatively larger, taller, more dominant buccal cusp.

南方古猿始祖种的描述

灵长目，林奈，1758 年
类人猿亚目，米瓦特，1864 年
人猿总科，格雷，1825 年

南方古猿属，达特，1925 年

南方古猿始祖种，新种

词源。按照居住在中阿瓦什研究地区并对那里的野外工作作出贡献的阿法尔人民的叫法而来。该名字来源于阿法尔语言。“Ramid”意思是“根”，既适用于人类也适用于植物。

正模标本。ARA-VP-6/1（图 1*a*）是来自同一个体的一套牙齿，包括：左上 I^1、$\underline{C}$、P^3、P^4，右上 I^1、$\underline{C}$(断裂)、P^4、M^2，以及右下 P_3 和 P_4。嘎达·哈米德于 1993 年 12 月 29 日星期三发现该样本。正模标本和副模标本系列都存放在位于亚的斯亚贝巴的埃塞俄比亚国家博物馆中。

副模标本。表 1 列出了正模标本和副模标本系列，它们都来自阿拉米斯。表中列出的标本包括相关联的颅后骨骼、两块头盖骨基底部分、一块儿童下颌骨，还有相关但游离的牙齿。

产地。阿拉米斯 1~7 地点位于阿瓦什河西边，在阿拉米斯与阿德甘托利流域的上游源头处，该地在埃塞俄比亚阿法尔凹地的中阿瓦什古人类学研究区 [10]。阿拉米斯 VP 地点 6 的地理坐标是北纬 10° 28.74′，东经 40° 26.26′；海拔约 625 米。

相关地层。所有原始人类标本都是在地表发现的，地层位于达阿姆–阿图玄武岩凝灰岩层 3 米内的剖面，其下覆的伽拉玻璃凝灰岩复合体地层年代为 439 万 ± 3 万年(参考文献 10)。

种征。南方古猿始祖种是南方古猿属中的一个种，它与包括南方古猿阿法种在内的其他原始人类物种有以下不同：相对于颊齿来说，上、下犬齿较大；下颌第一乳臼齿窄而斜向延长，有大的下原尖，小而分开很远的下后尖，没有前凹，小而低的跟座，小牙尖不发达；颞下颌关节没有可确定的关节界限；绝对及相对较薄的犬齿和臼齿釉质；下颌第三前臼齿非常不对称，具有突出的、高的颊侧牙尖，陡峭的后舌侧方向的横脊；上颌第三前臼齿更加不对称，具有相对较大、较高、较占优势的颊侧牙尖。

Table 1. Aramis fossil hominid specimens

Specimen number	Collection year	Element	Discoverer	Dental dimensions
ARA-VP-1/1	1992	RM^3	G. Suwa	RM^3: 10.2MD, 12.3BL
ARA-VP-1/2	1992	RI^1	A. Asfaw	RI^1: 8.2LL
ARA-VP-1/3	1992	L_c frag.	G. Suwa	
ARA-VP-1/4	1992	Right humerus, full shaft	S. Simpson	
ARA-VP-1/125	1992	Left temporal	S. Simpson	
ARA-VP-1/127	1992	L^c, RM^1, worn roots of incisors, canine and premolar	T. White	
ARA-VP-1/128	1992	Associated teeth	T. Assebework	L_c: 11.0LL; RP_3: 7.5Mn, (9.8)Mx; LP_3: 7.5 Mn, (9.9)Mx; RP_4: 7.3 (7.5)MD, 9.5BL; LP_4: 7.3 (7.5)MD; RM_1: 10.9 (11.2)MD, (10.3)BL; LM_1: 10.6 (11.0)MD, (10.1)BL; LM_2: 12.8 (13.0)MD, 11.9BL; RM_3: 12.7(MD),11.0(BL)
ARA-VP-1/129	1992	Right mandible (I_1, dm_1)	A. Asfaw	RI_1: 6.0MD; Rdm_1: 7.3MD, 4.9BL
ARA-VP-1/182	1992	RM3 fragment	Group	
ARA-VP-1/183	1992	UC fragment	Group	
ARA-VP-1/200	1993	LM_1	A. Ademassu	LM_1: 11.0MD, 10.3BL
ARA-VP-1/300	1993	R^c	Y. Haile-Selassie	R^c: (11.2)MD, 11.1LL, 14.3CH
ARA-VP-1/400	1993	LM^2	Y. Beyene	LM^2: (11.3–12.3)MD, (15.0)BL
ARA-VP-1/401	1993	LM_3 fragment	M. Feseha	
ARA-VP-1/500	1993	R.+L. temp.+occ.	T. White	
ARA-VP-6/1	1993	Associated teeth	G. Hamed	LI^1: 9.6 (10.0)MD, 7.5LL, 12.5swCH; R^c: 11.6LL, 14.5CH; L^c: 11.7LL, 14.6CH, 11.5MD; LP^3: 7.7MD, 8.4MxMD; 12.5BL; RP^4: 8.4MD, (11.3)BL; RM^2: 11.8MD, (14.1)BL; RP_3: (8.2) MnMD; (11.5)MxBL; RP_4: 8.9MD, 9.7BL
ARA-VP-7/2	1993	Left humerus, radius, ulna	A. Asfaw	

Fossil hominid specimens recovered from Aramis between December 1992 and December 1993. Dental dimensions are standard, estimates for breakage or interproximal attrition are shown in parentheses. BL, Buccolingual; LL, labiolingual; MD, mesiodistal; CH, distance from buccal enamel line to apex (canine height); Mn, minimum diameter; Mx, maximum diameter; sw, slightly worn. All measurements were taken on original specimens by the authors.

表 1. 阿拉米斯原始人类化石标本

标本号	采集年份	标本	发现者	牙齿尺寸
ARA-VP-1/1	1992	RM^3	诹访	RM^3: 10.2MD, 12.3BL
ARA-VP-1/2	1992	RI^1	阿斯富	RI^1: 8.2LL
ARA-VP-1/3	1992	L$_c$ 碎片	诹访	
ARA-VP-1/4	1992	右侧肱骨，全骨干	辛普森	
ARA-VP-1/125	1992	左侧颞	辛普森	
ARA-VP-1/127	1992	L^c, RM^1，切牙、尖牙和前磨牙的破损根	怀特	
ARA-VP-1/128	1992	相关的牙齿	阿塞贝沃克	L$_c$: 11.0LL; RP$_3$: 7.5Mn,(9.8)Mx; LP$_3$: 7.5 Mn,(9.9)Mx; RP$_4$: 7.3 (7.5)MD, 9.5BL; LP$_4$: 7.3 (7.5)MD; RM$_1$: 10.9 (11.2)MD,(10.3)BL; LM$_1$: 10.6 (11.0)MD,(10.1)BL; LM$_2$: 12.8 (13.0)MD, 11.9BL; RM$_3$: 12.7 (MD),11.0 (BL)
ARA-VP-1/129	1992	右下颌骨 (I$_1$, dm$_1$)	阿斯富	RI$_1$: 6.0MD; Rdm$_1$: 7.3MD, 4.9BL
ARA-VP-1/182	1992	RM3 碎片	团队	
ARA-VP-1/183	1992	UC 碎片	团队	
ARA-VP-1/200	1993	LM$_1$	阿德马苏	LM$_1$: 11.0MD, 10.3BL
ARA-VP-1/300	1993	R^c	海尔塞拉西	R^c:(11.2)MD, 11.1LL, 14.3CH
ARA-VP-1/400	1993	LM^2	贝耶内	LM^2:(11.3~12.3)MD,(15.0)BL
ARA-VP-1/401	1993	LM$_3$ 碎片	费塞哈	
ARA-VP-1/500	1993	R.+L. temp.+occ.	怀特	
ARA-VP-6/1	1993	相关的牙齿	弗拉梅德	LI^1: 9.6 (10.0)MD, 7.5LL, 12.5swCH; R^c: 11.6LL, 14.5CH; L^c: 11.7LL, 14.6CH, 11.5MD; LP^3: 7.7MD, 8.4MxMD; 12.5BL; RP^4: 8.4MD,(11.3)BL; RM^2: 11.8MD,(14.1)BL; RP$_3$:(8.2) MnMD;(11.5)MxBL; RP$_4$: 8.9MD, 9.7BL
ARA-VP-7/2	1993	左侧肱骨，桡骨，尺骨	阿斯富	

于 1992 年 12 月至 1993 年 12 月之间从阿拉米斯挖掘到的原始人类化石标本。牙齿尺寸标准，对破裂或磨损牙齿的估计值在圆括号中标明。BL，颊舌径；LL，唇舌径；MD，近中－远中径；CH，颊侧釉质线到牙尖点的距离（犬齿高度）；Mn，最小径；Mx，最大径；sw，轻微磨损。所有测量尺寸都是作者对原始标本测量而得。

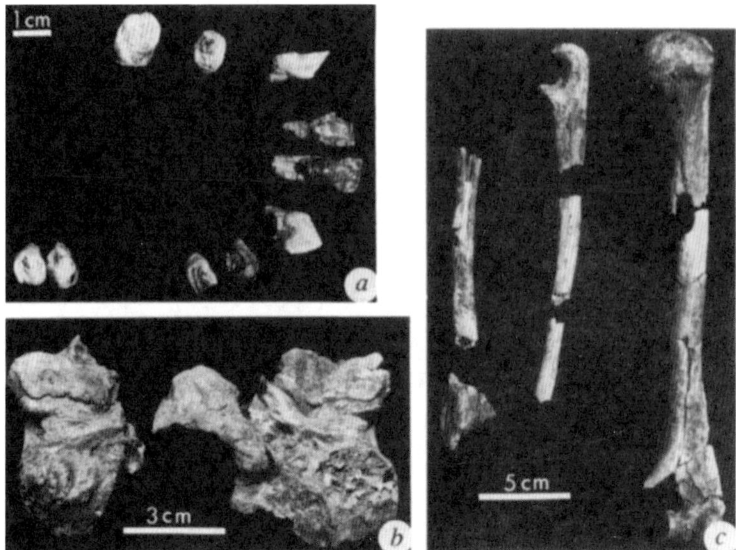

Fig. 1. *a*, Holotype specimen, ARA-VP-6/1 upper and lower dentition from a single individual; *b*, partial adult basicranium, ARA-VP-1/500; *c*, associated adult arm elements, ARA-VP-7/2. All alignments approximate. See text for descriptions.

A. ramidus is distinguished as a hominid from modern great apes and known elements of *Sivapithecus*, *Kenyapithecus*, *Ouranopithecus*, *Lufengpithecus* and *Dryopithecus* by the following: canine morphology more incisiform, crowns less projecting, with relatively higher crown shoulders; cupped distal wear pattern on lower canine; mandibular P_3 with weaker mesiobuccal projection of the crown base and without functional honing facet; modally relatively smaller mandibular P_3; modally relatively broader lower molars; foramen magnum anteriorly placed relative to carotid foramen; hypoglossal canal anteriorly placed relative to internal auditory meatus; carotid foramen placed posteromedial to tympanic angle.

A. ramidus is further distinguished from both *Pan* and *Gorilla* by the following: upper canine not mesiodistally elongate.

A. ramidus is further distinguished from *Pan troglodytes* and *Pan paniscus* by the following: upper central incisors small relative to postcanine teeth; lower third molars elongate and larger relative to other molars; molars not as crenulated, occlusal foveae not as broad.

A. ramidus is further distinguished from *Gorilla* by the following : smaller absolute tooth and upper limb size; flatter temporomandibular joint; lack of strong molar cusp relief; less sectorial first deciduous molar, dm_1.

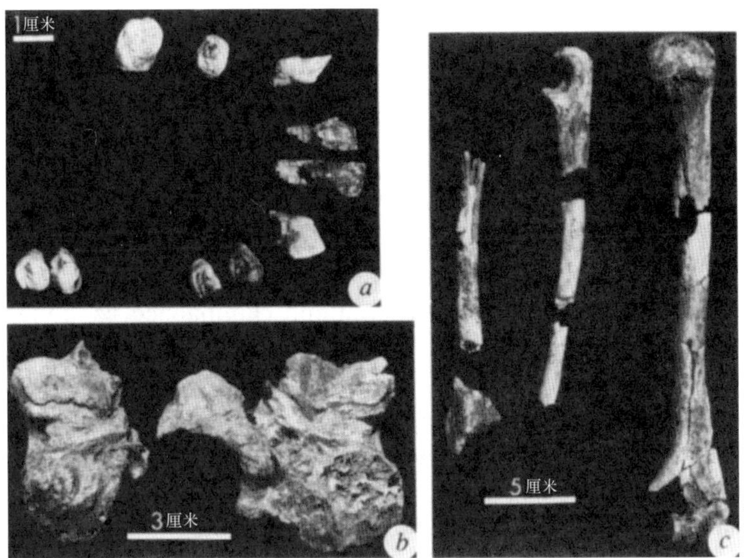

图 1. *a*，正模标本，来自同一个体的 ARA-VP-6/1 上、下齿系；*b*，部分成年颅底部，ARA-VP-1/500；*c*，相关的成年手臂部分，ARA-VP-7/2。所有比对都接近。描述见正文。

南方古猿始祖种被视为一种原始人类，与现代的大型猿类和已知的西瓦古猿、肯尼亚古猿、欧兰诺古猿、禄丰古猿和森林古猿在如下方面有所不同：犬齿门齿化，牙冠不太突出，具有相对较高的牙冠台肩；下颌犬齿是凹陷的远端磨损模式；下颌 P_3 在牙冠基部具有微弱的近中颊侧突出、没有功能性的研磨面；下颌 P_3 相对较小；下颌臼齿相对较宽；枕骨大孔相对于颈动脉孔的位置更靠前；舌下神经管相对于内耳道的位置更靠前；颈动脉孔位于鼓室倾角的中后部。

南方古猿始祖种还在如下方面与黑猩猩属和大猩猩属都不同：上颌犬齿不是在近中–远中方向上延伸。

南方古猿始祖种在如下方面与黑猩猩和倭黑猩猩都不同：相对于颊齿来说，上中门牙较小；下颌第三臼齿延伸，且相对于其他臼齿较大；臼齿皱纹少，咬合面凹也不宽。

南方古猿始祖种与大猩猩属还在如下方面有所不同：牙齿和上肢的绝对尺寸较小；颞下颌关节较平；缺乏强壮的臼齿齿尖突出；第一乳臼齿 dm_1 不太像扇形。

Dental Description

The ARA-VP-1/129 child's mandible retains a first deciduous molar (dm_1). The dm_1 has been crucially important in studies of *Australopithecus* since the discovery of the genus 70 years ago, and has been used frequently as a key character for sorting apes and hominids[11-13]. The Aramis dm_1 is morphologically far closer to that of a chimpanzee than to any known hominid (Fig. 2). It is very small—more than 4 s.d. units below the combined *A. afarensis*/*africanus* mean. It is at the low end of the common chimpanzee size range ($n = 29$) and comparable to the bonobo mean ($n = 21$) (Table 2). The apelike Aramis dm_1 lacks the apparently derived hominid features of buccolingual crown expansion, mesiolingually prominent metaconid, well-defined anterior fovea, and large talonid with well differentiated cusps. The only probable hominid derived feature shared with *A. afarensis* is an occlusally and mesiobuccally reduced protoconid, possibly associated with a loss of deciduous canine honing. The relative size of the talonid, whether judged by relative protoconid length or actual area ratios, lies at the chimpanzee means. The Aramis tooth stands farther in this feature from *A. afarensis* than *A. afarensis* is separated from robust *Australopithecus* homologues. The crown length to breadth ratio (1.49) shows a very narrow dm_1, surpassed in mean values only by the common chimpanzee (mean = 1.58) among fossil hominids and modern hominoids. The ratio between labiolingual breadth of the deciduous canine root and the square root of the computed dm_1 area shows a relatively large Aramis canine, nearly matching the *Pan paniscus* ratio mean and exceeding the *P. troglodytes* average and the *G. gorilla* range ($n = 20$). The only measurable *A. afarensis* specimen (L.H.-2) lies closer to the most extreme *A. boisei* condition (KNM-ER 1477) than it does to Aramis in this ratio.

Table 2. Lower first deciduous molar (dm_1) measurements

	Mesiodistal (MD) length	Buccolingual (BL) breadth	MD × BL area	Protoconid length	MD Length+protoconid length
Aramis ($n = 1$)	7.3	4.9	35.8	5.2	1.4
A. afarensis					
n	4	4	4	4	4
min	8.5	7.6	68.0	4.3	1.7
max	9.6	8.4	80.6	5.6	2.0
mean	**9.2**	**7.9**	**72.5**	**5.1**	**1.8**
s.d.	0.5	0.4	5.7	0.6	0.1
A. africanus					
n	7	5	5	3	3
min	8.4	7.1	59.6	5.2	1.6
max	9.1	8.1	73.7	5.3	1.7
mean	**8.8**	**7.6**	**66.6**	**5.2**	**1.7**
s.d.	0.2	0.4	5.5	0.1	0.1

牙齿描述

ARA-VP-1/129 儿童下颌骨保留一颗第一乳臼齿(dm_1)。自从 70 年前发现该属以来，该牙对于研究南方古猿属一直都意义重大，经常作为对猿类和原始人类进行分类的关键特征[11-13]。阿拉米斯 dm_1 在形态学上与黑猩猩的相似性比与其他已知原始人类的都更大(图 2)。这颗牙非常小——比南方古猿阿法种 / 非洲种的平均值要小 4 个标准偏差单位以上。该值处于普通黑猩猩尺寸范围的下限($n = 29$)，与倭黑猩猩的均值($n = 21$)相当(表 2)。似猿的阿拉米斯 dm_1 明显缺少人类的以下衍生特征：颊舌径扩展，近中舌侧突出的下后尖，界线明确的前凹以及具有分化良好的牙尖的大跟座。唯一与南方古猿阿法种可能共有的人类衍生特征就是下原尖在咬合面和近中颊侧减小，这可能与乳犬齿咬合功能缺失有关。无论是相对于下原尖长度还是面积，跟座的相对大小都与黑猩猩的均值相当。阿拉米斯牙齿在这个特征上与南方古猿阿法种的差别比南方古猿阿法种与同源的南方古猿粗壮种的差别更大。牙冠的长宽比(1.49)表明这是一个很窄的 dm_1，比化石人类和现代其他类人猿都大，仅低于普通黑猩猩的均值(均值 = 1.58)。乳犬齿牙根的唇舌宽度与 dm_1 牙根面积的平方根值相比，表明阿拉米斯的犬齿相对较大，几乎与倭黑猩猩的比值均值相匹配，而超过了黑猩猩的均值和大猩猩的范围($n = 20$)。与阿拉米斯样本的相比，唯一可测量的南方古猿阿法种标本(LH-2)的这一比值与南方古猿鲍氏种的极端情况(KNM-ER 1477)更接近。

表 2. 下颌第一乳臼齿(dm_1)的测量值

	近中–远中(MD)长度	颊舌(BL)宽度	MD × BL 面积	下原尖长度	MD 长度 + 下原尖长度
阿拉米斯样本($n = 1$)	7.3	4.9	35.8	5.2	1.4
南方古猿阿法种					
n	4	4	4	4	4
最小值	8.5	7.6	68.0	4.3	1.7
最大值	9.6	8.4	80.6	5.6	2.0
均值	**9.2**	**7.9**	**72.5**	**5.1**	**1.8**
标准差	0.5	0.4	5.7	0.6	0.1
南方古猿非洲种					
n	7	5	5	3	3
最小值	8.4	7.1	59.6	5.2	1.6
最大值	9.1	8.1	73.7	5.3	1.7
均值	**8.8**	**7.6**	**66.6**	**5.2**	**1.7**
标准差	0.2	0.4	5.5	0.1	0.1

Continued

	Mesiodistal (MD) length	Buccolingual (BL) breadth	MD × BL area	Protoconid length	MD Length+protoconid length
A. robustus					
n	8	8	8	8	8
min	9.0	7.7	71.0	4.3	1.8
max	10.8	9.7	101.9	5.8	2.3
mean	**10.1**	**8.3**	**83.7**	**4.9**	**2.1**
s.d.	0.5	0.6	9.5	0.5	0.1
P. paniscus					
n	21	21	21	20	20
min	6.3	4.4	27.7	4.3	1.3
max	8.8	5.5	48.4	6.0	1.6
mean	**7.4**	**5.1**	**37.6**	**5.0**	**1.5**
s.d.	0.6	0.31	4.7	0.5	0.1
P. troglodytes					
n	29	29	29	29	29
min	7.0	4.6	32.9	5.0	1.3
max	9.4	5.8	54.5	6.7	1.6
mean	**8.1**	**5.2**	**42.2**	**5.8**	**1.4**
s.d.	0.6	0.4	5.2	0.5	0.1
G. gorilla					
n	20	20	20	20	20
min	9.8	6.7	71.4	6.7	1.3
max	12.2	8.9	108.6	9.0	1.6
mean	**11.0**	**7.5**	**82.3**	**7.8**	**1.4**
s.d.	0.7	0.6	10.7	0.6	0.1
P. pygmaeus					
n	6	6	6	6	6
min	8.4	6.4	53.8	5.8	1.3
max	10.2	8.1	82.6	8.1	1.5
mean	**9.2**	**7.1**	**66.2**	**6.7**	**1.4**
s.d.	0.7	0.6	10.3	0.8	0.1
H. sapiens					
n	21	21	21	21	21
min	7.4	6.4	47.4	4. 0	1.4
max	9.2	8.1	69.9	5.7	2.1
mean	**8.4**	**7.2**	**60.4**	**4.9**	**1.7**
s.d.	0.5	0.4	6.1	0.5	0.2

Comparative metrics on deciduous lower first molars (dm_1) of various hominoid taxa. Abbreviations and conventions as in Table 1.

续表

	近中－远中（MD）长度	颊舌（BL）宽度	MD × BL 面积	下原尖长度	MD 长度 + 下原尖长度
南方古猿粗壮种					
n	8	8	8	8	8
最小值	9.0	7.7	71.0	4.3	1.8
最大值	10.8	9.7	101.9	5.8	2.3
均值	**10.1**	**8.3**	**83.7**	**4.9**	**2.1**
标准差	0.5	0.6	9.5	0.5	0.1
倭黑猩猩					
n	21	21	21	20	20
最小值	6.3	4.4	27.7	4.3	1.3
最大值	8.8	5.5	48.4	6.0	1.6
均值	**7.4**	**5.1**	**37.6**	**5.0**	**1.5**
标准差	0.6	0.31	4.7	0.5	0.1
黑猩猩					
n	29	29	29	29	29
最小值	7.0	4.6	32.9	5.0	1.3
最大值	9.4	5.8	54.5	6.7	1.6
均值	**8.1**	**5.2**	**42.2**	**5.8**	**1.4**
标准差	0.6	0.4	5.2	0.5	0.1
银背大猩猩					
n	20	20	20	20	20
最小值	9.8	6.7	71.4	6.7	1.3
最大值	12.2	8.9	108.6	9.0	1.6
均值	**11.0**	**7.5**	**82.3**	**7.8**	**1.4**
标准差	0.7	0.6	10.7	0.6	0.1
婆罗洲猩猩					
n	6	6	6	6	6
最小值	8.4	6.4	53.8	5.8	1.3
最大值	10.2	8.1	82.6	8.1	1.5
均值	**9.2**	**7.1**	**66.2**	**6.7**	**1.4**
标准差	0.7	0.6	10.3	0.8	0.1
智人					
n	21	21	21	21	21
最小值	7.4	6.4	47.4	4.0	1.4
最大值	9.2	8.1	69.9	5.7	2.1
均值	**8.4**	**7.2**	**60.4**	**4.9**	**1.7**
标准差	0.5	0.4	6.1	0.5	0.2

对各种人科动物类群的下颌第一乳臼齿（dm_1）的测量尺寸进行的比较。缩写及惯例同表 1。

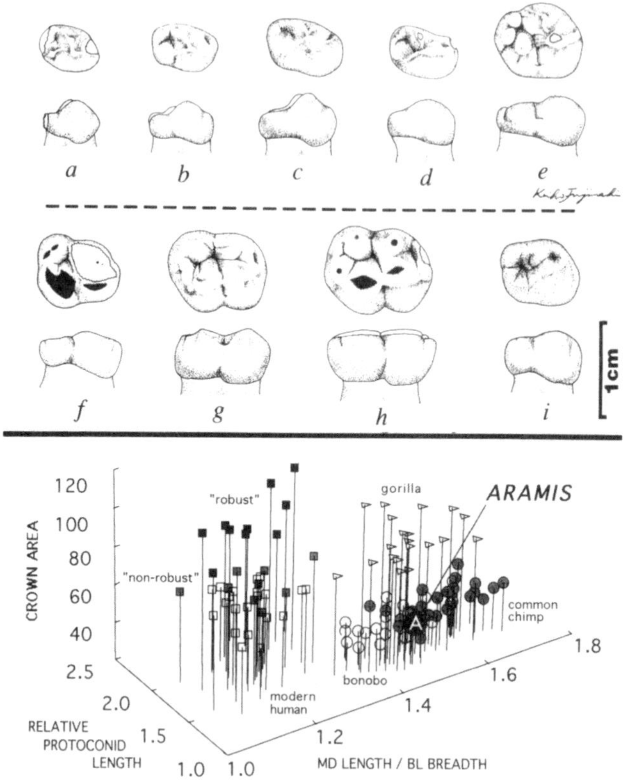

Fig. 2. Deciduous first molar comparisons. Metric and morphological comparisons show a wide separation between the dm_1 of Aramis and those of other early hominid species. *a*, *Dryopithecus* (IPS 42/1784); *b*, *Pan paniscus* (T-26992); *c*, *Pan troglodytes* (PRI 1372); *d*, *Australopithecus ramidus* (ARA-VP-1/129); *e*, *A. afarensis* (A.L. 333-43B); *f*, *A. africanus* (Taung); *g*, *A. robustus* (TM 1601); *h*, *A. boisei* (KNM ER-1477); *i*, *Homo sapiens* (modern). The three-dimensional plot shows dm_1 crown area (buccolingual (BL) multiplied by mesiodistal (MD)) in square mm on the vertical axis. MD length divided by total protoconid length is shown on the left depth axis. The third axis represents a measure of tooth crown narrowness, the MD length divided by the BL breadth. Individual specimens are shown. The "robust" sample includes *A. robustus*, *A. aethiopicus* (L704) and *A. boisei*. The "non-robust" sample includes *A. africanus*, *A. afarensis* and early *Homo* (KNM ER-1507; Omo 222). The new species *A. ramidus* is centred in the chimpanzee ranges for these measures. It represents a good ancestral morphotype for all later hominid species.

The *A. ramidus* permanent dentition is represented at most positions (Fig. 3; Table 3). Upper and lower incisors do not exhibit the large size typical of extant *Pan*. Upper and lower central incisor size relative to postcanine teeth is comparable to Miocene hominoids and gorillas. Of the five individuals for whom canine size is determinable, all five have crowns at or larger than the *A. afarensis* mean. Upper and lower canines are also large relative to postcanine teeth. ARA-VP-1/128 is over 5 s.d. units above the *A. afarensis* mean in measures of relative canine size within known individuals ($C \div P_4$; $C \div M_1$; and $C \div M_3$ ratios of maximum labiolingual canine crown breadth÷square root of computed molar or premolar crown area). In ARA-VP- 6/1 relative canine crown area is comparable to the female great ape condition. Morphology of the known Aramis canines, however, diverges from that of known apes (Fig. 3). The upper canines are slightly less incisiform than homologues of

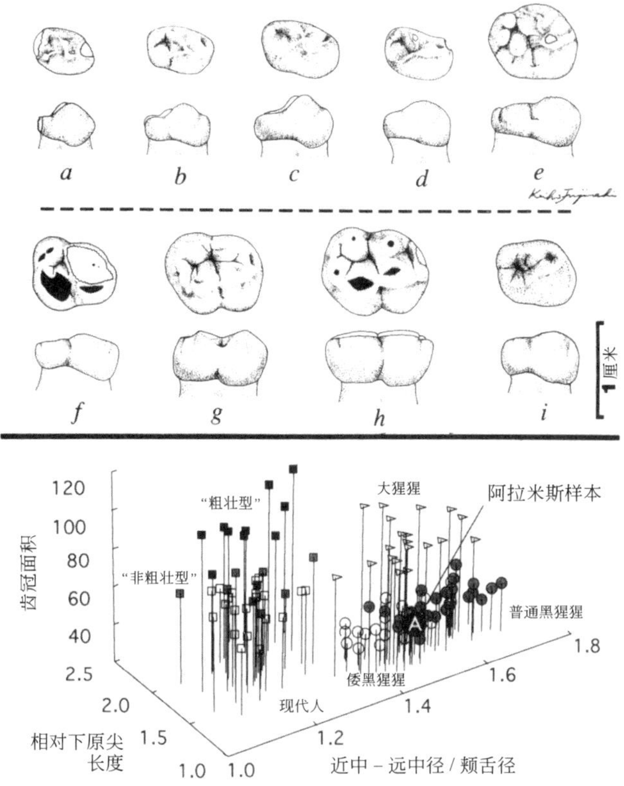

图 2. 第一乳臼齿的比较。度量及形态学比较表明阿拉米斯样本的 dm_1 和其他早期原始人类物种的 dm_1 之间具有很大的区别。*a*，森林古猿属(IPS 42/1784)；*b*，倭黑猩猩(T-26992)；*c*，黑猩猩(PRI 1372)；*d*，南方古猿始祖种(ARA-VP-1/129)；*e*，南方古猿阿法种(A.L. 333-43B)；*f*，南方古猿非洲种(汤恩)；*g*，南方古猿粗壮种(TM 1601)；*h*，南方古猿鲍氏种(KNM ER-1477)；*i*，智人(现代)。三维图的纵轴显示了以平方毫米表示的 dm_1 的牙冠面积(颊舌径(BL)乘以近中－远中径(MD))。MD 长度除以下原尖的全长的结果在左侧的深度轴线上表示出来了。第三条轴表示牙冠狭窄度的测量值，即 MD 长度除以 BL 宽度。各个标本都有显示。"粗壮"样本包括南方古猿粗壮种、南方古猿埃塞俄比亚种(L704)和南方古猿鲍氏种。"非粗壮"样本包括南方古猿非洲种、南方古猿阿法种和早期人属(KNM ER-1507；奥莫 222)。新物种——南方古猿始祖种的这些尺寸都集中在黑猩猩的范围内。其代表了一种较好的后来所有的原始人类物种的祖先形态型。

南方古猿始祖种的恒牙系在大部分位置都有出现(图 3；表 3)。上、下门齿没有显示出现存黑猩猩属所具有的典型的巨大尺寸。上、下中门齿相对于颊齿的大小与中新世人科动物和大猩猩的具有可比性。在可以确定犬齿尺寸的五个个体中，所有这五个样本的牙冠都与南方古猿阿法种的均值相同或更大。上、下犬齿相对于颊齿来说也较大。在已知的个体中，ARA-VP-1/128 犬齿相对大小的测量值超过南方古猿阿法种的均值 5 个标准偏差单位以上($C \div P_4$、$C \div M_1$ 和 $C \div M_3$ 最大唇舌犬齿牙冠宽度 ÷ 臼齿或前臼齿牙冠面积的平方根的比值)。ARA-VP-6/1 中，犬齿牙冠的相对面积与雌性大型猿类的情况相当。然而，已知的阿拉米斯犬齿形态学与已知猿类

A. afarensis but more incisiform than any ape counterpart, with occlusally placed terminations of the mesial and distal apical crests (Fig. 3*g*). The visual result of apically placed crown shoulders is a low, blunt canine tooth relative to more projecting ape canines, a morphological condition which may have important evolutionary implications. The Aramis upper canine is large buccolingually, forming a further contrast with mesiodistally elongate African ape canines. Wear pattern also differs significantly from the ape condition. Mandibular canine wear does not show the pattern typical of great apes. Some worn female *Pan* canines are superficially similar, but still lack the distal crown cupping seen on Aramis. Instead, they feature planar wear surfaces from contact with the upper canine, even on individuals with rounding (not honing) of the buccal P_3 face.

Table 3. Comparative dental metrics for permanent dentition

a, Upper dentition															
	Mesiodistal					Labio/buccolingual					Crown area (MD × BL)				
	n	Min	Max	Mean	s.d.	*n*	Min	Max	Mean	s.d.	*n*	Min	Max	Mean	s.d.
I^1															
A. afarensis	3	10.8	11.8	11.2	0.6	5	7.1	8.6	8.20	0.6	3	90.5	99.1	94.2	4.5
Aramis	1	—	—	(10.0)	—	2	7.5	8.2	—	—	1	—	—	(75.0)	—
C															
A. afarensis	9	8.9	11.6	10.0	0.8	10	9.3	12.5	10.9	1.1	9	82.8	145.0	109.9	18.9
Aramis	2	(11.2)	(11.5)	—	—	2	11.1	11.7	—	—	2	(124.3)	(134.5)	—	—
P^3															
A. afarensis	9	7.5	9.3	8.7	0.5	8	11.3	13.4	12.4	0.6	8	84.7	120.9	108	11.0
Aramis	1	—	—	7.7	—	1	—	—	12.5	—	1	—	—	96.3	—
P^4															
A. afarensis	10	7.6	9.7	9.0	0.6	6	11.1	12.6	12.1	0.6	6	84.4	119.7	106.8	12.6
Aramis	1	—	—	8.4	—	1	—	—	(11.3)	—	1	—	—	(94.9)	—
M^2															
A. afarensis	5	12.1	13.5	12.8	0.5	6	13.4	15.1	14.7	0.6	5	162.1	199.8	187.5	14.6
Aramis	2	(11.8)	11.8	—	—	2	(14.1)	(15.0)	—	—	2	(166.4)	(177.0)	—	—
M^3															
A. afarensis	8	10.5	14.3	11.9	1.4	8	13.0	15.5	13.8	1.0	8	136.5	215.9	165.1	30.9
Aramis	1	—	—	10.2	—	1	—	—	12.3	—	1	—	—	125.5	—
b, Lower dentition															
	Mesiodistal					Labio/buccolingual					Crown area (MD × BL)				
	n	Min	Max	Mean	s.d.	*n*	Min	Max	Mean	s.d.	*n*	Min	Max	Mean	s.d.
I_1															
A. afarensis	2	6.2	8.0	—	—	—	—	—	—	—	—	—	—	—	—
Aramis	1	—	—	6.0	—	—	—	—	—	—	—	—	—	—	—

的不同（图 3）。上颌犬齿的门齿化程度较同源的南方古猿阿法种低些，但是比其他猿类更明显，近中脊和远中脊顶点末端位于咬合面（图 3*g*）。可见的犬齿牙冠肩台现象说明它相比突出的猿类犬齿来说显得更低而钝，其形态学特征可能具有重要的进化含义。阿拉米斯上颌犬齿颊舌径较大，与近中–远中延伸的非洲猿的犬齿形成进一步的对比。磨耗模式也与猿的情况非常不同。下颌犬齿磨损情况并没表现出大型猿类的典型模式。有些雌性黑猩猩属犬齿表面的磨耗很相似，但是仍然缺乏在阿拉米斯标本中见到的杯状的远端牙冠。它们的特征性平面在与上颌犬齿的接触中相互磨损，甚至在 P_3 颊面圆滑（而非咬合式磨损）的个体中也是如此。

表 3. 恒齿系的牙齿测量尺寸的比较

a，上牙															
	近中–远中					唇舌 / 下原尖					牙冠面积（MD × BL）				
	n	最小值	最大值	均值	标准差	*n*	最小值	最大值	均值	标准差	*n*	最小值	最大值	均值	标准差
I^1															
南方古猿阿法种	3	10.8	11.8	11.2	0.6	5	7.1	8.6	8.20	0.6	3	90.5	99.1	94.2	4.5
阿拉米斯样本	1	—	—	(10.0)	—	2	7.5	8.2	—	—	1	—	—	(75.0)	—
C															
南方古猿阿法种	9	8.9	11.6	10.0	0.8	10	9.3	12.5	10.9	1.1	9	82.8	145.0	109.9	18.9
阿拉米斯样本	2	(11.2)	(11.5)	—	—	2	11.1	11.7	—	—	2	(124.3)	(134.5)	—	—
P^3															
南方古猿阿法种	9	7.5	9.3	8.7	0.5	8	11.3	13.4	12.4	0.6	8	84.7	120.9	108	11.0
阿拉米斯样本	1	—	—	7.7	—	1	—	—	12.5	—	1	—	—	96.3	—
P^4															
南方古猿阿法种	10	7.6	9.7	9.0	0.6	6	11.1	12.6	12.1	0.6	6	84.4	119.7	106.8	12.6
阿拉米斯样本	1	—	—	8.4	—	1	—	—	(11.3)	—	1	—	—	(94.9)	—
M^2															
南方古猿阿法种	5	12.1	13.5	12.8	0.5	6	13.4	15.1	14.7	0.6	5	162.1	199.8	187.5	14.6
阿拉米斯样本	2	(11.8)	11.8	—	—	2	(14.1)	(15.0)	—	—	2	(166.4)	(177.0)	—	—
M^3															
南方古猿阿法种	8	10.5	14.3	11.9	1.4	8	13.0	15.5	13.8	1.0	8	136.5	215.9	165.1	30.9
阿拉米斯样本	1	—	—	10.2	—	1	—	—	12.3	—	1	—	—	125.5	—
b，下牙															
	近中–远中					唇舌 / 下原尖					牙冠面积（MD × BL）				
	n	最小值	最大值	均值	标准差	*n*	最小值	最大值	均值	标准差	*n*	最小值	最大值	均值	标准差
I_1															
南方古猿阿法种	2	6.2	8.0	—	—	—	—	—	—	—	—	—	—	—	—
阿拉米斯样本	1	—	—	6.0	—	—	—	—	—	—	—	—	—	—	—

Continued

b, Lower dentition															
	Mesiodistal					Labio/buccolingual					Crown area (MD × BL)				
	n	Min	Max	Mean	s.d.	*n*	Min	Max	Mean	s.d.	*n*	Min	Max	Mean	s.d.
C															
A. afarenis						13	8.8	12.4	10.4	1.1	—	—	—	—	—
Aramis	—	—	—	—	—	1	—	—	11.0	—	—	—	—	—	—
P_3 (min/max)															
A. afarensis	19	6.5	9.8	8.6	1.1	19	9.7	13.3	11.6	1.1	19	63.1	127.7	99.7	20.4
Aramis	2	7.5	(8.2)	—	—	2	(9.9)	(11.5)	—	—	2	(74.2)	(94.3)	—	—
P_4															
A. afarensis	15	7.7	11.1	9.7	1.0	14	9.8	12.8	10.9	0.8	14	77.0	129.7	106.5	16.8
Aramis	2	7.5	8.9	—	—	2	9.5	9.7	—	—	2	71.2	86.3	—	—
M_1															
A. afarensis	17	11.2	14.0	13.0	0.6	16	11.0	13.9	12.6	0.8	16	124.3	194.6	164.9	17.1
Aramis	2	11.0	11.1	—	—	2	(10.2)	10.3	—	—	2	(113.2)	113.3	—	—
M_2															
A. afarensis	23	12.4	16.2	14.3	1.0	22	12.1	15.2	13.5	0.9	22	152.5	234.1	193.3	24.3
Aramis	1	—	—	(13.0)	—	1	—	—	11.9	—	1	—	—	(154.7)	—
M_3															
A. afarensis	14	13.7	16.3	14.8	0.8	14	12.1	14.9	13.3	0.8	13	172.0	231.5	195.7	17.7
Aramis	1	—	—	12.7	—	1	—	—	11.0	—	1	—	—	139.7	—

Comparative metrics for the permanent teeth of *A. afarensis* (comprises the Hadar pre-1990 sample and the full Laetoli and Maka samples) and *A. ramidus* (from Table 1). Data are shown only for tooth positions represented in both species. Measurements are standard and were taken by the authors on original specimens with conventions and abbreviations as in Table 1. There is considerable overlap between the known species ranges, as there is among other species in the genus. As documented in the text and illustrations, however, proportional differences within individual dentitions combine with other morphological considerations to warrant the recognition of *A. ramidus* as a species distinct from *A. afarensis.*

The broken canines and lower P_3 in ARA-VP-1 /128 and -6/1 exhibit thin enamel distinct from previously known hominid conditions. Canine enamel thickness approximates the chimpanzee condition, with a lack of apical thickening we observe in other hominids. The 1.0 mm buccal enamel thickness of the ARA-VP-6/1 broken upper right canine slightly exceeds the 0.9 mm maximum recorded in our small sample of broken female *P. troglodytes* upper canines ($n = 6$) and is approximately 2.4 s.d. units above our combined-sex chimpanzee mean of 0.65 mm ($n = 14$). The broadly constant enamel thickness of the Aramis maxillary canine above the midcrown height level contrasts with the *A. afarensis* condition in which buccal enamel thickens towards the apex, commonly reaching ~1.5 mm. The significance of maxillary canine enamel thickness can be evaluated in the light of proposed wear mechanics of the $\underline{C}/P_3$ complex[14]. The relatively thin enamel and large size

续表

b，下牙															
	近中—远中					唇舌 / 下原尖					牙冠面积（MD × BL）				
	n	最小值	最大值	均值	标准差	*n*	最小值	最大值	均值	标准差	*n*	最小值	最大值	均值	标准差
C															
南方古猿阿法种						13	8.8	12.4	10.4	1.1	—	—	—	—	—
阿拉米斯样本	—	—	—	—	—	1	—	—	11.0	—	—	—	—	—	—
P_3（最小值 / 最大值）															
南方古猿阿法种	19	6.5	9.8	8.6	1.1	19	9.7	13.3	11.6	1.1	19	63.1	127.7	99.7	20.4
阿拉米斯样本	2	7.5	(8.2)	—	—	2	(9.9)	(11.5)	—	—	2	(74.2)	(94.3)	—	—
P_4															
南方古猿阿法种	15	7.7	11.1	9.7	1.0	14	9.8	12.8	10.9	0.8	14	77.0	129.7	106.5	16.8
阿拉米斯样本	2	7.5	8.9	—	—	2	9.5	9.7	—	—	2	71.2	86.3	—	—
M_1															
南方古猿阿法种	17	11.2	14.0	13.0	0.6	16	11.0	13.9	12.6	0.8	16	124.3	194.6	164.9	17.1
阿拉米斯样本	2	11.0	11.1	—	—	2	(10.2)	10.3	—	—	2	(113.2)	113.3	—	—
M_2															
南方古猿阿法种	23	12.4	16.2	14.3	1.0	22	12.1	15.2	13.5	0.9	22	152.5	234.1	193.3	24.3
阿拉米斯样本	1	—	—	(13.0)	—	1	—	—	11.9	—	1	—	—	(154.7)	—
M_3															
南方古猿阿法种	14	13.7	16.3	14.8	0.8	14	12.1	14.9	13.3	0.8	13	172.0	231.5	195.7	17.7
阿拉米斯样本	1	—	—	12.7	—	1	—	—	11.0	—	1	—	—	139.7	—

对南方古猿阿法种(由哈达尔1990年之前发现的样本和所有莱托里与马卡的样本构成)和南方古猿始祖种(来自表1)的恒齿的测量尺寸进行的比较。表中数据仅展示了两个物种都保存了的牙位。测量尺寸标准，是作者遵守与表1相同的惯例与缩写对原始标本进行测量得到的。已知的物种范围间存在相当大的重叠区，正如该属中的其他物种一样。但是，就像正文及插图中所说明的，个体齿系内的比例差异与其他形态学因素共同证实南方古猿始祖种是与南方古猿阿法种不同的种。

ARA-VP-1/128和ARA-VP-6/1破裂的犬齿和下颌P_3展示出了与之前所知原始人类情况不同的薄型牙釉质。犬齿牙釉质的厚度接近于黑猩猩，同时缺少我们在其他原始人类中观察到的犬齿尖点加厚的现象。ARA-VP-6/1断裂的上颌右侧犬齿的牙釉质颊侧厚度为1.0毫米，稍微超过了我们的小型样本——雌性黑猩猩断裂的上颌犬齿所记录的最大值0.9毫米($n = 6$)，两种性别的黑猩猩的均值为0.65毫米，所以ARA-VP-6/1大约超过了黑猩猩均值2.4个标准差单位($n = 14$)。阿拉米斯上颌犬齿在牙冠中部之上的牙釉质厚度的恒定与南方古猿阿法种的情况形成了对比，后者颊侧到牙尖点的牙釉质厚度增加，通常约为1.5毫米。上颌犬齿牙釉质厚度的意义可

of the Aramis canine, together with its primitive P_3 morphology, suggest a $\underline{C}/P_3$ complex morphologically and functionally only slightly removed from the presumed ancestral ape condition.

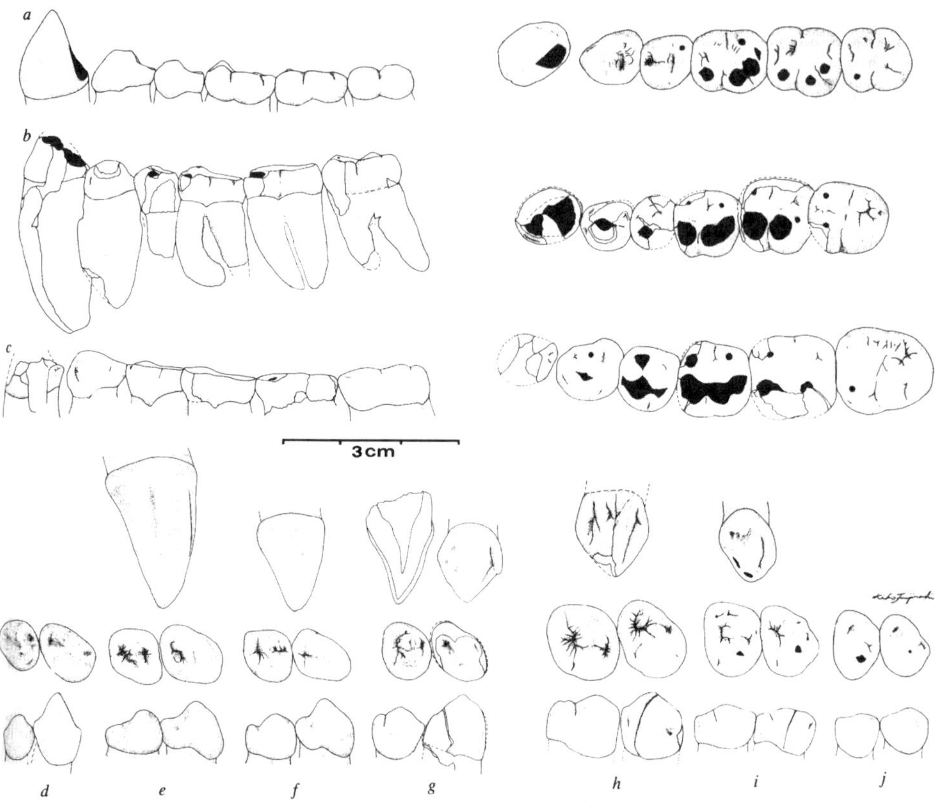

Fig. 3. Comparisons of upper canine/lower premolar complexes and tooth rows. Top three rows, Occlusal and lateral views of the lower tooth rows of: *a*, *Pan troglodytes* female (CMNH B1770); *b*, *A. ramidus* (ARA-VP-1/128); *c*, *A. afarensis* holotype (Laetoli Hominid 4). Bottom three rows, lingual views of upper canines and occlusal and buccal views of lower third and fourth premolars of: *d*, *Dryopithecus* (MNHNP); *e*, *Pan troglodytes* male (CMNH B1882); *f*, *P. troglodytes* female (CMNH B1721); *g*, *A. ramidus* holotype (ARA-VP-6/1; split right canine on the left); *h*, *A. afarensis* (LH-3); *i*, *A. afarensis* (A.L. 400); *j*, *A. afarensis* (A.L. 288-1, "Lucy"). *a*, *c* and *h* reversed for comparison.

The ARA-VP-6/1 P_3 is markedly more apelike than any *A. afarensis* homologue in its high protoconid with extensive buccal face and steep, distolingually directed transverse crest (Fig. 3*g*). In these features it is indistinguishable from ape homologues. The strong mesiobuccal protrusion of its crown base is also outside the known *A. afarensis* range. The Aramis P_3 deviates toward the *A. afarensis* condition in some details. These include a more occlusal termination of the mesial protoconid crest, weaker mesiobuccal protrusion of the crown base (especially ARA-VP-1/128), and a smaller size relative to P_4–M_3 although rare individual *Pan* specimens do approximate the Aramis condition in these features. The worn ARA-VP-1/128 P_3 lacks a honing facet but exhibits steep mesial and distal wear

以从 $\underline{C}/P_3$ 复合体的磨损机制中解读[14]。阿拉米斯相对薄的犬齿牙釉质和较大尺寸以及其 P_3 的原始形态，提示该 $\underline{C}/P_3$ 复合体在形态和功能上都仅仅能勉强将其从假定的祖先猿类中分离出来。

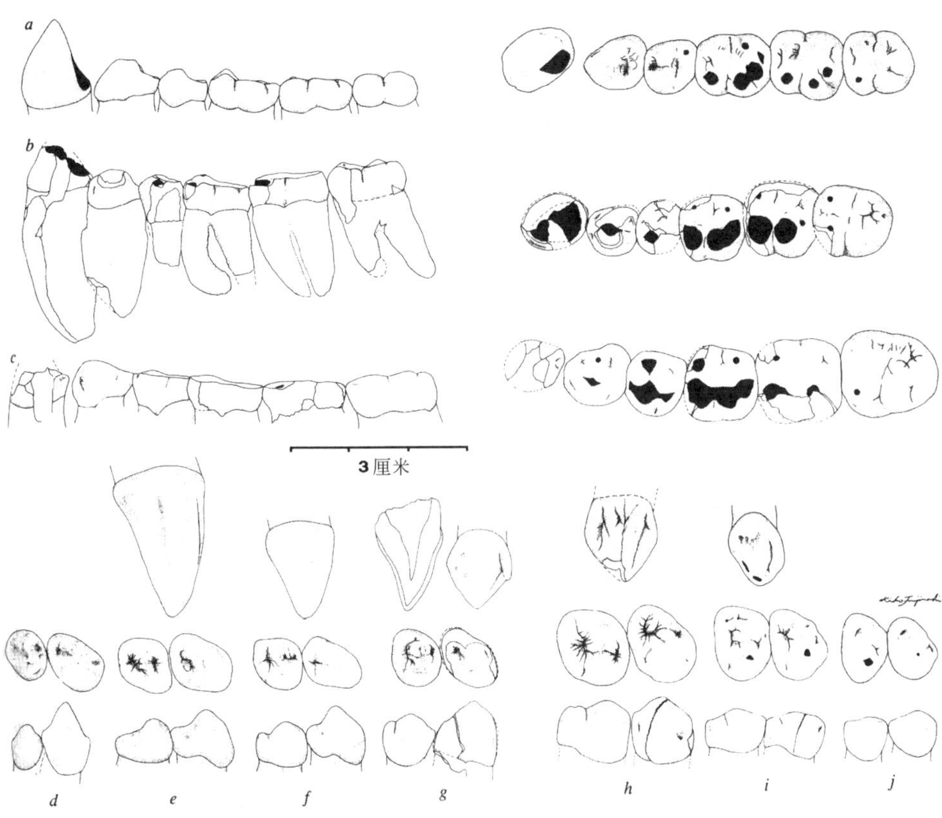

图 3. 上颌犬齿 / 下颌前臼齿复合体及齿列的比较。上面三排为如下物种的下颌齿列的咬合面和侧面视图：*a*，雌性黑猩猩（CMNH B1770）；*b*，南方古猿始祖种（ARA-VP-1/128）；*c*，南方古猿阿法种正模标本（莱托里原始人类 4）。下面三排为如下物种的上颌犬齿的舌面视图以及下颌第三和第四前臼齿的咬合面和颊面视图：*d*，森林古猿（MNHNP）；*e*，雄性黑猩猩（CMNH B1882）；*f*，雌性黑猩猩（CMNH B1721）；*g*，南方古猿始祖种正模标本（ARA-VP-6/1；左侧是有裂口的右犬齿）；*h*，南方古猿阿法种（LH-3）；*i*，南方古猿阿法种（A.L. 400）；*j*，南方古猿阿法种（A.L. 288-1，“露西”）。*a*、*c* 和 *h* 是反转后进行的比较。

高高的下原尖具有广阔的颊面和陡峭的远中舌侧方向的横脊，在这一点上，ARA-VP-6/1 的 P_3 比任何同源的南方古猿阿法种都更加像猿类（图 3*g*）。在这些特征上，难以将该样本与猿类的同源物种区别开来。牙冠基底强壮的近中颊侧突出也不属于已知的南方古猿阿法种的范围。阿拉米斯的 P_3 在某些细节方面偏离了南方古猿阿法种。这些细节包括近中下原尖脊的终端更加靠近咬合面、牙冠基底更微弱的近中颊侧突出（尤其是 ARA-VP-1/128）以及相对于 P_4~M_3 更小的尺寸，但是也很少有黑猩猩属的标本个体在这些特征上与阿拉米斯样本的情况接近。ARA-VP-1/128 的磨损的 P_3 缺少研磨面，但是展示出陡峭的近中和远中磨损斜面，这与南方古猿阿法

slopes not matched in *A. afarensis.*

The P^3 is distinctly primitive in its tall and mesiodistally elongate paracone. Both P^3 and P^4 exhibit a prominent anterior transverse crest. In the P^3 this crest defines an anteriorly facing triangular portion of the occlusal surface, as in apes. The lower P_4 exhibits a prominent transverse crest and minimal talonid development. The P4s from two known individuals are both single rooted.

Molar morphology resembles the *A. afarensis* condition, but lacks the extreme buccolingual breadth relative to mesiodistal length common in that species (Fig. 3*a–e*). The "serrate" root pattern and deep dentine wear on the buccal cusps described in *A. afarensis*, Tabarin, and Lothagam[15-17] also occur in Aramis specimens. All molars lack the extensive crenulation and broad occlusal foveae characteristic of modern chimpanzees, or the high cusp topography of gorillas. The Aramis lower third molar is rounded distally, like *A. afarensis* and Miocene hominoids. A great size discrepancy between M_1 and M_2 is seen in ARA-VP-1 / 128.

A distinct difference from known hominids occurs in molar enamel thickness. Maximum radial enamel thickness of crown faces can be measured in three fractured Aramis specimens and it ranges from 1.1 to 1.2 mm buccally, at or near the unworn cusp apex, perpendicular to the enamel–dentine junction. These values are comparable to the uppermost range of our homologous enamel thickness values measured on broken *P. troglodytes* molars ($n = 22$; M_1 through to M_3). Equivalent measures in *A. afarensis* range from 1.4 to 2.0 mm ($n = 5$). In one case (the ARA-VP-1 / 128 third molar) Aramis radial enamel thickness at the buccal protoconid face can be evaluated relative to cervical breadth. A comparison of this ratio of enamel thickness suggests that *A. ramidus* may be characterized as intermediate between the chimpanzee and the *A. afarensis* / *africanus* / early *Homo* conditions.

In postcanine size, the range of the available Aramis sample includes specimens smaller than known *A. afarensis* homologues (the two known M_1 teeth are both more than 3 s.d. units below the mean). Of the seven Aramis individuals for whom postcanine tooth size is determinable, all have crown sizes smaller than the *A. afarensis* mean. We interpret the limited morphology and metrical data available as indicating a single species with a postcanine dentition significantly smaller than in *A. afarensis.*

The postcanine mandibular row can be reconstructed for ARA-VP-1 / 128 by juxtaposing interproximal facets (Fig. 3*a–c*). This shows that the C to M_2 dental row is weakly concave buccally, as in modern and fossil apes and some *A. afarensis* specimens. The P_3 axis is less oblique than in most apes. The canine is positioned directly in line with the mesiodistal axis of the postcanine tooth row rather than being set mesiolingually to the postcanine axis as in the case for most *A. afarensis.* This is a more primitive arrangement shared with modern and Miocene great apes, and may suggest that the mesial part of the lower canine was not functionally incorporated into the incisal row as seen in *A. afarensis*[17].

种不相同。

P^3 非常原始，具有高的、近中 – 远中端延伸的上前尖。P^3 和 P^4 都展示出突出的前横脊。P^3 中，该脊决定了咬合面前部呈三角形，就像猿类一样。下 P_4 呈现出一条突出的横脊和不发达的跟座。两个已知个体的 P4 都是单牙根。

臼齿形态与南方古猿阿法种的情形相像，但是缺少后者常见的相对于近中 – 远中长度极端的颊舌宽度（图 3*a*~*e*）。在描述南方古猿阿法种、塔巴林样本和洛萨加姆标本 [15-17] 时提到的“锯齿状”牙根模式和颊侧牙尖上深的牙本质磨损在阿拉米斯标本中也有出现。所有臼齿都缺少现代黑猩猩特征性的广泛褶皱和宽阔的咬合窝，或者缺少大猩猩特征性的高牙尖形态。阿拉米斯的下颌第三下臼齿远端呈圆弧状，就像南方古猿阿法种和中新世人科动物的一样。ARA-VP-1/128 的 M_1 和 M_2 有很大的尺寸差异。

与已知原始人类的显著差异是在臼齿牙釉质厚度上。牙冠最大釉质辐射厚度可以在三个断裂的阿拉米斯标本中进行测量，其颊侧范围在 1.1 毫米到 1.2 毫米不等，测量位置位于或接近未磨损的牙尖顶点到牙釉质 – 牙本质结合处垂直位置。这些数值与我们得到的黑猩猩断裂的臼齿（$n = 22$；M_1 到 M_3）牙釉质厚度的最高范围相当。相应的南方古猿阿法种釉质厚度范围是从 1.4 毫米到 2.0 毫米（$n = 5$）。在一个例子中（ARA-VP-1/128 的第三臼齿），能估计阿拉米斯样本颊侧下原尖面的牙釉质辐射厚度相对于牙颈宽度的比值。这一牙釉质厚度比值的比较，提示南方古猿始祖种的特点介于黑猩猩和南方古猿阿法种 / 非洲种 / 早期人属之间。

在颊齿的大小方面，可用的阿拉米斯样本的分布范围包括比已知的同源的南方古猿阿法种小的标本（已知的两颗 M_1 牙齿都比平均值小了 3 个标准差单位以上）。在可以确定颊齿大小的 7 个阿拉米斯标本中，所有牙冠大小都比南方古猿阿法种的平均值还小。我们得到的有限的形态学和测量数据可认为它是比南方古猿阿法种的颊齿小得多的单一物种。

ARA-VP-1/128 的下颌颊齿列可以通过齿间接触面得以重建（图 3*a*~*c*）。重建结果表明该标本 C 到 M_2 的齿列在颊侧稍微凹陷些，正如在现代猿和化石猿以及一些南方古猿阿法种标本中看到的情况一样。P_3 轴线没有大部分猿类中的那么倾斜。犬齿的位置刚好与颊齿列的近中 – 远中轴线一致，而不是像大部分南方古猿阿法种那样位于颊齿轴线的近中舌侧。这是一种与现代和中新世的大型猿类相似的更加原始的排列方式，可能提示该个体下颌犬齿近中部在功能上没有并入门牙系统，这与南方古猿阿法种不同 [17]。

Cranial Description

The ARA-VP-1/125 and -1/500 specimens represent adult temporal and occipital regions (Fig. 1*b*). Both are smaller than their *A. afarensis* counterparts, but no female temporal is known for that species. The Aramis cranial fossils evince a strikingly chimpanzee-like morphology that includes marked pneumatization of the temporal squama which even invades the root of the zygoma. The occipital condyle is small, measuring 16 × 7.5 mm. The anterior border of the foramen magnum (basion) is intersected by a bicarotid chord connecting the centres of right and carotid foramina, and the endocranial opening of the hypoglossal canal is placed more anteriorly relative to the internal auditory meatus than in great apes. This condition, as in other fossil hominid taxa, reflects a shortened basioccipital component of the cranial base relative to modern African ape crania. The temporomandibular joint is very flat, with virtually no articular eminence and weak inferior projection of the entoglenoid process. The tympanic is tubular, bounded anteriorly and posteriorly by deep furrows, and the tube extends to the lateral edge of the postglenoid process in one individual and beyond it in the second. The mastoid process is a blunt eminence rather than the inflated, inflected pyramidal structure diagnostic of the chimpanzee. The digastric groove is distinctly deeper than in the chimpanzee.

Postcranial Description

The ARA-VP-7/2 specimen (Fig. l*c*) is a rare association of all three bones from the left arm of a single individual. In size the specimen indicates a hominid larger than the A.L. 288-1 *A. afarensis* from Hadar and smaller than other individuals of this species. Fracture of the specimen currently precludes length estimates for the three elements, but the humeral head is approximately 30% larger than the smallest (A.L. 288-1) *A. afarensis* specimen (breadth: Aramis = 34.6, A.L. 288-1 = 27.0; height: Aramis = 36.5, A.L. 288-1 = 28.1). The arm displays a mosaic of characters usually attributed to hominids and/or great apes. From proximal to distal, probable derived characters shared with other hominids include an elliptical humeral head; a blunt, proximally extended ulnar olecranon process surmounting a straight dorsal upper shaft profile; an anteriorly oriented trochlear notch; and, an anteriorly facing ulnar brachialis insertion. The specimen also shows a host of characters usually associated with modern apes, including a strong angulation of the distal radial articular surface due to a large styloid process, a strong lateral trochlear ridge on the distal humerus (also seen in some *A. afarensis*), and an elongate, superoposteriorly extended lateral humeral epicondyle. The Aramis arm diverges from the African ape condition in other features. The proximal humerus lacks the deep, tunnel-like bicipital groove often seen on African apes. Further studies will unravel the functional and phylogenetic significance (polarities) of these and other postcranial characters.

头骨描述

ARA-VP-1/125 和 ARA-VP-1/500 标本代表了成年个体的颞骨和枕骨区（图 1*b*）。二者相应地都比南方古猿阿法种要小，但是该物种没有发现任何雌性颞骨。阿拉米斯的头盖骨化石表明其具有明显似黑猩猩的形态，这些形态包括颞骨鳞部明显的气腔，该气腔甚至侵入到了颧骨根部。枕（骨）髁小，测量值为 16 毫米 × 7.5 毫米。枕骨大孔（颅底点）的前缘是颈动脉索相互交叉处，后者衔接着右边颈动脉孔，舌下神经管的颅内开口相对于中耳道的位置比大型猿类的更靠前。这种情况与其他原始人类类群化石一样，反映出相对于现代非洲猿类头盖骨，其颅底枕骨底部部分有所缩短。颞下颌关节很平坦，实际上并没有关节隆起，下颌窝内突出的下部突不明显。鼓室呈管状，前后以深沟为界，一个个体中的管状结构延伸至下颌窝后突出的侧边缘，另一个个体的则延伸至边缘之外。乳突是一个钝的突出而非黑猩猩特征性的膨胀的、弯曲的锥形结构。二腹肌沟比黑猩猩的要深得多。

颅后部分的描述

ARA-VP-7/2 标本（图 1*c*）是稀有的来自同一个体左臂的三块骨头。从大小上看，该标本比哈达尔 A.L. 288-1 南方古猿阿法种更大、而比该物种的其他个体小。由于标本的破碎，目前无法对三件标本的长度进行测量，但是肱骨头部比最小的（A.L. 288-1）南方古猿阿法种要大将近 30%（宽度：阿拉米斯 = 34.6，A.L. 288-1 = 27.0；高度：阿拉米斯 = 36.5，A.L. 288-1 = 28.1）。上臂显示出人类和（或）大型猿类的镶嵌特征。从近端到远端，可能与其他原始人类存在共同的衍生特征，包括椭圆形的肱骨头，一个钝的、近端延伸的尺骨鹰嘴（指肘部的骨性隆起）越过一个直的背侧骨干，向前的滑车切迹，朝前的尺骨肱肌附着部位。同时也显示出许多现代猿类通常具有的特征，包括因大的茎突而产生的尺骨远端关节面较大的角度，肱骨远端强壮的侧滑车脊（有些南方古猿阿法种中也见到过这种特征）和一个延伸的、向后上扩展的肱骨侧上髁。阿拉米斯标本的上肢与非洲猿在其他特征上有所区别。其近端肱骨缺少深的、管状的二头肌沟，而在非洲猿类中经常见到这种特征。进一步的研究将阐明这些特征和其他颅后特征的功能意义和系统发育意义（极性）。

Comparisons and Remarks

The pre-5 Myr record of hominid evolution is sparse. Although the Lothagam fragment has been attributed to *A.* cf. *afarensis*[18-20] this assignment was made mostly on the basis of primitive characters and in the absence of associated cranial, anterior dental or postcranial remains. Hominid remains from the period between 4 and 5 Myr are also few and poor, comprising a proximal humerus and jaw fragment from Baringo[16,19-22], and a distal humerus from Kanapoi of more uncertain age. These and the slightly younger Kubi Algi[23,24] and Fejej[25] specimens have all been attributed to *A. afarensis*.

To our knowledge, no fossils predating 4 Myr have been identified as representing taxa other than *Australopithecus afarensis* and *Australopithecus africanus*[16,19,20,23-26]. Assignment of the limited available > 4 Myr sample to *A. afarensis* was warranted for the comparatively undiagnostic Lothagam, Baringo, Kanapoi and Tabarin specimens from Kenya[16]. The discovery of more complete, more diagnostic specimens at Aramis allows a recognition of characters which distinguish them at (minimally) the species level from Hadar, Maka and Laetoli hominid fossils. The limited preserved morphology in the Lothagam, Tabarin and Baringo specimens broadly matches both the Aramis sample and *A. afarensis*. The discovery of the Aramis hominids demonstrates, however, that some of the suggested differences between Lothagam and *A. afarensis* (for example, enamel thickness[15]) are likely to be substantiated. However, the preserved anatomy of these Kenyan specimens may well reflect primitive character states for the basal hominid (and perhaps ancestral hominoids). Nothing available for these Kenyan specimens validates inclusion in the new Ethiopian taxon before the recovery of more diagnostic body parts.

We note that Ferguson, referring to casts and literature, has invented a plethora of new names for early African hominids (for instance, he divides *A. afarensis* into three species; an alleged dryopithecine ape[27], a diminutive early human[28], and a subspecies of *A. africanus*[29]). His invalid naming of the A.L. 288 specimen as "Homo antiquus" (in which he includes KNM-ER 1813)[28] was followed by his 1989 placement of the Baringo Tabarin specimen (which he incorrectly identified as "KNM-ER TI 13150") into a subspecies ("praegens") of that species[30]. Because of these problems, because Ferguson's diagnosis of that specimen did not differentiate it from *A. afarensis*, and because it lacks any characters that differentiate it from the latter species or unequivocally link it to the Aramis species[16], we consider Ferguson's subspecific nomen "*praegens*" to be a *nomen dubium* and propose that it be suppressed even in the event that the Tabarin specimen be shown conspecific with the Aramis series.

The 1992/93 Aramis hominids share a wide array of traits with *A. afarensis* but also depart anatomically from this species in lacking some of the key traits it possesses and which are shared exclusively among all later hominids. Because of relationships indicated by molecular studies, and because terminal Miocene to Pleistocene fossil African apes are

比较与总结

500 万年以前的人类进化记录只是零星的一点。尽管洛萨加姆碎块被归到南方古猿阿法种[18-20]，但是这主要是根据原始特征，而缺少相关的头盖骨、前牙或颅后遗骸标本。距今 400 万至 500 万年前的原始人类遗迹也很少，保存状况很差，包括来自巴林戈的一块近端肱骨和一块颌骨碎块[16,19-22]以及来自卡纳波伊的年代更加不确定的一块远端肱骨。这些标本以及稍微年轻点的库比阿尔及[23,24]和菲耶济[25]标本都被划分到了南方古猿阿法种中。

据我们所知，400 万年前的化石还没有被鉴定为代表南方古猿阿法种和南方古猿非洲种之外的种类[16,19,20,23-26]。有限可得的 400 万年以上的标本划分到南方古猿阿法种的是来自肯尼亚的相对无鉴别特征的洛萨加姆、巴林戈、卡纳波伊和塔巴林的标本[16]。在阿拉米斯发现的更加完整的、更加具有鉴别特征的标本能够在（至少）物种水平上与来自哈达尔、马卡和莱托里的原始人类化石区分开来。洛萨加姆、塔巴林和巴林戈的标本中保存下来的有限的形态学特征与阿拉米斯样本和南方古猿阿法种都能够匹配上。但是，阿拉米斯原始人类的发现表明洛萨加姆和南方古猿非洲种之间有些潜在的差异（例如，釉质厚度[15]）有可能得到证实。然而，这些肯尼亚标本的解剖学特征可能很好地反映了基底原始人类（可能是人科动物祖先）的原始特征状态。在发现更多具有鉴别性的身体骨骼标本之前，还没有证据可以证实这些肯尼亚标本属于埃塞俄比亚新类群。

我们注意到弗格森参考模型和文献发明了对早期非洲原始人类新命名的新模型（例如，他将南方古猿阿法种划分成了三个物种：所谓的森林古猿[27]、早期小型人类[28]和南方古猿非洲种的亚种[29]）。他将 AL 288 标本命名为“古人”（其中将 KNM-ER 1813 也包含了进去）[28]，这是一个无效的命名，后来他在 1989 年把巴林戈塔巴林标本（他误认为是“KNM-ER TI 13150”）放进了该物种的一个亚种（“普拉根”）[30]中。但是弗格森对那个标本的判断没有将其与南方古猿阿法种区别开来，并且它缺少可以将其与后一物种区分开来的任何特征或缺少将其与阿拉米斯物种清晰关联起来的特征[16]，基于这些问题，我们认为弗格森的亚种名字“普拉根”是一个疑难学名，并且提议，即使在塔巴林标本显示出与阿拉米斯系列属于同一物种的情况下，这个名字也应该被禁止使用。

1992/1993 的阿拉米斯原始人类与南方古猿阿法种具有许多相同的特征，但是在解剖学上也有所区别，例如缺少一些后者所具有的、与所有后来的原始人类共有的关键特征。由于分子学研究所暗示的关系，以及因为中新世末期到更新世时期的

unknown, comparison of the Aramis hominids and modern African apes is warranted. The Aramis remains evince significant cranial, dental and postcranial similarities to the chimpanzee condition, but some or all of these features may be primitive retentions. Only further discoveries and comparisons may elucidate which features actually define the chimp-human and/or African ape-human clades. Meanwhile, the modern African apes arc distinct in many dental features from both Aramis and middle to late Miocene hominoids, and thus probably do not represent the ancestral condition[8,9]. At the same time, the relatively thin Aramis molar enamel suggests that a simple "hard object feeder" model[7] is likely to be inaccurate for the ancestral African ape/hominid stock.

We have taken a conservative position here regarding placement of the Aramis fossils at the family and genus levels. The major anatomical/behavioural threshold between known great apes and Hominidae is widely recognized to be bipedality and its anatomical correlates. The two derived craniodental characters shared among all hominids are anterior placement of the occipital condyle/foramen magnum and a more incisiform canine with reduced sexual dimorphism. Acquisition of these states at Aramis may correlate with bipedality[31,32] although this remains to be demonstrated. The postcranial evidence available for *A. ramidus* is not definitive on the issue of locomotor pattern.

The anticipated recovery at Aramis of additional postcranial remains, particularly those of the lower limb and hip, may result in reassessment of these fossils at the genus and family level. Meanwhile, characters such as the modified C/P3 complex, an anterior foramen magnum, and proximal ulnar morphology (shared with later *Australopithecus* species) suggest that the Aramis fossils belong to the hominid clade. Similarity to the *A. afarensis* hypodigm warrants the inclusion of the Aramis fossils in the genus *Australopithecus*. At the same time, *A. ramidus* is the most apelike hominid ancestor known, and its remains suggest that modern apes are probably derived in many characters relative to the last common ancestor of apes and humans. More work at Aramis should further elucidate the sexual dimorphism, locomotion, diet and habitat of this species. The fossils already available indicate that a long-sought link in the evolutionary chain of species between humans and their African ape ancestors occupied the Horn of Africa during the early Pliocene.

(**371**, 306-312; 1994)

Tim D. White[*], **Gen Suwa**[†] **& Berhane Asfaw**[‡]

[*] Laboratory for Human Evolutionary Studies, University of California, Berkeley, California 94720, USA

[†] Department of Anthropology, University of Tokyo, Bunkyo-Ku, Hongo, Tokyo 113, Japan

[‡] Ethiopian Ministry of Culture and Sports Affairs, Paleoanthropology Laboratory, PO Box 5717, Addis Ababa, Ethiopia

Received 10 June; accepted 17 August 1994.

References:

1. Johanson, D. C. & White, T. D. *Science* **202**, 321-330 (1979).
2. Sarich, V. M. & Wilson, A. C. *Science* **158**, 1200-1203 (1967).
3. Hasegawa, M., Kishino, H. & Yano, T. *J. Molec. Evol.* **22**, 160-174 (1985).

非洲猿类化石都是未知的，所以其与阿拉米斯原始人类和现代非洲猿类的比较能够得以保证。阿拉米斯标本表明其与黑猩猩在头骨、牙齿和颅后骨骼方面具有显著的相似性，但是这些特征中有些或者全部可能都是原始特征的保留。只有进一步的发现和比较才可能解释清楚哪些特征可以最终确定黑猩猩－人类和（或）非洲猿－人类进化枝。同时，现代非洲猿在牙齿的许多特征上都与阿拉米斯样本、中新世中期到晚期的人科动物不同，因此可能并不代表祖先的情况[8,9]。与此同时，相对薄的阿拉米斯臼齿釉质提示：对非洲猿／原始人类祖先库来说，一种简单的“坚硬食物”模型[7]可能是不准确的。

对于在科和属水平将阿拉米斯化石置于何种地位，我们在这里采取了一个保守态度。已知的大型猿类和人科之间的主要解剖学／行为学界线被广泛认为是两足行走及其相关解剖学特征。所有原始人类共有的这两种衍生的颅骨、牙齿特征包括枕（骨）髁／枕骨大孔的前置以及具有减弱的两性异形的门齿化的犬齿。尽管还没有得到证实，不过阿拉米斯标本中这些性状的获得可能与两足行走有关[31,32]。可得到的南方古猿始祖种的颅后证据还不能确定行动模式这一问题。

我们期望在阿拉米斯发现其余颅后残骸，尤其是下肢和髋部遗骸，这样就可能在属和科的水平上重新对这些化石进行评价。同时，有些特征，例如改进的C/P3复合体、前置枕骨大孔和近侧尺骨形态学（后来的南方古猿样本也有此特征）提示阿拉米斯化石属于原始人类进化枝。与南方古猿阿法种的相似性证明了阿拉米斯化石是属于南方古猿属的。与此同时，南方古猿始祖种是已知的最像猿类的原始人类祖先，其遗迹提示现代猿类可能衍生了许多与猿类和人类的最终共同祖先相关的特征。在阿拉米斯进行的其他工作应该能进一步阐明该物种的两性异形、行动方式、食性和栖息环境等问题。已经得到的这些化石暗示：长久以来一直在寻找的人类与其非洲猿类祖先之间的进化链物种在早上新世时期生活在非洲好望角。

（刘皓芳 翻译；赵凌霞 审稿）

4. Pilbeam, D. R. *Am. Anthrop.* **88**, 295-312 (1986).
5. Sarich, V. M., Schmid, C. W. & Marks, J. *Cladistics* **5**, 3-32 (1989).
6. Andrews, P. & Martin, L. *Phil. Trans. R. Soc. Lond.* **334**, 199-209 (1991).
7. Andrews, P. *Nature* **360**, 641-646 (1992).
8. Begun, D. R. *Science* **257**, 1929-1933 (1992).
9. Dean, D. & Delson, E. *Nature* **359**, 676-677 (1992).
10. WoldeGabriel, G. *et al. Nature* **371**, 330-333 (1994).
11. Dart, R. A. *Fol. Anat. Jap.* **12**, 207-221 (1934).
12. Robinson, J. T. *Transv. Mus. Mem.* **9**, 1-179 (1956).
13. Le Gros Clark, W. E. *Q. J. Geol. Soc.* **105**, 225-264 (1950).
14. Walker, A. C. *Am. J. Phys. Anthrop.* **65**, 47-60 (1984).
15. White, T. D. *Anthropos* (Brno) **23**, 79-90 (1986).
16. Ward, S. C. & Hill, A. *Am. J. Phys. Anthrop.* **72**, 21-37 (1987).
17. White, T. D. *et al. Nature* **366**, 261-265 (1993).
18. Kramer, A. *Am. J. Phys. Anthrop.* **70**, 457-473 (1986).
19. Hill, A. & Ward, S. *Yearb. Phys. Anthropol.* **31**, 49-83 (1988).
20. Hill, A., Ward, S. & Brown, B. *J. Hum. Evol.* **22**, 439-451 (1992).
21. Hill, A. *Nature* **315**, 222-224 (1985).
22. Pickford, M., Johanson, D. C., Lovejoy, C. O., White, T. D. & Aronson, J. L. *Am. J. Phys. Anthrop.* **60**, 337-346 (1983).
23. Coffing, K., Feibel, C., Leakey, M. & Walker, A. *Am. J. Phys. Anthrop.* **93**, 55-65 (1994).
24. Heinrich, R. E., Rose, M. D., Leakey, R. E. & Walker, A. C. *Am. J. Phys. Anthrop.* **92**, 139- 148 (1993).
25. Fleagle, J. G., Rasmussen, D. T., Yirga, S., Bown, T. M. & Grine, F. E. *J. Hum. Evol.* **21**,145- 152 (1991).
26. Patterson, B., Beherensmeyer, A. K. & Sill, W. D. *Nature* **226**, 918-921 (1970).
27. Ferguson, W. W. *Primates* **24**, 397-409 (1983).
28. Ferguson, W. W. *Primates* **25**, 519-529 (1984).
29. Ferguson, W. W. *Primates* **28**, 258-265 (1987).
30. Ferguson, W. W. *Primates* **30**, 383-387 (1989).
31. Lovejoy, C. O. *Science* **211**, 341-350 (1981).
32. Lovejoy, C. O, in *The Origin and Evolution of Humans and Humanness* (ed. Rasmussen, D. T.) 1-28 (Jones and Bartlett, Boston, 1993).

Acknowledgements. We thank the Anthropology and Archaeometry programmes of the National Science Foundation, the University of California Collaborative Research Program of the Institute of Geophysics and Planetary Physics at Los Alamos National Laboratory, and the National Geographic Society for funding. This research was made possible by the Centre for Research and Conservation of the Cultural Heritage and the National Museum of Ethiopia in the Ethiopian Ministry of Culture and Sports Affairs, the Ethiopian Embassy to the USA, the Afar people, the American Embassy in Addis Ababa, and the Cleveland Museum of Natural History. Special thanks to L-S. Temamo, H. Ali-Mirah, M. Tahiro, Rep. N. Pelosi and M. Starr. K. Coffing, A. C. Walker and D. Begun showed us casts of East Turkana hominids and Miocene hominoids; Lyman Jellema facilitated comparative research on the Hamman-Todd collection. Thanks to Keiko Fujimaki for scientific illustrations. O. Lovejoy and B. Latimer provided assistance in postcranial interpretation and F. C. Howell provided comments, E. Kanazawa, H. Yamada and H. Ishida provided access to equipment and comparative collections in their care. Thanks to our colleagues in Middle Awash project geology and palaeontology for elucidating the environmental and chronostratigraphic placement of this new hominid species. A. Ademassu, A. Almquist, A. Asfaw, M. Asnake, Y. Beyene, J. D. Clark, M. Fisseha, A. Getty, Y. H.-Selassie, B. Latimer, K. Schick, S. Simpson, M. Tesfaye and S. Teshome contributed to the fieldwork. B. Wood, P. Andrews, E. Delson and F. C. Howell provided comments on the manuscript, Thanks to J. Desmond Clark for inviting us to participate in the Middle Awash research and for inspiring us in the search for human origins.

DNA Fingerprinting Dispute Laid to Rest

E. S. Lander and B. Budowle

Editor's Note

After its introduction to the US courts in 1988, DNA fingerprinting was criticized for its poorly defined procedures and inadequate scientific standards. But after hundreds of scientific papers, three sets of guidelines and 150 court decisions, its bad reputation remained difficult to shed, despite having become methodologically rigorous. Here Eric S. Lander, an early critic of the lack of standards, and Bruce Budowle, one of the principal architects of the FBI's DNA typing programme, declare that "the DNA fingerprinting wars are over." There is no scientific reason to doubt the accuracy of forensic DNA typing results, and the major hurdle now, they say, is to persuade the public that the DNA fingerprinting controversy has been resolved.

Two principals in the once-raging debate over forensic DNA typing conclude that the scientific issues have all been resolved.

THE US public, usually indifferent to matters scientific, has suddenly become obsessed with DNA. Nightly newscasts routinely refer to the polymerase chain reaction (PCR) and even the tabloids offer commentary on restriction fragment length polymorphisms (RFLPs). The new-found fascination with nucleic acids does not stem from recent breakthroughs in genetic screening for breast cancer susceptibility or progress in gene therapy—developments which will indeed affect the lives of millions. Rather, it focuses on the murder case against the former US football star, O. J. Simpson.

The Los Angeles trial, starting in November and to be broadcast live by several major television networks, will probably feature the most detailed course in molecular genetics ever taught to the US people. This bold experiment in public education should, in principle, be a cause for rejoicing among scientists. The catch is that the syllabus is being prepared by attorneys whose primary roles are as adversaries; the likely result is confusion. Already, the news weeklies are preparing the ground with warnings that DNA fingerprinting remains "controversial", being plagued by major unresolved scientific issues.

Forensic DNA typing certainly did provoke controversy soon after it was introduced into US courts in 1988. The technology itself represents perhaps the greatest advance in forensic science since the development of ordinary fingerprints in 1892, and is soundly rooted in molecular biology. The problem, however, stemmed from the manner of its introduction. Pioneered by biotech start-up companies with good intentions but no track

DNA 指纹分析争议的平息

兰德，布德沃

编者按

自 1988 年 DNA 指纹分析引入美国法庭以来，就因其缺少明确的步骤和充分的科学标准而饱受批评。虽然已有数百篇相关科学文章、3 组指南和 150 例法院裁决，在方法学上已非常缜密，但这项技术依然难以摆脱坏名声。在这篇文章中埃里克·兰德（早期关注标准缺失的评论家）和布鲁斯·布德沃（美国联邦调查局 DNA 分型项目的主要设计师之一）声称“DNA 指纹战争已经结束了”。他们说，没有科学理由怀疑法医 DNA 分型结果的准确性，现在主要的障碍是说服公众相信 DNA 指纹争议已经解决了。

法医 DNA 分型曾一度引发热议，争论的双方最终达成一致：已经解决所有的科学问题。

通常来说，美国民众对科学问题是漠不关心的，然而，近来他们突然对 DNA 产生了兴趣。晚间新闻广播常常会提及聚合酶链式反应（PCR），甚至连小报也开始出现关于限制性片段长度多态性（RFLP）的评论。这种新发现的对于核酸的痴迷，并非源于乳腺癌易感性遗传筛查方面的新突破，或是可以影响上百万人生命的基因治疗方面的进展，而是源于对美国前橄榄球明星辛普森谋杀案的关注。

多家主要的电视网将对始于 11 月份的洛杉矶审判进行现场直播，这一审判可能成为美国民众学习过的最为详细的分子遗传学课程。一般来说，科学家们应该会对这种公共教育方面的大胆尝试感到欣喜。然而这其中却隐藏着一个不利因素，即教学大纲由辩护律师准备，而律师的首要角色就是反对者，因此这一尝试的结果是很不确定的。在审判之前，新闻周刊就已经预留出了带有警告标语“DNA 指纹分析受困于主要的未解决的科学问题，仍然‘存在争议’”的版面。

法医 DNA 分型自 1988 年被引入美国法院后不久便引发了巨大的争议。这项完全基于分子生物学的技术，可能代表了自 1892 年普通指纹技术开始发展以来法庭科学的最大进步。但是，它的引进方式出现了问题。该技术由一些新创的生物技术公司开创，它们拥有不错的目的，然而在法庭科学方面却毫无业绩，并且在早期的几

record in forensic science, DNA typing was marred by several early cases involving poorly defined procedures and interpretation[1]. Standards were lacking for such crucial issues as: declaring a match between patterns; interpreting artefacts on gels; choosing probes; assembling databases; and computing genotype frequencies. There is broad agreement today that many of these early practices were unacceptable, and some indefensible. For its part, the US Federal Bureau of Investigation (FBI) moved much more deliberately in developing procedures, sought public comment and opted for conservative procedures.

As a result of these growing pains, forensic DNA typing was subjected to intense debate and scrutiny. When it first burst on the scene, the supporting scientific literature consisted of a mere handful of papers. By the middle of this year, there had been more than 400 scientific papers, 100 scientific conferences, 3 sets of guidelines from the Technical Working Group on DNA Analysis Methods (TWGDAM), 150 court decisions and, importantly, a 3-year study by a National Research Council (NRC) committee released in 1992 (ref. 2). In the light of this extraordinary scrutiny, it seems appropriate to ask whether there remains any important unresolved issue about DNA typing, or whether it is time to declare the great DNA fingerprinting controversy over.

As co-authors, we can address these questions in an even-handed manner. B.B. was one of the principal architects of the FBI's DNA typing programme, whereas E.S.L. was an early and vigorous critic of the lack of scientific standards, and served on the NRC committee. In a world of soundbites, we are often pegged as, respectively, a "proponent" and an "opponent" of DNA typing. Such labels greatly oversimply matters, but it is fair to say that we represent the range of scientific debate.

We recently discussed the current state of DNA typing, and could identify no remaining problem that should prevent the full use of DNA evidence in any court. What controversy existed seems to have been fully resolved by the NRC report, the TWGDAM guidelines and the extensive scientific literature. The DNA fingerprinting wars are over.

Our goal is to correct the lingering impression to the contrary. Our analysis below represents our unanimous opinions (apart from specific comments about the workings and intent of the NRC committee itself, which necessarily are based on E.S.L.'s recollection). We focus on the subject most often said to remain problematical: population genetics. Our main thesis is that the academic debate that continues to swirl about population genetic issues is rooted in a misunderstanding of the NRC report and is, in any case, of no practical consequence to the courts. We also touch on how the legal and scientific community should cope with the continuing evolution of DNA typing technology. In particular, we question whether a steady succession of *ad hoc* committees, however distinguished, is a wise solution.

例案件中，不明确的流程及解释[1]对 DNA 分型技术造成了一定的损害。缺乏标准的关键问题包括：如何判断图谱间相互匹配；如何解释凝胶上的人为假象；如何选择探针；如何整合数据库；如何计算基因型频率。如今，一个被普遍认可的观点是：许多的早期实践是不被认可的，并且有一些是站不住脚的。美国联邦调查局(FBI)方面为优化流程做了大量细致工作，并征求公众意见，最后选择了保守的程序。

作为发展初期必经的困难，法医 DNA 分型经受了激烈的争论和严格的检验。在它第一次出现在人们的面前时，支持性的科学文献是非常少的。到今年年中的时候，已经有 400 多篇科学论文、100 次科学会议、3 组来自 DNA 分析方法技术工作组(TWGDAM)的指南、150 例法院裁决，很重要的是，国家研究委员会(NRC)也在 1992 年发表了一篇用时 3 年的研究结果(参考文献 2)。在这异乎寻常的仔细检查之下，人们对于 DNA 分型是否存在其他未解决的重要问题，或者是否已经到了宣布结束 DNA 指纹争议的时刻的询问，似乎也都是合情合理的。

作为文章的共同作者，我们可以以公平的态度回答这些问题。布鲁斯·布德沃曾是美国联邦调查局 DNA 分型项目的主要设计师之一，而埃里克·兰德早期是一名活跃的关注科学标准缺失的批评家，并供职于国家研究委员会。在一个充满话语片段的世界里，人们通常认为我们分别是 DNA 分型的“支持者”和“反对者”。这样的标签极大地简化了存在的问题，但公平地说，我们确实代表了科学争论的幅度。

最近，我们讨论了 DNA 分型的现状，可以确定没有能阻止 DNA 证据全面应用于法庭的剩余问题。NRC 报告、TWGDAM 指南以及大量的科学文献似乎已经完全解决了存在的争议。这场 DNA 指纹分析战争结束了。

我们的目的是纠正挥之不去的反面印象。下文中的分析阐明了我们一致的观点(除了对于 NRC 委员会的工作内容和目的方面的具体评论，这部分内容必然是基于埃里克·兰德的回忆)。我们所关注的主题是群体遗传学，这也是最常被认为存在问题的方面。我们的主要论点是，围绕群体遗传学问题持续的学术争论是源于对 NRC 报告的错误理解，并且这一争论对于法院来说是没有实际意义的。我们也谈及了法律界和科学界应该如何应对持续发展的 DNA 分型技术。尤其是，我们考虑设立一个连续稳定且杰出的特设委员会是否是一个明智的解决方案。

Comparing autoradiographs from DNA samples at Cellmark Diagnostics.

Laboratory Practices

The initial outcry over DNA typing standards concerned laboratory problems: poorly defined rules for declaring a match; experiments without controls; contaminated probes and samples; and sloppy interpretation of autoradiograms[1]. Although there is no evidence that these technical failings resulted in any wrongful convictions, the lack of standards seemed to be a recipe for trouble. To address these problems, the NRC committee enunciated conservative standards for each laboratory step, based on more than a decade of experience with human DNA analysis. TWGDAM also developed guidelines along similar lines. Today, there is no doubt about the correct laboratory protocols to ensure reliable DNA typing results. Since the NRC report, US courts have unanimously accepted the technical reliability of DNA evidence, both in principle and in practice.

The NRC committee also highlighted the importance of laboratory accreditation, rigorous quality assurance and quality control (QA/QC) programs, and external blind proficiency tests (tests administered by persons outside the testing lab itself). The importance of these practices has been universally acknowledged, and most forensic labs follow TWGDAM's voluntary quality-control guidelines.

Population Genetics

The controversy over population genetics began as a secondary issue. If DNA analysis reveals that two samples match at the loci tested, the final step is to estimate the frequency of the shared genotype in the general population, which indicates the probability that a randomly chosen person would carry this genotype. Such estimates depend on surveys of

在塞尔马克诊断实验室对比 DNA 样品的放射自显影图像。

实验室规范

起初对 DNA 分型标准的强烈抗议是与实验室问题有关的：断定结果相互匹配的标准不明确；实验缺少对照；探针和样品受到污染；以及对于放射自显影图的草率解释[1]。虽然没有证据表明这些技术缺陷会导致错误定罪，但是标准的缺失很可能导致麻烦的产生。为了解决这些问题，NRC 委员会根据十多年对人类 DNA 进行分析的经验，明确阐述了每一项实验室步骤的保守性标准。TWGDAM 也制定了类似的指南。时至今日，这一能确保 DNA 分型结果可靠的正确实验室方案是毋庸置疑的。自 NRC 报告发布以来，美国法院已经在原则上和实践上一致承认 DNA 证据的技术可靠性。

NRC 委员会也强调了实验室认证、严格的质量保证和质量控制（QA/QC）程序，以及外盲水平测试（测试由该测试实验室之外的人执行）的重要性。这些操作的重要性已被广泛认可，并且多数法医实验室遵循了 TWGDAM 的非强制性质量控制指南。

群体遗传学

群体遗传学最初是作为一个次要问题而受争议的。如果 DNA 分析揭示两个样品在所检测的基因座处相互匹配，那么最后的一个步骤就是去评估该共享基因型在人群中的频率，这一结果将表明一个随机选择的个体携带该基因型的概率。然而，

the appropriate population.

In some early cases, the rarity of genotype frequencies was greatly overstated owing to a technical error: the calculations were based on overoptimistic assumptions about the precision with which genotypes could be measured. One commercial lab, for example, reported the astronomical frequency of 1 in 738,000,000,000,000, based on a four-locus match[1]. The NRC committee easily rectified these problems by requiring consistency between the measurement precision used for forensic analysis and for population genetic estimates (a practice that the FBI, in fact, had long followed).

IMAGE
UNAVAILABLE
FOR COPYRIGHT
REASONS

Opting for conservative procedures—FBI serology laboratory in Washington, DC.

A subtler but more challenging issue emerged in later cases, concerning the structure of human populations. The "product rule", used by forensic labs to calculate genotype frequencies, assumed that the individual alleles comprising a genotype could be treated as statistically independent, and their frequencies multiplied[2]. However, some population geneticists asserted that the assumption of independence was appropriate for well-mixed populations (technically, those at Hardy–Weinberg equilibrium and linkage equilibrium), but was not necessarily valid for populations with substructure. According to this argument, the frequency of a common Japanese genotype might be underestimated because the product rule ignored the fact that common Japanese alleles tend to occur together in the US Asian population. Moreover, the frequency of genotypes arising from mixed ethnic ancestry might be understated because the product rule was typically applied to separate racial databases (for the Caucasian, Black and Hispanic populations) and thus did not account for the presence of genotypes involving common alleles from different racial groups. The substructure argument became a *cause célèbre*, pitting such luminaries as Lewontin and Hartl[3] against Chakraborty and Kidd[4]. Both sides conceded that substructure could matter in principle, but many doubted that its effect could be significant in practice (see ref. 5).

这样的评估取决于对适宜人群的调查。

在早期的一些案例中，因为一个技术差错，基因型频率中的罕见情况被过分夸大了：计算结果是基于过于乐观的假设，即其准确性足以估测基因型。例如，一个商业性实验室报告了四个基因座相匹配的频率，得出了 1/738,000,000,000,000 这一天文数字级别的结果[1]。NRC 委员会则通过要求法医分析和群众遗传评估（事实上是一项 FBI 已经长期遵循的惯例）的测量准确性相一致，从而轻而易举地纠正了这些问题。

因为版权的原因
图像不可用

FBI 在位于华盛顿的血清学实验室选择保守程序。

在后来的一些案例中，浮现出了一个细微但是更具挑战性的问题，即人群的结构。在用于法医实验室计算基因型频率的“乘法定则”中，假设了包含一个基因型的个体等位基因可以被视为在统计学上是独立的，并且其频率是可以相乘的[2]。然而，一些人口遗传学家坚持认为，独立性的假设对于充分混合的人群而言是合理的（学术上称这些人群处于哈迪–温伯格平衡以及连锁平衡），但对于含有亚结构的人群而言未必成立。根据这一论点，因为乘法定则忽视了常见的日本人等位基因倾向于在美国亚裔人口中同时出现的事实，所以人们可能会低估常见的日本人基因型的频率。此外，由于该乘法定则适用于分离人种的数据库（白种人、黑种人以及西班牙裔美国人），不能解释其他种族的常见等位基因的出现，因此由不同种族混合产生的基因型，其频率可能会被低估。关于亚结构的争论成了轰动一时的事件，杰出人物们也开始相互辩论，就像路翁亭和哈特尔[3]对抗查克拉博蒂和基德[4]一样。争辩的双方都承认亚结构在理论上是能够产生影响的，但是许多人质疑其影响在实际操作中的重要性（见参考文献 5）。

The NRC committee at first attempted to settle the issue on its merits. The members agreed that the product rule was probably near the mark, but were hard pressed to say just how close. The committee considered applying formulas from theoretical population genetics based on the empirical measures of the degree of variation and admixture among and within populations. However, it concluded that there were, at the time, too few hard data about the loci used in forensic typing (most classical genetic surveys concerned protein polymorphisms, likely to be strongly influenced by natural selection) and about the precise structure of the US population. It would be too risky to base a recommendation on assumptions that might subsequently turn out to be faulty.

Thomas Caskey (Baylor College) eventually pointed the way out the quagmire when he asked, out of frustration, whether it was possible to ignore population substructure altogether. Taking up the notion, the NRC committee set out to fashion an extremely conservative rule having the virtue that it made virtually no assumptions.

The Ceiling Principle

The solution turned out to be quite simple. Suppose that the US population is descended from a collection of populations P_1, P_2, ..., P_n, each sufficiently old and well mixed to allow the product rule to be safely applied. Regardless of the population substructure, the multiplication rule requires only a slight modification to yield a strict upper bound on the frequency of any genotype *G*: for each allele in *G*, the allele frequency should be taken to be the maximum over the component subpopulations. In effect, the approach makes the worst-case assumption that the population may contain individuals who, for example, carry a common Caucasian allele at a locus on chromosome 2 and a common Black allele at a locus on chromosome 17. By assuming the worst, one is guaranteed to be conservative. Because it used the maximum frequency in any subpopulation, the method was dubbed the "ceiling principle"[2].

In practice, it is unnecessary to survey every possible subpopulation. The committee concluded that the likely variation in allele frequencies could be reckoned by conducting modest surveys of 100 individuals from each of 10–15 representative subpopulations spanning the range of ethnic groups represented in the United States—such as English, Germans, Italians, Russians, Navahos, Puerto Ricans, Chinese, Japanese, Vietnamese and west Africans. Each allele frequency could then be taken to be the maximum over these subpopulations, although never less than 5%. (The latter provision was designed to deal with unexamined populations. If an allele was rare in the 10–15 subpopulations surveyed, genetic drift was not likely to have caused its frequency to drift much above 5% in other significant subpopulations.) Even in advance of detailed data about ethnic groups, the committee felt that same principle could be applied to the available racial databases (Caucasian, Black, Hispanic, Asian), although it recommended a 10% floor on allele frequencies to reflect the greater uncertainty about subpopulation variation: this slightly amended form was called the "interim ceiling principle". (The choices of 5% and 10%

起初，NRC 委员会尝试按照实际情况解决这一问题。委员会成员一致认为该乘法定则基本接近标准，但是很难说究竟有多么接近。委员会考虑过使用理论性群体遗传学公式，这些公式是基于对不同群体间和同一群体内的变异度及混合度的经验测度的。然而，结果表明当时鲜有与法医分型相关的基因座（多数传统遗传学调查与蛋白质多态性有关，很有可能受到自然选择的显著影响）和美国人群精确结构的硬数据。将一项建议建立在可能随即会被证实有误的假设之上，是一件非常冒险的事情。

托马斯·卡斯基（来自贝勒医学院）最终给出了走出这个困境的方法，即是否有可能完全忽视人群亚结构。NRC 委员会采纳了这个提议，并开始设计一个非常保守的定则，这个定则的优点是几乎不存在假定条件。

上限原则

结果证明这个解决方法是十分简单的。假设美国人群是由许多不同种群的后代组成（P_1，P_2，…，P_n），且每一个种群都足够悠久且充分融合，以至于可以放心运用乘法定则。不考虑人群亚结构，只需要稍稍修改乘法法则就可以获得任何基因型 G 频率的严格上界：对于基因型 G 中的每个等位基因而言，其等位基因频率应被视为各亚人群组分中的最大值。实际上，这一方法假设了一个最糟糕的情况，即人群中可能存在 2 号染色体和 17 号染色体基因座上分别携带有一个常见的白种人等位基因和一个常见的黑种人等位基因的个体。对这种情况的假设保证了该方法的保守性。由于在任意亚人群中都使用了最大频率，因此该方法被命名为“上限原则”[2]。

事实上，调查每一个可能的亚人群是不必要的。美国种群中有 10~15 个具有代表性的亚人群，可以通过对来自每一个亚人群的 100 个个体进行合适的调查，从而估算在等位基因频率方面可能存在的变化，这些亚人群包括英国人、德国人、意大利人、俄罗斯人、纳瓦霍人、波多黎各人、中国人、日本人、越南人和西非人等。尽管这些等位基因频率的数值从未低于 5%，但是仍可将每一个等位基因频率视为这些亚人群的最大值。（其中，后一项规定是为处理未被调查的人群而设计的。如果某个等位基因在这 10~15 个被调查的亚人群中非常少见，那么不可能是遗传漂变导致其频率在其他重要亚人群中发生了明显超过 5% 的漂变。）甚至在拥有详细的种族群体数据之前，委员会就认为相同的原理可以应用于可获得的种族数据库（白种人、黑种人、西班牙裔美国人以及亚洲人），尽管等位基因频率上 10% 的最低值被建议用以反映更大的亚人群变化的不确定性：这一微小修正被称为“临时上限原则”。（对

were based on the quantitative effect of genetic drift on the match odds—that is, on the reciprocal of the allele frequency—although none of this reasoning survived into the text of the final report.) The practical effect of these rules was to limit the contribution of any single locus to a factor of 50:1 odds based only on aggregate data for racial classifications and 200:1 odds based on more detailed ethnic surveys.

The ceiling principle was unabashedly conservative. It gave the benefit of every conceivable doubt to the defendant, so that it could withstand attacks from the most stubborn and creative attorneys. Some of the statistical power was sacrificed to neutralize all possible worries about population substructure.

The committee was comfortable with such a lop-sided approach, because even these extreme assumptions did not undermine the practical use of DNA fingerprinting. A four-locus match performed by forensic labs could still provide odds of 6,250,000:1. If this were not enough, two additional loci could increase the odds to more than 15,000,000,000:1.

Finally, the ceiling principle was not intended to be exclusive. Expert witnesses were still free to provide their statistical "best estimate" of genotype frequencies based on the product rule. But if disagreement over such estimates arose, the ceiling principle provided an approach that all parties had to admit was biased to favour a defendant. By all rights, this seemingly solomonic solution should have ended the controversy over population genetics.

Hitting the Ceiling

Surprisingly, attacks came from an unexpected quarter. Some vocal theoretical population geneticists and statisticians concluded that the committee had been too conservative. They argued that the effect of population substructure was slight and that it would best be treated by using formulas from theoretical population genetics. The ceiling principle was accused of being clumsy and scientifically flawed. Suddenly, a new controversy over population genetics seemed to emerge[5-10].

The debate was based on a simple misunderstanding of the NRC Committee report but, with the committee disbanded, there was no easy way to address it. Moreover, the committee members had agreed to let the report speak for itself to avoid the emergence of conflicting gospels according to different members. In retrospect, this was probably an unwise decision because it has allowed a minor academic debate to snowball to the point that it threatens to undermine the use of DNA fingerprinting by suggesting that there is some problem with the use of population-genetic estimates in court.

Six objections have been raised to the ceiling principle, which are worth briefly refuting:

(1) The ceiling principle is premised on the flawed analysis of Lewontin and Hartl that there is significant population substructure[8]. On the contrary, the committee

于 5% 和 10% 的选择是根据遗传漂变对匹配概率的定量效应影响而做出的，即对等位基因频率的倒数的影响，虽然这些推论都未出现在最后的报告中。）这些规定的实际作用是限制任意单个基因座的贡献，在仅基于种族分类总数据时，将概率限制在 50∶1 之内，而当基于更详细的种族调查时，将概率限制在 200∶1 之内。

上限原则不掩饰自身的保守性。它将每一个可想到的疑点所带来的好处都提供给了被告，以便其能够抵抗来自最固执且富有创造力的律师的攻击。该方法就是通过牺牲一部分统计效能来消除所有关于人群亚结构的可能担忧。

委员会对这种偏向一方的解决方法感到满意，因为即使是这些极端的假设也不能从本质上对 DNA 指纹分析的实际应用产生损害。法医实验室进行了四个基因座的匹配，这一方法仍然可以提供 6,250,000∶1 的概率。如果这样仍不能满足要求，可以增加两个额外的基因座从而将概率提升至 15,000,000,000∶1 以上。

最后，上限原则不是排他性的。专家证人仍然可以自由地基于乘法定则，提供关于基因型频率的数据性“最佳估算”。但是，如果对于该估算的反对意见增多，那么上限原则就可以提供一个各方都认可且偏向于被告的解决方法。按理来说，这一看似智慧的解决方法本该结束关于群体遗传学的争议。

对上限原则的冲击

令人惊讶的是，攻击来自一个出乎意料的群体。一些理论群体遗传学家和统计学家直言不讳地指出委员会过于保守。他们反驳说人群亚结构的作用很微弱，而最佳的选择应该是使用来自理论群体遗传学的公式。上限原则被指责是笨拙并且存在科学性缺陷的。突然之间，似乎出现了一个围绕群体遗传学的新争论[5-10]。

这一争论起源于对 NRC 委员会报告的误解，然而，随着该委员会的解散，这一问题逐渐变得难以处理。此外，委员会成员已经同意让该报告为它自己正名，从而避免因成员们的不同说法而产生准则间自相矛盾的情况。回顾过去可以发现，这可能不是一个明智的决定，因为这使得一个小的学术争论，通过指出法庭使用群体遗传学评估存在的一些问题而迅速升级到可以威胁 DNA 指纹分析应用的程度。

反对者提出了六条关于上限原则的反对意见，而这些都是值得进行简略反驳的：

（1）**上限原则是以路翁亭和哈特尔提出的、存在重要群体亚结构的缺陷分析为前提的**[8]。事实恰恰相反，虽然委员会非常怀疑亚结构具有的重要影响，但是它仍

was quite dubious that substructure had significant effects, but felt that the possibility needed to be taken seriously rather than dismissed based on theoretical or indirect arguments. The NRC report cites the Lewontin–Hartl article[3] only twice, both times balanced against a longer list of opposing articles.

(2) **The ceiling principle is scientifically flawed because it is not used in population genetics[9,10]. Moreover, the plan to sample 10–15 representative populations is statistically unsound[8].** The choice of a statistical method necessarily depends on the dangers of overestimation versus underestimation. In forensics, there is strong agreement on the need to be conservative. By contrast, population geneticists do not need to be conservative in academic studies; they are content to err equally often on the high and low side, and thus ceiling approaches are unnecessary. However, ceiling approaches are common in a closely related genetic pursuit: the mapping of disease genes. To guard against falsely implicating or excluding a chromosomal region, human geneticists often analyse their data under worst-case conditions, such as using an unrealistically low ceiling on the penetrance of a disease gene or an unrealistically high frequency for a marker allele[11,12]. Also, some authors have complained that surveys of 100 individuals do not allow accurate estimates of the frequency of rare alleles[8]. The purpose of the suggested population studies, however, was not to estimate low frequencies, but rather to check that some alleles do not unexpectedly have extremely high frequencies (much more than 5–10%) in certain populations. For this purpose, 100 individuals is quite adequate.

(3) **The ceiling principle makes ludicrous assumptions about the possible substructure of a population[9,10].** Although the NRC report called for empirical studies of those groups that made significant contributions to the United States (such as English, Italians and Puerto Ricans), some commentators[9,10] were carried away by hyperbole—asserting that the ceiling principle assumes that "the culprit might be...a Lapp for one allele and a Hottentot for the other". However, if Lapp and Hottentot are replaced by Italians and Puerto Ricans, the assumption is perfectly reasonable. Indeed, it is unreasonable to assume that such genotypes don't occur in the population.

(4) **The ceiling principle is so conservative that it hampers the courtroom application of DNA fingerprinting.** Despite fears that an ultraconservative standard would clip the wings of the DNA fingerprinting, published analyses by several groups[13,14] agree that the effect is modest: Whereas the product rule typically gives four-locus genotype frequencies of about 10^{-8}–10^{-9}, the ceiling principle pares them back to about 10^{-6}–10^{-7}. That extreme assumptions have so little effect only underscores the power of DNA typing and the wisdom of taking a conservative approach.

(5) **The ceiling principle is not actually guaranteed to be conservative.** This concern appears in only a single paper[15], in which the authors point out that the conservativeness of the ceiling principle depends on the component populations P_1, ..., P_n being themselves well-mixed. The authors take great pains to construct counterexamples in the event that the populations are themselves substructured. In fact, the NRC committee considered this point

然认为需认真对待这一可能性，而不是根据理论性的或间接的论据而忽视它。NRC 的报告只引用了两次路翁亭–哈特尔的文章[3]，并且两次都用了更多相反的文章来维持平衡。

(2) **因为上限原则不用于群体遗传学，所以它具有科学性瑕疵[9,10]。此外，选取 10~15 个代表性群体的计划在统计学上也是不可靠的[8]。**统计学方法的选择必须基于对高估与低估所导致的危险的比较。法医方面对保守性的需求具有很强的一致性。相比之下，群体遗传学家在学术研究方面并不需要保守性；他们常常满足于在上限和下限同等地犯错，因此上限方法是非必需的。然而，上限方法却常见于与之密切相关的遗传寻踪之中：绘制疾病基因图谱。为了防止错误地引入或排除某个染色体区域，人类遗传学家常常在最坏情况下分析数据，例如将一个不合实际的较低上限值用于某个疾病基因的外显率，或者将一个不合实际的高频率用于某一标记等位基因[11,12]。与此同时，一些作者抱怨含有 100 个个体的调查并不能对罕见等位基因的基因频率进行精确估算[8]。但是，对于所建议的群体研究而言，其目的不是为了估算低频率，而是为了验证特定人群中的一些等位基因不具有出乎意料的极高频率(大大超过 5%~10%)。就这一目的而言，100 个个体已经是相当充足的。

(3) **上限原则中关于群体中可能的亚结构的假设是滑稽的[9,10]。**虽然 NRC 的报告需要那些对美国人群有显著影响的群体(例如英国人、意大利人以及波多黎各人)的经验性研究的支持，但是一些评论员[9,10]被夸张的说法冲昏了头脑——他们断言上限原则假定了“罪犯可能……根据一个等位基因而言是拉普人，根据另一个等位基因则是霍屯督人”。但是，如果拉普人和霍屯督人被替换为意大利人和波多黎各人，那么这个假设便变得极其合理了。事实上，假设这样的基因型不会在人群中发生是不合理的。

(4) **上限原则过于保守以至于阻碍了 DNA 指纹分析在法庭上的应用。**尽管对极端保守的标准可能会钳制住 DNA 指纹分析的翅膀的情况存在担忧，但是由多个团队[13,14]发表的分析结果表明，该影响是微弱的：乘法定则通常提供的四个基因座基因型频率在 10^{-8}~10^{-9} 左右，然而上限原则将其削减至 10^{-6}~10^{-7} 左右。前面所述的极端假设具有很弱的影响，它仅仅强调了 DNA 分型的效力以及实施保守性方法的智慧。

(5) **事实上，上限原则并未保证其保守性。**这一忧虑仅在一篇论文[15]中出现过，作者在文章中指出，上限原则的保守性依赖于自身充分混合的组分人群 P_1，…，P_n。该作者还努力构造了假如人群本身具有亚结构的反例。事实上，NRC 委员会认为，这一点是不证自明的(虽然未能足够清晰地阐述这一点)，因为从琐碎的观察中就可

to be self-evident (although failed to state it clearly enough), as can be seen from the trivial observation that the ceiling principle has no effect on genotype frequencies when one combines two identical, but substructured populations. In applying the ceiling principle in practice, the committee was confident that any residual substructure in the component subpopulations could safely be ignored in the context of such a conservative scheme.

(6) The NRC report is causing DNA fingerprinting cases to be thrown out of court[8]. To the contrary, most courts have used the NRC report as strong evidence that, notwithstanding disagreement over the best solution, there is at least one approach that is indisputably conservative. Of the few cases in which DNA evidence has ever been rejected on population-genetic grounds, virtually all involved evidence predating the NRC report. These courts have cited the report solely for its acknowledgement that a controversy existed and was a reason for constituting the committee.

The NRC report, to be sure, has important flaws. The ceiling principle was not elegant solution, but simply a practical way to sidestep a contentious and unproductive debate. The report had more than its share of miswordings, ambiguities and errors, many of which have been corrected by a vigilant commentator[16]. A few poorly worded sentences have been seized upon by lawyers trying to undermine the straightforward calculation of ceiling frequencies (although such arguments have not succeeded). Most important, the report failed to state clearly enough that the ceiling principle was intended as an ultra-conservative calculation, which did not bar experts from providing their own "best estimates" based on the product rule. This failure was responsible for the major misunderstanding of the report. Ironically, it would have been easy to correct.

A Law-enforcement Perspective

Even as academics debated fine points, forensic scientists got on with their business. The FBI and TWGDAM found the ceiling principle to be unnecessarily conservative, but nonetheless promptly adopted precise guidelines for implementing the ceiling principle, correctly clarifying a few ambiguous statements in the NRC report, such as which population databases to include and whether to sum adjacent bins in a frequency distribution. Forensic labs adopted a two-tiered approach, in which experts are prepared to quote both their best estimate and the conservative ceiling bound. As new population-genetic issues arise (such as how to modify or replace the ceiling principle to accommodate the less polymorphic PCR-based systems), the community is preparing to develop further guidelines. Overall, the system meets the spirit and the letter of the NRC report.

Conservative calculations have had no noticeable impact on the use of DNA evidence. In the vast majority of cases, a jury needs to know only that a particular DNA pattern is very rare to weigh it in the context of a case: the distinction between frequencies of 10^{-4}, 10^{-6} and 10^{-8} is irrelevant in the case of suspects identified by other means.

以得知，当一个人群中含有两个相同但具有亚结构的人群时，上限原则对基因型频率是无影响的。在上限原则的实际应用方面，委员会自信地认为基于这种保守性方案的前提可以安全地忽略组分亚人群中任何剩余的亚结构。

(6) **NRC 的报告将导致 DNA 指纹分析案件被踢出法庭**[8]。事实恰恰相反，尽管仍然存在对于这项最佳解决方法的异议，大多数法院已经将 NRC 报告作为一个强有力的证据来证明这个方法至少是一个无可争辩的保守性方法。在少数群体遗传学水平上 DNA 证据被弃用的案例中，几乎所有案例都包含在时间上早于 NRC 报告的证据。这些法院已经引用了这个报告，该报告承认争议的存在并以此作为建立该委员会的理由。

诚然，NRC 的报告存在重要的缺陷。虽然上限原则不是绝妙的解决办法，但的确是一个可以避开无休止、无结果的争论的实用方法。这个报告含有的不只是措辞不当、语意含糊以及谬误这些小瑕疵，其中多数不足已经被警惕的评论员修正了[16]。尝试损害上限频率这种简便计算方法的律师们已经抓住了少数措辞不当的语句(尽管这样的争论还未成功)。最重要的是，该报告未能明确阐述上限原则的目的是作为一个极端保守的计算方法，这一方法没有禁止专家们提出他们自己基于乘法定则的“最佳估算”。这一失误是引起对该报告产生严重误解的原因。讽刺的是，做出改正本应很容易。

执法视角

在学术界辩论细节问题的同时，法医学家继续着他们的工作。FBI 和 TWGDAM 认为上限原则具有的保守性是非必需的，尽管如此，它们也迅速采纳了精确的指南来履行上限原则，正确地阐明了 NRC 报告中的一些模糊表述，例如将哪一个人群数据库包含在内的问题以及是否将频率分布中相邻区间的数据进行相加。法医实验室采取了双层法，即专家们准备引用最佳估算结果以及保守的上限范围。随着新的群体遗传学问题的出现(例如如何修改或替换上限原则以适应基于 PCR 的具有更少多态性的系统)，科研团队正准备开发更进一步的指南。总的来说，该系统与 NRC 报告的精神和文字内容相符。

对于 DNA 证据的使用而言，保守性计算的影响已经不再明显。在绝大多数案例中，陪审团只需要了解在一种情况下，某个特定的 DNA 模式很难作为衡量案件的证据，即在用其他方法鉴别出嫌疑人的案件中，10^{-4}、10^{-6} 以及 10^{-8} 这三种频率间的差异是没有意义的 。

The FBI has also rapidly carried out population surveys, as recommended by the NRC committee. FBI scientists have studied more than 25 distinct subpopulations, as well as 50 separate samples from the US population[17-21]. The effort has yielded a remarkable database for examining allele frequency variation among ethnic groups. Reassuringly, the observed variation is modest for the loci used in forensic analysis and random matches are quite rare, supporting the notion that the FBI's implementation of the product rule is a reasonable best estimate. Nonetheless, the FBI has taken the scientifically sound position that it remains wiser to study new loci empirically than to assume that significant variation can never occur.

Most important, the admissibility of such DNA evidence prepared in accordance with the NRC recommendations is firmly established in virtually all US jurisdictions. In a few, the appellate courts have yet to rule formally, but there is little doubt that they will find such evidence acceptable.

Modest Proposals

Although the system ain't broke, there is no shortage of proposals about how to fix it. Some academic commentators advocate a return to the product rule; others propose an approach based on the kinship statistic F_{ST}; and still others recommend an approach involving likelihood ratios that combine gel electrophoresis artefacts and population-genetic considerations into a single statistic[9,10,22-24]. Some seek to determine genotype frequencies exactly, while others prefer conservative estimates. The NRC—at the urging of the National Institute of Justice, representing the academic wing of forensic scientists—has concluded that the best solution is to constitute another *ad hoc* committee on DNA fingerprinting, composed primarily of statisticians and population geneticists.

It is easy to forget that this new debate is purely academic. The most extreme positions range over a mere two orders of magnitude: whether the population frequency of a typical four-locus genotype should be stated, for example, as 10^{-5} or 10^{-7}. The distinction is irrelevant for courtroom use.

Rehashing issues may be a harmless pastime in the academic world, but not so in a legal system that lives by the dictum *stare decisis* (let the decision stand). From the standpoint of law enforcement, it is better to have a settled, if slightly imperfect, rule than ceaselessly to quest after perfection. Already the NRC's intention to re-examine forensic DNA typing has been seized upon by some lawyers as evidence that there remain fundamental problems.

Ad hoc committees typically imagine that they will be able to accomplish their mandate with speed and finality. The original NRC study was anticipated to take one year, but required three. The idea of a second NRC panel was first floated in June 1993 with the optimistic projection that it could report by the end of that year. In fact, the committee has only

就像 NRC 委员会建议的那样，FBI 也已经迅速地进行了群体调查。FBI 的科学家们已经研究了超过 25 个不同的亚人群，以及 50 个来自美国人群的独立样本 [17-21]。通过这些努力得到了一个可以检验种群间等位基因频率变化的非凡数据库。可以放心的是，观测到的变化对用于法医分析的基因座而言是轻微的，而且很少发生随机匹配，这一点支持了一个观点，即 FBI 对于乘法定则的应用是一个明智的最佳判断。尽管如此，FBI 已经采取了科学合理的态度，相对于假设从不发生显著变化的情况而言，根据经验研究新基因座仍然更为明智。

最重要的是，按照 NRC 的建议准备的 DNA 证据在几乎所有美国管辖的地区内都被认可了。但在少数地区，还有一些受理上诉的法院未能接受这类证据，但毫无疑问的是，他们终将发现这类证据是值得认可的。

小小的建议

虽然这个系统未被破坏，但是也不缺少关于如何修补它的建议。一些学术评论家主张重新启用乘法定则；另一些则提出了一个基于亲属关系统计值 F_{ST} 的解决方法；还有一些人建议使用涉及似然比的方法，即将凝胶电泳人为假象和群体遗传学方面要考虑的因素组合成一个单个的统计数据 [9,10,22-24]。一些人设法精确地确定基因型频率，然而其他人则更倾向于保守性的判断。国家司法研究所是法医学权威的代表，在它的督促下，NRC 最终做出了决定，即最佳的解决方法是建立另一个与 DNA 指纹印分析相关的特设委员会，该委员会主要由统计学家和群体遗传学家组成。

容易遗忘的一点是，这一新辩论是纯学术性的。最极端位置间的差异仅仅超过了两个数量级：例如，一个典型的四个基因座基因型的群体频率是否应被表述为 10^{-5} 或 10^{-7}。这个差异与法庭中的应用并不相干。

在学术界，对于一些问题的重新讨论也许是一种无恶意的娱乐，但是在以遵循先例（维持决议的效力）为信条的司法系统中，情况并非如此。从执法的角度来看，相比于不停地追求一个完美的规则，拥有一个即使存在些许不足的固定规则会更好一些 。一些律师已经抓住 NRC 打算重新检查法医 DNA 分型的意图，并将其作为支持 DNA 分型仍然存在严重问题的证据。

特设委员会曾设想自己能够迅速而彻底地完成使命。预期花费一年时间完成的初期 NRC 研究最终用了三年的时间才得以完成。1993 年 6 月，人们首次提出设立第二个 NRC 专家小组的想法，并且乐观地预期它能够在年末发表报告。事实上，委

just begun meeting and will probably not issue a report before late 1995. Even then, any recommendations will take 3 years to ripple through the legal system — guaranteeing that finality will not be achieved on these issues before early 1999. Despite the committee's best efforts, any new report will probably offer new opportunities for misunderstanding that will become apparent only after the panel is disbanded. And, if the new report endorses a different standard, some attorneys are sure to argue, rightly or wrongly, that differences between the reports demonstrate a lack of scientific consensus. These observations are not meant to dissuade the new NRC committee from its mission, but rather to point out the challenge facing any *ad hoc* group.

A Sounder Approach

The real solution is to recognize that forensic DNA typing has become a mature field and requires a more systematic approach. The NRC report anticipated that rapid evolution of technology would pose a steady succession of questions requiring attention. Its most important recommendation was the establishment of a permanent national committee on forensic DNA typing (NCFDT) to address issues as they arose. If such a committee had been appointed in 1992, it could have made short work of the population-genetics issues, by clarifying, changing or discarding the original NRC recommendations.

It is encouraging that this NRC recommendation has recently been adopted. The newly enacted DNA Identification Act of 1994 mandates the FBI to establish a DNA advisory board to recommend standards for laboratory procedures, quality assurance and proficiency testing. The act requires open meetings and broad representation, including molecular and population geneticists not affiliated with forensic laboratories, with board nominations to come from professional organizations including the National Academy of Sciences. Ideally, the panel will provide a forum for weighing important issues, including new laboratory techniques, population genetics and proficiency testing.

Most of all, the public needs to understand that the DNA fingerprinting controversy has been resolved. There is no scientific reason to doubt the accuracy of forensic DNA typing results, provided that the testing laboratory and the specific tests are on a par with currently practiced standards in the field. The scientific debates served a salutary purpose: standards were professionalized and research stimulated. But now it is time to move on.

(**371**, 735-738; 1994)

Eric S. Lander and Bruce Budowle

Eric S. Lander is in the Whitehead Institute for Biomedical Research, Nine Cambridge Center, Cambridge, Massachusetts 02142, USA and the Department of Biology, MIT, Cambridge, Massachusetts 02139, USA.

Bruce Budowle is in the Forensic Science Research and Training Center, FBI Laboratory, FBI Academy, Quantico, Virginia 22135, USA.

员会仅仅是开始进行了会议，并且不太可能在 1995 年年末前发表报告。即使这样，任何建议也都将需要三年的时间才能影响司法系统——确保了这些问题在 1999 年年初前都不会有最终的结果。虽然委员会在不懈努力，但是任何的新报道都有可能给误解的发生提供机会，并且这将在该专家小组解散后显现出来。此外，如果新报告支持一个不同的标准，那么无论正确与否一些律师一定会争辩说这些报告间的差异显示了它们之间缺乏科学共识。这些评论并非旨在劝阻这个新 NRC 委员会完成使命，而是为了指出任何特设委员会会面临的挑战。

一个更合理的方法

真正的解决办法是，认识到法医 DNA 分型已经成为一个成熟的技术领域并且需要一个系统性更强的方法。NRC 的报告做出了预测，技术的迅速发展将导致一连串需要注意的问题的出现。该报告中最重要的建议是建立一个持久的、与法医 DNA 分型相关的国家委员会（NCFDT），以便解决出现的问题。如果在 1992 年就成立了这样的委员会，那么它可能已经通过阐述、改变或弃用最初 NRC 的建议，从而迅速地解决群体遗传学的问题。

令人鼓舞的是，NRC 的这条建议最近已经被采纳了。最新制定的 1994 年 DNA 鉴定相关法案要求 FBI 组建一个 DNA 顾问委员会，用以为实验室程序、质量保证以及水平测试提供标准规范。该法案要求会议公开并且代表具有广泛性，这些代表包括不隶属于法医实验室的分子遗传学家和群体遗传学家，且都是由委员会提名的、来自包括美国国家科学院在内的专业组织机构。理想状况下，该委员会将提供一个权衡重要问题的平台，这些问题包括新的实验室技术、群体遗传学以及水平测试。

最重要的是，公众需要明白 DNA 指纹分析的争论已经结束。如果测试实验室和特定的测试与当前这一领域的标准规范是一致的，那么就不存在质疑法医 DNA 分型结果的科学理由。这些科学争论是为了一个有益的目的：使标准规范变得专业化并且调查研究能产生促进作用。现在，我们该继续前进了。

（高俊义 翻译；方向东 审稿）

References:

1. Lander, E. S. *Nature* **339**, 501-505 (1989).
2. National Research Council *DNA Technology in Forensic Science* (National Academy Press, 1992).
3. Lewontin, R. C. & Hartl, D. L. *Science* **254**, 1745-1750 (1991).
4. Chakraborty, R. & Kidd, K. K. *Science* **254**, 1735-1739 (1991).
5. *Science* **266**, 201-203 (1994).
6. Weir, B. S. *Proc. Natl. Acad. Sci. U.S.A.* **89**, 11654-11659 (1992).
7. Aldhous, P. *Science* **259**, 755-756 (1993).
8. Devlin, B., Risch, N. & Roeder, K. *Science* **259**, 748-749 (1993).
9. Morton, N. E., Collins, A. & Balasz, I. *Proc. Natl. Acad. Sci. U.S.A.* **90**, 1892-1896 (1993).
10. Morton, N. E. *Genetica* (in the press).
11. Schellenberger, G. D. *et al. Science* **258**, 668 (1992).
12. Terwlliger, J. D. & Ott, J. *Handbook of Human Genetic Linkage* (Johns Hopkins Univ. Press, 1994).
13. Duncan, G. T., Noppinger, K. & Tracy, M. *Genetica* **88**, 51-57 (1993).
14. Krane, D. E. *et al. Proc. Natl. Acad. Sci. U.S.A.* **89**, 10583-10587 (1992).
15. Slimowitz, J. R. & Cohen, J. E. *Am. J. Hum. Genet* **53**, 314-323 (1993).
16. Weir. B. S. *Am. J. Hum. Genet* **52**, 437-440 (1993).
17. Federal Bureau of Investigation *VNTR Population Data: A Worldwide Study* (1993).
18. Budowle, B., Monson, K. L., Giusti, A. M. & Brown. B. L. *J. Forensic Sci.* **39**, 319-352 (1994).
19. Budowle, B., Monson, K. L., Giusti. A. M. & Brown, B. L. *J. Forensic Sci.* **39**, 988-1008 (1994).
20. Budowle, B. *et al. J. Forensic Sci.* (in the press).
21. Huang, N. E. & Budowle, B. *J. Forensic Sci.* (in the press).
22. Morton, N. E. *Proc. Natl. Acad. Sci. U.S.A.* **89**, 2556-2560 (1992).
23. Evett, I. E., Scranage, J. & Pinchin, R. *Am. J. Hum. Genet.* **52**, 498-505 (1993).
24. Balding, D. J. & Nichols, R. A. *Forensic Sci. Int.* **64**, 125-140 (1994).

Positional Cloning of the Mouse *obese* Gene and Its Human Homologue

Yiying Zhang *et al.*

Editor's Note

The *obese* mouse carries a mutation that causes profound obesity and type II diabetes, but in the early 1990s the mechanism underlying this syndrome was far from clear. Here American biologist Jeffrey M. Friedman and colleagues describe the cloning and sequencing of the mouse *obese* gene and its human homologue, and speculate that the encoded fat-produced protein acts on the brain to inhibit food intake and/or regulate energy expenditure. The protein turned out to be the hormone leptin, which is released into the blood and transported to the hypothalamus, where it suppresses food intake. A mutation in the leptin gene has since been linked to obesity, and animal models of obesity continue to boost our understanding of weight control.

The mechanisms that balance food intake and energy expenditure determine who will be obese and who will be lean. One of the molecules that regulates energy balance in the mouse is the *obese* (*ob*) gene. Mutation of *ob* results in profound obesity and type II diabetes as part of a syndrome that resembles morbid obesity in humans. The *ob* gene product may function as part of a signalling pathway from adipose tissue that acts to regulate the size of the body fat depot.

OBESITY is the commonest nutritional disorder in Western societies. More than three in ten adult Americans weigh at least 20% in excess of their ideal body weight[1]. Increased body weight is an important public health problem because it is associated with type II diabetes, hypertension, hyperlipidaemia and certain cancers[2]. Although obesity is often considered to be a psychological problem, there is evidence that body weight is physiologically regulated[3].

The molecular pathogenesis of obesity is unknown. To identify components of the physiological system controlling body weight, we have applied positional cloning technologies in an attempt to isolate mouse obesity genes. Five single-gene mutations in mice that result in an obese phenotype have been described[3]. The first of the recessive obesity mutations, the *obese* mutation (*ob*), was identified in 1950[4]. *ob* is a single gene mutation that results in profound obesity and type II diabetes as part of a syndrome that resembles morbid obesity in humans[5]. Neither the primary defect nor the site of synthesis of the *ob* gene product is known. Cross-circulation experiments between mutant and wild-type mice suggest that *ob* mice are deficient for a blood-borne factor that regulates nutrient intake and metabolism[6], but the nature of this putative factor has not been determined.

小鼠肥胖基因及其人类同源基因的定位克隆

张一影（音译）等

编者按

肥胖小鼠携带一种基因突变体，该突变能引起严重的肥胖症和II型糖尿病，但是在20世纪90年代早期，人们并不清楚这种综合征的机制。在本文中，美国生物学家杰弗里·弗里德曼和同事描述了小鼠*obese*(肥胖)基因及其人类同源基因的克隆和测序，推测该基因所编码的脂生成蛋白作用于大脑来抑制食物摄入和(或)调控能量代谢。结果显示，该蛋白是种激素——瘦蛋白，它可以释放到血液中并运送到下丘脑，而下丘脑正是抑制进食的功能部位。瘦蛋白基因的突变与肥胖有关，肥胖动物模型有助于加深我们对于体重控制的理解。

平衡进食和能量消耗的机制决定了谁会肥胖而谁会纤瘦。在小鼠体内，调节能量平衡的其中一个分子就是肥胖(*ob*)基因。*ob* 基因突变会导致人类出现类似病态肥胖的综合征，其中包括严重的肥胖和II型糖尿病。*ob* 基因产物可能是脂肪组织中调节体脂存储量的信号通路中的一员。

在西方社会，肥胖是最普遍的营养失调疾病。超过三成的美国成年人体重至少超过理想体重的20%[1]。体重的增加是一个严重的公共健康问题，因为它总会伴随着II型糖尿病、高血压、高血脂和某些癌症[2]。虽然肥胖通常被认为是一种心理疾病，但是现在有证据表明体重是受生理调节的[3]。

肥胖的分子发病机理还不清楚。为了确定控制体重的生理系统的组成，我们尝试利用定位克隆技术来分离出小鼠的肥胖基因。现在已有报道显示，小鼠体内5种单基因突变可导致其肥胖表型[3]。第一个发现的隐性肥胖突变就是1950年报道的肥胖(*ob*)基因突变[4]。*ob* 是一个单基因突变，会导致人类出现类似于病态肥胖的综合征，其中包括严重的肥胖和II型糖尿病[5]。*ob* 基因产物的主要缺陷和合成位点均是未知的。突变型和野生型小鼠的交叉循环实验表明，*ob* 小鼠缺少一种调节营养摄入和代谢的血源性因子[6]，但是这种假定因子的性质仍不明确。

We report the cloning and sequencing of the mouse *ob* gene and its human homologue. *ob* encodes a 4.5-kilobase (kb) adipose tissue messenger RNA with a highly conserved 167-amino-acid open reading frame. The predicted amino-acid sequence is 84% identical between human and mouse and has features of a secreted protein. A nonsense mutation in codon 105 has been found in the original congenic C57BL/6J *ob/ob* mouse strain, which expresses a twentyfold increase in *ob* mRNA. A second *ob* mutant, the co-isogenic SM/Ckc-$+^{Dac}ob^{2J}/ob^{2J}$ strain, does not synthesize *ob* RNA. These data suggest that the *ob* gene product may function as part of a signalling pathway from adipose tissue that acts to regulate the size of the body fat depot.

For the positional cloning of mutant genes from mammals, it is necessary first to obtain genetic and physical maps, then to isolate the gene and detect the mutation. Here we describe the successful use of this approach to identify the *ob* gene.

Genetic and Physical Mapping of *ob*

The first *ob* mutation (carried on the congenic C57BL/6J *ob/ob* strain) was found proximal to the *Microphthalmia* (*Mi*) and *waved-1* (*wav-1*) loci on proximal mouse chromosome 6 (ref. 7); a second co-isogenic allele of *ob* has been identified in the SM/Ckc-Dac mouse strain (S. Lane, personal communication). We previously positioned *ob* relative to a series of molecular markers on mouse chromosome 6 and mapped the *ob* gene close to a restriction-fragment length polymorphism (RFLP) marker, D6Rck13, derived from chromosome microdissection[5,8]; we also found that *Pax4* in the proximal region of mouse chromosome 6 is tightly linked to *ob* (ref. 9). Both loci were initially used to type a total of 835 informative meioses derived from both interspecific and intersubspecific mouse crosses that were segregating *ob*. *Pax4* was mapped proximal to *ob* and was recombinant in two animals (111 and 420 in Fig. 1); no recombination between D6Rck13 and *ob* was detected among the first 835 meioses scored[8].

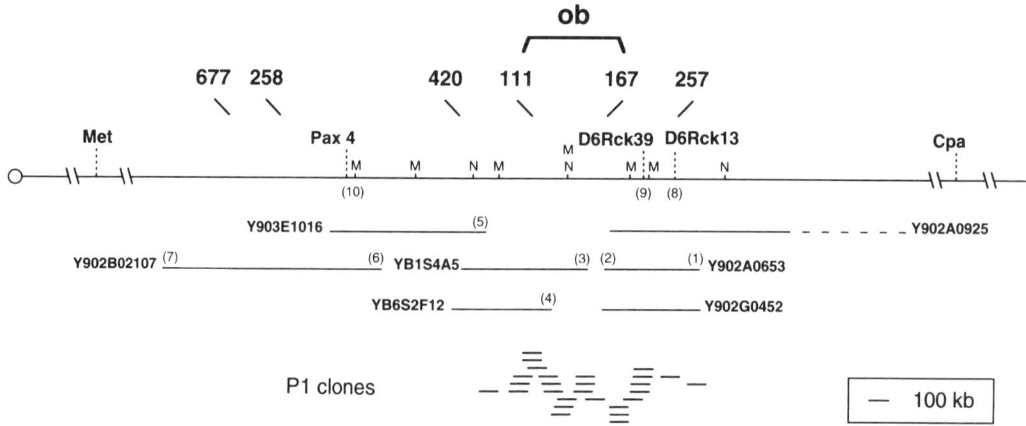

Fig. 1. Physical map of the mouse *ob* locus. A chromosome walk across the *ob* locus was completed in YACs and P1 clones using the flanking markers D6Rck13 and *Pax4* as starting points. The position of

在这篇文章里我们报道了小鼠 *ob* 基因及其人类同源基因的克隆与测序。*ob* 基因编码了一个 4.5 kb 的脂肪组织信使 RNA(mRNA)，该 mRNA 有一个高度保守的、由 167 个氨基酸组成的开放阅读框。预测的氨基酸序列在人与小鼠之间有 84% 的一致性，并且具有一些分泌蛋白的特性。在原始同类的 C57BL/6J *ob*/*ob* 小鼠品系里，我们发现了第 105 个密码子处的无义突变。这些突变小鼠中，*ob* 基因的 mRNA 表达量增加至原来的 20 倍。第二个 *ob* 突变体发现于同类的 SM/Ckc-$+^{Dac}$$ob^{2J}$/$ob^{2J}$ 品系的小鼠中，该小鼠不能合成 *ob* 基因的 RNA。这些数据提示了 *ob* 基因产物可能是脂肪组织中调节体脂存储量的信号通路中的一员。

为了从哺乳动物中定位克隆突变基因，我们必须先要得到遗传图谱和物理图谱，然后分离基因和检测突变。在这篇文章里面，我们成功应用这种方法确定了 *ob* 基因。

绘制 *ob* 基因突变的遗传图谱和物理图谱

在同类的 C57BL/6J *ob*/*ob* 的小鼠品系里，发现了第一个 *ob* 突变，它位于 6 号染色体的 *Microphthalmia*(*Mi*) 和 *waved-1* (*wav-1*) 两个基因座附近(参考文献 7)；随后，在 SM/Ckc-Dac 小鼠品系里发现了另外一个 *ob* 同源等位突变(莱恩，个人交流)。我们首先将 *ob* 基因的位置用一系列小鼠 6 号染色体的分子标记进行定位，并且确定 *ob* 基因位于一个限制性片段长度多态性(RFLP)标记——D6Rck13 附近，这个分子标记是通过染色体显微切割得到的 [5,8]。我们还发现在小鼠 6 号染色体近端区域的 *Pax4* 与 *ob* 基因紧密连锁(参考文献 9)。D6Rck13、*Pax4* 这两个基因座最初是用来鉴定 835 个种间和亚种间小鼠因杂交 *ob* 基因被隔离的减数分裂配子的。*Pax4* 邻近 *ob* 基因，杂交时在两个动物间可以发生重组(图 1 的 111 和 420)；而在第一轮得到的 835 个减数分裂里面没有检测到 D6Rck13 与 *ob* 基因之间发生重组 [8]。

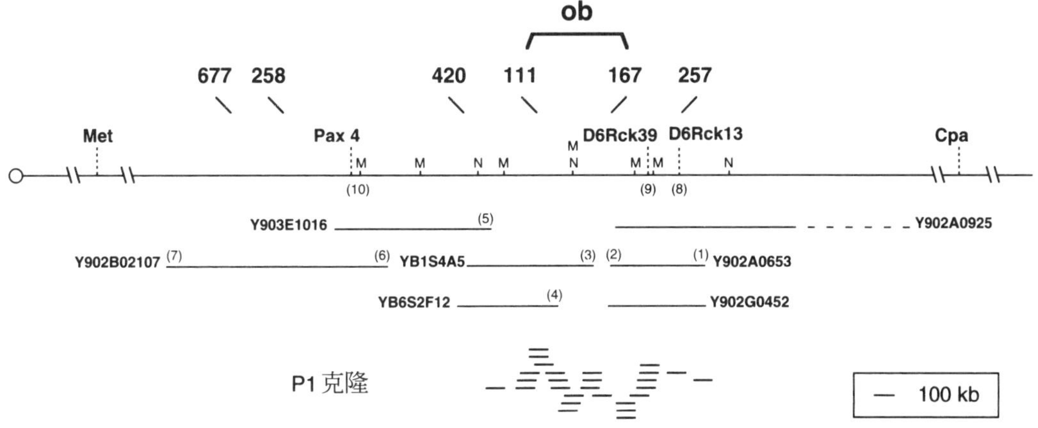

图 1. 小鼠 *ob* 基因座的物理图谱。以侧翼标记 D6Rck13 和 *Pax4* 作为起始位点通过一系列 YAC 和 P1 克隆完成了 *ob* 基因座的染色体步移。一些罕见的限制性酶切位点 *Mlu*1 (M) 和 *Not*1 (N) 的位置也被标记

the rare-cut restriction enzymes *Mlu*1 (M) and *Not*1 (N) are indicated. Numbers in bold type indicate designation of recombinant animals in the region of *ob* among the 1,606 meioses scored. Each of the ends of the YAC clones isolated with D6Rck13 and *Pax4* were recovered using vectorette PCR and/or plasmid end rescue and used to type the recombinant animals. These ends were sequenced and used in turn to isolate new YACs. The resulting YAC contig is shown on the middle panel. One of the YACs in this contig, y902A0925, was chimaeric. Each of the probes used to genotype the recombinant animals is indicated in parentheses. On this basis, *ob* was localized to the interval between the recombination events in animals 111 and 167, (between end (5) and D6Rck39). Selected probes were also used to isolate bacteriophage P1 clones across the non-recombinant region. The resulting P1 contig is shown in the bottom panel. These P1 bacteriophage spanned a ~70-kb interval between probes (2) and (3) which could not be recovered in a YAC clone. The *ob* gene was identified in a P1 clone isolated using the distal end of YAC yB6S2F12 (end 4).

Methods. YAC clones were isolated by PCR screening or hybridization to high-density filters of available mouse YAC libraries[38,40]. The YACs from ICRF begin with a 902 prefix. YAC clones were sized on a CHEF MAPPER (Bio-Rad). Restriction enzyme digestions were done according to the manufacturers recommendations. YAC ends were recovered using vectorette PCR and plasmid end rescue[10,11]. P1 clones were isolated by sending specific pairs of PCR primers to Genome Systems Inc (St Louis, MO) who provided single picks of individual mouse P1 clones[12]. P1 ends were recovered using vectorette PCR[13]. Cosmid subclones from YAC y902A0653 were isolated as described[41]. Primer selection and PCR amplification of simple sequence repeats were performed as described: initial denaturation at 94 °C for 3 min, 25 cycles of 94 °C for 1 min, 55 °C for 2 min and 72 °C for 3 min. Primer sequences for *Pax4: Pax4F*, 5′-GGAGGTAGAGATGGCAGCAG-3′; *Pax4R*, 5′-ACAGA- AAGCAAGGAGGATTTC-3′, with a product size of 126 bp. Primer sequences for D6Rck39 were: 39GTF, 5′-GCACACTGACAGTGCCCTTA-3′; 39GTR, 5′-TGTAACCTGGAATTGGGAGC-3′ with a product size of 128 bp. The breeding and maintenance of the various mouse crosses have been described[8,42].

To isolate the *ob* gene we cloned the DNA in the region of *Pax4* and D6Rck13 (Fig. 1), using both probes to start the construction of a physical map in the region of *ob*. Yeast artificial chromosomes (YACs) corresponding to *Pax4* and D6Rck13 were isolated and characterized. Centromeric and telomeric ends of each YAC were recovered, and ends mapping closer to *ob* used to screen for new YACs. YAC ends were recovered using either vectorette polymerase chain reaction (PCR) or the plasmid end-rescue technique[10,11]. One of the ends (labelled (1) in Fig. 1) of a D6Rck13 YAC, 902A0653, was recombinant in animal 257, positioning *ob* between this YAC end and *Pax4*. We were unable to recover any YACs linking the ends of YACs yBlS4A5 and 902A0653 (labelled (2) and (3) in Fig. 1). Pulsed-field gel electrophoresis (PFGE) indicated that there was a ~70-kb gap separating the two YAC ends. To bridge this gap, we used both YAC ends to isolate a set of mouse P1 clones[12,13]. Analysis of the ends of these Pl clones showed that they overlapped and that the gap in the YAC contig was ~70 kb. The size of the contig spanning the *ob* locus was 2.2 megabases (Mb) and *ob* was localized to the 900 kb between the distal end of YAC 903E1016 (labelled (5) in Fig. 1) and the distal end of YAC 902A0653 (labelled (1) in Fig. 1).

To position *ob* more precisely, we genotyped an additional 771 meioses derived from both a C57BL/6J *ob*/*ob* × DBA/2J intercross and backcross[5]. The typing of the intraspecific crosses required the development of informative single-strand length polymorphisms (SSLPs) for both D6Rck13 and *Pax4*. Sequencing of the *Pax4* gene itself revealed a microsatellite sequence, and an SSLP near D6Rck13, D6Rck39, was identified by sequencing cosmid subclones from YAC y902A0653 (a YAC isolated with D6Rck13).

出来了。粗体数字表示在记录的1,606次减数分裂中*ob*的位置上重组动物的名称。每个通过D6Rck13和*Pax4*分离得到的YAC克隆的末端都会用载体小件PCR和（或）质粒末端修复技术进行修复并且用于鉴定重组动物。这些末端序列先被测序然后被用于分离新的YAC。最终找到的YAC重叠群在图中间区域。这个重叠群中有一个标号为y902A0925的YAC，是嵌合的。每一个用于鉴定重组动物的探针都用圆括号在图中表示出来了。根据这个结果，*ob*基因位于111和167号动物发生重组处的中间（在(5)号末端和D6Rck39之间）。一些挑选的探针也被用来寻找位于非重组区的噬菌体P1克隆。得到的P1重叠群在图的底部展示。得到的P1克隆横跨了探针(2)和(3)之间约70 kb的区域，该区域不能通过YAC克隆进行修复。*ob*基因是通过标号yB6S2F12的YAC远末端探针((4)号末端)在一个P1克隆上找到的。

方法。YAC克隆是通过PCR筛选或者含有小鼠YAC文库的高密度滤纸杂交分离得到[38,40]。英国皇家癌症研究基金会(ICRF)来源的YAC都有一个902前缀。这些YAC克隆用CHEF MAPPER(伯乐公司)分出大小，根据制造商的推荐进行限制性酶切，然后用载体小件PCR和质粒末端修复对相应的YAC末端进行修复[10,11]。P1克隆则是将特定的PCR引物送到基因组系统公司(圣路易斯，密苏里)，它们会提供单个小鼠P1克隆的分离[12]。P1克隆末端用载体小件PCR修复[13]。来自y902A0653的YAC亚克隆黏粒根据文献描述的方法分离[41]。引物筛选和简单重复序列的PCR扩增根据以下描述进行：预变性用94 ℃持续3分钟，接下来25个循环是94 ℃持续1分钟，55 ℃持续2分钟，72 ℃持续3分钟。*Pax4*的引物序列：*Pax4F*，5′-GGAGGTAGAGATGGCAGCAG-3′；*Pax4R*，5′-ACAGAAAGCAAGGAGGATTTC-3′，产物大小是126 bp。D6Rck39的引物序列：39GTF，5′-GCACACTGACAGTGCCCTTA-3′；39GTR，5′-TGTAACCTGGAATTGGGAGC-3′，产物大小128 bp。小鼠品系的杂交和传代按照相关文献操作[8,42]。

为了分离*ob*基因，我们克隆了*Pax4*和D6Rck13之间的DNA（图1），并用这两个探针来构建*ob*基因区域的物理图谱。对应于*Pax4*和D6Rck13序列的酵母人工染色体（YAC）被分离出来并得到鉴定。对每一个YAC的着丝粒和端粒末端进行恢复，而靠近*ob*基因的末端则被用来筛选新的YAC。这些YAC末端序列的恢复是靠载体小件聚合酶链式反应（PCR）或质粒末端拯救技术[10,11]实现的。含有D6Rck13的YAC（编号902A0653）的一个末端（图1的标记(1)）在257号小鼠里发生重组，使得*ob*基因处于这个YAC末端和*Pax4*之间。我们得不到任何可以将编号为yB1S4A5和902A0653的两个YAC末端（图1的标记(2)和(3)）连接起来的YAC。脉冲场凝胶电泳（PFGE）结果显示有一个70 kb的间隙将这两个YAC末端隔开。为了桥连这个间隙，我们用这两个YAC末端的序列分离一系列小鼠P1克隆[12,13]。通过分析这些P1克隆末端，我们发现这些克隆之间有重叠，而且YAC重叠群中这个间隙的大小是70 kb左右。这个跨越*ob*基因基因座的YAC重叠群大小为2.2 Mb，并且*ob*基因就在YAC 903E1016的远末端（图1的标记(5)）和YAC 902A0653的远末端（图1的标记(1)）之间的900 kb区域内。

为了进一步精确定位*ob*基因，我们分析了C57BL/6J *ob*/*ob*品系和DBA/2J品系互交和回交得到的额外771个减数分裂后的基因型[5]。种内杂交的分型分析需要利用D6Rck13和*Pax4*的单链长度多态性（SSLP）技术的发展。通过单独对*Pax4*基因测序我们得到了一个微卫星DNA序列，而在D6Rck13标记附近，通过对来自YAC y902A0653（由D6Rck13分隔）的黏粒进行测序我们发现了另外一个SSLP，标

PCR amplification of genomic DNA with primers flanking these microsatellites revealed polymorphisms among the various progenitor strains of the genetic crosses. No additional recombinants between *ob* and *Pax4* were identified after genotyping the obese backcross and intercross progeny from the crosses to DBA mice. The genetic results indicated that D6Rck39 was distal to *ob* and recombined with *ob* in a single obese animal derived from the C57BL/6J *ob*/*ob* × DBA/2J backcross, animal 167. This recombination occurred between D6Rck39 and the distal end of YAC yB6S2F12 (end (4) in Fig. 1) because a B × D polymorphism defined by this end indicated that it was non-recombinant in animal 167 as well as animals 111 and 420. Thus the distal end of YAC yB6S2F12 (end (4) in Fig. 1) failed to recombine with *ob* among the 1,606 meioses scored. As recombination in animal 111 was localized between this end and the distal end of YAC y903E1016 (end (5), which was non-recombinant in animal 420), the combined data from animals 111 and 167 placed *ob* in, at most, a 650-kb interval between D6Rck39 and YAC end (5). The exact size of the critical region could not be determined until the points of recombination in these animals were precisely localized. For reasons discussed later, fine mapping of these recombination events was not necessary. There was a total of 6 recombination events in this 2.2-Mb contig among 1,606 meioses. Therefore in the region of *ob*, 1 cM of genetic distance corresponded to ~5.8 Mb, a rate of recombination nearly threefold lower than average for the entire mouse genome[14].

To facilitate the identification of genes in the region of *ob*, we isolated P1 clones using the ends of the YACs in this region as well as the ends derived from the initial set of P1 clones. Twenty-four P1 clones were used to construct a contig across most of the 650-kb critical region of *ob*.

Gene Identification

Genes from this 650-kb interval were isolated using the method of exon trapping[15]. DNA from individual or pools of P1 clones was subcloned as *Bam*H1/*Bgl*II digests into the pSPL3 exon trapping vector. Briefly, each exon trapping product was sequenced and compared to all sequences in Genbank using the BLAST computer programme[16]. Putative exons were screened for the presence of corresponding RNA from a variety of tissues using northern blots and reverse-transcription PCR. Six genes were identified: four mapped within the 650-kb critical region of *ob* and two to outside this region.

One of the trapped exons, 2G7, was hybridized to a northern blot of mouse tissues and detected a ~4.5-kb RNA only in white adipose tissue (Fig. 2*a*). This exon was derived from a P1 isolated with the distal end of YB6S2F12 YAC (labelled (4) in Fig. 1). Even when autoradiographs were exposed for up to a week, no signal was detected in any other tissue, but we cannot exclude the possibility that this gene may be expressed elsewhere or below

号 D6Rck39。通过这些微卫星两侧的引物对基因组 DNA 进行 PCR 扩增，我们发现在用来基因杂交的不同原始小鼠品系里面，这些微卫星 DNA 序列具有多态性。在分析与 DBA 品系的互交和回交后代中的肥胖小鼠的基因型后，我们没有发现 *ob* 基因和 *Pax4* 之间有其他的重组发生。这些遗传结果显示 D6Rck39 在 *ob* 基因的远端，并且与 C57BL/6J *ob*/*ob* × DBA/2J 回交得到的 167 号小鼠的 *ob* 基因发生重组。这个重组发生在 D6Rck39 与 YAC yB6S2F12 的末端之间（图 1 的（4）号末端）。因为我们分析了这个 YAC 末端的一个约 B × D 多态标记，并且发现它在 167 号小鼠以及 111 号和 420 号小鼠里面全都没有重组。因此在记录的 1,606 次减数分裂中，YAC yB6S2F12 的远末端（图 1 的（4）号末端）都没有与 *ob* 基因发生重组。由于 111 号小鼠的重组发生在 yB6S2F12 的末端与 YAC y903E1016 远末端之间（末端（5）在 420 号小鼠中未发生重组），所以从 111 号和 167 号小鼠重组的数据分析可知，*ob* 基因最有可能在 D6Rck39 标记和 YAC（5）号末端之间这 650 kb 间隙之内。更为精确的区域大小只有在准确定位这些小鼠体内的重组位点之后才能确定。后面会说到，这样精确的重组位置分析对我们这里来说是不必要的。我们分析的 1,606 次减数分裂中有 6 次重组发生在这个 2.2 Mb 的重叠群里。因此，在 *ob* 基因这个区域中 1 cM 遗传距离大致相当于 5.8 Mb，这大约是小鼠整个基因组的重组频率的 1/3[14]。

为了促进 *ob* 区域的基因鉴定，我们用该区域的 YAC 末端序列和先前得到的一些 P1 克隆末端序列分离 P1 克隆。我们总共用 24 个 P1 克隆来构建了一个涵盖了这 650 kb *ob* 基因临界区域的绝大部分的重叠群。

基 因 鉴 定

利用外显子捕获技术[15]，我们找到了这 650 kb 间隙里面的基因。单个 P1 克隆或者 P1 克隆库中提取的 DNA，用 *Bam*H1/*Bgl*II 消化并亚克隆进 pSPL3 外显子捕获载体。简而言之，就是将外显子捕获的产物测序，然后在 Genbank 数据库里用 BLAST 软件[16]和已知基因序列比对。假定的外显子再用 RNA 印迹或者逆转录 PCR（RT-PCR）技术检测是否在不同组织里有相应的 RNA。用这个方法找到 6 个基因：4 个在这 650 kb *ob* 基因临界区域内，另外 2 个在该区域外。

其中一个名为 2G7 的外显子与小鼠组织进行 RNA 印迹实验，只在白色脂肪组织中检测到一个约 4.5 kb 的 RNA（图 2*a*）。这个外显子是来自于 YB6S2F12 这个 YAC 远末端的一个 P1 克隆（图 1 的（4）号标记）。即使放射自显影的胶片曝光一周时间，也没有检测到来自其他组织的信号，但是我们不能排除这个基因在别处表达或

the level of detection in some tissues. Actin mRNA was detected in all samples (data not shown). Apparent adipose-tissue-specific expression was also seen after RT-PCR of RNA from a variety of tissues using specific primers from the 2G7 exon, with a strong signal being seen only in white fat and actin being detectable in all tissues tested (Fig. 2*b*). The high level of expression of 2G7 in adipose tissue suggested that this exon might be derived from the *ob* gene.

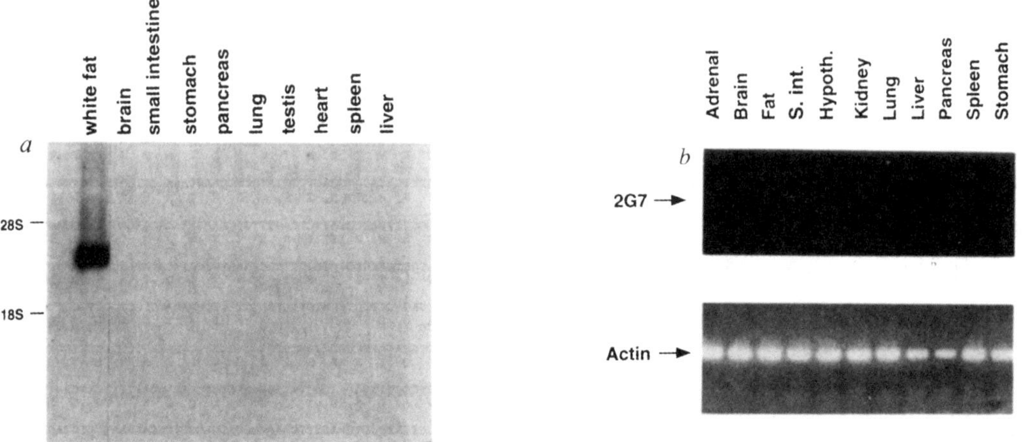

Fig. 2. Tissue distribution of the 2G7 transcript. *a*, Northern blot of total RNA (10 μg) from various tissues probed with labelled 2G7 exon. The 2G7 exon was identified using exon trapping with DNA from a pool of P1 clones in the region of *ob*. This probe hybridized specifically to RNA from white adipose tissue. Autoradiograph signals appeared after 1-h exposure (24-h exposure shown here). The transcript migrated between 28S and 18S ribosomal RNA markers and is estimated to be ~4.5 kb. *b*, Reverse transcription-PCR (RT-PCR) was performed with RNA from each of the tissue samples shown using primers specific for the 2G7 exon or actin. A positive signal was detectable only in white adipose tissue, even when PCR amplification was continued for 30 cycles.

Methods. *a*, Exon trapping was done by ligating *Bgl*II/*Bam*H1 digestion products of a pool of P1 clones into the pSPLIII vector as described[15]. Total RNA was prepared and electrophoresed in formaldehyde gels as in ref. 43. Northern blots were transferred to Immobilon and hybridized to radiolabelled probes as described[44]. *b*, Reverse-transcription PCR reactions were performed using 100 ng total RNA[45]. First-strand cDNA, prepared using random hexamers as primers, was PCR-amplified using primers derived from the 2G7 exon as well as mouse actin. The primers used to detect 2G7 were selected using the Primer program and were: 2G7F(5′CCAGGGCAGGAAAATGTG3′): 2G7R(5′CATCCTGGACTTTCTGGATAGG3′). The mouse actin primers were purchased from Clonetech. PCR amplification was performed for 30 cycles with 94 °C denaturation for 1 min 55 °C hybridization for 1 min, and 72 °C extensions for 2 min with a 1-s autoextension per cycle. RT-PCR products were resolved in a 1.5% low-melting-point agarose (Gibco/BRL) gel run in 0.5 × TBE buffer.

The level of expression of the 2G7 exon was then assayed in the two available *ob* strains by hybridization to northern blots as well as by RT-PCR. Northern blots showed that 2G7 RNA was absent in SM/Ckc-+^{Dac}ob^{2J}/ob^{2J} adipose tissue (Fig. 3*a*). This lack of 2G7 RNA in these animals was demonstrated by RT-PCR of the fat cell RNA, as shown by the absence of a signal after thirty cycles of amplification (Fig. 3*b*). As the ob^{2J} mutation is relatively recent and is maintained as a co-isogenic strain, these data indicate that 2G7

者以极低的水平在其他组织表达的可能性。肌动蛋白的 mRNA 可以在所有组织中被检测到（结果没有展示）。用 2G7 外显子的特异引物对来自不同组织的 RNA 进行 RT-PCR，我们同样可以明显检测到 2G7 在脂肪组织中特异性表达，只在白色脂肪组织中有强信号，并且肌动蛋白在所有组织中均被检测到（图 2*b*）。2G7 外显子在脂肪组织中的高表达提示它可能来自 *ob* 基因。

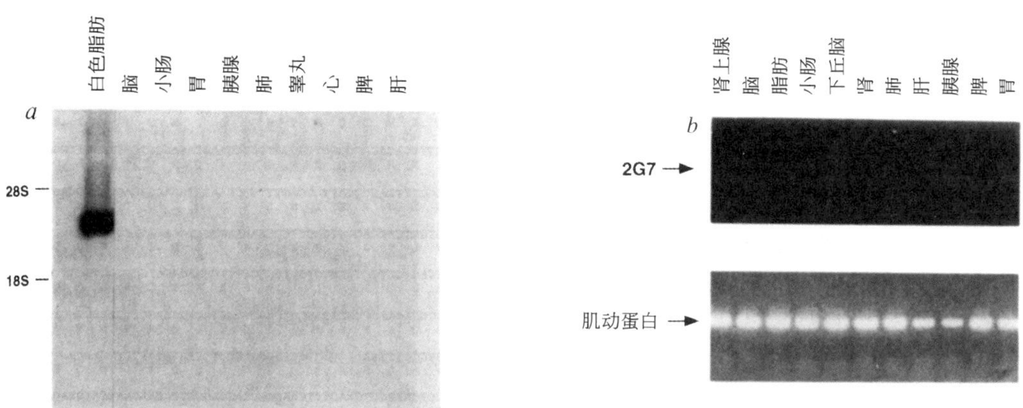

图 2. 2G7 转录产物的组织分布。*a*，不同组织来源的总 RNA（10 μg）用带标签的 2G7 外显子探针的 RNA 印迹结果。2G7 外显子是用外显子捕获技术从含有 *ob* 基因区域的 P1 克隆库的 DNA 序列中找到的。这个探针特异性地与白色脂肪组织的 RNA 杂交。放射自显影信号在曝光 1 小时后出现（这里显示的是 24 小时曝光的信号）。这个转录产物在 28S 和 18S 核糖体 RNA 之间，大小大概是 4.5 kb。*b*，利用特异的引物从各个组织的 RNA 里扩增 2G7 外显子或肌动蛋白基因的逆转录 PCR（RT-PCR）结果。即使是用 30 个循环的 PCR 扩增也只在白色脂肪组织检测到阳性信号。

方法。*a*，外显子捕获是按照文献描述操作，即将 P1 克隆库中 *Bgl*II/*Bam*H1 消化产物连接到 pSPLIII 载体里面 [15]。按照参考文献 43 提取总 RNA，并用甲醛胶进行电泳。按照文献描述的方法将 RNA 印迹转到 Immobilon 膜上，用同位素标记的探针进行 RNA 杂交 [44]。*b*，100 ng 总 RNA 用来进行逆转录 PCR[45]。单链 cDNA 使用随机六聚体引物进行合成，然后用 2G7 外显子和小鼠肌动蛋白的引物进行 PCR 扩增。用来扩增 2G7 外显子的引物是用 Primer 软件设计的，如下所示：2G7F（5′CCAGGGCAGGAAAATGTG3′），2G7R（5′CATCCTGGACTTTCTGGATAGG3′）。小鼠肌动蛋白引物购自 Clonetech 公司。PCR 扩增 30 个循环，94 ℃变性 1 分钟，55 ℃退火 1 分钟，72 ℃延伸 2 分钟加上每个循环有 1 秒的自动延伸。RT-PCR 产物用 1.5% 低熔点琼脂糖凝胶（Gibco/BRL 公司）在 0.5 倍 TBE 缓冲液中进行电泳分离。

接下来，用 RNA 杂交和 RT-PCR 检测 2G7 外显子在含有 *ob* 基因的两个小鼠品系中的表达水平。RNA 印迹显示 2G7 RNA 没有在 SM/Ckc-$+^{Dac}$$ob^{2J}$/$ob^{2J}$ 小鼠的脂肪组织中表达（图 3*a*）。这些小鼠缺少 2G7 RNA 表达的结论同时也被脂肪细胞 RNA 的 RT-PCR 结果证实，RT-PCR 显示 30 个扩增循环之后都没有任何信号（图 3*b*）。由于 ob^{2J} 突变是新近发现的并且一直作为同类品系保存，这些结果提示 2G7 确实是 *ob* 基

encodes an exon from the *ob* gene. This was confirmed by characterization of the mutation in C57BL/6J *ob*/*ob* mice. Northern blots of adipose tissue RNA from C57BL/6J *ob*/*ob* mice showed a ~20-fold increase in the level of *ob* RNA (Fig. 3*a*), suggesting that the original *ob* allele (C57BL/6J *ob*/*ob*) was associated with a non-functional gene product and that the mRNA was increased as part of a possible feedback loop. This turned out to be the case (Fig. 4*b*) as the C57BL/6J *ob*/*ob* phenotype is the result of a nonsense mutation.

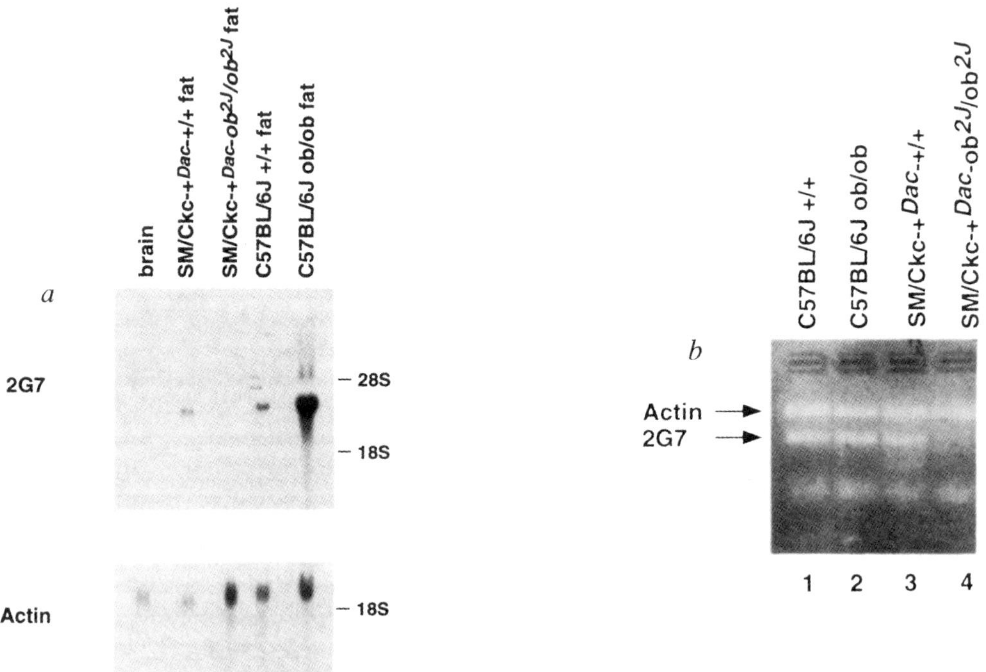

Fig. 3. 2G7 expression in mutant mice. *a*, Northern blot of fat cell RNA isolated from *obese* and lean mice. The 2G7 exon was hybridized to northern blots with 10 μg total RNA from white adipose tissue from each of the strains indicated. An approximately 20-fold increase in the level of 2G7 RNA was apparent in white fat RNA from the C57BL/6J *ob*/*ob* strain relative to lean littermates. There was no detectable signal in RNA from the SM/Ckc-+^{Dac}ob^{2J}/ob^{2J} mice even after a 2-week exposure. A 24-h autoradiographic exposure is shown. The same filter was hybridized to an actin probe (bottom panel). SM/Ckc-+^{Dac} and SM/Ckc-+^{Dac}ob^{2J}/ob^{2J} mice were provided by S. Lane and B. Paigen of the Jackson Laboratory. C57BL/6J *ob*/*ob* mice were purchased from the Jackson Laboratory. *b*, RT-PCR of mutant RNA. Reverse-transcription PCR was performed for 30 cycles of amplification using 100 ng total fat RNA from each of the samples shown. In this experiment both the actin and 2G7 primer pairs were included in the same PCR reaction. The complete absence of 2G7 RNA was demonstrated in the SM/Ckc-+^{Dac}ob^{2J}/ob^{2J} adipose tissue. For RT-PCR reactions, 100 ng total RNA was used[45].

因的一个外显子。这个猜想被 C57BL/6J *ob/ob* 小鼠的突变特征所证实。C57BL/6J *ob/ob* 小鼠脂肪组织的 RNA 印迹结果显示，*ob* 基因的 RNA 水平增加至原来的约 20 倍(图 3*a*)。这可能是由于原始 *ob* 基因的等位基因(C57BL/6J *ob/ob*)与一个非功能的基因产物有关，而 mRNA 作为可能的反馈环的一部分其表达量增加。图 4*b* 的结果正说明 C57BL/6J *ob/ob* 小鼠的表型是由一个无义突变导致的。

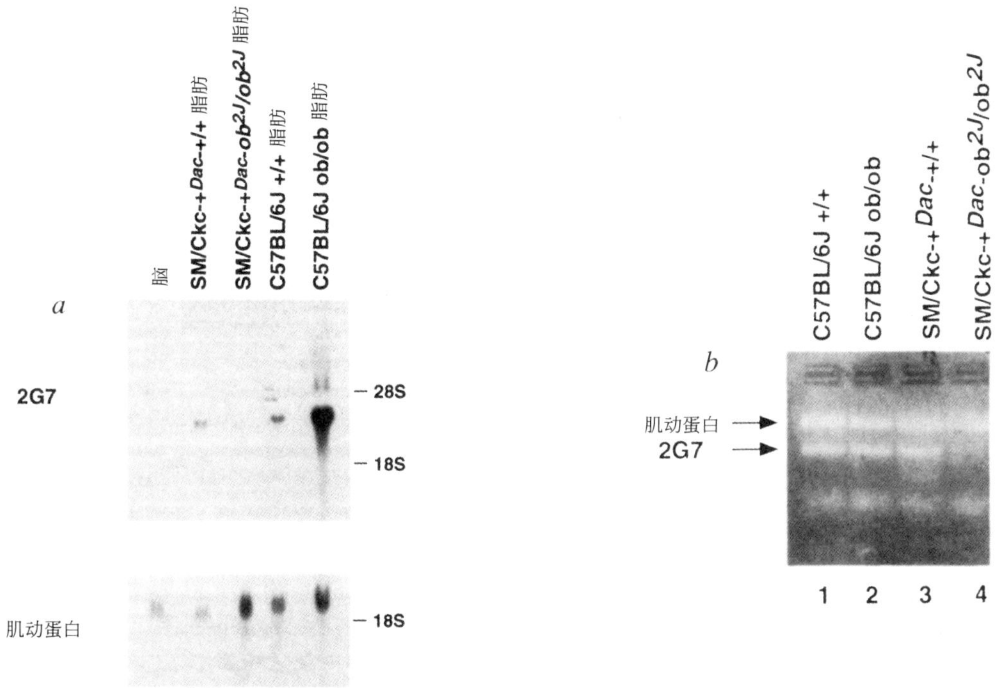

图 3. 2G7 在突变小鼠体内的表达。*a*，从肥胖和纤瘦小鼠脂肪细胞中提取的 RNA 的 RNA 印迹结果。如图所示，不同品系白色脂肪组织提取 10 μg 总 RNA 与 2G7 外显子杂交进行 RNA 印迹。从 C57BL/6J *ob/ob* 品系白色脂肪组织提取的 RNA，相对于同一窝较瘦小鼠的 RNA，2G7 外显子的表达水平大约增加至原来的 20 倍。即使是曝光两个星期，从 SM/Ckc-$+^{Dac}$$ob^{2J}/ob^{2J}$ 品系小鼠提取的 RNA 也检测不到信号。这里展示的是一个 24 小时放射自显影的图片。肌动蛋白的探针也是在同一张膜上杂交(底下一个图片)。SM/Ckc-$+^{Dac}$ 和 SM/Ckc-$+^{Dac}$$ob^{2J}/ob^{2J}$ 小鼠是杰克逊实验室的莱恩和帕伊根提供的。而 C57BL/6J *ob/ob* 小鼠则是从杰克逊实验室购买。*b*，突变 RNA 的 RT-PCR 结果。如图所示，不同小鼠的 100 ng 总脂肪 RNA 用来进行 30 个循环的逆转录 PCR。在这个实验中肌动蛋白和 2G7 外显子的引物是加到同一个 PCR 反应里面。这个结果证明了 2G7 RNA 在 SM/Ckc-$+^{Dac}$$ob^{2J}/ob^{2J}$ 品系小鼠的脂肪组织中完全没有表达。100 ng 总 RNA 被用来做 RT-PCR 反应[45]。

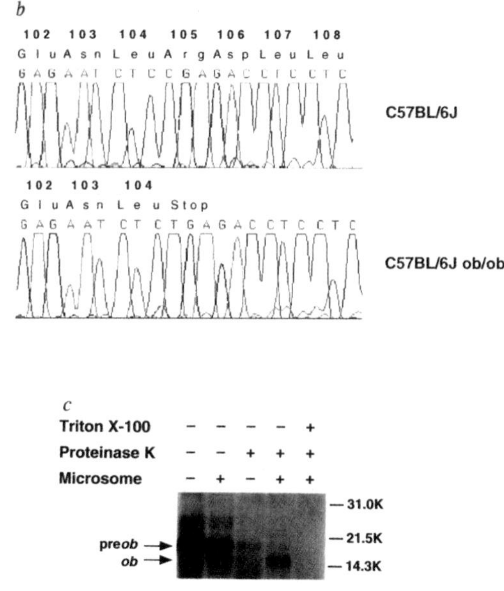

Fig. 4. *a*, Sequence of *ob* cDNA. 22 cDNA clones were isolated from a mouse white fat cDNA library and sequenced. A 97-bp 5′ leader was followed by a predicted 167-amino-acid ORF and a ~3,700-kb 3′-untranslated sequence. A total of ~2,500 base pairs of the 3′-untranslated sequence is shown. Analysis of the predicted protein sequence using the *sigseq* computer program suggests the presence of a signal sequence (underlined)[21]. Microheterogeneity of the cDNA was noted in that ~70% of the cDNAs had a glutamine codon at codon 49 and 30% did not. This amino acid is underlined as is the arginine codon that is mutated in C57BL/6J *ob/ob* mice (see *b*). After cleavage of the signal sequence, two cysteines remain at the C terminus of the molecule, possibly with formation of a disulphide bond. cDNA clones were isolated from a mouse adipose tissue library (Clonetech)[46]. Inserts were prepared directly from phage DNA or by PCR amplification of the insert using primers from the λgt10 cloning site (Clonetech). PCR products were prepared for sequencing after electrophoresis in low-melting-point agarose gels (Gibco/BRL) using agarase[47]. DNA was sequenced manually or using an ABI 373A automated sequencer[48,49]. 5′ RACE was done after dG tailing of first-strand cDNA using terminal transferase followed by PCR amplification with the reverse complement of the 2G7F primer and oligo (dC)[18]. *b*, The C57BL/6J *ob/ob* mutation. RT-PCR

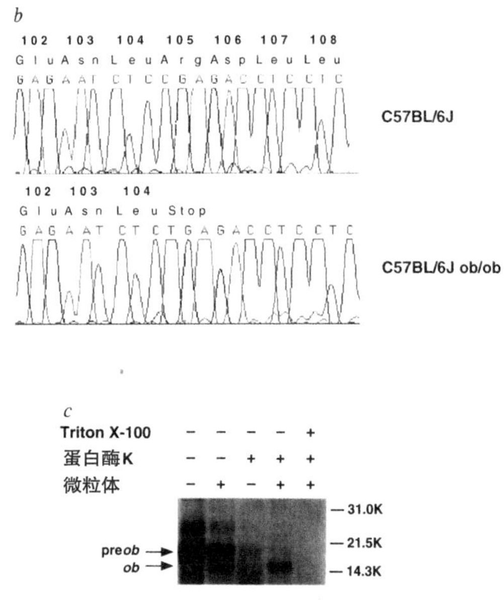

图 4. *a*，*ob* 基因 cDNA 序列。从小鼠白色脂肪组织 cDNA 文库中分离的 22 个 cDNA 克隆被用来测序。整个 cDNA 包括一个 97 bp 的 5′ 前导序列，然后是一个预测有 167 个氨基酸的开放阅读框和一个约 3,700 kb 的 3′ 非翻译序列。3′ 非翻译区的约 2,500 bp 在图中表示出来了。用 *sigseq* 软件分析预测的蛋白序列提示存在一个信号肽序列（图中下划线标注）[21]。cDNA 的微不均一性表现为，大约 70% cDNA 在 49 位密码子处有谷氨酰胺但是另外 30% 没有。该氨基酸用下划线标注，因为在 C57BL/6J *ob*/*ob* 小鼠体内突变为精氨酸密码子（见图 *b*）。信号肽被剪切以后，在蛋白的 C 端仍然有两个半胱氨酸可能会形成二硫键。cDNA 的克隆是从小鼠白色脂肪组织文库中筛选出来的（Clonetech 公司）[46]。插入片段直接是噬菌体 DNA 中得到或者用 λgt10 克隆位点的引物从插入片段中 PCR 扩增出来（Clonetech 公司）。PCR 产物在低熔点琼脂糖凝胶（Gibco/BRL 公司）中电泳，再用琼脂酶处理后进行测序 [47]。DNA 测序是用手工或者 ABI 373A 测序仪进行 [48,49]。5′ RACE 是用末端转移酶在单链 cDNA 上加脱氧鸟苷序列之后用 2G7F 引物和寡脱氧胞苷进行 PCR 扩增 [18]。*b*，C57BL/6J *ob*/*ob* 突变。用 C57BL/6J+/+ 和 C57BL/6J *ob*/*ob* 小鼠的白色脂肪组织 RNA 和 5′ 端以及 3′ 端非翻译区的引物进行 RT-PCR，扩增的产物经胶纯化后，用引物编码区的双链进行测序。C57BL/6J *ob*/*ob* 的小鼠有一个 C 到 T 的突变，导致第 105 位精

was carried out using white fat RNA from C57BL/6J+/+ and C57BL/6J *ob*/*ob* mice and primers from the 5′ and 3′ untranslated regions. PCR reaction products were gel-purified and sequenced using primers along both strands of the coding sequence. The C57BL/6J *ob*/*ob* mice had a C → T mutation which changed Arg 105 to a stop codon. This base change is shown as the output from the automated sequencer. The cDNA sequences of mutant and wild-type mice were otherwise identical. The wild-type cDNA sequence was also seen in genomic DNA from v/Le and C57BL/10 mice, the progenitor strains of the stock on which the C57BL/6J *ob*/*ob* mutation originally arose (data not shown). *c*, *In vitro* translation of *ob* RNA. A human *ob* cDNA was subcloned into the pGEM cloning vector (see Fig. 5*b*). Plus-strand RNA was synthesized using SP6 polymerase. The *in vitro* synthesized RNA was translated in the presence or absence of canine pancreatic microsomal membranes. An ~18K primary translation product was seen after *in vitro* translation. The addition of microsomal membranes (mb) to the reaction led to the appearance of a second translation product ~2K smaller than the primary translation product. The 16K product was resistant to proteinase K but was rendered protease-sensitive when the microsomal membranes were permeabilized by Triton X-100, indicating that a functional signal sequence was present. A human *ob* cDNA was cloned into the pGEM-3zf(+) plasmid. *In vitro* transcription was carried out using a Ribo MAX large-scale RNA production system with SP6 polymerase (Promega). The transcription mixture was used without further purification in a wheat-germ translation system (Promega) with or without 5 μg of a canine pancreas microsomal membrane preparation[50] at 27 °C for 2 h. Proteinase K (100 μg ml^{-1}) digests were performed on ice for 1 h with and without 0.1% Triton X-100. Translation products were analysed by SDS–PAGE.

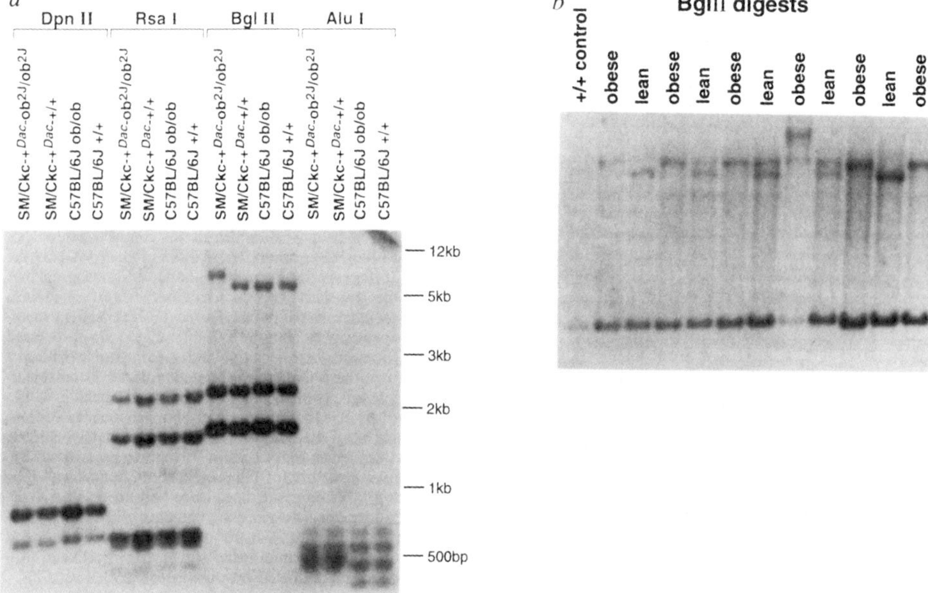

Fig. 5. The ob^{2J} mutation. *a*, Southern blots of genomic DNA from mutant animals. The 2G7 probe was hybridized to 5 μg of restriction-enzyme-digested genomic DNA from each strain. Restriction digestion with *Bgl*II revealed an increase in the size of a ~9 kb (the largest) *Bgl*II fragment in SM/Ckc-$+^{Dac}ob^{2J}/ob^{2J}$ DNA. RFLPs were not detectable with any other restriction enzymes. Preliminary restriction mapping of genomic DNA indicated that the polymorphic *Bgl*II site is ~7 kb upstream of the transcription start site (data not shown). None of the other enzymes tested extend past the mRNA start site. Genomic DNA preparation and Southern blotting have been described[8]. *b*, Segregation of a *Bgl*II polymorphism in the SM/Ckc-$+^{Dac}ob^{2J}/ob^{2J}$. Six obese and five lean progeny from the same generation of the coisogenic SM/Ckc-$+^{Dac}ob^{2J}/ob^{2J}$ colony were genotyped by scoring the *Bgl*II polymorphism in *a*. All of the phenotypically obese animals were homozygous for the larger allele of the polymorphic *Bgl*II fragment. The DNA in the control lane was prepared from an unrelated SM/Ckc-$Dac^{+/+}$ mouse, bred separately from the SM/Ckc-$+^{Dac}ob^{2J}/ob^{2J}$ colony.

氨酸密码子变成终止密码子。这个碱基突变在自动测序仪的测序结果中显示出来了。突变型和野生型小鼠的其他测序结果是一致的。野生型小鼠的 cDNA 序列也存在于 v/Le 和 C57BL/10 两个小鼠的基因组 DNA 中（结果没有展示）。这两个品系是最初出现 C57BL/6J *ob*/*ob* 突变的小鼠的原品系。*c*，*ob* 基因 RNA 的体外翻译。人 *ob* 基因 cDNA 被亚克隆到 pGEM 克隆载体里（见图 5*b*）。利用 SP6 聚合酶来合成正链 RNA。这个体外合成的 RNA 在犬胰腺微粒体膜系统存在或缺失的情况下被翻译。体外翻译得到一个约 18K 大小的初始翻译产物。反应中存在微粒体膜系统时会得到一个次生的、比原始产物小约 2K 的翻译产物。这个 16K 的蛋白对蛋白酶 K 不敏感，但是如果微粒体膜系统被 Triton X-100 透化以后这个蛋白对蛋白酶又再次敏感了。这个结果提示存在一个功能性的信号肽。人类的 *ob* 基因 cDNA 被克隆到 pGEM-3zf(+)质粒里面。使用 Ribo MAX 大规模 RNA 转录系统和 SP6 聚合酶（普洛麦格公司）进行体外转录。没有经过进一步纯化的转录混合物接下来被直接用在加或者不加 5 μg 犬胰腺微粒体膜系统的小麦胚芽翻译系统（普洛麦格公司）[50] 中，然后 27 ℃温浴 2 小时。在加或者不加 0.1% Triton X-100 的情况下，冰上进行蛋白酶 K(100 μg · ml^{-1}) 消化 1 小时。用十二烷基硫酸钠–聚丙烯酰胺凝胶电泳（SDS–PAGE）分析翻译产物。

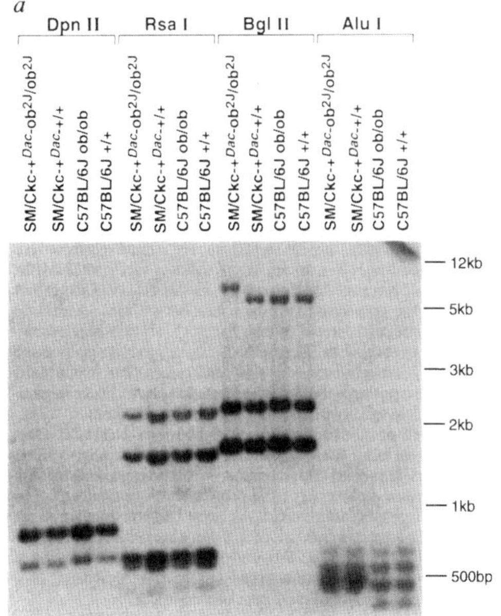

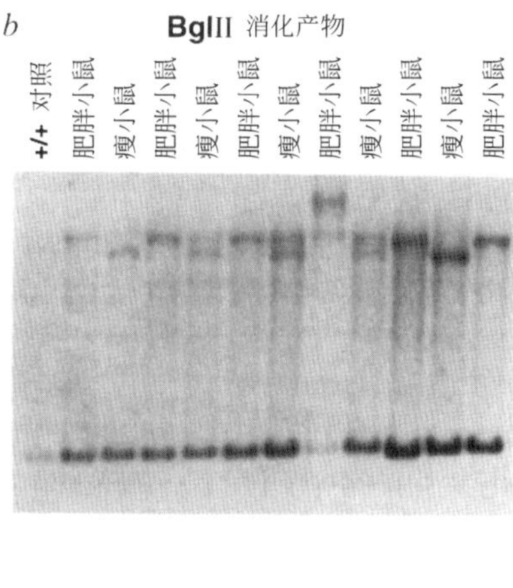

图 5. ob^{2J} 突变。*a*，突变小鼠基因组的 DNA 印迹结果。2G7 的探针与来自各个小鼠品系的经内切酶酶切后的 5 μg 基因组 DNA 进行杂交。*Bgl*II 酶切 SM/Ckc-+$^{Dac}ob^{2J}/ob^{2J}$ DNA 出现一个更大的约 9 kb（最大的）的产物。用其他的内切酶没有发现任何限制性片段长度多态性。基因组 DNA 的初步内切酶位点图谱显示这个 *Bgl*II 酶切位点的多态性位于转录起始位点上游大概 7 kb 的位置（结果没有展示）。在 mRNA 起始位点以后没有任何其他的酶切位点。基因组 DNA 的提取和 DNA 杂交的方法按照文献描述进行 [8]。*b*，*Bgl*II 位点多态性在 SM/Ckc-+$^{Dac}ob^{2J}/ob^{2J}$ 小鼠中得到分离。在图 *a* 中，SM/Ckc-+$^{Dac}ob^{2J}/ob^{2J}$ 品系同窝的六只肥胖小鼠和五只瘦小鼠的后代通过 *Bgl*II 多态性来进行基因型确定。所有肥胖表型的小鼠都是比较大的多态性 *Bgl*II 片段等位基因纯合子。图中对照组的 DNA 是从一个与 SM/Ckc-+$^{Dac}ob^{2J}/ob^{2J}$ 小鼠分开饲养的、无关的 SM/Ckc-Dac$^{+/+}$ 小鼠品系中提取的。

Sequence of the *ob* Gene

The 2G7 exon was used to isolate a total of 22 complementary DNA clones from a mouse adipose tissue cDNA library. None of the cDNA clones extended more than 97 base pairs upstream of 2G7, whereas each extended a variable distance downstream. Sequencing of the cDNA clones revealed a methionine initiation codon in the 2G7 exon, with a 167-amino-acid open reading frame followed by a long 3′-untranslated sequence (Fig. 4). A potential Kozak translation initiation consensus sequence was present with an adenosine residue three bases upstream of the ATG[17]. One of the cDNA clones extended to the 5′ end of the mRNA because its sequence was identical to that of the 5′ RACE (rapid amplification of cloned ends) products of adipose tissue RNA (data not shown)[18]. A total of 2.9 kb of cDNA sequence is shown, most of which is 3′-untranslated sequence. A search for internal homology within the cDNA revealed a 50-base-pair (bp) direct repeat in both the 5′ and 3′ untranslated sequence; this sequence was not found in Genbank and there were no additional segments of internal homology.

Two classes of cDNA were found which differed by inclusion or exclusion of a single glutamine codon (underlined in Fig. 4). This residue is found in a position immediately 3′ to the splice acceptor of the 2G7 exon. As the CAG codon of glutamine includes a possible AG splice-acceptor sequence, there appears to be slippage at the splice-acceptor site, with an apparent 3-bp deletion in a subset of the cDNAs[19]. This glutamine residue is located in a highly conserved region of the molecule but its significance is unknown.

To identify the mutation in C57B6/J *ob*/*ob* mice, RT-PCR products from the entire open reading frame (ORF) were prepared from adipose tissue RNA from this mutant and both strands sequenced. The coding sequences were identical apart from a C→T mutation in C57BL/6J *ob*/*ob* mice that results in a change of an arginine at position 105 to a stop condon (Fig. 4*b*). This amino acid is also underlined in Fig. 4*a*. This base change did not occur in genomic DNA from v/Le or C57BL/10 mice, the strains in which this mutation was carried before it was transferred to C57BL/6J (S. Lane, personal communication). DNA sequence changes from CGA to TGA are not uncommon as a result of the high mutation rate of methyl cytosine to thymidine[20]. This *ob* mutation is presumed to inactivate the protein as the phenotype in both C57BL/6J and SM/Ckc-$+^{\text{Dac}}ob^{2J}/ob^{2J}$ mice is identical (S. Lane, personal communication; data not shown).

A database search of the *ob* protein using the BLAST program[16] identified no significant homology to any sequences in Genbank. The predicted polypeptide was largely hydrophilic and had a putative N-terminal signal sequence[22]. Amino-acid sequence analysis using the *sigseq* computer program indicated a 98% likelihood that a signal sequence was present at the N terminus[21] (underlined in Fig. 4*a*). The predicted signal sequence cleavage site is C terminal to an alanine at position 21. Computer analysis of the human protein (Fig. 6*b*) also suggests that it has an N-terminal signal sequence, so human and mouse *ob* may both encode secreted proteins. There is some divergence in the mouse

ob 基因的序列

用2G7外显子从小鼠脂肪组织的cDNA文库里找到22个互补DNA(cDNA)克隆。但是没有任何一个cDNA克隆超过2G7上游97个碱基对，而且每一个克隆的下游序列长度都不一样。通过对cDNA克隆测序，我们在2G7外显子上找到一个甲硫氨酸的起始密码子和一个有很长的3′非翻译序列的167个氨基酸的开放阅读框(图4)。在起始位点ATG上游的3个碱基处是一个腺苷酸残基以及类似Kozak翻译起始的共有序列[17]。其中一个cDNA克隆其实是延伸到了mRNA的5′末端，因为它的序列和脂肪组织RNA的cDNA 5′ RACE(末端快速扩增法)得到的产物一致(结果没有展示)[18]。这个全长2.9 kb的cDNA序列，绝大部分是3′非翻译序列。对基因内部的同源性进行分析，我们在5′和3′非翻译序列各发现一个50碱基对(bp)的重复；这个重复序列在Genebank数据库没有找到，并且也没有其他的内部同源性片段。

研究发现两类cDNA，它们的不同之处在于是否包括单个谷氨酰胺密码子(图4下划线)。这个残基位于2G7外显子3′末端剪接受体处。由于谷氨酰胺的密码子CAG中包含一个可能的AG剪接受体序列，这就可能导致剪接位点有一个偏移进而导致后面的cDNA序列缺失3个碱基对[19]。这个谷氨酰胺残基位于整个分子的一个非常保守的区域，但是它的重要性还未知。

为了鉴定C57B6/J *ob*/*ob* 小鼠基因的突变，我们从突变小鼠的脂肪组织RNA里面扩增出基因的整个开放阅读框(ORF)cDNA并且对双链进行测序。所有编码序列都进行了鉴定，在C57BL/6J *ob*/*ob* 小鼠体内发现了一个会导致105位精氨酸变成终止密码子的一个C到T的突变(图4*b*)。这个氨基酸在图4*a*用下划线标记。这个碱基的改变并不存在于v/Le或C57BL/10两种小鼠的基因组DNA中，在这两种小鼠中，这种突变被转移到C57BL/6J品系之前已携带(莱恩，个人交流)。CGA到TGA的突变在DNA里面也很常见，因为甲基化的胞嘧啶变成胸腺嘧啶的突变率很高[20]。由于C57BL/6J和SM/Ckc-$+^{Dac}ob^{2J}/ob^{2J}$的表型一样，所以这个*ob*突变也被认为可以导致蛋白完全失活(莱恩，个人交流；结果没有展示)。

在Genebank数据库里面用BLAST软件[16]搜索不到任何高度同源的OB蛋白序列。预测的蛋白序列基本是亲水性的，而且N端可能有信号序列[22]。用*sigseq*软件分析氨基酸序列表明有98%的可能性蛋白N端有信号序列[21](图4*a*下划线标记)。C端的第21位的丙氨酸是预测的信号肽酶切位点。计算机分析人类的OB蛋白(图6*b*)也有一个N端信号序列，所以人和小鼠的*ob*基因可能都编码分泌蛋白。小鼠和人类的信号序列虽然有些差异，但是*sigseq*软件给出了同样高的分数。为了确认

and human signal sequences, but both give identical scores when analysed using *segseq*. To confirm the presence of a functional signal sequence, a human cDNA that included the entire ORF was subcloned into a pGEM vector (suitable mouse subclones were not recovered). Positive-strand human *ob* RNA was transcribed using SP6 polymerase and translated *in vitro* with and without canine pancreatic microsomal membranes[50]. The primary translation product migrated with an apparent relative molecular mass of ~18K, consistent with that predicted from the cDNA sequence. Inclusion of microsomal membranes in the reaction inhibited translation ~5-fold, but about half of the *ob* primary translation product was truncated by ~2K in the presence of microsomal membranes, which suggests that the signal sequence is functional (Fig. 4*c*). To confirm that the *ob* protein had been translocated, we treated *in vitro* translation products with proteinase K, which caused complete proteolysis of the 18K primary translation product whereas the 16K processed form was unaffected, indicating that the translation product had been translocated into the microsomal lumen. Permeabilization of the membranes with Triton X-100 rendered the processed form protease-sensitive and is compatible with the hypothesis that *ob* is a secreted molecule. After signal-sequence cleavage, two cysteine residues would remain within the predicted protein, suggesting that the molecule may contain the disulphide bond characteristic of other secreted polypeptides. Amino-acid sequence and secondary structure prediction support a globular structure for the protein (data not shown). The largely hydrophilic amino-acid sequence had no notable structural motifs or membrane-spanning domains other than the N-terminal signal sequence. We find no consensus for *N*-linked glycosylation or dibasic amino-acid sequences indicative of protein cleavage in the predicted processed protein[22].

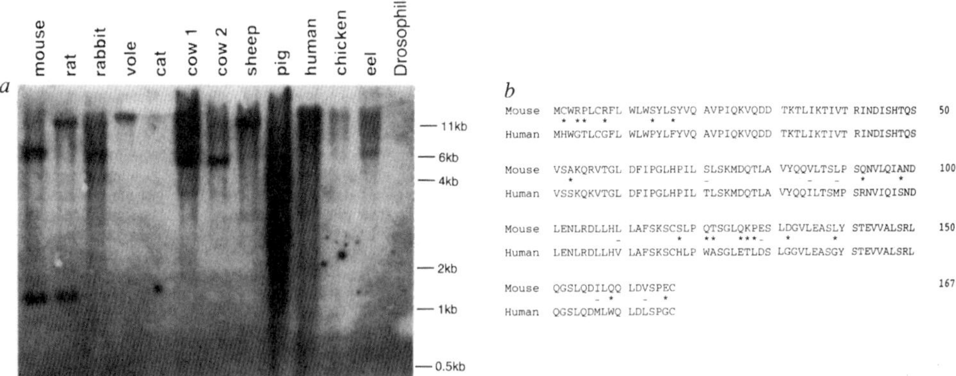

Fig. 6. Evolutionary conservation of *ob*. *a*, Cross-species hybridization of *ob*. Genomic DNA from each of the species shown was Southern blotted after *Eco*RI digestion and probed with *ob* cDNA. Hybridization signals were detectable in every vertebrate sample even after moderate-stringency hybridization. The cat DNA used in this experiment was slightly degraded. DNA was prepared from tissues or cell lines from the organisms shown. Southern blot hybridization was at 65 °C, with two washes in 2 × SSC/0.2% SDS at 65 °C for 20 min, and autoradiograph exposure was for 3 d using Kodak X-OMAT film. *b*, Sequence of human *ob* protein. The mouse *ob* gene was used as a probe to isolate human adipose tissue cDNA clones from a λgtII library. The sequence of human *ob* cDNA was highly homologous to that of the mouse cDNA in the predicted 167-amino-acid coding sequence. There was 30% homology in the 5′ and 3′ untranslated sequences in this tissue. Predicted amino-acid sequences of the human and mouse proteins have been

这个功能信号序列的存在，我们将一个含有整个开放阅读框的人类 cDNA 亚克隆到 pGEM 载体里（没有能够得到合适的小鼠亚克隆）。SP6 聚合酶用来反转录人 *ob* RNA 的正链，然后在有和没有犬胰腺微粒体膜系统[50]的两种情况下进行体外翻译。最初的翻译产物跑胶测量的相对分子质量大约 18K，这与 cDNA 序列预测的大小一致。如果翻译在微粒体膜系统进行，翻译效率大概下降为 1/5 左右，但是一半的翻译产物在这种情况下约被剪切了 2K。这个结果提示信号序列确实具有功能（图 4*c*）。为了确认 OB 蛋白确实可以易位，我们用蛋白酶 K 处理了体外翻译产物。结果 18K 的最初产物被全部酶解，但是剪切后的 16K 产物基本不受影响。这个结果提示我们翻译产物被转运进了微粒体腔内。用 Triton X-100 透化的微粒体膜可以让剪切后的产物对蛋白酶敏感，这个结果符合 OB 是分泌蛋白的假设。信号肽被剪切掉以后，OB 蛋白仍然有两个半胱氨酸残基，说明该分子可能像其他分泌蛋白一样具有二硫键。氨基酸序列和二级结构预测则支持这是一个球状蛋白（结果没有展示）。除了 N 端的信号序列，主要的亲水性氨基酸序列没有明显的结构模体或者跨膜结构域。在预测的剪切蛋白中，我们没有发现提示蛋白裂解的 *N* 连接糖基化或者预示蛋白酶位点的双碱基氨基酸序列[22]。

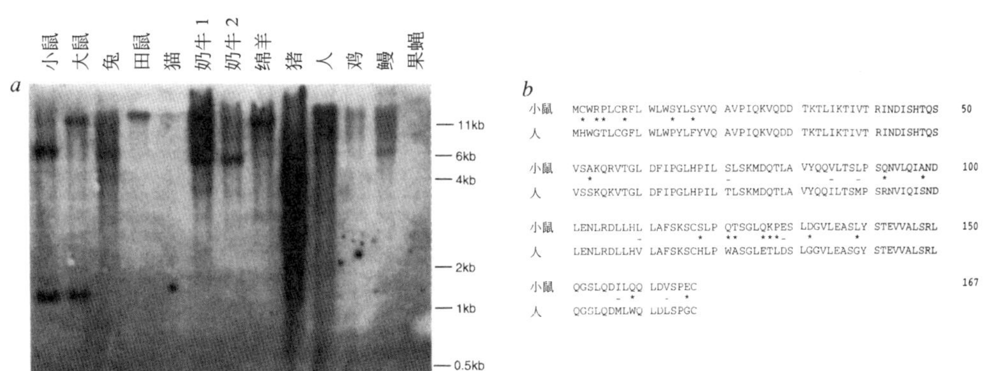

图 6. *ob* 基因的进化保守性。*a*，物种间 *ob* 基因的杂交。各个物种的基因组 DNA 都先用 *Eco*RI 酶切，然后与 *ob* 基因 cDNA 探针进行 DNA 印迹杂交。所有脊椎动物的基因组 DNA 即使在中等严格的杂交条件下都有信号显示。猫的 DNA 在这个实验中有少许降解。这些 DNA 是从图中所示的动物组织或者细胞系中提取。DNA 印迹杂交温度为 65 ℃，用 2×SSC/0.2% SDS 缓冲液在 65 ℃下洗脱 20 分钟，洗脱两次后用柯达 X-OMAT 显影膜放射自显影 3 天。*b*，人类 OB 蛋白的序列。小鼠的 *ob* 基因序列被用作探针从人类 λgtII 脂肪组织 cDNA 文库中分离人类 *ob* 基因。人类 *ob* 基因的 cDNA 序列在预测的 167 个氨基酸的编码区内与小鼠的同源性非常高。在该组织内 5′ 和 3′ 非翻译区域有 30% 的同源性。图示为预测的小鼠和人类蛋白序列比对，保守的氨基酸改变用虚线表示而不保守的改变为星号。那个可变的谷氨酰胺密码子用下划线标出，该位点同时是 C57BL/6J *ob*/*ob* 小鼠的无义突变位点。虽然从 22 位缬氨酸（紧

aligned. Conservative changes are noted by a dash and non-conservative changes by an asterisk. The variable glutamine codon is underlined, as is the position of the nonsense mutation in C57BL/6J *ob/ob* mice. Overall there is 84% identity at the amino-acid level, although only seven substitutions were found between the valine at codon 22 (immediately downstream of the signal sequence overage) and the cysteine at position 117. cDNA clones were isolated from a human adipose tissue cDNA library (Clonetech) as described for Fig. 4, except that PCR amplification of the phage insert was carried out using primers adjacent to the cloning site of λgtII. PCR products were sequenced directly.

The ob^{2J} Mutation

We next explored the molecular basis for the mutation in SM/Ckc-$+^{Dac}ob^{2J}/ob^{2J}$ mice. In these animals, the absence of 2G7 RNA was associated with an increase in the size of an ~9 kb *Bgl*II fragment of genomic DNA in affected animals (Fig. 5*a*). Although the precise nature of this polymorphism is not established, the altered *Bgl*II site maps ~7 kb upstream of the mRNA start site for the 4.5 kb *ob* RNA (data not shown), suggesting that this mutation may be the result of a structural alteration or sequence variation in the promoter. None of the other restriction fragments reach the promoter region. Nevertheless, this polymorphism is significant because it is always associated with the obese phenotype in a colony that segregates the *ob* 2J mutation (Fig. 5*b*). In this experiment, DNA from a total of six obese and five lean animals was Southern-blotted after *Bgl*II digestion and probed with 2G7. In each case homozygosity for the larger of the polymorphic *Bgl*II fragments was associated with an obese phenotype. Each of the lean animals was either homozygous for the smaller *Bgl*II allele (+/+) or heterozygous (*ob*/+). As reported, no overt phenotypic differences were apparent between +/+ and *ob*/+ mice[23].

ob Sequence Conservation

The coding sequence of the *ob* gene was hybridized to genomic Southern blots of vertebrate DNAs (Fig. 6*a*). Even at moderate stringency, there were detectable signals in all vertebrate DNAs tested, demonstrating that *ob* is highly conserved among vertebrates. To determine the extent of this *ob* sequence conservation, we isolated and sequenced cDNA clones hybridizing to *ob* from a human adipose tissue cDNA library. The nucleotide sequences from human and mouse were highly homologous in the predicted coding sequence, but had only 30% homology in the available 5′ and 3′ untranslated regions. Homology of the human polypeptide apparent N-terminal signal sequence to the mouse equivalent region was slightly lower than in the rest of the molecule. The signal sequence is seen to be functional from the truncation of the primary translation product in the presence of canine pancreas microsomal membranes (Fig. 4*c*). Alignment of the predicted human and mouse amino-acid sequences (Fig. 6*b*) shows that there is 84% overall identity and more extensive identity in the N terminus of the mature protein, with only four conservative and three non-conservative substitutions among the residues between the signal sequence cleavage site and the conserved cysteine at position 117. As in mouse, 30% of the clones were missing a glutamine at codon 49.

接着信号肽的下游）到 117 位半胱氨酸之间只有 7 个替换，但是氨基酸水平共有 84% 的一致性。如图 4 所示，除使用 λgtII 克隆位点附近的引物 PCR 扩增噬菌体插入片段以外，cDNA 克隆是从人类脂肪组织 cDNA 文库（Clonetech 公司）中分离的。PCR 产物直接用于测序。

ob^{2J} 的突变

我们接下来研究了 SM/Ckc-+$^{Dac}ob^{2J}/ob^{2J}$ 小鼠突变的分子机制。在这些突变体中，2G7 RNA 的缺少与患病小鼠基因组 DNA 中一个大约 9 kb 的 *Bgl*II 酶切片段的增加是相关的（图 5*a*）。虽然关于这个多态性的确切机制未知，但是改变的 *Bgl*II 内切酶位点在 4.5 kb *ob* RNA 的 mRNA 起始位点上游大约 7 kb 处（结果没有展示），由此提示这个突变可能是由启动子结构或者序列的改变导致的。在此启动子区域内没有任何其他酶切片段。然而这个多态位点还是非常重要，因为从群体中分离 *ob* 2J 突变的时候，它总是和肥胖表型联系在一起（图 5*b*）。在这个实验中，总共 6 只肥胖小鼠和 5 只瘦小鼠的 DNA 用 *Bgl*II 消化，然后用 2G7 探针进行 DNA 印迹杂交。在每个实验里面，多态性 *Bgl*II 大片段的纯合性总是和肥胖表型联系在一起；而瘦小鼠则可能是纯合小片段（+/+）或者是杂合子（*ob*/+）。与先前报道的结果一样，+/+ 和 *ob*/+ 小鼠并没有明显的表型不同 [23]。

ob 序列的保守性

ob 基因的编码区序列与其他脊椎动物的 DNA 进行 DNA 印迹杂交（图 6*a*）。即使是在中等严格的条件下，我们也能够在所有的脊椎动物 DNA 中检测到信号，这表明 *ob* 基因在所有脊椎动物里都非常保守。为了进一步确定 *ob* 基因序列保守性的程度，我们分离并且测定了从人类脂肪组织 cDNA 文库分离的与 *ob* 基因杂交的 cDNA 克隆。在编码区，人与小鼠的核酸序列是非常保守的，但是在得到的 5′ 和 3′ 非翻译区则只有 30% 的同源性。两个多肽的 N 端信号序列的同源性也稍微比其他区域低一些。而这些信号序列的功能性已经被含有犬胰腺微粒体膜系统的体外翻译实验证明（图 4*c*）。预测的人与小鼠氨基酸序列比对结果（图 6*b*）显示有 84% 的相同序列，并且成熟蛋白的 N 端相似性更高。在 N 端信号序列切割位点和第 117 位的保守半胱氨酸之间就只有 4 个保守和 3 个不保守的氨基酸残基替换。与小鼠体内一样，30% 的克隆缺少第 49 位密码子编码的谷氨酰胺。

Discussion

Obesity has been a focus of discussion[24], with a line of research dating back to Lavoisier and Laplace (1783) implying that energy balance—food intake versus energy output—is physiologically regulated[25-27]. The site of this regulation has been the subject of nearly one hundred years of debate[28-30]. Of the brain regions implicated in the regulation of feeding behaviour, the ventromedial nucleus of the hypothalamus (VMH) is considered to be the most important satiety centre in the central nervous system (CNS). The increase in body weight associated with VMH lesions is a result of both increased food intake and decreased energy expenditure[29].

The energy balance in mammals was therefore postulated to be controlled by a feedback loop in which the amount of stored energy is sensed by the hypothalamus, which adjusts food intake and energy expenditure to maintain a constant body weight[31,32]. The nature of the inputs to the hypothalamus was unclear[27]. According to the lipostasis theory[32], the size of the body fat depot is regulated by the CNS, with a product of fat metabolism circulating in plasma and affecting energy balance by interacting with the hypothalamus. The glucostasis theory[33] suggested that the plasma glucose level is a key signal regulating energy stores. A third possibility is that body temperature is an important input to the CNS centres controlling food intake[31]. The inability to identify the putative signal from fat has hindered the validation of the lipostasis theory. Moreover, neither the glucostasis theory nor theories on the thermal regulation of food intake fully account for the precision with which energy balance is regulated *in vivo*. The possibility that at least one component of the signalling system circulates in the bloodstream was first suggested by Hervey[34], who showed that the transfer of blood from an animal with a VMH lesion across a vascular graft to an untreated animal (a parabiosis experiment) resulted in a reduction of food intake in the intact animal. The biochemical nature of the putative signal has remained elusive, although it has been suggested that *ob* is responsible for the generation of this blood-borne factor[6].

Our results, particularly the evidence that the *ob* protein product is secreted, suggest that *ob* may encode this circulating factor. Evidence that the gene described here is *ob* includes the identification of a nonsense mutation in the C57BL/6J *ob*/*ob* strain and a genomic alteration in the SM/Ckc-$+^{Dac}ob^{2J}/ob^{2J}$ strain that is associated with an absence of RNA. The data showing that one strain overproduces *ob* RNA while the other fails to express this gene, make it unlikely that these changes are secondary. This conclusion can be confirmed using transgenic mice to complement the obese mutation. The nature of the mutation in the co-isogenic SM/Ckc-$+^{Dac}ob^{2J}/ob^{2J}$ mice is as yet unknown. Preliminary analysis of the gene structure has indicated that the polymorphic *Bgl*II site maps ~7 kb upstream of the promoter. As this mutation is co-isogenic, the data strongly suggest that this polymorphism is associated with the mutation, although it is not yet clear whether there is a structural alteration (insertion or deletion) or base change. Further studies will be required to identify the molecular basis for the absence of *ob* RNA expression in SM/Ckc-$+^{Dac}ob^{2J}/ob^{2J}$ mice, as well as the molecular mechanisms that result in the increased expression of *ob* RNA in C57BL/6J *ob*/*ob* mice.

讨　论

肥胖症已经是一个讨论的焦点[24]，追溯到拉瓦锡和拉普拉斯（1783年）时代的一系列研究已经暗示能量的平衡——食物的摄取与能量的消耗——是受生理因素调节的[25-27]。这种调节位点已经争论了上百年[28-30]。大脑的某个区域与调节进食行为有关，下丘脑腹内侧核（VMH）就是这样的一个区域，它被认为是中枢神经系统（CNS）最重要的饱中枢。与VMH的损伤相关的身体重量增加是由进食增加和能量代谢下降所致[29]。

因此，哺乳动物的能量平衡假定是由一个反馈环控制的，下丘脑能感受在反馈环中能量存储的数量，并以此调节进食和能量代谢来保持体重恒定[31,32]。输入到下丘脑的信息的本质并不清楚[27]。根据脂肪稳态学说[32]，体脂存量的体积是受中枢神经系统调节的。存在于血浆中的某种脂肪代谢产物会在下丘脑发生作用从而调节能量平衡。血糖稳态学说[33]则认为血糖水平是调节能量存储的重要信号分子。第三种可能性则是体温会作为中枢神经系统调节进食的重要信息[31]。不能从脂肪中找到假定的信号分子，这阻碍了脂肪稳态学说的证实。此外，血糖稳态学说和体温调节进食的学说都不能完全解释体内能量平衡的精确调节。赫维[34]通过血管嫁接手术将下丘脑腹内侧核受损动物的血液输入正常动物体内（连体共生的实验）导致正常动物进食减少，从而首次提出血液中至少存在一种信号分子调节进食。虽然OB被认为可能是这样一种血源性因子[6]，但是假定信号分子的生化性质仍然未知。

我们的结果，特别是OB作为一种分泌蛋白的证据，提示*ob*基因可能编码这样一种循环因子。文章中所说的基因就是*ob*基因的证据包括在C57BL/6J *ob*/*ob*小鼠中发现无义突变和*ob* RNA缺失的SM/Ckc-+^{Dac}*ob*^{2J}/*ob*^{2J}小鼠品系的基因组改变。数据显示，一个品系*ob* RNA表达增加而另外一个品系*ob*表达缺失的结果也不太可能是继发性效应。这个结论可以由转基因小鼠修复肥胖突变体来证实。在同类系的SM/Ckc-+^{Dac}*ob*^{2J}/*ob*^{2J}小鼠体内，基因突变的机制还是未知的。基因结构的初步分析结果显示，启动子上游约7 kb处有一个多态性*Bgl*II位点。由于这个突变是同类系的，虽然不知道是否有结构变化（插入或者缺失）或碱基改变，但是这个数据还是强烈暗示这个多态性与基因突变有关。进一步研究需要明确SM/Ckc-+^{Dac}*ob*^{2J}/*ob*^{2J}突变体中*ob* RNA表达缺失的分子基础，以及C57BL/6J *ob*/*ob*突变体中*ob* RNA表达增加的分子机制。

The increased level of *ob* RNA in the C57BL/6J mutant, which has greatly increased fat cell mass, raises the possibility that the level of expression of this gene signals the size of the adipose depot. This hypothesis is consistent with the lipostasis theory, implying that an increase in the level of the *ob* signal (as might occur after a prolonged binge of eating) may act directly or indirectly on the CNS to inhibit food intake and/or regulate energy expenditure as part of a homeostatic mechanism to maintain constancy of the adipose mass. Such effects could be explained if the *ob* protein activated the sympathetic nervous system via interactions with the VMH. Direct effects of *ob* on the CNS would require that mechanisms exist to allow its passage across the blood-brain barrier. The means by which the CNS effects changes in body weight remain to be elucidated and may include mechanisms independent of effects on food intake and energy expenditure. In addition, *ob* may have endocrine, paracrine and autocrine effects on different tissues. These and other possibilities will be testable if the active form of the *ob* protein can be shown to circulate in plasma.

The *ob* mutation is associated with a myriad of hormonal and metabolic alterations[35]. These include effects on thermo-regulation, fertility, adrenal and thyroid function as well as a wide range of biochemical abnormalities. Although we cannot exclude the possibility that many of these pleiotypic effects are the result of the expression of *ob* in sites not included in our initial survey of mouse tissues, the apparent expression of *ob* in only adipose tissue suggests that many of these changes are secondary. It has been suggested that the complex *ob* phenotype reflects an imbalance in the activity of the autonomic nervous system with low sympathetic and high parasympathetic tone[35]. If *ob* is a signal molecule that regulates body weight, it may act by interacting with CNS (and other) receptors to modulate food intake and the activity of the autonomic nervous system. Parabiosis experiments suggest that the *ob* receptor is encoded by the mouse *db* (diabetes) gene[6], a possibility that can be tested by positional cloning of *db* or by expression cloning and mapping of the *ob* receptor[36].

The extensive homology of the *ob* gene product among vertebrates suggests that its function is highly conserved. Now that the human homologue of *ob* is cloned, it will be possible to test for mutations in the human *ob* gene. As *ob* heterozygotes have an enhanced ability to survive a prolonged fast[23], heterozygous mutations at *ob* may provide a selective advantage in human populations subjected to caloric deprivation[37]. Identification of *ob* now offers an entry point into the pathways that regulate adiposity and body weight and should provide a fuller understanding of the pathogenesis of obesity.

(**372**, 425-432; 1994)

Yiying Zhang[*†], **Ricardo Proenca**[*†], **Margherita Maffei**[†], **Marisa Barone**[*†], **Lori Leopold**[*†] & **Jeffrey M. Friedman**[*†‡]
[*] Howard Hughes Medical Institute, †The Rockefeller University, 1230 York Avenue, New York, New York 10021, USA
‡ To whom correspondence should be addressed.

Received 17 October; accepted 4 November 1994.

C57BL/6J 突变小鼠的 *ob* RNA 表达增加同时脂肪细胞的质量也增加许多，这增加了这个基因表达水平是脂肪组织积累的一个信号的可能性。这个假设与脂肪稳态学说吻合，暗示增加 *ob* 信号分子的表达（可能发生在长期暴饮暴食之后）可能直接或者间接地影响中枢神经系统来抑制进食和（或）调节能量消耗，作为维持脂肪质量恒定的一个稳态机制。如果 OB 蛋白可以通过下丘脑腹内侧核激活交感神经系统，那么就能解释这个效果了。当然，OB 如果直接作用于中枢神经系统，它必须能够通过血脑屏障。中枢神经系统调节体重的方式仍然不清楚，可能包括不受进食和能量消耗影响的机制。另外，OB 在不同组织内可能起内分泌、旁分泌和自分泌作用。如果有活性的 OB 蛋白能在血液内循环，这些所有的可能性都要被考虑到。

ob 突变伴随着大量的激素和代谢水平的改变 [35]，这些改变包括体温调节，生育能力，肾上腺和甲状腺功能，以及大量的生化异常问题。虽然我们不能完全排除这些多效性是由 *ob* 在我们最初研究的小鼠组织之外的组织中表达造成的，但 *ob* 只在脂肪组织明显表达这一点提示所有这些变化都是继发效应。*ob* 突变体复杂的性状已经反映了自主神经系统活动的紊乱，即交感神经兴奋性低且副交感神经兴奋性高 [35]。如果 OB 真的可以作为一个信号分子调节体重，那么它可能与中枢神经系统（或者其他系统）的受体相互作用来调节进食和自主神经系统的活动。异种共生的实验提示 OB 的受体可能是由小鼠糖尿病（*db*）基因编码的 [6]。通过定点克隆 *db* 基因，或者表达性克隆 OB 受体和绘制 OB 受体的遗传图，我们可以验证这个假说 [36]。

脊椎动物 *ob* 基因产物的高度同源性提示它的功能是高度保守的。现在人类 *ob* 同源基因已经被克隆了，我们就可以检测人类 *ob* 基因里面的突变。由于 *ob* 杂合子能够活得更长 [23]，*ob* 基因的杂合突变可能在热量不足的人群中具有选择优势 [37]。*ob* 基因的鉴定会给脂肪和体重调节通路的研究提供一个突破口并让我们对肥胖的致病机制有更完善的了解。

（莫维克 翻译；焦炳华 审稿）

References:

1. Burros, M. *Despite awareness of risks, more in US are getting fat* in *The New York Times* 17 July 1994.
2. Grundy, S. M. & Barnett, J. P. *Disease-a-Month* **36**, 645-696 (1990).
3. Friedman, J. M. & Leibel, R. L. *Cell* **69**, 217-220 (1990).
4. Ingalls, A. M., Dickie, M. M. & Snell, G. D. *J. Hered.* **41**, 317-318 (1950).
5. Friedman, J. M. *et al. Genomics* **11**, 1054-1062 (1991).
6. Coleman, D. L. *Diabetologia* **14**, 141-148 (1978).
7. Dickie, M. M. & Lane, P. W. *Mouse News Lett.* **17**, 52 (1957).
8. Bahary, N. *et al. Mamm. Genome* **4**, 511-515 (1993).
9. Walther, C. *et al. Genomics* **11**, 424-434 (1991).
10. Riley, J. *et al. Nucleic Acids Res.* **18**, 2887-2890 (1990).
11. Hermanson, G. G. *et al. Nucleic Acids Res.* **19**, 4943-4948 (1991).
12. Sternberg, N. *Trends Genet.* **8**, 11-16 (1992).
13. Hartl, D. L. & Nurminsky, D. I. *BioTechniques.* **15**, 201-208 (1993).
14. Dietrich, W. *et al. Genetics* **131**, 423-447 (1992).
15. Church, D. M. *et al. Nature Genet.* **6**, 98-105 (1994).
16. Gish, W. & States, D. J. *Nature Genet.* **3**, 266-272 (1993).
17. Kozak, M. *Cell* **44**, 283 -292 (1986).
18. Frohman, M. A., Dush, M. K. & Martin, G. R. *Proc. Natl. Acad. Sci. U.S.A.* **85**, 8998-9002 (1988).
19. Padgett, R. A. *et al. A. Rev. Biochem.* **55**, 1119-1150 (1986).
20. Cooper, D. N. & Youssoufian, H. *Hum. Genet.* **78**,151-155 (1988).
21. Heijne, G. v. *Nucleic Acids Res.* **14**, 4683-4690 (1986).
22. Sabatini, D. D. & Adesnick, M. B. *The Metabolic Basis of Inherited Disease* (eds Scriver, C. R. *et al.*) 177-223 (McGraw-Hill, New York, 1989).
23. Coleman, D. L. *Science* **203**, 663-665 (1979).
24. Bray, G. A. *Int. J. Obesity* **14**, 909-926 (1990).
25. Rubner, M. *Ztschr. Biol.* **30**, 73-142 (1894).
26. Adolph, E. F. *Am. J. Physiol.* **151**, 110-125 (1947).
27. Hervey, G. R. *Nature* **222**, 629-631 (1969).
28. Cannon, W. B. & Washburn, A. L. *Am. J. Physiol.* **29**, 441-454 (1912).
29. Brobeck, J. R. *Physiol. Rev.* **25**, 541-559 (1946).
30. Hetherington, A. W. & Ranson, S. W. *Am. J. Physiol.* **136**, 609-617 (1942).
31. Brobeck, J. R. *Yale J. Biol. Med.* **20**, 545-552 (1948).
32. Kennedy, G. C. *Proc. R. Soc.* B**140**, 578-592 (1953).
33. Mayer, J. *Ann. N.Y. Acad. Sci.* **63**, 15-43 (1955).
34. Hervey, G. R. *J. Physiol.* **145**, 336-352 (1959).
35. Bray, G. A. *J. Nutrition* **121**, 1146-1162 (1991).
36. Bahary, N. *et al. Proc. Natl. Acad. Sci. U.S.A.* **87**, 8642-8646 (1990).
37. Neel, J. V. *Am. J. Hum. Genet.* **14**, 353-362 (1962).
38. Larin, Z., Monaco, A. P. & Lehrach, H. *Proc. Natl. Acad. Sci. U.S.A.* **88**, 4123-4127 (1991).
39. Green, E. D. & Olson, M. V. *Proc. Natl. Acad. Sci. U.S.A.* **87**, 1213-1217 (1990).
40. Kusuml, K. *et al. Mamm. Genome* **4**, 391-392 (1993).
41. Vidal, S. M. *et al. Cell* **73**, 469-485 (1993).
42. Friedman, J. M. *et al. Genomics* **11**, 1054-1062 (1991).
43. Chirgwin, J. M. *et al. Biochemistry* **18**, 5294-5299 (1979).
44. Friedman, J. M., Cheng, E. & Darnell, J. E. *J. Molec Biol.* **179**, 37-53 (1984).
45. Wang, A. M., Doyle, M. V. & Mark, D. F. *Proc. Natl. Acad. Sci. U.S.A.* **86**, 9717-9721 (1989).
46. Benton, W. D. & Davis, R. W. *Science* **195**, 180-182 (1977).
47. Burmeister, M. & Lehrach, H. *Trends. Genet.* **5**, 41 (1989).
48. Sanger, F., Nicklen, S. & Coulson, A. R. *Proc. Natl. Acad. Sci. U.S.A.* **74**, 5463-5467 (1977).
49. Venter, C. J. *Automated DNA Sequencing and Analysis* (Academic, New York, 1993).
50. Yu, Y. *et al. Proc. Natl. Acad. Sci. U.S.A.* **86**, 9931-9935 (1989).

Acknowledgements. We thank D. Coleman for prior genetic and physiological analysis; R. Leibel and N. Bahary for their important contributions to the early phases of this work; D. Koos and R. Cox for help in isolating the first set of YAC clones; A. Buckler, J. Rutter and C. Stotler for instruction in exon trapping; Y. Yu and B. Lauring for advice on *in vitro* translation and for canine pancreatic microsomal membranes; P. Gruss for providing the *Pax4* probe before publication and communicating its map position; J. E. Darnell, N. Heintz, R. Kucherlapati, S. Burley, J. Froude, D. Luck, C. Winestock and L. Safani for comments; S. Korres for preparing the manuscript; J. Sholtis for photography; F. Ilchert for help in preparing the figures; A. Popowicz, P. Dash and the staff of computing services for assistance; D. Sabatini for expert advice on the characterization of the *ob* polypeptide; and J. E. Darnell Jr, who provided the environment to initiate this work. J.M.F. is an investigator of the Howard Hughes Medical Institute. This work was supported by a grant from NIH / NIDDK.

Viral Dynamics in Human Immunodeficiency Virus Type 1 Infection

Xiping Wei *et al.*

Editor's Note

By 1995, experimental drugs that block the replication of the AIDS virus HIV-1 were being tested in patients. These studies provided an opportunity to study the dynamics of viral replication and of the CD4 immune-response cells (lymphocytes) that HIV attacks. In this paper, HIV researcher George M. Shaw and colleagues show that the composite lifespan of plasma virus and CD4 lymphocytes is remarkably short. The results suggested that the steady, persistent viral levels seen in patients actually represent a balance between the rapid destruction of infected cells and the ongoing infection of new cells. The findings surprised those involved in developing AIDS treatments, offering new understanding of how HIV works.

The dynamics of HIV-1 replication *in vivo* are largely unknown yet they are critical to our understanding of disease pathogenesis. Experimental drugs that are potent inhibitors of viral replication can be used to show that the composite lifespan of plasma virus and virus-producing cells is remarkably short (half-life ~2 days). Almost complete replacement of wild-type virus in plasma by drug-resistant variants occurs after fourteen days, indicating that HIV-1 viraemia is sustained primarily by a dynamic process involving continuous rounds of *de novo* virus infection and replication and rapid cell turnover.

THE natural history and pathogenesis of human immunodeficiency virus type-1 (HIV-1) infection are linked closely to the replication of virus *in vivo*[1-17]. Clinical stage is significantly associated with all measures of virus load, including infectious virus titres in blood, viral antigen levels in serum, and viral nucleic acid content of lymphoreticular tissues, peripheral blood mononuclear cells (PBMCs) and plasma (reviewed in ref. 18). Moreover, HIV-1 replication occurs preferentially and continuously in lymphoreticular tissues (lymph node, spleen, gut-associated lymphoid cells, and macrophages)[11,19,20]; virus is detectable in the plasma of virtually all patients regardless of clinical stage[6,10,13,21]; and changes in plasma viral RNA levels predict the clinical benefit of antiretroviral therapy (R. Coombs, unpublished results). These findings emphasize the central role of viral replication in disease pathogenesis.

Despite the obvious importance of viral replication in HIV-1 disease, relatively little quantitative information is available regarding the kinetics of virus production and clearance *in vivo*, the rapidity of virus and $CD4^+$ cell population turnover, and the

人免疫缺陷病毒 1 型感染的病毒动力学研究

魏西平（音译）等

编者按

截至 1995 年，多个用于阻断艾滋病病毒 HIV-1 复制的实验性药物用于临床测试。这些研究为研究病毒复制以及 HIV 攻击的 CD4 免疫应答细胞（淋巴细胞）的动力学提供了机会。在这篇文章中，HIV 研究者乔治·肖及其同事指出血浆病毒和 CD4 淋巴细胞的复合生存期非常短。研究结果表明，HIV 患者稳定且持续的病毒水平实际上是感染细胞迅速消亡与大量新感染细胞生成的平衡结果。这些发现使开发抗艾滋病疗法的研究者感到意外，为理解 HIV 发病机制提供新见解。

尽管 HIV-1 体内复制动力学对我们了解疾病的发病机制至关重要，但目前研究相对匮乏。有效抑制病毒复制的实验性药物结果显示，血浆病毒和产病毒细胞的复合生存周期非常短暂（半衰期大约是 2 天）。14 天治疗以后，血浆内野生型病毒几乎被耐药的变异体完全替代。这表明，HIV-1 病毒血症主要通过包括持续的新病毒感染、复制以及快速的细胞更替在内的动态过程维持。

人免疫缺陷病毒 1 型（HIV-1）感染的自然病史和发病机制都与病毒体内复制密切相关 [1-17]。临床分期与各种测量方式测定的病毒载量显著相关。这些测量方式包括血液感染性病毒滴度，血清病毒抗原水平，淋巴网状组织、外周血单个核细胞（PBMC）和血浆内的病毒核酸含量（参考文献 18 中综述）。此外，HIV-1 病毒优先并持续在淋巴网状组织（淋巴结、脾、肠相关淋巴样细胞和巨噬细胞）内复制 [11,19,20]；临床各期患者血浆中均能检测到病毒 [6,10,13,21]；而且血浆病毒 RNA 水平的改变可预测抗逆转录病毒治疗的临床疗效（库姆斯，未发表的结果）。这些结果强调了病毒复制在疾病发病中的关键作用。

尽管病毒复制在 HIV-1 型艾滋病中具有明显重要性，但是有关病毒在体内产生和清除、病毒及 $CD4^+$ 细胞群快速更替、生物学相关病毒突变固定速率的动力学定量信息相对较少 [22,23]。造成这种情况的原因，一方面是已有的抗逆转录病毒药物缺乏足够

fixation rates of biologically relevant viral mutations[22,23]. This circumstance is largely due to the fact that previously available antiretroviral agents lacked sufficient potency to abrogate HIV-1 replication, and methods to quantify virus and determine its genetic complexity were not sufficiently sensitive or accurate. We overcame these obstacles by treating subjects with new investigational agents which potently inhibit the HIV-1 reverse transcriptase (nevirapine, NVP)[24] and protease (ABT-538; L-735,524)[25,26]; by measuring viral load changes using sensitive new quantitative assays for plasma virus RNA[6,18,27]; and by quantifying changes in viral genotype and phenotype in uncultured plasma and PBMCs using automated DNA sequencing[28] and an *in situ* assay of RT function[29,30].

Virus Production and Clearance

Twenty-two HIV-1-infected subjects with $CD4^+$ lymphocyte counts between 18 and 251 per mm^3 (mean $\pm$ s.d., 102 ± 75 cells per mm^3) were treated with ABT-538 ($n = 10$), L-735,524 ($n = 8$) or NVP ($n = 4$) as part of phase I/IIA clinical studies. The design and clinical findings of those trials will be reported elsewhere (K. Squires *et al.*, and V.A.J. *et al.*, manuscripts in preparation). Plasma viral RNA levels in the 22 subjects at baseline ranged from $10^{4.6}$ to $10^{7.2}$ molecules per ml (geometric mean of $10^{5.5}$) and exhibited maximum declines generally within 2 to 4 weeks of initiating drug therapy (Figs 1 and 2*a*). For ABT-538- and L-735,524-treated patients, virus titres fell by as much as $10^{3.9}$-fold (mean decrease of $10^{1.9}$-fold) whereas for NVP-treated patients virus fell by as much as $10^{2.0}$-fold (mean decrease of $10^{1.6}$-fold). The overall kinetics of virus decline during the initial weeks of therapy with all three agents corresponded to an exponential decay process (Figs 1 and 2*a*).

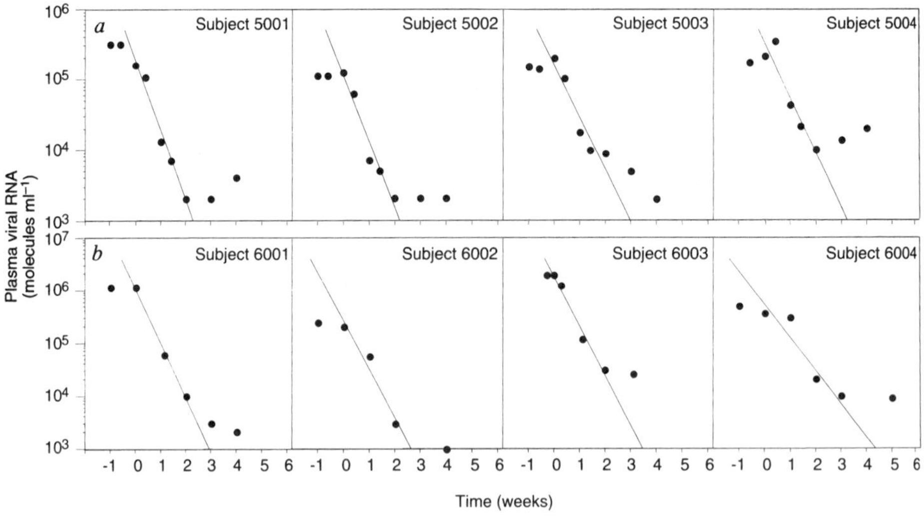

Fig. 1. Plasma viral RNA determinations in representative subjects treated with the HIV-1 protease inhibitors ABT-538 (*a*) and L-735,524 (*b*). Subjects had not received other anti-retroviral agents for at least 4 weeks before therapy. Treatment was initiated at week 0 with 400–1,200 mg d^{-1} of ABT-538 or

的效力抑制病毒 HIV-1 的复制，另一方面是量化病毒和测定病毒基因复杂性的方法不足够敏感或准确。我们通过多种手段克服这些障碍：使用新研发的高效 HIV-1 逆转录酶抑制剂（奈韦拉平，NVP）[24] 和蛋白酶抑制剂（ABT-538；L-735,524）[25,26] 治疗患者；使用新的敏感定量方法监测血浆中的病毒 RNA，进而评估病毒载量的改变 [6,18,27]；使用自动化的 DNA 测序 [28] 和原位 RT 功能检测方法来量化未培养的血浆和 PBMC 中病毒基因型和表型的变化 [29,30]。

病毒产生和清除

作为 I/IIA 期临床研究的一部分，22 名 HIV-1 感染受试者接受 ABT-538（$n = 10$）、L-735,524（$n = 8$）或者 NVP（$n = 4$）治疗。这些患者的 $CD4^+$ 淋巴细胞计数范围为每立方毫米 18~251 个（均值 ± 标准差：每立方毫米 102 ± 75 个）。相关临床研究的设计和结果将另有报道（斯夸尔斯等和约翰逊等，稿件准备中）。22 名受试者血浆病毒 RNA 的基线水平为每毫升 $10^{4.6}$~$10^{7.2}$ 分子（几何平均数是 $10^{5.5}$）。血浆病毒 RNA 水平通常在药物治疗后的 2~4 周内出现最大下降（图 1 和图 2*a*）。经 ABT-538 和 L-735,524 治疗的 HIV 感染患者，血内病毒滴度下降为 $1/10^{3.9}$（平均下降为 $1/10^{1.9}$），而经 NVP 治疗的患者病毒滴定最低降为 $1/10^{2.0}$（平均下降为 $1/10^{1.6}$）。在所有三种药物治疗的前几周，病毒减少的整体动力学呈现指数式衰减（图 1 和图 2*a*）。

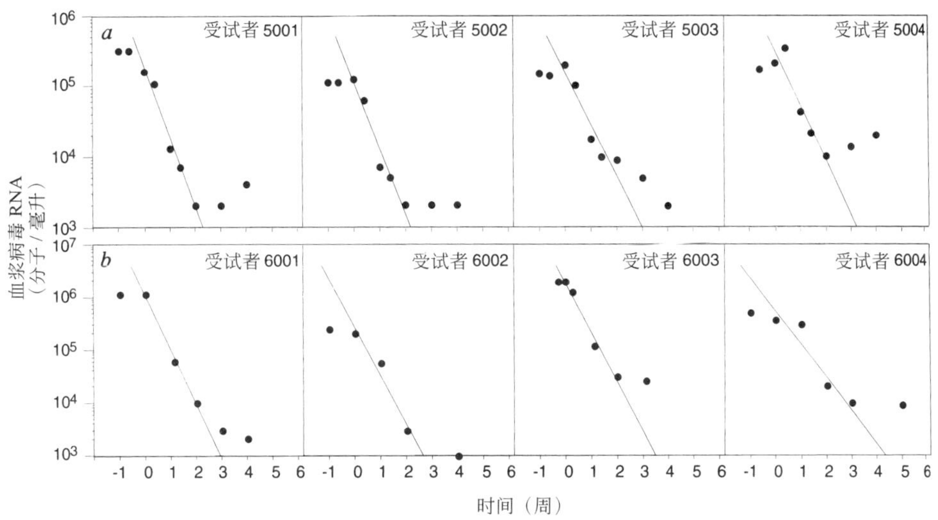

图 1. HIV-1 蛋白酶抑制剂 ABT-538（*a*）和 L-735,524（*b*）治疗后，代表性受试者血浆病毒 RNA 水平的测定。受试者开始此项研究前至少 4 周未接受其他抗逆转录病毒药物治疗。治疗开始后，患者全程接受

1,600–2,400 mg d^{-1} of L-735,524 and was continued throughout the study. Viral RNA was determined by modified branched DNA (bDNA)[18] (*a*) or RT-PCR[27] (*b*) assay and confirmed by QC-PCR[6]. Shown are the least-squares fit linear-regression curves for data points between days 0 and 14 indicating exponential (first-order) viral elimination.

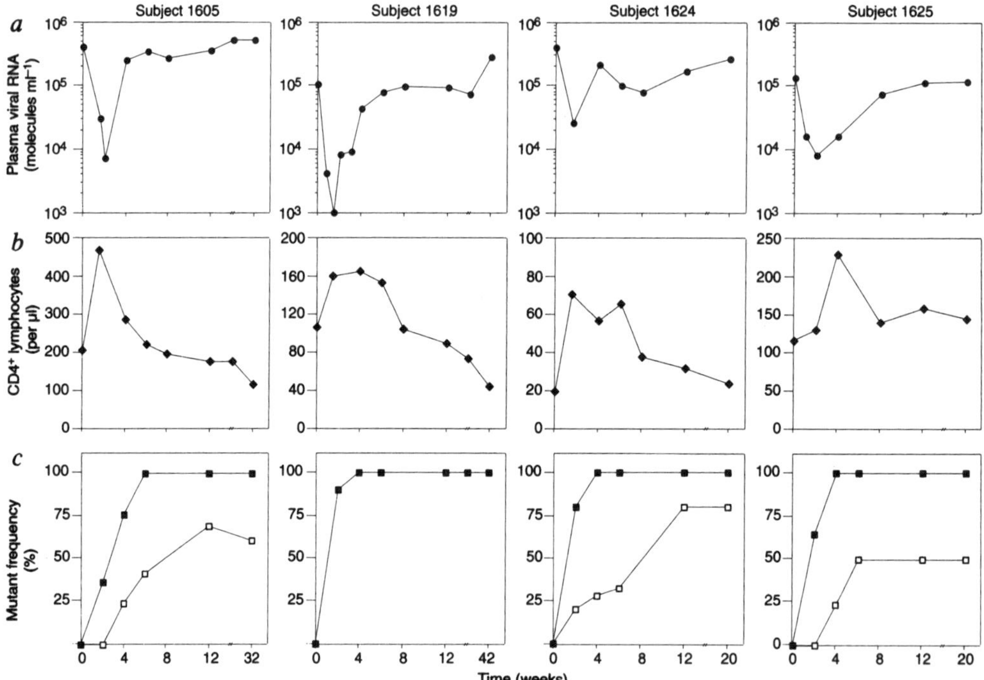

Fig. 2. Plasma viral RNA determinations (*a*), CD4⁺ lymphocyte counts (*b*), and percentages of mutant viral genomes in plasma and PBMCs (*c*) of subjects initiating treatment with NVP. Subjects were participants in a clinical protocol assessing the effects of NVP when added to existing treatment with ddI (subject 1605) or ddI plus zidovudine (subjects 1619, 1624, 1625). Treatment with NVP was initiated at week 0 using 200 mg per day and was increased to 400 mg per day after 2 weeks. ddI and zidovudine dosages were 400 mg per day and 300–600 mg per day, respectively. Viral RNA (●) was determined by QC-PCR assay[6]. CD4⁺ lymphocytes (◆) were quantified by flow cytometry. Frequencies of viral genomes containing NVP-resistance-associated mutations in plasma (■) and PBMCs (□) were determined by automated DNA sequence analysis (Fig.3, legend), with each data point representing the average of 3–6 independent PCR amplifications and sequence determinations.

The antiretroviral agents used in this study, despite their differing mechanisms of action, have a similar overall biological effect in that they block *de novo* infection of cells. Thus the rate of elimination of plasma virus that we measured following the initiation of therapy is actually determined by two factors: the clearance rate of plasma virus *per se* and the elimination (or suppression) rate of pre-existing, virus-producing cells. To a good approximation, we can assume that virus-producing cells decline exponentially according to $y(t) = y(0)e^{-\alpha t}$, where $y(t)$ denotes the concentration of virus-producing cells at time t after the initiation of treatment and α is the rate constant for the exponential decline. Similarily, we

400~1,200 mg · d^{-1} ABT-538 或 1,600~2,400 mg · d^{-1} L-735,524 治疗。病毒 RNA 通过改良的分支 DNA 测定法（bDNA）[18]（a）或者 RT-PCR 法 [27]（b）测定，并经 QC-PCR 确证 [6]。图中显示了第 0 天和第 14 天之间数据点的最小二乘法拟合的线性回归曲线，结果表明病毒呈指数式减少（一级动力学）。

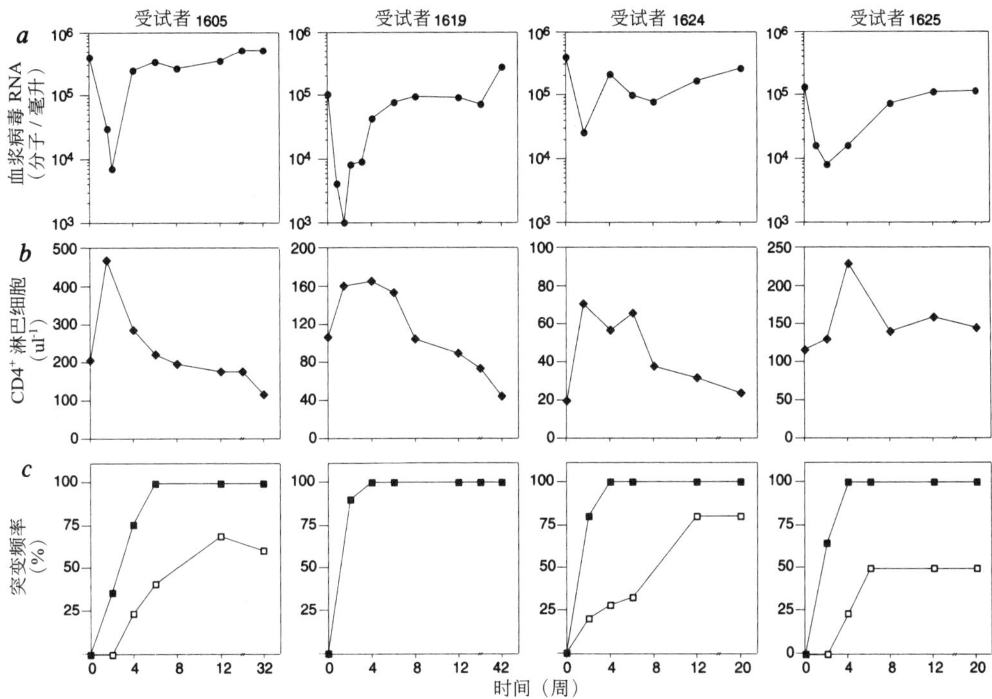

图 2. NVP 治疗组受试者的血浆病毒 RNA 水平（a），$CD4^+$ 淋巴细胞计数（b），血浆和 PBMC 中突变病毒基因组的比例（c）。NVP 治疗组受试者在接受 NVP 临床试验方案前已接受 ddI 治疗（受试者 1605）或者 ddI 复合齐多夫定治疗（受试者 1619、1624、1625）。初始 2 周内，NVP 剂量为 200 mg · d^{-1}，2 周后增加到 400 mg · d^{-1}。ddI 和齐多夫定的剂量分别是 400 mg · d^{-1} 及 300~600 mg · d^{-1}。病毒 RNA（●）经 QC-PCR 法测定 [6]。$CD4^+$ 淋巴细胞（◆）由流式细胞仪定量。血浆（■）和 PBMC（□）中 NVP 耐药相关病毒基因组突变频率由自动化 DNA 测序分析确定（图 3，图注），每一个数据点代表 3~6 个独立的 PCR 扩增和序列测量的均值。

尽管本研究使用的抗逆转录病毒药物作用机制不同，但是它们具有类似的生物学效果，主要表现在阻滞细胞重新感染。因此，药物治疗后我们测得的血浆病毒消除率实际上由两个因素共同决定：血浆病毒自身清除速率和早已存在的产病毒细胞的清除（或者抑制）速率。近似地，我们可以假设产病毒细胞按照 $y(t)=y(0)e^{-\alpha t}$ 呈指数式减少，其中 $y(t)$ 表示开始治疗后 t 时间的产病毒细胞数量，α 是指数式下降的速率常数。类似地，我们假设游离病毒 $v(t)$ 由产病毒细胞以速度 $ky(t)$ 产生，并

assume that free virus $v(t)$ is generated by virus-producing cells at the rate $ky(t)$ and declines exponentially with rate constant u. Thus, for the overall decline of free virus, we obtain $v(t) = v(0)[ue^{-\alpha t} - \alpha e^{-ut}]/(u-\alpha)$. The kinetics are largely determined by the slower of the two decay processes. As we have data only for the decline of free virus, and not for virus-producing cells, we cannot determine which of the two decay processes is rate-limiting. However, the half-life ($t_{1/2}$) of neither process can exceed that of the two combined. With these considerations in mind, we estimated the elimination rate of plasma virus and of virus-producing cells by three different methods: (1) first-order kinetic analysis of that segment of the viral elimination curve corresponding to the most rapid decline in plasma virus, generally somewhere between days 3 and 14; (2) fitting of a simple exponential decay curve to all viral RNA determinations between day 0 and the nadir or inflection point (Fig. 1); and (3) fitting of a compound decay curve that takes into account the two separate processes of elimination of free virus and virus-producing cells, as described. Method (1) gives a $t_{1/2}$ of 1.8 ± 0.9 days; method (2) gives a $t_{1/2}$ of 3.0 ± 1.7 days; and method (3) gives a $t_{1/2}$ of 2.0 ± 0.9 days for the slower of the two decay processes and a very similar value, 1.5 ± 0.5 days, for the faster one. These are averages ($\pm$ 1 s.d.) for all 22 patients. Method (3) arguably provides the most complete assessment of the data, whereas method (2) provides a simpler interpretation (but slightly slower estimate) for virus decline because it fails to distinguish the initial delay in onset of antiviral activity due to the drug accumulation phase, and the time required for very recently infected cells to initiate virus expression, from the subsequent phase of exponential virus decline. There were no significant differences in the viral clearance rates in subjects treated with ABT-538, L-735,524 or NVP, and there was also no correlation between the rate of virus clearance from plasma and either baseline $CD4^+$ lymphocyte count or baseline viral RNA level.

Virus Turnover

Direct population sequencing. As an independent approach for determining virus turnover and clearance of infected cells, we quantified serial changes in viral genotype and phenotype with respect to drug resistance in the plasma and PBMCs of four subjects treated with NVP (Fig. 2). NVP potently inhibits HIV-1 replication but selects for one or more codon substitutions in the reverse transcriptase (RT) gene[24,31,32]. These mutations result in dramatic decreases (up to 1,000-fold) in drug susceptibility and are associated with a corresponding loss of viral suppression *in vivo*[32]. Genetic changes resulting in NVP resistance can thus serve as a quantifiable molecular marker of virus turnover. A rapid decline in plasma viral RNA was observed following the institution of NVP therapy and this was associated with a reciprocal increase in $CD4^+$ lymphocyte counts (Fig. 2*a* and *b*). Both responses were of limited duration, returning to baseline within 6–20 weeks in these four patients. The proportion of virus in uncultured plasma and PBMCs that contain NVP-resistance-conferring mutations (Fig. 2*c*) was determined by direct automated nucleotide sequencing of viral nucleic acid (Fig. 3), as previously described[28]. We first validated this method by reconstitution experiments, confirming its sensitivity for detecting RT mutants that comprise as little as 10% of the overall virus population.

且以速率常数 u 呈指数式下降。因此，我们获得游离病毒总体下降速率 $v(t)=v(0)[ue^{-\alpha t}-\alpha e^{-ut}]/(u-\alpha)$。整体病毒动力学较大程度上由前述两项衰减中较慢的过程决定。由于我们仅有游离病毒减少的数据，而没有产病毒细胞的数据，所以我们不能决定哪一个是限速因素。但是，任何一个过程的半衰期($t_{1/2}$)都不超过两者联合的半衰期。在此设想下，我们通过三种方法估算血浆病毒和产病毒细胞的清除速率：(1)血浆病毒清除速率曲线下降最快区段进行一级动力学分析，此区段多发生在用药后第 3 天到第 14 天；(2)从第 0 天至最低点或拐点，拟合总病毒 RNA 测定水平的单一指数衰减曲线(图 1)；(3)综合考虑游离病毒清除和产病毒细胞清除这两个独立过程，拟合复合衰减曲线。方法(1)得出 $t_{1/2}$ 是 1.8±0.9 天；方法(2)得出 $t_{1/2}$ 是 3.0±1.7 天；方法(3)得出两种衰变 $t_{1/2}$ 慢速率 $t_{1/2}$ 是 2.0±0.9 天，快速率 $t_{1/2}$ 是 1.5±0.5 天。以上数据为 22 名患者的平均值(±1 标准差)。方法(3)可以说提供了最全面的数据分析，而方法(2)为病毒减少提供了一种相对简单直接的估算。方法(2)估算显得有些缓慢，这是由于没有把治疗早期由于药物蓄积而导致抗病毒起效滞后，以及最新感染的细胞产生病毒所需时间跟病毒指数减少后期时间区分开来。使用 ABT-538 或 L-735,524 或者 NVP 治疗的受试者病毒清除速率没有显著差异，而且血浆中病毒清除速率与基础 $CD4^+$ 淋巴细胞数或基础病毒 RNA 水平无相关性。

病毒更替

直接群体测序。作为一种独立确定病毒更替和感染细胞清除的方法，我们连续定量检测 4 名 NVP 治疗的受试者血浆和 PBMC 中与药物耐受相关的病毒基因型和表型的系列改变(图 2)。NVP 能高效抑制 HIV-1 复制，但是从逆转录酶(RT)基因中选择出一个或者多个密码子的替换 [24,31,32]。这些突变导致药物敏感性急剧下降(低至 1/1,000)，同时也与药物丧失体内病毒抑制效果相关 [32]。因此，导致 NVP 耐药的基因型改变可以作为定量病毒更替的分子标记。NVP 介入治疗能快速降低血浆病毒 RNA 水平，这与 $CD4^+$ 淋巴细胞数目增加相对应(图 2*a* 和图 2*b*)。两种反应均呈现时效性，4 名患者在 6~20 周治疗后均恢复到基础基线水平。正如之前文献报道 [28]，在未培养的血浆和 PBMC 中 NVP 耐药突变体的病毒比例(图 2*c*)通过自动化核酸测序对病毒核酸进行直接测定(图 3)。我们首先通过重建实验证实这种方法的有效性，结果发现此法灵敏度高，当病毒群体存在仅 10% RT 突变体时即可确认。我们对具有明确比例的野生型和突变型 HIV-1 RT cDNA 克隆(仅密码子 190 的第二位

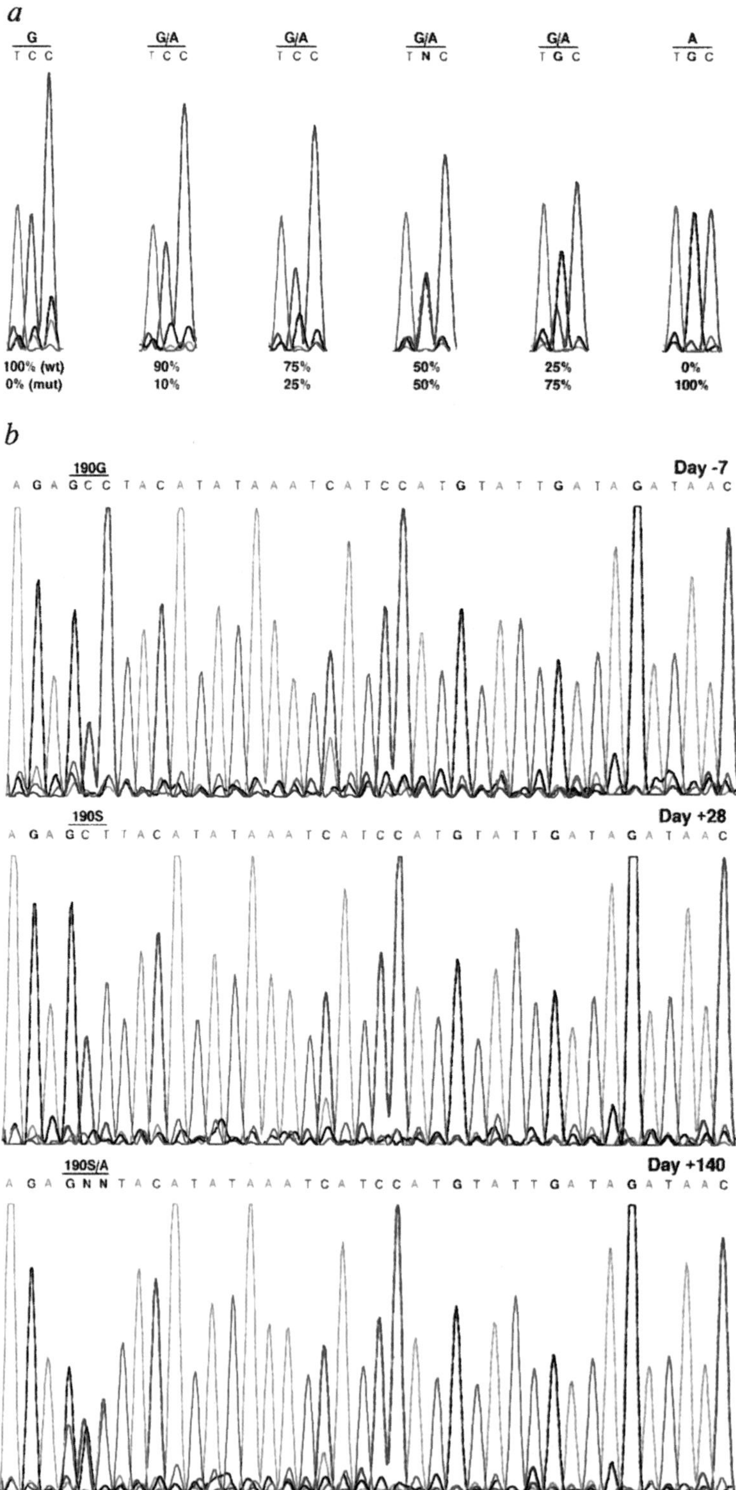

Fig. 3. Quantitative detection of HIV-1 drug-resistance mutations by automated DNA sequencing. *a*, DNA sequence chromatograms of RT codon 190 from a defined mixture of wild-type (wt) and mutant (mut)

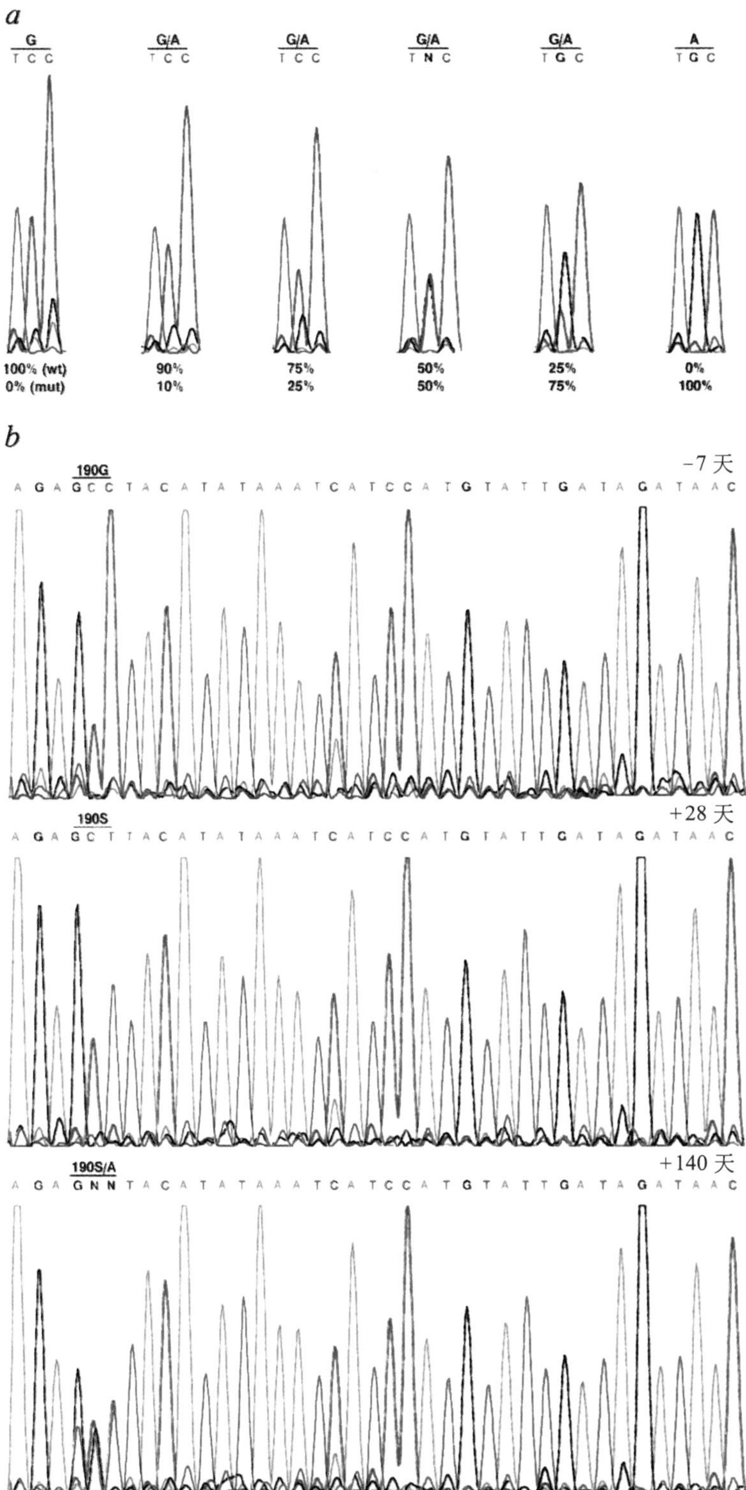

图 3. 自动化 DNA 测序定量检测 HIV-1 耐药突变体。*a*，特定比例的野生型(wt)和突变型(mut)HIV-1 cDNA 克隆混合物在 RT 密码子 190 区域的 DNA 序列层析谱，两者差别仅在于密码子第二位碱基。图

HIV-1 cDNA clones differing only at the second base position of the codon. Sequences shown were obtained from, and therefore are presented as, the minus (non-coding) DNA strand. For example, the minus-strand TCC sequence shown corresponds to the plus-strand codon GGA (glycine, G). Similarly, the minus-strand TGC sequence corresponds to the plus-strand codon GCA (alanine, A). The single-letter amino-acid code corresponds to the plus-strand DNA sequence. Mixed bases approximating a 50/50 ratio are denoted as N. *b*, DNA sequence chromatograms of RT codons 179–191 (again displayed as the minus-strand sequence) derived from plasma-virion-associated RNA of subject 1625 before (day −7) and after (days +28 and +140) starting NVP therapy. Codon changes resulting in amino-acid substitutions at position 190 are indicated for the plus strand. For example, the GCC minus-strand sequence at position 190 (day −7) corresponds to GGC (glycine, G), and the GCT minus-strand sequence at position 190 (day +28) corresponds to AGC (serine, S) in the respective plus strands.

Methods. Mixtures of wild-type and mutant cDNA clones (*a*) were prepared and diluted such that first-round PCR amplifications were done with 1,000 viral cDNA target molecules per reaction. HIV-1 RNA was isolated from virions pelleted from uncultured plasma specimens (*b*), as described[18]. cDNA was prepared using Moloney murine leukaemia virus reverse transcriptase (GIBCO BRL)[6] and an oligonucleotide primer corresponding to nucleotides 4,283 to 4,302 of the HXB2 sequence[43]. The full-length viral reverse transcriptase gene (1,680 bp) was amplified by means of a nested PCR using conditions and oligonucleotide primers (outer primers: nt 2,483–2,502 and 4,283–4,302; inner primers: nt 2,549–2,565 and 4,211–4,229), previously reported[30]. Subgenomic fragments of the RT gene were also amplified using combinations of the following oligonucleotide primers: (5′) 2,585–2,610; (5′) 2,712–2,733; (3′) 2,822–2,844; (3′) 3,005–3,028; (3′) 3,206–3,228; (3′) 3,299–3,324; (3′) 3,331–3,350; (3′) 3,552–3,572; and (3′) 3,904–3,921. All 3′ primers incorporated the universal primer sequence for subsequent dye-primer sequence analysis. The HIV-1 copy number in every PCR reaction was determined (100–10,000 copies). A total of three to six separate PCR amplifications of primary patient material was done on each sample using different combinations of primers, and representative chromatograms are shown. Rarely, codon interpretation was ambiguous. In the day +140 plasma sample from subject 1625 (bottom of panel *b*), the complementary (plus) strand could read: AGC (serine), GCN (alanine), ACN (threonine), AGA/AGG (arginine) or GGN (glycine). In this case, we sequenced 7 full-length RT molecular clones and found that they encoded only serine or alanine. For sequencing, an automated ABI 373A sequenator and the Taq Dye Primer Cycle Sequencing Kit (ABI) were used. Sequences were analysed using Sequencher (Gene Codes Corp.) and Microgenie (Beckman) software packages, and base-pair mixtures were quantified by measuring relative peak-on-peak heights[28].

Defined mixtures of wild-type and mutant HIV-1 RT cDNA clones (differing only at the second base position of codon 190) were amplified and sequenced (Fig. 3*a*). Varying proportions of wild-type and mutant viral sequences present in the original DNA mixtures (mutant composition: 0, 10, 25, 50, 75 and 100%) were faithfully represented in the relative peak-on-peak heights (and in the relative peak-on-peak areas) of cytosine (C) and guanine (G) residues at the second base position within this codon. Ratios of (mutant)/(mutant+wild type) nucleotide peak heights expressed in arbitrary fluorescence units were as follows (predicted/observed): 0/ < 10%; 10/18%; 25/29%; 50/49%; 75/71% and 100/94%.

We next determined the ability of direct population sequencing to quantify wild-type and mutant viral RNA genomes in clinical specimens. Figure 3*b* shows the sequence chromatograms of RT codons 179–191 from virions pelleted directly from uncultured plasma specimens of subject 1625 before (day −7) and after (days +28 and +140) the initiation of NVP therapy. At day −7, all codons within the amino-terminal half of the RT gene (codons 1–250), including those shown, were wild-type at positions associated with NVP resistance[31,32]. However, after only 28 days of NVP therapy, the wild-type plasma

中所显示的序列来自DNA负链(非编码链)。例如,显示的负链TCC序列对应于正链密码子GGA(甘氨酸,G)。类似地,负链 TGC 序列对应于正链密码子 GCA(丙氨酸,A)。这些单字母氨基酸编码对应于正链DNA 序列。当混合碱基比例大约为 50/50 时以 N 表示。*b*,受试者 1625 在 NVP 治疗前(治疗前的第 7 天)以及治疗后(第 28 天及第 140 天)血浆病毒粒子相关 RNA 在 RT 密码子 179~191 区域的 DNA 序列层析谱(仍以负链序列显示)。密码子的改变导致 190 位氨基酸替换(以正链显示)。例如,190 位的 GCC 负链序列(治疗前的第 7 天)对应于正链的 GGC(甘氨酸,G);而 190 位的 GCT 负链序列(治疗后的第 28 天)对应于正链的 AGC(丝氨酸,S)。

方法。野生型和突变型 cDNA 克隆的混合物(*a*)通过第一轮 PCR 扩增及稀释后制备,每次反应以 1,000 个病毒 cDNA 为目标分子。如参考文献所述 [18],从未培养血浆标本的病毒粒子中分离出 HIV-1 RNA(*b*)。cDNA 通过莫洛尼氏鼠白血病病毒逆转录酶(GIBCO BRL)[6] 和对应于 HXB2 序列的 4,283 到 4,302 位核苷酸的寡核苷酸引物制备 [43]。病毒逆转录酶全长基因(1,680 bp)使用参考文献 [30] 所述的条件和寡核苷酸引物(外部引物:nt 2,483~2,502 和 4,283~4,302;内部引物:nt 2,549~2,565 和 4,211~4,229)经巢式 PCR 扩增得到。RT 基因的亚基因组片段联合以下寡核苷酸引物进行扩增:(5′)2,585~2,610;(5′)2,712~2,733;(3′)2,822~2,844;(3′)3,005~3,028;(3′)3,206~3,228;(3′)3,299~3,324;(3′)3,331~3,350;(3′)3,552~3,572;(3′)3,904~3,921。所有的 3′ 端引物均整合用于随后染色引物序列分析的通用引物序列。每个 PCR 反应的 HIV-1 拷贝数是确定的(100~10,000 拷贝)。每份样本数据均使用原始样本通过不同引物组合经 3~6 次独立 PCR 扩增获得,图中显示典型 DNA 序列层析谱。极少情况下出现模糊解码,例如受试者 1625 第 140 天血浆样本(*b* 栏底部)互补链(正链)序列可解码为 AGC(丝氨酸),GCN(丙氨酸),ACN(苏氨酸),AGA(AGG)(精氨酸)或者 GGN(甘氨酸)。在此特例中,我们对 7 个全长 RT 分子克隆进行测序。结果发现,这些克隆仅编码丝氨酸或丙氨酸。使用自动化 ABI 373A 测序仪和 Taq Dye 引物循环测序试剂盒(ABI)进行测序。使用 Sequencher 软件(Gene Codes 公司)以及 Microgenie 软件包(贝克曼公司)进行序列分析,通过测量相对峰高对碱基对混合物进行定量 [28]。

碱基不同)混合物进行扩增并测序(图 3*a*)。原始 DNA 混合物中存在的不同比例野生型和突变型病毒序列(突变成分:0,10%,25%,50%,75% 和 100%)真实地被该密码子内第二位碱基的胞嘧啶(C)和鸟嘌呤(G)残基的相对峰高(以及相对峰面积)所展示。(突变型)/(突变型 + 野生型)核苷酸峰高度的比例通过荧光单位相对表达量(预测值 / 观察值)表示:0/ < 10%;10%/18%;25%/29%;50%/49%;75%/71% 和 100%/94%。

我们下一步确定是否可用直接群体测序来量化临床标本中的野生型和突变型病毒 RNA 基因组。图 3*b* 序列层析谱显示了受试者 1625 开始 NVP 治疗前(治疗前的第 7 天)以及治疗后(第 28 天和第 140 天)未培养血浆标本病毒粒子 RT 密码子 179~191 的测序结果。治疗前的第 7 天,RT 基因(密码子 1~250)的氨基末端所有 NVP 耐药相关密码子均为野生型 [31,32]。但是,仅经过 28 天 NVP 治疗,野生型血浆

virus population was completely replaced by a NVP-resistant mutant population differing from the wild-type at codon 190 (glycine-to-serine substitution). After 140 days of drug therapy, this codon had evolved further such that the plasma virus population consisted of an equal mixture of two drug-resistant strains, one containing G190S and the other containing G190A. There were no other NVP-resistance-conferring mutations detectable within the viral RT gene.

In all four subjects evaluated by direct viral population sequencing (Fig. 4), specific NVP-resistance-conferring mutations within the RT gene could be unambiguously identified and subsequently confirmed by molecular cloning, expression and drug susceptibility testing. In all cases, mutant virus increased rapidly in the plasma and virtually replaced wild-type virus after only 2–4 weeks of NVP therapy (Fig. 2*c*). By analysing the rate of accumulation of resistant mutants in the plasma population, we could obtain an independent estimate of the turnover rate of free virus. The rise of drug-resistant mutant virus is influenced substantially by the preceding increase in the $CD4^+$ cell population (which provides additional resources for virus production[33]) and therefore follows complex dynamics. However, we could obtain an estimate of these dynamics by making simplifying assumptions. We assume that wild-type virus declines exponentially with a decay rate α, and that the drug-resistant mutant increases exponentially with the rate β. Thus, the ratio of mutant to wild-type virus increases exponentially at the combined rate $\alpha+\beta$. Our genetic RNA (cDNA) data allow us to estimate this sum. Knowing α from our data on virus decline, we get $\beta \approx 0.27$, or a 32% daily virus production (average over 4 patients). Assuming that mutant virus rises exponentially, this corresponds to a doubling time of ~2 days, which is in excellent agreement with the measured elimination half-life of 2.0 ± 0.9 days for plasma virus (Figs 1 and 2*a*). Turnover of viral DNA from wild-type to drug-resistant mutant in PBMCs was delayed and less complete compared to plasma virus, reaching levels of only 50–80% of the total PBMC-associated viral DNA population by week 20 (Fig. 2*c*). Measurement of the time required for resistant virus to spread in the PBMC population allowed us also to estimate the half-life of infected PBMCs. After complete turnover of mutant virus in the plasma pool, we may assume that PBMCs infected with wild-type virus decline exponentially at a rate *d*, whereas cells infected by mutant virus are generated at a constant rate, but also decline exponentially at rate *d*. With these simplifying assumptions, the rate at which the frequency of resistant virus in the PBMC population increases provides an estimate for the parameter *d* and hence for the half-life of infected PBMCs. We obtained a half-life of ~50–100 days. This means that the average half-life of infected PBMCs is very long and of the same order of magnitude as the half-life of uninfected PBMCs[34,35]. Based on the long half-life of PBMCs, and the fact that these cells harbour predominantly wild-type virus at a time (days 14–28) when most virus in plasma is mutant, we conclude that most PBMCs contribute comparatively little to plasma virus load. Instead, other cell populations, most probably in the lymphoreticular system[11,19,20], must be the major source of virus production.

病毒群体完全被耐药突变型所替代。突变型仅在密码子 190 与野生型存在不同(甘氨酸替换为丝氨酸)。140 天的药物治疗后，此密码子发生进一步改变。血浆病毒群体由两种分别包括 G190S 和 G190A 的耐药病毒突变型等比例共存。在病毒 RT 基因内没有检测到其他 NVP 耐药相关的基因突变。

在经过直接病毒群体测序的四名受试者中(图 4)，通过分子克隆、表达和药物易感性测试可以清楚地鉴定并随后确认 RT 基因中的特异性 NVP 耐药性基因。在所有病例中，血浆中 NVP 耐药突变体可在治疗后 2~4 周迅速增加并取代野生型病毒(图 2*c*)。通过分析耐药突变体在血浆群体中的聚集速度，我们能够估算游离病毒更替速率。耐药突变型病毒的增加很大程度上受 $CD4^+$ 细胞增加的影响(这些细胞提供病毒生成的额外来源[33])，因此形成复杂的动力学变化。然而，我们可以通过简化的假设估算此动力学过程。我们假设野生型病毒以速率 α 呈指数式减少，而耐药突变型以速率 β 呈指数式增加。这样，突变型和野生型病毒的比例以速率 $\alpha+\beta$ 指数式增加。我们的遗传 RNA(cDNA) 数据帮助我们估计出总量。从病毒减少的数据中获得 α，同时我们得出 $\beta\approx0.27$，或者病毒以每天 32% 的速率增加(4 名患者的平均值)。假设突变型病毒以指数式增长，我们估算出病毒倍增时间约为 2 天。这与测得的血浆病毒的清除半衰期 2.0 ± 0.9 非常相符(图 1 和图 2*a*)。与血浆病毒相比，PBMC 中病毒 DNA 从野生型更替成突变型呈不完全、相对延迟等特性。在 20 周时，突变型仅占到总 PBMC 相关病毒 DNA 数量的 50%~80%(图 2*c*)。测定耐药病毒突变体在 PBMC 群体中扩散时间使得我们能够估计感染 PBMC 的半衰期。血浆中发生完全突变型病毒更替后，我们可以假设感染野生型病毒的 PBMC 以速率 d 呈指数式减少，而感染突变型病毒的细胞在以恒定速度增长的同时也以速率 d 呈指数式减少。经过这些简化假设，我们通过 PBMC 群体中耐药病毒增长频率估计参数 d，然后估算感染 PBMC 的半衰期。结果发现，感染 PBMC 的半衰期约 50~100 天。这意味着感染 PBMC 的平均半衰期非常长，与未感染 PBMC 的半衰期量级一致[34,35]。基于 PBMC 的长半衰期，以及这些细胞在一段时间内(第 14~28 天)主要产生野生型病毒，而此时血浆中的大部分病毒都是突变型，我们得出结论：大部分 PBMC 对血浆病毒载量的贡献相对很小，病毒主要来源于其他细胞群体。这些细胞群最有可能位于淋巴网状系统[11,19,20]。

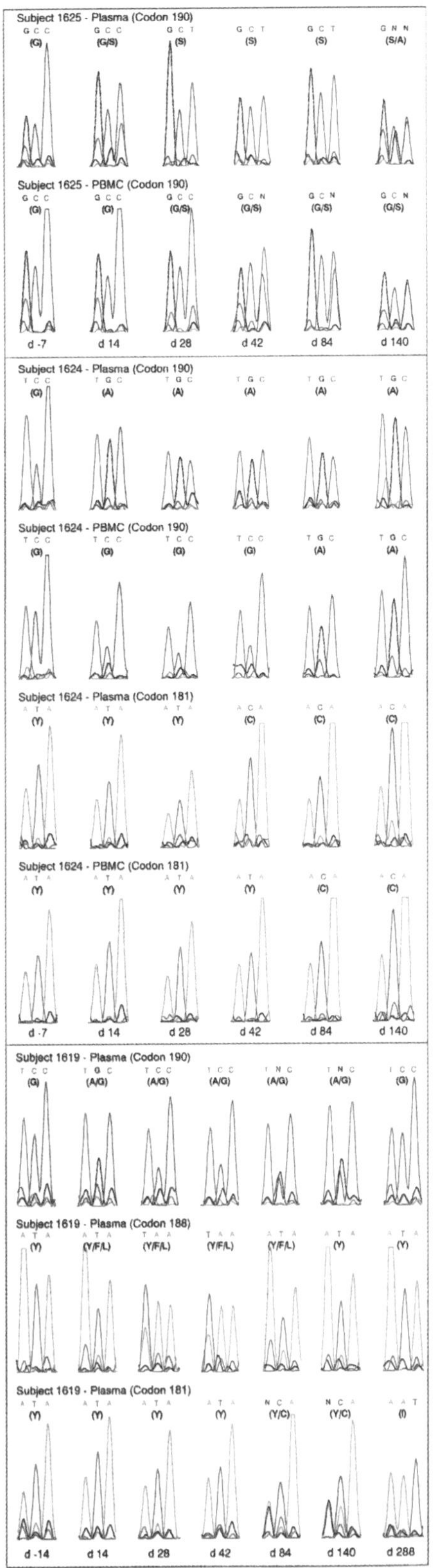

Fig. 4. Quantitative detection of HIV-1 drug resistance mutations by automated DNA sequencing in plasma viral RNA (cDNA) and PBMC-associated viral DNA populations before and after the initiation of NVP on day 0. As in Fig. 3, minus-strand sequences are shown together with single-letter amino-acid codes of the corresponding plus-strand sequence. Mixed bases approximating a 50/50 ratio are denoted as N.

Methods. HIV-1 cDNA was prepared from virions pelleted from uncultured plasma as described for Fig. 3. Viral DNA was isolated from uncultured PBMCs, as described[44]. The full-length viral reverse transcriptase genes as well as subgenomic fragments were amplified and sequenced as described for Fig. 3. The HIV-1 copy number in every PCR reaction was determined (100–10,000 copies). Some sequences were determined from both coding and non-coding DNA strands to ensure the accuracy of quantitative measurements.

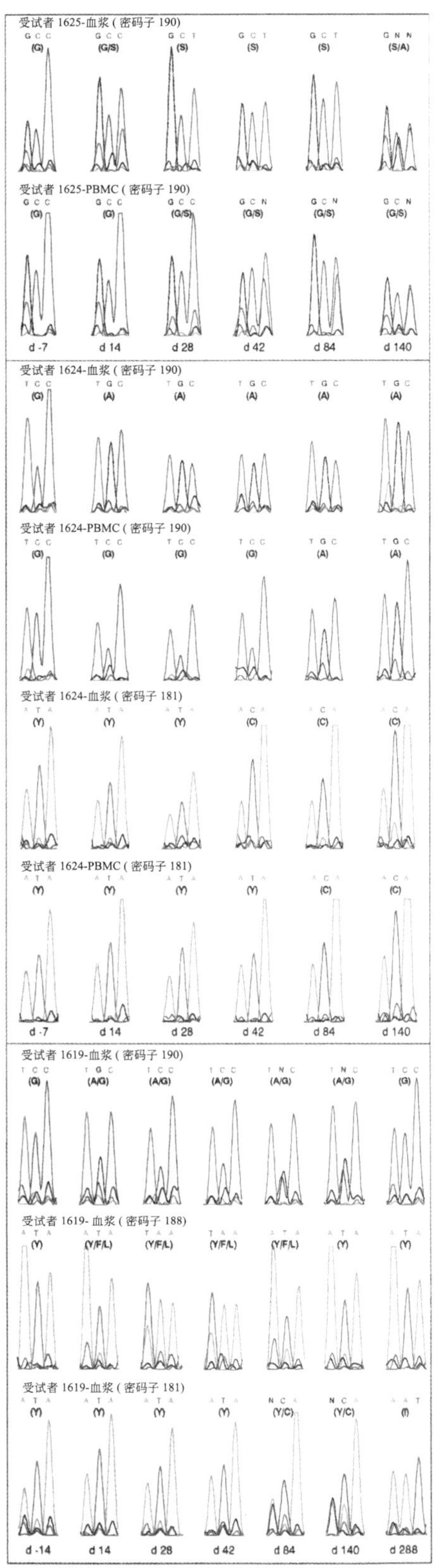

图 4. NVP 治疗前后经自动化 DNA 测序法定量检测血浆病毒 RNA（cDNA）和 PBMC 相关病毒 DNA 群体中的 HIV-1 耐药突变。如图 3 所示，负链序列与对应的正链序列的单字母氨基酸编码同时显示。当碱基大约以 50/50 比例混合时以 N 表示。

方法。如图 3 中所述，HIV-1 cDNA 从未培养血浆中分离的病毒粒子中制备。如参考文献所述 [44]，病毒 DNA 从未培养的 PBMC 中分离。按照图 3 所述方法扩增全长病毒逆转录酶基因和亚基因组片段并测序。每个 PCR 反应中的 HIV-1 拷贝数是确定的（100~10,000 拷贝）。为确保定量检测的准确性，一些序列同时通过编码和非编码 DNA 链确定。

Direct sequence analysis of viral nucleic acid revealed not only rapid initial turnover in viral populations but also continuing viral evolution with respect to drug resistance mutations. In subject 1625 (Fig. 4, top panel), wild-type virus in plasma was completely replaced after 28 days of NVP therapy by mutant virus (G190S), which in turn evolved by day 140 into a mixture of G190S and G190A. In subject 1624 (Fig. 4, middle panel), two codon changes conferring NVP resistance occurred. A G190A substitution appeared in plasma virus at day 14 and a Y181C appeared at day 42. Similarly, in subject 1605 (not shown), a Y181C mutation appeared in plasma at day 14 and a Y188L mutation at day 28. The sequential changes in plasma virus were mirrored by similar changes in PBMCs at later timepoints. In subject 1619, the pattern of resistance changes was even more complex (Fig. 4, bottom panel). By day 14, approximately 70% of plasma virus contained a G190A mutation. By day +28, this mutant population was largely replaced by virus containing a Y188F/L substitution. By day 84, still another major shift in the viral quasispecies occurred, this time resulting in a population of viruses containing mutations at both Y181C and G190A. Finally, by day 288 the viral population in plasma consisted exclusively of a mutant exhibiting a single tyrosine-to-isoleucine substitution at position 181 (Y181I); mutations at codons 188 and 190 were not present in this virus population. All of these amino-acid substitutions at RT codons 181, 188 and 190 were shown in our *in situ* expression studies and by others[31,32,36] to confer high-level NVP resistance. The direct sequence analyses thus demonstrate that major changes in the HIV-1 quasispecies occur quickly and continuously in response to selection pressures and that these changes are reflected first and most prominently in the plasma virus compartment.

***In situ* RT gene expression and drug susceptibility testing**. Because direct sequence analysis of viral mixtures provides only semiquantitative information and does not distinguish between viruses with functional rather than defective RT genes, we employed another method for quantifying virus turnover in uncultured plasma and PBMC compartments. Full-length RT genes were amplified by polymerase chain reaction (PCR), cloned into pLG18-l, expressed in *Escherichia coli*, and tested individually for enzymatic function and NVP susceptibility by *in situ* assay[29,30] (Table 1). For subject 1625 at day −7, 100% (80/80) of RT clones from plasma and 100% (163/163) of RT clones from PBMCs expressed enzyme that was sensitive to NVP inhibition. By day 14, however, 62% of plasma-derived clones expressed enzyme that was resistant to NVP, and by days 28, 84 and 140, 100% were resistant. Conversely, at day 14, 0% of PBMC-derived clones expressed NVP-resistant enzyme, and even after 28, 84 and 140 days, only 48–75% of clones were resistant. Similar results were obtained for the other study subjects (Table 1). Thus, the kinetics of virus population turnover determined by a quantitative RT *in situ* expression assay corresponded closely with those determined by direct population sequencing (Fig. 2*c*).

病毒核酸的直接序列分析不但显示了病毒群体的快速早期更替，同时也反映出病毒持续产生耐药突变。受试者 1625（图 4，上图）血浆中的野生型病毒在 NVP 治疗 28 天后完全被突变型（G190S）取代，进而在第 140 天进化成 G190S 和 G190A 突变体的混合物。受试者 1624（图 4，中图）发生了两个与 NVP 耐药相关的密码子改变。治疗后第 14 天血浆病毒出现 G190A 突变体，第 42 天出现 Y181C 突变体。与此类似，受试者 1605（未显示）第 14 天出现血浆病毒 Y181C 突变体，第 28 天出现 Y188L 突变体。血浆病毒的改变由后期 PBMC 内病毒类似的改变反映出来。受试者 1619 耐药突变的形式更加复杂（图 4，下图）。治疗后第 14 天，几乎 70% 的血浆病毒含有 G190A 突变体。治疗后第 28 天，这些突变体大部分被 Y188F 或 Y188L 突变体取代。治疗后第 84 天，病毒准种再次发生重大转变，导致含有 Y181C 和 G190A 两种突变体的群体占主导。最后，到第 288 天，血浆中的病毒完全由 Y181I（181 位的酪氨酸被异亮氨酸替换）突变体组成。密码子 188 和 190 的突变在此病毒群体中不存在。所有这些 RT 密码子 181、188 和 190 的氨基酸替换都在我们的原位表达研究和其他研究 [31,32,36] 中显示出高水平的 NVP 耐药性。直接序列分析显示 HIV-1 病毒准种应对选择压力时发生快速和持续的改变。这些变化最先且最明显地反映在血浆病毒成分中。

原位 RT 基因表达和药物敏感性测试。因为病毒混合物的直接测序分析仅仅提供半定量的信息，不能鉴别含功能性 RT 基因和缺陷 RT 基因的病毒。我们使用另一种方法来定量测定未培养血浆和 PBMC 中的病毒更替。用聚合酶链式反应（PCR）扩增全长的 RT 基因，克隆到 pLG18-1 载体并在大肠杆菌中表达。各种突变体分别用原位检测测定酶功能和 NVP 敏感性 [29,30]（表 1）。受试者 1625 治疗前的第 7 天，血浆中 100%（80/80）的 RT 克隆和 PBMC 中 100%（163/163）的 RT 克隆表达 NVP 敏感酶。但是治疗后第 14 天，62% 的血浆病毒来源的克隆表达 NVP 耐药酶。治疗后的第 28 天、84 天和 140 天，100% 克隆表达 NVP 耐药酶。与此相反，治疗后第 14 天，PBMC 来源的克隆没有表达耐药酶，即便是 28 天、84 天和 140 天后，仅有 48%~75% 的克隆表达耐药酶。其他研究对象得到类似的研究结果（表 1）。因此，通过定量 RT 原位表达与直接群体测序法（图 2*c*）得到的病毒群体更替动力学结果非常接近。

Table 1. *In situ* functional analysis of HIV-1 RT clones

Subject	Specimen		Functional clones	NVP-sensitive clones		NVP-resistant clones	
1625	Plasma	day -7	80	80	(100%)	0	(0%)
		+14	72	27	(38%)	45	(62%)
		+28	57	0	(0%)	57	(100%)
		+84	67	0	(0%)	67	(100%)
		+140	86	0	(0%)	86	(100%)
1625	PBMC	-7	163	163	(100%)	0	(0%)
		+14	121	121	(100%)	0	(0%)
		+28	258	134	(52%)	124	(48%)
		+84	133	43	(32%)	90	(68%)
		+140	261	65	(25%)	196	(75%)
1624	Plasma	-7	19	19	(100%)	0	(0%)
		+14	34	4	(12%)	30	(88%)
		+28	79	6	(8%)	73	(92%)
		+140	27	0	(0%)	27	(100%)
1624	PBMC	-7	24	24	(100%)	0	(0%)
		+14	34	29	(85%)	5	(15%)
		+28	52	42	(81%)	10	(19%)
		+140	87	26	(30%)	61	(70%)
1605	PBMC	-7	31	31	(100%)	0	(0%)
		+140	31	11	(35%)	20	(65%)
1619	Plasma	-14	79	79	(100%)	0	(0%)
		+28	41	0	(0%)	41	(100%)
		+140	38	0	(0%)	38	(100%)

Full-length RT genes were amplified by PCR from uncultured plasma and uncultured PBMCs as described in Fig. 3 legend. DNA products were cloned into the *Eco*RI and *Hin*dIII sites of the bacterial expression plasmid pLG18-1(refs 29, 30). The expression plasmids were screened for the presence of functional RT and tested *in situ* for susceptibility to NVP inhibition at 3,000 nM (~50–75 fold greater than the IC_{50})[24,31,32]. To ensure accuracy in distinguishing RT genes encoding NVP-resistant versus sensitive enzymes, and to confirm the identification of specific NVP-resistance-conferring RT mutations obtained by direct sequencing (Figs 3 and 4), we determined the complete nucleotide sequences of 21 cloned RT genes which had been phenotyped in the *in situ* assay (V.A.J. and G.M.S., submitted). There was complete concordance between the phenotypes and genotypes of these 21 clones with respect to NVP-resistance-conferring mutations, as well as complete concordance between direct viral population sequences and clone-derived sequences at NVP-resistance-conferring codons.

Infectious virus drug susceptibility testing. Plasma and PBMCs are known to harbour substantial proportions of defective or otherwise non-infectious virus[6,37]. To determine whether the viral genomes represented in total viral nucleic acid (Fig. 4 and Table 1) corresponded to infectious virus with respect to NVP-resistance-conferring mutations, we co-cultivated PBMCs from three of the study subjects (1605, 1624, 1625) with normal

表 1. HIV-1 RT 克隆的原位功能分析

受试者	标本		功能性克隆	NVP 敏感克隆		NVP 耐药克隆	
1625	血浆	天 −7	80	80	(100%)	0	(0%)
		+14	72	27	(38%)	45	(62%)
		+28	57	0	(0%)	57	(100%)
		+84	67	0	(0%)	67	(100%)
		+140	86	0	(0%)	86	(100%)
1625	PBMC	−7	163	163	(100%)	0	(0%)
		+14	121	121	(100%)	0	(0%)
		+28	258	134	(52%)	124	(48%)
		+84	133	43	(32%)	90	(68%)
		+140	261	65	(25%)	196	(75%)
1624	血浆	−7	19	19	(100%)	0	(0%)
		+14	34	4	(12%)	30	(88%)
		+28	79	6	(8%)	73	(92%)
		+140	27	0	(0%)	27	(100%)
1624	PBMC	−7	24	24	(100%)	0	(0%)
		+14	34	29	(85%)	5	(15%)
		+28	52	42	(81%)	10	(19%)
		+140	87	26	(30%)	61	(70%)
1605	PBMC	−7	31	31	(100%)	0	(0%)
		+140	31	11	(35%)	20	(65%)
1619	血浆	−14	79	79	(100%)	0	(0%)
		+28	41	0	(0%)	41	(100%)
		+140	38	0	(0%)	38	(100%)

如图 3 图注所述，通过 PCR 扩增来自未培养血浆和未培养 PBMC 标本的全长 RT 基因。DNA 产物克隆到细菌表达质粒 pLG18-1 的 *Eco*RI 和 *Hin*dIII 位点之间（参考文献 29，参考文献 30）。筛查表达质粒是否存在功能 RT，并原位检测对 3,000 nM（约 50~75 倍的 IC_{50} 浓度）NVP 的易感性[24,31,32]。为了确保准确鉴别特异性 NVP 耐药酶和敏感酶，并通过直接测序确认找到的特异性 NVP 耐药相关 RT 突变（图 3 和图 4），我们确定了原位实验中表达的所有 21 个克隆的 RT 基因完整核苷酸序列（约翰逊和肖，已投稿）。结果发现，这 21 个突变克隆在 NVP 耐药突变上的基因型和表型完全一致，而且直接病毒群体测序的结果和 NVP 耐药相关密码上克隆来源序列也完全一致。

感染性病毒药物敏感性检测。血浆和 PBMC 能够大比例容纳缺陷或非感染性病毒[6,37]。为了确定用整个病毒核苷酸（图 4 和表 1）表示的病毒基因组是否与 NVP 耐药突变相关的感染病毒的基因组一致，我们将三名研究对象（1605、1624、1625）的 PBMC 与正常供体淋巴母细胞共同培养，建立初始的病毒分离株。将治疗前和

donor lymphoblasts in order to establish primary virus isolates. The RT genes of these cultured viruses, obtained before and after therapy, were cloned (Fig. 3 and Table 1 legends) and sequenced in their entirety (V.A.J. and G.M.S., submitted). RT codons associated with NVP susceptibility were completely concordant in cultured and uncultured virus strains. Furthermore, the virus isolates exhibited NVP susceptibility profiles[38] consistent with their genotypes.

CD4$^+$ Lymphocyte Dynamics

Changes in CD4$^+$ lymphocyte counts during the first 28 days of therapy could be assessed in 17 of our patients (Fig. 2*b* and data not shown). CD4$^+$ cell numbers increased in every patient by between 41 and 830 cells per mm^3. For the entire group, the average increase was 186 ± 199 cells per mm^3 (mean ± s.d.), or $268 \pm 319\%$ from baseline. As CD4$^+$ lymphocytes increase in numbers because of (1) exponential proliferation of CD4$^+$ cells in peripheral tissue compartments, and/or (2) constant (linear) production of CD4$^+$ cells from a pool of precursors, we analysed our data based on each of these assumptions. The average percentage increase in cell number per day (assumption (1)) was $5.0 \pm 3.1\%$ (mean ± 1 s.d.). The average absolute increase in cell number per day (assumption (2)) was 8.0 ± 7.8 cells $mm^{-3}\ d^{-1}$. Given that peripheral blood contains only 2% of the total body lymphocytes[35] and that the average total blood volume is ~5 litres, an increase of 8 cells $mm^{-3}\ d^{-1}$ implies an overall steady-state CD4$^+$ cell turnover rate (where increases equal losses) of $(50) \times (5 \times 10^6\ mm^3) \times (8\ cells\ mm^{-3}\ d^{-1})$, or 2×10^9 CD4$^+$ cells produced and destroyed each day.

Discussion

Previously, it was shown that lymphoreticular tissues serve as the primary reservoir and site of replication for HIV-1 (refs 11, 19, 20) and that virtually all HIV-1-infected individuals, regardless of clinical stage, exhibit persistent plasma viraemia in the range of 10^2 to 10^7 virions per ml[6]. However, the dynamic contributions of virus production and clearance, and of CD4$^+$ cell infection and turnover, to the clinical "steady-state" were obscure, although not unanticipated[22,23,39]. We show by virus quantitation and mutation fixation rates that the composite lifespan of plasma virus and of virus-producing cells is remarkably short ($t_{1/2} = 2.0 \pm 0.9$ days). This holds true for patients with CD4$^+$ lymphocyte counts as low as 18 cells per mm^3 and as high as 355 cells per mm^3 (Figs 1 and 2; G.M.S., unpublished). These findings were made in patients treated with three different antiretroviral agents having two entirely different mechanisms of action and using three different experimental approaches for assessing virus turnover. The viral kinetics thus cannot be explained by a unique or unforeseen drug effect or a peculiarity of any particular virological assay method. Moreover, when new cycles of infection are interrupted by potent antiretroviral therapy, plasma virus levels fall abruptly by an average of 99%, and in some cases by as much as 99.99% (10,000-fold). This result indicates that the vast majority of circulating plasma virus derives from continuous rounds of *de*

治疗后获得的这些培养病毒的RT基因进行克隆（图3图注和表1表注）并全部测序（约翰逊和肖，已投稿）。与NVP敏感性相关的RT密码子在培养的和未培养的病毒株中完全一致。此外，这些病毒分离株呈现的NVP敏感性[38]与其基因型也完全一致。

$CD4^+$淋巴细胞动力学

我们有17位患者可以评价治疗初始28天内$CD4^+$淋巴细胞数目的变化（图2*b*，数据未显示）。患者每立方毫米血液中$CD4^+$淋巴细胞增加41到830个。在整个实验组中，患者每立方毫米血液平均增加186±199个$CD4^+$淋巴细胞（平均值±标准差），或者比基线水平增加268%±319%。$CD4^+$淋巴细胞数量增加存在多种原因，（1）外周组织中$CD4^+$淋巴细胞呈指数式增殖，和（或）（2）从前体细胞恒定（线性）地生成$CD4^+$淋巴细胞。我们基于以上假设分析数据。结果发现，每天细胞数量增加的平均百分比（假设（1））是5.0%±3.1%（平均值±1标准差）。每天细胞数目的绝对增加值（假设（2））为每立方毫米8.0±7.8个。考虑到外周血仅含有2%的全身淋巴细胞[35]，平均全血容量约5升，每天每立方毫米8个细胞意味着总体稳定的$CD4^+$淋巴细胞更替速度（增加与减少相互平衡）为（50）×（5×10^6立方毫米）×（每天每立方毫米8个细胞），相当于每天有2×10^9个$CD4^+$淋巴细胞生成和消亡。

讨　论

过去的研究发现，淋巴网状组织是HIV-1的主要储存地和复制区域（参考文献11，参考文献19，参考文献20）。同时，各期HIV-1感染者均存在持续的血浆病毒血症，血浆病毒粒子浓度为每毫升10^2~10^7个[6]。尽管有过预期[22,23,39]，但由于病毒产生和清除动态变化以及$CD4^+$淋巴细胞的感染和更替动力学过于复杂，目前对HIV-1患者出现的临床"稳态"机制尚不明确。我们通过病毒定量和固定突变速率发现血浆病毒以及产病毒细胞的复合生存周期非常短（$t_{1/2}=2.0\pm0.9$天）。此项发现在低$CD4^+$淋巴细胞数目（每立方毫米18个）患者和高$CD4^+$淋巴细胞数目（每立方毫米355个）患者中均成立（图1和图2；肖，未发表）。这些发现是通过三种具有两类完全不同作用机制的抗逆转录病毒药物，同时使用三种不同实验方法来评测HIV患者病毒的更替获得的。因此，特殊的或未知的药效以及特殊病毒学检测方法的特性均不能解释病毒动力学。此外，新的病毒感染周期被强效抗逆转录病毒治疗打断后，平均血浆病毒水平迅速下降了99%，某些特例甚至下降99.99%（1/10,000）。这个结果提示大部分的循环血浆病毒来自持续的新病毒感染、复制和细胞更替，而

novo virus infection, replication and cell turnover, and not from cells that produce virus chronically or are latently infected and become activated. The identity and location of this actively replicating cell population is not known, but appears not to reside in the PBMC pool, consistent with prior reports[11,19,20]. Nevertheless, PBMCs traffic through secondary lymphoid organs and to some extent are in equilibrium with these cells[35]. It is thus possible that a small fraction of PBMCs[8,9,14-17], like a small fraction of activated lymphoreticular cells[20], could make an important contribution to viraemia.

The magnitude of ongoing virus infection and production required to sustain steady-state levels of viraemia is extraordinary: based on a virus $t_{1/2}$ of 2.0 days and first-order clearance kinetics ($v(t) = v(0)e^{-\alpha t}$, where $\alpha = 0.693/t_{1/2}$), 30% or more of the total virus population in plasma must be replenished daily. For a typical HIV-1-infected individual with a plasma virus titre equalling the pretreatment geometric mean in this study ($10^{5.5}$ RNA molecules per ml/2 RNA molecules per virion = $10^{5.2}$ virions per ml) and a plasma volume of 3 litres, this amounts to $(0.30) \times (10^{5.2}) \times (3 \times 10^{3}) = 1.1 \times 10^{8}$ virions per day (range for all 22 subjects, 2×10^{7} to 7×10^{9}). Even this may be a substantial underestimate of virus expression because virions may be inefficiently transported from the interstitial extravascular spaces into the plasma compartment and viral protein expression alone (short of mature particle formation) may result in cytopathy or immune-mediated destruction. Because the half-life of cells producing the majority of plasma virus cannot exceed 2.0 days, at least 30% of these cells must also be replaced daily. In our patients, we estimated the rate of $CD4^+$ lymphocyte turnover to be, on average, 2×10^{9} cells per day, or about 5% of the total $CD4^+$ lymphocyte population, depending on clinical stage. This rapid and ongoing recruitment of $CD4^+$ cells into a short-lived virus-expressing pool probably explains the abrupt increase in $CD4^+$ lymphocyte numbers that is observed immediately following the initiation of potent antiretroviral therapy, and suggests the possibility of successful immunological reconstitution even in late-stage disease if effective control of viral replication can be sustained.

The kinetics of virus and $CD4^+$ lymphocyte production and clearance reported here have a number of biological and clinical implications. First, they are indicative of a dynamic process involving continuous rounds of *de novo* virus infection, replication and rapid cell turnover that probably represents a primary driving force underlying HIV-1 pathogenesis. Second, the demonstration of rapid and virtually complete replacement of wild-type virus by drug-resistant virus in plasma after only 14–28 days of drug therapy is a striking example of the capacity of the virus for biologically relevant change. In particular, this implies that HIV-1 must have enormous potential to evolve in response to selection pressures as exerted by the immune system[39]. Although other studies[40-42] have provided some evidence that virus turnover occurs sooner in plasma than in PBMCs, our data show this phenomenon most clearly. A similar experimental approach involving the genotypic and phenotypic analysis of plasma virus could be helpful in identifying viral mutations and selection pressures involved in resistance to other drugs, immune surveillance and viral pathogenicity. Third, the difference in lifespan between virus-producing cells and latently infected cells (PBMCs) suggests that virus expression *per se* is directly involved in $CD4^+$ cell destruction. The data do not suggest an "innocent bystander" mechanism of cell

不是感染细胞缓慢产生病毒或者感染细胞休眠再激活。与之前报告一致，这个活跃复制细胞群体的本质和位置尚不明确，但活跃复制中心可能不在 PBMC[11,19,20]。无论怎样，PBMC 经过次级淋巴器官在某种程度上与这些细胞达到动态平衡 [35]。这提示 PBMC 的小亚群 [8,9,14-17] 可能与这些活化的淋巴网状细胞 [20] 一样，对病毒血症起关键作用。

维持病毒血症稳态水平所需要的持续性病毒感染量和产生量非常可观。根据病毒 2.0 天的半衰期及其一级清除动力学（$v(t) = v(0)e^{-\alpha t}$，其中 $\alpha = 0.693/t_{1/2}$），每天血浆中总病毒群体的 30% 甚至更多都需要被重新补足。一个典型 HIV-1 患者血浆病毒滴度约为本研究治疗前的几何平均数（每毫升 $10^{5.5}$ 个 RNA 分子 / 每个病毒粒子 2 个 RNA 分子 = 每毫升 $10^{5.2}$ 个病毒粒子）。患者血浆体积约 3 升，每天需要的病毒粒子数为 $(0.30) \times (10^{5.2}) \times (3 \times 10^3) = 1.1 \times 10^8$ 个（所有 22 名患者每日所需病毒粒子数范围为 $2 \times 10^7 \sim 7 \times 10^9$）。尽管此数值可能低估了病毒的表达量，这是由于病毒粒子可能不能有效地从血管外间隙转运到血浆中，同时病毒蛋白表达本身（缺乏成熟颗粒）可能会导致细胞病变或者免疫介导损伤。由于产生大部分血浆病毒的产病毒细胞半衰期不会超过 2.0 天，所以每天至少 30% 的产病毒细胞要被替换。在目前研究的 HIV 感染患者群中，我们估算 $CD4^+$ 淋巴细胞更替速率平均是每天 2×10^9 个，或大约总 $CD4^+$ 淋巴细胞群体的 5%。这些估算会随着临床病期不同而有所差异。这种快速而持续的 $CD4^+$ 淋巴细胞补充至寿命短暂的病毒生成细胞群，可能是有效的抗逆转录病毒治疗后立即出现 $CD4^+$ 淋巴细胞数目增多的原因。同时提示只要有效地控制病毒复制，即便在 HIV 感染晚期也存在进行免疫重建的可能。

本文报道的病毒动力学和 $CD4^+$ 淋巴细胞产生及清除动力学具有多种生物学和临床应用前景。首先，这些结果提示持续的重复病毒感染、复制和快速细胞更替等动力学过程可能是 HIV-1 发病机制中的主要驱动力。其次，在药物治疗仅 14~28 天后就出现野生型血浆病毒几乎完全被耐药突变体替代，这是一个令人惊奇的病毒生物学相关变化。尤其是，这提示 HIV-1 病毒具有强大的潜力应对免疫系统的选择压力 [39]。尽管其他的研究 [40-42] 已经发现病毒更替在血浆中的发生比在 PBMC 中的发生更快，我们的数据更加清楚地阐释这种现象。类似的分析血浆病毒基因型和表型的实验方法可能对于发现病毒突变体或发现包括耐药性、免疫监视和病毒致病性等选择压力下病毒的改变是非常有帮助的。第三，产病毒细胞和潜伏感染细胞（PBMC）寿命不同表明病毒表达本身就直接导致 $CD4^+$ 淋巴细胞的损伤。这些研究结果并没有暗示在细胞杀伤机制中存在“旁观者效应”，也没有暗示未感染或处于潜伏期感染

killing whereby uninfected or latently infected cells are indirectly targeted for destruction by adsorption of viral proteins or by autoimmune reactivities.

Although we have emphasized that most virus in plasma derives from an actively replicating short-lived population of cells, latently infected cells that become activated or chronically producing cells that generate proportionately less virus (and thus do not contribute substantially to the plasma virus pool) may nonetheless be important in HIV-1 pathogenesis. Based on *in situ* analysis[20], these cells far outnumber the actively replicating pool and the diversity of their constituent viral genomes represents a potentially important source of clinically relevant variants, including those conferring drug resistance. In future studies, it will be important not only to discern the specific elimination rates of free virus and of the most actively producing cells, but also the dynamics of virus replication and cell turnover in other cell populations and in patients at earlier stages of infection. Such information will be essential to developing a better understanding of HIV-1 pathogenesis and a more rational approach to therapeutic intervention.

(**373**, 117-122; 1995)

Xiping Wei*, Sajal K. Ghosh*, Maria E. Taylor*, Victoria A. Johnson†, Emilio A. Emini‡, Paul Deutsch§, Jeffrey D. Lifson‖, Sebastian Bonhoeffer¶, Martin A. Nowak¶, Beatrice H. Hahn*, Michael S. Saag† & George M. Shaw*#

Divisions of * Hematology/Oncology and † Infectious Diseases, University of Alabama at Birmingham, 613 Lyons-Harrison Research Building, 701 South 19th Street, Birmingham, Alabama 35294, USA

Departments of ‡Antiviral Research and §Clinical Pharmacology, Merck Research Laboratories, West Point, Pennsylvania 19486, USA

‖ Division of HIV and Exploratory Research, Genelabs Technologies Inc., Redwood City, California 94063, USA

¶ Department of Zoology, University of Oxford, Oxford 0X1 3PS, UK

To whom correspondence should be addressed.

Received 22 November; accepted 16 December 1994.

References:

1. Ho, D. D., Moudgil, T. & Alam. M. *New Engl. J. Med.* **321**, 1621-1625 (1989).
2. Coombs, R. W. *et al. New Engl. J. Med.* **321**, 1626-1631 (1989).
3. Saag, M. S. *et al. J. Infect. Dis* **164**, 72-80 (1991).
4. Clark, S. J, *et al. New Engl. J. Med.* **324**, 954-960 (1991).
5. Daar, E. S., Mougdil, T., Meyer, R. D. & Ho, D.D. *New Engl. J. Med.* **324**, 961-964 (1991).
6. Piatak, M. Jr *et al. Science* **259**, 1749-1754 (1993).
7. Piatak, M. *et al. Lancet* **341**,1099 (1993).
8. Schnittman, S. M., Greenhouse, J. J., Lane, H. C., Pierce, P. F. & Fauci, A. S. *AIDS Res. Hum. Retrovir.* 7, 361-367 (1991).
9. Michael, N. L., Vahey, M., Burke, D. S. & Redfield, R. R. *J. Virol.* **66**, 310-316 (1992).
10. Winters, M. A., Tan, L. B., Katzenstein, D. A. & Merigan, T. C. *J. Clin. Microbiol.* **31**, 2960- 2966 (1993).
11. Pantaleo, G. *et al. Proc. Natl. Acad.* Sci. *U.S.A.* **88**, 9838-9842 (1991).
12. Connor, R. I., Mohri, H., Cao, Y. & Ho, D. D. *J. Virol.* **67**, 1772-1777 (1993).
13. Bagnarelli, P. *et al. J. Virol.* **66**, 7328-335 (1992).
14. Bagnarelli, P. *et al. J. Virol.* **68**, 2495-2502 (1994).
15. Graziosi, C. *et al. Proc. Natl. Acad. Sci. U.S.A.* **90**, 6405-6409 (1993).
16. Patterson, B. K. *et al. Science* **260**, 976-979 (1993).
17. Saksela, K., Stevens, C., Rubinstein, P. & Baltimore, D. *Proc. Natl. Acad. Sci. U.S.A.* **91**, 1104-1108 (1994).

细胞间接通过吸附病毒蛋白或者自身免疫反应造成损伤。

尽管我们强调血浆病毒大部分源于复制活跃且寿命短暂的细胞群体，但那些激活的潜伏感染细胞或者产生相对较少病毒的慢速产病毒细胞（因此不能对血浆病毒群产生较大的作用）仍然在 HIV-1 发病机制中发挥重要作用。根据原位分析[20]，这些细胞的数量远远超过复制活跃的细胞，而且这些病毒基因组构成的多样性是临床治疗过程中产生耐药性变异体的潜在重要来源。在今后的研究中，不仅要关注特异性游离病毒以及复制活跃的细胞的清除速率，而且要重视其他细胞群体以及早期感染阶段的患者的病毒复制和细胞更替动力学。这些信息对于更好地了解 HIV-1 发病机制以及开发更合理的治疗性干预艾滋病的方法至关重要。

（毛晨晖 翻译；胡卓伟 审稿）

18. Cao, Y. *et al. AIDS Res. Hum. Retrovir.* (in the press).

19. Pantaleo, G. *et al. Nature* **362**, 355-358 (1993).

20. Embretson, J. *et al. Nature* **362**, 359-362 (1993).

21. Aoki-Sei, S. *et al. AIDS Res. Hum. Retrovir.* **8**, 1263-1270 (1992).

22. Coffin, J. M. *Curr. Top. Microbiol. Immun.* **176**, 143-164 (1992).

23. Wain-Hobson, S. *Curr. Opin. Genet. Dev.* **3**, 878-883 (1993).

24. Meriuzzi, V. J. *et al. Science* **250**, 1411-1413 (1990).

25. Kempf, D. *et al. Proc. Natl. Acad. Sci. U.S.A.* (in the press).

26. Vacca, J. P. *et al. Proc. Natl. Acad. Sci. U.S.A* **91**, 4096-4100 (1994).

27. Mulder, J. *et al. J. Clin. Microbiol.* **32**, 292-300 (1994).

28. Larder, B. A. *et al. Nature* **365**, 671-675 (1993).

29. Prasad, V. R. & Goff, S. P. *J. Biol. Chem.* **264**, 16689-16693 (1989).

30. Saag, M. S. *et al. New Engl. J. Med.* **329**, 1065-1072 (1993).

31. Richman, D. D. *et al. Proc. Natl. Acad. Sci. U.S.A.* **88**, 11241-11245 (1991).

32. Richman, D. D. *et al. J. Virol.* **68**, 1660-1666 (1994).

33. McLean, A. R. & Nowak, M. A. *AIDS* **6**, 71-79 (1992).

34. Michie, C., McLean, A., Alcock, C. & Beverley, P. C. L. *Nature* **360**, 264-265 (1992).

35. Sprent, J. & Tough, D. F. *Science* **265**, 1395-1400 (1994).

36. Balzarini, J. *et al. Proc. Natl. Acad. Sci. U.S.A.* **91**, 6599-6603 (1994).

37. Meyerhans, A. *et al. Cell* **58**, 901-910 (1989).

38. Japour, A. J. *et al. Antimicrob. Agents Chemother.* **37**, 1095-1101 (1993).

39. Nowak, M. A. *et al. Science* **254**, 963-969 (1991).

40. Simmonds, P. *et al. J. Virol.* **65**, 6266-6276 (1991).

41. Smith, M. S., Koerber, K. L. & Pagano, J. S. *J. Infect. Dis.* **167**, 445-448 (1993).

42. Zhang, Y.-M., Dawson, S. C., Landsman, D., Lane, H. C. & Salzman, N. P. J. *Virol.* **68**, 425- 432 (1994).

43. Myers, G. K., Korber, B., Berzofsky, J. A. & Smith, R.F. *Human Retroviruses and AIDS 1993* (Los Alamos National Laboratory, New Mexico, 1993).

44. Shaw, G. M. *et al. Science* **226**, 1165-1171 (1984).

Acknowledgements. We thank the study participants; K. Squires, J. M. Kilby, M. Trechsel, L. DeLoach and the UAB 1917 Clinic staff; Abbott Laboratories, Merck & Co. and Boehringer Ingelheim Pharmaceuticals Inc. (BIPI); J. Coffin, R. May and F. Gao for discussion; J. Decker, S. Campbell-Hill, Y. Niu and S. Yin Jiang for technical assistance; and J. Wilson for artwork. This study was supported by the NIH, the US Army Medical Research Acquisition Activity, BIPI, the Wellcome Trust, Keble College and Boehringer Ingelheim Stiftung. Core research facilities were provided by the UAB Center for AIDS Research, the UAB AIDS Clinical Trials Unit and the Birmingham Veterans Administration Medical Center.

Rapid Turnover of Plasma Virions and CD4 Lymphocytes in HIV-1 Infection

D. D. Ho *et al.*

Editor's Note

This paper by David Ho of the New York University School of Medicine and coworkers, along with an accompanying paper by a British and American team reporting the same result, provided a fundamental insight into the behaviour of the AIDS virus HIV-1, which altered the thinking behind strategies to combat the virus. It had been widely thought that, after infection, HIV has an inactive period of "latency". Ho *et al.* show here that, on the contrary, the virus has a very rapid and active cycle of replication, which both exhausts the CD4 immune-response cells trying to combat it and enables the virus to rapidly evolve drug-resistance. This finding recommended an approach of hitting HIV with drugs "early and hard".

Treatment of infected patients with ABT-538, an inhibitor of the protease of human immunodeficiency virus type 1 (HIV-1), causes plasma HIV-1 levels to decrease exponentially (mean half-life, 2.1 ± 0.4 days) and CD4 lymphocyte counts to rise substantially. Minimum estimates of HIV-1 production and clearance and of CD4 lymphocyte turnover indicate that replication of HIV-1 *in vivo* is continuous and highly productive, driving the rapid turnover of CD4 lymphocytes.

IN HIV-1 pathogenesis, an increased viral load correlates with CD4 lymphocyte depletion and disease progression[1-9], but relatively little information is available on the kinetics of virus and CD4 lymphocyte turnover *in vivo*. Here we administer an inhibitor of HIV-1 protease, ABT-538 (refs 10, 11), to twenty infected patients in order to perturb the balance between virus production and clearance. From serial measurements of the subsequent changes in plasma viraemia and CD4 lymphocyte counts, we have been able to infer kinetic information about the pretreatment steady state.

ABT-538 has potent antiviral activity *in vitro* and favourable pharmacokinetic and safety profiles *in vivo*[10]. It was administered orally (600–1,200 mg per day) on day 1 and daily thereafter to twenty HIV-1-infected patients, whose pretreatment CD4 lymphocyte counts and plasma viral levels ranged from 36 to 490 per mm^3 and from 15×10^3 to 554×10^3 virions per ml, respectively (Table 1). Post-treatment CD4 lymphocyte counts were monitored sequentially, as were copy numbers of particle-associated HIV-1 RNA in plasma, using an ultrasensitive assay (Fig. 1 legend) based on a modification of the branched DNA signal-amplification technique[12,13]. The trial design and clinical findings of this study will be reported elsewhere (M.M. *et al.*, manuscript in preparation).

HIV-1 感染患者血浆内病毒粒子和 CD4 淋巴细胞的快速更新

何大一等

编者按

这篇由纽约大学医学院的何大一及其同事发表的文章提供了对艾滋病病毒 HIV-1 行为的基本认知。另一篇由英美团队发表的文章报道了相同的结果。这一结果改变了抗病毒的策略。人们广泛认为，HIV 病毒感染后有一个不活跃的“潜伏期”。何大一他们在这篇文章中指出，与这一认知相反，病毒有一个快速且活跃的复制期，这一时期不仅会耗尽试图对抗病毒的 CD4 免疫应答细胞而且会使病毒迅速出现耐药性。这一发现建议尽早用药物猛烈对抗 HIV。

使用 ABT-538（一种人免疫缺陷病毒 I 型（HIV-1）蛋白酶抑制剂）治疗 HIV-1 感染患者后，引起血浆内 HIV-1 水平指数式下降（平均半衰期为 2.1 ± 0.4 天）而 CD4 淋巴细胞数目显著升高。HIV-1 的产生和清除以及 CD4 淋巴细胞更新的最小估测值表明，HIV-1 在体内的复制是持续的、高产的，并可促使 CD4 淋巴细胞快速更新。

在 HIV-1 的发病机制中，体内病毒载量的增加与 CD4 淋巴细胞减少以及疾病进展密切相关[1-9]，但是目前关于体内病毒和 CD4 淋巴细胞更新的动力学信息相对较少。本文中我们在 20 名感染患者身上使用一种 HIV-1 蛋白酶抑制剂——ABT-538（参考文献 10，参考文献 11）以干预病毒产生和清除之间的平衡。通过检测血浆病毒血症和 CD4 淋巴细胞数目的后续改变，我们已经能够推断治疗前稳态的动力学信息。

ABT-538 在体外具有强大的抗病毒活性，而且在体内有良好的药动学特性和安全性[10]。20 名 HIV-1 感染患者每天口服药物（600~1,200 毫克 / 天），他们的 CD4 淋巴细胞数目和血浆内病毒水平在治疗前分别是每立方毫米 36~490 个细胞和每毫升 15×10^3~554×10^3 个病毒粒子（表 1）。随后使用基于改良的分支 DNA 信号扩增技术[12,13]的超敏感检测方法（图 1 图注）监测治疗后的 CD4 淋巴细胞数目以及血浆中 HIV-1 颗粒相关 RNA 的拷贝数。本研究的实验设计和临床结果将另有报道（马科维茨等，稿件准备中）。

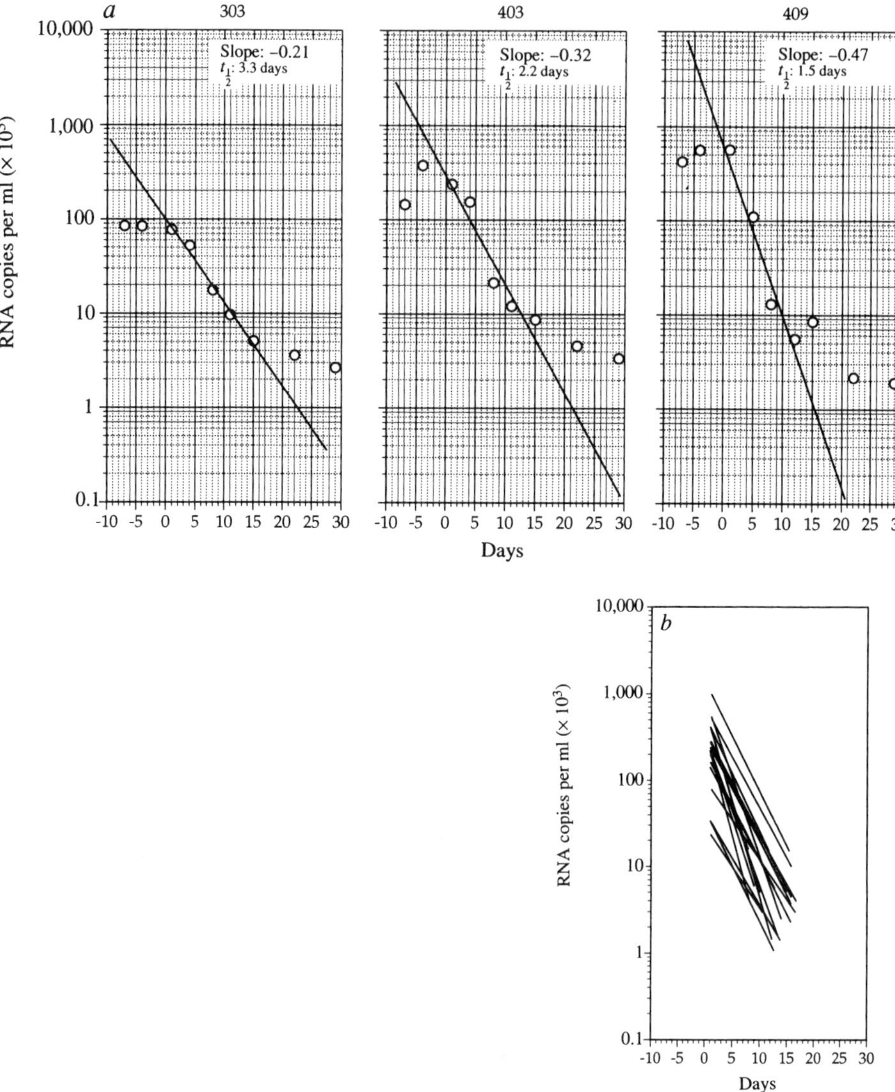

Fig. 1. *a*, Plasma viral load before and after ABT-538 treatment was begun on day 1 for three representative cases. Plasma samples were tested with the branched DNA signal-amplification assay as previously described[12,13]. Those samples with RNA levels below the detection sensitivity of 10,000 copies per ml were then tested using a modified assay differing from the original in two ways: hybridization of the bDNA amplification system is mediated by binding to overhangs on contiguous target probes; and the enzymatic amplification system has been enhanced by modification of wash buffers and the solution in which the alkaline phosphatase probe is diluted. The results of these changes are a diminution of background signals, an enhancement of alkaline phosphatase activity, and thus a greater detection sensitivity (500 copies per ml). Linear regression was used to obtain the best-fitting straight line for 3–5 data points between day 1 and the inflection point before the plateau of the new steady-state level. The slope, S, of each line represents the rate of exponential decrease; that is, the straight-line fit indicates that the viral load decreases according to $V(t) = V(0)\exp(-St)$. Given the exponential decay, $V(t_{1/2}) = V(0)/2 = V(0)\exp(-St_{1/2})$, and hence the viral half-life, $t_{1/2} = \ln(2)/S$. Before drug administration, the change in viral load with time can be expressed by the differential equation, $dV/dt = P - cV$, where P is the viral production rate, c is the viral clearance rate constant, and V is the number of plasma virions. During the pretreatment steady state, $dV/dt = 0$, and hence $P = cV$. We have also tested more intricate models, in

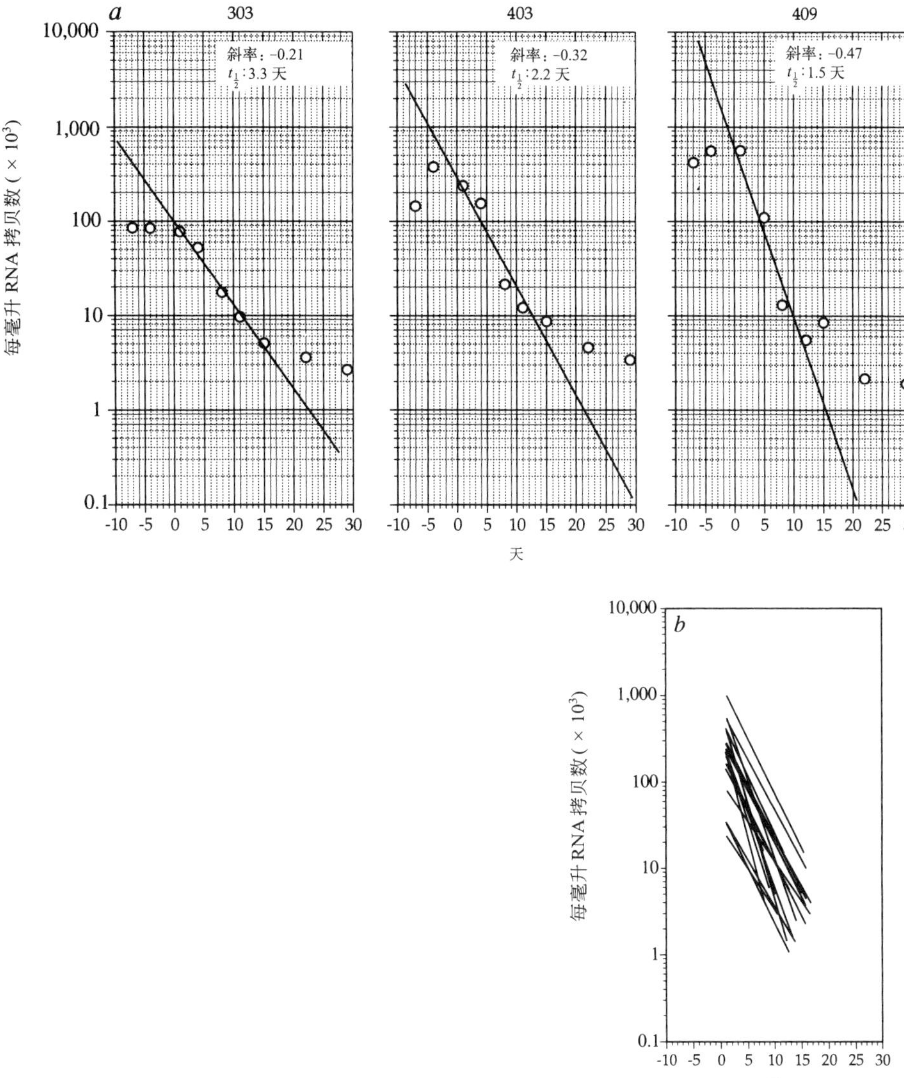

图 1. a，三个代表性病例开始 ABT-538 治疗前后的血浆内病毒载量。如参考文献所述 [12,13] 血浆样本用分支 DNA 信号扩增法检测。那些 RNA 水平处于检测敏感度（10,000 拷贝 / 毫升）以下的样本随后用改良的方法检测，该方法与前者有两方面的不同：bDNA 扩增系统的杂交是通过结合到邻近目标探针的突出部分上，而且酶扩增系统通过改良冲洗缓冲液和稀释碱性磷酸酶探针的溶液加以强化。这些改变可以减少背景信号，增强碱性磷酸酶活性从而提高检测敏感性（500 拷贝 / 毫升）。使用线性回归来获得第 1 天和新稳态水平平台期之前的拐点之间 3~5 个数据点的最合适直线。每一条直线的斜率 S 代表了指数式减少的速率。也就是说，直线拟合表示病毒载量的减少符合 $V(t)=V(0)\exp(-St)$。鉴于存在指数式衰减，$V(t_{1/2})=V(0)/2=V(0)\exp(-St_{1/2})$，因此病毒的半衰期 $t_{1/2}=\ln(2)/S$。药物使用前，病毒载量随时间的改变可以表达成微分方程 $\mathrm{d}V/\mathrm{d}t=P-cV$，其中 P 表示病毒产生速率，c 表示病毒清除速率常数，V 是血浆内病毒粒子的数量。在治疗前的稳态期，$\mathrm{d}V/\mathrm{d}t=0$，因此 $P=cV$。我们也检测了更多复杂的模型，其病毒减少由 2 到 3 种指数式速率决定，分别是病毒清除速率、产病毒细胞的减少速率和潜伏感染细胞的减少速率。但是这次没有足够的数据分别估计多个参数。然而，使用参考文献 14 中的模型，我们发现如果潜伏感染细胞的死亡速率相对于其他两个速率非常小的话，那么病毒的减少呈一种简单的指

which the viral decay is governed by two or three exponential rates, namely the viral clearance rate, the decay rate of virus-producing cells, and the decay rate of latently infected cells. But there are insufficient data at this time to estimate the multiple parameters separately. Nonetheless, using the model in ref. 14, we find that if the death rate of latently infected cells is very small compared with the other two rates, then viral decay follows a simple exponential decline, with $S=c$, because the slow activation of a large number of latently infected cells offsets the loss of actively infected cells. Irrespective of the model, on a log plot, $S=-\mathrm{d}(\ln V)/\mathrm{d}t=c-P/V$. If drug inhibition is complete and virus-producing cells are rapidly lost (so $P=0$), then $S=c$. If viral production continues, S is still $\leqslant c$, so the slope is a minimum estimate of the viral clearance efficiency, *b*, Decline of plasma viral load after ABT-538 treatment in all 20 patients. The slope for each case was obtained as already discussed, and the length of each line was determined by the initial viral load and the new steady-state level.

Table 1. Summary data of HIV-1 and CD4 lymphocyte turnover during the pretreatment steady state

Patient	Baseline values		Kinetics of HIV-1 turnover			Kinetics of CD4 lymphocyte turnover*		
	CD4 cell count (mm^{-3})	Plasma viraemia (virions per ml × 10^3) †	Slope	$t_{1/2}$ (days)	Minimum production and clearance‡ (virions per day × 10^9)	Slope	Minimum production and destruction: Blood§ (cells per day × 10^6)	Minimum production and destruction: Total‖ (cells per day × 10^9)
301	76	193	−0.30	2.3	0.56	0.070 (6.9)	21.7 (28.1)	1.1 (1.4)
302	209	80	−0.27	2.6	0.26	0.004 (0.5)	4.3 (2.7)	0.2 (0.1)
303	293	41	−0.21	3.3	0.11	0.005 (1.4)	9.9 (9.5)	0.5 (0.5)
304	174	121	−0.28	2.5	0.54	0.019 (1.9)	22.2 (13.0)	1.1 (0.6)
305	269	88	−0.33	2.1	0.50	0.055 (21.5)	108.0 (157.0)	5.4 (7.8)
306	312	175	−0.52	1.3	1.27	0.058 (25.7)	105.0 (150.0)	5.3 (7.5)
308	386	185	−0.46	1.5	1.48	0.020 (9.1)	55.9 (65.8)	2.8 (3.3)
309	49	554	−0.29	2.4	1.85	0.088 (11.8)	20.7 (56.6)	1.0 (2.8)
310	357	15	−0.26	2.7	0.05	0.038 (15.6)	71.0 (81.9)	3.6 (4.1)
311	107	130	−0.29	2.4	0.51	0.064 (11.0)	38.9 (62.8)	2.0 (3.1)
312	59	70	−0.30	2.3	0.30	0.048 (4.5)	17.0 (26.9)	0.8 (1.4)
313	47	100	−0.54	1.3	0.88	0.077 (5.9)	24.7 (40.5)	1.2 (2.0)
401	228	101	−0.40	1.7	0.47	NA	NA	NA
402	169	55	−0.28	2.5	0.21	0.014 (3.1)	13.4 (17.4)	0.7 (0.9)
403	120	126	−0.32	2.2	0.74	0.015 (2.4)	13.8 (18.7)	0.7 (0.9)
404	46	244	−0.27	2.6	1.06	0.080 (8.5)	24.6 (57.5)	1.2 (2.9)
406	490	18	−0.31	2.2	0.08	NA	NA	NA
408	36	23	−0.25	2.8	0.08	0.059 (3.4)	12.5 (19.7)	0.6 (1.0)
409	67	256	−0.47	1.5	2.07	0.073 (15.9)	35.3 (115.0)	1.8 (5.7)
410	103	99	−0.36	1.9	0.53	0.051 (5.6)	32.4 (34.5)	1.6 (1.7)
Range	36–490	15–554	−0.21 to −0.54	1.3–3.3	0.05–2.07	0.004–0.088 (0.5–25.7)	4.3–108.0 (2.7–157.0)	0.2–5.4 (0.1–7.8)
Mean	180±46	134±40	−0.34±0.06	2.1±0.4	0.68±0.13	0.047 (8.6)	35.1 (53.2)	1.8 (2.6)

* The results for the kinetics of CD4 lymphocyte turnover generated by an exponential growth model are shown without parentheses; results generated by a linear production model are shown in parentheses.

数式减少，其 $S=c$，因为大量潜伏感染细胞的缓慢激活抵消了活跃感染细胞的减少。不考虑这个模型，在对数图上，$S=-\mathrm{d}(\ln V)/\mathrm{d}t=c-P/V$。如果药物的抑制作用是完全的，而且产病毒细胞很快减少(那么 $P=0$)，那么 $S=c$。如果病毒持续产生，$S\leqslant c$，因此斜率是病毒清除效率的最小估测值。b，20 名患者 ABT-538 治疗后血浆内病毒载量的减少。每个病例的斜率是通过之前讨论的方法得到的，每条线的长度由初始病毒载量和新的稳态水平决定。

表 1. 在治疗前稳态期 HIV-1 和 CD4 淋巴细胞更新的数据总结

基线值			HIV-1 更新动力学			CD4 淋巴细胞更新动力学 *		
患者	CD4 细胞数目(每立方毫米)	血浆病毒血症(每毫升病毒粒子 $\times 10^3$)†	斜率	$t_{1/2}$(天)	最小生成和清除速率 ‡(每日病毒粒子 $\times 10^9$)	斜率	最小生成和破坏速率	
							血液 §(每日细胞 $\times 10^6$)	总计 ‖(每日细胞 $\times 10^9$)
301	76	193	−0.30	2.3	0.56	0.070 (6.9)	21.7 (28.1)	1.1 (1.4)
302	209	80	−0.27	2.6	0.26	0.004 (0.5)	4.3 (2.7)	0.2 (0.1)
303	293	41	−0.21	3.3	0.11	0.005 (1.4)	9.9 (9.5)	0.5 (0.5)
304	174	121	−0.28	2.5	0.54	0.019 (1.9)	22.2 (13.0)	1.1 (0.6)
305	269	88	−0.33	2.1	0.50	0.055 (21.5)	108.0 (157.0)	5.4 (7.8)
306	312	175	−0.52	1.3	1.27	0.058 (25.7)	105.0 (150.0)	5.3 (7.5)
308	286	185	−0.46	1.5	1.48	0.020 (9.1)	55.9 (65.8)	2.8 (3.3)
309	49	554	−0.29	2.4	1.85	0.088 (11.8)	20.7 (56.6)	1.0 (2.8)
310	357	15	−0.26	2.7	0.05	0.038 (15.6)	71.0 (81.9)	3.6 (4.1)
311	107	130	−0.29	2.4	0.51	0.064 (11.0)	38.9 (62.8)	2.0 (3.1)
312	59	70	−0.30	2.3	0.30	0.048 (4.5)	17.0 (26.9)	0.8 (1.4)
313	47	100	−0.54	1.3	0.88	0.077 (5.9)	24.7 (40.5)	1.2 (2.0)
401	228	101	−0.40	1.7	0.47	NA	NA	NA
402	169	55	−0.28	2.5	0.21	0.014 (3.1)	13.4 (17.4)	0.7 (0.9)
403	120	126	−0.32	2.2	0.74	0.015 (2.4)	13.8 (18.7)	0.7 (0.9)
404	46	244	−0.27	2.6	1.06	0.080 (8.5)	24.6 (57.5)	1.2 (2.9)
406	490	18	−0.31	2.2	0.08	NA	NA	NA
408	36	23	−0.25	2.8	0.08	0.059 (3.4)	12.5 (19.7)	0.6 (1.0)
409	67	256	−0.47	1.5	2.07	0.073 (15.9)	35.3 (115.0)	1.8 (5.7)
410	103	99	−0.36	1.9	0.53	0.051 (5.6)	32.4 (34.5)	1.6 (1.7)
范围	36~490	15~554	−0.21 至 −0.54	1.3~3.3	0.05~2.07	0.004~0.088 (0.5~25.7)	4.3~108.0 (2.7~157.0)	0.2~5.4 (0.1~7.8)
均值	180±46	134±40	−0.34±0.06	2.1±0.4	0.68±0.13	0.047 (8.6)	35.1 (53.2)	1.8 (2.6)

* 通过指数式增长模型得出的 CD4 淋巴细胞更新动力学的结果显示为不带括号的；通过线性模型得到的结果在括号内显示。

† Each virion contains two RNA copies.

‡ Calculated using plasma and extracellular fluid volumes estimated from body weights, and assuming that plasma and extracellular fluid compartments are in equilibrium.

§ Calculated using blood volumes estimated from body weights.

‖ Calculated on the assumption that the lymphocyte pool in blood represents 2% of the total population[16]. NA, not analysed owing to large fluctuations in CD4 cell counts.

Kinetics of HIV-1 Turnover

Following ABT-538 treatment, every patient had a rapid and dramatic decline in plasma viraemia over the first two weeks. As shown using three examples in Fig. l*a*, the initial decline in plasma viraemia was always exponential, demonstrated by a straight-line fit to the data on a log plot. The slope of this line, as defined by linear regression, permitted the half-life ($t_{1/2}$) of viral decay in plasma to be determined (Fig. 1 legend): for example, patient 409 was found to have a viral decay slope of −0.47 per day, yielding a $t_{1/2}$ of 1.5 days (Fig. l*a*). Hence the rate and extent of decay of plasma viraemia was determined for each patient. As summarized in Fig. 1*b*, in every case there was a rapid decline, the magnitude of which ranged from 11- to 275-fold, with a mean of 66-fold (equivalent to 98.5% inhibition). The residual viraemia may be attributable to inadequate drug concentration in certain tissues, drug resistance, persistence of a small long-lived virus-producing cell population (such as macrophages), and gradual activation of a latently infected pool of cells. As summarized in Table 1, the viral decay slopes varied from −0.21 to −0.54 per day, with a mean of -0.34 ± 0.06 per day; correspondingly, $t_{1/2}$ varied from 1.3 to 3.3 days, with a mean of 2.1 ± 0.4 days. The latter value indicates that, on average, half of the plasma virions turn over every two days, showing that HIV-1 replication *in vivo* must be highly productive.

The exponential decline in plasma viraemia following ABT-538 treatment reflects both the clearance of free virions and the loss of HIV-1-producing cells as the drug substantially blocks new rounds of infection. But although drug inhibition is probably incomplete and virus-producing cells are not lost immediately, a minimum value for viral clearance can still be determined (Fig. 1 legend) by multiplying the absolute value of the viral decay slope by the initial viral load. Assuming that ABT-538 administration does not affect viral clearance, this estimate is also valid before treatment. As the viral load varies little during the pretreatment phase (Fig. l*a*, and data not shown), we assume there exists a steady state and hence the calculated clearance rate is equal to the minimum virion production rate before drug therapy. Factoring in the patient's estimated plasma and extracellular fluid volumes based on body weight, we determined the minimum daily production and clearance rate of HIV-1 particles for each case (Table 1). These values ranged from 0.05 to 2.07×10^9 virions per day with a mean of $0.68 \pm 0.13 \times 10^9$ virions per day. Although these viral turnover rates are already high, true values may be up to a few-fold higher, depending on the $t_{1/2}$ of virus-producing lymphocytes. The precise kinetics of this additional parameter remains undefined. However, the mean $t_{1/2}$ of virus-producing cells is probably less, or in any case cannot be much larger, than the mean $t_{1/2}$ of 2.1 days observed for plasma virion elimination, demonstrating that turnover of actively infected cells is both

† 每个病毒粒子含有 2 个 RNA 拷贝。
‡ 用从体重估计的血浆和细胞外液计算得出，并假设血浆和细胞外液成分相同。
§ 用从体重估计的血容量计算得出。
‖ 假设血液中的淋巴细胞占总细胞数的 2% 得出的结果[16]。NA，由于 CD4 细胞数目波动太大而未分析。

HIV-1 更新的动力学

随着 ABT-538 治疗，每名患者在起始 2 周内血浆病毒血症出现快速且明显的降低。正如图 1*a* 中三个例子所示，血浆病毒血症的初始降低是指数式的，在以数据的对数值作出的图上显示为一条直线。其斜率可通过线性回归得出，并可用于确定血浆中病毒减少的半衰期($t_{1/2}$)(图 1 图注)。例如，患者 409 的病毒减少斜率是每天 −0.47，得出其半衰期是 1.5 天(图 1*a*)。这样就可以得到每名患者血浆病毒血症减少的速率和程度。如图 1*b* 总结的，每个病例中血浆内病毒载量都有快速的减少，其降低程度从 1/11 到 1/275 不等，平均 1/66 (相当于抑制了 98.5% 的病毒)。剩余的病毒血症可能是由某些组织中的药物浓度不够、耐药、存在一小群长寿命的产病毒细胞(比如巨噬细胞)以及潜伏的感染细胞池的逐步激活引起的。如表 1 所总结的，病毒减少的斜率从每天 −0.21 到 −0.54 不等，平均每天 −0.34 ± 0.06；相应地，$t_{1/2}$ 从 1.3 天到 3.3 天不等，平均 2.1 ± 0.4 天。后一个数值表明平均起来每两日就有一半的血浆病毒粒子发生更新，说明 HIV-1 在体内的复制肯定是高产量的。

ABT-538 治疗后血浆病毒血症的指数式减少说明同时发生了游离病毒粒子的清除和产 HIV-1 细胞的清除，因为该药物能有效地阻断进一步的感染。尽管药物的抑制可能是不完全的而且产病毒细胞并不是立即就被清除，但是仍能通过病毒减少斜率的绝对值与初始病毒载量相乘来确定病毒清除的最小数值(图 1 图注)。假设 ABT-538 的使用没有影响病毒的清除，这种治疗前的估计也仍然是有效的。因为在治疗前，病毒载量变化很小(图 1*a*，数据未显示)，我们假设存在一个稳定状态，因此在药物治疗前计算出来的清除速率与最小病毒粒子产生速率是相等的。将基于体重估计的患者血浆和细胞外液容积计算在内，我们得出每一例患者 HIV-1 颗粒的每日最小生产速率和清除速率(表 1)。这些数值从每天 0.05×10^9 到 2.07×10^9 个病毒粒子不等，平均每天 $(0.68 \pm 0.13) \times 10^9$ 个。尽管这些病毒更新速率已经非常快，根据产病毒淋巴细胞的半衰期，其真实数值可能会高出几倍。这些指标的准确动力学参数仍然未知。但是，产病毒细胞的平均半衰期很可能少于而不是大于血浆内病毒粒子清除的平均半衰期 2.1 天。这表明活跃感染细胞的更新是非常快速和持续的。从我们的

rapid and continuous. It could also be inferred from our data that nearly all (98.5%) of the plasma virus must come from recently infected cells.

Examination of Fig. 1*b* shows that the viral decay slopes (clearance rate constants) are independent of the initial viral loads. The slopes do not correlate with the initial CD4 lymphocyte counts (Fig. 2*a*), another indicator of the disease status of patients. Therefore these observations strongly suggest that the viral clearance rate constant is not dependent on the stage of HIV-1 infection. Instead, they indicate that viral load is largely a function of viral production, because clearance rate constants vary by about 2.5-fold whereas the initial loads vary by almost 40-fold (Table 1).

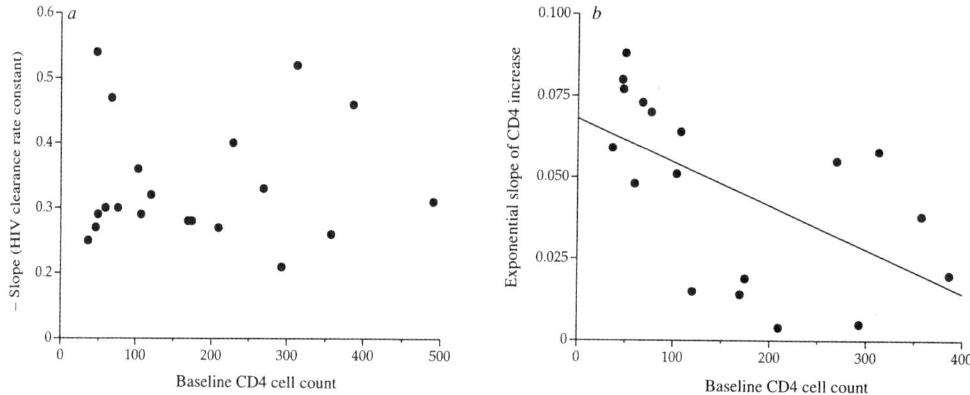

Fig. 2. *a*, Lack of correlation between viral decay slopes and disease status as indicated by baseline CD4 cell counts. Correlation coefficient = 0.05 (P value > 0.1). *b*, Inverse correlation between the exponential CD4 increase slopes and baseline CD4 cell counts. Correlation coefficient = −0.57 (P value < 0.01). Such an inverse correlation would be expected if T-cell proliferation were governed by a density-dependent growth function (logistic, for example), in which the growth rate decreases with increasing population level, if T cells were produced from precursors at a constant rate or from a combination of these two effects.

Kinetics of CD4 Lymphocyte Turnover

After ABT-538 treatment, CD4 lymphocyte counts rose in each of the 18 patients that could be evaluated. As shown in three examples in Fig. 3, some increases were dramatic (patient 409, for example) whereas others (such as patient 303) were modest. Based on the available data, it was not possible to determine with confidence whether the rise was strictly exponential (Fig. 3, top) or linear (Fig. 3, bottom). An exponential increase would be consistent with proliferation of CD4 lymphocytes in the periphery, particularly in secondary lymphoid organs, whereas a linear increase would indicate cellular production from a precursor source such as the thymus[14]. Given that the thymus involutes with age and becomes further depleted with HIV-1 infection[15], it is more likely that the rise in CD4 lymphocytes is largely due to proliferation. Nevertheless, as both components may contribute, we analysed the observed CD4 lymphocyte data by modelling both exponential and linear increases.

数据也能推断出几乎所有(98.5%)的血浆内病毒都来自新近感染的细胞。

图 1*b* 显示病毒减少的斜率(清除速率常数)与初始病毒载量无关，也与初始的 CD4 淋巴细胞数目(患者疾病状态的另一个指标)无关(图 2*a*)，提示患者处于疾病的稳态。因此这些结果显示病毒清除速率常数并不依赖于 HIV-1 感染的时期，并且说明病毒载量很大程度上是病毒产生引起的。因为清除速率常数变化了大约 2.5 倍，而初始载量则变化了几乎 40 倍(表 1)。

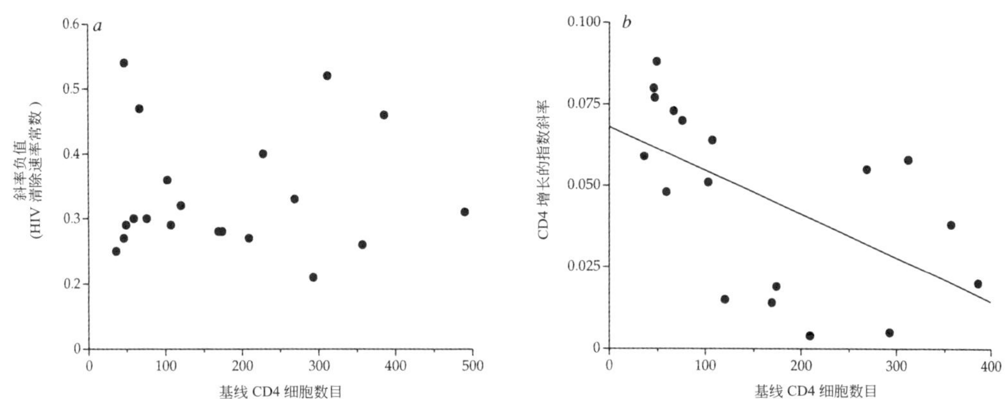

图 2. *a*，由基线 CD4 细胞数目所示，病毒减少斜率与疾病状态之间不存在相关性。相关系数 = 0.05 (P 值 > 0.1)。*b*，CD4 增长的指数斜率与基线 CD4 细胞数目之间存在负相关性。相关系数 = −0.57 (P 值 < 0.01)。出现这种负相关的情况如下：T 细胞的增殖是密度依赖的(比如对数式)，即增长速率随着群体量的增加而降低，或 T 细胞的增殖是以恒定的速率来源于前体，或两者的混合效果。

CD4 淋巴细胞更新动力学

经过 ABT-538 治疗后，所有可以评价的 18 名患者的 CD4 淋巴细胞数目都增加了。如图 3 中的三个例子所示，一些增加是非常显著的(比如患者 409)，而其他则相对适度(比如患者 303)。根据这些获得的数据，还不足以明确地确定这种增长是严格指数式的(图 3，上部)还是线性的(图 3，下部)。外周 CD4 淋巴细胞的增殖，尤其是次级淋巴器官中的增殖是指数式的增加，而线性增加则表明细胞来源于前体，比如胸腺 [14]。考虑到胸腺随年龄增长而退化并因 HIV-1 感染而进一步耗竭 [15]，很可能 CD4 淋巴细胞的增加大部分都来源于增殖。不管怎样，两种来源都可能发挥作用，我们分别用指数式和线性增长建模，分析观察到的 CD4 淋巴细胞数据。

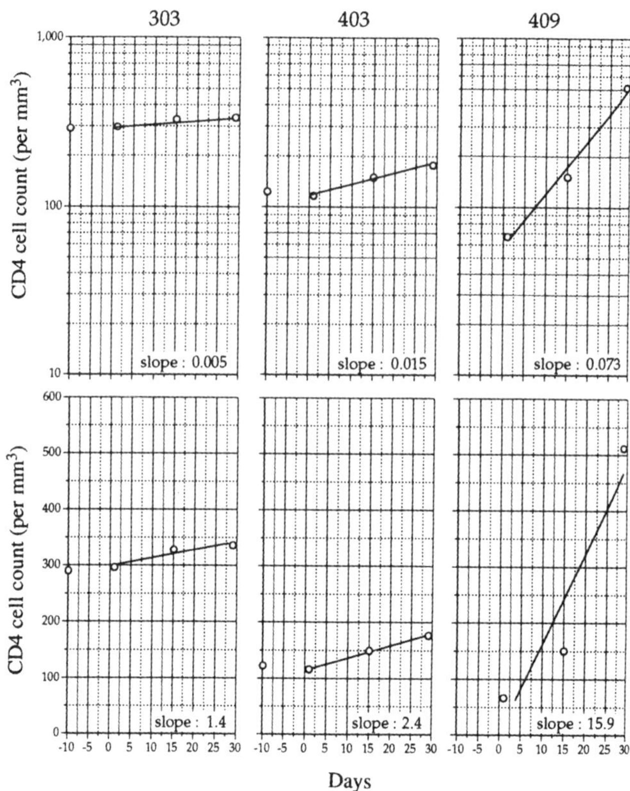

Fig. 3. Increase in CD4 cell counts after ABT-538 treatment plotted on a logarithmic (top) or linear (bottom) scale. Each slope was obtained from the best-fit line derived from linear regression on 2–4 data points. In the model for exponential increase, the doubling time was determined by dividing ln(2) by the slope. From the slope, we obtained minimum estimates of the CD4 lymphocyte production rate. The change in CD4 cell number over time can be described by the equation, $dT/dt = P - \mu T$, where T is the cell count, P is the cell production rate, and μ is the cell decay rate. The slope, S, on a log plot is thus $= d(\ln T)/dt = P/T - \mu$. Hence, $S \times T$ must be less than P, showing that our estimates indeed represent minimum CD4 lymphocyte production rates. Using a similar argument, slopes derived from a model of linear increases are also minimum estimates of CD4 lymphocyte production.

The slope of the line depicting the rise in CD4 lymphocyte counts on a log plot was determined for each case (Fig. 3, top). Individual slopes varied considerably, ranging from 0.004 to 0.088 per day, with a mean of 0.047 per day (Table 1), corresponding to a mean doubling time of ~15 days (Fig. 3 legend). On average, the entire population of peripheral CD4 lymphocytes was turning over every 15 days in our patients during the pretreatment steady state when CD4 lymphocyte production and destruction were balanced. Moreover, the slopes were inversely correlated with baseline CD4 lymphocyte counts (Fig. 2*b*) in that patients with lower initial CD4 cell counts had more prominent rises. This demonstrates convincingly that the CD4 lymphocyte depletion seen in AIDS is primarily a consequence of the destruction of these cells induced by HIV-1 not a lack of their production.

As ABT-538 treatment reduces virus-mediated destruction of CD4 lymphocytes, the observed increase in CD4 cells provides a minimum estimate (Fig. 3 legend) of the

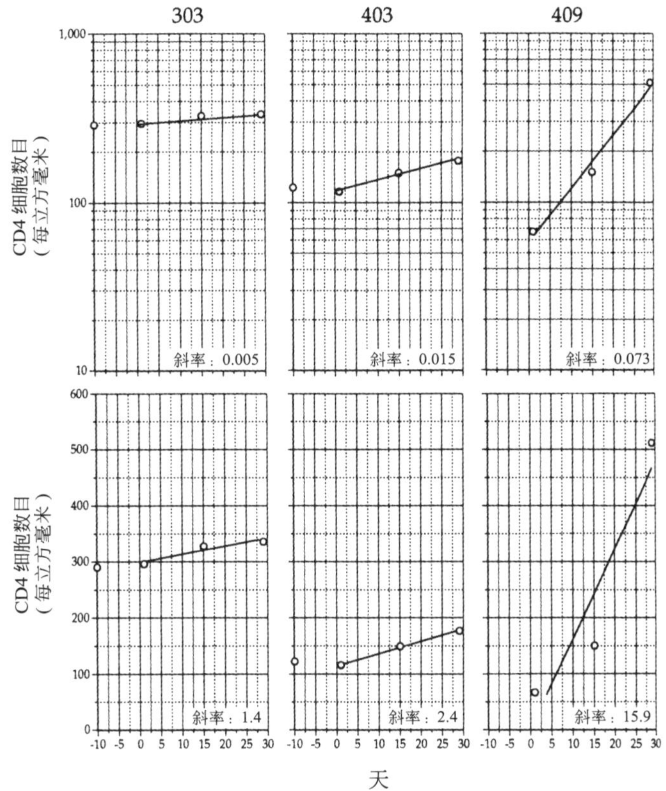

图 3. 在对数（上部）和线性（下部）坐标上绘出的 ABT-538 治疗后 CD4 淋巴细胞数目的增长。每一个斜率是通过 2~4 个数据点的线性回归得出的最适合直线获得的。在指数式增长模型上，倍增时间用 ln(2) 除以斜率得到。从斜率我们可以得到 CD4 淋巴细胞产生的速率的最小估测值。CD4 细胞数目随时间变化可以用以下公式描述 $dT/dt = P - \mu T$，其中 T 是细胞数目，P 是细胞生成速率，μ 是细胞减少速率。在对数图上斜率 $S = d(\ln T)/dt = P/T - \mu$。因此，$S \times T$ 必定小于 P，表明我们的估算确实代表了 CD4 淋巴细胞生成的最小速率。利用类似的论据，线性模型上得到的斜率也是 CD4 淋巴细胞生成的最小估测值。

我们确定了每个病例在对数图上描述 CD4 淋巴细胞数目增加的直线的斜率（图 3，上部）。各斜率差别很大，从每日 0.004 到 0.088 不等，平均每日 0.047（表 1），与平均倍增时间约是 15 天相对应（图 3 图注）。平均来说，在 CD4 淋巴细胞生成和破坏处于平衡状态的治疗前稳态阶段，患者整个外周 CD4 淋巴细胞群每 15 日更新一次。此外，CD4 增长的指数式斜率与基线 CD4 淋巴细胞数目（图 2*b*）呈负相关，即初始 CD4 淋巴细胞数目低的患者增加更明显。这就有力地证明了艾滋病患者体内 CD4 淋巴细胞的减少是 HIV-1 诱导的细胞破坏的结果，而不是产生不足的结果。

由于 ABT-538 的治疗减少了病毒介导的 CD4 淋巴细胞的破坏，观察到的 CD4 细胞增长值可以作为治疗前 CD4 淋巴细胞产生速率的最小估测值（图 3 图注），反

pretreatment CD4 lymphocyte production rate, which in turn equals the destruction rate during the steady state. Minimum production (destruction) rates were calculated for each case by multiplying the slope, the initial CD4 cell count, and the estimated blood volume. The minimum numbers of CD4 cells in blood produced or destroyed each day ranged from 4.3×10^6 to 108×10^6, with a mean of 35.1×10^6 (Table 1). Given that the blood lymphocyte pool is about 2% of the total population[16], the overall CD4 lymphocyte turnover in our patients was calculated to vary from 0.2×10^9 to 5.4×10^9 cells per day, with a mean of 1.8×10^9 cells per day.

The increase in CD4 lymphocyte counts following ABT-538 administration was also modelled linearly (Fig. 3, bottom). The slope of the line depicting the increase for each case was determined, and the values varied from 0.5 to 25.7 cells per mm^3 per day, with a mean of 8.6 cells per mm^3 per day (Table 1). Using the same argument as for the exponential case, minimum estimates of total CD4 production (or destruction) rates at baseline were determined to vary from 0.1×10^9 to 7.8×10^9 cells per day, with a mean of 2.6×10^9 cells per day.

Although our two sets of CD4 lymphocyte analyses do not yield identical numerical results, they are in close agreement and emphasize the same qualitative points about HIV-1 pathogenesis. The number of CD4 lymphocyte destroyed and replenished each day is of the order of 10^9, which is strikingly close to estimates of the total number of HIV-1 RNA-expressing lymphocytes in the body determined using *in situ* polymerase chain reaction and hybridization methods[8,17]. In addition, CD4 replenishment appears to be highly stressed in many patients in that the faster production rates are ~25–78-fold higher than the slowest rate (Table 1), which is presumably still higher than the as-yet-undefined normal CD4 turnover rate. The precise mechanisms of CD4 lymphocyte repopulation, however, will have to be addressed in the future by studies on phenotypic markers and functional status of the regenerating cells. Nonetheless, the rapid CD4 lymphocyte turnover has several implications. First, the apoptosis commonly observed in the setting of HIV-1 infection[18] may simply be an expected consequence of an active lymphocyte regenerative process. Second, the CD4 lymphocyte depletion seen in advanced HIV-1 infection may be likened to a sink containing a low water level, with the tap and drain both equally wide open. As the regenerative capacity of the immune system is not infinite, it is not difficult to see why the sink eventually empties. It is also evident from this analogy that our primary strategy to reverse the immunodeficiency ought to be to target virally mediated destruction (plug the drain) rather than to emphasize lymphocyte reconstitution (put in a second tap).

Discussion

We believe our new kinetic data have important implications for HIV-1 therapy and pathogenesis. It is self-evident that, with rapid turnover of HIV-1, generation of viral diversity and the attendant increased opportunities for viral escape from therapeutic agents are unavoidable sequelae[19,20]. Treatment strategies, if they are to have a dramatic clinical impact, must therefore be initiated as early in the infection course as possible, perhaps even

过来，这个值等于稳态期细胞的破坏速率。通过将斜率、初始 CD4 细胞数目和估计的血容量相乘计算出每个病例的最小产生(破坏)速率。每日血液中产生或者破坏的 CD4 细胞的最小数量是 4.3×10^6 到 108×10^6 不等，平均 35.1×10^6 (表 1)。考虑到血液淋巴细胞只占全体细胞的大约 2%[16]，计算出我们的患者中整体 CD4 淋巴细胞更新速率是每日 0.2×10^9 到 5.4×10^9 个细胞不等，平均每日 1.8×10^9 个细胞。

ABT-538 治疗后 CD4 淋巴细胞的增加也可以线性建模(图 3，下部)。直线的斜率描述了每个病例细胞的增加，其数值是确定的，从每日每立方毫米 0.5 到 25.7 个细胞不等，平均每日每立方毫米 8.6 个细胞(表 1)。使用相同的数据用于指数式模型，确定基线总 CD4 细胞产生速率(或者破坏速率)的最小估测值为每日 0.1×10^9 到 7.8×10^9 个细胞不等，平均每日 2.6×10^9 个细胞。

尽管我们的两种 CD4 淋巴细胞分析方法没有得出完全一致的结果，但它们是非常接近的，并进一步明确了 HIV-1 发病机制中病毒粒子与 CD4 淋巴细胞之间存在明确的反转关系。每日清除和补充的 CD4 淋巴细胞数量级是 10^9，这与使用原位聚合酶链式反应和杂交方法得到的体内表达 HIV-1 RNA 的淋巴细胞总数的估计值惊人地相近 [8,17]。此外，CD4 细胞补充在许多患者中存在较大差异，其产生速率快者是最慢者的 25~78 倍(表 1)，而且很有可能比目前未明确的正常 CD4 细胞更新速率更快。但是，CD4 淋巴细胞补充的精确机制需要通过进一步对再生细胞的表型标记和功能状态进行深入研究来确定。然而，这种快速的 CD4 淋巴细胞更新仍然具有研究价值。首先，在 HIV-1 感染中经常观察到的细胞凋亡现象 [18] 可能就是活跃的淋巴细胞再生引起的。其次，可以将 HIV-1 感染晚期出现的 CD4 淋巴细胞耗竭比喻成一个含有少量水的盆，其注入和排放水的速率相等。由于免疫系统的再生能力不是无限的，不难想象为什么最终水盆变空了。从这个类比中，我们也可以发现我们想要逆转免疫缺陷状态的主要措施应该是针对病毒介导的淋巴细胞破坏(堵住出水口)而不是强调淋巴细胞的再生(增加一个进水口)。

讨　论

我们相信我们新的动力学数据对 HIV-1 的治疗和发病机制探索具有重要的应用价值。显而易见，随着 HIV-1 的快速更新，必将引起病毒多样化、增加病毒逃逸治疗药物的机会等后遗症 [19,20]。因此治疗策略想要取得显著的临床效果，就必须尽可能在感染早期进行干预，甚至可能在血清转化时就开始。血浆中 HIV-1 的快速更新

during seroconversion. The rapid turnover of HIV-1 in plasma also suggests that current protocols for monitoring the acute antiviral activity of novel compounds must be modified to focus on the first few days following drug initiation. Our interventional approach to AIDS pathogenesis has shown that HIV-1 production and clearance are delicately balanced but highly dynamic processes. Taken together, our findings strongly support the view that AIDS is primarily a consequence of continuous, high-level replication of HIV-1, leading to virus- and immune-mediated killing of CD4 lymphocytes.

(**373**, 123-126; 1995)

David D. Ho, Avidan U. Neumann[*†], Alan S. Perelson[†], Wen Chen, John M. Leonard[‡] & Martin Markowitz
Aaron Diamond AIDS Research Center, NYU School of Medicine, 455 First Avenue, New York, New York 10016, USA
[*] Santa Fe Institute, Santa Fe, New Mexico 87501, USA
[†] Theoretical Division, Los Alamos National Laboratory, Los Alamos, New Mexico 87545, USA
[‡] Pharmaceutical Products Division, Abbott Laboratories, Abbott Park, Illinois 60064, USA

Received 16 November; accepted 15 December 1994.

References:

1. Weiss, R. A. *Science* **260**, 1273-1279 (1993).
2. Ho, D. D., Moudgil, T. & Alam, M. *New Engl. J. Med.* **321**, 1621-1625 (1989).
3. Simmonds, P. *et al. J. Virol.* **64**, 864-872 (1990).
4. Connor, R. l., Mohri, H., Cao, Y. & Ho, D. D. *J. Virol.* **67**, 1772-1777 (1993).
5. Patterson, B. K. *et al. Science* **260**, 976-979 (1993).
6. Piatak, M. *et al. Science* **259**, 1749-1754 (1993).
7. Pantaieo, G. *et al. Nature* **362**, 355-359 (1993).
8. Embretson, J. *et al. Nature* **362**, 359-362 (1993).
9. Wain-Hobson, S. *Nature* **366**, 22 (1993).
10. Kempf, D. *et al. Proc. Natl. Acad. Sci. U.S.A.* (in the press).
11. Markowitz, M. *et al. J. Virol.* (in the press).
12. Pachl, C. *et al. J. AIDS* (in the press).
13. Cao, Y. *et al. AIDS Res. Human Retroviruses* (in the press).
14. Perelson, A. S., Kirschner, D. E. & De Boer, R. *J. Math. Biosci.* **114**, 81-125 (1993).
15. Grody, W. W., Fligiel, S. & Naeim, F. *Am. J. Clin. Path.* **84**, 85-95 (1985).
16. Paul, W. E. *Fundamental Immunology* (Raven, New York, 1985).
17. Haase, A. T. *Ann. N. Y. Acad. Sci.* **724**, 75-86 (1994).
18. Ameisen, J. C. & Capron, A. *Immun. Today* **12**, 102-105 (1991).
19. Wain-Hobson, S. *Curr. Opin. Genet. Dev.* **3**, 878-883 (1993).
20. Coffin, J. M. *Curr. Top. Microbiol. Immun.* **176**, 143-164 (1992).

Acknowledgements. We thank the patients for their participation, Y. Cao and J. Wilbur for assistance with branched DNA assays, A. Hsu and J. Valdes for input on trial design, and J. Moore and R. Koup for helpful discussions. This work was supported by grants from Abbott Laboratories, the NIH and NYU Center for AIDS Research, the Joseph P. and Jeanne M. Sullivan Foundation, Los Alamos National Laboratory LRDR Program and The Aaron Diamond Foundation.

也提示目前监测新药的急性抗病毒活性的方法需要进行修改，特别是要着眼于开始治疗后前几日的检测。我们对艾滋病发病机制的研究显示 HIV-1 产生和清除存在微妙的平衡并且是高度动态的过程。综上所述，我们的发现强烈地支持这一观点，即艾滋病主要是持续的、高水平的 HIV-1 病毒复制以及病毒和免疫介导的 CD4 淋巴细胞死亡共同引起的。

（毛晨晖 翻译；胡卓伟 审稿）

A Plio-Pleistocene Hominid from Dmanisi, East Georgia, Caucasus

L. Gabunia and A. Vekua

Editor's Note

The conventional model of human evolution has it that the first hominid to leave the African homeland, around 1.8 million years ago, was *Homo erectus*, which migrated throughout Eurasia. Hence the interest of an excavation beneath a medieval monastery at Dmanisi in the Caucasus mountains, which has yielded primitive hominids and stone tools since the early 1990s. This paper—the description of a jawbone and its attribution to *Homo erectus*—was merely a harbinger. Dmanisi has since yielded several skulls and some skeletal material of a 1.8–1.6-million-year-old *erectus*-like creature, more like the primitive African form (known as *Homo ergaster*) than the more sophisticated East Asian forms. The migration from Africa was clearly a more complicated affair than had been thought.

Archaeological excavations at the mediaeval site of Dmanisi (East Georgia) revealed that the town was built on a series of deposits yielding Late Villafranchian mammalian fossils and led to the discovery in late 1991 of a well preserved early human mandible. Dmanisi, where excavations are being carried out by a joint expedition of the Archaeological Research Centre of the Georgian Academy of Sciences and the Römisch–Germanisches Zentralmuseum (Mainz, Germany), is located southwest of Tbilisi, at about 44° 20′ N, 41° 20′ E (Fig. 1). The fossils date to the latest Pliocene (or perhaps to the earliest Pleistocene), probably between 1.8 and 1.6 million years ago (Myr). Here we identify the mandible as belonging to the species *Homo erectus*, of which it is the earliest known representative in western Eurasia. It shows a number of similarities to the African and Chinese representatives of this species.

THE geology of the Dmanisi region and the stratigraphy of the continental deposits yielding the human jaw and the associated vertebrate fauna have been described in preliminary reports[1,2]. A basalt layer 80 m thick is overlain by about 4 m of alluvial (subaerial) deposits containing mammalian fossils and marked in places by soil horizons (Fig. 1*b*). The presence of *Kowalskia* sp., *Mimomys* sp. (*M. reidi-pitymyoides* group)[3], *Hypolagus brachygnathus* Kormos, *Pachycrocuta* cf *perrieri* (Croizet et Jobert), *Archidiskodon meridionalis* Nesti, *Equus* cf *stenonis* Cocchi, *Dicerorhinus etruscus etruscus* (Falconer), and *Cervus perrieri* Croizet et Jobert aligns the fauna with the Late Villafranchian of Europe, perhaps especially its earlier part[4]. A potassium-argon date of 1.8 ± 0.1 Myr was obtained on the basalt[1], which is in broad agreement with this age. Moreover, both the basalt and the bone-bearing layer

在高加索格鲁吉亚东部的德马尼西发现的上新世－更新世时期的原始人类

加布尼亚，韦夸

编者按

人类演化的传统模式认为，大约180万年前离开非洲家园的第一支人类是直立人，他们迁徙到整个欧亚大陆。因此大家关心在高加索山脉德马尼西的一个中世纪修道院地下的考古发掘，这里自20世纪90年代初以来就出土了原始人类化石和石器。本文描述了一个下颌骨并论及其归属于直立人，这只是一篇先期的报告。之后，德马尼西又出土了几个头骨和生活于180万～160万年前的一个类似直立人的生物的一些骨骼材料，相比进化得更为复杂的东亚化石类型，这种生物与被认为是匠人的非洲原始化石类型较为相似。从非洲迁徙出来显然是一个比早先认为的更复杂的事情。

在德马尼西（格鲁吉亚东部）中世纪遗址的考古发掘揭示出这个小镇是建立在一系列埋藏有维拉弗朗晚期哺乳类化石的堆积物之上的，并且1991年年末在这里发现了一块保存完好的早期人类下颌骨。由格鲁吉亚科学院考古研究中心和罗马－日耳曼中央博物馆（德国美因茨）联合组成的探险队正在德马尼西进行考古挖掘工作。德马尼西位于第比利斯西南，地理坐标大约为44° 20′ N，41° 20′ E（图1）。这些化石可以追溯到上新世最晚的一段时期（或者也可能是更新世最早的一段时期），大约是180万年前到160万年前。我们在本文中将该下颌骨鉴定为属于直立人这一物种，这是目前已知的欧亚大陆西部最早的直立人代表。该标本显示出很多与该物种的非洲代表和中国代表的相似性。

德马尼西地区的地质学和产出人类颌骨及相关的脊椎动物群的陆相堆积物的地层学特征在初步报告中已有所描述[1,2]。一层80米厚的玄武岩被约4米厚的包含哺乳动物化石的（接近地面的）冲积层所覆盖，玄武岩层中有些地方间杂有土壤层（图1*b*）。科氏仓鼠未定种、模鼠未定种（里德－小拟模鼠组）[3]、低颌次兔（科尔莫什）、佩里埃硕鬣狗（克鲁瓦泽和若贝尔）、南方原齿象（内斯蒂）、似古马（科基）、埃楚斯堪双角犀（福尔克纳）和佩里埃鹿（克鲁瓦泽和若贝尔）与欧洲的维拉弗朗晚期的动物群相匹配，尤其可能是该阶段的早一部分[4]。根据钾－氩法对玄武岩测得的年代为180万 ±10万年[1]，该年龄值与上述估计的年代大体一致。此外，玄武岩层和含骨

show normal geomagnetic polarity[1,5], which may indicate that they belong to the end of the Olduvai Subchron (1.95–1.77 Myr), although more detailed studies would clarify this point. Numerous stone tools of "Oldowan" appearance (Fig. 1*c*) also occur in the bone-bearing layers, at and above the level with the human mandible.

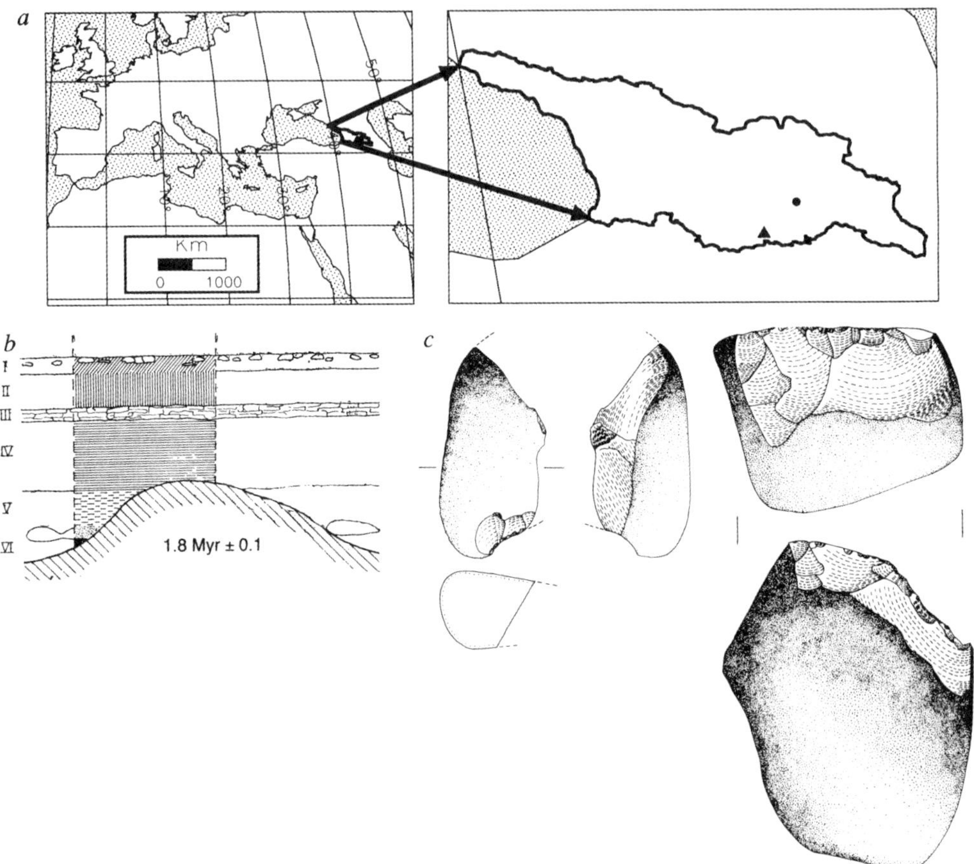

Fig. 1. *a*, Map showing rough location of Dmanisi (left, position of Georgia within Europe; right, Georgia, with Tbilisi (circle) and Dmanisi (triangle); *b*, simplified stratigraphic column of hominid locality; and *c*, two artefacts found in the fossiliferous horizons (*b* and *c* after refs 2 and 15 respectively). In *b*, the sequence includes, from top downward: (I) grey clays with limestone blocks, of which the top formed the habitation surface for the mediaeval town (0.4 m); (II) yellow-brown loam with many artefacts but few bones (0.5 m); (III) thin limestone crust (0.2 m); (IV) brown loam with many fossil vertebrates and artefacts (1.5 m); (V) blackish-brown loam with thin sandy layers, many fossil vertebrates including the hominid mandible and numerous artefacts (1.0 m); (VI) compact blackish sand and volcanic ash with rare mammalian fossils; basalt with unaltered surface. In *c*, distance between vertical lines is 9.5 cm.

The Dmanisi hominid mandible consists of an almost entirely preserved corpus with complete and little-worn dentition (Fig. 2). The base is damaged and both rami are broken away. The general appearance is of a jaw with small teeth but a thick corpus, especially below the first and second molars (see Table 1). It is not possible to determine its sex unambiguously.

层都显示出了地磁正极性[1,5]，这可能指示它们都属于奥杜威亚期末期(195 万至 177 万年前)，但是要确定这一点，还需要更详细的研究。大量属于“奥杜威文化”(图 1*c*)的石器也出现在含骨层中，与人类下颌骨位于同一层位或在其上位。

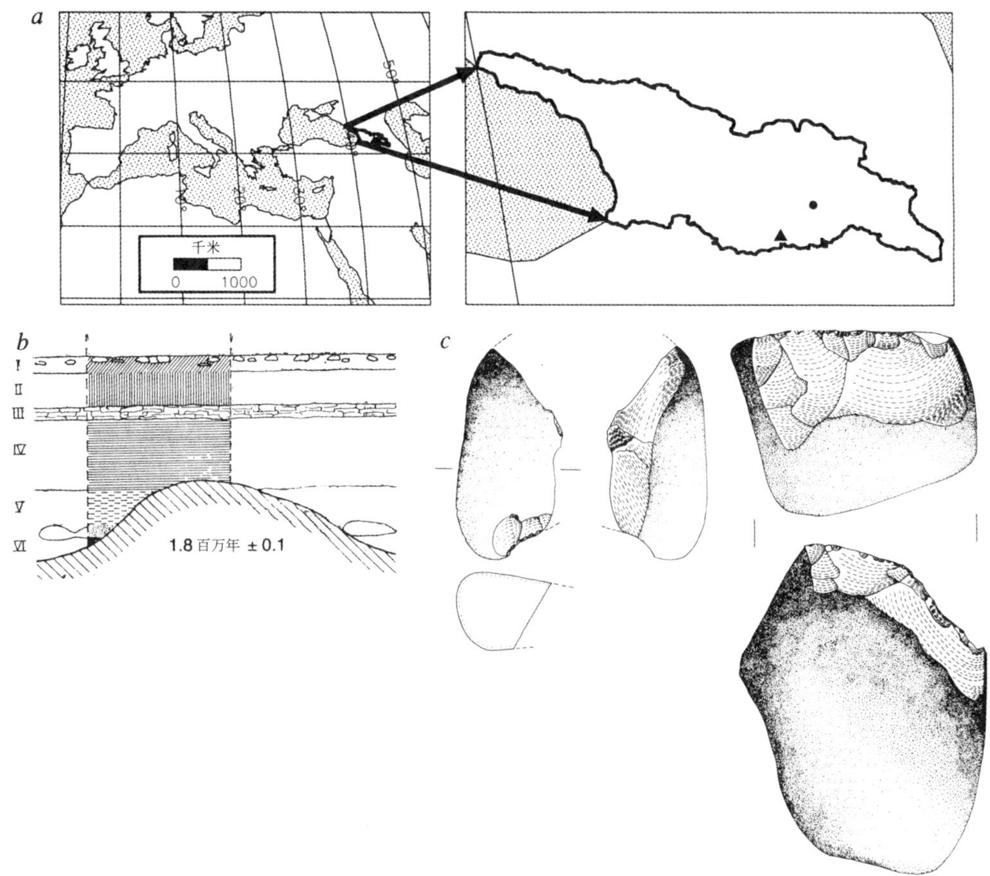

图 1. *a*，显示德马尼西大体位置的地图(左图：格鲁吉亚在欧洲的位置；右图，第比利斯(圆圈)和德马尼西(三角形)在格鲁吉亚的位置)；*b*，原始人类化石所在地点的地层剖面简图；*c*，在含有化石的地层发现的两件人工制品(在参考文献 2 和 15 中对 *b* 和 *c* 分别有介绍)。*b* 图中，地层顺序从上到下依次为：(I)含有石灰石块的灰色黏土，其顶层为这个中世纪小镇的居住地表面(0.4 米)；(II)含有许多人工制品但是骨骼少的黄棕色壤土(0.5 米)；(III)薄的石灰岩壳(0.2 米)；(IV)含有许多脊椎动物化石和人工制品的棕色壤土(1.5 米)；(V)含有薄砂层、许多脊椎动物化石(包括原始人类下颌骨)和许多人工制品的黑褐色壤土(1.0 米)；(VI)含有稀少的哺乳动物化石的密实的黑砂和火山灰；未改变表面的玄武岩。*c* 图中，垂线之间的距离是 9.5 厘米。

德马尼西原始人类下颌骨包括一块保存几乎完整的下颌体，其上带有完全的、磨损轻微的齿系(图 2)。下颌体基底损坏，两侧下颌支都脱落了。其整体外观就是一个带有小型牙齿和较厚下颌体的颌骨，尤其是第一、第二臼齿下的下颌体特别厚(见表 1)。不能明确地判断出它的性别。

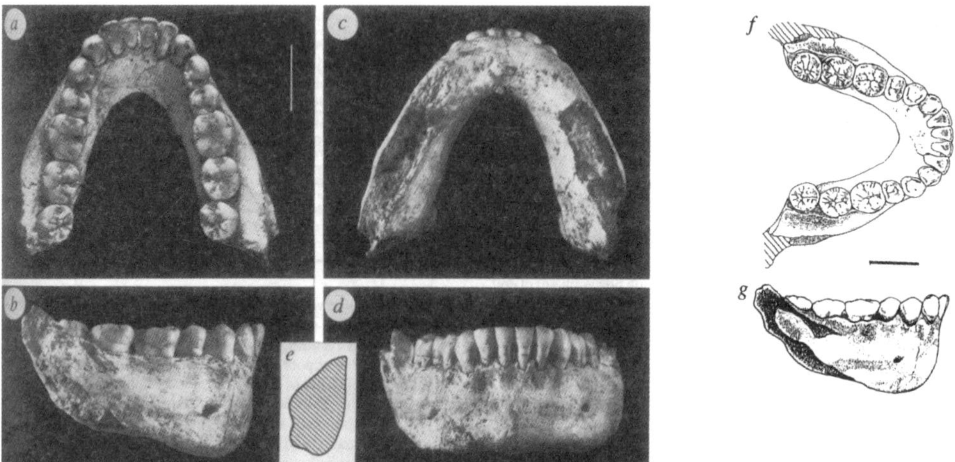

Fig. 2. Occlusal (*a*), right lateral (*b*), basal (*c*) and anterior (*d*) views of Dmanisi mandible, and symphyseal cross-section (*e*). Partially reconstructed drawing of Dmanisi mandible, in occlusal (*f*) and right lateral (g) views. Scale bars in *a* and *f*, 2 cm. The symphysis is nearly vertical in profile, with no sign of a mental protuberance, and it curves smoothly into the inferior border of the corpus. In anterior view, the shape appears subsquare. On each side a central swelling merges upward into the medioincisive and canine jugum, but does not extend anteriorly onto the symphysis; moreover, there is no incurvation below the incisors and thus no mental trigone. Internally, the laterally extensive alveolar planum slopes gently down to the superior transverse torus, but it does not extend posteriorly in a shelf. The genial fossa is distinct; the symphyseal base is flattened and thick anteroposteriorly, with weak digastric markings. The corpus is thick, and its superior and inferior borders may have been nearly parallel, although the base is broken below the molars (increasingly lacking posteriorly); anteriorly, the base is thick and slightly everted. A single large mental foramen is placed 14 mm below the mesial edge of P_4 on the right, whereas two smaller foramina occur on the left. A lateral prominence is developed below the anterior border of M_2, extending antero-inferiorly to merge with the well-defined superior lateral torus. The alveolar arcade is broadly U-shaped, each set of cheek teeth curving inward at both ends (the M_1s are most lateral). The retromolar space is small, whereas the extramolar sulcus is broad and hollowed. The anterior end of the buccinator groove lies at the level of mesial M_2, and the base of the triangular torus points superiorly, suggesting that the anterior border of the ramus might have been placed anteriorly and vertically. The slightly anteriorly inclined incisors show a small amount of flattened incisal edge wear. The canines are somewhat pointed, with a distinct external cingulum, but they do not project above the incisors. Both premolars bear distinct traces of cingulum and have a relatively large talonid. The P_3 is markedly asymmetric and rather triangular in outline, its buccal face being convex near the cervix. The anterior fovea is short and oriented nearly mesiodistally, whereas the posterior fovea is somewhat deeper and extends mesially to the very small lingual cusp. The crown of P_4 is almost quadrilateral in outline, wider than long and distinctly smaller (shorter) than that of P_3. The first and somewhat smaller second molars are quadrilateral in outline. The third molar is distinctly smaller, with a rather rounded outline. All three molars present a Y-6 cusp pattern, with the hypoconid surrounded by distinct grooves. The talonid of M_3 is somewhat longer than the trigonid and narrows distally; the occlusal surface of this tooth is tilted lingually by comparison to those of the other molars.

On the basis of the features detailed in Fig. 2, especially the overall size and robustness of the corpus and the teeth, shape of the symphysis, and dental proportions, the Dmanisi jaw represents an early form of *Homo*. Comparisons can be made with: east Asian populations of *H. erectus*, from Sangiran (Java) and Zhoukoudian (China); early African fossils assigned to that species or to *Homo ergaster* (mainly ER 730 and 992, WT 15000); later African

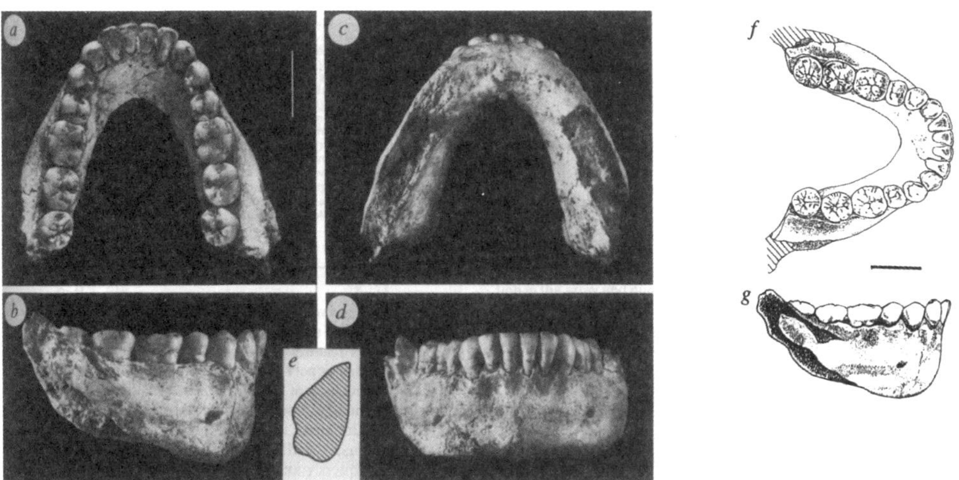

图 2. 德马尼西下颌骨。咬合面(*a*)、右侧面(*b*)、基底(*c*)和前面(*d*)视图,以及下颌联合部位的断面图(*e*)。德马尼西下颌骨部分复原的效果图，以咬合面(*f*)和右侧面(*g*)视图示出。*a* 和 *f* 中的比例尺是 2 厘米。下颌联合的侧面观轮廓接近垂直，没有颏隆凸的迹象，其平缓地弯曲过渡到下颌体的下边缘。从前面观察，其形状接近正方形。每侧都有一中央膨胀部，向上合并入内侧门齿和犬齿轭，但是并不向前延伸到下颌联合；此外，在门齿下没有内曲，所以没有颏三角。在下颌骨内侧面，向外侧延伸的牙槽平面平缓地向下倾斜至上横圆枕处，但是并没有向后延伸成类似猿板的构造。颏窝很明显；联合部位基底扁平，前后方向上很厚，具有微弱的二腹肌附着的痕迹。下颌体厚，虽然在臼齿以下的基底部分断裂了(越往后缺少的越多)，其上边缘和下缘可能近乎平行；在前面，基底厚且有些外翻。一个单一的大颏孔位于右侧 P_4 的近中侧边缘之下 14 毫米处，而两个较小的颏孔出现在左侧。在 M_2 前边缘的下方发育出了一个侧面隆起，其向前向下延伸与明显的上外侧圆枕融合在一起。齿槽弓呈现宽阔的 U 字形，每套颊齿都在两端向内弯曲(两侧的 M_1 位置在最外侧，最靠边)。臼齿后空间小，而臼齿外沟宽阔且中空。颊肌沟的前端位于 M_2 近中侧水平，三角形圆枕的基底指向上方，提示着下颌支的前端可能靠前并且是垂直的。稍微向前倾斜的门齿表现出轻微的扁平的切缘磨损迹象。犬齿有点尖，具有明显的外侧齿带，但是它们并不突出于门齿之上。两颗前臼齿都具有明确的齿带痕迹，并且具有相对较大的跟座。P_3 明显不对称，外廓有点儿呈三角形，其颊面在接近齿颈处呈凸形。前凹短，位置接近近中–远中方向，而后凹稍微深些，向近中侧延伸至很小的舌尖。P_4 的牙冠外形上几乎是四边形的，宽度比长度要大，比 P_3 明显要小(短)。第一和稍微有点小的第二臼齿外形上也是四边形的。第三臼齿明显较小，外廓有点儿接近圆形。三颗臼齿都具有 Y-6 牙尖的模式，具有被明显的沟槽围绕的下次尖。M_3 的跟座比三角座稍长，远侧端狭窄；与其他臼齿的相比，这颗牙的咬合面向舌侧倾斜。

根据图 2 中详细描述的特征，尤其是下颌体和牙齿的总体尺寸和粗壮性、下颌联合的形状、牙齿的比例等方面，可知德马尼西颌骨代表的是人属的一种早期类型。可以与如下几种原始人类进行比较：来自桑吉兰(爪哇岛)和周口店(中国)的东亚直立人群体；被归属于直立人或匠人的早期非洲化石(主要有 ER 730 和 ER 992、WT

H. erectus (OH 22, Tighenif (ex-Ternifine), Thomas Quarry 1); members of *Homo habilis* or *Homo rudolfensis* (such as OH 7 and 13, ER 1802); and early European fossils of archaic *Homo sapiens* (such as Mauer, Arago, Atapuerca). Selected comparative data on these fossils are included in Tables 1 and 2.

Table 1. Measurements (in mm) of the mandibular corpus in selected late Pliocene to late Middle Pleistocene *Homo* fossils

	Symphysis depth	Symphysis thickness	Symphysis ratio	$P_{(3-)4}$ depth	$P_{(3-)4}$ thickness	$P_{(3-)4}$ ratio	$M_{(1-)2}$ depth	$M_{(1-)2}$ thickness	$M_{(1-)2}$ ratio
Dmanisi 211 R	30.8	16.8	55%	26.8	18.5	69%	*24.7	18.4	74%
KNM ER 1802 R[10]	36.0	24.5	68%	40.0	20.0	50%	38.0	27.0	71%
OH 13 R[10]	25.0	18.0	72%	26.0	16.5	63%	28.5	22.5	79%
KNM ER 730[10]	32.5	17.5	54%	32.5	19.0	58%	31.5	19.0	60%
KNM ER 992 L[10]	37.0	21.0	57%	31.0	20.0	65%	33.0	22.0	67%
KNM WT 15000 L[11]				27.2	18.1	67%	24.8	21.0	85%
OH 22[10]	33.5	20.0	60%	29.0	20.5	71%	28.5	21.0	74%
Sangiran 1b[10]	32.0	17.0	53%	33.0	16.0	48%	33.0	17.0	52%
Sangiran 9 R[10]	41.0	19.0	46%	39.0	21.5	55%	32.0	23.0	72%
Zhoukoudian G 1-6[12]	40.0	13.7	34%				34.0	17.3	51%
Zhoukoudian H 1[12]	32.3	14.0	43%				26.0	14.9	57%
Zhoukoudian Mean[12]	34.5			28.5					
Tighenif 1[13]	37.5	18.5	49%	33.0	19.0	58%	35.0	22.0	63%
Tighenif 2[13]	31.0	17.0	55%	32.0	15.0	47%	31.5	17.5	56%
Tighenif 3[13]	35.0	19.0	54%	35.0	19.0	54%	38.0	21.0	55%
Thomas Quarry 1[13]				28.5	16.0	56%	26.5	*18.0	68%
Mauer	32.0	18.5	58%	32.5	18.5	57%	33.0	19.0	58%
Arago 2[10]	31.0	16.0	52%	30.5	17.0	56%	30.0	17.0	57%
Arago 13[10]	36.0	20.0	56%	31.0	22.5	73%	31.0	23.5	76%

Symphysis depth, minimum superoinferior distance between infradentale and base of symphysis (termed height in ref. 10, measurement no. 141); symphysis thickness, maximum mesiodistal (anteroposterior) distance perpendicular to preceding (termed depth, measurement no. 142); Symphysis ratio, percentage thickness/depth (rounded up); $P_{(3-)4}$ depth, minimum depth of corpus below septum between P_3 and P_4 or below middle of P_4 , depending on authority (ref. 10, measurement no. 147, at P_4, lingual side; termed minimum height in ref. 13); $P_{(3-)4}$ thickness, maximum thickness of corpus perpendicular to preceding (ref. 10, measurement no. 148; minimum corpus breadth[13]). Ratio as for symphysis. $M_{(1-)2}$, as $P_{(3-)4}$ (ref. 10, measurement nos 154–155, at M_2; at M_2 in ref. 13). L or R following specimen identification refers to side of mandible when both are present; superscript number gives source of data.

* Value estimated on Dmanisi mandible (see Fig. 2*g*) or roughly estimated by authors from data in cited source.

15000)；较晚的非洲直立人(OH 22、提盖尼夫(以前的特尼芬)、托马斯采石场 1)；能人或鲁道夫人的成员(例如 OH 7、OH 13、ER 1802)；以及早期的欧洲古老型智人化石(例如摩尔、阿拉戈、阿塔普埃尔卡)。对这些化石有选择地进行了比较，相关数据列于表 1 和表 2 中。

表 1. 选择的晚上新世到中更新世晚期的人属化石的下颌体的测量尺寸(单位：毫米)

	下颌联合深度	下颌联合厚度	下颌联合比例	$P_{(3-)4}$ 深度	$P_{(3-)4}$ 厚度	$P_{(3-)4}$ 比例	$M_{(1-)2}$ 深度	$M_{(1-)2}$ 厚度	$M_{(1-)2}$ 比例
德马尼西样本 211 R	30.8	16.8	55%	26.8	18.5	69%	*24.7	18.4	74%
KNM ER 1802 R[10]	36.0	24.5	68%	40.0	20.0	50%	38.0	27.0	71%
OH 13 R[10]	25.0	18.0	72%	26.0	16.5	63%	28.5	22.5	79%
KNM ER 730[10]	32.5	17.5	54%	32.5	19.0	58%	31.5	19.0	60%
KNM ER 992 L[10]	37.0	21.0	57%	31.0	20.0	65%	33.0	22.0	67%
KNM WT 15000 L[11]				27.2	18.1	67%	24.8	21.0	85%
OH 22[10]	33.5	20.0	60%	29.0	20.5	71%	28.5	21.0	74%
桑吉兰样本 1b[10]	32.0	17.0	53%	33.0	16.0	48%	33.0	17.0	52%
桑吉兰样本 9 R[10]	41.0	19.0	46%	39.0	21.5	55%	32.0	23.0	72%
周口店样本 G 1-6[12]	40.0	13.7	34%				34.0	17.3	51%
周口店样本 H 1[12]	32.3	14.0	43%				26.0	14.9	57%
周口店样本平均值 [12]	34.5			28.5					
提盖尼夫样本 1[13]	37.5	18.5	49%	33.0	19.0	58%	35.0	22.0	63%
提盖尼夫样本 2[13]	31.0	17.0	55%	32.0	15.0	47%	31.5	17.5	56%
提盖尼夫样本 3[13]	35.0	19.0	54%	35.0	19.0	54%	38.0	21.0	55%
托马斯采石场样本 1[13]				28.5	16.0	56%	26.5	*18.0	68%
摩尔样本	32.0	18.5	58%	32.5	18.5	57%	33.0	19.0	58%
阿拉戈样本 2[10]	31.0	16.0	52%	30.5	17.0	56%	30.0	17.0	57%
阿拉戈样本 13[10]	36.0	20.0	56%	31.0	22.5	73%	31.0	23.5	76%

下颌联合部的深度，即下颌左右内侧门齿间隔顶点和下颌联合基底之间的最短上下距离(参考文献 10 中称为高度，测量数值编号为 141)；下颌联合的厚度，即与前一侧径垂直的最大近中远中侧(前后)距离(称为深度，测量数值编号为 142)；下颌联合的比例，即厚度 / 深度百分比(取整)；$P_{(3-)4}$ 深度，即 P_3 和 P_4 之间的中隔之下或 P_4 中部之下的下颌体的最小深度，依研究者而定(参考文献 10，测量数值编号为 147，在 P_4 处测量，舌侧；参考文献 13 中称为最小高度)；$P_{(3-)4}$ 厚度，即与前一侧径垂直的下颌体的最大厚度(参考文献 10，测量数值编号为 148；下颌体最小宽度 [13])。比例同下颌联合。$M_{(1-)2}$ 的情况同 $P_{(3-)4}$ (参考文献 10 中的 M_2 处测量，测量数值编号为 154~155；参考文献 13 在 M_2 处测量)一样。当下颌左右侧都存在时，附在标本名称之后的 L 或 R 指的是左侧或右侧；上标的数字给出了数据出处。

* 数值是根据德马尼西下颌骨(见图 2*g*)估计出来的，或是作者根据引用来源中的数据粗略估算出来的。

Table 2. Measurements (in mm) of the mandibular dentition in selected late Pliocene to late Middle Pleistocene *Homo* fossils

	I_1 m-d	I_1 b-l	I_2 m-d	I_2 b-l	C_1 m-d	C_1 b-l	P_3 m-d	P_3 b-l	P_4 m-d	P_4 b-l	M_1 m-d	M_1 b-l	M_2 m-d	M_2 b-l	M_3 m-d	M_3 b-l	Module ratios		Molar Sizes
																	M_1/M_2	M_3/M_2	
Dmanisi 211 R	5.9	5.8	6.6	6.4	8.7	8.2	9.0	9.2	8.1	9.2	13.2	12.3	12.3	11.5	11.2	10.6	115%	84%	1 > 2 > 3
Dmanisi 211 L	6.2	5.9	6.4	6.3	8.6	7.9	8.9	9.6	8.0	9.6	13.0	12.5	11.5	11.6	10.7	10.6	122%	85%	1 > 2 > 3
KNM ER 1802 R[10]							10.7	11.5	11.4	12.0	14.8	13.0	16.6	14.2			82%		2 > 1
OH 7 L[10]	6.5	6.7	7.3	7.6	8.9	10.1	9.5	9.7	10.4	10.7	14.1	12.5	15.7	13.7			82%		2 > 1
OH 13 R[10]					7.6	7.9	9.0	8.7	9.0	9.9	13.0	11.6	14.2	12.0	14.8	12.4	88%	108%	3 > 2 > 1
KNM ER 820 L[10]	6.1	6.3	6.3	6.9							12.5	10.6							
KNM ER 992 L[10]			7.2	7.0	8.9	9.5	9.7	10.3	8.8	10.9	12.7	10.9	13.0	12.2	13.4	12.3	87%	104%	3 > 2 > 1
KNM WT 15000 L[14]	6.6	6.8	7.5	8.3	8.9	9.6	8.9	10.1	9.0	10.1	12.4	11.0	12.5	12.0			91%		2 > 1
OH 22[10]							10.1	9.2	9.0	10.0	13.4	12.0	13.0	11.7			106%		1 > 2
Sangiran 1b[13]									8.9	10.8	12.9	12.9	13.2	13.4	14.4	12.5	94%	102%	3 > 2 > 1
Sangiran 9 R[10]					7.0	8.8	9.1	11.3	9.3	11.6			14.1	12.7	13.8	12.7		98%	2 > 3
Zhoukoudian G1-6 L[12]	6.2	6.8		7.3	8.1	8.2	9.1	10.7	8.5	11.0	13.2	12.5	12.5	12.7	12.0	12.3	104%	93%	1 > 2 > 3
Zhoukoudian G1-7[12]													12.6	12.9	12.9	12.4		98%	2 > 3
Zhoukoudian Mean[12]	6.4	6.4	6.8	7.0	8.7	9.4	8.7	9.9	9.0	9.8	12.5	11.8	12.5	12.1	11.7	11.3	98%	87%	2 > 1 > 3
Tighenif 1[13]			6.0	8.0			8.5	10.2	8.3	10.1	13.0	12.5	13.0	13.0	12.0	12.2	96%	87%	2 > 1 > 3
Tighenif 2[13]							8.6	11.1	8.8	11.0	13.9	12.8	14.1	13.3	13.4	12.5	95%	89%	2 > 1 > 3
Tighenif 3[13]			5.0	9.0	7.7	10.7	8.0	10.2	8.2	10.0	12.4	12.0	12.0	12.2	12.0	11.5	102%	94%	1 > 2 > 3
Thomas Quarry 1[13]									9.0	10.5	14.0	12.8	14.8	13.0	12.8	11.7	93%	78%	2 > 1 > 3
Mauer	5.5	7.5	6.9	7.7	7.4	8.9	8.1	9.0	7.5	9.2	11.3	11.2	12.7	12.0	10.9	11.9	83%	85%	2 > 3 > 1
Arago 13							10.1	11.0	9.5	12.9	13.6	13.6	14.6	14.0	13.4	13.0	90%	85%	2 > 1 > 3

Maximum mesiodistal (m-d) and buccolingual (b-l) dimensions of lower teeth. L or R following specimen identification refers to side of mandible when both are present; superscript number gives source of data. Module ratios represent products of m-d and b-l diameters of first and third molars, respectively, divided by those of second molars. ER 1802 is identified as *Homo rudolfensis*, OH 7 and OH 13 as *H. habilis*, Thomas Quarry, Mauer and Arago as early *H. sapiens*, other specimens as *H. erectus*, although ER 820–OH 22 are sometimes considered *H. ergaster*.

表 2. 选择的晚上新世到中更新世晚期的人属化石的下颌骨齿系的测量尺寸（单位：毫米）

	I_1 m-d	I_1 b-l	I_2 m-d	I_2 b-l	C_1 m-d	C_1 b-l	P_3 m-d	P_3 b-l	P_4 m-d	P_4 b-l	M_1 m-d	M_1 b-l	M_2 m-d	M_2 b-l	M_3 m-d	M_3 b-l	模数比		臼齿尺寸
																	M_1/M_2	M_3/M_2	
德马尼西样本 211 R	5.9	5.8	6.6	6.4	8.7	8.2	9.0	9.2	8.1	9.2	13.2	12.3	12.3	11.5	11.2	10.6	115%	84%	1 > 2 > 3
德马尼西样本 211 L	6.2	5.9	6.4	6.3	8.6	7.9	8.9	9.6	8.0	9.6	13.0	12.5	11.5	11.6	10.7	10.6	122%	85%	1 > 2 > 3
KNM ER 1802 R[10]							10.7	11.5	11.4	12.0	14.8	13.0	16.6	14.2			82%		2 > 1
OH 7 L[10]	6.5	6.7	7.3	7.6	8.9	10.1	9.5	9.7	10.4	10.7	14.1	12.5	15.7	13.7			82%		2 > 1
OH 13 R[10]					7.6	7.9	9.0	8.7	9.0	9.9	13.0	11.6	14.2	12.0	14.8	12.4	88%	108%	3 > 2 > 1
KNM ER 820 L[10]	6.1	6.3	6.3	6.9							12.5	10.6							
KNM ER 992 L[10]			7.2	7.0	8.9	9.5	9.7	10.3	8.8	10.9	12.7	10.9	13.0	12.2	13.4	12.3	87%	104%	3 > 2 > 1
KNM WT 15000 L[14]	6.6	6.8	7.5	8.3	8.9	9.6	8.9	10.1	9.0	10.1	12.4	11.0	12.5	12.0			91%		2 > 1
OH 22[10]							10.1	9.2	9.0	10.0	13.4	12.0	13.0	11.7			106%		1 > 2
桑吉兰样本 1b[13]									8.9	10.8	12.9	12.9	13.2	13.4	14.4	12.5	94%	102%	3 > 2 > 1
桑吉兰样本 9 R[10]					7.0	8.8	9.1	11.3	9.3	11.6			14.1	12.7	13.8	12.7		98%	2 > 3
周口店样本 G1-6 L[12]	6.2	6.8		7.3	8.1	8.2	9.1	10.7	8.5	11.0	13.2	12.5	12.5	12.7	12.0	12.3	104%	93%	1 > 2 > 3
周口店样本 G1-7[12]													12.6	12.9	12.9	12.4		98%	2 > 3
周口店样本平均值 [12]	6.4	6.4	6.8	7.0	8.7	9.4	8.7	9.9	9.0	9.8	12.5	11.8	12.5	12.1	11.7	11.3	98%	87%	2 > 1 > 3
提盖尼夫样本 1[13]			6.0	8.0			8.5	10.2	8.3	10.1	13.0	12.5	13.0	13.0	12.0	12.2	96%	87%	2 > 1 > 3
提盖尼夫样本 2[13]							8.6	11.1	8.8	11.0	13.9	12.8	14.1	13.3	13.4	12.5	95%	89%	2 > 1 > 3
提盖尼夫样本 3[13]			5.0	9.0	7.7	10.7	8.0	10.2	8.2	10.0	12.4	12.0	12.0	12.2	12.0	11.5	102%	94%	1 > 2 > 3
托马斯采石场样本 1[13]									9.0	10.5	14.0	12.8	14.8	13.0	12.8	11.7	93%	78%	2 > 1 > 3
摩尔样本	5.5	7.5	6.9	7.7	7.4	8.9	8.1	9.0	7.5	9.2	11.3	11.2	12.7	12.0	10.9	11.9	83%	85%	2 > 3 > 1
阿拉戈样本 13							10.1	11.0	9.5	12.9	13.6	13.6	14.6	14.0	13.4	13.0	90%	85%	2 > 1 > 3

下牙的最大近中远中径（m-d）和颊舌径（b-l）。附于标本名称后面的 L 或 R 是指当下颌左右侧都保存时，进行测量的是下颌的左侧或右侧部分；上标的数字给出了数据出处。模数比代表第一和第三臼齿的近中远中径和颊舌径分别除以第二臼齿的相应测量值的结果。ER 1802 被鉴定为鲁道夫人，OH 7 和 OH 13 被鉴定为能人，托马斯采石场标本、摩尔和阿拉戈标本被鉴定为早期智人，其他标本被鉴定为直立人，而 ER 820~OH 22 有时被认为属于匠人。

The Dmanisi mandible appears to be most similar to African *H. erectus* fossils, with fewer similarities to those from Zhoukoudian. It is clearly smaller than mandibles of earliest *Homo*, especially in the P_4 and $M_{2,3}$ and in the corpus under the cheek teeth. The most striking feature of the tooth row is perhaps the great proportional decrease in tooth module ("area"), about 15–20% from M_1 to M_2 and from M_2 to M_3. No other specimen included in Table 2 presents such a strong distal reduction pattern, the closest approximations being OH 22 (from Bed IV, ~0.6 Myr, but lacking M_3 and all teeth mesial to P_3), Zhoukoudian Gl-6 (~0.4 Myr) and Tighenif 3 (~0.7 Myr). If the total Zhoukoudian and Tighenif samples are considered, however, the first two molars are nearly equal in module, with M_3 smaller. Other *H. erectus* jaws, such as ER 992, Sangiran 1 and 9, and Zhoukoudian Gl-7, have similar sized M_2 and M_3, usually with a smaller M_1, whereas the small M_1 of Mauer and Arago sets them apart from Dmanisi. The Thomas Quarry 1 mandible (~0.4 Myr) has very large mesial molars with a greatly reduced M_3. The Dmanisi M_3 (and to a lesser degree M_2) tilt lingually in their sockets, as seen also in ER 992 and WT 15000. The anterior teeth appear morphologically similar to those of African and Asian *H. erectus* (*sensu lato*).

In terms of tooth size, the Dmanisi M_2 is most comparable to those in the ER 992, WT 15000, OH 22, Zhoukoudian, Tighenif 3 and Mauer jaws. The Dmanisi premolars are small within the comparative sample, and especially narrow; they conform most closely to those of OH 13, the Zhoukoudian mean, Tighenif 3 and Mauer. The P_3 lingual cusp is strongly reduced, as seen also in the Zhoukoudian unworn specimen Uppsala M3549 described by Zdansky. The Dmanisi canine is small, with the buccolingual diameter only slightly smaller than the mesiodistal; its closest match is with those from Zhoukoudian and, surprisingly, OH 13. The incisors are close in size to those of the young juvenile ER 820 and the Zhoukoudian mean, lacking the buccolingual dominance of the North African and European specimens.

The Dmanisi corpus (Table 1) increases in relative thickness from the symphysis distally, although it must be emphasized that submolar depth is estimated. This pattern is also seen in ER 730 and 992, OH 22, Sangiran 9 (but not 1), possibly at Zhoukoudian, and in some but not all of the Tighenif and Arago jaws. The remaining Tighenif, Arago, Mauer and Sangiran specimens have a rather consistent relative thickness along the tooth row. The Dmanisi jaw is small; its symphysis is comparable in size, and also in overall morphology, to that of ER 730, as well as to Sangiran 1, Tighenif 2, Thomas Quarry, Arago 2 and perhaps Zhoukoudian H1. By contrast, the symphysis neither slopes backward as much as in the early African fossils, nor is there a well-developed trigone as in early European specimens. The corpus under the premolars is most similar in size to those of OH 13, OH 22 and Arago 2, as well as to the subadult WT 15000. Under the molars, size similarity is greatest to Zhoukoudian H1, Thomas Quarry and WT 15000, which probably would have presented a deeper (and thus far larger) corpus as an adult.

德马尼西下颌骨似乎与非洲直立人化石最相似，与周口店化石的相似性较少。该下颌骨比最早人属的下颌骨要小很多，尤其是 P_4 和 $M_{2,3}$ 以及颊齿以下的下颌体部分。齿列最明显的特征可能就是齿模数（“面积”）大比例减少，从 M_1 到 M_2 以及从 M_2 到 M_3 约减少了 15%~20%。表 2 列出的标本中没有其他标本表现出如此强烈的远端缩减模式，与该下颌骨具有最接近近似值的是 OH 22（来自奥杜威峡谷第 IV 层，约 60 万年前，但缺少 M_3 和 P_3 近中侧的所有牙齿）、周口店 G1-6（约 40 万年前）和提盖尼夫 3（约 70 万年前）。然而，如果我们将周口店和提盖尼夫的所有样本都纳入考虑范围内的话，那么前两颗臼齿在模数上几乎是相等的，而 M_3 稍小。其他的直立人颌骨，例如 ER 992、桑吉兰 1 和桑吉兰 9 以及周口店 G1-7 的 M_2 和 M_3 大小都相似，通常 M_1 较小，不过摩尔和阿拉戈的小型 M_1 与德马尼西不同。托马斯采石场 1 的下颌骨（约 40 万年前）具有非常大的近中侧臼齿和大大减小的 M_3。德马尼西样本的 M_3（以及程度稍小的 M_2）在其牙槽内朝向舌侧倾斜，就像在 ER 992 和 WT 15000 中看到的一样。前牙在形态上与非洲和亚洲的直立人（广义上）都很像。

从牙齿大小来看，德马尼西标本的 M_2 与 ER 992、WT 15000、OH 22、周口店、提盖尼夫 3 和摩尔颌骨的 M_2 非常相近。德马尼西标本的前臼齿在所有这些进行比较的样本中算是比较小的，并且格外狭窄；它们与 OH 13、周口店样本的平均值、提盖尼夫 3 和摩尔样本的最为接近。P_3 的舌侧尖减小明显，这种情况在师丹斯基描述的未磨损的周口店标本乌普萨拉 M3549 中也看到过。德马尼西标本的犬齿小，颊舌径只比近中远中径略小；与其最匹配的是周口店的标本以及 OH 13，后者与其惊人地相似。门齿大小与年轻的青少年标本 ER 820 的门齿和周口店标本的平均值很接近，缺少北非和欧洲标本所具有的颊舌径优势。

德马尼西标本的下颌体（表 1）在相对厚度上从联合部向远侧有所增加，但是这里必须强调的是，其臼齿以下的高度是估计出来的。这种模式在 ER 730 和 ER 992、OH 22、桑吉兰 9（而不是桑吉兰 1）中都有发现，可能在周口店标本中以及某些（但非全部）提盖尼夫和阿拉戈颌骨中也可见这种模式。其余的提盖尼夫、阿拉戈、摩尔和桑吉兰标本与整个齿列对应的相对厚度颇为一致。德马尼西的颌骨小；其下颌联合在尺寸以及总体形态学上与 ER 730 和桑吉兰 1、提盖尼夫 2、托马斯采石场标本、阿拉戈 2 相当，可能与周口店 H1 也是相当的。相比之下，下颌联合既不像早期非洲化石那么向后倾斜，也没有像早期欧洲标本具有非常发达的三角座。前臼齿下方的下颌体与 OH 13、OH 22 和阿拉戈 2 在大小上非常相似，也与亚成年的 WT 15000 的尺寸相仿。在臼齿下方的部分，其尺寸与周口店 H1、托马斯采石场标本和 WT 15000 的最为相似，这可能代表了一种成人所具有的较高（因此也更大）的下颌体。

There are thus both morphological and metrical similarities to fossils spanning a wide range of time and space. The Dmanisi jaw is distinctive in the great dental reduction distally, as well as in relatively small tooth and corpus size combined with apparently great corpus robusticity under the cheek teeth; in addition, the alveolar arcade is narrow and the beginning of the ramus is placed far anteriorly. Among the most similar specimens overall are certain Turkana Basin fossils (ER 730 and 992, WT 15000), OH 22 (lacking M_3) and Zhoukoudian Gl-6 (with a much larger corpus). Selected members of the Tighenif/Thomas, Arago/Mauer and Sangiran samples show specific similarities as well. The most reasonable interpretation of this jaw is that it belonged to a population of *H. erectus*, possibly foreshadowing (European) early *H. sapiens*.

The date provided by the faunal and geophysical evidence indicates that the Dmanisi population was one of the oldest human groups outside Africa, certainly the most ancient in western Eurasia. Dates for Chinese and Indonesian *H. erectus* have long been in doubt, but recent studies have suggested ages of ~1.1 Myr for the Lantian Gongwangling cranium[6] and ~1.7 Myr for the oldest Sangiran and Mojokerto fossils[7]. Although there are potential problems with these dates, the Indonesian evidence especially suggests hominid presence at least between 1.5–1.0 Myr. A similar age is estimated for the 'Ubeidiya archaeological site in Israel[8]. Ages between 2 and 1 Myr have also been suggested for a variety of localities in Europe, but the most recent review of this work[9] indicates that there is little support for demonstrated human occupation before ~0.5 Myr. The Dmanisi human mandible, attributed to *H. erectus* and dated around 1.8 Myr or slightly younger, might be the oldest evidence of humans outside Africa. Its presence in Georgia further suggests that humans either waited outside the "gates to Europe" for more than 1 Myr or inhabited the subcontinent at very low density during that interval.

(**373**, 509-512; 1995)

L. Gabunia & A. Vekua
Institute of Paleobiology, Georgian Academy of Sciences, 9 Sulkhan-Saba, Tbilisi 380007, Georgia

Received 11 August; accepted 1 December 1994.

References:

1. Majsuradze, G., Pavlenshvili, E., Schmincke, H. & Sologashvili, D. *Jahrb. Römisch-German. Zentralmus.* **36**, 74-76 (1991).
2. Bosinski, G., Lordkipanidze, D., Majsuradze, G. & Tvaicrelidze, M. *Jahrb. Römisch-German. Zentralmus.* **36**, 76-83 (1991).
3. Muskhelishvili, A. *Jahrb. Römisch-German. Zentralmus.* (in the press).
4. Gabunia, L. & Vekua, A. *Dmanisi Man and the Accompanying Vertebrate Fauna* 1-60 (Metsniereba, Tbilisi, 1993).
5. Sologashvili, D., Majsuradze, G., Pavlenshvili, E. & Klopotovskaja, N. *Jahrb. Römisch-German. Zentralmus.* (in the press).
6. An, Z. & Ho, C. K. *Quat. Res.* **32**, 213-221 (1989).
7. Swisher, C. C. *et al. Science* **263**, 1118-1121 (1994).
8. Verosub, K. L. & Tchernov, E. in *Les Premiers Européens* (eds Bonifay, E. & Vandermeersch, B.) 237-242 (Comité Travaux Historiques Scientifiques, Paris, 1991).
9. Roebroeks, W. *Curr. Anthropol.* **35**, 301-305 (1994).
10. Wood, B. A. *Koobi Fora Research Project, Vol. 4: Hominid Cranial Remains* (Clarendon, Oxford, 1991).
11. Walker, A. & Leakey, R. in *The Nariokotome* Homo erectus *Skeleton* (eds Walker, A. & Leakey, R.) 63-94 (Harvard Univ. Press, 1993).

因此该化石与横跨了大范围时间和空间的化石在形态学和测量方面都有相似性。德马尼西标本的颌骨的独特性表现在它的牙齿越向远端越表现出巨大缩减，它的牙齿和下颌体相对较小，却与粗壮度明显很大的颊齿下的下颌体相结合；另外，其齿弓狭，下颌支开始的部位很靠前。在所有标本中，最相似的当属图尔卡纳盆地的某些化石（ER 730、ER 992、WT 15000）、OH 22（缺少 M_3）和周口店 G1-6（具有大得多的下颌体）了。选择的提盖尼夫 / 托马斯、阿拉戈 / 摩尔和桑吉兰样本的成员也显示出了特定的相似性。对于该颌骨的最合理的解释就是它属于直立人群体，可能是（欧洲）早期智人的先驱。

动物群和地球物理学证据所提供的年代表明德马尼西群体是非洲之外的最古老的人类群体之一，他们肯定是欧亚大陆西部最古老的人群。长期以来一直有人对中国和印度尼西亚的直立人所处的年代表示怀疑，但是最近的研究显示公王岭蓝田人头盖骨[6]的年代为约 110 万年前，最古老的桑吉兰化石和莫佐克托化石的年代为约 170 万年前[7]。尽管这些年代存在着潜在的问题，但是印度尼西亚的证据显示了至少在 150 万～100 万年前就已经存在原始人类了。以色列的乌贝迪亚考古学遗址[8]估计出的年代也与此相似。欧洲的许多遗址也显示了原始人类早在 200 万～100 万年前就存在了，但是该项工作的最新综述[9]指出还没有多少证据支持在约 50 万年以前人类就在此居住了。德马尼西人类下颌骨被归属于直立人，年代为距今约 180 万年，也可能稍晚一些，它可能是非洲之外存在人类的最古老的证据。其在格鲁吉亚的出现进一步提示人类或者是在“通向欧洲的大门”之外等待了 100 多万年，或者在那段时间以非常低的密度居住在次大陆上。

（刘皓芳 翻译；吴新智 审稿）

12. Weidenreich, F. *Palaeontol. Sinica n.s.* **D1**, 1-180 (1937).

13. Rightmire, G. P. *The Evolution of* Homo erectus (Cambridge Univ. Press, 1990).

14. Brown, B. & Walker, A. in *The Nariokotome* Homo erectus *Skeleton* (eds Walker, A. & Leakey, R.) 161-192 (Harvard Univ. Press, 1993).

15. Bosinski, G., Bugianisvili, T., Mgeladze, N., Nioradze, M. & Tusabramisvili, D. *Jahrb. Römisch-German. Zentralmus.* **36**, 93-107 (1991).

Acknowledgements. We thank H. and M.-A. de Lumley for access to the rich comparative collections in the Laboratoire d'Anthropologie in Marseilles, G. Bosinski (Römisch-Germanisches Zentralmuseum, Neuwied), E. Delson (Anthropology, Lehman College of the City University of New York), J. L. Franzen (Forschungsinstitut Senckenberg, Frankfurt), F. C. Howell (Anthropology, University of California, Berkeley), G. P. Rightmire (Anthropology, State University of New York at Binghamton), A. Walker (Cell Biology and Anatomy, The Johns Hopkins University School of Medicine, Baltimore), T. White (Anthropology, University of California, Berkeley), D. Dean (Anatomy, Case Western Reserve University School of Medicine, Cleveland), H. Roth (Département d'Anthropologie et d'Ecologie, Université de Genève), B. Senut (Laboratoire d'Anthropologie du Muséum National d'Histoire Naturelle, Paris) for their help. E. Delson prepared the final versions of the manuscript and L. Meeker prepared Fig. 2, for which we are most grateful.

Sequence Variation of the Human Y Chromosome

L. S. Whitfield *et al.*

Editor's Note

The non-recombining region of the Y chromosome makes it an excellent tool for studying male human evolution. In this paper, clever comparisons and statistical analyses of Y chromosome DNA from five men of different geographical origin and one chimpanzee suggest that the men's most recent common ancestor lived somewhere between 37,000 and 49,000 years ago. Maternally derived mitochondrial DNA analyses, however, yielded a time of origin of modern humans between 120,000 and 474,000 years ago. The findings, presented by geneticists Simon Whitfield, John Sulston and Peter Goodfellow, fit with the idea of male migration. But an accompanying paper by evolutionary geneticist Michael Hammer places the common ancestral human Y chromosome at 188,000 years.

We have generated over 100 kilobases of sequence from the non-recombining portion of the Y chromosomes from five humans and one common chimpanzee. The human subjects were chosen to match the earliest branches of the human mitochondrial tree. The survey of 18.3 kilobases from each human detected only three sites at which substitutions were present, whereas the human and chimpanzee sequences showed 1.3% divergence. The coalescence time estimated from our Y chromosome sample is more recent than that of the mitochondrial genome. A recent coalescence time for the Y chromosome could have been caused by the selected sweep of an advantageous Y chromosome or extensive migration of human males.

IT has been suspected for some time that the non-recombining portion of the human Y chromosome might harbour a low level of variation compared with the autosomes and the X chromosome. However, studies of restriction-enzyme site polymorphism have not included determinations of the interspecies divergence needed to correct for the mutation rate of the region examined[1,2]. The amount of divergence between two sequences depends on the time since their last common ancestor and the rate of mutation. More recent comparisons of Y chromosome and autosomal or X chromosome DNA sequence have incorporated an estimate of mutation rate but have not allowed for the different modes of inheritance and levels of recombination[3,4].

The Y chromosome region that we have surveyed lies immediately proximal to the Yp pseudoautosomal boundary. It ends at the boundary and includes *SRY*, the mammalian testis-determining gene. This sequence has been extensively investigated to confirm

人类 Y 染色体的序列变异

惠特菲尔德等

编者按

Y 染色体的非重组区使其成为研究人类男性进化的绝佳工具。在这篇文章中，通过对五个不同地区起源的人类和一只黑猩猩的 Y 染色体 DNA 进行比较和统计分析，表明人类的最近共同祖先生活在 37,000 到 49,000 年前。然而，母系线粒体 DNA 分析得出现代人类起源于 120,000 到 474,000 年前。遗传学家西蒙·惠特菲尔德、约翰·萨尔斯顿和彼得·古德费洛的这些发现支持男性迁移的观点。但是进化遗传学家迈克尔·哈默在相应的一篇文章中将人类共同祖先的 Y 染色体定在 188,000 年前。

对来自五个人类个体和一只黑猩猩的 Y 染色体的非重组区，我们完成了大于 100 kb 序列的测定。人类样本的选择标准是他们的序列要与人类线粒体系统树最早期的分支相匹配。通过对每个人类样本 18.3 kb 序列进行分析，发现仅 3 个位点检测到碱基替换，而人类和黑猩猩碱基序列的差异为 1.3%。我们根据 Y 染色体样本估计的溯祖时间比由线粒体基因组估计得到的溯祖时间更为晚近。Y 染色体估计出了更近的溯祖时间可能是由两方面的原因引起的：一是优势 Y 染色体的选择性清除，二是人类男性的广泛迁移。

很长一段时间以来，人们就怀疑人类的 Y 染色体非重组区的变异水平与常染色体和 X 染色体相比较低。然而，目前在限制性酶切位点的多态性研究中，尚不包括可以用来确定种间差异的位点信息，而这些确定种间差异的信息对于校正我们所考察区域的突变速率来说是必需的 [1,2]。两条序列的差异程度依赖于他们的最近共祖时间以及突变速率。近来更多的研究在比较 Y 染色体与常染色体的 DNA 序列或 Y 染色体与 X 染色体的 DNA 序列时，都会将突变速率的估计纳入考虑范围之内，但尚未考虑到不同的遗传模式和重组水平的影响 [3,4]。

我们研究的 Y 染色体区域位于最接近 Y 染色体短臂假常染色体区的边界处。其终止于该边界处，且包含哺乳动物的睾丸决定基因——*SRY*。这一序列已被广泛地

that it contains no coding sequences other than *SRY*[5]. The sequences at the three sites that show substitutions in humans are presented in Table 1 and the genealogy that this suggests is depicted in Fig. 1.

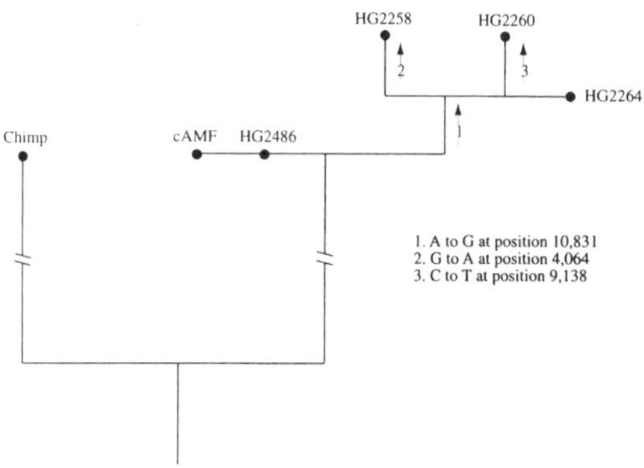

Fig. 1. Genealogy of Y chromosome sequences estimated using the principle of maximum parsimony.

Table 1. Nucleotides in human and chimpanzee subjects at three sites on the Y chromosome

Subjects	Origin	Language	Nucleotide position		
			4,064	9,138	10,831
cAMF	European	Italian	G	C	A
HG2260	Melanesian	Nasioi	G	T	G
HG2264	Rondonian surui	Tupi	G	C	G
HG2486	Tsumkwe san	!Kung	G	C	A
HG2258	Mbuti pygmy	Niger/Kordofanian	A	C	G
Chimpanzee			G	C	A

The names, origins and languages of the human subjects are given, together with the nucleotide that each human and the chimpanzee show at the three positions that are variable in the human sequences. Numbering refers to the sequence of cosmid cAMF3.1 (ref. 5). Sequences with names prefixed HG were obtained from direct sequencing of PCR products; HG numbers are accession numbers of lymphoblastoid cell lines (in the Y Chromosome Consortium repository) from which PCR-template DNA was prepared. cAMF is from the cosmid clone cAMF3.1, the sequencing of which is described in ref. 5. The sequence of cAMF3.1 was used to design PCR primers to produce amplification products covering the whole 18.3-kb region. All PCR primers were tailed at the 5′ end with M13 forward or reverse primer sequence. Products from PCR amplifications were purified on agarose gels, then electroeluted, precipitated and washed. Sequencing was by a linear amplification dideoxy termination method using fluorescent dye-labelled primers matching one of the tail sequences[5]. Candidate nucleotide variants were checked by hybridization of radiolabelled allele-specific oligonucleotides to PCR products fixed to nylon membranes (Hybond N+, Amersham). The same PCR primers and methods were used to obtain sequence from the chimpanzee. The lymphoblastoid cell line Colin (S. Marsh and P. Parham) was the source of genomic DNA. The numerous sites at which the chimpanzee sequence differed from the human sequence were not checked with allele-specific oligonucleotides. Some PCR primers did not readily produce template; although a complete sequence contig of the region was produced, only the 15.7 kb of sequence for which both DNA strands were sequenced at least once were used for analysis. This value represents the amount of sequence left after removal of insertions and deletions.

Interspecific comparison was made using 15,680 base pairs (bp) of sequence (after the removal of insertions and deletions) from the same region of the chimpanzee Y chromosome.

研究并证实该区域除了 *SRY* 外不再包含其他编码序列[5]。表 1 列出了在人类 Y 染色体中存在碱基替换的三个位点的序列，图 1 则描绘了不同序列间的谱系关系。

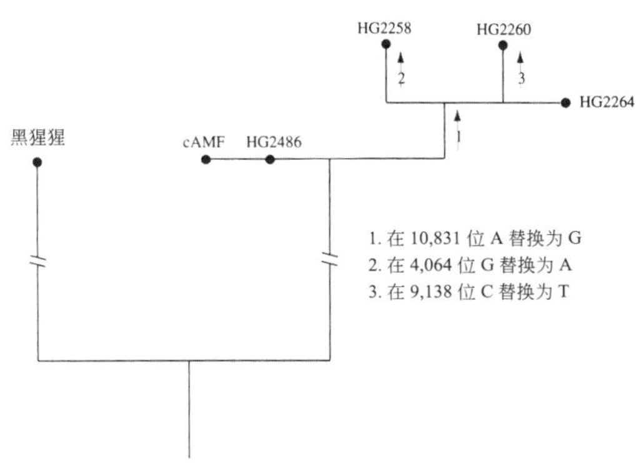

图 1. 使用最大简约性原则估计的 Y 染色体序列谱系关系

表 1. 人类和黑猩猩样本 Y 染色体三个位点的核苷酸

样本	来源	语言	核苷酸位置		
			4,064	9,138	10,831
cAMF	欧洲人	意大利语	G	C	A
HG2260	美拉尼西亚人	Nasioi	G	T	G
HG2264	朗多尼亚苏瑞族	图皮语	G	C	G
HG2486	楚姆奎桑人	!Kung	G	C	A
HG2258	姆布蒂俾格米人	尼日尔 / 科尔多凡语系	A	C	G
黑猩猩			G	C	A

该表给出了人类样本的名字、来源和语言，以及每个人和黑猩猩在人类序列中三个可变位置处对应的核苷酸。编号指的是黏粒 cAMF3.1（参考文献 5）的序列。名字前缀为“HG”的序列是由 PCR 产物直接测序得到的；HG 后的数字是类淋巴母细胞系在 Y 染色体联合资源库中的注册号，PCR 模板 DNA 是从该细胞系制备的。cAMF 来自黏粒克隆 cAMF3.1，其序列在参考文献 5 中有介绍。我们用 cAMF3.1 的序列设计了多对 PCR 引物，以产生涵盖整个 18.3 kb 区域的扩增产物。所有 PCR 引物都在 5′ 端加上了 M13 正向或反向引物序列的尾巴。PCR 扩增产物用琼脂糖凝胶进行纯化，然后电洗脱、沉淀和洗涤。使用荧光染料标记的引物与尾部序列之一配对，通过线性扩增双脱氧终止法进行测序[5]。候选核苷酸变异通过放射标记的等位基因特异性寡核苷酸与固定在尼龙膜上（尼龙膜 N+ 购自安马西亚公司）的 PCR 产物杂交的方法来检测。黑猩猩的序列使用同样的 PCR 引物和方法获得。基因组 DNA 来源于类淋巴母细胞系 Colin（马什和帕勒姆）。对于黑猩猩与人类序列中的众多碱基不同的位点，没有用等位基因特异性寡核苷酸进行检测。有些 PCR 引物并没有轻易地产生出模板；尽管该区域的全部序列重叠群都已经产生出来了，但只有 15.7 kb 的序列用于分析，这 15.7 kb 序列的 DNA 双链至少进行了一次测序。这一数值代表了去除了插入和缺失之后剩下的序列数量。

我们在去除了序列中的碱基插入和缺失情况后，选取了黑猩猩 Y 染色体上同一区域的长度为 15,680 bp 的序列进行种间比较。黑猩猩与人的 cAMF 序列在这

207 (1.32%) of these 15,680 bp differ by substitution between the chimpanzee and human sequence cAMF. The *SRY* coding sequence was removed before the calculation of the number of substitutions per site. This does not eliminate any of the three sites found to be variable in humans.

The non-recombining region of the Y chromosome has properties similar to those of the mitochondrial genome (mtDNA). Neither the mtDNA nor the Y-specific portion of the Y chromosome undergoes conventional meiotic recombination and both the Y chromosome and the mitochondrion are inherited from one sex only: the mtDNA is inherited matrilineally and the Y chromosome patrilineally. We have compared the variation of the human Y chromosome population with that of the mtDNA in a manner that corrects for the different respective mutation rates; a coalescence time for the Y chromosome sample has been calculated using the method of Tamura and Nei[6], and this has been compared to an estimate of a mtDNA coalescence time made using the same methodology (Table 2)[6]. The method of Tamura and Nei for estimating the numbers of substitutions between pairs of sequences allows for variation of the substitution rate over sites: ignoring such variation, where it exists, can drastically affect estimates of distance. Assuming that the Homo and Pan clades diverged 6 million years before present, the coalescence time of the mtDNA sample is estimated to be between 120,000 and 474,000 years, and the coalescence time of the Y chromosome sample is estimated to be 37,000 and 49,000 years. If the population samples are equivalent, this suggests a significant difference in the population genetics and demography of the Y chromosome and mtDNA.

Table 2. Estimates of numbers of nucleotide substitutions per site, rate of nucleotide substitution and time taken for human sequences to coalesce

	Y	mtDNA
$\hat{d}_{hh}$ (s.e.) Mean human-human distance via deepest fork (substitutions per site)	0.0000942 (0.0000732)	0.024 (0.006)
$\hat{d}_{ch}$ (s.e.) Mean chimp-human distance (substitutions per site)	0.0135 (0.000962)	0.752 (0.224)
$\hat{\lambda}_{max}$ Maximum substitution rate (substitutions per site per year)	1.284×10^{-9}	1.00×10^{-7}
$\hat{\lambda}_{min}$ Minimum substitution rate (substitutions per site per year)	9.631×10^{-10}	2.53×10^{-8}
Minimum age of coalescence ($\hat{d}_{hh}/2\hat{\lambda}_{max}$)	37,000 yr	120,000 yr
Maximum age of coalescence ($\hat{d}_{hh}/2\hat{\lambda}_{min}$)	49,000 yr	474,000 yr

The approach was as follows: (1) construction of a tree from human and chimpanzee sequences (Fig. 1), which indicates the location of the deepest human fork; (2) estimation of the average number of substitutions per site (using the method

15,680 bp 中有 207 个（1.32%）位点由于发生碱基替换而存在差异。我们在计算每个位点的替换数之前，会先将 *SRY* 的编码序列去掉，这样做并没有去除在人类中发现的那三个碱基可变的位点。

Y 染色体的非重组区与线粒体基因组（mtDNA）有着相似的特点。线粒体 DNA 和 Y 染色体的 Y 特异区段都不进行常规的减数分裂重组过程，并且 Y 染色体和线粒体都只能由一种性别遗传而来：线粒体是母系遗传，而 Y 染色体是父系遗传。我们将人类 Y 染色体和 mtDNA 各自的突变率进行修正后，再对二者的群体变异情况进行比较；此外，我们采用 Tamura–Nei 法[6]计算 Y 染色体样本的溯祖时间，并与用同样方法估算出的 mtDNA 溯祖时间进行比较（表 2）[6]。在估算成对序列间的碱基替换数时，使用 Tamura–Nei 法能够考虑到各位点替换率的变化：当该类变化存在时，如果将其忽略就会明显影响对遗传距离的估计。假设人属和黑猩猩属的进化枝在距今 600 万年前发生分歧，根据 mtDNA 样本估计出的溯祖时间在 120,000 年到 474,000 年之间，而根据 Y 染色体样本估计的溯祖时间为 37,000 年到 49,000 年之间。如果群体样本没有差异，那么这就提示：Y 染色体和 mtDNA 在群体遗传学和人口学方面存在显著差异。

表 2. 对每个位点的核苷酸替换数、核苷酸替换率及人类序列溯祖时间的估计

	Y	mtDNA
$\hat{d}_{hh}$(se) 通过最远的分叉测得的人与人之间的平均距离 （每个位点发生的替换）	0.0000942（0.0000732）	0.024（0.006）
$\hat{d}_{ch}$(se) 黑猩猩与人之间的平均距离 （每个位点发生的替换）	0.0135（0.000962）	0.752（0.224）
$\hat{\lambda}_{max}$ 最大替换率 （每年每个位点发生的替换）	1.284×10^{-9}	1.00×10^{-7}
$\hat{\lambda}_{min}$ 最小替换率 （每年每个位点发生的替换）	9.631×10^{-10}	2.53×10^{-8}
最小溯祖时间 ($\hat{d}_{hh}/2\hat{\lambda}_{max}$)	37,000 年	120,000 年
最大溯祖时间 ($\hat{d}_{hh}/2\hat{\lambda}_{min}$)	49,000 年	474,000 年

方法如下：（1）基于人类和黑猩猩序列构建系统树（图 1），该树表明人类分叉最远能到达的位置；（2）对人类系统树最深处的分叉上，处于相反分支上的个体进行两两比较后，估计每个位点的替换平均数（采用参考文献 6 描述

described in ref. 6) for all pairwise human comparisons involving individuals from opposite branches of the deepest fork of the human tree ($\hat{d}_{hh}$); (3) estimation of the average number of substitutions per site between chimpanzees and human ($\hat{d}_{ch}$) using the same method, and from this estimate the rate of substitution ($\hat{\lambda}$, substitutions per site per year) as a fraction of the time (T) of the separation of the Homo and Pan clades ($\hat{\lambda} = \hat{d}_{ch}/2T$); (4) use of the estimate of the rate of substitution, $\hat{\lambda}$, to calculate the age of the coalescence time of the human sequences (Age $= \hat{d}_{hh}/2\hat{\lambda}$). mtDNA values are taken from ref. 6 (but with $\hat{\lambda}$ recalculated using a single value of T). The method models variation of the substitution rate over sites on a γ distribution. Parameter α determines the shape of the γ distribution and hence the degree of variation. Estimation of α is difficult for the Y chromosome data. Yang[16] suggests that $\alpha = 0.2$ may represent extreme variation whereas $\alpha = 0.8$ may represent little variation. Tamura and Nei estimated and used $\alpha = 0.11$ for the mtDNA control region. $\alpha = 0.11$, $\alpha = 2$ and $\alpha = 100$ give very similar coalescence times for our Y data. All values given in the table and in the text are derived with $\alpha = 2$. The ranges of values given for coalescence times arise from the error for the substitution rate $\hat{\lambda}$. The mean of $\hat{d}_{hh}$ is used in this calculation rather than the minimum and maximum; the error for $\hat{d}_{hh}$ is not independent of the error for the substitution rate $\hat{\lambda}$ which has been considered[6].

We have tested three key aspects of the analysis. First, the positioning of the chimpanzee branchpoint during the construction of the human mtDNA tree determines where the human coalescence occurs. Rather than attempt to recover the true phylogeny of the mtDNA dataset, a range of tree reconstruction models were applied and the coalesence times of the trees that they suggested were assessed. The methods employed were maximum parsimony[7], neighbourhood joining and UPGMA[8], with several distance estimates, and maximum likelihood[7] with a variety of transition/transversion ratios. A total of 57 different candidate trees were obtained (after elimination of duplicates). Among these trees were three roots that were different from that used by Tamura and Nei[6]. None made a significant alteration to the date of the mtDNA coalescence.

Second, our Y chromosome sample is small and might be prone to sampling error. It is, however, not a random sample and it is difficult to say, therefore, what is the posterior probability of observing this level of Y chromosome variation. To estimate the number of new variable sites that would have to be observed in a human sample for the estimated Y chromosome coalescence time to overlap that of the mtDNA, we created a hypothetical sequence that has the ancestral sequence at the three known variable sites and then created "new" substitutions. Whereas we found only three variable sites in five sequences, six new substitutions are required before the resultant coalescence time approached that of the mtDNA (that is, the dataset of the five original human sequences plus the hypothetical sequence gave the Y coalescence time as 90,000–120,000 years).

Third, the method used assumes that the substitution rate is constant in all parts of the tree. This "molecular clock" assumption was tested for the mtDNA data set (we have insufficient data to test the Y clock) using the program BASEML from the PAML package[9]. A maximum-likelihood fit of the data to the range of reasonable trees was made under a general reversible "discrete gamma" model[10], with and without the assumption of a molecular clock. The likelihood ratio statistic ($2\Delta l$) from estimates with and without the clock assumption from each of several best trees could then be compared to a χ^2 distribution with 23 degrees of freedom[11,12]. This analysis indicated that the molecular clock was a reasonable assumption.

的方法）($\hat{d}_{hh}$)；(3) 采用相同方法对黑猩猩与人类之间各个位点发生的替换平均数的估计 ($\hat{d}_{ch}$)，以及基于此对人属和黑猩猩属进化枝发生分离的时间段 (T) 的分数——替换率 ($\hat{\lambda}$，每年每个位点发生的替换) ($\hat{\lambda}=\hat{d}_{ch}/2T$) 进行估计；(4) 使用替换率估计量 $\hat{\lambda}$ 计算人类序列溯祖时间发生的年代（年代 $=\hat{d}_{hh}/2\hat{\lambda}$)。mtDNA 值引自参考文献 6（但使用 T 的单一值重新计算了 $\hat{\lambda}$）。该方法模拟了 γ 分布中各位点的替换率变异情况。参数 α 决定了 γ 分布的形状以及变异程度。根据 Y 染色体数据很难估计出 α 值。杨子恒[16] 建议用 $\alpha = 0.2$ 代表极端变异情况，而 $\alpha = 0.8$ 代表几乎没有变异发生的情况。田村和根井估计出 mtDNA 控制区中 $\alpha = 0.11$，并且使用了这一数值。对于我们的 Y 染色体数据，$\alpha = 0.11$，$\alpha = 2$ 和 $\alpha = 100$ 给出的溯祖时间非常相似。表中及文中给出的所有数值都是从 $\alpha = 2$ 推导出来的。给出的溯祖时间的取值范围是由替换率 $\hat{\lambda}$ 的误差产生的。该计算中使用了 $\hat{d}_{hh}$ 的平均值而没有使用最小和最大值；$\hat{d}_{hh}$ 的误差并不独立于曾经认为的替换率 $\hat{\lambda}$ 的误差[6]。

我们从三个主要方面对这一分析进行了检验。首先，在构建人类 mtDNA 系统树的过程中将黑猩猩分支点定位在何处决定了人类溯祖发生的位置。我们并没有尝试从 mtDNA 数据库去复原真正的系统发生，而是采用了一系列系统树重建模型，并且对根据这些模型构建的进化树的溯祖时间进行评估。使用的方法包括基于数个距离估计值的最大简约法[7]、近邻结合法和非加权分组平均法（UPGMA）[8] 以及基于各种转换 / 颠换比值的最大似然法[7]。在去除了重复的系统树之后，我们共得到了 57 个不同的候选树。这些系统树中，有三个根与田村和根井使用的根[6] 不同。它们与 mtDNA 溯祖时间相比都没有显著变化。

其次，我们的 Y 染色体样本较少，因此易存在抽样误差。但是它并不是一个随机样本，所以很难说清楚观察到 Y 染色体这一变异水平的后验概率是多少。要使估计出的 Y 染色体溯祖时间和 mtDNA 的重叠，就必须观察人类样本中存在的新的可变位点，而为了对这些新可变位点的数目进行估计，我们创建了一条假定的序列，该序列上已知的三处可变位点具有祖先序列，然后又创建了“新”的替换位点。然而我们发现在五条序列中仅存在三个可变位点，而要使 Y 染色体的合成溯祖时间接近于 mtDNA 的溯祖时间，必须要有六个新的替换位点（即由原有的五条人类序列加上假定序列构成的数据集，共同给出的 Y 染色体的溯祖时间是 90,000 年到 120,000 年之间）。

再次，我们使用的方法假定整棵系统树的所有部分的替换率都是恒定的。我们使用 PAML 软件包[9] 里的 BASEML 程序来检验 mtDNA 数据集是否符合“分子钟”假设（我们没有足够的数据来检测 Y 染色体的分子钟）。无论是否符合分子钟假设，合理的进化树范围内的数据最大似然度是根据一个一般可逆的“离散 γ”模型[10] 确定的。因此，考虑或不考虑分子钟假设，我们得到的几棵最好的进化树的似然比统计量 ($2\Delta l$) 的估计值都可以与自由度为 23 的 χ^2 分布相比较[11,12]。这一分析暗示分子钟是一个合理的假设。

The "globalization" of the human population might be expected to have created isolated subpopulations. True isolation would prevent the homogenization of the whole population by hindering the complete fixation of individual variants. A post-globalization coalescence time for the Y chromosome sample would suggest that there has been recent replacement of Y chromosomes. The conclusion that our sample has a very recent coalescence may be a consequence of our sampling strategy, and the accuracy of the precise coalescence time for the Y chromosome also depends upon the accuracy of our date for the Homo-Pan divergence and upon undetected fluctuations in the molecular clock.

Y chromosome lineages are not shuffled by recombination. When selection acts upon a single Y-linked locus, the whole non-recombining region "hitchhikes" towards fixation as well[13-15]. The spread of an individual Y chromosome can occur purely by chance, but it is likely to be greatly influenced by selective advantage, male migration and even reproductive behaviour. For example the practice of polygyny may enable a small number of males to father a disproportionately large number of offspring, and their Y chromosomes could increase in frequency very rapidly. Nonetheless, widespread replacement of Y chromosomes is most readily explained by male migration and the occurrence of a selectively favoured Y chromosome.

(**378**, 379-380; 1995)

L. Simon Whitfield*, John E. Sulston† & Peter N. Goodfellow*
* Department of Genetics, University of Cambridge, Cambridge CB2 3EH, UK
† Sanger Centre, Hinxton Hall, Hinxton, Cambridgeshire CB10 1RQ, UK

Received 18 September; accepted 14 October 1995.

References:

1. Jakubiczka, S., Arnemann, J., Cooke, H. J., Krawczak, M. & Schmidtke, J. *Hum. Genet.* **84**, 86-88 (1989).
2. Malaspina, P. *et al. Ann. Hum. Genet.* **54**, 297-305 (1990).
3. Dorit, R. L., Akashi, H. & Gilbert, W. *Science* **268**, 1183-1185 (1995).
4. Ellis, N. *et al. Nature* **344**, 663-665 (1990).
5. Whitfield, L. S., Hawkins, T. L., Goodfellow, P. N. & Sulston, J. *Genomics* **27**, 306-311 (1995).
6. Tamura, K. & Nei, M. *Molec. Biol. Evol.* **10**, 512-526 (1993).
7. Felsenstein, J. *PHYLIP* (*Phylogeny Inference Package*) (University of Washington, Seattle, USA, 1993).
8. Kumar, S., Tamura, K. & Nei, M. *MEGA* (*Molecular Evolutionary Genetics Analysis*) (Pennsylvania State University, University Park, PA, 1993).
9. Yang, Z. *PAML* (*Phylogenetic Analysis by Maximum Likelihood*) (Inst. Molec. Evol. Genet., Pennsylvania State University, University Park, PA, 1995).
10. Yang, Z. B. *J. Molec. Evol.* **39**, 105-111 (1994).
11. Yang, Z., Goldman, N. & Friday, A. *Syst. Biol.* **44**, 385-400 (1995).
12. Felsenstein, J. *J. Molec. Evol.* **17**, 368-376 (1981).
13. Maynard-Smith, J. & Haigh, J. *Genet. Res.* **23**, 23-35 (1974).
14. Begun, D. J. & Aquadro, C. F. *Nature* **356**, 519-520 (1992).
15. Clark, A. G. *Genetics* **115**, 569-577 (1987).
16. Yang, Z. H. *J. Molec. Evol.* **40**, 689-697 (1995).

Acknowledgements. We thank N. Barton and J. Barrett for their comments on the manuscript, A. Friday for discussion, N. Ellis and M. Hammer for cells and DNA from the Y Chromosome Consortium repository, and T. Hawkins for his input to the design of the sequencing strategy. This work was supported by grants from the Wellcome Trust (to P.N.G. and J.E.S.). L.S.W. is supported by an Imperial Cancer Research Fund graduate bursary.

人们可能期望人类群体的“全球化”创造出了隔离的亚种群。真正的隔离会通过阻止个体变异的完全固定来阻止整个群体的均一化。Y 染色体样本后全球化的溯祖时间提示 Y 染色体在最近发生过替换。我们使用的样本得到了非常近的溯祖时间，这可能是因为我们的抽样策略，Y 染色体的准确溯祖时间的精确度也依赖于我们所选用的人属 – 黑猩猩属分歧时间的精确性以及未检测到分子钟的波动情况。

Y 染色体世系不会因为重组而发生重排。当选择作用于一个单独的 Y 连锁基因座时，整个非重组区也会通过“搭便车”而趋于固定 [13-15]。个体 Y 染色体的传播有可能只是偶然发生的，但也可能是选择优势、雄性迁移甚至繁殖行为对其产生了重大影响。例如一夫多妻制有可能使一小部分男性不成比例地生育了大量的后代，使得他们的 Y 染色体频率急速增加。然而，对 Y 染色体广泛替代最好的解释还是雄性迁移和选择上有利的 Y 染色体的出现。

（刘皓芳 翻译；文波 审稿）

Sheep Cloned by Nuclear Transfer from a Cultured Cell Line

K. H. S. Campbell *et al.*

Editor's Note

Here British biologist Ian Wilmut and colleagues announce the arrival of Morag and Megan, the first mammals cloned from an established cell line. The team injected DNA from cultured, differentiated sheep embryo cells into unfertilized sheep eggs lacking nuclei, then used an electrical prompt to trigger development. Several hundred attempted nuclear transfers yielded five live births and just two healthy, fertile adult sheep. The experiment confirmed that nuclear reprogramming was possible in mammals, but sceptics demanded the production of clones derived from adult donor DNA as final proof. One year later, that proof came with the birth of Dolly the sheep, and the prospect of therapeutic cloning, which aims to produce therapeutically useful, donor-matched stem cells, moved a step closer.

Nuclear transfer has been used in mammals as both a valuable tool in embryological studies[1] and as a method for the multiplication of "elite" embryos[2-4]. Offspring have only been reported when early embryos, or embryo-derived cells during primary culture, were used as nuclear donors[5,6]. Here we provide the first report, to our knowledge, of live mammalian offspring following nuclear transfer from an established cell line. Lambs were born after cells derived from sheep embryos, which had been cultured for 6 to 13 passages, were induced to quiesce by serum starvation before transfer of their nuclei into enucleated oocytes. Induction of quiescence in the donor cells may modify the donor chromatin structure to help nuclear reprogramming and allow development. This approach will provide the same powerful opportunities for analysis and modification of gene function in livestock species that are available in the mouse through the use of embryonic stem cells[7].

THE cells used in these experiments were isolated by microdissection and explantation of the embryonic disc (ED) of day 9 *in vivo* produced "Welsh mountain" sheep embryos. The line was established from early passage colonies with a morphology like that of embryonic stem (ES) cells. By the second and third passages, the cells had assumed a more epithelial, flattened morphology (Fig. 1*a*) which was maintained on further culture (to at least passage 25). At passage 6, unlike murine ES cells they expressed cytokeratin, and nuclear lamin A/C which are markers associated with differentiation[8]. This embryo-derived epithelial cell line has been designated TNT4 (for totipotent for nuclear transfer).

从体外培养的细胞系中通过核移植获得克隆羊的方法

坎贝尔等

编者按

在本文中英国生物学家伊恩·威尔穆特及其同事宣布了莫拉格和梅甘的诞生。它们是由已建立的细胞系克隆得到的首批哺乳动物。该团队将培养的已分化的绵羊胚胎细胞 DNA 注入未受精的去核绵羊卵母细胞中，然后用电脉冲诱导发育。在数百次核移植试验中得到五只羊羔，但仅得到两只健康的、有生殖能力的成年绵羊。这个实验证实了哺乳动物细胞核重编程的可行性，但持怀疑态度的人要求提供成年供体 DNA 的克隆产物作为最终的证据。一年后，作为证据的多莉羊诞生，而旨在产生有益于治疗的、与供体匹配的干细胞的治疗性克隆又向前迈进了一大步。

核移植是哺乳动物胚胎研究中一种很有价值的工具[1]，同时也是扩增优良胚胎的一种方法[2-4]。以往研究中，只有以早期胚胎或原代培养的胚胎来源细胞为细胞核供体得到后代的报道[5,6]。本研究首次报道了（据我们所知）对已建立的细胞系进行核移植并成功获得成活的哺乳类后代的方法。从绵羊胚胎分离细胞经传代培养 6~13 代后，采用血清饥饿法诱导其进入静止期，再将该细胞核移植到去核卵母细胞中，获得成活羔羊。诱导供体细胞进入静止期，这可能会修饰供体细胞的染色质结构以利于细胞核重编程并允许细胞发育。本方法可为分析和修饰家畜的基因功能提供强大的机遇，目前已经在小鼠上采用胚胎干细胞实现了这一点[7]。

这些实验中所使用的细胞，是利用显微切割术从体内生长到第 9 天的“威尔士山”绵羊的胚盘（ED）上分离出来并进行后续移植的。该细胞系来自传代早期的细胞克隆，其形态类似胚胎干（ES）细胞。到第 2 代和第 3 代，细胞呈现出上皮扁平的形态（图 1*a*），并在此后的培养中持续保持这一形态（至少到第 25 代）。在第 6 代，与鼠类 ES 细胞不同的是，这些细胞开始表达细胞角蛋白和核纤层蛋白 A/C（这是与分化有关的标记物）[8]。这一来源于胚胎的上皮细胞系被命名为 TNT4（即对核移植是全能的）。

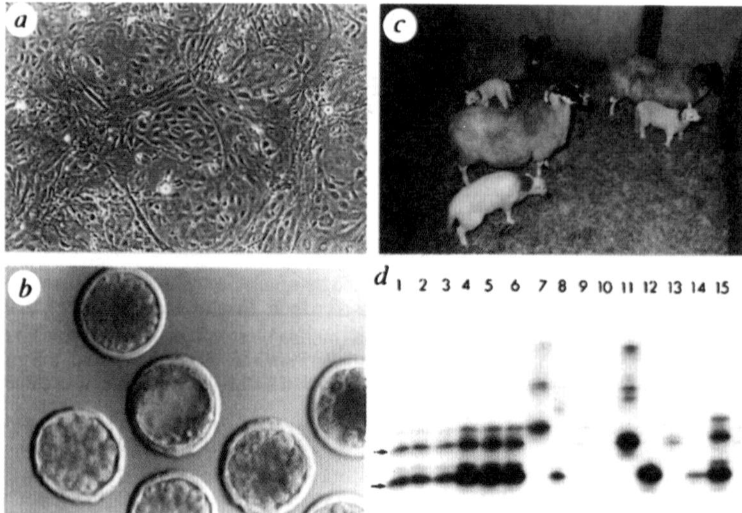

Fig. 1 . Production and characterization of the TNT4 cell line and the offspring produced by nuclear transfer from TNT4 cells. *a*, Morphology of the TNT4 cell line at passage 6. *b*, Group of embryos including a single blastocyst on day 7 after reconstruction. *c*, Group of three Welsh mountain lambs produced by nuclear transfer with surrogate Scottish blackface ewes. *d*, Autoradiogram showing the alleles generated following amplification of the microsatellite FCB266 (ref. 18). Lanes 1–6 are from, respectively, TNT4 cells and the five lambs generated by nuclear transfer. Both lambs and cells display an identical pattern, revealing 2 alleles (arrowed) at 114 and 125 bp. Lanes 7–15, nine randomly chosen Welsh mountain sheep, none of whom show an identical pattern to the nuclear transfer group. Lambs and TNT4 cells were also identical at six further microsatellite loci: MAF33, MAF48, MAF65, MAF209, OarFCB11, OarFCB128, OarRCB304 (data not shown). The nine unrelated random control animals showed extensive variation at all of these loci.

Methods. Groups of 4–6 microdissected embryonic discs were cultured on feeder layers of mitotically inactivated primary murine fibroblasts in Dulbeccos Modified Eagles medium (GIBCO) containing 10% fetal calf serum, 10% newborn serum and supplemented with recombinant human leukaemia inhibition factor (LIF). After 5–7 days of culture, expanding discs were treated with trypsin and passaged onto fresh feeders yielding 4 similar lines. At passage 12 of the $2n$ chromosome complement of 54 was observed in 31 of 50 spreads, the remaining aneuploid spreads are thought to be artefacts of preparation. For microsatellite analysis genomic DNA was extracted from whole blood, tissue culture cells or fetal tissues using a Puregene DNA isolation kit (Gentra Systems Inc., Minneapolis, USA). The PCR analysis of microsatellites was carried out using an end-labelled primer ($[\gamma\text{-}^{32}\mathrm{P}]$ATP). All other aspects of labelling and thermal cycling conditions were as described elsewhere[17].

The development of embryos reconstructed by nuclear transfer is dependent upon interactions between the donor nucleus and the recipient cytoplasm. We have previously reported the effects of the cytoplasmic kinase activity, maturation/mitosis/meiosis promoting factor (MPF), on the incidence of chromosomal damage and aneuploidy in reconstructed embryos and established two means of preventing such damage[9]. First, the effects of the donor cell-cycle stage can be overcome by transferring nuclei after the disappearance of MPF activity by prior activation of the recipient enucleated M II oocyte[9,10]. Using this approach we obtained the birth of lambs by nuclear transfer during establishment of the cell line (up to and including passage 3). On subsequent culture (passages 6 and 11) no development to term was obtained (see Table 1). From these numbers we cannot

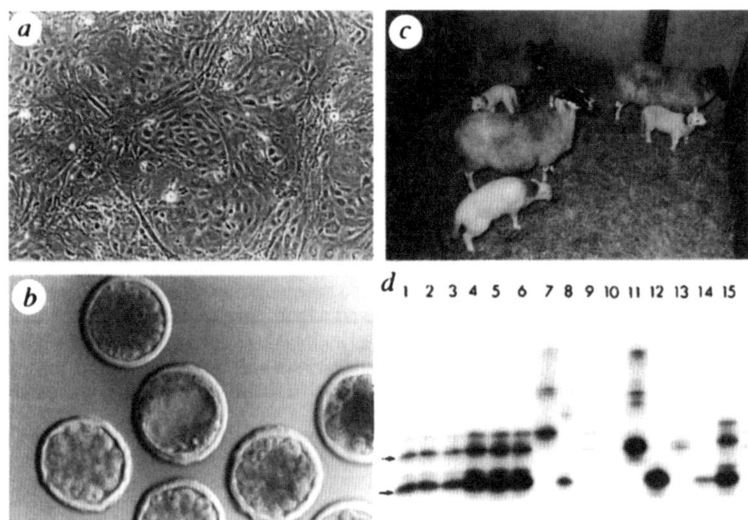

图 1. TNT4 细胞系的制备和鉴定，以及 TNT4 细胞核移植产生的羔羊后代。*a*，TNT4 细胞系传代培养至第 6 代的细胞形态。*b*，重组 7 天后含有单独胚泡的胚胎形态。*c*，以苏格兰黑脸母羊为受体代孕羊，通过核移植产生的三只威尔士山羊羔。*d*，通过扩增微卫星 FCB266 得到的等位基因的放射自显影图（参考文献 18）。泳道 1~6 分别代表来自 TNT4 细胞系以及通过核移植产生的五只羊羔的自显影结果。羊羔与 TNT4 细胞呈现一致的分布模式，显示了大小分别为 114 bp 和 125 bp 的两个等位基因（箭头所指的）的存在。泳道 7~15 代表九只随机挑选的威尔士山绵羊的自显影结果，没有一个与核移植组出现相同的模式。出生的羊羔与 TNT4 细胞在其他 6 个微卫星基因座上也显示出一致性：MAF33、MAF48、MAF65、MAF209、OarFCB11、OarFCB128、OarRCB304（数据未显示）。九只随机选择的对照组动物在这些基因座上显示了极高的变异性。

方法。将 4~6 个显微切割的胚盘在有丝分裂失活的原代鼠成纤维细胞饲养层上培养。培养体系为含有 10% 胎牛血清、10% 新生小牛血清，并添加重组人类白血病抑制因子（LIF）的 GIBCO 公司的 DMEM 培养基。培养 5~7 天之后，用胰蛋白酶处理膨大的胚盘并传代到新鲜的饲养层，产生 4 个相似的细胞系。当传代到第 12 代时，获得的 50 个细胞克隆中有 31 个含有 54 条染色体组成的 2 倍染色体组，其他非整倍的克隆则被认为是制备过程中的伪迹。为进行基因组 DNA 的微卫星分析，我们用 Puregene DNA 分离试剂盒（Gentra Systems 公司，明尼阿波利斯，美国）从全血、组织培养细胞或胎儿组织中提取基因组 DNA。微卫星 PCR 实验使用了 [γ-^{32}P]ATP 的末端标记引物。标记和热循环条件参考了其他研究中的方法 [17]。

通过核移植重新构建的胚胎，其发育过程依赖于供体细胞核与受体细胞质间的相互作用。我们以前报道过在重组的胚胎中细胞质激酶以及细胞成熟 / 有丝分裂 / 减数分裂促进因子（MPF）对染色体损伤以及染色体非整倍体发生率的影响，并建立了两种方法以防止这种损伤的发生 [9]。在第一种方法中，可以预先激活处于减数第二次分裂中期（MⅡ）的去核卵母细胞（受体）以使 MPF 活性消失，再移植细胞核，以消除供体细胞周期的影响 [9,10]。利用该手段，我们在细胞系建立阶段（第 3 代及以内）进行核移植并获得了新生羔羊。在随后的培养中（第 6 代和第 11 代），没有获得完成体内发育的胚胎（见表 1）。从这些数字中我们并不能得出这一方法不能获得完成体

conclude that development to term will not be obtained using this method. The lack of development of some control embryos is thought to relate to an infection in the oviduct of the temporary recipient ewe from which 6 were recovered.

Table 1. Development using unsynchronized TNT4 cells

Donor cell type	Number of morula and blastocysts/total embryos (%)	Number of lambs/ embryos transferred
	October 1993–February 1994	
16 cell	6/11 (27.3)	2/6
ED cell	1/15 (6.7)	0/1
ED P1	4/19 (21.0)	1/4
ED P2	1/11 (9.1)	1/1
ED P3	2/36 (5.5)	2/2
	October–December 1994	
16 cell	14/28 (50.0)	0/14
TNT4 P6	9/98 (9.2)	0*/9
TNT4 P11	10/92 (10.9)	0/10

Development of ovine embryos reconstructed by nuclear transfer of unsynchronized cells during isolation and after establishment of the TNT4 line to enucleated preactivated ovine oocytes. P, Passage number; ED, embryonic disc. For embryo reconstruction, donor oocytes were placed into calcium-free M2 containing 10% FCS, 7.5 μg ml^{-1} Cytochalasin B (Sigma) and 5.0 μg ml^{-1} Hoechst 33342 (Sigma) at 37 °C for 20 min to aspirate. A small amount of cytoplasm enclosed in plasma membrane was removed from directly beneath the 1st polar body using a glass pipette (~20 μm tip external diameter). Enucleation was confirmed by exposing this karyoplast to ultraviolet light and checking for the presence of a metaphase plate. At 34–36 h after GnRH injection enucleated oocytes were activated. Following further culture for 4–6 h in TC199, 10% FCS a single cell was fused. All activations and fusions were accomplished as previously described[10,17] in 0.3 M mannitol, 0.1 mM $MgSO_4$, 0.0005 mM $CaCl_2$[17]. For activation a single DC pulse of 1.25 kV cm^{-1} for 80 μs and for fusion an AC pulse of 3 V for 5 s followed by 3 d.c. pulses of 1.25 kV cm^{-1} for 80 μs were applied. All oocyte/cell couplets were cultured in TC199, 10% FCS 7.5 μg ml^{-1} Cytochalasin B (SIGMA) for 1 h following application of the fusion pulse and then in the same medium without Cytochalasin until transferred to temporary recipient ewes. Reconstructed embryos were cultured in the ligated oviduct of a recipient "blackface" ewe until day 7 after reconstruction. All morula and blastocyst stage embryos were transferred to synchronized recipient blackface ewes for development to term.

* A single pregnancy was established but subsequently lost at about 70–80 days.

An alternative means of avoiding damage due to the activity of MPF is to transfer diploid nuclei into metaphase II oocytes that have a high level of MPF activity[9]. The availability of TNT4 cells allows this approach to be used. In this study a synchronous population of diploid donor nuclei was produced by inducing the cells to exit the growth cycle and arrest in G0 in a state of quiescence. In the presence of a high level of MPF activity the transferred nucleus undergoes nuclear membrane breakdown and chromosome condensation. It has been argued[11] that the developmental potential of reconstructed embryos depends upon the "reprogramming of gene expression" by the action of cytoplasmic factors and that this might be enhanced by the prolongation of this period of exposure. To assess these effects donor cells were fused to oocytes either (1) 4–8 h before activation "post-activated" or (2) at the time of activation "fusion and activation" or (3) to preactivated oocytes "preactivated".

内发育的胚胎的结论。某些对照胚胎没有发育被认为是与临时受体绵羊的输卵管炎症有关，因为这些受体绵羊中有六只后来恢复了体内胚胎发育。

表 1. 采用未同步 TNT4 细胞的发育情况

供体细胞类型	桑椹胚和胚泡数 / 总胚胎数 (%)	羊羔数 / 移植的胚胎数
1993 年 10 月至 1994 年 2 月		
16 细胞	6/11 (27.3)	2/6
ED 细胞	1/15 (6.7)	0/1
ED P1	4/19 (21.0)	1/4
ED P2	1/11 (9.1)	1/1
ED P3	2/36 (5.5)	2/2
1994 年 10 月至 12 月		
16 细胞	14/28 (50.0)	0/14
TNT4 P6	9/98 (9.2)	0*/9
TNT4 P11	10/92 (10.9)	0/10

处于分离或已建立 TNT4 细胞系阶段的非同步细胞，与去核并事先激活的绵羊卵母细胞进行核移植后得到的重组绵羊胚胎的发育情况。P：传代数；ED：胚盘。为进行胚胎重组，将供体卵母细胞放入含 10% FCS、7.5 μg · ml^{-1} 细胞松弛素 B(西格玛公司) 和 5.0 μg · ml^{-1} Hoechst 33342 (西格玛公司) 的无钙离子 M2 培养基中，37 ℃放置 20 分钟以便吸入细胞。用一支玻璃移液管(尖端外径约 20 μm)，将少量贴附在细胞质膜上的细胞质直接从第一极体下方移走。将该细胞核暴露在紫外线下，并检查赤道板是否存在，以确认去核成功。在注射 GnRH 34~36 小时之后，去核卵母细胞即被激活，在含有 10% FCS 的 TC199 培养基中继续培养 4~6 小时，单细胞即可用于融合。按照参考文献所述 [10,17]，所有激活与融合过程是在含 0.3 M 甘露醇、0.1 mM $MgSO_4$、0.0005 mM $CaCl_2$[17] 的培养基中完成的。激活时，使用 1.25 kV · cm^{-1} 单次直流脉冲，持续 80 微秒。融合时先采用单次 3 V 交流脉冲，持续 5 秒，随后采用 3 次 1.25 kV · cm^{-1}，持续 80 微秒的直流脉冲。融合脉冲之后，需要在含 10% FCS、7.5 μg · ml^{-1} 细胞松弛素 B(西格玛公司) 的 TC199 培养基中培养所有卵母细胞 / 细胞组 1 小时，随后采用融合脉冲，然后在没有细胞松弛素的相同培养基中培养，直至移植到临时受体母绵羊体内。重组的胚胎于 7 天后移植到受体"黑脸"母绵羊结扎的输卵管内。所有桑椹胚和胚泡阶段的胚胎被移植到同步了的受体黑脸母绵羊体内完成体内发育。

* 一次成功妊娠，但在 70~80 天时流产。

另一种避免因 MPF 活性导致损伤发生的方法，是将二倍体细胞核移植到 MPF 活性很高的减数第二次分裂中期的卵母细胞中 [9]。TNT4 细胞的可用性使得这一方法很奏效。本研究中，通过诱导细胞退出生长周期，并锁定在静止期 G0 阶段，从而获得一群同步的二倍体供体细胞核。由于 MPF 活性处于高水平，移植后的细胞核会经历核膜破裂和染色体浓缩。有人争论说 [11] 重组胚胎的发育潜能依赖于细胞质因子活动所致的"基因表达重编程"，并可能通过延长这一暴露期而得以加强。为评估这些影响，我们将供体细胞与卵母细胞分别在以下时期融合进行研究：(1) 在"后激活"4~8 小时之前，或 (2) 在"融合与激活"期间，或 (3)"预先激活"卵母细胞。

During these studies *in vivo* ovulated metaphase arrested (M II) oocytes were flushed from the oviduct of "Scottish blackface" ewes. The methodology used was as previously described[10] with the following exceptions; oocytes were recovered 28–33 h after injection of gonadotropin-releasing hormone (GnRH), calcium/magnesium-free PBS containing 1.0% FCS was used for all flushing, and recovered oocytes were transferred to calcium-free M2 medium[12] containing 10% FCS and were maintained at 37 °C in 5% CO_2 in air until use. As soon as possible after recovery oocytes were enucleated and embryos reconstructed. At 50–54 h after GnRH injection, reconstructed embryos were embedded in agar and transferred to the ligated oviduct of dioestrus ewes. After 6 days the embryos were retrieved and development assessed microscopically (see Fig. 1*b*).

The development of embryos reconstructed using quiescent TNT4 cells and 3 different cytoplast recipients is summarized in Table 2. No significant difference was observed in the frequency of development with high and low passage number donor cells or with cytoplast recipient type used (results were analysed by the marginal model in ref. 13). All embryos that had developed to the morula/blastocyst stage were transferred as soon as possible to the uterine horn of synchronized final recipient ewes for development to term. Recipient ewes were monitored for pregnancy by ultrasonography. Ewes that were positive at day 35 were classified as pregnant (Table 3). A total of eight fetuses were detected in seven recipient ewes including a single twin pregnancy. A total of five phenotypically female Welsh mountain lambs were born from the Scottish blackface recipient ewes (Fig. 1*c*). Two of these lambs died within minutes of birth and a third at 10 days; the remaining two lambs are apparently normal and healthy (8–9 months old). Of the remaining 3 fetuses, one was lost at about 80 days of gestation, and a second was lost at 144 days of gestation. The third fetus was thought to be a twin pregnancy and was either misdiagnosed or lost at an unknown time. Microsatellite analysis of the cell line, fetuses and lambs showed that all of the female lambs were derived from a single cell population (Fig. 1*d*).

Table 2. Development to morula and blastocyst stage of ovine embryos reconstructed using quiescent TNT4 cells and 3 different cytoplast recipients (January–March 1995)

Cytoplast type		Number of morulae and blastocysts/total number of embryos recovered (%)		
Experiment number	TNT passage number	Post-activated	Activation and fusion	Preactivated
1	6	4/28	6/32*	–
2	7	1/10	1/26*	–
3	13	0/2	–	2/14
4	13	0/14	0/11	–
5	11	1/9	–	0/9
6	11	1/2	9/29***	–
7	12	–		6/45*
8	13	3/13*		–
Total		10/78(12.8%)	16/98 (16.3%)	8/68 (11.7%)

Development to the morula and blastocyst stage of ovine embryos recovered on day 7 after reconstruction by nuclear transfer of quiescent TNT4 cells at different passages into 3 cytoplast recipients. To induce quiescence, TNT4 cells

在这些研究中，体内处于 MⅡ 的卵母细胞从“苏格兰黑脸”母羊的输卵管中被冲洗出来。实验方法如参考文献所述 [10]，但有以下不同：卵母细胞被注射促性腺激素释放激素(GnRH)28~33 小时后恢复，整个实验过程中用含 1.0% FCS 的 PBS(不含钙离子和镁离子)进行漂洗，恢复后的卵母细胞被转移到含有 10% FCS 但不含钙离子的 M2 培养基 [12] 中，并在含 5% CO_2 的气体中 37 ℃保存待用。在卵母细胞恢复后尽快去核，并重组胚胎。在注射 GnRH 50~54 小时后，重组胚胎被埋植在琼脂中，并移植到间情期母羊被结扎的输卵管上。6 天后取回胚胎，并用显微镜评估其发育情况(见图 1*b*)。

使用静止期 TNT4 细胞和三种不同胞质体受体重组得到的胚胎的发育情况归纳为表 2。对于采用不同传代次数的供体细胞，或是不同类型的胞质体受体，其发育频率并未观察到显著差异(采用参考文献 13 的边际模型分析结果)。所有已经发育到桑椹胚 / 胚泡阶段的胚胎，都被迅速地移植到同步的受体母绵羊的子宫角内进行体内发育。研究中采用超声检查法来监控受体母绵羊的怀孕过程。第 35 天时呈现阳性的母绵羊即被归为怀孕组(表 3)。一共在七只受体母羊身上检测到 8 个胎儿，其中包括一例双胎妊娠。苏格兰黑脸受体母绵羊共生下了五只表型为雌性的威尔士山羔羊(图 1*c*)。其中两只在出生后几分钟即死亡，一只在第 10 天死亡，其他两只羔羊显然是正常且健康的(8 至 9 个月大)。其余 3 个胎儿，一只在妊娠 80 天时流产，另一只在妊娠 144 天时流产。而被认为是双胞胎的那例，也许是误诊，也许在某个未知的时间其中一只流产了。对细胞系、胎儿和羔羊的微卫星分析表明，所有雌性羔羊均来自一个单独的细胞群(图 1*d*)。

表 2. 静止期 TNT4 细胞和三种不同的胞质体受体重组得到的绵羊胚胎发育到桑椹胚和胚泡阶段的情况 (1995 年 1 月至 3 月)

胞质体类型		桑椹胚和胚泡数 / 恢复胚胎的总数 (%)		
试验编号	TNT 传代数	后激活	激活与融合	预先激活
1	6	4/28	6/32*	–
2	7	1/10	1/26*	–
3	13	0/2	–	2/14
4	13	0/14	0/11	–
5	11	1/9	–	0/9
6	11	1/2	9/29***	–
7	12	–		6/45*
8	13	3/13*		–
总计		10/78 (12.8%)	16/98 (16.3%)	8/68 (11.7%)

处于不同传代阶段的静止期 TNT4 细胞核移植到三种胞质体受体中进行重组，恢复 7 日后绵羊胚胎发育到桑椹胚和胚泡阶段的情况。为诱导细胞进入静止期，将 TNT4 细胞放置于预先铺好饲养层的 29 cm^2 培养瓶(GIBCO 公司)中

were plated into feeder layers in 29-cm^2 flasks (GIBCO) and cultured for 2 days, the semiconfluent exponentially growing cultures were then washed three times in medium containing 0.5% FCS and cultured in this low-serum medium for 5 days. Embryos were reconstructed using preactivated cytoplasts as previously described (Table 1) and by two other protocols. (1) post-activation, as soon as possible after enucleation a single cell was fused to the cytoplast in 0.3 M mannitol without calcium and magnesium, to prevent activation. Couplets were washed and cultured in calcium-free M2, 10% FCS at 37 °C, 5% CO_2 for 4–8 h. Thirty minutes before activation the couplets were transferred to M2 medium, 10% FCS containing 5 μM Nocodazole (SIGMA). Following activation the reconstructed zygotes were incubated in medium TC199, 10% FCS, 5.0 μM Nocodazole for a further 3 h. (2) Preactivation, at 34–36 h after GnRH injection a single cell was fused to an enucleated oocyte. The same pulse also induced activation of the recipient cytoplast. All activations and fusions were accomplished as described in Table 1 unless otherwise stated.

* Denotes number of pregnancies following transfer of morula and blastocyst stage embryos to synchronized final recipient ewes.

Table 3. Induction of pregnancy and further development following transfer of morula and blastocyst stage embryos reconstructed from quiescent TNT4 cells

Cytoplast type	Post-activated	Activation and fusion	Preactivated
Total number of morula and blastocyst stage embryos transferred	10	16	8
Total number of ewes	6	9	4
Number of pregnant ewes (%)	1 (16.7)	5 (55.5)	1 (25.0)
Number of fetuses / total embryos transferred (%)	2 / 10 (20.0)	5 / 16 (31.25)	1 / 8 (12.5)
Number of live births	1	3	1
Passage number of cells resulting in offspring	1 × P11	1 × P6, 2 × P11	1 × P13

Induction of pregnancy following transfer of all morula/blastocyst stage reconstructed embryos to the uterine horn of synchronized final recipient blackface ewes. The table shows the total number of embryos from each group transferred, the frequency of pregnancy in terms of ewes and embryos (in the majority of cases 2 embryos were transferred to each ewe and a single twin pregnancy was established (using the "post-activated" cytoplast)) and the number of live lambs obtained.

Because of the seasonality of sheep a direct comparison of all of these methods of embryo reconstruction has not yet been made. The success of the later studies may be due to a number of factors. First, quiescent nuclei are diploid and therefore the cell-cycle stages of the karyoplast and cytoplast in both the "post-activation" and "fusion and activation" methods of reconstruction are coordinated. The preactivated cytoplast will accept donor nuclei from G0, G1, S and G2 cell-cycle phases. Second, the G0 phase of the cell cycle has been implicated in the differentiation process and the chromatin of quiescent nuclei has been reported to undergo modification[14]. As a result the chromatin of quiescent donor nuclei may be more readily modified by oocyte cytoplasm. The TNT4 cells resemble several cell lines derived previously in sheep[15] and also pigs[16]. It remains to be determined whether comparable development is obtained with other such lines or other cell types. At the present time we are unable to differentiate the mechanisms involved and report that the combination of nuclear transfer and cell type described here support development to term of cloned ovine embryos from cells that had been in culture through up to 13 passages. As cell-cycle duration was about 24 h, this period of culture before nuclear transfer would be sufficient to allow genetic modification and selection if procedures comparable to those used in murine ES cells can be established.

培养 2 天，将半融合的呈指数型增长的培养物在含 0.5% FCS 的培养基中漂洗三次，然后在该低血清培养基中培养 5 天。采用以下两种方法，用此前描述的预先激活的胞质体（表 1）对胚胎进行重组。(1) 后激活方法：对单个细胞去核后，尽快在 0.3 M 的无钙镁甘露醇中将其融入胞质体，以防止激活。将细胞组在无钙离子但含有 10% FCS 的 M2 培养基中进行漂洗，并于 5% CO_2 环境中 37 ℃培养 4~8 小时。在激活之前 30 分钟，将细胞组转移到含 5 μM 诺考达唑（西格玛公司）以及 10% FCS 的 M2 培养基中。在激活之后，将重组的合子置于含有 10% FCS 和 5.0 μM 诺考达唑的 TC199 培养基中继续孵育 3 小时。(2) 预先激活方法：在注射 GnRH 34~36 小时之后，单个细胞融合到去核卵母细胞中。同样的脉冲也诱导激活了受体胞质体。除特别声明外，所有激活与融合操作均如表 1 所述。
* 指的是把桑椹胚和胚泡期的胚胎移植到同步化的最终受体母羊之后的受孕数量。

表 3. 处于静止期的 TNT4 细胞经重组获得的桑椹胚和胚泡移植后诱导妊娠和发育的情况

胞质体类型	后激活	激活与融合	预先激活
所移植的桑椹胚和胚泡期胚胎的总数	10	16	8
母绵羊总数	6	9	4
受孕绵羊数量 (%)	1 (16.7)	5 (55.5)	1 (25.0)
胎儿数量 / 移植的胚胎总数 (%)	2/10 (20.0)	5/16 (31.25)	1/8 (12.5)
生育数量	1	3	1
成功产生后代的细胞传代数	1 × P11	1 × P6, 2 × P11	1 × P13

把所有桑椹胚 / 胚泡期的重组胚胎移植到同步化的最终受体——黑脸母绵羊的子宫角后，诱导妊娠的情况。该表格给出了每个移植组得到的胚胎总数，与母绵羊以及胚胎有关的妊娠频率（大多数情况下，每只母羊移植两个胚胎，获得了一次双胎妊娠（采用“后激活”的胞质体）），以及得到的成活羊羔的数量。

由于绵羊的季节周期性，本研究尚无法对所有胚胎重组的方法进行直接比较。随后研究的成功可能与多种因素有关。首先，静止的细胞核为二倍体，“后激活”和“融合与激活”两种重组方法中的核体与胞质体的细胞周期阶段可以有效兼容。而“预先激活”的胞质体能接受来自细胞周期 G0、G1、S 和 G2 阶段的供体细胞核。其次，细胞周期的 G0 阶段与分化过程有关。另外，有报道指出处于休眠期的细胞核中的染色质会进行修饰 [14]。因此休眠期供体核中的染色质可能更容易被卵母细胞的细胞质所改变。TNT4 细胞与先前来源于绵羊 [15] 和猪 [16] 的几个细胞系类似。因此，类似的发育过程是否可利用其他细胞系或其他细胞类型得到，仍有待进一步证实。目前，我们尚不能区分相关分子机制，也无法表明本研究中所展示的核移植及其所涉及的细胞类型可以支持体外培养传代至 13 代以内的细胞系完成体内发育，并最终发育为成活个体。由于细胞周期大约为 24 小时，如果能够建立与鼠科 ES 细胞类似的程序，那么在核移植之前的培养阶段就可以充分地进行遗传修饰和选择。

The production of cloned offspring in farm animal species could provide enormous benefits in research, agriculture and biotechnology. The modification by gene targeting and selection of cell populations before embryo reconstruction coupled to the clonal origin of the whole animal provides a method for the dissemination of rapid genetic improvement and/or modification into the population.

(**380**, 64-66; 1996)

K. H. S. Campbell, J. McWhir, W. A. Ritchie & I. Wilmut
Roslin Institute (Edinburgh), Roslin, Midlothian EH25 9PS, UK

Received 27 November 1995; accepted 5 January 1996.

References:

1. McGrath, J. & Solter, D. *Science* **220**, 1300-1302 (1983).
2. Bondioli, K. R., Westhusin, M. E. & Looney, C. R. *Therio* **33**, 165-174 (1990).
3. Prather, R. S. & First, N. L. *Int. Rev. Cytol.* **120**, 169-190 (1990).
4. Chesne, P., Heyman, Y., Peynot, N. & Renard, J.-P. *C.R. Acad. Sci. Paris Life Sci.* **316**, 487-491 (1993).
5. Sims, M. & First, N. L. *Proc. Natl. Acad. Sci. U.S.A.* **91**, 6243-6147 (1994).
6. Collas, P. & Barnes, F. L. *Molec. Repr. Dev.* **38**, 264-267 (1994).
7. Hooper, M. L. *Embryonal Stem Cells: Introducing Planned Changes into the Germline* (ed. Evans, H. J.) (Harwood Academic, Switzerland, 1992).
8. Galli, C., Lazzari, G., Flechon, J. & Moor, R. M. *Zygote* **2**, 385-389 (1994).
9. Campbell, K. H. S., Ritchie, W. A. & Wilmut, I. *Biol. Reprod.* **49**, 933-942 (1993).
10. Campbell, K. H. S., Loi, P., Capai, P. & Wilmut, I. *Biol. Reprod.* **50**, 1385-1393 (1994).
11. Szollosi, D., Czolowska, R., Szollosi, M. S. & Tarkowski, A. K. *J. Cell Sci.* **91**, 603-613 (1988).
12. Whitten, W. K. & Biggers, J. D. *J. Reprod. Fertil.* **17**, 399-401 (1968).
13. Breslaw, N. E. & Clayton, D. G. *J. Am. Stat. Assoc.* **88**, 9-25 (1993).
14. Whitfield, J. F., Boynton, A. L., Rixon, R. H. & Youdale, T. *Control of Animal Cell Proliferation* Vol. 1 (eds Boynton, A. L. & Leffert, H. L.) 331-365 (Academic, London, 1985).
15. Piedrahita, J. A., Anderson, G. B. & Bon Durrant, R. H. *Therio* **34**, 879-901 (1990).
16. Gerfen, R. W. & Wheeler, M. B. *Anim. Biotechnol.* **6**, 1-14 (1995).
17. Willadsen, S. M. *Nature* **320**, 63-65 (1986).
18. Buchanan, F. C., Galloway, S. M. & Crawford, A. M. *Anim. Genet.* **25**, 60 (1994).

Acknowledgements. We thank M. Ritchie, J. Bracken, M. Malcolm-Smith and R. Ansell for technical assistance; D. Waddington for statistical analysis; and H. Bowran and his colleagues for their care of the animals. This research was supported by Ministry of Agriculture Fisheries and Food and the Roslin Institute.

农畜克隆后代的产生，将为科研、农业和生物技术领域带来巨大的研究与应用价值。在胚胎重组之前通过基因靶向修饰或在细胞群中选择优良细胞并克隆整只动物，将提供一种迅速将改良和(或)修饰基因分散到该种群中的有效方法。

(周志华 翻译；方向东 审稿)

A Comprehensive Genetic Map of the Mouse Genome

W. F. Dietrich *et al.*

Editor's Note

Linkage maps reveal the relative positions of genes and/or genetic markers in terms of recombination frequency, rather than distance along a chromosome. Here geneticist Eric S. Lander and colleagues report a complete genetic linkage map of the mouse genome, with over 7,000 genetic markers. The markers are sufficiently abundant, polymorphic and stable to enable mapping of simple and complex traits in different mouse crosses. The map also helped researchers positionally clone and pinpoint mouse mutations and provided a common framework for the mapping of mutations and cloned genes. The paper was published alongside a human genetic linkage map, marking the close of the first phase of the Human Genome Project: the construction of dense genetic maps of mouse and man.

The availability of dense genetic linkage maps of mammalian genomes makes feasible a wide range of studies, including positional cloning of monogenic traits, genetic dissection of polygenic traits, construction of genome-wide physical maps, rapid marker-assisted construction of congenic strains, and evolutionary comparisons[1,2]. We have been engaged for the past five years in a concerted effort to produce a dense genetic map of the laboratory mouse[3-6]. Here we present the final report of this project. The map contains 7,377 genetic markers, consisting of 6,580 highly informative simple sequence length polymorphisms integrated with 797 restriction fragment length polymorphisms in mouse genes. The average spacing between markers is about 0.2 centimorgans or 400 kilobases.

TO construct a simple sequence length polymorphism (SSLP) map, we identified more than 9,000 sequences from random genomic clones and public databases containing simple sequence repeats (mostly, $(CA)_n$-repeats), designed polymerase chain reaction (PCR) primers flanking the repeat, and tested each for polymorphism by measuring the allele sizes in 12 inbred mouse strains. Of the successful PCR assays, we genotyped the 90% of loci that revealed different alleles between the OB and CAST strains in an (OB × CAST) F_2 intercross with 46 progeny. These data were assembled into a map by performing genetic linkage analysis with the MAPMAKER computer package[7,8].

A total of 6,336 SSLP loci were scored in the F_2 intercross, with 6,111 derived from anonymous sequence and 225 from known genes (Table 1). Of these, 5,905 were scored as codominant markers and 431 as dominant markers (because the pattern of one allele

一张高精度的小鼠基因组遗传图谱

迪特里希等

编者按

连锁图显示的是根据重组率得到的基因和（或）遗传标记的相对位置，而不是在染色体上的距离。本文中，遗传学家埃里克·兰德及其同事报告了一个完整的小鼠基因组遗传连锁图，该图谱包含7,000多个遗传标记。这些标记具有较高的丰度、多态性和稳定性，能够用于不同小鼠杂交品系的单基因和多基因性状作图。此图还提供了一个突变和克隆基因作图的共同框架，能帮助研究人员进行定位克隆和小鼠突变位点定位。这篇文章与人类遗传连锁图谱同时发表，标志着人类基因组计划第一期工作——构建小鼠和人类的高密度遗传图谱的结束。

哺乳动物基因组的高密度遗传连锁图谱可用于广泛的研究领域，包括单基因性状的定位克隆、多基因性状的遗传解析、全基因组物理图谱的构建、快速标记辅助的同品系构建及进化比较[1,2]。过去五年里，我们共同致力于构建实验小鼠的高密度遗传图谱[3-6]。这里我们公布这一项目的最终结果。这张图谱包含7,377个遗传标记，由6,580个高信息量的简单序列长度多态性（SSLP）标记与797个小鼠基因内的限制性片段长度多态性（RFLP）标记组成。标记之间的平均间距约为0.2厘摩或400,000碱基。

为构建一个SSLP图谱，我们从随机基因组克隆和公共数据库中含有简单重复序列（多数是$(CA)_n$重复）的数据中鉴定出9,000多个序列，在重复序列两侧设计聚合酶链式反应（PCR）引物，通过在12个小鼠近交品系中鉴定每一个等位基因的大小来检验它们的多态性。对$(OB\times CAST)F_2$代互交的46个后代的成功PCR结果统计表明，我们可检出90%在OB品系和CAST品系间不同的等位基因的基因座。利用MAPMAKER程序包进行遗传连锁分析，将这些数据组装成遗传图谱[7,8]。

在F2代互交实验中共评价了6,336个SSLP基因座，其中6,111个来自未知序列，225个来自已知基因（表1）。它们中有5,905个是共显性标记，431个是显性标记

obscured the other). The map provides dense coverage of all 20 mouse chromosomes, with a total genetic length of 1,361 centimorgans (cM). Because the cross involves 92 meioses, the mean spacing between crossovers is 1.1 cM and thus loci can be mapped to "bins" of this average size. The map has 1,001 occupied bins (Table 3(*a*)), with an average of 6.3 markers per bin and an average spacing of 1.36 cM between consecutive bins.

Table 1. Genetic markers, genetic length and polymorphism* by chromosome

Chromosome	No. of markers	No. of random markers	No. from GENBANK	"Consensus" genetic length†	Observed genetic length‡	Polymorphism among lab strains (%)§‖	Lab strains versus SPR or CAST (%)‖
1	511	494	17	98	109.9	57	92
2	507	491	16	107	95.7	49	94
3	343	332	11	100	67.5	51	95
4	350	342	8	81	74.2	51	93
5	402	391	11	93	82.9	48	95
6	368	349	19	74	59.1	46	94
7	357	341	16	89	59.8	48	94
8	350	345	5	81	72.0	44	94
9	336	318	18	70	62.9	52	95
10	293	286	7	78	73.0	35	96
11	350	326	24	78	82.0	53	94
12	278	268	10	68	61.5	50	94
13	303	296	7	72	60.2	48	95
14	259	246	13	53	65.6	49	94
15	264	257	7	62	62.2	51	94
16	215	214	1	59	51.0	43	94
17	255	239	16	53	51.0	56	93
18	231	226	5	57	39.7	53	95
19	134	131	3	42	57.2	52	93
X	230	219	11	88	73.5	33	95
Total	6,336	6,111	225	1,503	1,360.9¶	48	94

* Polymorphism survey was based on visual comparisons of fragments across large acrylamide gels and was thus subject to mobility differences among lanes. To assess the accuracy of data in our database, 3,000 individual pairwise comparisons were repeated. Some 6% of reported polymorphic pairs turn out to be monomorphic upon careful comparison, while 4% of reported monomorphic pairs turn out to be polymorphic. The data are thus accurate enough to allow selection of markers for crosses, but geneticists wishing to know every polymorphic marker in a narrow region (for fine-structure genetic mapping and positional cloning, for example) are advised to recheck each locus.

† Based on "consensus" genetic map in Encyclopedia of the Mouse Genome, http://www.informatics.jax.org/encyclo.html (1993).

‡ Distance between most proximal and most distal markers in the map reported here.

（因为一个等位基因的模式会掩盖另一个等位基因）。这张高密度图谱覆盖了小鼠的20条染色体，总遗传长度达1,361厘摩。因为杂交涉及92次减数分裂，交换位点间的平均距离为1.1厘摩，因此可将这些基因座定位到以此平均值为单位的“箱子”中。这张图谱共覆盖1,001个“箱子”（表3（*a*）），平均每个“箱子”含有6.3个标记，连续“箱子”之间的平均距离为1.36厘摩。

表1. 染色体上的遗传标记、遗传长度和多态性统计*

染色体	标记数目	随机标记数目	来自GENBANK的数目	“共有”遗传长度†	观察的遗传长度‡	实验室品系间的多态性（%）§‖	实验室品系比SPR或CAST（%）‖
1	511	494	17	98	109.9	57	92
2	507	491	16	107	95.7	49	94
3	343	332	11	100	67.5	51	95
4	350	342	8	81	74.2	51	93
5	402	391	11	93	82.9	48	95
6	368	349	19	74	59.1	46	94
7	357	341	16	89	59.8	48	94
8	350	345	5	81	72	44	94
9	336	318	18	70	62.9	52	95
10	293	286	7	78	73	35	96
11	350	326	24	78	82	53	94
12	278	268	10	68	61.5	50	94
13	303	296	7	72	60.2	48	95
14	259	246	13	53	65.6	49	94
15	264	257	7	62	62.2	51	94
16	215	214	1	59	51	43	94
17	255	239	16	53	51	56	93
18	231	226	5	57	39.7	53	95
19	134	131	3	42	57.2	52	93
X	230	219	11	88	73.5	33	95
总计	6,336	6,111	225	1,503	1,360.9¶	48	94

* 多态性检测是通过比较PCR片段在长的聚丙烯酰胺凝胶泳道间迁移率的差异来实现。为评估我们数据库中数据的准确性，对3,000个标记对进行了重复比较。仔细比较后发现，已报道的多态性标记对中大约6%实际上是单态性的，而报道的单态性标记对中4%实际上是多态性的。因此这些数据足够准确，可用于筛选杂交标记。但是对于希望知道在一个很小的区域内每一个多态性标记（如用于精细结构遗传作图和定位克隆）的遗传学家，建议重新检查每个基因座。

† 基于小鼠基因组百科全书中的“共有”遗传图谱，http://www.informatics.jax.org/encyclo.html（1993）。

‡ 本图谱中最近端和最远端标记之间的距离。

§ Pairwise comparisons of OB, B6, DBA, A, C3H, BALB, AKR, NON, NOD and LP.
‖ Standard error of the mean for each chromosome depends on number of markers studied, but is < 1% in all cases.
¶ Distance is shorter than in previously published versions of this map (ref. 6) because final error checking reduced the number of apparent crossovers.

We next sought to integrate the map of largely anonymous SSLPs with the locations of known genes, because this information can suggest candidates for the genes underlying mouse mutations. We analysed a (B6 × SPRET) backcross that has been extensively used for restriction fragment length polymorphism (RFLP) mapping[9-11]. The backcross has been genotyped for 797 RFLPs. To integrate the maps, we genotyped 1,245 SSLPs from our map in 46 progeny from the SPRET backcross, providing a common reference point approximately every 1.1 cM. We also genotyped 244 additional SSLPs that were not polymorphic—and thus could not be mapped—in the (OB × CAST) intercross, but were polymorphic in the (B6 × SPRET) backcross. The SPRET cross was thus scored for a total of 1,543 SSLPs and 797 RFLPs.

The final map with 7,377 loci is shown in Fig. 1, with the SSLP map on the right and the integration with the RFLP map on the left. A full description of the markers—including primer sequences, locus sequence, genotypes in each cross, and allele sizes in the characterized strains—would require over 500 pages of this journal. The complete information is available electronically on the World Wide Web (see Fig. 1 legend).

The maps constructed in the CAST intercross and SPRET backcross maps have similar lengths (1,361 and 1,385 cM respectively), despite the fact that the intercross reflects sex-averaged recombination rates and the backcross reflects female recombination rates (because heterozygous mothers were used). Because there is typically about 80% more recombination in females than males, the SPRET backcross map might be expected to be about 40% longer. That it is not probably reflects recombinational suppression owing to structural heterogeneity (inasmuch as the laboratory mouse is evolutionarily twice as distant from SPRET as from CAST).

The SSLP map constructed in the cross was subjected to rigorous quality control and quality assessment[3,8]. All obligate double crossovers were identified and rechecked. The final data set contained no obligate double crossovers involving markers separated by less than 21 cM, indicating strong crossover interference in the mouse. (In the absence of interference, about 100 such events would be expected.) We also filled in any missing genotypes that could alter the position of a locus (by virtue of being adjacent to the site of a crossover). Despite our best efforts, some errors surely remain: in particular, an incorrect genotype adjacent to the site of a crossover would not necessarily produce a double crossover, and could shift a locus by 1.1 cM. Each chromosome is thus likely to contain a handful of loci that are slightly misplaced. The SSLPs used for integration with the SPRET backcross provided a different assessment of accuracy. We checked whether these 1,245 loci mapped to the same location in both crosses. There were ten apparent discrepancies. In five cases (*D5Mit198*, *D7Mit173*, *D9Mit132*, *D9Mit150* and *D19Mit61*), the loci were found to reproducibly amplify polymorphic fragments at

§ OB、B6、DBA、A、C3H、BALB、AKR、NON、NOD 和 LP 小鼠品系之间成对比较。
‖ 每条染色体的平均数标准误差依赖于所研究的标记数目，但在所有情况下都小于 1%。
¶ 因为在最终的错误检查时减少了明显交换的数目，距离较此图谱以前的版本（参考文献 6）短。

下一步，我们想将大量未命名的 SSLP 基因座与已知基因的位置整合在一起，因为这个信息可以提示携带潜在突变的小鼠候选基因。我们分析了一个被广泛用于 RFLP 作图的（B6 × SPRET）回交系 [9-11]。从这个回交系中已经鉴定出 797 个 RFLP。为了整合图谱，我们从 SPRET 的 46 个回交后代中鉴定了 1,245 个 SSLP，约每 1.1 厘摩可提供一个共同参考点。我们还鉴定了另外 244 个在（OB × CAST）互交系中不具多态性的 SSLP，因此这些基因座不能体现在（OB × CAST）互交系图谱中，但在（B6 × SPRET）回交系中有多态性。因此，在 SPRET 杂交系中共获得了 1,543 个 SSLP 和 797 个 RFLP。

包含 7,377 个基因座的最终图谱如图 1 所示，右侧为 SSLP 图谱，左侧为整合的 RFLP 图谱。全面描述这些标记——包括引物序列、基因座序列、每个杂交中的基因型和某些品系中的等位基因长度——将需要本杂志 500 多页的篇幅。完整的电子版信息可在万维网上获得（见图 1 图注）。

尽管互交反映的是性别的平均重组率，而回交反映的是雌性重组率（因为采用的是杂合子母本），但 CAST 互交系和 SPRET 回交系所构建的图谱具有相似的长度（分别为 1,361 厘摩和 1,385 厘摩）。因为典型的雌性重组比雄性多 80%，所以预期的 SPRET 回交图谱应大约比 CAST 互交图谱长 40%。因此它不太能反映出因为结构异质性所造成的重组抑制（因为实验室小鼠与 SPRET 的进化距离是它与 CAST 进化距离的两倍）。

杂交的 SSLP 图谱构建受到严格的质量控制和质量评价 [3,8]。所有必需双交换事件都被鉴定出来并重新检查。最终的数据集不包含任何涉及标记间距离小于 21 厘摩的必需双交换事件，这表明在小鼠中存在强交换干扰。（在无干扰条件下，预计将有 100 个这种事件发生。）我们还填补了任何可能改变一个基因座位置的缺失基因型（借助于与一个交换事件的位点毗邻）。虽然我们尽了最大努力，但肯定仍存在一些错误：特别是，一个与交换位点毗邻的错误基因型不一定会产生双交换，但是会造成基因座偏移 1.1 厘摩。每条染色体因此都可能包含少量稍有错位的基因座。用于与 SPRET 回交系整合的 SSLP 提供了一种不同的精度评估。我们检查了这 1,245 个基因座在两个杂交系中是否定位在相同的位置上，发现有 10 处明显的差异。我们发现有 5 处基因座（*D5Mit198*、*D7Mit173*、*D9Mit132*、*D9Mit150* 和 *D19Mit61*）在这两个

different chromosomal locations in the two crosses. This probably occurs because strain variation creates an alternative target for amplification, although the possibility that CAST and SPRET differ by small insertional translocations cannot be excluded. In remaining five cases, the results from the CAST cross were found not to be reproducible. These probably arose from laboratory errors that unfortunately cannot be identified in retrospect. These five loci were removed from the map. Based on the frequency (5 of 1,245), we would expect that 20 further erroneous loci remain, which corresponds to about one per chromosome.

We used several criteria to analyse the genomic distribution of loci. The spacing between SSLPs agrees reasonably well with expectation under a random distribution, although some deviation from randomness can be detected. The relative positions of markers and crossovers can be inferred completely in an experimental cross, and the entire data set can be reduced to a string of the form "mmcccmmmcccmcmcm...", with each m and c denoting the occurrence of a marker or a crossover, respectively. The hypothesis that markers are randomly distributed with respect to crossovers can be tested by comparing the observed clustering of consecutive markers and crossovers to that expected for tossing a biased coin with the probability of a marker being $p_m = M/(M+C)$, where M is the number of markers and C the number of crossovers[6]. There is some statistically significant evidence of clustering by this test (Table 3). The map contains an interval with eight consecutive crossovers (on chromosome 19) and a block of 54 recombinationally inseparable markers (on chromosome 2); the probability of such clusters of crossovers and markers occurring at random somewhere in the map is 0.5% and 3.4%, respectively. More generally, the frequency of both large and small clusters slightly exceeds expectation. Nonetheless, the distribution is not far from random expectation, at least at the level of resolution provided by the meioses studied here.

The chromosomal distribution of SSLPs among the autosomes agrees well with expectation under the assumption that loci are uniformly distributed with respect to cytogenetic length. (Chromosome 19 shows a slight deficit, which is not statistically significant after correction for multiple testing; it may reflect the unusually large proportion of heterochromatin on this chromosome.) In contrast, chromosome X shows a clear deficit, with only about 57% as many as expected (Table 2). This phenomenon appears to be general in mammalian genomes, as we have also found a similar deficit in an SSLP map of the rat[12] (62% of expectation), and Weissenbach and colleagues report a slightly less pronounced deficit in the human genome[13] (75% of expectation). In principle, the deficit of SSLPs on chromosome X could occur if $(CA)_n$-repeats were either less frequent on chromosome X, or were equally frequent but less polymorphic. The latter hypothesis would predict that the deficit of polymorphic loci on chromosome X would be offset by a great excess of non-polymorphic repeats. Of the SSLPs monomorphic between OB and CAST, 37% would have to lie to chromosome X to explain the observed data. We determined the chromosomal location of > 100 monomorphic loci (by genetic mapping for those that were polymorphic between B6 and SPRET and by somatic cell hybrid mapping for those that were not), but we found no significant excess on chromosome X.

杂交系中再扩增出位于染色体不同位置上的多态性片段。这可能是由品系间变异产生另外的扩增靶造成的，但不能排除小插入易位引起 CAST 和 SPRET 产生差别的可能性。而在剩余的 5 个基因座中，CAST 杂交系的结果不可重复。这可能是由实验错误引起的，遗憾的是无法追溯出原因，因此将这 5 个基因座从图谱中去除。基于这个频率（5/1,245），我们预计仍存在 20 个错误基因座，大约是每条染色体上一个。

我们用了几个标准来分析基因座在基因组上的分布。虽然可以检测到一些随机性偏差，但 SSLP 之间的距离与随机分布模式下的预测非常一致。标记与交换的相对位置完全可以通过一个实验性杂交推断出来，并且所有数据可以简化为一串“mmcccmmmcccmcmcm…”，每个 m 和 c 分别代表一个标记或交换的出现。标记相对于交换是随机分布的这一假设可以通过观察的连续标记簇和交换簇与期望值的比较来检验，犹如投掷硬币，一个标记的概率是 $p_m = M/(M+C)$，M 是标记数目，C 是交换数目 [6]。通过这个检验我们发现了一些具有显著统计学意义的簇（表 3）。图谱中含有一个由 8 个连续交换组成的区间（19 号染色体上）和一个由 54 个重组不分离标记组成的区间（2 号染色体上）；图谱上随机出现这种交换簇和标记簇的可能性分别为 0.5% 和 3.4%。更普遍的是，大、小簇的频率都略超预期。尽管如此，至少在本研究所提供的减数分裂这种分辨率水平上其分布仍接近随机分布预期。

SSLP 在常染色体上的分布与基因座相对于染色体细胞遗传学长度均匀分布的假设预期高度吻合。（经多重校验后，19 号染色体表现出略微差别，但不具有显著统计学意义；这可能反映了这条染色体上存在超高比例的异染色质区间。）相反，X 染色体表现出明显的差别，其仅为预期的 57%（表 2）。这一现象在哺乳动物基因组中似乎很普遍，我们在大鼠的 SSLP 图谱中也发现了相似的差别 [12]（预期的 62%），而韦森巴赫及其同事报道了在人类基因组中一处略小的差别 [13]（预期的 75%）。原则上，X 染色体上 SSLP 较少的原因可归结为 X 染色体上 $(CA)_n$ 重复频率低或频率相等但多态性低。后一种假设预期 X 染色体上多态性基因座的不足会被大量过剩的非多态性重复抵消。在 OB 和 CAST 间的单态 SSLP 中，必须有 37% 定位在 X 染色体上才可以解释所观察到的数据。我们鉴定出含有超过 100 个单态标记的基因座所在的染色体位置（通过遗传作图确定的在 B6 和 SPRET 之间存在多态性的基因座，通过体细胞杂交作图确定的不存在多态性的基因座），但并未在 X 染色体上观察到显著富

Accordingly, the deficit appears to be primarily due to an actual shortage of $(CA)_n$-repeats on chromosome X.

Table 2. Distribution of random markers based on cytogenetic length of chromosomes

Chromosome	No. of random markers†	Based on cytogenetic length*		
		Percentage of total length	Expected number of markers‡	*Z*-score§
Autosomes only				
1	494	7.68±.15	452.7±22.4	1.84
2	491	7.42±.15	437.0±21.9	2.47
3	332	6.39±.13	376.7±20.2	-2.20
4	342	6.29±.13	360.4±20.0	-1.41
5	391	6.06±.12	356.2±19.7	1.73
6	349	5.90±.12	347.7±19.4	0.07
7	341	5.54±.11	326.4±18.7	0.79
8	345	5.30±.11	312.5±18.3	1.78
9	318	5.11±.10	301.2±17.9	0.94
10	286	5.06±.10	298.1±17.8	-0.67
11	326	5.04±.10	296.8±17.8	1.65
12	268	5.21±.10	306.9±18.1	-2.14
13	293	4.67±.09	275.4±17.1	1.21
14	246	4.76±.10	280.5±17.3	-1.99
15	257	4.32±.09	254.7±16.4	0.15
16	214	4.07±.08	239.6±15.9	-1.60
17	239	4.12±.08	242.7±16.0	-0.22
18	226	4.14±.08	244.0±16.0	-1.11
19	131	2.91±.06	171.7±13.4	-3.04
Total	5,892	100	5,892.0	
Autosomes versus X chromosome				
Autosomes	5,892	93.76±.12	5,729.7±20.4	7.96
X	219	6.24±.12	381.3±20.4	-7.96
Total	6,111	100.0	6,111.0	

* Cytogenetic length taken from previous measurements[19]. Standard error of the mean was calculated directly from the raw data on chromosome measurements, generously provided by E. Evans.

† Only random markers are considered to avoid biases in chromosomal distribution of known genes.

‡ Mean ± standard deviation. Standard deviation in number of markers expected combines both standard error in the measurement of chromosome length and sampling error given to the total number of loci examined. Uncertainty in the precise length of chromosomes was not included in previous analyses[6], owing to its small magnitude, but it becomes relevant as the number of loci increases and sampling error correspondingly decreases. For comparison of autosomes to X chromosome, the expectation reflects the fact that 5% of the random markers were derived from male DNA (thus underrepresenting the X chromosome by a factor of two) and 95% from female DNA.

集。因此，造成这种差别的主要原因是 X 染色体上实际的 $(CA)_n$ 重复过少。

表 2. 基于染色体细胞遗传学长度的随机标记分布

染色体	随机标记数目 †	基于细胞遗传学长度 *		
		总长度的百分比	期望的标记数目 ‡	Z 值 §
仅常染色体				
1	494	7.68 ± .15	452.7 ± 22.4	1.84
2	491	7.42 ± .15	437.0 ± 21.9	2.47
3	332	6.39 ± .13	376.7 ± 20.2	−2.20
4	342	6.29 ± .13	360.4 ± 20.0	−1.41
5	391	6.06 ± .12	356.2 ± 19.7	1.73
6	349	5.90 ± .12	347.7 ± 19.4	0.07
7	341	5.54 ± .11	326.4 ± 18.7	0.79
8	345	5.30 ± .11	312.5 ± 18.3	1.78
9	318	5.11 ± .10	301.2 ± 17.9	0.94
10	286	5.06 ± .10	298.1 ± 17.8	−0.67
11	326	5.04 ± .10	296.8 ± 17.8	1.65
12	268	5.21 ± .10	306.9 ± 18.1	−2.14
13	293	4.67 ± .09	275.4 ± 17.1	1.21
14	246	4.76 ± .10	280.5 ± 17.3	−1.99
15	257	4.32 ± .09	254.7 ± 16.4	0.15
16	214	4.07 ± .08	239.6 ± 15.9	−1.60
17	239	4.12 ± .08	242.7 ± 16.0	−0.22
18	226	4.14 ± .08	244.0 ± 16.0	−1.11
19	131	2.91 ± .06	171.7 ± 13.4	−3.04
总计	5,892	100	5,892.0	
常染色体对 X 染色体				
常染色体	5,892	93.76 ± .12	5729.7 ± 20.4	7.96
X 染色体	219	6.24 ± .12	381.3 ± 20.4	−7.96
总计	6,111	100.0	6,111.0	

* 染色体细胞遗传学长度来自以前测量的结果 [19]。平均数标准误差直接根据染色体上测量的原始数据计算，由埃文斯慷慨提供。

† 只考虑随机标记，以避免已知基因在染色体上分布所造成的偏差。

‡ 平均值 ± 标准差。期望标记数目的标准差包括染色体长度测量的标准误差和检测基因座总数产生的取样误差。由于规模小，以前的分析不包括染色体精确长度所造成的不确定性 [6]，但其随着基因座数目的增加和取样偏差的相应减少而变得相关。对于常染色体和 X 染色体的比较，期望值反映的是 5% 的随机标记来自父本 DNA(因此代表 1/2 的 X 染色体）而 95% 来自母本 DNA 的事实。

§ Z-score = (observed−expected)/standard deviation. For the autosomes, none of the Z-scores are significant at the $P = 0.05$ level after Bonferroni correction for multiple testing. For the comparison of autosomes to X chromosome, the Z-score is significant at $P < 10^{-14}$.

The SSLPs show a polymorphism rate of about 50% among inbred laboratory strains surveyed and about 95% between laboratory strains and CAST or SPR (Table 1). The pairwise polymorphism rates among the 12 strains surveyed have not changed significantly from our previous report[6] and are not presented here. Interestingly, the distribution of polymorphism across the genome is not uniform[11]. The average polymorphism rate among the *Mus musculus* strains surveyed was just under 50%, but two chromosomes showed substantially lower polymorphism rates: chromosome X at 33%, and chromosome 10 at 35% (Table 1). Decreased polymorphism could reflect recent selection for specific ancestral chromosomes. For the X chromosome, it could also reflect a different mutation rate (inasmuch as each chromosome X resides in males only two-thirds as often each autosome, and most mutations are thought to occur in male germline) or different population genetic forces (with hemizygosity affecting selection and effective population size).

Table 3. Clusters of consecutive crossovers and markers

(*a*) Number of crossovers between consecutive random markers*					
No. of crossover per interval	Observed		Expected†		P (longest run ≥ n) (%)†
	No.	(percentage)	No.	(percentage)	
0	5,095	(83.85)	5,035.5 ± 29.6	(82.59)	
1	784	(12.90)	876.7 ± 27.4	(14.38)	100.0
2	151	(2.49)	152.6 ± 12.2	(2.50)	100.0
3	27	(0.44)	26.6 ± 5.1	(0.44)	100.0
4	14	(0.23)	4.6 ± 2.2	(0.08)	99.6
5	4	(0.07)	0.8 ± 0.9	(0.01)	62.2
6	0	(0.00)	0.1 ± 0.4	(< 0.01)	15.6
7	0	(0.00)	0.0 ± 0.2	(< 0.01)	2.9
8	1	(0.02)	0.0 ± 0.1	(< 0.01)	0.5
Total	6,076				
(*b*) Random markers occurring between consecutive crossovers‡					
No. of markers per block	Observed		Expected§		P (longest run ≥ n) (%)§
	No.	(percentage)	Number	(percentage)	
0	288	(22.3)	227.9 ± 13.7	(17.4)	100.0
1	208	(16.1)	188.2 ± 12.7	(14.4)	100.0
2	126	(9.8)	155.5 ± 11.7	(11.9)	100.0
3	111	(8.6)	128.4 ± 10.8	(9.8)	100.0
4	84	(6.5)	106.0 ± 9.9	(8.1)	100.0
5	73	(5.7)	87.6 ± 9.0	(6.7)	100.0

§ Z 值 =(观察的 − 期望的)/ 标准差。对于常染色体而言，在邦费罗尼校正多重检验后，所有 Z 值在 $P = 0.05$ 水平都不显著。常染色体和 X 染色体比较时，Z 值在 $P < 10^{-14}$ 时具有显著性。

SSLP 在所检测的近交实验室品系间的多态率约为 50%，在实验室品系和 CAST 或 SPR 间约为 95%(表 1)。在检测的 12 个品系间，成对多态率与我们先前的报道 [6] 没有明显的差别，这里未显示。有趣的是，多态性在基因组中的分布并不均一 [11]。在所检测的 *Mus musculus* 品系间，平均多态率刚好低于 50%，但是有两条染色体的多态率更低：X 染色体为 33%，10 号染色体为 35%(表 1)。多态性的降低可能反映了近期对特定祖先染色体的选择。对于 X 染色体而言，还可能反映了不同的突变率(因为每个雄性的 X 染色体只有常染色体的三分之二，而大多数突变被认为发生在雄性种系中)或不同的群体遗传压力(半合子状态会影响选择和有效群体大小)。

表 3. 连续交换簇和标记簇

(*a*) 连续随机标记之间的交换数目 *					
每个区间的交换数	观察		预期 †		P(最大游程 ≥ n)(%)†
	数量	(百分比)	数量	(百分比)	
0	5,095	(83.85)	5,035.5 ± 29.6	(82.59)	
1	784	(12.90)	876.7 ± 27.4	(14.38)	100.0
2	151	(2.49)	152.6 ± 12.2	(2.50)	100.0
3	27	(0.44)	26.6 ± 5.1	(0.44)	100.0
4	14	(0.23)	4.6 ± 2.2	(0.08)	99.6
5	4	(0.07)	0.8 ± 0.9	(0.01)	62.2
6	0	(0.00)	0.1 ± 0.4	(< 0.1)	15.6
7	0	(0.00)	0.0 ± 0.2	(< 0.1)	2.9
8	1	(0.02)	0.0 ± 0.1	(< 0.1)	0.5
总计	6,076				
(*b*) 连续交换之间的随机标记 ‡					
每个区间的标记数	观察		预期 §		P(最大游程 ≥ n)(%)§
	数量	(百分比)	数量	(百分比)	
0	288	(22.3)	227.9 ± 13.7	(17.4)	100.0
1	208	(16.1)	188.2 ± 12.7	(14.4)	100.0
2	126	(9.8)	155.5 ± 11.7	(11.9)	100.0
3	111	(8.6)	128.4 ± 10.8	(9.8)	100.0
4	84	(6.5)	106.0 ± 9.9	(8.1)	100.0
5	73	(5.7)	87.6 ± 9.0	(6.7)	100.0

Continued

(*b*) Random markers occurring between consecutive crossovers‡					
No. of markers per block	Observed		Expected§		*P* (longest run ≥ *n*) (%)§
	No.	(percentage)	Number	(percentage)	
6	62	(4.8)	72.3 ± 8.3	(5.5)	100.0
7	51	(4.0)	59.7 ± 7.6	(4.6)	100.0
8	36	(2.8)	49.3 ± 6.9	(3.8)	100.0
9	38	(2.9)	40.7 ± 6.3	(3.1)	100.0
10	32	(2.5)	33.7 ± 5.7	(2.6)	100.0
11	37	(2.9)	27.8 ± 5.2	(2.1)	100.0
12	19	(1.5)	23.0 ± 4.7	(1.8)	100.0
13	28	(2.2)	19.0 ± 4.3	(1.4)	100.0
14	18	(1.4)	15.7 ± 3.9	(1.2)	100.0
15	7	(0.5)	12.9 ± 3.6	(1.0)	100.0
16	12	(0.9)	10.7 ± 3.3	(0.8)	100.0
17	5	(0.4)	8.8 ± 3.0	(0.7)	100.0
18	5	(0.4)	7.3 ± 2.7	(0.6)	100.0
19	6	(0.5)	6.0 ± 2.4	(0.5)	100.0
20	10	(0.8)	5.0 ± 2.2	(0.4)	100.0
21	3	(0.2)	4.1 ± 2.0	(0.3)	100.0
22	5	(0.4)	3.4 ± 1.8	(0.3)	100.0
23	7	(0.5)	2.8 ± 1.7	(0.2)	100.0
24	4	(0.3)	2.3 ± 1.5	(0.2)	100.0
25	0	(0.0)	1.9 ± 1.4	(0.1)	100.0
26	5	(0.4)	1.6 ± 1.3	(0.1)	99.9
27	1	(0.1)	1.3 ± 1.1	(0.1)	99.8
28	1	(0.1)	1.1 ± 1.0	(0.1)	99.3
29	0	(0.0)	0.9 ± 0.9	(0.1)	98.4
30	1	(0.1)	0.7 ± 0.9	(0.1)	96.7
31	1	(0.1)	0.6 ± 0.8	(< 0.1)	94.0
32	0	(0.0)	0.5 ± 0.7	(< 0.1)	90.3
33	0	(0.0)	0.4 ± 0.6	(< 0.1)	85.4
34	1	(0.1)	0.3 ± 0.6	(< 0.1)	79.6
35	1	(0.1)	0.3 ± 0.5	(< 0.1)	73.1
38	1	(0.1)	0.2 ± 0.4	(< 0.1)	52.2
40	1	(0.1)	0.1 ± 0.3	(< 0.1)	39.6
54	1	(0.1)	< 0.1 ± 0.1	(< 0.1)	3.4
Total	1,289				

* The intervals with ≥ 1 crossover represent the 981 gaps between consecutive bins of recombinationally inseparable markers. Only random markers are considered to avoid biases in distribution of known genes.

续表

(b)连续交换之间的随机标记‡					
每个区间的标记数	观察		预期§		P(最大游程≥n)(%)§
	数量	(百分比)	数量	(百分比)	
6	62	(4.8)	72.3±8.3	(5.5)	100.0
7	51	(4.0)	59.7±7.6	(4.6)	100.0
8	36	(2.8)	49.3±6.9	(3.8)	100.0
9	38	(2.9)	40.7±6.3	(3.1)	100.0
10	32	(2.5)	33.7±5.7	(2.6)	100.0
11	37	(2.9)	27.8±5.2	(2.1)	100.0
12	19	(1.5)	23.0±4.7	(1.8)	100.0
13	28	(2.2)	19.0±4.3	(1.4)	100.0
14	18	(1.4)	15.7±3.9	(1.2)	100.0
15	7	(0.5)	12.9±3.6	(1.0)	100.0
16	12	(0.9)	10.7±3.3	(0.8)	100.0
17	5	(0.4)	8.8±3.0	(0.7)	100.0
18	5	(0.4)	7.3±2.7	(0.6)	100.0
19	6	(0.5)	6.0±2.4	(0.5)	100.0
20	10	(0.8)	5.0±2.2	(0.4)	100.0
21	3	(0.2)	4.1±2.0	(0.3)	100.0
22	5	(0.4)	3.4±1.8	(0.3)	100.0
23	7	(0.5)	2.8±1.7	(0.2)	100.0
24	4	(0.3)	2.3±1.5	(0.2)	100.0
25	0	(0.0)	1.9±1.4	(0.1)	100.0
26	5	(0.4)	1.6±1.3	(0.1)	99.9
27	1	(0.1)	1.3±1.1	(0.1)	99.8
28	1	(0.1)	1.1±1.0	(0.1)	99.3
29	0	(0.0)	0.9±0.9	(0.1)	98.4
30	1	(0.1)	0.7±0.9	(0.1)	96.7
31	1	(0.1)	0.6±0.8	(<0.1)	94.0
32	0	(0.0)	0.5±0.7	(<0.1)	90.3
33	0	(0.0)	0.4±0.6	(<0.1)	85.4
34	1	(0.1)	0.3±0.6	(<0.1)	79.6
35	1	(0.1)	0.3±0.5	(<0.1)	73.1
38	1	(0.1)	0.2±0.4	(<0.1)	52.2
40	1	(0.1)	0.1±0.3	(<0.1)	39.6
54	1	(0.1)	<0.1±0.1	(<0.1)	3.4
总计	1,289				

* ≥1 个交换的区间代表连续的重组不分离标记“箱子”间的 981 个间隙。只有随机标记被认为可避免因已知基因的分布而造成的偏差。

† The probability of the longest run is calculated in ref. 6. Briefly, if a coin with heads probability P is tossed n times, the length R_n of the longest head run has expected value $\mu = \log_{1/p}[(n-1)(1-p)+1]$ and the distribution of R_n is given approximately by Prob $(R_n - \mu > t) = 1 - \exp(-p^t)$. In this case, $p = 0.17$.

‡ The blocks with ≥1 marker represent the 1,001 bins of recombinationally separable markers. Only random markers are considered to avoid biases in distribution of known genes.

§ The probability of the longest head run is calculated with $p = 0.83$.

Our mouse genetic-mapping project is now at its conclusion. Although more SSLPs remain to be found (newly isolated repeats show < 10% overlap with our current set), we have reached the point of diminishing returns. The map covers the entire mouse genome, with the markers being sufficiently abundant, polymorphic and stable to allow the mapping of monogenic or polygenic traits in virtually any mouse cross of interest[5,8]. Moreover, the markers are sufficiently dense to facilitate positional cloning of most mouse mutations. With > 90% of the mouse genome being within 750 kb of a marker, and current mouse yeast artificial chromosome (YAC) libraries[14,15] having a mean insert size > 750 kb, the map affords ready access to the vast majority of the genome with little need for chromosomal walking, and provides a preliminary scaffold for constructing a genome-wide physical map[16].

The map also provides a common framework for the mapping of mutations and cloned genes. In addition to our integration with the Frederick cross, the SSLP map is being used as a framework for other mapping crosses, including public resources at the Jackson Laboratory[17] and the European Collaborative Interspecific Backcross (EUCIB)[18]. The EUCIB project (http://www.hgmp.mrc.ac.uk/MBx/MBxHomepage.html) is rescoring our SSLP markers in a cross with 1,000 meioses, which should yield finer resolution of order and correct remaining errors.

Together with the final report on the human genetic map[13], this paper marks the close of the first phase of the Human Genome Project: the construction of dense genetic maps of mouse and man.

(**380**, 149-152; 1996)

William F. Dietrich*, Joyce Miller*, Robert Steen*, Mark A. Merchant*, Deborah Damron-Boles*, Zeeshan Husain*, Robert Dredge*, Mark J. Daly*, Kimberly A. Ingalls*, Tara J. O'Connor*, Cheryl A. Evans*, Margaret M. DeAngelis*, David M. Levinson*, Leonid Kruglyak*, Nathan Goodman*, Neal G. Copeland†, Nancy A. Jenkins†, Trevor L. Hawkins*, Lincoln Stein*, David C. Page*‡§ & Eric S. Lander*‡‖

* Whitehead/MIT Center for Genome Research, Whitehead Institute for Biomedical Research, 9 Cambridge Center, Cambridge, Massachusetts 02142, USA

† Mammalian Genetics Laboratory, ABL-Basic Research Program, NCI-Frederick Cancer Research and Development Center, Frederick, Maryland 21702, USA

‡ Department of Biology, and § Howard Hughes Medical Institute, Massachusetts Institute of Technology, Cambridge, Massachusetts 02139, USA

‖ To whom correspondence should be addressed.

Received 23 October 1995; accepted 19 February 1996.

† 最大游程的概率按参考文献 6 计算。简单地讲，如果硬币正面的概率 P 是扔 n 次的结果，最大正面游程 R_n 的长度预期值 $\mu = \log_{1/p}[(n-1)(1-p)+1]$，$R_n$ 的分布可近似通过 $\text{Prob}(R_n-\mu>t)=1-\exp(-p^t)$ 得出。在本例中，$p=0.17$。

‡ ≥1 个标记的区间代表了重组可分离标记的 1,001 个“箱子”。只有随机标记被认为可避免因已知基因的分布而造成的偏差。

§ 最大正面游程的概率以 $p=0.83$ 计算。

我们的小鼠遗传图谱计划现在到了结局。虽然还有更多 SSLP 需要发现（新分离出的重复与我们现在的数据集之间的重叠小于 10%），但我们已经到达收益递减点了。这张图谱覆盖了整个小鼠基因组，标记十分丰富，具有多态性和稳定性，实际可用于任何感兴趣的小鼠杂交系的单基因和多基因性状作图 [5,8]。此外，标记密度非常高，有助于大多数小鼠突变的定位克隆。在这张图谱中，超过 90% 的小鼠基因组在 750 kb 之内都有一个标记，且目前小鼠酵母人工染色体（YAC）文库 [14,15] 的平均插入片段长度大于 750 kb，这意味着我们几乎不再需要利用染色体步移就可以获得基因组的绝大部分信息，该图谱为构建全基因组物理图提供了初步的框架 [16]。

这张图也为突变和克隆基因作图提供了一个共同框架。除了我们与弗雷德里克杂交系的整合外，SSLP 图谱也被用作其他杂交品系作图的框架，包括杰克逊实验室 [17] 和欧洲合作种间回交（EUCIB）中的公共资源 [18]。EUCIB 计划（http://www.hgmp.mrc.ac.uk/MBx/MBxHomepage.html）正在利用一个有 1,000 个减数分裂的杂交来重新评估我们的 SSLP 标记，这将产生更高精度的排序，并校正现存的错误。

这篇文章与人类遗传图谱最终结果 [13] 的共同发表，标志着人类基因组计划第一阶段——构建人类和小鼠高密度遗传图谱的结束。

（李梅 翻译；胡松年 审稿）

References:

1. Copeland, N. G. *et al. Science* **262**, 57-66 (1993).

2. Copeland, N.G. *et al. Science* **262**, 67-82 (1993).

3. Dietrich, W. F. *et al. Genetics* **131**, 423-447 (1992).

4. Dietrich, W. F. *et al.* in *Genetic Maps 1992* (ed. O'Brien, S.) 4.110-4.142 (Cold Spring Harbor Laboratory Press, NY, 1992).

5. Miller, J. C. *et al.* in *Genetic Variants and Strains of the Laboratory Mouse* 3rd Edn. (eds Lyons, M. F. & Searle, A.) (Oxford Univ. Press, New York, 1994).

6. Dietrich, W. F. *et al. Nature Genet.* 7, 220-245 (1994).

7. Lander, E. S. *et al. Genomics* **1**,174-181 (1987).

8. Lincoln, S. E. & Lander, E. S. *Genomics* **14**, 604-610 (1992).

9. Copeland, N. G. & Jenkins, N. A. *Trends Genet.* 7, 113 (1991).

10. Ceci, J. D. *et al. Genomics* **5**, 699-709 (1989).

11. Buchberg, A. M. *et al. Genetics* **122**, 153-161 (1989).

12. Jacob, H. J. *et al. Nature Genet.* **9**, 63-69 (1995).

13. Dib, C. *et al. Nature* **380**, 152-154 (1996).

14. Larin, Z., Monaco, A. P. & Lehrach, H. *Proc. Natl. Acad. Sci. U.S.A.* **88**, 4123 (1991).

15. Kusumi, K. *et al. Mamm. Genome* **4**, 391-392 (1993).

16. Hudson, T. *et al. Science* **270**, 1945-1955 (1995).

17. Rowe, L. B. *et al. Mamm. Genome* **5**, 253-274 (1994).

18. The European Backcross Collaborative Group. *Hum. Molec. Genet.* **3**, 621-627 (1994).

19. Evans, E. in *Genetic Variants and Strains of the Laboratory Mouse* 3rd Edn (eds Lyons, M. F. & Searle, A.) (Oxford Univ. Press, New York, 1994).

Acknowledgements. We thank L. Wangchuk, D. Tsering, G. Farino and K. Norbu for technical assistance; D. Gilbert and L. Maltais for help in ascertaining official nomenclature for gene loci; and Research Genetics Inc. for making SSLP primers available to the community. This work was supported in part by a grant from the National Center for Human Genome Research (to E.S. L.). L.K. was supported by a Special Emphasis Research Career Award from the National Center for Human Genome Research.

2.5-million-year-old Stone Tools from Gona, Ethiopia

S. Semaw *et al.*

Editor's Note

For many years, the earliest-known stone tools were from Olduvai, found alongside *Homo habilis* and *Zinjanthropus boisei*, and dated to around 1.8 million years ago. Further work in Ethiopia and Kenya pushed the earliest archaeology to around 2.3 million years. This report, from the Gona river drainage in Ethiopia, described primitive "Oldowan" (Olduvai-style) artefacts securely dated to 2.5 million years old, then the earliest known artefacts from anywhere in the world. But who made them? The world of the late Pliocene was occupied by australopithecines as well as early *Homo*. Louis Leakey's association of stone tools with *Homo*, so boldly made in 1964, had never looked so uncertain. Could the world's first tool-makers have been ape-men?

The Oldowan Stone tool industry was named for 1.8-million-year-old (Myr) artefacts found near the bottom of Olduvai Gorge, Tanzania. Subsequent archaeological research in the Omo (Ethiopia) and Turkana (Kenya) also yielded stone tools dated to 2.3 Myr. Palaeoanthropological investigations in the Hadar region of the Awash Valley of Ethiopia[1], revealed Oldowan assemblages in the adjacent Gona River drainage[2]. We conducted field work in the Gona study area of Ethiopia between 1992 and 1994 which resulted in additional archaeological discoveries as well as radioisotopic age control and a magnetic polarity stratigraphy of the Gona sequence. These occurrences are now securely dated between 2.6–2.5 Myr. The stone tools are thus the oldest known artefacts from anywhere in the world. The artefacts show surprisingly sophisticated control of stone fracture mechanics, equivalent to much younger Oldowan assemblages of Early Pleistocene age. This indicates an unexpectedly long period of technological stasis in the Oldowan.

IN 1976 the first archaeological occurrences of stone tools were identified in a fine-grained context at Gona[3], and two additional sites were reported later[4]. Fieldwork in 1992–94 increased the number of reported sites (Fig. 1a) and has provided the impetus for a reassessment of the geological context, age and character of the Gona archaeological concentrations.

埃塞俄比亚戈纳发现 250 万年前石制工具

塞马夫等

编者按

多年来，奥杜威地区与能人及鲍氏种东非人一起被发现的、距今约 180 万年前的石制品一直被学术界认为是最早的石制工具，随后在埃塞俄比亚和肯尼亚进一步的工作将这一年代向前推进到距今约 230 万年前。本文描述了在埃塞俄比亚戈纳河流域发现的、明确可以追溯到距今 250 万年前的远古“奥杜威文化”（奥杜威模式）的石制品，它们是世界上目前已知最早的人工制品。那么，这些石器的制造者究竟是何人呢？上新世晚期，南方古猿和早期人属成员同时生活在我们的星球上。路易斯·利基 1964 年提出制作石器工具是人属成员所独有的技能，这一激进观点从未面临如此严峻的挑战。难道最早的石制工具确是由猿人制造的吗？

坦桑尼亚奥杜威峡谷谷底附近发现的、距今 180 万年前的石制品被命名为奥杜威石器工业。随后奥莫（埃塞俄比亚）和图尔卡纳（肯尼亚）考古工作中所获得的石制品测年数据为距今 230 万年前。在埃塞俄比亚阿瓦什河谷哈达尔地区[1]进行的古人类调查显示，附近戈纳河流域[2]同样存在奥杜威石器工业制品。1992 年到 1994 年间，我们在埃塞俄比亚戈纳地区进行的田野工作中，发现了更为丰富的考古学遗存，并且获得了戈纳遗址地层的放射性同位素年龄和磁极性地层学信息。由于该地区石制品的年代可断定为距今 260 万年到 250 万年间，因此这些石器应是目前所知的、世界上最古老的人工制品。令人惊奇的是，这些石制品显示其生产者已经能熟练地掌握石料的断裂力学，具有与时代更晚的早更新世奥杜威石制品组合相同的加工工艺，这显示奥杜威石器技术的停滞期远比我们想象的久远。

1976 年，在戈纳的细粒地层沉积物中首次发现了石器考古遗存[3]。随后，另外两处遗址也陆续被报道出来[4]。1992~1994 年间的野外工作使所报道遗址的数目有所增加（图 1a），为重新认识戈纳地区考古遗址群的地质环境、年代和特征提供了契机。

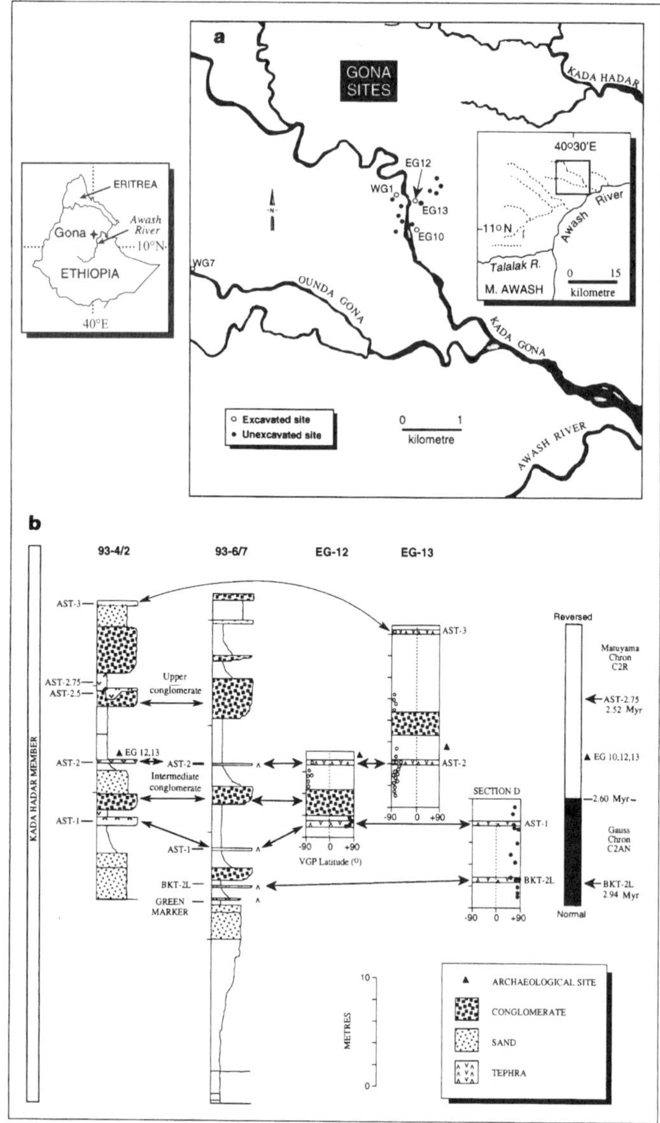

Fig. 1. **a**, Map of the Gona River drainage showing locations of the archaeological sites. **b**, Stratigraphic context of the Gona sites. Lithostratigraphy and markers from the East Gona exposures; composite sections (93-4/2 and 93-6/7) are correlated with results of magnetic polarity sampling from EG12/EG13 and section D (at 93-6/7). The profiles represent composite sections of the Gona stratigraphic sequence. Significant stratigraphic markers are listed next to the columns and correlations shown with solid lines. Filled circles indicate normal polarity, open circles reversed polarity. Chronostratigraphic context with units of the magnetic polarity timescale (MPTS) and isotopic age determinations are given on the right.

Strata associated with the Gona sites comprise three sedimentary intervals within the Kada Hadar Member of the Hadar Formation: an upward-coarsening interval of lacustrine and deltaic strata at the base of the sequence, five upward-fining fluvial cycles, and a predominantly fine-grained sequence of fluvial deposits capping the sequence. These sediments record

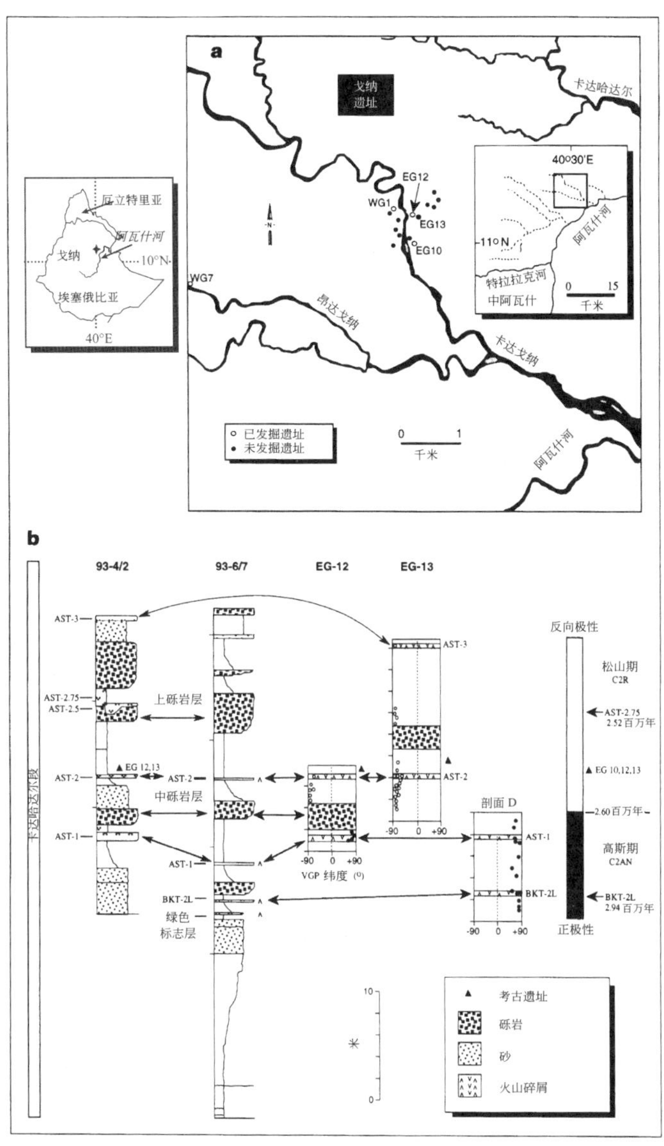

图 1. **a**，戈纳河流域考古遗址位置图。**b**，戈纳遗址地层堆积序列。戈纳东部岩相地层及标志层露头；组合剖面(93-4/2 和 93-6/7)与 EG12/EG13 以及剖面 D(93-6/7 位置)的地磁极性结果对比。该图为戈纳各地点的地层层序剖面集合。重要的地层标志在旁边纵列栏中列出，并以实线标示出它们之间的相互关系。实心代表正极性，空心代表反向极性。右侧标示的是包含地磁极性年代(MPTS)单位的年代地层环境和同位素年龄。

与戈纳旧石器时代考古遗址相关的地层包含了哈达尔组卡达哈达尔段中的三次沉积间断：地层层序底部沉积颗粒向上逐渐变粗的湖相和三角洲间隔层、五个颗粒向上逐渐变细的河流相旋回以及覆盖在此层序之上的一层主要由细微颗粒构成的河流沉积物。这些沉积物记录了整个遗址区湖盆被填充、随之而来的三角洲平原的进

the infilling of a lake basin, subsequent progradation of a delta plain, and cycling of fluvial channel and floodplain environments across the localities. The archaeological occurrences are found in floodplain environments, close to margins of channels that carried the volcanic cobbles used as raw materials for tool manufacture.

Relationships between Gona and Hadar strata are given in Fig. 1b. In the Gona, a series of marker beds consisting of vitric tephra and bentonites (altered vitric tephra) provide local correlation and tie the sequence to the Kada Hadar Member to the east. The lowest of these is a greenish bentonite in the Gona, the Green Marker. It correlates on lithologic and stratigraphic grounds with BKT-$2L_l$ in the Kada Hadar. A few metres above this is a discontinuous anorthoclase-phyric tephra, which correlates with BKT-2L. The next two markers, vitric tephra AST-1 and AST-2, have not yet been recognized outside the Gona. Between these tephra is a prominent conglomerate termed the Intermediate Conglomerate[5]. Another prominent conglomerate above AST-2 is the Upper Conglomerate, and a bentonite channelled into this conglomerate (cinerite III[5]) we refer to as AST-2.5. In outcrops immediately above this restricted exposure of AST-2.5 is plagioclase-phyric bentonite AST-2.75. This portion of the sequence is capped by a third vitric tephra, AST-3. None of the ASTs have been correlated into the Kada Hadar sequence. However, we have chronostratigraphic date on AST-2.75, and magnetic polarity stratigraphy through the sequence to provide additional control.

The age of the Gona artefact sites is constrained by a combination of radioisotopic dating and magnetic polarity stratigraphy (Fig. 1b). Magnetic stratigraphy revealed a transition from normally magnetized basal strata to reversed polarities at the level of the Intermediate Conglomerate, between tephra AST-1 and 2. Tuffs BKT-$2L_l$ and BKT-L have been identified near the bottom of the Gona sequence on the basis of lithologic similarity and outcrop tracing. In the adjacent Kada Hadar drainage, they underlie a third Bouroukie Tuff, BKT-2u. Ages of 2.95 Myr for BKT-2L and 2.92 Myr for BKT-2u have been recently cited[6], although no uncertainties were reported, nor have data been published. We report here new data documenting an age of 2.94 Myr for BKT-2L and 2.52 Myr for AST-2.75 (Fig. 2 and Table 1). This establishes the normal magnetozone at the base of the Gona sequence as the upper Gauss Chron, and the overlying reversed magnetozone as the lowermost Matuyama Chron. The Gauss–Matuyama Chron transition occurred at 2.6 Myr (ref. 7). The two excavated sites EG10 and EG12 directly overlie AST-2 and fall stratigraphically below AST-2.75. They are thus tightly bracketed between 2.6 and 2.5 Myr.

Table 1. Ar/Ar analytical data for AST-2.75 and BKT-2L

Lab no. (watts)	^{40}Ar(mol) ($\times 10^{-15}$)	^{40}Ar/^{39}Ar	^{37}Ar/^{39}Ar	^{36}Ar/^{39}Ar ($\times 10^{-3}$)	^{40}Ar*/^{39}Ar	%^{40}Ar*	Age (Myr)	$\pm\sigma$ (Myr)
AST-2.75 (Single crystal total fusion)								
8302-01	1.165	4.856	3.084	5.101	3.604	74.0	2.557	0.311
8302-02	2.014	7.017	2.485	12.385	3.562	50.7	2.527	0.260

积作用以及河道和河漫滩环境的循环过程。发现于河漫滩上的考古学遗存临近河道边缘，河流所携带的火山岩砾石可作为加工石制工具的原料。

戈纳地区和哈达尔地区地层之间的关系见图 1b。在戈纳地区，由火山玻璃碎屑和膨润土（蚀变的火山玻璃碎屑）所构成的一系列标志层提供了当地地层堆积之间的相互关系，并且将该地区的地层与卡达哈达尔段的东部区域联系起来。在戈纳地区的这些地层中，堆积物最下部是一个绿色的膨润土层，即绿色标志层，其岩性和地层层序与卡达哈达尔 BKT-2L$_1$ 相关。该层之上几米处与 BKT-2L 相关的是一层不连续的歪长石斑状火山碎屑。接下来的两处标志层是在戈纳以外的地方从未发现过的火山玻璃碎屑 AST-1 和 AST-2。这些火山玻璃碎屑之间有一层明显的砾岩，称为中砾岩层 [5]。在 AST-2 之上另一层明显的砾岩层是上砾岩层，还有我们称为 AST-2.5 的、一处嵌在该砾岩层中的膨润土（火山凝灰岩 Ⅲ[5]）。在 AST-2.5 暴露出来的有限区域之上，刚露头的部分是斜长石膨润土层 AST-2.75，该部分层序被第三层火山玻璃碎屑，即 AST-3 所覆盖。没有一个 AST 层位与卡达哈达尔的地层相关。但我们有 AST-2.75 的地层年代，磁极性地层学数据则给出了整个地层的年代控制。

戈纳遗址石制品的年代由放射性同位素年代测定法和磁极性地层学两种方法共同确定（图 1b）。磁性地层学数据揭示了在火山碎屑 AST-1 和 AST-2 之间发生的从下部正极性到中砾岩层反向极性的倒转。根据岩性地层的相似性和追踪露头，在接近戈纳层序的底部确定了凝灰岩层 BKT-2L$_1$ 和 BKT-L。在相邻的卡达哈达尔流域，它们下伏在第三种伯乌罗凯凝灰岩层 BKT-2u 之下。尽管没有年代数据不确定性的报道，结果也尚未发表，但最近已经有人引用 BKT-2L 的年代为 295 万年，BKT-2u 的年代为 292 万年 [6]。我们这里报道的新数据证实 BKT-2L 的年代为 294 万年，而 AST-2.75 的年代为 252 万年（图 2 和表 1）。这确立了以戈纳地层底部的正极性地层为高斯正极性时晚期，而覆盖在上面的反向极性地层为松山反向极性时的初期。高斯 – 松山极性倒转事件发生在距今 260 万年前（参考文献 7）。两处发掘地点 EG10 和 EG12 刚好处在 AST-2 地层之上、AST-2.75 地层之下，因此这两个遗址地层的年代恰好被严格控制在距今 260 万年到 250 万年之间。

表 1. AST-2.75 和 BKT-2L 的氩 – 氩法（Ar/Ar）分析数据

实验编号（watts）	^{40}Ar(mol)（$\times 10^{-15}$）	^{40}Ar/^{39}Ar	^{37}Ar/^{39}Ar	^{36}Ar/^{39}Ar（$\times 10^{-3}$）	^{40}Ar*/^{39}Ar	%^{40}Ar*	年代（百万年）	$\pm\sigma$（百万年）
AST-2.75（激光单颗粒全熔）								
8302-01	1.165	4.856	3.084	5.101	3.604	74.0	2.557	0.311
8302-02	2.014	7.017	2.485	12.385	3.562	50.7	2.527	0.260

Continued

Lab no. (watts)	^{40}Ar(mol) ($\times 10^{-15}$)	^{40}Ar/^{39}Ar	^{37}Ar/^{39}Ar	^{36}Ar/^{39}Ar ($\times 10^{-3}$)	^{40}Ar*/^{39}Ar	%^{40}Ar*	Age (Myr)	$\pm\sigma$ (Myr)
AST-2.75 (Single crystal total fusion)								
8302-03	3.106	6.054	3.030	9.675	3.445	56.8	2.444	0.163
8302-04	1.062	21.970	13.955	63.094	4.489	20.2	3.184	1.710
8302-05	0.651	5.269	2.812	8.941	2.858	54.1	2.028	0.597
8302-06	1.783	7.873	2.472	17.899	2.787	35.3	1.977	0.432
8302-07	0.533	6.698	2.934	12.699	3.187	47.5	2.261	0.857
8302-08	0.504	27.587	0.046	72.863	6.059	22.0	4.297	4.210
8302-09	0.577	4.694	3.120	7.706	2.672	56.8	1.896	0.480
8302-10	0.553	5.456	2.730	6.544	3.748	68.6	2.659	0.503
8302-11	0.341	5.942	3.197	9.146	3.503	58.8	2.485	0.926
8302-12	0.678	6.195	3.497	10.013	3.525	56.7	2.501	0.547
8302-13	2.049	5.610	0.004	6.637	3.649	65.0	2.589	0.181
8302-14	1.188	7.851	2.582	15.958	3.348	42.6	2.376	0.483
8302-15	0.754	19.284	1.172	56.896	2.567	13.3	1.821	1.971
8302-16	0.275	4.465	3.489	1.797	4.224	94.4	2.996	1.176
8302-20	82.311	33.003	0.033	70.249	12.247	37.1	*8.674*	*0.083*
8302-21	0.237	15.566	38.207	21.730	12.566	78.4	*8.899*	*1.505*
8302-22	0.282	4.903	3.050	4.711	3.763	76.6	2.670	0.474
8302-23	4.541	42.883	1.816	2.477	42.355	98.6	*29.823*	*0.296*
8302-24	6.624	35.672	0.816	8.843	33.144	92.9	*23.379*	*0.162*
8302-25	0.467	10.666	3.110	23.173	4.077	38.1	2.892	0.483
8302-26	0.704	10.816	4.077	24.397	3.945	36.4	2.799	0.318
8302-27	24.226	40.627	0.010	19.737	34.796	85.6	*24.536*	*0.087*
8302-28	0.785	6.732	4.952	12.390	3.480	51.5	2.469	0.198
8302-29	16.982	35.005	0.143	0.447	34.888	99.7	*24.601*	*0.075*
8302-30	0.454	5.000	3.201	4.392	3.967	79.2	2.814	0.303
8302-31	6.865	57.853	4.131	176.057	6.178	10.6	4.381	0.454
8302-32	0.258	5.389	3.075	6.967	3.585	66.4	2.543	0.629
Weighted mean								2.517 ± 0.075
BKT-2L (Single crystal total fusion)								
7201-01	2.099	3.495	0.633	0.344	3.438	98.5	2.936	0.070
7201-02	3.564	3.515	0.230	0.303	3.438	97.9	2.936	0.037
7201-03	2.195	3.541	0.857	0.490	3.458	97.7	2.953	0.057
7201-04	4.744	3.447	0.224	0.072	3.439	99.9	2.936	0.027
7201-05	2.825	3.507	0.231	0.163	3.472	99.1	2.965	0.054

续表

实验编号（watts）	^{40}Ar(mol)（$\times 10^{-15}$）	$^{40}Ar/^{39}Ar$	$^{37}Ar/^{39}Ar$	$^{36}Ar/^{39}Ar$（$\times 10^{-3}$）	$^{40}Ar^*/^{39}Ar$	$\%^{40}Ar^*$	年代（百万年）	$\pm\sigma$（百万年）
AST-2.75（激光单颗粒全熔）								
8302-03	3.106	6.054	3.030	9.675	3.445	56.8	2.444	0.163
8302-04	1.062	21.970	13.955	63.094	4.489	20.2	3.184	1.710
8302-05	0.651	5.269	2.812	8.941	2.858	54.1	2.028	0.597
8302-06	1.783	7.873	2.472	17.899	2.787	35.3	1.977	0.432
8302-07	0.533	6.698	2.934	12.699	3.187	47.5	2.261	0.857
8302-08	0.504	27.587	0.046	72.863	6.059	22.0	4.297	4.210
8302-09	0.577	4.694	3.120	7.706	2.672	56.8	1.896	0.480
8302-10	0.553	5.456	2.730	6.544	3.748	68.6	2.659	0.503
8302-11	0.341	5.942	3.197	9.146	3.503	58.8	2.485	0.926
8302-12	0.678	6.195	3.497	10.013	3.525	56.7	2.501	0.547
8302-13	2.049	5.610	0.004	6.637	3.649	65.0	2.589	0.181
8302-14	1.188	7.851	2.582	15.958	3.348	42.6	2.376	0.483
8302-15	0.754	19.284	1.172	56.896	2.567	13.3	1.821	1.971
8302-16	0.275	4.465	3.489	1.797	4.224	94.4	2.996	1.176
8302-20	82.311	33.003	0.033	70.249	12.247	37.1	*8.674*	*0.083*
8302-21	0.237	15.566	38.207	21.730	12.566	78.4	*8.899*	*1.505*
8302-22	0.282	4.903	3.050	4.711	3.763	76.6	2.670	0.474
8302-23	4.541	42.883	1.816	2.477	42.355	98.6	*29.823*	*0.296*
8302-24	6.624	35.672	0.816	8.843	33.144	92.9	*23.379*	*0.162*
8302-25	0.467	10.666	3.110	23.173	4.077	38.1	2.892	0.483
8302-26	0.704	10.816	4.077	24.397	3.945	36.4	2.799	0.318
8302-27	24.226	40.627	0.010	19.737	34.796	85.6	*24.536*	*0.087*
8302-28	0.785	6.732	4.952	12.390	3.480	51.5	2.469	0.198
8302-29	16.982	35.005	0.143	0.447	34.888	99.7	*24.601*	*0.075*
8302-30	0.454	5.000	3.201	4.392	3.967	79.2	2.814	0.303
8302-31	6.865	57.853	4.131	176.057	6.178	10.6	4.381	0.454
8302-32	0.258	5.389	3.075	6.967	3.585	66.4	2.543	0.629
加权平均值								2.517±0.075
BKT-2L（激光单颗粒全熔）								
7201-01	2.099	3.495	0.633	0.344	3.438	98.5	2.936	0.070
7201-02	3.564	3.515	0.230	0.303	3.438	97.9	2.936	0.037
7201-03	2.195	3.541	0.857	0.490	3.458	97.7	2.953	0.057
7201-04	4.744	3.447	0.224	0.072	3.439	99.9	2.936	0.027
7201-05	2.825	3.507	0.231	0.163	3.472	99.1	2.965	0.054

Continued

Lab no. (watts)	^{40}Ar(mol) ($\times 10^{-15}$)	$^{40}Ar/^{39}Ar$	$^{37}Ar/^{39}Ar$	$^{36}Ar/^{39}Ar$ ($\times 10^{-3}$)	$^{40}Ar^*/^{39}Ar$	$\%^{40}Ar^*$	Age (Myr)	$\pm\sigma$ (Myr)
BKT-2L (Single crystal total fusion)								
7201-06	4.568	3.469	0.233	0.196	3.424	98.8	2.924	0.033
7201-07	5.078	3.492	0.250	0.216	3.443	98.7	2.940	0.028
7201-08	4.639	3.462	0.228	0.017	3.470	100.4	2.963	0.025
7201-09	2.098	3.699	0.218	0.762	3.485	94.4	2.976	0.059
7201-10	3.603	3.481	0.224	0.125	3.457	99.4	2.952	0.035
7201-14	5.612	3.654	0.227	0.579	3.496	95.8	2.985	0.021
7201-15	5.201	3.552	0.233	0.385	3.451	97.3	2.947	0.020
7201-16	3.695	3.494	0.229	0.377	3.395	97.3	2.899	0.025
7201-17	2.412	3.511	0.227	0.336	3.424	97.7	2.924	0.039
7201-18	6.018	3.488	0.246	0.154	3.457	99.2	2.952	0.017
7201-19	4.206	3.472	0.230	0.106	3.453	99.6	2.949	0.024
7201-20	5.572	3.488	0.231	0.067	3.481	99.9	2.972	0.017
7201-21	2.002	3.574	0.734	0.510	3.475	97.3	2.968	0.045
7201-22	3.935	3.491	0.224	0.310	3.412	97.9	2.914	0.023
7201-24	0.513	3.800	0.205	1.104	3.484	91.8	2.976	0.175
7201-25	7.476	3.561	0.234	0.604	3.395	95.5	2.900	0.013
Weighted mean								2.940 ± 0.006
BKT-2L (Bulk-step heating)								
(2.1)	0.306	7.275	1.158	19.628	1.559	21.4	1.332	0.619
(2.5)	0.342	4.939	0.860	11.467	1.611	32.6	1.376	0.337
(2.8)	0.700	5.477	0.617	9.346	2.758	50.4	2.356	0.213
(3.0)	0.428	3.810	0.559	0.937	3.572	93.8	3.050	0.364
(3.3)	0.432	3.563	0.526	0.516	3.446	96.8	2.943	0.315
(3.6)	0.441	3.495	0.516	0.902	3.263	93.5	2.787	0.289
(3.9)	0.606	4.047	0.527	2.159	3.445	85.2	2.942	0.173
(4.4)	1.502	3.607	0.396	0.804	3.395	94.2	2.900	0.057
(4.7)	0.361	3.509	0.488	0.485	3.399	97.0	2.902	0.374
(5.1)	0.670	3.627	0.636	1.148	3.332	92.0	2.846	0.127
(5.6)	0.648	3.865	0.597	1.668	3.413	88.4	2.915	0.136
(6.0)	0.665	3.581	0.618	1.085	3.303	92.3	2.821	0.130
(6.5)	0.608	3.432	0.603	0.142	3.432	100.1	2.931	0.219
(7.0)	0.782	3.467	0.604	0.640	3.320	95.9	2.835	0.150
(7.5)	2.301	4.457	0.866	3.643	3.443	77.3	2.941	0.053
(8.0)	3.219	3.422	0.818	0.263	3.403	99.5	2.906	0.026

续表

实验编号(watts)	^{40}Ar(mol)($\times 10^{-15}$)	$^{40}Ar/^{39}Ar$	$^{37}Ar/^{39}Ar$	$^{36}Ar/^{39}Ar$($\times 10^{-3}$)	$^{40}Ar^*/^{39}Ar$	%$^{40}Ar^*$	年代(百万年)	$\pm\sigma$(百万年)
BKT-2L(激光单颗粒全熔)								
7201-06	4.568	3.469	0.233	0.196	3.424	98.8	2.924	0.033
7201-07	5.078	3.492	0.250	0.216	3.443	98.7	2.940	0.028
7201-08	4.639	3.462	0.228	0.017	3.470	100.4	2.963	0.025
7201-09	2.098	3.699	0.218	0.762	3.485	94.4	2.976	0.059
7201-10	3.603	3.481	0.224	0.125	3.457	99.4	2.952	0.035
7201-14	5.612	3.654	0.227	0.579	3.496	95.8	2.985	0.021
7201-15	5.201	3.552	0.233	0.385	3.451	97.3	2.947	0.020
7201-16	3.695	3.494	0.229	0.377	3.395	97.3	2.899	0.025
7201-17	2.412	3.511	0.227	0.336	3.424	97.7	2.924	0.039
7201-18	6.018	3.488	0.246	0.154	3.457	99.2	2.952	0.017
7201-19	4.206	3.472	0.230	0.106	3.453	99.6	2.949	0.024
7201-20	5.572	3.488	0.231	0.067	3.481	99.9	2.972	0.017
7201-21	2.002	3.574	0.734	0.510	3.475	97.3	2.968	0.045
7201-22	3.935	3.491	0.224	0.310	3.412	97.9	2.914	0.023
7201-24	0.513	3.800	0.205	1.104	3.484	91.8	2.976	0.175
7201-25	7.476	3.561	0.234	0.604	3.395	95.5	2.900	0.013
加权平均值								2.940±0.006
BKT-2L(块状逐步加热)								
(2.1)	0.306	7.275	1.158	19.628	1.559	21.4	1.332	0.619
(2.5)	0.342	4.939	0.860	11.467	1.611	32.6	1.376	0.337
(2.8)	0.700	5.477	0.617	9.346	2.758	50.4	2.356	0.213
(3.0)	0.428	3.810	0.559	0.937	3.572	93.8	3.050	0.364
(3.3)	0.432	3.563	0.526	0.516	3.446	96.8	2.943	0.315
(3.6)	0.441	3.495	0.516	0.902	3.263	93.5	2.787	0.289
(3.9)	0.606	4.047	0.527	2.159	3.445	85.2	2.942	0.173
(4.4)	1.502	3.607	0.396	0.804	3.395	94.2	2.900	0.057
(4.7)	0.361	3.509	0.488	0.485	3.399	97.0	2.902	0.374
(5.1)	0.670	3.627	0.636	1.148	3.332	92.0	2.846	0.127
(5.6)	0.648	3.865	0.597	1.668	3.413	88.4	2.915	0.136
(6.0)	0.665	3.581	0.618	1.085	3.303	92.3	2.821	0.130
(6.5)	0.608	3.432	0.603	0.142	3.432	100.1	2.931	0.219
(7.0)	0.782	3.467	0.604	0.640	3.320	95.9	2.835	0.150
(7.5)	2.301	4.457	0.866	3.643	3.443	77.3	2.941	0.053
(8.0)	3.219	3.422	0.818	0.263	3.403	99.5	2.906	0.026

Continued

Lab no. (watts)	^{40}Ar(mol) ($\times 10^{-15}$)	$^{40}Ar/^{39}Ar$	$^{37}Ar/^{39}Ar$	$^{36}Ar/^{39}Ar$ ($\times 10^{-3}$)	$^{40}Ar^{*}/^{39}Ar$	$\%^{40}Ar^{*}$	Age (Myr)	$\pm\sigma$ (Myr)
BKT-2L (Bulk-step heating)								
(8.5)	0.124	2.951	0.342	-1.199	3.327	112.9	2.841	0.804
(8.5)	0.201	3.491	0.359	0.383	3.401	97.5	2.904	0.565
(8.5)	5.930	3.461	0.458	0.152	3.447	99.7	2.943	0.010
Plateau age								2.936 ± 0.010

Methods. Palaeomagnetic sampling and analysis were done as described previously[23]. Ar/Ar dating used methods and facilities described in refs 24, 25. Samples were irradiated in the Oregon State University Triga reactor for 1.5 h using the cadmium-lined CLICKIT facility. Data for BKT-2L and AST-2.75 are presented in Table 1: isotope ratios are corrected for background, mass discrimination (1.00657 ± 0.00151 to 1.00873 ± 0.0010 per atomic mass unit, measured by routine automated analysis of air pipette samples) and radioactive decay. Sensitivities of the mass spectrometer–electron multiplier systems were (5–10) $\times 10^{13}$ nA per mol for argon isotopes. Nucleogenic interference corrections were made based on $(^{39}Ar/^{39}Ar)_{Ca} = 0.00764 \pm 0.0000557$, $(^{36}Ar/^{37}Ar)_{Ca} = 0.0002705 \pm 0.0000105$, $(^{40}Ar/^{39}Ar)_{Ca} = 0.00014 \pm 0.00060$. Calculated ages are based on J values (0.0003935 ± 0.0000003 for AST-2.75, 0.0004737 ± 0.0000003 for BKT-2L) derived from replicate analysis of the Fish Canyon sanidine standard, with an age of 27.84 Myr[24,25]. Age calculations are based on decay constants $\lambda = 5.543 \times 10^{-10}\ Y^{-1}$. Uncertainties in ages reflect uncertainties in isotope abundances and all corrections, but do not reflect uncertainty in the age of the standard. The arithmetic mean and standard error of the mean are 2.58 ± 0.11 Myr, whereas the inverse variance weighted mean is 2.52 ± 0.08 Myr; we adopt the latter as more appropriate because of the heterogeneous precision of individual analyses. Samples with ages shown in italics (AST-2.75) are contaminants and were excluded from the mean age calculations.

Figure 1a maps the Gona archaeological sites. Systematic excavations at EG10 (13 m^2) and at EG12 (9 m^2) yielded the largest and most informative artefact assemblages (n = 2,970 stone artefacts, 1,114 of these being *in situ*). The artefacts were restricted to a 10-cm interval of a 6-m-thick, well consolidated clay-rich palaeovertisol. An additional 43 surface and excavated artefacts were recovered in 1994 from WG7, increasing the geographical distribution of the archaeological occurrences to include the Ounda Gona catchment. The fine-grained context, the limited vertical dispersion, the fresh quality of the artefacts and the lack of artefact-size sorting suggests a low-energy depositional environment.

The excavated and surface artefact assemblages at EG10 and EG12 primarily comprise simple cores, whole flakes and flaking debris. Unifacially and bifacially flaked cores comprise the "flaked pieces" category. There are numerous examples of several generations of flake scars on the cores, indicating that Late Pliocene hominids had mastered the skills of basic stone knapping (Fig. 3). Moreover, the large number of well struck flakes with conspicuous bulbs of percussion indicate a clear understanding of conchoidal fracture mechanics. "Detached pieces" are numerically dominant, with values in the range of 75–95%. There are a few "pounded pieces", namely pieces modified or shaped by pounding or battering, like hammerstones, anvils or battered cobbles.

续表

实验编号 (watts)	^{40}Ar(mol) ($\times 10^{-15}$)	$^{40}Ar/^{39}Ar$	$^{37}Ar/^{39}Ar$	$^{36}Ar/^{39}Ar$ ($\times 10^{-3}$)	$^{40}Ar^*/^{39}Ar$	%$^{40}Ar^*$	年代 (百万年)	±σ (百万年)
BKT-2L(块状逐步加热)								
(8.5)	0.124	2.951	0.342	−1.199	3.327	112.9	2.841	0.804
(8.5)	0.201	3.491	0.359	0.383	3.401	97.5	2.904	0.565
(8.5)	5.930	3.461	0.458	0.152	3.447	99.7	2.943	0.010
坪年龄								2.936±0.010

方法。古地磁采样和分析的方法参见之前的描述[23]。Ar/Ar 年代测定使用的方法和设备在参考文献 24 和 25 中有描述。使用镉谱线 CLICKIT 设施，在俄勒冈州立大学的 Triga 反应堆中照射样本 1.5 小时。BKT-2L 和 AST-2.75 的数据在表 1 中列出：同位素比值经过背景校正、质量歧视校正(通过空气移液管样本的常规自动分析测定每原子质量单位 1.00657±0.00151 到 1.00873±0.0010)和放射性衰变校正。质谱仪电子倍增系统的敏感性对氩同位素而言是每摩尔(5~10)×10^{13} nA。元素的核源干扰修正是在($^{39}Ar/^{39}Ar$)$_{Ca}$ = 0.00764±0.0000557、($^{36}Ar/^{37}Ar$)$_{Ca}$ = 0.0002705±0.0000105、($^{40}Ar/^{39}Ar$)$_{Ca}$ = 0.00014±0.00060 的基础上进行的。计算出来的年代是根据菲什峡谷透长石标准的重复分析推出的 J 值(AST-2.75 的 J 值是 0.0003935±0.0000003，BKT-2L 的 J 值是 0.0004737±0.0000003)得到的，其年代为 2,784 万年[24,25]。年代的计算依据是衰变常数 $\lambda = 5.543 \times 10^{-10}$ Y^{-1}。年代的不确定性反映了同位素丰度和所有校正值的不确定性，但是并不反映标准年代的不确定性。算术平均值和均值的标准误差是(258±11)万年，而方差倒数加权均值是(252±8)万年；因为个体分析存在异质精确性，所以我们认为后者更准确。年代用斜体表示的样本(AST-2.75)是污染物，在计算均值时将它们排除在外。

图 1a 为戈纳考古遗址图。在 EG10 (13 平方米) 和 EG12 (9 平方米) 两处进行的系统发掘工作获取了数量最多、信息最为丰富的人工制品组合(石制品 2,970 件，其中 1,114 件为原位埋藏)。这些石制品夹杂在 6 米厚的、致密的、富含黏土的古土壤层的 10 厘米厚的文化层中。另外，1994 年在 WG7 地表采集和发掘出土了 43 件人工制品，它们将考古发现的地理分布区域扩展到了整个昂达戈纳流域。细微颗粒的沉积环境、分布在有限的垂直地层、人工制品的新鲜程度以及人工制品尺寸缺乏变化暗示着一种比较稳定的沉积环境。

在 EG10 和 EG12 发掘出土和地表采集的人工制品主要包括简单的石核、完整的石片类工具和剥片废屑。单面和两面剥片的石核类工具组成“剥片类”工具。石核体上有大量不同剥片阶段的石片疤，说明上新世晚期的人科成员已经掌握了基本的剥片技术(图 3)。此外，大量带有明显打击泡的石片说明制造者清楚地了解石料贝壳状断裂力学。“剥落石片类”数量居多，比例为 75%~95%。通过击打或敲打来进行剥片或修形的石锤、石砧或砾石等“打击类”工具数量较少。

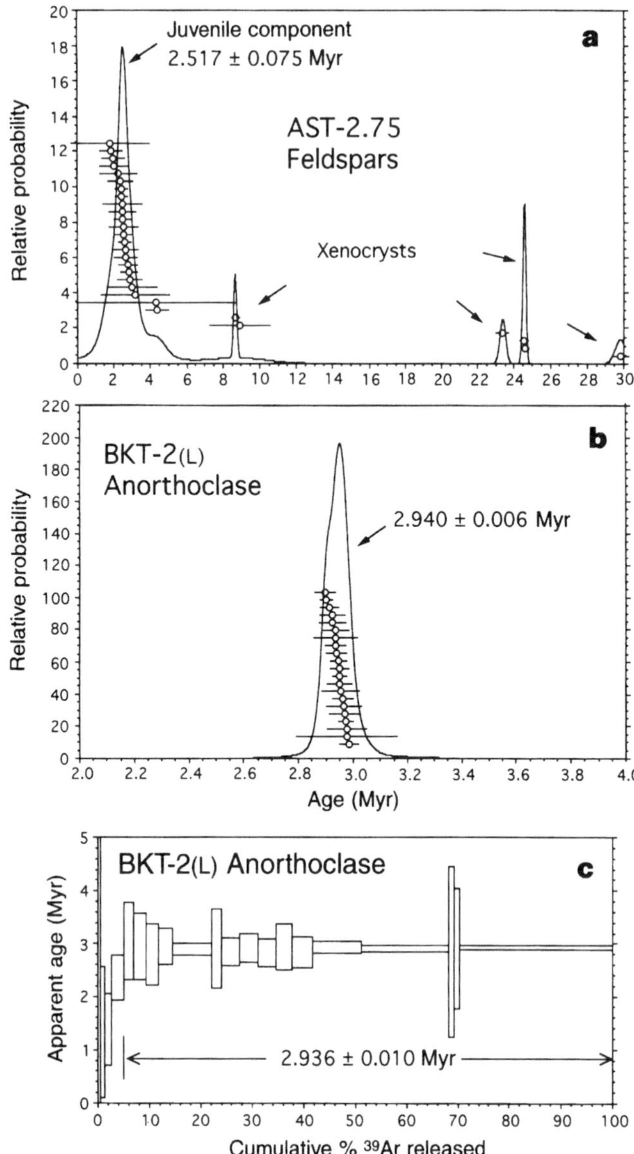

Fig. 2. Age-probability plots for **a,** AST-2.75, and **b,** BKT-2L for single-crystal laser fusion $^{40}Ar/^{39}Ar$ analyses. **c,** Apparent age spectrum for bulk argon-ion laser heating of BKT-2L.

Trachyte was the main raw material source, comprising > 70% of the artefacts. Other volcanics such as rhyolite and basalt were also used. The likely sources of raw materials were nearby stream conglomerates. Clasts from the level of the Intermediate Conglomerate show that trachyte cobbles were the most abundant, comprising about 50% of the total sample that also includes chalcedony, breccia and other lavas. The high proportion of trachyte artefacts could be taken to mean an appreciation of the flaking properties and selectivity for this particular raw material over others. There is a good correspondence between the sizes of the stone clasts and the sizes of the excavated stone artefacts. The

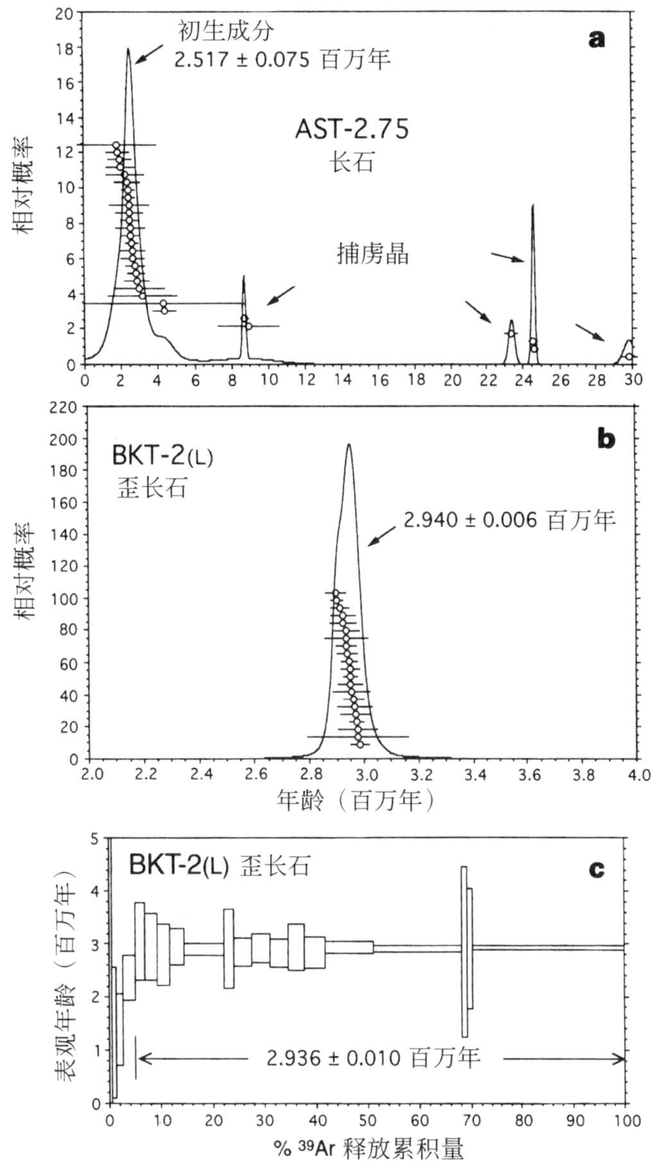

图 2. 年龄概率图：**a**，AST-2.75 和 **b**，BKT-2L 的激光单颗粒熔融 $^{40}Ar/^{39}Ar$ 分析。**c**，使用块状氩离子激光加热得到的 BKT-2L 表观年龄谱。

粗面岩是加工石制品的主要原料，占石制品的 70% 以上。其他的火山岩如流纹岩和玄武岩也有使用。这些原料可能出自附近溪流的砾岩层。从中砾岩层得到的岩块表明，粗面岩砾石最为丰富，约占标本总量的 50%。标本中也含玉髓、角砾岩和其他熔岩。粗面岩石制品的高比例可能意味着相较其他原料而言，古人类更倾向于选择这种熟悉的、便于剥片的原料。碎屑岩的大小和发掘出土的人工石器制品尺寸

close proximity of raw material sources may account for the less exhausted nature of the "flaked pieces" and the very low numbers of "unmodified pieces" (manuports) found at the archaeological sites.

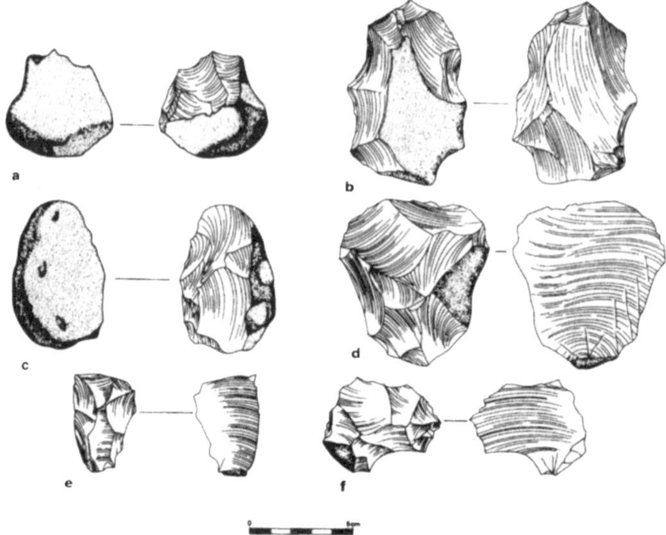

Fig. 3. Sketches of a sample of the excavated Gona artefacts. Flaked pieces: **a,** unifacial side chopper, EG12; **b,** discoid, EG10; **c,** unifacial side chopper, EG10. Detached pieces: **d–f,** whole flakes, EG10. Note that the maximum dimension of **d** is as large as some of the flaked pieces.

The artefact collections from Olduvai Gorge, Koobi Fora and the Omo[8-11] are the largest and best-studied samples of Oldowan artefacts. The oldest reliably dated among these are those from Lokalalei[12], which are constrained to be younger than the Kalochoro Tuff (= Tuff F of the Shungura Fm) at 2.36 ± 0.05 Myr. The Gona artefacts are thus at least 160 Kyr $\pm$ 90 Kyr older than these, the ages being distinguishable with greater than 92% confidence. All these stone assemblages are characterized by simple flaked cores made on cobbles and blocks, and associated flaking debris. The composition of the Gona assemblages is very similar to Plio-Pleistocene sites elsewhere, except for lower diversity of the cores from Gona and the high incidence of utilized pieces and manuports at Olduvai Gorge.

In addition to the lower diversity of the types seen in the Gona core forms, other contrasts exist between these and the Early Pleistocene assemblages. There are no retouched flakes such as light duty flake scrapers at Gona, which are present in small quantities in the Olduvai assemblages. In contrast to these minimal differences, we see strong parallels among all Plio-Pleistocene assemblages dated between 2.5–1.5 Myr. Assemblages from Olduvai and Koobi Fora show a greater degree of core reduction, but knowledge of fundamental conchoidal fracture mechanics suggests that the Gona artefacts are not significantly different. Indeed, the greater core reduction witnessed at Koobi Fora and Olduvai may be due to the considerable distances of these sites from the sources of raw materials compared to Gona.

之间有很好的一致性。原料来源近可以解释遗址中较少发现高损耗的“剥片类”制品和“未使用石料”的现象。

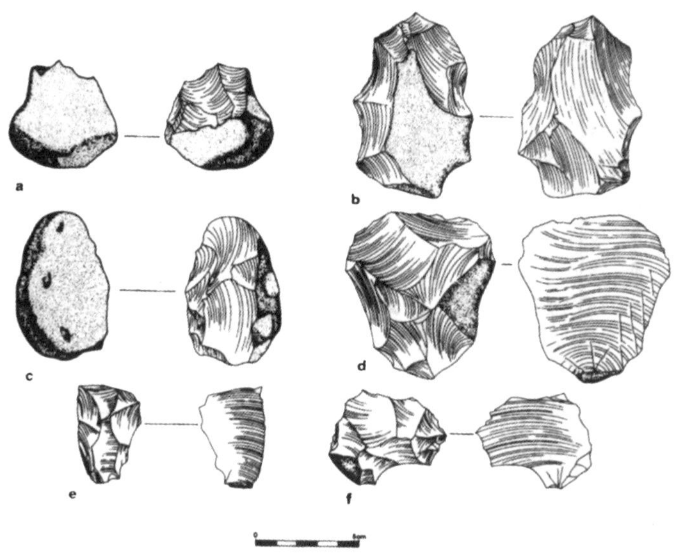

图 3. 戈纳发掘出土石制品线图。剥片类：**a**，单面砍砸器，EG12；**b**，饼状石核，EG10；**c**，单面砍砸器，EG10。剥落的石片类：**d~f**，完整石片，EG10。注意：**d** 的最大尺寸与一些剥片类的尺寸近似。

奥杜威工业数量最集中、研究最为充分的人工制品样本来自奥杜威峡谷、库比福勒和奥莫 [8-11]，其中最早的可靠年代数据来自罗卡拉雷 [12]，它们被认定比卡罗卓洛凝灰岩（= 上古拉地层的凝灰岩 F）更年轻，距今约 (236 ± 5) 万年。因此戈纳人工制品至少比这些要早 16 万年 ± 9 万年，其可辨年代的置信度大于 92%。所有这些石制品组合都是由毛坯为砾石或岩块的、简单剥片的石核和剥片废屑构成。戈纳遗址的石制品组合与其他地方上新世–更新世遗址的组合非常相似，只是戈纳遗址的石核变化少，而奥杜威峡谷遗址群中使用的片类工具和未使用石料比例较高。

除了戈纳石核形态多样性较低外，戈纳遗址的石制品与早更新世石制品组合之间还存在其他差异。戈纳无二次加工修理的工具如轻型刮削器，而这种工具在奥杜威石制品组合中少量存在。与这些细小的差异相比，我们发现所有距今 250 万年到 150 万年之间的上新世–更新世石制品组合都表现出高度的一致性。奥杜威和库比福勒发现的石制品组合显示出较高程度的石核剥片率，但是人类对贝壳状断裂力学最基本的了解表明，戈纳石制品并没有显著的不同之处。事实上，在库比福勒和奥杜威出现的石核高剥片率现象可能是由于与戈纳相比，这些地点离原料产地的距离更远。

The presence of large concentrations of stone artefacts at the early Gona sites shows that by 2.5 Myr some populations of Late Pliocene hominids had already mastered the basics of stone tool manufacture. The working edges of the majority of Gona artefacts are very fresh and sharp. Many of the cores show evidence of pitting and bruising. This suggests that in addition to being sources of sharp-edged flakes[13], the cores were used as multipurpose tools, for example as hammerstones and for other pounding activities.

The Gona evidence extends the known age of the Oldowan Industrial Complex to 2.5 Myr, and obviates the need to posit a "pre-Oldowan"[14] or an early facies of the Omo industrial complex[10,12]. Clearly, all the Oldowan assemblages group together because of the simplicity of craft practices in fashioning simple cores and the resultant flakes. These contrast with Acheulean assemblages that show elements of much more specific and preconceived designs[15]. The very early age of the Gona artefacts shows a technological stasis in the Oldowan industrial complex for over a period of 1 million years. The Acheulean appears abruptly at about 1.6–1.5 Myr with large bifacial tool forms such as handaxes and cleavers that were unknown in the Oldowan (*sensu stricto*)[16-18].

The hominids responsible for the manufacture of the Gona artefacts remain unidentified. There is, as yet, no evidence of stone artefacts in sediments older than 2.6 Myr within the Gona. Two contemporaneous hominid species *Homo* and *A. aethiopicus* are known elsewhere in eastern Africa from deposits that are comparable in age with the Gona[19,20]. However, no hominid specimens have yet been found associated with Late Pliocene stone tools.

The sophisticated understanding of conchoidal fracture evidence at Gona implies that the hominids that lived about 2.5 Myr ago were not novices to lithic technology. We predict that even older artefacts will be found. Future research in the Gona study area will be directed towards searching for archaeological evidence in the older deposits and towards assessing contrasting hypotheses linking global climatic changes, the origin of the genus *Homo*, and the beginning of stone-tool manufacture and use[21,22].

(**385**, 333-336; 1997)

S. Semaw[*], **P. Rennet**[†], **J. W. K. Harris**[*], **C. S. Feibel**[*], **R. L. Bernor**[‡], **N. Fesseha**[‡] **& K. Mowbray**[*]

[*] Department of Anthropology, Douglass Campus, Rutgers University, New Brunswick, New Jersey 08903, USA

[†] Berkeley Geochronology Center, 2455 Ridge Road, Berkeley, California 94709, USA

[‡] Laboratory of Paleobiology, Department of Anatomy, College of Medicine, Howard University, Washington DC 20059, USA

Received 18 June; accepted 25 November 1996.

References:

1. Johanson, D. C. *et al. Am. J. Phys. Anth.* **57**, 373-402 (1982).

2. Corvinus, G. & Roche, H. *L' Anthropologie* **80**, 315-324 (1976).

3. Harris, J. W. K. *Afr. Arch. Rev.* **1**, 3-31 (1983).

4. Harris, J. W. K. & Semaw, S. *Nyame Akuma* **31**, 19-21 (1989).

5. Roche, H. & Tiercelin, J. J. *C. r. hebd. Séanc. Acad. Sci. Paris* **284D**, 1871-1874 (1977).

在时代较早的戈纳遗址中，石制品的大量集中出现表明，在距今 250 万年前，一些上新世晚期的人科成员已经掌握了基本的石器制造技术。大部分戈纳遗址石制品的工作刃缘都非常新鲜且锐利。不少石核体上显示出挖掘和摩擦的痕迹，表明这些石核除了用于剥落刃缘锋利的石片之外 [13]，还被用作多用途的工具，如作为石锤、用于其他敲砸活动。

戈纳遗址将已知的奥杜威工业范畴的年代拓展到了距今 250 万年，从而排除了有“前奥杜威”[14] 或早期奥莫工业 [10,12] 存在的可能。很显然，因为石核－石片加工工艺十分简单，所以所有奥杜威工业的石制品组合都很类似，这与阿舍利石制品组合所呈现出的具有更加专业化趋向及预先设计的元素形成了鲜明的对比 [15]。戈纳石制品年代之早表明奥杜威石器工业技术曾历经 100 多万年的技术停滞期。阿舍利工业在距今约 160 万年到 150 万年间突然出现了大型的两面工具，如手斧和薄刃斧，这些在狭义的奥杜威工业中都是前所未见的 [16-18]。

现在尚无法确定戈纳遗址石制品的制造者究竟是哪一人科种群，也没有证据表明戈纳地区沉积物中存在早于距今 260 万年前的石制品。目前已知的、在东非与戈纳遗址时代相同的堆积物中发现的人科成员有两种——人属和南方古猿埃塞俄比亚种 [19,20]，但尚未发现与上新世晚期石器相伴随的人科化石标本。

在戈纳发现的人类熟练掌握石料贝壳状断裂力学的证据表明，生活在距今约 250 万年前的原始人类并不是制作石制工具的新手，我们预测未来将有更古老的石制品被发现。戈纳地区后续的研究工作将瞄准在更古老的地层堆积物中搜寻考古学证据以及甄别莫衷一是的全球气候变化、人属起源以及石制工具的出现与使用等方面关系的假说 [21,22] 上。

（刘皓芳 翻译；王社江 审稿）

6. Kimbel, W. H. *et al. Nature* **368**, 449-451 (1994).

7. McDougall, I. *et al. Geophys. Res. Lett.* **19**, 2349-2352 (1992).

8. Leakey, M. D. *Olduvai Gorge* Vol. 3 (Cambridge University Press, UK, 1971).

9. Isaac, G. L. in *Earliest Man and Environments in the Lake Rudolf Basin* (eds Coppens, Y., Howell, F. C., Isaac, G. L. & Leakey, R. E. F.) 552-564 (University of Chicago Press, Illinois, 1976).

10. Chavaillon, J. in *Earliest Man and Environments in the Lake Rudolf Basin* (eds Coppens, Y., Howell, F. C., Isaac, G. L. & Leakey, R. E. F.) 565-573 (University of Chicago Press, Illinois, 1976).

11. Merrick, H. V. & Merrick, J. P. S. in *Earliest Man and Environments in the Lake Rudolf Basin* (eds Coppens, Y., Howell, F. C., Isaac, G. L. & Leakey, R. E. F.) 574-589 (University of Chicago Press, Illinois, 1976).

12. Kibunjia, M. *J. Hum. Evol.* **27**, 159-171 (1994).

13. Toth, N. *J. Archaeol. Sci.* **12**, 101-120 (1985).

14. Piperno, M. in *Hominidae: Proc. 2nd Intern. Congr. Hum. Paleont.* 189-195 (Jaca Books, Milan, Italy, 1989).

15. Gowlett, J. A. J. in *Stone Age Prehistory* (eds Bailey, G. N. & Callow, P.) 243-260 (Cambridge University Press, UK, 1986).

16. Isaac, G. L. & Curtis, G. H. *Nature* **249**, 624-627 (1974).

17. Asfaw, B. *et al. Nature* **360**, 732-735 (1992).

18. Dominguez-Rodrigo, M. *Complutum* **7**, 7-15 (1996).

19. Walker, A. *et al. Nature* **322**, 517-522 (1986).

20. Hill, A. *et al. Nature* **355**, 719-722 (1992).

21. Vrba, E. S. in *Evolutionary History of the Robust Australopithecines* (ed. Grine, F.) 405-426 (de Gruyter, New York, 1988).

22. deMenocal, P. B. *Science* **270**, 53-59 (1995).

23. Renne, P. R. *et al. Geophys. Res. Lett.* **20**, 1067-1070 (1993).

24. Deino, A. & Potts, R. *J. Geophys. Res.* **95**, 8453-8470 (1990).

25. WoldeGabriel, G. *et al. Nature* **371**, 330-333 (1994).

Acknowledgements. Permission for this research was granted by the CRCCH and the National Museum of Ethiopia in the Ministry of Culture and Information. On behalf of members of the Gona research project, the two principal investigators (S.S. and J.W.K.H.) thank the Government of Ethiopia and the Afar People. We also thank J. D. Clark, T. White, and F. C. Howell and S. Asrat for their help; B. Asfaw, Y. Beyene and G. WoldeGabriel for encouragement, advice and laboratory assistance at the Museum in Addis; the NTO of Ethiopia and drivers; Y. Haile Selassie, A, Ademassu, J. Haile-Mariam and J. Wynn for overall support; M. Kahasai, A. Asfaw, A. Humet, A. Habib and the Afars at Eloha, Busidima and Talalak for support in the field; Y. Kanaa, M. Siegel and P. Jung for illustrations; T. Assebework and S. Eshete for fieldwork; J. Butterworth for palaeomagnetic analyses; and T. White, R. Blumenschine, D. Lieberman, M. Domínguez-Rodrigo and M. Rogers for comments on the manuscript. We thank the NSF, the L. S. B. Leakey Foundation, the Boise Fund, Ann and Gordon Getty Foundation, D. Holt, Z. Zelazo, and Rutgers University for financial support.

Correspondence and requests for materials should be addressed to S.S. (e-mail: semaw@eden.rutgers. edu).

Viable Offspring Derived from Fetal and Adult Mammalian Cells

I. Wilmut *et al.*

Editor's Note

Soon after Morag and Megan, the first mammals cloned from an established cell line, cloning pioneer Ian Wilmut and colleagues report the arrival of Dolly the sheep, the first mammal cloned from an adult cell. Dolly was proof that nuclear transfer could be used to reprogram adult DNA, raising the prospect of tailor-making patient-matched stem cells for use in regenerative medicine and of cloning other animals. Many species including dogs, horses and mice have been cloned subsequently, and nuclear transfer-derived human stem cells are being used to study development and disease. Dolly prompted researchers to devise new egg-free methods for making stem cells without destroying embryos, and it is believed these will greatly aid basic and applied biological research.

Fertilization of mammalian eggs is followed by successive cell divisions and progressive differentiation, first into the early embryo and subsequently into all of the cell types that make up the adult animal. Transfer of a single nucleus at a specific stage of development, to an enucleated unfertilized egg, provided an opportunity to investigate whether cellular differentiation to that stage involved irreversible genetic modification. The first offspring to develop from a differentiated cell were born after nuclear transfer from an embryo-derived cell line that had been induced to become quiescent[1]. Using the same procedure, we now report the birth of live lambs from three new cell populations established from adult mammary gland, fetus and embryo. The fact that a lamb was derived from an adult cell confirms that differentiation of that cell did not involve the irreversible modification of genetic material required for development to term. The birth of lambs from differentiated fetal and adult cells also reinforces previous speculation[1,2] that by inducing donor cells to become quiescent it will be possible to obtain normal development from a wide variety of differentiated cells.

IT has long been known that in amphibians, nuclei transferred from adult keratinocytes established in culture support development to the juvenile, tadpole stage[3]. Although this involves differentiation into complex tissues and organs, no development to the adult stage was reported, leaving open the question of whether a differentiated adult nucleus can be fully reprogrammed. Previously we reported the birth of live lambs after nuclear transfer from cultured embryonic cells that had been induced into quiescence. We suggested that inducing the donor cell to exit the growth phase causes changes in chromatin structure

胎儿和成年哺乳动物的细胞可以产生存活后代

威尔穆特等

编者按

由已建立的细胞系克隆得到的首批哺乳动物莫拉格和梅甘出生后不久，克隆先驱伊恩·威尔穆特和他的同事们报道了第一只由成年体细胞克隆得到的哺乳动物——多莉羊的诞生。多莉羊表明通过核移植可实现成体DNA的重编程，这为定制与患者相匹配的用于再生医学的干细胞以及克隆其他动物带来了希望。随后，包括狗、马和小鼠在内的其他物种也实现了克隆，源于核移植的人类干细胞被用来研究发育和疾病问题。多莉羊促使研究者设计新的无卵方法在不破坏胚胎的情况下获得干细胞，也会为基础生物研究和应用生物研究提供巨大帮助。

哺乳动物的卵细胞受精后出现连续的细胞分裂和渐进分化，最开始出现早期的胚胎，随后分化为组成成年动物所需的各类细胞。将发育特定阶段的单个细胞核移植到摘除细胞核的未受精的卵细胞内，为研究细胞分化到这个阶段是否发生了不可逆的遗传修饰提供了契机。由分化细胞发育而来的第一批后代是在转入已诱导为静止的胚胎来源细胞系的核后产生的[1]。采用相同的方法，我们从成年母羊乳腺、胎儿和胚胎建立的三个新细胞群中获得了活羊羔。从成体细胞发育出羊羔这个事实证实了这一细胞的分化并不包括完成体内发育所需遗传物质的不可逆修饰。这些来源于分化的胎儿细胞和成体细胞的小羊的出生也加强了先前的推测[1,2]，即通过诱导供体细胞进入静止期就可能实现不同分化阶段细胞的正常发育。

很久以前人们就知道，在两栖动物中，用培养的成体角质形成细胞的细胞核进行核移植，可以支持个体发育到幼年蝌蚪阶段[3]。尽管这一过程涉及复杂的组织和器官的分化，但是尚无报道称其能发育至成年阶段。因此，分化的成体细胞核能否完全重编程是一个有待解决的问题。先前我们报道了培养的胚胎细胞经诱导为静止状态后，经过核移植能够发育成活的小羊。我们认为诱导供体细胞退出其生长周期能引起染色质结构的改变，从而促进了基因表达的重编程；而且如果以类似的方式

that facilitate reprogramming of gene expression and that development would be normal if nuclei are used from a variety of differentiated donor cells in similar regimes. Here we investigate whether normal development to term is possible when donor cells derived from fetal or adult tissue are induced to exit the growth cycle and enter the G0 phase of the cell cycle before nuclear transfer.

Three new populations of cells were derived from (1) a day-9 embryo, (2) a day-26 fetus and (3) mammary gland of a 6-year-old ewe in the last trimester of pregnancy. Morphology of the embryo-derived cells (Fig. 1) is unlike both mouse embryonic stem (ES) cells and the embryo-derived cells used in our previous study. Nuclear transfer was carried out according to one of our established protocols[1] and reconstructed embryos transferred into recipient ewes. Ultrasound scanning detected 21 single fetuses on day 50–60 after oestrus (Table 1). On subsequent scanning at ~14-day intervals, fewer fetuses were observed, suggesting either mis-diagnosis or fetal loss. In total, 62% of fetuses were lost, a significantly greater proportion than the estimate of 6% after natural mating[4]. Increased prenatal loss has been reported after embryo manipulation or culture of unreconstructed embryos[5]. At about day 110 of pregnancy, four fetuses were dead, all from embryo-derived cells, and post-mortem analysis was possible after killing the ewes. Two fetuses had abnormal liver development, but no other abnormalities were detected and there was no evidence of infection.

Table1. Development of embryos reconstructed with three different cell types

Cell type	No. of fused couplets (%)*	No. recovered from oviduct (%)	No. cultured	No. of morula/ blastocyst (%)	No. of morula or blastocysts transferred†	No. of pregnancies/ no. of recipients (%)	No. of live lambs (%)‡
Mammary epithelium	277 (63.8)[a]	247 (89.2)	–	29 (11.7)[a]	29	1/13 (7.7)	1 (3.4%)
Fetal fibroblast	172 (84.7)[b]	124 (86.7)	–	34 (27.4)[b]	34	4/10 (40.0)	2 (5.9%)
			24	13 (54.2)[b]	6	1/6 (16.6)	1 (16.6%) §
Embryo-derived	385 (82.8)[b]	231 (85.3)	–	90 (39.0)[b]	72	14/27 (51.8)	4 (5.6%)
			92	36 (39.0)[b]	15	1 /5 (20.0)	0

* As assessed 1 h after fusion by examination on a dissecting microscope. Superscripts a or b within a column indicate a significant difference between donor cell types in the efficiency of fusion ($P < 0.001$) or the proportion of embryos that developed to morula or blastocyst ($P < 0.001$).

† It was not practicable to transfer all morulae/blastocysts.

‡ As a proportion of morulae or blastocysts transferred. Not all recipients were perfectly synchronized.

§ This lamb died within a few minutes of birth.

对不同分化程度的供体细胞进行处理后，其发育也会是正常的。这里我们研究了当来源于胎儿组织或者成体组织的供体细胞被诱导脱离生长周期并在核移植之前进入细胞周期的 G0 期后，是否还有可能完成正常的发育。

这三个新的细胞群分别来源于：(1) 第 9 天的胚胎；(2) 第 26 天的胎儿；(3)6 岁母羊孕晚期的乳腺。胚胎来源的细胞形态 (图 1) 不同于我们之前实验中使用过的小鼠胚胎干细胞 (ES) 和胚胎来源的细胞。根据我们建立的方案 [1] 进行核移植，然后将重组的胚胎移植到受体母羊中。发情期后第 50~60 天超声检测到了 21 个单独的胚胎 (表 1)。随后在大约以 14 天为间隔的扫描检测中发现，看到的胚胎数越来越少，暗示出现了错误诊断或者胎儿死亡。胎儿损失率共计 62%，显著高于自然交配的估计值 6%[4]。有研究报道，胚胎处理或者非重组胚胎的培养会增加胎儿损失率 [5]。大约在孕期的第 110 天，四个胎儿死亡，它们都来源于胚胎细胞核衍生的细胞，处死母羊后即可进行验尸分析。其中两个胎儿的肝脏发育异常，但是没有发现其他异常或者感染的证据。

表 1. 三种不同类型细胞重组胚胎的发育

细胞类型	融合胚胎的数量 (%)*	输卵管中收回的胚胎数量 (%)	培养的胚胎数量	桑椹胚 / 胚泡数量 (%)	移植的桑椹胚或胚泡数量 †	怀孕母羊数目 / 受体母羊数目 (%)	存活小羊数目 (%)‡
乳腺上皮细胞	277 (63.8)[a]	247 (89.2)	–	29 (11.7)[a]	29	1/13 (7.7)	1 (3.4%)
胎儿成纤维细胞	172 (84.7)[b]	124 (86.7)	–	34 (27.4)[b]	34	4/10 (40.0)	2 (5.9%)
			24	13 (54.2)[b]	6	1/6 (16.6)	1 (16.6%)§
胚胎来源细胞	385 (82.8)[b]	231 (85.3)	–	90 (39.0)[b]	72	14/27 (51.8)	4 (5.6%)
			92	36 (39.0)[b]	15	1/5 (20.0)	0

* 在融合 1 小时后用解剖显微镜进行检查。上标 a 或者 b 表示供体细胞融合的效率显著差异 ($P < 0.001$) 或者胚胎发育成桑椹胚或者胚泡的比例显著差异 ($P < 0.001$)。

† 要移植所有的桑椹胚或胚泡是不可行的。

‡ 占移植的桑椹胚或胚泡的比例。并不是所有的受体都能完全同步。

§ 该小羊出生后几分钟内死亡。

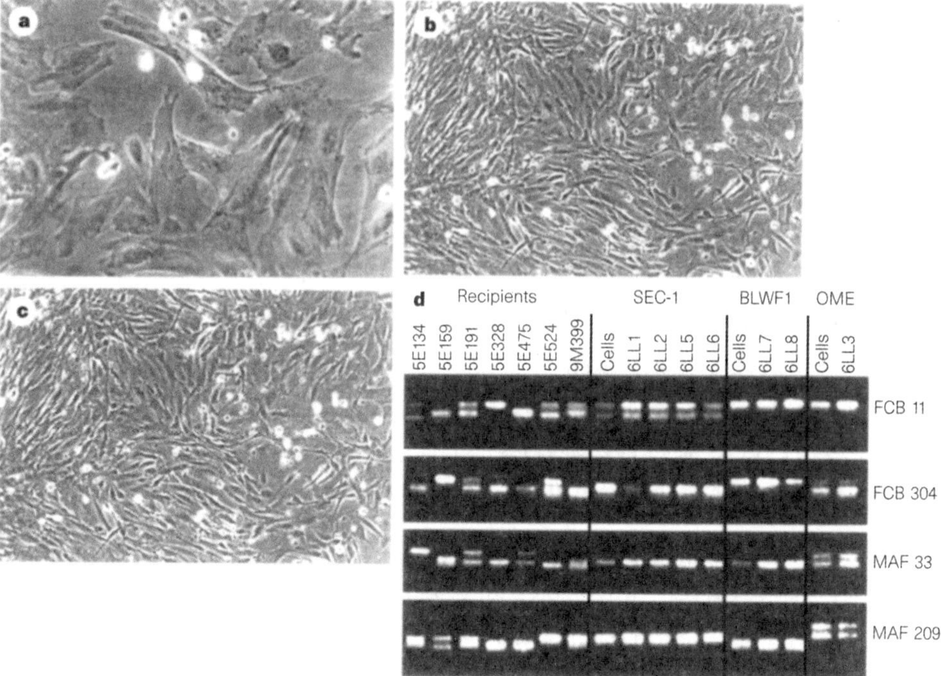

Fig. 1. Phase-contrast photomicrograph of donor-cell populations: **a**, Embryo-derived cells (SEC1); **b**, fetal fibroblasts (BLWF1); **c**, mammary-derived cells (OME). **d**, Microsatellite analysis of recipient ewes, nuclear donor cells and lambs using four polymorphic ovine markers[22]. The ewes are arranged from left to right in the same order as the lambs. Cell populations are embryo-derived (SEC1), fetal-derived (BLW1), and mammary-derived (OME), respectively. Lambs have the same genotype as the donor cells and differ from their recipient mothers.

Eight ewes gave birth to live lambs (Table 1, Fig. 2). All three cell populations were represented. One weak lamb, derived from the fetal fibroblasts, weighed 3.1 kg and died within a few minutes of birth, although post-mortem analysis failed to find any abnormality or infection. At 12.5%, perinatal loss was not dissimilar to that occurring in a large study of commercial sheep, when 8% of lambs died within 24 h of birth[6]. In all cases the lambs displayed the morphological characteristics of the breed used to derive the nucleus donors and not that of the oocyte donor (Table 2). This alone indicates that the lambs could not have been born after inadvertent mating of either the oocyte donor or recipient ewes. In addition, DNA microsatellite analysis of the cell populations and the lambs at four polymorphic loci confirmed that each lamb was derived from the cell population used as nuclear donor (Fig. 1). Duration of gestation is determined by fetal genotype[7], and in all cases gestation was longer than the breed mean (Table 2). By contrast, birth weight is influenced by both maternal and fetal genotype[8]. The birth weight of all lambs was within the range for single lambs born to Blackface ewes on our farm (up to 6.6 kg) and in most cases was within the range for the breed of the nuclear donor. There are no strict control observations for birth weight after embryo transfer between breeds, but the range in weight of lambs born to their own breed on our farm is 1.2–5.0 kg, 2–4.9 kg and 3–9 kg for the Finn Dorset, Welsh Mountain and Poll Dorset genotypes, respectively. The attainment of sexual maturity in the lambs is being monitored.

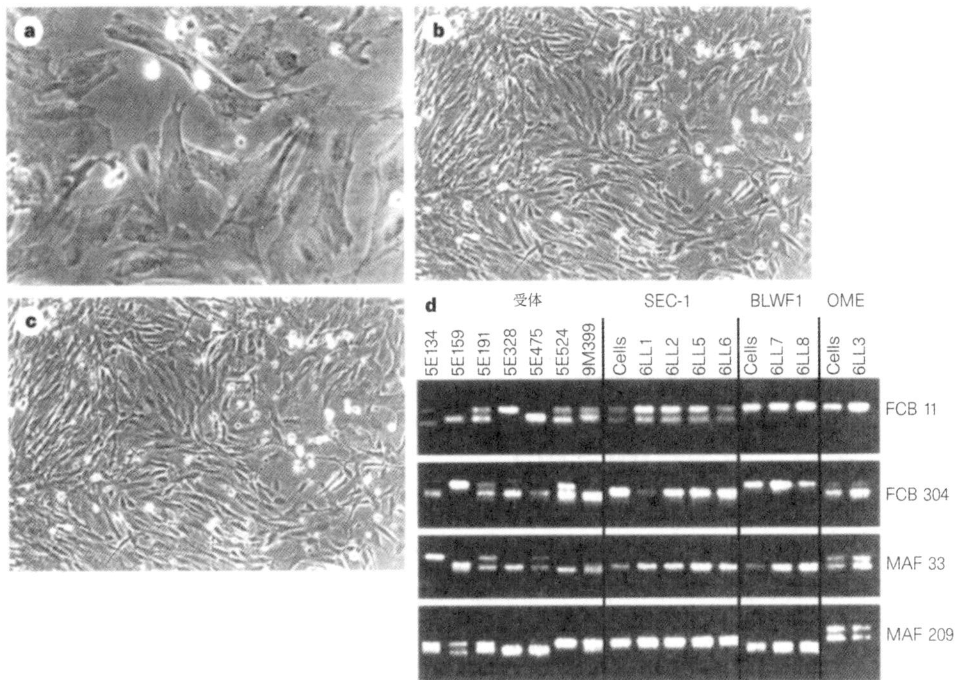

图 1. 供体细胞群的相差显微照片：**a**，胚胎来源的细胞（SEC1）；**b**，胎儿成纤维细胞（BLWF1）；**c**，乳腺来源的细胞（OME）。**d**，使用四种多态性绵羊标记物对受体母羊、细胞核供体细胞和小羊进行微卫星分析[22]。母羊从左到右排列的顺序和羊羔一致。细胞群分别是胚胎来源的细胞（SEC1）、胎儿来源的细胞（BLW1）和乳腺来源的细胞（OME）。小羊和供体细胞具有相同的基因型，而与受体母羊不同。

八只母羊生出了活的小羊（表 1，图 2），包含了所有三个细胞群来源。其中一只较弱的小羊来源于胎儿成纤维细胞，重 3.1 kg，并在出生后几分钟内死亡，然而验尸分析没有发现任何异常或者感染。12.5% 的围产期损失率与商品羊大型研究中大约 8% 的小羊在出生后 24 小时内死亡[6]的损失率相似。所有的小羊都表现出细胞核供体的形态特征，而不是卵细胞供体的特征（表 2）。仅仅这点就表明这些小羊的出生并非是卵细胞供体或者受体母羊偶然交配的结果。此外，利用 DNA 微卫星对细胞群和小羊的四个多态性基因座进行分析也证实了每只小羊均来自核供体细胞群（图 1）。妊娠期的长短由胎儿的基因型决定[7]，而且在所有例子中妊娠期都比品种平均妊娠期长（表 2）。相比之下，出生体重则受到母体和胎儿基因型的影响[8]。所有小羊的出生体重均在我们农场的黑面母羊生出的单胎羊羔的体重范围内（最重 6.6 kg），而且在大多数情况下都在核供体品种的体重范围内。对于不同品种间胚胎移植后的出生体重没有进行严格的对照观察，但是在我们农场中出生的芬兰多塞特、威尔士山和无角多塞特三个品种小羊的体重范围分别是 1.2~5.0 kg、2~4.9 kg 和 3~9 kg。小羊的性成熟程度也在监测之中。

Table 2. Delivery of lambs developing from embryos derived by nuclear transfer from three different donor cells types, showing gestation length and birth weight

Cell type	Breed of lamb	Lamb identity	Duration of pregnancy (days)*	Birth weight (kg)
Mammary epithelium	Finn Dorset	6LL3	148	6.6
Fetal fibroblast	Black Welsh	6LL7	152	5.6
	Black Welsh	6LL8	149	2.8
	Black Welsh	6LL9†	156	3.1
Embryo- derived	Poll Dorset	6LL1	149	6.5
	Poll Dorset	6LL2‡	152	6.2
	Poll Dorset	6LL5	148	4.2
	Poll Dorset	6LL6‡	152	5.3

* Breed averages are 143, 147 and 145 days, respectively for the three genotypes Finn Dorset, Black Welsh Mountain and Poll Dorset.

† This lamb died within a few minutes of birth.

‡ These lambs were delivered by caesarian section. Overall the nature of the assistance provided by the veterinary surgeon was similar to that expected in a commercial flock.

Fig. 2. Lamb number 6LL3 derived from the mammary gland of a Finn Dorset ewe with the Scottish Blackface ewe which was the recipient.

Development of embryos produced by nuclear transfer depends upon the maintenance of normal ploidy and creating the conditions for developmental regulation of gene expression. These responses are both influenced by the cell-cycle stage of donor and recipient cells and the interaction between them (reviewed in ref. 9). A comparison of development of mouse and cattle embryos produced by nuclear transfer to oocytes[10,11] or enucleated zygotes[12,13] suggests that a greater proportion develop if the recipient is an oocyte. This may be because factors that bring about reprogramming of gene expression in a transferred nucleus are required for early development and are taken up by the pronuclei during development of the zygote.

表 2. 由三种不同类型供体细胞核移植后得到的胚胎发育而成的小羊的妊娠期时间和出生体重

细胞类型	小羊品种	小羊代号	妊娠期长度(天)*	出生体重(kg)
乳腺上皮细胞	芬兰多塞特	6LL3	148	6.6
胎儿成纤维细胞	黑面威尔士	6LL7	152	5.6
	黑面威尔士	6LL8	149	2.8
	黑面威尔士	6LL9†	156	3.1
胚胎来源细胞	无角多塞特	6LL1	149	6.5
	无角多塞特	6LL2‡	152	6.2
	无角多塞特	6LL5	148	4.2
	无角多塞特	6LL6‡	152	5.3

* 三种基因型的羊——芬兰多塞特羊、黑面威尔士山羊和无角多塞特羊的品种平均妊娠期分别是 143 天、147 天和 145 天。

† 该小羊出生后几分钟内死亡。

‡ 这些羊出生都采用剖腹取胎术。兽医提供的帮助与商品羊中的相似。

图 2. 6LL3 号小羊，来源于芬兰多塞特母羊的乳腺细胞，受体母羊是苏格兰黑面羊。

通过核移植产生的胚胎的发育依赖于正常染色体倍性的保持以及基因表达的发育调控环境的创造。这些反应都受到供体细胞和受体细胞的细胞周期以及它们之间相互作用的影响(参考文献 9 中的综述)。将核移植到卵母细胞[10,11]或者无核的受精卵[12,13]中得到的小鼠和牛胚胎的发育情况的对比结果表明，如果受体是卵母细胞，那么大部分胚胎发育。这可能是因为在早期发育时就需要那些能引起移植的细胞核中基因表达重编程的因子，但是在受精卵的发育中这些因子已经被原核消耗了。

If the recipient cytoplasm is prepared by enucleation of an oocyte at metaphase II, it is only possible to avoid chromosomal damage and maintain normal ploidy by transfer of diploid nuclei[14,15], but further experiments are required to define the optimum cell-cycle stage. Our studies with cultured cells suggest that there is an advantage if cells are quiescent (ref. 1, and this work). In earlier studies, donor cells were embryonic blastomeres that had not been induced into quiescence. Comparisons of the phases of the growth cycle showed that development was greater if donor cells were in mitosis[16] or in the G1 (ref. 10) phase of the cycle, rather than in S or G2 phases. Increased development using donor cells in G0, G1 or mitosis may reflect greater access for reprogramming factors present in the oocyte cytoplasm, but a direct comparison of these phases in the same cell population is required for a clearer understanding of the underlying mechanisms.

Together these results indicate that nuclei from a wide range of cell types should prove to be totipotent after enhancing opportunities for reprogramming by using appropriate combinations of these cell-cycle stages. In turn, the dissemination of the genetic improvement obtained within elite selection herds will be enhanced by limited replication of animals with proven performance by nuclear transfer from cells derived from adult animals. In addition, gene targeting in livestock should now be feasible by nuclear transfer from modified cell populations and will offer new opportunities in biotechnology. The techniques described also offer an opportunity to study the possible persistence and impact of epigenetic changes, such as imprinting and telomere shortening, which are known to occur in somatic cells during development and senescence, respectively.

The lamb born after nuclear transfer from a mammary gland cell is, to our knowledge, the first mammal to develop from a cell derived from an adult tissue. The phenotype of the donor cell is unknown. The primary culture contains mainly mammary epithelial (over 90%) as well as other differentiated cell types, including myoepithelial cells and fibroblasts. We cannot exclude the possibility that there is a small proportion of relatively undifferentiated stem cells able to support regeneration of the mammary gland during pregnancy. Birth of the lamb shows that during the development of that mammary cell there was no irreversible modification of genetic information required for development to term. This is consistent with the generally accepted view that mammalian differentiation is almost all achieved by systematic, sequential changes in gene expression brought about by interactions between the nucleus and the changing cytoplasmic environment[17].

Methods. Embryo-derived cells were obtained from embryonic disc of a day-9 embryo from a Poll Dorset ewe cultured as described[1], with the following modifications. Stem-cell medium was supplemented with bovine DIA/LIF. After 8 days, the explanted disc was disaggregated by enzymatic digestion and cells replated onto fresh feeders. After a further 7 days, a single colony of large flattened cells was isolated and grown further in the absence of feeder cells. At passage 8, the modal chromosome number was 54. These cells were used as nuclear donors at passages 7–9. Fetal-derived cells were obtained from an eviscerated Black Welsh Mountain fetus recovered at autopsy on day 26 of pregnancy. The head was removed before tissues were cut into small pieces and the cells dispersed by exposure to trypsin. Culture was in BHK 21 (Glasgow MEM; Gibco Life Sciences) supplemented with

如果受体细胞质是通过将处于减数第二次分裂中期的卵母细胞去核制备的，这只可能通过移植二倍体细胞核来避免染色体的损伤并保持正常的倍性[14,15]，但是需要进一步的研究来确定最佳细胞周期。我们用培养的细胞进行的研究表明，如果细胞是静止的会有一些优势（参考文献 1 和本工作）。在早期的研究中，供体细胞是没有被诱导静止的胚胎卵裂球。各个细胞周期细胞的对比结果表明，供体细胞处于有丝分裂期[16]或者 G1 期（参考文献 10）比处于 S 期或者 G2 期的胚胎发育更好。使用 G0、G1 或有丝分裂期的供体细胞可以促进发育，这表明在此时卵母细胞的细胞质内可能存在更易获取的重编程因子。但是我们需要在同一个细胞群中直接比较不同时期的细胞，以便更加清楚地了解潜在的机制。

这些结果表明，不同细胞类型来源的细胞核通过合适的细胞周期组合提高重编程的机会之后应该具有全能性。反过来，用成年动物的细胞核进行移植可以确定基因优良的动物，在挑选好的良种动物群中通过有限的繁殖就可以达到基因改良的目的。此外，通过修饰后细胞群体的核移植可实现家畜的基因打靶，这将为生物技术提供新的机会。本文描述的这种技术也为研究表观遗传改变的持久性和影响提供了机会，比如基因印迹和端粒缩短，这两者被认为分别在体细胞的发育和衰老中发生。

据我们所知，这只由乳腺细胞核移植培育出的小羊是从成体组织中培养出的首个哺乳动物。供体细胞的表型不清楚。原代培养物中主要含有乳腺上皮细胞（超过 90%）以及其他分化的细胞类型，包括肌上皮细胞和成纤维细胞。我们不能排除存在小比例相对未分化的干细胞的可能，它们能在妊娠期支持乳腺的再生。小羊的出生表明在乳腺细胞的发育过程中没有完成体内发育所需遗传信息的不可逆修饰。这和普遍接受的观点一致，即哺乳动物的分化主要是通过细胞核和变化的细胞质环境相互作用引起的基因表达过程中系统且有序的改变实现的[17]。

方法。胚胎来源的细胞是按参考文献所述的方法[1]，从培养的无角多塞特母羊第 9 天胚胎的胎盘中获得的，并经过了以下处理。在干细胞基质中加入小牛 DIA/LIF。8 天后，用酶消化裂解移植的胎盘组织，将细胞置于新的饲养细胞上。再过 7 天，分离出一个大的扁平的细胞集落，然后在没有饲养细胞的情况下进一步培养。第 8 代的模式染色体数目是 54。第 7 代至第 9 代的细胞用作核供体。胎儿来源的细胞取自一只去除内脏的妊娠期 26 天的黑面威尔士山羊胎儿尸体。去除该尸体的头部，然后将组织切成碎片并用胰蛋白酶消化使细胞分散。培养基是加有 L– 谷氨酰胺（2 mM）、丙酮酸钠（1 mM）和 10% 胎牛血清的 BHK 21（GMEM

L-glutamine (2 mM), sodium pyruvate (1 mM) and 10% fetal calf serum. At 90% confluency, the cells were passaged with a 1:2 division. At passage 4, these fibroblast-like cells (Fig. 1) had modal chromosome number of 54. Fetal cells were used as nuclear donors at passages 4–6. Cells from mammary gland were obtained from a 6-year-old Finn Dorset ewe in the last trimester of pregnancy[18]. At passages 3 and 6, the modal chromosome number was 54 and these cells were used as nuclear donors at passage numbers 3–6.

Nuclear transfer was done according to a previous protocol[1]. Oocytes were recovered from Scottish Blackface ewes between 28 and 33 h after injection of gonadotropin-releasing hormone (GnRH), and enucleated as soon as possible. They were recovered in calcium- and magnesium-free PBS containing 1% FCS and transferred to calcium-free M2 medium[19] containing 10% FCS at 37 °C. Quiescent, diploid donor cells were produced by reducing the concentration of serum in the medium from 10 to 0.5% for 5 days, causing the cells to exit the growth cycle and arrest in G0. Confirmation that cells had left the cycle was obtained by staining with antiPCNA/cyclin antibody (Immuno Concepts), revealed by a second antibody conjugated with rhodamine (Dakopatts).

Fusion of the donor cell to the enucleated oocyte and activation of the oocyte were induced by the same electrical pulses, between 34 and 36 h after GnRH injection to donor ewes. The majority of reconstructed embryos were cultured in ligated oviducts of sheep as before, but some embryos produced by transfer from embryo-derived cells or fetal fibroblasts were cultured in a chemically defined medium[20]. Most embryos that developed to morula or blastocyst after 6 days of culture were transferred to recipients and allowed to develop to term (Table 1). One, two or three embryos were transferred to each ewe depending upon the availability of embryos. The effect of cell type upon fusion and development to morula or blastocyst was analysed using the marginal model of Breslow and Clayton[21]. No comparison was possible of development to term as it was not practicable to transfer all embryos developing to a suitable stage for transfer. When too many embryos were available, those having better morphology were selected.

Ultrasound scan was used for pregnancy diagnosis at around day 60 after oestrus and to monitor fetal development thereafter at 2-week intervals. Pregnant recipient ewes were monitored for nutritional status, body condition and signs of EAE, Q fever, border disease, louping ill and toxoplasmosis. As lambing approached, they were under constant observation and a veterinary surgeon called at the onset of parturition. Microsatellite analysis was carried out on DNA from the lambs and recipient ewes using four polymorphic ovine markers[22].

(**385**, 810-813; 1997)

I. Wilmut, A. E. Schnieke*, J. McWhir, A. J. Kind* & K. H. S. Campbell
Roslin Institute (Edinburgh), Roslin, Midlothian EH25 9PS, UK
* PPL Therapeutics, Roslin, Midlothian EH25 9PP, UK

Received 25 November 1996; accepted 10 January 1997.

References:

1. Campbell, K. H. S., McWhir, J., Ritchie, W. A. & Wilmut, I. Sheep cloned by nuclear transfer from a cultured cell line. *Nature* **380**, 64-66 (1996).

培养基，Gibco 生命科学公司）。这些细胞达到 90% 的覆盖率后，就按照 1∶2 的比例进行传代培养。到第 4 代时这些成纤维细胞样细胞（图 1）就有 54 个模式染色体。选用第 4 代至第 6 代的胎儿细胞作为核供体。乳腺细胞来自一只 6 岁的妊娠晚期芬兰多塞特母羊[18]。在第 3 代和第 6 代，模式染色体达到 54 个，因此第 3 代至第 6 代的细胞都被用作核供体。

根据参考文献的方法进行核移植[1]。注射促性腺激素释放激素（GnRH）后 28 到 33 小时之间从苏格兰黑面母羊中获得卵母细胞，并尽快去核。它们在含有 1% FCS 的无钙镁 PBS 溶液中恢复，然后转移到 37 ℃下含有 10% FCS 的无钙 M2 培养基中[19]。通过减少培养基中血清的浓度（5 天内从 10% 减到 0.5%）制备静止的二倍体供体细胞，这一方法使得细胞退出细胞周期并静止在 G0 期。利用抗 PCNA/ 细胞周期蛋白抗体（美国 Immuno Concepts 公司）进行染色，并用结合罗丹明的二抗（丹麦 Dakopatts 公司）进行显色，最终确定这些细胞退出细胞周期。

注射 GnRH 到供体母羊中 34~36 小时后，用同样的电脉冲诱导供体细胞和去核卵母细胞的融合以及卵母细胞的激活。大部分的重组胚胎如以前一样培养在羊结扎的输卵管中，但是一些通过胚胎来源细胞或者胎儿成纤维细胞核移植产生的胚胎培养在化学成分确定的培养基中[20]。经过 6 天培养后发育成桑椹胚或者胚泡的大部分胚胎都被移植到受体中并发育直到分娩（表 1）。根据胚胎的有效性，每只母羊中移植 1 个、2 个或 3 个胚胎。使用布雷斯洛和克莱顿[21]的边缘模型来分析细胞类型对细胞融合以及发育到桑椹胚或者胚泡的影响。发育到足月进行比较不太可能，因为将所有发育到一定阶段可以移植的胚胎都进行移植并不好操作。如果有太多的胚胎可供选择，那么形态较好的就被选中。

发情期后 60 天左右用超声进行妊娠诊断，此后每两周检测一次胎儿的发育情况。怀孕的受体母羊需要监测营养状况，身体状况，实验性变态反应性脑脊髓炎、Q 热、边界病、跳跃病和弓形体病的症状。产期临近时，进行实时监护，兽医随时准备接生。使用四个多态性羊标记物对小羊和受体母羊的 DNA 进行微卫星分析[22]。

（毛晨晖 翻译；方向东 审稿）

2. Solter, D. Lambing by nuclear transfer. *Nature* **380**, 24-25 (1996).

3. Gurdon, J. B., Laskey, R. A. & Reeves, O. R. The developmental capacity of nuclei transplanted from keratinized skin cells of adult frogs. *J. Embryol. Exp. Morph.* **34**, 93-112 (1975).

4. Quinlivan, T. D., Martin, C. A., Taylor, W. B. & Cairney, I. M. Pre- and perinatal mortality in those ewes that conceived to one service. *J. Reprod. Fert.* **11**, 379-390 (1966).

5. Walker, S. K., Heard, T. M. & Seamark, R. F. *In vitro* culture of sheep embryos without co-culture: successes and perspectives. *Therio* **37**, 111-126 (1992).

6. Nash, M. L., Hungerford, L. L., Nash, T. G. & Zinn, G. M. Risk factors for perinatal and postnatal mortality in lambs. *Vet. Rec.* **139**, 64-67 (1996).

7. Bradford, G. E., Hart, R., Quirke, J. F. & Land, R. B. Genetic control of the duration of gestation in sheep. *J. Reprod. Fert.* **30**, 459-463 (1972).

8. Walton, A. & Hammond, J. The maternal effects on growth and conformation in Shire horse–Shetland pony crosses. *Proc. R. Soc. B***125**, 311-335 (1938).

9. Campbell, K. H. S., Loi, P., Otaegui, P. J. & Wilmut, I. Cell cycle co-ordination in embryo cloning by nuclear transfer. *Rev. Reprod.* **1**, 40-46 (1996).

10. Cheong, H.-T., Takahashi, Y. & Kanagawa, H. Birth of mice after transplantation of early-cell-cycle-stage embryonic nuclei into enucleated oocytes. *Biol. Reprod.* **48**, 958-963 (1993).

11. Prather, R. S. *et al.* Nuclear transplantation in the bovine embryo. Assessment of donor nuclei and recipient oocyte. *Biol. Reprod.* **37**, 859-866 (1987).

12. McGrath, J. & Solter, D. Inability of mouse blastomere nuclei transferred to enucleated zygotes to support development *in vitro. Science* **226**, 1317-1318 (1984).

13. Robl, J. M. *et al.* Nuclear transplantation in bovine embryos. *J. Anim. Sci.* **64**, 642-647 (1987).

14. Campbell, K. H. S., Ritchie, W. A. & Wilmut, I. Nuclear-cytoplasmic interactions during the first cell cycle of nuclear transfer reconstructed bovine embryos: Implications for deoxyribonucleic acid replication and development. *Biol. Reprod.* **49**, 933-942 (1993).

15. Barnes, F. L. *et al.* Influence of recipient oocyte cell cycle stage on DNA synthesis, nuclear envelope breakdown, chromosome constitution, and development in nuclear transplant bovine embryos. *Mol. Reprod. Dev.* **36**, 33-41 (1993).

16. Kwon, O. Y. & Kono, T. Production of identical sextuplet mice by transferring metaphase nuclei from 4-cell embryos. *J. Reprod. Fert.* Abst. Ser. **17**, 30 (1996).

17. Gurdon, J. B. The control of gene expression in animal development (Oxford University Press, Oxford, 1974).

18. Finch, L. M. B. *et al.* Primary culture of ovine mammary epithelial cells. *Biochem. Soc. Trans.* **24**, 369S (1996).

19. Whitten, W. K. & Biggers, J. D. Complete development *in vitro* of the preimplantation stages of the mouse in a simple chemically defined medium. *J. Reprod. Fertil.* **17**, 399-401 (1968).

20. Gardner, D. K., Lane, M., Spitzer, A. & Batt, P. A. Enhanced rates of cleavage and development for sheep zygotes cultured to the blastocyst stage *in vitro* in the absence of serum and somatic cells. Amino acids, vitamins, and culturing embryos in groups stimulate development. *Biol. Reprod.* **50**, 390-400 (1994).

21. Breslow, N. E. & Clayton, D. G. Approximate inference in generalized linear mixed models. *J. Am. Stat. Assoc.* **88**, 9-25 (1993).

22. Buchanan, F. C., Littlejohn, R. P., Galloway, S. M. & Crawford, A. L. Microsatellites and associated repetitive elements in the sheep genome. *Mammal. Gen.* **4**, 258-264 (1993).

Acknowledgements. We thank A. Colman for his involvement throughout this experiment and for guidance during the preparation of this manuscript; C. Wilde for mammary-derived cells; M. Ritchie, J. Bracken, M. Malcolm-Smith, W. A. Ritchie, P. Ferrier and K. Mycock for technical assistance; D. Waddington for statistical analysis; and H. Bowran and his colleagues for care of the animals. This research was supported in part by the Ministry of Agriculture, Fisheries and Food. The experiments were conducted under the Animals (Scientific Procedures) Act 1986 and with the approval of the Roslin Institute Animal Welfare and Experiments Committee.

Correspondence should be addressed to I.W. (e-mail Ian.Wilmut@bbsrc.ac.uk).